SPECIES-AGE (DAYS)*

MONKEY	RAT	MOUSE	RABBIT	HAMSTER	GUINEA PIG	CHICK
166	21-22	18-20	31-34	15.5-16	64-68	20-21
4-9	3-5	3-6	2.6-6	3-4.5	4.8	
9	5-6	7	6	4.5-5	6	
18-20	9	8	6.5	7	13.5	0.25-0.75
19-21	9.5	8		7.5	13.5	1
20-21	10	8.3⁻		7.7 ⁻	14.5	1
	10	8.3		7.75	15.5 ⁻	1.5
	10.2			8. ⁻	16.5 ⁻	1.5 ⁻
	10				16.5	1.5
	10	9.1		8.5		2.75
23-24?	10.5	9	9.5	8.25	15.5	2.3 ⁻
	11-12	9.5		9	16.5	1.75
	11.3	9		8.5	16.5 ⁺	2.3 ⁻
23-24	10.5	8.6	9 ⁻	8	15.5	1.5
	11.5	9.6		8.25	16.5	2.3
	12	11			18.5	3
	10	8.5			16.5	1.8
25-26	10.5	9.3	10.5	8.25	16.5	2.2
	11.3	9.5		8.5	16.5?	3.75
1	12-13	8-10	6-9	10-14	2-3	
11	2, 5, 12, 16, 17, 18 30	4, 8, 20 32	4, 5	1, 6, 23	22	9, 17

* Midpoint of event except: ⁺ event beginning, ⁻ event finished (continued on the back cover)

CATALOG OF TERATOGENIC AGENTS

CATALOG OF TERATOGENIC AGENTS

Eighth Edition

Thomas H. Shepard, M.D.

Central Laboratory for Human Embryology
Department of Pediatrics
School of Medicine
University of Washington
Seattle, Washington

THE JOHNS HOPKINS UNIVERSITY PRESS

Baltimore and London

First edition (1973) sponsored by the National Institute of Child Health and Human Development, National Institutes of Health, Bethesda, Maryland 20892

The Johns Hopkins University Press
2715 North Charles Street
Baltimore, Maryland 21218-4319
The Johns Hopkins Press Ltd., London

The paper used in this book meets the minimum requirements of American National Standard for Information Sciences—Permanence of Paper for Printed Library Materials, ANSI Z39.48-1984.

Library of Congress Cataloging-in-Publication Data

Shepard, Thomas H., 1923–
  Catalog of teratogenic agents / Thomas H. Shepard — 8th ed.
    p. cm.
  Includes bibliographical references and indexes.
  ISBN 0-8018-5182-3 (hc : alk. paper)
  1. Teratogenic agents—Catalogs. I. Title
  [DNLM: 1. Teratogens—catalogs. 2. Abnormalities, Drug-Induced.
  3. Abnormalities, Radiation-Induced.   QS 626 S547c 1995]
  QM691.S53 1995
  616'.043—dc20
  DNLM/DLC
  for Library of Congress   95-8539
                                                                      CIP

This book is dedicated with appreciation to my past and present research fellows who stimulated its production

Prasanta K. Datta
Sorrel Waxman
Pierre Ferrier
Charles P. Mahoney
Ronald J. Lemire
Ralph R. Hollingsworth
Thomas Nelson
Takashi Tanimura
Alan G. Fantel
Godfrey P. Oakley
Gerald J. Bargman
Theodore H. Regimbal
Larklyn H. Fisher
Trent D. Stephens
Philip E. Mirkes
Kohei Shiota
Margo I. Van Allen
Marlisssa Campbell
John Rogers
Carol J. Quaife
Jack M. FitzSimmons
Masahiko Fujinaga
Raj P. Kapur
Hyoung W. Park

The eighth edition is dedicated to my wife, Alice.

CONTENTS

Endpapers: Comparative Time Periods of Embryonic and Fetal
Development in Humans and Experimental Animals

PREFACE TO THE EIGHTH EDITION

The eighth edition contains approximately 1500 additions, of which 270 are newly listed agents. A special effort has been made to obtain as much information as possible on drugs and other agents to which pregnant women may be exposed. Because of an increased awareness of the importance of negative published data, a considerable number of nonteratogenic agents have been included. Many such references are drawn from the Japanese literature and the author is, again, grateful for the help of Dr. Takashi Tanimura, Professor of Anatomy at Kinki University School of Medicine. Dr. Takahide Kihara of Kinki University also helped with the Japanese literature. Dr. Sonia Tabacova, Senior Research Scientist at the National Center of Hygiene, Ecology and Nutrition, Sofia, Bulgaria, helped with the Russian literature.

Some of the advantages of using a computer program for storage and production have been realized again with this edition. The originally published material theoretically does not need to be proofread; the names, references, and dosages cannot be mistakenly altered. When the program is completed, the entire final printout to be sent to the publisher can be run off within hours. Once in the hands of the publisher, the material does not need to be typeset, a consideration that reduces the cost of the book. Finally, the computer storage tape can be used by agencies such as the National Library of Medicine to extract specific data and references on various agents and supply their many clients rapidly. A week-to-week update of this data file is maintained on a mainframe computer at the University of Washington. Quarterly updates are available on CD-ROM through Micromedex, Inc., Denver, Colorado.

There are several other sources of teratology information. James G. Wilson's text (1973) remains one of the best introductions to teratology. A comprehensive 4-volume set of books on methods and principles in teratology has been edited by J. G. Wilson and F. C. Fraser (1977). Schardein (1993) has revised his *Chemically Induced Birth Defects*, a volume containing a large number of references. Heinonen et al. (1977) summarized drug exposures during more than 50,000 pregnancies and furnished useful tables of malformation-risk factors, nearly all of which are not significantly increased. The National Library of Medicine maintains a Developmental and Reproductive Toxicology (DART) database. The older references (before 1989) are available on ETICBACK, whose system of computerized information retrieval on teratology is available for those who require an in-depth listing of references. Their listing can be obtained directly through the National Library of Medicine (TOXNET).

Five especially useful publications on methods for detection of teratogenic agents and protection of the human population are available. The proceedings of a conference in Guadeloupe under the auspices of the Institute de la Vie address in vitro, pregnant animal testing, and epidemiologic methods (Shepard et al., 1975). Brent and Harris (1976) edited a volume emphasizing the scientific and epidemiologic techniques useful in preventing fetal and neonatal loss. A volume from a conference supported by the March of Dimes Birth Defects Foundation gives detailed guidelines for studies of human populations (Bloom, 1981). The U.S. Environmental Protection Agency published the proceedings of two conferences on techniques and risk assessment of substances that affect reproduction. Discussions of protocols for testing the effect of environmental substances on the male and female reproductive systems and the conceptus are given (Galbraith et al., 1982). Barlow and Sullivan (1982) summarized data on reproductive hazards in industry and this contains valuable information on a number of chemicals.

In the past few years there has been an increase in the demand for information concerning reproductive risk, and a number of delivery systems have been developed or expanded. TERIS is a database on teratologic information and risks and is available via modem, on hard disk, or via CD-ROM. This information has been reviewed by six nationally recognized clinical teratologists (Friedman et al., 1990) and is linked with an updated version of the *Catalog of Teratogenic Agents*. Anthony Scialli maintains another online database, REPROTOX (Scialli, 1990). A text by Briggs et al. (1994), entitled *Drugs in Pregnancy and Lactation*, gives clinical data on many drugs. The TERIS information has been published recently (Friedman and Polifka, 1994).

The eighth edition was formated with LaTeX at the University of Washington. This has allowed for greatly improved clarity and the use of Greek symbols, boldface type, and special punctuation. Entries containing a great deal of data have been reorganized using boldface headings. The section on occupational exposures has been organized by alphabetical listing of the type of work. A number of interesting defect syndromes have been created by transgenic insertions; to list these, the organ affected has been used (for example, an insertion that interferes with pancreatic development is listed under Pancreatic ablation).

REFERENCES

Barlow, S. M., and Sullivan, F. M. *Reproductive Hazards of Industrial Chemicals.* London: Academic Press, 1982.

Bloom, A. D., ed. *Guidelines for Studies of Human Populations Exposed to Mutagenic and Reproductive Hazards.* White Plains, N.Y.: March of Dimes Birth Defects Foundation, 1981.

Brent, R. L., and Harris, M., eds. *Prevention of Embryonic, Fetal, and Perinatal Disease.* Bethesda, Md.: DHEW (NIH) 76–853, 1976.

Briggs, G. G., Freeman, R. K., and Yaffe, S. J. *Drugs in Pregnancy and Lactation,* 4th ed. Baltimore: Williams and Wilkins, 1994.

Friedman, J. M., and Polifka, J. E. *Teratogenic Effects of Drugs: A Resource for Clin-*

icians (TERIS). Baltimore: Johns Hopkins University Press, 1994.

Friedman, J. M., Little, B. B., Brent, R. L., Cordero, J. F., Hanson, J. W., and Shepard, T. H. Potential human teratogenicity of frequenty prescribed drugs. *Obstet. Gynecol.* 75: 594–99, 1990.

Galbraith, W. M., Voytek, P., and Ryan, M. G., eds. *Assessment of Risks to Human Reproduction and to Development of the Human Conceptus from Exposure to Environmental Substances.* EPA-600/9-82-001. Springfield, Va.: National Technical Information Service, 1982.

Heinonen, O. P., Slone, D., and Shapiro, S. *Birth Defects and Drugs in Pregnancy.* Boston: John Wright PSG, Inc., 1977.

Schardein, J. L. *Chemically Induced Birth Defects*, 2nd ed. New York: Marcel Dekker, 1993.

Scialli, A. R. Making information work and Reprotox agent list additions. *Reprod. Toxicol.* 4: 249–50, 343–44, 1990.

Shepard, T. H., Miller, J. R., and Marois, M., eds. *Methods for Detection of Environmental Agents That Produce Congenital Defects.* New York: North Holland-American Elsevier, 1975.

Wilson, J. G. *Environment and Birth Defects.* New York: Academic Press, 1973.

Wilson, J. G., and Fraser, F. C., eds. *Handbook of Teratology*, vols. 1–4. New York: Plenum Press, 1977.

ACKNOWLEDGMENTS

I wish to thank the many unnamed colleagues who have supplied me with the information included in this catalog. My students and research fellows have contributed considerably, as have other teratologists. My associates, Drs. Alan Fantel and Philip Mirkes, were particularly helpful. Dr. Takashi Tanimura, of the Department of Anatomy at Kinki University School of Medicine (Osaka), aided with the selection and translation of the Japanese literature, and Dr. Sonia Tabacova, Senior Research Scientist at the National Center of Hygiene, Ecology and Nutrition, Sofia, Bulgaria, helped with the Russian literature.

The computerization of the book was simplified initially by the generous help of Dr. Victor A. McKusick, who allowed us to make use of a program similar to that employed for production of his catalog (*Mendelian Inheritance in Man* [8th ed.; Johns Hopkins University Press, 1988]). Dr. Richard Shepard and David Bolling, of the Johns Hopkins University School of Medicine, and James E. Peoples, of the University of Arizona School of Medicine, kindly helped to provide the edit-and-print programs for computerizing the body of the first edition of this book.

The computerization of the material and the processing of the catalog took place in the computer center of the University of Washington under the able direction of Jim Fox and Dr. John Jacobsen. Tom Stebbins designed the figures for the tables on the endpapers.

Patricia Stark entered my handwritten additions into the computer database. Dr. Janine E. Polifka, of the Department of Pediatrics, University of Washington, aided me by identifying typographical and other errors. Ms. Leanne Cornel efficiently retrieved library references.

The financial assistance of the National Institutes of Health (HD 00836) is appreciated.

INTRODUCTION

Defects existing at birth, irrespective of their cause, create societal problems of such magnitude that the subject needs little amplification. Approximately 3 percent of all human newborns have a congenital anomaly requiring medical attention, and approximately one-third of these conditions can be regarded as life threatening. With increasing age, more than twice as many congenital defects are detected. Close to one-half the number of children in hospital wards are there because of prenatally acquired malformations of one kind or another.

Our knowledge about the cause and prevention of these problems is extremely limited. About 20 percent are associated with gene mutations and another 5 percent with chromosomal aberrations. About 10 percent of the remaining anomalies are known to be due to a teratogenic agent. There are more than 2,500 agents listed in this catalog. About 1,200 can produce congenital anomalies in experimental animals, but only about forty of these are known to cause defects in the human. Therefore, there exists a wide difference between our knowledge of experimental teratology and the role that external agents play in producing human malformations.

A further problem is that the teratologic literature is dispersed throughout most of the biomedical publications rather than being confined to one or two of the specialized journals. Although a number of excellent review articles have been published, a catalog such as this has not been available. Harold Kalter's (1968) text deals with the teratology of the central nervous system in animals, and Josef Warkany's (1971) extensive treatise is concerned with congenital defects in the human. Jones's (1988) popular book on dysmorphic syndromes helps to define and extend our description of human defects. McKusick's (1994) Catalog of Mendelian inheritance in man includes many congenital syndromes produced by gene mutations.

The main purpose of this catalog is to help link the information on experimental teratogenic agents with the congenital defects in human beings. The catalog should provide a source of reference for teratologists wishing knowledge of the literature dealing with a particular teratogenic agent. For the obstetrician, pediatrician, and geneticist it should help answer the often-asked question: Does this agent produce congenital defects in the human or animal? Another function of this book may be to aid the scientists who protect us from our modified environment. Testing pharmaceutical products has become a major responsibility, and the environmental protectionist is also faced with the safety of food additives and household products. This catalog also may be useful to chemists working to develop new products. Unfortunately, because of species variability, at least in part, in teratogenic sensitivity, the ultimate testing of some products has been done in the human, with the alert clinician acting as monitor.

It is important that the presently accumulated information on teratogens be fully utilized to recognize potential teratogenic hazards and to prevent congenital malformations in humans. If it happens that many anomalies are produced by the interaction of genes with a teratogenic agent, it undoubtedly will be easier to remove the agent than to alter the action of the gene. The work of producing this book will be rewarded if it contributes to the prevention of any congenital defects.

The teratogenic agents listed in this book include chemicals, drugs, physical factors, and viruses. I have attempted to make a complete listing of all agents that have been studied for teratogenicity in animals or humans. The chemical names are, in most cases, those that appear in the Merck Index, but cross-indexing to alternate chemical and proprietary names has been done also. Studies carried out on species phylogenetically below the chick have been omitted. No attempt has been made to list agents that are teratogenic only when administered in combination with another agent. The presence of an agent in the catalog does not necessarily indicate that it is a teratogen because a number of compounds, with substantially negative teratogenic effects, have been included. When there is conflicting evidence of teratogenicity, equal representation has been attempted. An example of this is cigarette smoking.

The literature has been surveyed using the usual library aids. When possible, abstracts of work have not been utilized. Numerous excellent references dealing only with the mechanism of the defects' production have been omitted. Although some agents may have been omitted inadvertently from this listing, the method of production and printing of this type of book readily allows for easy revision. I hope that scientists in the field will feel free to send me new information or corrections for inclusion in future editions.

Each listing includes a main entry with synonyms. In the 8th edition, an attempt has been made to list any human data in the first part of the abstract. This is followed by a brief account of some of the work published including species, dose, gestational age at time of administration, and type of congenital defects produced. The references following each entry were chosen because of their review nature, originality, or because they are most current.

During the early planning of the book it appeared that the accumulation of the new teratologic literature could very rapidly outdate a text published in the usual manner. To obviate this partially, the catalog was constructed using a computerized system. The phrase computerized system has puzzled some of my associates, who initially concluded that I used some mechanical device for generating information. On the contrary, the text was compiled in the usual way and then transferred to the computer. This text was then stored on computer tape, from which a printout could be obtained. This computerized system for producing a catalog allows for easy insertion of new material or corrections during compilation and for easy production of future editions. It has the further advantage that, with the use of photocopying, the book can be printed in less than 3 months at a considerably reduced cost.

Since a teratogenic agent is defined by its ability to produce a congenital defect, it seems appropriate to provide a definition for "congenital defect" at this point. A congenital defect has its genesis during embryonic or fetal development and consists of a major or minor deviation from normal morphology or function. The border line between a minor congenital defect and normal variation is most difficult to define, and this accounts for a large difference in incidence rates. David Smith (1988) offered some criteria for distinguishing the two. In general, a minor defect should be present in less than 2 percent or 3 percent of the population. This small percentage could be defined in a statistical manner as the number of observations falling outside of three standard deviations from the mean. Large morphological types of congenital defects such as cleft palate or meningomyelocele may be called anomalies or malformations, but I feel the term *defect* also includes unknown or subtle structural defects that alter function. These functional changes also could be ascribed to molecular changes, many of which are still unknown. Particularly in the nervous and endocrine systems, changes in postnatal function have become an important aspect of both experimental and human teratology. An example of this is maternal hyperphenylalaninemia, which may lead to cerebral dysfunction and mental retardation in the offspring. Another new and important teratologic area requiring long term postnatal observation is prenatally induced oncogenesis, illustrated by vaginal carcinomas produced in grown girls whose mothers were treated with diethylstilbesterol during pregnancy.

A teratogenic agent acts during pregnancy to produce a physical or functional defect in the conceptus or offspring. The definition can be made more specific by using a modification of Koch's postulates in the following manner:

Koch's Postulates	*Application to Teratology*
1. A specific microorganism must be present in each case.	1. The agent must be present during the critical periods of development.
2. A pure culture of the organism should produce a similar disease in the experimental animal.	2. The agent should produce congenital defects in an experimental animal. The defect rate should be statistically higher in the treated group than in the control animals receiving the same vehicle or sham procedure.
3. Organisms from the experimental animal must be recovered and grown in pure culture.	3. Proof should be obtained that the agent in an unaltered state acts on the embryofetus either directly or indirectly through the placenta. In this area, biochemistry and in vitro culture are most often used instead of bacteriology.

The fulfillment of the first two conditions is sufficient to define a teratogenic agent. The third may be considered desirable but not essential. In this catalog, there are few teratogens that fit all three of these criteria. Surprisingly, teratogenic agents in the human generally do fill all three criteria—for instance, rubella virus, radiation, and androgens that masculinize the female fetus. Thalidomide, although accepted universally as a teratogen, does not fit the third criterion because the compound in its unaltered state has not been demonstrated to directly affect the conceptus. Although this third criterion may seem unnecessary to many teratologists, I believe that a more complete knowledge of these important molecular mechanisms can generate more rapidly the means for preventing malformations. An expanded discussion of the criteria for human teratogenicity can be found on page xxiii.

Several difficult problems face the teratologist in judging whether an agent is a human teratogen. When the dosage level of a compound must be raised to near-fatal levels for the mother before defects are produced in her fetus, most workers consider the agent weakly teratogenic. However, clinicians, the Federal Drug Administration, and the pharmaceutical industry encounter difficult decisions in applying these experimental findings to humans. Agents causing embryonic or fetal death in experimental animals often later prove to be teratogenic in humans but are not considered to be teratogens unless physical or functional defects are produced. Similarly, an agent that can cause fetal growth retardation does not necessarily qualify as a teratogen. Retardation of fetal skeletal maturation reported as decreased ossification centers of the manubrium or immaturity of the vertebral centra is another example of a change considered physiologic but not teratogenic.

REFERENCES

Jones, K. L. Smith's *Recognizable Patterns of Human Development*, 4th ed. Philadelphia: W. B. Saunders Company, 1988.

Kalter, H. *Teratology of the Central Nervous System*. Chicago: University of Chicago Press, 1968.

McKusick, V. A. *Mendelian Inheritance in Man: A Catalog of Human Genes and Genetic Disorders*, 11th ed. Baltimore: Johns Hopkins University Press, 1994.

Warkany, J. *Congenital Malformations: Notes and Comments*. Chicago: Year Book Medical Publishers, 1971.

NOTES FOR TABLE ON ENDPAPERS

Comparative Time Periods of Embryonic and Fetal Development in Humans and Experimental Animals

One of the main principles of teratology is that teratogenic activity is strongly dependent on exposure to an agent at a specific "sensitive" period in development. It seemed appropriate, therefore, to attempt an integration of embryonic and fetal events for the human and some common experimental animals. I have tabulated events that occur in the human, rhesus monkey, rat, mouse, rabbit, hamster, guinea pig, and chick. The times are given in days from copulation or ovulation for the monkey, mouse, rabbit, hamster, and guinea pig. An assumption has been made that ovulation time is approximately the same as copulation time. For humans, the time is derived by subtracting 14 days from the period between the onset of the last normal menses and time of delivery. For the rat, timing customarily has been from 9:00 A.M. following copulation, and I have chosen to follow this method. Timing in the chick is from the start of incubation.

The reader may note that the ages given for humans are 2-3 days greater than the dates originally given by G. L. Streeter, who based his timing on comparable stages in the monkey. This adjustment, based on last menstrual period dates, has been drawn largely from material collected in the Central Laboratory for Human Embryology at the University of Washington (Iffy et al., 1967; Shepard, 1975; and Shepard, 1988). Our data, based on North American and European specimens, agree quite well with the Japanese observations of Nishimura and Yamamura (1969). Exact dating of the early somite stages is still hazardous because of the meager number of available human embryos.

A good deal of confusion exists about what the days of the pregnancy should be called. In this volume, and, I believe, as a general practice among teratologists, the first 24 hours after fertilization is designated as the first day; however, if the word day precedes the cardinal number, this indicates that the age is at least that many days. Thus, a day-10 embryo is at least 10 days of age, but an embryo of the 10th day is between 9 and 10 days of age. Kalter (1968) eloquently discussed this problem.

Many definitions of the embryonic period have been offered. I define the end of the human embryonic period or start of the fetal period by these criteria: 54-60 gestational days, crown-rump length of 33 mm, end of Streeter's Horizon XXIII, and end of major organogenesis. Two other criteria that might help define the difference in some of the experimental animal models are the appearance of ossification and the presence of external characteristics specific for the species.

A definition of fetal growth in the human and monkey is only now beginning. Tanimura et al. (1971) designated fetal organ weights for the human, and Kerr et al. (1969). for the monkey. Gruenwald and Minh (1960), Potter (1961), and Shepard et al. (1988) gave organ weight standards for fetuses over the weight of 500 gm.

The references that follow were useful in constructing the tables appearing on the front and back inside covers.

REFERENCES

Gruenwald, P., and Minh, H. N. Evaluation of body and organ weights in perinatal pathology. I. Normal standards derived from autopsies. *Amer. J. Clin. Path.* 34: 247–53, 1960.

Kalter, H. How should times during pregnancy be called in teratology? *Teratology* 1: 231–34, 1968.

Kerr, G. R., Kennan, A. L., Waisman, H. A., and Allen, J. R. Growth and development of the fetal rhesus monkey. I. Physical growth. *Growth* 33: 201–13, 1969.

Long, J. A., and Burlingame, M. L. The development of the external form of the rat with some observations on the origin of the extraembryonic coelom and fetal membranes. *Univ. Calif. Mem. Zool.* 43: 143–83, 1938.

Potter, E. L. *Pathology of the Fetus and Infant.* Chicago: Year Book Medical Publishers, 1961. P. 14.

Shepard, T. H. Growth and development of the human embryo and fetus. In Gardner, L. I. (ed.), *Endocrine and Genetic Diseases of Childhood*, 2nd ed. Philadelphia: W. B. Saunders, 1975. Pp. 1–6.

Shepard, T. H., Barr, M., Fellingham, G. W., Fujinaga, M., Fitzsimmons, J., and Fantel, A. G. Organ weight standards for human fetuses. *Pediatric Pathology* 4: 513–22, 1988.

Tanimura, T., Nelson, T., Hollingsworth, R. R., and Shepard, T. H. Weight standards for organs from early human fetuses. *Anat. Rec.* 171: 227–36, 1971.

ENDPAPER REFERENCES

1. Boyer, C. C. Chronology of the development for the golden hamster. *J. Morph.* 92: 1–37, 1953.

2. Christie, G. A. Developmental stages in somite and post-somite rat embryos, based on external appearance and including some features of the macroscopic development of the oral cavity. *J. Morph.* 114: 263–86, 1964.

3. Davies, J. *Human Developmental Anatomy.* New York: Ronald Press, 1963.

4. ____, and Hesseldahl, H. Comparative embryology of mammalian blastocysts. In Blandau, R. J. (ed.), *Biology of the Blastocyst.* Chicago: University of Chicago Press, 1971. Pp. 27–48.

5. Edwards, J. A. The external development of the rabbit and rat embryo. In Woollam, D. H. M. (ed.), *Advances in Teratology*, Vol. 3. New York: Academic Press, 1968. Pp. 239–63.

6. Graves, A. P. Development of the golden hamster. *Cricetus auratus* waterhouse, during the first nine days. *Amer. J. Anat.* 77: 219–51, 1949.

7. Gruenwald, P., and Minh, H. N. Evaluation of body and organ weights in perinatal pathology. I. Normal standards derived from autopsies. *Amer. J. Clin. Path.* 34: 247–53, 1960.

8. Gruneberg, H. The development of some external features in mouse embryos. *J. Hered.* 34: 88–92, 1943.

9. Hamburger, V., and Hamilton, H. L. A series of normal stages in the development of the chick embryo. *J. Morph.* 88: 49-92, 1951.

10. Heuser, C. H., and Corner, G. W. Development horizons in embryos, description of age groups XIX, XX, XXI, XXII, and XXIII being the fifth issue of a survey of the Carnegic Collection. *Contrib. Embryol.* 34: 165–96, 1951.

11. Heuser, C. H., and Streeter, G. L. Development of the macaque embryo. *Contrib. Embryol.* 29: 15–55, 1941.

12. Huber, G. C. The development of the albino rat, *Mus norvegicus albinus*. *J. Morph.* 26: 246–386, 1915.

13. Iffy, L., Shepard, T. H., Jakobovits, A., Lemire, R. J., and Kerner, P. The rate of growth in young embryos of Streeter's horizons XIII to XXIII. *Acta Anat.* (Basel) 66: 178–86, 1967.

14. Kalter, H. How should times during pregnancy be called in teratology? *Teratology* 1: 231–34, 1968.

15. Kerr, G. R., Kennan, A. L., Waisman, H. A., and Allen, J. R. Growth and development of the fetal rhesus monkey. I. Physical growth. *Growth* 33: 201–13, 1969.

16. Long, J. A., and Burlingame, M. L. The development of the external form of the rat with some observations on the origin of the extraembryonic coelom and fetal membranes. *Univ. Calif. Mem. Zool.* 43: 143–83, 1938.

17. Monie, I. W. Comparative development of rat, chick and human embryos. In *Teratologic Workshop Manual* (supplement). Berkeley, Calif.: Pharmaceutical Manufacturers Associations, 1965. Pp.146–62.

18. Nicholas, J. S. Experimental methods and rat embryos. In Farris, E. J., and Griffith, J. Q. (eds.), *The Rat in Laboratory Investigation*. New York: Hafner Publishing, 1962. Pp. 51–67.

19. Nishimura, H., and Yamamura, H. Comparison between man and some other mammals of normal and abnormal developmental processes. In Nishimura, H., and Miller, J. R. (eds.). *Methods for Teratological Studies in Experimental Animals and Man.* Tokyo: Igaku Shoin, 1969. Pp. 223–40.

20. Otis, E. M., and Brent, R. Equivalent ages in mouse and human embryos. *Anat. Rec.* 120: 33–63, 1954.

21. Potter, E. L. *Pathology of the Fetus and Infant.* Chicago: Year Book Medical Publishers, 1961. P. 14.

22. Scott, J. P. The embryology of the guinea pig. I. Table of normal development.

Amer. J. Anat. 60: 397–432, 1937.

23. Shenefelt, R. E. Morphogenesis of malformations in hamsters caused by retinoic acid: Relation to dose and stage of treatment. *Teratology* 5: 103–18, 1972.

24. Shepard, T. H. Growth and development of the human embryo and fetus. In Gardner, L. I. (ed.), *Endocrine and Genetic Diseases of Childhood*, 2nd ed. Philadelphia: W. B. Saunders, 1975. Pp. 1–6.

25. Shepard, T. H., Barr, M., Fellingham, G. W., Fujinaga, M., Fitzsimmons, J., and Fantel, A. G. Organ weight standards for human fetuses. *Pediatric Pathology* 4: 513–22, 1988.

26. Streeter, G. L. Developmental horizons in human embryos, description of age group XI, 13 to 20 somites, and age group XII, 21 to 29 somites. *Contrib. Embryol.* 30: 211–45, 1942.

27. _____. Developmental horizons in human embryos, description of age group XIII embryos about 4 or 5 millimeters long, and age group XIV, period of indentation of lens vesicle. *Contrib. Embryol.* 31: 27–63, 1945.

28. _____. Developmental horizons in human embryos, description of age groups XV, XVI, XVII and XVIII, being the third issue of a survey of the Carnegie Collection. *Contrib. Embryol.* 32: 133–203, 1948.

29. Tanimura, T., Nelson, T., Hollingsworth, R. R., and Shepard, T. H. Weight standards for organs from early human fetuses. *Anat. Rec.* 171: 227–36, 1971.

30. Witschi, E. Development of the rat. In Altman, P., and Dittmer D. S. (eds.), *Growth Including Reproduction and Morphological Development*. Washington, D.C.: Federation of American Societies for Experimental Biology, 1962. Pp. 304–414.

31. O'Rahilly, R., and Muller, F. *Developmental Stages in Human Embryos*. Washington, D.C.: Carnegie Institution of Washington, Publication 637, 1987.

32. Theiler, K. *The House Mouse Development and Normal Stages from Fertilization to 4 Weeks of Age*. Berlin: Springer-Verlag, 1972.

HUMAN TERATOGENS

Proven, Possible, and Unlikely (January, 1995)

Several of my associates have encouraged me to include lists of agents that are considered, in human beings, to be proven, possible, or unlikely teratogens. This has been done with some trepidation, since I have attempted in the textual material to present the facts without comment. The substantiation for inclusion of an agent in one of these three lists has been drawn from data recorded in the text. The criteria used for proof of human teratogenicity are listed in table 1 and represent an amalgamation from writings by Wilson (1977), Brent (1978), Stein et al. (1984), Hemminki and Vineis (1985), and Shepard (1973–86).

In recent years, several papers on teratogenic risk have appeared. Jelovsek et al. (1989) analyzed the relationship between animal tests and the identification of human teratogens. In another work, Jelovsek et al. (1990) weighed the principles used by experts to identify and classify human teratogens. A paper by Kimmel and Gaylor (1988) discussed qualitative and quantitative evaluations of risk and how "acceptable" exposure levels are determined. A scheme for ranking chemicals by their potential risk has been detailed (Wang and Schwetz 1987).

The Teratology Society Public Affairs Committee (1994) has urged the FDA to abandon its teratogenic rating system because it may produce undue anxiety in the physcician and pregnant woman and contribute to unnecessary termination of wanted pregnancies.

REFERENCES

Brent, R. L. Editor's note. *Teratology* 17: 183, 1978.

Hemminki, K., and Vineis, P. Extrapolation of the evidence on teratogenicity of chemicals between humans and experimental animals: Chemicals other than drugs. *Teratogenesis, Carcinogenesis, and Mutagenesis* 5: 251–318, 1985.

Jelovsek, F. R., Manison, D. R., and Chen, J. J. Prediction of risk for human developmental toxicity: How important are animal studies for hazard identification? *Obstet. Gynecol.* 74: 624–36, 1989.

Jelovsek, F. R., Manison, D. R., and Young, J. F. Eliciting principles of hazard identification from experts. *Teratology* 42: 521–33, 1990.

Kimmel, C. A., and Gaylor, D. W. Issues in qualitative and quantitative risk analysis for developmental toxicity. *Risk Analysis* 8: 15–20, 1988.

Shepard, T. H. *Catalog of Teratogenic Agents*, 1st-7th eds. Baltimore: Johns Hopkins University Press, 1973–1992.

Shepard, T. H. Human teratogenicity. In Barness, L. (ed.), *Advances in Pediatrics*. Vol. 33. Chicago: Yearbook Publishers, 1986. Pp. 225–68.

Stein, Z., Kline, J., and Kharrazi, M. What is a teratogen? Epidemiologic criteria. In Kalter, H. (ed.), *Issues and Reviews in Teratology*, Vol. 2. New York: Plenum Press, 1984. Pp. 23–66.

Teratology Society Public Affairs Committee. FDA classification of drugs for teratogenic risk. *Teratology* 49: 446–447, 1994.

Wang, G. M., and Schwetz, B. A. An evaluation system for ranking chemicals with teratogenic potential. *Teratogenesis, Carcinogenesis and Mutagenesis* 7: 133–39, 1987.

Wilson, J. G. Embryotoxicity of drugs in man. In Wilson, J. G., and Fraser, F. C. (eds.), *Handoook of Teratology*, Vol. 1. New York: Plenum Press, 1977. Pp. 309–55.

TABLE 1. *Amalgamation of Criteria for Proof of Human Teratogenicity*

1. Proven exposure to agent at critical time(s) in prenatal development (prescriptions, physician's records, dates).
2. Consistent findings by two or more epidemiologic studies of high quality:
    (a) Control of confounding factors;
    (b) Sufficient numbers;
    (c) Exclusion of positive and negative bias factors;
    (d) Prospective studies, if possible; and
    (e) Relative risk of six or more (?).
3. Careful delineation of the clinical cases. A specific defect or syndrome, if present, is very helpful.
4. Rare environmental exposure associated with rare defect. Probably three or more cases (examples; oral anticoagulants and nasal hypoplasia, methimazole and scalp defects (?), and heart block and maternal rheumatism).
5. Teratogenicity in experimental animals important but not essential.
6. The association should make biologic sense.
7. Proof in an experimental system that the agent acts in an unaltered state. Important information for prevention.

Note: Items 1, 2, and 3 or 1, 3, and 4 are essential criteria. Items 5, 6, and 7 are helptul but not essential.

TABLE 2. *Teratogenic Agents in Human Beings*

Radiation	Drugs and Environmental Chemicals
Atomic weapons	Aminopterin and methylaminopterin
Radioiodine	Androgenic hormones
Therapeutic	Busulfan
	Captopril (renal failure)
Infections	Chlorobiphenyls
Cytomegalovirus (CMV)	Cocaine
Herpes virus hominis ? I and ll	Coumarin anticoagulants
Parvovirus B-19 (Erthema infectiosum)	Cyclophosphamide
Rubella virus	Diethylstilbestrol
Syphilis	Diphenylhydantoin
Toxoplasmosis	Enalapril (renal failure)
Varicella virus	Etretinate
Venezuelan equine encephalitis virus	Iodides and goiter
	Lithium
Maternal Metabolic Imbalance	Mercury, organic
Alcoholism	Methylene blue via intraamniotic
Chorionic villus sampling	injection
(before day 60)	Methimazole and scalp defects
Cretinism, endemic	Penicillamine
Diabetes	1 3-cis-Retinoic acid (Isotretinoin and
Folic acid deficiency	Accutane)
Hyperthermia	Tetracyclines
Phenylketonuria and Sjogren's	Thalidomide
syndrome	Toluene abuse
Rheumatic disease and congenital	Trimethadione
heart block	Valproic acid
Virilizing tumors	

TABLE 3. *Possible and Unlikely Teratogens*

Possible Teratogens	Unlikely Teratogens
?Binge drinking	Agent Orange
?Carbamazepine	Anesthetics
?Cigarette smoking	Aspartame
?Colchicine	Aspirin (but aspirin in the 2nd half of
?Disulfiram	pregnancy may increase cerebral
?Ergotamine	hemorrhage during delivery)
?High Vitamin A	Bendectin (antinauseants)
?Lead	Illicit drugs (marihuana, LSD)
?Primidone	Metronidazole (Flagyl)
?Quinine, suicidal doses	Oral contraceptives
?Streptomycin	Progesterone (Hydrogyprogesterone
?Zinc deficiency	and Medroxyprogesterone)
	Rubella vaccine
	Spermicides
	Video display terminals and
	electromagnetic waves
	Ultrasound

USE OF THE CATALOG

The text of the catalog consists of alphabetically organized entries of the agents. The chemical compounds are alphabetized without regard for single preceding letters, numbers, or Greek symbols. For instance, I-beta-D-arabinofuranosyl-5-fluorocytosine would be listed in the A's. The chemical name as listed in The Merck Index has been used most often as the main entry, with trade names as synonyms. Trade names are followed by ™. Chemical abstract numbers (CAS) have been added when possible to each agent entry. Following the catalog are the subject and author indexes for the reader's convenience. If there are various names for a single agent, the reader can locate the text entry more rapidly by consulting the subject index. For example, irradiation in the subject index refers the reader to the text entry under radiation. In both indexes, the numbers supplied refer to entries, not pages. Page numbers are given at the top of the page, near the inside margin; the entry numbers are in boldface, near the outside margin.

References added in the 7th or 8th edition are identified in the text by the use of boldface type for the year of publication.

CATALOG OF TERATOGENIC AGENTS

1   AA–2414 *(+-)-7–(3,5,8–Trimethyl–1–4–benzoquinone–2–yl)–7–phenylheptanoic*
Acid

Sugitari et al. (**1994**) gave up to 100 mg per kg orally to rats before and during pregnancy at doses of up to 100 mg per kg. No adverse reproductive, fetal or behavioral effects were found. Similar studies in the rabbit did not reveal teratogenicity (Nakatsu et al., **1993**).

Nakatsu, T.; Kanamori, H. and Yoshizaki, H.: Teratological study of AA–2414 in rabbits. Yakuri to Chiryo 21:s1781–s1787, 1993.

Sugitani, T.; Takatani, O.; Kanamori, H.; Nakatsu, T.; Ooshima, Y. and Yoshizaki, H.: Peri–and post–natal study of AA–2414 in rats. Yakuri to Chiryo 21:s1789–s1806, 1993.

Sugitani, T.; Nakatsu, T.; Kumada, S.; Yoshizaki, H.; Kanamori, H. and Takatani, O.: Reproduction study of AA–2414 in rats. Yakuri to Chiryo 21:s1765–s1780, 1993.

2   Abortion–Induced

The question, "Does an induced abortion affect subsequent pregnancies?" has been debated. Pantelakis et al. (1973) reported that women with previous induced abortions had double the rate of premature and stillborn children. Wright et al. (1972) found an increase in second trimester spontaneous abortions in women who had previously undergone cervical dilatation. The cervical dilatation was thought to produce cervical incompetence. Critical appraisal of these and other reports on the subject (Institute of Medicine, 1975; Lui et al., 1972) suggests that further matched control studies are needed to answer the question fully. One such study by Daling and Emanuel (1975), carefully matched for social and other contributing factors, could find no effect on subsequent pregnancies. The congenital defect rate was also unaltered.

Bracken and Holford (1979) studied 1,427 pregnancies with previous therapeutic abortions and could detect no significant increase in malformations. Hogue et al. (1982) reviewed published data on this subject in detail and found little indication of subsequent poor pregnancy outcome. Levin et al. (1982) studied 25 women with ectopic pregnancy and found that if they had one prior induced abortion, their relative risk after correction for confounding factors was increased to 1.3 and if there was more than one, the risk was 2.6. Neither risk was statistically significant.

Bracken, M.B. and Holford, T.R.: Induced abortion and congenital malformations in offspring of subsequent pregnancies. Am. J. Epidemiol. 109:425–432, 1979.

Daling, J.R. and Emanuel, I.: Induced abortion and subsequent outcome of pregnancy: A matched cohort study. Lancet 2:170–172, 1975.

Hogue, C.J.R.; Cates, W. and Tietze, C.: The effects of induced abortion on subsequent reproduction. Epidemiol. Rev. 4:66–94, 1982.

Institute of Medicine: Report of a Study on Legalized Abortion and the Public Health. Washington, D.C., Academy of Science. 47–71, 1975.

Levin, A.A.; Schoenbaum, S.C.; Stubblefield, P.G.; Zimicki, S.; Monson, R.R. and Ryan, K.R.: Ectopic pregnancy and prior induced abortion. Am. J. Public Health. 72:253–256, 1982.

Lui, D.T.Y.; Melville, H.A.H. and Martin, T.: Subsequent gestational morbidity after various types of abortion. Lancet 2:43 only, 1972.

Pantelakis, S.N.; Papadimitriou, G.C. and Doxiadis, S.A.: Influence of induced and spontaneous abortions on the outcome of subsequent pregnancies. J. Obstet. Gynecol. 116:799–805, 1973.

Wright, C.S.W.; Campbell, S. and Beazley, J.: Second trimester abortion after vaginal termination of pregnancy. Lancet 1:1278–1279, 1972.

3   Abrin

Extracts of this seed were gavaged or injected intraperitoneally into rats on days 7–14. The intraperitoneal dose one–tenth the maternal lethal dose produced decreased fetal weight and some unspecified defects.

El–Shabray, O.A.; El–Gengaihi, S. and Ibrahim, N.A.: Toxicity and teratogenicity of abrin. Egypt. J. Vet. Sci 24(2):135–142, 1987.

4   Abrobit
2-Hydroxyethyldimethyl-1-(10-phenothiazinylmethyl ethylammonium chloride)

West (1962) found fetal death in rat fetuses exposed to 5, 8 or 10 mg per kg of this antihistamine. No other fetal findings were included in the report.

West, G.B.: Drugs and rat pregnancy. J. Pharm. Pharmacol. 14:828–830, 1962.

5   Acebutolol Hydrochloride *dl-1(2-Acetyl-4-butyramido-phenoxy)–2–hydroxy–3–isopropyl aminopropane HCL CAS 37517–30–9*

Yokoi et al. (1978) gave up to 500 mg per kg orally or 25 mg per kg intravenously to pregnant rats on days 7–17. No adverse fetal or postnatal effects were found. Dubois et al. (1982) treated 56 hypertensive women at various times in pregnancy and found no increase in congenital defects. Neither bradycardia nor hypoglycemia were a problem at birth.

Dubois, D.; Petitcolas, J.; Temperville, B.; Klepper, A. and Catherine, P.: Treatment of hypertension in pregnancy with β–adrenoceptor antagonists. Br. J. Clin. Pharmac. 13:375S–378S, 1982.

Yokoi, Y.; Yoshida, H.; Hirano, K.; Masanori, N.; Okumura, M.; Nose, T.; Kawamoto, H.; Kaneko, K. and Hori, H.: Teratological studies of acebutolol hydrochloride in rats. Oyo Yakuri 15:885–904, 1978.

6   Acemetacin *CAS 53164–05–9*

Koga et al. (1981) studied the effect of this non–steroidal anti–inflammatory agent during organogenesis in the rat. Oral doses of up to 8.0 mg per kg gave maternal ulcers of the small intestine of the dams and fetal growth retardation was seen. Congenital defects were not increased. At 4.0 mg per kg during the prenatal period an increase in fetal mortaliy was seen (Aoki et al., 1981).

Aoki, Y.; Koga, T.; Ota, T.; Sugasawa, M. and Kobayashi, F.: Reproductive studies with acemetacin (K–708): Perinatal and postnatal study in rats. Oyo Yakuri 22:777–786, 1981.

Koga, T.; Ota, T.; Aoki, Y.; Sugasawa, M. and Kobayashi, F.: Reproductive studies with acemetacin (K–708): Teratological study in rats. Oyo Yakuri 22:765–776, 1981.

7 Acetaldehyde CAS 75–07–0

Veghelyi et al. (1978) first asked if acetaldehyde was the toxic agent in fetal alcohol syndrome. This metabolite of ethanol was administered to mice on days 7, 8 and 9 of gestation (O'Shea and Kaufman, 1979). A 1 or 2 percent solution of acetaldehyde in 0.1 ml of 0.9 percent saline per 25 gm of body weight was used. Increased resorptions and decreased embryonic protein and decreased fetal weights as compared to saline–injected controls occurred. No increase in defect rate was found in the treated fetuses on day 19, but on day 10 an increased number of embryos with open anterior or posterior neural tube was reported.

Campbell and Fantel (1983) studied the effect of acetaldehyde in whole embryo culture and found reduced growth at 25, 50 and 75 microM concentrations. Popov et al. (1981) studied the effect of ethanol (6.5–108 mM) and its metabolite, acetaldehyde (4.5–450 mM), in the culture of 9.5 day rat embryos. In both cases similar malformations were found. Skosyreva (1982) observed a high teratogenicity of acetaldehyde substance.

Webster et al. (1983) gave 4 percent acetaldehyde in arachis oil intraperitoneally to mice on single days from gestational day 7 through 10. In some experiments two doses either 30 minutes or 6 hours apart were given. A small increase in defects over control ingestions was found.

Ali and Persaud (1988) showed that embryonic defects and growth retardation occur when alcohol and an inhibitor of acetaldehyde (cyanamide) are coadministered on days 9–12. Neither chemical alone in the same dose caused defects.

Ali, F. and Persaud, T.V.N.: Mechanisms of fetal alcohol effects: role of acetaldehyde. Exp. Pathol. 33:17–21, 1988.

Campbell, M.A. and Fantel, A.G.: Teratogenicity of acetaldehyde in vitro: Relevance to the fetal alcohol syndrome. Life Sciences 32:2641–2647, 1983.

O'Shea, K.S. and Kaufman, M.H.: The teratogenic effect of acetaldehyde: Implications for study of the fetal alcohol syndrome. J. Anat. 128:65–76, 1979.

Popov, V.B.; Weisman, B.L.; Puchkov, V.F. and Ignatyeva, T.V.: Embryotoxic effect of ethanol and its biotransformation products in the culture of postimplantation rat embryos. Byull. Eksp. Biol. Med. (USSR) 12:725–728, 1981.

Skosyreva, A.M.: Comparative studies of embryotoxic effects of alcohol and its metabolite, acetaldehyde during organogenesis. Akush. Ginekol. (USSR) 1:49–50, 1982.

Veghelyi, P.V.; Osztovics, M.; Kardos, G.; Leisztner, L.; Szaszovszky, S.; Igali, S. and Imrei, J.: The fetal alcohol syndrome symptoms and pathogenesis. Acta Ped. Acad. Scient. Hung. 19:171–189, 1978.

Webster, W.S.; Walsh, D.A.; McEwen, S.E. and Lipson, A.H.: Some teratogenic properties of ethanol and acetaldehyde in C57BL6J mice: Implications for the study of fetal alcohol syndrome. Teratology 27:231–243, 1983.

8 Acetamide Carbonic Acid Hydrazides Hydroxamic Acids CAS 60–35–5

Thiersch (1962) reported that acetamide was non–toxic to rat litters, but N–methyl acetamide and N,N–dimethyl acetamide were lethal to the embryo–fetus when 2 gm per kg were given once intraperitoneally between the 4th and 14th days of gestation. Von Kreybig et al. (1969) found skeletal defects in fetuses from rats given 600 mg per kg of dimethyl or diethyl acetamide.

N–Methylforamide, N–methylacetamide, N–ethylacetamide, N–methylpropionamide and N–methylbutyramide were found by von Kreybig and Schmidt (1968) to be teratogenic in the rat. At individual daily subcutaneous doses of 600 to 1000 mg per kg extremity defects were frequent. Heptanoic acid and N–methylamide were not teratogenic. Formhydroxamic, methylhydroxamic and acethydroxamic acids were teratogenic in the range of 500 to 700 mg per kg given subcutaneously on day 13. Propionhydroxamic, hexanohydroxamic and phthalomonohydroxamic acids were not teratogenic.

Kennedy (1986) summarized the developmental toxicity of this chemical and its related congeners.

Kennedy, G.L.: Biological effects of acetamide, formamide and their monomethyl and dimethyl derivatives. CRC Crit. Rev. Toxicol. 17:129–182, 1986.

Thiersch, J.B.: Effects of acetamides and formamides on the rat litter in utero. J. Reprod. Fertil. 4:219–220, 1962.

von Kreybig, T.; Preussmann, R. and von Kreybig, I.: Chemische Konstitution und teratogene Wirkung bei der Ratte. II. N–alkylharnstoffe, N–alkylsulfonamide, N, N–dialkylacetamide, N–methylthioacetamid, Chloracetamid. Arzneim. Forsch. 19:1073–1076, 1969.

von Kreybig, T. and Schmidt, W.: Chemische Konstitution und Teratogene Wirkung bei der Ratte. Arzneim Forsch 18:645–657, 1968.

9 2–Acetamido–1,3,4–thiadiazole–5–(N–t–butylsulfonamide)

Maren and Ellison (1972) injected up to 600 mg per kg into rats on days 10 and 11. Resorption was increased at the 300 and 600 mg levels. At 200 mg and 100 mg per kg, low fetal weights and skeletal immaturity were observed.

Maren, T.H. and Ellison, A.C.: The teratological effect of certain thiadiazoles related to acetazolamide with a note on sulfanilamide and thiazide diuretics. Johns Hopkins Med. J. 130:95–104, 1972.

10 Acetaminophen CAS 103–90–2

Heinonen et al. (1977) reported on 226 infants born to mothers who took the drug during the first four months of pregnancy and found no increase in malformations. Aselton et al. (1985) studied the offspring of 347 women who took acetaminophen with codeine and found only 3 congenital defects. Char et al. (1975) reported an infant with polyhydramnios and renal failure. The nephrotoxicity was felt to be associated with acetaminophen which the mother ingested heavily during pregnancy.

Collins (1981) reviewed the subject and found no reported animal studies. In unpublished studies by S.C. Emerson and T.H. Shepard, rats given 750 mg per kg during organogenesis produced fetuses that were slightly underweight and had increased minor skeletal defects. No increase in defect rate or change in liver weight occurred. Lubawy and Garrett (1977) gavaged rats on days 8–19 with 250 mg per kg and produced no decrease in fetal weight. Ogawa et al. (1982) gave mice an oral preparation containing 43 percent acetaminophen along with ethenzamide, caffeine and a vehicle to mice. Treatment during organogenesis with approximately 387 mg of acetaminophen produced no ill effects in the fetuses.

Aselton, P.A.; Jick, H.; Milunsky, A.; Hunter, J.R. and Stergachis, A.: First trimester drug use and congenital disorders. Obstet. Gynecol. 65:451–455, 1985.

Char, V.C.; Chandra, R.; Fletcher, A.B. and Avery, G.B.: Polyhydramnios and neonatal renal failure: A possible association with maternal acetaminophen ingestion. J. Pediatr. 86:638–639, 1975.

Collins, E.: Maternal and fetal effects of acetaminophen and salicylate in pregnancy. Obstet. Gynecol. 58:57S–62S, 1981.

Heinonen, O.P.; Slone, D. and Shapiro, S.: Birth Defects and Drugs in Pregnancy. Publishing Sciences Group Inc., Littleton, Mass., 1977.

Lubawy, W.C. and Garrett, R.J.B.: Effect of aspirin and acetaminophen on fetal and placental growth in rats. J. Pharm. Sci. 66:111–113, 1977.

Ogawa, H.; Arakawa, E.; Morobushi, A.; Yamada, H. and Ito, H.: Reproductive studies of NB–6. Kiso to Rinsho 16:683–695, 1982.

11 Acetanilide CAS 103–84–4

Wright (1967) reported reduced fertility in mice maintained on 0.1 percent of the drug in the diet.

Wright, H.N.: Chronic toxicity studies of analgesic and antipyretic drugs and cogeners. Toxicol. Appl. Pharmacol. 11:280, 1967.

12 Acetazolamide *Diamox*™ CAS 59–66–5

Heinonen et al. (1977) reported 12 women who took the medication and no defects were observed in their offspring. Nakane et al. (1980) reported 5 defects among 55 epileptic women who took this medication along with other drugs. The malformations were dislocated hip (2), cleft lip and palate (2) and ventricular septal defect.

Layton and Hallesy (1965) produced post–axial forelimb defects in mice and rats by feeding a diet containing 0.6 percent acetazolamide. The defect was specifically localized to the forelimb and the right side was more often involved. Symmetric forelimb defects were produced in the hamster (Layton, 1971). In monkeys, administration of 600 mg per kg for 3 to 6 embryonic days gave negative results (Wilson, 1971). Nakatsuka et al. gave rabbits 50 to 150 mg per kg daily on days 6–18 of gestation and found a dose dependent increase in vertebral malformations. The treatments caused weight decreases in the dams.

Ellison and Maren (1972), in an attempt to understand the mechanism of action of this compound, produced respiratory acidosis, potassium depletion and metabolic acidosis in rats but found no forelimb defect. Maren and Ellison (1972a) did find that replacement of potassium during acetazolamide therapy reduced the teratogenic effect. Maren and Ellison (1972b) also published a study comparing teratogenicity of other related thiadiazoles.

Miller and Scott (1992)using a combination of acetazolamide and amiloride (500 mg per kg and 4 mg per kg respectively) produced abnormalities of the uteric buds in mice on days 9 and 10. Renal agenesis and short or displaced ureters were found.

Ellison, A.C. and Maren, T.H.: The effects of metabolic alterations on teratogenesis. Johns Hopkins Med. J. 130:87–94, 1972.

Heinonen, O.P.; Slone, D. and Shapiro, S.: Birth Defects and Drugs in Pregnancy. Publishing Sciences Group Inc., Littleton, Mass., 1977.

Layton, W.M.: Teratogenic action of acetazolamide in golden hamsters. Teratology 4:95–102, 1971.

Layton, W.M. and Hallesy, D.W.: Deformity of forelimb in rats: Association with high doses of acetazolamide. Science 149:306–308, 1965.

Maren, T.H. and Ellison, A.C.: The effect of potassium on acetazolamide–induced teratogenesis. Johns Hopkins Med. J. 130:105–115, 1972a.

Maren, T.H. and Ellison, A.C.: The teratological effect of certain thiadiazoles related to acetazolamide, with a note on sulfanilamide and thiazide diuretics. Johns Hopkins Med. J. 130:95–104, 1972b.

Miller T.A. and Scott, Jr., W.J.: Abnormalities in ureter and kidney development in mice give acetazolamide–amiloride or dimethadione (dmo) during embryogenesis. Teratology 46:541–550, 1992.

Nakane, Y.; Okuma, T.; Takahashi, R.; Sato, Y.; Wada, T.;Fukushima, Y.; Kumashiro, H.; Omo, T.; Takahashi, T.; Aoki, Y.; Kazamatsuri, H.; Inami, M.; Komai, S.; Seino, M.; Miyakoshi, M.; Tanimura, T.; Hazama, H.; Kawahara, R.; Otsuki, S.; Hosokawa, K.; Inanaga, K.; Nakazawa, Y. and Yamamoto, K.: Multi–institutional study on the teratogenicity and fetal toxicity of antiepileptic drugs: A report of a collaborative study group in Japan. Epilepsia 21:663–680, 1980.

Nakatsuka, T.; Komatsu, T. and Fujii, T.: Axial skeletal malformations induced by acetazolamide in rabbits. Teratology 45:629–636, 1992.

Wilson, J.G.: Use of rhesus monkeys in teratological studies. Fed. Proc. 30:104–109, 1971.

13 Acetoacetate

Zusman et al. (1987) found that mouse ova were inhibited in their development when 5 microgm per ml or more was added to the culture.

Zusman, I.; Yaffe, P. and Ornoy, A.: Effects of metabolic factors in the diabetic state on the in vitro development of preimplantation mouse embryos. Teratology 35:77–85, 1987.

14 Acetohexamide CAS 968–81–0

Bariljak (1965) gave pregnant rats 2000 mg per kg of body weight on days 9 and 10 and caused 46 percent embryonic death but no abnormalities.

Bariljak, I.R.: Comparison of teratogenic action of carbutamide and cyclamide in rat embryos. Pharmakol. Toxicol. (Russian) 5:616–621, 1965.

15 Acetohydroxamic Acid CAS 546–88–3

This organic acid is used to prevent renal calculi. There are no reports dealing with the drug in human pregnancy. Chaube and Murphy (1966) gave 750 to 1000 mg per kg to rats intraperitoneally on the 11th or 12th day and found exencephaly or encepholocele, cleft palate and seletal defects in the offspring. Bailie et al. (1986) gave 25 mg per kg orally to Beagles each day of pregnancy and found defects of the heart, skeletal system and ventral midline.

Bailie, N.C.; Osborne, C.A.; Leininger, J.R.; Fletcher, T.F.; Johnston, S.D.; Ogburn, P.N.; and Griffith, D.P.: Teratogenic effect of acetohydroxamic acid in clinically normal beagles. Am J Vet Res 47:2604–2611, 1986.

Chaube, S. and Murphy, M.L.: The effects of hydroxyurea and related compounds on the rat fetus. Cancer Res. 26:1448–1455, 1966.

16 Acetone Peroxides CAS 1336–17–0

Acetone peroxides are the reaction products formed by the addition of hydrogen peroxide to acetone in a mild acid solution and consist mainly of 2–2–dihydroproperoxy propane and bis(1,1–hydroperoxy–1,1–methyl) diethyl peroxide. Oser and Morgareidge (1967) fed bread containing up to 0.045 percent to rats for 24 months and found no decrease in numbers of offspring or weanlings.

Oser, B.L. and Morgareidge, K.: Safety evaluation of flour treated with acetone peroxides. Food Cosmet. Toxicol. 5:309–319, 1967.

17 Acetonitrile *Dichloroacetonitrile Trichloroacetonitrile Bromochloroacetonitrile* CAS 75–05–8

Willhite (1981) exposed hamsters to 1,800 to 8,000 ppm for 60 minutes during the early primitive streak stage. A concentration–dependent increase in exencephaly, encephaloceles and rib abnormalities was found. Thiocyanate prevented the teratogenicity. Johannsen et al. (1986) gavaged up to 275 mg per kg to rats on days 6–19 and found no teratogenicity but at high levels maternal and fetal toxicity were found.

Smith et al. (1986) studied various halogenated acetonitriles, including the mono, di and trichloride, bromochlorine and dibromide forms. At 55 mg per kg, some reduction in birth weight was observed but detailed autopsies were not performed on the offspring. They felt that increasing halide substitution at the α–carbon increased in utero toxicity. Smith et al. (1987) studied the developmental toxicity of this group of compounds in rat offspring. The di and

trichloronitrile and bromochloronitrile adversely impacted perinatal survival. Acetonitrile at maternally toxic dosages had no adverse postnatal effects.

Johannsen, F.R.; Levinskas, G.J.; Berteau, P.E. and Rodwell, D.E.: Evaluation of the teratogenic potential of three aliphatic nitriles in the rat. Fund. Appl. Toxicol. 7:33–40, 1986.

Smith, M.K.; Zenick, H. and George, E.L.: Reproductive toxicology of disinfection by–products. Environ. Health Perspect. 69:177–182, 1986.

Smith, M.K.; George, E.L.; Zenick, H.; Manson, J.M. and Stober, J.A.: Developmental toxicity of halogenated acetonitriles: Drinking water by–products of chlorine disinfection. Toxicology 46:83–93, 1987.

Willhite, C.C.: Malformations induced by inhalation of acetonitrile vapors in the golden hamster. (abs) Teratology 23:69A only, 1981.

18 Acetoxymethyl–methylnitrosamine CAS 329–83–8

Bochert et al. (1982) reported that this presumed reactive metabolite of dimethylnitrosamine produced digital defects in rat fetuses exposed to 5–16 mg per kg on day 11 of gestation.

Bochert, G.; Platzek, T. and Rahm, U.: Embryotoxicity induced by acetoxymethyl–methylnitrosamine: DNA methylation of embryonic tissues during organogenesis as compared to teratogenic effect in mice. (abs) Teratology 26:14A, 1982.

19 (2-Acetyllactoyloxyethyl) trimethylammonium 1,5-naphthalenedisulfonate

This cholinergic compound was tested in pregnant mice and rabbits by Takai et al. (1979). In mice doses of up to 7500 mg per kg orally before mating, during the first 6 days, during days 6–15 or from day 15 through the 21st day after birth had no effect on the fetus or offspring. In rabbit studies, 1000 mg per kg was given from the 6th through the 18th days without adverse fetal effects.

Takai, A.; Nakada, H.; Nakamura, S.; Inaba, J. and Orikawa, M.: Toxicity test of (2-acetyllactoyloxyethyl)trimethylammonium 1,5-naphthalenedisulfonate (TM 723). Reproductive tests in mice and rabbits. Oyo Yakuri 18:923–942, 1979.

20 2–Acetylaminofluorene

This carcinogen and mutagen was shown to be activated by a liver monooxygenase in an in vitro rat culture. Necrosis of the prosencephalon and open neural tubes were found when the embryos were exposed to 50 microgm per ml of medium (Faustman–Watts et al., 1983). Two metabolites N–hydroxy–2–acetylaminofluorene and N–acetoxy–2–acetylaminofluorene were teratogenic without bioactivation. Faustman–Watts et al. (1985) extended their work on the metabolites of the compound.

Faustman–Watts, E.M.; Greenaway, J.C.; Namkung, M.J.; Fantel, A.G. and Juchau, M.R.: Teratogenicity in vitro of 2–acetylaminofluorene: Role of biotransformation in the rat. Teratology 27:19–28, 1983.

Faustman–Watts, E.M.; Namkung, M.J.; Greenaway, J.C. and Juchau, M.R.: Analysis of metabolites of 2–acetylaminofluorene generated in an embryo culture system. Relationship of biotransformation to teratogenicity in vitro. Biochem. Pharmacol. 34(16):2953–2959, 1985.

21 4–Acetylaminophenylacetic Acid CAS 18699–02–0
MS–932

Toshida et al. (**1990**) gave rats orally up to 3000 mg per kg on days 7–17. At the highest dose some of the dams died and there was fetal growth retardation. At 300 mg per kg there were no fetal effects. Tateda et al. (**1990**) in rabbits gave up to 1000 mg per kg on days 6–18 and found no teratogenicity.

Tateda, C.; Ichikawa, K.; Ono, C.; Takehara, I.; Kiwai, S.; Oketani, Y.; Tanaka, E. and Sumi, N.: Reproductive and developmental toxicity studies of 4–acetylaminophenylacetic acid (ms–932) (iv) teratological study in rabbits by oral administration. Oyo Yakuri 40(3):305–310, 1990.

Toshida, K.; Tanaka, E.; Komatsu, K.; Umeshita, C.; Mizusawa, R. and Sumi, N.: Reproductive and developmental toxicity studies of 4–acetylaminophenylacetic acid (ms–932) (ii) study on oral administration during the period of organogenesis in rats. Oyo Yakuri 40(3):279–291, 1990.

22 Acetylcholine CAS 51–84–3

Beuker and Platner (1956) injected 1.2 to 2.85 mg of this compound into chick eggs during the interval 4 to 12 days of incubation and found no defects.

Beuker, E.D. and Platner, W.S.: Effect of cholinergic drugs on development of chick embryo. Proc. Soc. Exp. Biol. Med. 91:539–543, 1956.

23 1–α–Acetylmethadol

Kennedy et al. (1975) fed rabbits and rats up to 2.0 mg per kg during organogenesis and found no adverse fetal effects.

Kennedy, G.L.; Nuite, J.A.; Smith, S.; Keplinger, M.L.; Braude, M.C. and Calandra, J.C.: Teratogenic potential of methadone and 1–α–acetylmethadol (LAAM) in rats and rabbits. (abs) Toxicol. Appl. Pharmacol. 33:74 only, 1975.

24 N–(N–Acetyl–L–methionyl)–0,0–bis (ethoxy carbonyl) dopamine *TA–870*

Imado et al. (**1991**) gave rats and rabbits up to 2000 mg per kg orally during organogenesis and found no teratogenicity but at the highest dose the fetal weights were reduced. Prenatal and perinatal studies in the rat were done and only growth inhibition at 500 mg per kg or above and delay in vaginal opening were found.

Imado, N.; Asano, Y. and Ariyuki, F.: Reproductive and developmental toxicity studies of N–(N–acetyl–L–methionyl)–0,0–bis (ethoxycarbonyl) dopamine (TA–870) in rats and rabbits. The Clinical Report 25:2171–2179, 1991.

25 3–Acetylpyridine CAS 350–03–8

Landauer (1957) observed poorly developed musculature in 13–day chick embryos after injection of 750 microgm into the egg at 96 hours. Ackermann and Taylor (1948) found "undersized deformed legs" in chick embryos treated with the substance.

Ackermann, W.W. and Taylor, A.: Application of metabolic inhibitor to the developing chicken embryo. Proc. Soc. Exp. Biol. Med. 67:449–452, 1948.

Landauer, W.: Niacin antagonists and chick development. J. Exp. Zool. 136:506–530, 1957.

26 Acetylsalicylic Acid CAS 50–78–2

Discussed under Salicylate

27 Acetylsulfanilamide also see *Sulfonamides* CAS 121–61–9

Bariljak (1968) produced a 30 percent fetal mortality by giving rats 10 microgm per kg on the 10th day of gestation. No abnormalities were produced.

Bariljak, I.R.: Comparison of antithyroidal and teratogenic action of some hypoglycemic sulphanylamides. Prob. Endocrinol. (Russian) 14(6):89–94, 1968.

28 N–Acetyl–L–tryptophan CAS 1218–34–4

This parenteral substitute for l–tryptophan was studied in rats, rabbits and mice by Maruoka et al. (1980). Intraperitoneal doses of up to 600 mg per kg before conception and through day 7 were not associated with adverse effect in the rat. On days 7–17 the same intraperitoneal dose and up to 5.0 gm per kg daily was associated with some decrease in fetal weight but no increased malformations. Behavioral studies were carried out and no differences from controls were found. Intravenous doses in the rabbit on days 6–18 of up to 1000 mg per kg daily produced only fetal growth retardation at the highest dose. No effect was seen at 500 mg per kg.

Maruoka, H.; Kadota, Y.; Uesako, T.; Kakemoto, Y. and Ueshima, M.: Toxicological studies of N–acetyl–L–tryptophan 4–7. Iyakuhin Kenkyu 11:682–742, 1980.

29 Acidosis

Discussed under Acetazolamide and Salicylates

30 Aclacinomycin A CAS 57576–44–0

Kamata et al. (1980) studied this antineoplastic agent in rats and rabbits. Using a dose of 0.2–1.0 mg per kg per day on days 7 through 17 produced no ill effect in the offspring. The same doses on days 17–21 were found to have no effect on the performance of the offspring. Up to 2.0 mg per kg was given to the pregnant rabbit on days 6–18 and no adverse effects were found in the fetuses.

Kamata, K.; Tomizawa, S.; Sato, R. and Kashima, M.: Teratological studies on aclacinomycin A. Oyo Yakuri 19:783–794, 1980.

31 Acquired Immunodeficiency Syndrome *AIDS HIV*

Scott et al. (1984) described 14 patients with AIDS. Eight of them were hospitalized in the first 3 months of life and 5 had symptoms before 1 month of age. Failure to thrive, oral C. albicans and hepto–splenomegaly were the main presenting features. Transplacental or perinatal transmission was suspected. Jovaisas et al. (1985) cultured HTLV–3 virus from a 20 week fetus obtained after induced abortion of an infected mother. Marion et al. (1986) described a dysmorphic syndrome associated with the offspring with AIDS. Growth retardation, microcephaly (70 percent), bossing of the forehead, hypertelorism, flattened nasal bridge, well–formed philtrum, blue sclerae and patulous lips were the features. Qazi et al. (1988) in a study of 30 perinatally exposed New York infants could not identify a characteristic dysmorphism.No specific syndrome was found in a European collaborative study of 600 newborns (European Collaborative Study, **1991**).

Mok et al. (1987) in a preliminary follow–up of 71 infants born to sero–positive mothers found that 5 developed the clinical picture. The average follow–up time was 6 months. Their criteria for AIDS–related complex was interstitial pneumonia, persistent Candida, or parotid swelling for at least 2 months together with 2 or more of the following: Persistent general lymphadenopathy, recurrent bacterial infections, hepatomegaly or splenomegaly, chronic or reported episodes of diarrhea, failure to thrive or progressive encephalopathy. Krivine et al. (**1992**) measured the blood of 50 offspring from HIV–positive women with polymerase chain reaction, culture and p24 antigen measurements. At birth they failed to find evidence of infection in 70% of infants subsequently turning positive, a finding which supported transmission at the end of pregnancy or at delivery.

Falloon et al. (1988) summarized information on transmission from mother to infant. About 50 percent of the offspring of mothers with AIDS are infected. Caesarian section does not prevent infection. The virus was cultured from mid–term fetuses. Breast milk may also be a source of infection. The European collaborative study (**1991**) reported a vertical transmission rate of 13% somwehat lower than in other studies. Goedert et al. (**1991**) studied 22 twins from infected mothers and found most of the transmission was to the first born twin suggesting exposure to the cervix and birth canal was the main source of infection. Thirty–eight percent of the first twins were infected compared to 19% of the second born twins.

Using immunocytochemistry and in–situ hybridisation HIV–1 virus was identified in the decidual leucocytes, villous trophoblast, villous mesenchymal and embryonic blood cells from infected women at 8 weeks of gestation (Lewis et al. 1990).

European Collaborative Study: Children born to women with HIV-1 infection: natural history and risk of transmission. The Lancet 337:253–260, 1991.

Falloon, J.; Eddy, J.; Roper, M. and Pizzo, P.A.: AIDS in pediatric population. In: AIDS: Etiology, Diagnosis, Treatment Prevention, 2nd edition. J.B. Lippincott Comp., Philadelphia, 339–351, 1988.

Goedert, J.J.; Duliege, A.; Amos, C.I.; Felton, S.; Biggar, R.J. and International Registry of HIV–Exposed Twins: High risk of hiv–1 infection for first–born twins. The Lancet 338:1471–1475, 1991.

Jovaisas, E.; Koch, M.A.; Schafer, A.; Stauber, M. and Lowenthal, D.: LAVHTLV in a 20 week fetus. Lancet 2:1129, 1985.

Krivine, A.; Firtion, G.; Cao, L.; Francoual, C.; Henrion, R. and Lebon, P.: HIV replication during the first weeks of life. The Lancet 339:1187–1189, 1992.

Lewis, S.H.; Reynolds–Kohler, C.; Fox, H.E. and Nelson. J.A.: HIV–1 in trophoblastic and villous hofbauer cells, and haematological precursors in eight–week fetuses. The Lancet 335:565–568, 1990.

Marion, R.; Wiznia, A.; Hucheon, R.G. and Rubinstein, A.: AIDS embryopathy: A new dysmorphic syndrome in children with acquired immune deficiency syndrome (AIDS). (abs) Pediatr. Res. 20:339A, 1986.

Mok, J.Q.; De Rossi, A.; Ades, A.E.; Giaquinto, C.; Grosch–Worner, C. and Peckham, C.S.: Infants born to mothers seropositive for human immunodeficiency virus. Lancet 1:1164–1168, 1987.

Qazi, Q.H.; Sheikh, T.M.; Fikrig, S. and Menikoff, H.: Lack of evidence for craniofacial dysmorphism in perinatal human immunodeficiency virus infection. J. Pediatr. 112:7–11, 1988.

Scott, G.B.; Buck, B.E.; Leterman, J.G.; Bloom, F.L. and Parks, W.P.: Acquired immunodeficiency syndrome in infants. N. Eng. J. Med. 310:76–81, 1984.

32 Acriflavine *Trypaflavine* CAS 8048–52–0

Ancel (1946) used this local antiseptic dropped onto the chick embryo at 26 hours in a dose of 0.02 mg and found about 30 percent spina bifida. At 48 hours of incubation the development of the amnion was arrested by similar treatment.

Ancel, P.: Recherche experimentale sur le spina bifida. Arch. Anat. Microsc. Morphol. Exp. 36:45–68, 1946.

33 Acrolein CAS 107–02–8

This compound is one of the metabolites of cyclophosphamide. Mirkes et al. (1981) added this compound to an in vitro rat embryo culture at 5 microgm per ml (equimolar to a teratogenic dose of cyclophosphamide) and found no growth retardation or increase in defects. At twice the dose the compound was lethal. Schmid et al. (1981) found growth retardation but no structural defects at l00 and l50 microM concentrations when rat embryos were exposed in vitro. Parent et al. (1993) gave up to 2.0 mg per kg per day during rabbit organogenesis and found no teratogenicity at levels toxic to the does. Hales (1982) injected the compound intra–amniotically and at doses of l microgm or 0.0l8 micromoles per rat fetus found 85 percent defects which were similar to those produced by cyclophosphamide. Acrolein binds very strongly to serum proteins and this causes difficulty in assessing, either in vitro or in vivo, whether it is the final active metabolite in cyclophosphamide teratogenicity. Slott and Hales (1985) analyzed 5 structurally related compounds (acrylic acid, allyl alcohol, glycidol, glyceraldehyde and propionaldehyde) by intra–amniotic injection into rat embryos and concluded that acrolein itself was much more teratogenic.

Hales, B.F.: Comparison of the mutagenicity and teratogenicity of cyclophosphamide and its active metabolites, 4–hydroxycyclophosphamide, phosphoramide mustard and acrolein. Cancer Res. 42:3016–3021, 1982.

Mirkes, P.E.; Fantel, A.G.; Greenaway, J.C. and Shepard, T.H.: Teratogenicity of cyclophosphamide metabolites: phosphoramide mustard, acrolein and 4–ketocyclophosphamide in rat embryos cultured in vivo. Toxicol. Appl. Pharmacol. 58:322–330, 1981.

Parent, R.A.; Caravello, H.E.; Christian, M.S. and Hoberman, A.M.: Developmental toxicity of acrolein in New Zealand white rabbits. Fund. Appl. Toxicol. 20:248–256, 1993.

Schmid, B.P.; Goulding, E.; Kitchin, K.T. and Sanyal, M.K.: Assessment of the teratogenic potential of acrolein and cyclophosphamide in a rat embryo culture system. Toxicol. 22:235–243, 1981.

Slott, V.L. and Hales, B.F.: Teratogenicity and embryolethality of acrolein and structurally related compounds. Teratology 32:65–72, 1985.

34 Acrylamide

This widely used industrial chemical is known to produce adult toxicity as evidenced by peripheral neuropathy. Field et al (1990) gavaged mice and rats on days 6–17 and 6–20 respectively. Embryotoxicity was not observed in rats at doses of up to 15 mg per kg but the mouse fetuses were reduced in weight at 7.5 and 15 mg per kg. No evidence of teratogenicity was found. Postnatal studies of reproduction and neurobehavioral studies were performed by others and reviewed by Field et al. (1990). Wise et al. (**1995**) gave rats up to 20 mg per kg per day on days 6 to lactational day 10. Neurotoxicity was found starting at 5 mg per kg a dose level that produced maternal toxicity.

Field, E.A.; Price, C.J.; Sleet, R.B.; Marr, M.C.; Schwetz, B.A. and Morrissey, R.E.: Developmental toxicity evaluation of acrylamide in rats and mice. Fundam Appl Toxicol. 14:502–512, 1990.

Wise, L.D.; Gordon, L.R.; Soper, K.A.; Duchai, D.M. and Morrissey, R.E.: Developmental neurotoxicity evaluation of acrylamide in Sprague–Dawley rats. Neurotoxicology and Teratolocy 17:189–198, 1995.

35 Acrylic Acid CAS 79–10–7

Discussed under Methacrylate Esters

36 Acrylonitrile CAS 107–13–1

Murray et al. (1983) administered this compound found in plastics and fibers to rats on days 6 through 15 of gestation at doses of 25 mg per kg by mouth. Some maternal toxicity occurred and trunk and aortic anomalies were produced. Inhalation of 80 ppm produced teratogenic effects. Inhalation of 40 ppm or oral doses of 20 mg per kg caused no fetal effects. Willhite et al. (1981) produced encephaloceles and exencephaly in hamster fetuses by giving 80 mg per

kg intraperitoneally on day 8. Mehrota et al. (**1988**) found no postnatal ill effects in rats given 5mg per kg per day.

Murray, F.J.; Nitschke, K.D.; John, J.A.; Crawford, A.A.; McBride, J.G. and Schwetz, B.A.: Teratogenic potential of acrylonitrile given to rats by gavage or inhalation. Food Cosmet. Toxicol. 16:547–551, 1983.

Mehrota, J.; Khannal, V.K.; Hussain, R. and Seth, P.K.: Biochemical and developmental effects in rats following in utero exposure to acrylonitrile: a preliminary report. Industrial Health 26:251–25, 1988.

Willhite, C.C.; Ferm, V.H. and Smith, R.P.: Teratogenic effects of oliphatic nitriles. Teratology 23:317–323, 1981.

37 ACTH *Adrenocorticotrophic Hormone* CAS 9002–60–2

Cleft palate was produced in mice given 5 mg ACTH every 6 hours for 2 to 3 days beginning in the 13th day of gestation (Fraser et al., 1954). The subject is well summarized by Kalter and Warkany (1959).

Kotin et al. (1978) injected pregnant rats with 4 and 10 units of hormone from days 11 through 15 and observed alterations of DNA and protein biosynthesis in liver, brain and placenta of 15–day fetuses.

Fraser, F.C.; Kalter, H.; Walker, B.E. and Fainstat, T.D.: Experimental production of cleft palate with cortisone and other hormones. J. Cell Comp. Physiol. 43(1):237–259, 1954.

Kalter, H. and Warkany, J.: Experimental production of congenital malformations in mammals by metabolic procedure. Physiol. Rev. 39:69–115, 1959.

Kotin, A.M.; Chebotar, N.A. and Tishchenko, L.I.: Effect of adrenocorticotropic hormone on the synthesis and content of nucleic acids and proteins in the brain of rat embryos during early embryogenesis. Ontogenez (USSR) 9.1:70–77, 1978.

38 Actifed™

Discussed under Triprolidine HCl and Pseudoephedrine

39 Actinobolin Sulfate CAS 31327–55–6

Karnofsky and Lacon (1962) listed this substance as producing facial coloboma and cleft palate in the chick exposed to 5 to 10 mg. Edema and webbing of the toes also occurred.

Karnofsky, D.A. and Lacon, C.R.: Survey of cancer chemotherapy service center compounds for teratogenic effect in the chick embryo. Cancer Res. 22:84–86, 1962.

40 Actinomycin C *Cactinomycin* CAS 8052–16–2

Exencephaly, spina bifida and other defects were produced in rats injected on the 6th and 10th days of gestation (Takaya, 1963). The amount injected was not stated.

Takaya, M.: Teratogenic effect of the antitumor antibiotics. (abs) Proceedings of the Congenital Anomalies Research Association of Japan 3:47–48, 1963.

41 Actinomycin D CAS 50–76–0

Tuchmann–Duplessis and Mercier–Parot (1960), using 25 to 100 microgm per kg daily before the 10th day, found in the rat various degrees of craniorachischisis, other defects of the nervous system, and branchial arch malformations. The same authors (1960) found agenesis of the optic nerve, encephalocele and lumbosacral spina bifida in rabbits treated with 150 or 200 microgm per day during the 8th to 10th days of gestation. Wilson (1966) confirmed the interesting finding that after the 10th day defects cannot be produced in the rat. Dyban and Akimova (1967) showed that 50 to 250 microgm per kg had no effect given during the first 4 days, but given on the 9th day, the embryonic ectoderm and decidual cells were affected but not the yolk sac endoderm. Baranov et al. (1968) demonstrated radioautographically that the compound is localized in the embryo. Pierro (1961) produced rumplessness in chicks with 0.06 microgm injections at 48 hours of incubation.

Baranov, V.S.; Weisman, B.L.; Nikitina, S.S.; Repina, G.V. and Repina, V.S.: Differential inhibition of serum protein synthesis in mammalian embryos by actinomycin D in various stages of liver development. Biochemistry (Russian) 33(6):1174–1182, 1968.

Dyban, A.P. and Akimova, I.M.: Some peculiarities of embryogenesis; disturbances caused by blockade of RNA synthesis in mammals. Experiments with actinomycin D in rats. Arch. Anat. (Russian) 52(5):36–50, 1967.

Pierro, L.J.: Teratogenic action of Actinomycin D in the embryonic chick. J. Exp. Zool. 147:203–210, 1961.

Tuchmann–Duplessis, H. and Mercier–Parot, L.: The teratogenic action of the antibiotic Actinomycin D. In: Ciba Foundation Symposium on Congenital Malformations, G.E.W. Wolstenholme and C.M. O'Connor (eds). Boston: Little Brown and Co., 115–128, 1960.

Tuchmann–Duplessis, H. and Mercier–Parot, L.: Influence de l'actinomycine D sur la gestation et le developpement foetal du lapin. C. R. Soc. Biol. (Paris) 154:914–916, 1960.

Wilson, J.G.: Effects of acute and chronic treatment with Actinomycin D on pregnancy and the fetus in the rat. Harper Hospital Bulletin 24:109–118, 1966.

42 Acyclovir CAS 59277–89–3 *Zovirax*

Andrews et al. (1988) have described an ongoing epidemiology study and found among 49 women exposed in the first trimester and studied prospectively only one with a defect and this defect was attributed to an inherited familial disease. In an international study sponsored by Burroughs Wellcome, 466 women exposed in the first trimester to oral or intravenous therapy there were 14 infants with congenital defects. No specific syndrome was found and the types of defects included eye (1) hip displasia (1), limb deformity (1), cleft palate (1), pyloric stenosis (3) and brain defect (2). One infant had cardiac problems, hydrocephalus and calcified foci in the CNS. This prospective study is managed by R.R. Eldridge of Burroughs Wellcome Co. (800–722–9292, ext. 8465).

This antiviral drug was tested in a two–generational study of mice (450 mg per kg per day) and a study during organogenesis of rabbits and rats (to 50 mg per kg per day). No adverse effects were found (Moore et al., 1983). Postnatal

studies in rats were done using dosages up to 80 mg per kg at which dose minimal renal lesions were found. Klug et al. (1985) found in whole embryo culture that at 50 microM concentrations and above, embryonic differentiation was disturbed. Growth parameters were altered at 25 but not at 10 microM. Twenty–five microM is equivalent to 5.6 microgm per ml.

Stahlmann et al. (1987) reported that subcutaneous doses of 200–300 mg on day 10 produced malformations in rats of the tail, eye, ear and mandible. No adverse effects were found with two 25 mg per kg doses daily on days 6–15. Further details were reported by Stahlmann et al. (1988).

Andrews, E.B.; Tilson, H.H.; Hurn, B.A.L. and Cordero, J.F.: Acyclovir in pregnancy registry an observational epidemiologic approach. Am J Medicine 85:123–128, 1988.

Klug, S.; Lewandowski, C.; Blankenburg, G.; Merker, H–J. and Neubert, D.: Effect of acyclovir on mammalian embryonic development in culture. Arch. Toxicol. 58:89–96, 1985.

Moore, H.L.; Szczech, G.M.; Rodwell, D.E.; Kapp, R.W.; De Miranda, P. and Tucker, W.E.: Preclinical toxicology studies with acyclovir. Teratology, reproductive and neonatal tests. Fund. Appl. Toxicol. 3:560–568, 1983.

Stahlmann, R.; Klug, S.; Lewandowski, C.; Chahoud, I.; Bochert, G.; Merker, H–J. and Neubert, D.: Teratogenicity of acyclovir in rats. Infection 15(4):261–262, 1987.

Stahlmann, R.; Klug, S.; Lewandowski, C.; Bochert, G.; Chahoud, I.; Rahm, U.; Merker, H–J. and Neubert, D.: Prenatal toxicity of acyclovir in rats. Arch Toxicol. 61(6):468-479, 1988.

43 Adenine also see 2–*Deoxyadenosine* CAS 73–24–5

Fujii and Nishimura (1970) gave 250 and 300 mg of adenine per kg to mice on the 12th gestational day and produced cleft palate, digital defects and subcutaneous hematomas in 13 to 20 percent of the fetuses.

Fujii, T. and Nishimura, H.: Teratogenic action of adenine on mouse embryos. Jpn. J. Pharmacol. 20:445–447, 1970.

44 Adenosine CAS 58–61–7

Discussed under 2–Deoxyadenosine

45 Adenosine–5–Monophosphate CAS 61–19–8

Hashimoto et al. (1970) gave mice and rats up to 400 mg per kg intraperitoneally on days 7–13 and produced only slight growth inhibition at the highest dose.

Hashimoto, Y.; Nobuo, T. and Nomura, M.: Studies on teratogenic action of adenosine–5'–monophosphate in mice and rats. Oyo Yakuri 4:625–633, 1970.

46 Adenosine Triphosphate CAS 56–65–5

Gordon et al. (1963) studied the effect of adenosine triphosphate and its decomposition products on cortisone–induced defects in the mouse fetus. Adenosine monophosphate (29 microgm) increased the number of fetuses with hydrocephalus, spinal and eye deformity but adenosine triphosphate decreased the incidence. The mice received

2.5 mg of cortisone acetate along with these substances from days 11–14 of gestation. Hashimoto et al. (1970) injected this compound intraperitoneally into pregnant rats and mice on days 7–13 of gestation. Doses of 400 mg per kg per day caused slight fetal growth retardation but no teratogenicity.

Gordon, H.W.; Traczyk, W.; Peer, L.A. and Bernhard, W.G.: The effect of adenosine triphosphate and its decomposition products on cortisone induced teratology. J. Embryol. Exp. Morphol. 11:475–482, 1963.

Hashimoto, Y.; Toshioka, N. and Nomura, M.: Studies on teratogenic action of adenosine–5–monophosphate in mice and rats. Oyo Yakuri 4:625–633, 1970.

47 Adhesives, Spray

Discussed under Spray Adhesives

48 Adiponitrile

Johannsen et al. (1986) gavaged up to 80 mg per kg to rats on days 6–19 and found no teratogenicity. At high levels, however, maternal and fetal toxicity were found.

Johannsen, F.R.; Levinskas, G.J.; Berteau, P.E. and Rodwell, D.E.: Evaluation of the teratogenic potential of three aliphatic nitriles in the rat. Fund. Appl. Toxicol. 7:33–40, 1986.

49 Adriamycin CAS 23214–92–8 *Doxorubicin™*

Oguro et al. (1973) gave this antineoplastic medication intraperitoneally to pregnant rats and intravenously to mice on days 7 through 13 in amounts up to 1,000 microgm per day and observed no teratogenicity. Using doses of 2 mg per kg intraperitoneally and longer treatment, Thompson et al. (1978) produced atresia of the esophagus and intestine and cardiovascular anomalies in rat fetuses. Intravenous treatment of the rabbit with up to 0.6 mg per kg on days 6–18 produced no fetal defects but abortions occurred. Kavlock et al. (1986) produced postnatal renal damage in rats exposed to 1.0 or 1.5 mg during days 6–9 or 10–12. Gililland and Weinstein(1983) reported 6 pregnancies where treatment was after the 1st trimes Murray et al. (1984) in a case report found an imperorate anus and rectovaginal fistula in a fetus exposed to intravenous doxorubicin (325 mg) and oral cyclophosphamide during the first trimester. premature.

Gililland, J. and Weinstein, L.: The effects of cancer chemotherapeutic agents on the developing fetus. Obstet Gynecol Surv 38(1):6–13, 1983.

Kavlock, R.J.; Rogers, E.H. and Rehnberg, B.F.: Renal functional teratogenesis resulting from adriamycin exposure.

Murray, C.L.; Reichert, J.A.; Anderson, J. and Twiggs, L.B.: Multimodal cancer therapy for breast cancer in the first trimester of pregnancy. JAMA 252:2607–2608, 1984. Teratology 33:213–220, 1986.

Oguro, T.; Hatano, M.; Imamura, T. and Shimizu, M.: Toxicological studies on adriamycin–HCl. 4. Teratological study. (Japanese) Yakubutsu Ryoho (Medical Treatment) 6:1152–1164, 1973.

Thompson, D.J.; Molello, J.A.; Strebing, R.J. and Dyke, I.L.: Teratogenicity of adriamyin and daunomycin in the rat and rabbit. Teratology 17:151–158, 1978.

50 Aflatoxin B$_1$ CAS 1402–68–2

Aflatoxins are products of aspergillus flavus and were shown to be carcinogenic as well as teratogenic. Elis and Dipaolo (1967) injected 4 mg per kg intraperitoneally into hamsters on the 8th day of pregnancy. DeVries found that 54% of 125 pregnant women in Kenya had aflatoxin in their blood and those with levels had infants with reduced birth rates (by 255g). It was also found in 37% of the cord bloods. No increase in defects was mentioned, but two with aflatoxin levels had stillbirths.Early embryos were malformed in about 30 percent, whereas surviving fetuses showed a defect rate of 6 percent. Neural tube closure defects, microcephaly, umbilical hernia and cleft palate were reported. Fetal liver cell necrosis was noted. Premixing the aflatoxin B$_1$ with DNA reduced the teratogenic effect. Aflatoxin feeding at 10 ppm during the last half of gestation produced nodular hyperplasia of the liver in four and cholangiocarcinoma in one of 79 rat offspring (Grice et al., 1973).

Yamamoto et al. (1981) treated mice intraperitoneally with 16 or 32 mg per kg for various two day consecutive periods during organogenesis. Cleft palate, open eyelid, wavy ribs and bent long bones were found after the 32 mg dose. Geissler and Faustman (1988) studied rat embryos grown in vitro in concentrations of 0–150 mM solutions. At 10 and 20 mM concentrations, neural tube defects occurred. S9 fractions from rats treated with phenobarbitol or 3–methylcholanthrene did not enhance the effect.

DeVries, H.R.; Maxwell, S.M. and Hendrickse, R.G.: Foetal and neonatal exposure to aflatoxins. Acta Paediatr Scand 78:373–378, 1989.

Elis, J. and Dipaolo, J.A.: Aflatoxin B$_1$ induction of malformations. Arch. Pathol. 83:53–57, 1967.

Geissler, F. and Faustman, E.M.: Developmental toxicity of aflatoxin B$_1$ in the rodent embryo in vitro: Contribution of exogenous biotransformation systems to toxicity. Teratology 37:101–111, 1988.

Grice, H.C.; Moodie, C.A. and Smith, D.C.: The carcinogenic potential of aflatoxin or its metabolites in rats from dams fed aflatoxin pre– and post–partum. Cancer Res. 33:262–268, 1973.

Yamamoto, Y.; Kihara, Y. and Tanimura, T.: Effects of aflatoxin B$_1$ on teratogenicity of mice. (abs) Teratology 24:25A, 1981.

51 Afridol Blue CAS 3627–06–3

This azo dye produced malformations in rats (review by Beck and Lloyd, 1966). Also see the heading, Trypan Blue.

Beck, F. and Lloyd, J.B.: The teratogenic effects of azo dyes. In: Advances in Teratology, D.H.M. Woollam (ed.), New York, Logos and Academic Press, 1:131–193, 1966.

52 Agent Orange also see *2,4-Dichlorophenoxyacetic Acid;* also see *2,4,5-Trichlorophenoxyacetic Acid (2,4,5–T);*

also see *Tetrachlorodibenzo–p–dioxin; Dioxins; TCDD* CAS 39277–47–9

Agent Orange was composed of about equal parts of the n–butyl esters of 2,4 dichloroacetic acid (2,4–D) and 2,4,5–trichlorophenoxyacetic acid (2,4,5–T) and was used widely in agriculture and forestry in the United States before the Vietnam War. Between 1962 and 1971, Agent Orange was sprayed over about one–tenth of the Vietnamese countryside. 2,4,5–T was regularly contaminated with trace amounts of highly toxic dioxins, the most prevalent of which was 2,3,7,8–tetrachlorodibenzo–p–doxin (TCDD).

In this section of the Catalog, selected epidemiologic literature of humans exposed to mixtures of the above components will be summarized. Animal studies, except for those dealing with mixtures, will be covered elsewhere under the specific chemical.

Nelson et al. (1979) found no correlation between agricultural use of 2,4,5–T in Arkansas and facial clefts observed over a six to seven year period. Field and Kerr (1979) found a positive correlation between yearly neural tube defects and 2,4,5–T usage in Australia. In a report which never appeared in a peer reviewed journal in the U.S., the Environmental Protection Agency (1979) found a correlation between abortion frequency and 2,4,5–T spraying. The U.S. Department of Defense (1970) studied births among nearly one–half million South Vietnamese during the Vietnam War and found no increase in defects, stillbirths or hydatidiform moles during the period of large scale military use of herbicides.

Thomas and Czeizel (1982) found no association between yearly use of 2,4,5–T in Hungary and stillbirths or births of children with neural tube defects, oral clefts or cystic renal disease. Hanify et al. (1981) studied the malformations in Northland, New Zealand over 17 years and found a correlation with aerial 2,4,5–T applications. They did not find an association with neural tube defects and cleft palate. Smith et al. (1982) studied three groups of male chemical applicators and found no increase in malformations among 486 pregnancies in which the father sprayed 2,4,5–T. The two control groups were composed of workers who did not spray 2,4,5–T during the previous year and workers who used other types of pesticides. A very useful summary of the above reports and others was given in tabular form by Friedman (1984). Summaries of the reports following the Seveso industrial accident with TCDD are reported under tetrachlorodibenzo–p–dioxin.

Comprehensive case–control studies of Vietnam veterans' offspring by Donovan et al. (1984) and Erickson et al. (1984) were reported. Donovan et al. (1984) matched 8,517 Australian infants born with defects with normal infants born to mothers of similar age. The father was then identified as to whether he was a Vietnam veteran. One hundred and twenty–seven fathers of defective infants were veterans as compared to 123 of the controls. They concluded that there was no evidence of increased risk.

Erickson et al. (1984) interviewed approximately 5,000 families of infants with major defects in the Atlanta area. Controls were matched with respect to race, year of birth and hospital of birth. A multiple interviewer technique was used to decrease the opportunity for interviewer to recognize case from control. An exposure opportunity index for Agent Orange was made by the armed service for each Vietnam veteran. There was no evidence that Vietnam veterans fathered more babies with defects. The Agent Orange index of exposure was not associated with an increased risk although in a few defect categories there was an increased risk which could have been by chance according to the author's detailed analysis. The odds ratio for spina bifida among all veterans and Vietnam veterans was 1.25 and 1.05. The ratios for anencephaly which is believed to be associated with spina bifida, were significantly lower, namely 0.99 and 0.89. Additionally, the odds ratio for other neoplasms was 1.80 among Vietnam veterans' offspring. In the total group of veterans the odds ratio was 0.83. The group of other neoplasms included dermoid and epidermoid cysts (26 cases), benign tumors (24 cases) and a variety of malignant tumors (41 cases). The Vietnam veteran's odds ratio for Down's Syndrome was 0.95. Both of these case control studies are available in greater detail by request to the authors.

Sterling and Arundel (1986) attempted to analyze Vietnamese studies. Odds ratios for hydatidiform moles were reported to be 4.6 and 12 in two separate studies. The odds ratio for congenital anomalies was 2.5, but the epidemiologic resources were felt to be limited.

Lamb et al. (1981) formulated Agent Orange from its constituents and gave it orally to mice for 8 weeks. The males were subsequently mated to untreated females. No evidence of dominant lethality was found and the embryos had no excess malformations.

Donovan, J.W.; Maclennan, R. and Adena, M.: Vietnam service and the risk of congenital anomalies a case–control study. Med. J. Australia 140:394–397, 1984.

Erickson, J.D.; Mulinare, J.; McClain, P.W.; Fitch, T.G.; James, L.M.; McClearn, A.B. and Adams, M.J.: Vietnam veterans' risk for fathering babies with birth defects. JAMA 252:903–912, 1984.

Field, B. and Kerr, C.: Herbicide use and incidence of neural–tube defects. Lancet 1:1341–1342, 1979.

Friedman, J.M.: Does Agent Orange cause birth defects? Teratology 29:193–221, 1984.

Hanify, J.A.; Metcalf, P.; Nobbs, C.L. and Worsley, K.J.: Aerial spraying of 2,4,5–T and human birth malformations. An epidemiologic investigation. Science 212:349–351, 1981.

Lamb, J.C.; Moore, J.A.; Marks, T.A. and Haseman, J.K.: Development and viability of offspring of male mice treated with chlorinated phenoxy acids and 2,3,7,8–tetrachlorodibenzo–p–dioxin. J. Toxicol. Environ. Health 8:835–844, 1981.

Nelson, C.J.; Holson, J.F.; Green, H.G. and Gaylor, D.W.: Retrospective study of the relationships between agricultural use of 2,4,5–T and cleft palate occurrence in Arkansas. Teratology 19:377–384, 1979.

Smith, A.H.; Fisher, D.O.; Pearce, N. and Chapman, C.J.: Congenital defects and miscarriages among New Zealand 2,4,5–T sprayers. Arch. Environ. Health 37:197–200, 1982.

Sterling, T.D. and Arundel, A.: Review of recent Vietnamese studies on the carcinogenic and teratogenic effects of phenoxy herbicide exposure. Int. J. Health Services 16:265–278, 1986.

Thomas, H.F. and Czeizel, A.: Safe as 2,4,5–T. Nature 295:276 only, 1982.

U.S. Department of Defense: Congenital Malformations, Hydatidiform moles and stillbirths and the Republic of Vietnam 1960–1969. U.S. Government Printing Office (No. 903–233), Washington, D.C., 1970.

U.S. Environmental Protection Agency: Preliminary report of assessment of a field investigation six–year spontaneous abortion rates in three Oregon areas in relation to forest 2,4,5–T spray practices. Washington, D.C., 1979.

53 AH–8165

This muscle relaxant was shown not to pass the placental barrier significantly in women undergoing termination of pregnancy by hysterotomy (Blogg and Simpson, 1973).

Blogg, C.E. and Simpson, B.R.: Placental transfer of AH–8165. Br. J. Anaesth. 45:438–439, 1973.

54 Akabane Virus

This virus has been associated with increased rates of abortion, stillbirth and congenital malformations in cattle, sheep and goats (Kurogi et al., 1977). The major amount of study has been done in Japan. Kurogi et al. (1977) isolated the virus and inoculated pregnant cattle intravenously. The two fetuses examined had myositis and some of the full term animals had the typical syndrome with cerebral defects, hydroanencephaly and arthrogryposis.

Kurogi, H.; Inaba, Y.; Takahashi, E.; Sato, K.; Satoda, K.; Goto, Y.; Omori, T. and Matumoto, M.: Congenital abnormalities in newborn calves after inoculation of pregnant cows with akabane virus. Infection and Immunity 17:338–343, 1977.

55 Alazopeptin A CAS 1397–84–8

Thiersch (1958) injected this peptide derived from streptomyces into rats on the 7th and 8th or 11th and 12th days (0.3 to 0.5 mg per kg) and produced a high litter resorption rate but no defects.

Thiersch, J.B.: Effect of alazopeptin (A) on litter and fetus of the rat in utero. Proc. Soc. Exp. Biol. Med. 97:888–889, 1958.

56 Albendazole *Methyl*
(5–propylthio)–1H–benzmidazol–2–yl) carbamate CAS 54965–21–8

Delatour et al. (1981) fed this antihelminthic to rats and found teratogenicity at 8.8 mg per kg. Treatment was from days 8–15. In some experiments, they used the liver of cattle which had been fed the compound. A number of metabolites of the drug were studied and methyl(5–(propylsulfinyl)–1H–benzimidazol) carbamate was teratogenic.

Delatour, P.; Parish, R.C. and Gyurik, R.J.: Albendazole: A comparison of relay embryotoxicity with embryotoxicity of individual products. Ann. Rech. Vet. 12:159–167, 1981.

57 Alcide™

Skowronski et al. (1985) gavaged rats and mice on days 6–15 with 1 ml or 0.1 ml and found no adverse reproductive effects.

Skowronski, G.A.; Abdel–Rahman, M.S.; Gerges, S.E. and Klein, K.M.: Teratologic evaluation of Alcide™ liquid in rats and mice. J. Appl. Toxicol. 5:97–103, 1985.

58 Alclofenae *4–Allyloxy–3–chlorophenylacetic Acid Mervan™*

Lambelin et al. (**1970**) gave rats 150 mg or 100 mg orally or intraperitoneally respectively on days 6–15 and produced no increase in defects. The fetal weights in both groups were decreased.

Lambelin, G.; Roba, J.; Gillet, C.; Gautier, M. and Buu–Hoi, N.P.: Toxicity studies of 4–allyloxy–3–chlorophenylacetic acid, a new analgesic, antipyretic, and anti–inflammatory agent. Arzneim–Forsch. 20:618–630, 1970.

59 Alcohol *Alcoholism Ethanol Fetal Alcohol Syndrome*
CAS 64–17–5

A fetal alcohol or alcoholism syndrome has been delineated in the human. Jones et al. (1973) described 8 children from mothers with chronic alcoholism. The children were characterized by small birth weight, microcephaly, reduction in width of the palpebral fissures and maxillary hypoplasia. Five had cardiac anomalies. The facial features of 21 children with fetal alcohol syndrome were compared with controls and 6 of 7 experts were able to differentiate them accurately by using AP and lateral photographs (Clarren et al., 1987). Streissguth et al (**1991**) have described 61 adolescents and adults with fetal alcoholism syndrome and point out that although mental deficit remains, the facial changes are less distinctive. Detailed psychologic studies at age 14 years found fluctuating attention states, problems with response inhibition and spatial learning to be correlated with maternal drinking (Streissguth et al, **1994** A,B). Spohr et al. (**1993**) in a 10 year follow–up of 60 patients with fetal alcohol syndrome found that the growth and mental retardation persisted.

In a subsequent report, Jones and Smith (1973) estimated that about one–third of the offspring of chronic alcoholic mothers had the syndrome. Two of their children had cleft palates. All the infants had some form of developmental delay (Jones et al., 1974). One autopsy report showing malorientation of the brain appeared (Clarren et al., 1978). Lemoine et al. (1968) in a previously obscure paper described 127 children from alcoholic mothers and found a very similar clinical picture with a 25 percent malformation rate which included, in particular, cleft palate and cardiac anomalies. Other reports of the syndrome appeared (Palmer et al., 1975; Ferrier et al., 1973; Manzke and Grosse, 1975; Majewski, 1981). Little and Sing (1987) reported a decrease of 137 gm in birth weight among offspring whose fathers drank regularly during the month of conception. The lowered weights could not be explained by 21 other measured variables. Vertebral defects have been described in children with the fetal alcohol syndrome (de Cornulier et al., 1991). Stromland and Pinazo–Duran (1994) reported optic nerve hypoplasia in children with FAS and in treated rats.

In a study of the frequency of the syndrome made in a group of 1,500 infants from middle class families, 11 newborns were judged without the examiner's knowledge to have signs compatible with the prenatal effect of alcohol (Hanson et al., 1978). Two of these had fetal alcohol syndrome and were offspring of heavy drinkers, and 9 were from the high risk group who took over two ounces of ethanol per day. Behavior studies of the at–risk newborns showed increased tremors, decreased vigorous activity and increased non–alert wake states (Landesman–Dwyer et al., 1978). In a large prospective study, Mills et al. (1984) found that one to two drinks per day was associated with a substantial increased risk for producing a growth–retarded infant. Their results were drawn after factoring out 15 other variables. The cardiac defects associated with the syndrome were detailed (Sandor et al., 1981).

Roman et al. (1988) reviewed the human data and raised the question of a dosage response between alcohol consumption and intelligence. A review of the clinical and animal work (Streissguth et al., 1980) indicates that the fetal alcohol syndrome may be the most commonly recognized cause of mental retardation. These authors felt that adverse fetal effects are not as likely when the mother drinks less than two drinks per day (1980). Streissguth et al. (1984) summarized their prospective study of 1529 pregnancies and found evidence for a dose–response effect of alcohol on adverse behavior; they did not find an increase in clinical fetal alcohol syndrome below two ounces of alcohol consumption per day. Streissguth et al. (1986) detailed these studies in a group of seven–year–olds. In a subsequent study of four–year–olds, Streissguth et al. (1989) found a significant drop in IQ of the offspring of mothers who drank 1.5 ounces (approximately 3 drinks) a day during pregnancy. The drop was 5 IQ points after adjustment for paternal and maternal education and numerous other variables. Day et al. (1990) interviewed 461 mothers during pregnancy and at 8 months of age found a significant reduction in growth (including head circumference) associated with women who took one or more drinks daily.

Mills and Graubard (1987) did not find an increase in defect rate among 793 mothers who drank 1–2 drinks per day. Harlap and Shiono (1980), Kline et al. (1980) and Streissguth et al. (1981) published data suggesting that even moderate drinking may be associated with increased spontaneous abortion in the first and/or second trimester. Pauli and Feldman (1986) reported two skeletal defects occurring in newborns exposed to ethanol and they reviewed other such cases. Stromland (1985) described eye defects in children exposed to alcoholism. Tikkanen and Heinonen (1991) compared 573 cases with congenital heart disease to 1,055 controls and found that 14.1% of the defect group had mothers who took 2–3 drinks per day while in the control there were 6.9% (P less than 0.001). No difference was found in the group that took one drink per day.

Autti–Ramo et al. (1992) compared the mental development of 20 infants exposed to alcohol consumption in the 1st trimester to 20 exposed in the 1st and 2nd trimester and 20 exposed in all three trimesters. Those exposed only in the 1st trimester had Bayley scores -0.54 standard deviations below the control while those exposed in the 1st and 2nd or all trimesters were reduced by -1.02 and -1.81 standard deviations respectively. Streissguth et al. (1994) have studied short term memory and attention longitudinally in 462 adolescents exposed to prenatal alcohol.

Paternal alcohol use among 10,232 women was shown by Savitz et al. (1992) to have no association with gestational age or birth weight.

The human syndrome may not be due primarily to alcohol. Other factors such as poor protein intake, pyridoxine or other vitamin B deficiency, as well as alcohol contaminants such as lead or genetic predisposition may play an important etiologic role. Veghelyi et al. (1978) presented the hypothesis that the syndrome is related to elevated acetaldehyde levels secondary to a defect in mitochondrial aldehyde dehydrogenase (see Acetaldehyde). Streissguth and Dehaene (1993) studied 16 twin pairs and found alcohol teratogenesis was more uniformly expressed in momozygotic than dizygotic twins.

Clarren et al. (1987) and Clarren and Astley (1992) studied pregnancy rates in monkeys gavaged weekly with alcohol (2.5 or 1.8 g per kg) and found increased abortion in those dosed after the first 30 days of pregnancy. Clarren et al. (1988) found physical and developmental anomalies in monkey offspring exposed to once a week gavage doses of alcohol from 0.3 to 4.1 gm per kg. Treatment was started in the first week of gestation and continued until birth. Facial anomalies and nervous system dysfunction were found in 57 percent of the 33 exposed. Some minor neurologic dysfunctions occurred in the groups treated with 0.3 and 0.6 gm per kg. Extreme neurologic deviation occurred in the groups receiving 1.2 gm and above. Neurologic dysfunction occurred often without any malformations. Only one offspring had microcephaly. More detailed studies of monkeys exposed weekly for the first 3,6 or 24 weeks of pregnancy to 1.8 gm of alcohol per kg indicated behavioral and cognitive dysfunction in the 6 and 24 week exposure periods. The results in the 3 weeks exposure group were equivocal. Scott and Fradkin (1984) gavaged cynomologus monkeys with up to 5 gm per day from days 20 to 150 and found a decrease in birth weights but no increase in defects. The weight of the eyes and of the brains in relation to the body were not decreased.

Sandor (1968) injected the equivalent of 0.3 ml of ethanol into the chick air sac at 23 hours of incubation and produced neural tube and cerebral vesicle deformities along with some mesodermal defects. Sandor and Amels (1971) injected pregnant rats at 6 and 7 days of gestation with 1.0 to 2.0 gm of ethanol per kg and found embryolethality but no defects in the surviving fetuses. Kawashima et al. (1985) gavage–fed rats on days 7–17 of gestation with up to 6 gm per kg of ethanol. Decreased fetal weight and skeletal malformations and dilated renal pelves were found at the 4 and 6 gm dose levels. Gage and Sulik (1991) studying the pathogenesis of hydroureter and hydronephiosis seen with alcohol observed alterations at the ureterovesicle junction and in neural crest cells proximal to the posterior neuropore. Chernoff (1977) fed pregnant mice before and during pregnancy with a Metracal diet containing 15 to 30 percent ethanol derived calories. At higher concentrations, resorptions occurred frequently. Neural defects and cardiac malformations were found in a significant number of offspring. Yanai and Ginsburg (1976) found increased susceptibility to audiogenic seizures in offspring of mice given alcohol. Learning is impaired in rats exposed during intrauterine life (Martin et al., 1977). Phillips et al. (1991) carried out detailed controlled measurements of myelin and axons in the rat optic nerve after prenatal and postnatal exposure to alcohol. Myelin acquisition was delayed and the myelin thickness was premanently reduced.

Intraperitoneal or gavage treatment of preovulatory mice produced parthenogenesis and after ovulation, nondysjunction (Dyban and Baranov, 1987).

Randall and Taylor (1979) showed a dose response curve between defects and dietary content of ethanol. Maternal weight changes between the sucrose–fed control and ethanol groups of mice were not different and the ethanol group was supplemented with vitamins. Abel et al. (1987) found reduced life span of rats from mothers treated with 3.5 gm per kg on days 11–21. These offspring were raised by foster dams. The females died 20 weeks and the males 2.5–7 weeks earlier than pair fed controls. The cause was not determined but the incidence of tumors did not differ.

Arishima et al. (1993) found that the ductus arteriosis of day 20 rat fetuses was constricted following administration of alcohol (20 ml of 30% alcohol per kg).

Boggan et al. (1989) produced hydronephrosis in mice with 0.3 mg per 10 g of body weight on day 10 + 4 hours of pregnancy. Studies of day 18 fetuses using indigo carmine injected into the bladder suggested that both reflux into and obstruction of the ureters was related to the defect.

Further studies seeking to understand the molecular pathogenesis of the syndrome were reported by Chernoff (1980) in three mouse strains. Maternal alcohol dehydrogenase levels in the three strains were related inversely to the maternal alcohol level and the number of congenital defects. Sulik and Johnston (1982) carried out detailed embryologic studies after exposing the mouse to acute doses of alcohol. The morphologic features followed a deficiency in the neural plate. Diaz and Samson (1980) studied the effect on brain growth of alcohol administration through intragastric cannulas to newborn rats. Kotch et al. (1992) found increased cell death in limb buds of mouse embryos with alcohol–induced limb defects.

Mankes et al. (1982) fed male rats 20 percent ethanol for 60 days and after discontinuing for three weeks, bred them. There was an increase in congenital malformations in their offspring. Fifty–five percent of the offspring had microcephalus, cranial fissures and hydronephrosis. Kaufman (1983) reported production of meiotic nondysjunction in the female pronucleus of mice given ethanol 1–2 hours

after ovulation. Webster et al. (1983) studied the pathogenesis of fetal alcohol syndrome in the mouse. Randall et al. (1991) found in mice that aspirin reduced the frequency of alcohol–induced birth defects in a dose–dependent manner. The aspirin also reduced prostoglandin E levels in day 10 uterine/embryo tissue.

The experimental animal teratology of ethanol was reviewed by Blakely (1988). The direct effects of ethanol were studied in in–vitro culures of rat embryos (Brown et al., 1979; Campbell and Fantel, 1983; Popov et al., 1981). A growth retarding effect, especially of the central nervous system, became apparent at concentrations of 300 mg per 100 ml. Ali and Persaud (1988) showed that embryonic defects and growth retardation occur when alcohol and an inhibitor of acetaldehyde (cyanamide) are coadministered on days 9–12. Neither chemical alone in the same dose caused defects. Zajac et al. (1989) found reduced density of neurons and neuronal nuclear volume in the rostral red nucleus of intrauterine exposed rat fetuses. This nucleus is an integrative center in the brain. Bonthius and West (1991)studied in detail the anatomy and histology of fetal rat brains after exposure to exact amounts of alcohol in the perinatal period. The literature on animal models was summarized by Streissguth et al., 1980.

Russian Literature on Teratogenicity of Alcoholism

Kiryushenkov (1987) reported that the critical level of ethanol responsible for fetal alcohol syndrome (FAS) is a daily dose 60 g calculated for 100 percent of ethanol. Mastyukova et al. (1987) studied etiological factors in 270 children with hearing defects and showed that 4.7 percent were children of mothers, or both parents, having several drinks per day. The main symptoms of FAS were present in all children: Growth retardation in the prenatal period (7.8 percent), retardation of physical development (80 percent), skeletal–facial anomalies (70 percent) and CNS lesion (65 percent). In addition to hearing defects, 70 percent of FAS children had clear symptoms of mental retardation.

Mirovich et al. (1987) reported the anomalies found in the offspring of parents with chronic alcoholism. The number of such children from alcoholic mothers exceeded that from alcoholic fathers by 3 to 3.5. Narzullaeva et al. (1987) reported that 29 percent of pregnant women who drank alcohol had spontaneous abortions. The perinatal death of fetuses was 12–25 percent, premature birth was 22 percent and offspring with signs of FAS, 0.1–0.4 percent. Between 30–50 percent of the offspring of alcoholic mothers had FAS. Special tests showed that 74 percent of the offspring had physical and mental abnormalities. Quite often offspring from alcoholic mothers with no FAS had oligophrenia, epilepsy, deteriorated vision and hearing, speech disturbance and neuroses. With the addition of nicotine, the teratologic effect of alcohol was found to be more pronounced.

Ormission et al. (1987) studied 16 cases of FAS in newborns, 12 being premature. All newborns had a delay in intrauterine development and birth weight, length and circumference of head were all reduced. Some metabolic disorders were evident. All children had a disturbance of CNS (anxiety, alteration of sleep–wakefulness and appetite). Severe congenital heart disease and CNS occurred in one-fifth of all children.

Skosyreva et al. (1987) made a study of 43 newborns with FAS. All had hypotrophy, signs of prematurity, prenatal lesions of CNS, typical cranio–facial anomalies and defects of extremities. Teiverlaur (1987) reported the results of a questionaire of 46 women who had children with mental delay (group 1) and 54 who had moron children (group 2). Women of group 2 drank more alcohol and smoked more tobacco. Their husbands were charaterized by more aggressive behavior and were heavy drinkers.

Ternovskaya et al. (1987) made a study of 68 children from infant and children's homes and found that every third child had a full symptom–complex of the fetal alcohol syndrome (FAS). Major anomalies (congenital heart disease and developmental CNS defects) were found in 3, but stigma of disembryogenesis were in 98 percent of all children. They all had disturbance of psychic development. Out of 143 children, 15.4 percent were not even able to attend special schools for mentally retarded children.

Tikha et al. (1987) made a study of 60 children with symptoms of FAS, 50 of them were studied in dynamics. All children were characterized by pre- and/or postnatal retardation of growth, CNS lesion and facial anomalies. All newborns exhibited nervous irritation and sleep disturbances. Children under school age showed hyperactivity and school children had symptoms of weak brain dysfunction. In newborns, the content of all amino acids was abnormal. A decreased content of proline was associated with growth retardation and disturbance of skeleton function.

Usov et al. (1987) made a study of 150 children (group 1) from alcoholic and healthy (group 2) mothers. Attention was focused on psychic and physical development, developmental anomalies and the frequency of somatic and infectious diseases. Prematurely born and children with congenital hypotrophy consisted in group 1 of 18.7 percent and 31.5 percent, and in group 2, 4 percent and 5 percent, respectively. The birth weight and length of 50 percent of the children in group 1 were taken as low or not exceeding the mean values. Microcephaly and oligophrenia were observed in 20.5 percent and 27.3 percent, respectively, psychomotor and speech retardation in 41.3 percent and 24 percent, respectively. On the whole, in group one there were three times more cases of developmental defects.

Utin et al. (1987) summarized the data obtained in the study of offspring from 118 healthy (control) and 112 alcoholic parents. Among infants of parents with chronic alcoholism, especially when accompanied by epilepsy or oligophrenia, the frequency of somatic and psychical pathology was 3 to 14–fold that of control families. If both parents, or only the mother drank alcohol, the risk for abnormal offspring was higher than in a case of an alcoholic father.

Voinilo (1987) made a study of 67 infants (aged 4–8 years) from mothers with chronic alcoholism. Twenty–nine infants had clear symptoms of FAS. Zeits et al. studied a group of 35 mature and premature newborns from women who were heavy drinkers. Premature offspring (more than 25 percent) showed predisposition to infectious and allergic diseases, rickets and hypotrophy. They often had encephalopathy and congenital developmental defects. Mature offspring showed psychomotor retardation (50 percent), as well as retardation in birth weight and length, or both. Among premature offspring, the percentage of retardation was even higher.

Popov et al. (1985) studied the embryotoxicity of different concentrations of ethanol (17–108 mM) and acetaldehyde (4.5–450 microM) on the cultures of postimplantation rat embryos. Exposure time required for ethanol at concentrations of 65–108mM to exert the embryolethal and teratogenic action was 24–28 hours. At a concentration of 110 mM, acetaldehyde had a marked teratogenic effect after a three hour exposure. At higher concentrations, the effect was present after one hour. Semyonov et al. (1987) reported that one of the mechanisms likely to be responsible for harmful influence of alcohol on offspring is a stable disturbance of zinc and copper metabolism.

Chebotar et al. (1987) administered i.g. ethanol to pregnant rats at a dose of 2.4 g per kg, 2–3 times per day, or at a total dose of 2.4, 4.8, 7.2 g per kg per day on days 7 to 15. Phthalimide (500 mg per kg) and ethanol (2.4 g per kg) were administered simultaneously on various days between 9 and 13 of pregnancy. The total dose of ethanol 4.8 g per kg led to a decrease of the fetus weight. At a dose 7.2 g per kg, there was an insignificant increase of embryolethal effect. On day 8 of pregnancy, teratogenic action of alcohol reached maximum. Single injections intraperitoneally of ethanol (4.5 g per kg) on days 10, 11 or 12 led to embryonic death (64 percent). Administration of phthalimide to pregnant rats induced single cases of embryonic death (3 percent), no developmental defects were observed. Simultaneous administration of phthalimide and ethanol prevented embryolethal and nephrotoxic action of ethanol.

Drzhevetskaya et al. (1985) studied the function of hypothalamo–hypophyseal adrenocortical system (HHAS) in rat offspring which received 20 percent ethanol solution instead of fresh water from the first 30 days of lactation. Ethanol was shown to induce a stable inhibition of HHAS. Weak adaptation to muscle load was rather stable too.

Popov et al. (1985) described embryolethal and teratogenic effects of ethanol before and after biotransformation, as was evidenced by the number of sister chromatid exchanges (SCE) in the cells of cultured rat embryos. Ethanol (concentration 4.5–225 mM) was added to the culture medium. The minimal concentration which induced embryolethal or teratogenic effect was 65 mM. However, even at concentrations of 108 mM, some embryos remained alive. Concentrations of 65 and 106 mM did not lead to an increase of SCE. Acetylaldehyde at concentrations of 45–225 mM had a marked embryolethal and teratogenic effect. At a concentration of 45 mM, the number of SCE exceeded control 2.6-fold. At concentrations of 112 and 225 mM, it was 6, 7 and 9 times that of the controls.

Skalny et al. (1987) reported results of single intragastric administration of 20 percent ethanol solution in rats during 50–100 days. Females received ethanol during the pregnancy and lactation period. In the offspring, the absorption of ^{65}Zn by the liver epididymides and muscle tissue was decreased. The same was found in bone tissue, skin and ovaries. The absorption of ^{65}Zn by the kidney was decreased, while for adrenal glands, it was increased.

Abel, E.L.; Church, M.W. and Dintcheff, B.A.: Prenatal alcohol exposure shortens life span in rats. Teratology 36:217–220, 1987.

Ali, F. and Persaud, T.V.N.: Mechanisms of fetal alcohol effects: role of acetaldehyde. Exp. Pathol. 33:17–21, 1988.

Arishima, K.; Yamamoto, M.; Takizawa, T.; Sohmiya, H.; Eguchi, Y. and Shiota, K.: Effect of acute maternal alcohol consumption on the fetal ductus arteriosus in the rat. Biol Neonate 63:40–43, 1993.

Autti–Ramo, I.; Korkman, M.; Hilakivi–Clarke, L.; Lehtonen, M.; Halmesmaki, E. and Granstrom, M.L.: Mental development of 2–year–old children exposed to alcohol in utero. The Journal of Pediatrics 120:740–746, 1992.

Blakely, P.M.: Experimental teratology of ethanol. In: Issues and Reviews in Teratology, H. Kalter (ed.), Plenum Press, New York. 4:237–282, 1988.

Boggan, W.O.; Monroe, B.; Turner, Jr., W.R.; Upshur, J. and Middaugh, L.D.: Effect of prenatal ethanol administration on the urogenital system of mice. Alcohol.: Clin. Exp. Res. 13:206–208, 1989.

Bonthius, D.J. and West, J.R.: Permanent deficits in rats exposed to alcohol during the brain growth spurt. Teratology 44:147–163, 1991.

Brown, N.A.; Goulding, E.H. and Fabro, S.: Ethanol embryotoxiciy: Direct effects on mammalian embryos in vitro. Science 206:573–575, 1979.

Campbell, M.A. and Fantel, A.G.: Teratogenicity of formaldehyde in vitro: Relevance to the fetal alcohol syndrome. Life Sciences 32:2641–2647, 1983.

Chebotar, N.A.; Dzhandzhaliya, A.M.; Konopistseva, L.A. and Puchkov, V.F.: Teratogenic action of alcohol in rats and its prevention. Alcoholism and Heredity. Inter. Symposium, Leningrad, 1986 (USSR), Moscow, 171–175, 1987.

Chernoff, G.F.: The fetal alcohol syndrome in mice. An animal model. Teratology 15:223–230, 1977.

Chernoff, G.F.: The fetal alcohol syndrome in mice: Maternal variables. Teratology 22:71–75, 1980.

Clarren, S.K.; Alvord, E.C.; Sumi, M.; Streissguth, A.P. and Smith, D.W.: Brain malformations related to prenatal exposure to ethanol. J. Pediatr. 92:64–67, 1978.

Clarren, S.K. and Astley, S.J.: Pregnancy outcomes after weekly oral administration of ethanol during gestation in the pig–tailed macaque: comparing early gestational exposure to full gestational exposure. Teratology 45:1–9, 1992.

Clarren, S.K.; Astley, S.J. and Bowden, D.M.: Physical anomalies and developmental delays in nonhuman primate infants exposed to weekly doses of ethanol during gestation. Teratology 37:561–569, 1988.

Clarren, S.K.; Astley, S.J.; Gunderson, V.M. and Spellman, D.: Congitive and behavioral deficits in nonhuman primates associated with very early embryonic binge exposures to ethanol. The Journal of Pediatrics 121:789–796, 1992.

Clarren, S.K.; Bowden, D.M. and Astley, S.J.: Pregnancy outcomes after weekly oral administration of ethanol during gestation in the Pig–tailed Macaque (Macaca nemestrina). Teratology 35:345–354, 1987.

Clarren, S.K.; Sampson, P.D.; Larsen, J.; Donnell, D.J.; Barr, H.M.; Bookstein, F.L.; Martin, D.C. and Streissguth, A.P.: Facial effects of fetal alcohol exposure: Assessment by photographs and morphometric analysis. Am. J. Med. Genetics 26:651–656, 1987.

Day, N.L.; Richardson, G.; Robles, N.; Sambooth, U.; Taylor, P.; Scher, M.; Stoffer, D.; Jasperse, D. and Cornelius, M.: Effect of prenatal alcohol exposure on growth and morphology of offspring at 8 months of age. Pediatrics 85(5):748–752, 1990.

de Cornulier, M.; de Lacour, F.; Avet–Loiseau, H.; Passuit, N.; Branger, B.; Lemoine, P. and Picherot, G.: Atteintes vertebrales et syndrome d'alcoolisme foetal. Pediatrie 46:685–689, 1991.

Diaz, J. and Samson, H.H.: Impaired brain growth in neonatal rats exposed to ethanol. Science 208:751–753, 1980.

Drzhevetskaya, I.A. and Butova, O.A.: The influence of alcohol intoxication of rats during lactation on the function of hypothalamo–hypophyseal adrenocortical system of offspring. Farmakologia i Toxikologia (USSR) 6:93–96, 1985

Dyban, A.P. and Baranov, V.S.: Cytogenetics of Mammalian Embryonic Development. Oxford Science Publications, Chapter 1, 24–25, 1987.

Ferrier, P.E.; Nicod, I. and Ferrier, S.: Fetal alcohol syndrome. Lancet 2:1496 only, 1973.

Gage, J.C. and Sulik, K.K.: Pathogenesis of ethanol-induced hydronephrosis and hydroureter as demonstrated following in vivo exposure of mouse embryos. Teratology 44:299–312, 1991.

Hanson, J.W.; Streissguth, A.P. and Smith, D.W.: The effects of moderate alcohol consumption during pregnancy on fetal growth and morphogenesis. J. Pediatr. 92:457–460, 1978.

Harlap, S. and Shiono, P.H.: Alcohol, smoking and incidence of spontaneous abortions in the first and second trimester. Lancet 2:173–178, 1980.

Jones, K.L. and Smith, D.W.: Recognition of the fetal alcohol syndrome in early infancy. Lancet 2:999–1101, 1973.

Jones, K.L. and Smith, D.W.: The fetal alcohol syndrome. Teratology 12:1–10, 1975.

Jones, K.L.; Smith, D.W.; Streissguth, A.P. and Myrianthopoulos, N.C.: Outcome in offspring of chronic alcoholic women. Lancet 1:1076–1078, 1974.

Jones, K.L.; Smith, D.W.; Ulleland, C.N. and Streissguth, A.P.: Pattern of malformation in offspring of chronic alcoholic mothers. Lancet 1:1267–1271, 1973.

Kotch, L.E.; Dehart, D.B.; Alles, A.J.; Chernoff, N. and Sulik, K.K.: Pathogenesis of ethanol–induced limb reduction defects in mice. Teratology 46:323–332, 1992.

Kaufman, M.H.: Ethanol–induced chromosomal abnormalities at conception. Nature 302:255–260, 1983.

Kawashima, K.; Tanaka, S.; Nakaura, S.; Takanaka, A. and Omori, Y.: Effect of ethanol on prenatal developments of rats. Eisei Shikenjo Hokoku 103:10–14, 1985.

Kiryushenkov, A.P.: Fetal alcohol syndrome. Alcoholism and Heredity. Proceedings of inter. symposium, Leningrad, 1986 (USSR). Moscow, 79–83, 1987.

Kline, J.; Shroat, P.; Stein, Z.A.; Susser, M. and Warburton, D.: Drinking during pregnancy and spontaneous abortion. Lancet 2:176–180, 1980.

Landesman–Dwyer, S.; Keller, L.S. and Streissguth, A.P.: Naturalistic observations of newborns: Effects of maternal alcohol intake. Alcoholism: Clin. Exper. Res. 2:171, 1978.

Lemoine, P.; Harousseau, H.; Borteyru, J.P. and Menvet, J.C.: Les enfants de parents alcooliques: anomalies observees, a propos de 127 cas. Arch. Franc. Pediatr. 25:830–832, 1968.

Little, R.E. and Sing, C.F.: Father's drinking and infant birth weight: report of an association. Teratology 36:59–65, 1987.

Majewski, F.: Alcohol embryopathy: Some facts and speculations about pathogenesis. Neurobehav. Toxicol. Teratol. 3:129–144, 1981.

Mankes, R.F.; LeFevre, R.; Benitz, K.F.; Rosenblum, I.; Bates, H.; Walker, A.T. and Abraham, R.: Paternal effects of ethanol in the Long–Evans rat. J. Toxicol. Environ. Health 10:871–878, 1982.

Manzke, H. and Grosse, F.R.: Incomplettes und Komplettes fetal Alkoholsyndrom bei drei Kindern einer Trinkerin. Med. Welt. 26:709–712, 1975.

Martin, J.C.; Martin, D.C.; Sigman, G. and Radow, B.: Offspring survival, development and operant performance following maternal ethanol administration. Dev. Psychobiol. 10:435–446, 1977.

Mastyukova, E.M. and Pevzner, M.S.: Fetal alcohol syndrome and neuromental defects in children. Alcoholism and Heredity. Proceedings of inter. symposium, Leningrad, 1986 (USSR). Moscow, 107–108, 1987.

Mills, J.L.; Graubard, B.I.; Harley, E.E.; Rhodes, G.G. and Berendes, H.W.: Maternal alcohol consumption and birth weight. JAMA 253:1875–1879, 1984.

Mills, J.L. and Graubard, B.I.: Is moderate drinking during pregnancy associated with an increased risk for malformations? Pediatrics 80:309–314, 1987.

Mirovich, D.Yu. and Barabadze, E.V.: The influence of alcohol on reproductive function and the offspring. Akusherstvo i Gigiena (USSR) 10:10–11, 1987.

Narzullaeva, M.S.; Nikitin, A.I. and Slozina, N.M.: The influence of alcohol on the female reproductive function and the fetus. Zdravookhranenie Tajikistana (USSR) 6:14–18, 1987.

Ormission, A. and Ginter, L.: Conference on the results of campaign for sober life (theses), Tartu, USSR. Nov 25–26, 1987. Tartu 106–107, 1987.

Palmer, R.H.; Ovellete, E.M.; Warner, L. and Leichtman, S.R.: Congenital malformations in offspring of a chronic alcoholic mother. Pediatrics 53:490–494, 1975.

Pauli, R.M. and Feldman, P.F.: Major limb malformations following intrauterine exposure to ethanol: Two additional cases and literature review. Teratology 33:273–280, 1986.

Phillips, D.E. and Rydquist, J.E.: Short and long–term effects of combined pre and postnatal ethanol exposure (three trimester equivalency) on the cevelopment of myein and axons in rat optic nerve. Int. J. Devl. Neuroscience 0:000–000, 1991.

Popov, B.V. and Bichevaya, N.K.: The embryotoxic action of ethanol and acetaldehyde on the rat embryonic cultures. In: Main Trends and Controlling Mechanisms of the Early Normal and Abnormal Mammalian Embryogenesis. (USSR) Leningrad, Meditsina 77–84, 1985.

Popov, B.V. and Pyatkin, E.L.: The dependence of teratogenic effect on the number of sister chromatid exchanges in biotransformation of ethanol in the rat embryo culture. In: Main trends and controlling mechanisms of the early normal and abnormal mammalian embryogenesis. (USSR) Leningrad, Meditsina 85–90, 1985.

Popov, V.G.; Weisman, B.L.; Puchkov, V.F. and Ignatyeva, T.V.: Embryotoxic effect of ethanol and biotransformation products in the culture of postimplantation rat embryos. Byull. Eksp. Biol. Med. (USSR) 12:725–728, 1981.

Randall, C.L.; Anton, R.F.; Becker, H.C.; Hale, R.L. and Ekblad, U.: Aspirin dose–dependently reduces alcohol–induced birth defects and prostaglandin E levels in mice. Teratology 44:521–529, 1991.

Randall, C.L. and Taylor, W.J.: Prenatal ethanol exposure in mice. Teratogenic effects. Teratology 19:305–312, 1979.

Roman, E.; Beral, V. and Zuckerman, B.: The relation between alcohol consumption and pregnancy outcome in humans, a critique. In: Issues and Reviews in Teratology, H. Kalter (ed.), Plenum Press, New York 4:205–235, 1988.

Sandor, G.G.S.; Smith, D.F. and MacLeod, P.M.: Cardiac malformations in the fetal alcohol syndrome. J. Pediatr. 98:771–773, 1981.

Sandor, S.: The influence of aethyl alcohol on the developing chick embryo. Rev. Roum. Embryol. Cytol. Ser. Embryol. 5:167–171, 1968.

Sandor, S. and Amels, D.: The action of aethanol on the prenatal development of albino rats. Rev. Roum. Embryol. Cytol. Ser. Embryol. 8:105–118, 1971.

Savitz, D.A.; Zhang, J.; Schwingl, P. and John, E.M.: Association of paternal alcohol use with gestational age and birth weight. Teratology 46:465–471, 1992.

Scott, W.J. and Fradkin, R.: The effects of prenatal alcohol on cynomolgus monkeys. Teratology 29:49–56, 1984.

Semyonov, A.S.; Seranov, T.I.; Skalny, A.V. and Eldelman, T.N.: Pediatria (USSR) 4:49–54, 1987.

Skalny, A.V. and Skosyreva, A.M.: Zn deficiency due to alcohol in mother, fetus and offspring. Akusherstvo i Ginekologia (USSR) 4:6–8, 1987.

Skosyreva, A.M.; Kurilo, L.F.; Skalny, A.V.; Tatarinskaya, E.V. and Mihailova, G.A.: Alcoholism and Heredity. Proceedings of inter. symposium, Leningrad, 1986, (USSR), 134–138, 1987.

Spohr, H.L.; Willims, J.; Steinhausen, H.C.: Prenatal alcohol exposure and long–term developmental consequences. The Lancet 34–910, 1993.

Streissguth, A.P.; Barr, H.M. and Martin, D.C.: Alcohol

exposure in utero and functional deficits in children during the first four years of life in mechanisms of alcohol damage in utero, Ciba Foundation Symposium 105, Pitman, London, 1984.

Streissguth, A.P.; Barr, H.M.; Sampson, P.D.; Parrish–Johnson, J.C.; Kirchner, G.L. and Martin, D.C.: Attention, distraction and reaction time at age 7 years and prenatal alcohol exposure. Neurobehav. Toxicol. Teratol. 8:717–725, 1986.

Streissguth, A.P.; Barr, H.M.; Sampson, P.D.; Darby, B.L. and Martin, D.C.: IQ at age 4 in relation to maternal alcohol use and smoking during pregnancy. Dev. Psychol. 25:3–11, 1989.

Streissguth, A.P. and Dehaene, P.: Fetal alcohol syndrome in twins of alcoholic mothers: concordance of diagnosis and IQ. American Journal of Medical Genetics 47:857–861, 1993.

Streissguth, A.P.; Landesman–Dwyer, S.; Martin, J.C. and Smith, D.W.: Teratogenic effects of alcohol in humans and animals. Science 209:353–361, 1980.

Streissguth, A.P.; Martin, D.C.; Martin, J.C. and Barr, H.M.: The Seattle longitudinal prospective study on alcohol and pregnancy. Neurobehav. Toxicol. Teratol. 3:223–233, 1981.

Streissguth, A.P.; Aase, J.M.; Clarren, S.K.; Randels, S.P.; LaDue, R.A. and Smith, D.F.: Fetal alcohol syndrome in prospective study. Alcohol Clin. Exp. Res. 18:202–218, 1994.

Streissguth, A.P.; Sampson, P.D.; Olson, H.C.; Bookstein, F.L.; Barr, H.M.; Scott, M.; Feldman, J. and Mirsky, A.F.: Maternal drinking during pregnancy: Attention and short–term memory in 14 year old offspring–a longitudinal prospective study. Alcoholosm: Clinical and Experimental Research 18:202–218, 1994.

Streissguth, A.P.; Barr, H.M.; Olson, H.C.; Sampson, P.D.; Bookstein, F.L. and Burgess, D.M.: Drinking during pregnancy decreases word attack and arithmetic scores on standardized tests: Adolescent data from a population-based prospective study. Alcoholism: Clinical and Experimental Research 18:248–254, 1994.

Stromland, K.: Ocular abnormalities in the fetal alcohol syndrome. Acta Ophthalmo Logica 63(171):5–50, 1985.

Stromland, K. and Pinazo–Duran, M.D.: Optic nerve hypoplasia: Comparative effects in children and rats exposed to alcohol during pregnancy. Teratology 50:100–111, 1994.

Sulik, K.K. and Johnston, M.C.: Acute ethanol administration in an animal model results in craniofacial features characteristic of the fetal alcohol syndrome. Science 214:936–938, 1982.

Teiverlaur, M.P.: Moron and mentally retarded children from alcoholic parents (theses). Conference on the results of the campaign for sober life, Tartu, USSR, November 25–26, 1987. Tartu 68–69, 1987.

Ternovskaya, V.A.; Dregalo, A.A. and Ternovsky, L.N.: Alcoholism and congenital defects. Zdravookhranenie Rossiyskoi Federatsiy (USSR) 12:35–38, 1987.

Tikha, R. and Matlokha, Z.: Free aminoacids in blood plasma of children with fetal alcohol syndrome. Alcoholism

and Heredity. Proceedings of inter. symposium, Leningrad, 1986. (USSR). Moscow 147–151, 1987.

Tikkanen, J. and Heinonen, O.P.: Maternal exposure to chemical and physical factors during pregnancy and cardiovascular malformations in the offspring. Teratology 43:591–600, 1991.

Usov, I.N.; Skripchenko, T.A. and Efimenko, G.N.: The development of children from alcoholic mothers. Pediatria (USSR) 7:77–81, 1987.

Utin, A.V.; Nazirov, B.D. and Danielov, M.B.: Offspring from alcoholic parents with psychical pathology. Alcoholism and Heredity. Proceedings of inter. symposium, Leningrad, 1986 (USSR). Moscow 159–162, 1987.

Veghelyi, P.V.; Osztovics, M.; Kardos, G.; Leisztner, L.; Szaszovszky, E.; Igali, S. and Imrei, J.: The fetal alcohol syndrome: symptoms and pathogenesis. Acta Paediat. Acad. Scient. Hungaricas 19:171–189, 1978.

Voinilo, O.A.: The importance of early detection and urgent correction of the pathology in offspring of mothers with chronic alcoholism (theses). Regional conference on urgent narcology, Sept. 17, 1987. (USSR) Kharkov 66–67, 1987.

Webster, W.S.; Walsh, D.A.; McEwen, S.E. and Lipson, A.H.: Some teratogenic properties of ethanol and acetaldehyde in C57BL6J mice. Implications for the study of fetal alcohol syndrome. Teratology 27:231–243, 1983.

Yanai, J. and Ginsburg, B.E.: Audiogenic seizures in mice whose parents drank alcohol. J. Stud. Alcohol 37:1564–1571, 1976.

Zajac, C.S.; Bunger, P.C. and Moore, J.C.: Changes in red nucleus neuronal development following maternal alcohol exposure. Teratology 40:567–570, 1989.

Zeits, R.I.; Dmitrieva, O.M. and Kolodkina, E.S.: Development and predisposition to disease of children of alcoholic mothers. Pediatria (USSR) 4:54–55, 1987.

60 Alcohols, Industrial *1–Octanol 1–Nonanol 1–Decanol*

Nelson et al. (**1990**) exposed rats for 7 hours on days 1–19 to 1–octanon (400 mg per m^3), 1–nonanol (150 mg per m^3), or 1–deconol (100 mg per m^3). No maternal or fetal ill effects were found.

Nelson, B.K.; Brightwell, W.S.; Khan, A.; Krieg Jr., E.F. and Hoberman, A.M.: Developmental toxicology assessment of 1–octanol, 1–nonanol, and 1–decanol administered by inhalation to rats. Journal of the American College of Toxicology 9:93–97, 1990.

61 Alcohol Sulfate

Discussed under Surfactants

62 Aldicarb *2-Methyl-2(methylthio)propionaldehyde O-(methylcarbamoyloxime)* CAS 116–06–3

Reduced acetylcholinesterase activity was shown in rat fetal brain and liver for as long as 24 hours after oral administration of 0.0l or 0.l0 mg per kg to the dams (Cambon et al., 1979).

Cambon, C.; Declume, C. and Derache, R.: Effect of insecticidal carbamate derivatives (carbofuran, pirimicarb, aldicarb) on the activity of acetylcholinesterase in tissues from pregnant rats and fetuses. Toxicol. Appl. Pharmacol. 49:203–208, 1979.

63 Aldosterone CAS 52–39–1

Grollman and Grollman (1962) gave 0.05 mg daily to rats during their gestation and observed increased blood pressure in the offspring at one year of age.

Grollman, A. and Grollman, E.F.: Teratogenic induction of hypertension. J. Clin. Invest. 41:710–714, 1962.

64 Alfentanil CAS 71195–58–9

Fujinaga et al. (1988) administered this narcotic to rats giving 8 mg per kg on days 5–20 via implanted osmotic minipumps. No adverse fetal effects were found.

Fujinaga, M.; Mazze, R.I.; Jackson, E.C. and Baden, J.M.: Reproductive and teratogenic effects of sufentanil and alfentanil in Sprague–Dawley rats. Anesth. Analg. 67:166–169, 1988.

65 Alinidine Hydrobromide CAS 33178–86–8
2–[N–allyl–N–(2,6–dichlorophenyl)–amino]–2–inidazoline hydrobromide.

This bradycardic agent was given intravenously to rats and rabbits in fertility, teratogenicity and during perinatal studies. Doses of up to 5–10 mg per kg were used. Maternal toxicity occurred at 1.5 mg per kg in rats and 3 mg per kg in rabbits. No reproductive or developmental effects were found in either species. (Matsuo et al., **1989**)

Matsuo et al. (**1989**) gave this bradycardic agent in up to 20 mg per kg to rabbits and rats during organogenesis. No teratogenicity was found but postnatal survival of the rat pups was decreased by maternal behavior and oligolactia. Postnatal development and reproductive function was uneffected.

Matsuo, A.; Kast, A. and Tsunenari, Y.: Reproductive toxicology of intravenous alinidine hydrobromide. Iyakuhin Kenkyu 20:318–337, 1989.

Matsuo, A.; Ida, H.; Kast, A. and Tsunenari, Y.: Oral reproduction toxicology of adinidine hydrobromide. Iyakuhin Kenkyu 20:153–170, 1989.

66 Alkaloid Q

Keeler and Binns (1968) fed 1.0–1.5 gm to 5 ewes on the 14th day of gestation and found no cyclopia in the offspring.

Keeler, R.F. and Binns, W.: Teratogenic compounds of Veratrum californicum (Durand). Teratology 1:5–10, 1968.

67 Alkylbenzene Sulfonate also see *Surfactants*

Mikami et al. (1973) applied this detergent containing 13 mg of alkylbenzene sulfonate per 0.5 ml to the bare skin of pregnant rats from the first to the 13th day of pregnancy. The fetuses had subcutaneous bleeding, edema, cleft palate and musculo–skeletal defects. Other workers in the Japanese Ministry of Health and Welfare were not able yet to reproduce these findings (Masuda et al., 1973).

Masuda, F.; Okamoto, K. and Inoue, K.: Effects of linear alkylbenzene sulfonate spread on mouse skin during pregnancy on fetuses. J. Food Hyg. Soc. Jpn. 14:580–582, 1973.

Mikami, Y.; Sakae, Y. and Miyamoto, I.: Anomalies induced by ABS applied to the skin. (abs) Teratology 8:98 only, 1973.

68 Allergy Shots *Desensitization*

Heinonen et al. (1977) reported no increase in defects among 9,222 women given immunizing agents during the first 4 lunar months. Fourteen women were given desensitization and no malformed offspring were found in the group showing uniform rates by hospital.

Heinonen, O.P.; Slone, D. and Shapiro, S.: Birth Defects and Drugs in Pregnancy. John Wright Publishing Sciences Group, Inc., Littleton, Mass. 1977.

69 Allopurinol *4–Hydroxypyrazolo–(3,4–d)–pyrimidine* CAS 315–30–0

Chaube and Murphy (1968) gave 50 to 500 mg per kg to pregnant rats on the 11th or 12th gestational day and found no defects in fetuses surviving to the 21st day. Spezia et al. (**1992**) found in whole rat embryo culture that concentration of 0.66 mM or above interfered with development. Rat and human S–9 activation enhanced the toxicity.

Chaube, S. and Murphy, M.L.: The teratogenic effects of the recent drugs active in cancer chemotherapy. In: Advances in Teratology, D.H.M. Woollam (ed.), New York, Academic Press, 3:181–237, 1968.

Spezia, F.; Fournex, R. and Vannier, B.: Action of allopurinol and aspirin on rat whole–embryo cultures. Toxicology 72:239–250, 1992.

70 Alloxan CAS 50–71–5

Takano and Nishimura (1967) reported a 6 to 7 percent incidence of defects in the offspring of rats and mice made diabetic by alloxan. The defects were microphthalmia, exencephaly, cleft lip and palate and spina bifida. These authors in discussion of their work and that of others concluded that the diabetic state and not the alloxan was responsible for the defects. Kalter (1968) summarized in detail the experimental literature on defects found after use of alloxan. More recent Japanese work was summarized by Shimada et al. (1975).

Kalter, H.: Teratology of the Central Nervous System. Chicago: University of Chicago Press, 150–152, 1968.

Shimada, T.; Endo, A. and Watanabe, G.: Lactic dehydrogenase isoenzymes in embryos of alloxan diabetic mice. Cong. Anom. 15:127–132, 1975.

Takano, K. and Nishimura, H.: Congenital malformations induced by alloxan diabetes in mice and rats. Anat. Rec. 158:303–312, 1967.

71 Allyl Alcohol CAS 107–18–6

Slott and Hales (1985) injected 10, 100 or 1000 microgm of this chemical directly into the amniotic cavity of day 13 rat embryos. Embryo lethality was increased at doses of 100 microgm and above. Limb defects and other sorts of deformities were found in the 1000 microgm group

Slott, V.L. and Hales, B.F.: Teratogenicity and embryolethality of acrolein and structurally related compounds in rats. Teratology 32:65–72, 1985.

72 Allyl Chloride *3–Chloro–1–propene* CAS 107–05–1

John et al. (1983) exposed rats and rabbits to vapor concentrations of 0, 30 and 300 ppm for 7 hours daily during active organogenesis. At the highest level, maternal toxicity occurred and a slight delay in fetal skeletal development was observed.

John, J.A.; Gushow, T.S.; Ayres, J.A.; Hanley, T.R., Jr.; Quast, J.F. and Rao, K.S.: Teratologic evaluation of inhaled epichlorhydrin and allyl chloride in rats and rabbits. Fund. Appl. Toxicol. 3:437–442, 1983.

73 Allylestrenol

Czeizel and Huiskes (1988) did a case control study in Hungary at a time when 24 percent of the women were being supported with the hormone during pregnancy. The treated group consisted of 2,645 and the control group, 7,248. No associated significant increase in hypospadias was observed in the treated group that took the medication during the critical fourth month. No other congenital defects were increased.

Czeizel, A. and Huiskes, N.: A case–control study to evaluate the risk of congenital anomalies as a result of allylestrenol therapy during pregnancy. Clin. Ther. 10(6):725–739, 1988.

74 Allylisothiocynate CAS 57–06–7

Discussed under Ethylenethiourea

75 17–α-Allyl–4–oestrene–17–β-ol *Gestanon™*

Jost and Moreau–Stinnakre (1970) administered 10 to 20 mg of this synthetic progestin to rats by mouth or subcutaneously from the 13th through the 21st day of gestation. Some persistence of wolffian ducts and increase in the ano-rectal distance was found in female fetuses. The males were normal. The trans–intestinal route and presence of maternal ovaries were shown experimentally to favor the occurrence of the anomalies. Konstantinova et al. (**1975**) gave women volunteers 30 mg per day orally for 15 days before induced abortion. No effects on organogenesis or chromosomal structure were found.

Jost, A. and Moreau–Stinnakre, M.: Action d'une substance progestive synthetique (17–α–allyl–4–oestrene–17 β–ol) sur la differenciation sexuelle des foetus de rat. Acta Endocrinol. 65:29–49, 1970.

Konstantinova, B.; Deespodova. C.V. and Yankov, D.: Study of the influence of gestanon on the early development of human embryo. Akush. GInekol. (Sofia) 14:251–254, 1975.

76 Allylthiourea

Discussed under Ethylenethiourea

77 Aloe

Nath et al. (**1992**) fed dried aloe leaves to rats and found increased fetal loss and defects at a dose of 125 mg per kg. Kinky tails, clubbing and wrist drop were found increased.

Nath, D.; Sethi, M.; Singh, R.K. and Jain, A.K.: Commonly used Indian abortifacient plants with special reference to their teratologic effects in rats. J Ethnopharmacol 36:147–154, 1992.

78 Alosenn

Matsumoto et al. (1981) gave rats this laxative composed of senna leaves and other vegetables during organogenesis. Oral doses of up to 2 gm per kg daily caused no adverse fetal effects.

Matsumoto, T.; Tsugitami, M.; Ouchi, M.; Tomizawa, S. and Kamata, K.: Teratological study of alosenn in rats. Kiso to Rinsho 15:36–53, 1981.

79 Alphaxalone CAS 23930–19–0

Esaki et al. (1976) studied this anesthetic component in pregnant rats and mice. Doses of up to 30 mg per kg in mice and 12 mg in rats were given intravenously on the 8th or 12th day in mice and 9th or 13th day in rats. No adverse fetal effects occurred.

CT-1341, which is a 3 to 1 mixture of alphaxalone with alphadolone, was given subcutaneously to mice and rats during active organogenesis in doses of up to 5 ml per kg and no anomalies resulted (Esaki et al., 1975). Some postnatal growth retardation was found in rats. Negative studies in the monkey were reported (Tanioka et al., 1977).

Esaki, K.; Oshio, K. and Yoshikawa, K.: Effects of intravenous administration of Alphaxalone on the fetuses of the mouse and rat (Japanese). CIEA Preclin. Rep. 2:229–236, 1976.

Esaki, K.; Tsukada, M.; Izumiyama, K. and Oshio, K.: Influence of CT–1341 on the fetuses of the mouse and rat (Japanese) CIEA Preclin. Rep. 1:165–172, 1975.

Tanioka, Y.; Koizumi, H. and Inaba, K.: Teratogenicity test by intravenous administration of CT–1341 in rhesus monkeys. (Japanese) CIEA Preclin. Rep. 3:35–45, 1977.

80 Alprazolam *Xanax™* CAS 28981–97–7

Schick–Boschetto and Zuber (**1992**) followed 184 pregnancies exposed during the first trimester to therapeutic doses. Five infants (2.94 percent) had malformations. No pattern of defects was identified.

This benzodiazepine tranquilizer was given to pregnant rats orally at doses of 0.5 to 50 mg per kg. Dosing was on days 7–17. At the highest dose, some increase in thoracic vertebral anomalies occurred along with increased fetal death. No adverse effects were found at the other dosage levels (Esaki et al., 1981).

Esaki, K.; Oshio, K. and Yanagita, J.: Effects of oral administration of alprazolam (TUS–1) on the rat fetus: Experiment on drug administration during the organogenesis period. Preclin. Rep. Cent. Inst. Exp. Anim. 7:65–77, 1981.

Schick–Boschetto, B. and Zuber, C.: Alprazolam exposure during early human pregnancy (abst.). Teratology 45:460 only, 1992.

81 Alprenolol

Chimura (**1985**) used this β–adrenoceptor blocker in pregnant rat studies. Increased fatal death and decreased weight was found when 0.8 mg was given to the dam for 5 days during gestation. Hypoglycemia also occurred in the newborns. Parturition time was increased.

Chimuar, T.: Effects of β–adrenoceptor blockade on parturition and fetal carrdiovascular and metabolic system. Act. Gynec Jpn 37:691–695, 1985.

82 Alternariol Monomethyl Ether CAS 641–38–3

Pero et al. (1973) administered this mold toxin subcutaneously to pregnant mice on days 13 through 16 and produced an increase in unspecified malformations at a dose of 100 mg per kg. Resorptions and runts were increased at the same dose but not at 50 mg per kg. The monomethyl ether also produced congenital defects.

Pero, R.W.; Posner, H.S.; Blois, M.; Harran, D. and Spalding, J.W.: Toxicity of metabolites produced by "alternaria". Environ. Health Perspect. 4:87–94, 1973.

83 Alum *Aluminum Potassium Sulfate* CAS 10043–01–3

Kanoh et al. (1982) fed pregnant rats for 7 days during mid–pregnancy with up to 10 percent of this compound in the diet. No adverse effects were noted in day 20 fetuses.

Kanoh, S.; Ema, M. and Kawasaki, H.: Studies on the toxicity of alum. Oyo Yakuri 24:65–69, 1982.

84 Aluminum CAS 7429–90–5

Ridgway and Karnofsky (1952) were unable to find defects in chick embryos after injecting 15 mg of $AlCl_3$ into the yolk sac on the 4th day. McCormack et al. (1978) fed rats 0.1 percent aluminum trichloride in the drinking water from days 6–19 of gestation and found no adverse fetal effects. Golub et al. (1985) added up to 1000 mg per kg to the diet of mice during gestation and studied the offspring. Although there were no neurobehavioral changes, weight gain was decreased and the spleen weights reduced as compared to pair–fed controls. Cranmer et al. (1986) found a reduction in fetal weight and an increase in resorptions in mice treated with 100 mg per kg intraperitoneally or 200 mg per kg orally. Misawa and Shigeta (**1993**) gave a single dose of aluminum chloride of up to 1,800 mg per kg orally to rats on day 15. Postnatal development and learning acquisition was slower in the group exposed to the highest dose.

Wide (1984) found no impairment of implantation in mice after giving 100 millimoles on day 3. Fetal hemorrhage was increased when treatment was on day 8. Yokel (1985) found reduced conditioned reflexes and memory in rabbit offspring exposed by injection during most of gestation to 100 or 400 micromoles of aluminum. Muller et al. (**1990**) gave rats aluminum lactate (400 mg Al per kg) and although there was no effect or mortality, litter size or fetal weight significant decrease in geotaxis was seen following treatment in the 2nd and 3rd weeks. Locomotion coordination and operant conditioning were also reduced.

Bernuzzi et al. (1989) gave oral aluminum lactate or chloride to rats on days 1–21 at doses of 272 (chloride) and 378 (lactate) mg per kg per day, an increased mortality was found postnatally. Neuromotor development was delayed at these high doses.

Bernuzzi, V.; Desor, D. and Lehr, P.R.: Developmental alterations in offspring of female rats orally intoxicated by aluminum chloride or lactate during gestation. Teratology 40:21–27, 1989.

Cranmer, J.M.; Wilkins, J.D.; Cannon, D.J. and Smith, L.: Fetal–placental–maternal uptake of aluminum in mice following gestational exposure: effect of dose and route of administration. NeuroToxicol. 7:601–608, 1986.

Golub, M.S.; Negri, S.R.; Keen, C.L. and Gershwin, M.E.: Developmental toxicity of chronic oral aluminum exposure in mice. (abs) Teratolgy 31:64A only, 1985.

McCormack, K.M.; Ottosen, L.O.; Mayor, G.H.; Sanger, V.L. and Hook, J.B.: The teratogenic effects of aluminum in rats. (abs) Teratology 17:50A only, 1978.

Misawa, T. and Shigeta, S.: Effects of prenatal aluminum treatment on development and behavior in the rat. J. Tox. Sc. 18:43–48, 1993.

Muller, G.; Bernuzzi, V.; Desor, D.; Hutin, M.F.; Burnel, D. and Lehr, P.R.: Developmental alterations in offspring of female rats orally intoxicated by aluminum lactate at different festation periods. Teratology 42:253–261, 1990.

Ridgway, L.P. and Karnofsky, D.A.: The effects of metals on the chick embryo: toxicity and production of abnormalities in development. Ann. N.Y. Acad. Sci. 55:203–215, 1952.

Wide, M.: Effect of short-term exposure to five industrial metals on the embryonic and fetal development of the mouse. Environ. Res. 33:47–53, 1984

Yokel, R.A.: Toxicity of gestational aluminum exposure to the maternal rabbit and offspring. Toxicol. Appl. Pharmacol. 79:121–133, 1985.

85 Aluminum Magnesium Silicate *Magnesium Aluminosilicate* CAS 12511–31–8

Sakai and Moriguchi (1975) administered up to 6000 mg per kg to mice from days 7–12 of gestation and found no fetal changes or postnatal effect.

Sakai, K. and Moriguchi, K.: Effect of magnesium aluminosilicate administered to pregnant mice on pre– and postnatal development of offsprings. (Japanese) Oyo Yakuri 9:703–714, 1975.

86 Aluminum Nitrate

Paternain et al. (1988) gave rats by gavage 180, 360 or 720 mg per kg on days 6–14 of gestation. At all levels the maternal and fetal weight were reduced. Hypoplastic deformed fetal ribs occurred at all dose levels and micrognathia was increased at 360 mg per kg. Other visceral defects appeared in small numbers among fetuses exposed to 360 and 720 mg per kg.

Paternain, J.L.; Domingo, J.L.; Llobet, J.M. and Corbella, J.: Embryotoxic and teratogenic effects of aluminum nitrate in rats upon oral administration. Teratology 38:253–257, 1988.

87

6–Amidino–2–napthyl4–[(4,5–dihydro–1H–imidazol–2–yl) amino] benzoate Dimethanesulfonate *FUT–187*

This enzyme inhibits the decomposition of serum protein. Furuhashi et al. (**1992**) found no ill effects of oral doses up to 720 mg per kg on fertility. Shimamura et al. (**1992**) and Kawanishi et al. (**1992 A**) found no ill effects on fetuses and postnatal development after giving 800 mg per kg during organogenesis in the rat. After 800 mg per kg perinatally Kawanishi et al. (**1992 B**) found no abnormal behavior. Decreased fetal survival was found in rabbit offspring exposed to 100 mg per kg during organogenesis but no teratogenicity was reported.

Furuhashi, T.; Uehara, M.; Kodama, R.; Yoshida, R.; Maruden, A. and Shimamura, K.: Reproductive and developmental toxicity studies of FUT–187 (I)–fertility study in rats with oral administration of FUT–187. The Journal of Toxicological Sciences 17:201–219, 1992.

Furuhashi, T.; Uehara, M.; Takei, A.; Kurihara, H.; Yoshida, R.; Maruden, A. and Shimamura, K.: Reproductive and developmental toxicity studies of FUT–187 (V)–perinatal and postnatal study in rats with oral administration of FUT–187. The Journal of Toxicological Sciences 17:263–286, 1992.

Kawanishi, H.; Shiraaishi, M.; Igarashi, Y.; Takeshima, T.; Toyohara, S.; Imai, S.; Shimamura, K. and Maruden, A.: Reproductive and developmental toxicity studies of FUT–187 (III)–postnatal study in rat f1 offspring from dams treated orally with FUT–187 during the period of fetal organogenesis. The Journal of Toxicological Sciences 17:231–252, 1992.

Shimamura, K.; Terabayashi, M.; Kuramoto, S.; Hasegawa, N.; Terazawa, K. and Maruden, A.: Reproductive and developmental toxicity studies of 6–amidino–2–naphthyl 4–[4, 5–dihydro–1h–imidazol–2–yl) amino] benzoate dimethanesulfonate (FUT–187) (IV)–oral administration of New Zealand white rabbits during the period of fetal organogenesis. The Journal of Toxicological Sciences 17:253–261, 1992.

Shimamura, K.; Terazawa, K.; Terabayashi, M.; Watanabe, K.; Hasegawa, N.; Kuramoto, S.; Yokomoto, Y.; Otani, K. and Maruden, A.: Reproductive and developmental toxicity study of 6–amidino–2–napthyl 4–[4, 5–dihydro–1h–imidazol–2–yl) amino] benzoate dimethanesulfonate (FUT–187) (II) - oral administration to rats during the period of fetal oraganogenesis (prenatal examination). The Journal of Toxicological Sciences 17:221–230, 1992.

88 AMI–25

Ishida et al. (**1994 A,B**) gave this contrast media containing iron oxide to rats before and in the first week of pergnancy and in the peri and postnatal period. In maximum intravenous doses of 17.9 and 28 mg per kg respectively. Some decrease in fetal weight in the perinatal study was found but no decreased fertility or adverse fetal effects were found.

Ishida, S.; Matsuoka, T.; Sugisawa, K.; Ogawa, J.; Fuseya, H. and Watanabe, S.: Fertility study in rats treated intravenously with AMI–25. Yakuri to Chiryo 22:819–830, 1994.

Ishida, S.; Ikeya, M.; Sugisawa, K.; Ogawa, J.; Fuseya, H. and Watanabe, S.: Peri–and post–natal study in rats treated intravenously with AMI–25: Yakuri to Chiryo 22:831–843, 1994.

89 Amantadine Hydrochloride *1–Adamantanamine Hydrochloride* CAS 768–94–5

Nora et al. (1975) reported that a woman taking 100 mg per day during the first trimester gave birth to a child with a single ventricle with pulmonary atresia. Levy et al. (**1991**) report one women who gave birth to 2 normal infants while on treatment. Pandit et al. (**1994**) report an infant with tetralogy of Fallot and tibial hemimelia. Four other exposed mothers had normal offspring.

Vernier et al. (1969) gave this antiviral agent to rats and rabbits and found no evidence of teratogenicity. The rats were maintained for 3 litters on 10 mg per kg and during 1 pregnancy period were given 32 mg per kg per day orally. The rabbits received up to 32 mg per kg from days 6–16 of gestation. Lamar et al. (1970) reported malformations in rat fetuses exposed to 50 and 100 mg per kg. These defects included edema, malrotated hindlimbs and other skeletal defects.

Lamar, J.K.; Calhoun, F.J. and Darr, A.G.: Effects of amantadine hydrochloride on cleavage and embryonic development in the rat and rabbit. (abs) Toxicol. Appl. Pharmacol. 17:272 only, 1970.

Levy, M.; Pastuszak, A. and Koren, G.: Fetal outcome following intrauterine amantadine exposure. Reproductive Toxicology 5:79–81, 1991.

Nora, J.J.; Nora, A.H. and Way, G.L.: Cardiovascular maldevelopment associated with maternal exposure to amantadine. Lancet 2:607 only, 1975.

Pandit, P.B.; Chitayat, D.; Jefferies, A.L.; Landes, A.; Qamar, I.U. and Koren, G.: Tibial hemimelia and tetralogy of fallot associated with first trimester exposure to amantadine. Reproductive Toxicology 8:89–92, 1994.

Vernier, V.G.; harmon, J.B.; Stump, J.M.; Lynes, T.E.;

Vernier, V.G.; Harmon, J.B.; Stump, J.M.; Lynes, T.E.; Marvel, J.P. and Smith, D.H.: The toxicologic and pharmacologic properties of amantadine hydrochloride. Toxicol. Appl. Pharmacol. 15:642–665, 1969.

90 Amaranth *Food Dye and Coloring Red No. 2 3–Hydroxy–4–[(4–sulfo–1–naphthalenyl)azo]–2,7–naphtha lene disulfonic Acid Trisodium Salt* CAS 915–67–3

Shtenberg and Gavrilenko (1970) treated three generations of rats with this food dye in amounts of 1.5 and 15 mg per kg of body weight. The mode of oral administration was not clear. They reported decreased fertility and increased numbers of stillborn in the treated group and an unspecified number of deformed and macerated fetuses. Collins et al. (1972) administered 200 mg per kg to the rat during pregnancy and reported no increase in malformations but a small decrease in viable implants with the larger doses.

Collins and McLaughlin (1973) studied the metabolites of amaranth and found an increase in sternal malformations using sodium naphthionate (100 mg per kg) and skeletal defects using the R–amino salt (200 mg per kg) in rats fed by stomach tube during gestation. The dye used by the Russian workers was obtained from Imperial Chemical Industries and that used by the American workers from H. Kohnstamm and Co., New York.

Holson et al. (1975) were unable to show fetotoxicity in a multilaboratory study. In rat offspring exposed in utero to up to 1250 mg per kg daily, no carcinogenic effect was found (Clode et al., 1987). Keplinger et al. (1974) gavaged rats and fed capsules to rabbits during their organogenesis periods. No adverse fetal or reproductive findings were seen at maximum doses of 150 mg per kg in rats and 15 mg per kg in rabbits.

Clode, S.A.; Hooson, J.; Grant, D. and Butler, W.H.: Long–term toxicity study of amaranth in rats using animals exposed in utero. Food Chem. Toxicol. 25(12):937–946, 1987.

Collins, T.F.X.; McLaughlin, J. and Gray, G.C.: Teratology studies on food colorings. Part I. Embryotoxicity of amaranth (FD and C red No. 2) in rats. Food Cosmet. Toxicol. 10:619–624, 1972.

Collins, T.F.X. and McLaughlin, J.: Teratology studies on food colourings. Part II. Embryotoxicity of R salt and metabolites of amaranth (FD and C red No. 2) in rats. Food Cosmet. Toxicol. 11:355–365, 1973.

Holson, J.F.; Gaines, T.B.; Schumacher, H.J. and Cranmer, M.F.: Is red dye No. 2 Teratogenic? A joint government–industry approach to a toxicological problem. (abs) Toxicol. Appl. Pharmacol. 33:122 only, 1975.

Keplinger, M.L.; Wright, P.L.; Plank, J.B. and Calandra, J.C.: Teratologic studies with FD and C red No. 2 in rats and rabbits. Toxicol. Appl. Pharmacol. 28:209–215, 1974.

Shtenberg, A.I. and Gavrilenko, E.V.: Influence of the food dye amaranth upon the reproductive function and development of progeny on albino rats. (Russian) Vopr. Pitan. 29:66–73, 1970.

91 Ambazone CAS 539–21–9

Rottkay (1987) studied the effect of gavaging 350 mg per kg in mice. Treatment during preimplantation or embryogenesis did not produce malformation increases or weight reduction and postnatal development was normal.

Rottkay, F.V.: Prenatal toxicity of ambazone. Studia Biophysica 117:193–198, 1987.

92 Ambroxol

Trans-4-[2–amino–3,5(dibromo–benzyl)amino] cyclohexanol hydrochloride CAS 18683–91–5

Iida et al. (1981) gave this mucolytic agent orally to rats and rabbits during organogenesis and found no evidence of teratogenesis. At the highest dose (3000 mg per kg, rats; 200 mg per kg, rabbits), maternal toxicity was present and fetal growth retardation was found.

Matsuzawa et al. (1981) studied the effect of this expectorant on rats before conception, during the first 7 days of gestation and on days 17–20 of gestation. Maternal and fetal weight gain was decreased with doses of over 500 mg per kg but no adverse reproductive or postnatal changes were found.

Iida, H.; Kast, A. and Tsunenari, Y.: Teratology studies with ambroxol (NA 872) in rats and rabbits. Oyo Yakuri 21:271–279, 1981.

Matsuzawa, K.; Tanaka, T.; Enjo, H.; Mikita, T. and Hashimoto, Y.: Reproductive studies on ambroxol (NA 872) 1 and 2. Iyakuhin Kenkyu 12:358–387, 1981.

93 Ameltolide *LY201116*

Higdon et al. (**1991**) gave rats and rabbits oral doses of this new anticonvulsant during organogenesis. In the rat maternal and fetal weight gain was reduced at the highest dose (50 mg per kg). In rabbits maternal toxicity was found at 50 and 100 mg per kg. Incomplete digit ossification occurred at the highest level. Teratologic effects were seen in neither species. Pohland and Vavrek (**1991**) found that significant amounts of a radioactive labelled drug appeared in the rat fetus.

Higdon, G.L.; McKinley, E.R. and Markham, J.K.: Ameltolide. I: Developmental toxicology studies of a novel anticonvulsant. Teratology 44:37–44, 1991.

Pohland, R.C. and Vavrek, M.T.: Ameltolide. II: Placental transfer of radiocarbon following the oral administration of a novel anticonvulsant in rats. Teratology 44:45–49, 1991.

94 Americium CAS 7440–35–9

Sikov and Mahlum (1975) injected rats on day 9 with doses of up to 60 microcuries per kg and found no adverse effect. Most of the radioactivity was trapped in the maternal liver.

Sikov, M.R. and Mahlum, D.D.: Toxicity of [241]Am and [244]Cm after administration at nine days of gestation in the rat. Radiation Res. 62:565, 1975.

95 Amesergide

Seyler et al. (**1993**) studied this serotonin antigonist in pregnant rats giving up to 30 mg per kg orally from before pregnancy through postpartum day 21. The pups were smaller but otherwise there were no adverse reproductive effects. The decrease in weight gain was thought to be due to reduced milk production.

Seyler, D.E.; Cohen, I.R. and Sauter, S.: Effects of the serotonin antagonist amesergide on reproduction in female rats. Reproductive Toxicology 7:607–612, 1993.

96 Ametantrone Acetate

1,4–Bis–[[2–[2–hydroxyethyl]amino]ethyl]amino] 9–10–anthracene–dione

Petrere et al. (1986) found no teratogenicity of this antineoplastic in rats using doses of up to 6.0 mg per kg during organogenesis. In rabbits, at doses of 0.4 mg but not 0.2 mg per kg, teratogenicity was observed. Kim et al. (1984) found facial clefts, omphalocele and skeletal defects were found and at the highest dose, 12 percent of the rabbit fetuses were malformed.

Petrere, J.A.; Kim, S.N.; Anderson, J.A.; Fitzgerald, J.E.; De La Iglesia, F.A. and Schardein, J.L.: Teratology studies of ametantrone acetate in rats and rabbits. Teratology 34:271–278, 1986.

Kim, S.N.; Petrere, J.A.; Anderson, J.A.; Fitzgerald, J.E. and De La Iglesia, F.A.: Teratology studies of ametantrone acetate in pregnant rats and rabbits. (abs) Teratology 29:41A only, 1984.

97 Amezinium Methylsulfate CAS 30578–37–1

Satoh and Narama (**1988** A&B) gavaged rats and rabbits with up to 180 and 125 mg per kg respectively. Maternal and fetal toxicity were observed at the highest doses but no evidence of teratogenicity was found. This anti–hypertensive agent was given during organogenesis. Post–natal studies in rats revealed no adverse effects.

Satoh, T. and Narama, I.: Reproduction studies of amezimium emtilsulfate(3)–teratogenicity study in rats. Yakuri to Chiryo 16:1557–1572, 1988.

Satoh, T. and Narama, I.: Reproduction studies of Amezinium metisulfate(4)–teratogenicity study in rabbits. Yakuri to Chiryo 16:1573–1583, 1988.

98 Amfepramone 2–Diethylaminopropiophenone CAS 90–84–6

Cahen et al. (1964) reported that this drug was non–teratogenic in the rat and mouse. The treatment period was during the entire gestation and the dose in the mouse was 16 to 50 mg per kg and in the rat was 7 to 20 mg per kg.

Cahen, R.L.; Sautai, M.; Montagne, J. and Pessonnier, J.: Recherche de l'effet teratogen de la 2–diethylaminopropiophenone. Medicine Experimentale 10:201–224, 1964.

99 Amifostine YM–08310 Ethiofos

Sato et al. (1984) tested this radiation protection agent in pregnant rats and rabbits. A dose of up to 100 mg per kg was given intravenously during organogenesis in the rat and rabbit and in the perinatal period in the rat. Before fertilization in the rat, 50 mg per kg was used. No teratogenicity was found although some fetal weight decrease was found.

Sato, T.; Kaneko, Y. and Narama, I.: Reproduction studies of amifostine (YM 08310) in rabbits and rats. Oyo Yakuri 27:847–884, 1984.

100 Amikacin CAS 37517–28–5

Akutsu et al. (1982) gave up to 200 mg per kg intraperitoneally to rats before fertilization and for seven days afterwards. No evidence of teratogenicity or embryotoxic effects was seen.

Akutsu, S.; Katoh, M.; Kawana, K.; Shinamura, T.; Fukushima, Y. and Matsukaki, M.: Effect of amikacin on reproduction fertility study. Jpn. J. Antibiotics 35:194–204, 1982.

101 Amiloride

Miller and Scott (**1991**) gave mice acetazolamide (500

mg per kg) and amiloride (4 mg and 8 mg) on day 9.5 and produced 77% and 94% supernumerary kidneys.

Miller, T.A. and Scott, W.J.: Pathogenesis of supernumerary kidneys and renal agenesis in mice. (Abs) Teratology 43:429–430, 1991.

102 Aminacrine CAS 90–45–9

This acridine derivative is used as a vaginal topical disinfectant. Heinonen et al. (1977) reported no increase in defects of offspring among 59 women treated in the first 4 lunar months. In a study by Rosa et al. (1987), 113 women reported using aminacrine among a group of 6,564. There was no evidence of an association with heart defects, oral clefts or neural tube defects.

Heinonen, O.P.; Slone, D. and Shapiro, S.: Birth Defects and Drugs in Pregnancy. Publishing Sciences Group Inc., Littleton, Mass., 300–302, 1977.

Rosa, F.W.; Baum, C. and Shaw, M.: Pregnancy outcomes after first-trimester vaginitis drug therapy. Obstet. Gynecol. 69:751–755, 1987.

103 Amine Fluorides Hexadeclamine Hydrofluoride 9–Octadecenylamine Hydrofluoride 2,2-([3–(2 Octadecylamino)propyl]iminodiethanol Dihydro flouride Hydroxyethyl); Hetaflur Dectaflur Olaflur

Smith et al. (1974) fed up to 30 mg per day to rats and rabbits and found no adverse reproductive or fetal effects. At the highest dose, maternal weight gain was reduced in the rabbit.

Smith, J.M.; Rapp, W.R.; Strauss, W.F.; Dolan, M.M. and Yankell, S.L.: Reproductive, subacute and chronic safety evaluation studies of amine fluorides. Toxicol. Appl. Pharmacol. 29:85, 1974.

104 5-[2-Aminoacetamido)methyl]-1-[4-chloro-2-(o-chlorobenzoyl) chlorobenzoyl) phenyl]–N,N–dimethyl–1H–s–triazole–3–carboxamide HCl dihydrate

Hasegawa et al. (1985a) studied this triazolam–like sleep inducer in pregnant rats by gavage. In rats, doses of 200 or 400 mg per kg on days 7–17 produced decreased weights in the dam and growth retardation with cleft palates and extremity anomalies in the fetuses. No adverse effects occurred at 50 or 100 mg per kg. Postnatal behavior studies revealed no differences at 400 mg per kg. In the rabbit, treated at levels 3, 7 or 15 mg per kg orally, fetal weight was reduced but no increase in defects was found (Hasegawa et al., 1985b).

Kobayashi and Itoh (1985) fed up to 400 mg per kg from day 17 through 20 to rats. At 100 and 400 mg per kg per day, maternal weight gain was inhibited and reduced fetal weight and survival was found. No effect was seen at 25 mg per kg.

Hasegawa, Y.; Yoshida, T.; Takegawa, Y.; Miyago, M. and Fukiishi, Y.: Reproduction studies of 5-((2-aminoacetamido) methyl)-1-(4-chloro-2-(o–chlorobenzoyl)phenyl)–N, N–dimethyl–1H–s–triazole–3–carboxamide hydrochloride dihydrate. Oral administration during the period of fetal organogenesis in rats. Oyo Yakuri 30(5):765–791, 1985a.

Hasegawa, Y.; Yamagata, H.; Hirashiba, M.; Yoshida, T. and Fukiishi, Y.: Reproduction studies of 5–((2–aminoacetamido) methyl)–1–(4–chloro–2–(o–chlorobenzoyl)phenyl)–N, N–dimethyl–1H–s–triazole–3–carboxamide hydrochloride dihydrate. Oral administration during the period of fetal organogenesis in rabbits. Oyo Yakuri 30(5):819–833, 1985b.

Kobayashi, F. and Itoh, M.: Reproduction studies of 5–[(2–aminoacetamido)methyl]–1–[4–chloro–1– (o–chlorobenzoyl) phenyl]–N,N–dimethyl–1H–s–triazole–3–carboxamide hydrochloride dihydrate. Perinatal and postnatal study in rats. Oyo Yakuri 30:793–818, 1985.

105 Aminoacetonitrile also see *Lathyrism* CAS 540–61–4

This nitrile compound produces the lathyrism syndrome in fetuses (Stamler, 1955).

Stamler, F.W.: Reproduction in rats fed lathyrus peas or aminonitriles. Proc. Soc. Exp. Biol. Med. 90:294–298, 1955.

106 2–Aminoanthracene CAS 613–13–8

Martin and Erickson (1982) fed this compound to mice throughout pregnancy in amounts of 5 or 10 mg per kg. Skeletal anomalies and reduced spermatogonia and oocytes were found postnatally.

Martin, P.G. and Erickson, B.H.: Teratological and reproductive effects of ingested 2–aminoanthracene in CD–1 mice. (abs) Teratology 25:61A only, 1982.

107 Aminoazobenzene and Derivatives CAS 60–09–3

Sugiyama et al., (1960) used the following derivatives: monomethyl M–fluorodimethyl, p–chlorodimethyl and dimethyl. In all cases, the derivatives were teratogenic in mice producing skeletal defects with a few cleft palates. Dosages of 200 to 700 mg per kg were used on single days during embryogenesis.

Sugiyama, T.; Nishimura, H. and Fukui, K.: Abnormalities in mouse embryos induced by several aminoazobenzene derivatives. Okajimas Folia Anat. Jpn. 36:195–206, 1960.

108 Aminoazotoluene CAS 97–56–3

Kolesnichenko et al. (1978) administered this chemical to CBA mice intragastrically, thrice, during 4–5 days before delivery in sublethal doses (12 mg per kg). Numerous benign and malignant tumors of the offspring's liver were found.

Kolesnichenko, T.S.; Popova, N.V. and Shabad, L.M.: Tumors of the liver in mice induced by prenatal and postnatal administration of orthoaminoazotoluol. Byull. Eksper. Med. (USSR) (85)2:199, 1978.

109 2–p–Aminobenzene–sulfonamido–1,3,4 thiadiazole–5–sulfonamide

Maren (1972) injected 600 mg per kg subcutaneously on days 10 and 11 of rat pregnancy and did not find increased malformations.

Maren, T.H. and Ellison, A.C.: The teratological effect of certain thiadiazoles related to acetazolamide with a note on sulfanilamide and thiazide diuretics. Johns Hopkins Med J. 130:95–104, 1972.

110 p–Aminobenzoic Acid CAS 150–13–0

Ershoff (1946) fed 1 percent in the diet before and during pregnancy and found no adverse reproductive effects or malformations in rats. Telford et al. (1962) fed about 25 mg per kg to rats and reported a 23 percent resorption rate. Kato (1973) using 50 mg per kg on days 9–11, found no teratogenicity and no increase in resorptions.

Ershoff, B.H.: Effects of massive doses of p–aminobenzoic acid and inositol on reproduction in the rat. Soc. Exper. Biol. Med. 63:479–480, 1946.

Kato, T.: Effect of folate metabolism–related factors on the teratogenic action of sulfonamide in mice. Cong. Anom. 13:85–92, 1973.

Telford, I.R.; Woodruff, C.S. and Linford, R.H.: Fetal resorption in the rat as influenced by certain antioxidants. Am. J. Anatomy 110:29–36, 1962.

111 p–Aminobenzoic acid–N–xyloside–Na Salt CAS 72782–43–5

This antineoplastic agent was given orally by Matsumoto et al. (1982) to mice and rabbits in doses of up to 1250 and 5000 mg per kg per day. No effect on fertility or postnatal function was seen and in neither species was there any adverse fetal effect.

Matsumoto, M.; Nishiyama, M.; Hayashi, H.; Ikuzawa, M.; Seto, T. and Tanaka, O.: Reproductive studies of p–aminobenzoic acid–N–xyloside Na salt. Kiso to Rinsho 16:4029–4036,4037–4066,4067–4073,4074–4099, 1982.

112 2–Aminoethyl–isothiuronium bromide–hydrobromide

Mazur (1985) injected mice with 40 mg per kg intraperitoneally on day zero and found reduced implantation and increased resorptions.

Mazur, L.: Embryonic survival in mice treated with AET, MEA or 5–HT on the first day of pregnancy. Acta Physiologica

113 Aminobutyric Acid

Kohler et al. (1973) gave mice 400 or 800 mg per kg on day 9 and found no adverse reproduction.

Kohler, F.; Ockenfels, H. and Meise, W.: Teratogenicity of n–phthalylglycine and 4–phthalimidobutyric acid. Pharmazie 28: 680–681, 1985.

114 4–Amino Folic Acid

Discussed under Aminopterin

115 Aminoguanidine CAS 79–17–4

Neuman and McCoy (1955) produced inhibition of chick liver development by yolk sac injection with 2 to 20 mg of aminoguanidine on the 2nd through the 4th day of incubation. West (1962) gave daily subcutaneous injections to pregnant rats using 20 mg per kg and found no increase in fetal death.

Neuman, R.E. and McCoy, T.A.: Inhibition of development of chick embryo liver by aminoguanidine. Proc. Soc. Exp. Biol. Med. 90:339–342, 1955.

West, G.B.: Drugs and rat pregnancy. Letter to the editor. J. Pharm. Pharmacol. 14:828–830, 1962.

116   Amino–mercaptopurine

Discussed under 6–Mercaptopurine

117   2–Amino–3–methylpyridine

Morimoto and Bederka (1974) gave mice intraperitoneal doses of an unspecified amount on days 7, 9, 11, 13 and 15 and found neural tube defects increased.

Morimoto, R.I. and Bederka, J.P.: Teratogenesis of nicotinamide analogues in the 1CR mouse. (abs) Teratology 9:A29–A30, 1974.

118   6–Amino–3–methylpyridine

Morimoto and Bederka (1974) gave mice intraperitoneal doses of an unspecified amount on days 7, 9, 11, 13 and 15 and found no adverse fetal effects.

Morimoto, R.I. and Bederka, J.P.: Teratogenesis of nicotinamide analogues in the ICR mouse. (abs) Teratology 9:29A–30A, 1974.

119   6–Aminonicotinamide CAS 329–89–5

Landauer (1957) injected 10 microgm of this niacin antagonist into chick eggs at 96 hours and observed that nearly all survivors had micromelia. This effect could be prevented by concomitant niacin administration. Since Dr. Landauer's work, 6–aminonicotinamide became a common model for producing hydrocephalus, cleft palate and skeletal defects in the mouse (Pinsky and Fraser, 1959; Curley et al., 1968) and rat (Chamberlain and Nelson, 1963; Chamberlain, 1966). Chamberlain (1966) reported production of cleft palate in 100 percent of rat fetuses when the mother was given 8 mg per kg on the 15th gestational day. Defects of the urogenital and cardiovascular systems were produced by similar injections on the 10th day (Chamberlain and Nelson, 1963). Schardein et al. (1967) produced defects in the rabbit fetus. Courtney and Valerio (1968) treated monkeys on day 26 to 29 and produced four abortions and three normal fetuses.

The mechanism of action of this teratogen was suggested by the work of Dietrich et al. (1958) who demonstrated in mouse liver insertion of 6–aminonicotinamide into the analogues of nicotinamide adenine dinucleotide (NAD) and nicotinamide adenine dinucleotide phosphate (NADP). These analogues were non–physiologic under in vitro conditions. Re et al. (1984) fed this compound to rats in amounts up to 1.0 percent of the diet. No adverse effects occurred even though the weight gain of the dams was reduced.

Chamberlain, J.G.: Development of cleft palate induced by 6–aminonicotinamide late in rat gestation. Anat. Rec. 156:31–40, 1966.

Chamberlain, J.G. and Nelson, M.M.: Congenital abnormalities in the rat resulting from single injections of 6–aminonicotinamide during pregnancy. J. Exp. Zool. 153:285–299, 1963.

Courtney, K.D. and Valerio, D.A.: Teratology of the macaca mulatta. Teratology 1:163–172, 1968.

Curley, F.J.; Ingalls, T.H. and Zappasodi, P.: 6–aminonicotinamide–induced skeletal malformations in mice. Arch. Environ. Health 16:309–315, 1968.

Dietrich, L.S.; Friedland, I.M. and Kaplan, L.A.: Pyridine nucleotide metabolism: mechanism of action of niacin antagonist, 6–aminonicotinamide. J. Biol. Chem. 233:964–968, 1958.

Landauer, W.: Niacin antagonists and chick development. J. Exp. Zool. 136:509–530, 1957.

Pinsky, L. and Fraser, F.C.: Production of skeletal malformations in the offspring of pregnant mice treated with 6–aminonicotinamide. Biol. Neonate 1:106–111, 1959.

Re, T.A.; Loehr, R.F.; Rodriguez, S.C.; Rodwell, D.G. and Burnett, C.M.: Results of teratogenicity testing of M–aminophenol in Sprague–Dawley rats. Fund. Appl. Toxicol. 4:98–104, 1984.

Schardein, J.L.; Woosley, E.T.; Peltzer, M.A. and Kaup, D.H.: Congenital malformations induced by 6–aminonicotinamide in rabbit kits. Exp. Mol. Pathol. 6:335–346, 1967.

120   p–Aminophenol CAS 23–30–8

p–Aminophenol, a metabolic product of acetaminophen, was administered intraperitoneally or intravenously to hamsters on day 8 of gestation. At doses of 100–250 mg, neural tube, eye and skeletal defects were increased. The ortho form was also teratogenic but no evidence for teratogenicity of the meta form was obtained (Rutkowski and Ferm, 1982).

Spengler et al. (1986) gave rats up to 250 mg per kg orally on days 6–15. At the highest dose, maternal toxicity and teratogenicity was found, but at 85 mg per kg and lower, no adverse effects were found.

Rutkowski, J.V. and Ferm, V.H.: Comparison of the teratogenic effects of isomeric forms of aminophenol in the Syrian Golden Hamster. Toxicol. Appl. Pharmacol. 63:264–269, 1982.

Spengler, J.; Osterburg, I. and Korte, R.: Teratogenic evaluation of p–toluenediamine sulphate, resorcinol and p–aminophenol in rats and rabbits. (abs) Teratology 33:31A, 1986.

121   Aminophyllin also see *Theophylline* CAS 317–34–0

Georges and Denef (1968) reported that 100 to 200 mg per kg given subcutaneously to the rat fetus on days 1–17 produced a low but significant number of digital defects, most of which were localized to the posterior left limb.

Georges, A. and Denef, J.: Les anomalies digitales: manifestations teratogeniques des derives xanthiques chez le rat. Arch. Int. Pharmacodyn. Ther. 172:219–222, 1968.

122 β–**Aminopropionitrile** also see *Lathyrism* CAS 151–18–8

Stamler (1955) produced the lathyrus syndrome in rat fetuses by feeding diets containing 0.025 percent after the 17th day of gestation. Rosenberg (1957) injected 0.125 to 0.5 mg into the yolk sac of eggs on the 7th day and produced bowing of the long bones and scoliosis.

Rosenberg, E.E.: Teratogenic effects of β–aminopropionitrile in the chick embryo. Nature 180:706–707, 1957.

Stamler, F.W.: Reproduction in rats fed lathyrus peas or aminonitriles. Proc. Soc. Exp. Biol. Med. 90:294–298, 1955.

123 Aminopterin *4–Aminopteroylglutamic Acid* CAS 54–62–6

Since Thiersch and Phillips (1950) first observed the embryocidal effect of this folic acid antagonist, a large body of information became available on the subject (see Folic Acid Deficiency).

Thiersch (1952) induced therapeutic abortions in 12 patients in whom surgical interruption was not indicated and detected malformations (hydrocephalus, cleft palate and meningomyelocele) in three fetuses . The total dosage was 6 to 12 mg given over several days. There are 3 case reports of malformed children born to mothers ingesting the drug. Small stature, abnormal cranial ossification, arched palate and reduction in derivatives of the first branchial arch were observed (Meltzer, 1956; Shaw and Steinbach, 1968; Warkany et al., 1959). Shaw (1972) gave a nine–year follow–up on the aminopterin–affected child he previously described.

Baranov (1966a) gave rats 0.1 mg per kg of body weight on the 6th day and observed defects of the eye, face, skull, brain, extremities, abdominal wall and tail. A complete mitotic block was noted in the ectotrophoblast, but not in the embryoblast (Baranov, 1966b). Dyban et al. (1977) cultured eight–cell–blastomeres of rat embryos in Bigger's medium with the addition of 20 percent rat serum from animals treated with different doses of this drug. Retardation of cleavage rate, abnormalities of blastocyst formation and damage of inner cell mass were observed. Puchkov (1967) applied aminopterin locally to different parts of the chick embryo and produced abnormalities of the head, eye, limbs and trunk.

Baranov, V.S.: The specificity of the teratogenic effect of aminopterin as compared to other teratogenic agents. Bull. Exp. Biol. (Russian) 1:77–82, 1966a.

Baranov, V.S.: Mechanism of aminopterin pathogenic effect upon embryogenesis in the albino rat. Arch. Anat. (Russian) 51(8):17–28, 1966b.

Dyban, A.P.; Sekirina, G.G. and Golinsky, G.F.: The effect of aminopterin on the preimplantation rat embryos cultivated in vitro. Ontogenez, 8.2:121–127, 1977.

Meltzer, H.J.: Congenital anomalies due to attempted abortion with 4–aminopteroglutamic acid. JAMA 161:1253, 1956.

Puchkov, V.F.: Teratogenic action of aminopterin and 5–

fluorouracil on 4 to 23 somite chick embryos after application in ovo. Bull. Exp. Biol. (Russian) 7:99–102, 1967.

Shaw, E.B.: Fetal damage due to maternal aminopterin ingestion, follow–up at age 9 years. Am. J. Dis. Child. 124:93–94, 1972.

Shaw, E.B. and Steinbach, H.L.: Aminopterin–induced fetal malformation: survival of infant after attempted abortion. Am. J. Dis. Child. 115:477–482, 1968.

Thiersch, J.B.: Therapeutic abortions with folic acid antagonist 4–aminopteroylglutamic acid (4–amino P.G.A.) administered by oral route. Am. J. Obstet. Gynecol. 63:1298–1304, 1952.

Thiersch, J.B. and Phillips, F.S.: Effect of 4–amino–pteroylglutamic acid (aminopterin) on early pregnancy. Proc. Soc. Exp. Biol. Med. 74:204–208, 1950.

Warkany, J.; Beaudry, P.H. and Hornstein, S.: Attempted abortion with aminopterin (4–aminopteroylglutamic acid). A.M.A. J. Dis. Child. 97:274–281, 1959.

124 4–Amino–pteroylaspartic Acid CAS 25312–31–6

This folic acid antagonist was used to produce defects in the mouse (Tuchmann–Duplessis and Mercier–Parot, 1957).

Tuchmann–Duplessis, H. and Mercier–Parot, L.: Production de malformations chez la souris par administration d'acide x–methylfolique. C. R. Soc. Biol. 151:1855–1857, 1957.

125 Aminopyrine CAS 58–15–1

Loosli et al. (1964) found no deleterious effects in pregnant rats (150 mg per kg), rabbits (90 mg per kg) and mice (180 mg per kg) treated daily during most of their gestation. Nomura et al. (1977) injected mice subcutaneously on days 9–11 and produced defective fetuses with doses of 0.2 mg per gm of body weight. The defects were mostly ruptured omphaloceles. Sanyal et al. (1981) could not demonstrate teratogenicity in vitro in the rat embryo at concentrations of 80 microgms per ml medium.

This drug was tested in pregnant rats and mice with oral dosages of 150–200 mg per kg together with sodium nitrate on day 21 or 16, 17 and 18, respectively. Fetal pulmonary tumors, leukemia and mammary gland tumors were observed but there were no congenital malformations. Carcinogenic effect was eliminated by simultaneous injection of ascorbic acid (Alexandrov and Napalkov, 1979).

Alexandrov, V.A. and Napalkov, N.P.: Transplacental carcinogenesis effect as a result of a combined injection of aminopyrine and nitrite in mice. Vopr. Onkol. (USSR) 25.7:48–52, 1979.

Loosli, R.; Loustalot, P.; Schalch, W.R.; Sievers, K. and Stenger, E.G.: Joint study in teratogenicity research in some factors affecting drug toxicity. Proceedings of the European Society for the Study of Drug Toxicity. Cambridge: England, 214–216, 1964.

Nomura, T.; Isa, Y.; Tanaka, H.; Kanzaki, T.; Kimura, S. and Sakamoto, Y.: Teratogenicity of aminopyrine and its molecular compound with barbital. (abs) Teratology 16:118 only, 1977.

Sanyal, M.K.; Kitchin, K.T. and Dixon, R.L.: Rat conceptus development in vitro: Comparative effects of alkalating agents. Toxicol. Appl. Pharmacol. 57:14–19, 1981.

126 p–Aminosalicylic Acid CAS 65–49–6

Heinonen et al. (1977) reported that among 43 women who took this drug in the first 4 lunar months, 5 had children with congenital defects.

Heinonen, O.P.; Slone, D. and Shapiro, S.: Birth Defects and Drugs in Pregnancy. Publishing Sciences Group Inc., Littleton, Mass., 1977.

127 2–Amino–1,3,4–thiadiazole CAS 4005–51–0

Beaudoin (1973) produced congenital defects in rats injected peritoneally with 25 to 200 mg per kg. The most sensitive period for defect production was during days 9 through 13. Eye defects, hydrocephalus, skeletal reduction defects and cleft palate were found. Nicotinamide supplementation diminished the teratogenic action.

Beaudoin, A.R.: Teratogenic activity of 2–amino–1,3,4–thiadiazole hydrochloride in Wistar rats and the protection afforded by nicotinamide. Teratology 7:65–72, 1973.

128 2–Amino–1,3,4–thiadiazole–5–sulfonamide

Discussed under Acetazolamide

129 3–Amino 1,2,4–triazole Amitrole™ CAS 61–82–5

Landauer et al. (1971) injected 20 to 40 mg of this herbicide into chick yolk sacs at 0 to 96 hours of incubation. Abnormalities of the beak and occasionally bent tibias were produced.

Landauer, W.; Salam, N. and Sopher, D.: The herbicide 3–amino–1,2,4–triazole (Amitrole™) as teratogen. Environ. Res. 4:539–543, 1971.

130 Amiodarone™

This anti–arrhythmic drug contains 39 percent iodine. Laurent et al. (1987) and De Wolf et al. (1988) reported congenital hypothyroidism in newborns born to mothers treated with 200 mg per day. The diffuse goiters and hypothyroidism subsequently subsided with thyroid replacement. Ovadia et al. (**1994**) reported 3 pregnancies where 200 to 600 mg was used daily and found one normal newborn, one incompletely autopsied normal fetus and an infant with a large ventricular defect. Yamada and Collin (**1992**) treated rats with up to 100 mg per kg during organogenesis. At the highest dose reduced survival and slight edema were found. No teratogenic findings were identified. Perinatal and prenatal studies were also performed. In rabbits doses of up to 90 mg per kg did not produce an increase in defects but the fetal weight was reduced.

De Wolf, D.; De Schepper, J.; Verhaaren, H.; Deneyer, M.; Smitz, J. and Sacre–Smits, L.: Congenital hypothyroid goiter and amiodarone. Acta Paediatr. Scand. 77:616–618, 1988.

Laurent, M.; Betremieux, P.; Biron, Y. and LeHelloco, A.: Neonatal hypothyroidism after treatment by amiodarone during pregnancy. Am. J. Cardiol. 60:942, 1987

Ovadia, M.; Brito, M.; Hoyerr, G. and Marcus, F.: Human experience with amiodarone in the embryonic period. The American Journal of Cardiology 73:316–317, 1994.

Yamada, T. and Collin, A.: Reproductive and developmental toxicity studies of amiodarone. The Clinical Report 26:3871–3885, 1992.

131 Amitraz

Palermo–Neto et al. (**1994**) gavaged rats every 3rd day with 20 mg per kg. The pups were crossfostered and no significant differences were found in open–field behaviors. Some transient delays were found.

Palermo–Neto, J.; Florio, J.C. and Sakate, M.: Developmental and behavioral effects of prenatal amitraz exposure in rats. Neurotoxicology and Teratology 16:65–70, 1994.

132 Amitriptyline Elavil™ CAS 50–48–6

Heinonen et al. (1977) reported that among 21 women taking this drug in the first 4 lunar months, there was no increase in offspring with defects. Imipramine, a closely related antidepressant, is discussed under its separate heading.

Beyer et al. (1984) gave hamsters 70 mg per kg intraperitoneally on day 8 of pregnancy and found a small but significant increase in encephaloceles and bent tails in the fetuses. Briggs et al. (1983) summarized 86 pregnancies with first trimester exposures and found no increase in defects.

Beyer, B.K.; Guram, M.S. and Geber, W.F.: Incidence and potentiation of external and internal fetal anomalies resulting from chlordiazepoxide and amitriptyline alone and in combination. Teratology 30:39–45, 1984.

Briggs, G.G.; Bodendorfer, T.W.; Freeman, R.K. and Yaffe, S.J.: Drugs in Pregnancy and Lactation. Williams and Wilkins, Baltimore, 1983.

Heinonen, O.P.; Slone, D. and Shapiro, S.: Birth Defects and Drugs in Pregnancy. Publishing Sciences Group Inc., Littleton, Mass. 1977.

133 Amlodipine

This calcium channel blocker was studied in rats and rabbits by Horimoto et al. (**1991**). Using oral doses of up to 25 mg per kg in the rat no adverse reproductive or fetal effects were found when treatment was before and during early prergnancy or during organogenesis. Perinatal doses of 10 mg per kg caused dystocia and an increase in stillbirths but no postnatal behavioral changes.

Horimoto, M.; Takeuchi, K.; Iijima, M. and Tachibana, M.: Reprroductive and developmental toxicity studies with amlodipine in rats and rabits. Oyo Yakuri 42:167–176, 1991.

134 Ammonium Chloride CAS 12125–02–9

Goldman and Yakovac (1964) gave one–sixth molar ammonium chloride to mice orally in the drinking water after day 7 during pregnancy and although the offspring were small sized, no congenital defects were found.

Weaver and Scott (1984) gave mice 600 mg per kg orally at 8 and 10 am and 12 and 1 pm on day 10 of gestation and produced 7 percent ectrodactyly in the offspring.

Goldman, A.S. and Yakovac, W.C.: Salicylate intoxication and congenital anomalies. Arch. Environ. Health 8:648–656, 1964.

Weaver, T.E. and Scott, W.J.: Acetazolamide teratogenesis: Interaction of maternal metabolic and respiratory acidosis in the induction of ectrodactyly in C57BL–6J mice. Teratology 30:195–202, 1984.

135 Ammonium Perfluorooctanoate

Unpublished reports by the EPA found discoloration of embryonic lens eye nucleus in rats. Staples et al. (1984) were unable to confirm the studies.

Staples, R.E.; Burgess, B.A. and Kerns, W.D.: The embryo–fetal toxicity and teratogenic potential of ammonium perfluorooctanoate (APFO) in the rat. Fund. Appl. Toxicol. 4:429–440, 1984.

136 Ammonium–o–sulfobenzoic Acid

Discussed under Saccharin

137 Amniocentesis

One might expect a higher incidence of cleft palate in infants developing in utero with reduced amounts of amniotic fluid (i.e. Potter's syndrome). The absence of this association may be due to differences in fetal posture between the rodent and human. Congenital defects in humans from amniocentesis performed after the first trimester were not found (Parrish et al., 1957). In a multicenter study, the incidence of unexplained respiratory difficulties in newborns subjected to amniocentesis was 1.3 percent as compared to 0.4 percent in matched controls. Postural deformities were 1.0 percent as compared to 0.2 percent (Anonymous, 1978).

Hislop and Fairweather (1982) found significant decreases in the number of alveoli in lungs from monkey fetuses, after amniocentesis, of an unspecified amount of amniotic fluid. In their preliminary findings the birth weight and fixed lung volume were reduced in the fetuses exposed to amniocentesis on days 47–64 (equivalent of 14–17 weeks in the human).

Kennedy and Persaud (1977) punctured the amniotic sac of the rat on day 17 and produced fetal edema and hemorrhage which led to limb reduction and other compression defects. Trasler et al., (1956) and Jost (1956) observed defects in mice and rats, respectively, after amniocentesis. The reduction in the amount of amniotic fluid results in contraction of the uterus with compression of the head on the chest. This in turn lodges the tongue between the palatine shelves and prevents closure of the palate. In addition to cleft palate, stiff extremities, club foot, adactyly, microstomia and short umbilical cords resulted from amniocentesis of the rat on days 14.5, 15.5 or 16.5. The subject was reviewed by Kendrick and Feild (1967) and Demyer and Baird (1969).

Anonymous: Medical research working party. An as-

sessment of the hazards of amniocentesis. Br. J. Obstet. Gynaecol. Suppl. 2:85, 1978.

Demyer, W. and Baird, I.: Mortality and skeletal malformations from amniocentesis and oligohydramnios in rats: cleft palate, club foot, microstomia and adactyly. Teratology 2:33–38, 1969.

Hislop, A. and Fairweather, D.V.I.: Amniocentesis and lung growth: An animal experiment with clinical implications. Lancet 2:1271–1272, 1982.

Jost, A.: The age factor in some prenatal endocrine events. Ciba Found. Coll. Ageing 2:18–27, 1956.

Kendrick, F.J. and Feild, L.E.: Congenital anomalies induced in normal and adrenalectomized rats by amniocentesis. Anat. Rec. 159:353–356, 1967.

Kennedy, L.A. and Persaud, T.V.N.: Pathogenesis of developmental defects induced in the rat by amniotic sac puncture. Acta Anat. 97:23–25, 1977.

Parrish, H.M.; Lock, F.R. and Rountree, M.E.: Lack of congenital malformations in normal human pregnancies after transabdominal amniocentesis. Science 126:77 only, 1957.

Trasler, D.G.; Walker, B.E. and Fraser, F.C.: Congenital malformations produced by amniotic–sac puncture. Science 124:439 only, 1956.

138 Amobarbital CAS 57–43–2

Discussed under Barbituric Acid

139 Amosulalol HCl

Schardein et al. (1988) studied α and β–adrenoceptor blocker in rats using up to 1,000 mg per kg daily. Gavage feedings were made before fertilization and continued through day 7 of gestation. Fertility was decreased at 300 and 1000 mg daily doses but no effect was found at 30 mg per day. No fetal toxicity or teratogenicity were found.

Schardein, J.L.; Schwartz, C.; Miller, J.M. and Shibata, M.: Effects on reproduction: Study by oral administration of amosulalol hydrochloride before and in the early stage of gestation. Pharmacometrics 35:151–158, 1988.

140 Amoxanox
2–Amino–7–isopropyl–5–oxo–5H-(1)benzopyrano(2,3-b) pyridine–3–carboxylic Acid

Ihara et al. (1985A, 1985B) administered this antiallergic drug to rats and rabbits by gavage in doses of up to 300 mg per kg daily on days 6–17 or 6–18, respectively, and found no adverse fetal effects.

Ihara, T.; Sugitani, T.; Yoshida, T. and Negishi, R.: Teratogenic effects of amoxanox (AA–673) in the rat. Yakuri to Chiryo 13(9):4885–4900, 1985A.

Ihara, T.; Nakatsu, T.; Kanamori, H. and Hisada, K.: Teratogenic effects of Amoxanox (AA–673) in the rabbit. Yakuri to Chiryo 13(9):4901–4908, 1985B.

141 Amphetamine Sulfate

Discussed under Dextroamphetamine Sulfate

142 **Amphotericin B** CAS 1397–89–3

Ismail and Lerner (1982) reported a single case and reviewed the available data on 21 other cases. There was no evidence of teratogenicity but reversible renal damage and hypokalemia were found in some infants after birth.

Ismail, M.A. and Lerner, S.A.: Disseminated blastomycosis in a pregnant woman. Am. Rev. Respir. Dis. 126:350–353, 1982.

143 **Ampicillin** CAS 69–53–4

Bachev et al. (1974) could not produce malformations in rats given 100 mg per kg during pregnancy. Korzhova et al. (1981) fed rats 250 mg per kg on the 4th through the 13th or 15th through the 20th days of gestation. No teratogenic effect was found but the fetuses were smaller than the controls. Pregnancy was prolonged.

Bachev, S.; Petrova, L.; Voicheva, V.; Shishkova, M. and Kolev, N.: Experimental studies on the teratogenic effect, acute and chronic toxicity of ampicillin. Suvrem. Med. 25:28–32, 1974.

Korzhova, V.V.; Lisitsyna, N.T. and Mikhailova, E.G.: Effect of ampicillin and oxacillin on fetal and neonatal development. Bull Exp. Med (USSR) 91:169–171, 1981.

144 **Ampiroxicam**

Horimoto et al. (**1991**) gave rats this new non–steroidal anti–inflammatory agent orally in the pre and early stages of pregnancy and during organogenesis. The maximum dose was 3.5 mg per kg in the early studies and 7.0 mg per kg during pregnancy. No adverse effects were found on the fetus or postnatal development. In rabbits doses of up to 75 mg were given without teratogenicity.

Horimoto, M.; Takatsu, S.; Takeuchi, K.; Iijima, M. and Tachibana, M.: Reproductive and developmental toxicity studies with amipiroxicam in rats and rabbits. Oyo Yakuri 42:559–569, 1991.

145 **Amrinone**

Nishikawa et al. (1989) applied 0.1 to 2 micromoles to chick embryos and found increased heart defects. Komai et al. (**1990** A) gave rats up to 50 mg per kg intravenously on days 7–17. At 50 mg per kg there was an increase in skeletal changes and growth retardation. In rabbits Komai et al. (**1990** B) gave up to 15 mg per kg intravenously on days 6–18 and found some increase in fetal mortality at the highest dose but no teratogenicity.

Komai, Y.; Iriyama, ; Itoh, I.; Hattori, M.; Inoue, S.; Ishimura, KK. and Shibata, M.: Teratology study or amrinone in rats by subcutaneous treatment. Kiso to Rinsho 24:1883–1892, 1990.

Komai, Y.; Hattori, M.; Inoue, S.; Hibino, H.; Ishimura, K. and Shibata, M.: Teratology study of amrinone in rabbits. iso to Rinsho 24:415–421, 1990.

Nishiawa, T.; Kasajima, T. and Kanai, T.: Cardiovascular malformations associated with administration of amrinone to young chick embryos. Teratology 40:685, 1989.

146 **M–AMSA** *4'–(9–acridinylamino)* *methane–sulfon–m–anisidide*

Anderson et al. (**1986**) found embryolethality at 2.0 mg per kg on days 6 to 9 in the rat. Embryotoxicity but no teratogenicity was found at dose levels of 0.5 and 1.0 mg per kg. This antineoplastic agent which targets DNA topoisomerase II was studied in whole rat embryo cultures by Mirkes and Zwelling (1990). Abnormal embryos and growth decrease were seen at 100mM. As a control o–AMSA was used and at 100 micromolar abnormals appeared. The abnormalities were hypoplasia of the prosencephalon, microophthalmia and edema of the rhombencephalon.

Anderson, J.A.; Petrere, J.A.; Sakowski, R.; Fitzgerald, J.E. and DelaIglesia, F.A.: Teratology study in rats with amsacine, an antineoplastic agent. Fund. Appl. Toxicol. 7:214–220, 1986.

Mirkes, P.E. and Zwelling, L.A.: Embryotoxicity of the intercalating agents–m–AMSA and o–AMSA and the epipodophyllotoxin vp–16 in postimplantation rat embryos in vitro. Teratology 41:679–688, 1990.

147 **Amsacrine**
N–(4–(9–Acridinylamino)-3-methoxyphenyl) *methanesulfonamide* CAS 51264–14–3

Anderson et al. (1986) studied this antineoplastic agent in rats giving intraperitoneal doses of 0.5, 1.0 or 2.0 mg per kg on days 6–9. Decreased litter size and fetal weight occurred at the highest dose but teratogenicity was not found. Ng et al. (1987) injected rats intraperitoneally with this antitumor agent on days 6–15 at 0.5 mg per kg and produced eye, jaw and other skeletal defects.

Anderson, J.A.; Petrere, J.A.; Sakowski, R.; Fitzgerald, J.E. and De La Iglesia, F.A.: Teratology study in rats with amsacrine, an antineoplastic agent. Fund. Appl. Toxicol. 7:214–220, 1986.

Ng, W.W.; Anderson, J.A. and Sakowski, R.: Teratogenicity of amsacrine lactate given IP to rats during the entire organogenesis period. Teratology 35:76A, 1987.

148 **Amsulsin** *YM617)*

This phenylethylamine derivative was given orally to rats in doses up to 300 mg per kg on days 7–17 of gestation and no fetal ill effects were found (Watanabe and Fujiwara, **1991**). No ill effects were found in rabit fetuses exposed to up to 50 mg per kg during organogenesis (Itou and Fujiwara, **1991**).

Itou, S. and Fujiwara, M.: Teratology study of orally administered amusulosin hydrochloride (YM617) in rabbits. Oyo Yakuri 41:39–42, 1991.

Watanabe, T. and Fujiwara, M.: Teratology study of orally administered amsulosin hydrochloride (YM617) in rats. Oyo Yakuri 41:31–37, 1991.

149 **Anabasin Hydrochloride**
α-*Piperidile*–β-*pyridine-HCl* CAS 15251–47–5

Keeler et al. (1984) isolated this compound from wild tree tobacco (Nicotiana glauca) and showed that 2.6 mg per kg given orally to swine produced arthrogryposis and cleft palate in the offspring. Ryabchenko et al. (1982) administered this compound to rats at doses up to 15 mg per kg on days 1–16 of gestation and found slight embryotoxic effect. In lower doses (5 mg per kg), no teratogenic or embryotoxic activity was found. Doses of 3 mg per kg produced no damage to rabbit embryos and did not influence postnatal development.

Keeler, R.F.; Crowe, M.W. and Lambert, E.A.: Teratogenicity in swine of tobacco alkaloid anabasine isolated from Nicotiana glauca. Teratology 30:61–69, 1984.

Ryabchenko, V.P.: Influence of anabasin hydrochloride on the embryogenesis of albino rats and rabbits. Farmakol. Toksikol. (USSR) 1:87–90, 1982.

150 Anagyrine *Lupin Alkaloids* CAS 486–89–5

Kilgore et al. (1981) reported a child with deformities of the legs and hands and radial aplasia. The mother had ingested goat's milk contaminated with anagyrine. At the same time a dog and a goat on the farm gave birth to offspring with similar leg deformities. The dog had ingested goat's milk.

Keeler (1973) implicated this alkaloid found in lupins to be the compound which produces crooked calf disease, a congenital syndrome of calves which includes scoliosis, arthrogryposis and/or cleft palate. The compound was found as a major alkaloid in the plant samples that were associated with the defects.

Keeler, R.F.: Lupin alkaloids from teratogenic and nonteratogenic lupins. Teratology 7:23–30, 1973.

Kilgore, W.W.; Crosby, D.G.; Craigmill, A.L. and Poppen, N.K.: Toxic plants as possible human teratogens. Calif. Agricult. 35:6 only, 1981.

151 Anatoxin–a

Astrachan et al. (1980) injected hamsters intraperitoneally on the 8th day of gestation and on subsequent days, with up to 0.6 mg per kg per day and found no significant increase in defects.

Astrachan, N.B.; Archer, B.G. and Hilbelink, D.R.: Evaluation of the subacute toxicity and teratogenicity of anatoxin–a. Toxicon. 18:684–688, 1980.

152 3 α,17–β–Androstanediol

Schultz and Wilson (**1974**) injected rats on days 14–21 with 4 mg and the offspring with 0.16 mg for 3 days and produced masculinization of the female internal genitalia.

Schultz, F.M. and Wilson, J.D.: Virlization of the wolffian duct in the rat fetus by various androgens. Endo. 94:979–986, 1994.

153 Androstanedione

Schultz and Wilson (**1974**) injected rats with 4 mg on days 14–21 and produced no masculinization of the female fetuses internal genitalia.

Schultz, F.M. and Wilson, J.D.: Virlization of the wolffian duct in the rat fetus by varirous androgens. Endo. 94:979–986, 1994.

154 Androstenedione CAS 846–46–8

Greene et al. (1939) treated rats from before the 16th day through the 19th day with total doses of 87 to 137 mg and obtained masculinization of the external genitalia and persistence of the wolffian ducts in the female fetuses. These authors provide detailed descriptions of their experimental findings with this androgen as well as with androsterone and testosterone.

Schultz and Wilson (1974) gave pregnant rats 4 mg subcutaneously on days 14 through 21 and did not produce masculinization of female fetuses.

Greene, R.R.; Burrill, M.W. and Ivy, A.C.: Experimental intersexuality. The effect of antenatal androgens on sexual development of female rats. Am. J. Anat. 65:415–470, 1939.

Schultz, F.M. and Wilson, J.D.: Virilization of the Wolffian duct in the rat fetus by various androgens. Endocrinology 94:979–986, 1974.

155 Androsterone CAS 53–41–8

Greene et al. (1939), using 250 to 800 mg during the latter half of pregnancy in the rat, produced masculinization of the external genitalia and persistence of the wolffian ducts in female fetuses.

Greene, R.R.; Burrill, M.W. and Ivy, A.C.: Experimental intersexuality. The effect of antenatal androgens on sexual development of female rat. Am. J. Anat. 65:415–470, 1939.

156 Anemia, Hemorrhagic *Iron Deficiency*

Wilson (1953) and Moscarella et al., (1962) were unable to produce fetal defects by hemorrhagic anemia in rats and mice. Severe iron deficiency in the rat was associated with a 75 percent resorption rate. Most embryos died on or about day 12 (Shepard et al., 1980). There were twice as many female fetal survivors than male and no significant increase in defects was found. Grote (1969a and 1969b) produced skeletal defects in rabbits whose anesthetized mothers were bled. Immediate treatment with salt solution reduced the number of defects. Clark et al. (1984) gave rabbits 180 mg per kg on day 5 and produced as anemia which lasted through day 15. The axial skeletal defects were believed to be related to the anemia since the drug was cleared by day 9.

Clark, R.; Robertson, R.T.; Minsker, D.H.; Cohen, S.M.; Tocco, D.J.; Allen, H.L.; James, M.L. and Bokelman, D.L.: Diflunisal-induced maternal anemia as a cause of teratogenicity in rabbits. Teratology 30:319–332, 1984.

Grote, W.: Embryonale Fehlentwicklungen bei Kaninchen nach mutterlichem Blutverlust. Deutsch Med. Wochenschrift 94:1120 only, 1969a.

Grote, W.: Verhutung von Skelettfehlbildungen nach mutterlichem Blutverlust durch Elektrolytersatz. Deutsch Med. Wochenschrift 94:2342–2344, 1969b.

Moscarella, A.A.; Stark, R.B. and De Forest, M.: Anemia, cortisone and maternal stress as teratogenic factors in mice. Surg. Forum 13:469–471, 1962.

Shepard, T.H.; Mackler, B. and Finch, C.A.: Reproductive studies in the iron–deficient rat. Teratology 22:329–334, 1980.

Wilson, J.G.: Influence of severe hemorrhagic anemia during pregnancy on development of the offspring in the rat. Proc. Soc. Exp. Biol. Med. 84:66–69, 1953.

157 Anesthetics also see *Halothane, Enflurane and Nitrous Oxide*

Since the preliminary observations of increased spontaneous abortion rates in anesthesiologists in Denmark and Russia, others in the United States confirmed the observation. Cohen et al. (1971), in a controlled study, found a 38 percent abortion rate among anesthetists and a 30 percent rate among operating room nurses with only a 10 percent rate for the control group of general nurses. In view of our lack of information on the true incidence of spontaneous abortion, it is difficult to assess reports of increased frequency of abortion (Corbett, 1972).

Smith (1975) and Fink and Cullen (1976) reviewed the general subject of teratology as it relates to anesthesia. Corbett et al. (1974) reported a higher than expected rate of malformations among the offspring of female anesthetists but Cote (1975) criticized their definitions of anomalies and methods of clinical examination. Spence et al. (1977) reviewed the combined findings in Britain and the United States and believe that the increased risk of abortion in anesthetists and their wives is significantly increased over physician controls. From the combined answers of women, the malformation rate in anesthetists' offspring was 5.5 percent as compared to 4.4 percent in the controls (p = to 0.04). Tomlin (1978) found 4 families of 75 anesthetists to have cerebral defects in their offspring. Three had obstructive lesions of the CNS. Erickson and Kallen (1979) found no increase in risk among Swedish workers in operating rooms.

Cohen et al. (1980) collected questionnaires from 30,650 dentists and from 30,547 of their chairside assistants. The abortion rates among the chairside assistants were 14 for light users and 19 for heavy users while rates for the non-users were 8. A heavy user was defined as being exposed for over 3,000 hours during a ten–year period. Malformation rates in the same group were 5.7 for light users, 5.2 for heavy users and 3.6 for non–users. Musculo–skeletal and nervous system malformations were the categories increased. The wives of dentists did not have increased congenital defect rates but their abortion rates were 10.2 as compared to 6.7 in wives of non–users. Other disease categories among the chairside assistants of heavy users were significantly increased also. Tannenbaum and Goldberg (1984) reviewed the epidemiologic studies and conclude that because of flaws in the studies it is not possible to conclude that anesthetics increase the number of spontaneous abortions in exposed anesthetists.

Brackbill (1977) in studies of infants whose mothers received general anesthesia or no anesthesia during delivery found that auditory–invoked heart rate changes in the two groups differed. With repeated stimulation, infants born without anesthesia developed heart deceleration whereas those born with anesthesia shifted from initial deceleration to acceleration. Further studies are needed.

Friedman (1988) reviewed the subject including local anesthetics.

Brackbill, Y.: Long–term effects of obstetrical anesthesia on infant autonomic function. Dev. Psychobiol. 10:529–535, 1977.

Cohen, E.N.; Brown, B.W.; Wu, M.L.; Whitcher, C.E.; Brodsky, J.B.; Gift, H.C.; Greenfield, W.; Jones, T.W. and Driscoll, E.J.: Occupational disease in dentistry and chronic exposures to trace anesthetic gases. J. Am. Dent. Assoc. 101:21–31, 1980.

Cohen, E.N.; Bellville, J.W. and Brown, B.W.: Anesthesia, pregnancy and miscarriage. Anesthesiology 35:343–347, 1971.

Cote, C.J.: Birth defects among infants of nurse anesthetists. Anesthesiology 42:514–515, 1975.

Corbett, T.H.: Anesthetics as a cause of abortion. Fertil. Steril. 23:866–869, 1972.

Corbett, T.H.; Cornell, R.G.; Endres, J.L. and Leiding, K.: Birth defects among children of nurse anesthetists. Anesthesiology 41:341–344, 1974.

Erickson, A. and Kallen, B.: Survey of infants born in 1973 and 1975 to Swedish women working in operating rooms during their pregnancies. Anesth. Analg 58:302–305, 1979.

Fink, B.R. and Cullen, B.F.: Anesthetic pollution: What is happening to us? Anesthesiology 45:79–83, 1976.

Friedman, J.M.: Teratogen update: Anesthetic agents. Teratology 37:69–77, 1988.

Smith, B.E.: Teratology in anesthesia. Clin. Obstet. Gynecol. 17:145–163, 1975.

Spence, A.A.; Cohen, E.N.; Brown, B.W.; Knell–Jones, R.P. and Himmelberger, D.U.: Occupational hazards for operating room–based physicians. Analysis to date from the United States and United Kingdom. JAMA 238:955–959, 1977.

Tannenbaum, T.N. and Goldberg, R.J.: Exposure to anesthetic gases and reproductive outcome. J. Occup. Med. 27:659–668, 1984.

Tomlin, P.T.: Teratogenic effects of waste anesthetic gases. Br. Med. J. 1:108 only, 1978.

158 Angiotensin CAS 1407–47–2

Thompson and Gautieri (1969) gave angiotensin II (valyl–5) subcutaneously or intravenously to pregnant mice and produced no increase in defects. A dose of 10 mg per kg was given once on days 7 through 12 of gestation. Geber (1969) injected guinea pigs intravenously on the 8th day with 0.02 to 1.7 mg per kg and obtained 3.9 to 9.9 percent defective fetuses. The anomalies included hydrocephalus, myelocele, general and local edema and micrognathia. Hiramatsu et al. (**1990**) gave rats up to 173.5 mg per kg intravenously on days 7–17 and found no adverse fetal effects. Kato et al. (**1990**) gave rabbits intervenously up to 2,000 mg per kg daily on days 6–18 and found no adverse fetal effects.

Geber, W.F.: Angiotensin teratogenicity in the fetal hamster. Life Sci. 8:525–531, 1969.

Hiramatsu, Y.; Shimizu, M.; Suzuki, T.; Odo, K.; Iwasaki, S.; Katoh, M.; Koike, T.; Asano, M.; Suzuki, K.; Kato, M.; Konishi, K. and Kimura, T.: Study by intravenous administration of angiotensin ii human during the period of fetal organogenesis in rats. Kiso to Rinsko 24:6109–6127, 1990.

Kato, M.; Shishido, M.; Sakuma, T.; Mindmidake, J. and Konishi, K.: Study by intravenous administration of angiotensin ii human during the period of organogenesis in rabbits. Kiso to Rinsko 24:6129–6138, 1990.

Thompson, R.S. and Gautieri, R.F.: Comparison and analysis of the teratogenic effects of serotonin, angiotensin II and bradykinin in mice. J. Pharm. Sci. 58:406–412, 1969.

159 Aniline Hydrochloride *Aminobenzene* CAS 62–53–3

Price et al. (1985) gavaged pregnant rats with doses up to 100 mg per kg on days 7–20 and found no evidence of teratogenicity. Nakatsu et al. (**1993**) found an increase in rat fetal ventricular septal defects when the dams were given 260 mg per kg on days 12–14 or 15–17.

Nakatsu, T.; Kanamori, H.; Matsumoto, K. and Kusanagi, T.: Cardiovascular malformations in rat retuses induced by administration of aniline hydrochloride. (abs) Cong. Anomal. 33:278–279, 1993.

Price, C.J.; Tyl, R.W.; Marks, T.A.; Paschke, L.A.; Ledoux, T.A. and Reel, J.R.: Teratologic and postnatal evaluation of aniline hydrochloride in the Fischer 344 rat. Toxicol. Appl. Pharmacol. 77:465–478, 1985.

160 Anoxia

Discussed under Hypoxia

161 Antibodies

The extensive literature on this group of agents which can be teratogenic in many species was reviewed by Brent (1966, 1971). Ebert (1950) studied the effects of antiserum in the pre–differentiated chick embryo. McCallion and Langman (1964) reported that a soluble antigen from adult brain tissue caused abnormalities of the brain and eye of the chick embryo. In 1961, Brent et al. reported production of a wide spectrum of congenital defects in the rat receiving rabbit anti–kidney antiserum on the 8th day of gestation. These defects included anencephaly and other nervous system defects, anophthalmia and defects of the ventral body wall, heart, lip and urologic tract. Tuchmann–Duplessis et al., (1966) described the facial defects found in the rat fetus when the mother received heterologous kidney antibodies on the 7th through the 11th gestational day. By careful extension of this work by Brent et al. (1961), an anti–placental and an anti–yolk sac antibody were shown to produce defects. They showed that these antibodies become localized in the visceral and parietal yolk sac (Brent, 1971). This localization, coupled with the finding that these antibodies are particularly teratogenic during active yolk sac nutrition, may suggest that their mechanism of action is by interruption of some function played by the yolk sac.

Thyroid antibodies have been implicated in congenital athyrotic cretinism (see Thyroid Antibodies). Nora et al. (1974) produced heart and other defects by treating the pregnant mouse with antiheart antibody.

New and Brent (1972) injected yolk sac antibody inside the yolk sac of explanted rat embryos and found the teratogenic effect was removed. This suggested that an interaction of the yolk sac with the antibody was necessary to produce teratogenic action.

Antibodies in a test ribosomal vaccine from Klebsiella, streptococcus pneumoniae and pyogenes and H. influenza were injected subcutaneously into rats, mice and rabbits from 8 days before until delivery. No adverse fetal or reproductive effects were found (Labie, 1980).

Brent, R.L.: Immunologic aspects of developmental biology. In: Advances in Teratology, D.H.M. Woollam (ed.), London, Logos, Academic Press, 1:81–129, 1966.

Brent, R.L.: Antibodies and malformations. In: Malformations Congenitales Des Mammiferes. H. Tuchmann–Duplessis (ed.), Paris: Masson and Cie, 187–220, 1971.

Brent, R.L.; Averich, E. and Drapiewski, V.A.: Production of congenital malformations using tissue antibodies. 1. Kidney antisera. Proc. Soc. Exp. Biol. Med. 106:523–526, 1961.

Ebert, J.D.: An analysis of the effect of anti–organ sera on the development in vitro of the early chick blastoderm. J. Exp. Zool. 115:351–377, 1950.

Labie, Ch.: Study of the teratogenic activity of ribsomal vaccine. Arzneim–Forsch Drug Res. 30(1):173–181, 1980.

McCallion, D.J. and Langman, J.: An immunological study on the effect of brain extract on the developing nervous tissue in the chick embryo. J. Embryol. Exp. Morphol. 12:77–88, 1964.

New, D.A.T. and Brent, R.L.: Effect of yolk sac antibody on rat embryos grown in culture. J. Exp. Embryol. Morph. 27:543–553, 1972.

Nora, J.J.; Miles, V.N.; Morriss, J.H.; Weishuhn, E.J. and Nihill, M.R.: Antiheart antibody production of cardiovascular malformations in the mouse: A preliminary study. Teratology 9:143–150, 1974.

Tuchmann–Duplessis, H.; David, G. and Mercier–Parot, L.: Malformations de la face produites par un serum antitissulaire. Bulletin de l'Association des Anatomistes, 51st Reunion, 1000–1004, 1966.

162 Anticoagulant Therapy CAS 7440–36–0

Discussed under Coumarin

163 Anticonvulsants

Discussed under Diphenylhydantoin and Trimethadione

164 Antifoam A

Siddiqui (**1994**) fed rabbits with up to 2.5% of their diet on days 6–19 and found no adverse reproductive effects.

Siddiqui, W.H.: Developmental toxicity evaluation of Dow Corning antifoam A compound, food grade in rabbits. (abs) Teratology 49:397, 1994.

165 Antimony CAS 7440–36–0

James et al. (1966) gave 2 mg per kg of antimony potassium tartrate to 4 sheep during the major portion of gestation and found no fetal changes. Belyayeva (1967) exposed rats by inhalation to a dose of 50 mg per kg antimony trioxide. There were fewer than expected offspring but no morphologic alterations were seen. Ridway and Karnofsky (1952) produced no defects in chicks given 0.10 mg on the 4th day.

Belyayeva, A.P.: The effect of antimony on reproduction. Gig. Tr. Prof. Zabol. 11:32–37, 1967.

James, L.F.; Lazar, V.A. and Binns, W.: Effects of sublethal doses of certain minerals on pregnant ewes and fetal development. Am. J. Vet. Res. 27:132–135, 1966.

Ridgway, L.P. and Karnofsky, D.A.: The effects of metals on the chick embryo: Toxicity and production of abnormalities in development. Ann. N.Y. Acad. Sci. 55:203–215, 1952.

166 Antimycin A$_1$ CAS 642–15–9

Duffy and Ebert (1957) and Reporter and Ebert (1965) observed heart–specific effects and anomalous development in chick embryos grown on medium containing 0.03 microgm per ml of medium. Shepard (unpublished data) was unable to produce defects in rat embryos by intravenous injection of antimycin A$_1$.

Duffy, L.M. and Ebert, J.D.: Metabolic characteristics of the heart–forming areas of the early chick embryo. J. Embryol. Exp. Morphol. 5:324–339, 1957.

Reporter, M.C. and Ebert, J.D.: A mitochondrial factor that prevents the effects of antimycin A on myogenesis. Dev. Biol. 12:154–184, 1965.

167 Apholate *2,2,3,3,6,6–Hexahydro–2,2,4,4,6,6–hexakis(1–aziridinyl)1,3,5, 2,4,6 triazatriphosphorine* CAS 52–46–0

Younger (1965) reported multiple congenital defects in a lamb exposed to 1.0 mg per kg during the entire gestation. Anophthalmia, ectopic liver and defects of the skull were found.

Younger, R.L.: Probable induction of congenital anomalies in a lamb by apholate. Am. J. Vet. Res. 26:991–995, 1965.

168 Apomorphine CAS 58–00–4

Ancel and Scheiner (1951) reported that 0.025 mg added to 2 to 3 day chick embryos produced abnormalities of the limbs and body.

Ancel, P. and Scheiner, H.: Sur le pouvoir teratogene de certaines substances chimiques et en particulier de l' apomorphine. Arch. Anat. Histol. Embryol. (Strasb.) 34:19–25, 1951.

169 Aprindine HCl

Komai et al. (1987a,b,c) studied this antiarrhythmic agent in rats during organogenesis and the perinatal period and in the rabbit during organogenesis. The medication was given intravenously in both species and the daily maximum

dose in rats was 10 mg per kg and 6 mg in rabbits. No adverse effects were seen in the term rabbits. No malformations were produced in the rat but the adrenal glands of the rat dams were increased in weight. A decrease in fetal survival after perinatal treatment of the rat was found.

Komai, Y.; Fukuda, T.; Hattori, M.; Ishimura, K. and Hatano, M.: Reproduction study of aprindine hydrochloride. Teratogenicity study in rats by intravenous administration. Jpn. Pharmacol. Ther. 15:135–152, 1987a.

Komai, Y.; Fukuda, T.; Hattori, M.; Ishimura, K. and Hatano, M.: Reproduction study of aprindine hydrochloride. Teratogenicity study in rabbits by intravenous administration. Jpn. Pharmacol. Ther. 15:153–162, 1987b.

Komai, Y.; Fukuda, T.; Hattori, M.; Ishimura, K. and Hatano, M.: Reproduction study of aprindine hydrochloride. Peri–and postnatal study in rats by intravenous administration. Jpn. Pharmacol. Ther. 15:163–176, 1987c.

170 Aprotinin CAS 9087–70–1

This trypsin inhibitor was given to rats intravenously on days 7 through 17. At a dose of 100,000 IU per kg, maternal body weight was reduced and at 200,000, some fetal death and decreased weight occurred with hypoplasia of the sternebrae (Toyoshima et al., 1976).

Toyoshima, S.; Sato, H. and Sato, R.: Effects of trasylol (aprotinin) administered to the pregnant rat during the organogenetic period (7–17 days of gestation) on the pre–and postnatal development of their offspring. (Japanese) Clinical Report 10:2291–2307, 1976.

171 1–β–D–Arabinofuranosylcytosine

Discussed under Cytosine

172 1–β–D–Arabinofuranosyl–5–fluorocytosine

Discussed under Cytosine Arabinoside

173 Aranidpine *MPC–1304*

Yamakita et al. (**1993** A,B,C)gavaged rats with this calcium antagonist before pregnancy, during organogenesis and in the perinatal and via the dam in the postnatal period. At 7.5 mg per kg no adverse fetal effects were found but at higher doses fetal weight was reduced and postnatal viability declined. Similar studies in mice at doses of up to 50 mg per kg during organogenesis were not associated with teratogenicity (Aso et al., **1993**). Some decrease in fetal weight was found at the highest dose as well as an increawse in skeletal defects. In rabbits treated during organogenesis at 120 and 180 mg per kg the fetuses were lighter and had retarded bone age (Umemura et al., **1993**).

Aso, S.; Sueta, S.; Kajiwara, Y.; Ehara, H.; Kikuchi, Y.; Yamakita, O. and Yamamoto, H.: Reproductive and developmental toxicity studies of MPC–1304 (I): Fertility study in rats by oral administration. Yakuri to Chiryo 21:s1083–s1093, 1993.

Shimazu, H.; Ishida, S.; Yamazaki, E.; Fujioka, M.; Yamakita, O. and Yamamoto, H.: Reproductive and developmental toxicity strudies of MPC–1304 (III): Teratological study in mice by oral administration. Yakuri to Chiryo 21:s1115–s1124, 1993.

Umemura, T.; Katsumata, Y.; Takigami, H.; Yamakita, O. and Yamamoto, H.:Reproductive and developmental toxicity studies of MPC–1304 (IV): Teratogenicity study in rabbits by oral administration. Yakuri to Chiryo 21:s1125–s1137, 1993.

Yamakita, O.; Koida, M.; Shinomiya, M.; Katayama, S. and Yoshida, R.: Reproductive and developmental toxicity studies of MPC–1304 (II): Teratological study in rats by oral administration. Yakuri to Chiryo 21:s1095–s1113, 1993.

Yamakita, O.; Koida, M.; Shinomiya, M.; Ikebuchi, K. and Yoshida, R.: Reproductive and developmental toxicity studies of MPC–1304 (V): Peri- and postnatal study in rats by oral administration. Yakuri to Chiryo 21:s1139–s1157, 1993.

174 Arbutin CAS 497–76–7
Hydroquinone–B–D–glucopyranoside

Itabashi et al. (**1988**) gave rats subcutaneously up to 400 mg per kg before fertilization and during pregnancy and lactation. This drug is the major component of bearberry leaves. No adverse fetal or reproductive effects were seen except for reduction in the fetal weight of the female offspring. Postnatal development was not altered.

Itabashi, M.; Aihara, H.; Inoue, T.; Yamate, J.; Sannai, S.; Tajima, M.; Tanaka, C. and Wakisaka, Y.: Reproduction study of arbutin in rats by subcutaneous administration. Iyakuhin Kenkyu 19(2):282–297, 1988.

175 Areca catechu

The nuts of this Indian plant were extracted and then fed on days 1–7 to rats in an infertility study (Garg and Garg, 1970). Given in doses of 500 mg per kg implantation was partially prevented.

Garg, S.K. and Garg, G.P.: Antifertility screening of plants. VII. Effect of five indigenous plants on early pregnancy in albino rats. Indian J. Med. Res. 59:302–306, 1970.

176 Arotinoid Ethyl Ester *RO 13–6298*

Zimmermann et al. (1985) used this very potent retinoid to study pregnant mice. Doses of 10 microgm per kg daily on days 9–11 or 12–14 produced a spectrum of malformations similar to that seen with other retinoids. Flanagan et al. (1987) used the methanol and methyl derivatives and compared them to the acid using oral dosage in hamsters on the 8th day. The acid was 375 times as potent as all–trans retinoic acid and a defect pattern similar to retinoic acid was found. This retinoid was applied to the skin of hamsters in amounts of 0.01–1.0 mg per kg and at 0.05 mg per kg crooked tail and craniofacial defects were increased. (Willhite et al. **1990**)

Flanagan, J.L.; Willhite, C.C. and Ferm, V.H.: Comparative teratogenic activity of cancer chemopreventive retinoidal benzoic acid cogeners (arotinoids). JNCI 78(3):533–538, 1987.

Willhite, C.C.; Sharma, R.P.; Allen, P.V. and Berry, D.L.: Percutaneous retinoid absorption and embryotoxicity. The Journal of Investigative Dermatology 95:523–529, 1990.

Zimmermann, B.; Tsambaos, D. and Sturje, H.: Teratogenicity of arotinoid ethyl ester (RO 13–6298) in mice. Teratogenesis, Carcinogenesis, and Mutagenesis 5:415–431, 1985.

177 Arotinolol

Shimakoshi and Kato (**1984**) gavaged this β–blocker in studies of rabbits during organogenesis. Some increase in fetal mortality was found at the highest dose (100 mg per kg).

Shimakoshi, Y. and Kato, T.: Teratology study of arotinolol in rabbits. Clinical Reports 18:109–114, 1984.

178 Arsenic CAS 7440–38–2

Five women with arsenic poisoning during pregnancy had normal offspring (Kantor and Levin, 1948). Willhite and Ferm (1984) reviewed the subject and pointed out that the methylated metabolites are less toxic. Workers exposed to arsenic and other heavy metals in Sweden were found to have increased malformations and spontaneous abortions and decreased birth weights (Nordstrom et al., 1979). This work is discussed in more detail under the heading, Occupation.

Hood and Bishop (1972) injected mice with a single dose of 45 mg of arsenate per kg on gestational days 6 through 11 and produced fetuses with exencephaly, agnathia, anophthalmos and a few cleft palates. Skeletal defects were also present. In preliminary experiments with sodium arsenite, they were able to produce the same spectrum of defects with 10 mg per kg. More detailed studies of arsenate and arsenite teratogenicity in mice were reported by Hood et al. (1978) and Baxley et al. (1981). Hood et al. (1987) reported details on the distribution in pregnant mice. The teratogenicity of arsenate was reported in the rat (Beaudoin, 1974). Hood et al. (1987) reported more details of distribution and metabolism of pentavalent arsenic in the mouse and found large amounts of the mono and dimethyl arsenic in the fetus.

Ridgway and Karnofsky (1952) administered sodium ortho arsenate (0.20 mg) to 4–day chick embryos and found stunting, mild micromelia, impaired feather growth and swelling of the abdomen in the resulting 18–day–old chicks. Potassium arsenate given to ewes at 0.5 mg per kg during most of pregnancy caused no defects in 4 fetuses (James et al., 1966). Ferm et al. (1971) found that hamsters treated with 15 to 25 mg per kg of disodium arsenate on the 8th day produced fetuses with a high incidence of anencephaly and also other defects. Ferm and Hanlon (1986) could not decrease the rate of neural tube defects from arsenate in hamsters by constant infusions of folic acid. Golub (**1994**) has reviewed the animal teratogenicity and concluded that the conceptus was more sensitive to toxicity than the maternal subject.

Baxley, M.N.; Hood, R.D.; Vedel, G.C.; Harrison, W.P. and Szczech, G.M.: Prenatal toxicity of orally administered sodium arsenate in mice. Bull. Environ. Contam. Toxicol. 26:749–756, 1981.

Beaudoin, A.R.: Teratogenicity of sodium arsenate in rats. Teratology 10:153–158, 1974.

Ferm, V.H.; Saxon, A. and Smith, B.M.: The teratogenic profile of sodium arsenate in the golden hamster. Arch. Environ. Health 22:557–560, 1971.

Ferm, V.H. and Hanlon, D.P.: Arsenate–induced neural tube defects not influenced by constant rate administration of folic acid. Pediatr. Res. 20:761–767, 1986.

Golub, M.S.: Maternal toxicity and the identification or inorganic arsenic as a developmental toxicant. Reproductive Toxicology 8:283–295, 1994.

Hood, R.D. and Bishop, S.L.: Teratogenic effects of sodium arsenate in mice. Arch Environ. Health 24:62–65, 1972.

Hood, R.D.; Thacker, G.T. and Patterson, B.L.: Prenatal effects of oral versus intraperitoneal sodium arsenate in mice. J. Environ. Path. Toxicol. 1: 857–864, 1978.

Hood, R.D.; Vedel–Macrander, G.C.; Zaworotko, M.J.; Tatum, F.M. and Meeks, R.G.: Distribution, metabolism and fetal uptake of pentavalent arsenic in pregnant mice following oral or intraperitoneal administration. Teratology 35:19–25. 1987.

James, L.F.; Lazar, V.A. and Binns, W.: Effects of sublethal doses of certain minerals on pregnant ewes and fetal development. Am. J. Vet. Res. 27:132–135, 1966.

Kantor, H.I. and Levin, P.M.: Arsenic encephalopathy in pregnancy with recovery. Am. J. Obstet. Gynecol. 56:370–374, 1948.

Nordstrom, S.; Beckman, L. and Nordenson, I.: Occupational and environmental risks in and around a smelter in northern Sweden. VI. Congenital malformations. Hereditas 90:297–302, 1978.

Ridgway, L.P. and Karnofsky, D.A.: The effects of metals on the chick embryo: Toxicity and production of abnormalities in development. Ann. N.Y. Acad. Sci. 55:203–215, 1952.

Willhite, C.C. and Ferm, V.H.: Prenatal and developmental toxicology of arsenicals. Adv. Exp. Med. Biol. 177:205–228, 1984.

179 Arsine Gas

Morrissey et al. (**1990**) exposed rats and mice to concentrations of up to 2.5 ppm during organogenesis and found no reproductive toxicity. The maternal rats had increased spleen weights.

Morrissey, R.E.; Fowler, B.A.; Harris, M.W.; Moorman, M.P.; Jameson, C.W. and Schwetz, B.A.: Arsine: Absence of developmental toxicity in rats and mice. Fundamental and Applied Toxicology 15:250–356, 1990.

180 Artificial Insemination

Forse et al. (1982), in a multicenter study of 374 liveborn children conceived by artificial insemination, found 21 with major malformations and 5 with trisomies. The malformations included neural tube defects (1), congenital heart disease (5) and multiple defects (8).

Forse, R.A.; Ackman, D.G. and Fraser, F.C.: Is artificial insemination (AI) teratogenic? (abs) Teratology 25:40A–41A, 1982.

181 Artinolol 2,–(3–t–Butylamino–2–hydroxypropyl–thio)–4–(5–carbamoyl–2–thieny–thiazole HCl

Shimakoshi and Kato (1984) studied this β adrenergic blocker in rabbits using oral doses up to 200 mg per kg daily during organogenesis. No malformation increase was seen but embryolethality was increased at 30 and 200 mg per kg.

Shimakoshi, Y. and Kato, T.: Teratology study of artinolol in rabbits. Clin. Report 18:3613–3618, 1984.

182 Arvin™ CAS 9046–56–4

Penn et al. (1971) administered this defibrinating fraction of snake venom to pregnant mice and rabbits and found no significant increase in defects. A high death and resorption rate was associated at dosage ranges of 10 to 25 units per kg. Hemorrhagic placentas were observed.

Penn, G.B.; Ross, J.W. and Ashford, A.: The effects of arvin on pregnancy in the mouse and the rabbit. Toxicol. Appl. Pharmacol. 20:460–473, 1971.

183 α–Asarone

This chemical from a Mexican plant (Gautterin gaumeri) was given orally to rats on days 6–15. Fetal resorption was increased at the highest dose (75 mg per kg) but no significant increase in defects occurred.

Jimenez, L.; Chamorro, G.: Salazar, M. and Pages, N.: Evaluation teratologique de l'α–asarone chez le rat. Ann. Pharmaceutiques Francaises 46:179–183, 1988.

184 Asbestos CAS 1332–21–4

Schneider and Maurer (1977) fed mice up to 143 micrograms of chrysotile asbestos per ml of drinking water and found no teratogenic effects. Studies carried out with blastocyst exposure to the material found some decrease in postimplantation survival.

Schneider, U. and Maurer, R.R.: Asbestos and embryonic development. Teratology 15:273–280, 1977.

185 Ascorbic Acid Vitamin C CAS 50–81–7

Sharma (1988) found lowered blood ascorbic acid in seven mothers with abruptio placentae. The mean concentration was 1.17 as compared to controls with 1.62 mg per deciliter but the difference was not statistically significant. Eight women with placenta previa did not have reduced levels. Martin et al. (1957) studied a large group of women during pregnancy and did not find increased malformation rates in those with low serum and intake levels. The lowest group had the highest premature rate.

Frohberg et al. (1973) gave up to 1000 mg of ascorbic acid by mouth to pregnant mice and rats from day 6 through day 15 and found no adverse effects on the conceptus.

Kalter and Warkany (1959) reviewed work which has been done on scorbutic guinea pigs during pregnancy. The animals did not become pregnant on a standard vitamin–C deficient diet. With daily supplements of 5 ml of orange juice, abortions or resorptions occurred (Kramer et al.,

1933). Harman and Warren (1951) reported that when the pregnant guinea pigs were started on a vitamin–C deficient diet on the 21st day of gestation, the fetuses were smaller than controls and retarded in development of skin and muscle.

Palludan and Wegger (1989) studied a Danish mutant pig that was unable to synthesize ascorbic acid. Fetal hemorrhages, edema and irregular ossification were seen in the fetuses. The finding was similar to scurvy.

Frohberg, H.; Gleich, J. and Kieser, H.: Reproduktionstoxikologische Studien mit ascorbinsaure an Mausen und Ratten. Arzneim–Forsch 23:1081–1082, 1973.

Harman, M.T. and Warren, L.E.: Some embryologic aspects of vitamin–C deficiency in the guinea pig (Cavia cobaya). Trans. Kans. Acad. Sci. 54:42–57, 1951.

Kalter, H. and Warkany, J.: Experimental production of congenital malformations in mammals by metabolic procedure. Physiol. Rev. 39:69–115, 1959.

Kramer, M.M.; Harman, M.T. and Brill, A.K.: Disturbances of reproduction and ovarian changes in guinea pig in relation to vitamin C deficiency. Am. J. Physiol. 106:611–622, 1933.

Martin, M.P.; Bridgforth, E.; McGanity, W.J. and Darby, W.J.: The Vanderbilt cooperative study of maternal and infant nutrition 10. Ascorbic acid. J. Nutrit. 62:201–224, 1957.

Palludan, B. and Wegger, I.: The influence of ascorbic acid deficiency on fetal development in swine. Teratology 39:152, 1989.

Sharma, S.C.: Evaluation of antenatal and perinatal vitamin C status in pregnancies threatened with prematurity. In: Vitamins and Minerals in Pregnancy and Lactation, H. Berger (ed). Vol. 16, Nestec Ltd., Vevey/Raven Press, Ltd., New York, 1988.

186 L–Asparaginase CAS 9015–68–3

Blatt et al. (**1980**) reported two normal newborns exposed to this drug along with multiple other chemotherapeutic agents. Adamson et al. (1970) administered intravenously 50 IU per kg to rabbits on days 8 and 9 of gestation and observed 6 to 11 percent fetal defects. The defects consisted of abdominal extrusion, spina bifida, missing tail and defects of the lung, kidney or skeleton. No defects were produced in the rat. The enzyme crossed the rabbit placenta and entered the fetus. Ohguro et al. (1969) gave this compound intravenously to mice and intraperitoneally to rats in doses up to 4,000 IU per kg from the days 7–13 of gestation. Exencephaly and skeletal abnormalities occurred in fetuses from both species when the dosage was 1000 IU or more per day.

Adamson, R.H.; Fabro, S.; Hahn, M.A.; Creech, C.E. and Whang–Peng, J.: Evaluation of the embryotoxic activity of L–asparaginase. Arch. Int. Pharmacodyn. Ther. 186:310–320, 1970.

Blatt, J.; Mulvihill, J.J.; Ziegler, J.L.; Young, R.C. and Poplack, D.G.: Pregnancy outcome following cancer chemotherapy. Am. J. Med. 69:828–832, 1980.

Ohguro, Y.; Imamura, S.; Koyama, K.; Hara, T.; Miyagawa, A.; Hatano, M. and Kanda, K.: Toxicological studies on L–asparaginase. (Japanese) Yamaguchi Igaku 18:271–292, 1969.

187 Aspartame CAS 22839–47–0

The effect of this chemical sweetner on phenylalanine levels of normal and phenylketonuric heterozygotes was determined by Stegink et al. (1980) to be of no clinical significance even at doses as high as 49 to 58 mg per kg. Mahalik and Gautieri (1984) gavaged mice on days 15–18 of gestation with 1 or 4 gm per kg of aspartame. Controls were gavaged with normal saline. Visual placing was delayed by 2 and 4 days in the postnatal mice exposed to 1 and 4 gm per kg, respectively. Surface righting, air righting and negative geotaxis were not altered. Parameters such as maternal weight gain, fetal weight and litter size were not included in the report. This work was repeated by McAnulty et al. (1989) who gavaged up to 4000 mg per kg on days 15 through 18 and they found no differences in development including the visual system. Lederer (1985) fed aspartame as 1 percent of the diet to rats during gestation and found no ill effects.

Lederer, J.; Bodin, J. and Colson, A.: L'Aspartame et son effet sur la gestation chez le rat. J. de Toxicologie Clinique et Experimentale 7–14, 1985.

Mahalik, M.P. and Gautieri, R.F.: Research Committee. Psychol. Psych. Behavior 9:385–403, 1984.

McAnulty, P.A.; Collier, M.J.; Enticott, J.; Tesh, J.M.; Mayhew, D.A.; Comer, C.P.; Hjelle, J.J. and Kotsonis, F.N.: Absence of developmental effects in cf–1 mice exposed to aspartame in utero. Fundamental and Applied Toxicology 13:296–302, 1989.

Stegink, L.D.; Filer, L.J.; Baker, G.L. and McDonnell, J.E.: Effect of an abuse dose of aspartame upon plasma and erythrocyte levels of amino acids in phenylketonuria heterozygous and normal adults. J. Nutr. 110:2216–2224, 1980.

188 L–Aspartate, Monosodium CAS 56–84–8

Inouye and Murakami (1973) injected 6 mM per kg into pregnant mice on day 16, 17 or 18 of gestation. Pyknosis and edema were found in the arcuate and ventromedial nuclei of the fetal brains for 24 hours after treatment.

Inouye, M. and Murakami, U.: Brain lesions in mouse infants and fetuses induced by monosodium aspartate. Cong. Anom. 13:235–244, 1973.

189 Astemizole

Rats and rabbits were treated orally during organogenesis with up to 160 and 10 mg per kg respectively. No increase in defects among the fetuses was found but there was increased fetal death at 40 and 160 mg per kg, doses which were toxic to the mother.

Whyatt, P.L.: Astemizole in pregnancy. Aust. Fam. Physician 15:382–384, 1986.

190 Asthma

Gordon et al. reviewed 277 records of mothers with asthma during pregnancy. These patients were part of a large prospective collaborative study sponsored by the National Institute of Health and carried out on 50,000 women. There were 16 perinatal deaths about twice the expected rate. Sixteen mothers with severe asthma had 3 intrauterine deaths and two had neurologic problems. One had a congenital defect (polydactyly).

Gordon, M.; Niswander, K.R.; Berendes, H. and Kantor, A.G.: Fetal morbidity following potentially anoxigenic obstetric conditions. Am. J. Obst. Gynecol. 106:421–429, 1970.

191   Astromycin CAS 55779–06–1

This aminoglycoside antibiotic was studied in rats and rabbits by Nishikawa et al. (1981). The route was intramuscular and maximum daily doses were 500 mg per kg in the rat and 300 mg per kg in the rabbit. No adverse fetal changes were observed in either species. Fertility studies and perinatal studies in the rat failed to produce any alterations in the treated group. Some renal damage occurred in the dams.

Nishikawa, S.; Hara, T.; Miyazaki, H. and Ohguro, Y.: Safety evaluation of KW–1070. Chemotherapy 29:167–175, 1981.

192   Asverin AT–327

Kowa (1972) gave orally 20 or 70 mg per kg daily to rats in the 7–13 days of gestation. No adverse effects were found.

Kowa, Y.: Studio teratologico di un antitussigeno non narcotico l'asverin. Boll. Chim. Farm. 111:47–51, 1972.

193   Atabrine

Discussed under Quinacrine

194   Atenolol CAS 29122–68–7

Rubin et al. (1983) performed a double–blind study on 120 women with pregnancy–associated hypertension in the last trimester. Bradycardia (less than 120) occurred in 18 treated fetuses and 4 of the controls. No other ill effects in the fetus were noted. Esaki and Imai (1980) treated rats by gavage with 20 to 2000 mg per kg daily. This β–adrenergic blocker was associated with some fetal loss at the highest doses but no increase in congenital defects was found. Similar results were found in pregnant rabbits using up to 1600 mg per kg (Esaki, 1980).

Esaki, K.: Effects of oral administration on the rabbit fetus. Preclin. Rep. Cent. Inst. Exp. Animal 6:259–264, 1980.
Esaki, K. and Imai, K.: Effects of oral atenolol on reproduction in rats. Preclin. Rep. Cent. Inst. Exp. Animal 6:239–258, 1980.
Rubin, P.C.; Clark, D.M.; Sumner, D.J.; Low, R.A.; Butters, L.; Reynolds, B.; Steedman, D. and Reid, J.L.: Placebo-controlled trial of atenolol in treatment of pregnancy-associated hypertension. Lancet 1:431-434, 1983.

195   Atorvastatin

This inhibitor of microsomal 3–hydroxy–3–methylglutaryl–coenzyme A reductase was given by gavage to rats and rabbits during organogenesis. Maximum doses of 300 mg per kg in rats and 100 mg in rabbits were used. Maternal toxicity at the higher doses was associated with fetal toxicity but no evidence of teratogenicity was found. (Dostal et al., **1994**)

Dostal, L.A.; Schardein, J.L. and Anderson, J.A.: Developmental toxicity of the HMG–CoA reductase inhibitor, atorvastatin, in rats and rabbits. Teratology 50:387–394, 1994.

196   Atracurium

Skarpa et al. (1983) gave rabbits 0.15 mg per kg daily or 0.10 mg per kg twice daily by subcutaneous route on days 6-18 of pregnancy. No teratogenicity was found. No transfer of the drug was detected across the cat placenta.

Skarpa, M.; Dayan, A.D.; Follenfant, M.; James, D.A.; Moore, W.B.; Thomson, P.M.; Lucke, J.N.; Morgan, M.; Lovell, R. and Medd, R.: Toxicity testing of atracurium. Br. J. Anaesth. 55:27S–29S, 1983.

197   Atrazine

Yau et al. (1989) gavaged rats with up to 100 mg per kg daily during organogenesis. At maternally toxic doses, there was fetal osseous retardation but no other adverse effects.

Peters and Cook (1973) injected up to 2000 mg per kg on days 3 6 or 9 of rat pregnancy and found no adverse reproductive effect in the small groups studied.

Peters, J.W. and Cook, R.M.: Effects of atrazine on reproduction in rats. Bulletin of Environ Contamination & Toxicology 9:301–304, 1973.
Yau, E.T.; Amemiya, K.; Lindsay, L.A.; Wimbert, K.V.; Giknis, M.L.A.; Wetzel, L.; Stevens, J.; Traina, V.M. and Breckenridge, C.: The effect of the triazine herbicide, atrazine on embryonic and fetal development. Teratology 39:P81, 1989.

198   Atropine CAS 51–55–8

Beuker and Platner (1956) injected 0.6 to 1.5 mg of atropine into chick eggs during the interval of 4 to 12 days incubation and produced no defects. Arcuri and Gautieri (1973) injected pregnant mice on day 8 or 9 of gestation with 50 mg per kg. No increase in soft tissue anomalies occurred, but skeletal abnormalities were increased when treatment was given on day 9.

Arcuri, P.A. and Gautieri, R.F.: Morphine–induced fetal malformations. III. Possible mechanisms of action. J. Pharm. Sci. 62:1616–1634, 1973.
Beuker, E.D. and Platner, W.S.: Effect of cholinergic drugs on development of chick embryo. Proc. Soc. Exp. Biol. Med. 91:539–543, 1956.

199   Augmentin™ Amoxicillin Clavulanic Acid BRL 25000

Baldwin et al. (1983) administered this antibiotic to rats along with its β–lactamase inhibitor in maximum doses of 400 mg and 1,200 mg per day on days 6–15 and 15–21, respectively. No adverse fetal effects were found. Hirakawa et al. (1983) gave the same amount to males and females before fertilization and to females during the first 7 days and found no adverse reproductive effects. James et al. (1983) gavaged BRL 25000 in doses of 600 mg per kg to pigs on days 12–42 and observed no adverse fetal effects.

Baldwin, J.A.; Schardein, J.L. and Koshima, Y.: Reproduction studies of BRL 14151K and BRL 25000. Teratology and peri–and postnatal studies in rats. Chemotherapy 31:238–262, 1983.

Hirakawa, T.; Suzuki, T.; Sano, Y.; Tamura, K.; Koshima, Y.; Hiura, K–I.; Fujita, K. and Hardy, T.L.: Reproduction studies in rats. Chemotherapy. 31:263–272, 1983.

James, P.A.; Hardy, T.L. and Koshima, Y.: Reproduction studies of BRL 25000. Teratology studies in the pig. Chemotherapy 31:274–279, 1983.

200 Auranofin

Szabo et al. (**1978** A) gavaged rats with 5 ml per kg on days 6–15 and found no fetal ill effects. In rabbits, Szabo et al. (**1978** B) given 2 mg and 8 mg per kg subcutaneously gastroschisis, umbilical hernia and lung abnormalities were found.

Szabo, K.T.; Difebbo, M. and Phelan, D.G.: The effects of gold–containing compounds on pregnant rabbits and their fetuses. Vet Pathol 15:97–102, 1978a.

Szabo, K.T.; Guerriero, F.J. and Kang, Y.J.: The effects of gold–containing compounds on pregnant rats and their fetuses. Vet Pathol 15:89–96, 1978b.

201 Avidin CAS 1405–69–2

Discussed under Biotin Deficiency

202 N—(1–Azabicyclo[2.2.2]oct–3–yl)–6–chloro–4–methyl–3–oxo3, 4–dihydro–2H–1,4–benzoxazine–8–carboxamide monohydrochloride Y–25130

Aso et al. (**1992** A,B,C) studied this antiemetic in rats before and in early pregnancy, during embryogenesis and in the peinatal period. Intravenous dosage up to 60 mg per kg was used prenatally and 100 mg per kg during organogenesis and perinatally. No treatment related changes were found in the dams or fetuses and postnatal development was normal.

Aso, S.; Kajiwara, Y.; Kikuchi, Y. and Horiwaki, S.: Fertility study of (+)–N–(1–azaabicyclo[2.2.2]oct–3–yl)––6–chloro–4–methyl–3–oxo–3,4–dihydro–2h–1,4–benzoxazine–8–carboxamide monohydrochloride (Y–25130) in rats. Oyo Yakuri 44:53–60, 1992A.

Aso, S.; Kajiwara, Y.; Kikuchi, Y. and Horiwaki, S.: Teratogenicity study of (+)–n–(1–azabicyclo[2.2.2]oct–3–yl)–6–chloro–4–methyl–3–oxo–3,4–dihydro–2h–1,4–benzoxazine–8–carboxamide monohydrochloride (Y–35130) in rats. Oyo Yakuri 44:61–77, 1992B.

Aso, S.; Sueta, S.; Nakao, H. and Horiwaki, S.: Perinatal and postnatal study of (+)–n–(1–azabicyclo[2.2.2]oct–)–3–yl)–6–chloro–4–methyl–3–oxo–3,4–dihydro–2h–1.4–benzoxazine–8–carboxamide monohydrochloride (Y–25130) in rats. Oyo Yakuri 44:79–91, 1992C.

203 Azachlorozine

Smolnikova and Strekalova found no adverse embryogenesis using 35 mg per kg at varying times during gestation in the rat.

Smolnikova, N.M. and Strekalova, S.N.: Results of studies on embryotoxic and teratogenic action of nonachlazine. Akash. Ginekal (Mosk)46:70–71, 1970.

204 5–Azacytidine CAS 320–67–2

Matsuda et al. (1988) administered 0.1 and 1.0 mg intraperitoneally on day 7.5 to a strain of mice genetically susceptible to exencephaly (MT/HokIdr). At 1.0 mg, total resorption occurred and at 0.1 mg, they found 50 percent exencephaly as compared to controls with 10 percent. In another strain, Slc:ICR given 1.0 mg, exencephaly was high in the offspring in a similar way. Matsuda (1990) studied the embryos from these two strains in vitro and found that the differential in sensitivity persisited.

This RNA inhibitor damaged mouse fetal brain when given intraperitoneally to the mother in doses of 2 mg per kg (Schmahl, 1979). Takeuchi and Murakami (1978) produced exencephaly, encephalocele and eye defects in rat fetuses whose mothers were given 1 mg per kg intraperitoneally on day 8.

Matsuda, M.; Kimura, R.; Matsui, F.; Oohira, A. and Shoji, R.: Exencephaly induced by 5–azacytidine in MT/HokIdr strain mice. (abs) Teratology 38:512 only, 1988.

Matsuda, M.: Comparison of the incidence of 5–azacytidine–induced exencephaly between mt/hokldr and slc:icr mice. Teratology 41:147–154, 1990.

Schmahl, W.: Different teratogenic efficacy to mouse fetal CNS of 5–azacytidine in combination with x–irradiation depends on the sequence of successive application. Teratology 19:63–70, 1979.

Takeuchi, I. and Murakami, U.: Influence of cysteamine on the teratogenic action of 5–azacytidine. (abs) Teratology 18:143 only, 1978.

205 N-(4-Aza-endo-tricyclo[5.2.1.5]–decan–4–yl)–4–chloro–3–sulfamoylbenzamide
TDS

Osumi et al. (1979) administered this diuretic orally to rats on days 8 through 17. An increase in cleft palate occurred at 500 mg per kg per day, but not at 2000 and 4000 mg per kg. At 4000 mg per kg, there was some reduction in fetal weight. In the rabbit, given up to 2000 mg per kg daily, no harmful effects were noted in the fetuses (Tagaya et al., 1979).

Osumi, I.; Okada, F.; Mikami, T. and Suzuki, Y.: Effects of N–(4–aza–endo–tricyclo[5.2.1.5]–decan–4–yl) 4–chloro–3–sulfamoylbenzamide (TDS) orally administered to rats in the period of organogenesis upon the pre– and postnatal development. (Japanese) Yakubutsu Ryoho 12:651–667, 1979.

Tagaya, O.; Matsubara, T.; Goto, R. and Suzuki, Y.: Effects of N–(4–aza–endo–tricyclo[5.2.1.5]–decan–4–yl) 4—chloro–3–sulfamoylbenzamide (TDS) administered to rabbits in the period of organogenesis upon the pre– and postnatal development. (Japanese) Yakubutsu Ryoho 12:669–674, 1979.

206 Azaguanine *Guanazolo* CAS 134–58–7

Thiersch (1960) briefly reported that giving 500 mg per kg on days 7 and 8 caused exencephaly and stunting in rats. Waddington and Perry (1958) observed embryonic degeneration in chicks exposed in ova or in culture to the substance.

Thiersch, J.B.: In discussion of paper by M.L. Murphy. C.M. O'Connor and G.E.W. Wolstenholme (eds), Ciba Foundation Symposium on Congenital Malformations. Boston: Little Brown and Co., 111 only, 1960.

Waddington, C.H. and Perry, M.M.: Effects of some amino–acid and purine antagonists on chick embryos. J. Embryol. Exp. Morphol. 6:365–372, 1958.

207 Azalomycin F CAS 11003–24–0

This antibiotic effective against monilia and trichomoniasis was found non–teratogenic in pregnant rats and mice (Arai, 1968). The rats received 20 mg per kg and the mice received 100 mg per kg daily during the active period of embryogenesis.

Arai, M.: Azalomycin F, an antibiotic against fungi and trichomonas. Arzneim. Forsch. 18:1396–1399, 1968.

208 Azanidazole

(4–[(E)]–2–Methyl–5–nitro–1H–imidazole–2–yl) ethenyl)–2–pyrimidine) CAS 62973–76–6

Tammiso et al. (1978) studies this antitrichomonas agent in rabbits and rats. Oral daily doses of up to 300 mg per kg in rats and 110 mg per kg in rabbits were not associated with adverse fetal effects.

Tammiso, R.; Olivari, G.; Coccoli, C.; Garzia, G. and Vittadini, G.: Toxicological and teratological studies of azanidazole. Arzneim–Forsch. Drug Res. 28(II):2251–2256, 1978.

209 β–Azaserine CAS 115–02–6

Dagg and Karnofsky (1955) found defects of the appendicular skeleton of the chick after injecting 0.150 to 2.4 mg into the yolk on the 4th day or after. Blattner et al. (1958) studied the effect of this teratogen in the chick. D–Azaserine was non–teratogenic. Murphy and Karnofsky (1956) produced skeletal and palate defects in rats receiving 2.5 mg per kg from the 8th to 12th gestational days.

Blattner, R.J.; Williamson, A.P. and Simonsen, L.: Teratogenic changes in early chick embryos following administration of antitumor agent (azaserine). Proc. Soc. Exp. Biol. Med. 97:560–564, 1958.

Dagg, C.P. and Karnofsky, D.A.: Teratogenic effects of azaserine on the chick embryo. J. Exp. Zool. 130:555–572, 1955.

Murphy, M.L. and Karnofsky, D.A.: Effect of azaserine and other growth–inhibiting agents on fetal development of the rat. Cancer 9:955–962, 1956.

210 Azathioprine *Imuran*[TM] CAS 446–86–6

Golby (1970) studied 25 pregnancies of mothers with renal transplants treated with azathioprine and other drugs.

Eighteen infants were normal, 5 women had early abortions and 2 had premature babies. Nishimura and Tanimura (1976) reviewed reports on 26 exposed human fetuses and reported 1 with a defect (pulmonary valvular stenosis). Pirson et al. reported on 20 pregnancies treated with 1–2 mg per kg and prednisolone. Forty percent were below the 10th percentile for birth weight and obstetrical complications such as oligohydramnios, hypoxia and meconium staining were more common.

Tuchmann–Duplessis and Mercier–Parot (1968) were unable to produce defects in rats and mice but observed limb reduction deformities in rabbits given as little as 5 mg per kg per day from the 6th to 14th day. Scott (1977) found an increase in resorptions and fetal growth retardation in rats given 5 mg per kg. No defects were mentioned.

Golby, M.: Fertility after transplantation. Transplantation 10:201, 1970

Nishimura, H. and Tanimura, T.: Clinical Aspects of the Teratogenicity of Drugs. New York, American Elsevier, 106–107, 1976.

Pirson, Y.; VanLierde, M.; Ghysen, J.; Squifflet, J.P.; Alexandre, G.P.J. and DeStrihou, C.V.: Retardation of fetal growth in patients receiving immunosuppressive therapy. N. Eng. J. Med. 313:328, 1985.

Scott, J.R.: Fetal growth retardation associated with maternal administration of immunosuppressives. Am. J. Obstet. Gynecol. 128:668–676, 1977.

Tuchmann–Duplessis, H. and Mercier–Parot, L.: Foetopathes therapeutiques: production experimentale de malformations des membres. Union Med. Can. 97:283–288, 1968.

211 5–Azauracil CAS 71–33–0

Kosmachevskaya (1968) using doses of 0.5 and 4.0 mg per chick egg found 32 and 61 percent mortality, respectively. Twelve of 14 survivors were defective. Anophthalmia, microphthalmia, coloboma and ectopia of the lens were observed.

Kosmachevskaya, E.A.: Comparison of pathogenic activity of5–and 6–azauracil in chick embryos. Arch. Anat. (Russian) 3:85–92, 1968.

212 6–Azauracil CAS 461–89–2

Kosmachevskaya (1968) found that doses of 0.5 and 4.0 mg injected on the first day of incubation caused 35 and 77 percent mortality, respectively. All surviving chick embryos had eye deformities indicating anophthalmia, microphthalmia, coloboma and ectopia of the lens.

Kosmachevskaya, E.A.: Comparison of pathogenic activity of5–and 6–azauracil in chick embryos. Arch. Anat. (Russian) 3:85–92, 1968.

213 6–Azauridine CAS 54–25–1

Vojta and Jirasek (1966) administered 7.5 to 15 gm to women who were 1.5 to 2.5 months pregnant and interrupted the pregnancy within an eight–day period. Although spontaneous abortions and congenital defects did not occur, microM degeneration of the chorionic villi was seen.

Sanders et al. (1961) reported that single doses of 1 mg per gm over 3 days from the 4th to the 6th day of gestation interrupted all mouse pregnancies. Van Wagenen et al. (1970), utilizing the triacetyl derivative, produced abortions and one fetus with skeletal, palate and renal defects in rhesus monkeys.

Sanders, M.A.; Wiesner, B.P. and Yudkin, J.: Control of fertility by 6–azauridine. Nature 189:1015–1016, 1961.

Van Wagenen, G.; Deconti, R.C.; Handschumacher, R.E. and Wade, M.E.: Abortifacient and teratogenic effects of triacetyl–6–azauridine in the monkey. Am. J. Obstet. Gynecol. 108:272–281, 1970.

Vojta, M. and Jirasek, J.: 6–Azauridine–induced changes of the trophoblast in early human pregnancy. Clin. Pharmacol. Ther. 7:162–165, 1966.

214 Azelastine

Suzuki et al. (1981) fed rats on days 7–17 up to 100 mg per kg. The maternal toxic dose was 65 mg per kg. At that dose increased resorption and growth retardation occurred but no increase in major defects. Rabbits received up to 30 mg per kg daily during organogenesis and no teratologic findings were observed.

Suzuki, Y.; Okada, F.; Mikami, T.; Goto, M.; Hasegawa, H. and Chiba, R.: Teratology and reproduction studies of azelastine, a novel antiallergic agent, in rats and rabbits. Arzneim Forsch. Drug Res. 31:1225–1230, 1981.

215 Azetidine–2–carboxylic Acid CAS 2517–04–6

Alescio (1973) showed that this proline analogue, which inhibits collagen accumulation in the chick embryo, produces a decrease in lung epithelial budding. Nagai et al. (1978) produced fusions of the vertebral bodies of rat fetuses whose mothers were given 300 mg per kg, intraperitoneally, on day 8. The use of 100 mg per day for 3–day periods during organogenesis caused no ill effects in the fetuses.

Alescio, T.: Effect of a proline analogue, azetidine–2–carboxylic acid, on the morphogenesis in vitro of mouse embryonic lung. J. Embryol. Exp. Morph. 29:439–451, 1973.

Nagai, H.; Kambara, K.; Sudo, H.; Yokoyama, S.; Tatsuya, T. and Nagai, N.: Teratogenic effect of proline analogue L–azetidine 2 carboxylic acid on the skeletal system of rats. Lung Anom. 18:19–23, 1978.

216 Azide Sodium CAS 26628–22–8

Sana et al. (**1990**) infused hamsters subcutaneously with 6×10^{-2} millimoles per kg per hour on days 7 through 9. Although resorptions were increased no teratogenicitty was found. At this dose maternal toxicity occurred. It is of interest that this electron transport particle poison as well as carbon monoxide has little effect on the early embryo which is dependent on aerobic glycolysis. Spratt (1950) found in explanted chick embryos that sodium azide above a concentration of $10^{-4}\mu$ caused degeneration. All tissues of the embryo seemed to be equally affected.

Sana, T.R.; Ferm, V.H.; Smith, R.P.; Kruszyna, R.; Kruszyna, H. and Wilcox, D.E.: Embryotoxic effects of sodium azide infusions in the syrian hamster. Fundamental and Applied Toxicology 15:754–759, 1990.

Spratt, N.T.: Nutritional requirements of the early chick embryo. III. The metabolic basis of morphogenesis and differentiation as revealed by the use of inhibitors. Biol. Bull. 99:120–135, 1950.

217 Azo–bis (cyclohexanecarbonitrile)

DeSesso et al. (**1993**) injected this free radical initiator intracoelomically into day 12 rabbit embryos. Five hours after 165 micrograms the embryos had increased mitotic figures and moderate cell necrosis. Coadministration with the antioxidant propyl gallate delayed the appearance of cellular changes.

DeSesso, J.M.; Scialli, A.R. and Goeringer, G.C.: Azobix (cyclohexanecarbonitrile), a free radical initiator, causes early cellular changes in rabit embryos. (abs) Teratology 47:417, 1993.

218 Azoethane CAS 821–14–7

Druckrey (1973) reported the transplacental production of postnatal brain tumors in rats by inhalation exposure of this gas in doses of 37 to 150 mg per kg on the 15th or 22nd day of gestation. Azoxyethane given intravenously on the 15th day (50 mg per kg) also produced a high incidence of postnatal brain tumors.

Druckrey, H.: Specific carcinogenic and teratogenic effects of indirect alkylating methyl and ethyl compounds, and their dependency on stages of ontogenic developments. Xenobiotica 3:271–303, 1973.

219 Azoniaspiro
3–α–Benziloyloxynortropane–8,1–pyrrolidine Chloride

Antweiler et al. (1966) gave up to 20 mg per kg on days 8–12 to rats and found no adverse effects.

Antweiler, V.H.; Lauterbach, F.; Lehmann, H–D.; Uebel, H. and Vogel, G.: Zur Pharmakologie und Toxikologie von Azoniaspiranene in der Nortropin–bzw. Pseudonortropin–Reihe. Arzneim. Forsch. 16:1581–1591, 1966.

220 Azosemide CAS 27589–33–9

Hayasaka et al. (1984) studied the effect of this diuretic on mice, rats and rabbits. In rats, oral doses of 90 mg per kg per day were associated with wavy ribs, bent scapulae and bent humeri, but these deformations were only temporary. In mice, similar types of skeletal changes occurred at 120 mg per kg. No changes occurred in the rabbit fetuses whose mothers received 6 mg per kg close to the lethal dose. No other teratogenicity was observed.

This loop diuretic, which is structurally related to furosemide, was tested in rats before pregnancy and during the first 7 days and from day 17 through 3 weeks after birth (Hayasaka et al., 1984). Oral doses up to 90 mg per kg did not produce any adverse reproductive effects.

Hayasaka, I.; Ichiyama, K.; Murakami, K.; Kato, Z.; Tamaki, F.; Shibata, T.; Sugawara, T. and Hayashi, M.: Teratogenicity of azosemide a loop diuretic in rats, mice and rabbits. Cong. Anom. 24:111–121, 1984.

Hayasaka, I.; Murakami, K.; Kato, Z. and Uchiyama, K.: Reproduction study of azosemide in rats. Fertility and peri– and postnatal study. Clin. Report. 18:5176–5195, 1984.

221 Azoxyethane CAS 16301–26–1

Griesbach (1973), using 30 or 50 mg per kg subcutaneously on days 8 through 10, produced eye defects and internal hydrocephalus in a high proportion of the rat offspring.

Griesbach, U.: Selektive erzengung von Missbildungen durch Azoxyathan wahrend der Fruhenwicklung der Ratte. Naturwissenschaften 60:555 only, 1973.

222 Azthreonam CAS 78110–38–0

Furuhashi et al. (1985) administered this antibiotic intravenously to rats on days 0–7, 7–17 and 17–20 of gestation. At doses of up to 750 mg per kg, no adverse effects on reproduction were found.

Furuhashi, T.; Kato, I.; Igarashi, Y. and Nakayoshi, H.: Toxicity study of azthreonam fertility study in rats. Chemotherapy 33:190–202, 1985.

Furuhashi, T.; Ushida, K.; Sato, K. and Nakayoshi, H.: Toxicity study of azthreonam teratology study in rats. Chemotherapy 33:203–218, 1985.

Furuhashi, T.; Ushida, K.; Kakei, A. and Nakayoshi, H.: Toxicity study on azthreonam perinatal and postnatal study in rats. Chemotherapy. 33:219–231, 1985.

223 Azuletil

This antiulcer drug was studied by Nemec et al. (**1991**) and Tesh et al. (**1991** A,B) in pregnant rats by gavage before and during early pregnancy, during organogenesis and in the peri–and postnatal period. Adult toxicity was found at the highest dose 200 mg per kg, but no reproductive toxicity. At 600 mg per kg no teratogenicity or behavioral changes occurred. In the rabbit Tesh et al. (**1991**) gave up to 125 mg per kg without adverse fetal effects.

Nemec, M.D.; Holson, J.F.; Tasker, E.J.; Oberholtzer, K.D.; Knapp, J.F.; Tomiyama, A.; Shimada, T. and Matsumoto, M.: Assesement of the effect of the antiulcer agent, azuletil sodium (KT1–32)f on reproduction and prenatal development in Sprague–Dawley crl: CD BR rats. Japan Pharm. & Therap. 19:1317–1333, 1991.

Tesh, J.M.; McAnulty, P.A.; Deans, C.F.; Tesh, S.A.; Matsumoto, M. and Tomiyama, A.: The effects of oral administration of azuletil sodium (KT1–32) upon pregnancy in the rat. Japan Pharm. & Therap. 19:1335–1357, 1991A.

Tesh, J.M.; McAnulty, P.A.; Higgins, C.; Courtney, S.; Matsumoto, M. and Tomiyama, A.: The effects of oral administration or azuletil sodium (KT1–32) upon peri–and postnatal development in the rat. Japan Pharm. & Therap. 19:1371–1389, 1991B.

Tesh, J.M.; Ross, F.W.; Wilby, O.K.; Tesh, S.A.; Matsumoto, M. and Tomiyama, A.: The effects of oral adminis-tration of azuletil sodium (KT1–32) upon pregnancy in the rabbit. Japan Pharm. & Therap. 19:1359–1369, 1991.

224 Bacampicillin Hydrochloride CAS 37661–08–8

Noguchi and Ohwaki (1979) gavage fed this antibiotic to rats and rabbits at maximum doses of 3000 and 250 mg per kg, respectively. Treatment in both sexes before mating and during the first week of gestation had no adverse effects. No increase in defects occurred in rat or rabbit fetuses. Abortion and maternal death occurred in some of the rabbit dams at the higher doses. Treatment on days 17–21 in the rat at 3000 mg per kg, produced some maternal deaths and an increase in stillbirths.

Noguchi, Y. and Ohwaki, Y.: Reproductive and teratologic studies of bacampicillin hydrochloride in rats and rabbits. Chemotherapy 27:30–35, 1979.

225 Bacitracin

Heinonen et al. (1977) included 18 pregnancies exposed to this cutaneously administered antibiotic among a group of 79 women taking other antibiotics. There was no increase in defect rates in the group.

Heinonen, O.P.; Slone, D. and Shapiro, S.: Birth Defects and Drugs in Pregnancy. Publishing Sciences Group Inc., Littleton, Mass., 1977.

226 Baclofen CAS 1134–47–0

This muscle relaxant was tested in pregnant mice, rats and rabbits with negative teratogenic findings. The test was done orally during the active periods of organogenesis. Maximum daily dose per kg was 15, 12.5 and 4.55 mg respectively in the mouse, rat and rabbit (Hirooka, 1976A, B; Hirooka et al., 1976).

Hirooka, T.: Effects of baclofen (CIBA 34,647–Ba) administered orally to pregnant rats upon pre– and postnatal development of their offspring. Osaka Daigaku Igaku Zasshi 28:181–194, 1976A.

Hirooka, T.: Effects of baclofen (CIBA 34,647–Ba) administered orally to pregnant mice upon pre– and postnatal development of the offspring. Osaka Daigaku Igaku Zasshi 28:195–203, 1976B.

Hirooka, T.; Morimoto, K.; Tadokoro, T.; Takahashi, S.; Ikemori, M.; Hirano, Y. and Miyaji, T.: Effects of baclofen (CIBA 34647–BA) administered orally to pregnant rabbits upon pre– and postnatal development of their offspring. Osaka Daigaku Igaku Zasshi 28:257–264, 1976.

227 Bacterial Toxins Staphlococcal Toxin

Elis et al. (1973) reported that the administration of either Shigella or Staphlococcal toxin intravenously to mice during the later part of gestation was associated with increased postnatal death and decreased development of motor and other activities.

Elis, J.; Kpsiak, M. and Gutova, M.: The effect of prenatal bacterial toxins on postnatal development. (abs) Teratology 8:220 only, 1973.

228 Balagrin *Balan–N–butyl–N–ethyl–2,6 dinitro–4 Trifluoromethyl Analine Xylene*

Bulgarin is a herbicide mixture (20:80) of the two chemicals listed as synonyms. Ivanova–Chemishanska et al. (1979) gavaged rats 55 mg per kg daily or 110 mg per kg every second week for three generations. Reduced fertility and increased embryonic loss were found in the first generation but not in the second generation. Dosing throughout pregnancy at 55 or 110 mg per kg was associated with increased post–implantation loss and reduced fetal size. Increased micro ophthalmia, mild hydrocephalus and intracerebal hematomas were observed at 22 mg per kg or above. Single dosing on day 13 was associated with meningomyelocele.

Ivanova–Chemishanska, L.; Vashakidze, V.; Mirkova, E. and Antov, G.: Study of remote effects of the herbicide balagrin. Khig. Zdraveopaz. 22:552–560, 1979.

229 Bamifylline CAS 2016–63–9

Georges and Denef (1968) were unable to produce defects in rats using up to 1000 mg per kg from the 10th to 12th days.

Georges, A. and Denef, J.: Les anomalies digitales manifestations teratogeniques des derives xanthiques chez le rat. Arch. Int. Pharmacodyn. Ther. 172:219–222, 1968.

230 Barbital

Discussed under Barbituric Acid

231 Barbituric Acid and Derivatives *Pentabarbital Hexabarbital Heptabarbital Phenobarbital* CAS 57–44–3

In conjunction with studies of other antiepileptic drugs and several tranquilizers, prospective studies of the effect of barbiturates on pregnancy outcome were published. Milkovich and Van den Berg (1974), in a group of about 325 women taking barbiturates during pregnancy, found no increase in the defect rate. Shapiro et al. (1976) studied 8,000 mothers who took phenobarbital during their pregnancies and found no evidence of fetal damage when the drug was taken for indications other than epilepsy. Fedrick (1973) studied 41 epileptic mothers taking only phenobarbital during the first trimester and found no increase in congenital defects. Bethenod and Frederich (1975) studied epileptic mothers and noted 1 abnormal facies of 6 offspring exposed only to phenobarbital. Seip (1976) described 2 siblings with fetal hydantoin–like syndrome from a mother treated with high doses of phenobarbital.

Goldhaber et al. (1990) found no association between prenatal barbiturate therapy and 86 children who had cord or brain tumors.

Setala and Nyyssonen (1964) briefly reported the production of malformations of the head and extremities in mice given pentabarbital, but details including controls were lacking. McColl et al. (1963) found double vertebral centra in the offspring of mice fed 0.16 percent phenobarbital. This skeletal finding could be due to a nutritional effect with delayed ossification rather than representing a true

congenital defect. McColl et al. (1967) reported skeletal and aortic arch defects in the offspring of rabbits treated with 50 mg of phenobarbitol per kg from days 8 through 16. In the rat, Persaud (1965) found a low incidence of limb defects with hexobarbital and barbital given on the 4th and 8th day of gestation.

Bethenod, M. and Frederich, A.: Les enfants des antiepileptiques. (French) Pediatrie 30:227–248, 1975.

Fedrick, J.: Epilepsy and pregnancy. A report from the Oxford record linkage study. Br. Med. J. 2:442–448, 1973.

Goldhaber, M.K.; Selby, J.V.; Hiatt, R.A. and Quesenberry, C.P.: Exposure to barbiturates in utero and during childhood and risk of intracranial and spinal cord tumors. Cancer Research 50:4600–4603, 1990.

McColl, J.D.; Globus, M. and Robinson, S.: Drug–induced skeletal malformations in the rat. Experientia 19:183–184, 1963.

McColl, J.D.; Robinson, S. and Globus, M.: Effect of some therapeutic agents on the rat fetus. Toxicol. Appl. Pharmacol. 10:244–252, 1967.

Milkovich, L. and Van den Berg, B.J.: Effects of prenatal meprobamate and chlordiazepoxide hydrochloride on human embryonic and fetal development. New Eng. J. Med. 291:1268–1271, 1974.

Persaud, T.V.N.: Tierexperimentelle Untersuchungen zur Frage der teratogenen Wirkung von Barbituraten. Acta Biol. Med. Ger. 14:89–90, 1965.

Seip, M.: Growth retardation, dysmorphic facies and minor malformations following massive exposure to phenobarbitone in utero. Acta Paediatr. Scand. 65:617–621, 1976.

Setala, K. and Nyyssonen, O.: Hypnotic sodium pentobarbital as a teratogen in mice. Naturwissenschaften 51:412 only, 1964.

Shapiro, S.; Hartz, S.C.; Siskind, V.; Mitchell, A.A.; Slone, D.; Rosenberg, L.; Monson, R.R.; Heinonen, O.P.; Idanpaan, J.; Haro, S. and Saxen, L.: Anticonvulsants and prenatal epilepsy in the development of birth defects. Lancet 1:272–275, 1976.

232 Barium CAS 7440–39–3

Ridgway and Karnofsky (1952) injected 20 mg of BaCl(2) into the chick yolk sac on the 8th day and observed curled toes in about 50 percent of survivors. Earlier treatment had no teratogenic effect.

Ridgway, L.P. and Karnofsky, D.A.: The effects of metals on the chick embryo: Toxicity and production of abnormalities in development. Ann. N.Y. Acad. Sci. 55:203–215, 1952.

233 Batroxobin *Defibrase*™

Ozaki et al. (1983) studied this defibrinogenic agent in rats and dogs. In beagle dogs, up to 16 BU per kg were given intravenously daily during organogenesis and no evidence of teratogenicity was found. In rats, treated before fertilization, during organogenesis and in the perinatal period, no adverse fetal effects were found except for growth retardation at all doses (4,15 and 60 BU per kg). Maternal lethality was increased at 60 BU dose level and some increase in fetal death was seen.

Ozaki, M.; Sato, S.; Hiyama, T.; Kunikane, K.; Tsushima, K. and Mori, N.: Reproduction studies of batroxobin (Defibrase™) obtained from snake venom in dogs and rats. Oyo Yakuri 25:519–576, 1983.

234 Baygon *Propoxur Bayer™*

Rosenstein and Chernoff (1976) gave rats 1000 ppm in the diet from day 2 through gestation and weaning. The neonates were lighter and spontaneous and evoked ECG changes were found. Courtney et al. (1985) found no adverse effects of gavage treatment in rats during days 7–19 or 6–16 with up to 50 mg per kg.

Courtney, K.D.; Andrews, J.E.; Springer, J. and Dalley, L.: Teratogenic evaluation of the pesticides baygon, carbofuran, dimethoate and EPN. J. Environ. Sci. Health B20:373–406, 1985.

Rosenstein, L. and Chernoff, N.: Spontaneous and evoked ECG changes observed in neonatal rats following in utero exposure to baygon: a preliminary investigation. Toxicol. Appl. Pharmacol. 37:130, 1976.

235 BCG Vaccine

Takashima et al. (**1994**) gave this subcutaneously to rats before pregnancy and during the first 7 days of pregnancy. The dams had granulomas and polyarthritis but doses to 5 mg per kg did not alter fetal development.

Takashima, H.; Kaneko, Y.; Azegami, J.; Inada, H.; Yoshimura, S.; Nagao, T. and Mizutani, M.: Fertility study of dried BCG vaccine by subcutaneous administration in rats. Iyakuhin Kenkyu 25:25–36, 1994.

236 BD40A

Saegusa et al. (1980) studied this bronchodilator in pregnant rats using treatments on days 7–21. The amount and route were not specified. Wavy ribs occurred when single–day treatments were given on days 18 through 21.

Saegusa, T.; Kaneko, Y.; Sato, T.; Narama, I. and Segima, Y.: BD40A induced wavy ribs in rats. (abs) Teratology 22:14A only, 1980.

237 Beclomethasone Dipropionate CAS 5534–09–8

Greenberger and Patterson (1983) used the medication in 45 pregnancies (4 to 16 inhalations per day). The only malformation was a child with congenital heart disease which occurred in the offspring of women who had diabetes as well as asthma.

Esaki et al. (1976) administered this steroidal anti-inflammatory to mice by inhalation for up to 10 minutes daily, from day 6 through 15. The concentration in the chamber was 23 microgm per liter. The fetuses had approximately 50 percent malformations which were mainly cleft palate but vertebral defects also occurred. Tanioka (1976) administered up to 200 microgm per kg to monkeys on days 23 through 35 and produced no malformations in 7 fetuses.

Oral administration of up to 16 mg per kg to rats during the first week of gestation, days 7 through 17 or days 17 through 28 after delivery was studied by Sudon et al. (1979), Furuhashi et al. (1979) and Hasegawa et al. (1979).

In the early treatment group, a decrease in fetal survival and weight was associated with some maternal toxicity. During organogenesis, the 1.6 and 16 mg per kg dose decreased fetal growth but no abnormalities occurred. In late gestation, the treatment prolonged gestation and increased the number of dead fetuses when 1.6 or 16 mg per kg was given.

Esaki, K.; Izumiyama, K. and Yasuda, Y.: Effects of inhalant administration of beclomethasone dipropionate on the reproduction in mice. (Japanese) CIEA Preclin. Rep. 2:213–222, 1976.

Furuhashi, T.; Nomura, A.; Miyoshi, K.; Ikeya, E. and Nakayoshi, H.: Teratologic and fertility studies on beclomethasone dipropionate. 2. Teratological studies by oral administration. Oyo Yakuri 18:1021–1038, 1979.

Greenberger, P.A. and Patterson, W.: Beclomethasone dipropionate for severe asthma during pregnancy. Annals of Internal Medicine 98:478–480, 1983.

Hasegawa, T.; Nomura, A. and Nakayoshi, H.: Teratological and fertility studies on beclomethasone dipropionate. 3. Perinatal and postnatal study in rats by oral administration. Oyo Yakuri 18:1039–1054, 1979.

Sudou, S.; Sendota, H.; Yamamoto, K.; Miyoshi, K. and Nakayoshi, H.: Teratological and fertility studies on beclomethasone dipropionate. 1. Fertility study in rats by oral administration. Oyo Yakuri 18:1003–1019, 1979.

Tanioka, Y.: Teratogenicity test on beclomethasone dipropionate by inhalation in rhesus monkeys. (Japanese) CIEA Preclin. Rep. 2:155–164, 1976.

238 Befunolol CAS 39552–01–7

Yoshinka et al. (1979) and Nakamura et al. (1979) and others studied this β blocker in mice, rats and rabbits using maximal doses of 500, 200 and 100 mg per kg daily, respectively. At the highest doses, some delay in ossification of the fetuses was found. Stillbirths were increased in the rabbits at 100 mg per kg. Fertility and perinatal studies did not show significant differences from control in the mice.

Nakamura, K.; Yoshida, J.; Aoyama, I.; Moritoka, M. and Moritoki, H.: Teratological studies in mice. Kiso to Rinsho 13:4161–4177, 1979.

Nakamura, K.; Aoyama, I. and Moritoki, H.: Teratological study in rabbits. Kiso to Rinsho 13:3715–3739, 1979.

Okuda, T.; Matubara, T.; Moritoka, M. and Moritoki, H.: Perinatal and postnatal studies in mice. Kiso to Rinsho 13:3740–3759, 1979.

Yoshida, J.; Ohtani, A. and Moritoka, M.: Fertility study in mice. Kiso to Rinsho 13:3725–3739, 1979.

Yoshinka, I.; Saito, K.; Hikita, S.; Komori, A.; Iizuka, H. and Moritoki, H.: Toxicological studies of befunolol hydrochloride (BFE–60) Kiso to Rinsho 13:3678–3714, 1979.

239 Benazepril Hydrochloride

Takahashi et al. (**1991** A,B) gavaged rats before and during early pregnancy and during organogenesis with doses up to 1000 mg per kg. At the highest dose of this ACE inhibitor maternal food intake was decreased and the number of surviving fetuses was lower. In the fetuses at term there was an increase in dilated renal pelves and ureters. Reproductive function was normal in the offspring after exposure during organogenesis but decreased when the medication was given perinatally and in postnatal life. In the perinatal and postnatal studies dilatation of the ureters and pelves was found at the 100 and 1000 mg per kg but not at 1 mg per kg (Nagae et al., **1991**).

Nagae, Y.; Takahashi, S.; Deguchi, J.; Morimoto, K.; Takahashi, H.; Kawaharra, M.; Moori, T. and Miyamoto, M.: Reproductive and developmental toxicity studies of CGS 1482A (benazepril hudrochloride) (3): Segment III study in rats. Jpn. Pharm. Therap. 9:3471–3490, 1991.

Takahashi, S.; Nagae, Y.; Takahashi, H.; Deguchi, J.; Harada, T. and Miyamoto, M.: Reproductive and developmental toxicity studies of CGS 1482A (benazepril hydrochloride) (I): Segment I study in rats. Jpn. Pharm. Therap. 9:2445–2451, 1991A.

Takahashi, S.; Nagae, Y.; Takahashi, H.; Harada, T. and Miyamoto, M.: Reproductive and developmental toxicity studies of CGS1482A (benazepril hudrochloride) (2): segment II study in rats. Jpn. Pharm. Therap. 9:3453–3470, 1991B.

240 Bencyclane *N–[3–(1–Benzyl–cycloheptyl–oxy) propyl]-N,N-dimethyl-ammonium–hydrogen fumarate* CAS 2179–37–5

Boissier (1970) gave rabbits, mice and rats oral daily doses of 5 to 100 mg per kg and found no increase in defects in the offspring.

Boissier, J.R.: Untersuchungen ueber eine Maegliche Teratogene Wirkung von Bencyclan. Arzneim. Forsch. 20:1399–1402, 1970.

241 Bendazole *Tromasedan*

Shimada et al. (**1970**) gave mice and rats orally up to 100 and 200 mg per kg respectively during organogenesis. No ill effects were found in the fetuses or pups followed up to the time of weaning.

Shimada, T.; Endo, A. and Ichikari, I.: Effects of tromasdean upon the fetal and post–natal developments in mice and rats. Niigata Igakkai Zasshi 84:347–352, 1970.

242 Bendectin™ *Debendox™* CAS 8064–77–5

Numerous epidemiologic studies on Bendectin have been published. Milkovich and Van den Berg (1976) reported no increase in malformation rate in the offspring of 628 mothers taking the drug. Shapiro et al. (1977), in over 1,000 exposed mothers, found no increase in malformation rate and the offspring's intelligence at 4 years was not different from controls. Bunde et al. (1963) found no increase in congenital defects among 2000 women using Bendectin™. Cordero et al. (1981) determined the rate of Bendectin use in the first trimester among over 1200 pregnancies with birth defects. No significant differences in drug exposure were found in any of the 12 defect categories. Borderline increases were found among the groups with amniotic band defects, encephalocele and esophageal atresia. Mitchell et

al. (1981), using a case control study method, analyzed 343 infants with facial clefts or congenital heart disease and found no association with Bendectin. In addition to the numerous epidemiological studies (see below) indicating that Bendectin is not a human teratogen, a number of panels appointed by different countries have concurred. A recent article (Anonymous, 1990) chronicles the historical aspects of this unfortunate saga. The Bendectin data has been utilized to illustrate how multiple studies may be combined objectively to assess safety (Einarson et al., 1988). McKeitue et al. (**1994**) conducted a meta–analysis of the 16 cohort and 11 case–control studies and found no increase in risk for birth defects.

Official statements affirming the safety of benedectin have been issued by the governments of the United Kingdom, Canada, Switzerland, West Germany and Australia. Ornstein et al. (**1995**) reviewed the contrast between the handling of the problem in the United States and Canada. Drug availability in Canada was not limited by the combined action of the media and litigation but scientific review prevailed.

Morelock et al. (1982) analyzed 1,690 mother–infant pairs and among the 375 exposed to Bendectin and no adverse fetal outcome was detected. Jick et al. (1981) linked computer stored prescriptions for Bendectin with subsequent malformations and found no association among over 6,800 pregnancies. MacMahon (1981), in an editorial, pointed out that among the 9 studies on Bendectin a few significant associations inevitably appeared but no consistent excess of any particular malformation was found. Michaelis et al. (1983) studied 874 matched pairs and found no increase in defects among Bendectin takers. If Bendectin is teratogenic at all, it is only under very rare circumstances. Holmes (1983) summarized the epidemiologic data. Mitchell et al. (1983) were unable to show an association between pyloric stenosis and the use of Bendectin. They compared 325 infants with pyloric stenosis to 2 control groups. Shiono & Klebanoff (1989) analyzed defect rates of 58 different types of congenital malformations occurring among 31,564 newborns. The odds ratio for all defects was 1.0. Odds ratios were elevated for microcephaly, cataract and lung malformations. This was "exactly the number of associations that would be expected by chance."

Bendectin, consisting of equal parts doxylamine succinate, dicyclomine HCl and pyridoxine HCl, was given by Gibson et al. (1968). They used doses of 30 mg per kg per day in the rabbit and 60 mg per kg per day in the rat during organogenesis and no increase in congenital defects was found. In 1976, dicyclomine was removed from the formulation. Tyl et al. (1988) gave up to 800 mg per day of Bendectin to rats on days 6–15 and found only short 13th ribs and reduced ossification in the fetuses. No increase in malformations was found. The fetal and maternal weights were reduced at 800 mg per kg doses.

Hendrickx et al. (1985A) administered Bendectin in doses of 10 to 20 times the human dose during days 22 to 50 of gestation in cynomolgus and rhesus monkeys and baboons. The autopsies at 100 days showed an increase in ventricular septal defects. However, in a double–blind study of term cynomolgus monkeys exposed to the same treatment, no increase in heart or other defects was found among 69 autopsies (Hendrickx et al., 1985B).

Anonymous: Key witness against morning sickness drug faces scientific fraud charges. JAMA 263:1468–1473, 1990.

Bunde, C.A. and Bowles, D.M.: A technique for controlled survey of case records. Curr. Ther. Res. 5:245–248, 1963.

Cordero, J.F.; Oakley, G.P.; Greenberg, F. and James, L.M.: Is Bendectin a teratogen? JAMA 245:2307–2310, 1981.

Einarson, T.R.; Leeder J.S. and Korenn, G.: A method for meta–analysis of epidemiological studies. Pharmacoepidemiology 22:813–824, 1988.

Gibson, J.P.; Staples, R.E.; Larson, E.J.; Kuhn, W.L.; Holtkamp, D.E. and Newberne, J.W.: Teratology and reproduction studies with an antinauseant. Toxicol. Appl. Pharmacol. 13:439–447, 1968.

Hendrickx, A.G.; Cukierski, M.; Prahalada, S.; Janos, G. and Rowland, J.: Evaluation of Bendectin embryotoxicity in non–human primates. Ventricular septal defects in prenatal macaques and baboons. Teratology 32:179–189, 1985A.

Hendrickx, A.G.; Cukierski, M.; Prahalada, S.; Janos, G.; Booher, S. and Nyland, T.: Evaluation of Bendectin embryotoxicity in nonhuman primates. 2. Double–blind study in term cynomolgus monkeys. Teratology 32:191–194, 1985B.

Holmes, L.B.: Teratogen update: Bendectin. Teratology 27:277–281, 1983.

Jick, H.; Holmes, L.B.; Hunter, J.R.; Madsen, S. and Stergachis, A.: First trimester drug use and congenital defects. JAMA 246:343–346, 1981.

MacMahon, B.: More on Bendectin. JAMA 246:371–372, 1981.

McKiegue, P.M.; Lamm, S.H.; Linn, S. and Kutcher, J.S.: Bendectin and birth defects: I.A meta–analysis of epidemiologic studies. Teratology 50:27–27, 1994.

Michaelis, J.; Michaelis, H.; Gluck, E. and Koller, S.: Prospective study of suspected associations between certain drugs administered during early pregnancy and congenital defects. Teratology 27:57–64, 1983.

Mitchell, A.A.; Rosenberg, L.; Shapiro, S. and Slone, D.: Birth defects related to Bendectin use in pregnancy I: Oral clefts and cardiac defects. J. Am. Med. Assoc. 245:2311–2314, 1981.

Mitchell, A.A.; Schwingl, P.J.; Rosenberg, L.; Louik, C. and Shapiro, S.: Birth defects in relation to Bendectin use in pregnancy II: Pyloric stenosis. Obstetrics 147:737–742, 1983.

Milkovich, L. and Van den Berg, B.J.: An evaluation of the teratogenicty of certain antinauseant drugs. Am. J. Obstet. Gynecol. 125:244–248, 1976.

Morelock, S.; Hingson, R.; Kayne, H.; Dooling, E.; Zuckerman, B.; Day, N.; Alpert, J.J. and Flowerdew, G.: Bendectin and fetal development. Am. J. Obstet. Gynecol.

142:209–213, 1982.

Ornstein, M.; Einarson, A. and Koren, G.: Bendectin/diclectin for morning sickness: a canadian follow–up of an American tragedy. Reproductive Toxicology 9:1–6, 1995.

Shapiro, S.; Heinonen, O.P.; Siskind, V.; Kaufman, D.W.; Monson, R.R. and Slone, D.: Antenatal exposures to perinatal mortality rate, birth weight and intelligence quotient score. Am. J. Obstet. Gynecol. 128:480–485, 1977.

Shiono, P.H. and Klebanoff, M.A.: Bendectin and human congenital malformations. Teratology 40:151–155, 1989.

Tyl, R.W.; Price, C.J.; Marr, M.C. and Kimmel, C.A.: Developmental toxicity evaluation of Bendectin in CD rats. Teratology 37:539–552, 1988.

243 Benidipene KW–3049

Naya et al. (**1989** A&B) treated rats and rabbits orally during organogenesis with up to 35 and 50 mg respectively. In both species at the highest dose fetal deaths were increased but no teratogenicity or decrease in fetal weight was found.

Naya, M.; Fujita, T.; Sakuma, T. and Deguchi, T.: Reproduction study of kw–3049–teratogenicity study in rats. Kiso to Rinsho 23:6747–6757, 1989.

Naya, M.; Waki, Y.; Fujita, T.; Nishikawa, S. and Deguchi, T.: Toxicological study of kw–3049 teratogenicity study p.o. in rabbits. Kiso to Rinsho 23:6759–6767, 1989.

244 Benomyl
Methyl–1–(butylcarbamoyl)–2–benzimidazolecarbamate
CAS 17804–35–2

Spagnole et al. (**1994**) studied a population of over 900,000 for any association between Benomyl and eye defects. There were 33 cases of anophtholmia and 78 cases of microphthalmia (.035 and 0.83 per 10,000 respectively). Parental occupation in agriculture was not associated with the eye defects. Additionally, there was no association with the areas which had high use of Benomyl.

This fungicide was studied in pregnant rats and when a diet containing 0.50 percent benomyl was fed on days 6–15, no fetal effects or malformations resulted (Sherman et al., 1975). When the material was given by gavage (Kavlock et al., 1982), malformations were produced in both rats and mice. The teratogenic dose in the rat was 62.5 mg per kg and in the mouse, 100 mg per kg. Hydrocephalus, cleft palates, hydronephrosis and skeletal defects were found. Fetotoxicity but not teratogenicity was found when 62.5 mg per kg was administered via the diet. Shtenberg and Torchinsky (1972) reported teratogenicity in rats. The pathogenesis of periventricular overgrowth in the rat was studied by Ellis et al. (**1988**).

Ellis, W.G.; De Roos, F.; Kavlock, R.J. and Zeman, F.J.: Relationship of periventricular overgrowth to hydrocephalus in brains of fetal rats exposed to benomyl. Teratogenesis, Carcinogenesis, and Mutagenesis 8:377–391, 1988.

Kavlock, R.J.; Chernoff, N.; Gray, L.E., Jr.; Gray, J.A. and Whitehouse, D.: Teratogenic effects of Benomyl in wistar rat and CD–1 mouse with emphasis on rate of administration. Toxicol. Appl. Pharmacol. 62:44–54, 1982.

Sherman, H.; Culik, R. and Jackson, R.A.: Reproduction, teratogenic and mutagenic studies with benomyl. Toxicol. Appl. Pharmacol. 32:305–315, 1975.

Shtenberg, A.I. and Torchinsky, A.M.: On the interrelationship of general toxic, embryotoxic and teratogenic action of chemicals. Vestnik Akad. Nauk SSR 3:39–46, 1972.

Spagnolo, A.; Bianchi, F.; Calabro, A.; Calzolari, E.; Clementi, M.; Mastroiacovo, P.; Meli, P.; Petrelli, G. and Tenconi, R.: Anophthalmia and Benomyl in Italy: A multicenter study based on 940,615 newborns. Reproductive Toxicology 8:397–403, 1994.

245 Benz–a–anthracene CAS 56–55–3

Discussed under Dimethylbenz–a–anthracene

246 Benzalkonium Chloride

Buttar (1985) studied the effect of vaginal application on day 1 of the rat gestation. Doses of up to 200 mg per kg were used. At 143 times the human contraceptive dose, there was an increase in resorptions and fetal death. No congenital visceral defects occurred. Momma et al. (1987) gave mice up to 100 mg per kg on days 0–6 or 0–18 and found no adverse effects on implantation or fetal development. At doses of 30 mg per kg on days 0–6, there was a tendency to a "slight decrease in the pregnancy rate" and increase in resorptions.

Buttar, H.S.: Embryotoxicity of benzalkonium chloride in vaginally treated rats. J. Appl. Toxicol. 5:398–401, 1985.

Momma, J.; Takada, K.; Aida, Y.; Takagi, A.; Yoshimoto, H.; Suzuki, Y.; Nakaji, Y.; Kurokawa, Y. and Tobe, M.: Effects of benzalkonium chloride on pregnant mice. Bull. Natl. Inst. Hyg. Sci. (Tokyo) 105:20–25, 1987.

247 Benzbromarone CAS 3562–84–3

Aoyama et al. (1979) gave this drug to mice, rats and rabbits during organogenesis. The maximum oral doses were 220, 80 and 220 mg per kg in mice, rats and rabbits, respectively. At the highest dose in the mouse, there was embryo lethality, but no adverse effects were found in the rabbit fetus. At the 80 mg per kg dose in the rat, 14 percent of the fetuses were abnormal. The defects included skeletal reductions, club hands, curled tails and cleft lips.

Aoyama, T.; Terabayashi, M.; Konatru, S.; Hasegawa, T.; Shibutani, N. and Shimimura, K.: Teratologic study on benzobromarone. 1. Experiments in mice, rats and rabbits. Shinryo to Shinaku 16:1521–1545, 1979.

248 Benzene CAS 71–43–2

Savitz et al. (1989) studied 150 women occupationally exposed to benzene and found an adjusted odds ratio for stillbirth of 1.3 (CI 1.0–1.8) but for the 335 fathers had odds ratio of 1.0. The odds ratio for a small for gestational age and preterm delivery were not significantly increased.

Pushkina et al. (1968) exposed pregnant rats continuously during pregnancy to benzene vapors (1 to 670 mg per cubic meter) and found no developmental malformations.

The number of fetuses per liter was reduced with exposure to the higher doses. Some reduction in ascorbic acid content of the fetuses was found. Watanabe and Yoshida (1970) injected 3.0 ml per kg of this solvent into pregnant mice on single days of gestation. Injections on the 13th day resulted in an increased incidence of cleft palate and mandible reduction. Hudak and Ungvary (1978) exposed rats to 3.3 and 1500 mg per cubic meter from days 9 through 14 and found no increase in malformations although fetal growth retardation and skeletal anomalies occurred along with some maternal mortality. Ungvary and Tatrai (1985) exposed rats, mice and rabbits to 100 mg per square meter during embryogenesis and found growth and skeletal delay but no increase in defects.

Nawrot and Staples (1979) demonstrated embryolethality in the mouse gavaged with as little as 0.3 ml per kg, but no teratogenicity at 1.0 mg per kg. Green et al. (1978) exposed pregnant rats to 2200 ppm of the vapor and found no increase in fetal malformations. Some skeletal growth retardation was reported. Murray et al. (1979) exposed rabbits and mice to 500 ppm during organogenesis and found no teratogenicity. They summarized the predominantly negative findings of other workers.

Kitaev et al. (1979) exposed female rats (for 4 hours daily during 30–45 days) to benzene vapors (300–1000 mg per cubic meter) and observed abnormalities in cleaving embryos.

Chapman et al. (1994) added up to 1.6 mM to whole rat embryo cultures and produced no statistical effects.

Chapman, D.E.; Namkung, M.J. and Juchau, M.R.: Benzene and benzene metabolites as embryotoxic agents: Effects on cultured rat embryos/. Toxicology and Applied Pharmacology 128: 129–137, 1994.

Green, J.D.; Leong, B.K.J. and Laskin, S.: Inhaled benzene fetoxicity in rats. Toxicol. Appl. Pharmacol. 46:9–18, 1978.

Hudak, A. and Ungvary, G.: Embryotoxic effects of benzene and its methyl derivatives. Toxicology 11:55–63, 1978.

Kitaev, E.M.; Nikitin, A.I.; Lipovsky, S.M. and Louchikov, V.A.: Effect of benzene on some indexes of early rat embryogenesis. Akush. Ginekol. (USSR) 4:51–53, 1979.

Murray, F.J.; John, J.A.; Rampy, L.W.; Kuna, R.A. and Schwetz, B.A.: Embryotoxicity of inhaled benzene in mice and rabbits. Am. Ind. Hyg. Assoc. J. 40:993–998, 1979.

Nawrot, P.S. and Staples, R.E.: Embryo fetal toxicity and teratogenicity of benzene and toluene in the mouse. (abs) Teratology 19:41A only, 1979.

Pushkina, N.N.; Gofmekler, V.A. and Klevtsoua, G.N.: Changes in content of ascorbic acid and nucleic acids produced by benzene

Savitz, D.A.; Whelan, E.A. and Kleckner, R.C.: Effect of parents' occupational exposures on risk of stillbirth, preterm delivery, and small–for–gestational–age infants. American Journal of Epidemiology 129:1201-1218, 1989.

Ungvary, G. and Tatrai, E.: On the embryotoxic effects of benzene and its alkyl derivatives in mice, rats and rabbits. Arch. Toxicol. Suppl. 8:425–430, 1985.

Watanabe, G. and Yoshida, S.: The teratogenic effect of benzene in pregnant mice. Acta Medica Biol. 17:285–291, 1970.

249 Benzene Hexachloride

Discussed under Lindane

250
2–Benzenesulfonamido–5–tertiobutyl–1–thia–3,4–dia zole (1395–TH)

Koyama et al. (1969) gave this oral hypoglycemic to pregnant rats, mice and rabbits during organogenesis. In the mouse, cleft palate, exencephaly and vertebral defects were seen following 150, 300, or 600 mg per kg. Resorptions and fetal growth defects were increased in rats. The defects in the mice were attributed to hypoglycemia.

Koyama, K.; Imamura, S.; Ohguro, Y. and Hatano, M.: Toxicological studies on 2–benzenesulfonamido–5–tertiobutyl–1–thia–3, 4–diazole (1395–TH). Yamaguchi Igaku 20–27, 1969.

251 Benzenesulfonic Acid Hydrazide CAS 80–17–1

Matschke and Fagerstone (1977) gavage fed this rodenticide to pregnant mice on days 8–13. Resorptions occurred at maternal toxic levels (62 mg per kg), but only delayed ossification was found in the fetuses.

Matschke, G.H. and Fagerstone, K.A.: Effects of a new rodenticide, benzensulfonic acid hydrazide, on prenatal mice. J. Toxicol. Environ. Health 3: 407–411, 1977.

252 Benzhydrylpiperazines

Discussed under Meclizine™

253 Benzidine CAS 531–86–2

Mice receiving 150 mg per kg from day 12 to delivery, had offspring with increased rates of liver–cell tumors (Vesselinovitch et al., 1979).

Vesselinovitch, S.D.; Rao, K.V.N. and Mihailovich, N.: Neoplastic response of mouse tissues during perinatal age periods and its significance in chemical carcinogenesis. Natl. Cancer Inst. Monogr. 51:239–250.

254 Benzilic Acid
N,N–Dimethyl–2–hydroxy–methyl–piperidinum

Osterloh et al. (1966) treated rats and rabbits orally or subcutaneously with up to 200 mg per kg daily during gestation. An increase in defects among the rabbit fetuses was not significantly greater than the controls.

Osterloh, V.G.; Lagler, F.; Staemmler, M. and Helm, F.: Pharmakologische und toxikologische Untersuchungen uber Benzilsaure–(N,N–dimethyl–2–hydroxymethyl–piperidinium)–ester methylsulfat)–ein neues Spasmolyticum. Arzneim–Forsch 16:901–909, 1966.

255 Benzimidazole *2–Ethyl–5–methylbenzimidazole*
CAS 51–17–2

Waddington et al. (1955) injected 0.5 ml of 0.5 percent solution in eggs and found a significant incidence of omphalocephalics (head reduced in size and protruding through endoderm). The teratogenic effects of three analogues (2, 5–dimethyl, 2–ethyl–5–methyl and 2–hepta–5––methylbenzimidazole) were studied by Blackwood (1960). Doses of 0.2 to 0.7 mg injected into the egg albumen produced 53 percent defective embryos by the 14th day. The defects consisted of extruded brain and viscera and abnormal beaks.

Delatour and Richard (1976) studied this compound and 24 related chemicals in the pregnant rat. Oral doses of up to 53 mg per kg on days 8 through 12 caused neither fetal weight reduction nor malformations. Parabenzimidazole was teratogenic at doses of 10 mg per kg. The N–benzimidazolyl–2 and N–benzimidazolyl–5 carbamates were highly teratogenic. All types of defects occurred including cleft lip and palate, neural tube closure and skeletal types.

Blackwood, U.B.: Selective inhibitory and teratogenic effects of 2, 5–alkylbenzimidazole homologues on chick embryonic development. Proc. Soc. Exp. Biol. Med. 104:373–378, 1960.

Delatour, P. and Richard, Y.: Proprietes embryotoxiques et antimitotiques en serie benzimidazole. Therapie 31:505–515, 1976.

Waddington, C.H.; Feldman, M. and Perry, M.M.: Some specific developmental effects of purine antagonists. Exp. Cell Res. 3:366–380, 1955.

256 Benzo[a]pyrene and Derivatives CAS 50–32–8

Rigdon and Rennels (1964) fed rats this substance (1 mg per gm of diet) during pregnancy and found many resorptions and dead fetuses, but only 1 malformed fetus from 7 litters. MacKenzie et al. (1979) observed sterility in female mice exposed to 40–160 mg per kg on gestation days 7–16.

Shum et al. (1979) studied the in–utero toxicity in relation to the allelic difference at the Ah locus in mice. A dose of 50–300 mg per kg was given intraperitoneally on day 7 or 10. They identified the Ah genotype of individual fetuses by measurement of the AHH inducibility and reported that when the mothers were Ah nonresponsive, the fetuses with Ah responsive genotype showed decreased body weight and higher resorption and malformation rates, while the Ah nonresponsive fetuses in the same uterus did not. The type of defect included mainly club foot, hemangioma, cleft lip and cleft palate. All of these defects tend to be associated with late organogenesis. Hoshino et al. (1981) used the same general protocol and 150 or 300 mg per kg on day 8 and confirmed the findings of Shum et al. (1979) for toxicity (reduced fetal weight and increased resorptions). They did not find the same increase in malformations. They found only an increase in cervical ribs and this occurred among fetuses from Ah responsive mothers. Borodin et al. (**1989**) studied postnatal immune function in rats exposed in utero and found suppression of the B–cell system.

Barbieri et al. (1986) studied some synthetic derivatives in mice using an intraembryonal injection on day 10, 12 or 14. 7–β,8–α–Dihydroxy–9–α, 10–α–epoxy–7,8,9,10 tetrahydro was the most potent embryotoxic compound causing 100 percent malformations when injected on day 10. Benzo(a)pyrene–4,5–oxide was not teratogenic, but 6–methylbenzo(a)pyrene induced multiple malformations. The defects included exencephaly, thoraco–and gastroschisis, phocomelia and edema. Benzo(a)pyrene by intraamniotic or transplacental route was not teratogenic. For these experiments, 47 micromoles per kg was used on day 10, 12 or 14.

Barbieri, O.; Ognio, E.; Rossi, O.; Astigiano, S. and Rossi, L.: Embryotoxicity of benzo(a)pyrene and some of its synthetic derivatives in swiss mice. Cancer Res. 46:94–98, 1986.

Borodin, Y.L.; Slyanova, N.A.; Ivanov, V.V.: Sklyanov, Y.L. and Semenyuk, A.V.: Effect of low–molecular–weight toxicants on the immune system of the mother and offspring. Gistol. Embriol. 96:41–45, 1989.

Hoshino, K.; Hayashi, Y.; Takehira, Y. and Kameyama, Y.: Influences of genetic factors on the teratogenicity of environmental pollutants: Teratogenic susceptibility to benzo(a)pyrene and Ah locus in mice. Congenital Anomalies 21:97–103, 1981.

MacKenzie, K.M.; Lucier, E.W. and McLachlan, J.A.: Infertility in mice exposed prenatally to benzo–a–pyrene B.P. (abs) Teratology 19:37A only, 1979.

Rigdon, R.H. and Rennels, E.G.: Effect of feeding benz(a)pyrene on reproduction in the rat. Experientia 20:224–226, 1964.

Shum, S.; Jensen, N.M. and Nebert, D.W.: The murine Ah locus: In utero toxicity and teratogenesis associated with genetic differences in benzo(a)pyrene metabolism. Teratology 20:365–376, 1979.

257 Benzoate, Sodium

Minor and Becker (1971) gave rats up to 1000 mg per kg intraperitoneally on days 9–11. Gross anomalies were observed and fetal weight was reduced.

Minor, J.L. and Becker, B.A.: A comparison of teratogenic properties of sodium salicylate, sodium benzoate and phenol. Toxicol. Appl. Pharmacol. 19:373, 1971.

258 1,2–Benzofluorene

Keeler and Binns (1968) gave 1.2–1.5 gm on the 14th day of gestation to four ewes and found no craniofacial defects in the offspring.

Keeler, R.F. and Binns, W.: Teratogenic compounds of veratrum californicum (Durand). Teratology 1:5–10, 1968.

259 Benzolamide CAS 3368–13–6

Maren and Ellison injected rats with 600 mg per kg every six hours starting at 4 p.m. on day 10 and ending at 10 a.m. on day 11. They found that these very large doses produced the postaxial limb defects typical of acetazolamide. Theophylline or potassium deficiency potentiated this drug's teratogenicity.

Maren, T.H. and Ellison, A.C.: The teratological effect of benzolamide, a new carbonic anhydrase inhibitor. Johns Hopkins Med. J. 130:116–123, 1972.

260 5–Benzoyl–α–methyl–2–thiophene Acetic Acid

Hiramatsu et al. (1980) gave this antiflammatory agent orally to rabbits during organogenesis. At the highest dose (75 mg per kg), there was maternal toxicity, implantations were reduced and fetal ossification was delayed.

Hiramatsu, Y.; Tamura, Y. and Koniba, S.: Teratological study of RU–15060 (5-benzoyl-α-methyl-2-thiophene). Yakuri to Chiryo 8:1773–1776, 1980.

261 Benztropine Mesylate Cogentin™ CAS 132–17–2

This drug given in conjunction with phenothiazines or other drugs with anticholinergic activity may be associated with small left colon syndrome. Falterman and Richardson (1980) reported two newborns with small left colon syndrome. Both were exposed to maternal benztropine (2 mg or 6 mg daily) and a psychotropic drug.

Falterman, C.G. and Richardson, C.J.: Small left colon syndrome associated with maternal ingestion of psychotropic drugs. J. Pediatr. 97:308–310, 1980.

262 Benzydamine HCl

1–Benzyl–3–γ–dimethylamino–propoxy–1 H–indazole HCl CAS 132–69–4

Namba and Hamada (1969) gave this compound orally and subcutaneously to pregnant mice and rats during organogenesis and some growth retardation was found but no teratogenicity. The maximum oral doses were 200 mg per kg and for the subcutaneous, they were 100 (mice) and 150 mg per kg per day (rats). Silvestrini et al. (1967) found no fetal effects in long term studies in the mouse.

Namba, T. and Hamada, Y.: Teratogenic tests with benzydamine hydrochloride. Oyo Yakuri 3:271–281, 1969.

Silvestrini, B.; Barcellona, P.S.; Garau, A. and Catanese, B.: Toxicology of benzydamine. Toxicol. Appl. Pharmacol. 10:148–159, 1967.

263 Benzyl Alcohol CAS 100–51–6

Duraiswami (1954) injected 0.01 or 0.02 ml of benzyl alcohol into the yolk sac of the chick from before incubation up to the 7th day. Meningoceles and skeletal defects were produced.

Duraiswami, P.K.: Experimental teratogenesis with benzyl alcohol. Johns Hopkins Hospital Bull. 95:57–67, 1954.

264 Benzyl Benzoate CAS 120–51–4

Eibs et al. (1982) found no increase in malformations among the offspring of mice injected with 10 percent benzyl benzoate in castor oil on days 1–12.

Eibs, H.G.; Spielmann, H. and Hagele, M.: Teratogenic effects of cyproterone acetate and medroxyprogesterone treatment during the pre– and postimplantation period of mouse embryos. Teratology 25:27–36, 1982.

265 Benzyl Chloride CAS 100–44–7

Skowronski and Abdel–Rahman (1986) gave rats up to 100 mg per kg orally on days 6–15 and at the highest dose, produced a decrease in fetal length but no significant increase in defects.

Skowronski, G. and Abdel–Rahman, M.S.: Teratogenicity of benzyl chloride in the rat. J. Toxicol. Environ. Health 17:51–56, 1986.

266 Benzyloxycarbonyl–phenylalanine–alanine–diazomethane

Z–Phe–Ala–CHN₂

This inhibitor of lysosomal cysteine proteinases was added to rat whole embryo culture by Ambroso & Harris (**1994**). The day 10 or 11 embryos evidenced embryotoxicity in a dose dependent manner starting at 2.5 micromolar.

Ambroso, J.L. and Harris, C.: In vitro embryotoxicity of the cysteine proteinase inhibitors benzyloxycarbonyl–phenylalanine–alanine–diazomethane (Z–Phe–Ala–CHN₂) and benzyloxycarbonyl–phenylalanine–phenylalanine–diazomethane (Z–Phe–Phe–CHN₂). Teratology 50:214–228, 1994.

267 Benzyloxycarbonyl–phenylalanine–phenylalanine–diazomethane

Z–Phe–Phe–CHN₂

This inhibitor of lysosomal cysteine proteinases was added to rat whole embryo cultures by Ambroso and Harris (**1994**). Embryotoxicity occurred in a doser dependent manner starting at 2.5 micromolar concentration.

Ambroso, J.L. and Harris, C.: In vitro embryotoxicity of the cysteine proteinase inhibitors benzyloxycarbonyl–phenylalanine–alanine–diazomethane (Z–Phe–Ala–CHN₂) and benzyloxycarbonyl–phenylalanine–phenylalanine–diazomethane (Z–Phe–Phe–CHN₂). Teratology 50:214–228, 1994.

268 Bepridil Hydrochloride

Furuhashi et al. (**1991 A&B**) gave up to 200 mg per kg to rats during organogenesis and no ill effects were found except the fetal body weight was reduced at the highest dose. In the perinatal period survival and growth were decreased in the fetuses at the highest dose.

Furuhashi, T.; Ushida, K.; Kodama, R.; Ishikawa, M. and Sawamura, K.: Study of oral administration of bepridil HCl during the period of organogenesis in rats. The Clinical Report 25:3130–3152, 1991.

Furuhashi, T.; Ushida, K.; Kodama, R.; Ishikawa, M. and Sawamura, K.: Study by oral administration of bepridil HCl during the perinatal and lactating period in rats. The Clinical Report 25:3155–3182, 1991.

269 Beraprost Sodium

Nakamura et al. (**1989**) gave rats up to 2 mg per kg orally on days 7–17 and found no adverse fetal effects. Postnatal growth was normal.

Nakamura, K.; Matsubara, T.; Ohtami, A.; Yoshida, J.; Tanaka, M.; Nagao, H. and Yoshinaka, I.: Teratogenicity study of beraprost sodium in rats. Kiso to Rinsho 23:3613–3631, 1989.

270 Bermoprofen

Satoh et al. (**1988**) gavaged rats with up to 6 mg per kg on days 7–17. At 3 mg per kg and above the offspring had delay in tooth eruption and righting in air. No other reproductive effects were seen.

Satoh, K.; Imura, Y.; Mukumoto, K.; Terada, Y.; Shigematsu, K.; Yoshioka, M.; Nishimura, K. and Ohnishi, K.: Reproduction studies of bermoprofen(2)–teratogenicity study in rats. Yakuri to Chiryo 17:2797–2814, 1988.

271 Beryllium CAS 7440–41–7

No defects were produced in the chick embryo by administration of up to 113 micromoles of BE(SO)₄ by Ridgway and Karnofsky (1952). Selivanova and Savinova (1986) reported the results of administration of beryllium chloride solution and beryllium oxide suspension (0.19 and 50 mg per m³, respectively) in pregnant rats on various days between 3 and 20. There was higher embryonic death and the number of embryos with edema of the internal organs increased.

Savitz et al. (**1989**) found no increased risk for stillbirth, preterm or small for gestational age among the offspring of fathers exposed to beryllium.

Ridgway, L.P. and Karnofsky, D.A.: The effect of metals on the chick embryo: Toxicity and production of abnormalities in development. Ann. N.Y. Acad. Sci. 55:203–215, 1952.

Savitz, D.A.; Whelan, E.A. and Kleckner, R.C.: Effect of parents' occupational exposures on risk of stillbirth, preterm delivery, and small-for-gestational-age infants. American Journal of Epidemiology 129: 1201–1218, 1989.

Selivanova, L.N. and Savinova, T.B.: The influence of chloride and beryllium oxide on the sexual function of female rats and offspring. Gigiena i Sanitaria. (USSR) 28:44–46, 1986.

272 Betamethasone

9–α–Fluoro–16–β–methylprednisolone CAS 378–44–9

This steroid is a glucocorticoid with a prominent sodium retaining ability. Walker (1971) produced cleft palate in the rat and mouse using 0.05–0.3 mg daily for 3 to 4 days during organogenesis.

Walker, B.E.: Induction of cleft palate in rats with antiinflammatory drugs. Teratology 4:39–42, 1971.

273 Betamethasone Butyrate Propionate

Takeshima et al. (**1990**) gave this steriod to rats subcutaneously on days 7–17. Growth retardation and fetal death were increased at 0.4mg per kg but no defects increase was seen at 3.2 mg per kg. In rabbits given up to 3.10 microgm subcutaneously Saijo et al. (**1990**) found a tendency for cleft palate and joint contracture but these were not statistically increased.

Saijo, T.: Fujita, T.; Sadanaga, O. and Deguchi, T.: Reproduction study of betamethasone butyrate propionate (bbp) (4)–teratogenicity study in rabbits by subcutaneous administration. Kiso to Rinsho 24:5779–5786, 1990.

Takeshima, T.; Tauchi, K. and Imai, S.: Reproduction study of betamethasone butyrate propionate (bbp) (2)–teratogenicity stuy in rats by subcutaneous administration. Kiso to Rinsho 24:5747–5763, 1990.

274 Betaxolol CAS 63659–188–7

Tateda et al. (**1990**) gave rats up to 200 mg per kg orally on days 7–17 and found no teratogenicity or postnatal growth or developmental changes. Tesh et al. (**1990**) gave rabbits up to 36 mg per kg orally during organogenesis. Decreased litter size was seen at the highest dose but no evidence of teratogenicity was found.

Tateda, C.; Ichikawa, K.; Ono, C.; Kiwai, S.; Oketani, Y.; Tanaka, E. and Toshida, K.: Reproduction study of betaxolol–teratogenicity study in rats. Yakuri to Chiryo 18:s–1719–s–1739, 1990.

Tesh, J.M.; Ross, F.W.; Tesh, S.A.; Davies, W.; Durand, A.; Tanaka, E. and Toshida, K.: Reproduction study of betaxolol–teratogenicity study in rabbits. Yakuri to Chiryo 18:s–1741–s–1752, 1990.

275 Bezafibrate

Naitoh et al. (**1988**) gave rats orally up to 800 mg per kg on days 7–17. This anticholesterolemic had no adverse effects on the fetuses.

Naitoh, J.; Misawa, T.; Shibata, N.; Nishigaki, T.; Ozawa, S. and Ohba, M.: Reproduction studies of bezafibrate (2nd report)–teratological study in rats. Kiso to Rinsho 22:4415–4431, 1988.

276 Bhopal Disaster

Bhandari et al. (**1990**) reported details of a survey of the population of Bhopal exposed to toxic gas containing cyanide. An abortion rate (spontaneous loss before 20 weeks) was 24% among 1468 exposed as compared to 5.6% in the control group of 485. Stillbirths were not increased. There were 31 infants with defects (13 with club foot) among 2117 live births while among controls of 1160 there were 15 with defects. The defect rates were 14.2 vs 12.6 per 1000. No pattern of defect was identified.

Bhandari, N.R.; Syal, A.K.; Kambo, I.; Nair, A.; Beohar, V.; Saxena, N.C.; Dabke, A.T.; Agarwal, S.S. and Saxena, B.N.: Pregnancy outcome in women exposed to toxic gas at bhopal. Indian J. Med. Res. B:28–33, 1990.

277 Bifenox 2,4–Dichlorophenyl 3'–carboxymethyl–4–nitrophenyl Ether

Francis (1986) gave mice up to 100 mg per kg orally on days 5–14 and found no teratogenicity.

Francis, B.M.: Teratogenicity of bifenox and nitrofen in rodents. Environ. Sci. Health B21(4):303–317, 1986.

278 Bifonazole (1–[(4–biphenyly0phenylmethyl]1–H–Imidazole

Schluter (1983) gavaged rats before and then during the first 7 days of pregnancy with up to 40 mg per kg or during days 6–15 with up to 100 mg per kg. Fetal growth retardation was found at maternal toxic levels but no defects

increase was observed. Studies in the rabbit found no teratogenicity.

Schluter, G.: The toxicology of bifonazole. Arzneim.–Forsch./Drug Res 33:739–743, 1983.

279 Bilirubin CAS 635–65–4

Yeary (1977) injected 25 mg per kg on days 9–15 into heterozygous Gunn rats and found embryo lethality. The homozygous Gunn rat also had infertility.

Yeary, R.A.: Embryotoxicity of bilirubin. Am. J. Obstet. Gynecol. 127:497–498, 1977.

280 Biodiastase

Tsutsumi et al. (1979) administered up to 2500 mg per kg orally to mice and rats during organogenesis and produced no ill effects in the fetuses.

Tsutsumi, S.; Sakuma, N. and Fukiage, S.: Investigations on the possible teratogenicity of biodiastase in mice and rats. (Japanese) Clinical Report 11:1335–1343, 1979.

281 Biotin Deficiency CAS 58–85–5

Giroud et al. (1956) were unable to produce malformations in rat fetuses after maternal dietary deficiency of this substance. The dietary deficiency was augmented either by oral succinyl sulfathiazole or by mixture with the biotin antagonist avidin. Watanabe and Endo (1984) found cleft palates and micromelia in the offspring of mice maintained on a diet of egg whites. Simultaneous feeding of 10 mg of biotin prevented the defects. Watanabe and Endo (**1990**) have extended their studies in deficient mice and report exencephaly in addition to skeletal defects.

Giroud, A.; Lefebvres, J. and Dupuis, R.: Carence en biotine et reproduction chez la ratte. C. R. Soc. Biol. (Paris) 150:2066–2067, 1956.

Watanabe, T. and Endo, A.: Teratogenic effects of avidin–induced biotin deficiency in mice. Teratology 30:91–94, 1984.

Watanabe, T. and Endo, A.: Teratogenic effects of maternal biotin deficiency on mouse embryos examined at midgestation. Teratology 42:295–300, 1990.

282 [1α (Z),2β,5α]-(+)-7-[5-[[(1,1'-Biphenyl) 4-yl]mehtoxy]-2-(4-morpholinyl)-3-oxocyclopentyl]-4-heptenoic acid AH23848 4–Biphenylmethanol

This thromboxane A$_2$–receptor blocker was studied in the rat during organogenesis by Sutherland et al. (1989). Diaphragmatic hernia was increased at doses of 95 mg per kg and above. A breakdown product, 4–biphenylmethanol was associated with production of the defect.

Sutherland, M.F.; Parkinson, M.M. and Hallett, P.: Teratogenicity of three substituted 4–biphenyls in the rat as a result of the chemical breakdown and possible metabolism of a thromboxane A$_2$–receptor blocker. Teratology 39:537–545, 1989.

283 Birth Control Pills

Discussed under Oral Contraceptives

284 4–Bis(2-chlorobenzyl aminomethyl) Cyclohexane
AY 9944™ CAS 366–93–8

Roux et al. (1979) fed this anticholesterolemic to rats on days 2 through 4 of gestation in amounts of 50 mg per kg. Holoprosencephaly, ocular defects, uterohydronephrosis and testicular ectopies occurred in a high proportion of the fetuses. A hypercholesterolemic diet reduced the incidence of malformations. Repetto et al. (**1991**) showed that the chemical could directly effect culture rat embryos. The ratio of 7–dehydrocholesterol to cholesterol in the embryos was altered with in vitro or invivo treatment.

This inhibitor of cholesterol synthesis was studied by Roux et al. (1972) in rats, giving 50 mg per kg on the second through fourth gestational days. Ocular anomalies were found in 42 percent of the Wistar rat fetuses, but none were seen in the Sprague-Dawley strain. Both groups of fetuses had hydronephrosis and ectopia of the testes.

Repetto, M.; Maziere, J.C.; Citadelle, D.; Dupuis, R.; Meier, M.; Biade, S.; Quiec, D. and Roux, C.: Teratogenic effect of the cholesterol synthesis inhibitor ay 9944 on rat embryos in vitro. Teratology 42:611–618, 1991.

Roux, C.; Horvath, C. and Dupuis, R.: Teratogenic action and embryolethality of AY–9944™ : Prevention by a hypercholesterolemia provoking diet. Teratology 19:35–38, 1979.

Roux, C.; Taillemite, J.L.; Aubry, M. and Dupuis, R.: Effets teratogenes compares du chlorhydrate du [trans–1, 4–bis (2-chlorobenzyl aminomethyl) cyclohexane] (AY-9944) chez le rat Wistar et le rat Sprague-Dawley. C. R. Soc. Biol. 166:1233-1236, 1972.

285 1,3–Bis(2–chloroethyl)–1–nitrosourea *Carmustine*
CAS 154–93–8

Thompson et al. (1974) treated rats with up to 4.0 mg per kg intraperitoneally. At doses of 1.5 per kg and higher abdominal wall defects, ectopia cordis, encephaloceles and skeletal defects were increased. Renal and eye defects and hydrocephalus also occurred. Treatment of rabbits with up to 4.0 mg per kg on days 6–18 did not produce defects at maternal doses that were lethal.

Thompson, D.J.; Molello, J.A.; Strebing, R.J.; Dyke, I.L. and Robinson, V.B.: Reproduction and teratology studies with oncolytic agents in the rat and rabbit. 1. 1,3–Bis(2–chloroethyl)–1–nitrosourea (BCNU). Toxicol. Appl. Pharmacol. 30:422–439, 1974.

286 2,2–Bis(bromomethyl)–1,3–propanediol

The chemical is a flame retardant. Treinen et al. (1989) fed mice up to 0.4% of the diet and observed decreased litters and litter size due to an effect in the female. Two generations were studied. At the highest dose the weight of the dam was decreased.

Treinen, K.A.; Chapin, R.E.; Gulati, D.K.; Mounce, R.; Morris, L.Z. and Lamb IV, J.C.: Reproductive toxicity of 2,2–bis(bromomethyl)–1,3–propanediol in a continuous breeding protocol in swiss (cd–1) mice. Fundamental and Applied Teratology 13:245–255, 1989.

287 Bisdiamine

Okamoto et al. (1980) gave rats 150–200 mg per kg orally on days 9 and 10. Cardiovascular defects and diaphragmatic hernias were found in the fetuses. Sumida (1988) studied the pathogenesis of left and right aortic arches with aberrant subclavian arteries in rats. He administered a total dose of 400 mg on days 9 and 10 after conception and suggested that the drug might inhibit the migration of neural crest cells. Tasaka et al. (**1991**) studied the effect on the cardiovascular system and noted aplasia of the 6th aortic arch.

Okamoto, N.; Miyabara, S.; Satow, Y. and Hidaka, N.: Persistent truncus arteriosus, hypoplasia of the pulmonary trunk and aortic arch abnormalities in the rat induced by bisdiamine. (abs) Teratology 22:18A only, 1980.

Sumida, H.: Study of abnormal formation of the aortic arch in rats. By methacrylate casts method and by immunohistochemistry for appearance and distribution of desmin, myosin and fibronectin in the tunica media. Hiroshima J. Med. Sci. 37(1):19–36, March, 1988.

Tasaka, H.; Takenaka, H.; Okamoto, N.; Onitsuka, T.; Koga, Y. and Hamada, M.: Abnormal development of cardiovascular systems in rat embryos treated with bisdiamine. Teratology 43:191–200, 1991.

288 Bis–dibenzyloxyoxopropylindanpropionic Acid

This F–2 α–prostaglandin inhibitor was given by Del Vecchio and Rahwan (1984) intramuscularly to mice in amounts of 50 mg per kg daily during gestation or with 50 to 200 mg per kg twice daily on day 15. Neither teratogenic nor behavioral alterations were found.

Del Vecchio, F.R. and Rahwan, R.G.: Teratological evaluation of a novel anti–abortifacient dibenzyloxyindanpropionic acid. Drug Chem. Toxicol. 7:357-381, 1984.

289 N,N-Bis(dichloroacetyl)-1,8-octamethylenediamine
Bis-dichloroacetyl–amine

Oster et al. (1974) studied this male antifertility agent in pregnant rats and found the drug to be highly embryocidal. When they used 200 mg per kg per day on days 11 and 12, a high incidence of fetal edema, cleft lip and microcephaly was found. In a later publication, these authors also detailed septal heart defects, diaphragmatic hernias and small irregular thymuses (Taleporos et al., 1978). More detailed studies were published by Kilburn et al. (1982). Hayakawa et al. (1983) studied the congenital heart defects found in rats after administration of this compound. Porter and Schmidt (1986) gave mice 2,000 microgm by gavage on day 10 and found hypoplasia of the thymus in the newborn. Also, the morphology of the spleen was altered and the postnatal immune function was decreased.

Hayakawa, K.; Okishima, T.; Ohdo, S. and Okamoto, N.: Experimental study on pathogenesis and morphogenesis of congenital heart defects: Conotruncal abnormality produced by bis(dichloroacetyl)diamine. Cong. Anom. 23: 267–277, 1983.

Kilburn, K.H.; Hess, R.A.; Lesser, M. and Oster, G.: Perinatal death and respiratory apparatus dysgenesis due to bis(dichloroacetyl)diamine. Teratology 26:155–162, 1982.

Oster, G.; Salgo, M.P. and Taleporos, P.: Embryocidal action of bis(dichloroacetyl)diamine: An oral abortifacient in rats. Am. J. Obstet. Gynecol. 119:583–588, 1974.

Porter, J.F. and Schmidt, R.R.: Developmental toxicity of N,N-bis(dichloroacetyl)–1,8–octamethylene diamine. Effects of in utero exposure on the postnatal murine immune system. Biol. Neonate 50:221–230, 1986.

Taleporos, P.; Salgo, M.P. and Oster, G.: Teratogenic action of a bis(dichloroacetyl)diamine on rats. Patterns of malformations produced in high incidence at time–limited periods of development. Teratology 18:5–16, 1978.

290 1,2–Bis(3,5–dioxopiperazine–1–yl)propane
1CRF–159

This anticancer drug was given orally to pregnant mice, rats and rabbits (Duke, 1975). Embryolethality was found in each species and the margin between no activity and total embryolethality was narrow. No clear defect increase was found. The most sensitive time in gestation was early embryogenesis. For mice and rats, the lethality started between 5–10 mg per kg daily while for rabbits, it was between 20–35 mg per kg. The most sensitive time of administration in the rat was day 6 through 8. The offspring of the mice were retarded in growth.

Duke, D.I.: Prenatal effects of the cancer chemotherapeutic drug 1CRF–159 in mice, rats and rabbits. Teratology 11:119–126, 1975.

291 N,N–Bis–[2–hydroxyethyl)–p–phenylenediamine
Sulfate CAS 58262–44–5

Burnett et al. (1986) did not find teratogenicity or dominant lethal effects in rats fed up to 0.3 percent in their diet.

Burnett, C.M.; Re, T.A.; Loehr, R.F.; Rodriguez, S.C. and Corbett, J.F.: Evaluation of the teratological and dominant lethal potential of N,N–bis–(2–hydroxyethyl)–p–phenylenediamine sulphate in a 6–month feeding study in rats. Food Chem. Toxicol. 24:875–880, 1986.

292 Bismark Brown CAS 8005–78–5

Gillman et al. (1951) could find no congenital defects after treating the pregnant rat with this substance.

Gillman, J.; Gilbert, C.; Spence, I. and Gillman, T.: A further report on congenital anomalies in the rat produced by trypan blue. S. Afr. J. Med. Sci. 16:125–135, 1951.

293 Bismuth CAS 7440–69–9

Heinonen et al. (1977) reported no increase in congenital defects among 144 women who were treated with bismuth subgallate during pregnancy. Only 13 of these women received treatment in the first 4 lunar months.

Ridgway and Karnofsky (1952) found no defects in chick embryos receiving 20 mg of $BiCl_3$ via yolk sac on either the

4th or 8th day. James et al. (1966) fed 5 mg per kg to sheep daily for extended periods of gestation and found 1 of the 4 fetuses to be stunted, hairless and exophthalmic.

Heinonen, O.P.; Slone, D. and Shapiro, S.: Birth Defects and Drugs in Pregnancy. Publishing Sciences Group Inc., Littleton, Mass., 1977.

James, L.F.; Lazar, V.A. and Binns, W.: Effects of sublethal doses of certain minerals on pregnant ewes and fetal development. Am. J. Vet. Res. 27:132–135, 1966.

Ridgway, L.P. and Karnofsky, D.A.: The effects of metals on the chick embryo. Toxicity and production of abnormalities in development. Ann. N.Y. Acad. Sci. 55:203–215, 1952.

294 Bisoprolol Fumarate

Suzuki et al. (**1989**) gave this β adrenergic receptor blocker orally to rats and rabbits during organogenesis in doses of up to 100 and 50 mg per kg respectively. In the rat fetal weight and survival was decreased at the highest dose. Left umbilical arteries were common in the treated but no teratogenicity was found. In the rabbit fetal mortality was increased at the highest dose.

Suzuki, T.; Naitoh, Y.; Narama, I.; Ariyuki, F. and Noguchi, Y.: Reproductive studies of bisoprolol in rats and rabbits. Kiso to Rinsho 23:768–778, 1989.

295 Bisphenol A, Diglycidyl Ether *DGEBPA* CAS 80–05–7

Hardin et al. (1981) carried out preliminary studies injecting 85 or 125 mg per kg intraperitoneally into rats on days 0 through 14 of gestation. At the higher dose very few rats became pregnant. At the lower dose there was a significant increase in dilated cerebral ventricles and growth retardation.

Morrissey et al. (1987) gavaged rats and mice with up to 1250 mg per kg and 640 mg per kg during organogenesis. No effects were found in the rat. In the mouse, maternal weight was reduced and resorptions were increased. This epoxy resin was studied in pregnant rabbits by Breslin et al. (1988) using 300 mg per kg on days 6–18. The material was applied to clipped skin. No fetal toxicity or increase in defects occurred.

Breslin, W.J.; Kirk, H.D. and Johnson, K.A.: Teratogenic evaluation of diglycidyl ether of bisphenol A (DGEBPA) in New Zealand white rabbits following dermal exposure. Fund. Appl. Toxicol. 10:736–743, 1988.

Hardin, B.D.; Bond, G.P.; Sikov, M.R.; Andrew, F.D.; Beliles, R.P. and Niemeier, R.W.: Testing of selected workplace chemicals for teratogenic potential. Scand. J. Work Environ. Health 7(4):66–75, 1981.

Morrissey, R.E.; George, J.D.; Price, C.J.; Tyl, R.W.; Marr, M.C. and Kimmel, C.A.: The developmental toxicity of bisphenol A in rats and mice. Fund. Appl. Toxicol. 8:571–582, 1987.

296 1–[3,5–Bis-trifluoromethyl phenyl]–4–methyl
Thiosemicarbazide *Ciba 2696GO*

Rao et al. (1973) gave 2.5 to 10 mg per day by gavage to rats from day 6 through day 15. A dose dependent increased incidence of exencephaly, skeletal defects and resorptions were found.

Rao, R.R.; Bhat, N.G.; Nair, T.B. and Shukla, R.G.: Toxicological and teratological studies with 1–[3,5–bis–trifluoromethyl)phenyl]–4–methylthiosemicarbazide (Ciba 2696GO). Arzneim–Forsch 23:797–800, 1973.

297 Bis(tri–n–butyltin) Oxide

Davis et al. (**1987**) gavaged mice on days 6–15 with 6, 11.7, 23.4 or 35 mg per kg. At 11.7 and above cleft palates were found and maternal toxicity was present. No teratogenicity was found at 6 mg per kg. Crofton et al. studies postnatal development in rats exposed in utero to 0–16 mg per kg on days 6–20. Persistent behavioral effects were found at levels which produced maternal toxicity or death (10 mg per kg).

Crofton, K.M.; Boncek, V.M.; Rosen, M.B.; Sheets, L.P.; Chernoff, N. and Reiter, L.W.: Prenatal or postnatal exposure to bis(tri–n–butyltin)oxide in the rat: Postnatal evaluation of teratology and behavior. Toxicology and Applied Pharmacology 97:113–123, 1989.

Davis, A.; Barale, R.; Brun, G.; Forster, R.; Gunther, T.; Hautefeuille, H.; van der Heijden, C.A.; Knaap, A.G.A.C.; Krowke, R.; Kuroki, T.; Loprieno, N.; Malaveille, C.; Merker, H.J.; Monaco, M.: Mosesso, P.; Neubert, D.; Norppa, H.; Sorsa, M.; Vogel, E.; Voogd, C.E.; Umeda, M and Bartsch, H.: Evaluation of the genetic and embryotoxic effects of bis(tri–n–butyltin)oxide (TBTO), a broad–spectrum pesticide, in multiple in vivo and in vitro short–term tests. Mutation Research 188:65–95, 1987.

298 Blastokinin

Administration of rabbit antiblastokinin to rabbits during and after blastocyst formation reduced the number of implantations in the rabbit (Krishnan, 1971).

Krishnan, R.S.: Effect of passive administration of antiblastokinin on blastocyst development and maintenance of pregnancy in rabbits. Experientia 27:955–956, 1971.

299 Bleeding also see *Anemia*

Vaginal bleeding during early pregnancy with its multitude of causes is considered by most obstetricians to be associated with a higher than normal abnormality rate in the conceptus. However, Nishimura et al. (1966), in comparing 2,328 embryos obtained by therapeutic abortion from mothers without vaginal bleeding, could not show a significant difference in malformation rate as compared to 427 embryos from mothers who did have vaginal bleeding before therapeutic abortion.

For the effect of major maternal hemorrhage on the embryo, see the heading, Anemia.

Nishimura, H.; Takano, K.; Tanimura, T.; Yasuda, M. and Uchida, T.: High incidence of several malformations in the early human embryos as compared with infants. Biol. Neonate 10:93–107, 1966.

300 Bleomycin

Briggs et al. (1986) were able to find two pregnant women treated in the second and third trimesters and both delivered newborns without defects.

Nishimura and Tanimura (1976) reported limb and tail defects in the rat and mouse fetus treated with 0.05–1 or 3–5 mg per kg, respectively, daily during organogenesis. No teratogenicity was found in the rabbit.

Briggs, G.G.; Freeman, R.K. and Yaffe, S.J.: Drugs in Pregnancy and Lactation, 2nd edition. Williams and Wilkins, Baltimore, 1986.

Nishimura, H. and Tanimura, T.: Clinical Aspects of Teratogenicity of Drugs. Excerpta Med., New York, 1976.

301 Bluetongue Vaccine Virus

This virus can produce encephalopathy followed by cystic cavities in the subcortical tissues of the cerebrum and cerebellum in lambs after the mother is injected with live virus around the 40th day (Young and Cordy, 1964). The fetuses exposed after 70 gestational days are less susceptible to encephalopathy. The subject is reviewed briefly by Kalter (1968).

Kalter, H.: Teratology of the Central Nervous System. Chicago: University of Chicago Press, 177 only, 1968.

Young, S. and Cordy, D.R.: An ovine fetal encephalopathy caused by bluetongue vaccine virus. J. Neuropathol. Exp. Neurol. 23:635–659, 1964.

302 Bonaphthon CAS 6954–48–9

Proinova et al. (1980) treated rats with 400 mg per kg of this antiviral drug on days 10–13 of pregnancy and found neither embryotoxic and teratogenic effect nor any postnatal morphological and functional deviations.

Proinova, V.A.; Pershin, G.N.; Shashkina, L.F.; Nechushkina, L.V. and Fedorova, E.A.: Postnatal development of rats after exposure to bonaphthon. Farmakol. Toksikol. (USSR) 4:404–408, 1980.

303 Bopindolol

Hamada et al. (**1989**) gave this β adrenergic receptor blocking agent to rats orally on days 7–17 in doses up to 108 mg per kg. At the highest dose fetal weight and ossification were reduced but no teratogenicity was found. Schon et al. (**1992**) gave rats and rabbits oral doses of up to 30 or 100 mg per kg respectively. In the rat no teratogenic effect was seen but at the highest dose embryonic death was increased and fetal weight decreased. No dose–related teratogenic effect was found in the rabbit. In rats treated perinatally body weight increase was inhibited at 10 and 20 mg per kg (Suter et al., **1992**).

Hamada, M.; Watanabe, M. and Nakazima, Y.: Teratological study of bopindolol in rats. Kiso to Rinsho 23:4373–4384, 1989.

Schon, H.; Bruggemann, S. and Sato, N.: Teratological studies of bopindolol in rats and rabbits. The Clinical Report 26:115–121, 1992.

Suter, K.E.; Schon, H. and Sato, N.: Peri–and postnatal study of bopindolol in rats. The Clinical Report 26:13–128, 1992.

304   Boric Acid CAS 10043–35–3

Heinonen et al. (1977) reported on 253 offspring of mothers using boric acid in the first 4 lunar months and observed 19 major malformations when only 7 were expected. The increase was still within the 95 percent confidence levels.

Landauer (1952) produced rumplessness, curled toe and facial palate defects in chicks using injections of 2.5 mg. Rumplessness was common following treatment at 24 hours incubation while the facial and palate defects appeared after treatment at 4 days.

Dousset (1971) reported that boric acid is metabolized differently when the rat becomes pregnant. The chemical begins to appear first in the spinal fluid. Weir and Fisher (1972) gave boric acid or borax to pregnant rats at 350 ppm given during pregnancy. No reduction in liveborn or physical defects were found. Heindel et al. (**1992**) fed rats and mice up to 330 or 1003 mg per kg respectively. At the highest doses there was lethality in the dams. In the rat at 0.2 percent in the diet (about 163 mg per kg) hydrocephalus and skeletal defects were found to be increased. In the mouse fetuses no increase in visceral defects occurred. Treatment was either during embryogenesis or during the entire gestation. Heindl et al (**1994**) published more details on the developmental toxicity in rats and mice and in addition reported increased malformations in the rabbit at of 250 mg per kg. No effect in the rabbit was found at 125 mg per kg by gavage. The malformations in the rabbit were mainly cardiovascular.

Dousett, G.: Penetration du bore organique dans le liquide cephalorhachildien de la ratte gestante. Compt. Rendu (Paris) 165:722–724, 1971.

Heindel, J.J.; Price, C.J.; Field, E.A.; Marr, M.C.; Myers, C.B.; Morrissey, R.C. and Schwetz, B.A.: Developmental toxicity of boric acid in mice and rats. Fund. Appl. Toxicol 18:266–277, 1992.

Heindel, J.J.; Price, C.J. and Schwetz, B.A.: The developmental toxocity of boric acid in mice, rats, and rabbits. Environmental Health Perspectives 102:107–112, 1994.

Heinonen, O.P.; Slone, D. and Shapiro, S.: Birth Defects and Drugs In Pregnancy. Publishing Sciences Group Inc., Littleton, Mass., 1977.

Landauer, W.: Malformations of chicken embryos produced by boric acid and the probable role of riboflavin in their origin. J. Exp. Zool. 120:469–508, 1952.

Weir, R.J. and Fisher, R.S.: Toxicologic study on borax and boric acid. Toxicol. Appl. Pharmacol. 23:351–364, 1972.

305   Botulinum Toxin

Drachman and Sokoloff (1966) paralyzed chick embryos with this material and inhibited joint cavity production.

Drachman, D.B. and Sokoloff, L.: The role of movement in embryonic joint development. Dev. Biol. 14:401–420, 1966.

306   Bovine Albumin Anaphylaxis

Takayama (1981) immunized mice to purified bovine albumin and then administered 0.05 to 0.5 mg intraperi-

toneally or intravenously on day 8.5 of gestation. A significant increase in defects mainly exencephaly and omphalocele occurred in about 3 percent of the fetuses. No defects occurred in the non–immunized controls receiving the same dose. Bovine serum albumin intraperitoneally in a dose of 50 mg did produce similar defects.

Takayama, Y.: Teratogenic effects of anaphylactic immune reaction in mice. Cong. Anom. 21:175–186, 1981.

307   Bovine Serum Albumin

Takayama (1981) immunized mice with bovine serum albumin. The fetuses from mothers with anaphylactoid shock had 3.2 percent malformations, mostly exencephaly. This was increased over the control.

Takayama, Y.: Teratogenic effects of anaphylactic immune reaction in mice. Cong. Anom. 21:175–186, 1981.

308   Bovine Viral Diarrheal–Mucosol Disease *BVD–MD*

This mild diarrheal disease occurring in pregnant cattle was associated with ataxia in the newborn calf. Kahrs et al. (1970) injected the attenuated virus into pregnant cows between 100 to 200 days of gestation and observed ataxia with cerebellar hypoplasia in the newborn calves. Optic atrophy and lenticular opacities were found in some.

Kahrs, R.F.; Scott, F.W. and De Lahunta, A.: Congenital cerebellar hypoplasia and ocular defects in calves following bovine viral diarrhea–mucosol disease infection in pregnant cattle. Am. Vet. Med. Assoc. 156:1443–1450, 1970.

309   Bradykinin CAS 58–82–2

Thompson and Gautieri (1969) injected synthetic bradykinin into pregnant mice once on gestational days 7 through 12. When they used 25 microgm intravenously, they found an increased incidence of fetal hydronephrosis. Fetal growth retardation and defects in ossification were also found. An increased transfer rate of radioactive sodium from the mother to the placentas was shown to occur during the treatment.

Thompson, R.S. and Gautieri, R.F.: Comparison and analysis of the teratogenic effects of serotonin, angiotensin II and bradykinin in mice. J. Pharm. Sci. 58:406–412, 1969.

310   Bretylium CAS 61–75–6

West (1962) gave daily subcutaneous injections to pregnant rats using 15 mg per kg and found no increase in fetal death.

West, G.B.: Drugs and rat pregnancy. Letter to the Editor. J. Pharm. Pharmacol. 14:828–830, 1962.

311   BRL 16644, 16657 CAS 59257–18–0; CAS 59257–24–8

These two structurally related dopaminergic agonists were studied by Baldwin and Ridings (1986) in rats using 10 to 35 or 40 mg per kg daily on days 6–15 by the oral route. At the intermediate (20 mg per kg) and highest doses, anasarca, brachygnathia, cleft palate and undescended testes were found in the offspring. Some increase in defects was found at 10 mg per kg doses. No maternal toxicity was seen at this dose level.

Baldwin, H.L. and Ridings, J.: Teratogenicity in rats of two dopaminergic agonists. Toxicology 42:291–302, 1986.

312 BRL38227

Baldwin et al. (**1994**) treated rats orally before pregnancy and during the first week, during organogenesis and in the peri and postnatal periods with this potassium channel opener. The number of implantations was reduced at 1.25 mg per kg in the reproductive study and in the organogenesis study. In the organogenesis study at 100 mg per kg 6.8 percent had an extra vessel exiting the outflow tracts. Postnatal development was uneffected. In the rabbit Baldwin et al. (**1994**) found an increase in minor cardiovascular variations at 1.0 and 4.0 mg per kg given during organogenesis.

Baldwin, H.A.; Ridings, J.E.; Davidson, E.J.; Pritchard, A.L.: Ishii, R.; Tanaka, K.; Uemura, S. and Nishioka, Y.: Oral reproductive and developmental toxicity studies of BRL38227 in rats. Yakuri to Chiryo 22:s1467–s1500, 1994.

Baldwin, J.A.; Ridings, J.E.; Davidson, E.J.; Pritchard, A.L.; Ishii, R.; Tanaka, K.; Uemura, S. and Nishioka, Y.: Oral reproductive and developmental toxicity study of BRL38227 in rabbits: Administration during the period of organogenesis. Yakuri to Chiryo 22:s1501–s1511, 1994.

313 Bromacil

Dilley et al. (1977) exposed rats to up to 165 mg per sq. meter during days 7 through 14 and produced no prenatal fetal change.

Dilley, J.V.; Chernoff, N.; Kay, D.; Winslow, N. and Newell, G.W.: Inhalation studies of five chemicals in rats. Toxicol. Appl. Pharmacol. 41:196, 1977.

314 Bromazepam CAS 1812–30–2

Oketani et al. (1983) studied this tranquilizer in pregnant rats and rabbits. In the rabbit, at doses of 10 and 40 mg per kg during organogenesis, the dams were depressed in their activity and there was increased fetal mortality. No adverse fetal effects occurred in the rat after 30 mg doses given before fertilization and for 7 days afterwards, or after 40 mg on days 7–17. Perinatal and postnatal studies with 30 mg per kg caused no significant behavioral changes.

Oketani, Y.; Ichikawa, K.; Ono, C.; Gofuku, M.; Banba, I.; Tsugawa, R.; Takehara, I.; Shishido, H.; Sannohe, H. and Iriya, S.: Reproductive studies in rats and rabbits. Oyo Yakuri 26:99–135 and 199–232, 1983.

315 Bromide CAS 24959–67–9

Heinonen et al. (**1977**) studied 986 women exposed during the first 4 months to bromide and found 54 with malformations (RR=1.13). The relative risk for musculoskeletal defects was 1.64 (95% confidence 0.90–2.74). Opitz et al. (1972) reported short stature and small cranium in 2 children born to a mother ingesting large amounts of bromide. One of these children had congenital heart disease.

Mangurten and Kaye (1982) reported a neonate with hypotonia and neurologic depression and a serum bromide of 33 mg per 100 ml. The symptoms cleared and development at 9 months was normal. The mother was employed in a photographic laboratory and mixed chemicals until 5 weeks before delivery.

Heinonen, O.P.; Slone, D. and Shapiro, S.: Birth Defects and Drugs in Pregnancy. John Wright Publishing Sciences Group, Inc., Littleton, Mass. 1977.

Mangurten, H.M. and Kaye, C.I.: Neonatal bromidism secondary to maternal exposure in a photographic laboratory. J. Pediatr. 100:596–598, 1982.

Opitz, J.M.; Grosse, F.R. and Haneberg, B.: Congenital effects of bromism. Lancet 1:91–92, 1972.

316 Brominated Soybean Vegetable Oil CAS 8016–94–2

Vorhees et al. (1983) found impaired reproduction at 1.0 and 2.0 percent addition to the rat's diet. At 0.5 percent in the diet during the rat gestation they found associated behavioral defects in the offspring. The maximum allowable diet in humans is 0.01 percent.

Vorhees, C.V.; Butcher, R.E.; Wooten, V. and Brunner, R.L.: Behavioral and reproductive effects of chronic developmental exposure to brominated vegetable oil in rats. Teratology 28:309–318, 1983.

317 2–Bromoaceticacid

Randall et al. (**1991**) gavaged rats with up to 100 mg per kg daily during days 6–15. At the highest dose maternal and fetal weight gain was reduced and at the highest dose craniofacial and heart defects were increased.

Randall, J.L.; Christ, S.A.; Perez, P.H.; Nolen, G.A.; Read, E.J. and Smith, M.K.: Developmental effects of 2–bromoaceticacid in the Long Evans rat. (Abs) Teratology 43:454, 1991.

318 Bromochlorodifluoromethane

This fire–extinguisher chemical was tested in rats during organogenesis or before fertilization in males and females. For organogenesis, exposured to 50,000 ppm and for prefertilization 5,000 ppm had no adverse effect on reproduction, or fetal development.

Wicramaratne,G.A.D.S.; Tinston, D.J.; Kinsey, D.L. and Doe, J.E.: Assessment of the reproductive toxicology of bromochlorodifluoromethant (bcf,halon 1211) in the rat. British Journal of Industrial Medicine 45:755–760, 1988.

319 Bromocriptine Methanesulfonate CAS 25614–03–3

In over 1,400 pregnancies of women in whom the drug was given primarily in the early weeks of pregnancy, spontaneous abortions were found in 11 percent, minor defects in 2.5 percent and major defects in 1.0 percent of the offspring (Turkalj et al., 1982). These rates were not increased as compared to normal populations. In another exposed population of 448 pregnancies, only 11 congenital malformations were found (Griffith et al., 1978). Czeizel et al. (1989) found no aneuploidy in lumphocytes of 31 children exposed in utero.

Schardein (1985) cites animal studies in the rat and rabbit.

Czeizel, A.; Kiss, R,; Racz, K.; Mohori, K. and Glaz, E.: Case–control cytogenetic study in offspring of mothers treated with bromocriptine during early pregnancy. Mutation Research 210:23–27, 1989.

Griffith, R.W.; Turkalj, I. and Braun, P.: Outcome of pregnancy in mothers given bromocriptine. Br. J. Clin. Pharmacol. 5:227–231, 1978.

Schardein, J.L.: Chemically Induced Birth Defects. Marcel Dekker, Inc., New York, 1985.

Turkalj, I.; Braun, P. and Krupp, P.: Surveillance of bromocriptine in pregnancy. Am. J. Med. Assoc. 247:1589–1591, 1982.

320 Bromodichloromethane

Ruddick et al. (1980) gavaged rats on days 6 through 15 with up to 200 mg per kg. Some maternal and fetal toxicity was seen at the highest dose. Narotsky et al. (**1993**) reported full liter resorption in F–344 rats receiving 75 mg per kg by gavage during organogenesis.

Nartosky, M.G.; Hamby, B.T.; Mitchell, D.S. and Kavlock, R.J.: Evaluation of the critical period of bromodichloromethane–induced full–litter resorption in F–344 rats. (abs) Teratology 47:429, 1993.

Ruddick, J.A.; Villeneuve, D.C.; Chuand, I. and Valli, V.E.: Teratogenicity assessment of 4 trihalomethanes. Teratology 21:66A only, 1980.

321 5–Bromo–1–(2–deoxy–β–d–ribofuranosyl) 5–fluoro–6–methoxy–o Hydrouracil CAS 1031–61–4

Chaube and Murphy (1968) reported production of brain, palate and skeletal defects in the rat with 75 to 300 mg per kg injected on the 11th or 12th day.

Chaube, S. and Murphy, M.L.: The teratogenic effects of the recent drugs active in cancer chemotherapy. In: Advances in Teratology, D.H.M. Woollam (ed.), New York: Academic Press, 3:204–205, 1968.

322 5–Bromodeoxyuridine CAS 59–14–3

Dipaolo (1964) produced polydactylous offspring by injecting pregnant mice with 124 mg per kg. Chaube and Murphy (1968) produced defects of the nervous system, palate and skeleton in rat fetuses whose mothers were injected with 100 to 800 mg per kg on the 11th or 12th day. In the hamster, Ruffolo and Ferm (1965) produced mostly nervous system defects with 400 mg per kg given on the 8th day. No deleterious effects on 6.5 day rabbit blastocysts from mothers treated with 50 to 80 mg per kg were noted by Adams et al. (1961). Biggers et al. (1987), using tissues from in vitro exposed rat embryos, showed reduced neurite outgrowth from neuroblasts.

Adams, C.E.; Hay, M.F. and Lutwak–Mann, C.: The action of various agents upon the rabbit embryo. J. Embryol. Exp. Morphol. 9:468–491, 1961.

Biggers, W.J.; Barnea, E.R. and Sanyal, M.K.: Anomalous neural differentiation induced by 5–bromo–2'–deoxyuridine during organogenesis in the rat. Teratology 35:63–75, 1987.

Chaube, S. and Murphy, M.L.: The teratogenic effects of the recent drugs active in cancer chemotherapy. In: Advances in Teratology, D.H.M. Woollam (ed.), New York: Academic Press, 3:204–205, 1968.

Dipaolo, J.A.: Polydactylism in the offspring of mice injected with 5–bromodeoxyuridine. Science 145:501–503, 1964.

Ruffolo, P.R. and Ferm, V.H.: The embryocidal and teratogenic effects of 5–bromodeoxyuridine in the pregnant hamster. Lab. Invest. 14:1547–1553, 1965.

323 Bromofenofos CAS 21466–07–9

Yoshimura and Delatur (1986) administered this organophosphorous anthelmintic to rats by gavage on days 8–15. At 20 mg per kg, 90 percent resorptions occurred and survivors were malformed. At 10 mg per kg the resorptions were not increased and no teratogenicity was identified.

Yoshimura (1987) expanded these studies and found teratogenicity on days 8, 9, and 10 using 50 mg per kg. The defects included displaced testes, cleft lip, nasal atresia, short tail and anal atresia. The dephosphate form was also found to be teratogenic.

Yoshimura, H. and Delatour, P.: Embryolethality of bromofenofos in rats. Toxicology Letters 31:243–247, 1986.

Yoshimura, H.: Embryolethal and teratogenic effects of bromofenofos in rats. Arch Toxicol 60:319–324, 1987.

324 Bromoform CAS 75–25–2

Ruddick et al. (1980) gave up to 200 mg per kg by gavage to rats on days 6 through 15. No overt effect in the mother or fetus was noted.

Ruddick, J.A.; Villeneuve, D.C.; Chuand, I. and Valli, V.E.: Teratogenicity assessment of 4 trihalomethanes. Teratology 21:66A only, 1980.

325 Bromolysergic Acid Diethylamide

West (1962) gave daily subcutaneous injections to pregnant rats using 4 mg per kg and found no increase in fetal death.

West, G.B.: Drugs and rat pregnancy. Letter to the Editor. J. Pharm. Pharmacol. 14:828–830, 1962.

326 Bromophos

Nehez et al. (1986) studied this alkylating agent in two strains of mice using oral or intraperitoneal administration. At doses of 73 mg per kg on days 6, 8, 10 and 12, by the intraperitoneal route, a significant postimplantation loss was increased in one strain. With oral administration in another strain at the same dose, postimplantation loss was increased. Examination for defects was not mentioned. The tetramethylammonium and sodium salts of demethylbromophos were also studied.

Nehez, M.; Huszta, E.; Mazzag, E.; Scheufler, H.; Fischer, G.W. and Desi, I.: Cytogenetic and embryotoxic effects of bromophos and demethylbromophos. Regulatory Toxicology and Pharmacology 6:416–421, 1986.

327 Bromperidol CAS 10457–90–6

Imai et al. (1984) administered this neuroleptic orally to rats in doses up to 10 mg per kg per day on days 7 through 17 of gestation. No malformations were increased but maternal food intake was decreased and fetal growth retardation was found.

Imai, S.; Tauchi, K.; Huang, K–J.; Takeshima, T. and Sudo, T.: Teratogenicity study on bromperidol in rats. J. Toxicol. Sci. 9:109–126, 1984.

328 Brompheniramine CAS 86–22–6

In a study of 65 women exposed during the first 4 lunar months to this antihistamine, there were 10 offspring with defects (relative risk: 2.34, p < 0.05). Three of these children had syndactyly (Heinonen et al., 1977).

Heinonen, O.P.; Slone, D. and Shapiro, S.: Birth Defects and Drugs in Pregnancy. Publishing Sciences Group Inc., Littleton, Mass., 1977.

329 Bropirimine

Marks and Poppe (1988) gavaged rats with up to 100 mg per kg on days 7–15. The weight increase of the dams was lowered at 25 mg per kg and above. Some skeletal variations of the fetuses occurred in the highest dose level but no defect increase was found. Poppe et al. (1989) gave 100 to 400 mg per kg to rats at various times and recorded maternal toxicity which was associated with increased fetal death. Marks et al. (1990) studied the embryotoxicity of maternally toxic doses and observed a relationship between the effects and progesterone levels.

Marks, T.A.; Black, D.L.; Terry, R.D.; Branstetter, D.G. and Kirton, K.T.: Variability in the developmental toxicity of bropirimine with the day of administration. Teratology 42:55–66, 1990.

Marks, T.A. and Poppe, S.M.: Developmental toxicity of bropirimine in rats after oral administration. Teratology 38:7–14, 1988.

Poppe, S.M.; Marks, T.A.; Renis, H.E.: Broprimine–induced embryolethality after oral administration to the pregnant rat. Fundamental and Applied Toxicology 13:87–101, 1989.

330 Brotizolam

Matsuo et al. (1985) studied this sleep producing medication in pregnant rats and rabbits. Adult toxicity occurred at doses of 2.5 mg per kg in rats and 0.05 mg for rabbits. No adverse effects on reproduction or on the fetus were found. At dose levels (oral) of 2.5 mg per kg during organogenesis, some of the open field behavioral studies were found to be abnormal in the postnatal rat. Hewett et al. (1986) found no teratogenic effect of 30 mg per kg given to rats and rabbits.

Hewett, C.; Ellenberger, J.; Kollmer, H.; Kreuzer, H.; Niggeschulze, A. and Stotzer, H.: Safety assessment of brotizolam. Arzneim Forsch. Drug Res. 36(1):592–596, 1986.

Matsuo, A.; Kast, A. and Tsunenari, Y.: Reproduction studies of brotizolam in rats and rabbits. Iyakuhin Kenkyu 16:818–838, 1985.

331 Brovanexine Hydrochloride CAS 54340–60–2

This expectorant was given by Tsuruzaki et al. (1982) to rats orally in maximum doses of 5000 mg per kg during organogenesis and no adverse fetal effects were found.

Tsuruzaki, T.; Naki, H.; Shimo, T.; Noguchi, Y.; Kato, H.; Ito, Y. and Yamamoto, M.: Reproductive studies of brovanexine hydrochloride in rats. Kiso to Rinsho 16:7179–7195, 1982.

332 Brown HT

Mangham et al. (1987) fed up to 500 mg per kg to rats in a three–generation study. No adverse effects were found.

Mangham, B.A.; Moorhouse, S.R.; Grant, D.; Brantom, P.G. and Gaunt, I.F.: Three–generation toxicity study of rats ingesting brown HT in the diet. Food Chem. Toxicol. 25(12):999–1007, 1987.

333 Bucladesine DBcAMP CAS 16980–89–5

Akiyama et al. (1984) found no adverse reproductive effects from this drug developed for acute circulatory insufficiency. Dosage was given intravenously to rats for fertility studies (120 mg per kg) and perinatal studies (90 mg per kg). The higher doses were lethal to some of the dams.

Akiyama, Y.; Yamashita, N.; Harada, S.; Takayama, S.; Matsuhashi, K.; Mochida, K. and Ohura, K.: Reproduction studies of bucladesine (DBcAMP) in rats. Oyo Yakuri 27:571–610, 1984.

334 Buclizine CAS 82–95–1

This antihistamine belongs to the benzhydrylpiperazine series of compounds and was used to produce cleft palate in the rat. It is more fully described under Meclizine.

335 Bucloxic Acid 4–(3–Chloro–4–cyclohexyl–phenyl) 4–oxo–butyric Acid CAS 32808–51–8

Mazue et al. (1974) gave rabbits, rats and mice 25 mg per kg during active organogenesis and found no fetal damage.

Mazue, G.; Landsmann, F. and Brunaud, M.: Study of possible teratogenicity of bucloxic acid (804 CB). Arzneim Forsch 24:1413–1425, 1974.

336 Budesonide CAS 51333–22–3

Kihlstrom and Lundberg (1987) gave rabbits subcutaneously up to 0.29 micromoles per kg daily during organogenesis. At the highest dose, abortions occurred and at 0.01 and 0.06, fetal weight was reduced and delayed ossification was found.

Kihlstrom, I. and Lundberg, C.: Teratogenicity study of the new glucocorticosteroid budesonide in rabbits. Arzneim–Forsch Drug Research 37:43–46, 1987.

337 Budralazine 1–(2–(1,3–dimethyl–2–butenylidene)–hydrazino) phthalazine) CAS 36798–79–5

Shimada et al. (1981) carried out fertility and teratogenicity studies in rats with this antihypertensive. Fertility was not altered by dietary doses of 80 mg per kg daily.

At 100 mg per kg during days 7–17, no adverse fetal effects occurred. Perinatal and postnatal studies by Watanabe et al. (1981) found some weight reduction, but development and function were not affected. Studies in the rabbit during organogenesis with doses of 80 mg per kg did not produce teratogenic effects (Nagaoka et al., 1980).

Nagaoka, T.; Narama, I. and Oshima, Y.: Reproductive studies of budralazine; perinatal and postnatal studies in rats. Oyo Yakuri 21:343–350, 1980.
Shimada, H.; Ohura, K.; Mochida, K.; Arauchi, T. and Morita, H.: Reproductive studies of budralazine in rats. Oyo Yakuri 21:313–330, 1981.
Watanabe, T.; Ohura, K.; Mochida, K.; Arauchi, T. and Morita, H.: Reproductive studies of budralazine, perinatal and postnatal studies in rats. Oyo Yakuri 21:331–341, 1981.

338 Bufexamac *p–Butoxyphenylacethydroxamic Acid Droxaryl™* CAS 2438–72–4

Roba et al. (1970) studied this non–steroidal anti–inflammatory agent in pregnant rats and rabbits. At the highest dose in rats (750 mg per kg) there was fetal growth retardation but no increase in defects. Doses of up to 500 mg per kg in the rabbit were not associated with adverse fetal effects. The medication was given by gavage.

Roba, J.; Lambelin, G. and Buu–Hoi, N.P.: Teratological studies of p–butoxyphenylacethydroxamic acid (CP 1044J3) in rats and rabbits. Arzneim–Forsch. 20:565–569, 1970.

339 Buflomedil

Fukushima et al. (**1988** A) gave rats up to 250 mg per kg on days 7–17. No fetal ill effects were found. At the highest dose level F1 and F2 offspring were heavier than controls. The weight increases in the F2 generation were statistically significant. Fukushima et al. (**1988** B) gave rabbits orally up to 100 mg per kg on days 6–18 and found no adverse fetal effects.

Fukushima, T.; Ishihara, M.; Okuyama, K.; Igarashi, E.; Watanabe, Y.; Funahashi, N. and Nakazawa, M.: Teratogenicity test of buflomedil given during the period of fetal organogenesis in rats. Kiso to Rinsho 22:447–464, 1988.
Fukushima, T.; Yokomori, S.; Inoue, S.; Takiuchi, H.; Funahashi, N. and Nakazawa, M.: Teratogenicity test of buflomedil given during the fetal organogenesis period in rabbits. Kiso to Rinsho 22:465–472, 1988.

340 Bumadizon *Butyl–malonic Acid–mono-(1,2-diphenylhydrazine) Calcium Eumotol™* CAS 3583–64–0

Konig et al. (1973) found no teratogenicity in mice, rats and rabbits. Oral doses of 80 (mouse) or 100 mg per kg (rat and rabbit) were used before and during gestation. A two–generation study was done.

Konig, J.; Knoche, Ch. and Schafer, H.: Der Einfluss von Butyl–malonsaure–mono–(1,2–diphenyl–hydrazid–calcium) (bumadizon–calcium) auf die Reproduktion und die Praenatale Entwicklung. Arzneim–Forsch 23:1246–1251, 1973.

341 Bumetanide

McClain and Dammers (1981) found no teratogenic effects in rabbits, rats, mice or hamsters. Doses up to 100 mg per kg were used during organogenesis in rats and mice.

McClain, R.M. and Dammers, K.D.: Toxicologic evaluation of bumetanide, a potent diuretic agent. J. Clin. Pharmacol. 21:543–554, 1981.

342 Bunitrolol Hydrochloride
o–[–2–Hydroxy–3–(tert–butylamine)propyl] benzonitrile hydrochloride CAS 34915–68–9

Matsuo et al. (1981) gave rats up to 300 mg per kg and rabbits up to 50 mg per kg daily during organogenesis. Although there was maternal toxicity in both species at the higher doses, no adverse effects on reproduction were found. Behavioral studies in the offspring were not different from controls. This drug is a β–adrenoceptor blocker. Studies before breeding and during the postnatal period were also done. Only an increase in neonatal death was found at 300 mg per kg dose levels (Matsuo et al., 1981).

Matsuo, A.; Kast, A. and Tsunenari, Y.: Fertility and reproduction studies with bunitrolol hydrochloride (KOE 1366 Cl) in rats. Iyakuhin Kenkyu 12:976–987, 1981.
Matsuo, A.; Kast, A. and Tsunenari, Y.: Teratological studies on bunitrolol hydrochloride (KOE 1366 Cl) in rats and rabbits. (Japanese) Iyakuhin Kenkyu 12:12–24, 1981.

343 Bupranolol Hydrochloride *1-(tert-Butylamino)–3–(2-chloro-5-methylphenoxy)-propanol Hydrochloride* CAS 15148–80–8

Kagiwada et al. (1973) gave this β blocker to pregnant mice and rats during organogenesis in maximum doses of 100 and 150 mg per kg, respectively. At the highest doses, fetuses of both species had retarded bone development.

Kagiwada, K.; Ishizaki, O. and Saito, G.: Effects of β–receptor blocking agent, 1–(tert-butylamino)–3–(2–chloro–5–methylphenoxy)–2–propanol hydrochloride (KL–255), on pre– and postnatal development of the offsprings in pregnant mice and rats. Oyo Yakuri 7:65–74, 1973.

344 Buprenorphine CAS 52485–79–7

Mori et al. (1982) studied the effect of this analgesic in rats on days 7–17 and 17–21. With doses over 0.05 mg per kg per day, fetal growth retardation was found but no defects were produced up to 5 mg per kg. In the late treatment group at 5 mg per kg, the offspring had a reduced survival rate. Heel et al. (1979) reported briefly that intramuscular doses of up to 5 mg per kg during organogenesis in rats and rabbits had no adverse fetal effects. Tiong and Olley (**1988**) studied the effect of this chemical on enkephalin levels in the brain of the rat fetus. At doses of 1 and 2 mg per kg no change occurred but with 8 mg per kg of methadone a drop was seen.

Heel, R.C.; Brogden, R.N.; Speight, T.M. and Avery, G.S.: Buprenorphine: a review of its pharmacological properties and therapeutic efficiency. Drugs 17:81–110, 1979.
Mori, N.; Sakanoue, M.; Kamata, S.; Takeuchi, M.; Shimpo, K. and Tamagawa, M.: Toxicological studies of

buprenorphine teratogenicity, perinatal and postnatal studies in the rat. Iyakuhin Kenkyu 13:509–544, 1982.

Tiong, G.K.L. and Olley, J.E.: Effects of exposure in utero to methadone and buprenorphine on enkephalin levels in the developing rat brain. Neuroscience Letters 93:101–106, 1988.

345 Buserelin Acetate CAS 68630–75–1; CAS 57982–77–1

This LH–releasing hormone analogue was given by Akaike et al. (1987a) to mice in doses of 0.01 to 10,000 microgm per kg per day on days 6–15. No defects resulted, but the duration of gestation was prolonged.

Akaike et al. (1987b) studied the LH–RH analogue in mice using up to 1000 microgm per kg subcutaneously on day 15 of gestation through day 21 of lactation. At 1 microgm dose levels and above, there was prolonged parturition with increased stillbirths. No morphological or behavioral changes were noted in the survivors.

Akaike, M.; Takayama, K.; Ohno, H.; Kobayashi, T. and Sakaguchi, T.: Teratogenicity study of subcutaneously administered buserelin acetate in mice. Oyo Yakuri 33:631–646, 1987a.

Akaike, M.; Takayama, K.; Ohno, H.; Kobayashi, T. and Sakaguchi, T.: Perinatal and postnatal study of subcutaneously administered buserelin acetate in mice. Oyo Yakuri 33:647–663, 1987b.

346 Buspirone Hydrochloride CAS 33386–08–2

Kai et al. (1990 A) gave up to 75 mg per kg orally to rats during organogenesis. At the highest dose the dams had decreased activity and weight gain and the fetuses were smaller but without congenital defect increases. Behavior was normal postnatally but the 10 week pups had decreased lung and brain weights. Kai et al. (1990, B) in perinatal studies (d17–20) found that 12 and 75 mg per kg decreased maternal activity and survival of the newborns. At the highest dose the 10 week old pups had decreased brain weight but learning behavior and reproduction were not altered.

Kai, S.; Kohmura, H.; Ishikawa, K.; Ohta, S.; Kuroyanagi, K.; Kawano, S.; Kadota, T.; Chikazawa, H.; Kondo, H. and Takahashi, N.: Reproductive and developmental toxicity studies of buspirone hydrochloride (i)–oral administration to rats during the period of fetal organogenesis. The Journal of Toxicological Sciences 15:31–60, 1990.

Kai, S.; Kohmura, H.; Ishikawa, K.; Ohta, S.; Kuroyanagi, K.; Kawano, S.; Kadota, T.; Chikazawa, H.; Kondo, H. and Takahashi, N.: Reprodductive and developmental roxicity studies of buspirone hydrochloride (ii)–oral administration to rats during perinatal and lactation periods. The Journal of Toxicological Sciences 15:61–84, 1990.

347 Busulfan 1–4–Butanedioldimethanesulfonate Myleran™ CAS 55–98–1

Diamond et al. (1960) reported a stunted infant with cleft palate, eye defects and generalized cytomegaly born to a pregnant mother who received 6–mercaptopurine, X–ray

treatments and busulfan. These authors attributed the effects to busulfan because of a previous pregnancy in which 6–mercaptopurine and X–rays alone caused no fetal defects.

Murphy et al. (1958) produced cleft palate, stunting and digital defects in the fetuses of rats treated on the 12th gestational day with 18–34 mg per kg. Heller and Jones (1964) produced ovarian dysgenesis in the rat fetus by treating the mother intraperitoneally on day 13 of gestation with 10 mg per kg. Vanhems and Bousquet (1972) gave 10 mg per kg intraperitoneally to rats on the 13th day of gestation and found destruction of the seminiferous tubules of the surviving male fetuses.

Diamond, I.; Anderson, M.M. and McCreadie, S.R.: Transplacental transmission of busulfan (Myleran) in a mother with leukemia. Production of fetal malformations and cytomegaly. Pediatrics 25:85–90, 1960.

Heller, R.H. and Jones, H.W.: Production of ovarian dysgenesis in the rat and human by busulphan. Am. J. Obstet. Gynecol. 89:414–420, 1964.

Murphy, M.L.; Delmoro, A. and Lacon, C.R.: The comparative effects of polyfunctional alkylating agents in the rat fetus with additional notes on the chick embryo. Ann. N.Y. Acad. Sci. 68:762–781, 1958.

Vanhems, E. and Bousquet, J.: Influence du busulphan sur le Developpement du testicule du rat. Ann. Endocrin. 33:119–128, 1972.

348 Butabarbital

Heinonen et al. (1977) found no increase in defects among 305 women who took this sedative at any time during pregnancy.

Heinonen, O.P.; Slone, D. and Shapiro, S.: Birth Defects and Drugs in Pregnancy. Publishing Sciences Group Inc., Littleton, Mass., 1977.

349 Butadiene 1,3–Butadiene CAS 106–99–0

Hackett et al. (1987) studied pregnant mice exposed to up to 1000 ppm for six hours daily on day 6–15. At 200 and 1000 ppm, maternal and fetal toxicity occurred but no evidence of teratogenicity was found. Morrissey et al. (1990) exposed rats and mice to atmospheric concentrations of up to 1,000 ppm for 6 hours daily on days 6–15. At the highest concentrations the rat dams failed to gain weight normally but no developmental toxicity was found. Mice at 40 ppm had maternal toxicity. Neither species evidenced teratogenicity.

Carpenter et al. (1944) found no decrease in offspring of rats after exposure up to 6700 ppm for 7.5 hours per day, 6 days a week during pregnancy. No evidence for infertility was found in rabbits and guinea pigs exposed to the same concentrations.

Carpenter, C.P.; Shaffer, C.B.; Weil, C.S. and Smyth, H.F.: Studies on the inhalation of 1:3–butadiene with a comparison of its narcotic effect with benzol, toluol, and styrene, and a note on the elimination of styrene by the human. J. Ind. Hyg. Toxicol. 26:69–78, 1944.

Hackett, P.L.; Sikov, M.R.; Mast, T.J.; Brown, M.G.; Buschbom, R.L.; Clark, M.L.; Decker, J.R.; Evanoff, J.J.;

Rommereim, R.L.; Rowe, S.E. and Westerberg, R.B.: Inhalation developmental toxicology studies: Teratology study of 1,3–butadiene in mice. U.S. Dept. Energy Contract DE–AC06–76RLO 1830, Pacific Northwest Laboratory–6412, pp 1–21, 1987.

Morrissey, R.E.; Schwetz, B.A.; Hackett, P.L.; Sikov, M.R.; Hardin, B.D.; McClanahan, B.J.; Decker, J.R. and Mast, T.J.: Overview of reproductive and developmental toxicity studies of 1,3–butadiene in rodents. Environmental Health Perspectives 86:79–84, 1990.

350 Butalamine

3-Phenyl–5–dibutylaminoethylamino–1,2,4–oxadiazole HCl
CAS 56974–46–0

This vasodilator was tested in pregnant rats by Fujimura et al. (1975) in oral doses as high as 300 mg per kg during the 7th to the 14th day. No teratogenicity was found.

Fujimura, H.; Hiramatsu, Y.; Tamura, Y. and Suzuki, T.: Effect of butalamine administered to pregnant rats on pre– and postnatal development of their offspring. Oyo Yakuri 9:727–731, 1975.

351 Butalbital

Heinonen et al. (1977) reported 112 pregnancies in which first trimester exposure occurred. No increase in defects was found.

Heinonen, O.P.; Slone, D. and Shapiro, S.: Birth Defects and Drugs in Pregnancy. Publishing Sciences Group Inc., Littleton, Mass., 1977.

352 Butanol

Brightwell et al. (1987) exposed rats by inhalation to n–butanol 8,000 ppm, sec–butanol (5000 ppm) or t–butanol (5000 ppm) for 7 hours daily during gestation. At these high doses, maternal and fetal weights were reduced, but no increase in defects was found. Daniel and Evans (1982) gave up to 1.0 percent in the diet of mice on days 6–20. A 50 percent reduction in weight of the offspring was associated with reduced behavioral performances.

Nelson et al. (1989) exposed rats by inhalation to 1–butanol 2–butanol and t–butanol in concentrations of up to 8,000, 7,000, or 5,000 ppm respectively. Maternal toxicity occurred at the highest doses but no teratogenicity except for a slight

Brightwell, W.S.; Nelson, B.K.; Mackenzie-Taylor, D.R.; Burg, J.R.; Khan, A. and Goad, P.T.: Lack of teratogenicity of three butanol isomers administered by inhalation to rats. Teratology 56A, 1987.

Daniel, M.A. and Evans, M.A.: Quantitative comparison of maternal ethanol and maternal tertiary butanol diet on postnatal development. J. Pharmacol. Exp. Ther. 222:294–300, 1982.

Nelson, B.K.; Brightwell, W.S.; Khan, A.; Burg, J.R. and Goad, P.T.: Lack of selective developmental toxicity of three butanol isomers administered by inhalation to rats. Fundamental and Applied Toxicology 12:469–479, 1989.

353 Butaperazine

Szabo and Brent (**1974**) found cleft palates in mice exposed to oral doses of 30 mg per kg or more. Since limitation of food and water also produced 15% of fetuses with cleft palates the authors suggested that the effect might be due to nutritional or water deprivation.

Szabo, K.T. and Brent, R.L.: Species differences in experimental teratogenesis by tranquillising agents. The Lancet 1:565, 1974.

354 Butenafine Hydrochloride

Shimomura and Hatakeyama (**1990**) treated rabbits percutaneously with up to 50 mg per kg during days 6–18. No reproduction ill effects were found. Shibuya et al. (**1990**) gave up to 25 mg per kg to rats subcutaneously on days 7–17 and found no adverse fetal ill effects.

Shibuya, K.; Daidohji, S.; Yasui, A.; Okamura, H. and Iwamoto, S.: Teratogenicity study of butenafine hydrochloride in rats. Kiso to Rinsho 24:1709–1723, 1990.

Shimomura, K. and Hatakeyama, Y.: Reproduction study by percutaneous administration of butenafine hydrochloride during the fetal organogenesis period in rabbits. Preclin. Rep. Cent. Inst. Exp. Anim. 16(2):1–12, 1990.

355 L–Buthionine–(S,R)–sulfoximine *BSO*

Wong et al. (1989) found that this glutothione synthesis inhibitor enhanced phenytoin teratogenicity in the rat. The inhibitor was given in a dose of 1,800 mg per kg and followed by 55 mg per kg. Study days were 12 and 13.

Wong, M.; Helston, L.M.J. and Wells, P.G.: Enhancement of murine phenytoin teratogenicity by the gamma-glutamylcysteine synthetase inhibitor l–buthionine–(S,R)–sulfoximine and by the glutathione depletor diethyl maleate. Teratology 40:127–141, 1989.

356 Butoctamide semisuccinate

N–(2–Ethylhexyl)–3–hydroxybutyramide hydrogen succinate M–2H

Kuraishi et al. (1979) gave this hypnotic orally to rats during organogenesis in doses up to 1500 mg per kg daily. Although maternal weight gain was less than controls, no adverse fetal effects were found.

Kuraishi, K.; Nabeshima, J.; Haresaku, M. and Inoue, S.: Teratologic study with N–(2–ethylhexyl)–3–hydrobutyramide hydrogen succinate (M–2H) in rats. Oyo Yakuri 17:315–324, 1979.

357 Butorphanol CAS 42408–82–2

In a series of papers, Takahashi et al. (1982) studied the effect of this analgesic on rat and rabbit reproduction. In the rat up to 25 mg per kg daily, subcutaneously before mating, during organogenesis or late in gestation caused no reproductive problems, teratogenicity or postnatal behavioral defects. Some decrease in viability was found 3 days postpartum after treatment on days 17–20 of gestation. Intravenously administered doses of up to 5 mg on days 7–17 of gestation in the rat produced some increase in fetal deaths, but no increase in defects. Doses of up to 10 mg per kg subcutaneously in the rabbit on days 6–18 of gestation produced no teratogenic effect.

Takahashi, N.; Kai, S.; Kohmura, H.; Ishikawa, K.; Karoyangai, K.; Hamajimma, Y.; Kadota, T.; Kawano, S.; Yamada, K. and Koike, M.: Reproductive studies of butorphanol tartrate 1–5.

358 Butoxamine

Saksena and Gokhale (**1972**) fed rats 10 or 20 mg per kg during days 1–5 and found no effect on reproduction.

Saksena, S.K. and Gokhale, S.V.: Effect of some adrenergic receptor blocking drugs on early pregnancy in rats. Indian J. Med. Res. 60:281–283, 1972.

359 n–Butoxyacetic Acid

Rawlings et al. (1985) exposed explanted day 9.5 rat embryos to 5 mM and found only minimal embryotoxicity.

Rawlings, S.J.; Shuker, D.E.G.; Webb, M. and Brown, N.A.: The teratogenic potential of alkoxy acids in post–implantation rat embryo culture. Structure–activity relationships. Toxicol. Lett. 1472:49–58, 1985.

360 2–Butoxyethanol

Sleet et al. (**1991**) exposed rats by gavage to 200 or 300 mg per kg during 3 day periods of organogenesis and were not able to confirm the marginal effects on the cardiovascular system observed with inhilation exposure.

Sleet, R.B.; Price, C.J.; Marr, M.C.; Morrissey, R.M. and Schwetz, B.A.: Cardiovascular development (CVD) in f–344 rats following phase–specific exposure to 2–butoxyethanol. (Abs) Teratology 43:466–467, 1991.

361 Butriptyline CAS 35941–65–2

Di Carlo et al. (1971) gave this medication orally to rats in doses up to 25 mg per kg on the 4th through the 18th day and found no adverse fetal effects.

Di Carlo, R.; Paghini, G. and Pelagalli, G.V.: Comparative action of amitriptyline and butriptyline on skeletal development in the rat embryo. (abs) Teratology 4:486 only, 1971.

362 Butropium Bromide Butoxybenzyl Hyoscyamine Bromide CAS 29025–14–7

Suzuki et al. (1974) studied this drug in pregnant mice and rats with maximum oral doses of 500 mg per kg per day and maximum intraperitoneal doses of 40 mg per kg. The dams evidenced autonomic reactions. No increase in fetal death or congenital defects were found in fetuses exposed during active organogenesis.

Suzuki, Y.; Okada, F.; Kondo, S.; Asano, F.; Chiba, T. and Wakabayashi, T.: Teratological study with butoxybenzyl hyoscyamine bromide in rats and mice. Oyo Yakuri 8:319–337, 1974.

363 Butyl Benzyl Phthalate

Ema et al. (**1990**) fed 0.25, 0.5, 1.0 or 2.0 percent of this plasticizer in the diet to rats from day 0–20 of gestation

at the two highest doses (654 mg per kg) maternal and fetal weight was reduced and at 2.0% complete resorption occurred. No teratogenic activity was found. Ema et al. (**1991**) demonstrated by pair feeding that the resorptions were not due to decreased maternal nutrition.

Ema et al. (**1992**) gave this plasticizer in the diet of rats at a level of 2.0 percent. Maternal toxicity and complete resorption was found when it was given throughout pregnancy. Treatment on days 6–16 was associated with cleft palates and fusion of the sternebrae. Ema et al. (**1991**) found cleft palates and fused sternobrae in rats treated with 0.75g per kg on days 7–16.

Ema et al. (**1992**) fed rats a 2.0% diet of butyl benzyl phthalate on days 0–11 and found complete resorption and on days 11–20 feeding produced cleft palate and skeletal defects. Ema et al. (**1994**) presented data suggesting that lowered plasma progesterone accounted for the postimplantation loss.

Ema, M.; Itami, T. and Kawasaki, H. : Embryolethality and teratogenicity of butyl benzyl phthalate. J. Appl. Tox. 12:179–183, 1992.

Ema, M.; Itami, T. and Kawasaki, H.: Teratogenicity of butyl benzyl phthalate in rats (A). Cong. Anomal 31:230, 1991.

Ema, M.; Itami, T. and Kawasaki, H.: Evaluation of the embryolethality of butyl benzyl phthalate by conventional and pair–feeding studies in rats. Journal of Applied Toxicology 11:39–42, 1991.

Ema, M.; Murai, T.; Takafumi, I. and Hironoshin K.: Evaluation of the teratogenic potential of the plasticizer butyl benzyl phthalate in rats. Journal of Applied Toxicology 10(5):339–343, 1990.

Ema, M.; Itami, T. and Kawasaki, H.: Effect of period of exposure on the developmental toxicity of butyl benzyl phthalate in rats. Journal of Applied Toxicology 12:57–61, 1992.

Ema, M.; Kurosaka, R.; Amano, H. and Ogawa, Y.: Embryolethality of butyl benzyl phthalate during earrly pregnancy in rats. Reproductive Toxicology 8:231–236, 1994.

364 Butyl Acrylate CAS 141–32–2

Discussed under Methacrylate

365
4–(3–tert–Butylamino–2–hydroxypropoxy)–2–methyl–1 (2H)–isoquinolinone hydrochloride CAS 62774–96–3
N–696

Saegusa et al. (1988) gave rabbits up to 100 mg per kg on days 6–18 of pregnancy. Some increase in supernumerary coronary orifices was found at 20 mg per kg but not at 100 mg per kg. Other defects were not increased. At 100 mg per kg, resorptions were increased and fetal weight reduced. Sato et al. (**1988** A) gave rats orally up to 50 mg per kg of this β adrenergic receptor blocker daily during organogenesis. No adverse fetal or postnatal development was found. In rabbits given up to 100 mg per kg during organogenesis increased lethality of fetuses was found but no teratogenicity.

In perinatal studies Sato et al. (**1988** B) found no developmental ill effects in the offspring.

Saegusa, T.; Naito, Y.; Narama, I. and Kawase, S.: Reproduction study of 4–(3–tert–butylamino–2–hydroxypropoxy)–2–methyl–1(2H)–isoquinolinone hydrochloride (N–696), a new β adrenergic blocking agent. (4) Teratology study in rabbits. Pharmacometrics (Oyo Yakuri) 35:49–58, 1988.

Sato, T.; Narama, I. and Kawase, S.: Reproduction study of (+)(-)-4–(3–tert–butylamino–2–hydroxypropoxy)–2–methyl–1(2h)–isoquinolinone hyddrochloride (n–696), a new β–adrenergic blocking agent (2) teratogenicity study in rats. Oyo Yakuri 35(1):11–30, 1988.

Sato, T.; Narama, I. and Kawase, S.: Reproduction study of (+)(-)-4–(3–tert–buthylamino–2–hydroxypropoxy)–2–methyl–1(2H)–isoquinolinone hydrochloride (N–696), a new β adrenergic blocking agent (3) peri–and postnatal study in rats. Oyo Yakuri 35(1):31–48, 1988.

366
1–tert–Butylamino–3–(2–nitrilophenoxy)–2–propanol
KO–1366–Cl CAS 23093–74–5

Fuyuta et al. (1973) injected this β adrenergic blocker into pregnant rats from the 7th to the 13th day of pregnancy and found no increase in malformation rate. Doses up to 7.5 mg per kg were given orally and to 10 mg per kg per day intraperitoneally.

Fuyuta, M.; Fujimoto, T. and Kaihara, N.: Examination of the teratogenic effect of an adrenergic β receptor blocking agent KO–1366–Cl in rats. (abs) Teratology 8:92 only, 1973.

367
5–n–Butyl–1–cyclohexyl–2,4,6–trioxoperhydropyrimidine

Tanabe (1967) gave up to 400 mg per kg to rats orally on days 7–13 and found no adverse effects in the offspring.

Tanabe, K.: Hypersensitive toxicity of 5–n–butyl–1–cyclohexyl–2,4,6–trioxoperhydropyrimidine in the pregnant rat. Jpn. J. Pharmacol. 17:381–392, 1967.

368   Butylated Hydoxyanisole *3–BHA 2–BHA*
3–tert–Butyl–4–hydroxyanisole CAS 25013–16–5

Clegg (1965) administered, by mouth, 500 to 1000 mg per kg to several strains of mice and rats for various periods of time during gestation and observed no increase in fetal congenital malformations. Reproductive toxicity in other studies including rabbits, pigs and monkeys was reviewed by the International Agency for Research in Cancer (Anonymous, 1986)

Anonymous: Butylated hydoxyanisole. In: Vol. 40, International Agency for Research in Cancer. Anonymous, 1986. eb
Clegg, D.J.: Absence of teratogenic effect of butylated hydoxyanisole (BHA) and butylated hydroxytoluene (BHT) in rats and mice. Food Cosmet. Toxicol. 3:387–403, 1965.

369   Butylated Hydroxytoluene
3–tert–Butyl–4–hydroxyanisole

2,6–Bis(1,1–dimethylethyl)–4–methyl–phenol CAS 128–37–0

Brown et al. (1959) reported that 3 out of 30 litters from rats receiving this material at levels of 0.5 percent of the diet contained fetuses with anophthalmia. The dose was approximately 250 mg per kg per day. Clegg (1965) was unable to confirm their findings using 500 to 1000 mg per kg in several strains of mice and rats. Further reproductive studies are summarized by IARC (1986).

Anonymous: Butylated hydroxytoluene (BHT), International Agency for Research in Cancer, 40:177–179, 1986.
Brown, W.D.; Johnson, A.R. and Halloran, M.W.: The effect of the level of dietary fat on toxicity of phenolic antioxidants. Aust. J. Exp. Biol. Med. Sci. 37:533–548, 1959.
Clegg, D.J.: Absence of teratogenic effect of butylated hydroxyanisole (BHA) and butylated hydroxytoluene (BHT) in rats and mice. Food Cosmet. Toxicol. 3:387–403, 1965.

370   2–sec–Butyl–4,6–dinitrophenol

Discussed under Dinoseb

371   Butylene Glycol Adipic Acid Polyester

Fancher et al. (1973) found no adverse effects of a diet with 10,000 PPM fed to 3 generations of rats.

Fancher, O.E.; Kennedy, G.L.; Plank, J.B.; Lindberg, D.C.; Hunt, W.H. and Calandra, J.C.: Toxicology of a butylene glycol adipic acid polyester. Toxicol. Appl. Pharmacol. 26:58–62, 1973.

372   Butylene Oxide CAS 26249–20–7

Kimmel et al. (1984) reviewed studies done on pregnant rats and rabbits exposed to 1000 ppm for 7 hour periods during organogenesis. No developmental toxicity was observed. No toxicity in the rabbit or rat mother or fetus was found at inhalation doses of 250 and 1000 ppm (Hardin et al., 1981). The inhalation exposure was for 6 hours daily and on days 1–15 or 1–19 in the rat and rabbit, respectively. No teratogenicity was found.

Hardin, B.D.; Bond, G.P.; Sikov, M.R.; Andrew, F.D.; Beliles, R.P. and Niemeier, R.W.: Testing of selected workplace chemicals for teratogenic potential. Scand J. Work Environ Health 7(4):66–75, 1981.
Kimmel, C.A.; LaBorde, J.B. and Hardin, B.D.: Reproductive and developmental toxicology of selected epoxides. In: Toxicology and the Newborn, S. Kacew and M.J. Reasor (eds.), Elsevier Sc. Publ., Amsterdam, Chapter 13, 270–287, 1984.

373   Butylhydroquinone *TBHQ* CAS 1948–33–0

Tertiary butylhydroquinone was fed to pregnant rats during active organogenesis and no deleterious fetal effects were found (Krasavage, 1977). The highest dietary dose was 0.5 percent.

Krasavage, W.J.: Evaluation of the teratogenic potential of tertiary butylhydroquinone (TBHQ) in the rat. Teratology 16:31–34, 1977.

374 N–Butyl Ketone CAS 591–78–6

Bus et al. (1979) exposed rats to 1000 ppm of n–hexane on days 8–12, 12–16 or 8–16 of gestation. The chemical is broken down into methyl n–butyl ketone and 2,5–hexanedione and the parent and breakdown compounds were found in equal concentrations in the mother and fetus. No adverse fetal effects were found.

Bus, J.S.; White, E.L.; Tyl, R.W. and Barrow, C.S.: Perinatal toxicity and metabolism of n–hexane in Fischer–344 rats after inhalation exposure during gestation. Toxicol. Appl. Pharmacol. 51:295–303, 1979.

375 9–Butyl–6–mercaptopurine CAS 6165–01–1

Discussed under 6–Mercaptopurine

376 Butyric Acid

Dawson (**1991**) using a frog embryo treatogenesis assay found that 50% of the offspring were malformed at 400 mg per liter. Microencephaly, eye defects edema and gut defects were most commonly seen.

Dawson, D.A.: Additive incidence of developmental malformation for xenopus embryos exposed to a mixture of ten aliphatic carboxylic acids. Teratology 44:531–546, 1991.

377 p–tert–Butylphenol Formaldehyde Resin

This adhesive was studied by Tanaka et al. (**1992**) in rats which were gavaged with up to 1,000 mg per kg during organogenesis. No adverse fetal effects were found. Itami et al. (**1993**) fed rats up to 10% of their diet with chemical used in making adhesives. No adverse fetal effects were found after treatment on days 6–15.

Itami, T.; Ema, M. and Kawasaki, H.: Teratogenic evaluation of p–tert–butylphenol formaldehyde resin (novolak type) in rats following oral exposure. Drug and Chemical Toxicology 16:369–382, 1993.

Tanaka, R.; Usami, M.; Kawashima, K. and Takanaka, A.: Studies of the teratogenic potential of p–tert–butylphenolformaldehyde resin in rats. Bulletin of National Institute of Hygienic Sciences 110:22–26, 1992.

378 1–Butyryl–4–cinnamylpiperazine HCl CAS 17730–82–4

This analgesic was studied in pregnant rabbits, mice and rats by Irikura et al. (1972) using oral or subcutaneous routes of up to 180 mg per kg daily during organogenesis. No congenital defects resulted but some reduction in fetal weight was found in all species at the highest doses.

Irikura, T.; Sugimoto, T.; Suzuki, H. and Hosomi, J.: Studies on analgesic agents, part 2. Teratogenic studies of 1–butyryl–4–cinnamylpiperazine hydrochloride (AP237). Oyo Yakuri 6:271–277, 1972.

379 BV–26–723 *Methyl–11–bromo–14,15–dihydro–14–β–hydroxy–(3d,16d)eburna–menine 14–carboxylate hydrogen fumarate*

This drug used for cerebral circulation disorders was tested by oral dosage in pregnant rats and rabbits (Nakashima et al., 1983). Doses up to 250 mg per kg given before fertilization and during the first 7 days, during organogenesis or in the perinatal period, did not produce adverse reproductive effects in rats. The rabbits received up to 60 mg per kg during days 6–18. Resorptions were increased at 60 mg and minor skeletal variations were seen.

Nikashima, T.; Ishizaka, K.; Hamada, M. and Matsuda, K.: Reproduction studies of BV–26–723 in rats and rabbits. Clin. Report 17:1565–1592, 1983.

380 Cacodylic Acid CAS 75–60–5

This organic arsenical herbicide was tested in rats and mice on days 7–16 (Rogers et al., 1981). Gavage feedings of 200–600 mg per kg and 7.5–60 mg per kg were administered to mice and rats, respectively. At doses of 400 mg or more, cleft palate occurred in mice. This dose was above the maternal toxic dose. Irregular palatine rugae were found in rat fetuses at doses of 15 mg per kg.

Rogers, E.H.; Chernoff, N. and Kavlock, R.J.: The teratogenic potential of cacodylic acid in the rat and mouse. Drug Chem. Toxicol. 4:49–61, 1981.

381 Cadmium CAS 7440–43–9

Cvetkova (1970) reported that the weights of newborns of women employed in the cadmium industry were significantly lower than controls. Nomiyama in a personal communication to Thuerauf et al. (1975), stated that no fetal damage could be detected in Itai–itai disease, a syndrome associated with high cadmium exposure. The reproductive effects of cadmium in animals and humans was reviewed in an IARC monograph (**1993**).

Ferm and Carpenter (1967) reported mid–line facial and palate defects in hamsters after maternal intravenous injection of 2 mg per kg of cadmium sulfate on the 8th day of gestation. Mulvihill et al. (1970) described the embryonic events contributing to these defects. Barr (1973), using 16 micromoles of cadmium per kg intraperitoneally in rats on days 9, 10 or 11 of gestation, produced eye defects, hydrocephaly, thinning of the abdominal wall with undescended testes, and other defects. Subcutaneous administration was not teratogenic. Chernoff (1973) produced weight reduction in the developing fetal lungs of maternally treated rats. Dencker (1975) showed that CD–109 enters the hamster and mouse embryo via the yolk sac and then through the primitive gut. The pharmacokinetics in the rat fetus was described (Levin et al., 1987). Saltzman et al. (1989) studied the dynamics of placental and uterine blood flow in the rat after cadmium exposure. Padmanabhan & Hameed (**1990**) described limb malformation produced in mice.

Christley and Webster (1983) studied the distribution of cadmium during organogenesis in the mouse. Padmanabhan (1987) described changes in the external and internal ear of the mouse fetus. Golden (1975 PhD thesis, University of Michigan), using 100 ppm in the drinking water during pregnancy, found an increase in blood pressure in the postnatal period. Levin and Miller (1981) found a decrease in placental blood flow associated with placental necrosis and fetal death after administration of 40 micromoles of cadmium chloride subcutaneously in rats. The acute distribution of cadmium in the rat was studied (Samarawickrama and Webb, 1981). Cadmium chloride embryotoxicity was evaluated in rats following the single i.p. administration of 2.5, 10 and 20 mg/kg doses (calculated per Cd^{2+}) on 1–20 days of pregnancy. The dose 2.5 mg/kg gave the lowest effect: the body weight of fetuses was decreased and the number of resorptions was increased. At the dose 20mg/kg the toxic effect was maximal and by the 21st day of pregnancy the weight of females was two times less, compared to control. No teratogenic effect was observed. Wier et al. (**1990**) perfused human placentae with medium containing concentrations of 10–100 nanomoles of cadmium. Although there was no change in oxygen, glucose or aminoacid utilization the synthesis and release of hCG was reduced. Kreis et al. (**1993**) studied reproduction by cows in an area of the Netherlands where cadmium was increased (0.1–2.5 microgram per liter of ground water). Lower twinning rates and more birth complications occurred among the exposed animals.

Yu et al. (1985) studied the effect of 5 and 10 microgm per ml on 4 cell and morula stage mouse embryos and found that most of the embryos degenerated. At an exposure of 1.0 microgm per ml, a decrease in implantation of transferred embryos was observed. Yu and Chan (1987) found that concentrations of 0.5 and 1.0 microgm per ml did not affect trophoblastic invasiveness of mouse embryos treated for 24 hours at the four–cell stage. Ali et al. (1986) observed rats given 4.2 or 8.4 microgm per ml in the drinking water during gestation. Hyperactivity and delay in cliff aversion and swimming behavior was noted in both groups.

Nayak et al. (1989) studied sister chromatid exchanges, nucleolar organizing regions and chromosomes in the livers of mouse fetuses at doses that caused resorptions and found no changes in their rates.

Chen and Hales (**1994**) studied cadmium in whole embryo cultures and their finding of increased E–cadherin suggested to them that the mechanism of toxicity was related to cytoskeletal protein changes in the yolk sac.

Ali, M.M.; Murthy, R.C. and Chandra, S.V.: Developmental and longterm neurobehavioral toxicity of low level in utero cadmium exposure in rats. Neurobehav. Toxicol. Teratol. 8:463–468, 1986.

Barr, M.: The teratogenicity of cadmium chloride in two stocks of Wistar rats. Teratology 7:237–242, 1973.

Chen, B. and Hales, B.F.: Cadmium–induced rat embryotoxicity in vitro is associated with an increased abundance of E–cadherin protein in the yolk sac. Toxicol. & Appl. Pharm. 128:293–301, 1994.

Chernoff, N.: Teratogenic effects of cadmium in rats. Teratology 8:29–32, 1973.

Christley, J. and Webster, W.S.: Cadmium uptake and distribution in mouse embryos following maternal exposure during the organogenic period. Teratology 27:305–312, 1983.

Cvetkova, R.P.: Materials on the study of the influence of cadmium compounds on the generative function. Gig. Tr. Prof. Zabol. 14:31, 1970.

Dencker, L.: Possible mechanisms of cadmium fetotoxicity in golden hamsters and mice. Uptake by the embryo, placenta and ovary. J. Reprod. Fertil. 44:461–471, 1975.

Ferm, V.H. and Carpenter, S.J.: Teratogenic effect of cadmium and its inhibition by zinc. Nature 216:1123 only, 1967.

IARC.: Iarc monographs on the evaluation of carcinogenic risks in humans. Vol 58, IARC, Lyon, France 1993, 184–188.

Kreis, I.A.; DeDoes, M.; Hoekstra, J.A.;Coulander, C.L.; Peters, P.W.J. and Wentink, G.H.: Effects of cadmium on reproduction, an epizootologic study. Teratology 48:189–196, 1993.

Levin, A.A. and Miller, R.K.: Fetal toxicity of cadmium in the rat. Decreased utero–placental blood flow. Toxicol. Appl. Pharmacol. 58:297–306, 1981.

Levin, A.A.; Kilpper, R.W. and Miller, R.K.: Fetal toxicity of cadmium chloride. The pharmacokinetics in the pregnant Wistar rat. Teratology 36:163–170, 1987.

Litvinov, N.N.; Kazachkov, V.L.; Astakhova, L.F.: The problem of dose–effect relations of cadmium chloride embryotoxic effects. Gig. Sanit. 4:86–88, 1989.

Mulvihill, J.E.; Gamm, S.H. and Ferm, V.H.: Facial formation in normal and cadmium–treated golden hamsters. J. Embryol. Exp. Morphol. 24:393–403, 1970.

Nayak, B.N.; Ray, M.; Persaud, T.V.N. and Nigli, M.: Embryotoxicity and in vivo cytogenetic changes following maternal exposure to cadmium chloride in mice. Exp. Pathol. 36:75–80, 1989.

Padmanabhan, R.: Abnormalities of the ear associated with exencephaly in mouse fetuses induced by maternal exposure to cadmium. Teratology 35:9–18, 1987.

Padmanabhan, R. and Hameed, M.S.: Characteristics of the limb malformations induced by maternal exposure to cadmium in the mouse. Reproductive Toxicology 4:291–304, 1990.

Saltzman, R.A.; Miller, R.K. and Di Sant' Agnese, P.A.: Cadmium exposure on day 12 of gestation in the Wistar rat: Distribution, uteroplacental blood flow, and fetal viability. Teratology 39:19–30, 1989.

Samarawickrama, G.P. and Webb, M.: The acute toxicity and teratogenicity of cadmium in the pregnant rat. J. Appl. Toxicol. 1:264–269, 1981.

Thuerauf, J.; Schaller, K.H.; Engelhardt, E. and Gossler, K.: The cadmium content of the human placenta. Int. Arch. Occup. Environ. Health 36:19–27, 1975.

Wier, P.J.; Miller, R.K.; Maulik, D. and DiSant'Agnes, P.A.: Toxicity of cadmium in the perfused human placenta. Toxicology and Applied Pharmacology 105:156–171, 1990.

Yu, H.S.; Tam, P.P.L. and Chan, S.T.H.: Effects of cadmium on preimplantation mouse embryos in vitro with special reference to their implantation capacity and subsequent development. Teratology 32:347–353, 1985.

Yu, H.S. and Chan, S.T.H.: Effects of cadmium on preimplantation and early postimplantation mouse embryos in vitro with special reference to their trophoblastic invasiveness. Pharmacol. Toxicol. 60:129–134, 1987.

382 **Ca–DTPA** *Calcium Trisodium Diethylenetriaminepentacetate* CAS 12111–24–9

Fisher et al. (1976) injected 1,440 micromoles subcutaneously into mice daily from 2 to 6 or from 7 to 11 days and produced neural tube closure defects in the fetuses.

Fisher, D.R.; Calder, S.E.; Mays, C.W. and Taylor, G.N.: Ca–DTPA–induced fetal death and malformations in mice. Teratology 14:123–128, 1976.

383 **Caerulein** CAS 17650–98–5

This decapeptide with gastrin–like activity was not teratogenic in rabbits and rats. Doses of 25 (rabbit) and 50 mg (rat) per kg were given subcutaneously during pregnancy (Chieli et al., 1972).

Chieli, T.; Bertazzoli, C.; Ferni, F.; Delloro, I.; Capella, C. and Solci, E.: Experimental toxicology of caerulein. Toxicol. Appl. Pharmacol. 23:480–491, 1972.

384 **Caffeine** *Coffee* CAS 58–08–2

Although initial reports suggested that caffeine might be associated with adverse fetal outcome, a number of more carefully done studies indicated that there is no increased risk for congenital defects and only a possible risk for miscarriage at amounts over 150 mg per day. Barr and Streissguth (**1991**) in a prospective study of 500 mothers could not associate pregnancy outcome with caffeine intake. Studies of I.Q. were done at age 7 years. Nehlig and Debry (**1994**) have reviewed the animal and human prenatal toxicity of caffeine. From their comprehensive report they concluded that congenital defects were unlikely to be associated with coffee consumption and that the data on spontaneous abortion and prematurity were not consistent. They did suggest that an additive role by ethanol and tobacco might be present.

Weathersbee et al. (1977) and Borlee et al. (1978) carried out studies on caffeine and coffee in pregnancy. Weathersbee et al. studied 489 households in the Utah area and in 16, where the mothers had ingested 600 mg or more per day, it was found that 15 had fetal loss or prematurity. A coffee serving is estimated at 75 mg of caffeine. In the study by Borlee et al., 202 malformed children and 175 controls were included. In the group of mothers of malformed offspring, the over–consumers (more than 8 cups per day) were 44 while in the control, there were only 21 (p < 0.05). Hydrocephalus, interventricular septal defect and cleft lip were the most commonly found malformations. Both studies are subject to criticism since the matching of controls was either not detailed or not performed. Sociological variables might have accounted for the differences. Heinonen et al. (1977) listed 5,000 pregnancies where caffeine exposure was present and did not find an increase in defect rate over matched controls.

Linn et al. (1982) published a carefully controlled analysis of the coffee drinking (and smoking) habits of over 12,000 pregnant women. After controlling for smoking, other habits, demographic characteristics and medical history, no relationship was found between coffee consumption

and low birth weight or malformation in the offspring. In a case–control study of 2,030 women giving birth to offspring with selected congenital defects, no malformation group was found in which caffeine consumption was significantly increased (Rosenberg et al., 1982).

Srisuphan and Bracken (1986) interviewed 3,135 pregnant women and found after fitting into a multiple logistic regression model of other confounders, that the relative risk for spontaneous abortion for women taking over 150 mg of caffeine was 1.97 (p = 0.07). Wilcox et al. (1988) studied 221 healthy women who planned their pregnancies. A significant reduction in fecundity occurred in those who took the equivalent of 5–30 cups of coffee per month (501–3000 mg). A dose response was found in their studies. Those with consumption equivalents of 30–50 cups per month had about a doubling of the decrease in fecundity. Other confounding associations were studied and did not account for the change. Christianson et al. (1989) found that women who drank 7 cups or more of coffee per day reported "difficulty in becoming pregnant" twice as often as those who drank little or no coffee. In a retrospective study of 2817 fertile women, Joesoef et al. (1990) could find Infante–Rivard et al. (**1993**) found significantly increased risks for fetal loss among women taking 163–321 or greater than 321 mg per day. The odds ratios were 1.95 (CI, 1.29–2.93) and 2.62 (CI, 1.38–5.01) respectively. For women taking 48–162 or less than 48 mg per day the odds ratios were not significantly increased. Mills et al. (**1993**) could not associate mild coffee drinking with increased spontaneous abortion or low birth weight.

Martin and Bracken (1987) reported reduced birth weight (120 g) in women who took 300 mg or more of caffeine per day. The reduction in weight was 105 g when other variables were controlled. Kurpa et al. (1983) paired 706 pregnancies with defects to the pregnancy immediately preceding in the same maternity ward. They found no increase in coffee consumption and there was no general sub–group of defects that had a significantly higher consumption.

Nishimura and Nakai (1960), using single injections of 250 mg per kg to mice on days 10 through 14, found fetuses with predominantly digital defects but also with subcutaneous hemorrhage and cleft palate. Knoche and Konig (1964) found only 7 different defects among over 700 mouse fetuses whose mothers received 50 mg per day during gestation. Loosli et al. (1964) reported no defects in rats, mice or rabbits treated with 100 mg per kg during organogenesis. Bertrand et al. (1965) could not produce significant numbers of defects in the mouse with 100 mg per kg, but in rats gavaged with 75 to 150 mg per kg per day during gestation, they found digital defects and associated hemorrhages. The rate of ectrodactyly in the fetuses was dose dependent and there was a predilection for the left posterior extremity. They estimate that the amount of caffeine that could be ingested by a person drinking 1 liter per day would be approximately 20 to 25 mg per kg. Beck and Urbano (**1991**) showed that a subteratogenic dose of caffeine could alter the skeletal development of mice given acetazolamide.

Collins et al. (1983) administered caffeine in drinking

water during the entire rat gestation. At 161 and 205 mg per kg, decreased implantation and increased resorptions were found. Fetal body weight reduction and edema occurred at 161 and 205 mg per kg, but no dose–related increase in gross anomalies occurred after ad lib exposure in drinking water. Coffee was administered to pregnant rats as their sole source of water throughout pregnancy. Intakes of 38 mg per kg (equivalent to 32 cups of coffee per day) did not produce an increase in defects or postnatal development or reproductive capacity (Palm et al., 1976).

Bertrand et al. (1970) produced ectrodactyly in rabbit fetuses by administration of doses of 100 mg per kg. Fujii and Nishimura (1974) blocked the teratogenic effects of caffeine by preadministration of the β–adrenergic blocker, propranolol. Caffeine administered directly to the chick embryo produced aortic arch anomalies (Gilbert et al., 1977). Renwick (1988) commented on the similarity of the limb defects in mice and those that were anecdotally reported in humans. He plotted the animal effects against the cubed dose of caffeine and found no evidence for threshold.

Nolen (1988) comprehensively reviewed the findings from animal studies of teratogenicity and postnatal development. Mulvihill (1973) reviewed the teratogenicity and mutagenicity of this chemical. Brown and Scialli (1987) reviewed the animal and human data on caffeine and concluded that although there is evidence that very large bolus doses produce defects in experimental animals, no relationship was found between coffee ingestion and congenital defects in the human. Nash and Persaud (1988) reviewed animal and human teratogenic data.

Barr, H.M. and Streissguth, A.P.: Caffeine use during pregnancy and child outcome: A 7–year prospective study. Neurotoxicology and Teratology 13:441–448, 1991.

Beck, S.L. and Urbano, C.M.: Potentiating effect on caffeine on the teratogenicity of acetazolamide in c57bl/6j mice. Teratology 44:241–250, 1991.

Bertrand, M.; Giroud, J. and Rigaud, M.F.: Ectrodachtylie provoquee par la cafeine chez les rongeurs. Role des facteurs specifiques et genetiques. C. R. Soc. Biol. (Paris) 164:1488–1489, 1970.

Bertrand, M.; Schwam, E.; Frandon, A.; Vagne, A. and Alary, J.: Sur un effet teratogene systematique et specifique de la cafeine chez les rongeurs. C. R. Soc. Biol. (Paris) 159:2199–2202, 1965.

Borlee, I.; Lechat, M.F.; Bouckaert, A. and Misson, C.: Coffee, risk factor during pregnancy? (French) Louvain Med. 97:279–284, 1978.

Brown, N.A. and Scialli, A.R.: Update on caffeine. A medical letter on environmental hazards to reproduction. Reproductive Toxicology. Vol. 1, No. 3, 1987.

Christianson, R.E.; Oechsli, F.W. and Van den Berg, B.J.: Caffeinated beverages and decreased fertility. Lancet 1:378, 1989.

Collins, T.F.X.; Welsh, J.J.; Black, T.N. and Ruggles, D.L.: A study of teratogenic potential of caffeine ingested in drinking water. Food Chem. Toxicol. 21:763–777, 1983.

Fujii, T. and Nishimura, H.: Reduction in frequency of fetopathic effects of caffeine in mice by pretreatment with propranolol. Teratology 10:149–152, 1974.

Gilbert, E.F.; Bruyere, H.J.; Ishikawa, S.; Cheung, M.O. and Hodach, R.J.: The effects of methylxanthines on catecholamine–stimulated and normal chick embryos. Teratology 16:47–52, 1977.

Heinonen, O.P.; Slone, D. and Shapiro, S.: Birth Defects and Drugs in Pregnancy. Publishing Science Group Inc., Littleton, Mass. 1977.

Infante-Rivard, C.; Fernandez, A.; Gauthier, R.; David, M. and Rivard, G.: Fetal loss associated with caffeine intake before and during pregnancy. JAMA 270:2940–2943, 1993.

Joesoef, M.R.; Beral, V.; Rolfs, R.T.; Aral, S.O. and Cramer, D.W.: Are caffeinated beverages risk factors for delayed conception? The Lancet 335:136-137, 1990.

Knoche, Ch. and Konig, J.: Zur prenatelen Toxizitat von Diphenylpyralin-8-theophyllinat unter Berucksichtigung von Erfahrungen mit Thalidomid und Coffein. Arzneim. Forsch. 14:415–424, 1964.

Kurppa, K.; Holmberg, P.C.; Kuosma, E. and Saxen, L.: Coffee consumption during pregnancy and selected congenital malformations: A nationwide case–control study. Am. J. Pub. He. 73:1397–1399, 1983.

Linn, S.; Schoenbaum, S.C.; Monson, R.R.; Rosner, B.; Stabblefield, P.G. and Ryan, K.J.: No association between coffee consumption and adverse outcomes of pregnancy. N. Eng. J. Med. 306:141–145, 1982.

Loosli, R.; Loustalot, P.; Schalch, W.R.; Sievers, K. and Steager, E.G.: Proceedings European Society for Study of Drug Toxicity 4:214–216, 1964.

Martin, T.R. and Bracken, M.B.: The association between low birth weight and caffeine consumption during pregnancy. Am. J. Epidemiol. 126(5):813–821, 1987.

Mills, J.L.; Holmes, L.B.; Aarons, J.H.; Simpson, J.L.; Brown, Z.A.; Jovanovic–Peterson, L.G.; Conley, M.R.; Graubard, B.I.; Knopp, R.H. and Metzger, B.E.: Moderate caffeine use and the risk of spontaneous abortion and intrauterine growth retardation. JAMA 269:593–597, 1993.

Mulvihill, J.J.: Caffeine as teratogen and mutagen. Teratology 8:69–72, 1973.

Nash, J. and Persaud, T.V.N.: Reproductive and teratological risks of caffeine. Anat. Anz. Jena 167:265–270, 1988.

Nishimura, H. and Nakai, K.: Congenital malformations in offspring treated with caffeine. Proc. Soc. Exp. Biol. Med. 104:140–142, 1960.

Nehlig, A. and Debry, G.: Potential teratogenic and neurodevelopmental consequences of coffee and caffeine exposure: A review on human and animal data. Neurotoxicology and Teratology 16:531–543, 1994.

Nolen, G.A.: The developmental toxicology of caffeine. In: Issues and Reviews in Teratology, H. Kalter (ed.), Plenum Press, New York 4:305–350, 1988.

Palm, P.E.; Arnold, E.P.; Rockwall, P.C.; Teyczek, J.C.; Teague, K.W. and Kensler, C.J.: Evaluation of the teratogenic potential of fresh–brewed coffee, caffeine, and aspirin in rats. Arthur D. Little, Inc., Cambridge, Mass. Toxicol. Appl. Pharmacol. 37:125–126, 1976.

Renwick, J.H.: Caffeine cubes. Lancet 1:821–822, 1988.

Rosenberg, L.; Mitchell, A.A.; Shapiro, S. and Slone, D.:

Selected birth defects in relation to caffeine–containing beverages. JAMA 247:1429–1432, 1982.

Srisuphan, W. and Bracken, M.B.: Caffeine consumption during pregnancy and association with late spontaneous abortion. Am. J. Obstet. Gynecol. 154:14–20, 1986.

Weathersbee, P.S.; Olsen, L.K. and Lodge, J.R.: Caffeine and pregnancy. Postgraduate Med. 62:64–69, 1977.

Wilcox, A.; Weinberg, C. and Baird, D.: Caffeinated beverages and decreased fertility. Lancet 2:1453–1455, 1988.

385 Calatropis gigantea

Bhima Rao et al. (1974) smeared one–half of the pregnant rat uterus at day 8 with the latex of this plant or with the dialyzed fraction from the homogenate. Fetal loss was 100 percent in the first group and 61 percent with the second treatment. Feeding on day 8 with 1.0 ml of the latex produced 30 percent fetal loss rate.

Bhima Rao, B.S.; Devaraj Sarkar, H.B. and Sheshadri, H.S.: Effect of latex of calatropis gigantea on pregnancy in the albino rat. J. Reprod. Fertil. 38:234, 1974.

386 Calcitonin, Synthetic *YM–11221 Elcatonin*

Ito et al. (**1988**) injected this synthetic calcitonin from salmon subcutaneously into rats from day 17 through the 20th postnatal day. No effects were seen on the pups with 1.0, 0.6, or 0.2 iu per kg. Takahashi et al. (**1993 A,B**) gave this calcitonin product from eels intravenously to rats before and in early pregnancy and during organogenesis in doses of up to 80 units per kg. No adverse reproductive or fetal effects were found. In a preliminary report Takahashi et al. (**1993 B**) indicated that perinatal administration was associated with severe hypocalcemia and death in the dams. Takahashi et al. (**1993 C**) gave up to 80 units intravenously to rabbits during organogenesis and found no ill effects.

Ito, I.; Hibino, H.; Iriyama, K.; Ishimura, K.; Fujiwara, M.; Shibata, M.; Yoshida, T. and Miki, T.: Peri- and postnatal study in rats treated subcutaneously with ym–11221. Kiso to Rinsho 22:5667–5674, 1988.

Takahashi, M.; Sakura, T.; Karasawa, N.; Motoyama, M. and Kobayashi, Y.: Fertility and reproductive toxicity study in rats administered intravenously with elcatonin. Yakuri to Chiryo 21:4503–4511, 1993A.

Takahashi, M.; Sakurai, T.; Karasawa, N.; Motoyama, M. and Kobayashi, Y.: Teratogenicity study in rats administered intravenously with elcatonin. Yakuri to Chiryo 21:4513–4530, 1993B.

Takahashi, M.; Sakurai, T.; Ohsuka, Y. and Kobayashi, Y.: Terratogenicity study in rabbits administered intravenously with elcatonin. Yakuri to Chiryo 21:4531–4537, 1993C.

387 Calcium CAS 7440–70–2

Heinonen et al. (1977) in a survey of 1007 women taking calcium compounds in the first trimester found 10 defects of the central nervous system while 4.7 were expected. The type of defect was not confined to any specific deformity. Shackelford et al. (**1993**) fed rats up to 1.25 % dietary calcium carbonate for 6 weeks before mating and during gestation and found no adverse reproductive effects.

Heinonen, O.P.; Slone, D. and Shapiro, S.: Birth Defects and Drugs in Pregnancy. Publishing Sciences Group Inc., Littleton, Mass., 1977.

Shackelford, M.E.; Collins, T.F.X.; Welsh, J.J.; Black, T.N.; Ames, M.J.; Chi, R.K. and O'Donnell, M.W.: Foetal development in rats fed ain–76A diets supplemented with excess calcium. Fd. Chem. Toxic. 31:953–961, 1993.

388 Calcium Cyclamate CAS 139–06–0

Discussed under Cyclamate

389 Californium [252]

Satow et al. (**1989**) exposed rats on days 8 or 9 to 20–120 rad and found that the radiation effect on the fetus was stronger than Co^{60}. Cardiovascular defects were the most common type produced.

Satow, Y.; Lee, J.Y.; Hori, H; Okuda, H.; Tsuchimoto, S.; Sawada, S. and Yokoro, K.: Teratogenic effect of californium [252] irradiation in rats. J. Radiat. Res. 30(2):155–163, 1989.

390 Camphor

No increase in congenital malformations was found among 168 pregnancies where topical application occurred in the first four lunar months (Heinonen et al., 1977).

Heinonen, O.P.; Slone, D. and Shapiro, S.: Birth Defects and Drugs in Pregnancy. Publishing Sciences Group Inc., Littleton, Mass., 1977.

391 Canadine

Canadine administered intraperitoneally at a dose of 500 mg per kg in rats on the 8th day of pregnancy led to the death of animals and increased postimplantation death of fetuses up to 28%. A single 50 mg per kg dose (1–17 days of pregnancy) increased preimplantation death of embryos up to 32%, especially if administered in the early period of pregnancy (1–7 days). The 10 mg per kg dose showed no embryotoxic effect. In none of the above cases there was a teratogenic effect.

Akhmedkhodzhaeva, K.S.: Teratogenic and embryotoxic action of canadine. Dokl. Akad. Nauk UzSSR 5:50–52, 1989.

392 Cancer Chemotherapy

Li et al. (1979) reviewed pregnancy outcome of 84 women and 62 men who had cancer treatment before conceiving. There were 242 live births and no increase in defect rate. One offspring had retinoblastoma and one, whose mother was treated by X–ray for brain tumor, had myelocytic leukemia. Seiber and Adamson (1975) reviewed the subject comprehensively. McKeen et al. (1979) analyzed the reproductive histories of 37 women with Hodgkins disease. Of 17 whose pregnancies occurred before treatment, one had a minor abnormality. Of 14 treated during pregnancy, five had elective abortions, one had a spontaneous abortion and four had normal newborns. Four of the 14 offspring had defects (cleft palate, hydrocephalus, inner ear defect, learning disability). Forty–four had pregnancies after treatment; two offspring had serious defects (hydrocephalus, tracheomalacia) and three had minor defects. Green et al. (**1991**) studied 202 pregnancies occurring after chemotherapy in childhood or adolescence and concluded that the incidence of congenital anomalies was not increased. Two of 20 patients who received dactinomycin had infants with congenital heart diseases.

Blatt et al. (1980) reviewed 40 pregnancies in 30 women who received aggressive chemotherapy. There were 10 elective abortions and 2 spontaneous abortions. No major malformations occurred in the 28 newborns, but 83 percent were conceived after therapy. Murray et al. (1984) summarized the effect of various antineoplastic drugs in 164 women treated during the first trimester and 76 treated in the second and third trimesters. No abnormalities were reported in the women treated in the second and third trimesters, but 19 of the 164 fetuses from the first trimester therapy were malformed. Cyclophosphamide, busulfan and aminopterin accounted for most of the fetuses with malformations. Green and Hall (1988) followed the reproduction of 93 former male or female patients with Hodgkins' disease. Among 33 pregnancies, there was no increase in malformations. Saurel–Cubizolles et al. (**1993**) reported that nurses exposed to chemotherapy had a higher risk for ectopic pregnancy (RR=10.0 CI 2.1–56.2). They regarded their findings as preliminary.

Senturi et al. (1985) studied the offspring of 27 fathers treated before conception for testicular cancer with radiotherapy and 25 treated with chemotherapy. No increase in adverse reproductive outcome was detected. The chemotherapy varied but vinblastine, bleomycin and cis-platin were often used.

Gulati et al. (1986) studied 18 children who were born after chemotherapy. Of 14 children born to parents treated before chemotherapy, one had trisomy 13–15 and another was stillborn with multiple congenital defects. Two offspring of women treated during pregnancy failed to thrive after birth. Mulvihill et al. (1987) studied 2,308 survivors of cancer therapy during childhood or adolescence and found only seven had other cancers later. This did not differ from the rate in siblings. Follow–up time was generally to 20 years. Dr. Mulvihill maintains an informational database on this subject at the National Cancer Institute. Doll et al. (1989) have summarized the outcome of 184 pregnancies exposed to chemotherapy. Dividing the agents into alkalating, antimetabolite, miscellaneous with plant alkaloids and antibiotics they found malformation rtes of 14–17%. Pregnancies with combined treatment had a malformation rate of 16%.

Blatt, J.; Mulvihill, J.J.; Ziegler, J.L.; Young, R.C. and Poplack, D.G.: Pregnancy outcome following cancer chemotherapy. Am. J. Med. 69:828–832, 1980.

Doll, D.C.; Ringenberg, Q.S. and Yarbro, J.W.: Antineoplastic agents and pregnancy. Seminars in Oncology 16:337–346, 1989.

Green, D.M. and Hall, B.: Pregnancy outcome following treatment during childhood or adolescence for Hodgkin's disease. Pediatr. Hematol. Oncol. 5:269–277, 1988.

Green, D.M.; Zevon, M.A.; Lowrie, G.; Siegelstein, N. and Hall, B.: Congenital anomalies in children of patients who received chemotherapy for cancer in childhood and adolescence. The New England Journal of Medicine 325:141–146, 1991.

Gulati, S.K.; Vega, R.; Gee, T.; Koziner, B. and Clarkson, B.: Growth and development of children born to patients after cancer therapy. Cancer Investigation 4:197–205, 1986.

Li, F.P.; Fine, W.; Jaffe, N.; Holmes, G.E. and Holmes, F.F.: Offspring of patients treated for cancer in childhood. J. Nat. Cancer Inst. 62:1193–1197, 1979.

McKeen, E.A.; Mulvihill, J.J.; Rosner, F. and Zarrabi, M.H.: Pregnancy outcome in Hodgkins disease. Lancet 2:590 only, 1979.

Mulvihill, J.J.; Connelly, R.R.; Austin, D.F.; Cook, J.W.; Holmes, F.F.; Krauss, M.R.; Meigs, J.W.; Steinhorn, S.C.; Teta, M.J.; Myers, M.H.; Byrne, J.; Bragg, K.; Hassinger, D.W.; Homes, G.F.; Latourette, H.B.; Naughton, M.D.; Strong, L.C. and Weyer, P.J.: Cancer in offspring of long–term survivors of childhood and adolescent cancer. Lancet 2:813–817, 1987.

Murray, C.L.; Reichert, J.A.; Anderson, J. and Twiggs, L.B.: Multimodal cancer therapy for breast cancer in the first trimester of pregnancy. JAMA 252:2607–2608, 1984.

Saurel–Cubizolles, M.J.; Job–Spira, N. and Estryn–Behar, M.: Ectopic pregnancy and occupational exposure to antineoplastic drugs. The Lancet 341:1169–1171, 1993.

Seiber, S.M. and Adamson, R.H.: Toxicity of Neoplastic Agents in Advances in Cancer Research, Academic Press Inc., New York 2:57–155, 1975.

Senturi, Y.D.; Peckham, C.S. and Peckham, M.J.: Children fathered by men treated for testicular cancer. Lancet 2:766–769, 1985.

393 Canreonate *Spironolactone Aldactone*™ SC–14266 CAS 2181–04–6

This diuretic and aldosterone antagonist was studied in pregnant rats and mice at intraperitoneal doses of up to 80 mg per kg during days 8 to 14 (rats) and 7 to 13 (mice). No defects were produced, but at 80 mg per kg some resorptions occurred in mice (Miyakubo et al., 1977). Hecker et al. (1976) gave 40 mg to rats on days 13–21 and found shortening of the anogenital distances in the males.

Hecker, A.; Hasan, S.H.; Neuman, F. and Schering, A.G.: Effect of spironolactone on the sexual differentiation of rat fetuses. Acta Endocrinologica 215(5):32–78, 1976.

Miyakubo, T.; Saito, S.; Tokunaga, U.; Ando, H. and Namba, H.: Toxicological studies of SC–14266. 5. Effects on the pre– and postnatal development in mice and rats. (Japanese) Nichidai Igaku Zasshi 36:261–282, 1977.

394 Capobenate

Ariano et al. (**1971**) studied this drug in the rabbit using 1000 mg per kg intravenously and found no adverse fetal effects.

Ariano, M.; Cappellini, V.; Lucca, L.; Naimzada, M.K.;

Pe, A.; Sgorbati, M.; Tammiso, R. and Rognoni, F.: Tossicologia e teratogenesi. Riv. Farmarol. Ter. 2(Suppl):17–43, 1971.

395 Caprolactam CAS 105–60–2

Gad et al. (1987) treated pregnant rats on days 6–15 with 100, 500 and 1000 mg per kg orally. Some decrease in viability and skeletal variants were found at the highest dose. Rabbits treated on days 6–28 with up to 250 mg per kg did not have adverse reproductive outcomes.

Gad, S.C.; Robinson, K.; Serota, D.G. and Colpean, B.R.: Developmental toxicity studies of caprolactam in the rat and rabbit. J. Appl. Toxicol. 7(5):317–326, 1987.

396 Captan
N–Trichloromethylthio–4–cyclohexane–1–2–dicarboximide
CAS 133–06–2

A study of insecticide workers exposed mainly to Captan is reviewed under occupation–insecticide workers. Fabro et al. (1966) fed 80 mg per kg daily to rabbits on days 7 through 12 and found no congenital defects in the fetuses. Kennedy et al. (1968) found no teratogenicity using rats and hamsters. Verrett et al. (1969) injected 3 to 20 mg per kg into the air cell or yolk of chick eggs before incubation and observed defects of the nervous or skeletal system in 7.8 percent. Phocomelia and amelia were especially noteworthy.

Negative studies with this compound were reported in the pregnant beagle (Kennedy et al., 1975) and non–human primate (Vondruska et al., 1971). Vondruska et al. (1971) gave monkeys up to 75 mg per kg from the 21st through the 34th gestational day and found no adverse fetal effects.

Fabro, S.; Smith, R.L. and Williams, R.T.: Embryotoxic activity of some pesticides and drugs related to phthalimide. Food Cosmet. Toxicol. 3:587–590, 1966.

Kennedy, G.L.; Fancher, O.E. and Calandra, J.C.: An investigation of the teratogenic potential of captan, folpet and difolatan. Toxicol. Appl. Pharmacol. 13:420–430, 1968.

Kennedy, G.L.; Fancher, O.E. and Calandra, J.C.: Non-teratogenicity of captan in beagles. Teratology 11:223–226, 1975.

Verrett, M.J.; Matchler, M.K.; Scott, W.F.; Reynaldo, E.F. and McLaughlin, J.: Teratogenic effects of captan and related compounds in the developing chicken embryo. Ann. N.Y. Acad. Sci. 160:334–343, 1969.

Vondruska, J.F.; Fancher, O.E. and Calandra, J.C.: An investigation into the teratogenic potential of captan, folpet and difolatan in non–human primates. Toxicol Appl. Pharmacol. 18:619–624, 1971.

397 Captopril CAS 62571–86–2 *ACE Inhibitors*

Oligohydramnios and/or neonatal anuria, pulmonary hypoplasia, mild or severe intrauterine growth failure, persistent patent ductus arteriosus and fetal death have been associated with treatment with captopril or enalopril. Hypoplasia of the skull bones is a frequent finding. A summary and recommendations has been published by Brent and Beckman (**1991**). Barr (**1994**) has reviewed the subject and recommends the ACE inhibitors not be used in pregnant women. Bontroy et al. (1984) reported a pregnancy in which 100 mg per day was given for nephrotic symdrome and hypertension. Oligohydramnios was found at 32 weeks and at 34 weeks. The infant was delivered and had bradycardia and hypotension with patent ductus arteriosus. The serum potassium level was elevated. Another inhibitor of angiotensin conversion enzyme, enalapril, was also associated with renal failure in the newborn. Knott et al. (1989) reported an ultrasound diagnoses of renal dysplasia in a woman receiving 75 mg of captopril and other agents for renal transplantation. Autopsy of the intrauterine dead fetus did not confirm renal changes.

Barr and Cohen (**1991**) have detailed the changes in skull and renal pathology. The hypocalvaria involves the membrane bones in particular. The renal histology is characterized by dilated glomerular spaces and no differentiation between the proximal and distal convoluted tubules.

Pipkin et al. (1980) noted increased stillbirths in lambs exposed late in pregnancy. Rabbits treated orally from day 24 to 28 with 3.3 mg gave birth to 75 fetuses of which 47 were liveborn and 28 stillborn. Only five of the 86 control fetuses were stillborn. In rabbits treated on days 24–28 with 3.3 mg orally, gestation length and neonatal death were significantly increased (Pipkin et al., 1982). Al–Shabanah et al. (**1991**) gave rats 3,100 30 mg per kg on days 6–15. Food intake of the dam was decreased at all doses. At the highest dose partially ossified skulls were found in 23% of the fetuses. Growth retardation occurred in the fetuses but no increase in defects were identified.

Al–Shabanah, O.A.; Al–Harbi, M.M.; AlGharably, N.M.A. and Islan, M.W.: The effect of maternal administration of captopril on fetal development in rat. Research Communications in Chemical Pathology and Pharmacology 73:221–229, 1991.

Barr Jr., M.: Teratogen update: angiotensin–converting enzyme inhibitors. Teratology 50:399–409, 1994.

Barr Jr., M. and Cohen Jr, M.: ACE Inhibitor fetopathy and hypocalvaria: the kidney–skull connection. Teratology 44:485–495, 1991.

Bontroy, M–J.; Vert, P.; Hurault de Ligny, B. and Miton, A.: Captopril administration in pregnancy impairs fetal angiotensin converting enzyme activity and neonatal adaption. Lancet 2:935–936, 1984.

Brent, R.L. and Beckman, D.A.: Angiotensin–converting enzyme inhibitors, an embryopathic class of drugs with unique properties: Information for clinical teratology counselors. Teratology 43:543–546, 1991.

Knott, P.D.; Thorpe, S.S. and Lamont, C.A.R.: Congenital renal dysgenesis possibly due to captopril. Lancet 1:451, 1989.

Pipkin, F.B.; Turner, S.R. and Symonds, E.M.: Possible risk with captopril in pregnancy: Some animal data. Lancet 1:1257, 1980.

Pipkin, F.B.; Symonds, E.M. and Turner, S.R.: The effects of captopril (SQ 14225) upon mother and fetus in the chronically cannulated ewe and in the pregnant rabbit. J. Physiol. 323:415–422, 1982.

398 **Carbamates** *Ethyl Carbamate n–Propyl Carbamate Allyl Carbamate n–Butyl Carbamate β–Hydroxyethyl Carbamate Ethyl M–Methylcarbamate Ethyl N,N–Dimethylcarbamate Ethyl N–Hydroxycarbamate Diethyl Carbamate Urethan*

DiPaolo and Elis (1967) administered various carbamate compounds intraperitoneally to hamsters on the 8th day of pregnancy. The ethyl form (urethan) produced defects of the palate, brain, eye, extremities and omphalocele at microM doses of 5 to 17 per gm of body weight. The propyl form produced growth retardation at 4.8 micromoles per gm. No defects were produced with the allyl (5.0 micromoles), butyl (2 micromoles), β–hydroxyethyl (4.8 micromoles) forms. The ethyl N–methylcarbamate at 4 and 10 micromoles produced various defects including exencephaly and omphalocele. The ethyl N,N–dimethyl form produced some defects at 4.3 but not at 2.1 micromoles per gm. The ethyl N–hydroxy form produced defects of most organ systems at 2.4 micromoles per gm as did the diethyl form at 4.2 micromoles per gm.

Dipaolo, J.A. and Elis, J.: The comparison of teratogenic and carcinogenic effects of some carbamate compounds. Cancer Res. 27:1696–1701, 1967.

399 **Carbamazepine** *Tegretol*™ CAS 298–46–4

Rosa (**1991**) drawing on an eight year period of Medicaid recipients in Michigan found two offspring with spina bifida among 107 women who took carbamazepine alone as an antiepileptic. He estimated from these data and other cohorts that the risk of spina bifida is about one percent. In a preliminary report, Van Allen et al. (1988) reported that 10 of 21 infants exposed alone (15) or in combination with other medications (6), had a distinct phenotype which included round facies, upslanting palpebral fissures, hypertelorism, hypoplastic nasal bridge, short upturned nose, flamus nevus, large anterior fontanelle and variable nail hypoplasia. Two infants had microcephaly which developed postnatally.

Hiilesmaa et al. (1981) measured head circumference of the offspring of 139 women with epilepsy. In those exposed to carbamazepine or carbamazepine with phenobarbitone, there was a 7 mm and 6 mm reduction, respectively. This change persisted at 18 months but no congenital defect increase or mental retardation was reported. In the 55 infants exposed to phenytoin, there was no head circumference reduction. Bertollini et al. (1987) studied 70 pregnancies in which monotherapy was employed. Reduced body length, weight and head circumference was found in spite of normal gestational length. Vorhees er al (1990) gavaged 200,400 and 600 mg per kg on days 7–18 in the rat. At the two highest doses there was maternal toxicity at all doses and fetal weight was reduced. Malformations were increased at 400 and 600 mg ker kg but not at 200 mg per kg. The defects were generalized edema and skeletal in type.

Jones et al. (1989) identified prospectively 35 women who were on carbamazepine alone during pregnancy. Craniofacial defects consisting of upslanting eyes, long philtrum and short nose were present in 11 percent. Fingernail hypoplasia was found in 26 percent and developmental delay in 20 percent. One had congenital heart disease. There was an increase in infants with three or more minor congenital defects as reported by an examiner who did not know the prenatal exposure. Keller (1989), Scialli and Lione (1989),and Bortnichak and Wetter (1989) have pointed out lack of validity in the new behavior studies by Jones et al. because of lack of controls and a too narrow definition of delayed development. Gladstone et al. (**1992**) found in a prospective study that of 23 women treated with carbamazepine one gave birth to an infant with myleomeningocele. Kallen (**1994**) found an increase in spina bifida among exposed neonates O.R. 6.0 (CI 0.9–56.9) and recommended fetal ultrasound to exclude the risk of around 1%.

Kaneko et al. (1988) studied 172 pregnancies in which epileptic drugs were used. For the 31 in which only one drug was used, there was 6.5 percent malformations while among the remaining 141 patients which were treated with more than one drug, the rate was 15.6 percent. By Wilcoxon rank–sum tests, valproate use was significantly associated with a defect increase and carbamazepine was almost significantly associated (p = 0.053). Among the polydrug users who used carbamazepine, there were 11 malformations among 105. When primidone was used in polydrug use, there were 8 malformations among 80 cases. In this study of Japanese children, 4 facial clefts and 3 heart defects occurred among 172.

Briggs et al. (1983) listed publications giving the pregnancy outcome from 531 pregnancies. Kaneko et al. (1986) found seven malformations among 45 patients treated with carbamazepine and another drug (except valproic acid). This was a statistically significant increase. Millar and Nevin (1973) report one infant exposed to phenobarbital and carbamazepine with myelomenigocele. Schardein (1977) listed references which reported malformed offspring generally exposed to several anticonvulsants.

The pharmaceutical company distributing this compound reported two malformations in 135 rat fetuses whose mothers were exposed to 250 mg per kg daily (Paulson and Paulson, 1981).

Bertollini, R.; Kallen, B.; Mastroiacovo, P. and Robert, E.: Anticonvulsant drugs in monotherapy. Effect on the fetus. Eur. J. Epidemiol. 3:164–171, 1987.

Briggs, G.G.; Bodendorfer, T.W.; Freeman, R.K. and Yaffe, S.J.: Drugs in Pregnancy and Lactation. Williams and Wilkins, Baltimore, 1983.

Gladstone, D.J.; Bologa, M.; Maguire, C.; Pastuszak, A. and Koren, G.: Course of pregnancy and fetal outcome following maternal exposure to carbamazepine and phenytoin: a prospective study. Reproductive Toxicology 6:257–261, 1992.

Hiilesmaa, V.K.; Teramo, K.; Granstrom, M–L. and Bardy, A.H.: Fetal head growth retardation associated with maternal antiepiletic drugs. Lancet 2:165–167, 1981.

Jones, K.L.; Lacro, R.V.; Johnson, K.A. and Adams, J.: Pattern of malformations in the children of women treated with carbamazepine during pregnancy. N. Eng. J. Med. 320:1661–1666, 1989.

Kallen, A.J.B.: Maternal carbamazepine and infant spina bifida. Reproductive Toxicology 8:203–205, 1994.

Kaneko, S.; Fukushima, Y.; Sato, T.; Ogawa, Y.; Nomura, Y.; Ono, T.; Kan, R.; Nakane, Y.; Teranishi, T. and Goto, M.: Teratogenicity of antiepileptic drugs: A prospective study. Jpn J. Psych. Neurol. 40:447–450, 1986.

Kaneko, S.; Otani, K.; Fukushima, Y.; Ogawa, Y.; Nomura, Y.; Ono, T.; Nakane, Y.; Teranishi, T. and Goto, M.: Teratogenicity of antiepileptic drugs: Analysis of possible risk factors. Epilepsia 29:459–467, 1988.

Keller, D.M.; Scialli, A.R.; Lione, A.; Bortnichak, E.A. and Wetter, M.S.: Teratogenic effects of carbamazepine. New England Journal of Medicine 321:1480, 1989.

Millar, J.H.D. and Nevin, N.C.: Congenital malformations and anticonvulsant drugs. Lancet 1:328 only, 1973.

Paulson, G.W. and Paulson, R.B.: Teratogenic effects of anticonvulsants. Arch. Neurol. 38:140–143, 1981.

Schardein, J.L.: Drugs as Teratogens, CRC Press, Cleveland, 116, 1976.

Rosa, F.W.: Spina bifida in infants of women treated with carbamazepine during pregnancy. The New England Journal of Medicine 324:674–677, 1991.

Van Allen, M.I.; Yerby, M.; Leavitt, A.; McCormick, K.B. and Loewenson, R.B.: Increased major and minor malformations in infants of epileptic mothers: Preliminary results of the pregnancy and epilepsy study. Am. J. Hum. Gen. 43:A73, 1988.

Vorhees, C.V.; Acuff, K.D.; Weisenburger, W.P. and Minck, D.R.: Teratogenicity of carbamazepine in rats. Teratology 41:311–317, 1990.

400 Carbaryl 1–Naphthyl N–Methylcarbamate CAS 63–25–2

This carbamate cholinesterase inhibitor used as a pesticide was studied by Robens (1969) in the guinea pig, hamster and rabbit. Only in the guinea pig were defects produced. Using a near lethal dose for the mother (300 mg per kg) from day 11 to day 20, she found vertebral anomalies in about one–half of the surviving fetuses. Weil et al. (1973) found no teratogenicity in the rat (200 mg per kg orally).

Smalley et al. (1968) fed 3.1 to 50 mg per kg to beagles during pregnancy. At levels of 6.25 and above, the defect rate was increased in the offspring. Midline abdominal wall defects and skeletal defects were the most common type. Reduced conception was found at the 50 mg per kg level.

Robens, J.F.: Teratologic studies of carbaryl, diazinon, norea, disulfiram and thiram in small laboratory animals. Toxicol. Appl. Pharmacol. 15:152–163, 1969.

Smalley, H.E.; Curtis, J.M. and Earl, F.L.: Teratogenic action of carbaryl on beagle dogs. Toxicol. Appl. Pharmacol. 13:392–403, 1968.

Weil, C.S.; Woodside, M.D.; Bernard, J.B.; Condra, N.I.; King, J.M. and Carpenter, C.P.: Comparative effect of carbaryl on rat reproduction and guinea pig teratology when fed either in the diet or by stomach intubation. Toxicol Appl. Pharmacol. 26:621–638, 1973.

401 Carbazochrome Sodium Sulfonate

1–Methyl–5–semicarbazo–ono–6–oxo–2,

3,5,6–tetrahydroindole–3–sulfonate sodium CAS 51460–26–5

Fujii and Kowa (1970) gave this adrenochrome derivative to pregnant mice and rats on the 7th through the 14th days of gestation. Both the intraperitoneal and oral routes were used and the dosages were as high as 3,000 mg per kg, orally and 1,600 mg per kg, intraperitoneally. No increase in malformations was found.

Fujii, T.M. and Kowa, Y.: Teratological studies of carbazochrome sodium sulfonate in mice and rats. Oyo Yakuri 4:39–46, 1970.

402 Carbendazim CAS 10605–21–7

Methyl benzimidazole carbamate, a precursor of benomyl, was gavage fed to rats and rabbits in amounts of up to 80 and 160 mg per kg, respectively. The rats were treated on days 5–15 and the rabbits on days 6–18. No increase in defects was observed at 20 mg per kg in the rat and 40 mg in the rabbit (Janardhan et al., 1984).

Janardhan, A.; Sattur, P.B. and Sisodia, P.: Teratogenicity of methyl benzimidazole carbamate in rats and rabbits. Bull. Environ. Contam. Toxicol. 33:257–263, 1984.

403 Carbenoxolone CAS 5697–56–3

Pinder et al. (1976) reported briefly that this anti–ulcer medication given in amounts of up to 125 mg per kg at various times in gestation in mice, rabbits and rats produced no adverse fetal effects in mice, rabbits and rats.

Pinder, R.M.; Brogden, R.N.; Sawyer, P.R.; Speight, T.M.; Spencer, R. and Avery, G.S.: Carbenoxolone: A review of its pharmacological properties. Drugs 11:245–307, 1976.

404 Carbinoxamine Diphenyldisulfonate

Maruyama and Yoshida (1968) administered this antihistamine orally to rats and mice at maximum doses of 80 and 100 mg per kg during organogenesis. No teratologic changes occurred, but there was some decrease in surviving mouse fetuses and a reduction of fetal weight in the rat offspring.

Maruyama, H. and Yoshida, S.: Pharmacology of a new antihistamine, carbinoxamine diphenyldisulfonate. 2. Toxicity and influence on fetuses. J. Med. Soc. Toho Univ. 15:367–374, 1968.

405 Carbocysteine S–Carboxymethyl Cysteine CAS 638–23–3

Ito et al. (1977A,B) studied this expectorant in pregnant rats and rabbits during active organogenesis. No harmful effects were noted with maximum daily doses of 500 mg and 250 mg per kg in the rat and rabbit. Kawabata and Sugimoto (1979) gave this expectorant by mouth to rats on days 17–21 of gestation and for 21 days after birth. No adverse effects on postnatal development were found.

Ito, R.; Toida, S.; Matsuura, S.; Tanihata, T.; Hidano, T.; Miyamoto, K.; Matsuura, M. and Nakai, S.: A safety study in reproduction and teratology of S–carboxymethyl cysteine: A new organogenesis, perinatal and lactation in rats. (Japanese) J. Med. Soc. Toho Univ. 24:62–66, 1977A.

Ito, R.; Toida, S.; Tanihata, T.; Matsuura, S.; Miyamoto, K.; Matsuura, M. and Nakai, S.: A safety study of teratogenicity in organogenesis in rabbits of S–carboxymethyl cysteine: A new expectorant. (Japanese) J. Med. Soc. Toho Univ. 24:663–666, 1977B.

Kawabata, R. and Sugimoto, T.: Perinatal and postnatal study of S–CMC in rats. Effect on function reproductive function of the offspring. Kiso to Rinsho 18:1311–1317, 1979.

406 Carbofuran *Dihydro–2, 2–dimethyl–7–benzofuranol– methylcarbamate* CAS 1563–66–2

McCarthy et al. (1971) reported that although no increase in defects occurred in the offspring of treated rats, rabbits and dogs, a reduced survival rate was seen in rat offspring exposed to 100 ppm in the diet. Wolfe and Esher (1980) repeated this dietary dose level in mice and found a reduced survival rate but the difference was not significant. Fetal weights were not reduced. Postnatal plasma corticosterone levels were elevated in the offspring of mice exposed in utero to 0.01 or 0.50 mg per kg per day (Cranmer et al., 1978). A transplacental effect on activity of acetylcholinesterase of the day 18 rat fetus was shown (Cambon et al., 1979). Courtney et al. (1985) gavaged rats with up to 20 mg per kg and found no adverse effects.

Cambon, C.; Declume, C. and Derache, R.: Effect of the insecticidal carbamate derivatives (carbofuran, pirimicard, aldicarb) on the activity of acetylcholinesterase in tissues from pregnant rats and fetuses. Toxicol. Appl. Pharmacol. 49:203–208, 1979.

Courtney, K.D.; Andrews, J.E.; Springer, J. and Dalley, L.: Teratogenic evaluation of the pesticides baygon, carbofuran, dimethoate and EPN. J. Environ. Sci. Health B20:373–406, 1985.

Cranmer, J.P.; Avery, D.L.; Grudy, R.R. and Kitay, J.I.: Endocrine dysfunction resulting from prenatal exposure to carbofuran, diazinon and chlordane. J. Environ. Path. Toxicol. 2:357–369, 1978.

McCarthy, J.F.; Fancher, O.E.; Kennedy, G.L.; Keplinger, M.L. and Calandra, J.C.: Reproduction and teratology studies with the insecticide carbofuran. (abs) Toxicol. Appl. Pharmacol. 19:370 only, 1971.

Wolfe, J.L. and Esher, R.J.: Toxicity of carbofuran and lindane to the old–field mouse (P. gossypinus). Bull. Environ. Contam. Toxicol. 24:894–902, 1980.

407 Carbomethoxythiazide

6–Chloro–3–carbomethoxy–3, 4–dihydro–2–methyl– 2H–1,2,4–benzothia diazine–7–sulfonamide–1–dioxide CAS 42583–55–1

Takai et al. (1973) gave this antihypertensive drug to pregnant mice and rats orally in amounts up to 3,000 mg per kg per day during active organogenesis. Although some delay in fetal ossification was seen, no increase in gross malformations occurred. The neonates were also followed for three weeks after birth without appearance of significant differences from the control group.

Takai, A.; Nakada, H. and Yoneda, T.: A pharmacological study on 6–chloro–3–carbomethoxy-3,4–dihydro–2–methyl–2H–1,2, 4–benzothiadiazine–7–sulfonamide–1,1–fiocifr (DU 5747), a new antihypertensive agent. 5. Teratological studies of DU 5747. (Japanese) Oyo Yakuri 7:267–274, 1973.

408 Carbon Dioxide CAS 124–38–9

Haring (1960) exposed rats during a single gestational day to 6 percent CO_2, 20 percent oxygen and 74 percent nitrogen and observed 23 percent cardiac malformations in the offspring as compared to 6.8 percent in controls. The highest incidence of the defect occurred on the 10th day. The cardiac lesions were characterized as due to overgrowth. The majority of these lesions were partial transposition or ventricular outflow stenosis, but interventricular septal defects also were observed.

Grote (1965) administered 10–13 percent CO_2 to rabbits between the 7th and 12th days of gestation and found 16 fetuses of 67 to have defects of the vertebral column. Only one abnormal animal was found in the 30 controls. He was not able to determine the role of indirect O_2 deficiency in this experiment. Storch and Layton (1971) were unable to mimic the teratogenic effect of acetazolamide in hamsters by giving 10 percent carbon dioxide in the atmosphere.

Weaver and Scott (1984) found ectrodactyly in mice of the C57BL–6J strain following exposure to 20 percent carbon dioxide on day 10. No such changes occurred in SWV mice. Metabolic acidosis with ammonium chloride produced a lower increased frequency.

Grote, W.: Stoerung der Embryonalentwicklung bei erhoetem CO_2 und O_2. Partialdruck und bie Unterdruck. 2. Morphol. Anthropol. 56:165–194, 1965.

Haring, O.M.: Cardiac malformations in rats induced by exposure of the mother to carbon dioxide during pregnancy. Circ. Res. 8:1218–1227, 1960.

Storch, T.G. and Layton, W.M.: The role of hypercapnia in acetazolamide teratogenesis. Experientia 27:534–535, 1971.

Weaver, T.E. and Scott, W.J.: Acetazolamide teratogenesis. Interaction of maternal metabolic and respiratory acidosis in the induction of ectrodactyly in C57BL–6J mice. Teratology 30:195–202, 1984.

409 Carbon Disulfide CAS 75–15–0

Tabacova et al. (1978) exposed rats to 50, 100 and 200 mg per square meter throughout pregnancy and found increased hydrocephalus and club foot at 50 mg and above. In a later paper, Tabacova et al. (1981) found permanent postnatal changes in offspring exposed prenatally to 10 mg per square meter. Hexobarbital sleeping times were increased. Activity was increased and T–maze learning decreased. Reproduction was not affected. At a level of 0.03 mg per square meter only a transient mild motor hyperactivity was found.

Tabacova, S.; Hinkova, L. and Balabaeva, L.: Carbon

disulfide teratogenicity and postnatal effects in rat. Toxicology Letters 2:129–133, 1978.

Tabacova, S.; Hinkova, L. and Balabaeva, L.: Hazards for the progeny after maternal exposure to low carbondisulfide concentrations. G. Ital. Med. Lav. 3:121–125, 1981.

410 Carbon Monoxide CAS 630–08–0

Koren et al. (**1991**) in a multicenter prospective study reported no adverse effects among 31 fetuses exposed to mild or moderate poisoning (i.e. mother was alert). In 3 of 5 women with severe toxicity two stillbirths and one infant with cerebral palsy occurred. Two mothers with severe poisoning were treated with hyperbaric oxygen and had normal infants.

Carbon monoxide poisoning of the mother resulted in impaired development of the central nervous system of the human fetus. Longo (1977) reviewed the biologic effects of carbon monoxide on pregnancy. Warkany (1971) in his review of the subject could not detect any clear correlation between time of exposure and symptomatology of the children. Beaudoin et al. (1969) also reviewed the subject.

Schwetz et al. (1979) found no teratogenicity from 250 ppm exposure for 7 of 24 hours between days 6 through 15 in mice and 6 through 18 in rabbits. Some increase in minor skeletal anomalies was associated in mice with the exposure. Astrup et al. (1972) exposed rabbits during pregnancy to 180 ppm carbon monoxide. Perinatal death occurred in 43 of 123 treated offspring but in only one of a comparable control group. The birth weight was approximately 10 gm less in the treated group and 3 had defects of their extremities. A group of rabbits exposed to 90 ppm gave birth to smaller offspring but the increase in perinatal death was less dramatic. Fechter and Annau (1977) reported behavioral changes in the offspring of rats exposed to concentrations of carbon monoxide producing only 15 percent maternal carboxyhemoglobin concentrations. Singh and Moore–Cheatum (**1993**) found that protein deficiency in mice enhanced fetotoxicity of carbon monoxide.

Ginsberg and Meyers (1976) studied the effect of carbon monoxide on the fetal monkey brain. Brain swelling and hemorrhagic necrosis were found. Acute exposure studies in late fetal monkeys produced hemorrhagic necrosis of the fetal cerebral hemispheres at levels which were tolerated by the mothers (Ginsberg and Myers, 1974).

Robkin et al. (1976) could not produce alteration in fetal heart or viability in explanted rat embryos exposed to as much as 10 percent carbon monoxide. Robkin and Cockroft (1978) showed that carbon monoxide causes a shift toward anerobic glucose metabolism in exposed rat embryos. Mactutus and Fechter (1985) exposed rats to 150 ppm during their entire gestation and observed permanent memory deficits in the offspring. Singh (1986) exposed mice to 65 and 125 ppm from days 7–18 and found that the offspring had significant deficits in three different tests for righting. Singh and Scott (1984) found reduced fetal weight in mice after exposures of 150 ppm and higher during organogenesis. Fechter (1984) reviewed the subject. Singh et al. (**1993**) found that protein deficiency in mice increased teratogenic-

ity and developmental toxicity of carbon monoxide.

Astrup, P.; Trolle, D.; Olsen, H.M. and Kjeldsen, K.: Effect of moderate carbon–monoxide exposure on fetal development. Lancet 2:1220–1222, 1972.

Beaudoin, A.; Gachon, J.; Butin, L–P. and Bost, M.: Les consequence foetales de l'intoxication oxycarbonee de la mere. Pediatrie 24:459–461, 1969.

Fechter, L.D.: Toxicity of carbon dioxide exposure in early development. In: Toxicology and the Newborn, S. Kacew and M.J. Reasoner (eds.), Elsevier Sc. Publ., Amsterdam, 122–140, 1984.

Fechter, L.D. and Annau, Z.: Toxicity of mild prenatal carbon monoxide exposure. Science 197:680–682, 1977.

Ginsberg, M.D. and Myers, R.E.: Fetal brain damage following maternal carbon monoxide intoxication. An experimental study. Acta Obstet. Gynecol. Scand. 53:309–317, 1974.

Ginsberg, M.D. and Myers, R.E.: Fetal brain injury after maternal carbon monoxide intoxication. Neurology 26:15–23, 1976.

Koren, G.; Sharav, T.; Pastuszak, A.; Garrettson, L.K.; Hill, K.; Samson, I.; Rorem, M.; King, A. and Dolgin, J.E.: A multicenter, prospective study of fetal outcome following accidental carbon monoxide poisoning in pregnancy. Reproductive Toxicology 5:397–403, 1991.

Longo, L.D.: The biological effects of carbon monoxide in the pregnant woman, fetus and newborn infant. Am. J. Obstet. Gynecol. 129:69–103, 1977.

Mactutus, C.F. and Fechter, L.D.: Moderate prenatal carbon monoxide produces persistent, and apparently permanent, memory deficits in rats. Teratology 31:1–12, 1985.

Robkin, M.A.; Beachler, D.W. and Shepard, T.H.: Effect of carbon monoxide on early rat embryo heart rates. Environ. Res. 12:32–37, 1976.

Robkin, M.A. and Cockroft, D.L.: The effects of carbon monoxide on glucose metabolism and growth of rat embryos. Teratology 18:337–342, 1978.

Schwetz, B.A.; Smith, F.A.; Leong, B.K.J. and Staples, R.E.: Teratogenic potential of inhaled carbon monoxide in mice and rabbits. Teratology 19:385–392, 1979.

Singh, J. and Scott, L.H.: Threshold of carbon monoxide induced fetotoxicity. Teratology 30:253–259, 1984.

Singh, J.: Early behavioral alterations in mice following prenatal carbon monoxide exposure. Neurotoxicol. 7:475–482, 1986.

Singh, J.; Aggison, Jr., L. and Moore-Cheatum, L.: Teratogenicity and developmental toxicity of carbon monoxide in protein–deficient mice. Teratology 48:149–159, 1993.

Singh, J. and Moore–Cheatum, L.: Gestational protein deficiency enhances fetotoxicity of carbon monoxide. Maternal Nutrition and Pregnancy Outcome. Ann. N.Y. Acad. Sci. 678:366–368, 1993.

Warkany, J.: Congenital Malformations Notes and Comments. Chicago: Year Book Medical Publishers, 128 only, 1971.

411 Carbon Tetrachloride CAS 56–23–5

Wilson (1954) administered carbon tetrachloride on 2

or 3 days of gestation by mouth (0.3cc) or subcutaneously (0.8cc) and caused no congenital defects in the rat offspring. Adams et al. (1961) noted some degeneration of embryonic discs from rabbit blastocysts exposed in vivo to 1.01 ml per kg. Schwetz et al. (1974) exposed pregnant rats to 300 and 1000 ppm for 7 hours on days 6 through 15 of gestation. Fetal size was reduced but neither resorptions nor malformations were increased over the control.

Transplacental passage was shown in the pregnant mouse (Roschlau and Rodenkirchen, 1969), rat (Tsirel'nikov and Dobrovol'ska, 1973) and human (Dowty and Laseter, 1976).

Adams, C.E.; Hay, M.F. and Lutwak–Mann, C.: The action of various agents upon the rabbit embryo. J. Embryol. Exp. Morphol. 9:468–491, 1961.

Dowty, B.J. and Laseter, J.L.: The transplacental migration and accumulation in blood of volatile organic constituents. Pediatr. Res. 10:696–701, 1976.

Roschlau, G. and Rodenkirchen, H.: Histologische Untersuchungen uber die diaplazentare Wirkung von Tetrachlorkohlenstaff und Ally–alkohol ieber auf Mauesefeten. Exp. Path. 3:255–263, 1969.

Schwetz, B.A.; Leong, B.K.J. and Gehring, P.J.: Embryo and fetotoxicity of inhaled carbon tetrachloride, 1,1–dichloroethane and methyl ethyl ketone in rats. Toxicol. Appl. Pharmacol. 28:452–464, 1974.

Tsirel'nikov, N.I. and Dobrovol'ska, S.G.: Morphohistochemical investigation of the embryonic liver after CC14 administration at various stages of ontogeny. Bull. Exp. Biol. Med. 76:1467–1469, 1973.

Wilson, J.G.: Influence on the offspring of altered physiologic states during pregnancy in the rat. Ann. N.Y. Acad. Sci. 57:517–525, 1954.

412 Carbophos

Carbophos is widely used phosphoorganic pesticide. Female rats were subjected to inhalation of carbophos (0.35-1.8 mg/m^3) and formaldehyde (0.4-2.4 mg/m^3) 4 hours a day during the whole pregnancy. Another group of rats received the same mixture at +35 degrees C and air humidity 85–95%. Postimplantation resorptions of fetuses was observed and the rate of skeletal bone ossification was retarded. Embryotoxic effects increasd and reached maximum when the temperature and air humidity was elevated.

Korshunova, E.P.: Hygienic evaluation of embryotoxic effects of carbophos and formaldehyde mixture separated and combined with elevated temperature and air humidity. Gig. Sanit. 10:13–15, 1988.

413 Carboplatin CAS 41575–94–4

Kai et al. (1988) gave this oncostatic drug intervenously to rats on days 7–17. At the highest dose (4 mg per kg) maternal toxicity and lowered fetal weight occurred. Postnatal studies of behavior indicated no adverse effects.

Kai, S. ; Kohmura, H.; Ishikawa, K.; Takeuchi, Y.; Ohta, S.; Kuroyanagi, K.; Kadota, T.; Kawano, S.; Chikazawa, H.; Konda, H.; Sakai, A. and Takahashi, N.: Reproductions

studies ofcarboplatin (ii) intravenous administration to rats during the period of fetal organogenesis. The Journal of Toxicological Sciences 13:35–61, 1988.

414 Carboxyethylgermanium Sesquioxide GE–132 CAS 12758–40–6

Nagari et al. (1980) gave this compound intraperitoneally at 500 and 1000 mg per kg per day in a three generation study and found no adverse effects. Shimpo and Mori (1980) gave rabbits intravenously up to 1000 mg per kg during organogenesis without fetal ill effects.

Nagari, H.; Hasegawa, K. and Shimpo, K.: Reproductive study of rats intraperitoneally treated with carboxyethylgermanium sesquioxide (GE–132). Oyo Yakuri 20:271–280, 1980.

Shimpo, K. and Mori, N.: Teratogenicity test of carboxyethylgermanium sesquioxide (GE–132) given during the period of organogenesis in rabbits. Oyo Yakuri 20:675–679, 1980.

415 Carboxymethyl Ethylcellulose

Ohkuma et al. (1985) administered up to 5000 mg per kg orally to rats on days 7–17 and found no adverse fetal effects.

Ohkuma, H.; Tanabe, M. and Maehashi, H.: Teratogenicity study of carboxymethyl ethylcellulose in rats. Oyo Yakuri 30(4):677–685, 1985.

416 N–[N–(4–Carboxyphenyl)glycyl]aminoacetonitrile CAS 19065–92–0

This drug used for chronic hepatitis was given by Irikura et al. (1975) orally to mice and rabbits during organogenesis and no teratologic effects were seen. The maximum dose in the mouse was 2,500 mg per kg and 1,500 in the rabbit.

Irikura, T.; Sugimoto, T.; Hosomi, J. and Suzuki, H.: Teratological studies on N–[N–(4–carboxyphenyl)glycyl]aminoacetonitrile in mice and rabbits. Oyo Yakuri 9:523–534, 1975.

417 Carbutamide 1–Butyl–3–sulfanilylurea CAS 339–43–5

Demeyer and Isaac–Mathy (1958) found cleft palate, eye defects and generalized edema in rat fetuses whose mothers received 500 mg per day for 1, 2 or 3 days during the embryonic period. Tuchmann–Duplessis and Mercier–Parot (1959a) using 800 mg per kg from the 1st through 12th day in rats produced 10 to 33 percent malformed fetuses. The majority of the defects involved the eye and included absence of the optic nerve. Some nervous system defects were reported. The same authors (1959b) noted that carbutamide sulfylurea was more teratogenic in rats than other hypoglycemia sulfylureas. Demeyer (1961) caused cleft palates and hydrocephalus with this substance in rats. Bariljak (1967) found that sodium carbonate reduced this drug's teratogenic and lethal action in the rat.

Bariljak, I.R.: The influence of natrium hydrocarbonate on teratogenic activity of oranil (carbutamide). Pharmacol. Toxicol. (Russian) 5:631–633, 1967.

Demeyer, R.: Etude experimentale de la glycoregulation gravidique et de l'action teratogene des perturbations du metabolisme glucidique. Arcia, Bruxelles: Paris, Masson Et Cie, 175–183, 1961.

Demeyer, R. and Isaac–Mathy, M.: A propos de l'action teratogene d'un sulfamide hypoglycemiant. Ann. Endocrinol. 19:167–172, 1958.

Tuchmann–Duplessis, H. and Mercier–Parot, L.: Sur l'action teratogene d'un sulfamide hypoglycemiant. Etude experimentale chez la rat. J. Physiol. 51:65–83, 1959a.

Tuchmann–Duplessis, H. and Mercier–Parot, L.: Influence de divers sulfamides hypoglycemiants sur le developpement de l'embryon. Etude experimentale chez le rat. Acad. Nat. Med. (Paris) 143:238–241, 1959b.

418 Carcinogens

There are a large number of carcinogenic chemicals which are also teratogenic. Dipaolo and Koten (1966) reviewed this subject comprehensively. Several reviews on the subject of transplacental carcinogenesis appeared (Alexandrov, 1973; Druckery, 1973; Rice, 1973).

Alexandrov, V.A.: Embryotoxic and teratogenic effects of chemical carcinogens. In: Transplacental Carcinogenesis. L. Tomatis and U. Mohr (eds.), International Agency For Research in Cancer Scientific Publication (4), Lyons, 112–126, 1973.

Dipaolo, J.A. and Kotin, P.: Teratogenesis–oncogenesis: A study of possible relationships. Arch. Pathol. 81:3–23, 1966.

Druckrey, H.: Chemical structure and action in transplacental carcinogenesis and teratogenesis. In: Transplacental Carcinogenesis. L. Tomatis and U. Mohr (eds.), International Agency For Research in Cancer Scientific Publication (4), Lyons, 45–57, 1973.

Rice, J.M.: An overview of transplacental chemical carcinogenesis. Teratology 8:113–126, 1973.

419 Carcinolipin *Cholesteryl–14–methylhexadecanoate* CAS 19477–24–8

Shabad et al. (1973) gave this compound intraperitoneally during the last one–third of rat pregnancy in total doses of 5 to 18 mg. A fourfold increase in lung tumors was found in the postnatal animals.

Shabad, L.M.; Kolesnichenko, T.S. and Savluchinskaya, L.A.: Transplacental effect of carcinolipin in mice. Neoplasma 20:347–348, 1973.

420 Cargutocin *Deamino–dicarba–(gly–7)oxytocin* CAS 33605–67–3

This oxytocin–like drug was given intravenously to pregnant rats and rabbits in maximum daily doses of 100 and 10 units in the rat and rabbit, respectively. Treatment during major organogenesis was not associated with any fetal ill effects (Hamada et al., 1979).

Hamada, Y.; Imanishi, M.; Onishi, R. and Hashiguchi, M.: Teratogenicity study of cargutocin in rats and rabbits. Iyakuhin Kenkyu 10:26–40, 1979.

421 Carica papaya *Papain Papaya*

Schmidt (**1995**) gave rats in the first week or during organogenesis up to 800 mg per kg orally and found no maternal toxicity or reproductive ill effects. A relatively pure and standardized form of the compound was used. Garg and Garg (1970) tested extracts of the pulp of the unripe fruit and found that at 500 mg per kg on days 1–7 the implantation of rats was decreased 60 percent.

Garg, S.K. and Garg, G.P.: Antifertility screening of plants. part VII. Effect of five indigenous plants on early pregnancy in albino rats. Ind. J. Med. Res. 59:302–306, 1970.

Schmidt, H.: Effect of papain on different phases of prenatal ontogenesis in rats. Reproductive Toxicology 9:49–55, 1995.

422 Carmine *Cochineal* CAS 1390–65–4

Grant et al. (1987) gave rats up to 1000 mg per kg by gavage of carmine of cochineal with no untoward fetal results. Schluter (1970) injected lithium or sodium carmine into mice on the 8th gestational day and produced fetuses with malformation of the ribs and vertebral column as well as exencephalies. He demonstrated that the carmine was localized in the visceral yolk sac epithelium.

Grant, D.; Gaunt, I.F. and Carpanini, F.M.B.: Teratogenicity and embryotoxicity study of carmine of cochineal in the rat. Food Chem. Toxicol. 25(12):913–917, 1987.

Schluter, G.: Ueber die embryotoxische Wirkung von Carmin bei der Maus. Z. Anat. Entivicklungsgesch. 131:228–235, 1970.

423 Carminomycin

Damjanov and Celluzzi (1980) treated mice on days 8–11 with up to 3.5 mg per kg subcutaneously or 0.5 mg per kg intraperitoneally. Fetal survival was reduced along with fetal weight but no teratogenicity was clearly identified.

Damjanov, I. and Celluzzi, A.: Embryotoxicity and teratogenicity of the anthracycline antibiotic carminomycin in mice. Res. Commun. Chem. Pathol. Pharmacol. 28:497–504, 1980.

424 Carmoisine *Azorubine Food Dye and Coloring Red No. 3*

Ford et al. (1987) gave up to 1200 mg per kg per day orally to male and female rats before and during pregnancy, lactation and weaning. Four phaeochromocytomas occurred in 53 exposed males at the highest dose. They cite negative teratogenicity studies in rabbits and rats at maximum oral doses of 400 and 1000 mg per kg daily.

Ford, G.P.; Stevenson, B.I. and Evans, J.G.: Long–term toxicity study of carmoisine in rats using animals exposed in utero. Food Chem. Toxicol. 25(12):919–925, 1987.

425 I–Carnitine Chloride

Toteno et al. (1988) administered up to 1000 mg per kg to rabbits on days 6–18 and found no adverse fetal effects.

Toteno, I.; Furukawa, S.; Haguro, S.; Matsushima, T.; Awazu, K.; Nirubagam T,; Fujii, S.; Terada, T. and Wada, Y.: Teratogenicity study of l–carnitine chloride in rabbits. Iyakuhin Kenkyu 19(3):510–521, 1988.

426 L–Carnosine CAS 305–84–0

Akatsuka et al. (1974) gave this chemical to pregnant rats intraperitoneally on the 8th through the 14th day and to mice intravenously on the 7th through the 13th day of gestation. Maximum doses of 3,000 mg per kg gave some increases in resorptions but no increase in defects. Postnatal studies were negative.

Akatsuka, K.; Hashimoto, T.; Takeuchi, K.; Miyame, Y. and Horisaka, K.: Effects of L–carnosine on the development of their fetuses and offspring in the pregnant mice and rats. Oyo Yakuri 8:1219–1228, 1974.

427 Carotene CAS 36–88–4

No case reports of excess carotene intake associated with congenital defects have been found. Symptoms of hypervitaminosis A were not found with excess carotene intake (Underwood, 1984).

Underwood, B.A.: Vitamin A in animal and human nutrition. In: The Retinoids. M.B. Sporn; A.B. Roberts and D.S. Goodman (eds.), New York, Academic Press, 1:282–377, 1984.

428 Carperitide α–Human Atrial Natriuretic Polypeptide SUN4936

Shikuma et al. (**1992** A&B) studied this compound intravenously in rats before pregnancy and in early gestation and during organogenesis at doses up to 10 mg per kg. Some decreased motor activity occurred at the highest dose but there were no adverse effects on the fetuses. Shimuza et al. (**1992**) studied rats during the perinatal period at a dose of 10 mg per kg and found no changes in the behavior or reproduction of the fetuses. Sakai et al (**1992**) gave 10 mg per kg intravenously during organogenesis to rabbits and found no ill effects in the dams or fetuses.

Sakai, Y.; Kinoshita, K.; Sugiyama, K. and Otaka, T.: Reproductive and developmental toxicity studies of α–human atrial natriuretic polypeptide (Carpertide, SUN4936) in rabbits teratogenicity study. Oyo Yakuri 44:465–471, 1992.

Shikuma, H.; Ishihara, M.; Arai, C.; Nakazawa, M.; Sugiyama, K.; Okamoto, M.; Ohnishi, S. and Otaka, T.: Reproductive and developmental toxicity studies of α–human atrial naturiuretic polypeptide (Carpertide, SUN 4936) in rats (1) fertility study. Oyo Yakuri 44:485–494, 1992A.

Shikuma, H.; Ishihara, M.; Sakaguchi, Y.; Nakazawa, M.; Sugiyama, K.; Okamoto, M.; Ohnishi, S. and Otaka, T.: Reproductive and developmental toxicity studies on f α–human atrial natriuretic polypeptide (Carperitide, SUN

4936) in rats (2) teratogenicity study. Oyo Yakuri 44:495–510, 1992B.

Shimizu, M.; Katoh, M.; Kobayashi, Y.; Yamashita, Y.; Shinoda, A.; Sugiyama, K.; Okamoto, M.; Ohnishi, S. and Otaka, T.: Perinatal and postnatal studies of intravenously administered α–human atrial natriurretic polypeptide (SUN4936) in rats. Oyo Yakuri 44:123-134, 1992.

429 Carprofen CAS 53716–49–7

This non–steroidal anti–inflammatory agent was given in amounts of up to 20 mg per kg orally to rats on day 7 through 15 of gestation. No adverse fetal effects were noted, but the agent slightly prolonged gestation which was associated with an increased number of dead pups at birth.

McClain, R.M. and Hoar, R.M.: Reproduction studies with carprofen a nonsteroidal antiinflammatory agent in rats. Toxicol. Appl. Pharmacol. 56:376–382, 1980.

430 Cartap HCl

3–Bis(carbamoylthio)–2–(N,N–dimethylamino)–propane Hydrochloride CAS 15263–52–2

Mizutani et al. (1971) studied this insecticide in pregnant mice, rats and hamsters. Doses up to 100 mg per kg given during organogenesis caused only minor skeletal changes which were ascribed to the maternal toxicity which occurred.

Mizutani, M.; Ihara, T.; Kanamori, H.; Takatani, D.; Matsokawa, J.; Amano, T. and Kaziwara, K.: Teratogenesis studies with cartap hydrochloride in the mouse, rat and hamster. J. Takeda Res. Lab. 30:776–785, 1971.

431 Carteolol CAS 51781–06–7

This β–adrenergic blocker was given orally to pregnant mice in doses of up to 150 mg per kg daily. Resorptions were increased at 75 and 150 mg levels, but the rate of congenital defects was not increased (Tanaka et al., 1979). Perinatal and postnatal studies were also negative (Tamagawa et al., 1979).

Tamagawa, M.; Namoto, T.; Tanaka, N. and Hishino, H.: Reproduction study of carteolol hydrochloride in mice, part 2. Perinatal and postnatal toxicity. J. Toxicol. Sci. 4:59–78, 1979.

Tanaka, N.; Shingai, F.; Tamagawa, M. and Nakatsu, I.: Reproductive study of carteolol hydrochloride in mice, part 1. Fertility and reproductive performance. J. Toxicol. Sci. 4:47–58, 1979.

432 Carumonam CAS 86832–68–0; CAS 87638–04–8

Ihara et al. (1986) studied this antibiotic in the pregnant rabbit using up to 1000 mg per kg intravenously on days 6–18. At the highest dose, 3 of 13 dams aborted but no fetal ill effects resulted in the remaining litters.

Ihara, T.; Ooshima, Y.; Nakamura, H. and Sugitani, T.: Teratogenic effects of Carumonam in the rabbit. Jpn. Pharmacol. Ther. 14:173–180, 1986.

433 Carvedilol

Bode et al. (**1991** A,B,C) studied this β–blocker in rats before and in early pregnancy, during organogenesis and in the perinatal and postnatal period. At the highest doses (60 or 200 mg per kg orally) there was some decreased birth weight and delayed postnatal physical development. The weight of the dams was also reduced at 200 mg per kg. Bode et al. (1991 D) found no teratogenicity in a second species rabbits after oral doses of up to 75 mg per kg orally.

Bode, G.; Lexa, P.; Sterz, H.; Ohura, K.; Harada, S.; Watanabe, T. and Takayama, S.: Reproductive toxicology study (fertility) with carvedilol in rats drug administration per os. The Clinical Report 25:3071–3099, 1991A.

Bode, G.; Lexa, P.; Sterz, H.; Ohura, K.; Harada, S.; Watanabe, T. and Takayama, S.: Reproductive toxicology study (teratogenicity) with carvedilol in rats drug administration per os. The Clinical Report 25:3101–3107, 1991B.

Bode, G.; Lexa, P.; Sterz, H.; Ohura, K.; Harada, S.; Watanabe, T. and Takayama, S.: Peri–and postnatal toxicity study with carvedilol in rats drug administration per os. The Clinical Report 25:3109–3129, 1991C.

Bode, G.; Lexa, P.; Sterz, H.; Ohura, K.; Harada, S.; Watanabe, T. and Takayama, S.: Reproductive toxicology study (teratogenicity) with carvedilol in rabbits drug administration per os. The Clinical Report 25:127–133, 1991D.

434 Carzinophilin CAS 1403–28–7

Takaya (1965) injected this antibiotic into rats at doses of 0.01 to 0.03 mg per kg on the 6th through 10th day and produced 10 percent defective offspring. The defects were micro(an)ophthalmia and hydronephrosis.

Takaya, M.: Teratogenic effects of antibiotics. J. Osaka City Med. Cent. 14:107–115, 1965.

435 Cassava Root

Rats fed the powder of this root as 80 percent of their diet during the first 15 days of gestation had fetuses with limb reduction defects and microcephaly (Singh, 1981). Cyanide is known to be released from this food by hydrolysase.

The active ingredient, linamarin, was fed to pregnant guinea pigs and at 120 and 140 mg per kg axial skeletal defects and encephaloceles occurred in the offspring (Frakes et al., 1985).

Frakes, R.A.; Sharma, R.P. and Willhite, C.C.: Developmental toxicity of the cyanogenic glycoside linamarin in the golden hamster. Teratology 31:241–246, 1985.

Singh, J.D.: The teratogenic effects of dietary cassava on the albino rat: A preliminary report. Teratology 24:289–291, 1981.

436 Castor Oil Seed

Mauhoub et al. (1983) reported a three month old infant with ectrodactyly, vertebral defects and growth retardation. The distal phalanges and nails were absent in the feet and hands. The mother had taken one castor oil seed during each of the first three months of pregnancy.

Mauhoub, M.E.; Khalifa, M.M.; Jaswal, O.B. and Garrah, M.S.: Ricin syndrome, a possible new syndrome associated with ingestion of castor–oil seed in early pregnancy: A case report. Ann. Trop. Paediat. 3:57–61, 1983.

437 Catena–(S)–[u–[Nα–(3–aminopropionyl) histidinato(2–)–N¹,N², O:Nψ]–zinc] Z–103

Matsuda et al. (1991) studied this antiulcer medication in pregnant rats and rabbits. At 1200 mg per kg orally fetal weights were reduced but no teratogenicity was found. Fertilization and perinatal studies were also done and no ill effects identified. Rabbits exposed to 300 mg per kg had increased resorptions and fetal weight reduction but no teratogenicity.

Matsuda, K.; Nishi, N.; Hiramatsu, Y.; Shimizu, M.; Ohta, T. and Kato, M.: Reproductive and developmental toxicity studies on catena–9s)–[u–[Nα–(3–aminopropionyl) histidinato(2–)–N¹,N²,O:Nψ]–zinc]. Arzneim Forsch/Drug Res. 41:1042–1048, 1991.

438 Cefaclor CAS 53994–73–3

This cephalosporin antibiotic was given orally by Nomura et al. (1979) to mice, rats and rabbits. No increase in defects was found among the rat and mouse fetuses exposed daily to up to 2000 mg per kg daily during organogenesis and no adverse fetal effects observed. Prenatal or perinatal administration of up to 2000 mg per kg to rats was not associated with reproductive or postnatal changes (Furuhashi et al., 1979).

Furuhashi, T.; Nomura, A.; Uehara, M.; Komuro, E. and Nakayoshi, H.: Fertility and perinatal study of cefaclor in rats. Chemotherapy 27:865–880, 1979.

Nomura, A.; Furuhashi, T.; Ikega, E.; Sawaki, A. and Nakayoshi, H.: Fertility and perinatal study of cefaclor in mice, rats and rabbits. Chemotherapy 27:846–864, 1979.

439 Cefadroxil CAS 50370–12–2

Tauchi et al. (1981) gave this antibiotic to rats and rabbits orally in amounts of up to 2300 and 100 mg per kg daily, respectively. No significant increase in defects was seen in either species. Fetal weight was reduced in the rabbit when 100 mg per kg was used. Postnatal studies in the rat were done and no differences from controls were found.

Tauchi, K.; Kawanishi, H. and Igarashi, T.: Studies of the reproductive toxicity of cefradroxil. Jpn. J. Antibiot. 33:478–486, 487–496, 497–502, 503–509, 1981.

440 Cefatrizine S–640–P CAS 51627–14–6

Matsuzaki et al. (1976A) carried out teratogenicity studies in the rat and mouse at 3,200 mg per kg orally during active organogenesis. Except for weight reduction in the treated rat fetuses, no ill effects were observed. In the rabbit at doses of 400 and 800 mg per kg, an increase in fetal death and reduction in fetal weight was found (Matsuzaki et al., 1976B).

Matsuzaki, A.; Akutsu, S.; Shimamura, T. and Honda, H.: Studies on the possible teratogenicity of cefatrizine (S–640–P). Jpn J. Antibiot. 29:812–825, 1976A.

Matsuzaki, A.; Akutsu, M.; Mukaikawa, H. and Kobayashi, K.: Studies of the possible teratogenicity of cefatrizine (S–640–P). 2. Effects on the embryos when administered to pregnant rabbit. Jpn J. Antibiot. 29:144–152, 1976B.

441 Cefazolin

Hasegawa et al. (**1987**) gave rats intravenously up to 800 mg per kg on days 7–17 of gestation and found no adverse effects. A slight reduction in fetal weight was found. Postnatal development was normal.

Hasegawa, Y.; Takegawa, Y.; Yoshida, Y.: Reproduction of rats under 6351s (flomoxef). 2. Intravenous administration during fetal organogenesis. Chemotherapy (Tokyo) (Nippon Kagaku Ryoho Gakkai Zasshi) 35 (Suppl 1):370–403, 1987.

442 Cefbuperazone *T–1982* CAS 76610–84–9

This antibiotic agent was given subcutaneously to male and female rats at doses of 2000 mg per kg and no adverse effects on fertility were found. Intravenous dosage (2000 mg per kg) during organogenesis and the perinatal period produced no ill effects except for fetal weight reduction (Nakada et al., 1982).

Nakada, H.; Nakamura, S.; Komae, N.; Takimoto, Y. and Takai, A.: Toxicity test of T–1982. Chemotherapy 30:319–344, 1982.

443 Cefcapene *S–1108*

This oral cephem antibiotic was studied in rats before fertilization and during the first 7 days of gestation (Hara et al, **1993** A) and no change in reproduction was found at oral doses up to 1000 mg per kg. Similar studies during organogenesis showed only slight fetal growth retatdation (Hasegawa et al., **1993**). At the same dosage perinatal studies were performed and except for a delay in weight increase behavior and reproduction were normal (Komai et al., **1993** B). Rabbits were given up to 20 mg per kg during organogenesis and no fetal changes occurred.

Hara, K.; Yoshizaki, T. and Nara, H.: Reproductive and developmental toxicity studies of S–1108, a new ester–type oral cephem antibiotic (I) a study on oral administration in rats prior to and in the early stages of pregnancy. Iyakuhin Kenkyu 24:21–38, 1993.
Hasegawa, Y.; Yoshida, T.; Hirata, M. and Nara. H.: Reproductive and developmental toxicity studies of S–1108, a new ester–type oral cephem antibiotic (II) a study on oral administration in rats during the period of fetal organogenesis Iyakuhin Kenkyu 24:39–78, 1993.
Hasegawa, Y.; Kanamori, S.; Hirata, M. and Nara, H.: Reproductive and developmental toxicity studies of S–1108, a new ester–type oral cephem antibiotic (III) a study on oral administration in rabbits during the period of fetal organogenesis. Iyakuhin Kenkyu 24:79–95, 1993.
Komai, Y.; Ogura, H.; Inoue, S.; Katano, T.; Hattori, M.; Isowa, K.; Ishimura, K. and Hasegawa, Y.: Reproductive and developmental toxicity studies of S–1108, a new ester–type

oral cephem antibiotic (IV) a study on oral administration in rats during the prenatal and lactation periods. Iyakuhin Kenkyu 24:878–915, 1993.

444 Cefclidin

Goto et al. (**1992**) tested this cephalosporin antibiotic in rats giving up to 1,000 mg per kg on days 7–17 or gestation. No adverse effects were found.

Gogo, M.; Nishimura, O.; Okada, F.; Osumi, I. and Matsubara, Y.: Reproductive study of cefclidin by intravenous administration during the period of fetal organogenesis in rats. Chemotherapy 40:117–153, 1992.

445 Cefdinir

Shimazu et al. (**1989**) gave this antibiotic to rats and rabbits during organogenesis in oral doses of up to 1,000 mg per kg. At 320 mg per kg and higher in rats fetal and maternal weight was reduced. No increase in malformations was found.

Shimazu, H.; Ishikawa, Y.; Fujioka, M.; Matsuoka, T.; Shiota, Y.; Kadoh, Y.; Shimizu, I. and Noguchi, H.: Toxicity study of cefdinir (3rd report)–reproduction study. Kiso to Rinsho 23:5833–5843, 1989.

446 Cefditoren pivoxil

Ito et al. (**1992**) studied this oral cephem antibiotic orally in rabbits using up to 30 mg per kg during organogenesis. The number of fetuses deceased at 7.5 mg per kg and higher and fetal weight was reduced in the 30 mg per kg group. No congenital defect rate increase was found.

Ito, M.; Hasunuma, K.; Okano, K.; Izawa, M.; Fijii, C.; Kosugi, I. and Fujita, M.: Toxicity studies of cefditoren pivoxil, a new oral cephem antibiotic. Chemotherapy 40:1397–1408, 1992.

447 Cefepime

This cephem antibiotic was given subcutaneously to rats before and during gestation in amounts of up to 1000 mg per kg. Softening of the adult stools occurred but no decrease in fertility or postnatal behavior was found.

Kai, S.; Kohmura, H.; Ishikawa, K.; Kawano, S.; Sakai, A.; Kuroyanagi, K.; Kadota, T. and Takahashi, N.: Reproductive and developmental toxicity studies of cefepime dihydrochloride administered subcutaneously to rats during the premating, gestation and lactation periods. The Japanese Journal of Antibiotics 45:643–660, 1992.

448 Cefetamet Pivoxil

Hayashi et al. (**1990**) gave up to 500 mg per kg orally on days 7–17 to rats. No adverse effects were found at term or through weaning.

Hayashi, M,; Kurihara, M.; Takahashi, M,; Mineshima, H. and Horii, I.: Reproduction segment ii study of cefetamet pivoxil in rats. Yakuri to Chiryo 18:s–2981–s–3003, 1990.

449 Cefmetazole CAS 56796–20–4

Esaki et al. (1980) administered this antibiotic to pregnant beagle dogs on the 18th through the 35th day in doses of up to 1000 mg per kg. No adverse effects were noted. Masuda et al. (1978) gave this antibiotic in doses of up to 1000 mg per kg to rats for 14 days after mating and found no adverse reproductive effects. On days 6–15 up to 2000 mg per kg was given intravenously without teratogenic changes. Perinatal and postnatal studies using 2000 mg per kg intravenously in rats on days 17–21 were carried out and no changes were found in behavior or reproductive ability of the offspring. Reversible weight reduction was found in testicular of treated infants receiving 100 mg or more per kg daily.

Esaki, K.; Nomura, G. and Iwaki, T.: Effects of intravenous administration of cefmetazole (CS–1170) on the fetus of the beagle dog. Preclin. Rep. Cent. Inst. Exp. Animal 6:289–292, 1980.

Masuda, H.; Kimura, K. and Hirose, K.: Toxicological studies of CS1170. 2. Reproductive studies of CS 1170 in mice and rats. Sankyo Kenkyusho Nenpo (Ann. Rep. Sankyo Res. Lab) 30:148–167, 1978.

Moe, J.B.; Manabe, J.M.; Ikegami, N.; Tanase, H.; Lohrberg, S.M.; Larsen, E.R. and Piper, R.C.: Differential effects of cefmetazole sodium on the reproductive system of infant and pubertal male rats. Fundamental and Applied Toxicology 13:146–155, 1989.

450 Cefodizime Sodium CAS 86329–79–5

Kitatani et al. (**1988**) treated mice intravenously with up to 3000 mg per kg on days 6–15. No teratogenic effects were found although fetal weight reduction was found at the highest dose. In perinatal studies of mice Yamakita et al. (**1988** A) used similar treatments and found no significant effects except a reduction in the spleen weights of 10 week pups exposed to the highest dose. Fertility was normal in pretreated and early gestational treated animals given up to 3000 mg per kg intravenously. (Akaike et al., **1988**)

Akaike, M.; Kitatani, T.; Takayama, K. and Kobayashi, T.: Fertility study of cefodizime sodium in mice–intravenous administration from pre–conceptional period through early period of gestation. The Journal of Toxicological Sciences 13:177–190, 1988.

Kitatani, T.; Akaike, M.; Takayama, K. and Kobayashi, T.: Teratological study of cefodizime sodium in mice–intravenous administration during period of organogenesis. The Journal of Toxicological Sciences 13:191–214, 1988.

Yamakita, O.; Koida, M.; Shinomiya, M.; Katayama, S.; Ikebuchi, K.; Ohuchida, A.; Kouchi, Y. and Yoshida, R.: Peri- and postnatal study of cefodizime sodium in mice–late gestation through period of lactation. The Journal of Toxicological Sciences 13:25–229, 1988.

451 Cefoperazone CAS 62893–19–0

Tanioka and Koizumi (1979) injected this antibiotic intravenously into pregnant monkeys on days 23–47 at doses of up to 400 mg per kg per day. No teratogenic effect was found in nine pregnancies.

Nakada et al. (1980) administered this drug subcutaneously to rats in amounts of up to 1000 mg per kg daily on days 17–21 or 7–17 of gestation. No adverse fetal effects were seen and postnatal function was not changed. Administration of 500 mg per kg before mating in both sexes and during the first 7 days was done with no adverse effect.

Nakada, H.; Nakamura, S.; Inaba, J.; Komae, N. and Takai, A.: Toxicity test of cefoperazone (T–1551). Reproductive study in rats. Chemotherapy 28:268–291, 1980.

Tanioka, Y. and Koizumi, H.: Effects of T–1551 on the fetus of rhesus monkeys. Preclin. Rept. Cent. Inst. Exp. Animal 5:145–156, 1979.

452 Cefotaxime CAS 63527–52–6

This cephalosporin antibiotic was studied in mice and rabbits by Sugisaki et al. (1981A). Intravenous doses of up to 2000 mg per kg were used prenatally and perinatally and subcutaneous doses of up to 6000 mg per kg were used for studies during organogenesis. No adverse effects on fertility, the fetuses or the behavior of the offspring were found. Pregnant rabbits received up to 50 mg intravenously on days 6–18 of gestation and the fetuses were not different from controls (Sugisaki et al., 1981B).

Sugisaki, T.; Kitatani, T.; Takagi, S.; Akaike, M. and Hayashi, S.: Reproductive studies of cefotaxime in mice. Oyo Yakuri 21:351–373, 1981A.

Sugisaki, T.; Akaike, M. and Hayashi, S.: Teratological study of cefotaxime given intravenously in rabbits. Oyo Yakuri 21:375–384, 1981B.

453 Cefotetan CAS 69712–56–7

This antibiotic was given intraperitoneally to rats before mating and during the first 7 days of gestation in doses of up to 2000 mg per kg and no adverse reproductive changes were found (Odani and Nakata, 1982). Uchida and Odani (1982) gave rats intravenously doses of up to 2000 mg per kg on days 17–21. Some growth retardation was found but no behavioral effects were observed. Shibata and Tamada (1982) gave up to 2000 mg per kg intravenously to rats on days 7–17 and produced no adverse fetal effects.

Allen et al. (1982) gave monkeys 600 mg intravenously on days 20–45 and produced no embryopathic activity.

Allen, D.G.; Clark, R.; Palmer, A.K. and Heywood, R.: An embryotoxicity study in macaca fascicularis with cefotetan disodium (a cephamycin antibiotic). Toxicol. Lett. 11:43–48, 1982.

Odani, Y. and Nakata, M.: Reproduction study of cefotetan (YM 09330). Fertility study in rats when administered intraperitoneally. Oyo Yakuri 23:641–649, 1982.

Shibata, M. and Tamada, H.: Teratological evaluation of cefotetan (YM 09330) administered intravenously to rats. Chemotherapy 30:278–294, 1982.

Uchida, T. and Odani, Y.: Reproductive studies of cefotetan (YM 09330). Perinatal and postnatal study in rats by intravenous administration. Oyo Yakuri 23:767–783, 1982.

454 Cefotiam Hexetil Hydrochloride CAS 66309–69–1

Mizutani et al. (**1988**) carried out fertility studies using gavage and up to 1,000 mg per kg and found no adverse effects. In studies during organogenesis and in the perinatal period at doses of up to 3,000 mg there was some reduced fetal weight but no teratogenicity. At the highest dose some of the dams died of gastric ulcers. In monkeys Korte et al. (**1988**) found no adverse fetal effects from 1,000 mg per kg on days 25–50.

Korte, R.; Osterburg, I.; Vogel, F. and Ihara, T.: Teratogenicity study with cefotiam hexetil hydrochloride (ctm-he) in the cynomolgus monkey. Oyo Yakuri 36(1):83–90, 1988.

Mizutani, M.; Hashimoto, Y.; Shirota, M.; Morita, Y.; Shimizu, Y. and Ihara, T.: Reproduction studies of cefotiam hexetil hydrochloride (ctm-he) in rats. Oyo Yakuri 36(1):73–82, 1988.

455 Cefoxitin

This cephalosporin antibiotic was tested by Watanabe et al. (1978) in pregnant rats and mice. No effects on fertility were found with doses of 400 mg per kg. Intravenous doses of up to 900 mg per kg did not produce adverse effects in the offspring of either species. Prenatal studies did not produce changes in postnatal function.

Watanabe, T.; Ohura, K.; Morita, H. and Akimoto, T.: Toxicological studies on cefoxitin. IV. Influence of cefoxitin on reproduction in mice and rats. Chemotherapy 26:205–227, 1978.

456 Cefpirome Sulfate

Sugiyama et al. (**1990** A,B and C) treated rats intramuscularly or intraperitoneally on days 7–17 with up to 800 mg per kg. At 400 and 800 mg per kg there was toxicity seen in the dams and reduced fetal weight. No teratogenicity was found. Fertility and reproduction were not hampered by intraperitoneal doses similar to the above. Administration on days 17 through day 21 of lactation reduced fetal weight but postnatal development was normal.

Sugiyama, O.; Tanaka, K.; Toya, M.; Igarashi, S.; Watanabe, K.; Watanabe, S.; Tsuji K. and Kumagai, Y.: Teratological study of cefpirome sulfate in rats. The Journal of Toxicological Sciences 15(III):65–89, 1990.

Sugiyama, O.; Tanaka, K.; Toya, M.; Igarashi, S.; Watanabe, K.; Watanabe, S.; Tsuji, K. and Kumagai, Y.: Fertility study of cefpirome sulfate in rats. The Journal of Toxicological Sciences 15:(III):53–64, 1990.

Sugiyama, O.; Watanabe, S.; Tanaka, K.; Toya, M.; Igarashi, S.; Watanabe, K.; Tsuji, K. and Kumagai, Y.: Perinatal and postnatal study of cefpirome sulfate in rats. The Journal of Toxicological Sciences 15(III):91–107, 1990.

457 Cefpoxine Proxetil

Tanase & Hirose (**1988**) gave rats orally up to 500 mg per kg on days 7–17. Maternal weight loss and diarrhea in the dams was seen at 100 mg per kg or above. No teratogenicity or postnatal development delay was seen although ossification was retarded at the highest dose.

Tanase, H. and Hirose, K.: Reproduction study of cs–807 administered during the period of fetal organogenesis in rats. Chemotherapy 36:320–328, 1988.

458 Cefroxadin CAS 51762–05–1

This antibiotic was given to rabbits orally in doses of up to 10 mg per kg on days 6–18 of gestation and no significant fetal changes were found (Hirooka et al., 1980). With oral doses of up to 1000 mg per kg in the rat no adverse effects on reproduction were reported (Hirooka et al., 1979).

Hirooka, T.; Tadokoro, T.; Takahashi, S.; Katagawa, K.K. and Kitagawa, S.: Reproductive tests of cefroxadin (CGP–9000) in rats. Iyakuchin Kenkyu 10:802–824, 1979.

Hirooka, T.; Takahashi, S.; Tadokoro, T. and Kitagawa, S.: Reproductive studies of cefroxadin (CGP–9000) a new oral antibiotic. 4. Teratological tests in rabbits. Oyo Yakuri 19:669–679, 1980.

459 Ceftazidime CAS 72558–82–8

Furuhashi et al. (1983A) gave this cephalosporin antibiotic to rats intravenously before and during the first 7 days, on days 7–17 and days 17–21 of gestation. At dosages of 1.0 gm per kg no adverse fertility or postnatal effects were observed. During organogenesis a dose of 2.0 gm per kg was associated with osseous retardation but no defects. Rabbits were treated with 200 mg per kg during days 6–18. No adverse fetal effects were seen (Furuhashi et al., 1983B).

Furuhashi, T.; Veharsa, M.; Takai, A.; Honda, T.; Nomura, A. and Nakayoshi, H.: Safety study on ceftazidime, fertility study in rats, teratological study in rats and perinatal–postnatal study in rats. Chemotherapy 31:968–986, 1983A.

Furuhashi, T.; Kato, I. and Nakayoshi, H.: Safety study on cephazidime: Teratological study in rabbits. Chemotherapy 31:961–967, 1983B.

460 Ceftezole CAS 41136–22–5

Niki et al. (1976) studied this antibiotic in mice and rats using the intravenous and subcutaneous routes and found no increase in fetal defects. The mice received 4000 and 2000 mg per kg by the subcutaneous and intravenous routes, respectively, and the rats received 1000 mg by each route. Treatments were given during mid–gestation.

Niki, R.; Shiota, S.; Usami, M.; Noguchi, G.; Sugiyama, O.; Ohkawa, H. and Takagaki, Y.: Studies of toxicity and teratogenicity of ceftezole. Chemotherapy 24:671–702, 1976.

461 Ceftibuten

Hasegawa and Takegawa (**1989**) gavaged rats with up to 4,000 mg per kg on days 7–17. Fetal growth retardation was found at 4,000 mg per kg. Postnatally behavior and reproduction were normal except growth retardation was found at 1,000 and 4,000 mg per kg doses. No teratogenicity was found. Rabbits given up to 40 mg per kg had no increase in offspring with congenital defects but growth retardation occurred at the highest doses. (Hasegawa & Fukiishi, **1989**) Treatment of both sexes before pregnancy and females during the first 7 days after fertilization did not cause any reproductive ill effects.

Hasegawa, Y. and Fukiishi, Y.: Reproduction studies on 7432–s 3 teratology study in rabbits. Chemotherapy 37:1026–1039, 1989.

Hasegawa, Y. and Takegawa, Y.: Reproduction studies on 7432–s 2. teratology study in rats. Chemotherapy 37:990–1025, 1989.

Hasegawa, Y. and Takegawa, Y.: Reproduction studies on 7432–s 1. fertility study in rats. Chemotherapy 37:972–989, 1989.

462 Ceftizoxime Sodium CAS 68401–82–1

This cephalosporin antibiotic was given to rats subcutaneously by Fukuhara et al. (1981) in amounts of up to 1000 mg per kg daily. Dosing was either before mating and during the first 17 days, during days 7–17 or on days 17–21. No adverse effects were found in the fetuses or offspring.

Fukuhara, K.; Fujii, T.; Kado, Y. and Watanabe, N.: Reproductive studies on ceftizoxime sodium in rats. Jpn. J. Antibiotics 34:466–476, 1981.

463 Ceftriaxone CAS 73384–59–5

Shimizu et al. (1984) gave rats up to 750 mg per kg intravenously before pregnancy and during the first 7 days and on days 7–17 and from day 17 through 3 weeks, postnatally. No adverse reproductive or fetal effects were found.

Shimizu, M.; Noda, K.; Honma, M. and Udaka, K.: Reproductive studies of ceftriaxone (RO 13–9904) in the rat. Clin. Report 18:1891–1922, 1984.

464 Cefuroxime CAS 55268–75–2

Furuhashi et al. (1979) administered this cephalosporin antibiotic to rabbits subcutaneously or intravenously on days 6 through 18 of gestation. Doses of up to 150 mg per kg had no adverse effects on the fetuses.

Furuhashi, T.; Nomura, A.; Ikeya, E. and Nakazawa, M.: Teratological studies on cefuroxime in rabbits. Chemotherapy 27:245–292, 1979.

465 Celiprolol CAS 56980–93–9

Ninomiya et al. (1989,A,B,C) studied this B–adrenergic blocker in rats before fertilization, during organogenesis and in the perinatal period. The maximum dose was 320 mg per kg. No teratogenicity or adverse postnatal behavior was found.

Ninomiya, H.; Akitsuki, S.; Kondo, J.; Nishikawa, K.; Yamashita, Y.; Watanabe, M.; Nagasawa, H.; Sumi, N. and Nomura, A.: Reproduction study of celiprolol (2) teratogenicity study in rats. Oyo Yakuri 37(2)215–229, 1989.

Ninomiya, H.; Akitsuki, S.: Kondo, J.; Nishiawa, .; Yamashita, Y.; Watanabe, M.; Nagasawa, H.; Sumi, N. and Nomura, A.: Reproduction study of celiprolol (1) fertility study in rats. Oyo Yakuri 37(2):201–213, 1989.

Ninomiya, H.; Akitsui, S.; Kondo, J.; Nishikawa, K.; Watanabe, M.; Nagasawa, H.; Sumi, N. and Nomura, A.: Reproduction study of celiprolol (3) peri–and postnatal study in rats. Oyo Yakuri 37(2):231–242, 1989.

466 Cellryl

Sukegawa et al. (1976) gave up to 10 mg subcutaneously to rats on days 8–17. No adverse fetal effects was found.

Sukegawa, M.; Makino, M.; Kusunoki, F.; Eguchi, K.; Yamamoto, T.; Ito, K. and Suzuki, M.R.: Acute toxicity and reproduction studies of cellryl. Yakuri to Chiryo 4:74–82, 1976.

467 Cellulose Acetate Phthalate

Suzuki et al. (1975) fed up to 1500 mg per kg per day to rats on days 6–9 and found no adverse fetal effects.

Suzuki, Y.; Hirose, K.; Takahashi, A.; Takayanagi, M.; Maita, K.; Yamashita, K. and Masuda, H.: Teratological study of cellulose acetate in rats. Iyakuhin Kenkyu 6:41–48, 1975.

468 Centalun™
3–Methyl–3,4,dihydroxy–4–phenyl–butin–1

Kuhn and Wick (1963) tested rats and mice orally during organogenesis with maximum amounts of 600 and 900 mg per kg, respectively, and found no adverse effects. Rabbits were given up to 1000 mg per kg without teratogenicity.

Kuhn, F–J. and Wick, H.: Pharmakologische eigen Schaften des 3–Methyl–3,4–dihydroxy–4–phenyl–butin–1. Arzneim–Forsch. 13:728–734, 1963.

469 Centazolone *3–Aminobenzo–6,1–quinazolone–4–one*

Sethi and Mukherjee (1978) gavaged mice with up to 200 mg per kg and rabbits with up to 160 mg per kg during organogenesis and found no evidence of teratogenicity.

Sethi, N. and Mukherjee, S.K.: Studies of 3–aminobenzo–6–l–quinazolone–4–one (centazolone, compound 65469): A new tranquillosedative compound. Ind. J. Exp. Biol. 16:206–207, 1978.

470 Centchroman *3, 4–Trans–2, 2–dimethyl–3–phenyl–4–p–(β–pyrrolidin–oethoxy)–phenyl–7–methoxychromano*

Sethi (1977) gave up to 100 mg and 80 mg per kg to pregnant mice and rabbits, respectively. Daily treatment during organogenesis did not produce adverse prenatal development in the fetuses.

Sethi, N.: Influence of centchroman on prenatal development in mice and rabbits. Ind. J. Exp. Biol. 15:1182–1183, 1977.

471 Cephacetrile *Celospor™*

Esposti et al. (1986) injected up to 500 mg per kg daily during organogenesis to pregnant rats and rabbits. No reproductive ill effects were found. Postnatal survival was normal.

Esposti, G.; Esposti, D.; Guerino, G.; Bichisao, E. and Fraschini, F.: Effects of cephacetrile on reproduction cycle. Arzneim–Forsch. 36:484–486, 1986.

472 Cephalexin CAS 15686–71–2

Aoyama et al. (1969) found no teratogenicity in the offspring of mice and rats given oral doses of up to 800 (mice) and 4,000 mg per kg (rats) during organogenesis.

Briggs et al. (1986) reviewed several reports of pregnant patients receiving the medication and found no increase in defects or toxicity of the newborn.

Aoyama, T.; Furuoka, R.; Hasegawa, N. and Nemoto, K.: Teratologic studies of cephalexin in mice and rats. Oyo Yakuri 3:249–263, 1969.

Briggs, G.G.; Freeman, R.F. and Yaffe, S.J.: Drugs in Pregnancy and Lactation. Williams and Wilkins, Baltimore, 2nd ed., 1986.

473   Cephamycin *MT141* CAS 84305–41–9

Kurebe et al. (1984) studied the effect of this antibiotic in pregnant rats and rabbits. In rats treated intramuscularly with up to 1,600 mg per day during the first 7 days, 7–17 days and 7–21 days, the only adverse effect was in fetal osseous development which was delayed when the females were treated for 63 days before mating and for 7 days after fertilization. In rabbits given 40 mg per kg on days 6–18, there was an increase in dead or resorbed fetuses but no congenital defects.

Kurebe, M.; Asaoka, H.; Hata, T.; Watanabe, T. and Hirota, C.: Toxicological studies of a new cephamycin, MT141. Fertility in the rat, teratogenicity in the rat and rabbit, perinatal studies and postnatal study in rats. Jpn. J. Antibiot. 37:218–269, 1984.

474   Cephem *7432–S (6R–7R)–7–(2)–2–(2–amino–4–5hiazolyl)–4–carboxy–2–butenoylamino) 8–oxo–5–thia–1–azabicyclo [4.2.0] oct–2–end–2 carboxylic Acid*

Hasegawa and Itoh (**1991**) gave rats orally up to 2,000 mg per kg in the perinatal and postnatal period. At 500 and 2,000 mg per kg the cecal weights were reduced in the offspring.

Hasegawa, Y. and Itoh, M.: Reproduction studies on 7432–S–Supplement to perinatal and postnatal study in rats. Chemotherapy 39:253–257, 1991.

475   Cephradine CAS 38821–53–3

This cephalosporin antibiotic was given orally andce and intraperitoneally to mice and rats before, during and afters were gestation. No adverse effects were detected at 300 mg per kg daily (Hassert et al., 1973).

Hassert, G.L.; DeBaecke, P.J.; Kulesza, J.S.; Traina, V.M.; Sinha, D.P. and Bernal, E.: Toxicological, pathological and teratological studies in animals with cephradine. Antimicrob. Agents Chemother. 3:682–685, 1973.

476   Cerium CAS 7440–45–1

Ridgway and Karnofsky (1952) injected 8 mg of Ce(NO₃)₃ into chick eggs on the 4th day and found no defects.

D'Agostino et al. (1982) injected cerium chloride subcutaneously into mice on day 7 or 12 of gestation at a dose of 80 mg (Ce) per kg. Except for an increased frequency of rearings in the open field of the adult offspring exposed in utero, there was no apparent effect on behavioral parameters. Other reproductive effects were not reported in the paper.

D'Agostino, R.B.; Lown, B.A.; Morganti, J.B. and Massaro, E.J.: Effects of in utero or suckling exposure to cerium (citrate) on the postnatal development of the mouse. J. Toxicol. Environ. Health 10:449–458, 1982.

Ridgway, L.P. and Karnofsky, D.A.: The effects of metals on the chick embryo: Toxicity and production of abnormalities in development. Ann. N.Y. Acad. Sci. 55:203–215, 1952.

477   Cerium [144]

McFee (1964) treated rats orally with 0.5 or 1.0 mc on days 4–14. Resorption rates were increased. Treatment on days 7, 8 or 9 was associated with eye defects.

McFee, A.F.: Effects of [144]Ce administered to pregnant rats. P.S.E.B.M. 116:712–715, 1964.

478   Ceruletide Diethylamine CAS 71–247–25–1

Hattori et al. (**1989**) and Hasegawa et al. (**1989**) gave rats and rabbits doses of up to 300 micro gm per day during organogenesis and found no teratogenicity. The rats received the drug subcutaneously and the rabbits intramuscularly. Administration before fertilization in rats did not alter reproduction. Perinatally administered drug did not effect postnatal development.

Hasegawa, Y.; Yamagata, H.; Hirashiba, M.; Fukiishi, Y.; Yoshida, T.; Takegawa, Y.; Miyago, M.; Ochida, H. and Yokozawa, Y.: Reproduction studies on 883–s (ceruletide diethylamine)–a teratology study in rabbits by the intramuscular administration. Kiso to Rinsho 23:1677–1691, 1989.

Hattori, M.; Ogura, H.; Inoue, S.; Isowa, K.; Komai, Y.; Ishimura, K. and Kobayashi, F.: Reproduction study of 883–s (ceruletide diethylamine)–teratogenicity study in rats by subcutaneous administration. Kiso to Rinsho 23:1637–1676, 1989.

479   Cesium CAS 7440–46–2

Lemeshevskaya and Silaev (1979) exposed rats by inhalation with concentrations of 4.6 and 0.43 mg per m squared during the entire gestation and observed different types of congenital malformations. The non–toxic dose for females (0.003 mg per m squared) produced no embryotoxic activity. The radioactive form (^{137}Cs) was studied by Hirata (1964) in the mouse. At doses of 60 and 120 rads, malformations were found. Intravenous doses were given on days 4 and 10, or 8 and 10 of pregnancy.

Hirata, M.: Experimental studies on embryotoxicity of radioactive cesium in mice. II. Embryotoxicity by ^{137}Cs internal radiation. Hiroshima Sanfujinka Gakkai–Kashi 3:224–242, 1964.

Lemeshevskaya, E.M. and Silaev, A.A.: Embryotoxic studies of cesium arsenate at the inhalation action. Gig. Tr. Prof. Zabol. (USSR) 9:56, 1979.

480   Cesium [137] CAS 10045–97–3

Hirata (1964) injected mice intravenously with 2 to 4 microcuries per gm of body weight on day 4, 8, 10, 12 or 15. A small number of cleft palates or tail anomalies were detected. Treatment on day 8 for instance produced 7 abnormal fetuses of the 157 examined.

Hirata, M.: Experimental study on fetal disorders in mice due to radioactive cesium. Fetal disorders due to ^{137}Cs irradiation. J. Hiroshima Obstet. Gynecol. Soc. 3:224, 1964.

481 Cetirizine

Kamijima et al. (**1994**) gave rats oral doses of up to 200 mg per kg before conception and during the first week of pregnancy, during organogenesis or during the peri and postnatal period. Estrous cycles were lengthened at 200 mg per kg but no decrease in reproduction or adverse effect on the fetuses were found. At the highest dose the fetal weight increase was reduced and postnatal viability tended to decrease. Rabbits were treated during organogenesis and no teratogenic effect was found.

Kamijima, M.; Sakai, Y.; Kinoshita, K.; Aruga, K.; Morimoto, K. and Kawakami, T.: Reproductive and developmental toxicity studies of cetirizine in rats and rabbits. The Clinical Report 28:1877–1903, 1994.

482 Cetylpyridinium Chloride *Cepacol*™ CAS 123–03–5

Aselton et al. (1985) studied the offspring of 292 women using this medication in the first trimester and found nine with malformations. This rate was 31 congenital defects per 1000 while the mean rate was 16 per 1,000. This increase was not statistically significant. No increase in defects was observed among 326 women who took the medication in the first four lunar months (Heinonen et al., 1977).

Aselton, P.A.; Jick, H.; Milunsky, A.; Hunter, J.R. and Stergachis, A.: First trimester drug use and congenital disorders. Obstet. Gynecol. 65:451–455, 1985.

Heinonen, O.P.; Slone, D. and Shapiro, S.: Birth Defects and Drugs in Pregnancy. Publishing Sciences Group Inc., Littleton, Mass., 1977.

483 CGA–24704

Amemiya et al. (1989) fed this herbicide in amounts up to 1000 ppm for two generations and found no adverse reproductive effects.

Amemiya, K.; Giknis, M.; Jacobs, M.; Youreneff, M.; Yau, E.; Traina, V.; Breckenridge, C.; Stevens, J. and Tisdel, M.: Two generation study on the effect of CGA 24704 technical on reproduction in rats. Teratology 39:P55, 1989.

484 CGS–15529

This potential anti–inflammatory agent was studied by Giknis et al. (1989) who gavaged rats throughout pregnancy with up to 15 mg per kg daily. Delay in partuition was associated with increased stillborns and decreased viability and newborn weight.

Giknis, M.L.A.; Wimbert, K.V.; Mainiero, J.; Yau, E.T. and Traina, V.M.: The reproductive effects of CGS 15529, a potential anti–inflammatory agent, in the rat. Teratology 39:P58, 1989.

485 α–Chaconine CAS 20562–03–2

This glycoalkaloid constituent of potato tubers given in doses of 16 mg per kg to mice on days 8 or 9 produced skeletal defects and midline facial defects (Pierro et al., 1977).

Pierro, L.J.; Haines, J.S. and Osman, S.F.: Teratogenicity and toxicity of purified α–chaconine and α–solanine. (abs) Teratology 15:31A only, 1977.

486 Chaetochromin Mycotoxin

Ito and Ohtsubo (1987) studied the effect of this mycotoxin on pregnant mice. They fed 10 or 30 ppm in the diet during the entire gestation or on days 7–9 or 10–12. From the short duration therapy, no adverse effects were seen in the fetuses. In those exposed to 30 ppm on days 0–18, 10 of the 97 survivors had exencephaly.

Ito, Y. and Ohtsubo, K.: Teratogenicity of oral chaetochromin, a polyphenolic mycotoxin produced by chaetomium spp., to mice embryo. Bull. Environ. Contam. Toxicol. 39:299–303, 1987.

487 Chaetoglobosin A

Ohtsubo (1980) injected this chemical intraperitoneally, on days 7–9 or 7–12 in the mouse. At doses of 0.5 mg per kg of body weight, exencephaly and fused ribs were increased.

Ohtsubo, K.: Teratogenicity of chaetoglobosin A in mice. Maikolokishin 10:17–18, 1980.

488 Chenodiol *Chenodeoxycolic Acid* CAS 474–25–9

Kitao et al. (1982) gave rats orally up to 360 mg per kg on days 7–17 and found no fetal ill effects.

Kitao, T.; Kamishita, S.; Yoshikawa, H. and Sakaguchi, M.: Teratogenicity studies of chenodeoxycolic acid in rats. Yakuri to Chiryo 10:3887–3901, 1982.

489 Cherry, Wild Black

Selby et al. (1971) reported an epidemic of six malformed pigs from nine mothers who fed on wild black cherries during the 16th to 28th day of gestation. Defects did not occur in sows pastured nearby in a field without black cherries. The one defective piglet described in detail was very similar to the human form of sirenomelia with absence of the anus, plantar–flexed hind legs, rudimentary external genitalia and a blindly–ending colon.

Selby, L.A.; Menges, R.W.; Houser, E.C.; Flatt, R.E. and Chase, A.C.: An outbreak of swine malformations associated with wild black cherry, Prunus serotina. Arch. Environ. Health 22:496–501, 1971.

490 Chicken Pox

Discussed under Varicella

491 Chloral Hydrate CAS 302–17–0

Heinonen et al. (1977) found no increase in the offspring of 71 women who took the drug in their first four lunar months.

Heinonen, O.P.; Slone, D. and Shapiro, S.: Birth Defects and Drugs in Pregnancy. Publishing Sciences Group Inc., Littleton, Mass., 1977.

492 Chlorambucil

α–(N–N–Di–2–Chlorethyl)aminophenylbutyric Acid CAS
305–03–3

Shotton and Monie (1963) described a human fetus of
120 mm crown–rump length with unilateral absence of a
ureter and kidney. The mother received six mg of chlorambucil daily during early pregnancy.

This alkylating agent used in treating malignancy produces defects of the nervous system, palate and skeleton in
rat fetuses when given to the mother in doses of 8 to 12
mg per kg on the 11th or 12th day. Chaube and Murphy
(1968) summarized the effects of these agents. Monie (1961)
provided a detailed account of the defects this substance
produces in the urogenital system of the rat. Soukup et al.
(1967) found about 10 percent abnormal metaphase figures
in rat embryos after maternal treatment with 8 mg per kg.

Mirkes and Greenaway (1982) produced defects of the
prosencephalon in rat embryos grown in vitro with as little
as 1.5 microgm per ml of medium. The toxic activity was enhanced by the presence of a liver monooxygenase system.

Chaube, S. and Murphy, M.L.: The teratogenic effects of
the recent drugs in cancer chemotherapy. In: Advances in
Teratology, III., D.H.M. Woollam (ed.), Academic Press, New
York, 194–196, 1968.

Mirkes, P.E. and Greenaway, J.C.: Teratogenicity of
chlorambucil in rat embryos in vitro. Teratology 26:135–
143, 1982.

Monie, I.W.: Chlorambucil–induced abnormalities of
the urogenital system of rat fetuses. Anat. Rec. 139:145–
153, 1961.

Shotton, D. and Monie, I.W.: Possible teratogenic effect
of chlorambucil on a human fetus. JAMA 186:74–75, 1963.

Soukup, S.; Takacs, E. and Warkany, J.: Chromosome
changes in embryos treated with various teratogens. J. Embryol. Exp. Morphol. 18:215–226, 1967.

493 Chloramine

George et al. (1987) treated male and female rats orally
before fertilization, during the entire gestation and for 10
days postnatally with 10 mg per kg of body weight. They
found no adverse fetal or postnatal growth problems.

George, E.L.; Carlton, B.D.; Barasan, A.H. and Smith,
M.K.: Developmental effects of drinking water disinfectants. Teratology 36:42A, 1987.

494 Chloramphenicol Chloromycetin™ CAS 56–75–7; CAS 530–43–8

Heinonen et al. (1977) reported no increase in defects
among 98 women treated during the first four lunar months.
Three hundred, forty–eight women treated any time in
pregnancy had no increase in defects.

Generalized edema was found in the rat in 12 to 71 percent of fetuses whose mothers received 2–4 percent chloramphenicol in the diet during the latter half of gestation (Mackler et al., 1975). Prochazka et al. (1964) reported no defects
in the rat fetuses whose mothers received 100 to 200 mg
per kg during pregnancy. Fritz and Hess (1971) using the

rat, mouse and rabbit, found little evidence of teratogenicity.
Dyban and Chebotar (1971) administered very large doses
(2500 mg per kg) by stomach tube on the 9th through the
11th day and produced eventration, hydrocephalus, cleft lip
and defects of the diaphragm.

In chick embryos, chloramphenicol inhibited protein
synthesis and growth but only infrequent defects of the
splanchnopleure and neural tube were reported (Blackwood, 1962; Billett et al., 1965). In the rabbit (Brown et al.,
1968) and monkey (Courtney and Valerio, 1968), no congenital defects were produced.

Billet, F.S.; Collini, R. and Hamilton, L.: The effects of
D and L threo–chloramphenicol in the early development of
the chick embryo. J. Embryol. Exp. Morphol. 13:341–356,
1965.

Blackwood, U.B.: The changing inhibition of early differentiation and general development in the chick embryo
by 2–ethyl–5–methyl–benzimidazole and chloramphenicol.
J. Embryol. Exp. Morphol. 10:315–336, 1962.

Brown, D.M.; Harper, K.H.; Palmer, A.K. and Tesh, S.A.:
Effect of antibiotics upon the rabbit. (abs) Toxicol. Appl.
Pharmacol. 12:295 only, 1968.

Courtney, K.D. and Valerio, D.A.: Teratology in the
Macaca mulatta. Teratology 1:163–172, 1968.

Dyban, A.P. and Chebotar, N.A.: Can cleft palate be induced by chloramphenicol? Arch. Anat. (Russian) 60(5):25–
29, 1971.

Heinonen, O.P.; Slone, D. and Shapiro, S.: Birth Defects
and Drugs in Pregnancy. Publishing Sciences Group Inc.,
Littleton, Mass, 1977.

Fritz, H. and Hess, R.: The effect of chloramphenicol on
the prenatal development of rats, mice and rabbits. Toxicol.
Appl. Pharmacol. 19:667–674, 1971.

Mackler, B.; Grace, R.; Tippit, D.F.; Lemire, R.J.; Shepard, T.H. and Kelley, V.C.: Studies of the development of congenital anomalies in rats. 3. Effects of inhibition of mitochondrial energy systems on embryonic development. Teratology 12:391–396, 1975.

Prochazka, J.; Simkova, V.; Havelka, J.; Hejzlar, M.;
Viklicky, J.; Kargerova, A. and Kubikova, M.: Concerning
the penetration of the placenta by chloramphenicol. (Russian) Pediatriia 19:311–314, 1964.

495 2–Chlor–4–azaphenthiazine Cloxypendyl™ CAS 15311–77–0

Gross et al. (1968) gave 5 mg per kg to rats from the 8th
through the 20th day and found no evidence of teratogenicity.

Gross, A.; Thiele, K.; Schuler, W.A. and Schlichtegroll,
A.: 2–chloro–4–azaphenthiazine Synthese und Pharmakologische Eigenschaften von Cloxypendyl. Arzneim. Forsch.
18:435–442, 1968.

496 Chlorazepate, Dipotassium

Dipotassium–7–chloro–3–carboxy–1,3–dihydro–2, 2–
dihydroxy–5–phenyl–2h–1,4–benzodiazepine–chlorazepate

Brunaud et al. (1970) studied this tranquillizer in pregnant rats, mice and rabbits during organogenesis. The rats and mice received up to 50 mg per kg subcutaneously. No adverse fetal effects were reported.

Brunaud, M.; Navarro, J.; Salle, J. and Sciou, G.: Pharmacological, toxicological, and teratological studies on dipotassium–7–chloro–3–carboxy–1,3–dihydro–2, 2-dihydroxy–5–phenyl–2H–1,4–benzodiazepine-chloroazepate (dipotassium chlorazepate, 4306 CB), a new tranquilizer. Arzneim–Forsch. 20:123–125, 1970.

497 Chlorcyclamide™

This oral hypoglycemic used in a dose of 1000 mg per kg in rats on the 9th and 10th days of pregnancy resulted in 29 percent abnormal fetuses. The defects included hydrocephalus, microcephaly, hydronephrosis and heart malformations (Bariljak, 1968).

Bariljak, I.R.: Comparison of antithyroidal and teratogenic activity of some hypoglycemic sulphanylamides. (Russian) Problems of Endocrinology 14(6):89–94, 1968.

498 Chlorcyclizine

1–(p–Chloro–α–phenylbenzyl)–4–methyl–piperizine CAS 82–93–9

This antihistamine belongs to the benzhydrylpiperazine series of compounds and has been used to produce cleft palate in the rat. It is more fully discussed under the heading, Meclizine.

499 Chlordane CAS 57–74–9

1,2,4,5,6,7,8,8–Octachloro–4–7–methano–3α, 4,7,7,α–tetrahydroindane

Ingle (1952) found normal offspring of rats fed 150 to 300 PPM chlordan during and after gestation. Maintained with their lactating mothers, these young developed excitability and tremors. If they were transferred to foster mothers on normal diets, their development was normal. Cranmer et al. (1979) treated mice throughout pregnancy and with 8.0 mg per kg per day the offspring were shown to have a defect in cell–mediated immune response. Usami et al. (1986) gavaged rats on days 7–17 with up to 80 gm per kg and found no increase in fetal death or malformations.

Cranmer, J.S.; Avery, D.L. and Barnett, J.B.: Altered immune competence of offspring exposed during development to the chlorinated hydrocarbon pesticide chlordane. (abs) Teratology 19:23A only, 1979.

Ingle, L.: Chronic oral toxicity of chlordan to rats. Arch. Ind. Hyg. Occup. Med. 6:357–367, 1952.

Usami, M.; Kawashima, K.; Nakaura, S.; Yamaguchi, M.; Tanaka, S.; Takanaka, A. and Omori, Y.: Effect of chlordane on prenatal development of rats. Eisei Shikenjo Hokoku 104:68–73, 1986.

500 Chlordantoin

Heinonen et al. (1977) did not find an increase in defects among the offspring of 24 women who used the drug in the first four lunar months.

Heinonen, O.P.; Slone, D. and Shapiro, S.: Birth Defects and Drugs in Pregnancy. Publishing Sciences Group Inc., Littleton, Mass., 1977.

501 Chlordecone *Kepone* CAS 143–50–0

This polychlorinated organic insecticide and fungicide was fed to mice in amounts of 40 ppm (Huber, 1965). Reproduction failure related to constant estrus was observed. The average young per litter was reduced by 30 ppm. Although the material was detected in fetuses, no congenital defects were reported. Cannon and Kimbrough (1979) fed rats 25 ppm for three months and completely inhibited reproduction. Chernoff and Rogers (1976) reported enlarged cerebral ventricles and dilated renal pelves in rat fetuses exposed by gavage to 10 and 6 mg per kg on days 7–16. In the mouse, clubfeet were increased over controls at 12 mg per kg.

Canon, S.B. and Kimbrough, R.D.: Short–term chlordecone toxicity in rats including effects on reproduction, pathological organ changes, and their reversibility. Toxicol. Appl. Pharmacol. 47:469–476, 1979.

Chernoff, N. and Rogers, E.H.: Fetal toxicity of kepone in rats and mice. Toxicol. Appl. Pharmacol. 38:189–194, 1979.

Huber, J.J.: Some physiologic effects of the insecticide kepone in the laboratory mouse. Toxicol. Appl. Pharmacol. 17:516–524, 1965.

502 Chlordiazepoxide *Librium*™ CAS 58–25–3

In a study of 136 women taking this drug in the first 13 weeks of pregnancy, Crombie et al. (1975) found only three offspring with congenital defects. Hartz et al. (1975) in a followup of 50,000 pregnancies could not associate an increased defect rate in the 501 pregnancies where the drug was taken in the first trimester. Milkovich and Van den Berg (1974), using a computer linkage between prescriptions given to mothers and the outcome of their pregnancies, found four congenital defects among 35 women taking the drug in the first 42 days.

Crombie, D.L.; Pinsent, R.J.; Fleming, D.M.; Rumeau–Rouquette, C.; Goujard, J. and Huel, G.: Fetal effects of tranquilizers in pregnancy. N. Eng. J. Med. 293:198–199, 1975.

Hartz, S.C.; Heinonen, O.P.; Shapiro, S.; Siskind, V. and Slone, D.: Antenatal exposure to meprobamate and chlordiazepoxide in relation to malformations, mental development and childhood mortality. N. Eng. J. Med. 292:726–728, 1975.

Milkovich, L. and Van den Berg, B.J.: Effects of prenatal meprobamate and chlordiazepoxide hydrochloride on human embryonic and fetal development. N. Eng. J. Med. 291:1268–1271, 1974.

503 Chlorfenvinphos

Diethyl–1–(2–4–dichlorophenyl)–2–chlorovinyl Phosphate CAS 470–90–6

Ambrose et al. (1970) fed rats diets with up to 300 PPM of this organophosphate ester pesticide during several generations and found no increase in congenital defects.

Ambrose, A.M.; Larson, P.S.; Borzelleca, J.F. and Hennigar, G.R.: Toxicologic studies on diethyl–1–(2,4–dichlorophenyl) 2–chlorovinyl phosphate. Toxicol. Appl. Pharmacol. 17:323–336, 1970.

504 Chlorguanide *Paludrine*™ CAS 500–92–5

Dyban et al. (1966) using 5 to 30 mg of paludrine between days 8 to 13 detected no toxic or teratogenic effect in the offspring of rats.

Dyban, A.P.; Udalova, L.D. and Akimova, I.M.: Correlation between teratogenic activity and chemical structure of drugs. Experiments with chloridine and paludrine. Dokl. Acad. Sci. U.S.S.R. (Russian) 167(1):228–231, 1966.

505 Chlorinated Humic Substances

Smith et al. (1986) reviewed studies done with artificially chlorinated humics. The concentration in the drinking water used in the tests was up to 0.8 gm per liter. No prenatal effects were found but a slight postnatal growth retardation was found in rat pups.

Smith, M.K.; Zenick, H. and George, E.L.: Reproductive toxicology of disinfection by–products. Environ. Health Perspect. 69:177–182, 1986.

506 Chlorine Dioxide

George et al. (1987) treated male and female rats orally before fertilization, during the entire gestation and for 10 days postnatally with 10 mg per kg of body weight. They found no adverse fetal or postnatal growth problems.

George, E.L.; Carlton, B.D.; Basaran, A.H. and Smith, M.K.: Developmental effects of drinking water disinfectants. Teratology 43A, 1987.

507 Chlorisopropamide

Bariljak (1968) using 1000 mg per kg on the 9th and 10th gestational days of the rat produced 30 to 45 percent fetal mortality but no congenital defects were found.

Bariljak, I.R.: Comparison of antithyroidal and teratogenic activity of some hypoglycemic sulphanylamides. Problems of Endocrinology (Russian) 14(6):89–94, 1968.

508 Chlorite, Sodium

This substance has been proposed for water purification. Moore et al. (**1980**) added this chemical to the water of pregnant and lactating mice in amounts up to 100 mg per liter. The only adverse effect was a reduction in the offspring's weight gain. Couri et al. (**1982**) studied rats subjected to drinking up to 2% in the water or up to 50 mg per kg intraperitoneal injections on gestational days 8–15. At 10 and 20 mg per kg intraperitoneally the fetal loss was increased and the growth decreased but no increase in congenital defects occurred. With .01 and .05% in the drinking water fetal weight was reduced but no congenital defect increase was seen.

Couri, D.; Miller, Jr., C.H.; Bull, R.J.; Delphia, J.M. and Ammar, E.M.: Assessment of maternal toxicity, embryotoxicity and teratogenic potential of sodium chlorite in Sprague–Dawley rats. Environmental Health Perspectives 46:25–29, 1982.

Moore, G.S.; Calabrese, E.J. and Leonard, D.A.: Effects of chlorite exposure on conception rate and litters of A/J strain mice. Bull. Environ. Contam. Toxicol. 25:689–696, 1980.[A

509 Chlormadinone Acetate *17 α–Acetoxy–6–chloro–4, 6–pregnadiene–3, 20–dione* CAS 302–00–7

Kawashima et al. (1977) administered 5 mg orally to rats on days 17–20 and produced a reduction in urovaginal septum length in the female fetuses.

Takano et al. (1966) reported that this progestin produced defects in mice and rabbits. The mice were given 3–50 mg per kg orally during the latter two–thirds of pregnancy. Cleft palate and club foot were the main defects observed. In rabbits with administration of 10 mg per day from the 8th through the 17th gestational day skeletal and palate defects occurred along with a relatively high incidence of abdominal wall defect. Masculinization of the female mouse and rabbit fetuses was not produced. Chambon et al. (1966) reported a similar type of teratogenicity in the rabbit and rat, but the dose in the rabbit was 0.5 mg per day, considerably less than that used by Takano.

Chambon, Y.; Depagne, A. and Leveve, Y.: Malformations et deformations foetales par insuffisance hormonale gestative chez le lapin et chez le rat. Comptes Rendus De l'association Des Anatomistes 131:270–279, 1966.

Kawashima, K.; Nakaura, S.; Nagao, S.; Tanaka, S.; Kuwamura, T. and Omori, Y.: Virilizing activities of various steroids in female rat fetuses. Endocrinol. Japon. 24(1):77–81, 1977.

Takano, K.; Yamamura, H.; Suzuki, M. and Nishimura, H.: Teratogenic effect of chlormadinone acetate in mice and rabbits. Proc. Soc. Exp. Biol. Med. 121:455–457, 1966.

510 Chlormequat *Chlorocholine Chloride* CAS 999–81–5

Juszkiewicz et al. (1970) administered 25–400 mg per kg orally to hamsters on the 8th gestational day. With doses of 300 and 400 mg, 5.9 and 8.8 percent of the fetuses, respectively, were abnormal. The defects were polymorphic but included encephaloceles, anophthalmia and microphthalmia. Ackermann et al. (1970) found no teratogenicity in the rat.

Ackermann, H.; Proll, J. and Luder, W.: Untersuchungen zur Toxikologischen Beurteilung von Chlorcholinchlorid. Arch. Experiment. Veterin. 24:1049–1057, 1970.

Juszkiewicz, T.; Rakalska, Z. and Dzierzawski, A.: Effet embryopathie du chlorure de chlorocholine (CCC) chez le hamster dore. Eur. J. Toxicol. 3:265–270, 1970.

511 Chlormezanone CAS 80–77–3

Heinonen et al. (1977) studied the offspring of 26 women who took this tranquilizer anytime during pregnancy. Three had malformations but the type was not described.

Heinonen, O.P.; Slone, D. and Shapiro, S.: Birth Defects and Drugs in Pregnancy. Publishing Sciences Group Inc., Littleton, Mass, 1977.

512 Chloroacetamide CAS 79–07–2

Thiersch (1971) gave pregnant rats 20 mg per kg on single days (7th, 11th or 12th) and found no effects on litter size or fetuses.

Thiersch, J.B.: Investigations into the differential effect of compounds on rat litter and mother. In: Malformations Congenitales Des Mammiferes, H. Tuchmann–Duplessis (ed.), Paris: Masson et Cie, 95–113, 1971.

513 Chloroacetic Acid

Smith et al. (1990) gavaged rats during organogenesis with up to 140 mg per kg daily. The highest dose was maternally toxic. At the highest dose there was an increase in cardiovascular malformations primarily levocardia.

Smith, M.K.; Randall, E.J.; Stober, R and Strober J.A.: Developmental effects of chloroacetic acid in the Long–Evans rat. (abst)Teratology 41:593, 1990.

514 Chlorobenzene CAS 108–90–7

John et al. (1984) exposed rats and rabbits for six hours daily during organogenesis to concentrations of up to 590 ppm. At the highest dose in rabbits a few fetuses with ablepharia and fore–limb flexure were found but the total malformations were not increased significantly. No adverse fetal effects in the rats. Black et al. (1988) studied three congeners (1,2,3; 1,2,4 and 1,3,5) in pregnant rats at up to 600 mg per kg daily by gavage on days 6–15. No feto–toxicity or teratogenicity was found.

Black, W.D.; Valli, V.E.O.; Ruddick, J.A. and Villeneuve, D.C.: Assessment of teratogenic potential of 1,2,3- 1,2,4- and 1,3,5–trichlorobenzenes in rats. Bull. Environ. Contam. Toxicol. 41:719–726, 1988.

John, J.A.; Hayes, W.C.; Hanley, T.R., Jr.; Johnson, K.A.; Gushow, T.S. and Rao, K.S.: Inhalation teratology study of monochlorobenzene in rats and rabbits. Toxicol. Appl. Pharmacol. 76:365–373, 1984.

515 o–Chlorobenzylidene Malononitrile CS CAS 2698–41–1

Upshall (1973) studied this irritant which is used for dispersing unruly crowds. Five minute exposures of up to 60 mg per cubic meter were used on days 6 through 15 in rats and day 6 through 18 in rabbits. No adverse fetal effects were found. Rats given 20 mg per kg intraperitoneally on days 6, 8, 10, 12 and 14 gave birth to normal fetuses.

Upshall, D.G.: Effects of o–chlorobenzylidene malononitrile (CS) and the stress of aerosol inhalation upon rat and rabbit embryonic development. Toxicol. Appl. Pharmacol. 24:45–59, 1973.

516 Chlorobiphenyls PCB

Maternal ingestion of chlorobiphenyls from contamination of a brand of cooking oil in Japan was responsible for dark–brown staining of the skin of the newborn babies (Miller, 1971). This rice oil disease is called "Yusho" in Japanese and "Yucheng" in Chinese. This syndrome called the cola–colored baby was associated frequently with intrauterine growth retardation but no defects were recorded and the skin discoloration faded in a few months (Taki et al., 1969). Funatsu et al. (1972) reported four newborns with the syndrome and noted that three had exophthalmus and two had teeth present at birth. Shiota (1976) gave some data on

polychlorinated biphenyl concentrations in human fetuses. Rogan (1982) reviewed the clinical aspects. Sunahara et al. (1987) found that the birth weight was proportional to the decrease in EGF–stimulated receptor autophosphorylation of the placental EGF receptor in the exposed pregnancies. Yu et al. (**1991**) followed 128 children exposed to contaminated cooking oil prenatally or via breast feeding. By most developmental tests the exposed group was decreased significantly. The children who were smaller at birth or who had signs of intoxication were more severely effected. Lione (1988) reviewed the effects of PCB's on reproduction with particular emphasis on the human.

Fein et al. (1984) studied the offspring of mothers eating contaminated fish from the Great Lakes and reported that exposed infants were 160 to 190 gm lighter than controls and their heads were 0.6 to 0.7 cm smaller. The exposed newborn was defined as having 3.0 or greater nanogms per ml in the cord blood. None of the confounding variables were found to explain this finding. Taylor et al. (1984) found a birth weight reduction of 153 gm in the offspring of 51 women exposed to PCB's in the workplace. Their gestation was also reduced by 6.6 days as compared to a low exposure group. Rogan et al. (1986) found no association between birth weight or head circumference and the levels of PCB's in cord blood, placenta and milk among 912 infants. In postnatal functional studies, higher levels were associated with hypotonicity and hyporeflexia.

Kato et al. (1972) demonstrated that the compound crosses the rat placenta but found no defects in the fetuses. Marks et al. (1981) gavaged hexachlorobiphenyl on days 6 through 15 of pregnancy in the mouse. At doses of 0.1 and 1 mg per kg, no significant increase in defects was produced but at 2, 4, 8 and 16 mg per kg one was found. Cleft palate and hydronephrosis were the most common types found. Maternal toxicity was found at 8 mg per kg. From study of other isomers they concluded that the 3, 3–prime, 4 and 4–prime positions seemed essential for teratogenicity.

Tilson et al. (1990) summarized data on the effect of biphenyls on the developing nervous system of animals and humans. They conclude that exposed humans have detectable effects on motor maturation and some evidence of impaired infant learning.

Fein, G.G.; Jacobson, J.L.; Jacobson, S.W.; Schwartz, P.M. and Dowler, J.K.: Prenatal exposure to polychlorinated biphenyls effects on birth size and gestational age. J. Pediatr. 105:315–320, 1984.

Funatsu, H.; Yamashita, F.; Ito, Y.; Tsugawa, S.; Funatsu, T.; Yoshikane, T.; Hayoshi, M.; Kato, T.; Yakashiji, M.; Okamoto, G.; Yamasaki, S.; Arima, T.; Kuno, T.; Ioe, H. and Ide, I.: Polychlorbiphenyls (PCB) induced fetopathy. I. Clinical observation. Kurume M. J. 19:43–50, 1972.

Kato, T.; Yakushiji, M.; Tuda, H.; Arimi, A.; Takahashi, K.; Shimomura, M.; Miyahara, M.; Adachi, M.; Tashiro, Y.; Matsumoto, M.; Funatsu, I.; Yamashita, F.; Ito, Y.; Tsugawa, S.; Funatsu, T.; Yoshikane, T. and Hayashi, M.: Polychlorbiphenyls (PCB) induced fetopathy. II. Experimental studies: Possible placental transfer of polychlorobiphenyls in rats. Kurume M. J. 19:53–59, 1972.

Lione, A.: Polychlorinated biphenyls and reproduction. Reprod. Toxicol. 2:83–90, 1988.

Marks, T.A.; Kimmel, G.L. and Staples, R.E.: Influence of symmetrical polychlorinated biphenyl isomers on embryo and fetal development. Toxicol. Appl. Pharmacol. 61:269–276, 1981.

Miller, R.W.: Cola–colored babies. Chlorobiphenyl poisoning in Japan. Teratology 4:211–212, 1971.

Rogan, W.J.: PCBs and cola–colored babies: Japan, 1968, and Taiwan, 1979. Teratology 26:259–261, 1982.

Rogan, W.J.; Gladen, B.C.; McKinney, J.D.; Carreras, N.; Hardy, P.; Thullen, J.; Tinglestad, J. and Tully, M.: Neonatal effects of transplacental exposure to PCBs and DDE. J. Pediatr. 109:335–341, 1986.

Shiota, K.: Fetal accumulation of polychlorinated biphenyls (PCB's). A consideration on the problem of human pollutants. Cong. Anom. 16:9–16, 1976.

Sunahara, G.I.; Nelson, K.G.; Wong, T.K. and Lucier, G.W.: Decreased human birth weights after in utero exposure to PCB's and PCDF's are associated with decreased placental EGF-stimulated receptor autophosphorylation capacity. Molecular Pharmacology 32:572–578, 1987.

Taki, I.; Hisanaga, S. and Amagase, Y.: Report on Yusho (chlorobiphenyls poisoning). Pregnant women and their fetuses. Fukuoka Acta Med. 60:471–474, 1969.

Taylor, P.R.; Lawrence, C.E.; Hwang, H–L. and Paulson, A.S.: Polychlorinated biphenyls influence on birthweight and gestation. Am. J. Publ. Health 74:1153–1154, 1984.

Tilson, H.A.; Jacobson, J.L. and Rogan, W.J.: Polychlorinated biphenyls and the developing nervous system: cross–species comparisons. Neurotoxicology and Teratology 12:239–248, 1990.

Yu, M.; Hsu, C.; Gladen, B.C. and Rogan, W.J.: In utero pcb/pcdf exposure: Relation of developmental delay to dysmorphology and dose. Neurotoxicology and Teratology 13:195–202, 1991.

517 Chlorobutanol

Smoak (**1993**) studied the effects of this preservative on whole embryo culture of mice. Early somite embryos were malformed and growth retarded at 25 microgm per ml. Defects of the neural tube and heart occurred. No effect was found at the 10 microgm level. Plasma levels in chronically treated patients have been up to 85 microgm per ml.

Smoak, I.W.: Embryotoxic effects of chlorobutanol in cultured mouse embryos. Teratology 47:203–208, 1993.

518
3–Chloro–5–[3–(4–carbamyl–4–piperidinopiperidino) propyl]–10,11–dihydro–5H–dibenz [b,f] Azepine Dihydrochloride Monohydrate CAS 65016–29–7

Hamada (1970) gave this compound orally to pregnant rats and mice on several or more days during midgestation. At the highest dose in the mouse (200 mg per day) fetal growth retardation and cleft palates were increased. In the rats at doses up to 250 mg per kg, no fetal changes occurred.

Hamada, Y.: Studies on psychotropic drugs. 14. Toxicological studies of 3–chloro–5–[3–(4–carbamyl–4–piperidinopiperidino) propyl]–10,11–dihydro–5H–dibenz [b,f] azepine dihydrochloride monohydrate. Oyo Yakuri 5:663–668, 1970.

519 5–Chloro–2–deoxyuridine CAS 50–90–8

Chaube and Murphy (1964) produced defects of the nervous system, palate and skeleton in rats treated with 125 to 1000 mg per kg on the 12th day. Thymidine protected the fetus against the teratogenic effects of this compound.

Chaube, S. and Murphy, M.L.: Teratogenic effects of 5–chlorodeoxyuridine on the rat fetus: Protection by physiological pyrimidines. Cancer Res. 24:1986–1989, 1964.

520 2–Chlorodibenzofuran

Usami et al. (**1993**) gavaged mice with the drinking water contaminant in doses of up to 500 mg per kg during organogenesis. The liver weights of the dams were increased at doses of 12.5 mg per kg and above and an increase in the number of fetuses with a skeletal variation in the suboccipital bone was found at the highest dose. No other adverse effects occurred in the fetuses.

Usami, M.; Sakemi, K.; Tabata, H.; Kawashima, K. and Takanaka, A.: Effects of 2–chlorodibenzofuran on fetal development in mice. Bull. Environ. Contam. Toxicol. 51:748–755, 1993.

521 Chlorodibromomethane

Ruddick et al. (1980) gave chlorodibromomethane by gavage to rats on days 6–15. Maternal and fetal toxicity were noted at the highest level (200 mg per kg).

Ruddick, J.A.; Villeneuve, D.C.; Chuand, I. and Valli, V.E.: Teratogenicity assessment of four trihalomethanes. Teratology 21:66A only, 1980.

522 Chlorodifluoromethane CAS 75–45–6

Litchfield and Longstaff (1984) summarized reproductive tests in animals. Exposure of rats to 50,000 ppm for six hours daily on days 6–15 was associated with an increase in anophthalmia. At this dose maternal toxicity was present. At 1000 ppm no defects were found. No adverse effects were found in rabbits exposed to similar doses during organogenesis.

Litchfield, M.H. and Longstaff, E.: The toxicological evaluation of chlorofluorocarbon 22 (CFC 22). Food Chem. Toxicol. 22:465–475, 1984.

523 p–Chlorodimethylaminobenzene

Discussed under Aminoazobenzene

524 1–(2–Chloroethyl)–3–cyclohexyl–1–nitrosourea
Lomustine

Thompson et al. (1975) gave rats intraperitoneal doses of 2 to 8 mg per kg during 4–day periods of organogenesis. At 2 mg and above omphalocele, ectopia cordis, hydrocephalus, syndactyly, anophthalmia and aortic arch anomalies occurred. Doses of up to 3.0 mg per kg in the rabbit caused resorptions but no defects.

Thompson, D.J.; Molello, J.A.; Strebing, R.J. and Dyke, I.L.: Reproduction and teratological studies with 1–(2–chloroethyl)–3–cyclohexyl–1–nitrosurea (CCNU) in the rat and rabbit. Toxicol. Appl. Pharmacol. 34:456–466, 1975.

525 Chlorofluoromethane CAS 593–70–4

Coate et al. (1979) treated rats for 6 hours daily on days 6–15 using 0.1 percent and found only a low incidence of cervical ribs (8 out of 208).

Coate, W.B.; Voelker, R. and Kapp, R.W., Jr.: Inhalation toxicity of monochloromonofluoromethane. Toxicol. Appl. Pharmacol. 48:A109, 1979.

526 Chloroform CAS 67–66–3

Ruddick et al. (1980) gavaged rats with 100 mg per kg on days 6 through 15. At all dose levels maternal toxicity occurred and this was associated with fetal toxicity. Smith et al. (1986) summarized animal tests and at the highest inhalation doses found some maternal toxicity which was associated with fetotoxic effects.

Smith, M.K.; Zenick, H. and George, E.L.: Reproductive toxicology of disinfection by–products. Environ. Health Perspect. 69:177–182, 1986.

Ruddick, J.A.; Villeneuve, D.C.; Chuand, I. and Valli, V.E.: Teratogenicity assessment of four trihalomethanes. Teratology 21:66A, 1980.

527 Chloro(triethyl–phosphine) gold SK&F D–39162

Szabo et al. (1978) gave rabbits up to 8 mg per kg during organogenesis. Oral doses of 2 and 8 mg per kg were associated with lung defects and umbilical wall defects.

Szabo, K.T.; DiFebbo, M.E. and Phelan, D.G.: The effects of gold–containing compounds on pregnant rabbits and their fetuses. Vet. Path. 15(Suppl.5):97–102, 1978.

528 Chloromethylenecycline

Discussed under Chlorotetracycline

529 6–Chloro–δ[6]–1, 17α–methylene–17–α–hydroxyprogesterone

Junkmann and Neumann (1964) found that this compound produced hypospadias in male rats when the mother was treated with 0.1 to 10 mg per animal per day on days 16–19.

Junkmann J. and Neumann, F.: M Wirkungsmechanismus von an feten antimaskulin wirksamen gestagenen. Acta Endocrinologica 90:139–154, 1964.

530 4–Chloro–2–methylphenoxyacetic Acid Ethylester
MCPEE CAS 2698–38–6

Yasuda and Maeda (1972) tested this herbicide in pregnant rats. It was given (with the diet at levels of 40, 500, 1000 and 2000 PPM) on days 8 through 15 of gestation. At the highest dose which was equivalent to approximately 100 mg per kg per day, congenital malformations including cleft palate, ventricular septal defect and kidney defects were found. At this high level there was maternal weight loss.

Their report contains a summary of teratologic testing of other phenoxyalkanoic acid herbicides.

Yasuda, M. and Maeda, H.: Teratogenic effects of 4–chloro–2–methylphenoxyacetic acid ethylester (MCPEE) in rats. Toxicol. Appl. Pharmacol. 23:326–333, 1972.

531 6–Chloro–4–nitro–2–aminophenol

This ham colorant was given to rats in amounts of up to 350 mg per kg orally on days 6–15 and no adverse reproductive effects were found (Picciano et al., 1984).

Picciano, J.C.; Morris, W.E. and Wolf, B.A.: Evaluation of the teratogenic potential of the oxidative dyes 6–chloro–4–nitro–2–aminophenol and o–chloro–p–phenylenediamine. Food Chem. Toxicol. 22:147–149, 1984.

532 3–Chloro–5–[3–(2–oxo–1,2,3,5,6,7,8a–octahydro–imidazo[1,2a]pyridind–3–spiro–4'–piperidino) propyl]–10,11–dehudro–5H' sibenz(b,f) azepine Dihydrochloride CAS 89419–40–8 Y–516

Yoneyama et al. (**1989**) administered orally up to 100 mg per kg to male and female rats before fertilization and for 7 days of pregnancy in the female. Food consumption was decreased at the highest dose and diestrus was increased at 1, 10 and 100 mg per kg but not at 1.15 mg per kg. No skeletal anomalies occurred. Imanishi et al. (**1989**) gave rats and rabbits up to 30 mg and 100 mg per kg respectively during organogenesis. No teratogenicity was found but at the highest dose maternal sedation and fetal mortality was increased in the rabbits.

Imanishi, M.; Yoneyama, M.; Hashiguchi, M.; Takeuchi, M. and Maruyama, Y.: Teratogenicity study of (+)(-)–3–chloro–5–[3–(2–oxo–1,2,3,5,6,7,8,8a–octahydroimidazo[1,2–a]pyridine–3–spiro–4"–piprtifino)–propyl]–10,11–dihydro–5h–dibenz[b,f] azepine dihydrochloride y–516 in rats and rabbits. Oyo Yakuri 38(2):91–107, 1989.

Yoneyama, M.; Imanishi, M.; Taeuchi, M. and Maruyama, Y.: (+)(-)–3–chloro–5–[3–(2–oxo–1,2,3,5,6,7,8,8a–octahydroimidazo[1,2–a]pyridine–3–spiro–f N–dimridino) propyl]–10,11–dihydro–5h–dibenz[b,f]azepine dihydrochloride (y–516) in rats. Oyo Yakuri 38(2):79–90, 1989.

533 Chlorophenols *Tetrachlorophenol* *Pentachlorophenol*

Schwetz et al. (1974a) report studies with 2,3,4,6–tetrachlorophenol in the rat using 30 mg and 50 mg per kg, respectively, orally on days 6 through 15 of gestation. Although some fetal edema and delay in ossification occurred no defects were produced. For pentachlorophenol, Schwetz et al. (1974b) found embryotoxicity when oral doses above 5 mg per kg were given. No teratogenic activity was found.

Schwetz, B.A.; Keeler, P.A. and Gehring, P.J.: Effect of purified and commercial grade tetrachlorophenol on rat embryonal and fetal development. Toxicol. Appl. Pharmacol. 28:146–150, 1974a.

Schwetz, B.A.; Keeler, P.A. and Gehring, P.J.: The effect of purified and commercial grade pentachlorophenol on rat embryonal and fetal development. Toxicol. Appl. Pharmacol. 28:151–161, 1974b.

534 o–Chloro–p–phenylenediamine

Picciano et al. (1984) gavaged up to 40 mg per kg in the pregnant rat on days 6–15 and observed decreased fetal weight and resorptions, but no increase in fetal malformations.

Picciano, J.C.; Morris, W.E. and Wolf, B.A.: Evaluation of the teratogenic potential of the oxidative dyes 6–chloro–4–nitro–2–aminophenol and o–chloro–p–phenylenediamine. Food Chem. Toxicol. 22:147–149, 1984.

535 (E)–[4–(2–Chlorophenyl)–1,3–diathiolan–2–ylidene]–1–imidazolylacetonitrile
NND–318

Mayfield et al. (1992 A,B,C) studied this fungicide in rats at doses of up to 125 mg per kg subcutaneously before pregnancy, during organogenesis and in the perinatal period. At the highest dose there were no pregnant animals. At 25 mg pe kg there were increases in the number of fetuses with dilated renal calyces. No significant increase in defects was found at a dose level of 7 mg per kg and postnatal development and reproduction was not affected.

Mayfield, R.; Jones, K.; John, D.M.; Parker, C.A. and Konaka, S.: Fertility and reproductive performance of the rat following subcutaneous administration of (+)–(E)–[4–(2–chlorrophenyl0–1,2–dithiolan–2–ylidene]–1–imidazolylacetontrile (NND–318) prior to mating and up to day 7 of pregnancy. Oyo Yakuri 43:353–367, 1992A.

Mayfield, R.; Jones, K.; Parker, C.A.; Hughes, E.W. and Konaka, S.: Pre– and post natal development of the rat following subcutaneous administration of (+)–(E)–[4–2–chlorophenyl)–1–dithiolan–2–ylidene]–1–imidazolylacetonitril (NND–318) during organogenesis. Oyo Yakuri 43:369–382, 1992B.

Mayfield, R.; Jones, K.; Parker, C.A.; Hughes, E.W. and Konaka, S.: Pre–post natal developmental of the rat following subcutaneous administration of (+)–(E)–[4–(2–Chlorophenyl)–1,   3–dithiolan–2–ylidene]–1–imidazolylacetonitrile (NND–318) during the peri–and post natal period. Oyo Yakuri 43:383–393, 1992C.

Mayfield, R.; John, D.M. and Konaka, S.: Pre–natal development of the rabbit following subcutaneous administration of (+)–(E)–[4(2–chlorophenyl)–1,3–dithiolan–2–ylidene]–1–imidazolylacetonitrile (NND–318). Oyo Yakuri 43:395–400, 1992.

536 1–(3–Chlorophenyl)–3–N,N–dimethylcarbamoyl–5–methoxyprazole
PZ–177

Yamamoto Et al (1978 A) treated rats with up to 250 mg per kg on days 7–17 and found no adverse fetal effects frrom this anti-inflammatory agent. Rabbits were given up to 400 mg per kg on days 6–18. At 400 but not 200 mg per kg there was decxreased maternal weight gain and increased fetal resorption.

Yamamoto, H.; Miyake, J.; Miyake, H.; Nakamura, Y.; Kawase, Y.; Ohhata, H. and Yamada, S.: The effects of 1–(3–chlorophenyl)–3–N,N–dimethylcarbamoyl–5–methoxypryazole (PZ–177) on reproduction in rats. Iyakuhin Kenkyu 9:538–548, 1978.

Yamamoto, H.; Miyake, J.; Miyake, H. and Asada, M.: A terratological study of 1–(3–chlorophenyl)–3–N, N–dimethylcarbamoyl–5–methoxypyrazole (PZ–177) in rabbits: Its administration during the period of majorr organogenesis of the progeny. Iyakuhin Kenkyu 9:558–562, 1978.

537 o–Chloro–β–phenyl–ethylhydrazine Dihydrogen Sulfate *WL–27*

Poulson and Robson (1964) found that this amineoxidase inhibitor prevented implantation in mice. When they gave 20 mg per kg per day on the 6th through 11th days, five litters were examined and no defects noted. Rabbits receiving 27 mg per kg on day 7 to 11 of pregnancy gave birth to 30 fetuses, one of which exhibited unusual facial, skull and limb defects.

Poulson, E. and Robson, J.M.: Effect of phenelzine and some related compounds on pregnancy and sexual development. J. Endocrinol. 30:205–215, 1964.

538 Chloroprene

Salnikova and Fomenko (1973) exposed rats for 4 hours daily through gestation to up to 4.0 mg per cubic meter and found weight and skeletal reduction at 3 and 4 mg per cubic meter. Decreased postnatal viability was found at the 0.6 and 0.13 mg per cubic meter levels.

Salnikova, L.S. and Fomenko, V.N.: Experimental study of the effect of chloroprene on embryogenesis (Russian). Gig. Tr. Prof. Zabol. 17:23–26, 1973.

539 β–Chloroprene *2–Chlorobutadiene–1,3* CAS 126–99–8

Culik et al. (1978) exposed rats by inhalation to 25 ppm for four hours daily on days 3 through 20 and found no adverse fetal effects.

Culik, R.; Kelley, D.P. and Clary, J.J.: Inhalation studies to evaluate the teratogenic and embryotoxic potential of β–chloroprene (2–chlorobutadiene–1,3). Toxicol. Appl. Pharmacol. 44:81–88, 1978.

540 6–Chloropurine CAS 87–42–3

Tuchmann–Duplessis and Mercier–Parot (1959) produced circular body constrictions of rat fetuses treated with 5 mg doses on the 9th and 10th gestational day of the rat. Amniotic bands were present also. Murphy and Chaube (1962) gave 100 to 400 mg per kg to rats and produced cleft palate and skeletal defects. Thiersch (1957) used 100 mg per kg in rats and found that this dose on the 11th and 12th day caused defects in 53 percent of the fetuses. He commented that the placenta was more resistant than the fetus to this drug.

Murphy, M.L. and Chaube, S.: Teratogenic effects of abnormal purines and their ribosides in the rat. (abs) Proc. Amer. Ass. Cancer Res. 3:347 only, 1962.

Thiersch, J.B.: Effect of 2–6 diaminopurine (2–6 DP): 6–Chloropurine (CLP) and thioguanine (THG) on rat litter in utero. Proc. Soc. Exp. Biol. Med. 94:40–43, 1957.

Tuchmann–Duplessis, H. and Mercier–Parot, L.: Sur l'action abortive et teratogene de la 6–chloropurine. C. R. Soc. Biol. (Paris) 153:1133–1136, 1959.

541 Chloroquine CAS 54–05–7

Smith (1966) cites the case of a woman who during four of her eight pregnancies was given chloroquine (250 mg daily from the 6th week after conception). Two of these children were congenitally deaf with instability of gait. One of these children had chorioretinitis of the type associated with chloroquine toxicity in the adult. A third exposed child had hemihypertrophy and developed a Wilm's tumor. The family was reported by Hart and Naunton (1964).

Wolfe and Cordero (1985) studied 169 infants from pregnancies where the mother took the medication once weekly. Two infants had defects (tetralogy of Fallot and congenital hypothyroidism) but there was no increase over the rate in a control group of 454 unexposed. Cot et al. (1992) found no decrease in birth weight of women treated in a randomize trial.

Udalova (1967), using 1000 mg per kg on the 9th day, induced embryonic death in 27 percent and eye abnormalities in 47 percent of the surviving rat fetuses. Anophthalmia and microphthalmia were the only defects found. Dencker et al. (1975) using an I–125 labeled chloroquine analogue showed transplacental passage and concentration in the monkey fetus. The adrenal cortex and retina were found to concentrate the isotope. Ambroso and Harris (1993) found embryotoxicity and dysmorphogenesis in rat embryos exposed to 10, 20 or 30 micromolar concentrations in whole embryo culture.

Ambroso, J.L. and Harris, C.: Chloroquine embryotoxicity in the postimplantation rat conceptus in vitro. Teratology 48:213–226, 1993.

Cot, M.; Roisin, A.; Barro, D.; Yada, A.; Verhave, J.P.; Carnevale, P. and Breart, G.: Effect of chloroquine chemoprophylaxis during pregnancy on birth weight: Results of a randomized trial. Am J Trop Med Hyg 46(1):21–27, 1992.

Dencker, L.; Lindquist, N.G. and Ulberg, S.: Distribution of an I–125 labeled chloroquine analogue in a pregnant macaca monkey. Toxicology 5:255–264, 1975.

Hart, C.W. and Naunton, R.F.: The ototoxicity of chloroquine phosphate. Arch. Otolaryngol. 80:407–412, 1964.

Smith, D.W.: Dysmorphology (teratology). J. Pediatr. 69:1150–1169, 1966.

Udalova, L.D.: The effect of chloroquine on the embryonal development of rats. Pharmacol. Toxicol. (Russian) 2:226–228, 1967.

Wolfe, M.S. and Cordero, J.F.: Safety of chloroquine in chemosuppression of malaria during pregnancy. Br. Med. J. 290:1466–1467, 1985.

542 Chlorosil

Boikova et al. (1981) studied this agent in pregnant rats. At doses of up to 100 mg taken orally during 14 days of gestation, no adverse fetal effects occurred.

Boikova, V.V.; Golikov, S.N.; Korkhov, V.V. and Mots, M.N.: Reproductive studies with a new cholinolytic chlorosil in rats. Farmakol. Toksikol. (USSR) XLIV, 1:03–95, 1981.

543 Chlorotetracycline Aureomycin™ CAS 57–62–5

Tubaro (1964) produced skeletal defects in chicks receiving one mg via the yolk sac on the 8th day. Chlormethylenecycline was non–teratogenic. This author reports negative findings in pregnant rabbits and rats receiving ten mg per kg per day of chlorotetracycline.

Tubaro, E.: Possible relationship between tetracycline stability and effect on foetal skeleton. Br. J. Pharmacol. 23:445–448, 1964.

544 8–Chlorotheophylline

This drug constitutes about 45 percent of the commonly used anti–motion sickness compound Dramamine™. In 319 exposed pregnancies, Heinonen et al. (1977) found no general increase in malformations. Five children with congenital heart disease were identified but this was not statistically significant.

Heinonen, O.P.; Slone, D. and Shapiro, S.: Birth Defects and Drugs in Pregnancy. Publishing Sciences Group Inc., Littleton, Mass., 1977.

545 Chlorothiazide CAS 58–94–6

Heinonen et al. (1977) reported no increase in defect rate among 5,283 exposed infants.

Grollman and Grollman (1962) reported persistent hypertension in rat offspring after maternal treatment with 166 mg daily from day 15 until birth. Chlorothiazide and hydrochlorothiazide were injected into the pregnant rat in amounts of 250 mg per kg on day 10 and 11 and no defects resulted (Maren and Ellison, 1972).

Grollman, A. and Grollman, E.F.: Teratogenic induction of hypertension. J. Clin. Invest. 41:710–714, 1962.

Heinonen, O.P.; Slone, D. and Shapiro, S.: Birth Defects and Drugs in Pregnancy. Publishing Sciences Group Inc., Littleton, Mass., 1977.

Maren, T.H. and Ellison, A.C.: The teratogenic effect of certain thiadiazoles related to acetazolamide with a note on sulfanilamide and thiazide diuretics. Johns Hopkins Med. J. 130:95–104, 1972.

546 (4–Chloro–o–toloxy)acetic Acid
2–Methyl–4–chlorophenoxyacetic Acid MCPA CAS 94–74–6

Buslovich et al. (1979) observed single administration one–half LD–50 on the 9th and 10th day of pregnancy induced teratogenic and embryotoxic effects in rat embryos. Phenobarbital (80 mg per kg) administered to pregnant rats before the agent decreased the embryotoxic action.

Buslovich, S.Y.; Aleksashina, Z.A. and Kolosovskaya, V.M.: Effect of phenobarbital on the embryotoxic action of 2–methyl–4–chlorophenoxyacetic acid. Farmakol. Toksicol. (USSR) 42, 2:167–170, 1979.

547 Chlorphenesin Carbamate CAS 886–74–8

Jacobs (1971) produced profound muscular flaccidity in rat fetuses given 500 mg per kg on day 15.3 of gestation.

This flaccidity did not interfere with normal closure of the palate.

Jacobs, R.M.: Failure of muscle relaxants to produce cleft palate in mice. Teratology 4:25–30, 1971.

548 Chlorpheniramine CAS 132–22–9

Heinonen et al. (1977) reported that 1,070 women were exposed to this antihistamine in the first four lunar months. Only 58 malformations were observed and this was not a significant increase.

Heinonen, O.P.; Slone, D. and Shapiro, S.: Birth Defects and Drugs in Pregnancy. Publishing Sciences Group Inc., Littleton, Mass., 1977.

549 Chlorphentermine CAS 461–78–9

Lullmann–Rauch (1973) reported that injection of 40 mg per kg of body weight given to pregnant rats from day 10 through day 21 produced a lipidosis characterized by cytoplasmic inclusions in the fetal cardiac and pulmonary cells. This drug is used as an anorexigenic in humans.

Lullmann–Rauch, R.: Chlophentermine-induced ultrastructural alterations in foetal tissues. Virchows Arch. Abt. B. Zellpath. 12:295–302, 1973.

550 Chlorproguanil CAS 537–21–3

Discussed under Cycloguanil

551 Chlorpromazine *Thorazine*[TM] CAS 50–53–3

This phenothiazine compound is of doubtful teratogenicity. The literature on its teratog7nicity is critically surveyed by Kalter (1968).

Sobel (1960) reviewed 52 instances where the mother was treated and found no increase in defects or wastage, but he did observe respiratory distress in newborns from three women receiving very high doses (500 to 600 mg daily) up to parturition. Vince (1969) commented on the reports of others indicating that phenothiazines were not teratogenic in humans. His experience with one patient who had transposition of the great vessels and the reported myocardial damage from phenothiazines raised his question about embryonic or fetal myocardial toxicity.

Chambon (1955) found increased fetal mortality when rats were given 10 mg per kg, but no defects. Radioactive ^{35}S from the drug accumulates in the fetal mouse and monkey retina (Lindquist et al., 1970). Clark et al. (1970) injected 3 mg per kg into rats from days 12 through 15 of gestation and then subjected the offspring to several types of behavior tests. Although the latencies were shorter than controls, the treated group made significantly more errors in the maze and took longer to learn the bar–press response. Reversal in the T–maze or extinction of operant response were not altered. Umemura et al. (1983) studied learning behavior in rat offspring exposed to 2 mg per kg on days 17 to postpartum day 21. Most of the learning tasks were not decreased but reverse learning of light–dark discrimination was impaired.

Chambon, Y.: Action de la chlorpromazine sur l'evolution et l'avenir de la gestation chez la ratte. Ann. Endocrinol. 16:912–922, 1955.

Clark, C.V.H.; Gorman, D. and Vernadakis, A.: Effects of prenatal administration of psychotropic drugs on behavior of developing rats. Dev. Psychobiol. 34:225–235, 1970.

Kalter, H.: Teratology of the Central Nervous System. Chicago: University of Chicago Press, 164 only, 1968.

Lindquist, N.G.; Sjostrand, S.E. and Ullberg, S.: Accumulation of chorio–retinotoxic drugs in the foetal eye. Acta. Pharmacol. Toxicol. 28:64 only, 1970.

Sobel, D.E.: Fetal damage due to ECT, insulin, coma, chlorpromazine or reserpine. A.M.A. Gen. Psych. 2:606–611, 1960.

Umemura, T.; Hironaka, N.; Takada, K.; Sasa, H. and Yanagita, T.: Influence of chlorpromazine administered to rat dams in the peripartum and nursing periods on the learning behavior of the second generation. J. Toxicol. Sci. 8:105–118, 1983.

Vince, D.J.: Congenital malformations following phenothiazine administration during pregnancy. Canad. Med. Assoc. J. 100:223 only, 1969.

552 Chlorpropamide CAS 94–20–2

This oral hypoglycemic sulfylurea compound was found not to be teratogenic in rats when Tuchmann–Duplessis and Mercier–Parot (1959) gave 200 mg per day from the 1st through 12th day of gestation. Brock and von Kreybig (1965) found 'disrupted' 13–day embryos in one strain of rats after injecting 10 mg per kg on the 4th day.

Although Jackson et al. (1962) initially reported an unusually high fetal mortality, subsequent studies and review of other work have tended to indicate that the poor fetal outcome was due to the diabetes rather than the drug (Sutherland et al., 1974).

Brock, N. and von Kreybig, T.: Experimenteller Beitrag zur prufung Teratogener Wirkungen von Arzneimitteln an der Laboratoriumratte. Naunyn–Schmiedebergs Arch. Pharmakol. Exp. Pathol. 249:117–145, 1964.

Jackson, W.P.U.; Campbell, G.D.; Notelovitz, M. and Blumsohn, D.: Tolbutamide and chlorpropamide during pregnancy in humans. Diabetes 11: 98–101, 1962.

Sutherland, H.W.; Brewsher, P.D.; Cormack, J.D.; Hughes, C.R.T.; Reid, A.; Russell, G. and Stowers, J.M.: Effect of moderate dosage of chlorpropamide in pregnancy on fetal outcome. Arch. Dis. Child. 49:283–291, 1974.

Tuchmann–Duplessis, H. and Mercier–Parot, L.: Action de la chlorpropamide, sulfamide hypoglycemiant, sur la gestation et le developpement foetal du rat. C. R. Acad. Sci. (Paris) 249:1160–1162, 1959.

553 3–Chlor–pyridazinyl–(6)–thio–acetic acid–diethylamide

Lindner et al. (1964) gave mice this choleretic agent orally from day 8 in doses of up to 900 mg per kg. No increase in defects was found even at the highest dose which was lethal to a number of the dams.

Lindner, V.I.; Stormann, H. and Wendtlandt, W.:

Toxikologische Untersuchungen mit [3-Chlor–pyridazinly–(6)-thio]-essigaure-diathylamid und dessen Wirkungen auf Verschiedene Korperfermente. Arzneim Forsch. 14:271–279, 1964.

554 Chlorpyrifos *Dursban*[TM]

Deacon et al. (1980) gavaged mice with doses of up to 25 mg per kg on days 6–15 and produced delayed ossification but no increase in defects.

Deacon, M.; Murray, J.; Pilny, M.; Rao, K.S.; Dittenber, D.; Hanley, T.R., Jr. and John, J.A.: Embryotoxicity and fetotoxicity of orally administered chlorpyrifos in mice. Toxicol. Appl. Pharmacol. 54:31–40, 1980.

555 Chlorthalidone *Phthalamudine*

Fabro et al. (1966) found no teratogenic action of this compound when it was fed to rabbits in amounts of 150 mg per kg per day during days 7 through 12 of gestation.

Fabro, S.; Smith, R.L. and Williams, R.T.: Embryotoxic activity of some pesticides and drugs related to phthalimide. Food Cosmet. Toxicol. 3:587–590, 1966.

556 Cholecalciferol *Vitamin D Hypervitaminosis D* CAS 1406–16–2

Friedman (1968) reviewed the evidence that vitamin D may produce the syndrome of supravalvular aortic stenosis with elfin facies and mental retardation (Garcia et al., 1964). The virtual absence of this cardiac lesion before the years of routine institution of vitamin D prophylaxis and the high incidence of the syndrome in Gottingen, Germany where rickets prophylaxis during pregnancy consisted of huge doses of vitamin D, may be circumstantial evidence of the vitamin's role in supravalvular aortic stenosis. The great difficulty in measuring vitamin D in serum and the frequent absence of hypercalcemia in patients with supravalvular aortic stenosis have made it impossible to definitely implicate vitamin D in the etiology of supravalvular aortic stenosis. Forbes (1979) commented on the lack of evidence that vitamin D is related to Williams syndrome. O'Brien et al (**1993**) studied women exposed to 100 to 600 times the recommended daily allowance and found no increase in defects among 36 offspring. Thirty–two of the women had elevated vitamin D levels but none had elevated serum calcium.

Friedman and Roberts (1966) showed in rabbits that 1.5 million units to the mother produced abnormalities of the aorta in 14 of 24 offspring. Many of these aortic lesions resembled the form seen in children. In addition to the aortic lesion, the rabbits exhibited a reduction defect in dentition which was similar to the maxillary changes in patients with the syndrome (Friedman and Mills, 1969).

Ornoy et al. (1972) gave 20,000 or 40,000 IU of vitamin D_2 to rats by gavage from day 10 through 21 of gestation and produced growth retardation, neonatal death, and postnatal impairment of ossification with fractures. Later, Ornoy (1981) reported that 1,2,5 OH_2D_3 was the metabolite causing the adverse effects.

Makita et al. (1977) administered 1–α–hydroxy cholecalciferol (D_3) to rabbits on days 6 through 18. Embryolethality was increased at doses of 0.8 microgm per kg. Three malformations (cleft palates and absence of gall bladder) occurred at 0.02 and 0.2 microgm per kg and one grossly malformed fetus at 0.5 microgm per kg. The total number of fetuses examined in the three dose level groups was 134.

Friedman, W.F.: Vitamin D and the supravalvular aortic stenosis syndrome. In: Advances in Teratology, D.H.M. Woollam (ed.), New York: Academic Press, 3:83–96, 1968.

Friedman, W.F. and Mills, L.F.: The relationship between vitamin D and the craniofacial and dental anomalies of the supravalvular aortic stenosis syndrome. Pediatrics 43:12–18, 1969.

Friedman, W.F. and Roberts, W.C.: Vitamin D and the supravalvular aortic stenosis syndrome: The transplacental effects of vitamin D on the aorta of the rabbit. Circulation 34:77–86, 1966.

Forbes, G.B.: Letter to the editor: Vitamin D in pregnancy and the infantile hypercalcemic syndrome. Pediatr. Res. 13:1382, 1979.

Garcia, R.E.; Friedman, W.F.; Kaback, M.M. and Rowe, R.D.: Idiopathic hypercalcemia and supravalvular aortic stenosis: Documentation of a new syndrome. N. Eng. J. Med. 271:117–120, 1964.

Makita, T.; Kato, M.; Matuzawa, K.I.; Ojima, N.; Hashimoto, Y. and Noguchi, T.: Safety evaluation studies on the hormonal form of vitamin D_3. III. Teratogenicity in rabbits by oral administration. (Japanese) Iyakuhin Kenkyu 8:615–624, 1977.

O'Brien, J.; Rosenwasser, S.; Feingold, M. and Lin, A.: Prenatal exposure to milk with excessive vitamin D suplementation. (abs) Teratology 47:387, 1993.

Ornoy, A.; Kaspi, T. and Nebel, L.: Persistent defects of bone formation in young rats following maternal hypervitaminosis D_2. Israel J. Med. Sci. 8:943–949, 1972.

Ornoy, A.; Zusman, I. and Hirsh, B.E.: Transplacental effects of vitamin D_3 metabolites on the skeleton of rat fetuses. (abs) Teratology 23:55A, 1981.

557 Cholecalciferol Deficiency *Vitamin D Deficiency*

Warkany (1943) produced a congenital malformation consisting of curvatures of the long bones in rat offspring after chronic feeding of a rachitogenic diet. The histologic appearance of the bones was similar to that seen in rickets except for the absence of large amounts of osteoid tissue. Defective mandibles were produced in chicks whose mothers were maintained on vitamin D deficient diets. 25–Hydroxy–vitamin D_3 corrected the defect, but feeding of 1,25–dihydroxyvitamin D_3 did not (Sunde et al., 1978).

Sunde, M.L.; Turk, C.M. and DeLuca, H.F.: The essentiality of vitamin D metabolites for embryonic chick development. Science 200:1067–1069, 1978.

Warkany, J.: Effect of maternal rachitogenic diet on skeletal development of young rat. Am. J. Dis. Child. 66:511–516, 1943.

558 Cholestyramine CAS 11041–12–6

Rosa (**1994**) found no offspring with defects among 37 women taking this medication. Only 4 took it in the first trimester.

Koda et al. (1982) gave this anion exchange resin by mouth to rats and rabbits in amounts of up to 2 gms per kg daily. No adverse fetal or fertility effects were found. Postnatal studies in rats were done and no changes were found in function.

Koda, S.; Anabuki, K.; Miki, T.; Kahi, S. and Takahashi, N.: Reproductive studies on cholestyramine. Kiso to Rinsho 16:2040–2049, 2050–2069, 2070–2077, 2078–2094, 1982.

Rosa, F.: Anti–cholesterol agent pregnancy exposure outcomes. (abs) Reproductive Toxicology 8:445–446, 1994.

559 Choline CAS 62–49–7

A deficiency of choline in the pregnant mouse did not produce congenital defects (Meader and Williams, 1957).

Meader, R.D. and Williams, W.L.: Choline deficiency in the mouse. Am. J. Anat. 100:167–204, 1957.

560 Chondroitin Polysulfate

Hamada (1972) administered orally up to 5000 mg per kg to mice and rats during organogenesis and found no adverse effects.

Hamada, Y.: Studies on anti–atherosclerotic agents. IX. Teratological studies of sodium chondroitin polysulfate. Oyo Yakuri 6:589–594, 1972.

561 Chondroitin Sulfate CAS 9007–28–7

Kamei (1961) injected 1 cc of a 2 percent solution into mice on the 9th, 10th and 11th gestational days. Cleft palates or kinky tails were produced in some of the offspring.

Kamei, T.: The teratogenic effect of excessive chondroitin sulfate in the DDN strain of mice. Med. Biol. 60:126–129, 1961.

562 Chorionic Villus Sampling CVS

Firth et al. (**1991**) reported 5 limb reduction defects among 289 pregnancies where chorionic villus sampling was performed between 56–66 days of gestation (42 to 52 days from estimated ovulation). Terminal digits were effected in all cases and micrognathia was also present in 4. These findings are in contrast to the lack of reported findings in over 3,400 cases (Rhoads et al. **1989** and Anonymous, **1989**) where sampling was done predominantly after 52 days from ovulation. The presence of an extremely sticky substance in the exocelom which becomes obliterated at approximately 52 days is an association that is probably of significance. The inadvertent rupture of the amniotic sac could expose the developing embryonic digits to this very sticky substance and lead to immobilization and their maldevelopment. An alternate hypothesis which fits the animal teratologic work is that oligohydraminios occurs and leads to pressure necrosis of the extremities and abnormal craniofacial molding.

Mastroiacovo and Cavalcanti (**1991**) reported that among 118 childrn with transverse limb defects there were 4 exposed to chorionic villus sampling while among the control (malformed without limb reduction) there were 15 among 6,486. The odds ration was 15.1 (confidence interval 4.1118–49.65). Smidt–Jensen et al. (**1992**) compared abdominal CVS, cervical CVS and amniocentesis in a randomized study totaling about 4,000 patients. The CVS sampling was performed at 9–11 menstrual weeks. The fetal loss rate was 6.3%, 6.4% and 10.9% for transabdominal CVS, cervical VCS and amniocentesis. No oromandibular–limb abnormality occurred in the CVS group but one infant with aplasia of the hand occurred in the amniocentesis group. Burton et al. (**1992**) found four transverse limb defects among 394 offspring of women with CVS. These defects followed CVS at 9.5, 9.5, 10.5 and 11 weeks of gestation. Three of the transverse defects included more than one extremity. In three other cases there was cleft lip or palate.

Holmes (**1993**) has summarized a report of a National Institutes of Health on CVS. Rodeck (**1993**) summarized an international conference recommendation that chorionic villus sampling should not be done before 9 1/2 menstrual weeks. Rodeck et al. (**1993**) reported a higher rise in alpha feto protein after transabdominal sampling as compared to the transcervical approach which produces less placental trauma. Gruber and Burton (**1994**) have presented an example of the syndrome and summarized other reported cases.

Olney et al. (**1994**) reported based on over 1,400 cases that the relative risk for limb defect was 6.2 for pregnancies less than 70 gestional days, 3.2 for those 70–76 days and 1.8 for those over 76 days.

Anonymous: Multicentre randomised clinical trial of chorion villus sampling and amniocentesis. The Lancet 1:1–6, 1989.

Burton, B.K.; Schulz, C.J. and Burd, L.I.: Limb anomalies associated with chorionic villus sampling. Obstetrics & Gynecology 79:726–730, 1992.

Firth, H.V.; Boyd, P.A.; Chamberlain, P.; Macenzie, I.Z.; Lindenbaum, R.H. and Huson, S.M.: Severe limb abnormalities after chorion villus sampling at 56–66 days' gestation. The Lancet 337:762–763, 1991.

Gruber, B. and Burton, B.K.: Oromandibular–limb hypogenesis syndrome following chorionic villus sampling. International Journal of Pediatric Otorhinolaryngology 29:59–63, 1993.

Holmes, L.B.: Report of national institute of child health and human development workshop on chorionic villus sampling and limb and other defects, October 20, 1992. Teratology, 48:7–13, 1993.

Mastroiacovo, P. and Cavalcanti, D.P.: Limb–reduction defects and chorion villus sampling. The Lancet 337:1091 only, 1991.

Olney, R.S.; Khoury, M.J.;Botto, L.D. and Mastrioacova, P.: Limb defects and gestational age at chorionic villus sampling The Lancet 344:476, 1994.

Rhoades, G.G.; Jacson, L.G.; Schlesselman, S.E.; De La Cruz, F.F.; Desnic, R.J.; Golbus, M.S.; Ledbetter, D.H.; Lubs,

H.A.; Mahoney, M.J.; Pergament, E.; Simpson, J.L.; Carpenter, R.J.; Elias, S.; Ginsberg, N.A.; Goldberg, J.D.; Hobbins, J.C.; Lynch, L.L.; Shiono, P.H.; Wapner, R.J. and Zachary, J.M.: The safety and efficacy of chorionic villus sampling for early prenatal diagnosis of cytogenetic abnormalities. The New England Journal of Medicine 30:609–617, 1989.

Rodeck, C.H.: Fetal development after chorionic villus sampling. The Lancet 341:468–469, 1993.

Rodeck, C.H.; Sheldrake, A.; Beattie, B. and Whittle, M.J.: Maternal serum alphafetoprotein after placental damage in chorionic villus sampling. The Lancet 341:500, 1993.

Smidt–Jensen, S.; Permin, M.; Philip, J.; Lundsteen, C.; Zachary, J.M.; Fowler, S.E. and Gruning, K.: Randomised comparison of amniocentesis and transabdominal and transcervical chorionic villus sampling. The Lancet 340:1237–1244, 1992.

563 Chondroitinase ABC

Morris–Kay and Tuckett (1989) added chondroitinase ABC to whole rat embryo culture and observed arrest and neural crest migration and inhibited neural tube closure.

Morriss–Kay, G. and Tuckett, F.: Immunihistochemical localisation of chondroitin sulphate proteoglycans and the effects of chondroitinase ABC in 9–to 11–day rat embryos. Development 106:787–798, 1989.

564 Chromium CAS 7440–47–3

Ridgway and Karnofsky (1952) reported mild achondroplasia in chicks after adding 2.5 mg sodium dichromate to the chorioallantoic membrane on the 8th day. Gale and Bunch (1979) injected hamsters intravenously on day 7, 8 or 9 with 8 mg per kg of chromium trioxide and produced cleft palates and resorptions.

Gale, T.F. and Bunch, J.D.: The effect of the time of administration of chromium trioxide on the embryotoxic response in hamsters. Teratology 19:81–86, 1979.

Ridgway, L.P. and Karnofsky, D.A.: The effects of metals on the chick embryo: Toxicity and production of abnormalities in development. Ann. N.Y. Acad. Sci. 55:203–215, 1952.

565 Chromomycin A$_3$ CAS 7059–24–7

This substance has been reported to be teratogenic in rats (Takaya, 1963). Tanimura and Nishimura (1963) injected 0.5 to 2.0 mg per kg into mice on day 10, 11 or 12 and found 36 to 51 percent of the fetuses to have kinky tails, hydrops or abnormal head shapes.

Takaya, M.: Teratogenic effect of the antitumor antibiotics. (abs) Proceedings of the Congenital Anomalies Research Association of Japan 3:47–48, 1963.

Tanimura, T. and Nishimura, H.: Effects of antineoplastic agents especially chromomycin A$_3$ administered to pregnant mice upon the development of their offspring. Acta Anat. Nippon 38:1 only, 1963.

566 CI–921

Beyer et al. (1989) tested this anticancer agent in pregnant rats and rabbits. Maximum intervenous doses during organogenesis were 1.0 and 2.0 mg per kg in the rat and rabbit, repectively. Hydrocephalus and vertebral defects were found at the highest dose in rat fetuses but no increase was seen in the rabbit.

Henck et al. (**1992**) gave this antitumor agent intravenously to rats and rabbits on days 6–15 and 6–18 respectively. In the rat there was maternal toxicity and an increase in defects (microphthalmia, hydronephrosis) at a dose level of 1.0 mg per kg. Growth retardation and skeletal variations were found in rabbit fetuses at a non–toxic maternal dose of 2.0 mg per kg.

Beyer, B.K.; Brown, S. and Anderson, J.A.: Teratology studies in rats and rabbits with the anticancer agent CI–921. Teratology 39:P63, 1989.

Henck, J.W.; Brown, S.L. and Anderson, J.A.: Developmental toxicity of CI–921, an anilinoacridine antitumor agent. Fund. Appl. Toxicol. 18:211–220, 1992.

567 CI–943

(+)8–Ethyl–7,8–dihydro–1,3,5–trimethyl–1H–imidazo [1,2–c]–Pyrazolo[3,4–e]pyrimidine

Henck et al. (**1995**) administered this antipsychiotic agent in the diet to rats in amounts up to 75 mg per kg before and during gestation or in the last week of gestation and through lactation. At the highest dose there was decreased maternal weight gain. Survival, growth acquisition of developmental landmarks and reproductive potential of the offspring were not affected. In the treated perinatal–postnatal group vertical movement was decreased and the acoustic startle learning and memory were diminished.

Hench, J.W.; Petrere, J.A. and Anderson, J.A.: Developmental neurotoxicity of CI–943: a novel antipsychotic. Neurotoxicology and Teratology 17:13–24, 1995.

568 Ciafos *O, O–Dimethyl–O–(4–cyanophenyl) phosphorothioate; Cyanox*^{TM} CAS 2636–26–2

This insecticide was given in doses of up to 10 mg per kg to pregnant rats on days 9 through 14 and no teratogenic activity was detected (Yamamoto et al., 1972).

Yamamoto, H.; Yano, I.; Nishino, H.; Furuta, H. and Masuda, M.: Teratological studies on Cyanox^{TM} in rats. Oyo Yakuri 6:523–528, 1972.

569 Cianidanol

Yokoi et al. (1982) gave up to 5000 mg per kg orally before and during the first 7 days of gestation or on days 7–17 of gestation in the rat. Some fetal growth inhibition was found at the highest dose but no increased defect rate. Rabbits received up to 500 mg per kg during gestation with no adverse fetal effects.

Yokoi, Y.; Yoshida, H.; Mitsumori, T.; Nagano, M.; Hirano, K.; Terasaki, M. and Nose, T.: Reproductive studies of cianidanol (KB–53). Oyo Yakuri 24:383–390, 495–507, 521–529, 1982.

570 Cibenzoline Succinate

Kadoh et al. (**1988**) gave rats orally up to 125 mg per kg on days7–17. Fetal weight and postnatal growth was retarded at the highest dose. No effect was found at 50 mg per kg.

Shimazu et al. (**1991**) gave rats up to 30 mg per kg intravenously before or during organogenesis and found no teratogenicity but fetal death and growth retardation was found at the highest dose. Perinatal studies at 20 mg per kg found no ill effects. In rabbits doses of up to 10 mg per kg during organogenesis did not produce fetal ill effects.

Kadoh, Y.; Fujii, T. and Tenshoh, A.: Toxicity study of cibenzoline succinate (3rd report)–reproduction studies. Kiso to Rinsho 22:4527–4539, 1988.

Shimazu, H.; Nishimuro, N.; Ishida, S.; Ikeyo, M.; Matsuoka, T.; Serizawa, K.; Saegusa, T.; Shimizu, I. and Noguchi, H.: Toxicity study of cibenzoline succinate (7th report)–reproductive and developmental toxicity studies (intravenous dosing). The Clinical Report 25:3394–3410, 1991.

571 **Ciclopirox** CAS 29342–05–0 *HOE 296*
6–Cyclohexyl–1–hydroxy–4–methyl–2(h)–pyridone ethanolamine Salt CAS 41621–49–2

Miyamoto et al. (1975) studied this new antifungal agent in mice and rats. The maximal subcutaneous dose was 10 mg per kg and for the oral route it was 100 mg per kg. No teratogenic effects were seen with treatment during organogenesis.

Miyamoto, M.; Ohtsu, M.; Sugisaki, T. and Takayama, K.: Teratological studies of 6–cyclohexyl–1–hydroxy–4–methyl–2(1h)–pyridone ethanolamine salt (HOE 296) in mice and rats. Oyo Yakuri 9:97–108, 1975.

572 **Cigarette Smoking** *Tobacco*

Low–Birth Weight

Since Simpson (1957) identified increased prematurity rates in fetuses from smoking mothers, several reports documenting reduced birth size appeared (Lowe, 1959; Herriot et al., 1962; Ravenholt and Levinski, 1965). Rubin et al. (1986) presented evidence that passive exposure to cigarette smoke reduces the weight of a non–smoking mother's fetus. There were reports that the proportion of male newborns was reduced (Ravenholt and Levinski, 1965). Miller and Hassanein (1974) reported evidence that the reduced birth weight in smokers' offspring was not due to decreased nutritional intake. MacArthur and Knox (1988) analyzed the effect of stopping smoking at different stages of pregnancy. Based on 1,235 smokers and controls, those that stopped before 6 weeks and between 6 and 16 weeks, had infants 217 and 213 gm, respectively, heavier than the persistent smokers and similar to the non–smokers. Day et al. (**1992**) in a study of smoking among 763 women found that the growth deficit in the smoker's offspring disappeared after 18 months of age.

Martin and Braken (1986) summarized the literature and in a study of 3,891 prenatal, identified 853 with exposure to passive smoking via the father in most cases. After logistic regression analysis using 15 other risk factors, a decrease in term newborn weight of 25 g was found. This was not statistically significant. The relative risk of a birth weight below 2500 gm was 2.17 (p = 0.037) after correcting for other risk factors. Makin et al. (**1991**) carefully analyzed 91 children exposed to in utero and postnatally to active or passive smoking or to non–smoking. Their analysis suggested that those passively exposed were intermediate between those directly exposed and the unexposed. Among Indian tobacco chewers the offspring were 395 gm lighter at birth than controls (Verma, et al., **1983**).

Alderman et al. (**1994**) found that smoking was associated with a relative odds of craniosynostosis of 1.7 (95% conf. 1.2–2.6). For smoking over one pack per day, the relative odds were 3.5 (1.5–8.4).

Paulson et al. (**1993**) gave rats an extract of smokeless tobaco by gavage on days 6–20. Doses were 1.33 and 4.0 mg per kg of nicotine. For the higher dose the maternal and fetal weights were reduced. Cross-fostering was done. No consistent behavior changes were noted.

Intelligence Levels

The important question of whether or not cigarette smoking causes a change in intellectual function in the intrauterine exposed child has been studied in detail by Davie et al. (1972), who reported a three to four month delay in reading achievement after carefully factoring out the many other related variables. Hardy and Mellitis (1972) could find no intellectual impairment in another study of the offspring of heavy smokers.

Dunn et al. (1977) found reduced neurological and intellectual maturation in six–year–olds whose mothers smoked during pregnancy. Some increase in the incidence of "minimal brain dysfunction" was also found. The authors give a balanced discussion of the other characteristics of smokers that might contribute to differences in their offspring. Drawing on studies of seven–year–olds from the Collaborative Perinatal Project, Naeye and Tafari (1983) reported an increase in behavioral abnormalities (hyperactivity, shortened attention span) in the offspring of smoking mothers. Impairment in language, reading and writing skills were small. Freid (1989) found lower mental scores at 12 months and 36 months but not at 24 months in over 100 children exposed prenatally to cigarette smoking.

Congenital Defects

The neonatal mortality rate and incidence of congenital defects are not increased by smoking (Lowe, 1959; Underwood et al., 1965; Frazier et al., 1961; Yerushalmy, 1964). Malloy et al (**1989**) studied a population of 288,067 with 10,223 congenitally deformed infants and found no association.

Evans et al. (1979) found no general increase in congenital defects among offspring of smokers in a group of 67,609 pregnancies. However, Evans did find the rate of anencephaly was 1.7, 1.7, 2.5 and 3.0 per 1,000 women who did not smoke, smoked lightly, moderately or heavily (greater than 20 per day), respectively. Ericson et al. (1979) found a significant increase in smokers among 66 women who gave birth to infants with cleft lip and/or cleft palate.

Naeye (1978) reported that white cigarette smokers in a large collaborative study gave birth to anencephalics with a frequency of 1.72 per 1000 while white non–smokers had a rate of 0.1 per 1000. A similar change was not found among Blacks. Smoking is more prevalent in the lower social classes which is associated with higher incidence of anencephaly.

Himmelberger et al. (1978), using a mail survey of operating room personnel, reported significant increases in spontaneous abortion and congenital defects in the offspring of smoking mothers. Although all systems were involved, the cardiovascular and urogenital were most often affected. Hook and Cross (1985) reported that smoking decreased the risk ratio for Down's Syndrome in older mothers.

Pregnancy Complications

Underwood et al. (1965) reported that premature ruptured membranes were significantly increased in heavy smokers. Naeye (1979) reported that abruptio placentae, placenta previa and amniotic infections were positively associated with the number of cigarettes smoked and the duration of smoking. For instance, there were 17 non–smokers as compared to 36 long term heavy smokers among a group of 359 abruptio placentae patients. Miller and Gardner (1981) reported that the cadmium content of placentas from smokers was nearly three times that of non–smokers.

Sheveleva et al. (1986) made a study of 500 women (aged 18–45) composed of two groups, smokers (187) and control or nonsmokers (313). The study showed smoking to be a risk factor. It was associated with a disturbance of ovarian cycles, inflammatory gynecological diseases, spontaneous abortion, toxicosis of pregnancy, premature delivery, chronic hypoxia of fetuses and hypotrophy of newborns.

Sanyal et al. (**1994**) studied the placentas from smoking women and found an increase in the metabolism of polynuclear aromatic hydrocarbon (PAH) however DNA adducts of PAH were not increased.

Spontaneous Abortions

Boue et al. (1975) studied a large number of spontaneous abortions and found a significant reduction in chromosome aberrations among women who inhaled cigarettes. This rate was 50 percent as compared to the noninhalers' rate of 62 percent. This increase in abortuses with normal karyotype may be explained by an increase of the incidence of abortions in chromosomally normal embryos as a result of cigarette smoking. Kline et al. (1977) studied smoking habits among 574 women who aborted spontaneously and 320 women who delivered after the 28th week. Forty–one percent of those aborting smoked as compared to 28 percent of those reaching 28 weeks. The association did not vary with age or previous obstetrical events. Harlap and Shiono (1980) found an increase in second trimester abortion in smokers but the increase was not significant. Wilcox et al. (1989) reported that cigarette smoking was associated with increased fertility in one perspective and one retrospective study. They comment that the finding lacks biologic plausibility. Gindoff and Tidey (**1989**) summarized published data on human fecundity and spontaneous abortions.

Reproductive Function

Reduced fecundability women whose mothers smoked during pregnancy has been reported by Weinberg et al. (1989). They studied the number of cycles that a woman experienced before the onset of a planned pregnancy. The fecundability ratio (per cycle probability of conception) was 0.5. Correcting for other variables (age, frequency of intercourse, age of menarche) did not appreciably change the ratio. The authors viewed the findings as preliminary. Studies of exposure to household smoking after birth showed increased fecundability, a finding which was not biologically plausible (Wilcox et al., 1989).

Cancer Risk

In a case control study, the rate of cancer in the offspring of smokers was found by Stjernfeldt et al. (1986) to be 50 percent higher than controls. The rate was doubled for Hodgkin lymphoma, acute lymphoblastic leukemia and Wilms tumor. McKinney et al. (1986) vigorously responded to the work of Stjernfeldt et al. with data and comments that fail to show an association between cigarette smoking and cancer in the offspring.

Animal Studies

Nicotine and cigarette smoking in animals has also been studied and reduced fetal weight was recorded (see Nicotine). In rats injected with 3 mg of nicotine daily during pregnancy, a learning deficit was found in the offspring (Martin and Becker, 1971). Paulson et al. (**1994**) found increased mortality and reduced weight gain in rat offspring exposed to smokeless tobacco extracts with 4.0 or 6.0 mg of nicotine per kg.

Schoeneck (1941) showed that rabbits exposed to the equivalent of 20 cigarettes per day gave birth to fetuses that were 17 percent less in weight than controls.

Reviews

Nash and Persaud (1988) reviewed the various risks associated with human pregnancy outcome. Stillman et al. (1986) reviewed the reproductive effects of smoking. The subject was also reviewed by Landesman–Dwyer and Emanuel (1979). Streissguth (1986) reviewed the data of smoking and learning disabilities.

Alderman, B.W.; Bradley, C.M.; Greene, C.; Fernbach, S.K. and Baron, A.E.: Increased risk of craniosynostosis with maternal cigarette smoking during pregnancy. Teratology 50:13–18, 1994.

Boue, J.; Boue, A. and Lazar, P.: Retrospective and prospective epidemiological studies of 1500 karyotyped spontaneous human abortions. Teratology 12:11–26, 1975.

Davie, R.; Butler, N. and Goldstein, H.: From Birth to Seven: A report of the National Child Development Study, London: Longman and the National Children's Bureau, 175–177, 1972.

Day, N.; Cornelius, M.; Goldschmidt, L.; Richardson, G.; Robles, N. and Taylor, P.: The effects of prenatal tobacco and marijuana use on offspring growth from birth through 3 years of age. Neurotoxicology and Teratology 14:407–414, 1992.

Dunn, H.G.; Karaa, A.; Ingram, S. and Hunter, C.M.: Maternal cigarette smoking during pregnancy and the child's subsequent development. II. Neurological and intellectual maturation to the age of six and one–half years. Canad. J. Public Health 68:43–49, 1977.

Ericson, A.; Kallen, B. and Westerholm, P.: Cigarette

smoking as an etiologic factor in cleft lip and palate. Am. J. Obstet. Gynecol. 135:348–351, 1979.

Evans, D.R.; Newcombe, R.G. and Campbell, H.: Maternal smoking habits and congenital malformations: A population study. Br. Med. J. 2:171–173, 1979.

Frazier, T.M.; Davis, G.H.; Goldstein, H. and Goldberg, I.D.: Cigarette smoking and prematurity: A prospective study. Am. J. Obstet. Gynecol. 81:988–996, 1961.

Freid, P.A.: Cigarettes and marijuana: are there measurable long–term neurobehavioral teratogenic effects. NeuroToxicology 10:577–584, 1989.

Gindoff, P.R. and Tidey, G.F.: Effects of smoking on female fecundity and early pregnancy outcome. Sem. Rep. End. 7(4):305–313, 1989.

Hardy, J.B. and Mellitis, E.D.: Does maternal smoking have a long–term effect on the child? Lancet 2:1332–1336, 1972.

Harlap, S. and Shiono, P.H.: Alcohol, smoking and incidence of spontaneous abortions in the first and second trimester. Lancet 2:173–178, 1980.

Herriot, A.; Billewicz, W.Z. and Hytten, F.E.: Cigarette smoking in pregnancy. Lancet 1:771–773, 1962.

Himmelberger, D.U.; Brown, B.W. and Cohen, W.N.: Cigarette smoking during pregnancy and the occurrence of spontaneous abortion and congenital abnormality. Am. J. Epidemiol. 108:470–479, 1978.

Hook, E.B. and Cross, P.K.: Cigarette smoking and Down syndrome. Am. J. Hum. Genet. 37:1216–1224, 1985.

Kline, J.; Stein, Z.A.; Susser, M. and Warburton, D.: Smoking: A risk factor for spontaneous abortion. N. Eng. J. Med. 297:793–796, 1977.

Kullander, S. and Kallen, B.: A prospective study of smoking and pregnancy. Acta. Obstet. Gynecol. Scand. 50:83–94, 1971.

Landesman–Dwyer, S. and Emanuel, I.: Smoking during pregnancy. Teratology 19:119–126, 1979.

Lowe, C.R.: Effect of mothers smoking habits on birth weight of their children. Br. Med. J. 2:673–676, 1959.

MacArthur, C. and Knox, E.G.: Smoking in pregnancy: Effects of stopping at different stages. Br. J. Obstet. Gynecol. 95:551–555, 1988.

Makin, J. Fried, P.A. and Watkinson, B.: A comparison of active and passive smoking during pregnancy: Long–term effects. Neurotoxicology and Teratology 13:5–12, 1991.

Malloy, M.H.; Kleinman, J.C.; Baewell, J.M.; Schramm, W.F. and Land, G.H.: Maternal smoking during pregnancy; no association with congenital malformations in Missouri 1980–83. AJPH 79:1243–1246, 1989.

Martin, J.C. and Becker, R.F.: Effects of maternal nicotine absorption or hypopoxic episodes upon appetitive behavior of rat offspring. Dev. Psychobiol. 4:133–147, 1971.

Martin, T.R. and Bracken, M.B.: Association of low–birth weight with passive smoke exposure in pregnancy. Am. J. Epidemiol. 124:633–642, 1986.

McKinney, P.A. and Stiller, C.A.: Maternal smoking during pregnancy and the risk of childhood cancer. Lancet 2:519 only, 1986.

Miller, H.C. and Hassanein, K.: Maternal smoking and fetal growth of full term infants. Pediatr. Res. 8:960–963, 1974.

Miller, R.K. and Gardner, K.A.: Cadmium in the human placenta: Relationship to smoking. (abs) Teratology 23:51A only, 1981.

Naeye, R.L.: Relationship of cigarette smoking to congenital anomalies and perinatal death. Am. J. Path. 90:289–293, 1978.

Naeye, R.L. and Tafari, N.: Risk Factors in Pregnancy and Diseases of the Fetus and Newborn, Williams and Wilkins. Baltimore, 180–183, 1983.

Naeye, R.L.: The duration of maternal cigarette smoking, fetal and placental disorders. Human Devel. 3:229–237, 1979.

Nash, J.E. and Persaud, T.V.N.: Embryopathic risks of cigarette smoking. Exp. Pathol. 33:65–73, 1988.

Paulson, R.B.; Shanfeld, J.; Vorhees, C.V.; Cole, J.; Sweazy, A. and Paulson, J.O.: Behavioral effects of smokeless tobacco on the neonate and young Sprague Dawley rat. Teratology 49:293–305, 1994.

Paulson, R.B.; Shanfeld, J.; Vorhees, C.V.; Sweazy, A.; Gagni, S.; Smith, A.R. and Paulson, J.O.: Behavioral effects of prenatally administered smokeless tobacco on rat offspring. Neurotoxicology and Teratology 15:183–192, 1993.

Ravenholt, R.T. and Levinski, M.J.: Smoking during pregnancy. Lancet 1:961 only, 1965.

Rubin, D.H.; Krasil–Nikoff, P.A.; Leventhal, J.M. and Weile, B.: Effect of passive smoking on birth–weight. Lancet 23:415–417, 1986.

Sanyal, M.K.; Li, Y. and Belanger, K.: Metabolism of polynuclear aromatic hydrocarbon in human term placenta influenced by cigarette smoke exposure. Reproductive Toxicology 8:411–418, 1994.

Schoeneck, F.J.: Cigarette smoking in pregnancy. N.Y. State J. Med. 41:1945–1948, 1941.

Sheveleva, G.A.; Karacharova, L.F.; Frolova, O.G. and Filimonov, V.G.: The influence of smoking on the function of female organism. In: Protection of Maternity and Childhood. (USSR) Meditsina, 11:48–50, 1986.

Simpson, W.J.: A preliminary report in cigarette smoking and the incidence of prematurity. Am. J. Obstet. Gynecol. 73:808–815, 1957.

Stillman, R.J.; Rosenberg, M.J. and Sachs, B.P.: Smoking and reproduction. Fertil. Steril. 46:545–566, 1986.

Stjernfeldt, M.; Berglund, K.; Lindsten, J. and Ludvigsson, J.: Maternal smoking during pregnancy and risk of childhood cancer. Lancet 1:1350–1352, 1986.

Streissguth, A.P.: Smoking and drinking during pregnancy and offspring learning disabilities: A review of the literature and development of a research strategy. In: Learning Disabilities and Prenatal Risk, M. Lewis (ed.), Urbana–Champaign, Ill., University of Illinois Press, 28–67, 1986.

Underwood, P.; Hester, L.L.; Tucker, L. and Gregg, K.U.: The relationship of smoking to the outcome of pregnancy. Am. J. Obstet. Gynecol. 91:270–276, 1965.

Verma, R.C.; Chansoriya, M. and Kaul, K.K.: Effect of tobacco chewing by mothers on fetal outcome. Indian Pediatr. 20:105–111, 1983.

Wilcox, A.J.; Weinberg, C.R. and Baird, D.R.: Reduced fecundability in women with prenatal exposure to cigarette smoking. Am. J. Epidemiology. 129:1079–1083, 1989.

Wilcox, A.J.; Baird, D.D. and Weinberg, C.R.: Do women with childhood exposure to cigarette smoking have increased fecundability? Amer Journal of Epidemiology 129:1079–1083, 1989.

Yerushalmy, J.: Mother's cigarette smoking and survival of infant. Am. J. Obstet. Gynecol. 88:505–518, 1964.

573 Cimetidine *Tagamet*™ CAS 51481–61–9

Jones et al. (**1985**) reported that of 22 women taking the medication during pregnancy there was one offspring with Down's syndrome and a second with hypoxea–induced brain damage. A case report of liver impairment has been reported in the offspring of a mother treated in late pregnancy (Glade et al., 1980).

Hirakawa et al. (1980) gave this anti–ulcer agent to rats and rabbits by mouth in maximum daily doses of 2000 and 100 mg per kg, respectively. No adverse fertility effects were found in rats. The fetuses from both species were not adversely affected and postnatal development and function in the rat were not altered. Brimblecombe et al. (1985) found no adverse effects of oral doses of 950 mg per kg in the pregnant rat, rabbit and mouse. Parker et al. (1984) found reduced sexual performance in the offspring of rats treated on days 12–21. Shapiro et al. (**1988**) studied post natal rat male and female reproduction. They administered 4.0 mg per ml of drinking water on days 17 through the first post natal week. No changes in the offspring's function was observed.

Leslie and Walker (1977) gave oral doses of 950 mg per kg daily to rabbits, rats and mice during organogenesis. They found no adverse fetal effects. Postnatal studies in rats indicated some delay in pinna opening and onset of nocturnal activity. In the F_2 generation, females had lower numbers of corpora lutea of pregnancy. No effect on the male phallus was found after in utero exposure but some delay in development of the prostate and seminal vesicles was observed.

Brimblecombe, R.W.; Leslie, G.B. and Walker, T.F.: Toxicology of cimetidine. Hum. Toxicol. 4:13–25, 1985.

Glade, G.; Saccar, C.L. and Pereira, G.R.: Cimetidine in pregnancy: Apparent transient liver impairment in the newborn. Am. J. Dis. Child. 134:87–88, 1980.

Hirakawa, T.; Suzuki, T.; Hayashizaki, A.; Nishimura, N.; Sano, Y.; Nishikawa, M.; Nagashima, Y.; Seki, K. and Kihara, T.: Reproductive studies of cimetidine. Kiso to Rinsho 14:2819–2831, 1980.

Jones, D.G.C.; Langman, M.J.S.; Lawson, D.H. and Vessey, M.P.: Post–marketing surveillance of the safety of cimetidine: twelve–month morbidity report. Quarterly Journal of Medicine 215:253–268, 1985.

Leslie, G.B. and Walker, T.F.: Toxicological profile of cimetidine. In: Cimetidine, W.L. Burland and M.A. Simkins (eds.), Proceedings of the Second International Symposium of Histamine H–2 Receptor Antagonists, Excerpta Medica Amsterdam, 30–31, 1977.

Parker, S.; Udani, M.; Gavaler, J.S. and Van Thiel, D.H.: Pre– and neonatal exposure to cimetidine but not ranitidine adversely affects adult sexual functioning of male rats. Neurobehav. Toxicol. Teratol. 6:313–318, 1984.

Shapiro, B.H.; Hirst, S.H.; Babalola, G.O. and Bitar, M.S.: Prospective study on the sexual development of male and female rats perinatally exposed to maternally administered cimetidine. Toxicology Letters 44:315–329, 1988.

574 Cinepazide Maleate *[1–(1–Pyrrolidinylcarbonyl) methyl]–4–(3,4,5– trimethoxy–cinnamoyl) piperazine maleate* CAS 26328–04–1

Ino et al. (1979) gave up to 1,060 mg per kg daily by mouth to mice on days 6 through 15 and observed no malformations. Some reduction in fetal weight occurred at the highest dose. Shimada et al. (1979) gave 200 mg orally to rabbits on days 6 through 18 and found no alterations in the fetuses.

Ino, T.; Kobayashi, H. and Morita, H.: Reproduction studies of cinepazide maleate. (2) Teratogenicity study in mice. Iyakuhin Kenkyu 10:546–558, 1979.

Shimada, H.; Tashiro, K.; Morita, H. and Akimoto, T.: Reproduction studies of cinepazide maleate. (4) Teratogenicity study in rabbits. (Japanese) Iyakuhin Kenkyu 10:572–578, 1979.

575 Cinoxacin CAS 28657–80–9

Sato et al. (1980) gave rats orally up to 200 mg per kg daily before mating, during organogenesis or on days 17–21. Increased mortaly of the dams was seen at 200 and 300 mg per kg. No effects on fertility were found and no increase in defects occurred. In the perinatal studies at 200 mg per kg there was an increase in stillbirths and neonatal deaths. Rabbit studies during organogenesis using 800 mg per kg did not show adverse effects (Sato and Kobayashi, 1980).

Sato, T.; Keneko, Y.; Saegusa, T. and Kobayashi, F.: Reproductive studies of cinoxacin in rats. Chemotherapy 28:484–507, 1980.

Sato, T. and Kobayashi, F.: Teratological study on cinoxacin in rabbits. Chemotherapy 28:508–575, 1980.

576 Cisplatin *Platinum, Diamminedichloro* CAS 15663–27–1

Anabuki et al. (1982) administered this antineoplastic platinum compound intraperitoneally to rats and rabbits in maximum daily doses of 0.5 mg per kg. With rats treated before mating and during the first 7 days of gestation, resorptions were increased at 0.25 mg per kg and above. At the highest dose, some lethality and growth retardation was seen when rats were treated during organogenesis. Postnatal studies showed decreased viability and exploratory behavior in rat offspring exposed to 0.25 and 0.5 mg per kg. Teratologic studies were also negative in the rabbit except for increased embryolethality over 0.125 mg per kg.

Kopf–Maier et al. (1985) gave mice up to 10 mg intraperitoneally on individual days during organogenesis and produced growth and osseous retardation but no gross malformations. Some increase in slight hydrocephalus was seen but no dose response was present. At doses of 20 mg per kg the embryos were resorbed. Nagaoka et al. (1981) gave this antineoplastic agent intravenously to rats and rabbits in maximum doses of 0.54 and 0.3 mg per kg, respectively. No ill effects were found in the rabbit fetuses exposed during organogenesis. In the rat, some embryolethality and decreased body weight occurred at 0.375 mg per kg during organogenesis and in the perinatal period.

Anabuki, K.; Kitazima, S.; Koda, S. and Takahashi, N.: Reproductive studies on cisplatin. Yakuri to Chiryo 10:659–671, 673–694, 695–701, 1982.

Kopf–Maier, P.; Erkenswick, P. and Merker, H–J.: Lack of severe malformations versus occurrence of marked embryotoxic effects with treatment of pregnant mice with cis–platinum. Toxicology 34:321–331, 1985.

Nagaoka, T.; Nanama, I. and Konoha, N.: Reproductive studies on cisplatin. Kiso to Rinsho 15:5769–5781, 5782–5800, 5801–5808, 1981.

577 Cisplatinum–2–thymine

Beaudoin and Connelly (1978) gave rats single intraperitoneal doses of 20 to 80 mg per kg. The most sensitive days were 6 and 7. At doses of 60 to 80 mg per kg an increase in eye defects and hydrocephalus was found.

Beaudoin, A.R. and Connelly, T.G.: Teratogenic studies with platinum thymine blue. (abs) Teratology 17:46–47A, 1978.

578 Citral

Gaworski et al. (**1992**) exposed rats to up to 34 ppm of the agent for 6 hours daily on days 6–15 and found no reproductive ill effects. At 68 ppm there was maternal toxicity and slight decrease in fetal weight.

Gaworski, C.L.; Vollmuth, T.A.; York, R.G.; Heck, J.D. and Aranyi, C.: Developmental toxicity evaluation of inhaled citral in Sprague–Dawley rats. Fd. Chem. Toxic. 30:269–275, 1992.

579 Citrate

No teratogenic effect was observed among offspring of mice, rats, hamsters or rabbits treated with citric acid during pregnancy at doses of 2.4–241, 3.0–295, 2.7–272 and 4.2–425 mg per kg per day respectively (Anonymous 1973).

Anonymous: Food and Drug Research Labs, Inc.: Teratologic evaluation of fda 71–54 (citric acid). NTIS Report PB–223 814, 1973.

580 Citreoviridin

Morrissey and Vesonder (1986) gavaged rats with this product of penicillium mold in doses up to 15 mg per kg on days 8–11 or 12–15. They found retardation of skeletal development, embryolethality (days 8–11), maternal lethality and decreased fetal weight but no teratogenicity.

Morrissey, R.E. and Vesonder, R.F.: Teratogenic potential of the mycotoxin, citreoviridin, in rats. Food Chem. Toxicol. 24:1315–1320, 1986.

581 Citrinin CAS 518–75–2

Hood et al. (1976) injected mice intraperitoneally with up to 40 mg per kg on days 7 through 10 and produced no malformations. Prenatal survival was decreased as was birth weight. Reddy et al. (1982) gave rats 35 mg per kg subcutaneously on single days during organogenesis and produced fetuses that were small but not deformed.

Hood, R.D.; Hayes, A.W. and Cammell, S.: Effects of prenatal administration of citrinin and viriditoxin in mice. Food Cosmet. Toxicol. 14:175–178, 1976.

Reddy, R.V.; Mayura, K.; Hayes, A.W. and Berndt, W.O.: Embryocidal, teratogenic and fetotoxic effects of citrinin in rats. Toxicology 25:151–180, 1982.

582 CL 115,347 *Prostaglandin E₂ Analogue*

Angelov et al. (1989) studied the effect of large doses (9 mg per kg per day) on the ovulation, fertilization and implantation in the mouse. Reduced corpora lutea and implantations as well as increases in preimplantation losses were found.

Angelov, G.A.; Tesh, J.M. and Willoughby, C.R.: Prostaglandin E_2: Effects on ovulation, fertilization and early pre–implantation development of fertilized ova in the rat. Teratology 39:P56, 1989.

583 Clebopride Malate

Kawana et al. (1982) gave this anti–ulcer agent orally to rats and rabbits in amounts up to 100 mg per kg daily. In the rats there was no effect on fertility or postnatal function. No adverse fetal effect was seen in either species.

Kawana, K.; Katoh, M.; Akutsu, S.; Simamura, T.; Komatsu, H.; Matsuyama, K. and Matsuzaki, M.: Effect of clebopride malate (LAs) on reproduction. Kiso to Rinsho 16:5649–5660, 5661–5669, 5670–5680, 5681–5687, 1982.

584 Clenbuterol CAS 37148–27–9

Matsuzawa et al. (1984) studied this β2–adrenergic drug in pregnant rats and rabbits. At doses of up to 50 mg per kg before fertilization, and for one week after, no abnormal fetal effects occurred but at the highest dose, the dams were less active and had piloerection and increased salivation. On gestational days 7–17 at doses up to 20 mg per kg flexure of the ribs was significantly increased. Body weight of the fetuses was reduced at 2.0 mg per kg and above. Rabbits were given up to 20 mg per kg on days 6–18 of gestation and no malformations occurred. Fetal weight gain was decreased and mortality increased at 2.0 mg per kg and above.

Matsuzawa, K.; Tanaka, T.; Enjo, H.; Kawamura, H. and Makita, T.: Reproduction studies on clenbuterol. Iyakuhin Kenkyu 15:564–596, 1984.

585 Clentiazem

Asano et al. (**1994**) gave this calcium antagonist orally in doses of up to 50 mg per kg before and during early pregnancy and in the perinatal period. No teratogenicity was found but diestrus and dystocia were found. The perinatal mortality was increased when 25 or 50 mg per kg doses were used perinatally and during lactation. Postnatal behavior was unchanged but weight gain was slower.

Imahie et al. (**1992**) gave this antihypertensive to rats during organogenesis at oral doses of 3, 10 and 30 mg per kg. At 80 mg per kg there was maternal toxicity and dystocia. At 10 and 30 mg per kg ventrricular septal defects and coccygeal defects were increased. The postnatal development and reproduction were not altered. Rabbits given up to 30 mg per kg during organogenesis had maternal deaths and the fetuses had an increase in ventricular septal defects. No adverse effects were seen in the fetuses at 10 mg per kg.

Asano, Y.; Imahie, H. and Ariyuki, F.: Reproductive and developmental toxicity studies in rats given clentiazen (TA–3090) orally before mating and in the early stage of pregnancy (segment I) or during the perinatal and lactation periods (segment III). Oyo Yakuri 47:165–175, 1994.

Imahie, H.; Asano, Y. and Ariyuki, F.: Reproductive and developmental toxicity studies on oral administration of clentiazem (TA–3090) during the perirod of organogenesis in rats and rabbits. Oyo Yakuri 44:511–522, 1992.

586   3 CL(4) hydroxyethyl piperazinyl 1–1, propylphenothiazine–3, 4",5" trimethoxybenzoic Acid

Horvath et al. (**1976**) gave rats 2,000 mg per kg orally on day 11 or 12 and found an increase in orofacial defects, skeletal defects and hydronephrosis. This drug is a phenothiazine derivative.

Horvath, C.; Szony, L. and Mold, K.: Preventive effect of riboflavin and ATP on the teratogenic effects of the phenothiazine derivative T82. Teratology 14:167–170, 1976.

587   Clidanac *TA1–284* CAS 34148–01–1

Kusanagi et al. (**1977**) gavaged pregnant mice with 10,30 and 90 mg per kg per day from day 7 through day 15. At 90 mg per kg skeletal defects involving fused ribs and irregular vertebral bodies occurred. Other defects were not increased over the controls.

Kusanagi, T.; Ihara, T. and Mizutani, M.: Teratogenic effects of non–steroidal anti–inflammatory agents in mice. (Japanese) Cong. Anom. 17:177–185, 1977.

588   Clindamycin CAS 18323–44–9

Bollert et al. (**1974**) administered this antibiotic to rats and mice on days 6 through 15 and observed no adverse fetal effects. Doses up to 180 mg per kg per day were given subcutaneously.

Bollert, J.A.; Gray, J.E.; Highstrete, J.D.; Moran, J.; Purmalis, B.P. and Weaver, R.N.: Teratogenicity and neonatal toxicity of clindamycin 2–phosphate in laboratory animals. Toxicol. Appl. Pharmacol. 27:322–329, 1974.

589   Clobazam

7–Chloro–1–methyl–5–phenyl–1H–1,5–benzodiazepine–2, 4(3H, 5H–dione) CAS 22316–47–8

Fuchigami et al. (**1983**) studied the effect of this benzodiazepine in pregnant rats and rabbits. At 750 mg per kg given orally before pregnancy, during organogenesis and in the perinatal period, no adverse fetal effects occurred. In rabbits, at 50 mg per kg, the mothers' weights were less than the controls and the fetal mortality was higher but no defects were increased.

Fuchigami, K.; Komai, Y.; Ito, K.; Ishimura, K.; Hatano, M.; Kitatani, T.; Miyamoto, M.; Hobino, H. and Yokota, F.: Reproductive studies of clobazam in the rat and rabbit. Oyo Yakuri 25:907–929, 1039–1064, 1983.

590   Clobetasone CAS 54063–32–0

Shinpo et al. (1980) studied this topical steroid in rats giving up to 3.0 mg per kg by subcutaneous route. Cleft palates and omphaloceles were produced at the higher doses (1.0 and 3.0 mg). Some decrease in pregnancy rate and resorptions was found in females treated before pregnancy and during the first week of gestation. Pregnancy was prolonged when 1.0 mg per kg was given on days 17–21. Cleft palates and joint contractures were increased in rabbits at 60 microgms per kg during organogenesis.

Shinpo, D.; Mori, N.; Takahashi, M.; Togashi, H. and Tanabe, T.: Reproductive studies of clobetasone 17–butyrate (SN–203). Kiso to Rinsho 14:333–342, 343–358, 359–372, 373–379, 1980.

591   Clobutinol *1–(4–Chlorophenyl)–2, 3–dimethyl –4–dimethylamino–2–butanol HCl CAS 14860–49–2*

This antitussive was studied in rats and mice and no fetal or postnatal effects were observed (Kataoka et al., 1970). Treatment was from the 8th through the 14th day and oral doses to 75 mg per kg per day were used.

Kataoka, M.; Yuizono, T.; Kase, Y.; Miyata, T.; Kito, G. and Ishihara, T.: Teratological studies of clobutinol in mice and rats. (Japanese) Oyo Yakuri 4:981–989, 1970.

592   Clofazimine *Lampren* CAS 2030–63–9

Red–brown skin discoloration has been reported in the offspring of mothers who took this drug (Farb et al., 1982).

Stenger et al. (1970) gave this antibiotic orally to mice, rats and rabbits during their gestational periods and found no evidence of teratogenic activity. The rats and mice received up to 50 mg per kg and rabbits, 15 mg per kg daily.

Farb, H.; West, D.P. and Pedvis–Leftick, A.: Clofazimine in pregnancy complicated by leprosy. Obstet. Gynecol. 59:122–123, 1982.

Stenger, E.G.; Aeppli, L.; Peheim, E. and Thomann, P.E.: Zur Toxikologie des Leprostaticums 3–(p–chloranilino)–10–(p–chlorophenyl)–2–10–dihydro 2(isopropyl–amino)phenazin (G–30320). Arzneim Forsch. 20:794–799, 1970.

593   Clofezone CAS 17449–96–6

This anti–inflammatory agent was given orally to mice and rats during organogenesis in maximum doses of 600 mg and 240 mg, respectively. No teratogenic effects were observed but the rat fetuses at the maximal dose were retarded in growth (Kamada and Tomizawa, 1979).

Kamada, K. and Tomizawa, S.: Influences of clofezone on fetuses of mice and rats. Oyo Yakuri 18:235–246, 1979.

594 Clofibrate *Clofibric Acid* CAS 637–07–0

This antilipoproteinemic was studied in pregnant rats at up to 750 mg per day orally on days 5 through 15 (Sterner and Korn, 1980). Some minor skeletal variations and fetal weight reduction were found at the highest dose but no increase in malformations. In the rabbit, doses up to 600 mg per kg were used without evidence of teratogenicity. At the highest dose the weight of the pups was lighter.

Sterner, W. and Korn, W.D.: Zur Pharmakologie und Toxikologie von Etofyllinclofibrat. Arzneim–Forsch. 30:2023–2031, 1980.

595 Clofibride *4–Hydroxy–N–dimethylbutyramide* *4–chlorophenoxy–isobutyrate*

Da Lage et al. (1972) fed this cholesterol reducing agent to mice orally during the entire gestation in doses up to 400 mg per kg and no malformations resulted.

Da Lage, C.; Labie, Ch.; Loiseau, G.; Lohier, G.; Marquet, J.P. and Trichand, M.: Etud toxicologique et teratologique d'un hypocholesterolemiant (mg 46). Eur. J. Toxicol. 4:239–253, 1972.

596 Clomiphene *Clomid™* *2-[p-(2–Chloro–1,2–diphenylvinyl)phenoxyl]–triethylamine* CAS 911–45–5

Mili et al. (**1991**) studied 4,904 women exposed to clomiphene and after adjusting for a confounding factor (subfertility) found no increase in the odds ration for congenital defects (OR=0.97). In 66 types of defects analyzed microcephaly hydrocephaly, anophthalmia/microophthalmia and defects of lower GI tract were increased but this did include neural tube defects.

Oakley and Flynt (1972) reported that in a group of women receiving clomiphene or menotropins, a total of six Down's syndrome infants occurred when the corrected age expectation was only two. They proposed that the increased risk might be the underlying abnormal physiology of the women rather than the medication. Nevin and Harley (1976) and James (1977) noted an increase in neural tube defects among the offspring of women using the drug. They note that these women might have been subjected to a higher neural tube defect risk for other reasons. Cornel et al. (1989) studied 970 live birth pregnancies exposed to clomiphene and found three with anencephaly (about a four–fold increase in risk). Cornel et al. (**1990**) review studies on the outcome of ovulation stimulation (mostly clomiphene) and find a small increase in nural tube defects. The relative risks varied from 1.6 to 5. There is difficulty in matching these closely watched pregnancies with control pregnancies under similar surveillence. Surveillance factors could increase the rate. Cuckle and Ward (**1989**) summarized their own and other reports and found in prospective studies one pregnancy with neural tube defect per 633 births which was a relative risk of 2.44 which they judged too slight to judge the association. Milunsky et al. (**1990**) briefly summarized 5 studies of women with ovulation induction and noted that the relative risk was increased to

neural tube closure defects (RR = 2.2–2.5). Is it possible that heightened prenatal observation of these critical pregnancies could allow more abnormal fetuses to be discovered or reported? Mills (**1990**) discusses other bias factors that could contribute to the increased relative risk. They include the underlying infertility of the couple and publication and submission of reports. He points out that the exposure to clomiphene was usually not concurrent with neural tube closure and biologic plausability is absent. In Mills study of 5712 NTD case mothers no increase in ovulation induction was found.

Ahlgren et al. (1976) found 8 out of 148 (5.4 percent) exposed pregnancies had major malformations and this was not statistically increased over the controls (3.2 percent). Kurachi et al. (1983) studied 1034 pregnancies after clomiphene citrate treatment and found an abortion rate of 14 percent, a stillbirth rate of 1.6 percent and a malformation rate of 2.3 percent.

In a pre and post–ovulatory study in mice Dziadek (**1993**) found reduced implantation, fetal survival and fetal weight in those treated 1–5 days before ovulation. A slight but statistically significant increase in exencephaly was found but no dose response was found at 1.25 to 5.0 mg per gm of body weight.

Eneroth et al. (1971) produced hydramnios and cataracts in rat fetuses by injecting maternal rats with this substance from the 6th to the 14th day of gestation. The hydramnios was found from dose levels of 2 mg per kg and the cataracts appeared at 50 mg per kg. Courtney and Valerio (1968) treated 18 monkeys with 1 to 4 mg per kg per day for several days during the embryonic period and found no defects in the fetuses. Suzuki (1970a,b) has treated pregnant rats and mice with amounts up to 128 mg per kg orally and in both species found a dose–dependent increase in fetal mortality but no gross malformations.

McCormack and Clark (1979) injected 2 mg per kg into rats on day 5 or 12 of pregnancy and produced multiple abnormalities in the genital tract of the female offspring. Some of the changes were reminiscent of the vaginal adenosis seen in young girls whose mothers had received diethylstilbestrol.

Schardein (1980) reviewed the clinical aspects.

Motta and Hutchinson (**1991**) studied guinea–pig pregnancies after giving 2 mg per kg on the day of mating. Only 25% of the females continued pregnancy but their fetuses were large and normal. An effect on progesterone concentration and endometrium was associated with the loss of pregnancies.

Ahlgren, M.; Kallen, B. and Rahnevik, G.: Outcome of pregnancy after clomiphene therapy. Acta Obstet. Gynecol. Scand. 55:371–375, 1976.

Cuckle, H. and Ward, N.: Ovulation induction and neural tube defects. The Lancet 1281 only, 1989.

Cornel, M.C.; Kate, L.P.T.; Dukes, M.N.G.; De Jong–Van Den Berg, L.T.W.; Meyboom, R.H.B.; Garbis, H. and Peters, P.W.J.: Ovulation induction and neural tube defects. Lancet 2:1386, 1989.

Cornel, M.C.; Ten Kate, L.P. and te Meerman, G.J.: Association between ovulation stimulation, in vitro fertilization, and neural tube defects? Teratology 42:201–203, 1990.

Courtney, K.D. and Valerio, D.A.: Teratology in the Macaca mulatta. Teratology 1:163–172, 1968.

Dziadek, M.: Preovulatory administration of clomiphene citrate to mice causes fetal growth rtardation and neural tube defects (exencephaly) by an indirect maternal effect. Teratology 46:263–273, 1993.

Eneroth, G.; Eneroth, V.; Forsberg, U.; Grant, C.A. and Gustafsson, J.A.: Clomiphene–induced hydramnios and fetal cataracts in rats inhibited by progesterone. (abs) Teratology

James, W.H.: Clomiphene, anencephaly and spina bifida. Lancet 1:603 only, 1977.

Kurachi, K.; Aono, T.; Minagawa, J. and Miyake, A.: Congenital malformations of newborn infants after clomiphene–induced ovulation. Fertil. Steril. 40:187–189, 1983.

McCormack, S. and Clark, J.H.: Clomid administration to pregnant rats causes abnormalities of the reproductive tract in offspring and mothers. Science 204:629–631, 1979.

Mili, F.; Khoury, M.J. and Lu, X.: Clomiphene citrate use and the risk of birth defects: a population–based case–control study. (Abs) Teratology 43:422–423, 1991.

Mills, J.L.: Fertility drugs and neural tube defects: why can't we make up our minds? Teratology 42:595–596, 1990.

Mills, J.L.; Simpson, J.L.; Rhoads, G.G.; Granbard, B.I.; Hoffman, H.J.; Conley, M.R.; Lassman, M. and Cunningham, G.: The risk of neural tube defects in relation to maternal infertility and fertility drug use. Lancet 336:;103–104, 1990.

Milunsky, A.; Derby, L.E. and Jick, H.: Ovulation induction and neural tube defects. Teratology 42:593, 1990.

Motta, C.M. and Hutchinson, J.S.M.: Effects of clomiphene citrate on early pregnancy in guinea–pigs. J. Reprod. Fert. 92:65–73, 1991.

Nevin, N.C. and Harley, J.M.G.: Clomiphene and neural tube defects. Ulster Med. J. 45:59–64, 1976.

Oakley, G.P. and Flynt, J.W.: Increased prevalence of Down's syndrome (mongolism) among the offspring of women treated with ovulation–inducing agents. (abs) Am. J. Hum. Gen. 24:20a only, 1972.

Schardein, J.L.: Congenital abnormalities and hormones during pregnancy: A clinical review. Teratology 22:251–270, 1980.

Suzuki, M.R.: Effects of oral cyclofenil and clomiphene, ovulation inducing agents on pregnancy and fetuses in rats. (Japanese) Oyo Yakuri 4:635–644, 1970a.

Suzuki, M.R.: Effects of oral cyclofenil and clomiphene, ovulation inducing agents on pregnancy and fetuses in mice. Oyo Yakuri 4:645–651, 1970b.

597 Clomipramine

Neonatal convulsions were reported after maternal use (Cowe et al., 1982).

At doses of 110 mg per kg on day 9 in the mouse, teratogenicity was found but the maternal LD–50 was 120 mg per kg (Jarand, 1980). Behavioral changes in rats given 10 mg per kg on days 10–21 were reported. Conditioned avoidance was reduced (Drago et al., 1985).

Watanabe et al. (1970) studied pregnant rats, mice and rabbits during organogenesis. Oral and subcutaneous doses were used in the rat; oral, subcutaneous and intravenous dosing in the mouse and i.v. dosing in the rabbit. Oral doses to 100 mg per kg and intravenous doses of 2.5 mg per kg (rabbit) did not produce teratogenic effects.

Cowe, L.; Lloyd, D.J. and Dawling, S.: Neonatal convulsions caused by withdrawal from maternal clomipramine. Br. Med. J. 284:1837–1838, 1982.

Drago, F.; Continella, G.; Alloro, M.C. and Scapagnini, U.: Behavioral effects of perinatal administration of antidepressant drugs in the rat. Neurobehav. Toxicol. Teratol. 7:493–497, 1985.

Jurand, A.: Malformations of the central nervous system induced by neurotropic drugs in mouse embryos. Dev. Growth Differ. 22:61–78, 1980.

Watanabe, N.; Nakai, T.; Iwanami, K. and Fujii, T.: Toxicological studies of clomipramine hydrochloride. Kiso to Rinsho 4:2105–2124, 1970.

598 Clonazepam CAS 1622–61–3

Takeuchi et al. (1977) administered this benzodiazepine derivative to rabbits orally on days 6 through 18 in doses up to 20 mg per kg and found no fetal ill effects. No epidemiologic studies of pregnant women taking clonazepam have been located, but studies on a related drug diazepam do not indicate human teratogenicity.

Takeuchi, Y.; Shiozaki, U.; Noda, A.; Shimizu, M. and Udaka, K.: Studies on the toxicity of clonazepam. Part 3. Teratogenicity tests in rabbits. (Japanese) Yakuri to Chiryo 5:2457–2466, 1977.

599 Clonidine CAS 4205–90–7

Chahoud et al. (1985) gave this antihypertensive orally to mice on day 11 and found fetal growth retardation and cleft palates increased at doses of 10 to 40 mg per kg levels at which maternal lethality was present. At one mg per kg no fetal weight reduction was found. Shimizu et al. (1993) found no adverse effects of subcutaneous doses of up to 0.1 mg per kg in rats given before and during early pregnancy. No teratogenicity was found when 0.2 mg per kg was given during organogenesis (Shirota et al., 1993). At o.1 and 0.03 emg per kg the spontaneous movement of the dams was reduced and increased perinatal fetal deaths were found (Kaneko et al., 1993). In rabbits receiving up to 0.5 mg per kg no teratogenicity was found (Wada et al., 1993).

Chahoud, I.; Platzek, T. and Neubert, D.: The maternal and embryotoxicity of clonidine in mice. (abs) Teratology 32:19A, 1985.

Kaneko, Y.; Hashimoto, Y.; Shirota, M.; Nagao, T.; Mizutani, M.; Namba, K. and Asano, N.: Reproductive and developmental toxicity studies of clonidine (III) perinatal and postnatal study by subcutaneous administration in rats. Iyakuhin Kenkyu 24:953–968, 1993.

Shimizu, Y.; Sekino, S.; Shirota, M.; Nagao, T.; Mizutani, M.; Namba, K. and Asano, N.: Reproductive and developmental toxicity studies of clonidine (I) fertility and general reproductive performance study by subcutaneous administration in rats. Iyakuhin Kenkyu 24:926–934, 1993.

Shirota, M.; Watanabe, C.; Nagao, T.; Mizutani, M.; Namba, K. and Asano, N.: Reproductive and developmental toxicity studies of clonidine (II) teratogenicity study by subcutaneous administration in rats. Iyakuhin Kenkyu 24:935–952, 1993.

Wada, K.; Hashimoto, Y.; Shirota, M.; Nagao, T.; Mizutani, M.; Namba, K. and Asano, N.: Reproductive and developmental toxicity studies of clonidine (IV) teratogenicity study by subcutaneous administration in rabbits. Iyakuhin Kenkyu 24:969–976, 1993.

600 Clopyralid *3,6–Dichloropicolinic Acid*

This herbicide was given by gavage to rats and rabbits during active organogenesis and no adverse effects were noted at doses up to 250 mg per kg. At the highest dose there was a decrease in body weight gain in the dams (Hayes et al., 1984).

Hayes, W.C.; Smith, F.A.; John, J.A. and Rao, K.S.: Teratological evaluation of 3,6–dichloropicolinic acid in rats and rabbits. Fund. Appl. Toxicol. 4:91–97, 1984.

601 Clorazepate CAS 20432–69–3 *Tranxene* ᵀᴹ

Bavoux et al. (**1981**) reported neurologic depression in 4 newborns and that the half–life in serum was 152 hours.

Bavous, F.; Lafranchi, C.; Olive, G.; Asensi, D.; Dulac, O.; Francoual, C.; Huault, G.; Olivier, C.; Seilaniantz, M.; Castot, A.; Medernach, C.; Benzaken, C. and Efthymiou, M.L.: Grossesse et psychotropes:effects indesirables chez le nouveau–ne. Therapie 36:305–312, 1981.

602 Clorotepine *Octoclothepine*

Jelinek et al. (**1967**) gave rats up to 60 mg orally on days 10–17 and found no increase in defects. Fetal weight was reduced at the highest level.

Jelinek, P.V.; Zikmund, E.; and Reichlova, R.: L'influence de quelques medicaments psychotropes sur le developpement du foetus chez le rat. Therapie 23:1429–1433, 1967.

603 Clospirazine *8–[3–(2–Chloro–10–phenothiazinyl) propyl]–3–oxo–1–thia–4, 8–diazaspiro [4,5] decane HCl*

Hamada et al. (1970) studied the effect of this drug on the pregnant rat and mouse. They gave up to 100 mg per kg per day to mice and 200 mg per kg per day to rats during organogeneseis. Some minor abnormalities of the cervical vertebrae and sternum were found in fetuses from both species but no gross defects or change in survival occurred.

Hamada, Y.; Namba, T.; Okada, T. and Izaki, K.: Studies on psychotropic drugs. 9. Teratological studies of APY–606. Oyo Yakuri 4:497–504, 1970.

604 Clostridium Perfringens

This topical wound promoting agent was given subcutaneously to rats and rabbits in amounts of up to 50 and 1

mg per kg respectively (Miyazaki et al.,**1994**). Studies during organogenesis did not alter fetal development in either species. Local tissue reactions were found in the dams and fertility and perinatal survival of fetuses of both species was decreased.

Miyazaki, Y.; Shibano, T.; Sasaki, M.; Koyama, A.; Sato, S.; Kamijima, M.; Adachi, Y.; Yoneyama, S.; Sakai, Y.; Nakagawa, H.; Uchiyama, K.; Murakami, K. and Kondo, Y.: Reproductive and developmental toxicity studies of active protein derived from clostridium perfringens (SNK–863) by subcutaneous administration to rats and rabbits. Oyo Yakuri 48:135–150, 1994.

605 Clothiapine

2-Chloro-11-(4–methyl)piperazinodibenzo(b,f)[1,4] thiazepine CAS 2058–52–8

Kohn et al. (1969) fed rabbits and guinea pigs 6 and 15 mg per kg during gestation and reported no harmful effects.

Kohn, F.E.; Kay, D.L.; Cervenka, H.; Kay, J.H. and Schultz, F.H.: Reproduction studies in guinea pigs and rabbits following clothiapine administration. (abs) Toxicol. Appl. Pharmacol. 14:641 only, 1969.

606 Clotrimazole

Rosa et al. (1987) studied the spontaneous abortion risk by record linkage and found that the relative risk was 1.34. They warn that since many other associations were being tested in the study no firm conclusions could be drawn. No increase in usage was found among 6564 infants with birth defects.

Rosa, F.W.; Baum, C. and Shaw, M.: Pregnancy outcomes after first–trimester vaginitis drug therapy. Obstet. Gynecol. 69:751–755, 1987.

607 Cloxacillin CAS 61–72–3

Brown et al. (1968) reported negative teratogenic results after giving 100 mg per kg to rabbits during pregnancy.

Brown, D.M.; Harper, K.H.; Palmer, A.K. and Tesh, S.A.: Effect of antibiotics upon pregnancy in the rabbit. (abs) Toxicol. Appl. Pharmacol. 12:295 only, 1968.

608 Cloxazolam

10–Chloro–11b–(2–chlorophenyl)–2,3,5,6,7, 11b–hexahydrobenzo[6,7]–1, 4–dibenzepino[5,4–b]oxazol–6–one

Tanase et al. (1971) treated mice and rats with this tranquilizer during active organogenesis with up to 600 mg per kg per day by mouth. No fetal defects or postnatal effects were found.

Tanase, H.; Hirose, K. and Suzuki, Y.: 2. Effects of CS–370 upon the development of pre– and post–natal offspring of experimental animals. (Japanese) Ann. Sankyo Res. Lab. 23:180–191, 1971.

609 Clozapine

*8–Chloro–11–(4–methyl–1–piperazinyl)5–H,dibenzo- [1,4
-diazepine* CAS 5786–21–0

Lindt et al. (1971) gave up to 40 mg per kg daily to pregnant rabbits and rats during active organogenesis and observed no deleterious effects in the offspring.

Lindt, S.; Lauener, E. and Eichenberger, E.: The toxicology of 8–chloro–11–(4-methyl-1-piperazinyl)-5H–dibenzo [1,4]–diazepine (clozapine). Farmaco (Sci.) 26:585–602, 1971.

610 CN–88,823–2 *6,7–di*

methoxy–2–(4–thiomorpholinyl)–4–quinazolinamine hydrochloride

Schardein et al. (**1981**) gavaged up to 600 mg per kg on days 6–15 to rats. At the highest dose maternal toxicity was present and skeletal defects and dysplastic eyes were increased.

Schardein, J.L.; Lake, R.S.; Brusick, D.; Fitzgerald, J.E. and de la Iglesia, F.A.: Genotoxic and teratogenic potential of cn–88,823–2, a new antihypertensive agent. Toxicologist 1:140, 1981.

611 CPT–11 *+)–(4S)–4,11–diethyl–f–dihydroxy–9–[(4– piperidinopiperidino)–carbonyloxy]–1H–pyrano[3',4':6,7] indolizino[1,2–6]quinoline–3,14–(4H 12H)–dione Hydrochloride Trihydrate*

Itabashi et al. (**1990** A) gave rats this antineoplastic intravenously on days 7–17. At the highest dose (6.0 mg per kg) the dams did not gain weight normally and teratogenicity was found. The malformations were hydrocephalas, eye and skeletal defects as well as spina bifida at 1.2 mg per kg. Placental weights were decreased when 1.2 mg per kg was given. Fetuses exposed to 6.0 mg per kg had behavior and reproduction abnormalities. In rabbits treated during organogenesis (Itabashi et al. (**1990** B)) intravenously with up to 6.0 mg per kg a wide range of malformations occurred including nasal atresia, heart and skeletal systems. Studies before fertilization in rats (Itabashi et al., **1990** C) found copulation and fertility rates to be unchanged at doses of up to 6.0 mg per kg.

Itabashi, M.; Inoue, T.; Fujii, T.; Aihara, H. and Sannai, S.: Reproduction and developmental toxicity studies of ctp–11 (2nd report)–study on administration of the test sybstance during the period of organogenesis in rats. Kiso to Rinsho 24:7275–7304, 1990.

Itabashi, M.; Yoshida, K.; Suzuki, K.; Aihara, H. and Sannai, S.: Reproduction and developmental toxicity study of cpt–11 (4th report)–study on administration of the test substance during the period of organogenesis in rabbits. Kiso to Rinsho 24:7335–7337, 1990.

Itabashi, M.; Inoue, T.; Fujii, T.; Sato, K.; Aihara, H. and Sannani, S.: Reproduction and developmental toxicity studies of cpt–11 (3rd report)–study on administration of the test substance during the perinatal and lactation periods in rats. Kiso to Rinsho 24:7305–7323, 1990.

612 Coal Products *Fuel Oil*

Andrew et al. (1982) used solvent, refined coal solvent and heavy distillate by gavage on days 12–16 of rat gestation and observed increased perinatal mortality. Springer et al. (1982) exposed rats to an aerosol of heavy distillate from solvent refined coal. At the highest dose cleft palate, small lungs and growth retardation were found at 0.66 mg per liter. At levels of 0.084 and 0.017 mg per liter no effect was found. McKee et al. (1987) found increased lethality in rats exposed to EDS recycle solvent or fuel oil. No increase in malformations were found.

Khan et al. (1987) studied the effect of oral administration of crude oil (Prudhoe Bay) on rat pregnancy. Reduced maternal weight and resorptions were found.

Andrew, F.D.; Lytz, P.S.; Buschbom, R.L. and Springer, D.L.: Postnatal effects following prenatal exposure of rats to solvent refined coal (SRC) hydrocarbons. (abs) Teratology 25:26A, 1982.

Khan, S.; Martin, M.; Payne, J.F. and Rahimtula, A.D.: Embryotoxic evaluation of a prudhoe bay crude coal products (fuel oil) in rats. Toxicology Letters 38:109–114, 1987.

McKee, R.H.; Pasternak, S.J. and Traul, K.A.: Developmental toxicity of EDS recycle solvent and fuel oil. Toxicology 46:205–215, 1987.

Springer, D.L.; Poston, K.A.; Mahlum, D.D. and Sikov, M.R.: Teratogenicity following inhalation exposure of rats to a high boiling coal liquid. J. Appl. Toxicol. 2:260–264, 1982.

613 Cobalt CAS 7440–48–4

Kury and Crosby (1968) injected the chick yolk sac on the 4th day with 0.4 to 0.5 mg of cobaltous chloride and observed anemia in the surviving 20 day chicks. Some thyroid epithelial hyperplasia was observed in three embryos.

Nadeenko et al. (1980) demonstrated that cobaltous chloride was embryotoxic to rat fetuses when it was administered during the entire gestation (dose of 0.05 mg per kg). The dose of 0.005 mg per kg was non–toxic to females; however, the progeny of treated females had reduced weight. Paternain and Domingo (1988) gavaged rats with up to 100 mg per kg on days 6–15. At the highest dose, red blood cell indices were increased in the fetuses. Some stunted fetuses occured at 50 and 100 mg per kg but no evidence for teratogenicity or fetotoxicity was found.

Kury, G. and Crosby, R.J.: Studies on the development of chicken embryos exposed to cobaltous chloride. Toxicol. Appl. Pharmacol. 13:199–206, 1968.

Nadeenko, V.G.; Lenchenko, V.G.; Saichenko, S.P.; Arkhipenko, T.A. and Radovskaya, T.L.: Embryotoxic action of cobalt administered per os. Gigiena i Sanitariya (USSR) 2:6–8, 1980.

Paternain, J.L. and Domingo, J.L.: Developmental toxicity of cobalt in the rat. J. Toxicol. Environ. Health 24:193–200, 1988.

614 Cobalt[60]

Angleton et al. (1988) used exposure doses at 0.16 to 0.83 Gy on single days to beagles. Litter sizes were reduced with treatment on gestational days 8 and 28 and an excess of male offspring occurred. Wang et al. (**1993**) exposed mice to 0.1, 0.2 and 0.4 Gy on days 13–18. At 0.2 Gy or more behavioral deficits or changes were noted.

Angleton, G.M.; Benjamin, S.A. and Lee, A.C.: Health effects of low–level irradiation during development: Experimental design and prenatal and early neonatal mortality in beagles exposed to ^{60}cobalt γ rays. Radiation Research 115:70–83, 1988.

Wang, H.; Chen, D.; Gao, C. and Zhou, X.: Effects of low level prenatal ^{60}CO gamma–irradiation on postnatal growth and behavior in mice. Teratology 48:451–457, 1993.

615 Cocaine CAS 50–36–2 *Crack*

In the past 5 years a significant reproductive impact of cocaine has been observed particularly in the United States. The number of women using cocaine during pregnancy varies between 20 per cent in large city hospitals to 1.4% in residental areas (Streissguth et al., **1991**; Matera et al., **1990**; Chasnoff et al., (**1989**). Of the exposed fetuses, between 2 to 37 per cent exhibit significant nervous system problems (Fantel and Shepard, **1990**). Based on these data, the potentially impaired children born in one year in the United States is between 3,200 and 296,000.

Lutinger et al. (**1991**) performed a meta–analysis on cocaine literature published up to August 1989. When studies of polydrug/cocaine users were compared to polydrug (no cocaine) users head circumference, gestational age and birth weight were not significantly different. The odds ratio for abruptio was 5.40 (CL 0.97–30.1) and genitourinary tract malformations were significantly increased. An analysis with multiple linear regression of cocaine exposed (361 infants) and non exposed infants was done by Bateman et al. (**1993**) who found head circumference and duration of gestation reduced. Slutsker (**1992**) has summarized publications dealing with cocaine risks. Hutchings (**1993**) in a commentary suggested that teratogenic effects were found in only those exposed to the highest doses and in these studies one could not be cetain that the effects of concommitant alcohol and smoking were contributory.

Forman et al. (**1993**) found that 6.25% of neonates in Toronto tested positive for cocaine and that 52 percent required initial medical support (vs 30% who tested negative). Other factors such as smoking and hepatitis carrier state contributed to the finding.

Abruptio placenta is increased ten–fold to about 10% of all exposed pregnancies based on 9 publications. This pathology in addition to other vascular effects may account for a major portion of the damage to the central nervous system. Although there may be an increase in congenital defects this agent has the potential of damaging development any time during fetal life.

Effects on the Nervous System

Chasnoff et al. (1985) studied the offspring of 23 women and found 4 who had the onset of labor with abruptio placenta closely associated with an intravenous dose. A depression of the newborns interactive behavior was found. LeBlanc et al. (1987) studied 38 infants whose mothers took "crack". Five required sedation for tremulousness and irri-

tability during the first week. The average birth weight was 2.69 kg. No congenital defects were found but two had muscular rigidity that persisted past the neonatal period but cleared by the fourth month of age. Acker et al. (1983) reported two women who had onset of abruptio shortly after an intravenous and an intranasal administration of cocaine. Bingol et al. (1987) compared 50 women who used cocaine alone to a group of polyusers and drug free women. The cocaine group of 50 had no increase in abortions but the stillbirth rate was increased to 8. All stillbirths were associated with abruptio placentae. Some increase in defects occurred among the cocaine users offspring and these included exencephaly, interparietal encephalocele and one newborn with parietal bone defect.

Chouteau et al. (1988) studied 343 women who delivered without prior obstetrical care and found 124 who had evidence for cocaine use by analysis of the urine. There was significant reduction in birth weight but no correlation with abruptio or maternal hypertension.

Chasnoff et al. (1987) compared 57 cocaine using pregnant women to 73 women who were former addicts but on methadone. A significantly higher rate of premature labor, precipitous labor, abruptio placentae and fetal meconium staining was found but birth weight and head circumference were not different from the control. By the Brazelton Neonatal Assessment, a significant depression of organizational response to environmental stimuli was present. SIDS occurred in 15 per cent as compared to 4 percent in the control.

Doberczak et al. (1988) studied 39 exposed newborns and found irritability in 34 and EEG abnormalities in 17 of 38. These abnormalities were transient based on further studies at 3 to 12 months.

Dixon and Bejar (1989) studied 32 infants greater than 32 weeks gestation and found abnormal echoencephalography in 41% while in the control group of asphyxiated and at–risk infants only 27% were abnormal. The changes included white matter cavities, acute infarction, hemorrhages and ventricular enlargement. Hadeed and Siegel (1989) found microcephaly (21%) in the offspring of cocaine users. Singer et al. (**1994**) found an increased incidence of low grade intraventricular hemorrhage among very small neonates whose mother used cocaine. The rate was 42% as compared to 20% in an identical but not exposed group.

Brain hemorrhages involving the germinal matrix were reported in three of 4 exposed fetuses autposied after spontaneous abortion (Kapur et al., **1991**). Gilbert et al. (**1990**) could find no specific morphologic change in the placentas of exposed fetuses. Webster et al. (**1991**) found brain hemorrhage and cavitation of the brain in rat fetuses receiving 50–70 mg per kg of cocaine on day 16. Since these findings were similar to those seen two days after gentle handling of the uterus, they proposed that constriction/occlusion of the uterine vessels was the cause.

Scanlon (**1991**) has reviewed the neuroteratology of cocaine. There is a need for information on the later intellectual function of these children.

Congenital Malformations

Chasnoff et al. (1988) compared the pregnancies of 50 women using cocaine when pregnancy began with a group of 30 pregnant women who were polydrug, non–cocaine users. Nine of the cocaine users had congenital defects, of which seven had genitourinary tract defects and two had ileal atresia. One of these patients had prune belly syndrome which has been reported in another study of cocaine exposed infants (Bingol et al., 1987). In the group of 30 offspring from the non–cocaine, polydrug users, there was one malformed fetus with stigmata of fetal alcohol syndrome. Fries et al. (1993) have described the clinical facies they observed in 14 newborns exposed only to cocaine. Sixty–four percent had low nasal bridges with transverse nasal creases and 57% had periorbital edema. Irritability was present in 71%. The authors state that a further blinded prospective study are needed to establish their findings.

Chavez et al. (1989) found a statistically significant increase in urinary tract defects (odds ratio = 4.39) among women who took cocaine one month before pregnancy or in the first 3 months. Genital defects were not increased significantly. There were 276 infants with urinary tract defects and 2,835 controls. Various renal defects occurred in the exposed group and they included hydroephrosis and renal agenesis. Rosenstein et al. (1990) examined 100 consecutive newborns exposed to cocaine and found two with murmurs consistent with ventricular septal defect and two with renal defects (unilateral cystic kidney and abnormally small kidneys).

Congenital heart disease has been associated with cocaine use. Neerhof et al. (1989) found a 7% congenital defect rate and four congenital heart cases among 138 exposed cases. Little et al. (1989) reported four heart defects among 35 exposed infants and Bingol et al. (1987) found two heart defects among 50 offspring of women who abused cocaine. Lipshultz et al. (1991) found a 3.7 relative risk for congenital heart disease among a large group with positive urine screens in a retrospective study. Shepard et al (1991) described an autopsy on an exposed fetus that had a coronary occlusion which had presumably converted a normal heart into a single ventricular heart. Martin and Khoury (1992) studied 27 infants with single ventricle and found no increase over controls in cocaine use. No increase in single ventricles occurred in Atlanta during the years 1968 to 1990. Gilloteaux and Dalbec (1991) described endothelial sloughing with thrombi in the atria of newborn hamsters whose mothers were given 3–7 mg per kg intraperitoneally on days 6, 7 and 9.

Dominquez et al. (1991) have described the eye and brain abnormalities in 7 infants exposed to cocaine in utero. Nystagmus, strabismus and hypoplasia of the optic discs were encountered.Good et al. (1992) described 13 newborns exposed in utero to cocaine and found optic nerve abnormalities delayed visual maturation and in 3 persistent eyelid edema.

Hoyme et al. (1990) have described several limb reduction defects in the offspring of cocaine users.

Callahan et al. (1992) reported that radioimmune assay of infants hair and gc mass spectrometry of meconium were more sensitive than immune assay of urine for cocaine.

Animal Studies

In the mouse and rat, 60 mg and 50 mg per kg, respectively, given intraperitoneally caused no congenital defects (Fantel and MacPhail, 1982). Treatment days were 8 through 12 for the rat and days 7 through 16 for the mouse. In both species fetal weight was reduced as compared to pair–fed controls. Edema was found in the treated rat fetuses. Church et al. (1988) gave rats up to 90 mg subcutaneously on days 7–19 of gestation. Fetal edema and abruptio placentae occurred at 60 mg per kg and above. Fetal hemorrhage was also increased but defects were not. Hayasaka et al. (1976) reported that cocaine (20 mg per kg) potentiated the teratogenic effects of caffeine in mice. Mahalik et al. (1980, 1984) injected mice subcutaneously with single doses of 60 mg per kg on days 7–12. The fetal weight was not reduced but a low, but significant increase in eye defects and skeletal defects was reported. Pair–feeding was not carried out. Webster et al. (1990) found hemorrhage and edema of tail, nose and extremities followed by reduction defects when rats were treated after day 15. The no effect dose on day 16 was 40 mg per kg. Meyer et al. (1992) treated rats on days 11–20 with two injections of 20 mg per kg daily. Reduced sensitivity to cocaine injeced on postnatal day 11 was found as well as increased mortality. Kaurmann and Armant (1992) studied the effect of cocaine and benzoylecogonine on preimplantation mouse embryos at 25–400 mg of cocaine per ml. Blastocyst formation was impaired. Concentrations of 100–400 mg per ml of benzoylecogonine inhibited blastocyst formation.

Fantel et al. (1990) studied the effect of cocaine on isolated mitochondria and found reduced oxidative metabolism. In isolated embryos and hearts the heart rate decreased at or above 5 micromolar concentrations. Fantel et al. (1992 A) have proposed that the vasoconstriction and hypoprofusion after cocaine may increase reactive oxygen species which damage tissue. By exposing day 14 or 15 rat embryos in vitro to hypoxia or cocaine limb defects were produced (Fantel et al., 1992 B). Evidence has been presented that the initiating event in embryogenesis is produced by free radicals produced in the uterus (Fisher et al. 1994 and Zimmerman et al. 1994). The vasodiliation and hemorrhage seen in mouse embryos was reversed by two antioxidants (2–oxothiazolidine 4–carboxylate and α–phenyl–N–t–butylnitrone) given to the dam.

Woods and Plessinger (1990) have studied cardiovascular toxicity Woods and Plessinger (1990) have studied cardiovascular toxicity in pregnant ewes and found that pregnancy increased the sensitivity. Woods et al. (1987) produced increase in uterine vascular resistance and fetal hypoxemia, hypertension and tachycardia in the ewe given 0.5 to 2.0 mg per kg intravenously. Plessinger and Woods (1991) have reviewed the cardiovascular effects of cocaine in pregnancy. Duhart et al. (1993) studied pharmacokinetic parameters of cocaine and benzoylecgonine in pregnant and non–pregnant monkeys and found few differences.

El–Bizri et al. (1991) reported hydroephrosis in rat fetuses exposed intraperitoneally on days 0–19 to 6.25 umoles

(1.8 mg) per kg or more. A few cerebral hemorrhages were found at 25 and 50 umoles per kg. Concentrations of dopamine, epinephrine, and norepinephrine in fetal brain was not different from controls.

Hutchings et al. (1989) studied locomotion in foster parented rats exposed during the last 2 weeks of intrauterine life to maternal cocaine (30 or 60 mg per kg). A significant increase in activity was found on postnatal days 20 to 23 in the group exposed to 60 mg per kg. Barron et al. (1991) reported significant shortening of the rat umbilical cord in fetuses exposed orally to 60 mg per kg on days 14–21. Using a similar regeme Riley and Foss (1991) were unable to find changes in exploratory behavior or locomotor activity on postnatal day 21 or as adults. Riley and Foss (1991) studied active and passive avoidance and spatial navigation in rat offspring exposed to 60 mg of oral cocaine on days 14–21 and found no significant differences from controls. Foss and Riley (1991) tested the effect of prenatal cocaine exposure on the acoustic and open field behavior of the offspring. No evidence was found that the postnatal responses were altered.

Disposition of cocaine in pregnant sheep has been described by DeVane et al. (1991). Rapid equilibration with the fetus and rapid metabolism by the mother and fetus was reported. The serum half–lives were 4–5 minutes. Burchfield et al. (1991) found an increase in fetal sheep blood pressure and a fall in arterial oxygen tension when cocaine was administered to the dam.

Acker, D.; Sachs, B.P.; Tracey, K.J. and Wise, W.E.: Abruptio placentae associated with cocaine use. Am. J. Obstet. Gynecol. 146:220–225, 1983.

Barron, S.; Foss, J.A. and Riley, E.P.: The effect of prenatal cocaine exposure on umbilical cord length in fetal rats. Neurotoxicology and Teratology 13:503–506, 1991.

Bateman, D.A.; Ng, S.K.C.; Hansen, C.A. and Heagarty, M.C.: The effects of intrauterine cocaine exposure in newborns. (abs) American Journal of Public Health 83:190–193, 1993.

Bingol, N.; Fuchs, M.; Diaz, V.; Stone, R.K. and Gromisch, D.S.: Teratogenicity of cocaine in humans. J. Pediatr 110:93–96, 1987.

Burchfield, D.J.; Abrams, R.M.; Miller, R. and DeVane, C.L.: Disposition of cocaine in pregnant sheep II. Physiological responses. Developmental Pharmacology and Therapeutics 16:130–138, 1991.

Callahan, C.M.; Grant, T.M.; Phipps, P.; Clark, G.; Novack, A.H.; Streissguth, A.P. and Raisys, V.A.: Measurement of gestational cocaine exposure: sensitivity of infants hair, meconium, urine. The Journal of Pediatrics 120:763–768, 1992.

Chasnoff, I.J.; Burns, W.J.; Schnoll, S.H. and Burns, K.A.: Cocaine use in pregnancy. N. Eng. J. Med. 313:666–669, 1985.

Chasnoff, I.J.; Burns, K.A. and Burns, W.J.: Cocaine use in pregnancy: Perinatal morbidity and mortality. Neurotoxicol. Teratol. 9:291–293, 1987.

Chasnoff, I.J.; Chisum, G.M. and Kaplan, W.E.: Maternal cocaine use and genitourinary tract malformations. Teratology 37:201–204, 1988.

Chasnoff, I.J.; Landress, H.J.; Barrett, M.E.: The prevalence of illicit–drug or alcohol use during pregnancy and discrepancies in mandatory reporting in Pinellas county, Florida. NEJM 322:1202–1206, 1990.

Chavez, G.G; Mulinare, J and Cordero, J.F.: Maternal cocaine use during early pregnancy as a risk factor for congenital urogenital anomalies. JAMA 262:795–799, 1989.

Chouteau, M.; Namerow, P.B. and Leppert, P.: The effect of cocaine abuse on birth weight and gestational age. Obstet. Gynecol. 72:351–354, 1988.

Church, M.W.; Dintcheff, B.A. and Gessner, P.K.: Dose-dependent consequences of cocaine on pregnancy outcome in the Long-Evans rat. Neurotoxicol. Teratol. 10:51–58, 1988.

Church, M.W. and Overbeck, G.W.: Prenatal cocaine exposure in the Long–Evans rat. II. Dose–dependent effects on the brainstem auditory–evoked potential. Neurotoxicol. Teratol. 12:345–351, 1990.

DeVane, C.L.; Burchfield, D.J.; Abrams, R.M.; Miller, R.L. and Braun, S.B.: Disposition of cocaine in pregnant sheep I. Pharmacokinetics. Dev. Pharmacol. Ther. 16:123–129, 1991.

Dixon, S.D. and Bejar, R.: Echoencephalographic findings in neonates associated with maternal cocaine and methamphetamine use: Incidence and clinical correlates. J. Pediatrics 115:770–779, 1989.

Doberczak, T.M.; Shanzer, S.; Senle, R.T. and Kandall, S.R.: Neonatal neurologic and electroencephalographic effects of intrauterine cocaine exposure. Journal of Pediatrics 113:354–358, 1988.

Dominguez, R.; Vila–Coro, A.A.; Slopis, J.M. and Bohan T.P.: Brain and ocular abnormalities in infants with in utero exposure to cocaine and other street drugs. Am. J. Dis. Child. 145:688–695, 1991.

Duhart, H.M.; Fogle, C.M.; Gillam, M.P.; Bailey, J.R.; Slikker Jr., W. and Paule, M.G.: Pharmacokinetics of cocaine in pregnant and nonpregnant rhesus monkeys. Reproductive Toxicology 7:429–437, 1993.

El–Bizri, H.; Guest, I. and Varma, D.R.: Effects of cocaine on rat embryo development in vivo and in cultures. Pediat. on rat embryo development in vivo and in cultures. Pediat. Research 29:187–190, 1991.

El–Bizri, H.; Guest, I. and Varma, D.R.: Effects of cocaine on rat embryo development in vivo and in cultures. Pediat. Res. 29:187–190, 1991.

Fantel, A.G.; Barber, C.V.; Carda, M.B.; Tumbic, R.W. and Mackler, B.: Studies of the role of ischemia/reperfusion and superoxide anion radical production in the teratogenicity of cocaine. Teratology 46:293–300, 1992.

Fantel, A.G.; Barber, C.V. and Mackler, B.: Ischemia/reperfusion: a new hypothesis for the developmental toxicity of cocaine. Teratology 46:285–292, 1992.

Fantel, A.G. and MacPhail, B.J.: The teratogenicity of cocaine. Teratology 26:17–19, 1982.

Fantel, A.G.; Person, R.E.; Burroughs–Gleim, C.J. and Mackler, B.: Direct embryotoxicity of cocaine in rats: effects on mitochondrial activity, cardiac function, and growth and development in vitro. Teratology 42:25–3, 1990.

Fisher, J.E.; Potturi, R.B.; Collins, M.; Resnick, E. and

Zimmerman, E.F.: Cocaine–induced embryonic cardiovascular disruption in mice. Teratology 49:182–191, 1994.

Forman, R.; Klein, J.; Meta, D.; Barks, J.; Greenwald, M. and Koren, G.: Maternal and neonatal characteristics following exposure to cocaine in Toronto. Reproductive Toxicology 7:619–622, 1993.

Foss, J.A. and Riley, E.P.: Failure of acute cocaine administration to differentially affect acoustic startle and activity in arts prenatally exposed to cocaine. Neurotoxicology and Teratology 13:547–551, 1991.

Fries, M.H.; Kuller, J.A.; Norton, M.E.; Yankowitz, J.; Kobori, J.; Good, W.V.; Ferriero, D.; Cox, V.; Donlin, S.S. and Golabi, M.: Facial features of infants exposed prenatally to cocaine. Teratology 48:413–420, 1993.

Gilbert, W.M.; Lafferty, C.M.; Benirschke, K. and Resnik, R.: Lack of specific placental abnormality associated with cocaine use. Am. J. Obstet. Bynecol. 163:998–999, 1990.

Gilloteaux, J. and Dalbec, J.P.: Transplacental cardiotoxicity of cocaine: Atrial damage following treatment in early pregnancy. Scanning Microscopy 5:519–531, 1991.

Good, W.V.; Ferriero, D.M.; Golabi, M. and Kobori, J.A.: Abnormalities of the visual system in infants exposed to cocaine. Opthamology 99:341–346, 1992.

Hadeed, A.J.; Siegel, S.R.: Maternal cocaine use during pregnancy effect on newborn infant. Abruptio occurred in 11% of their 57 exposed patients and 1.7% of their controls. Pediatrics 84:205–210, 1989.

Hayasaka, I.; Sasaki, H. and Fujii, T.: Potentiation of the fetopathic effects of caffeine in mice by modification of catecholamine turnover. 2. By cocaine, an uptake inhibitor. (abs) Teratology 14:239–240, 1976.

Hoyme, H.E.; Jones, K.L.; Dixon, S.D.; Jewett, T.; Hanson, J.W.; Robinson, L.K.; Msall, M.E. and Allanson, J.E.: Prenatal cocaine exposure and fetal vascular disruption. Pediatrics 85:743–747, 1990.

Hutchings, D.E.: The puzzle of cocaine's effects following maternal use during pregnancy: are there reconcilable differences? Neurotoxicology and Teratology 15:281–286, 1993.

Hutchings, D.E.; Fico, T.A. and Dow–Edwards, D.L.: Prenatal cocaine: Maternal toxicity, fetal effects and locomoter activity in rat offspring. Neurotoxicol. Teratol. 11:65–69, 1989.

Kapur, R.P.; Shaw, C.M.; Shepard, T.H.: Brain hemorrhages in cocaine–exposed fetuses. Teratology 44:11–18, 1991.

Kaufmann, R.A. and Armant, D.R.: In vitro exposure of preimplantation mouse embryos to cocaine and benzoylecgonine inhibits subsequent development. Teratology 46:85–89, 1992.

LeBlanc, P.E.; Parekh, A.J.; Naso, B. and Glass, L.: Effects of intrauterine exposure to alkaloidal cocaine ("crack"). Am. J. Dis. Child. 141:937–938, 1987.

Little, B.B.; Snell, L.M.; Klein, V.R. and Gilstrap III, L.C.: Cocaine abuse during pregnancy: Maternal and fetal implications. Obstet. Gynecol. 73:157–160, 1989.

Lipshultz, S.E.; Frassica, J.J. and Orav, E.J.: Cardio-vascular abnormalities in infants prenatally exposed to cocaine. The Journal of Pediatrics 118:44–51, 1991.

Lutiger, B.; Graham, K.; Einarson, T.R. and Koren, G.: Relationship between gestational cocaine use and pregnancy outcome: A meta–analysis. Teratology 44:405–414, 1991.

Mahalik, M.P.; Gautieri, R.F. and Mann, D.E.: Teratogenic potential of cocaine hydrochloride in CF_1 mice. J. Pharm. Sci. 69:703–706, 1980.

Mahalik, M.P.; Gautieri, R.F. and Mann, D.E.: Mechanisms of cocaine–induced teratogenesis. Res. Commun. Substances of Abuse 5:279–302, 1984.

Martin, M.L. and Khoury, M.J.: Cocaine and single ventricle: a population study. Teratology 46:267–270, 1992.

Matera, C.; Warren, W.B.; Moomjy, M.; Fink, D.J. and Fox, H.E.: Prevalance of use of cocaine and other substances in an obstetric population. Am. J. Obstet. Gynecol. 163:797–801, 1990.

Meyer, J.S.; Sherlock, J.D. and Macdonald, N.R.: Effects of prenatal cocaine on behavioral responses to a cocaine challenge on postnatal day 11. Neurotoxicology and Teratology 14:183–189, 1992.

Neerhof, M.G.; MacGregor, S.N.; Retzky, S.S. and Sullivan, T.P.: Cocaine abuse during pregnancy: Peripartum prevalance and perinatal outcome. Am. J. Obstet. Gynecol. 161:633–638, 1989.

Plessinger, M.A. and Woods, Jr., J.R.: The cardiovascular effects of cocaine use in pregnancy. Reproductive Toxicology 5:99–113, 1991.

Riley, E.P. and Foss, J.A.: The acquisition of passive avoidance, active avoidance, and spatial navigation tasks by animals prenatally exposed to cocaine. Neurotoxicology and Teratology 13:559–564, 1991.

Riley, E.P. and Foss, J.A.: Exploratory behavior and locomotor activity: a failure to find effects in animals prenatally exposed to cocaine. Neurotoxicology and Teratology 13:553–558, 1991.

Rosenstein, B.J.; Wheeler, J.S. and Heid, P.L.: Congenital renal abnormalities in infants with in utero cocaine exposure. The Journal of Urology 144:110–112, 1990.

Scanlon, J.W.: The neuroteratology of cocaine: background, theory and clinical implications. Reproductive Toxicology 5:89–98, 1991.

Shepard, T.H.; Fantel, A.G. and Kapur, R.P.: Fetal coronary thrombobis as a cause of single ventricular heart. Teratology 43:113–117, 1991.

Singer, L.T.; Yamashita, T.S.; Hawkins, S.; Cairns, D.; Bailey, J. and Kliegman, R.: Increased incidence of intraventricular hemorrhage and developmental delay in cocaine exposed very low birth weight infants. J. Pediat. 124:765–771, 1994.

Slutsker, L.: Risks associated with cocaine use during pregnancy. Obstetrics & Gynecology 79:778–789, 1992.

Streissguth, A.P.; Grant, T.M.; Barr, H.M.; Brown, J.A.; Martin, J.C.; Mayock, D.E.; Ramey, S.L. and Moore, L.: Cocaine and the use of alcohol and other drugs during pregnancy. Am. J. Obstet. Gynecol. 164(5):1239–1243, 1991.

Webster, W. and Brown–Woodman, P.D.C.: Cocaine as a

cause of congenital malformations of vascular origin: experimental evidence in the rat. Teratology 41:689–697, 1990.

Webster, W.; Brown–Woodman, P.D.C.; Lipson, A.H. and Ritchie, H.E.: Fetal brain damage in the rat following prenatal exposure to cocaine. Neurotoxicology and Teratology 13:621–626, 1991.

Woods, Jr. J.R. and Plessinger, M.A.: Effect of cocaine on uterine blood flow and fetal oxygenation. JAMA 257:957–961, 1987.

Woods, Jr., J.R. and Plessinger, M.A.: Pregnancy increases cardiovascular toxicity to cocaine. Am. J. Obstet. Gynecol. 162:529–533, 1990.

Zimmerman, E.F.; Potturi, R.B.; Resnick, E. and Fisher, J.E.: Role of oxygen free radicals in cocaine–induced vascular disruption in mice. Teratology 49:192–201, 1994.

616 Cocoa Powder

Tarka et al. (1986) fed rats 2.5, 5.0 or 7.5 cocoa powder in their diets. This is equivalent to up to 10 pounds of milk chocolate per day in the human. No adverse effects were found with treatment either in perinatal or organogenesis periods.

Tarka, S.M., Jr.; Applebaum, R.S. and Borzelleca, J.F.: Evaluation of the perinatal, postnatal and teratogenic effects of cocoa powder and theobromine in Sprague–Dawley CD rats. Food Chem. Toxicol. 24:375–382, 1986.

Webster, W.S. and Brown–Woodman, P.D.C.: Cocaine as a cause of congenital malformations of vascular ofigin: experimental evidence in the rat. Teratology, 41:689–697, 1990.

617 Codeine *Hydrocodone* CAS 76–57–3

Heinonen et al. (1977) reported on 563 pregnancies where codeine was used during the first four lunar months and found no increase in defect rates. Eight respiratory malformations were reported while only 3 were expected. This increase was not statistically significant. Heinonen et al. (1977) found no increase in malformations among the offspring of 60 women who took the dihydro form (hydrocodone).

Geber and Schramm (1975) injected hamsters on day 8 subcutaneously with 73 to 360 mg per kg and found 6 to 8 percent cranioschisis in the offspring. No dose response was found and not a single malformation occurred in over 1400 saline injected controls. Williams et al. (**1991**) studied pregnant mice and hamsters at 150 mg per kg found an increase in meningoencephaloceles. No increase in defects was found in the mouse at doses up to 300 mg per kg. Williams et al. (**1991**) gavaged hamsters and mice with up to 300 or 150 mg per kg twice daily respectively. Decreased fetal weight was found at maternally toxic doses but no teratogenicity.

Geber, W.F. and Schramm, L.C.: Congenital malformations of the central nervous system produced by narcotic analgesics in the hamster. Am. J. Obstet. Gynecol. 123:705–713, 1975.

Heinonen, O.P.; Slone, D. and Shapiro, S.: Birth Defects and Drugs in Pregnancy. Publishing Sciences Group Inc.,

Littleton, Mass., 1977.

Williams, J.; Price, C.J.; Sleet, R.B.; George, J.D.; Marr, M.C.; Kimmel, C.A. and Morrissey, R.E.: Codeine: developmental toxicity in hamsters and mice. Fundamental and Applied Toxicology 16:401–413, 1991.

618 Coitus

Naeye and Tafari (1983) presented data suggesting that amniotic fluid infections and premature rupture of the membranes were significantly more common when coitus occurred between the last clinic visit and delivery. The mechanism may involve vaginal introduction of microorganisms or the proteolytic activity of seminal fluid. The effect of coitus was less significant when condoms were used.

Naeye, R.L. and Tafari, N.: Risk Factors in Pregnancy and Diseases of the Fetus and Newborn. Williams and Wilkins, Baltimore, 113–116, 200–203, 1983.

619 Colchicine CAS 64–86–8

Ferreira and Frota–Pessoa (1969) reported that two persons under colchicine therapy were found among 54 parents of Down's syndrome patients. Zemer et al. (1976) reported three normal offspring from fathers on therapy and one normal from a mother treated during the entire pregnancy. Of four mothers who discontinued the medication when pregnancy was detected, one aborted and the other three gave birth to normal infants. Ehrenfeld et al. (1987) reviewed histories on 36 women who took long term treatment for Mediterranean fever and found no increase in abortions due to the medication. Sixteen normal infants were born to these women. Rabinovitch et al. (**1992**) found no increase in abnormalities among the offspring of 131 women taking the drug for Familial Mediterranean Fever. The two cases of Down Syndrome in younger mothers in this group they felt were probably coincidence since a nationwide survey of 430 similarly treated women yielded no further cases.

In most experiments, demecolcine was employed. Morris et al. (1967) in rabbits using 0.1 to 5.0 mg per kg, found the higher dosages after the 9th day to be highly lethal to fetuses. With dosage levels of 0.1 to 0.5 mg per kg a small incidence of gastroschisis and failure of neural tube closure was found. The same authors used 1 to 2 mg per kg in monkeys on single days over a wide gestational period and obtained four normal fetuses. Embryocidal effects of the compound were shown in rabbits (Didcock et al., 1965) and in rats (Tuchmann–Duplessis and Mercier–Parot, 1958; Thiersch, 1958).

Ferm (1963) gave 10 or 20 mg per kg intravenously to hamsters on the 8th day and found microphthalmia, anophthalmia, umbilical hernia exencephaly and skeletal defects.

Adams et al. (1961) showed arrested cleavage in ova and degeneration of rabbit blastocysts after administering 2 to 8 mg per kg. Sieber et al. (1978) injected mice with 0.5 or 1.0 mg per kg intraperitoneally on day 6, 7 or 8 and produced a low but significant increase in defective fetuses. These defects included microphthalmia, microtia, exencephaly and other defects. Didcock et al. (1956) gave 1.5 to 18 mg per kg by oral and subcutaneous routes on various single days during organogenesis of the mouse and produced interruption of pregnancy. Similar findings were observed in rabbits.

An unusual experiment was performed by Chang (1944) who suspended sperm in 0.1 percent colchicine and artificially inseminated female rabbits. Thirty–three young resulted; one had an open fontanelle and very small philtrum. One was otocephalic and the third had microcephaly with enlarged eyes. Does of similar origin were artifically inseminated and gave 425 normal young.

Adams, C.E.; Hay, M.F. and Lutwak–Mann, C.: The action of various agents on the rabbit embryo. J. Embryol. Exp. Morphol. 9:468–491, 1961.

Chang, M.C.: Artificial production of monstrosities in the rabbit. Nature 154:150 only, 1944.

Didcock, K.; Jackson, D. and Robson, J.M.: The action of some nucleotoxic substances on pregnancy. Br. J. Pharmacol. 11:437–441, 1956.

Ehrenfeld, M.; Brzezinski, A.; Levy, M. and Eliakim. M.: Fertility and obstetric history in patients with familial Mediterranean fever on long–term colchicine therapy. Br J Obstet Gynaecol. 94:1186–1191, 1987.

Ferm, V.H.: Colchicine teratogenesis in hamster embryos. Proceedings of the Society for Experimental Biology and Medicine 112:775–778, 1963.

Ferreira, N.R. and Frota–Pessoa, O.: Trisomy after colchicine therapy. Lancet 1:1160–1161, 1969.

Morris, J.M.; Van Wagenen, G.; Hurteau, G.D.; Johnston, D.W. and Carlsen, R.A.: Compounds interfering with ovum implantation and development: Alkaloids and antimetabolites. Fertil. Steril. 18:7–17, 1967.

Rabinovitch, O.; Zemer, D.; Kukia, E.; Sohar, E. and Mashiach, S.: Colchicine treatment in conception and pregnancy: two hundred thirty–one pregnancies in patients with Familial Mediterranean Fever. Am. J. Reprod. Immunol. 28:245–246, 1992.

Sieber, S.M.; Whang–Peng, J.; Botkin, C. and Knutsen, T.: Teratogenic and cytogenetic effects of some plant–derived antitumor agents (vincristine, colchicine, maytansine, VP 16–213 and VM–26) in mice. Teratology 18:31–48, 1978.

Thiersch, J.B.: Effect of N–desacetyl–thio–colchicine (TC) and N–desacetyl–methyl–colchicine (MC) on the rat fetus and litter in utero. Proc. Soc. Exp. Biol. Med. 98:479–485, 1958.

Tuchmann–Duplessis, H. and Mercier–Parot, L.: Sur l'action teratogene de quelques substances antimitotiques chez le rat. C. R. Acad. Sci. (Paris) 247:152–154, 1958.

Zemer, D.; Pras, M.; Sohar, E. and Gofni, J.: Colchicine in familial Mediterranean fever. N. Eng. J. Med. 294:170–171, 1976.

620 Cold, Common *Coryza*

Kurppa et al. (**1991**) made a case control study of 393 pairs where one had anencephaly. Cold–like illness with fever occurred in 26 vs 6 controls (OR=7; CI=2.5-19.8) and in women with colds but no fever in 44 vs 11 controls (OR=5.1 CI=2.6–10.1). The odds ratio increases in the second trimester were not significantly increased. They did not think that memory bias played an important role and medications did not show a significant association.

Kurppa, K.; Holmberg, P.C.; Kuosma, E.; Aro, T. and Saxen, L.: Anencephaly and maternal common cold. Teratology 44:51–55, 1991.

621 Colestipol HCl *1–Chloro 2,3–epoxypropane Hydrochloride* CAS 26658–42–4

Webster and Bollert (1974) did reproduction studies in rats and studies of organogenesis in rats and rabbits. The doses were up to 1000 mg per kg daily. No adverse effects on reproduction were found.

Webster, H.D. and Bollert, J.A.: Toxicologic, reproductive and teratologic studies of colestipol hydrochloride: A new bile acid sequestrant. Toxicol. Appl. Pharmacol. 28:57–65, 1974.

622 Colistimethate Sodium *Colistin Sodium Methanesulfonate* CAS 8068–28–8

Tomizawa and Kamada (1973) administered this compound intraperitoneally to pregnant mice and rats during active organogenesis. The maximum dose in the mouse was 150 mg per kg per day and 40 mg in the rat. At the highest doses some fetal deaths occurred in the mouse fetuses but no gross or skeletal defects occurred in either species.

Saitoh et al. (1981) gave mice intravenously up to 500 mg per kg before mating and during the first 7 days of gestation. No changes in fertility, defect rate or postnatal behavior were found. Rats were given up to 25 mg per kg daily before mating and during the first 7 days, on days 7–17 or on days 17–21. No changes in fertility, defect rate or postnatal behavior were reported (Tsujitani et al., 1981A). Rabbits received up to 80 mg per kg on days 6–18 and no adverse fetal effects were found (Tsujitani et al., 1981B).

Saitoh, T.; Tsujitani, M.; Ohuchi, M. and Matsumoto, T.: Reproductive studies of sodium colistin methanesulfonate. Chemotherapy 29:887–890, 1044–1050, 1051–1061, 1981.

Tomizawa, S. and Kamada, K.: Colistin sodium methanesulfonate on fetuses of mice and rats. Oyo Yakuri 7:1047–1060, 1973.

Tsujitani, M.; Ohuchi, M.; Saitoh, T. and Matsumoto, T.: Reproductive studies of sodium colistin methanesulfonate in rabbits. Chemotherapy 29:300–305, 1981B.

Tsujitani, M.; Oyama, M.; Kadoya, K.; Ohuchi, M.; Saitoh, T. and Matsumoto, T.: Reproductive study of sodium colistin methane sulfonate in rats. Chemotherapy 29:149–163, 1981A.

623 Collagen, Bovine

Sato and Narama (1984) tested this topical hemostatic in pregnant rats and rabbits. The material was injected subcutaneously in maximum doses of 2.25 gm per animal in fertility and perinatal studies and up to 500 mg per kg in studies of rats during organogenesis. No adverse effects were noted. In the rabbit no defect increase was seen after 200 mg per kg on days 6–18.

Sato, T. and Narama, I.: Reproductive studies of mirofibrillar collagen hemostat (ZA 552) in rats and rabbits. Oyo Yakuri 27:639–687, 1984.

624 Collagen, Gene Mutant

Stacey et al. (1988) produced an osteogenesis–imperfecta–like syndrome in newborn mice by adding a mutant gene for pro–α 1 (I) collagen to the pronucleus of mouse ova.

Stacey, A.; Bateman, J.; Choi, T.; Mascara, T.; Cole, W. and Jaenisch, R.: Perinatal lethal osteogenesis imperfecta in transgenic mice bearing an engineered mutant pro–α 1(I) collagen gene. Nature 332:131–136, 1988.

625 Collagen, Wound Dressing

Purified collagen from bovine skin was injected subcutaneously in amounts of 10 ml per kg of a 1 to 10 diluted sample. Injection on days 6–15 resulted in no adverse reproductive effects (Takada et al., 1982).

Takada, U.; Moriguchi, M.; Hata, T. and Yamamoto, A.: Teratological studies on collagen wound dressing (CAS) in mice. J. Toxicol. Sci. 7:57–61, 1982.

626 Compazine™

Discussed under Prochlorperazine

627 Concanavalin A CAS 11028–71–0

Desesso (1979) gave 4 microgm of this material intraperitoneally to rabbit embryos on days 12, 13, 14 or 15 and produced craniofacial trunk and limb anomalies. In some cases the limbs were fused to the head or body. Hayasaka and Hoshino (1979) injected 3 and 5 mg per kg intravenously into mice and on day 7 produced exencephaly. They postulated that the mechanism was via an effect on the surface coat material of the neural tube.

DeSesso, J.M.: Lectin teratogenesis. Defects produced by concanavalin A in fetal rabbits. Teratology 19:15–26, 1979.

Hayasaka, I. and Hoshino, K.: Teratogenicity of concanavalin A in the mouse. Cong. Anom. 19:125–128, 1979.

628 Congo Red CAS 573–58–0

Beaudoin (1964) reported that a single 20 mg dose of this substance on the 8th day of gestation produced hydronephrosis, hydrocephalus or microphthalmia in about 15 percent of rat offspring. Wilson (1955), using 10 mg over 3 days, found only one rat fetus with hydrocephalus out of more than 100 examined specimens.

Beaudoin, A.R.: The teratogenicity of congo red in rats. Proc. Soc. Exp. Biol. Med. 117:176–179, 1964.

Wilson, J.G.: Teratogenic acitivity of several azo dyes chemically related to trypan blue. Anat. Rec. 123:313–334, 1955.

629 Coniine CAS 458–88–8

Keeler and Balls (1978) produced arthrogryposis and spinal curvature in calves whose mothers were gavaged with the plant, Conium maculatum, between days 50–75 of gestation. The active principle was 15 coniine. A number of analogs of coniine were not teratogenic. Forsyth & Frank (**1993**) gavaged rats and rabbits with 25 mg or 40 mg per kg respectively. An increase in arthrogryposis was found in the rabbit offspring.

Forsyth, C.S. and Frank, A.A.: Evaluation of Developmental toxicity of coniine to rats and rabbits. Teratology 48:59–64, 1993.

Keeler, R.F. and Balls, L.O.: Teratogenic effects in cattle of conium maculatum, conium alkaloids and analogs. Clinical Toxicol. 12:2 only, 1978.

630 Contraceptives, Oral

Discussed under Oral Contraceptives

631 Copper Deficiency and Excess CAS 7440–50–8

Buamah et al. (1984) found that maternal serum copper was reduced in seven of nine anencephalic pregnancies.

Copper deficiency during pregnancy has been associated with swayback, a degeneration of the nervous system in lambs (Innes and Saunders, 1962). Kalter (1968) summarized the relatively negative teratogenic findings in the rat and the cerebellar dysgenesis in guinea pig fetuses (Everson et al., 1968). O'dell et al. (1961) reported edema, hemorrhage and abdominal hernias in offspring of rats on a copper–deficient diet.

Ridgway and Karnofsky (1952) administered 0.6 mg of $CuCl_2$ to chicks on the 4th day and observed no abnormalities. James et al. (1966) treated four sheep with 10 mg per kg daily throughout gestation and found three normal fetuses and one abortion. DiCarlo (1980) administered 2.7 mg per kg intraperitoneally to hamsters on the morning of the eighth day of gestation. A high percentage of the embryos were edematous and had various cardiovascular defects. Grin et al. (1986) studied the results of 24–hour exposure of pregnant white rats to copper chloride (concentration 0.020 mg per m^3) during 21 days and found an increased number of intrauterine deaths. At a concentration of 0.003 mg per m^3, copper chloride had no adverse effect on the reproduction processes in female rats.

To study the mechanism of action of copper intrauterine devices, Brinster and Cross (1972) carried out in vitro tests on mouse ova and found that copper wire or exposure to concentrations of $CuCl_2$ above 2.5×10^{-4} M were lethal at the blastocyst stage. Webb (1973) found embryo lethality in rats during the late preimplantation period. This could be due to either a direct action of copper or a secondary effect from the inflammatory reaction of endometrial tissue. Mieden et al. (1986) reported the effect of copper deficient serum on the in vitro growth of explanted somite stage rat embryos. The associated growth and brain retardation was reversed by adding copper to the medium.

Brinster, R.L. and Cross, P.C.: Effect of copper on the preimplantation mouse embryo. Nature 238:398–399, 1972.

Buamah, P.K.; Russell, M.; Millford–Ward, A.; Taylor, P. and Roberts, D.F.: Serum copper concentration significantly less in abnormal pregnancies. Clin. Chem. 30:1676–1677, 1984.

DiCarlo, F.J., Jr.: Syndromes of cardiovascular malformations induced by copper citrate in hamsters. Teratology 21:89–101, 1980.

Everson, G.J.; Shrader, R.E. and Wang, T.: Chemical and morphological changes in brains of copper–deficient guinea pigs. J. Nutr. 96:115–125, 1968.

Grin, N.V. and Govorunova, N.N.: The experimental study of embryotoxic action of copper chloride. Gigiena i Sanitaria (USSR) 28:38–39, 1986.

Innes, J.R.M. and Saunders, L.Z.: Comparative Neuropathology. New York: Academic Press, 577–590, 1962.

James, L.F.; Lazar, V.A. and Binns, W.: Effects of sublethal doses of certain minerals on pregnant ewes and fetal development. Am. J. Vet. Res. 27:132–135, 1966.

Kalter, H.: Teratology of the Central Nervous System. Chicago: University of Chicago Press, 45–46, 1968.

Mieden, G.D.; Keen, C.L.; Hurley, L.S. and Klein, N.: Effects of whole rat embryos cultured on serum from zinc–and copper–deficient rats. J. Nutr. 116:2424–2431, 1986.

O'dell, B.L.; Hardwick, B.C. and Reynolds, G.: Mineral deficiencies of milk and congenital malformations in the rat. J. Nutr. 73:151–157, 1961.

Ridgway, L.P. and Karnofsky, D.A.: The effects of metals on the chick embryo: Toxicity and production of abnormalities in development. Ann. N.Y. Acad. Sci. 55:203–215, 1952.

Webb, F.T.G.: Contraceptive action of the copper IUD in the rat. J. Reprod. Fertil. 32:429–439, 1973.

632 Copper Naphthenate

Hardin et al. (1981) injected 10 mg per kg into rats on days 1–15 and found no maternal or fetal toxicity or teratogenicity.

Hardin, B.D.; Bond, G.P.; Sikov, M.R.; Andrew, F.D.; Beliles, R.P. and Niemeier, R.W.: Testing of selected workplace chemicals for teratogenic potential. Scand. J. Work Environ. Health 7(4):66–75, 1981.

633 Cordemcura™

Frosch and Manuel (1986) gave rats up to 100 mg per kg on days 6–15 of gestation. Skeletal defects were increased at 100 mg per kg but not at 30 mg per kg.

Frosch, I. and Mannel, S.: Zur Wirkung von Cordemcura auf die pranatale Entwicklung der Ratte. Pharmazie 41:214–216, 1986.

634 Corticotropin–Releasing Hormone *hCRH*

Takeda et al. (1992) gave this hormone to male and female rats before copulation and found no ill effects in the fetuses or in fertility with doses given intravenously up to 51 microgm per kg. No teratogenic or behavior changes were found after giving 100 mg intravenously during organogenesis (Iwase et al., 1992 A). Similar studies perinatally and during lactation were associated with decreased weight gain but the pups were able to reproduce normally (Iwase et al., 1992 B). Studies during organogenesis in the raabbit at the same dose did not interfer with fetal development (Suzuki et al., 1992).

Iwase, T.; Ohyama, N.; Umeshita, C.; Inazawa, K.; Namiki, M. and Ikeda, Y.: Reproductive and developmental toxicity studies of hCRH [corticotropin–releasing hormone (human)] (II): Study on intravenous administration of hCRH during the period of organogenesis in rats. Yakuri to Chiryo 20:s1325–s1338, 1992A.

Iwase, T.; Ohyama, N.; Inazawa, K.; Umeshita, C.; Namiki, M. and Ikeda, Y.: Reproductive and developmental toxicity studies of hCRH [corticotropin–releasing hormone(human)] (IV): Study on intravenous administration of hCRH during the perinatal and lactation period in rats. Yakuri to Chiryo 20:s1349–s1361, 1992B.

Suzuki, A.; Hoshino, N.; Ohwada, K.; Hayashi, K.; Ohyama, N. and Ikeda, Y.: Reproductive and developmental toxicity studies of hCRH [corticotropin–releasing hormone (human)] (III): Study on intravenous administration of hCRH during the period of organogenesis in rabbits. Yakuri to Chiryo 20:s1339–s1347, 1992.

Takeda, K.; Matsuura, K.; Isono, T.; Kaneshima, M.; Iwai, M.; Ohyama, N. and Ikeda, Y.: Reproductive and developmental toxicity studies of hCRH [corticotropin–releasing hormone (human)] (I): Study on intravenous administration of hCRH prior to and in the early stages of pregnancy in rats. Yakuri to CHiryo 20:s1313–S1323, 1992.

635 Cortisone CAS 53–06–5

In 260 pregnant women treated with corticoids during pregnancy, Bongiovanni and McPadden (1960) found reports of only two newborns with cleft palate and Serment and Ruf (1968) identified among 428 treated cases, three clefts of the palate or lip and a small number of defects of other systems. Although this did not exclude the possibility of human teratogenicity, it indicated that this compound is not highly dangerous.

This compound has been used extensively by teratologists as a tool for the study of mechanisms which produce cleft lip and palate in experimental animals (see reviews by Fraser et al., 1954; Giroud and Tuchmann–Duplessis, 1962). Cortisone has an effect on the growth and intrinsic movement of the palatine shelves (Larsson, 1962) and an additional action possibly through reduction of the amount of amniotic fluid resulting in contraction of the uterus with compression of the head on the chest, thus lodging the tongue between the palatine shelves and preventing their closure. Walker (1965), using 2.5 mg of the acetate form on the 11th through 14th days, could not associate the amount of amniotic fluid with production of cleft palate. Fainstat (1954) produced cleft palate in the rabbit. Grollman and Grollman (1962) produced hypertension in the offspring of rats treated with cortisone.

Cortisone acts on the multifactorial system which was shown by Fraser and colleagues to produce cleft palate in the mouse (Fraser, 1969). Walker (1967, 1971) compared a number of the glucocorticoid steroids for their ability to produce cleft palate in the mouse, rat and rabbit. In the rat, high doses of methylprednisolone, prednisolone and cortisone produced no cleft palate while triamcinolone, betamethasone and dexamethasone produced clefts. Methylprednisolone, betamethasone, dexamethasone and triamcinolone produced cleft palate in the mouse. Walker compared the same six drugs in pregnant rabbits and found methylprednisolone ineffective in producing cleft palate.

Bongiovanni, A.M. and McPadden, A.J.: Steroids during pregnancy and possible fetal consequences. Fertil. Steril. 11:181–186, 1960.

Fainstat, T.D.: Cortisone–induced congenital cleft palate in rabbits. Endocrinology 55:502–508, 1954.

Fraser, F.C.: Gene–environment interactions in the production of cleft palate. In: Methods for Teratological Studies in Animals and Man, edited by H. Nishimura and J.R. Miller. Tokyo: Igaku Shoin Ltd, 34–49, 1969.

Fraser, F.C.; Kalter, H.; Walker, B.E. and Fainstat, T.D.: The experimental production of cleft palate with cortisone and other hormones. J. Cell. Comp. Physiol. Suppl. 43:237–259, 1954.

Giroud, A. and Tuchmann–Duplessis, H.: Malformations congenitales: Role des facteurs exogenes. Pathol. Biol. (Paris) 10:119–151, 1962.

Grollman, A. and Grollman, E.F.: Teratogenic induction of hypertension. J. Clin. Invest. 41:710–714, 1962.

Kalter, H. and Warkany, J.: Experimental production of congenital malformations in mammals by metabolic procedure. Physiol. Rev. 39:69–115, 1959.

Larsson, K.S.: Studies on the closure of the secondary palate. IV. Autoradiographic and histochemical studies of mouse embryos from cortisone–treated mothers. Acta Morphol. Neerl. Scand. 4:369–386, 1962.

Serment, H. and Ruf, H.: Corticotherapie et grossesse. Bull. Fed. Soc. Gynecol. Obstet. Lang. Fr. 20:77–85, 1968.

Walker, B.E.: Amniotic fluid measurement in cortisone–treated and x–irradiated mice. Proc. Soc. Exp. Biol. Med. 118:606–609, 1965.

Walker, B.E.: Induction of cleft palate in rabbits by several glucocorticoids. Proc. Soc. Exp. Biol. Med. 125:1281–1284, 1967.

Walker, B.E.: Induction of cleft palate in rats with anti-inflammatory drugs. Teratology 4:39–42, 1971.

636 Coumaphos O,O–Diethyl
O–(3–chloro–4–methyl–2–oxo–2H–1 benzopyran–7–yl) CAS 56–72–4

Bellows et al. (1975) applied this insecticide dermally to heifers for various 10 day treatments from 10 days to 100 days after breeding. Doses of up to 28 gm per 45 kg of body weight produced no adverse fetal effects.

Bellows, R.A.; Rumsey, T.S.; Kasson, C.W.; Bond, J.; Warwick, E.J. and Pahnish, O.F.: Effects of organic phosphate systemic insecticides on bovine embryonic survival and development. Am. J. Vet. Res. 36(8):1133–1140, 1975.

637 Coumadin Derivatives *Dicumarol™ Coumadin Sodium–Warfarin Phenindione Bishydroxycoumarin* CAS 66–76–2

A very uncommon but strikingly similar pattern of con-

genital anomalies has been observed in over 50 children born to women treated with warfarin or its derivatives. The warfarin syndrome consists of nasal hypoplasia, stippled epiphyses and growth retinal-optic atrophy and central nervous system anomalies may occur. The subject has been reviewed by Schardein (1993) and Freidman & Polifka (1994).

Sbarouni and Oakley (1994) reported on 140 pregnancies of women with heart valve prostheses. Of the 66 treated for the first trimester with heparin followed by warfarin 92% had healthy babies without embryopathy. For the 36 with warfarin throughout pregnancy there were 83% healthy babies and for continual heparin there were 73% healthy infants. (The decreased rate was caused by stillbirths) In an editorial Ginsburg and Barron (94) recommend not giving warfarin between 6 and 12 weeks of gestation.

Since Kerber et al. (1968) reported nasal hypoplasia in the offspring of a woman treated with warfarin during pregnancy, an association has been made between the use of coumarin derivatives and Conradi syndrome which consists of nasal hypoplasia with calcific stippling of the secondary epiphyses (Becker et al., 1975; Holmes et al., 1972; Pettiflor and Benson, 1975 and Shaul et al., 1975). The dose given to the mother was from 2.5 to 10 mg per day during the first trimester. One patient received coumadin during only the first 8 weeks and another took phenindione. Nasal obstruction complicated the neonatal course of these infants who were of low birth weight. Three of the children were reported to be blind but their long term development is still unknown. Eleven reported cases appeared in the literature (see summaries by Shaul and Hall, 1977; Pauli et al., 1976). Kaplan (1985) reported an infant with Dandy Walker malformation and dysgenesis of the anterior eye segment. He summarized the 14 reported patients with CNS defects after coumarin.

Quaini et al. (1986) studied 105 pregnancies in patients with heart valves and most were treated with dicumerol. One newborn had the Warfarin syndrome and one had cleft palate. Four infants had minor hemorrhagic complications. Fetal intraventricular hemorrhage has been observed (Ville et al. 1993).

Warkany (1976) summarized the clinical data and cites five offspring with central nervous system problems including hydrocephalus. Hall et al. (1980) reported 13 patients with nervous system defects after second and third trimester treatment. Hall et al. (1980) summarized the results of 418 pregnancies in which coumarin derivatives were used. Where treatment was during the first trimester 69 percent of the pregnancies were abnormal (mostly prematurity and miscarriage). Thirty–eight percent of the 418 took coumarin during all three trimesters and in this group 75 percent of the newborns were judged to be normal. Chong et al. (1984) followed up on 22 of 42 patients exposed in utero and all were normal (only two were treated in the first trimester though). The pathology of a human fetus exposed to warfarin was reported (Barr and Burdi, 1976). Congenital defects following maternal use of heparin were not reported. In 1965, Villasanta reviewed 38 pregnancies where heparin was employed and reported no malformations, but the pregnancy loss rate with heparin is on the order of 20–

30 percent (Pauli and Hall, 1980). Pauli (1988) reported two infants with inborn errors giving the Warfarin Syndrome. They had defects vitamin K epoxide reductase.

In experimental animals, Kronick et al. (1974) found no significant malformations in mice whose mothers were given up to 4 mg per kg on days 8 through 11. On days 3 through 11, placental hemorrhage and subsequent fetal loss occurred. Grote and Weinmann (1973) treated rabbits intravenously on the 6th through the 18th day with up to 100 times the therapeutic dose and found no effect on resorption rate or the fetus (including skeletal studies). Howe and Webster (1992) describe a model in the rat. They gave warfarin and vitamin K–1 and produced cerebral hemorrhage in 25% of the offspring. In the rat fetus, maxillonasal hypoplasia and abnormal calcification of the nasal septal cartilage was found. The authors have also described cerebral hemorrhages (Howe & Webster, 1990).

Barr, M. and Burdi, A.R.: Warfarin–associated embryopathy in a 17 week abortus. Teratology 14:129–134, 1976.

Becker, M.H.; Genieser, N.B.; Finegold, M.; Miranda, D. and Spackman, T.: Chondrodysplasia punctata: Is warfarin therapy a factor? Am. J. Dis. Child. 129:365–369, 1975.

Chong, M.K.B.; Harvey, D. and Deswiet, M.: Follow-up study of children whose mothers were treated with warfarin during pregnancy. Br. J. Obstet. Gynecol. 91:1070–1073, 1984.

Freidman, J.M. and Polifka, J.E.: Teratogenic effects of drugs: a resource for clinicians (TERIS). The Johns Hopkins Press, Baltimore, 1994. p658.

Ginsburg, J.S. and Barron, W.M.: Pregnancy and prosthetic heart valves. The Lancet 344:1170–1171, 1994.

Grote, W. and Weinmann, I.: Uberprufung der Wirkstoffe Cumarin und Rutin im teratologischen Versuch an Kaninchen. Arzneim–Forsch 23:1319–1320, 1973.

Hall, J.G.; Pauli, R.M. and Wilson, K.M.: Maternal and fetal sequelae of anticoagulants during pregnancy. Am. J. Med. 68:122–144, 1980.

Holmes, B.; Moser, H.W.; Halldorsson, S.; Mack, C.; Pant, S.S. and Matzilevich, B.: Mental Retardation: An Atlas of Disease with Associated Physical Abnormalities. New York, Macmillan Co., 136–137, 1972.

Howe, A.M. and Webster W.S.: Exposure of the pregnant rat to warfarin and vitamin k1: an animal model of intraventricular hemorrhage in the fetus. Teratology 42:413–420, 1990.

Howe, A.M. and Webster, W.S.: The warfarin embryopathy: A rat model showing maxillonasal hypoplasia and other skeletal disturbances. Teratology 46:379–390, 1992.

Kaplan, L.C.: Congenital Dandy Walker malformation associated with first trimester Warfarin: A case report and literature review. Teratology 32:333–337, 1985.

Kerber, I.J.; Warr, O.S. and Richardson, C.: Pregnancy in a patient with prosthetic mitral valve. JAMA 203:223–225, 1968.

Kronick, J.; Phelps, N.E.; MCallion, D.J. and Hirsh, J.: Effects of sodium warfarin administered during pregnancy in mice. Am. J. Obstet. Gynecol. 118:819–823, 1974.

Pauli, R.M.: Mechanism of bone and cartilage maldevelopment in the warfarin embryopathy. Pathol. Immunopathol. Res. 7:107–112, 1988.

Pauli, R.M. and Hall, J.G.: Warfarin embryopathy. Am. J. Med. 68:122–144, 1980.

Pauli, R.M.; Madden, J.D.; Kranzler, K.J.; Culpepper, W. and Port, R.: Warfarin therapy initiated during pregnancy and phenotypic chondrodysplasia punctata. J. Pediatr. 88:506–508, 1976.

Pettiflor, J.M. and Benson, R.: Congenital malformations associated with the administration of oral anticoagulants during pregnancy. J. Pediatr. 459–462, 1975.

Shaul, W.L.; Emery, H. and Hall, J.G.: Chondrodysplasia punctata and maternal warfarin during pregnancy. Am. J. Dis. Child. 129:360–362, 1975.

Quaini, E.; Vitali, E.; Colombo, T. et al.: Complicanze materne e fetali in 105 gravidanze di portatrici di protesi valvolari cardiache. Minerva Ginecol 38:217–224, 1986.

Sbarouni, E. and Oakley, C.M.: Outcome of pregnancy in women with valve prostheses. Br. Heart J. 71:196–201, 1994.

Schardein, J.L.: Chemically induced birth defects 2nd ed. Marcel Dekker, Inc. New York, 1993 p. 108.

Shaul, W.L. and Hall, J.G.: Multiple congenital anomalies associated with oral anticoagulants. Am. J. Obstet Gynecol. 127:191–198, 1977.

Villasanta, U.: Thromboembolic disease in pregnancy. Am. J. Obstet. Gynecol. 93:142–160, 1965.

Ville, Y.; Jenkins, E.; Shearer, M.J.; Hemley, H.; Vasey, D.P.; Layton, M. and Nicolaides, K.H.: Fetal intraventricular haemorrhage and maternal warfarin. The Lancet 341:1211, 1993.

Warkany, J.: Warfarin embryopathy. Teratology 14:205–209, 1976.

638 Coumestrol *Phytoestrogen*

Fredricks et al. (1981) fed this phytoestrogen to mice in amounts of 100 or more parts per billion and found decreased ovulation and embryo degeneration. This chemical is found in many fruits, grams and in coffee. Later fetal stages were not studied.

Fredricks, G.R.; Kincaid, R.L.; Boniioli, K.R. and Wright, Jr., R.W.: Ovulation rates and embryo degeneracy in female mice fed the phytoestrogen coumestrol. Proc. Soc. Exp. Biol. Med. 167:237–241, 1981.

639 Coxiella burnetii *Q Fever agent*

Giroud and Giroud (1964) reviewed the serological studies of women with spontaneous abortions and defective offspring. Positive serologic values for different rickettsial antigens were found in eight of 16 mothers who gave birth to anencephalics. In three there was a positive reaction to C. burnetii. A one percent incidence of positive rickettsial serums was reported in controls.

Giroud et al. (1968) administered this live agent to rats on the 9th gestational day and found cataracts in 46 of 133 fetuses. Histologic changes were present in the retina or optic nerves of 105 of these fetuses. The maternal rats had no evidence of illness except for antibody rises.

Giroud, P. and Giroud, A.: Anencephalie et rickettsioses maladies inapparentes. Bull. Acad. Nat. Med. (Paris) 148:621–626, 1964.

Giroud, A.; Giroud, M.; Martinet, M. and Deluchat, C.H.: Inapparent maternal infection by Coxiella burnetii and fetal repercussions. Teratology 1:257–262, 1968.

640 Coxsackie Virus B

Although this virus was reported to produce meningoencephalitis and carditis in newborn infants (Kibrick and Benirschke, 1958) no clear–cut association with teratogenicity are known to the author. Evans and Brown (1963) carried out a prospective study of antibodies to four coxsackie strains during pregnancy and found a statistically significant increase in the overall congenital defect rate in those mothers who had a rise in titer.

Evans, T.N. and Brown, G.C.: Congenital anomalies and virus infections. Am. J. Obstet. Gynecol. 87:749–758, 1963.

Kibrick, S. and Benirschke, K.: Severe generalized disease encephalohepatomyocarditis occurring in the newborn period and due to infection with coxsackie virus, group B: Evidence of intrauterine infection with this agent. Pediatrics 22:857–875, 1958.

641 Cromolyn

Spector (1984) cites a report containing over 300 pregnancies in which cromolyn was used in combination with isoproterenol in most cases. No link with congenital defects was found.

In animals, Cox et al. (1970) found no teratogenicity in rats given up to 90 mg subcutaneously during pregnancy. In the rabbit, 500 mg given daily intravenously did not produce deformities but was lethal to some of the dams. Studies in the mouse were also non–teratogenic at 540 mg per kg.

Cox, J.S.G.; Beach, J.E.; Blair, A.M.J.N. and Clarke, A.J.: Disodium cromoglycate (Intal.). Advances in Drug Research 5:135–136, 1970.

642 Crooked Calf Disease *Piperidine alkaloids Ammodendrine*

Crooked calf disease a congenital defect of cattle has been associated with ingestion of anagyrine and γ–coniceine. Keeler and Panter (1989) extracted Lupinus formosus and fed it to cattle from the 40th to the 70th day. At doses of above 5.6 mg per kg spinal curvature (scoleosis) limb defects and cleft palate resulted. Ammodendrine was felt to be the causative agent.

Keeler, R.F. and Panter, K.E.: Piperdine alkaloid composition and relation to crooked calf disease–inducing potential of lupinus formosus. Teratology 40:423–432. 1989.

643 Crufomate *4–tert–Butyl–2–chlorophenyl methyl methylphosphoramidate* CAS 299–86–5

Bellows et al. (1975) treated heifers with this insecticide by applying for various 10 day periods from the 10th through 100th day after breeding. Up to 28.5 gm was applied dermally per 45 kg of body weight. No adverse fetal effects were found.

Bellows, R.A.; Rumsey, T.S.; Kasson, C.W.; Bond, J.; Warwick, E.J. and Pahnish, O.F.: Effects of organic phosphate systemic insecticides on bovine embryonic survival and development. Am. J. Vet. Res. 36:1133–1140, 1975.

644 CS–1 *9 α–Fluoro–11–β, 17–dihydroxy 3–oxo–4 androstene–17 α–proprionic acid*

Tache et al. (1974) decreased implantation with 5 mg per 100 gm twice daily in rats. Dosing after day 13 until parturition had no effect on the number of live fetuses. The medication was given orally.

Tache, Y.; Tache, J. and Selye, H.: Antifertility effect of CS–1 in the rat. J. Reprod. Fertil. 37:257–262, 1974.

645 CS–439 *ACNU* *3–[(4–Amino–2–methyl–5–pyrmidinyl) methyl]–1–(2–chloroethyl)–1–nitrosourea hydrochloride*

Masuda et al. (1977) used this nitrosourea–like compound intravenously to study pregnant rats and rabbits. In the rat, 0.5 and 0.1 mg per kg produced malformations when given during organogenesis. Skeletal defects and eye defects were observed at 0.5 mg per kg. No malformations increase was seen in rabbits tested with 1 mg per kg.

Masuda, H.; Kimura, K.; Okada, T.; Matsunuma, N.; Maita, K. and Akuzawa, M.: Toxicological studies of CS–439 (ACNU). Ann. Rep. Sankyo Res. Lab. 29:118–137, 1977.

646 CT–1341 CAS 8067–82–1

Discussed under Alphaxalone

647 Cuprizone *Biscyclohexanone Oxaldihydrazone*

Carlton (1966) fed this chelating agent to mice from the third day of pregnancy to term. Levels of 0.1 percent had no effect but there were no live young in litters receiving 0.3 and 0.5 percent.

Carlton, W.W.: Response of mice to chelating agents sodium diethyldithiocarbamate, α–benzoinoxime, and bicyclohexanone oxaldihydrazone. Toxicol. Appl. Pharmacol. 8:512–521, 1966.

648 Curare CAS 8063–06–7

Discussed under D–Tubocurarine

649 Curium244 CAS 7440–51–9

Sikov and Mahlum (1975) injected up to 60 microcuries intravenously in day 9 rats and found no adverse effects. Most of the radioactivity was trapped in the maternal liver.

Sikov, M.R. and Mahlum, D.D.: Toxicity of [241]Am and [244]Cm after administration at nine days of gestation in the rat. Radiation Res. 62:565, 1975.

650 Cyanamide CAS 420–04–2

Ali and Persaud (1988) administered this aldehyde dehydrogenase inhibitor intraperitoneally to rats on days 9–12 in amounts up to 100 mg per kg daily. No changes occurred in the day 12 embryos but when ethanol was coadministered, both defects and growth retardation occurred.

116

Ali, F. and Persaud, T.V.N.: Mechanisms of fetal alcohol effects: Role of acetaldehyde. Exp. Pathol. 33:17–21, 1988.

651 Cyanazine *Bladex™* CAS 21725–46–2

Lu et al. (1982) gavaged pregnant rats on days 12 and 15 with 25 mg per kg and found maternal and fetal weight decrease along with some minor skeletal variations.

Lu, C.C.; Tang, B.S. and Chai, E.Y.: Teratogenicity evaluations of technical Bladex™ in Fishers–344 rats. Teratology 25:59A–60A, 1982.

652 Cyanide also see *Cassava Root and Laetrile; Bhopal Disaster* CAS 57–12–5

Spratt (1950) found in explanted chick embryos that sodium cyanide above a concentration of 5×10^{-3} M inhibited development of the central nervous system with less effect on heart development. Doherty et al. (1982) administered 0.126 to 0.1295 millimoles per kg per hour intravenously to hamsters between days 6 and 9 of gestation. Neural tube, heart and limb defects occurred. Sodium thiosulfate given concurrently prevented the defects.

Doherty, P.A.; Ferm, V.H. and Smith, R.P.: Congenital malformations induced by infusion of sodium cyanide in the golden hamster. Toxicol. Appl. Pharmacol. 64:456–464, 1982.

Spratt, N.T.: Nutritional requirements of the early chick embryo. III. The metabolic basis of morphogenesis and differentiation as revealed by the use of inhibitors. Biol. Bull. 99:120–135, 1950.

653 2α–Cyano–4,4,17α–trimethylandrost–5–en–17 β–ol–3–one

Goldman (1969) injected this inhibitor of 3–β–hydrosteroid dehydrogenase into rats in amounts of 60 mg per kg on individual days before fertilization or during the first week of gestation and produced a slight but highly sifnificant degree of hypospadias in the male fetuses and clitoral hypertrophy in the female fetuses.

Goldman, A.S.: Congenital effectiveness of an inhibitor of 3–β–hydroxysteroid dehydrogenase administered before implantation of the rat blastula. Endocrinology 84:1206–1212, 1969.

654 17–α–Cyanomethyl–17–β–hydroxy–estra–4, 9–dien–3–one *STS 557*

Heinecke and Kohler (1983) gave this steroid to rabbits orally on days 5 to 25 and found no evidence of teratogenicity at doses up to 0.5 per kg daily.

Heinecke, H. and Kohler, D.: Prenatal toxic effects of STS 557. Exper. Clin. Endocrinol. 81:206–209, 1983.

655 p–Cyanophenol

Copeland et al. (1989) gavaged rats with up to 1000 mg per kg on day 11 and found little potential to induce developmental toxicity. In the in vitro studies of whole rat embryos (25 and 50 microg per ml), a concentration dependent increase in abnormality was observed. The presence of hepatocytes in the system reduced the effect.

Copeland, M.F.; Kavlock, R.J.; Oglesby, L.A.; Hall, L.L.; Beyers, P.E.; Ebron–McCoy, M.T. and Shrivastava, S.P.: Embryonic dosimetry of p–cyanophenol in in vivo and in vitro test systems. Teratology 39:P89, 1989.

656 Cyanuric Acid CAS 108–80–5

Canelli (1974) gave 500 mg per kg with or without calcium hypochlorite to rats on days 6–15. No increase in skeletal or external defects were found.

Canelli, E.: Chemical, bacteriological and toxicological properties of cyanuric acid and chlorinated isocyanurates as applied to swimming pool disinfection. Am. J. Pub. Health 64:155–162, 1974.

657 Cyclacillin

Mizutani et al. (1970) treated mice and rats with up to 200 mg per kg orally during organogenesis and found no adverse effects.

Mizutani, M.; Iharra, T.; Tanaka, S.; Kanamori. H.: Fetotoxic studies of amino–cyclohexyl penicillin in mice and rats. Takdea Kenkysuho HO (J Takeda Res Lab) 29:124–133, 1970.

658 Cyclamic Acid *Cyclamate* CAS 100–88–9

Tuchmann–Duplessis and Mercier–Parot (1970) administered 100 to 500 mg per kg by stomach tube to rats on the 12th or else 6th through 14th day of pregnancy and found neither an increase in resorptions nor congenital defects in the offspring. Perinatal and postnatal studies were done and no effect of early cyclamate treatment was found. Negative teratologic studies in the rat and rabbit were reported in an abstract by Vogin and Oser (1969). Long term studies using 5 percent sodium cyclamate for the diet of mice showed no teratogenicity (Kroes et al., 1977).

Although there are verbal reports of teratogenic findings in the chick (Taylor, 1969), these publications have eluded the author.

Kroes, R.; Peter, D.W.J.; Berkvens, J.M.; Verschuuren, T.D. and Van Esch, G.J.: Long term toxicity and reproduction study (including a teratogenicity study) with cyclamate, saccharin and cyclohexylamine. Toxicology 8:285–300, 1977.

Taylor, G.: Cyclamates. Lancet 2:1189 only, 1969.

Tuchmann–Duplessis, H. and Mercier–Parot, L.: Influence du cyclamate de sodium sur la fertilite et le developpement pre– et post–natal du rat. Therapie 25:915–928, 1970.

Vogin, E.E. and Oser, B.L.: Effects of cyclamate: Saccharin mixture on reproduction and organogenesis in rats and rabbits. Fed. Proc. 28:2709 only, 1969.

659 Cyclazocine CAS 3572–80–3

This narcotic antagonist has been given to pregnant rats and rabbits in maximal oral doses of 30 and 10 mg per kg, respectively, during active organogenesis and no fetal effects or anomalies were found (Nuite et al., 1975).

Nuite, J.A.; Smith, S.; Kennedy, G.L.; Keplinger, M.L. and Calandra, J.C.: Reproductive, teratogenic, perinatal and postnatal studies with cyclazocine. Toxicol. Appl. Pharmacol. 31:534–543, 1975.

660   Cyclizine CAS 82–92–8

This antihistamine belongs to the benzhydrylpiperazine series of compounds. It has been used as a teratogen in the rat. The details are discussed under meclizine. Two chemically similar hypolipemic agents (M and B 30227 and 31426) were shown to produce teratogenicity in rat embryo culture (Steele et al., 1983). Milkovich and Van den Berg (1976) reported no increased malformation rate among 111 offspring of mothers taking the drug during the first 84 days of pregnancy.

Milkovich, L. and Van den Berg, B.J.: An evaluation of the teratogenicity of certain antinauseant drugs. Am. J. Obstet. Gynecol. 125:244–248, 1976.

Steele, C.E.; New, D.A.T.; Ashford, A. and Copping, G.P.: Teratogenic action of hypolipemic agents: An in vitro study with postimplantation rat embryos. Teratology 28:229–236, 1983.

661   Cyclo(I–amino–cyclopentanecarbonyl–alanyl) *Cyclo(Acp–Ala)*

This melanocyte stimulating hormone inhibitor was injected into chick eggs on days 1.5, 2, 3, or 4 in a dose of 10 3^{v-1} moles and defects were produced in the group treated on days 3 and 4 (Kosar and Vanzura, 1988). The defect was of body walls.

Kosar, K. and Vanzura, J.: Embryotoxicity of L–prolyl–Lleucyl–glycinamide, cyclo(1–amino–cyclopentanecarbonyl-alanyl) and cyclo(glycyl–leucyl), new potential neuropeptides in chick embryos. Pharmazie 43:715–716, 1988.

662   Cyclo(glycyl–L–leucyl) *Cyclo(gly–Leu)*

Kosar and Vanzura (88) injected this neuropeptide into chick eggs on days 1.5–4. On day 4 embryotoxicity and malformations (body wall defects) were found. An effect was found at 10^{-4} and above.

Kosar, K. and Vanzura, J.: Embryotoxicity of L–prolyl–Lleucyl–glycinamide, cyclo(1–amino–cyclopentanecarbonyl-alanyl) and cyclo(glycyl–leucyl), new potential neuropeptides in chick embryos. Pharmazie 43:715–716, 1988.

663   Cyclocytidine *2,2-Anhydro–1–β–d–arabinofuranosylcytosine HCl CAS 31698–14–3*

Ohkuma et al. (1974) studied this antitumor agent in rats and rabbits. A dose of 50 mg per kg intraperitoneally in the rat from days 7 through 14 of pregnancy produced reduction defects of the extremities. In the rabbit, an intravenous dose of 6.0 mg per kg on days 7 through 16 resulted in cleft palate, open eye and reduction defects of the extremities.

Ohkuma, H.; Hikita, J.; Kiyota, K.; Tsuyama, S. and Hirayama, H.: Teratogenic evaluation of cyclocytidine, a new antitumor agent in rat and rabbit. Oyo Yakuri 8:1681–1691, 1974.

664   Cyclodiene Pesticides *Aldrin Dieldrin Endrin*

This group of chlorinated cyclodiene pesticides was studied by Ottolenghi et al. (1974) in hamsters and mice. Single oral doses of approximately one–half the respective LD–50 doses were given on days 7, 8 or 9 in the hamster and on day 9 in the mouse. A significant number of defects were produced in both species on all the days treated. The malformations in both species were open eye, webbed feet and cleft palate. For further information, see heading, Endrin.

Ottolenghi, A.D.; Haseman, J.K. and Suggs, F.: Teratogenic effects of aldrin, dieldrin, and endrin in hamsters and mice. Teratology 9:11–16, 1974.

665   Cyclofenil *Bis–(p–acetoxyphenyl)cyclohexylidenemethane CAS 2624–43–3*

Suzuki (1970a) gave pregnant rats this ovulation inducing agent during the middle stages of gestation (days 9 through 14) in amounts up to 128 mg per kg orally. An increase in fetal death occurred but no teratogenic action was detected. The same type of result was found using clomiphene. In the mouse the findings for both drugs were essentially the same (Suzuki, 1970b).

Einer–Jensen (**1968**) injected rats with this estrogenic compound on days 11–12 in dosage of 50 mg per kg and found only 55% normal fetuses with an increase in resorptions. Treatment from day 7 to 5 days after birth with oral doses of up to 16 mg per kg was not associated with any changes in external genitalia. There was no evidence of teratogenicity.

Einer–Jensen, N.: Antifertility properties of two diphenylethenes. Acta Pharmacol. (Copenh.) 26(suppl. 1):1–97, 1968.

Suzuki, M.R.: Effects of oral cyclofenil and clomiphene, ovulation inducing agents on pregnancy and fetuses in rats. Oyo Yakuri 4:635–644, 1970a.

Suzuki, M.R.: Effects of cyclofenil and clomiphene, ovulation inducing agents on mouse fetuses. Oyo Yakuri 4:645–651, 1970b.

666   Cycloguanil CAS 516–21–2

This antimalarial drug given by gavage at 4 hour intervals at 30 mg per kg to pregnant rats on the first day of gestation caused 90 percent of the embryos to die. Treatment on the 9th and 13th days had no harmful effect on the fetuses. Two analogues chlorproguanil and proguanil in similar doses and periods of gestation had no effect on the segmenting embryo or fetus.

Chebotar, N.A.: Embryotoxic and teratogenic action of proguanil, chlorproguanil and cycloguanil on albino rats. (Russian) Byulleten Eksperimentalnoi Biologii Meditsing 77:56–57, 1974.

667 **Cyclohexanone** CAS 108–94–1

No adverse effects were found in two studies of mice treated intraperitoneally with 50 mg per kg throughout pregnancy (Hall et al, 1974) or orally with up to 800 mg per kg of body weight (Chernoff and Kavlock, 1983).

Chernoff, N. and Kavlock, R.J.: A teratology test system which utilizes postnatal growth and viabilitity in the mouse. Short–term Bioassays in the Analysis of Complex Environ. Mixtures, 417–427, 1983.

Hall, I.H.; Carlson, G.L.; Abernethy, G.S. and Piantadosi, C.: Cycloalkanones 4 antiferility activity. J. Med. Chem. 17:1253–1257, 1974.

668 **Cycloheximide** *Actidione* CAS 66–81–9

Zimmerman et al. (1970) used this inhibitor of protein synthesis to study glucocorticoid teratogenicity. In experiments with 10 mg per kg in mice on day 11.5, they found 29 percent resorptions but no cleft palates. Eighty mg per kg was embryolethal. Lary et al. (1982) produced skeletal defects including polydactyly and oligodactyly in the offspring of mice treated with 30 mg per kg intraperitoneally on day 9.

Thiersch (1971) reported that 1 mg given intraperitoneally on the 11th day was embryocidal to the rat embryo and given on the 6th day caused stunting in 24 percent of the survivors.

Lary, J.M.; Hood, R.D. and Lindahl, R.: Interactions between cycloheximide and T–locus alleles during mouse embryogenesis. Teratology 25:345–349, 1982.

Thiersch, J.B.: Investigations into the differential effect of compounds on the rat litter and mother. In: Malformations Congenitales des Mammiferes. H. Tuchmann–Duplessis (ed.), Paris: Masson Cie, 95–113, 1971.

Zimmerman, E.F.; Andrew, F.D. and Kalter, H.: Glucocorticoid inhibition of RNA synthesis responsible for cleft palate in mice: A model. Proc. Nat. Acad. Sci. 67:779–785, 1970.

669 **Cyclohexylamine** CAS 108–91–8

Becker and Gibson (1970) injected pregnant mice intraperitoneally with 61 to 122 mg per kg on day 12 and produced no increase in congenital defects. Tanaka et al. (1973) gave up to 36 mg per kg on days 7 through 13 of gestation and observed no fetal changes. Takano and Suzuki (1971) found no teratogenic effect after feeding mice 100 mg per kg per day on days 6 through 11 of gestation.

Becker, B.A. and Gibson, J.E.: Teratogenicity of cyclohexylamine in mammals. Toxicol. Appl. Pharmacol. 17:551–552, 1970.

Takano, K. and Suzuki, M.: Cyclohexylamine, a chromosome–aberration producing substance: No teratogenicity in the mouse. (Japanese) Cong. Anomalies 11:51–57, 1971.

Tanaka, S.; Nakaura, S.; Kawashima, K.; Nagao, S.; Kuwamura, T. and Omori, Y.: Teratogenicity of food additives. 2. Effect of cyclohexylamine and cyclohexylamine sulfate on fetal development in rats. Shokuhin Eiseigaku Zasshi 14:542, 1973.

670 **N–Cyclohexyl–2–benzothiazylsulfenamide** *CBS*

Ema et al. (**1989**) fed up to 288 mg per kg on days 0–20 during rat pregnancy. Maternal food consumption was decreased at the highest dose as was fetal weight but no fetal ill effects were found.

Ema, M.; Murai, T.; Itami, T.; Kawasaki, H. and Kanoh, S.: Evaluation of the teratogenic potential of the rubber accelerator n–cyclohgexyl–2–benzothiazylsulfenamide in rats. Journal of Applied Toxicology 9:1887–190, 1989.

671 **N–Cyclohexyl–2–pyrrolidone**

Using up to 500 mg per kg in rats and 300 mg per kg in rabbits by gavage during organogenesis, Becci et al. (1984) found no adverse fetal changes as compared to vehicle controls.

Becci, P.J.; Reagan, E.L.; Wedig, J.H. and Barbee, S.J.: Teratogenesis study of N–cyclohexyl–2–pyrrolidone in rats and rabbits. Fund. Appl. Toxicol. 4:587–593, 1984.

672 **Cyclopamine** *Veratramine Rubijervine* CAS 4449–51–8

This compound derived from the plant Veratrum californicum has been shown to be teratogenic in ruminants (Binns et al., 1963). Natural epidemics of cyclopian sheep with cleft palates and cerebral defects were observed when the ewes grazed on Veratrum californicum. Dosages of 0.8 to 2.0 gm of the purified substance on the 8th gestational day produced the cyclopian defects in newborn sheep. Jervine and veratrosine, two alkaloids prepared from Veratrum californicum, were similarly teratogenic (Keeler and Binns, 1968). By buffering the material to prevent its conversion to the inactive form veratramine, Keeler (1970) produced similar defects in rabbit offspring. Keeler (1968) found rubijervine to be non–teratogenic in rabbits treated orally with 300 mg daily during organogenesis.

Binns, W.; James, L.F.; Shupe, J.L. and Everett, G.: A congenital cyclopian–type malformation in lambs induced by a maternal ingestion of a range plant, Veratrum californicum. Am. J. Vet. Res. 24:1164–1175, 1963.

Keeler, R.F.: Teratogenic compounds of Veratrum californicum (Durand): X. Cyclopia in rabbits produced by cyclopamine. Teratology 3:175–180, 1970.

Keeler, R.F. and Binns, W.: Teratogenic compounds of Veratrum californicum (Durand) v. Comparison of cyclopian effects of steroidal alkaloids from plant and structurally related compounds from other sources. Teratology 1:5–10, 1968.

673 **Cyclophosphamide and Related Analogues** *Cytoxan*™ CAS 50–18–0

Greenberg and Tanaka (1964) reported that a treated mother gave birth to a 1900 gm infant with multiple anomalies including 4 toes on each foot, flattening of the nasal bridge and a hypoplastic 5th finger. The clinical picture was compatible with fetal injury occurring during an intensive course of intravenous therapy (1,800 mg) given at about the 77th to 82nd day of gestation. Scott (1977) summarized some human exposure experience. Kirshon et al. (1988) reported that a woman treated on two days (gestational age 15 and 45) with 200 mg intravenously gave birth to an infant with absent thumbs, cleft palate and multiple eye defects (blepharophimosis and microphthalmos). The infants nails were distrophic at 10 months. Hypotonia and possible developmental delay were present.

Chaube et al. (1967) administered 7 to 10 mg per kg to rats and with treatment on the 11th or 12th day, the fetuses developed skeletal defects, cleft palates and exencephaly or encephalocele. This compound was shown to be relatively more embryolethal than chloroambucil when the fetal–maternal toxicity ratios of the two were compared. Brock and von Kreybig (1964) examined the early embryonic effects of the teratogen in the rat. Gibson and Becker (1968) showed the compound to be teratogenic in mice. In the rabbit, Fritz and Hess (1971) found a high incidence of cleft lip and/or palate and reduction defects of the extremities using intravenously 30 mg per kg on single days 11, 12 or 13. In the rhesus monkey, 10 mg per kg on days 27 through 29 produced facial clefts and when given on days 32 through 40, meningoencephalocele was observed (Wilk et al., 1978).

A number of studies of the metabolism of cyclophosphamide and its products in in vitro cultures with rat embryos showed that the compound must be bioactivated by a liver monofunctional oxygenase system in order to be teratogenic (Fantel et al., 1979; Popov, 1981; Popov et al., 1981; Kitchin et al., 1981). The morphologic changes found in vitro were very similar to those seen in vivo (Greenaway et al., 1982). Phosphoramide mustard in equimolar doses caused effects similar to those of bioactivated cyclophosphamide in vitro (Mirkes et al., 1982) and when given intraamniotically (Hales, 1982). Acrolein was toxic but its effect was difficult to assess because of protein binding (see Acrolein). The other stable metabolite, 4–ketocyclophosphoramide, was only weakly teratogenic in vitro (Mirkes et al., 1982). Little and Mirkes (**1992**) have shown in rats that the preactivated forms of cyclophosphamide bind to DNA and effect cell cycle by changes in the S phase followed by G_2/M arrest.

Spielmann and Jacob–Muller (1981) concluded that phosphoramide mustard was the active teratogenic metabolite in a mouse blastocyst system. Mirkes (1985) reviewed the subject comprehensively. The monofunctional form of phosphoramide mustard (with only one chloroethyl side chain) was shown to have the same embryotoxicity as cyclophosphamide (Mirkes et al., 1985). Mirkes et al. (1989) demonstrated that phosphoramide mustard lengthened S phase in neuroepithelial cells and slowed or arrested their G_2 phase.

Ignatieva (1985) studied in vitro action of cyclophosphamide on mouse embryos (8.5 days of development). The preparation was added to the culture medium or administered to females. At doses of 0.5 and 0.25 mg per ml added to the culture medium, 7.1 percent and 15.4 percent of embryos had developmental defects. Used with a liver microsome fraction, the preparation increased teratogenic effect.

The dose of 30 mg per kg administered in pregnant mice led to the death of all embryos and the dose of 15 mg per kg was lethal for 39 percent. The rest had no signs of teratogenic effect.

Trasler et al. (1985) administered 1.4, 3.4 or 5.1 mg per kg daily before mating to male rats. The males were not treated on the day of mating. Minimal changes in the male reproductive tract were found but malformations and retardation of growth were found in the offspring of the untreated females with whom they were bred. There was a dose–dependent increase in resorptions and fetal deaths. In the offspring of males treated at 7–9 weeks there were 4 defects in 57, compared to one in 254 of the controls. The defects were hydrocephalus, micrognathia and edema. Growth retardation occurred in 7 percent of the fetuses in this group. Kelly et al. (**1992**) found that paternal treatment in the rat caused inner cell mass damage. Trophectoderm was not affected.

Jarrell et al. (**1991**) using a single dose have found short term ovarian toxicity with loss of growing follicles in the adult rat.

Hales (1983) found that 5,5–dimethylcyclophosphamide was not teratogenic when injected intraamniotically into day 13 rat sites in doses of up to 1.0 mg. Diethylcyclophosphamide injected in the same mode and amount produced the same defects as cyclophosphamide. The latter chemical does not liberate acrolein after metabolism. Hales (1982) found that 4–hydroperoxycyclophosphamide injected intraamniotically on day 13 produced defects at a dose of one microgm.

Brock, N. and von Kreybig, T.: Experimenteller Beitrag zur Pruefung teratogener Wirkungen von Arzneimitteln an der Laboratoriumsratte. Naunym–Schmiedelbergs Arch. Pharmakol. Exp. Pathol. 249:117–145, 1964.

Chaube, S.; Kury, G. and Murphy, M.L.: Teratogenic effects of cyclophosphamide (NCA–26271) in the rat. Cancer Chemother. Rep. 51:363–376, 1967.

Fantel, A.G.; Greenaway, J.C.; Juchau, M.R. and Shepard, T.H.: Teratogenic bioactivation of cyclophosphamide in vitro. Life Sciences 25:67–72, 1979.

Fritz, H. and Hess, R.: Effects of cyclophosphamide on embryonic development in the rabbit. Agents and Actions 2:83–86, 1971.

Gibson, J.E. and Becker, B.A.: Teratogenicity of cyclophosphamide in mice. Cancer Res. 28:475–480, 1968.

Greenaway, J.C.; Fantel, A.G.; Shepard, T.H. and Juchau, M.R.: The in vitro teratogenicity of cyclophosphamide in the rat. Teratology 25:335–342, 1982.

Greenberg, L.H. and Tanaka, K.R.: Congenital anomalies probably induced by cyclophosphamide. JAMA 188:423–426, 1964.

Hales, B.F.: Comparison of the mutagenicity and teratogenicity of cyclophosphamide and its active metabolites, 4–hydroxycyclophosphamide, phosphoramide mustard and acrolein. Cancer Res. 42:3016–3021, 1982.

Hales, B.F.: Relative mutagenicity and teratogenicity of cyclophosphamide and two of its structural analogs. Biochem. Pharmacol. 32:3791–1795, 1983.

Ignatieva, T.V.: The action of cyclophosphamide in vivo and in vitro on postimplantation development of mice. In: Main Trends and Controlling Mechanisms of the Early Normal and Abnormal Mammalian Embryogenesis. (USSR) Leningrad, Meditsina 70–76, 1985.

Jarrell, J.F.; Bodo, L.; YoungLai, E.V.; Barr, R.D. and O'Connell, G.J.: The short–term reproductive toxicity of cyclophosphamide in the female rat. Reproductive Toxicology 5:481–485, 1991.

Kelly, S.M.; Robaire, B. and Hales, B.F.: Paternal cyclophosphamide treatment causes postimplantation loss via inner cell mass–specific cell death. Teratology 45:313–318, 1992.

Kirshon, B.; Wasserstrum, N.; Willis, R.; Herman, G.E. and McCabe, E.R.B.: Teratogenic effects of first trimester cyclophosphamide therapy. Obstet. Gynecol. 72:462 1988.

Kitchin, K.T.; Schmid, B.P. and Sanyal, M.K.: Teratogenicity of cyclophosphamide in a coupled microsomal activating embryo culture system. Biochem. Pharmacol. 30:59–64, 1981.

Little, S.A. and Mirkes, P.E.: Effects of 4–hydroperoxycyclophosphamide (4–ooh–cp) and 4–hydroperoxydechlorocyclophosphamide (4–ooh–decicp) on the cell cycle of post implantation rat embryos. Teratology 45:163–173, 1992.

Mirkes, P.E.: Cyclophosphamide teratogenesis: A review. Teratogenesis, Carcinogenesis and Mutagenesis 5:75–88, 1985.

Mirkes, P.E.; Fantel, A.G.; Greenaway, J.C. and Shepard, T.H.: Teratogenicity of cyclophosphamide metabolites: phosphoramide mustard, acrolein and 4–ketocyclophosphamide in rat embryos cultured in vitro. Toxicol. Appl. Pharmacol. 58:322–330, 1981.

Mirkes, P.E.; Greenaway, J.C.; Hilton, J. and Brundrett, R.: Morphological and biochemical aspects of monofunctional phosphoramide mustard teratogenicity in rat embryos cultured in vitro. Teratology 32:241–249, 1985.

Mirkes, P.E.; Ricks, J.L.; Pascoe–Mason, J.M.: Cell cycle analysis in the cardiac and neuroepithelial tissues of day 10 rat embryos and the effects of phosphoramide mustard, the major teratogenic metabolite of cyclophosphamide. Teratology 39:115–120, 1989.

Popov, V.B.: A study of the effect of cyclophosphamide in the culture of postimplantation rat embryos. Ontogenez (USSR) 12.3:251–256, 1981.

Popov, V.B.; Weisman, B.L. and Puchkov, V.F.: Embryotoxic effect of cyclophosphamide after biotransformation in the culture of postimplantation rat embryos. Byull. Eksper. Biol. Med. (USSR) 5:613–615, 1981.

Scott, J.R.: Fetal growth retardation associated with maternal administration of immunosuppressive drugs. Am. J. Obstet. Gynecol. 128:668–676, 1977.

Spielmann, H. and Jacob–Muller, U.: Investigations on cyclophosphamide treatment during the preimplantation period. 2. In vitro studies on the effects of cyclophosphamide and its metabolites 4–OH–cyclophosphamide, phosphoramide mustard, and acrolein on blastulation of four–cell and eight–cell mouse embryos and on their subsequent development during implantation. Teratology 23:7–13, 1981.

Trasler, B.F. and Robaire, B.: Paternal cyclophosphamide treatment of rats causes fetal loss and malformations without affecting male fertility. Nature 316:144–146, 1985.

Wilk, A.L.; McClure, H.M. and Horigan, E.A.: Induction of craniofacial malformations in the rhesus monkey with cyclophosphamide. (abs) Teratology 17:24A only, 1978.

674 Cyclopiazonic Acid

Khera et al. (1985) studied this mold metabolite in mice gavaging up to 16 mg per kg on days 9 through 12. This dose is calculated to be 1000 times human exposure. No adverse fetal effects were found.

Khera, K.S.; Cole, R.J.; Whalen, C. and Dorner, J.W.: Embryotoxicity study on cyclopiazonic acid in mice. Bull. Environ. Contam. Toxicol. 34:423–426, 1985.

675 Cyclosiloxanes

Le Feure et al. (1972) administered an equilibrated copolymer of mixed cyclosiloxanes represented by cylic [(Phmesio)2] (PMXMMY) orally to rabbits and rats. Before implantation in both species it proved lethal to zygotes or prevented implantation. Maternal rats given 220 mg per kg per day on day 16 gave birth to female pups with a urogenital structural anomaly associated with urinary incontinence.

Le Feure, R.; Coulston, F. and Goldberg, L.: Action of a copolymer of mixed phenylmethylcyclosiloxanes on reproduction in rats and rabbits. Toxicol. Appl. Pharmacol. 21:29–44, 1972.

676 Cyclosporin

In a national registry of transplanted mothers treated during pregnancy there was no pattern of increased malformations among 338 live borns (Armenti et al., **1993**). There was a 58 percent rate of prematurity (less than 37 weeks) among those women who were receiving other immunosuppressives. In another series Shaheen et al. (**1993**) followed renal function in 22 children and found no adverse effects. Ryffel et al. (1983) studied rats and rabbits with oral doses of up to 300 mg per kg during organogenesis. No increase in malformations was found but at maternal toxic doses (30 mg per kg), fetal death and growth retardation occurred.

Armenti, V.T.; Ahlswede, K.M.; Bhlswede, B.A.; Jarrel, B.E. and Moritz, M.J.: The national transplantation pregnancy registry: outcomes of 414 pregnancies in female transplant recipients. Teratology 47(5):393, 1993.

Ryffel, B.; Donatsch, P.; Madorin, M.; Matter, B.E.; Ruttimann, G.; Schon, H.; Stoll, R. and Wilson, J.: Toxicological evaluation of cyclosporin A. Arch. Toxicol. 53:107–141, 1983.

Shaheen, F.A.M.; Al–Suhaiman, M.H.; and Al–Khader, A.A.: Long–term nephrotoxicity after exposure to cyclosporine in utero. Transplantation 56(1):224–225, 1993.

677 Cyproheptadine CAS 129–03–3

This compound is thought to inhibit the contractile effect of serotonin on smooth muscles by competing with serotonin for its receptor. Sadowsky et al. (1972) treated 41 pregnancies with 4 to 16 mg and no teratogenic effects were observed.

Pfeifer et al. (1969) gave rats 5 mg per kg subcutaneously on days 5–7, 8–14 or 15–20. In the combined fetuses (140), one was dead and 9 resorbed. None were malformed.

Fuente and Alia (**1982**) injected rats intraperitoneally on 4 days during organogenesis with 50 mg per kg. Skeletal retardation, liver and brain damage were increased in day 21 fetuses. Hydronephrosis was also increased. The no effect dose was 10 mg per kg. Mortality of newborns was increased in a group receiving 2 mg per kg during gestation as well as those receiving higher doses during organgogenesis.

Fuente, M. and Alia, M.: The teratogenicity of cyproheptadine in two generations of wistar rats. Atch. Int. Pharmacodyn 257:168–176, 1982.

Pfeifer, Y.; Sadowsky, E. and Sulman, F.G.: Prevention of serotonin abortion in pregnant rats by five serotonin antagonists. Obstet. Gynecol. 33:709–14, 1969.

Sadowsky, E.; Pfeifer, Y.; Polishuk, W.Z. and Sulman, F.: A trial of cyproheptadine in habitual aborters. Israel Med. J. Sci. 8:623–625, 1972.

678 Cyproterone Acetate 1,2 α–Methylene–chlor δ(4,6)–pregnadeine–17–α–01–3, 20–dione–acetate CAS 427–51–0

Neumann et al. (1966) produced vaginas in male rat fetuses using 10 mg per day from the 13th to the 22nd gestational day. Forsberg and Jacobsohn (1969) reproduced this work. Similar results were described in the rabbit (Elger, 1966; Jost, 1966). Jost emphasized that although this antiandrogenic compound suppressed the masculinizing effects of the fetal testis, it did not prevent the inhibitory action on the mullerian ducts. Early differentiation of the testes was not prevented by this compound (Jost, 1966). Eibs et al. (1982A) showed in the mouse that the time of treatment determines whether the malformations are in the urinary tract, lung or palate. During preimplantation 30 microgm was lethal and 3 microgm inhibited inner cell mass growth in vitro (Eibs et al., 1982B).

Goldman (1973) and Neumann et al. (1970) reviewed the enzymatic mechanism of action of this and related compounds. A non–competitive binding of androgen biosynthesizing enzymes was reported.

Eibs, H.G.; Spielmann, H. and Hagele, M.: Teratogenic effects of cyproterone acetate and medroxyprogesterone treatment during the pre– and postimplantation period of mouse embryos. Teratology 25:27–36, 1982A.

Eibs, H.G.; Spielmann, H.; Jacob–Muller, U. and Klose, J.: Teratogenic effects of cyproterone acetate and medroxyprogesterone treatment during the pre– and postimplantation period of mouse embryos. II Cyproterone acetate and medroxyprogesterone acetate treatment before implantation in vivo. Teratology 25:291–299, 1982B.

Elger, W.: Die Rolle der fetalen Androgene in der Sexualdifferenzierung des Kaninchens und ihre Abgrenzung gegen andere Hormonale und somatische Faktoren durch Anwendung eines starken Antiandrogens. Arch. Anat. Microsc. Morphol. Exp. 55:658–743, 1966.

Forsberg, J.G. and Jacobsohn, D.: The reproductive tract of males delivered by rats given cyproterone acetate from days 7 to 21 of pregnancy. J. Endocrinol. 44:461–462, 1969.

Goldman, A.S.: Sexual programming of the rat fetus and neonate studied by selective biochemical testosterone–depriving agents. In: Advances in the Biosciences, G. Raspe and S. Berhnard (eds), 13, Oxford: Pergamon Press, 17–40, 1973.

Jost, A.: Steroids and sex differentiation of the mammalian foetus. Proceedings of the Second International Congress on Hormonal Steroids. Milan: Excerpta Medica International Congress Series 132:74–81, 1966.

Neumann, F.; Elger, W. and Kramer, M.M.: Development of a vagina in male rats by inhibiting androgen receptors through an anti–androgen during the critical phase of organogenesis. Endocrinology 78:628–632, 1966.

Neumann, F.; Von Berswordt–Wallrabe, R.; Elger, W.; Steinbeck, H.; Hahn, J.D. and Kramer, M.M.: Aspects of androgen–dependent events as studied by antiandrogens. Recent Prog. Horm. Res. 26:337–410, 1970.

679 Cysteamine Hydrochloride CAS 156–57–0

Adams et al. (1961) exposed rabbits to 70 to 160 mg per kg during preimplantation and recovered normal 6–day blastocysts. Mazur (1985) gave mice 40 mg per kg intraperitoneally on the first day of pregnancy and found reduced implantations and increased resorptions.

Adams, C.E.; Hay, M.F. and Lutwak–Mann, C.: The action of various agents upon the rabbit embryo. J. Embryol. Exp. Morphol. 9:468–491, 1961.

Mazur, L.: Embryonic survival in mice treated with AET, MEA or 5–HT on the first day of pregnancy. Acta Physiologica Hungarica 65:227–231, 1985.

680 Cysteinamine CAS 60–23–1

Rugh and Clugston (1956) gave 3 mg intraperitoneally to mice on days 14.5 or 17.5 before exposure to 300 or 7000 R x–rays. A protective effort was evident in survival and weight increase of the newborns. The cysteinamine alone had no adverse fetal effects.

Rugh, R. and Clugston, H.: Protection of mouse fetus against x–irradiation death. Science 123:28–29, 1956.

681 Cysteine CAS 52–90–4

Olney et al. (1972) treated pregnant mice and rats subcutaneously with 1.2 mg per gm on the last day of pregnancy and observed brain degeneration one day later in the fetus. Oral dosing did not produce lesions.

Olney, J.W.; Ho, O.L.; Rhee, V. and Schainker, B.: Cysteine–induced brain damage in infant and fetal rodents. Brain Res. 45:309–313, 1972.

682 Cytarabine Cytosine Arabinoside 1–β–D–Arabinofuranosylcytosine CAS 147–94–4

Wagner et al. (1980) reported that limb reduction and small ears with atresia of the canals occurred in the offspring of a leukemia–affected mother treated with this drug alone.

Karnofsky and Lacon (1966) produced facial and skeletal defects in the chick; Chaube and Murphy (1965), skeletal and palate defects in the rat and Fischer and Jones (1965), cerebellar hypoplasia in the hamster. Chaube and Murphy in their review (1968) reported that 20–800 mg per kg to rats or mice produced cleft palate and skeletal defects in the offspring and that the analogue, 1–β–D–arabinofuranosyl–5–fluorocytosine was teratogenic at 50–200 mg per kg given to the rat on the 12th day. Chaube and Murphy (1968) nullified the teratogenic effect of cytosine arabinoside by simultaneous injection of deoxycytidine.

Percy (1975), treating rats and mice for three days toward the end of gestation, produced cerebellar hypoplasia, microcystic renal changes and retinal dysplasia. Dosages of 50 mg per kg produced retinal dysplasia. Nomura et al. (1969) treated rats and mice with this anti–tumor drug using 1.5 to 15 mg per kg on the 7th to 12th days intravenously in the mouse and 15 to 60 mg on the 9th through 14th days intravenously in the rat. In the mouse, defects were increased at 3.0 mg per kg and above. In the rat, doses of 15 mg or more were associated with defects. Cleft palates and skeletal defects were frequent in both species.

Tesh et al. (**1990**) gave this antileukemic drug orally to rats on days 7–17 in doses of up to 10 mg per kg. Fetal toxicity was found after 10 mg per kg and digital defects were present after exposure to 4 or 10 mg per kg. No fetal effects were seen at 1.6 mg per kg.

Chaube, S. and Murphy, M.L.: Teratogenic effects of cytosine arabinoside (CA) in the rat fetus. (abs) Proc. Amer. Ass. Cancer Res. 6:11 only, 1965.

Chaube, S. and Murphy, M.L.: The teratogenic effects of the recent drugs active in cancer chemotherapy. In: Advances in Teratology, D.H.M. Woollam (ed.), New York: Academic Press, 3:204–205, 1968.

Fischer, D.S. and Jones, A.M.: Cerebellar hypoplasia resulting from cytosine arabinoside treatment in the neonatal hamster. (abs) Clin. Res. 13:540 only, 1965.

Karnofsky, D.A. and Lacon, C.R.: The effects of 1–β–d–arabinofuranosylcytosine on the developing chick embryo. Biochem. Pharmacol. 15:1435–1442, 1966.

Nomura, A.; Watanabe, M.; Yamagata, H. and Ohata, K.: 1–β–D–Arabinofuranosylcytosine (AC–1075). Gendai no Rinsho 3:758–775, 1969.

Percy, D.H.: Teratogenic effects of pyrimidine analogues 5–iododeoxyuridine and cytosine arabinoside in late fetal mice and rats. Teratology 11:103–118, 1975.

Tesh, J.M.; Wilby, O.K.; Tesh, S.A. and Handa, J.: Toxicological studies on ynk01 (vii)–study for effects of ynk01 treated orally upon organogenesis and development in rats. Kiso to Rinsho 23:2397–2418, 1990.

Wagner, V.W.; Hill, J.S.; Weaver, D. and Baehner, R.L.: Congenital abnormalities in a baby born to cytarabine treated mother. Lancet 2:98–100, 1980.

683 Cytochalasin B CAS 14930–96–2

Linville and Shepard (1972) exposed explanted pre-somite chick embryos to 0.5–1.0 microgm of this naturally–occurring fungal metabolite and observed a high incidence of neural tube closure defects.

Greenaway et al. (1977) found cytochalasins A,D and E to prevent neural tube closure in the chick but at different concentrations. Ruddick et al. (1974) fed 1 mg per kg to rats on days 6 through 15 and produced no defects. Wiley (1979) produced exencephaly in hamsters using 7.5 mg per kg intraperitoneally. Snow (1973) and Niemierko (1975) produced polyploidy in the mouse embryo by blocking polar body formation or later cleavage.

Greenaway, J.C.; Shepard, T.H. and Kuc, J.: Comparison of cytochalasins (A,B,D and E) in chick explant teratogenicity and tissue culture systems. Proc. Soc. Exp. Biol. Med. 155:239–242, 1977.

Linville, G.P. and Shepard, T.H.: Neural tube closure defects due to cytochalasin B. Nature 236:246–247, 1972.

Niemierko, A.: Induction of triploidy in the mouse by cytochalasin B. J. Embryol. Exp. Morph. 34:279–289, 1975.

Ruddick, J.A.; Harwig, J. and Scott, P.M.: Non–teratogenicity in rats of blighted potatoes and compounds contained in them. Teratology 9:165–168, 1974.

Snow, M.H.L.: Tetraploid mouse embryos produced by cytochalasin B during cleavage. Nature 244:513–515, 1973.

Wiley, M.J.: Ultrastructural analysis of cytochalasin–induced neural tube defects in vivo. (abs) Teratology 19:53A only, 1979.

684 Cytochalasin D CAS 22144–77–0

Shepard and Greenaway (1977) injected intraperitoneally 0.4 to 0.9 mg per kg on days 7 through 11 and produced exencephaly, hypognathia and skeletal reduction defects in C–57 and BALB–C strain mice but not in Swiss Webster mice. Wiley (1980) produced exencephaly in hamsters treated on the 8th day with 1.5 mg per kg intraperitoneally. Although non–teratogenic in vivo in the rat, three nanogm per ml of medium in vitro is associated with open neural tubes (Fantel et al., 1981).

Fantel, A.G.; Greenaway, J.C.; Shepard, T.H.; Juchau, M.R. and Selleck, S.B.: The teratogenicity of cytochalasin D and its inhibition by drug metabolism. Teratology 23:223–231, 1981.

Shepard, T.H. and Greenaway, J.C.: Teratogenicity of cytochalasin D in the mouse. Teratology 16:131–136, 1977.

Wiley, M.J.: The effect of cytochalasms on the ultrastructure on neurulating hamster embryos in vivo. Teratology 22:59–69, 1980.

685 Cytochalasin E CAS 36011–19–5

Austin et al. (1982) using 0.9 to 2.0 mg per kg intraperitoneally in two strains of mice produced neural tube defects, cleft lip and palate and other skeletal defects.

Austin, W.L.; Wind, M. and Brown, K.S.: Differences in toxicity and teratogenicity of cytochalasins D and E in various mouse strains. Teratology 25:11–18, 1982.

686 Cytomegalic Monoclonal Antibodies TI–23

Ikegawa et al. (**1994**) gave this antibody intravenously

to pregnant rats and rabbits in doses of up to 20 mg per kg daily. No adverse teratogenic activity was found in either species. Prenatal administration did not decrease fertility and perinatal administration was without ill effects. Postnatal behavior and fertility were not changes.

Ikegawa, S.; Sugawara, S.; Matsuzawa, K.; Nishizawa, S.; Hoshi, Y. and Izawa, Y.: Toxicity studies of TI–23 (4th report) - reproductive and developmental toxicity studies in rats and rabbits. The Clinical Report 28:1199–1221, 1994.

687 Cytomegalovirus *CMV*

This virus with a DNA core and estimated molecular weight of 65 million is classified as a herpes virus. The entire subject was comprehensively reviewed by Hanshaw (1970) and Weller (1971).

Yow et al. (1988) followed 4,578 pregnancies and found 2.2 percent experienced a primary infection. Twenty–four percent of these transmitted the infection to the infant. Of 22 infants, two had symptomatic disease at birth. Three of 16 nonsymptomatic had deafness and one had developmental delay. There is evidence that about five percent of pregnant women become infected with CMV virus (Sever et al., 1962) and approximately 1.3 percent of newborns have positive urine cultures for the virus (Hanshaw, 1970).

The virus may cross the placenta during the last two trimesters or the fetus could be infected while in the birth canal (Alexander, 1967). The virus has not been shown to cause spontaneous abortions. The syndrome produced by the virus, cytomegalic inclusion disease (CID), is characterized by intrauterine growth retardation associated with or followed by microcephaly. The ependymal area of the nervous system is primarily affected and may contain calcified areas. Chorioretinitis, seizures, blindness and optic atrophy may be associated with the syndrome.

In the neonatal period, hepatosplenomegaly with jaundice and thrombocytopenia can be the presenting symptoms. Although deafness may occur it is not a common feature. Serological or virus culture methods for diagnosis are available. Although this infection plays a significant role in about 25 percent of children with microcephaly, the true spectrum of the clinical syndrome is only beginning to appear (Emanuel and Kenny, 1966). Pass et al. (1980) gave results of a follow–up of 34 symptomatic newborns.

Melish and Hanshaw (1973) surveyed 3,800 newborns and found 20 with positive cultures. Seventeen of the 20 were asymptomatic and of the remaining three, one had deafness, one possible brain damage and the third had slow development following premature birth. Stagno and Whitley (1985) summarized the problems and pointed out that only 11 percent of women who seroconvert during pregnancy have infants with clinical manifestations. Benirschke et al. (1974) described the varied fetal and placental pathology.

London et al. (1986) used rhesus cytomegalovirus infections to infect rhesus fetuses and produce ventricular dilatation and chronic leptomeningitis.

Studies of the Towne virus an attenuated cytomegalovirus are progressing but not complete (Plotkin et al., **1989**).

Alexander, E.R.: Maternal and neonatal infection with cytomegalovirus in Taiwan. (abs) Pediatr. Res. 1:210 only, 1967.

Benirschke, K.; Mendoza, G.R. and Bazeley, P.L.: Placental and fetal manifestations of cytomegalovirus infection. Virchows Arch. B Cell Path. 16:121–139, 1974.

Emanuel, I. and Kenny, G.E.: Cytomegalic inclusion disease in infancy. Pediatrics 38:957–965, 1966.

Hanshaw, J.B.: Developmental abnormalities associated with congenital cytomegalovirus infection. In: Advances in Teratology, D.H.M. Woollam (ed.), New York: Academic Press, 4:62–93, 1970.

London, W.T.; Martinez, A.J.; Houff, S.A.; Wallen, W.C.; Curfman, B.L.; Traub, R.G. and Sever, J.L.: Experimental congenital disease with simian cytomegalovirus in rhesus monkeys. Teratology 33:323–331, 1986.

Melish, M.E. and Hanshaw, J.B.: Congenital cytomegalovirus infection: Developmental progress of infants detected by routine screening. Am. J. Dis. Child. 126:190–194, 1973.

Pass, R.F.; Stagno, S.; Myers, G.J.; Alford, C.A.: Outcome of symptomatic congenital cytomegalovirus infection: Results of long–term longitudinal follow–up. Pediatrics 66:758–762, 1980.

Plotkin, S.A.; Starr, S.E.; Friedman, H.M.; Gonczol, E. and Weibel, R.E.: Protective effects of towne cytomegalovirus vaccine against low–passage cytomegalovirus administered as a challenge. The Journal of Infectious Diseases 159:860–865, 1989.

Sever, J.L.; Huebner, R.J.; Castellano, G.A. and Bell, J.A.: Serological diagnosis 'en masse' with multiple antigens. Am. Rev. Resp. Dis. Suppl. 88:342–359, 1962.

Stagno, S. and Whitley, R.J.: Herpes virus infections of pregnancy. Part 1: Cytomegalovirus and Epstein–Barr virus infections. N. Eng. J. Med. 313:1270–1274, 1985.

Weller, T.H.: The cytomegaloviruses: Ubiquitous agents with protean clinical manifestations (second of two parts). N. Eng. J. Med. 285:267–274, 1971.

Yow, M.D.; Williamson, D.W.; Leeds, L.J.; Thompson, P.; Woodward, R.M.; Walmas, B.F.; Lester, J.W.; Six, H.R. and Griffiths, P.D.: Epidemiologic charateristics of cytomegalovirus infection in mothers and their infants. Am. J. Obstet. Gynecol. 158:1189–1195, 1988.

688 2,4–D

Discussed under Dichlorophenoxyacetic Acid

689 Dacarbazine *5–(3, 3–dimethyl–1–triazeno) imidazole–4–carboxamide* CAS 4342–03–6

Chaube (1973) studied rats by giving this antineoplastic compound intraperitoneally in doses of 200–1000 mg per kg. On days 11 or 12 a single dose of 800 or 1000 mg per kg produced skeletal reduction defects and some cleft palates and encephaloceles.

Chaube, S.: Protective effects of thymidine, and 5–aminoimidazolecarboxamide and riboflavin against fetal abnormalities produced in rats by 5–(3,3–dimethyl–1–triazeno)imidazole–4–carboxamide. Cancer Res. 33:2231–

2240, 1973.

690 Dactinomycin

Discussed under Actinomycin D

691 Danazol CAS 17230–88–5

This steroid which is closely related to 17 α–ethinyl testosterone was associated with masculinization of the female external genitalia (Kingsburg, 1985). Brunskill (**1992**) studied adverse drug reports of 94 women taking the drug during pregnancy. The males were normal but 23 of the 57 females had masculinized external genitalia. Although masculinization was more common with the higher doses one case was found at 200 mg per day. No cases were reported where the medication was discontinued before the 8th week of pregnancy.

Brunskill, P.J.: The effects of fetal exposure to danazol. British Journal of Obstetrics and Gynecology 99:212–215, 1992.

Kingsburg, A.C.: Danazol and fetal masculinization: A warning. Med. J. Aust. 143:410–411, 1985.

692 Dantrolene CAS 7261–97–4

Nagaoka et al. (1977A,B) gave up to 60 mg per kg daily by mouth to pregnant rats and rabbits during days 7–17 and 6–18, respectively. At the highest dose, minor skeletal variations occurred in both species and in the rat, both maternal and fetal weights were reduced. Shime er al (1988) studied 20 women who were given this drug to prevent malignant hyperthermia and found no adverse effects.

Nagaoka, T.; Osuka, F.; Shigemura, T. and Hatano, M.: Reproductive test of dantrolene. Teratogenicity test on rats. (Japanese) Clinical Report 11:2218–2230, 1977A.

Nagaoka, T.; Osuka, F. and Hatano, M.: Reproductive studies of dantrolene. Teratogenicity study in rabbits. Clinical Report 11:2212–2217, 1977B.

Shime, J.; Gare, D.; Andrews, J. and Britt, B.: Dantrolene in pregnancy: lack of adverse effects on the fetus and newborn infant. Am J Obstet Gynecol 159:831–834, 1988.

693 Daptomycin

Liu et al. (1988) injected rats and rabbits intravenously with this lipopeptide antibiotic. Treatment with up to 75 mg per kg per day during organogenesis caused no ill effects in either species.

Liu, S.L.; Howard, L.C.; Van Lier, R.B.L. and Markham, J.K.: Teratology studies with daptomycin administered intravenously (iv) to rats and rabbits. Teratology 37:475 only, 1988.

694 Daunorubicin *Daunomycin Rubidomycine* CAS 20830–81–3

No malformations were found among a total of 15 offspring of mothers treated with this drug along with other chemotherapeutic agents, (Alegre et al., 1982; Tobias and Bloom, 1980). One stillborn of a treated mother exhibited myocardial necrosis (Schaisen et al., 1979).

Chaube and Murphy (1968), injected this antibiotic into pregnant rats once on days 5–12 and produced no congenital defects in survivors. Maternal doses of 20 mg per kg were not lethal to the fetus. Thompson et al. (1978), using 1–4 mg per kg in rats for various periods during organogenesis, produced atresia of the alimentary canal, urogenital defects and cardiovascular anomalies. Studies in the rabbit did not produce defects.

Roux and Taillemite (1969) injected 1 to 3 mg per kg of this antibiotic into rats starting on the 7th day of gestation. With the higher dose which was given for three days by intraperitoneal route, they found that 45 percent of the fetuses were defective. Ocular anomalies were the most common but defects of the heart, kidney and brain were present also. Using one mg daily from days 7 through 14, they obtained a malformation rate of 16 percent. Julou et al. (1967) did not produce congenital defects in the chick, mouse and rabbit. Although their dose in the rabbit was 0.25 mg per kg, the mice received subcutaneous doses of 1.25 mg per kg during pregnancy.

Alegre, A.; Chunchurreta, R.; Rodriguez–Alarcon, J. and Cruz, E.: Successful pregnancy in acute promyelocytic leukemia. Cancer 49:152–153, 1982.

Chaube, S. and Murphy, M.L.: The teratogenic effects of the recent drugs active in cancer chemotherapy. In: Advances in Teratology, D.H.M. Woollam (ed.), New York: Logos and Academic Press, 3:181–237, 1968.

Julou, L.; Ducrot, R.; Fournel, J.; Ganter, P.; Maral, R.; Populaire, P.; Koenig, F.; Myon, J.; Pascal, S. and Pasquet, J.: Un nouvel antibiotique doue d/activite antitumorale: La rubidomycine (13.057 R.P.). Arzneim. Forsch. 17:948–954, 1967.

Roux, C. and Taillemite, J.L.: Action teratogene de la rubidomycine chez le rat. C.R. SOc. Biol. 163:1299–1302, 1969.

Schaison, G.; Jacquillat, C.; Auclerc, G. and Weil, M.: Les risques foetoembryonnaires des chimiotherapies. Bull. Cancer 66:165, 1979.

Thompson, D.J.; Molello, J.A.; Strebing, R.J. and Dyke, I.L.: Teratogenicity of adriamycin and daunomycin in the rat and rabbit. Teratology 17:151–158, 1978.

Tobias, J.S. and Bloom, H.J.G.: Doxorubicin in pregnancy. Lancet 1:776, 1980.

695 DDT *1,1,1–Trichloro–2,2–bis(p–chlorophenyl)ethane Chlorophenothane* CAS 50–29–3

O'Leary et al. (1970) measured the DDT and DDE levels in the serum of patients having spontaneous abortions and found no difference from serums drawn from women with normal pregnancies.

Although the thickness of egg shells of birds is reduced by chlorinated hydrocarbons (Hickey and Anderson, 1968), no reports of teratogenicity in the chick, mouse (Ware and Good, 1967) or rat (Ottoboni, 1969) were identified. In Ottoboni's publication a significant increase in ringtail, a constriction of the tail followed by amputation, occurred in the offspring of mothers whose diets contained 200 p.p.m. Ware and Good (1967) showed reduced fertility in mice maintained for long periods on diets of 7 P.P.M. Hart et al. (1971), using 50 mg per kg in the rabbit on day 7, 8 and 9 of gestation, noted premature delivery, increased fetal resorptions and reduced intrauterine growth but no congenital defects were produced.

Fabro (1973) reported a reduction in the brain weights of rabbit fetuses given 1 mg orally on days 4 through 7. DDT may disrupt the early developing hypothalamic–gonado–tropin mechanism according to the experiments by Heinrichs et al. (1971), who showed persistent estrus in rats following injection of 1 mg on the first three postnatal days. Ware (**1975**) has reviewed the subject.

Fabro, S.: Passage of Drugs and Other Chemicals into Uterine Fluids and Preimplantation Blastocyst, L. Boreus (ed), Fetal Pharmacology Raven Press, New York, 443–461, 1973.

Hart, M.M.; Adamson, R.H. and Fabro, S.: Prematurity and intrauterine growth retardation induced by DDT in the rabbit. Arch. Int. Pharmacodyn. Ther. 192:286–290, 1971.

Heinrichs, W.L.; Gellert, R.J.; Bakke, J.L. and Lawrence, N.L.: DDT administered to neonatal rats induces persistent estrus syndrome. Science 173:642–643, 1971.

Hickey, J.J. and Anderson, D.W.: Chlorinated hydrocarbons and eggshell changes in raptorial and fish–eating birds. Science 162:271–273, 1968.

O'Leary, J.A.; Davies, J.E. and Feldman, M.: Spontaneous abortion and human pesticide residues of DDT and DDE. Am. J. Obstet. Gynecol. 108: 1291–1292, 1970.

Ottoboni, A.: Effect of DDT on reproduction in the rat. Toxicol. Appl. Pharmacol. 14:74–81, 1969.

Ware, G.W.: Effects of DDT on reproduction in higher animals. Residue Rev 59:119–140, 1975.

Ware, G.W. and Good, E.E.: Effects of insecticides on reproduction in the laboratory mouse II mirex, telodrin and DDT. Toxicol. Appl. Pharmacol. 10:54–64, 1967.

696 Decabromodiphenyl Oxide

Norris et al. (**1975**) reported no adverse effects from administration of up to 1,000 mg per kg to rats on days 6–15. Reproductive capacity at 100 mg per kg was not effected.

Norris, J.M.; Kociba, R.J.; Schwetz, B.A.; Rose, J.Q.; Humiston, C.G.; Jewett, G.L.; Gehring, P.J. and Mailhes, J.B.: Toxicology of octabromobiphenyl and decabromodiphenyl oxide. Environmental Health Perspectives 11:153–161, 1975.

697 Decachlorooctahydro–1–3–4–metheno–2h–cycylobuta–(6d)–pentalen–one 2–one

Discussed under Kepone

698 Decamethonium Bromide CAS 541–22–0

Drachman and Sokoloff (1966) paralyzed chick embryos with intravenous infusions of this material after the 9th day and for several days joint cavity formation was inhibited.

Drachman, D.B. and Sokoloff, L.: The role of movement in embryonic joint development. Dev. Biol. 146:401–420, 1966.

699 Decloxizine

Giurgea et al. (**1968**) gavaged rats and mice with up to 105 mg per kg during organogenesis and found no reproductive toxicity or teratogenicity.

Giurgea, C.; Vanremoortere, E.; Giurgea, M.; Greindl, M.G.; Puigdevall, J. and Wellens, D.: General pharmacology of a new bronchodilator, decloxizine. Arzneim–Forsch 18:1002–1008, 1968.

700 Deferoxamine

Of 25 women given this drug for iron overdose only one malformed infant (Webbed fingers on one hand) could be associated with the treatment (McElhatton et al., **1991**).

Gower et al. (**1990**) used this iron chelator to reduce resorptions in mice given endotoxinon the 13th day of gestation. Details of fetal results were not given.

Gower, J.D.; Baldock. R.J.; O'Sullivan, A.M.; Dore, C.R.; Coid, C.R. and Green, C.J.: Protection against endotoxin-induced foetal resorption in mice by desferrioxamine and ebselen. Int. J. Exp. Path. 71:433–440, 1990.

McElhatton, P.R.; Roberts, J.C.; Sullivan, F.M.: The consequences of iron overdose and its treatment with desferrioxamine in pregnancy. Hum. Exp. Toxicol. 10:251–259, 1990.

701 Dehydration

Brown et al. (1974) produced isolated cleft palate in A–J mice by depriving them of water for 72 hours or by depriving them for 48 hours in the presence of dehumidified air. This treatment was started during day 12 of gestation. Schwetz et al. (1977) thirsted CF_1 mice for 48 hours during periods of organogenesis and found up to 28 percent of the litter contained fetuses with cleft palate. The water deprivation on days 12 and 13 caused a reduction in food intake which by itself produced malformations. Twenty–five percent of the litters had fetuses with cleft palate and 5 percent of the litters had a fetus with exencephaly.

Brown, K.S.; Johnston, M.C. and Murphy, P.F.: Isolated cleft palate in A–J mice after transitory exposure to drinking–water deprivation and low humidity in pregnancy. Teratology 9:151–158, 1974.

Schwetz, B.A.; Nitschke, K.D. and Staples, R.E.: Cleft palates in CF_1 mice after deprivation of water during pregnancy. Toxicol. Appl. Pharmacol. 40:307–315, 1977.

702 Dehydroacetate Sodium CAS 4418–26–2

Shiobara (1980) gavaged this food preservative into pregnant mice on days 6–15 and found fetal lethality increased at 200 mg per kg per day. An increase in 14th rib was found in the 50, 100 and 200 mg per kg groups but no other defects were significantly increased. Tanaka et al. (**1988**) gavaged this food preservative into rats on days 6–17 in doses of 0, 25, 50 and 100 mg per kg. Maternal and fetal weight gain was suppressed at 50 and 100 mg per kg and some skeletal variations occurred in the fetuses. No teratogenic action was found.

Shiobara, S.: Effect of sodium dehydroacetate (DHA–NA) orally administered to pregnant mice on the pregnants and their fetuses. Nippon Koshu Eisei Zasshi 27:91–97, 1980.

Tanaka, S.; Kawashima, K.; Nakaura, S.; Djajalaksana, S. and Takanaka, A.: Studies on the teratogenic potential of sodium dehydroacetate in rats. Eiseishikenjo Hokoku (Bulletin of National Institute of Hygienic Sciences) 106:54–61, 1988.

703 Dehydroepiandrosterone CAS 53–43–0

Greene et al. (1939) produced masculinization of female fetuses by treatment of the maternal rat with a total dose of 100 to 280 mg. They found regression of the wolffian ducts was inhibited by treatment. Before the 16th day, masculinization of the external genitalia was associated.

Greene, R.R.; Burrill, M.W. and Ivy, A.C.: Experimental intersexuality. The effect of antenatal androgens on sexual development of female rats. Am. J. Anat. 65:415–470, 1939.

704 Dehydroepiandrosterone Sulfate *Prasterone DHA–S*

Ryle et al. (**1994**) administered this steriod intravaginally from day 17 until before delivery in amounts up to 100 mg per kg. No adverse fetal effects were found and the behavior and sexual reproducdtion of the offspring was not changed.

Ryle, P.R.; Masters, R.E.; Parker, C.A.; Cadel, S. and Watanabe, I.: Reproductive and developmental toxicity study in rats given sodium prasterone sulfate (DHA–S) intravaginally during the perinatal period. Yakuri to Chiryo 48:201-211, 1994.

705 Deladroxone

This acetophenone derivative of 16–α, 17 α–dihydroxy progesterone was given by Munshi and Rao (1969) to mice subcutaneously in doses of up to 1.0 mg per kg from the 6th day until the end of pregnancy. At doses of 0.6 mg or less the litter sizes were unaffected.

Munshi, S.R. and Rao, S.S.: Effect of the acetophenone derivative of 16 α, 17 α–dihydroxyprogesterone on the organ weights and reproduction of female mice. Ind. Jour. Med. Res. 57:8 only, 1969.

706 Delayed Fertilization *Fertilization*

Since the early work of Blandau and Young (1939), there has been little doubt that a delay in fertilization of an egg once ovulated may lead to abnormal development of the ovum and embryo.

In the human, Jongbloet (1971) collected information on conditions which might lead to delayed fertilization of the ovum. He reported the incidence of pathologic progeny to be increased to 50 percent when conception occurred in the first month after marriage of couples without prenuptial contact. Based on small numbers, he found that couples conceiving while practicing the safe period of birth control experienced a higher incidence of abortion and pathologic offspring when their planned period of abstinence was increased.

Guerrero and Rojas (1975) studied 965 women who recorded their basal temperatures and coital history and observed that the probability of abortion diminished significantly at the time of ovulation. At ovulation, the chance was 7.5 percent while three days later, it was 24 percent. Boue et al. (1975) found that polyploid human abortuses were more common when the mother's ovulation occurred after the 14th day. The average presumed day of ovulation determined by temperature curves was 15 for trisomic and normal abortuses and 17 for those with polyploidy.

There is evidence that the delayed fertilization leads to abnormalities in chromosome complement which in some cases may be the result of polyspermy (Vickers, 1969; Shaver and Carr, 1967). Triploidy is the most common abnormality found. The mouse, rat, guinea pig, rabbit, pig and frog were studied and in general, after fertilization delay of 4 to 7 hours, abnormal blastocysts appear and with further delay the recovery rate of embryos drops sharply.

Simpson et al. (1988) reviewed the experimental and clinical findings of studies on the effects of aging sperm and ova and delayed fertilization. Several summaries of this experimental field were published (Austin, 1970; Carr, 1971; Witschi, 1970).

Austin, C.R.: Aging and reproduction: Post–ovulatory deterioration of the egg. J. Reprod. Fertil. Suppl. 12:39–53, 1970.

Boue, J.; Boue, A. and Lazar, P.: Retrospective and prospective epidemiological studies of 1500 karyotyped spontaneous human abortions. Teratology 12:11–26, 1975.

Blandau, R.J. and Young, W.C.: The effects of delayed fertilization on the development of the guinea pig ovum. Am. J. Anat. 64:303–329, 1939.

Carr, D.H.: Chromosome abnormalities in the preimplanting ovum. In: The Biology of the Blastocyst, R.J. Blandau (ed.), Chicago: University of Chicago Press, 355–357, 1971.

Guerrero, R. and Rojas, O.I.: Spontaneous abortion and aging of human ova and spermatozoa. N. Eng. J. Med. 293:573–575, 1975.

Jongbloet, P.H.: Mental and Physical Handicaps in Connection with Overripeness Ovopathy. H.E. Stenfert Kroese N.V.–Leiden, 22–61, 1971.

Shaver, E.L. and Carr, D.H.: Chromosome abnormalities in rabbit blastocysts following delayed fertilization. J. Reprod. Fertil. 14:415–420, 1967.

Simpson, J.L.; Gray, R.H.; Queenan, J.T.; Mena, P.; Perez, A.; Kambic, R.T.; Tagliabue, G.; Pardo, F.; Stevenson, W.S.; Barbato, M.; Jennings, V.H.; Zinaman, M.J. and Spieler, J.M.: Pregnancy outcome associated with natural family planning (NFP): Scientific basis and experimental design for an international cohort study. Adv. Contracept. 4:247–264, 1988.

Vickers, A.D.: Delayed fertilization and chromosomal anomalies in mouse embryos. J. Reprod. Fertil. 20:69–76, 1969.

Witschi, E.: Teratogenic effects from overripeness of the egg. In: Congenital Malformations, F.C. Fraser and V.A. McKusick (ed.), Amsterdam: Excerpta Medica, 157–169,

1970.

707 Demeclocycline

No increase in defects was observed by Heinonen et al. (1977) among 90 women who took this drug in the first 4 lunar months. Dental staining has been reported in the offspring of mothers taking the medication during the third trimester.

Mangi et al. (1976) reported growth retardation and defects of the mouth and limbs in mouse fetuses from mothers treated with up to 600 mg per kg on days 8–13.

Heinonen, O.P.; Slone, D. and Shapio, S.: Birth Defects and Drugs in Pregnancy. John Wright Publishing Sciences Group, Inc., Littleton, Mass. 1977.

Mangi, Y.; Muzutani, M. and Kaziwara R.: Effects of demethylchortetrayline hydrochloride on the fetuses of experimental animals. I observation by cesarean section of CF. mice at term (Personal communication) in Nishimura, H. and Tanimura T. eds Clinical Aspects of the Teratogenicity of Drugs American Elsevier Publishing Comp. New York. 1976 p125.

708 Demecolcine *Colcemid*™ CAS 477–30–5

Sokal and Lessmann (1960) cite two reports where pregnant women given 1.5 to 7.5 mg per day from early in gestation produced normal infants.

Sokal, J.E. and Lessmann, E.M.: Effects of cancer chemotherapeutic agents on the human fetus. JAMA 172:1765–1771, 1960.

709 Demeton™ CAS 8065–48–3

This organophosphorus insecticide was given intraperitoneally to mice on a single day between days 7 and 12 in doses up to 14 mg per kg. The higher doses produced fetal growth retardation (Budreau and Singh, 1973). A few minor skeletal malformations were found when dosing was carried out daily during 3 day periods of organogenesis.

Budreau, C.H. and Singh, R.P.: Teratogenicity and embryotoxicity of demeton and fenthion in CF No. 1 mouse embryos. Toxicol. Appl. Pharmacol. 24:324–332, 1973.

710 Dengue II Virus

London et al. (1979) injected an attenuated form of this virus intraamniotically on day 100 of the rhesus pregnancy and found hemorrhagic necrosis and hydrocephalus in 3 of 20 exposed fetal brains.

London, W.T.; Levitt, H.H.; Martinez, A.J.; Palmer, A.E.; Curfman, B.L. and Sever, J.L.: Hydrocephalus and necrotic encephalopathy in rhesus fetuses caused by Dengue II virus. (abs) Fed. Proc. 38:911 only, 1979.

711 2–Deoxyadenosine CAS 958–09–8

Karnofsky and Lacon (1961) reported that chick embryos injected on the 4th day with 0.5 to 4 mg of this substance developed skeletal and palate defects. Seven percent of embryos given adenine (2 to 8 mg) were abnormal but none receiving adenosine (2 to 8 mg) were defective.

Karnofsky, D.A. and Lacon, C.R.: Effects of physiological purines on the development of the chick embryo. Biochem. Pharmacol. 7:154–158, 1961.

712 Deoxycholic Acid *Bile Acids*

Deoxycholic acid or lithocolic acid was injected intraperitoneal into rats on days 6–15. (Zimber and Zusman, 1990) Resorptions, growth retardation and malformations were increased. Two ml of 5 mmolar or 2 mmolar solutions were used respectively. Ectrodachtyly, hydroamnion and umbilical hernias were observed. Intrauterine injection produced decrease in the microvilli of the yolk sac.

Zimber, A. and Zusman, I.: Effects of secondary bile acids on the intrauterine development in rats. Teratology 42:215–224, 1990.

713 2'–Deoxycoformycin *Pentostatin YK–176*

This inhibitor of adenosine deaminese was studied in the mouse and on days 7 and 8 resorptions were increased by intraperitoneal injections of 5 mg per kg. Treatment on days 6, 9, 10 or 11 produced no increase in resorption (Knudsen et al., 1989). The toxicity was associated with reduced adenasine deaminase activity in the antimesometrial portion of the decidua on day 7. Dostal et al. (1991) administered it to rats and rabbits intravenously during organogenesis. In the rabbit fetal toxicity occurred but no teratogenicity at doses of up to 0.02 mg per kg daily. In rats given 0.75 mg per kg fetal resorptions were increased and body weight reduced along with skeletal variations and defects. Airhart et al. (1993) extended their studies in the mouse using 5 mg per kg on day 7. Defects of neural tube closure and arch development were found on day 10 and followed by a high embryonic loss rate. Fuchigami et al. (1991) gave mice up to 1.0 mg per kg intravenously during the perinatal period and found increased death postnatally and physical retardation at the highest dose. In rabbits Fuchigami et al. (1991) gave up to 0.1 mg per kg intravenously and found fetal death and toxicity at 0.02 mg per kg but no teratogenicity.

Airhart, M.J.; Robbins, C.M.; Knudsen, T.B.; Church, J.K. and Skalko, R.G.: Occurrence of embryotoxicity in mouse embryos following in utero exposure to 2'–deoxycoformycin (pentostatin). Teratology 47:17–27, 1993.

Dostal, L.A.; Brown, S.; Bleck, J. and Anderson, J.A.: Developmental toxicity of pentostatin (2'–deoxycoformycin) in rats and rabbits. Teratology 44:325–334, 1991.

Fuchigami, K.; Sameshima, K.; Izumi, H.; Otsuka, T.; Ogami, C.; Yoshinaga, K.; Inoue, H.; Kuwata, S.; Umehashi, M. and Shigaki, T.: Study by intravenous administration of YK–276 during the perinatal and lactation periods in mice. The Clinical Report 25:4329–4347, 1991.

Fuchigami, K.; Sameshima, K.; Izumi, H.; Honda, H.; Yamauchi, M.; Hayashida, M.; Inoue, H.; Kuwata, S.; Umehashi, M. and Shigaki, T.: Study by intravenous administration of YK–176 during the period of organogenesis in rabbits. The Clinical Report 25:4319–4327, 1991.

Knudsen, T.B.; Gray, M.K.; Church, J.K.; Blackburn, M.R.; Airhart, M.J.; Kellems, R.E. and Skalko, R.G.: Early postimplantation embryolethality in mice following in utero inhibition of adenosine deaminase with 2'–deoxycoformycin. Teratology 40:615–626, 1989.

714 Deoxycorticosterone Acetate *DOCA* CAS 56–47–3

Grollman and Grollman (1962) administered one mg per day from day four through pregnancy and found that the rat offspring had hypertension which persisted for at least one year. Walker (1965) injecting 0.1 to 1.25 mg of this compound in mice on days 11 through 14 found no cleft palates and only one spina bifida out of 69 fetuses.

Grollman, A. and Grollman, E.F.: Teratogenic induction of hypertension. J. Clin. Invest. 41:710–714, 1962.

Walker, B.E.: Cleft palate produced in mice by human–equivalent dosage with triamcinolone. Science 149:862–863, 1965.

715 2–Deoxyglucose CAS 61–58–5

Demeyer (1961) gave 120 mg per day from the 9th through the 20th gestational day of the rat. Resorptions were 69 percent and the surviving fetuses were all malformed. Anophthalmia, cleft lip and palate and lesions of the extremities were observed. Spielmann et al. (1973) gave 1 gm per kg on days 8, 9, 10 or 11 and found no malformations in the surviving rat fetuses.

Demeyer, R.: Etude Experimentale de la Glycoregulation Gravidigne et de L'action Teratogene des Perturbations du Metabolisme Glucidigne. Bruxelles: Editions Arscia S.A., 184–189, 1961.

Spielmann, H.; Meyer–Wendecker, R. and Spielmann, F.: Influence of 2–deoxy–D–glucose and sodium fluoroacetate on respiratory metabolism of rat embryos during organogenesis. Teratology 7:127–134, 1973.

716 2–Deoxyguanosine CAS 961–07–9

Karnofsky and Lacon (1961) reported coloboma, cleft palate and skeletal defects in chicks receiving 0.25 to 1.0 mg via the yolk sac on the 4th day. Guanine had no effects but a few defects were found when guanosine was used.

Karnofsky, D.A. and Lacon, C.R.: Effects of physiological purines on the development of the chick embryo. Biochecm. Pharmacol. 7:154–158, 1961.

717 2–Deoxyinosine

Karnofsky and Lacon (1961) found some facial colobomas and skeletal defects in chicks after injecting 0.5 to 4 mg into the yolk sac on the 4th day. Hypoxanthine was non–teratogenic in doses of 2 to 8 mg per egg. An occasional defect was seen in chicks receiving 2 to 8 mg of inosine.

Karnofsky, D.A. and Lacon, C.R.: Effects of physiological purines on the development of the chick embryo. Biochem. Pharmacol 7:154–158, 1961.

718 4–Deoxynivalenol *Vomitoxin*

Khera et al. (1984) fed pregnant mice (2.0 mg per kg) and observed no significant adverse reproductive effects.

Khera, K.S.; Arnold, D.L.; Whalen, C.; Angels, G. and Scott, P.M.: Vomitoxin (4–deoxynivalenol) effects on reproduction of mice and rats. Toxicol. Appl. Pharmac. 74:345–356, 1984.

719 Deprodone Propionate

Ito et al. (**1990** A) gave rats subcutaneously up to 10 mg per kg during organogenesis. At doses of 10 mg per kg or more there was some maternal lethality and at 2 mg per kg the fetal weights were decreased. No teratogenicity was seen but umbilical hernias were increased at 2 mg per kg. Ito et al. (**1990** B) treated rabbits on days 6–18 with 2.0, 0.4, 0.08 mg per kg subcutaneously. At 0.4 mg per kg cleft palate exencephaly and umbilical hernias were increased as well as fetal deaths. The maximum non–effective dose was 0.08 mg per kg. Ito et al. (**1990** C) gave the steriod before fertilization and found reproduction inhibited in the male at 0.6 mg per kg and at 0.08 mg per kg in the female.

Ito, R.; Tanihata, T.; Onda, T.; Oshima, M. and Iizua, T.: Teratological studies on deprodone propionate s.c. on prematal, early gravidity (seg i) in rats. Kiso to Rinsho 24:3055–3076, 1990.

Ito, R.; Tanihata, T.; Onda, T.; Mori, S. and Miyamoto, K.: Teratological studies on deprodone propionate s.c. on organogenesis (seg ii) in sd rats. Kiso to Rinsho 24:3077–3083, 1990.

Ito, R.; Suzui, Y.; Miyagi, H.; Miyasaka, M.; Mori, K.; Kamiya, T.; Kitajima, S.; Hashimoto, M.; Niimi, J.; Mitsumata, H. and Miura, M.: Subchronic toxicity study on deprodone propionate for 15 weeks and 5 weeks recovery in rats. Kiso to Rinsho 24:3043–3054, 1990.

720 Deptropine Citrate *3*
β(10,11–Dihydro–5H–dibenzo[A,D] cyclohepten–5–yloxy)Tropane Citrate

Van Eeken and Mulder (**1966**) gave rats 10.5 mg by gavage for varying periods after day 4 and found no adverse fetal effects. Rabbits similarly treated did not produce deformed fetuses.

Van Eeken, C.J. and Mulder, D.: Studies on the influence on reproduction and fertility of 3β(10,11–dihydro–5h–dibenzo[a,d] cyclohpeten–5–yloxy0tropane citrate (deptropine citrate). Arch. Int. Pharmacodyn. Ther. 159:240–248, 1966.

721 Deserpidine *11–Desmethoxyreserpine* CAS 131–01–1

Tuchmann–Duplessis and Mercier–Parot (1961) gave large doses (1.8 mg per kg) from the 6th through the 16th day of gestation and found rat fetuses with reduced weight and multiple subcutaneous hemorrhages associated with some necrosis of the extremities and skeletal defects.

Tuchmann–Duplessis, H. and Mercier–Parot, L.: Malformations foetales chez le rat traite par de fortes doses de deserpidine. C. R. Soc. Biol. 155:2291–2293, 1961.

722 Desipramine CAS 50–47–5

This chemical is an active metabolite of imipramine. Neonatal withdrawal symptoms have been reported.

723 Desoximetasone

9–Fluoro–11–β–21–dihydroxy–16–α–methylpregna–1,
4–diene–3, 20–dione Desoxymethasone

This anti–inflammatory agent was tested in pregnant rats and mice (Miyamoto et al., 1975). Cleft palate and edema was found at 1,600 microgm per kg per day when given to mice during organogenesis.

Miyamoto, M.; Ohtsu, M.; Sugisaki, T. and Sakaguchi, T.: Teratogenic effect of 9–fluoro–11–β,21–dihydroxy–16–α–methylpregna–1, 4–diene,3,20–dione (A41304) a new antiinflammatory agent, and of dexamethasone in rats and mice. (Japanese) Folia Pharmacol. Japon 71:367–378, 1975.

724 Desoxycorticosterone

Desoxycorticosterone acatate (DOCA) (at a dose 0.8–1.0 mcg/100g) was injected intramuscularly into pregnant rats. DOCA induced hyperplasia and high proliferative activity of epidermis and inhibition in proliferation of fibroblasts in fetuses and newborn rats.

Pavlova, I.G.: Morphofunctional characteristics of the newborn rat skin under prenatal effects of desoxycorticosterone acetate. Arkh. Anat. Gistol. Embriol. 93(12):65–72, 1987.

725 4–Desoxypyridoxine Hydrochloride also see
Pyridoxine Deficiency CAS 61–67–6

This antagonist to pyridoxine was used to augment a pyridoxine deficient diet and produced anomalies in rats (Davis et al., 1970). Fetuses from mothers maintained during pregnancy on a B_6 deficient diet and 2.3 mg of desoxypyridoxine developed digital defects, cleft palates and omphaloceles and occasionally, exencephaly. Based on splenic and thymus hypoplasia, the authors raised the question of defective development of the immune system. The syndrome could be prevented by supplementation with pyridoxine.

Davis, S.D.; Nelson, T. and Shepard, T.H.: Teratogenicity of vitamin B_6 deficiency: Omphalocele, skeletal and neural defects with splenic hypoplasia. Science 169:1329–1330, 1970.

726 Detralfate CAS 37209–31–7

Towizawa et al. (1972) gave this anti–ulcer agent orally to pregnant mice and rats during active organogenesis in amounts up to 4.0 gm per day and observed no teratogenicity or postnatal effects.

Towizawa, S.; Kamata, K. and Yoshimari, M.: Effects of detralfate administered to pregnant mice and rats on pre– and post–natal development of their offsprings. Oyo Yakuri 6:599–611, 1972.

727 Dexamethasone also see *Cortisone* CAS 50–02–2

This potent glucocorticoid has the same general teratogenic action as cortisone. Buck et al. (1962) produced neural tube closure defects in rabbits. Pinsky and DiGeorge (1965) found that dexamethasone, when compared to hydrocortisone, had a much greater cleft palate producing activity than would be expected by comparison of their glucocorticoid activities. Esaki et al. (1981) applied the 17–valerate form to rabbits dermally at doses of up to 0.1 mg per kg and no adverse fetal effects were found. Umemura et al. (1982) studied the 17–valerate form in the postnatal and perinatal period in the rat giving up to 0.06 mg per kg daily by dermal application. Jerome and Hendrickx (1988) found cranium bifidum (1) and aplasia cutis congenita (3) among 6 monkeys treated with 10 mg per kg daily between days 22–50. No adverse changes in behavior or reproduction were observed.

Buck, P.; Clavert, J. and Rumpler, Y.: Action teratogenique des corticoides chez la lapine. Ann. Chir. Infant 3:73–87, 1962.

Esaki, K.; Shikata, Y. and Yanagita, T.: Effect of dermal administration of dexamethason–17–valerate in rabbit fetuses. Preclin. Rep. Cent. Inst. Exp. Anim. 7:245–256, 1981.

Jerome, C.P. and Hendrickx, A.G.: Comparative teratogenicity of triamcinolone acetonide and dexamethasone in the rhesus monkey (Macaca mulatta). J. Med. Primatol. 17:195–203, 1988.

Pinsky, L. and DiGeorge, A.M.: Cleft palate in the mouse: A teratogenic index of glucocorticoid potency. Science 147:402–403, 1965.

Umemura, T.; Sasa, H.; Takada, N.; Esaki, K. and Yanagita, T.: Effect of dexamethasone 17–valerate on reproduction in rats: Administration during the peripartum and nursing periods. Preclin. Rep. Cent. Exp. Anim. 8:235–246, 1982.

728 Dextroamphetamine Sulfate *Amphetamine* CAS 51–63–8

Some significant increase in the incidence of congenital heart disease following amphetamine ingestion was reported by Nora et al. (1970). They found 20 of 184 mothers of children with congenital heart disease had taken the drug during the vulnerable period as opposed to 3 of 108 control mothers (p < 0.025). Levin (1971) reported that among eleven mothers of children with biliary atresia, four took amphetamines during the second or third month of gestation. Milkovich and Van den Berg (1977) in a prospective study found no increase in the malformation rate among 1824 exposed children followed until their fifth birthday. Facial clefts occurred in three of 175 offspring exposed during the first 56 days after the last menses. The authors state that this could have occurred by chance. They did not find an increase in congenital heart defects.

Heinonen et al. (1977) studied 367 exposed mother–child pairs and found no increase in defect rate. Naeye (1983) analyzed weights of newborns whose mothers took the drug and found a 4 percent (144 gm) reduction when the drug was continued beyond the 28th week. The head circumferences were not decreased. In a control group of 20 asphyxiated newborns, only one had evidence for hemorrhagic infarction. Ultrasound studies follosed by CT or MRI were used. Little et al. (**1988**) compared the offspring of 52 women who abused methamphetamines with 52 non–drug–abusing women's offspring. No increase in defects was found but the birth weight and head circumferences were less in the exposed group. (2957 gms and 33.2cm)

Nora et al. (1968) using 50 mg per kg of body weight on the 8th day in two strains of mice produced a significant increase in ventricular and atrial septal defects. Yasuda et al. (1967) gave amphetamine sulfate (50 mg per kg) from gestational day 0 to 17 in the mouse and found five cleft palates in 274 fetuses as compared to 1 in 127 controls. Fetal weight was not reduced in the treated group. Kasirsky and Tansy (1971) using methamphetamine, a derivative of dexamphetamine, treated pregnant mice and rabbits. In the mice with a dose of 10 mg per kg on days 9 through 15, 13.6 percent of the fetuses had exencephaly, cleft palate or eye defects. In the rabbit treated with 1.5 mg per kg for 3 or 18 days during organogenesis, a significant increase in brain and eye defects was found.

Clark et al. (1970) injected rats with 1 mg per kg on days 12 through 15 of gestation and performed behavior tests on the offspring. The treated animals had low activity test scores early in testing and were delayed in reaching adult activity levels. No differences were found in T–maze learning or reversal or in performance or extinction of the operant response. Seliger (1973) found that 5 mg doses during day 5 through 9 in rat gestation was associated with higher activity in the offspring. Adams et al. (1982) reported deficits of acquisition of escape in rat offspring whose mothers received 0.5 to 2.0 mg per kg on days 12–15 of gestation. Increased sensitivity to postnatal d–amphetamine challenge was also seen in the females.

Yamamoto et al. (**1992**) gave mice up to 21 mg of methamphetamine intraperitoneally on day 8 of gestation. Exencephaly was increased at 13 mg per kg and above and cleft palate at 19 mg per kg. Skeletal defects were also found to be increased.

Adams, J.; Buelke–Sam, J.B.; Kimmel, C.A. and LaBorde, J.B.: Behavioral alterations in rats prenatally exposed to low doses of d–amphetamine. Neurobehav. Toxicol. Teratol. 4:63–70, 1982.

Clark, C.V.H.; Gorman, D. and Vernadakis, A.: Effects of prenatal administration of psychotropic drugs on behavior of developing rats. Dev. Psychobiol. 34:225–235, 1970.

Heinonen, O.P.; Slone, D. and Shapiro, S.: Birth Defects and Drugs in Pregnancy. Publishing Sciences Group Inc., Littleton, Mass., 1977.

Kasirsky, G. and Tansy, M.F.: Teratogenic effects of methamphetamine in mice and rabbits. Teratology 4:131–134, 1971.

Levin, J.N.: Amphetamine ingestion with biliary atresia. J. Pediatr. 79: 130–131, 1971.

Little, B.B.; Snell, L.M. and Gilstrap, L.C.: Methamphetamine abuse during pregnancy: Outcome and fetal effects. Obstet Gynecol 72(4):541–544, 1988.

Milkovich, L. and Van den Berg, B.J.: Effects of antenatal exposure to anorectic drugs. Am. J. Obstet. Gynecol. 129:637–642, 1977.

Naeye, R.L.: Maternal use of dextroamphetamine and growth of the fetus. Pharmacology. 26:1l7–120, 1983.

Nora, J.J.; Sommerville, R.J. and Fraser, F.C.: Homolo-

gies for congenital heart diseases: Murine models, influenced by dextroamphetamine. Teratology 1:413–416, 1968.

Nora, J.J.; Vargo, T.A.; Nora, A.H.; Love, K.E. and McNamara, D.G.: Dexamphetamine: A possible environmental trigger in cardiovascular malformations (letter). Lancet 1:1290 only, 1970.

Seliger, D.L.: Effect of prenatal maternal administration of D–amphetamine on rat offspring activity and passive avoidance learning. Physiol. Psych. 1:273–280, 1973.

Yamamoto, Y.; Yamamoto, K.; Fukui, Y. and Kurishita, A.: Teratogenic effects of methamphetamine in mice. Jpn. J. Legal Med. 46:126–131, 1992.

Yasuda, M.; Ariyuki, F. and Nishimura, H.: Effect of successive administration of amphetamine to pregnant mice upon the susceptibility of the offspring to the teratogenicity of thio–tepa. Congenital Anomalies 7:66–73, 1967.

729 Dextromethorphan *Racemethorphan*

Heinonen et al. (1976) reported 17 congenital defects among 300 women taking this antitussive in the first 4 lunar months.

Heinonen, O.P.; Slone, D. and Shapiro, S.: Birth Defects and Drugs in Pregnancy. John Wright Publishig Sciences Group, Inc., Littleton, Mass. 1977.

730 Dextromoramide

Jurand and Martin (1990) injected mice intraperitoneally on the 9th day of gestation. At 25 mg per kg exencephaly and dilated ventricles were observed. Maternal mortality was observed at 28 mg per kg.

Jurand, A. and Martin, L.V.H.: Teratogenic potential of two neurotropic drugs, haloperidol and dextromoramide, tested on mouse embryos. Teratology 42:45–54, 1990.

731 Diabetes

Malformations

There is controversy about whether diabetes contributes to a general increase in human congenital defects. Rubin and Murphy (1958) showed that the increased incidence of defects in the diabetic offspring can be explained by intensified scrutiny of these children either during hospitalization or at autopsy examination. Farquhar (1965) was unable to show an increased rate of defects but Hagbard (1961) and Pedersen et al. (1964) gave reports with small increases. Comess et al. (1969), in an Indian population, reported a 38 percent incidence of anomaly in the offspring of diabetics diagnosed before 25 years of age. McCarter et al. (1987) found a doubling of multi–organ defects in the offspring of diabetics. The study group contained 2,639 women. Simpson et al. (1983) found 9 major defects among 106 class B to F diabetics in a prospective study. Becerra et al. (1990) list increased relative risks for various defects among 47 insulin using mothers. Major CNS and cardiovascular defects were 15 and 18 times more common respectively. Rosa (**1994**) in a study of medicaid records reported a heart defect rate ratio of 6.6 (66 versus 10) for insulin users and 5.0 (5 versus 1) for the offspring of oral hypoglycemics.

Although there is doubt about the overall incidence of congenital defects, most authors have observed a significant number of offspring with caudal dysplasia or caudal regression syndrome. The defect may appear in as high as 0.2 to 1 percent of diabetic offspring and consists of varying amounts of sacral and femoral agenesis sometimes associated with defects of the palate and branchial arches (Kucera et al., 1965; Passarge, 1965; Rusnak and Driscoll, 1965; Kucera, 1971). It is of interest that an early report of 439 diabetic offspring by White (1949) cited only one patient with congenital skeletal defects. A small left colon syndrome was also described in children (Davis and Campbell, 1975). Milunsky et al. (1982), using serum α-fetoprotein for diagnosis, found a ten–fold increase in neural tube defects among 411 pregnant diabetics. Vtorova (1986) reported the results of a clinical–statistical analysis of 695 women with insulin–dependent diabetes. A parallel study of the hormonal profile of offspring with birth weight 3000–3900 and more than 4000 g was carried out. The results suggest a polycausal character of diabetic fetopathy. Ferencz et al. (1990) found increased odds ratios for the offspring of overt diabetics to be 21.3 for double outlet right ventricle and 12.8 for truncus arteriosus. They pointed out that these types of defects were thought to be related to neural crest derived tissues.

Stehbens et al. (1977) studied the intellectual function of diabetic offspring and found that the presence of acetone in the urine during gestation was associated with a significant reduction in IQ at three and five years of age.

Maternal Treatment and Severity

Programs which control the diabetic state starting before conception may be successful in reduction of the number of defects (Fuhrmann et al., 1983, White, 1949). Pedersen and Pedersen (1985) studied 2,587 diabetic offspring over a period of 56 years. Among the mothers with severe forms of the disease, a decrease in those with congenital defects was observed in the last four years of observation. The ultrasound measured crown–rump was decreased in the complicated cases and in the group with congenital defects.

The role of insulin, oral hypoglycemics and other physiologic changes of diabetes is difficult to evaluate. In a preliminary report from a large collaborative study of 1015 women, Mills (1987) reported major malformations to be 4.9 percent among well–controlled patients as compared to 2.1 percent in controls. Diabetic women who registered late had a rate of 9.0 percent. Glycolysated hemoglobin was not increased among women who gave birth to deformed infants (Mills et al., 1988). A review editorial (Anonymous, 1988) discussed the various studies that have correlated the glucose levels with diabetic pregnancy outcome. Their conclusion is that in spite of the data of Mills et al. (1988), "high blood glucose in early pregnancy contributes to an increased risk of fetal malformation."

Spontaneous Abortion

Kalter (1987), in an extensive review of the literature, could not find an increase in abortion rate among insulin dependent mothers or those with gestational diabetes. Crane and Wahl (1981) studied spontaneous abortion in 154 diabetics and found no increase. Greene et al. (1989) studied 303 diabetics and did not find increased spontaneous abortions until hemoglobin A_1 levels were very high (greater than 9 percent). Major malformations were increased among the group with hemoglobin A_1 above 12.1 percent. Mills et al. (**1988**) found no increase in spontaneous abortions among 386 pregnant diabetics under good control but an increase was found among a small group with elevated glycosylated hemoglobin levels.

Animal Studies

An interesting model using streptozotocin–treated rats was used to show that with insulin treatment the incidence of caudal dysplasia is reduced (Eriksson et al., 1982; Eriksson, 1984; Baker et al., 1981). Sadler (1980) studied the effect of serum from streptozotocin–treated rats on mouse embryos grown in vitro. Neural closure defects were found. Zusman et al. (1987), using preimplantation mouse eggs, studied the effect of insulin, glucagon, β–hydroxybutyrate and acetoacetate. All substances inhibited growth at higher concentrations. Zusman et al. (1989) subsequently studied the effect of human diabetic serum and its contents on growth of preimplantation mouse fetuses. Johansson et al. **1991** found transient behavioral effects in the offspring of streptozotocin diabetic rats. Eriksson et al. (1989,A) have summarized the effect of streptozotocin diabetes on reproduction in two rat strains. In the same model Erickson et al. (1989,B) interrupted insulin treatment on various days and found the highest malformation rates in those animals with the highest β–hydroxybatyrate and triglycerides. Styrud and Eriksson (**1992**) studied various concentrations of normal and diabetic rat serum. The diabetic serum enhanced the growth retardation of high glucose levels. Ooshima and Ihara (**1986**) found increased fetal death in yellow KK mice with non–insulin dependent diabetes. The fetal death was reduced by treatment with ciglitazone an oral hypoglycemic agent.

Summaries

Warkany (1971) carefully reviewed the evidence for teratogenicity of diabetes.

Anonymous: Congenital abnormalities in infants of diabetic mothers. Lancet 1:1313–1314, 1988.

Baker, L.; Egler, J.M.; Klein, S.H. and Goldman, A.S.: Meticulous control of diabetes during of anogenesis prevents congenital lumbosacral defects in rats. Diabetes 30:955–959, 1981.

Becerra, J.E.; Khoury, M.J.; Cordero, J.F. and Erickson, J.D.: Diabetes mellitus during pregnancy and the risks for specific birth defects: a population–based case–control study. Pediatrics 85:1–9, 1990.

Comess, L.J.; Bennett, P.H.; Man, M.B.; Burch, T.A. and Miller, M.: Congenital anomalies and diabetes in the Pima indians of Arizona. Diabetes 18:471–477, 1969.

Crane, J.P. and Wahl, N.: The role of maternal diabetes in reproductive spontaneous abortion. Fertil. Steril. 36:477–479, 1981.

Davis, W.S. and Campbell, J.B.: Neonatal small left colon syndrome. Am. J. Dis. Child 129:1024–1027, 1975.

Eriksson, U.J.: Congenital malformations in diabetic

animal models: A review. Diabetes Res. 1:57–66, 1984.

Eriksson, U.J.; Dahlstrom, E.; Larsson, K.S. and Hellerstrom, C.: Increased incidence of congenital malformations in the offspring of diabetic rats and their prevention by maternal insulin therapy. Diabetes 31:1–6, 1982.

Eriksson, R.S.M.: Thunberg, L. and Eriksson, U.J.: Effects of interrupted insulin treatment on fetal outcome of pregnant diabetic rats. Diabetes 38:764–772, 1989.

Eriksson, U.J.; Styrud, J. and Eriksson, R.S.M.: Carbohydrate Metabolism in Pregnancy and the Newborn IV. Springer–Verlag, New York, 1989.

Farquhar, J.W.: The influence of maternal diabetes on the fetus and child. In: Recent Advances in Pediatrics, D. Gairdner (ed.), Boston: Little Brown and Company, (3rd Ed) 126–129, 1965.

Ferencz, C.; Rubin, J.D.; NcCarter, R.J. and Clark, E.B.: Maternal diabetes and cardiovascular malformations: predominance of double outlet right ventricle and truncus arteriosus. Teratology 41:319–326, 1990.

Fuhrmann, K.; Reiher, H.; Semmler, K.; Fischer, F.; Fischer, M. and Glockner, F.: Prevention of congenital malformations in infants of insulin–dependent diabetic mothers. Diabetes Care 6:219:223, 1983.

Greene, M.F.; Hare, J.W.; Cloherty, J.P.; Benacerraf, B.R. and Soeldner, J.S.: First–trimester hemoglobin A_1 and risk for major malformation and spontaneous abortion in diabetic pregnancy. Teratology 39:225–231, 1989.

Hagbard, L.: Pregnancy and Diabetes. Springfield, Ill.: Thomas, Charles C., 21–25, 1961.

Johansson, B.; Meyerson, B. and Eriksson, U.J.: Behavioral effects of an intrauterine or neonatal diabetic environment in the rat. Biology of the Neonate 59:226–235, 1991.

Kalter, H.: Diabetes and spontaneous abortion: A historical review. Am. J. Obstet. Gynecol. 156:1243–53, 1987.

Kucera, J.: Rate and type of congenital anomalies among offspring of diabetic women. J. Reprod. Med. 7:61–70, 1971.

Kucera, J.; Lenz, W. and Maier, W.: Missbildungen der Beine und der kaudalen Wirbelsaule bei Kindern diabetischer Muetter. Dtsch. Med. Wochenschr. 90:901–905, 1965.

McCarter, R.J.; Kessler, I.I. and Comstock, G.W.: Is diabetes mellitus a teratogen or a coteratogen? Am. J. Epidemiol. 125:195–205, 1987.

Mills, J.L.: Findings from NICHD diabetes in early pregnancy study. Teratology 35:7A only, 1987.

Mills, J.L.; Knopp, R.H.; Simpson, J.L.; Jovanovic–Peterson, L.; Metzger, B.E.; Holmes, L.B.; Aarons, J.H.; Brown, Z.; Reed, G.F.; Bieber, F.R.; Van Allen, M.; Holzman, I.; Ober, C.; Peterson, C.M.; Withiam, M.J.; Duckles, A.; Mueller–Heubach, E. and Polk, B.F.: The National Institute of Child Health and Human Development Diabetes. Early pregnancy study: Lack of relation of increased malformation rates in infants of diabetic mothers to glycemic control during organogenesis. N. Eng. J. Med. 318:671–676, 1988.

Mills, J.L.; Simpson, J.L.; Driscoll, S.G.; Jovanovic–Peterson, L.; Van Allen, M.; Aarons, J.H.; Metzger, B.; Bieber, F.R.; Knopp, R.H.; Holmes, L.B.; Peterson, C.M.;

Withiam–Wilson, M.; Brown, Z.; Ober, C.; Harley, E.; Macpherson, T.A.; Ducles, A.; Mueller–Heubach, E. and The National Institute of Child Health and Human Development–Diabetes in Early Pregnancy Study. Incidence of spontaneous abortion among normal women and insulin–dependent diabetic women whose pregnancies were identified within 21 days of conception. New England Journal of Medicine 319:1617–1623, 1988.

Milunsky, A.; Alpert, E.; Kitzmiller, J.L.; Younge, M.D. and Neff, R.K.: Prenatal diagnosis of neural tube defects. The importance of serum αfetoprotein in diabetic pregnant women. Am. J. Obstet. Gynecol. 142:1030–1032, 1982.

Passarge, E.: Congenital malformations and maternal diabetes. Lancet 1:324–325, 1965.

Ooshima, Y. and Ihara, T.: Embryopathy in genetically diabetic mice, yellow kk, and its prevention by maternal therapy. Congenital Anomalies 26:169–177, 1986.

Pedersen, L.M. and Pedersen, J.F.: Congenital malformations in diabetic pregnancies. Acta Paediatr. Scand. Suppl. 320:79–84, 1985.

Pedersen, L.M.; Tygstrup, I. and Pedersen, J.: Congenital malformations in newborn infants of diabetic women. Correlation with maternal diabetic vascular complications. Lancet 1:1124–1126, 1964.

Rosa, F.: Casdiovascular defect (CVD) diagnoses with 1st trimester prescriptions. (abs) Teratology 49:373, 1994.

Rubin, A. and Murphy, D.P.: Studies in human reproduction. 3. The frequency of congenital malformations in the offspring of nondiabetic and diabetic individuals. J. Pediatr. 53:579–585, 1958.

Rusnak, S.L. and Driscoll, S.G.: Congenital spinal anomalies in infants of diabetic mothers. Pediatrics 35:989–995, 1965.

Sadler, T.W.: Effects of maternal diabetes on early embryogenesis: I. The teratogenic potential of diabetic serum. Teratology 21:339–347, 1980.

Stehbens, J.A.; Baker, G.I. and Mitchell, M.: Outcome at ages 1, 2 and 5 years of children born to diabetic women. Am. J. Obstet. Gynecol. 127:408, 1977.

Simpson, J.L.; Elias, S.; Martin, A.O.; Palmer, M.S.; Orgata, E.S. and Radvany, R.A.: Diabetes in pregnancy, Northwestern University series (1977–1981). I. Prospective study of anomalies in offspring of mothers with diabetes mellitus. Am. J. Obstet. Gynecol. 146:263–270, 1983.

Styrud, J. and Eriksson, U.J.: Development of rat embryos in culture media containing different concentrations of normal and diabetic rat serum. Teratology 46:473–483, 1992.

Zusman, I.; Yaffe, P. and Ornoy, A.: Effects of human diabetic serum on the in vitro development of mouse preimplantation embryos. Teratology 39:581–590, 1989.

Vtorova, V.G.: Factors exerting influence on the formation of diabetic fetopathy. Voprosy Okhrany Materinstva (USSR) 2:61–65, 1986.

Warkany, J.: Congenital Malformations: Notes and Comments. Chicago: Year Book Publishers, 124–125, 1971.

White, P.: Pregnancy complicating diabetes. Am. J. Med. 7:609–616, 1949.

Zusman, I.; Yaffe, P. and Ornoy, A.: Effects of metabolic factors in the diabetic state on the in vitro development of preimplantation mouse embryos. Teratology 35:77–85, 1987.

732 Diacetoxyscirpenol *Trichothecone™*

3–Hydroxy–4,15–diacetoxy–12,13–epoxytrichothec–9–ene; DAS

Mayura et al. (1987) injected this mycotoxin intraperitoneally in mice at doses of up to 6.0 mg per kg on days 7–11 of pregnancy. Fetal death was increased at 1.0 mg per kg and above. Neural tube closure defects, hydrocephalus, microcephaly, microphthalmia, omphalocele and skeletal defects occurred at 1 mg per kg given on days 7, 8 or 9.

Mayura, K.; Smith, E.E.; Clement, B.A.; Harvey, R.B.; Kubena, L.F. and Phillips, T.D.: Developmental toxicity of diacetoxyscirpenol in the mouse. Toxicology 45:245–255, 1987.

733 Diamide

Hiranruengchok and Harris (**1993**) studied this thiol oxidant in whole rat embryo culture and found abnormal rotation and growth decrease at 75 and 100 uM. Inhibition of glutathione disulfide reductase with 1,2–bid (2–chloroethyl)–1–nitroxourea potentiated the action of diamide.

Hiranruengchok, R. and Harris, C.: Glutathione oxidation and embryotoxicity elicited by diamide in the developing rat conceptus in vitro. Toxicology and Applied Pharmacology 120:52–71, 1993.

734 4,4–Diaminodiphenylenesulfate

Discussed under Hair Dyes

735 1,7–Diamino–8–naphthol–3,6–disulphonic Acid

Christie (1965) injected pregnant rats subcutaneously with 5 mg on day 8.5. Maternal mortality was high and resorptions were increased. No teratogenic findings were mentioned.

Christie, G.A.: Teratogenic effects of synthetic compounds related to trypan blue: The effect of 1,7–diamino–8–naphthol–3,6–disulphonic acid on pregnancy in the rat. Nature 208:1219–1220, 1965.

736 2–6–Diaminopurine CAS 1904–98–9

Chaube and Murphy (1968) report that this purine analogue has a fetal LD–100 of 200 mg per kg of maternal weight. No congenital defects were found. Thiersch (1957), using 50 mg per kg, found a high incidence of resorption when rats were treated on the 4th and 5th days and a few stunted fetuses when the same dose was injected on the 7th and 8th days.

Chaube, S. and Murphy, M.L.: The teratogenic effects of the recent drugs active in cancer chemotherapy. In: Advances in Teratology, D.H.M. Woollam (ed.), New York: Logos and Academic Press, 3:181–237, 1968.

Thiersch, J.B.: Effect of 2–6 diaminopurine (2–6 DPP), 6–chlorpurine (CLP) and thioguanine (THG) on rat litter in utero. Proc. Soc. Exp. Biol. Med. 94:40–43, 1957.

737 2,5–Diaminotoluene *Hair Dyes*

This hair–dye constituent was given in single dose subcutaneously to mice on days 7 through 14. Using 50 mg per kg on day 8, eighteen percent of the fetuses were malformed with vertebral and rib abnormalities most common but exencephaly and prosoposchisis was also seen (Inouye and Murakami, 1977). Reproductive studies in rats gave negative studies (Burnett et al., 1976; Wernick et al., 1975).

Burnett, C.; Goldenthal, E.I.; Harris, S.B.; Wazeter, F.X.; Strausburg, J.; Kopp, R. and Voelker, P.: Teratology and percutaneous toxicity studies on hair dyes. J. Toxicol. Environ. Health 1:1027, 1976.

Inouye, M. and Murakami, U.: Teratogenicity of 2,5–diaminotoluene, a hair–dye constituent in mice. Food Cosmet. Toxicol. 15:447–451, 1977.

Wernick, T.; Lanman, B.M. and Fraux, J.L.: Chronic toxicity, teratologic and reproductive studies with hair dyes. Toxicol. Appl. Pharmacol. 32:450 only, 1975.

738 cis–Diammine(glycolato)platinum *254–S*

Hara et al. (**1990**) administered this antineoplastic drug intravenously to rats on days 7–18 in doses of up to 540 microgm per kg. At the highest dose the weight of the dams was reduced. Except for fetal weight reduction at the highest dose no adverse teratogenic or postnatal behavioral changes were noted. Hasegawa et al. (**1990**) treated rabbits with up to 500 microgm per kg on days 6–18. At the highest dose fetal survival and weight was decreased but no teratogenicity was reported. Komai et al. (**1994**) gave up to 540 microgm per day intravenously before and during the early days of pregnancy in the rat. Increased resorptions were found at the highest dose. Decreased weight of the fetuses occurred when the same dose was given perinatally but behavior and reproduction of the offspring was not altered (Komai et al.,**1991** B).

Hara, K.; Muranaka, R.; Andou, M.; Itou, M.; Takegawa, Y.; Uchida, H.; Hasegawa, Y.; Yoshizaki, T. and Muraoka, Y.: A teratological study of a new platinum complex, cis–diammine(glycolato)platinum (254–s), by intravenous administration in rats. Iyakuhin Kenkyu 21:1233–1269, 1990.

Hasegawa, Y.; Kanamori, S.; Fukiishi, Y.; Hirashiba, M.; Yoshizaki, T. and Muraoka, Y.: A teratological study of a new platinum complex, cis–diammine(glycolato)platinum (254–s), by intravenous administration in rabbits. Iyakuhin Kenkyu 21:1215–1232, 1990.

Komai, Y.; Iriyama, K..; Kumada, J.; Hibino, H.; Itou, I.; Isowa, K. and Kobayashi, F.: A fertility study of a new platinum complex, cis–diammine(glycolato)platinum (354–S), by intravenous administration in rats. Iyakuhin Kenkyu 22:714–731, 1991A.

Komai, Y.; Nishiwaki, K.; Kumada, J.; Hibino, H.; Furutaki, K.; Isowa, K. and Kobayashi, F.: A perinatal and postnatal study of a new platinum complex, cis–

diammine(glycolato)platinum (254–S), by intravenous administration in rats. Iyakuhin Kenkyu 22:732–759, 1991.

739 Diathermy

Discussed under Microwave Radiations

740 Diazepam *Benzodiazepine Valium*™ CAS 439–14–5

Although initially there was some evidence that facial clefts might be increased in the offspring of exposed mothers, more detailed studies summarized below failed to confirm this.McElhatton (**1994**) has summarized the literature on benzodiazopine effects on pregnancy and lactation.

Aarskog (1975) questioned 130 mothers of children with oral clefts and found that seven of 111 had first trimester exposure. Only four out of 362 women controls questioned by another method had exposures. Czeizel (1988) found that 15 percent of mothers with children having facial clefts had taken the drug and this rate was the same for the mothers of infants with neural tube closure defects.

Saxen (1975) and Saxen and Saxen (1975) reported a significant increase in incidence of cleft palate or cleft lip and palate among mothers ingesting antineurotics (mostly diazepam) during the first trimester. She viewed these observations as mostly prospective because prescriptions were required for the use of these drugs. Safra and Oakley (1975) also found a four–fold increase in the ingestion of this drug among mothers giving birth to children with cleft lip or cleft lip and palate. Safra and Oakley (1976) pointed out that a four–fold increase in oral clefts if confirmed would imply only a 0.4 percent risk for cleft lip with or without cleft palate and a 0.2 percent risk for having a child with cleft palate. Crombie et al. (1975) in a study of 64 women who took this drug in pregnancy found no offspring with defects. Case reports of children with eye defects after benzodazepine exposures have included Moebius syndrome with nystagmus, tortuosity of retinal arteries, hypoplasia of optic cup (one with Warfarin exposure) and coloboma (Stromland, 1988).

Rosenberg et al. (1983) compared diazepam exposure during the first four lunar months among 611 pregnancies associated with oral clefts with 2,498 pregnancies associated with other types of serious defects. No increase was found. The relative risks were 1.0 for cleft lip with or without cleft palate and 0.8 for cleft palate alone. Shiono and Mills (1984) found no significant increase in oral clefts among 854 women who were prospectively determined to have taken diazepam in the first trimester. Braken and Holford (1981), in a retrospective study of diazepam use, found an odds ratio of 2.8 (p < .0001) for malformed offspring. The type of defects which were increased included polydachtyly, heart and circulatory defects and hemangiomas but did not include oral clefts.

Erkkola et al. (1974) showed that this medication is transferred easily to the early human fetus. Cerqueira et al. (1988) reported that among five pregnant women poisoned with 10 gm of benzodiazepine in mid or late pregnancy, no permanent adverse effects occurred in the offspring as observed at six months of age. Cardiac rhythm changes in the fetal heart were observed in one case during the acute phase.

Czeizel (1988) studied 11,073 pregnancies and found that 14.9 percent took benzodiazepines (diazepam, chlordiazepoxide or nitrazepam). Among 630 cases with cleft lip with or without cleft palate and 179 with isolated cleft palate and 392 with multiple congenital defects, there was no increase in use during pregnancy. Laegreid et al. (1987) described what they claim is a specific benzodiazepine syndrome in seven infants. Evidence of maternal diazepam ingestion was reported in five, and two mothers had taken oxazepam. The clinical findings included Moebius syndrome, Dandy Walker malformation with lissencephaly, and polycystic kidneys, (2) submucous clefts of the hard palate and (7) varying degrees of mental retardation. These authors did not cite a single previous study of exposed newborns including the complete reports given above. In response to this report, a number of investigators have criticized the accuracy of the diagnoses as well as the validity of the conclusions (Winter, 1987; Czeizel and Lendvay, 1987; Gerhardsson and Alfredsson, 1987).

The work of Laegreid et al. was reported in more detail (Laegreid et al., 1989). The methods for selection of cases and the denominator of the group studied were not given and this made the significance of the study difficult to interpret. The syndrome was charaterized by hypotonia at birth, delayed motor development with slanted eyes, epicantal folds and other facial changes. Bergman et al. (**1990**) in a prospective study of over 4,000 pregnancies were unable to detect any adverse effect of benzodiazopines given at somewhat lower doses. Bergman et al (**1992**) studied 80 pregnant women who had 10 or more benzodiazepine prescriptions over a 4 year period. Records were located in 64 cases and 6 had congenital defects (heart (2), undescended testes (2), talipes (2), syndactyly (1) and limb anomaly (1)). Eight had various forms of neurologic deficit. The authors felt the increased rates might be due to multiple alcohol exposure.

Beall (1972) reported no teratogenicity in the rat exposed to as much as 200 mg per kg given orally from day 6 through 15. Diazepam and N–demethyl–diazepam produced cleft palate in mice treated with 40 mg per kg by mouth on day 14 (Miller and Becker, 1973). Spielmann et al. (1986) injected mice with 100 mg per kg on day 2 and found no ill effects on day 17. Miller and Becker (1975) treated pregnant mice with diazepam orally on days 11, 12 and 13 and at the dosage level of 140 mg per kg found an increase in cleft palates in the fetuses. Kellog et al. (1980) reported that rats exposed prenatally on days 13–20 to 2.5–10 mg per kg of maternal weight exhibited delayed locomotion and acoustic startle reflexes that normally appear in the third postnatal week.

Weber (1985) reviewed animal and human studies on the teratogenicity of benzodiazepines. Livezey et al. (1986) reported that treatment of the rat on days 5–20 was associated with poor focus of attention and decreased benzodiazepine receptors postnatally. Livezey et al. (1986) treated rats subcutaneously with 6 mg per kg during the last 5 days of gestation and found 13 postnatal malignancies in 52 exposed rats and none in the 44 controls. Plasma immune globulin G was also increased in the exposed offspring. Cagiano et al. (1990) administered 0.1 or 1.0 mg per kg to rat dams on days 14–20. Subtle but statistically significance changes in locomotion activity and ejaculating behavior were reported.

Aarskog, D.: Association between maternal intake of diazepam and oral clefts. Lancet 2:29 only, 1975.

Beall, J.R.: Study of the teratogenic potential of diazepam and SCH 12041. Can. Med. Assoc. J. 106:1061 only, 1972.

Bergman, U.; Boethius, G.; Swartling, P.G. and Isacson. D. and Smedby, B.: Teratogenic effects of benzodiazepine use during pregnancy. The Journal of Pediatrics 116:490–491, 1990.

Bergman, U.; Rosa, F.W.; Baum, C.; Wiholm, B. and Faich, G.A.: Effects of exposure to benzodiazepine during fetal life. The Lancet 240:694–696, 1992.

Bracken, M.B. and Holford, T.R.: Exposure of prescribed drugs in pregnancy and association with congenital malformations. Obstet. Gynecol. 58:336-344, 1981.

Cagiano, R.; DeSalvia, M.A.; Persichella, M.; Renna, G.; Tattoli, M. and Cuomo, V.: Behavioural changes in the offspring of rats exposed to diazepam during gestation. European Journal of Pharmacology 177:67–74, 1990.

Cerqueira, M.J.; Olle, C.; Bellart, J.; Baro, F.; Cabero, I.; Queralto, J.M. and Espinosa, J.R.: Intoxication by benzodiazepines during pregnancy. Lancet 1:1341, 1988.

Crombie, D.L.; Pinsent, R.J.; Fleming, D.M.; Rumeau–Rouquette, C.; Goujard, J. and Huel, G.: Fetal effects of tranquilizers in pregnancy. N. Eng. J. Med. 293:198–199, 1975.

Czeizel, A.: Diazepam, phenytoin and etiology of cleft lip and=or cleft palate. Lancet 1:810 only, 1976.

Czeizel, A.: Lack of evidence of teratogenicity of benzodiazepine drugs in Hungary. Reproductive Toxicology 1(3):183–188, 1988.

Czeizel, A. and Lendvay, A.: In–utero exposure to benzodiazepines, Letter to the Editor, Lancet 1:627, 1987.

Erkkola, R.; Kanto, J. and Sellman, R.: Diazepam in early human pregnancy. Acta Obstet. Gynecol. Scand. 53:135–138, 1974.

Gerhardsson, M. and Alfredsson, L.: In–utero exposure to benzodiazepines, Letter to the Editor, Lancet 1:628, 1987.

Kellogg, C.; Tervo, D.; Ison, J.; Parisi, T. and Miller, R.K.: Prenatal exposure to diazepam alters behavioral development in rats. Science 207:205–207, 1980.

Laegreid, L.; Olegard, R.; Walstrom, J. and Conradi, N.: Teratogenic effects of benzodiazepine use during pregnancy. J. Pediatr. 114:126–131, 1989.

Laegreid, L.; Walstrom, J. and Conradi, N.: Abnormalities in children exposed to benzodiazepines in utero. Lancet 1:108–109, 1987.

Livezey, G.T.; Marczynski, T.J. and Isaac, L.: Receptors in mature rat progeny. Neurobehav. Toxicol. Teratol. 8:425–432, 1986.

Livezey, G.T.; Marczynski, T.J.; McGrew, E.A. and Beluhan, F.Z.: Prenatal exposure to diazepam: Late postnatal teratogenic effect. Neurobehav. Toxicol. Teratol. 8:433–440, 1986.

McElhatton, P.R.: The effects of benzodiazepine use during pregnancy and lactation. Reproductive Toxicology 8:461–475, 1994.

Miller, R.P. and Becker, B.A.: The teratogenicity of diazepam metabolites in Swiss–Webster mice. (abs) Toxicol. Appl. Pharmacol. 25:453 only, 1973.

Miller, R.P. and Becker, B.A.: Teratogenicity of oral diazepam and diphenylhydantoin in mice. Toxicol. Appl. Pharmacol. 32:53–61, 1975.

Rosenberg, L.; Mitchell, A.A.; Parsells, J.L.; Pashayan, H.; Louik, C. and Shapiro, S.: Lack of correlation of oral clefts to diazepam use during pregnancy. N. Eng. J. Med. 309:1282–1285, 1983.

Safra, M.J. and Oakley, G.P.: An association of cleft lip with or without cleft palate and prenatal exposure to valium. Lancet 2:478–479, 1975.

Safra, M.J. and Oakley, G.P.: Valium: An oral cleft teratogen? Cleft Palate Journal 13:198–200, 1976.

Saxen, I.: Associations between oral clefts and drugs taken during pregnancy. Int. J. Epidemiol. 4:37–44, 1975.

Saxen, I. and Saxen, L.: Association between maternal intake of diazepam and oral clefts. Lancet 2:498 only, 1975.

Shiono, P.H. and Mills, J.L.: Oral clefts and diazepam use during pregnancy. N. Eng. J. Med. 311:919–920, 1984.

Spielmann, H.; Kruger, C.; Tenschert, B. and Vogel, R.: Studies on the embryotoxic risk of drug treatment during the preimplantation period in the mouse. Arzneim Forsch. 36:219–223, 1986.

Stromland, K.: Ocular malformations in children exposed to drugs during gestation. Clinical Pediatrics 27:257, 1988.

Weber, L.W.D.: Benzodiazepines in pregnancy-academical debate or teratogenic risk? Biol. Res. Pregnancy 16:151–167, 1985.

Winter, R.M.: In–utero exposure to benzodiazepines. Letter to the Editor, Lancet 1:627, 1987..

741 Diazinon Dimpylate O, O–DiethylO–[2–isopropyl–6–methyl–4–pyrimidinyl] phosphorothioate CAS 333–41–5

This organo phosphorus pesticide was given to rats in doses of 100 to 200 mg per kg intraperitoneally on the 11th day of gestation (Kimbrough and Gaines, 1968). At 200 mg per kg some maternal mortality occurred along with limb reduction and hydrocephalus in surviving fetuses. At 150 mg per kg only fetal weight reduction was found. Robens (1969) studied rabbits and hamsters at doses up to 30 and 0.25 mg per kg, respectively, and reported no malformations.

Kimbrough, R.D. and Gaines, T.B.: Effect of organic phosphorus compounds and alkalating agents on the rat fetus. Arch. Environ. Health 16:805–808, 1968.

Robens, J.F.: Teratologic studies of carbaryl, diazinon, Norea, disulfiram and thiram in small laboratory animals. Toxicol. Appl. Pharmacol. 15:152–163, 1969.

742 O–Diazoacetyl–l–serine

Discussed under Azaserine

743 Diazocine *2,8–Dichloro–6,12–diphenyl–dibenzo (b,f) diazocine*

Duncan et al. (1965) studied this non–steroid estrogenic drug in rats and found that as little as 0.5 mg subcutaneously or orally for 7 days at the time of mating inhibited implantation. Treatment with oral doses of 1.0 mg on days 8 through 15 led to fetal death of many of the fetuses. No effect was seen when treatment was given on days 15–20.

Duncan, G.W.; Lyster, S.C. and Wright, J.B.: Reproductive mechanisms influenced by a diazocine. Proc. Soc. Exp. Biol. Med. 120:725–728, 1965.

744 6–Diazo–5–oxo–l–norleucine *DON* CAS 157–03–9

Chaube and Murphy (1968) reviewed the teratogenic information on this tumor inhibiting glutamine analogue. Murphy (1960) produced lip or palate defects in the rat by using a lethal dose (0.5 mg per kg) of DON modified by addition of adenine or guanine (100 mg per kg). Friedman (1975) produced cleft palate in the dog fetus by giving DON in amounts of 0.15 mg per kg on days 20, 21 and 22. Greene and Kochhar (1975) gave mice intramuscularly 0.5 mg per kg on day 10 or 11 of gestation and produced median cleft lip and skeletal defects in the offspring.

Chaube, S. and Murphy, M.L.: The teratogenic effects of the recent drugs active in cancer chemotherapy. In: Advances in Teratology, D.H.M. Woollam (ed.), New York: Logos and Academic Press, 3:181–237, 1968.

Friedman, M.H.: The effect of O–diazoacetyl–l–serine (azaserine) on the pregnancy of the dog: A preliminary report. J. Am. Vet. Med. Assoc. 130:159–162, 1975.

Greene, R.M. and Kochhar, D.M.: Limb development in mouse embryos: Protection against teratogenic effects of 6–diazo–5–o]o–norleucine (DON) in vivo and in vitro. J. Embryol. Exp. Morph. 33:355–370, 1975.

Murphy, M.L.: Teratogenic effects of tumor inhibiting chemicals in the rat. In: A Ciba Foundation Symposium on Congenital Malformations, G.E.W. Wolstenholme and C.M. O'Connor (eds.), London: J. and A. Churchill Ltd., 92–95, 1960.

745 5–Diazouracil CAS 2435–76–9

Skalko et al. (1973) reported that 5–20 mg per kg on day 10 was embryolethal but not teratogenic in the mouse embryo.

Skalko, R.G.; Caniano, D.A. and Packard, D.S.: The teratogenic interaction of 5–diazouracil and 5–iododeoxyuridine in the mouse embryo. Toxicol. Appl. Pharmacol. 25:453–454, 1973.

746 Diazoxide

Perinatal complications (maternal hypotension and hyperglycemia) have been reported with the use of this drug (Neumann et al., 1979). Alopecia or hypertrichosis has also been found in the offspring (Milner and Chonksey, 1972).

Milner, R.D.G. and Chonksey, S.K.: Effects of fetal exposure to diazoxide in man. Arch. Dis. Childhood 47:537–543, 1972.

Neumann, J.; Weiss, B.; Rabello, Y.; Cabal, L. and Freeman, R.K.: Diazoxide for the acute control of severe hypertension complicating pregnancy: A pilot study. Obstet. Gynecol. 53:52S–55S, 1979.

747 Dibekacin *3,4–Dideoxykanomycin B* CAS 34493–98–6

Koeda and Moriguchi (1973) gave rats and mice doses of 400 and 300 mg per kg, respectively, and produced no gross or skeletal defects. The doses were given daily during organogenesis by intraperitoneal or intramuscular route.

Koeda, T. and Moriguchi, M.: Teratological studies of 3,4–dideoxykanomycin B (DKB) in rats and mice. Jpn. J. Antibiot. 26:40–48, 1973.

748 2–[5H–Dibenzo[a,d]cyclohepten–5–one]–2–Propionic Acid

This anti–inflammatory was administered orally to rabbits in doses of up to 8.0 mg per kg during organogenesis. Acrania, gastroschsis and limb defects were increased at the highest doses. (Stevens et al. **1981**)

Stevens, T.L.; Thacker, G.T. and Parker, J.: Fetal anomalies associated with administration of 2–[5H–dibenzo[a,d] cyclohepten–5–one] 2–proprionic acid. Teratology 23:64A, 1981.

749 Dibenzofurans *1,2,3,7,8–Pentachlorodibenzofuran 2,3,4,7,8–Pentachlorodibenzofuran 1,2,3,4,7,8–Hexachlorobenzofuran*

Birnbaum et al. (1987A) gave these three dibenzofurans to mice on days 10–13. Hydronephrosis and cleft palate occurred 30 to 100 microgm per kg daily, doses which were not toxic to the dams. They also studied the combination of the three chemicals (Birnbaum et al., 1987B). Couture et al. (**1989**) gavaged rats on days 8, 10, or 12 with up to 200 microgm per kg. Maternal toxicity was observed and at the highest dose an increase in cleft palate occurred.

Birnbaum, L.S.; Harris, M.W.; Barnhart, E.R. and Morrissey, R.E.: Teratogenicity of three polychlorinated dibenzofurans in C57BL6N mice. Toxicol. Appl. Pharmacol. 90:206–216, 1987A.

Birnbaum, L.S.; Harris, M.W.; Barnhart, E.R. and Morrissey, R.E.: Teratogenic effects of polychlorinated dibenzofurans in combination in C57BL6N mice. Toxicol. Appl. Pharmacol. 91:246–255, 1987B.

Couture, L.A.; Harris, M.W. and Birnbaum, L.S.: Developmental toxicity of 2,3,4,7,8–pentachlorodibenzofuran in the Fischer 344 rat. Fundamental and Applied Toxicology 12:358–366, 1989.

750 Dibenz[b,f][1,4]oxazepine

Upshall (1974) exposed rabbits during organogenesis to 5–7 minutes of aerosol containing up to 200 mg per meter squared. Intragastric dosing with up to 100 mg per kg on alternate days of organogenesis was also performed. No adverse fetal effects were noted. Rabbits receiving 0.33 of the maternal LD–50 intravenously had increased fetal mortality.

Upshall, D.G.: The effects of dibenz[b,f][1,4]oxazepine (CR) upon rat and rabbit embryonic development. Toxicol. Appl. Pharmacol. 29:301–311, 1974.

751 Dibenzthiazyl Disulphide

Ema et al. (**1989**) fed rats up to 596 mg per kg on days 0–20 of gestation. No fetal ill effects were found and postnatal growth and survival were normal.

Ema, M.; Sakamot, J.; Murai, T. and Kawasaki, H.: Evaluation of the teratogenic potential of the rubber accelerator dibenzthiazyl disulphide in rats. Journal of Applied Toxicology 9:413–417, 1989.

752 Dibenzyltoluene

Kurosaki et al. (**1988**) gavaged rats on days 7–17 with up to 100 mg per kg. At the highest dose which was toxic to the mother some reduction in fetal weight occurred but no teratogenicity.

Kurosaki, T.; Kawashima, K.; Nakaura, S.; Tanaka, S.; Djajalaksana, S. and Takanaka, A.: Effects of dibenzyltoluene on fetal developments of rats. Eisei Shiensho Hokoku 106:61–66, 1988.

753 1,2–Dibromo–3–chloropropane *DBCP* CAS 96–12–8

Goldsmith et al. (1984) followed reproduction in 30 exposed workers for 5 years. There were no miscarriages, fetal deaths or malformations in 12 pregnancies. Ten females and only two males were born, a circumstance with a probability equal to 0.0l5. Whorton and Foliart (1988) summarized the studies of male infertility in exposed workers. The herbicide was banned in 1985. Potashnik and Phillip (1988) studied 34 offspring from fathers who had azo or oligospermia and found no increase in defects as compared to fifty-four siblings born before exposure. One child with extrophy of the bladder and two with minor anomolies occurred in the exposure period group. Ruddick and Newsome (1979) gave up to 50 mg per kg daily to rats by gavage on days 6–15 of gestation. Some maternal toxicity, fetal weight decrease and decreased fetal viability was found but no increase in malformations. Warren et al. (1988) treated pregnant rats with 25 mg per kg on various days after day 16 and found that the testicle size was significantly reduced in adult life. Sexual behavior was abnormal and in some cases, seminiferous tubules were absent.

Goldsmith, J.R.; Potashnik, G. and Israeli, R.: Reproductive outcomes in families of DBCP–exposed men. Arch. Environ. Health 39:85–89, 1984.

Potashnik, G. and Phillip. M.: Lack of birth defects among offspring conceived during or after paternal exposure to dibromochloropropane (DBCP). Andrologia 20(1):90–94, 1988.

Ruddick, J.A. and Newsome, W.H.: A teratogenicity and tissue distribution study on dibromochloropropane in the rat. Bull. Environ. Contam. Toxicol. 21:483–487, 1979.

Warren, D.W.; Ahmad, N. and Rudeen, P.K.: The effects of fetal exposure to 1,2–dibromo–3–chloropropane on adult male reproductive function. Biol. Reprod. 39:707–716, 1988.

Whorton, D. and Foliart, D.: DBCP: Eleven years later. Reprod. Toxicol. 2:155–161, 1988.

754 Dibromomannitol *DBM*

Hosomi et al. (1972) gave by gavage up to 150 mg per kg during organogenesis to mice and rats. Skeletal defects and cleft palates were increased in the mice treated on single days. In the rat, doses of over 100 mg per kg produced meningo and encephaloceles. Exencephaly and cleft palate also occurred.

Hosomi, J.; Suzuki, H.; Ishiyama, N.; Sano, Y. and Irikura, T.: Teratological studies on DBM in mice and rats. Kysoto Rynsho 6:30–42, 1972.

755 Di–n–Butyltin Dichloride

Ema et al. (**1991**) gavaged rats with up to 10 mg per kg daily on days 7–15. At doses of 5.0, 7.5 and 10 mg congenital defects were increased. At 5.0 mg per kg no maternal toxicity was found. The type of defects found were cleft jaw, ankyloglossia, and other skeletal defects. Ema et al. (**1992**) determined that day 8 was the most sensitive period for teratogenicity.

Ema, M.; Itami, T. and Kawasaki, H.: Teratogenicity of di–n–butyltin dichloride in rats. Toxicology Letters 58:347–356, 1991.

Ema, M.; Itami, T. and Kawasaki, H.: Susceptible period for the teratogenicity of di–n–butyltin dichloride in rats. Toxicology 73:81–92, 1992.

756 Di–n–butyl Phthalate

Discussed under Phthalate Esters

757 Dichloralphenazone CAS 480–30–8

McColl et al. (1965) reported no teratogenic effect of 500 mg per kg fed daily during the entire pregnancy of the rat.

McColl, J.D.; Globus, M. and Robinson, S.: Effect of some therapeutic agents on the developing rat fetus. Toxicol. Appl. Pharmacol. 7:409–417, 1965.

758 Dichloroacetic Acid

Smith et al. (**1992**) gavaged rats on days 6–15 with up to 400 mg per kg. At 14 mg per kg adverse maternal health occurred. At 400 mg per kg (and some higher levels) cardiovascular defects were found. Epstein et al.(**1992**) detailed the cardiac defects found.

Epstein, D.L.; Nolen, G.A.; Randall, J.L.; Christ, S.A.; Read, E.J.; Stober, J.A. and Smith, M.K.: Cardiopathic effects of dichloroacetate in the fetal long–evans rat. Teratology 46:225–235, 1992.

Smith, M.K.; Randall,J.L.; Read, E.J. and Stober, J.A.: Development toxicity of dichloroacetate in the rat. Teratology 46:217–223, 1992.

759 Dichloroacetonitrile *DCan*

Smith et al. (1989) gavaged rats on days 6–18 with 5 to 45 mg per kg. The highest dose was lethal to 9% of the dams.

At the highest dose heart and renal defects were increased. The no–observed–adverse effect was 15 mg per kg.

Smith, M.K.; Randall, J.R.; Stober, J.A. and Read, E.J.: Developmental toxicity of dichloroacetonitrile: a by—product of drinking water disinfection. Fundamental and Applied Toxicology 12:765–772, 1989.

760 Dichlorobenzenes

Hayes et al. (1985) studied ortho and para dichlorobenzene in pregnant rats and rabbits exposing for 6 hours a day during organogenesis to maximum concentrations of 400 and 800 ppm, respectively. Maternal toxicity occurred at the higher concentrations but neither fetotoxicity nor teratogenicity was found in either species. Giavini et al. (1986) gavaged rats with up to 1000 mg per kg per day on days 6–15. Using para–dichlorobenzene, no teratogenic effect was seen although at the maternally toxic levels some delay in ossification was seen in the fetuses.

Giavini, E.; Broccia, M.L.; Prati, M. and Vismara, C.: Teratologic evaluation of p–dichlorobenzene in the rat. Bull. Environ. Contam. Toxicol. 37:164–168, 1986.

Hayes, W.C.; Hanley, T.R., Jr.; Gushow, T.S.; Johnson, K.A. and John, J.A.: Teratogenic potential of inhaled dichlorobenzenes in rats and rabbits. Fund. Appl. Toxicol. 5:190–202, 1985.

761 1,4–Dichlorobutene–2 *1,4–DCB* CAS 764–41–0

This chemical, an intermediate in the synthesis of β–chloroprene, was given by inhalation to rats on days 6–15 in concentrations of 0.5 and 5.0 ppm for six hours per day. No embryotoxic or teratogenic effects were found (Kennedy et al., 1982).

Kennedy, G.L.; Culik, R. and Trochimowicz, H.J.: Teratologic evaluation of 1,4–dichlorobutene–2 in the rat following inhalation exposure. Toxicol. Appl. Pharmacol. 64:125–130, 1982.

762 Dichlorodiphenyl Dichloroethane *DDE*

Rogan et al. (1986) measured levels in human newborn blood, placenta and maternal milk and did not find an association with birth weight or head circumference. In a follow-up of 912 infants, hyporeflexia was found.

Rogan, W.J.; Gladen, B.C.; McKinney, J.D.; Carreras, N.; Hardy, P.; Thullen, J.; Tinglestad, J. and Tully, M.: Neonatal effects of transplacental exposure to PCB's and DDE. J. Pediatr. 109:335–341, 1986.

763 1,2–Dichloroethane *Ethylene Chloride Ethylene Dichloride* CAS 107–06–2

Alumot et al. (1976) administered 250 or 500 ppm in feed mash to rats for a two–year period. Approximately 60–70 percent of the dose was consumed. No significant decrease in fertility, litter size or fetal weight was observed. Rao et al. (1980) exposed rats to vapor at 100 and 300 ppm for seven hours daily during days 6–15 of gestation. Ten of the 16 rats at 300 ppm died and only one rat had an implanted pregnancy with total resorption. With 100 ppm,

there was neither increased resorption nor decrease in fetal weight. Lane et al. (1982) administered up to 1,000 mg per kg in drinking water and found no adverse reproductive effects. Vozovaya (1976) and Vozovaya and Malyarova (1975) reported preimplantation reproductive failure and accumulation of the material in fetal rat liver.

Alumot, E.; Nachtomi, E.; Mandel, E. and Holstein, P.: Tolerance and acceptable daily intake of chlorinated fumigants in the rat diet. Food Cosmet. Toxicol. 14:105–110, 1976.

Lane, R.W.; Riddle, B.L. and Borzelleca, J.F.: Effect of dichloroethane 1,1,1–trichloroethane in drinking water on reproduction and development in mice. Toxicol. Appl. Pharmacol. 63:409–421, 1982.

Rao, K.S.; Murray, J.S.; Deacon, M.M.; John, J.A.; Calhoun, L.L. and Young, J.T.: Teratogenicity and reproduction studies in animals inhaling ethylene dichloride. In: Ethylene Dichloride: A Potential Health Risk, Banbury Report 5, B. Ames, P. Infante and R. Reitz (eds.), Cold Spring Harbor Laboratory 149–161, 1980.

Vozovaya, M.A.: The effect of small concentrations of benzene and dichloroethane separately and combined on the reproductive function of animals. Gig. Sanit. 6:l00, 1976.

Vozovaya, M.A. and Malyarova, L.K.: Mechanism of action of ethylene dichloride on the fetus of experimental animals. Gig. Sanit. 6:94, 1975.

764 p–Di–2–Chloroethylamine Phenyl Butyric Acid

Didcock et al. (1956) gave mice 10 to 40 mg per kg by intraperitoneal and intravenous routes on days 12–13 and caused increased lethality of the fetuses. Cleft palates, enlarged spleen, astomia and hypoplasia of the hind limbs was found.

Didcock, K.; Jackson, D. and Robson, J.M.: The action of some nucleotoxic substances on pregnancy. Br. J. Pharmacol. 11:437–441, 1956.

765 Dichloroethylene

Dawson et al. (**1990**) continuously perfused the rat uterine cavity from day 7 to term with 1.5 and 1,500 ppm of dichloroethylene in saline. The control treated fetuses had 1.5% heart defects while at the treatment levels 17 and 29% had heart defects. Defects of the valves and septa were noted and all examinations were carried out on coded specimens. Goldberg et al. (**1992**) inoculated the chemical above the chick embryo and found increases in congenital heart defects when 30 microliters of 5 or 20 micromolar solution was used. At 25 micromolar concentration no heart defects were found. Non-cardiac defects were not increased.

Dawson, B.V.; Johnson, P.D.; Goldberg, S.J. and Ulreich, J.B.: Cardiac teratogenesis of trichloroethylene and dichloroethylene in a mammalian model. J. Am. Coll. Cardiol. 16:1304–1309, 1990.

Goldberg, S.J.; Dawson, B.V.; Johnson, P.D.; Hoyme, H.E. and Ulreich, J.B.: Cardiac teratogenicity of dichloroethylene in a chick model. Pediatric Research 32:23–26, 1992.

766 **1,1–Dichloroethylene** *Vinylidene chloride* CAS 75–35–4

Murray et al. (1979) exposed rats and rabbits by inhalation to concentrations of up to 160 ppm for a seven–hour period during days 6 through 18 and 6 through 15 of gestation. Maternal toxicity was demonstrated in both species. No increased general malformation rate was found but wavy ribs and delayed ossification were increased in rat fetuses exposed to 80 and 160 ppm. Rats were fed 200 ppm in their drinking water on days 6 through 15 of gestation and no differences from the fetal controls were found.

Alumot et al. (1976) fed rats up to 500 ppm over a two year period and found no alteration in fetal mortality or weight. Approximately 60–70 percent of the substance was actually consumed. Anderson et al. (1977) exposed male mice to 50 ppm, six hours daily for five days and found no evidence of dominant lethality. Short et al. (1977) had similar findings in the rat.

Alumot, E.; Nachtomi, E.; Mandel, E. and Holstein, P.: Tolerance and acceptable daily intake of chlorinated fumigants in the rat diet. Fd. Cosmet. Toxicol. 14:105–110, 1976.

Anderson, D.; Hodge, M.C.E. and Purchase, I.F.H.: Dominant lethal studies with halogenated olefins vinyl chloride and vinylidene dichloride in male CD–1 mice. Environ. Health Perspect. 21:71–78, 1977.

Murray, F.J.; Nitschke, K.D.; Rampy, L.W. and Schwetz, B.A.: Embryotoxicity and fetotoxicity of inhaled or ingested vinylidene chloride in rats and rabbits. Toxicol. Appl. Pharmacol. 49:189–202, 1979.

Short, R.D.; Minor, J.L.; Winston, J.M. and Lee, C.C.: A dominant lethal study in male rats after repeated exposure to vinyl chloride or vinylidene chloride. J. Toxicol. Environ. Health 3:965–968, 1977.

767 **Dichloroisocyanurate Sodium** *Sodium 3,5-chloro–1,3,5–trizime 2,4,6-trione*

Tani et al. (1981) gave mice up to 400 mg per kg on days 6–15 of gestation. No adverse fetal effects were found.

Tani, I.; Shibata, H.; Ninomiya, M.; Taniguchi, J. and Fuyita, I.: Effect of SD (SDIC) on the embryonic development and newborns given orally in the period of organogenesis in mice. Yakubutsu Ryoho Medical Treatment 13:353–363, 1981.

768 **Dichloromethotrexate** CAS 528–74–5

Discussed under Methotrexate

769 **2,4–Dichlorophenoxyacetic Acid** *2,4–D* CAS 94–75–7

Mutagenicity studies and cytogenetic effects in cultured human lymphocytes and in somatic cells from humans were positive in some experiments and negative in others. Friedman (1984) reviewed this large volume of data.

To date, studies of workers and Vietnam war veterans have not demonstrated any convincing human teratogenicity. These studies are discussed under Agent Orange. Casey and Collie (1984) reported that the offspring of a

mother who sprayed 2,4–D as part of her work was mentally retarded and had facial dysmorphology including a small nasal bridge and frontal bossing.

Schwetz et al. (1971) gave rats up to 87.5 mg per kg on days 6 through 15 of gestation. With doses above 25 mg per kg subcutaneous edema, wavy ribs, delayed ossification and lumbar ribs were observed. Similar findings were reported following administration of the butyl ether and isooctyl esters of the compound. Courtney (1977) studied the effect on mice of several forms of 2,4–D and reported growth retardation and cleft palate.

Konstantinova et al. (1978) studied the action of this agent in rats at the doses of 0.5–1.0 mg per kg and found that this herbicide is highly embryotoxic. Aleksashina et al. (1979) injected rats with one–twentieth LD–50 on the 4,5,6,9,10,11,12,13,14 and 15th day of pregnancy and noted strong embryotoxic effects. Khera and Mckinley (1972) studied the effects on the rat of the following derivatives: Butyl, isooctyl, butoxyethanol and dimethylamine. Minor skeletal defects were observed in dosages of 150 mg per kg on single days during organogenesis.

Aleksashina, Z.A.; Buslovich, S.Y. and Kolosovskaya, V.M.: Embryotoxical activity of herbicide derivatives 2,4–dichlorophenoxyacetic acid. Gigiena i Sanitaria (USSR) 4:70–71, 1979.

Casey, P.H. and Collie, W.R.: Severe mental retardation and multiple anomalies of uncertain cause after extreme parental exposure to 2,4–D. J. Pediatr. 104:313–315, 1984.

Courtney, K.D.: Prenatal effects of herbicides. Arch. Environ. Contam. Toxicol. 6:33–46, 1977.

Friedman, J.M.: Does agent orange cause birth defects? Teratology 29: 193–221, 1984.

Khera, K.S. and McKinley, W.P.: Pre–and postnatal studies on 2,4,5–trichlorophenoxyacetic acid, 2,4–dichlorophenoxyacetic acid and their derivatives in rats. Toxicol. Appl. Pharmacol. 22:14–28, 1972.

Konstantinova, T.K.; Efimenko, L.P.; Antonenko, T.A.; Nechkina, M.A. and Shilov, V.N.: Data for hygienic standardization of herbicides 2,4–D in the environmental objects. Gigiena i Sanitariya (USSR) 11:9–13, 1978.

Schwetz, B.A.; Sparschu, G.L. and Gehring, P.J.: The effect of 2,4–dichlorophenoxyacetic acid (2,4–D) and esters of 2,4–D on rat embryonal, foetal and neonatal growth and development. Food Cosmet. Toxicol. 9:801–817, 1971.

770 **2,2–Dichlorphenoxy–γ–butyric Acid** *2,4–DM*

Sokolova (1976) observed increased fetal mortality, growth retardation and behavioral abnormalities in the offspring of rats receiving herbicide in amounts of 0.1 mg per kg on days 4,5,7,9,10,11 and 14 or during the entire gestation (0.1–3 mg per kg).

Sokolova, L.A.: Studies of herbicide 2,4-D effect on gestation, embryogenesis and gonads of white rats. Gigiena i Sanitariya (USSR) 7:20–23, 1976.

771 **1,3–Dichloropropene**

Hanley et al. (1987) exposed rats and rabbits to up to 120 ppm for 6 hours daily during organogenesis. In the rat at the highest dose, a delay in ossification was the only positive change.

Hanley, T.R., Jr.; John–Greene, J.A.; Young, J.T.; Calhoun, L.L. and Rao, K.S.: Evaluation of the effects of inhalation exposure to 1,3–dichloropropene on fetal development in rats and rabbits. Fund. Appl. Toxicol. 8:562–570, 1987.

772 Dichlorphenamide CAS 120–97–8

Hallesy and Layton (1967) reported that this carbonic anhydrase inhibitor produced postaxial defects of the right upper extremity in rats. The drug was given in the diet providing 350 mg per kg per day. The defect appeared to be the same as that caused by acetazolamide, a compound of quite different chemical structure. Purichia and Erway (1972) produced fetal otolith deficits by injecting pregnant mice with a single dose (2.4–3.6 mg) between days 13 to 18 of pregnancy.

Landauer and Wakasugi (1968) found that ADP reduced the teratogenic effect of this compound in the chick.

Hallesy, D.W. and Layton, W.M.: Forelimb deformity of offspring of rats given dichlorphenamide during pregnancy. Proc. Soc. Exp. Biol. Med. 126:6–8, 1967.

Landauer, W. and Wakasugi, N.: Teratological studies with sulphonamides and their implications. J. Embryol. Exp. Morphol. 20:261–284, 1968.

Purichia, N. and Erway, L.C.: Effect of dichlorophenamide, zinc and manganese on otolith development in mice. Dev. Biol. 27:395–405, 1972.

773 Dichloroprop

Buschmann et al. (1986A) gavaged rats on days 4, 10, 13, and 18 after conception with 200 mg per kg of dichloroprop and produced no adverse fetal effects. Mecoprop administered in a similar regime at 330 mg per kg was associated with an increase in hydroureters. Postnatal function was retarded in the offspring exposed to dichloroprop (Buschmann et al., 1986B).

Buschmann, J.; Clausing, P.; Banasiak, U. and Grundel, D.: Comparative prenatal toxicity of phenoxyalcanic herbicides. (abs) Teratology 33:11A, 1986A.

Buschmann, J.; Clausing, P.; Solecki, E.; Fischer, B. and Peetz, U.: Comparative prenatal toxicity in phenoxyalcanic herbicides: Effects on postnatal development and behavior in rats. (abs) Teratology 33:11A–12A, 1986B.

774 Dichlorvos Dimethyl–2,2–dichlorovinyl PO4 CAS 61–73–7

Thorpe et al. (1972) exposed pregnant rats and rabbits to this active principle of various vapor insecticides and found no teratogenic effect even at levels toxic to the mother. The highest dose used was 6.25 microgm per liter of air during the entire gestation. Schwetz et al. (1979) reported lack of teratogenic effect in the mouse and rabbit given 60 and 5 mg per kg per day, respectively. Inhalation studies were also negative at 4 microgm per liter.

Schwetz, B.A.; Ioset, H.D.; Leong, B.K.J. and Staples, R.E.: Teratogenic potential of dichlorvos given by inhalation and gavage to mice and rabbits. Teratology 20:383–388, 1979.

Thorpe, E.; Wilson, A.B.; Dix, K.M. and Blair, D.: Teratologic studies with dichlorvos vapour in rabbits and rats. Arch. Toxikol. 30:29–38, 1972.

775 Dichromate

Discussed under Chromium

776 Diclofenac

Carp et al. (1988) studied rat implantation after exposure of ova to this prostaglandin inhibitor. Exposure in vitro for one hour to 75 microgm per ml resulted in no survivors among those transfered to the uterus but with 40 micrograms per ml, only 35 percent of the embryos survived to day 12. About two–thirds of these embryos were retarded. In the control group, 72 percent survived and only 15 percent were retarded. After injecting dams one hour before transplantation (0.3 mg per 100 gm), only 41 percent developed into embryos.

Carp, H.J.A.; Fein, A. and Nebel, L.: Effect of diclofenac on implantation and embryonic development in the rat. Eur. J. Obstet. Gynecol. Reprod. Biol. 28:273–277, 1988.

777 Dicloxacillin CAS 3116–76–5

Brown et al. (1968) reported briefly that giving 100 mg per kg to rabbits during pregnancy caused no fetal loss or congenital defects.

Brown, D.M.; Harper, K.H.; Palmer, A.K. and Tesh, S.A.: Effect of antibiotics upon the rabbit. Toxicol. Appl. Pharmacol. /2:295 only, 1968.

778 Dicofol

Lemonica et al. (1992) gave 10 mg to rats on days 4–15 and studied postnatal behavior. Open field testing results were decreased in the exposed offspring.

Lemonica, I.P.; Dos Santos, G. and Bernardi, M.M.: Effect of administration of organochloride pesticide (dicofol) during gestation of neuro–behavioral development of rats (Abst). Teratology 46:25A.

779 Dicrotophos
3–Hydroxy–N,N–dimethyl–cis–crotonamide Dimethyl Phosphate CAS 141–66–2

Roger et al. (1969) injected this organophosphate insecticide into chick eggs and amounts of 30 microgm or more per egg produced parrot beaks, micromelia, straight legs, abnormal feathering and edema. Nicotinic acid, nicotinamide and certain of their precursors alleviated this teratogenic action. Injection on the 4th day of incubation gave the maximum response.

Roger, J.C.; Upshall, D.G. and Casida, J.E.: Structure-activity and metabolism studies on organophosphate teratogens and their alleviating agents in developing hen eggs with special emphasis on bidrin. Biochem. Pharmacol. 18:373–392, 1969.

780 Dicyclomine also see *Bendectin*^{TM} CAS 77–19–0

Bendectin^{TM}, which consists of equal parts doxylamine succinate, dicyclomine HCl and pyridoxine HCl, was given by Gibson et al. (1968) in doses of 30 mg per kg per day in the rabbit and 60 mg per kg per day in the rat during organogenesis and no increase in congenital defects was found. In 1976, dicyclomine was removed from the formulation. Numerous epidemiologic studies on Bendectin have been published.

Gibson, J.P.; Staples, R.E.; Larson, E.J.; Kuhn, W.L.; Holtkamp, D.E. and Newberne, J.W.: Teratology and reproduction studies with an antinauseant. Toxicol. Appl. Pharmacol. 13:439–447, 1968.

781 Dicynone

Tuchmann–Duplessis (**1967**) treatd rats by gavage with up to 300 mg per kg during organogenesis and produced no adverse fetal effects.

Tuchmann–Duplessis, H.: Influenza di un nuovo farmaco il (Dicynone) (141 M.D.) sulla gestazione e lo sviluppo prenatale del roditore. Gaz. Med. It. 126:5–9, 1967.

782 2',3'–Dideoxycytidine *DDC*

Lindstrom et al. (**1990**) gavage mice twice a day during organogenesis. At the highest does (2000 mg per day) maternal weight loss and resorptions were observed. At 1000 and 2000 mg per day there were increased skeletal defects (small mandible and bent long bones). No statistical increase in defects occurred at the 200 and 400 mg per kg levels.

Lindstrom, P.; Harris, M.; Hoberman, A.M.; Dunnick, J.K. and Morrissey, R.E.: Developmental toxicity of orally administered 2',3'–dideoxycytidine in mice. Teratology 42:131–136, 1990.

783 Dieldrin also see *Cyclodiene Pesticides* CAS 60–57–1

This chlorinated hydrocarbon pesticide was administered orally to pregnant sows during periods in the last month of gestation. Doses up to 15 mg per kg produced no fetal changes but the compound was detectable in the fetal tissues (Uzoukwu and Sleight, 1972).

Oral doses of 4.0 mg per kg per day of this active organochlorine produced no teratogenic effects in mice (Dix et al., 1977).

Dix, K.M.; Van der Pauw, C.L. and McCarthy, W.V.: Toxicity studies with dieldrin: Teratological studies in mice dosed orally with HEOD. Teratology 16:57–62, 1977.

Uzoukwu, M. and Sleight, S.D.: Effects of dieldrin on pregnant sows. J.A.V.M.A. 160:1641–1643, 1972.

784 Dienestrol

Einer–Jensen (**1968**) found that doses of 0.3–0.6 mg per kg given orally during two–day periods of organogenesis produced an increase in resorptions in the rat. At 0.3 mg per on days 17 and 18 there was an increase in dead newborns. For rats and mice treated on days 3–4 a dose of 0.003 mg per kg was the ED–50 for abortion.

Einer–Jensen, N.: Antifertility properties of two diphenylethenes. Acta Pharmacol. (Copenh.) 26(suppl. 1):1–97, 1968.

785 Diesel Fuel

Shreiner (1983) reported "negative" results after exposing rats to 100 or 400 ppm from day 6 through 15. Exposures were for 6 hours daily.

Schreiner, C.A.: Petroleum and petroleum products: A brief review of studies to evaluate reproductive effects. Environmental and Health Sciences Laboratory, Mobil Oil Corporation. In: Assessment of Reproductive and Teratogenic Hazards, volume III. M.S. Christian, W.M. Galbraith, P. Voytek, M.A. Mehlman (eds), Princeton Scientific Publishers Inc., 29–46, 1983.

786 o–Diethylaminoethoxy–benzanilide

Kienel (1968) tested this compound and its chlorinated derivatives in the aniline radical as well as those methylated in the 5–position of the salicyl ring. Two halogenated derivatives were also studied. The mouse fetuses were exposed to doses of up to 75 mg per kg and for the parent compound, cleft palate and defects of the terminal digits were seen starting at 50 mg per kg. The three compounds chlorinated on various positions of the aniline radical were not teratogenic at 50 mg per kg. Those derivatives chlorinated, methylated or brominated at the 5–position of the salicyl ring were teratogenic. The chlorinated form was the most teratogenic (30 mg per kg).

Kienel, G.: Halogenierung und Embryotoxische Wirkung. Arzneim-Forsch. 18:658–661, 1968.

787 1–(2–Diethylaminoethyl)phenobarbital

Discussed under Barbituric Acid

788 N–N–Diethylbenzene Sulfonamide CAS 1709–50–8

Leland et al. (1972) administered this proposed mosquito repellant to pregnant rats on days 5 through 10 or 13 of gestation in doses up to 500 mg per kg. The material was given by stomach tube or intraperitoneally. Tail defects, umbilical hernias and reduction defects of the extremities were found, especially following treatment on day 9.

Leland, T.M.; Mendelson, G.F.; Steinberg, M. and Weeks, M.: Studies on the prenatal toxicity and teratology of N–N diethylbenzene sulfonamide. (abs) Toxicol. Appl. Pharmacol. 22:315 only, 1972.

789 Diethylcarbamazine CAS 90–89–1

This antifilariasis drug was tested orally in pregnant rats at 100 mg per kg from day 8 through 16 and in rabbits at 200 mg per kg on days 8 through 16 and no adverse fetal effects were found (Fraser, 1972).

Fraser, P.J.: Diethylcarbamazine: Lack of teratogenic and abortifacient action in rats and rabbits. Ind. J. Med. Res. 60:1529–1532, 1972.

790 **1,4 Diethylene Dioxide** *Dioxane* CAS 123–91–1

Giavini et al. (1985) gavaged rats on days 6–15 with up to 1.0 ml per kg and found only maternal and fetal weight reduction at the highest dose.

Giavini, E.; Vismara, C. and Broccia, M.L.: Teratogenesis study of dioxane in rats. Toxicology Letters 26:85–88, 1985.

791 **1–[p–(β–Diethylaminoethoxy)phenyl–21–(p–methoxyphenyl)–ethane**
MRL–37

Barnes and Meyer (1962) found that 20 mg per kg given orally inhibited implantation in rats. Some decrease in fetal survival was found when the treatment was given on day 18 through the day of parturition. This later effect could have been due to surgery on day 18.

Barnes, L.E. and Meyer, R.K.: Effects of ethamoxytriphetol, MRL–37, and clomiphene on reproduction in rats. Fertility and Sterility 13:472–480, 1962.

792 **Diethylene Glycol** CAS 111–46–6

Kawasaki et al. (1984) studied the effect of this solvent on the rat pregnancy giving up to 5 percent of the diet orally on days 0–20 of gestation. No increase in defects was seen but the weight of the neonates was slightly but not statistically reduced.

Kawasaki, K.; Murai, T. and Kanoh, S.: Fetal toxicity of diethylene glycol. Oyo Yakuri 27:801–807, 1984.

793 **Diethylene Glycol Dimethyl Ether**

Schwetz et al. (1992) gavaged rabbits on days 6–19 with 175 and 100 mg kg per day and found increased maternal mortality with an increase in skeletal defects and hydronephrosis among the fetuses. At 25 and 50 mg per kg no significant increase in defects was found.

Schwetz, B.A.; Price, C.J.; George, J.D.; Kimmel, C.A.; Morrissey, R.E. and Marr, M.C.: The developmental toxicity of diethylene and triethylene glycol dimethyl ethers in rabbits. Fundamental and Applied Toxicology 19:238–245, 1992.

794 **Diethylene Glycol Monomethyl Ether**

Hardin et al. (1986) dosed rats orally on days 7–16 of gestation with 720 and 2165 mg per kg. Neither dose was maternally toxic but rib and heart malformations were increased. Ventricular septal defects were the most common type of heart defect.

Price et al. (1987) administered up to 500 mg per kg to mice on days 6–15 orally. At doses of 150 mg per kg or more, malformations of most of the organ systems were found. These included the neural tube, heart, renal and skeletal systems.

Hardin, B.D.; Goad, P.T. and Burg, J.R.: Developmental toxicity of diethylene glycol monomethyl ether (diEGME). Fund. Appl. Toxicol. 6:430–439, 1986.

Price, C.J.; Kimmel, C.A.; George, J.D. and Marr, M.C.: The developmental toxicity of diethylene glycol dimethyl ether in mice. Fund. Appl. Toxicol. 8:115–126, 1987.

795 **Di(2–ethylhexyl) Adipate** CAS 103–23–1

Singh et al. (1973) gave rats up to 10 ml per kg intraperitoneally on days 5, 10 and 15. Only reduced fetal weight was found at the highest doses.

Singh, A.R.; Lawrence, W.H. and Autian, J.: Embryonic–fetal toxicity and teratogenic effects of adipic acid esters in rats. J. Pharm. Sci. 62:1596–1600, 1973.

796 **Di(2–ethylhexyl) Phthalate** CAS 117–81–7

Discussed under Phthalate Esters

797 **1,2–Diethylhydrazine** *Hydrazines* CAS 1615–80–1

Druckrey (1973) reviewed the extensive work with the transplacental carcinogenicity of the hydrazines. Twelve to 33 percent of the LD–50 of diethylhydrazine on the 15th day by any route produced postnatally a very high incidence of tumors of the nervous system. Administration before the 11th day produced no tumors. The doses used ranged from 10 to more than 50 mg per kg. On the 10th day, the development of the eyes and optic nerves was inhibited completely. Later in gestation, hydrocephalus and malformations of the paws was produced. 1–Methyl–2–benzyl hydrazine also had transplacental carcinogenicity (15–20 mg subcutaneously on the 15th and 21st days).

Druckrey, H.: Specific carcinogenic and teratogenic effects of indirect alkylating methyl and ethyl compounds, and their dependency on stages of ontogenic developments. Xenobiotica 3:271–303, 1973.

798 **Diethyl Maleate**

Wong et al. (1989) found that this glutathione depletor enhanced the teratogenicity of phenytoin in the rat. The chemical was given intraperitoneally 150 or 300 mg per kg and followed by 55 mg per kg of phenytoin.

Wong, M.; Helston, L.M.J. and Wells, P.G.: Enhancement of murine phenytoin teratogenicity by the gamma–glutamylcysteine synthetase inhibitor l–buthionine–(S,R)–sulfoximine and by the glutathione depletor diethyl maleate. Teratology 40:127–141, 1989.

799 **Diethyl Phthalate**

Field et al. (1993) fed rats to doses up to 5.0% of their diet on days 6 to 15 and found some maternal weight restriction but no effect on fetal development. These studies along with those on dimethyl phthalate suggested to the authors that the short–chain phthalic acid esters were less toxic than those with long chains.

Field, E.A.; Price, C.J.; Sleet, R.B.; George, J. D. and Marr, M.C.: Developmental toxicity evaluation of diethyl and dimethyl phthalate in rats. Teratology 48:33–44, 1993.

800 **Diethylpropion** *Tenuate*™

Heinonen et al. (1977) reported no significant increase in defects following maternal use by 40 women in the first four lunar months and 225 women who took it at any time.

Bunde and Leyland (**1965**) in a controlled retrospective survey by practioners of 1232 exposure control pairs found no increase in malformations in the treated group.

Bunde, C.A. and Leyland, H.M.: A controlled retrospective survey in evaluation of teratogenicity. J. of New Drugs 5:193–198, 1965.

Heinonen, O.P.; Slone, D. and Shapiro, S.: Birth Defects and Drugs in Pregnancy. Publishing Sciences Group Inc., Littleton, Mass., 1977.

801 Diethylstilbestrol *DES Stilbestrol™* CAS 56–53–1

Cancer Risk

Herbst et al. (1971) reported that adenocarcinoma of the vagina occurring in relatively young women was associated in seven out of eight instances with Stilbestrol™ treatment of the mother of the patients during the first trimester of pregnancy. A high incidence of benign adenosis of the vagina was found in these girls. The authors comment that their prenatal exposure to the synthetic estrogen may have increased the amount of benign tissue subsequently at risk for malignant change. This is a remarkable delay in teratogenic response and has obvious implications.

Further studies of this problem (Herbst et al., 1975A,B) based on many more cases indicate that the precancerous vaginal adenosis occurs in 73 percent of women exposed before the 9th week of gestation but in only 7 percent of those exposed after the 17th week. Clear–cell carcinoma has not occurred in women who were exposed in utero after the 18th week. Over 90 percent of the cancers occurred after age 14 years. The total dose of diethylstilbestrol received by cancer patients had varied from 1.5 mg to 225 mg. Herbst et al. (1977) pointed out that the risk of vaginal carcinoma was not related to the maternal dose. Two other non–steroidal estrogens (dienestrol and hexestrol) may have the same effect. It is estimated that the risk of development of adenocarcinoma after in utero exposure is exceedingly small at least in the population under 25 years. Haney et al. (1986) have found paraovarian cysts in 8 of 9 women who were exposed in utero to DES.

O'Brien et al. (1979), in a prospective study of 2,940 intrauterine exposed women, found that both time in gestation and amount of administered drug were related to incidence of vaginal adenosis. Exposure starting at 4 weeks was associated with adenosis in 56 percent of the offspring while at 16 and 20 weeks, the figure was 30 and 10 percent, respectively. The vaginal changes from DES decreased as the age of the examined woman increased. Not a single subject with clear cell carcinoma was identified from 1,340 participants identified from record review. Herbst et al. (1975B) suggested that the drug produced adenosis by either inhibiting upward growth of the vaginal plate or stimulating mullerian epithelium resulting in its persistence in the developing vagina.

Reproduction by Adult Female Offspring Exposed in Utero

The reproductive history of intrauterine exposed women was studied by Barnes et al. (1980). Miscarriages and absence of full term infants were significantly more common. Nineteen percent of these women had no full term infants as compared to a matched control of 4.9 percent. Kaufman et al. (1984) studied upper genital tract abnormalities in exposed women and the relationship of these changes to the increase in poor pregnancy outcome. Mittendorf and Williams (**1995**) found an increased odds ratio (2.4, 95 CI 1.2–4.5) for toxemia of pregnancy among women who had been exposed in utero.

Schechter et al. (**1991**) in a study of 20 DES and 20 controls could find no differences in menstrual cycle functioning or pain.

Gustavson et al. (**1991**) surveyed 1,711 women exposed to diethylstilbestrol in utero and found that 18.7 per 1,000 subsequently had a loss to 80% of their expected weight as compared to a control rate of 3.3 per 1,000. It was not known that this increase was due to anorexia nervosa.

Malformations

Bongiovanni et al. (1959) reported clitoromegaly in female newborns whose mothers were treated with diethylstilbestrol (5 to 50 mg per day during the first trimester).

Gill et al. (1976) studied 134 males exposed in utero and found that 27 percent had genital lesions (epididymal cysts, hypotrophic testes or capsular induration of the testes). In 29 percent, pathologic changes were found in spermatozoa. No signs of malignancy were found.

Experimental and Animal Studies

Newbold et al. (1987) reported that among 277 prenatally exposed male mice, testicular tumors were observed in 8 percent. Giannina et al. (1971) compared a large number of related chemicals for their antifertility activity in rats and hamsters. These compounds included ORF 3858, F 6066, CN 55, 945–27, U11, 555A, Su 13320 and Mer–25. Greene et al. (1940) used this compound and produced feminization of male rat fetuses but partial retention of wolffian bodies in female fetuses. The maternal rats were treated with a total of 10 to 42 mg from the 12th to 19th gestational day. Beyer et al. (1987) studied the biochemical mechanism in vitro using whole embryo culture. Cunha et al. (1987) transplanted human genital tracts to beneath the renal capsules of nude mice and then treated the mice with 20 mg pellets subcutaneously. Abnormal condensation and segregation of uterine mesenchyme was found. Newbold et al. (1987) studied the persistence of Mullerian duct remnants in male mice exposed in utero to DES. Hendrickx et al. (1988) made a study of prenatally exposed rhesus monkeys and found that 5 out of 8 had gross and histologic evidence of vaginal adenosis which reverted to squamous epithelium. Bullock et al. (1988) also reported cryptorchidism in mice exposed prenatally. Walker (**1990**) found that a high fat diet during pregnancy increased the postnatal tumor rate in female mice.

Summaries

A comprehensive review of the experimental models and the human clinical experience is given by McLachlan and Dixon (1977). Haney (1987) summarized the human anomalies and clinical associations induced by prenatal diethylstilbestrol. Walker (**1984**) has critically reviewed the subject.

Barnes, A.B.; Colton, T.; Gundersen, J.; Noller, K.L.; Tilley, B.C.; Strama, T.; Townsend, D.E.; Hatab, P. and O'Brien, P.C.: Fertility and outcome of pregnancy in women exposed in utero to diethylstilbestrol. N. Eng. J. Med. 302:609–613, 1980.

Beyer, B.K.; Greenaway, J.C.; Fantel, A.G. and Juchau, M.R.: Embryotoxicity induced by diethylstilbestrol in vitro. J. Biochem. Toxicol. 2:77–92, 1987.

Bongiovanni, A.M.; DiGeorge, A.M. and Grumbach, M.M.: Masculinization of the female infant associated with estrogenic therapy alone during gestation. Four cases. J. Clin. Endocrinol. Metab. 19:1004–1011, 1959.

Bullock, B.C.; Newbold, R.R. and McLachlan, J.A.: Lesions of testis and epididymis associated with prenatal diethylstilbestrol exposure. Environ. Health Perspect. 77:29–31, 1988.

Cunha, G.R.; Taguchi, O.; Namikawa, R.; Nishizuka, Y. and Robboy, S.J.: Teratogenic effects of clomiphene, tamoxifen and diethylstilbestrol on the developing human female genital tract. Human Pathology 18(11):1132–1143, 1987

Giannina, T.; Butler, M.; Popick, F. and Steinetz, B.: Comparative effects of some steroidal and nonsteroidal antifertility agents in rats and hamsters. Contraception 3:347–359, 1971.

Gill, W.B.; Schumacher, G.F.B. and Bibbo, M.: Structural and functional abnormalities in the sex organs of male offspring of mothers treated with diethylstilbestrol (DES). J. Reprod. Med. 16:147–153, 1976.

Greene, R.R.; Burrill, M.W. and Ivy, A.C.: Experimental intersexuality. The effects of estrogens on the antenatal sexual development of the rat. Am. J. Anat. 67:305–345, 1940.

Gustavson. C.R.; Gustavson, J.C.; Noller, .L.; O'Brien, P.C.; Melton, L.J.; Pumariega, A.J.; Kaufman, R.H. and Colton, T.: Increased risk of profound weight loss among women exposed to diethylstilbestrol in utero. Behavioral and Neural Biology 55:307–312, 1991.

Haney, A.F.: Structural and functional consequences of prenatal exposure to diethylstilbestrol in women. In: Developmental Toxicology: Mechanisms and Risks. J.A. McLachlan, R.M. Pratt and C.L. Market (eds.), Banbury report No. 26. Cold Spring Harbor Laboratory, Cold Spring, N.Y., 271–285, 1987.

Haney, A.F.; Newbold, R.R.; Fetter, B. and McLachlan, J.A.: Paraovarian cysts associated with prenatal diethylstilbestrol exposure. Comparison of the human with a mouse model. Am. J. Pathol. 124:405–411, 1986.

Hendrickx, A.G.; Prahalada, S. and Binkerd, P.E.: Long–term evaluation of the diethylstilbestrol (DES) syndrome in adult female rhesus monkeys (Macaca mulatta). Reproductive Toxicology 1:253–261, 1988.

Herbst, A.L.; Cole, P.; Colton, T.; Robboy, S.J. and Scully, R.E.: Age incidence and risk of DES–related clear cell adenocarcinoma. Am. J. Obstet. Gynecol. 128:43–48, 1977.

Herbst, A.L.; Poskanzer, D.C.; Robboy, S.J.; Friedlander, L. and Scully, R.E.: Prenatal exposure to stilbestrol: A prospective comparison of exposed female offspring with unexposed controls. N. Eng. J. Med. 292:334–339, 1975A.

Herbst, A.L.; Scully, R.E. and Robboy, S.J.: Effects of maternal DES ingestion on the female genital tract. Hospital Practice, October, 51–57, 1975B.

Herbst, A.L.; Ulfelder, H. and Poskanzer, D.C.: Adenocarcinoma of the vagina: Association of maternal stilbestrol therapy with tumor appearance in young women. N. Eng. J. Med. 284:878–881, 1971.

Kaufman, R.H.; Noller, K.; Adam, E; Irwin, J.; Gray, M.; Jeffries, J.A. and Hilton, J.: Upper genital tract abnormalities and pregnancy outcome in diethystilbestrol–exposed progeny. Am. J. Obstet. Gynecol. 48:973–984, 1984.

McLachlan, J.A. and Dixon, R.L.: Teratologic comparisons of experimental and clinical exposures to diethylstilbestrol during gestation. In: Regulatory Mechanisms Affecting Research, J.A. Thomas and R.L. Singhal (eds.), University Park Press, Baltimore, 309–336, 1977.

Mittendorf, R. and Williams, M.A.: Stilboestrol exposure in utero and risk of pre–eclampsia. The Lancet 245:265–266, 1995.

Newbold, R.R.; Bullock, B.C. and McLachlan, J.A.: Mullerian remnants of male mice exposed prenatally to diethylbestrol. Teratogenesis, Carcinogenesis and Mutagenesis 7:377–389, 1987.

Newbold, R.R.; Bullock, B.C. and Mclachan, J.A.: Testicular tumors in mice exposed in utero to diethylstilbestrol. J. Urology 138:1446–1449, 1987.

O'Brien, P.C.; Noller, K.L.; Robboy, S.J.; Barnes, A.B.; Kaufman, R.H.; Tilley, B.C. and Townsend, D.E.: Vaginal epithelial changes in young women enrolled in the National Cooperative Diethylstilbestrol Adenosis (DESAD) Project. Obstet. Gynecol. 53:300–308, 1979.

Schechter, D.; Ehrhardt, A.A.; Endicott, E.J.; Meyer–Bahlburg, H.F.L.; Nee, J. and Veridiano, N.P.: Menstrual cycle functioning in women with a history of prenatal diethylstilbestrol exposure. J. Psychosom. Obstet. Gynaecol. 12:51–66, 1991.

Walker, B.E.: Tumors in female offspring of control and diethylstilbestrol–exposed mice fed high–fat diets. The Journal of the National Cancer Institute 82:50–54, 1990.

Walker, B.E.: Transplacental exposure to diethylstilbestrol. Issues and Reviews in Teratology 2:157–187, 1984.

802 Diethylsulfoxide

Caujolle et al. (1965) were unable to demonstrate teratogenicity in mice, rats or rabbits. The mice were injected daily during gestation with 2 gm per kg and the rabbits 4 gm per kg. Celosomia and limb defects were produced in the chick embryo with as little as 0.1 percent of the LD–100.

Caujolle, F.M.E.; Caujolle, D.H.; Cros, S.B.; Calvet, M.J. and Tollon, Y.: Pouvoir teratogene du dimethylsulfoxyde et du diethylsulfoxyde. C. R. Acad. Sci. (d) (Paris) 260:327–330, 1965.

803 N,N–Diethyl–m–toluamide *DEET m–DET*

Robbins and Chernack (1986) reviewed the toxicology of this insect repellant. They cited unpublished work by B. Hardin. He treated rats subcutaneously with 0.3 ml per kg of 97–98 percent solution on days 6–15 of gestation and found no adverse reproductive effects. A single report of a mentally retarded offspring followed daily application of this insecticide has appeared (Schaefer and Peters, **1992**). Muscular hypotonia, hearing loss and strabimus were present.

Robbins, P.J. and Chernack, M.G.: Review of the biodistribution and toxicity of the insect repellant N,N–diethyl–m–toluamide (DEET). J. Toxicol. Environ. Health 18:503–525, 1986.

Schaefer, C. and Peters, P.W.J.: Intrauterine diethyltoluamide exposure and fetal outcome. Reproductive Toxicology 6:175–176, 1992.

804 Diethyl–triazene CAS 63980–20–1

Discussed under 1–Phenyl–3,3–dimethyl–triazine

805 Diflorasone CAS 2557–49–5

Suzuki and Narama (1984) and Satoh (1984) studied this anti-inflammatory steroid after subcutaneous administration in the rat before fertilization and during the first 7 days, during organogenesis and in the perinatal period. Corpora lutea were reduced at the highest dose (0.45 mg per day). No defects were produced but fetuses whose dams were treated with 0.45 mg per kg showed depressed weights of thymus, liver, spleen and adrenals. Fetal organ weight decrease was not seen at the 0.045 mg dose level. No congenital defect was increased. In the rabbit treated percutaneously with 0.016 mg per kg but not at 0.002 mg per kg, congenital defects were increased (cleft palate, congenital heart defects, encephalocele and omphalocele), (Narama, 1984).

Narama, I.: Reproduction studies of diflorasone diacetate (DDA). 4. Teratogenicity study in rabbits by percutaneous administration. Oyo Yakuri 28:241–250, 1984.

Satoh, T.; Narama, I. and Odani, Y.: Reproduction studies of diflorasone diacetate (DDA). 2. Teratogenicity study in rats by subcutaneous administration. Oyo Yakuri 28:207–224, 1984.

Suzuki, T. and Narama, I.: Reproduction studies of diflorasone diacetate (DDA). 1. Fertility study in rats by subcutaneous administration. Oyo Yakuri 28:195–205, 1984.

806 Diflucortolone Valerate 6–α

9–Difluor–11–β–hydroxy–16–α–methyl–21–valeryloxy–1,4–pregnadien–3,20–dione; Nerisona™ CAS 59198–70–8

Ezumi et al. (1977A,B and 1978A,B) made studies of this steroid in mice, rats and rabbits. No defects were produced in the rat at 0.1 mg per kg but cleft palate occurred in the mouse and rabbit fetuses at subcutaneous doses of 0.01 and 0.05 mg per kg given during organogenesis. In all three species, the fetal weight decreased and fetal mortality increased.

Daily dermal application of this steroid ointment during organogenesis produced tail aplasia, omphalocele and growth retardation in rat fetuses at 500 mg per kg and vertebral defects, omphalocele and cleft palate in rabbit fetuses at 250 mg per kg (Gunzel et al., 1976).

Ezumi, Y.; Tomoyama, J. and Kodama, N.: Effects of diflucortolone valerate subcutaneously injected into rats in mid gestation (period of organogenesis) or late gestation on the pre and postnatal development of their offspring. (Japanese) Yakubutsu Ryoho 10:1357–1365, 1977A.

Ezumi, Y.; Tomoyama, J. and Kodama, N.: Teratogenicity especially on the formation of cleft palate in mouse embryos of diflucortolone valerate by a single administration. (Japanese) Yakubutsu Ryoho 10:1585–1594, 1977B.

Ezumi, Y.; Tomoyama, J. and Kodama, N.: Effects of diflucortolone valerate subcutaneously injected to mice in mid gestation (period of organogenesis) or late gestation on the pre and postnatal development of the offspring. Yakubutsu Ryoho 11:237–256, 1978A.

Ezumi, Y.; Tomoyama, J. and Kodama, N.: Effects of diflucortolone valerate subcutaneously injected to rabbits in midgestation on the prenatal development of their offspring. (Japanese) Yakubutsu Ryoho 11:229–236, 1978B.

Gunzel, P.; El Etreby, M.F.; Bhargava, A.S.; Poggel, H.A.; Schobel, C.; Schuppler, J.; Siegmund, F. and Staben, P.: Tier experimentelle Vertraglichkeitsprufung von Diflucortolonvalerianat als reiner Wirkstoff und als Salbe, Fettsalbe und Creme. Arzneim. Forsch. 26:1476–1479, 1976.

807 Diflunisal *5–(2,4 Difluorophenyl) Salicylic Acid* CAS 22494–42–4

Nakatsuka and Fujii (1979) treated rats orally with up to 100 mg per kg on days 7 through 17. The dams gained less weight than controls but no adverse fetal effects were found. Rowland et al. (1987) gave up to 80 mg per kg to monkeys during organogenesis and found no evidence of teratogenicity.

Clarke et al. (1984) gave rabbits 180 mg per kg on day 5 and produced an anemia which lasted through day 15. The axial skeletal defects were believed to be related to the anemia since the drug was cleared by day 9.

Clark, R.; Robertson, R.T.; Minsker, D.H.; Cohen, S.M.; Tocco, D.J.; Allen, H.L.; James, M.L. and Bokelman, D.L.: Diflunisal-induced maternal anemia as a cause of teratogenicity in rabbits. Teratology 30:319–332, 1984.

Nakatsuka, T. and Fujii, T.: Comparative teratogenicity study of diflunisal (MK–647) and aspirin in the rat. Oyo Yakuri 17:551–557, 1979.

Rowland, J.M.; Robertson, R.T.; Cukierski, M.; Prahalada, S.; Tocco, D. and Hendrickx, A.G.: Evaluation of the teratogenicity and pharmacokinetics of diflunisal in cynomolgus monkeys. Fund. Appl. Toxicol. 8:51–58, 1987.

808 α–Difluoromethylornithine

Gray et al. (1989) studied the effects of this irreversible ornithine decarboxylase inhibitor in the pregnant rat. They gave 500 mg twice daily subcutaneously for 5 doses beginning on day 11, 14 or 17. Renal morphologic changes were studied posnatally and those treated on days 14–16 had permanent changes which included elevation of serum creatinine.

Gray, J.A.; Rehnberg, B.F.; Rogers, E.H.; Lau, C.; Slotkin, T.A. and Kavlock, R.J.: Prenatal β–

difluoromethylornithine treatment: effects on postnatal renal growth and function in the rat. Teratology 40:105–111, 1989.

809 Difluprednate CAS 23674–86–4

Ikeda et al. (1984A) gave 0.1–10 microgm subcutaneously on days 6 through 18 to rabbits. Above 1.0 microgm cleft palate, cerebral hernia and hypogenesis of the first digit were found. Ikeda et al. (1984B) found no perinatal or postnatal adverse effects when doses of up to 10 microgm were given on day 17 to day 21.

Ikeda, Y.; Iwase, T.; Sukegawa, J. and Osamu, F.: Reproductive studies on difluprednate. 4. Teratogenicity study in rabbits. Iyakuhin Kenkyu 15:1055–1060, 1984A.

Ikeda, Y.; Sukegawa, J.; Iwase, T. and Osamu, F.: Reproduction studies on difluprednate. 3. Perinatal and postnatal study in rats. Iyakuhin Kenkyu 15:1046–1054, 1984B.

810 Difolatan™

N–Tetrachloroethylthio–4–cyclohexene–1–2– dicarboximide
Captafol

This fungicide with a molecular structure similar to thalidomide was tested in rabbits by Kennedy et al. (**1968**). At up to 150 mg daily orally no adverse reproductive effects were found. In rats up to 500 mg per kg daily produced no adverse fetal effects. In three generation studies in rats there was slight reduction in weight in the first and third generation 21–day olds at 0.1% in the diet. Verrett et al. (1969) injected 3–20 mg per kg into the air cell or yolk sac of the chick egg before incubation and found 6.7 percent abnormal embryos. The defects included those of the central nervous and skeletal system with phocomelia and amelia being particularly noteworthy. The epoxide form was even more teratogenic.

Kennedy, G.; Fancher, O.E. and Calandra, J.C.: An investigation of the teratogenic potential of captan, folpet, and difolatan. Toxicology and Applied Pharmacology 13:420–430, 1968.

Verrett, M.J.; Matchler, M.K.; Scott, W.F.; Reynaldo, E.F. and McLaughlin, J.: Teratogenic effects of captan and related compounds in the developing chicken embryo. Ann. N.Y. Acad. Sci. 160:334–343, 1969.

811 Digitalis

Many human pregnancies in which digitalis was used were complicated by heart failure and the use of other drugs, particularly warfarin. Laros et al. (1970) reviewed 24 pregnancies associated with heart prostheses and except for the presence of two infants with warfarin embryopathy, the offspring were healthy.

Laros, R.K.; Hage, M.L. and Hayashi, R.H.: Pregnancy and heart valve prostheses. Obstet. Gynecol. 35:241–247, 1970.

812 Digoxin CAS 20830–75–5

Aselton et al. (1985) reported no congenital malformations among the offspring of 142 women who took the medication during the first trimester.

Singh and Mirkin (1978) demonstrated that digoxin crosses the rat fetal placenta. Hatano et al. (1976) gave up to 1.75 mg of methyldigoxin orally to rats on days 7–17. Except for variation in lumbar ribs at 0.5 mg per kg, no fetal changes occurred. In the rabbit, the same workers (Nagaoka et al., 1976), found an increase in fetal lethality at 10 mg per kg when the drug was given orally. There were no other adverse fetal changes. Hamamoto er al (1990) demonstrated in the fetal sheep fetus that prompt and near complete uptake of intraamniotic digoxxin occurring. The transfer from maternal to fetal circulation was much reduced.

Aselton, P.A.; Jick, H.; Milunsky, A.; Hunter, J.R. and Stergachis, A.: First–trimester drug use and congenital disorders. Obstet. Gynecol. 65:451–455, 1985.

Hamamoto, K.; Iwamoto, H.S.; Roman, C.M.; Benet, L.Z. and Rudolph, A.M.: Fetal uptake of intraamniotic digoxin in sheep. Pediatric Research 27:288–285, 1990.

Hatano, M.: Reproduction studies of β–methyldigoxin. 1. Teratogenicity study in rats. (Japanese) Clinical Report 10:579–593, 1976.

Nagaoka, T.; Osuka, F.; Shigemura, T. and Hatano, M.: Teratogenicity test of β–methyldigoxin (β–MD). (Japanese) Clinical Report 10:405–411, 1976.

Singh, S. and Mirkin, B.L.: Placental transfer and tissue localization of digoxin in the pregnant rat. Toxicol. Appl. Pharmacol. 46:395–403, 1978.

813 Dihydrostreptomycin also see *Streptomycin* CAS 128–46–1

Varpela et al. (1969) found normal hearing in 50 children who were exposed in utero to dihydrostreptomycin or streptomycin.

The dihydro form of streptomycin has been injected into rats and mice at 200 to 800 mg per kg daily during pregnancy (Suzuki and Takeuchi, 1969). The 800 mg dose level was lethal in all the mothers and lower doses caused reduction in fetal size. A large number of surviving fetuses were tested and responded to sound.

Suzuki, Y. and Takeuchi, S.: Etude experimentale sur l'influence de la streptomycine sur l'appareil audit du foetus apres administration de doses variees a la mere enceinte. Keio J. Med. 10:31–41, 1969.

Varpela, E.; Hietalahti, J. and Aro, M.J.T.: Streptomycin and dihydrostreptomycin medication during pregnancy and their effect on the child's inner ear. Scand. J. Resp. Dis. 50:101–109, 1969.

814 Dihydrotestosterone

Schultz and Wilson (**1974**) gave rats subcutaneous injections of 4 mg on days 14–21 and 0.16 mg to the newborns for the first 3 days. Masculinization of the females internal genitalia occurred.

Schultz, F.M. and Wilson, J.D.: Virilization of the wolffian duct in the rat fetus. Endo. 94:979–986, 1974.

815 5–α–Dihydrotestosterone

Schultz and Wilson (**1974**) injected rats with 4 mg on

days 14–21 and produced masculinization of the female fetal internal genitalia.

Schultz, F.M. and Wilson, J.D.: Virilization of the wolffian duct in the rat fetus. Endo. 94:979–986, 1974.

816 2,3–Dihydrooxynaphthalene

Discussed under Hair Dyes

817 5–β–Dihydrotestosterone

Schultz and Wilson (1974) gave pregnant rats four mg subcutaneously on days 14–21 and did not produce masculinization of female fetuses.

Schultz, F.M. and Wilson, J.D.: Virilization of the wolffian duct in the rat fetus by various androgens. Endocrinology 94:979–986, 1974.

818 Dihydroxymorphine

Yeh and Woods (**1970**) found that 15 or 28 mg per kg to rats during gestation did not reduce the number of fetuses or their weight.

Yeh, S.Y. and Woods, L.A.: Maternal and fetal distribution of H³–dihydromorphine in the tolerant and nontolerant rat. The Journal of Pharmacology and Experimental Therapeutics 174:9–13, 1970.

819 α,24(R)–Dihydroxyvitamin D₃ *TV–02 Vitamin D3*

Matsuzawa et al. (**1989**) administered up to 0.10 microgm per kg to rats on days 7–17. At o.10 and 0.50 microgm per kg maternal toxicity occurred. No teratogenicity or postnatal behavioral changes were found. At the highest dose decreased placental and fetal weight was found.

Matsuzawa, K.; Enjyo, H.; Nishizawa, S.; Hisada, H.; Oguri, M. and Makita, T.: Reproductive toxicity studies of tv–02 (2)–teratogenicity study in rats.Yakuri to Chiryo 17:4829–4849, 1989.

820 Diiodomethyl p–Tolyl

Ema et al. (**1992**) found no adverse effects in fetuses whose mothers received up to 1.0% of this antimicrobial agent in the diet during days 6–15 of gestation.

Ema, M.; Itami, T.and Kawasaki, H.: Teratological assessment of diiodomethyl p–tolyl sulfone in rats. Toxicology Letters 62:45–52, 1992.

821 Dikurin

Dirukin, a pesticide, was administered in the food at 2 or 20 mg per kg of feed during the pregnancy and no adverse fetal effects were found.

Shepel'skaya, N.R.: Gonadotoxic effect of dikurin under intragastric administration by probes and with food. Gig. Sanit. 11:78–79, 1988.

822 Dilazep CAS 35898–87–4 *K–285*

Abel et al. (1972) gave rats orally up to 100 mg per kg on days 7–16 and found no adverse effects on reproduction. Nishigaki et al. (**1990**) gave up to 12 mg per kg daily to rats

intravenously on days 7–17 and found no effects on the fetuses or offspring. Similar treatment to rabbits using up to 6 mg per kg on days 6–18 resulted in no adverse fetal effects. (Sakai et al, **1990**)

Abel, H.H.; Brock, N. and Lenke, D.: Zur Toxikologie von Dilazep, einer neuen Koronarakfiven Substanz. Arzneim–Forsch 22:667–674, 1972.

Nishigaki, K.; Ohta, T.; Moriwaki, T.; Koga, T. and Takimoto, M.: Teratogenicity study of k–285 (dilazep dihydrochloride) by intravenous administration during the period of fetal organogenesis in rats. Yakuri to Chiryo 18:s3241–s3256, 1990.

Sakai, Y.; Sasaki, M.; Kawakami, T.; Kinoshita, K.; Nishigaki, K. and Akiba, T.: Teratogenicity study of k–285 (dilazep dihydrochloride) by intravenous administration during the period of fetal organogenesis in rabbits. Yakuri to Chiryo 18:s3257–s3267, 1990.

823 Diltiazem Hydrochloride CAS 33286–22–5

This coronary dilator was injected intraperitoneally during organogenesis in pregnant mice, rats and rabbits (Ariyuki, 1975). At 80 mg and 12.5 mg per kg, respectively, in rats and rabbits, malformations of the limbs and tail were found. No malformations occurred in the mouse.

Ariyuki, F.: Effects of diltiazem hydrochloride on embryonic development: Species differences in the susceptibility and stage specificity in mice, rats and rabbits. Okajimas Fol. Anat. Jpn. 52:103–107, 1975.

824 Dimenhydrinate CAS 523–87–5

McColl et al. (1965) reported no teratogenic effect of 75 mg per kg per day of this material fed during the entire pregnancy of the rat.

McColl, J.D.; Globus, M. and Robinson, S.: Effect of some therapeutic agents in the developing rat fetus. Toxicol. Appl. Pharmacol. 7:409–417, 1965.

825 Dimercaprol *2,3–Dimercapto–l–propanol BAL* CAS 59–52–9

Nishimura and Takagaki (1959) injected mice subcutaneously with 0.001 ml per gm of body weight once on the 9th, 11th or 12th day of gestation. Reduction defects of the fetal extremities occurred in many of the fetuses with the highest incidence (69 percent) following treatment on the 12th day. A few cases of cleft palate and brain hernia were reported. Hood et al. (**1984**) were able with BAL to decrease the fetal deaths and skeletal defects produced by 12 mg per kg intraperitoneal doses.

Hood, R.D.; Vedel–Macrander, G.C.: Evaluation of the effect of BAL (2,3–dimoptopanol) on arsenite–induced teratogenesis in mice. Toxicol Appl Pharmacol 73:1–7, 1984.

Nishimura, H. and Takagaki, S.: Developmental anomalies in mice induced by 2,3–dimercaptopropanol (BAL). Anat. Rec. 135:261–268, 1959.

826 Dimethisterone

Kawashima et al. (**1977**) masculinization of female rats

exposed to 5 mg per kg. No effect was observed at 1 mg per kg.

Kawashima, K.; Nakaura, S.; Nagao, S.; Tanaka, S.; Kuwamura, T. and Omori, Y.: Virilizing activities of various steroids in female rat fetuses. Endocrinol. Japon. 24:77–81, 1977.

827 meso–2–3–Dimercaptosuccinic Acid *DMSA*

Domingo et al. (1988) gave this heavy metal antidote to mice subcutaneously during days 6–15 in doses of 410, 820 and 1640 mg per kg per day. At the highest dose, resorptions and the incidence of stunting was increased. Congenital defects were increased at 820 and 1640 mg per kg but not at 410 mg per kg. Cleft palate, eye, skeletal and neural tube closure defects were found. Maternal toxicity also occurred.

Domingo et al. (**1990**) gavaged rats on days 6–15 with 100, 300 and 1,000 mg per kg daily. Maternal and fetal toxicity occurred at all doses and resorptions were increased. No increase in malformations occurred. Domingo et al. (**1991**) were able to counteract the teratogenicity of sodium arsenite in mice by administering DMSA within an hour of the arsenite.

Domingo, J.L.; Ortega, A.; Paternain, J.L. and Llobet, J.M.: Oral mesoooo–2,3–dimercaptosuccinic acid in pregnant sprague–dawley rats: teratogenicity and alterations in mineral metaolism i teratological evaluation. Journal of Toxicology and Environmental Health 30:181–190, 1990.

Domingo, J.L.; Paternain, J.L.; Llobet, J.M. and Corbella, J.: Developmental toxicity of subcutaneously administered meso–2, 3–dimercaptosuccinic acid in mice. Fund. Appl. Toxicol. 11:715–722, 1988.

Domingo, J.L.; Bosque, M.A.; and Piera, V.: Meso–2, 3–dimercaptosuccinic acid and prevention of arsenite embryotoxicity and teratogenicity in the mouse. Fundamental and Applied Toxicology 17:314–320, 1991.

828 Dimetacrine *Istonyl*™ CAS 4757–55–5

Aoyama et al. (1970) administered this compound orally to rats in amounts up to 100 mg per kg and found no teratogenicity. Taoka et al. (1971) gave the compound intraperitoneally to rats and mice in maximum doses of 60 and 90 mg per kg, respectively, and found no adverse fetal effects.

Aoyama, J.; Nakai, K.; Ogura, M.; Saito, K. and Iwaki, R.: Acute toxicity and teratological studies of dimetacrine in mice and rats. Oyo Yakuri 4:855–869, 1970.

Taoka, K.; Nakamura, T.; Mitarai, H.; Tsuchiya, S. and Honda, K.: Teratological studies on demetracine. Oyo Yakuri 5:129–139, 1971.

829 Dimethadione

Miller and Scott (**1992**) gave this active metabolite of trimethadione to mice on days 10–10.5. At a dose of 800 mg per kg renal agenesis was found in a high percentage of the fetuses.

Miller, T.A. and Scott Jr., W.J.: Abnormalities in ureter and kidney development in mice given acetazolamide–

amiloride or dimethadione (dmo) during embryogenesis. Teratology 46:541–550, 1992.

830 Dimethoate

Courtney et al. (1985) gave this pesticide intragastrically to mice in doses of up to 80 mg per kg on days 6–16 and found no adverse fetal effects.

Courtney, K.D.; Andrews, J.E.; Springer, J. and Dalley, L.: Teratogenic evaluation of the pesticides baygon, carbofuran, dimethoate and EPN. J. Environ. Sci. Health B20(4):373–406, 1985.

831 2–(3,4–Dimethoxyphenyl)–5–methylthiazolidine

Ito et al. (1984) studied this antiulcer medication in pregnant rats and rabbits. No adverse effects on the fetus were found when as much as 300 mg per kg was given before pregnancy and during the first 7 days or in the perinatal period. During the organogenesis period 150 mg per kg did not produce an increase in defects. Rabbits received 300 mg per kg during organogenesis and no adverse fetal effects occurred.

Ito, R.; Kajiwara, S.; Mori, S.; Ouda, T.; Miyamoto, K.; Toida, S. and Tanihata, T.: Reproduction studies on 2–(3,4–dimethoxyphenyl)–5–methylthiazolidine. Oyo Yakuri 28:251–305, 1984.

832 Dimethylacetamide CAS 127–19–5

Solomon et al. (**1991**) exposed rats to 32,100 or 282 ppm for 6 hours daily on days 6–15. At 282 ppm fetal and maternal toxicity occurred but no increase in defects was found at any level. Thiersch (**1971**) found that giving pregnant rats 1.5 ml per kg on gestational day 4 or 7 led to over 60 % resorptions. The survivors were stunted but no defects were found. Painting the tails for weekly periods caused no effect on the litters. Ferenz and Kennedy (1986) exposed rats to up to 300 ppm for six hours daily before and during gestation. The exposed weanlings were lower in body weight than controls.

Ferenz, R.L. and Kennedy, G.L.: Reproduction study of dimethylacetamide following inhalation in the rat. Fund. Appl. Toxicol. 7:132–137, 1986.

Solomon, H.M.; Ferenz, R.L.; Kennedy, Jr., G.L. and Staples, R.E.: Developmental toxicity of dimethylacetamide by inhalation in the rat. Fundamental and Applied Toxicology 14:414–422, 1991.

Thiersch, J.B.: Investigations into the differential effect of compounds on rat litter and mother. In: Malformations Congenitales Des Mammiferes. H. Tuchmann–Duplessis (ed), Paris: Masson et Cie, 95–113, 1971.

833 N,N–Dimethylacetamide

Stula and Krauss (1977) exposed rats and rabbits cutaneously to 600 mg per kg on days 10, 11 or 12 and produced no significant fetal changes. Johannsen et al. (1987) gave up to 400 mg per kg by gavage to rats on days 6–19. At this high dose, oral and cardiovascular defects were found as well as anasarca.

Johannsen, F.R.; Levinskas, G.J. and Schardein, J.L.: Teratogenic response of dimethylacetamide in rats. Fund. Appl. Toxicol. 9:550–556, 1987.

Stula, E.F. and Krauss, W.C.: Embryotoxicity in rats and rabbits from cutaneous application of amide–type solvents and substituted ureas. Toxicol. Appl. Pharmacol. 41:35–55, 1977.

834 O,O–Dimethyl–s–(2–acetylaminoethyl) Dithiophosphate CAS 13265–60–6

Hashimoto et al. (1972) gave this pesticide to mice orally in maximal daily doses of 40 mg per kg from day zero through day 14 of gestation and found no increase in defects but some reduction in fetal survival was seen at the highest dose.

Hashimoto, Y.; Makita, T. and Noguchi, T.: Teratogenic studies of O,O–dimethyl–S–(2–acetylaminoethyl) dithiophosphate (DAEP) in ICR strain mice. Oyo Yakuri 6:621–626, 1972.

835 Dimethylamine

Guest and Varma (**1991**) gave 2.5 or 5.0 millimoles per kg intraperitoneally on days 1–17 in mice and found no maternal or fetal effects.

Guest, I. and Varma, D.R.: Developmental toxicity of methylamines in mice. Journal of Toxicology and Environmental Health 32:319–330, 1991.

836 Dimethylaminoazobenzene CAS 60–11–7

Discussed under Aminoazobenzene

837 7–12–Dimethylbenz–(a)–anthracene *DMBA* CAS 57–97–6

This potent carcinogen which has adrenolytic properties was shown to be teratogenic in the rat (Currie et al., 1970). Injection of 25 mg per kg on day 8 or 13 produced a high incidence of incomplete neural tube closure and other defects in the fetuses. Also, cleft palate, eye and kidney defects were found. A monohydroxymethyl product of metabolism of DMBA, 7–hydroxymethyl–12–methylbenz(a)anthracene was teratogenic, but two other products lacking adrenolytic acitivity were non–teratogenic. For adrenolytic and teratogenic activity benz(a)anthracene must possess two active side chains situated at C–7 and C–12. An active methyl group at C–12 was mandatory and at C–7 one of several substitutions was possible.

Currie et al. (1970) reported that if the maternal adrenal is inhibited at the time of exposure to these compounds, teratogenicity is abolished. Lambert and Nebert (1977) used this compound and 3–methylcholanthrene at doses of 25–50 mg and 70 mg per kg, respectively, on single days between day 7 and 13 in the mouse. Stillbirths and resorptions in mouse strains were produced whose bioactivating liver systems were inducible. A few malformations, mostly club foot, were found.

Currie, A.R.; Bird, C.C.; Crawford, A.M. and Sims, P.: Embryopathic effects of 7,12–dimethylbenz(a)anthracene

and its hydroxymethyl derivatives in the Sprague–Dawley rat. Nature 226:911–914, 1970.

Lambert, G.H. and Nebert, D.W.: Genetically mediated induction of drug–metabolized enzymes associated with congenital defects in the mouse. Teratology 16:147–154, 1977.

838 2,5–Dimethylbenzimidazole CAS 1792–41–2

Discussed under Benzimidazole

839 Dimethyldioxane CAS 25136–55–4

Smirnov et al. (1978) treated rats for 4 months with 0.01 and 0.004 mg per cubic meter of this substance and found a prolongated oestrus phase and slight embryolethal effect. Treatment during gestation with the same doses caused both pre–and postimplantation embryonic death.

Smirnov, V.T.; Dubrovskaya, F.I. and Kiseleva, T.I.: Action of dimethyldioxane on the generative function of experimental animals. Gigiena i Sanitariya (USSR) 9:16–18, 1978.

840 N,N–Dimethylformamide *Dimethylformamide* CAS 68–12–2

Stula and Krauss (1977) exposed rats cutaneously to 600 mg per kg on days 6, 7, or 8. Subcutaneous hemorrhage was present in fetuses exposed on day 8. Thiersch (1971) reported that unlike methylformamide, 0.5 to 2.0 ml of dimethylformamide per kg had no litter effects in the rat. Hellwig et al. (**1991**) treated rabbits, rats and mice by gavage, dermal, inhalation or intraperitoneal routes during organogenesis. In rats treated dermally with 94 mg per kg or more teratogenicity was increased without maternal toxicity. Teratogenicity in mice was found at 944 mg per kg intraperitoneally. In rabbits exposed dermally to 400 mg per kg teratogenicity was found. The defects were generally changes in the ribs and vertebrae. Lewis et al. (**1992**) exposed rats 6 hours daily to up to 300 ppm on days 6–15 and found maternal and fetal toxicity but no increase in defects.

Hellwig, J.; Merkle, J.; Klimisch, H.J. and Jackh, R.: Studies on the prenatal toxicity of n,n–dimethylformamide in mice, rats and rabbits. Fd. Chem. Toxic. 29:193–201, 1991.

Lewis, S.C.; Schroeder, R.E. and Kennedy, Jr., G.L.: Developmental toxicity of dimethylformamide in rat following inhalation exposure. Drug and Chemical Toxicology 15:1–14, 1992.

Stula, E.F. and Krauss, W.C.: Embryotoxicity in rats and rabbits from cutaneous application of amide–type solvents and substituted ureas. Toxicol. Appl. Pharmacol. 41:35–55, 1977.

Thiersch, J.B.: Investigations into the differential effect of compounds on rat litter and mother. In: Malformations Congenitales Des Mammiferes. H. Tuchmann–Duplessis (ed.), Paris: Masson and Cie, 95–113, 1971.

841 N,N–Di(n–butyl)foramide

Stula and Krauss (1977) exposed rats cutaneously to up to 1200 mg per kg on day 10 and did not produce significant fetal changes.

Stula, E.F. and Krauss, W.C.: Embryotoxicity in rats and rabbits from cutaneous application of amide–type solvents and substituted ureas. Toxicol. Appl. Pharmacol. 41:35–55, 1977.

842 3,7–Dimethyl–1–(5–oxo–hexyl)xanthine

Schultes et al. (1971) gave 57, 170 and 570 mg per kg to rats during organogenesis and found no effects on the offspring except at 570 mg per kg where there was a reduction in number of live offspring.

Schultes, E.; Popendiker, K.; Doerr, D.I. and Leuschner, F.: Die Vertraglichkeit von 3,7–Dimethyl–1–(5–oxo–hexyl)–xanthin im Tierversuch. Arzneim–Forsch. 21:1446–1453, 1971.

843 3,3–Dimethyl–1–phenyltriazene *DMPT*

Frank et at (1989) studied the pathology produced by this methylating agent in rat embryos whose mothers received 30 mg on day 12. Malformations of the palate, digits, limbs and brain have been previously reported. Frank et al. (**1991**) studied the distribution of the radioactively labelled compound in rodent embryos.

Frank, A.A.; Kazacos, E.A.; Hullinger, R.L. and Thompson, D.J.: Teratogenicity of 3,3–dimethyl–1–phenyltriazene in the rat: histopathologic and ultrastructural alterations. Teratology 40:495–504, 1989.

Frank, A.A.; Kazacos, E.A. and Thompson, D.J.: Teratogenicity of 3,3–dimethyl–l–phenyltriazene: distribution in rats and rat embryos. Toxicology Letters 55:55–884, 1991.

844 N,N–Dimethyl–p–phenylenediamine CAS 99–98–9

Discussed under Hair Dyes

845 3–3–Dimethyl–l–phenyl–triazene *Triazene* CAS 15056–34–5

Murphy et al. (1957) summarized their work with this compound in the chick and the rat. Skeletal defects in the chick were reversed partially by nicotinamide. Treatment at 11 days in the rat produced major skeletal defects and at 12 days, cleft palate and absence of the mandible became predominant. The single daily dose was 30 mg per kg. Rats exposed to 0.2 or 2 mg per square meter during pregnancy gave birth to fetuses with hydrocephalous, hematomas, bone malformations or infarct–like lesions of the liver (Mirkova, **1981**).

Mirkova, E.: A propos of the embryotoxic effect of triazine herbicide polyzin 50 (Russ). IARC Monographs 53:511, 1981.

Murphy, J.L.; Dagg, C.P. and Karnofsky, D.A.: Comparison of teratogenic chemicals in the rat and chick embryos: An exhibit with additions for publication. Pediatrics 19:701–714, 1957.

846 N,N–Dimethylphosphoramidocyanidate CAS 77–81–6 *Tabun*[TM]

Denny et al. (1989) gave up to 300 microg orally per day during rat embryogenesis. Maternal weight gain was reduced but no adverse fetal effects were found. Denny et al. (1990) found no developmental toxicity in rabbits given 317 microgm per kg during organogenesis.

Denny, K.H.; Parker, R.M.; Bucci, T.J. and Dacre, J.D.: Developmental toxicity of Tabun (GA) in the CD rat. Teratology 39: P72, 1989.

Denny, K.H.; Parer, R.M.; Divine, B.L. and Bucci, T.J.: Negative developmental toxicity test of tabun (GA) in the new zealand white rabbit. (abst)Teratology 41:549, 1990.

847 Dimethyl Phthalate CAS 131–11–3

Plasterer et al. (1985) gavage fed mice on days 7–14 with a dose considered to be just below adult lethality (3500 mg per kg). No gross congenital defects were detected in neonates and there was no increase in lethality.

Plasterer, M.R.; Bradshaw, W.S.; Booth, G.M.; Carter, M.W.; Schuler, R.L. and Hardin, B.D.: Developmental toxicity of nine selected compounds following prenatal exposure in the mouse: Naphthalene, p–nitrophenyl, sodium selenite, dimethyl phthalate, ethylene–thiourea and four glycol ether derivatives. J. Toxicol. Environ. Health 15:25–38, 1985.

848 Dimethylstilbestrol

Martin et al. (1960) studied the ability of this estrogen to interrupt pregnancy. Intramuscular doses of up to 64 microgm had no effect on litter size when admininstered on days 10, 11 and 12. Doses of 16, 32 and 64 microgm given on days 4, 5 and 6 did cause reduction in litter size.

Martin, L.; Emmens, C.W. and Cox, R.I.: The effects of oestrogens and anti–oestrogens in early pregnancy in mice. J. Endocrinol. 20:299–306, 1960.

849 Dimethyl Sulfoxide *DMSO* CAS 67–68–5

Ferm (1966) reported that DMSO injected in hamsters on the 8th day (0.5 to 8.0 gm per kg) of gestation produced exencephaly in a high proportion of fetuses. Rib fusions, microphthalmia, limb abnormalities and cleft lip were found, also. Thiersch (1971), using 4.0 ml per kg on days 4 and 5 or tail painting on days 11 through 18, caused no defects in rat fetuses. Caujolle et al. (1967) could produce only a few defects using 8 gm per kg in the mouse and rat. When they used 10 to 12 mg per egg, a 22 percent incidence of limb defects was seen in chicks. At doses between 10 and 30 gm per kg, Juma and Staples (1967) found increased resorptions in the rat.

Bariljak et al. (1978) showed that the peritoneal injection of 5.0 ml per kg on 13–day rat pregnancy reduced teratogenic and embryolethal action of pyrimethamine (50 mg per kg).

Bariljak, I.R.; Neumerzhitskaja, L.V. and Turkevich, A.N.: Studies on the antimutagenic and antiteratogenic characteristics of dimexide (DMSO). Tsitol. Genet. (USSR) 12:50–56, 1978.

Caujolle, F.M.E.; Caujolle, D.H.; Cros, S.B. and Calvet, M.J.: Limits of toxic and teratogenic tolerance of dimethyl sulfoxide. Ann. N.Y. Acad. Sci. 141:110–125, 1967.

Ferm, V.H.: Congenital malformations induced by dimethyl sulphoxide in the golden hamster. J. Embryol. Exp. Morphol. 16:49–54, 1966.

Juma, M.B. and Staples, R.E.: Effect of maternal administration of dimethylsulfoxide on the development of rat fetuses. Proc. Soc. Exp. Biol. Med. 125:567–569,1967.

Thiersch, J.B.: Investigations into the differential effect of compounds on rat litter and mother. In: Malformations Congenitales Des Mammiferes. H. Tuchmann–Duplessis (ed.), Paris: Masson et Cie, 95–113, 1971.

850   Dimethyl Terephthalate

Krotov and Chebotar (**1972**) exposed rats to 1 mg per cubic meter of air continuously during pregnancy and although there was embryolethality no teratogenicity was found. This exposure is about 110 times higher than the maximum allowable exposure in the Soviet Union.

Krotov, J.A. and Chebotar, N.A.: Embryotoxic and teratogenic action of some industrial substances formed during production of dimethyl terephthalate. Gig. Tr. Prof. Zabol. 16:40–43, 1972.

851   N,N–Dimethylthiourea

Discussed under Ethylenethiourea

852   Dimethylthiourea

Teramoto et al. (1981) gave 2000 mg per kg orally to rats on day 12 and to mice on day 10 and found no increase in the defect rate or resorptions in the mice. In rats resorptions were increased as were defects. Micrognathia, cleft palate and skeletal defects were found. This chemical is also discussed under ethylenethiourea.

Teramoto, S.; Kaneda, M.; Aoyama, H. and Shirasu, Y.: Correlation between the molecular structure of N–alkylureas and N–alkylthioureas and their teratogenic properties. Teratology 23:335–342, 1981.

853   Dimethyl Tubocurarine Chloride CAS 35–57–6

Discussed under Tubocurarine

854   1,3 Dimethylurea

Teramoto et al. (1981) gave 2000 mg per kg orally to rats on day 12 and to mice on day 10 and found no increase in the defect rate or resorptions in the rat. In the mice resorptions were increased and cleft palate and skeletal defects were increased.

Teramoto, S.; Kaneda, M.; Aoyama, H. and Shirasu, Y.: Correlation between the molecular structure of N–alkylureas and N–alkylthioureas and their teratogenic properties. Teratology 23:335–342, 1981.

855   Diminazene Diaceturate

Yoshimura (**1990**) gave rats up to 1000 mg per kg orally on days 8–15. At the highest dose resorptions and decrease in fetal weight was seen but no teratogenicity.

Yoshimura, H.: Teratological assessment of the antiprotozoal, diminazene diaceturate, in rats. Toxicology Letters 54:55–59, 1990.

856   Dinitrophenol CAS 25550–58–7

Goldman and Yakovac (1964) produced fetal growth inhibition but no defects in rat fetuses after maternal doses of 8 to 40 mg per kg on gestational days 9, 10 or 11. Gibson (1973) used the compound orally (38.3 mg per kg) and intraperitoneally (13.6 mg per kg) on days 10 through 12 in the mouse and found no teratogenic action.

Klein–Obbink et al. (**1964**) administered dinitrophenol intraperitoneally on four days of pregnancy in amounts of up to 30 mg per kg. Fourteen offspring (of 389) had anophthalmia or microoophthalmia. In another group fed DNP 192 offspring did not have eye defects.

Gibson, J.E.: Teratology studies in mice with 2-sec-butyl-4, 6-dinitrophenol (dinoseb). Food Cosmet. Toxicol. 11:31–43, 1973.
Goldman, A.S. and Yakovac, W.C.: Salicylate intoxication and congenital anomalies. Arch. Environ. Health 8:648–656, 1964.
Klein–Obbink, H.J. and Dalderup, L.M.: Effect of acetylsalicylic acid on fetal mice and rats. Lancet 2:152, 1964.

857   Dinitrotoluene

Wolkowski–Tyl et al. (1981) gavaged rats on days 7–20 with up to 100 mg per kg and found no adverse effects in the fetuses. Price et al. (1985) reported these experiments in more detail.

Price, C.J.; Tyl, R.W.; Marks, T.A.; Paschke, L.L.; Ledoux, T.A. and Reel, J.R.: Teratologic evaluation of dinitrotoluene in the Fischer 344 rat. Fund. Appl. Toxicol. 5:948–961, 1985.
Wolkowski–Tyl, R.; Jones–Price, C.; Ledoux, T.A.; Marks, T.A. and Langhoff–Paschke, L.: Teratogenicity evaluation of technical grade dinitrotoluene in the Fischer 344 rat. (abs) Teratology 23:70A only, 1981.

858   Dinocap *2,4–Dinitro–6–(1–methylheptyl) phenyl crotonate 2,6–Nitro–4–(1–methylheptyl) phenyl crotonate* CAS 39300–45–3

This fungicide, a mixture of crotonates and phenols, was studied in mice, rats and hamsters. Mice exposed to 25 mg per kg orally on days 7–16 of gestation had increased mortality and cleft palates. At 12 mg per kg, the offspring had torticollis. No effects were seen in hamster fetuses exposed to up to 200 mg per kg. No adverse effects were found when the rats were given 50 mg per kg. Gray et al. (1986) fed this fungicide to mice and at 25 mg per kg on days 7–16 found increased postnatal death, cleft palates and head tilting in the offspring. Hamsters received up to 200 mg per kg without postnatal behavioral changes. Rats receiving 100 mg per kg on days 7–10 had offspring with normal development.

Rogers et al. (1989) showed that otolith development in the mouse and hamster was abnormal after treatment during organogenesis. The two isomers 2,4–dinitro–6–(1–methylheptyl) phenyl crotonate that account for dinocap's activity were studied in mice during organogenesis and cleft palates and abnormal otoliths were found (Rogers et al., 1987).

Gray, L.E., Jr.; Rogers, J.M.; Kavlock, R.J.; Ostby, J.S.; Ferrell, J.M. and Gray, K.L.: Prenatal exposure to the fungicide dinocap causes behavioral torticollis, ballooning and cleft palate in mice, but not rats or hamsters. Teratogenesis, Carcinogenesis, and Mutagenesis 6:33, 1986.

Rogers, J.M.; Gray, L.E., Jr.; Carver, B.D. and Kavlock, R.J.: The developmental toxicity of dinocap in the mouse is not due to two isomers of the major active ingredients. Teratogenesis, Carcinogenesis and Mutagenesis 7:341–346, 1987.

Rogers, J.M.; Burkhead, L.M. and Barbee, B.D.: Effects of dinocap on otolith development: Evaluation of mouse and hamster fetuses at term. Teratology 39:515–524, 1989.

859 Dioctyl Sodium Sulfosuccinate *Colace™*
1,8–Dihydroxy Anthraquinone Solven Docusate Danthron

Among the 792 women who took this stool softener during pregnancy there was no increase in congenital defects. (Jick et al **1981** and Aselton et al. **1985**)

Ichikawa and Yamamoto (1980) fed this laxative combination to rats on days 7–17 in doses of up to 800 mg per kg. At the highest dose viability of the fetuses was decreased. No other ill effects were found.

Aselton, P.; Jick, H.; Milunsky, A.; Hunter, J.R. and Stergachis, A.: First–trimester drug use and congenital disorders. Obstet Gynecol 65(4):4511–455, 1985.

Ichikawa, Y. and Yamamoto, Y.: Teratology study of solvents in rats. Gendai Iryo 12:819–831, 1980.

Jick, H.; Holmes, L.B.; Hunter, J.R.; Madsen, S. and Steragachis, A.: First–trimester drug use and congenital disorders. JAMA 246(4):343–346, 1981.

860 Dinoseb *Dibutox™ 2–sec–Butyl–4,6–dinitrophenol* CAS 88–85–7

Gibson (1973) administered this herbicide orally, subcutaneously or intraperitoneally to mice in doses of up to 20 mg per kg per day. Doses over 17.7 to 20 mg per kg were toxic to the mother. Several time schedules during organogenesis were employed. At 17.7 mg per kg doses by the intraperitoneal and subcutaneous route, malformations were found. These included skeletal defects, cleft palate, hydrocephalus and adrenal agenesis. Oral administration produced no gross or soft tissue defects but some skeletal defects were seen at maternally toxic levels. Gibson and Rao (1973) studied the herbicides distribution in the pregnant mouse.

McCormack et al. (1980) studied postnatal renal function in rats exposed on days 10–12 of gestation by maternal intraperitoneal injection of 16 mg per kg. The dilated tubules and pelves seen in the neonatal period were not found or reduced by 42 days postnatally and renal function was normal. Beaudoin and Fisher (1981) administered 10 mg per kg intraperitoneally on day 9 of rat pregnancy and studied the embryos in vitro after day 10. Little or no effect was observed on growth and development of the embryos. Oral doses of 10 mg per kg on days 6 through 18 in the rabbit were associated with brain, diaphragmatic and skeletal defects (Anonymous, 1986). Fetal weight reduction was found

in rats exposed to 10 mg per kg on days 6–15.

Anonymous: Cited in Federal Register, 51(198):36637 only, 1986.

Beaudoin, A.R. and Fisher, D.L.: An in vivo=in vitro evaluation of teratogenic action. Teratology 23:57–61, 1981.

Gibson, J.E.: Teratology studies in mice with 2–sec–butyl-4, 6-dinitrophenol (dinoseb). Food Cosmet. Toxicol. 11:31–34, 1973.

Gibson, J.E. and Rao, K.S.: Disposition of 2–sec–butyl-4, dinitro–phenol (dinoseb). Food Cosmet. Toxicol. 11:45–52, 1973.

McCormack, K.M.; Abuelgasim, A.; Sanger, V.L. and Hook, J.B.: Postnatal morphology and functional capacity of the kidney following prenatal treatment with dinoseb in rats. J. Toxicol. Environ. Health 6:633–643, 1980.

861 Diosgenin CAS 512–04–9

Keeler and Binns (1968) gave 1.2–1.5 gm on the 14th day of gestation to 4 ewes and found no craniofacial defects in the offspring.

Keeler, R.F. and Binns, W.: Teratogenic compounds of veratrum californicum (Durand). Teratology 1:5–10, 1968.

862 Dioxin

Discussed under 2,3,7,8–Tetrachlorodibenzo–p–dioxin

863 Diphenhydramine HCl CAS 147–24–0

Heinonen et al. (1977) reported no significant increase in congenital defects among the offspring of 595 women who used the drug in the first 4 lunar months of their gestation. Aselton et al. (1985) studied the offspring of 270 women taking the medication in the first trimester and reported only four malformations.

This antihistamine was given to rats and rabbits and no defects were produced (Schardein et al., 1971). The compound was given orally in amounts of 15 to 20 mg per kg during active organogenesis. Saxen (1975) reported that a concentration of one microgm per ml delayed fusion of explanted mouse palatal shelves. Naranjo and Narango (1968) fed mice water containing 10–4 or 10–3 mg per kg during pregnancy. The fetal weight and survival were slightly reduced as was the day for vaginal and external ear opening.

Aselton, P.A.; Jick, H.; Milunsky, A.; Hunter, J.R. and Stergachis, A.: First–trimester drug use and congenital disorders. Obstet. Gynecol. 65:451–455, 1985.

Heinonen, O.P.; Slone, D. and Shapiro, S.: Birth Defects and Drugs in Pregnancy. Publishing Sciences Group Inc., Littleton, Mass., 1977.

Naranjo, P. and Naranjo, E.: Embryotoxic effects of antihistamines. Arzneim Forsch. 18:188–195, 1968.

Saxen, I.: Etiological variables in oral clefts. Proceedings of the Finnish Dental Society. 71(3):3–40, 1975.

Schardein, J.L.; Hentz, D.L.; Petrere, J.A. and Kurtz, S.M.: Teratogenesis studies with diphenhydramine HCl. Toxicol. Appl. Pharmacol. 18:971–976, 1971.

864 Diphenoxylate *Lomotil* CAS 915–30–0

Heinonen et al. (1977) reported no defects among seven mothers exposed during the first trimester.

Heinonen, O.P.; Slone, D. and Shapiro, S.: Birth Defects and Drugs in Pregnancy. Publishing Sciences Group Inc., Littleton, Mass., 1977.

865 Diphenyl CAS 92–52–4

Khera et al. (1979) administered this fungicide by gavage to rats at doses up to 500 mg per kg on days 6–15 and found no teratogenicity.

Khera, K.S.; Whalen, C.; Angers, G. and Trivett, G.: Assessment of the teratogenic potential of piperonyl butoxide, biphenyl and phosalone in the rat. Toxicol. Appl. Pharmacol. 47:353–358, 1979.

866 Diphenylamine CAS 122–39–4

Crocker and Vernier (1970) fed a diet of 2.5 percent diphenylamine to rats during the last six days of pregnancy and produced cystic dilatation of the renal collecting ducts and degeneration of the proximal tubules in the fetuses. Tube feeding of two ml of a one percent solution also produced the fetal kidney changes.

Crocker, J.F.S. and Vernier, R.L.: Chemically induced polycystic disease in the newborn. (abs) Pediatr. Res. 4:448 only, 1970.

867 2,6–cis–Diphenylhexamethylcyclotetrasiloxane

Levier and Jankowiak (1972) gave this antiestrogen to pregnant rats and with 0.33 mg per kg on days 1–5 or 3.0 mg per kg on day 1 prevented implantation. Resorptions were increased when 10 mg per kg was given on days 6–8.

Levier, R.R. and Jankowiak, M.E.: The hormonal and antifertility activity of 2,6–cis–diphenylhexamethylcyclotetrasiloxane in the female rat. Biology of Reproduction 7:260–266, 1972.

868 Diphenylhydantoin *Dilantin™ Phenytoin* CAS 57–41–0

Malformation Studies

Although initial studies suggested that diphenylhydantoin was teratogenic in humans, subsequent studies have cast doubt on this association. The various studies are detailed below.

The offspring of epileptic mothers treated with diphenylhydantoin were reported to have about twice the expected rate of congenital defects (Janz and Fuchs, 1964; Speidel and Meadow, 1972; Fedrick, 1973; Monson et al., 1973). The type of defect encountered included cleft lip and palate, congenital heart disease and microcephaly. Czeizel (1976) studied medications ingested by mothers of babies with cleft lip and=or palate and found 11 out of 413 (2.7 percent) had taken phenytoins. Hypoplasia of the nails and distal phalanges may represent a specific malformation in as many as 18 percent of pregnancies associated with this drug. (Barr et al., 1974; Hill et al., 1974). This type of defect tends to become less noticeable with age and it could be easily overlooked in the newborn (Hanson and Smith, 1975).

Some reports indicated no increase in malformation after dilantin exposure and these studies tended to be more prospective rather than retrospective in nature. Hanson

(1986) reviewed the clinical effects and believes that not more than 5–10 percent of exposed infants have the hydantoin syndrome and that subtle changes can be found in 30 percent. Goujard et al. (1974) in a study of their own and a review of five prospective studies, found no increase in malformations. Meyer (1973) studied retrospectively 199 children born of mothers treated with antiepileptics and could not establish any proof of teratogenic activity. Livingston et al. (1973) also had negative findings. Other factors associated with chronic seizure conditions may contribute to a slight increase in malformations. Bertolini et al. (1987) studied 153 pregnancies in which monotherapy was employed; they found 7 defects when 5 were expected.

In a detailed study of twenty children exposed to phenytoin with or without another drug Vanoverloop et al. (1992) found about a 10 percent decrease in intellectual function tests. The authors review other studies dealing with behavior and comment on the finding that if other drugs are used the deficit may be increased.

Strickler et al. (1985) studied 24 children exposed in utero to phenytoin using an in vitro assay (lymphocyte killing). When phenytoin was added to the lymphocytes along with a microsomal P–450 system, 14 children had abnormal tests and 10 were normal. Major defects were present in 12 of the 14 positive but only 2 of the 10 negatives. Positive tests were found in about 50 percent of the parents. Koren et al. (1989) found that none of the 188 cases of neuroblastoma they studied was associated with prenatal phenytoin exposure. Van Dyke et al. (1988) found that mothers who had one affected child are at a much higher risk for having a second one affected (9 of 10). If the first was not affected only 5 of 52 had an affected second child.

Cancers derived from neural crest cells appear to be more common than expected in children exposed to diphenylhydantoin. At least 4 cases are reported in the literature (Ehrenbard and Chaganti, 1981).

Many malformations probably have a multifactorial etiology; the role of folic acid deficiency and genetic predisposition need to be more thoroughly studied here. Dansky et al. (1987) found that twice as many women on diphenylhydantoin had depressed red cell folic acid levels as those on phenobarbitol.

Shapiro et al. (1976) drawing on the Collaborative Perinatal Project in the United States and a large Finnish registry, recorded an increase in malformations in the offspring when either the mother or father was epileptic and suggested that the increased rate was not due only to drug exposure. Based on 701 at–risk parents, the defect rate was 10.5 percent when the mother was epileptic as compared to 8.3 percent when the father was affected and 6.4 when neither parent was affected. The Committee on Drugs of the American Academy of Pediatrics (1979) in their review gave sound recommendations useful to those advising pregnant women with epilepsy.

Mirkin (1971) showed that diphenylhydantoin equilibrates across the human placenta at term and when the fetus is 8 to 10 cm in crown–rump length. Kelly (1984) reviewed the literature on anticonvulsants and teratogenicity.

Intellectual Function

Hanson et al. (1976) drawing on a Seattle study and the records of the Collaborative Perinatal Project of the National Institute of Neurological and Communication Disorders and Stroke, emphasize that while only 11 percent of exposed offspring have the hydantoin syndrome, three times as many have impairment of performance, especially intellectual.

Kelly et al. (1984) studied prospectively 89 mothers who took the drug during pregnancy and they reported that 30 percent of the offspring had minor craniofacial and digital changes but they did not find a significant increase in growth retardation, mental retardation or major malformations.

Gaily et al. (1988) examined blindly 82 offspring of mothers taking phenytoin. The examinations performed at age five and one–half years found an excess of minor anomalies but there was evidence that many of these were genetically related and only hypertelorism and digital hypoplasia were associated with phenytoin exposure. None of the patients had mental retardation.

Adams et al. (1990) summarized the neurobehavioral effects in animals and humans. They conclude that intellectual delay is present with the Fetal Hydantoin syndrome but in those without the syndrome an increased risk for mental retardation is not substantiated. Dessens et al. (**1994**) reviewed studies done on the cognitive and endocrine function of children who were exposed prenatally to anticonvulsants. The studies which have been reported generally show a lowered IQ level.

Animal Studies

Harbinson and Becker (1969) used 75 to 150 mg per kg in mice on days 8–13 and produced cleft lip and palate, ectrodactyly and minor skeletal defects. Hydronephrosis and internal hydrocephalus were found in those fetuses treated on day 9. Open eye was common in fetuses exposed on day eight. Elshove (1969) produced cleft palate in mice also and wondered if the mechanism of action was through folic acid deficiency induced by the drug. Schardein et al. (1973) could not prevent cleft palate in diphenylhydantoin–treated mice by giving folinic acid to 100 mg per kg. Khera (1979) found embryolethality in the cat given 2 mg per kg.

Finnell (1981) studied several strains of mice including one type with seizures (quaker) and found that the malformation rate correlated with both the dose and blood level of the drug. In the rabbit, doses of 75 and 100 mg per kg per day were teratogenic (McClain and Langhoff, 1980). Cleft palate and limb defects were found. Mullenix et al. (1983) studied behavior in rat fetuses exposed prenatally. Vorhees (1987) reported that at gavage doses of 100, 150 and 200 mg per kg on days 7–18 in the rat, dose–related postnatal changes occurred. Roberts et al. (**1991**) gave evidence that the maternal plasma level determined the amount of fetal toxicity observed in different strains of mice.

Shanks et al. (1989) found that diphenylhydantoin embryotoxicity was enhanced by adding a source of cytochrome P–450 to an in vitro mouse culture system. Since covalent binding of the parent compound was observed in embryos cultured without bioactivation the authors suggested that the embryo itself was capable of bioactivation. Finnell et al. (**1993**) found that stiripentol, a cytochrome P–450–inhibiting drug reduced the teratogenicity of dilantin in 3 mouse strains.

Danielson et al. (**1992**) found that reduction of the terminal digits of expose rabbit fetuses was associated with fetal hypoxia followed by vascular disruption and necrosis.

Wong and Wells (1988) have shown that prostoglandin synthetase and glutathione modulate teratogenicity in the mouse. Coadministration of N–acetylcysteine reduced fetotoxicity. Agents affecting prostoglandin synthetase (acetylsalicylic acid, α–phenyl–N–t–butylnitrone and caffeic acid) decreased the incidence of hydantoin induced cleft palate in the mouse.

Adams, J.; Vorhees, C.V. and Middaugh, L.D.: Developmental neurotoxicity of anticonvulsants: human and animal evidence on phenytoin. Neurotoxicology and Teratology 12:203–214, 1990.

Barr, M.; Pozanski, A.K. and Schmickel, R.D.: Digital hypoplasia and anticonvulsants during gestation, a teratogenic syndrome. J. Pediatr. 4:254–256, 1974.

Bertollini, R.; Kallen, B.; Mastroiacovo, P. and Robert, E.: Anticonvulsant drugs in monotherapy. Effect on the fetus. Eur. J. Epidemiol. 3:164–171, 1987.

Committee on Drugs, American Academy of Pediatrics: Anticonvulsants and pregnancy. Pediatrics 63:331–333,1979.

Czeizel, A.: Diazepam, phenytoin and etiology of cleft lip and/or palate. Lancet 1:81 only, 1976.

Danielson, M.K.; Danielsson, B.R.G.; Marchner, H.; Lundin, M.; Rundqvist, E. and Reiland, S.: Histopathological and hemodynamic studies supporting hypoxia and vascular disruption as explanation to phenytoin teratogenicity. Teratology 46:485–497, 1992.

Dansky, L.V.; Andermann, E.; Rosenblatt, D.; Sherwin, A.L. and Andermann, F.: Anticonvulsants, folate levels, and pregnancy outcome: A prospective study. Annals of Neurology 21:176–182, 1987.

Dessens, A.B.; Boer, K.; Koppe, J.G.; van de Poll, N.E. and Cohen–Kettenis, P.T.: Studies on long–lasting consequences of prenatal exposure to anticonvulsant drugs. Acta Paediatr. Suppl. 404:54–64, 1994.

Ehrenbard, L.T. and Chaganti, R.S.K.: Cancer in the fetal hydantoin syndrome. Lancet 2:97 only, 1981.

Elshove, J.: Cleft palate in the offspring of female mice treated with phenytoin. Lancet 2:1074 only, 1969.

Fedrick, J.: Epilepsy and pregnancy: A report from Oxford record linkage study. Br. Med. J. 2:442–448, 1973.

Finnell, R.H.: Phenytoin–induced teratogenesis: A mouse model. Science 211:483–484, 1981.

Finnell, R.H.; Van Waes, M.; Musselman, A.; Kerr, B.M. and Levy, R.H.: Differences in the patterns of phenytoin–induced malformations following stiripentol coadministration in three inbred mouse strains. Reproductive Toxicology 7:439–448, 1993.

Gaily, E.; Granstrom, M–L.; Hilesmaa, V. and Bardy, A.: Minor anomalies in offspring of epileptic mothers. J. Pediatr. 112:520–529, 1988.

Goujard, J.; Huel, G. and Rumeau–Rouquette, C.: Antiepileptiques et malformations congenitales. J. Gynecol. Obstet. Biol. Reprod. 8:831–842, 1974.

Hanson, J.W.: Teratogen update: Fetal hydantoin effects. Teratology 33:349–353, 1986.

Hanson, J.W.; Myrianthopoulos, N.C.; Harvey, M.A.S. and Smith, D.W.: Risks to the offspring of women treated with hydantoin anticonvulsants, with emphasis on the fetal hydantoin syndrome. J. Pediatr. 89:662–668, 1976.

Hanson, J.W. and Smith, D.W.: The fetal hydantoin syndrome. J. Pediatr. 87:285–290, 1975.

Harbinson, R.D. and Becker, B.A.: Relation of dosage and time of administration of diphenylhydantoin to its teratogenic effect in mice. Teratology 2:305–312, 1969.

Hill, R.M.; Verniaud, W.M.; Horning, M.G.; McCulley, L.B. and Morgan, N.F.: Infants exposed in utero to antiepileptic drugs. Am. J. Dis. Child. 127:645–653, 1974.

Janz, D. and Fuchs, U.: Sind Antiepileptische Medikamente waehrend der Schwangerschaft schaedlich. Dtsch. Med. Wochenschr. 89:241–243, 1964.

Kelly, T.E.: Teratogenicity of anticonvulsant drugs. 1. Review of the literature. Am. J. Med. Genet. 19:413–434, 1984.

Kelly, T.E.; Edwards, P.; Rein, M.; Miller, J.Q. and Dreifuss, F.E.: Teratogenicity of anticonvulsant drugs. 2. A prospective study. Am. J. Med. Genet. 19:435–443, 1984.

Khera, K.S.: Teratogenicity study of hydroxyurea and diphenylhydantoin in the cat. Teratology 20:447–452, 1979.

Koren, G.; Demitraoudis, D.; Weksberg, R.; Rieder, M.; Shear, N.H.; Sonely, M.; Shandling, b. and Spielberg, S.P.: Neuroblastoma after prenatal exposure to phenytoin: cause and effect. Teratology 40:157–162, 1989.

Livingston, S.; Berman, W. and Pauli, L.: Maternal epilepsy and abnormalities of the fetus and newborn. Lancet 2:1265 only, 1973.

McClain, R.M. and Langhoff, L.: Teratogenicity of diphenylhydantoin in the New Zealand white rabbit. Teratology 21:371–379, 1980.

Meadow, S.R.: Congenital abnormalities and anticonvulsant drugs. Proc. R. Soc. Med. 63:48–49, 1970.

Meyer, J.G.: The teratological effects of anticonvulsants and the effects on pregnancy and birth. Europ. Neurol. 10:179–190, 1973.

Mirkin, B.L.: Diphenylhydantoin: Placental transport, fetal localization, neonatal metabolism and possible teratogenic effects. J. Pediatr. 78:329–337, 1971.

Monson, R.R.; Rosenberg, L.; Hartz, S.C.; Shapiro, S.; Heinonen, O.P. and Slone, D.: Diphenylhydantoin and selected malformations. N. Eng. J. Med. 289:1049–1052, 1973.

Mullenix, P.; Tassinari, M.S. and Keith, D.A.: Behavioral outcome after prenatal exposure to phenytoin in rats. Teratology 27:149–157, 1983.

Roberts, L.G.; Laborde, J.B. and Slikker Jr., W.S.: Phenytoin teratogenicity and midgestational pharmacokinetics in mice. Teratology 44:497–505, 1991.

Schardein, J.L.; Dresher, A.J.; Hentz, D.L.; Petrere, J.A.; Fitzgerald, J.E. and Kurtz, S.M.: The modifying effect of folinic acid on diphenylhydantoin–induced teratogenicity in mice. Toxicol. Appl. Pharmacol. 24:150–158, 1973.

Shanks, M.J.; Wiley, M.J.; Kubow, S. and Wells, P.G.: Phenytoin embryotoxicity: role of enzymatic bioactivation in a murine culture model. Teratology 40:311–320, 1989.

Shapiro, S.; Hartz, S.C.; Siskind, V.; Mitchell, A.A.; Slone, D.; Rosenberg, L.; Monson, R.R.; Heinonen, O.P.; Idanpaan–Heikkila, J.; Haro, S. and Saxen, L.: Anticonvulsants and parental epilepsy in the development of birth defects. Lancet 1:272–275, 1976.

Speidel, B.D. and Meadow, S.R.: Maternal epilepsy and abnormalities of the fetus and newborn. Lancet 2:839–843, 1972.

Strickler, S.M.; Miller, M.A.; Andermann, E.; Dansky, L.V.; Seni, M–H. and Spielberg, S.P.: Genetic predisposition to phenytoin–induced birth defects. Lancet 2:746–749, 1985.

Van Dyke, D.C.; Hodge, S.E.; Heide, F. and Hill, L.R.: Family studies in fetal phenytoin exposure. Journal of Pediatrics 113:301–306, 1988.

Vanoverloop, D.; Schnell, R.R.; Harvey, E.A. and Holmes, L.B.: The effects of prenatal exposure to phenytoin and other anticonvulsants on intellectual function at 4 to 8 years of age. Neurotoxicology and Teratology 14:329–335, 1992.

Vorhees, C.V.: Fetal hydantoin syndrome in rats: Dose–effect relationships of prenatal phenytoin on postnatal development and behavior. Teratology 35:287–303, 1987.

Wong, M and Wells, P.G.: Effects of n–acetylcysteine on fetal development and on phenytoin teratogenicity in mice. Teratogenesis, Carcinogenesis, and Mutagenesis 8:65–79, 1988.

869 Diphenylpyraline

King et al. (**1965**) studied teratogenicity antihistamine of this in the rat. They used 2.5 to 37.5 mg per kg on days 1–16. No adverse effects were found.

King, C.T.G.; Weaver, S.A. and Narrod, S.A.: Antihistimines and teratogenicity in the rat. J. Pharmacol. Exp. Therapeut. 147:391–398, 1965.

870 Diphenylpyraline–8–chlorotheophyllinate

Knoche and Konig (1964) gave mice 4 mg per kg in drinking water from the 3rd to 5th day of gestation until term and found no increase in defective fetuses.

Knoche, Ch. and Konig, J.: Zur pranatalen Toxizitat von Diphenylpyralin-8-chlor–theophyllinat unter Berucksichtigung von Erfahrungen mit Thalidomid und Coffein. Arzneim Forsch. 14:415–424, 1964.

871 Dipyridamole *Perdcantin*™ CAS 58–32–2

Uzan et al. (**1991**) used 225 mg daily in conjunction with aspirin 150 mg daily to reduce fetal growth retardation and abruptio in at–risk mothers. Unfavorable features of newborns was not increased in 119 studied. Chen et al. (1982) reported 10 pregnancies in women taking this medication. There were three spontaneous abortions and seven newborns without serious defects. These ten pregnancies were complicated by anticoagulant therapy for prosthetic heart valves.

Chen, W.W.C.; Chan, C.S.; Lee, P.K.; Wang, R.Y.C. and Wong, V.C.W.: Pregnancy in patients with prosthetic heart valves: An experience with 45 pregnancies. Quart. J. Med. 51:358–365, 1982.

Uzan, S.; Beaufils, M.; Breart, G.; Uzan, M. and Paris, J.: Prevention of fetal growth retardation with low–dose aspirin: Findings of the EPREDA trial. Lancet 337(8755):1427–1431, 1991.

872 2,2 Dipyridyl CAS 366–18–7

This chelator for ferrous iron was given intraperitoneally to rats on days 11.5 through 14.5 in doses of 60–75 mg per kg (Oohira and Nogami, 1978). Skeletal defects especially of the limbs were found.

Oohira, A. and Nogami, H.: Limb anomalies produced by 2,2 dipyridyl in rats. Teratology 18:63–70, 1978.

873 Dipyrithione *Omadine MDS* CAS 3696–28–4

Heinonen et al. (1977) found no increase in defects among the offspring of 141 women treated in pregnancy but only 6 were tested in the first four lunar months.

This antimicrobial antifungal chemical was applied dermally to pregnant swine to give a dose of 10, 30 and 100 mg per kg. Crooked tail occurred in some of the fetuses. In the rat receiving 30 or 100 mg per kg on days 6 through 15 of gestation, maternal toxicity, increased resorption sites and rib fusions were found (Wedig et al., 1977). Johnson et al. (1984) studied this broad spectrum antibacterial and antifungal agent in pregnant rats and rabbits. Maternal mortality was high at 7.5 mg per kg orally, but doses to 30.0 mg per kg applied dermally during organogenesis did not cause an increase in malformations.

Heinonen, O.P.; Slone, D. and Shapiro, S.: Birth Defects and Drugs in Pregnancy, Publishing Sciences Group, Inc., 1977.

Johnson, D.E.; Schardein, J.L.; Mitoma, C. and Wedig, J.H.: Reproductive toxicological evaluation of omadine MDS. Fund. Appl. Toxicol. 4:81–90, 1984.

Wedig, J.H.; Kennedy, G.L.; Jenkins, D.H. and Keplinger, M.L.: Teratological evaluation of magnesium sulfate adduct of [2,d–dithio–bis(pyridine–1–oxide)] in swine and rats. Toxicol. Appl. Pharmacol. 42:561–570, 1977.

874 Dipyrone *Methampyrone* CAS 5907–38–0

Ungthavorn et al. (1970) injected this antipyretic into pregnant mice in doses up to 1000 mg per kg on days 8, 9 or 10. A low incidence of defects resulted and the incidence was not dose dependent. Six out of 68 fetuses receiving 750 mg per kg on day 9 had exencephaly or encephaloceles

Ungthavorn, S.; Chiamsawatphan, S.; Chatsanga, C.; Tangsanga, K.; Limpongsanuruk, S. and Jeyasak, N.: Studies on sulpyrin–induced teratogenesis in mice. J. Med. Assoc. Thailand 53:550–557, 1970.

875 Diquat *1,1–Ethylene–2,2–dipyridilium Dibromide*

CAS 2764–72–9

Khera et al. (1970) reported studies on this herbicide. Pregnant rats were injected with 7 or 14 mg per kg on one of several days during organogenesis. Maternal death occurred often with the higher dose. Skeletal defect of the sternum and lack of ossification or absence of one of the auditory ossicles were detected in some fetuses. Using repeated intraperitoneal injections of 0.5 mg per kg daily no teratogenic effects were found.

Khera, K.S.; Whitta, L.L. and Clegg, D.J.: Embryopathic effects of diquat and paraquat in rats. Pesticides Symposia. Inter–American Conference on Toxicology and Occupational Medicine, Eds: Deichmann, W.B.; Radomski, J.L. and Penalver, R.A.. Miami, Halos and Assoc. Inc. 257--261, 1970.

876 Direct Black 38

Wilson (1955) injected about 40 mg on days 7–10 of rat gestation. Of the 16 dams, 4 resorbed completely but no malformations occurred.

Wilson, J.G.: Teratogenic activity of several azo dyes chemically related to trypan blue. Anat. Rec. 123:313–333, 1955.

877 Disopyramide Phosphate CAS 22059–60–5

This drug was given intravenously to rats before, during or at the end of pregnancy in doses of up to 30 mg per kg and no adverse effects were observed (Umemura et al., 1981). In the rabbit intravenous doses of up to 11 mg per kg had no adverse effect during organogenesis (Esaki and Yanagita, 1981). Umemura et al. (1984) found no adverse effects on the rat fetus or neonate after giving up to 150 mg by gavage on days 7–17.

Esaki, K. and Yanagita, T.: Effects of intravenous administration of disopyramide phosphate on the rabbit fetus. Preclin. Rep. Cent Inst. Exp. Animal. 7:189–198, 1981.

Umemura, T.; Sasa, H.; Esaki, K.; Suzuki, S.; Unagami, S. and Yanagita, T.: Effects of disopyramide phosphate on reproduction in rats. Preclin. Rep. Cent. Inst. Exp. Animal 7:157–173, 1981.

Umemura, T.; Esaki, K.; Ando, K.; Sasa, H. and Yanagita, T.: Effects of intragastric administration of disopyramide phosphate on reproduction in rats. III. Experiment on drug administration during the organogenesis period. Preclin. Rep. Cent. Inst. Exp. Animal 10:87–110, 1984.

878 Disulfiram *Antabuse*™ CAS 97–77–8

Jones et al. (1989) studied 13 newborns whose mothers took disulfiram. Among 4 women who ingested alcohol while on treatment three had infants with fetal alcohol syndrome. The other 9 infants were normal except for one with fetal hydantoin syndrome. Based upon one patient with elevated acetaldehyde, Veghelyi et al. (1978) postulated that the fetal alcohol syndrome might occur in women with acetaldehyde levels over 30 micromolar. Nora et al. (1977) observed 2 infants with severe limb reduction in pregnancies whose exposure was in the first trimester. Gardner et al. (1981) reported that the offspring of a mother on disulfiram (denying alcohol use) had an infant with fetal alcohol syndrome. Jones et al. (**1991**) followed prospectively 21 women using disulfiram during pregnancy. Ten or the mothers took alcohol in addition and among the six offspring were three with the fetal alcohol syndrome. Seven of the women who took only disulfiram gave birth to normal infants.

Favre–Tissot and Delatour (1965) reported that among five women treated with disulfiram and tranquilizers during pregnancy, there were two offspring with club foot and one spontaneous miscarriage. Harding and Edwards (**1993**) gave 125 mg per kg orally to guinea pigs on days 17–19 or 19–21 of gestation. Brain weight was reduced in the female offspring exposed on days 19–21.

Rats fed 100 mg daily from day 3 were found to have 88 percent resorptions on day 13 (Salgo and Oster 1974). Surviving fetuses had no gross malformations as also reported by Robens (1969). The mechanism of embryotoxicity is believed to be via copper chelation but excess dietary copper was not used to reverse the resorption effect. Veghelyi et al. (1978) found that ethanol (4 ml) followed by disulfiram (150 mg per kg) in the pregnant rat on day 4 through day 13 was associated with a 65 percent resorption rate, reduced fetal weight and skeletal retardation. No comment on defects in the offspring was given.

Favre–Tissot and Delatour (1965) also administered up to 400 mg per kg to rats on days 5 through 10 and produced no increase in fetal anomalies. Thompson and Foeb (1985) gave mice 1 or 10 mg per kg orally through pregnancy and except for resorptions at the higher dose found no adverse fetal effects. In vitro studies of day 8 or 9 embryos at 10 and 100 microgram per ml of medium were toxic and the central nervous system was abnormal with exposures of 0.1 microgm per ml.

Favre–Tissot, M. and Delatour, P.: Psycopharmacologie et teratogenese a propos du sulfirame: Essai experimental. (French) Annales Medico–psychogiques 1:735–740, 1965.

Gardner, R.J.M. and Clarkson, J.E.: A malformed child whose previously alcoholic mother had taken disulfiram. The New Zealand Med. J. 680:184–186, 1981.

Harding, A.J. and Edwards, M.J.: Retardation of prenatal brain growth of guinea pigs by sidulfiram. Cong. Anom 33:197–202, 1993. f Finasterride, K.L.; Johnson, K.A.; Chambers, C.C. and Cooper, J.: Effect of disulfiram on the unborn baby: Implications relative to the mechanism of alcohol teratogenesis. Proceedings of the Greenwood Genetic Center 9:70(only), 1990.

Jones, K.L.; Chambers, C.C. and Johnson, K.A.: The effect of disulfiram on the unborn baby. (Abs) Teratology 43:438, 1991.

Nora, A.H.; Nora, J.J. and Blu, J.: Limb reduction anomalies in infants born to disulfiram–treated alcoholic mothers. Lancet 2:664 only, 1977.

Robens, J.F.: Teratologic studies of carbaryl diazionin, norea, disulfiram and thiram in small laboratory animals. Toxicol. Appl. Pharmacol. 15:152–163, 1969.

Salgo, M.P. and Oster, G.: Fetal resorption induced by disulfiram in rats. J. Reprod. Fertil. 39:375–377, 1974.

Thompson, P.A.C. and Foeb, P.I.: The effects of disulfiram on the experimental C–3H mouse embryo. J. Appl. Toxicol. 5:1–10, 1985.

Veghelyi, P.V.; Osztovics, M.; Kardos, G.; Leisztner, L.; Szaszovszky, S. and Imrei, J.: The fetal alcohol syndrome: Symptoms and pathogenesis. Acta Paed. Acad. Scient. Hung. 19:171–189, 1978.

879 Ditazol

4,5–Diphenyl–2–bis(2–hydroxy–ethyl)aminoxazol CAS 18471–20–0

Caprino et al. (1973) gave this anti–inflammatory compound to pregnant mice, rats and rabbits during active organogenesis and except for mild growth failure in the rat, found no fetal changes.

Caprino, L.; Borrelli, F. and Falchetti, R.: Toxicological investigation of 4,5–diphenyl–2–bis (2–hydroxyethyl–)–aminoxazol (ditazol or S222). Arzneim Forsch 23:1287–1291, 1973.

880 2,2–Dithio–bis–(pyridine–1–oxide) CAS 3696–28–4

Discussed under Dipyrithione

881 Dithiocarbamate *Maneb*[TM] *Zineb*[TM]

This fungicide in its manganese or zinc form was given orally to rats in doses of 2–4 gm per kg on day 11 or 13 and a high incidence of neural tube closure defects, cleft lip and palate and skeletal defects resulted (Petrova–Vergieva and Ivanova–Tchemishanska, 1973). These dose levels are at least one–half the LD–50 maternal dose. Below 0.5 gm per kg no defects occurred. They also studied the effect of inhalation of 100 mg of zineb per cubic meter for 4 hours per day after implantation through the end of gestation and found no adverse fetal effects.

Larsson et al. (1976) found teratogenic effects in the rat treated by gavage with 770 mg per kg on day 11 but in the mouse treated on day 9 or 13 with up to 1420 mg per kg, no fetal changes occurred. They studied two other related pesticides, mancozeb and propineb, in mice and rats. No teratogenicity was found in mice but at 1320 mg per kg for mancozeb and 230 mg per kg for propineb, teratogenic activity was found.

Larsson, K.S.; Arnander, C.; Cvekanova, E. and Kjellberg, M.: Studies of teratogenic effects of the dithiocarbamates maneb, mancozeb and propineb. Teratology 14:171–184, 1976.

Petrova–Vergieva, T. and Ivanova–Tchemishanska, L.: Assessment of the teratogenic activity of dithiocarbamate fungicides. Food Cosmet. Toxicol. 11:239–244, 1973.

882 DL111–IT

3–2(ethylphenyl)–5–(3–methoxyphenyl)–1H–1,2,4–triazole

Zhou et al. (**1991**) terminated approximately one-half of rat and mouse pregnancies with a single subcutaneous dose of 11.5 mg per kg on day 6 or 3.0 mg per kg for 5 days starting on day 4. The rodents were then allowed a second pregnancy after 30, 45 or 90 days. No increase in fetal loss or malformation was found but in the group treated after 30 days nephrohydrosis was 8.5% compared to 2.5 and 2.4% at 45 and 90 days.

Zhou, H.; Fang, R.; Yang, B. and Zhang, Y.: Embryotoxicity and teratogenicity of dl111–it, an early pregnancy–terminating agent, in the subsequent gestation following administration in rats. Contraception 43:287–293, 1991.

883 Dobutamine Hydrochloride

Nagaoka et al. (1979) gave rats up to 50 mg per kg daily intravenously before mating and during the first 7 days of gestation. During organogenesis up to 70 mg per kg was given and perinatally 50 mg per kg was used. At 70 mg per kg there was some delay in ossification of the fetus. In the perinatal studies, growth retardation of the offspring occurred to 35 days of age. In the rabbit intravenous doses of up to 30 mg per kg produced no fetal ill effects.

Nagaoka, T.; Fuchigami, K.; Shigemura, T.; Takatouka, K.; Osuga, F. and Hatano, M.: Reproductive studies on S–1000 (dobutamine hydrochloride). Yakuri to Chiryo 7:1691–1706, 1707–1730, 1731–1741, 1742–1763, 1979.

884 Dodecylethoxysulphate, Sodium

Discussed under Surfactants

885 Dofetilide *L–691,121*

Ban et al. (**1992**) treated rats with 1 mg per kg daily. The drug is a class III antiarrhythmic. On individual days 10–13 a high embryonic eath rate occurred. Surviving fetuses had enlarged slow beating hearts. Embryo culture in 0.05 microgm per ml was associated with bradycardia which could be reversed by changing to fresh media. This class III antiarrhythmic was studied in whole rat embryo cultures on day 11 and 14 by Spence et al. (**1994**). At concentrations of 0.01 microgm per ml or above both types of embryo had significant heart rate drops.

Ban, Y.; Konishi, R.; Kawana, K.; Nataksuka, T. and Fujii, T.: Embryotoxic effects of a class III antiarrhythmic agent, l–691,121, in rats (A). Teratology 45:462, 1992.

Spence, S.G.; Vetter, C. and Hoe, C.: Effects of the class iii antiarrhythmic, dofetilide (UK–68,798) on the heart rate of midgestation rat embryos, in vitro. Teratology 49:282–292, 1994.

886 Domperidone *KW–5338* CAS 57808–66–9

Hara et al. (1980) gave this antiemetic orally to mice, rats and rabbits in maximum doses of 120, 200 and 120 mg per kg, respectively. The percent of successful mating was decreased at one mg per kg. Maternal and fetal weight was reduced and fetal survival was lower at 70 and 200 mg per kg treatment during organogenesis. Displacement of the subclavian artery, skeletal defects and eye defects occurred at 200 mg per kg. No postnatal effects were found except for retarded genital development at doses over 70 mg per kg. The same authors studied the effects of intravenous and intraperitoneal dosing at up to 25 and 30 mg per kg, respectively. Fertility was reduced when the dose exceeded 0.2 mg

per kg. Following treatment with 15 or 30 mg per kg during organogenesis, growth retardation and skeletal delay were found. Postnatal studies were not remarkable except for growth retardation. Rabbits treated during organogenesis had fetuses with reduced survival (25 mg per kg) but no significant increase in defects.

Hara, T.; Nishikawa, S.; Miyazaki, E. and Ogura, T.: Toxicologic studies on KW–5338 reproductive studies. Yakuri to Chiryo 8:4045–4060, 4125–4136, 1980.

887 DON

Discussed under 6–Diazo–5–oxo–l–norleucine

888 L–Dopa *3–(3,4–Dihydroxyphenyl)–L–alanine Dopamine CAS 59–92–7*

Staples and Mattis (1973) gave this medication by intubation to rabbits in doses up to 250 mg per kg per day from days 8 through 15 of gestation. At doses of 125 and 250 mg per kg malformations of the circulatory system were found. In mice treated with up to 500 mg per kg per day on days 6 through 15 only fetal stunting was significant.

Samojlik et al. (1969) administered 10 mg per kg daily or for 5 days from day 10 or 15 of the pregnancy of the rat. Many of the offspring died during suckling. Of 216 fetuses exposed during the entire gestation period, there were two with cataracts, 17 with suppurative inflammation of the eyes and one with polydactyly. Kitchin and DiStefano (1976) found hemorrhages in the brown fat of rat offspring after giving L–Dopa orally or Dopamine intraperitoneally. Treatment with L–Dopa prenatally produced the same effect in up to 9 percent of the offspring.

Kitchin, K.T. and DiStefano, V.: L–Dopa and brown fat hemorrhage in the rat pup. Toxicol. Appl. Pharmacol. 38:251–263, 1976.

Samojlik, E.; Khing, O.J. and Chang, M.C.: Effects of dopamine on reproductive processes and fetal development in rats. Am. J. Obstet. Gynecol. 104:578–585, 1969.

Staples, R.E. and Mattis, P.A.: Teratology of L–dopa. (abs) Teratology 8:238 only, 1973.

889 Dorzolamide Hydrochloride *MK–0507*

Nakatsuka et al. (**1994** A) gave this glaucoma medication to rabbits orally during organogenesis in doses of 2.5 to 10 mg per kg. Axial skeletal and costal defects occurred at all dose levels and acidosis was found in the dams. In rats Nakatsuka et al (**1994** B) gave up to 7.5 mg per kg before and during early gestation, during organogenesis and in the perinatal period. At 10 mg per kg there was fetal weight reduction but no teratogenicity or behavioral deficits were found.

Nakatsuka, T.; Komatsu, T.; Akutsu, S. and Matsumoto, H.: Topical carbonic anhydrase inhibitoe, dorzolamide hydrochloride (MK–0507); toxicity study (5th report)–developmental toxicity studies in rabbits. The Clinical Report 28:1331–1344, 1994A.

Nakatsuka, T.; Komatsu, T.; Ban, Y.; Katoh, M. and Matsumoto, H.: Topical carbonic anhydrase inhibitor, dorzolamide hydrochloride (MK–0507); toxocity study (4th report)–reproductive and developmental toxicity studies in rats. The Clinical Report 28:1301–1330, 1994B.

890 Dosulepin Hydrochloride CAS 897–15–4

Nakamura et al. (1983) gave this tricyclic antidepressant orally to rats in doses up to 40 mg per kg per day before fertilization and for the first 7 days of gestation. They found no adverse reproductive effects. Similar doses from day 17 to day 21 post partum did not affect behavior. Rabbits were given 80 mg per kg on days 6–18 and no adverse fetal effects were found. Nakamura et al. (1984) gave this tricyclic antidepressant to rats orally on days 7–17 of gestation. Maximum doses of 40 mg per kg produced no adverse fetal effects.

Nakamura, K.; Ohtani, A.; Kitagawa, N. and Tanaka, M.: Reproduction studies of dosulepin hydrochloride. Iyakuhin Kenkyu 14:571–601, 1983.

Nakamura, K.; Hashimoto, Y.; Nakamura, A. and Ichikawa, N.: Teratogenicity studies of dosulepin hydrochloride in rats. Oyo Yakuri 27:1103–1117, 1984.

891 Doxapram CAS 309–29–5

Imai (1974a) studied this respiratory stimulant in pregnant mice using an intraperitoneal maximum dose of 144 mg per kg per day of major organogenesis. At the highest dose some fetal death and growth retardation occurred but no gross defects. One treated pregnancy gave fetuses with minor skeletal defects. Post–natal studies for 6 weeks did not show any differences between the treated and control group. No teratogenicity was found in rats (Imai 1974b).

Imai, K.: Effect of doxapram hydrochloride administered to pregnant mice on pre– and post–natal development of their offsprings. Oyo Yakuri 8:229–236, 1974a.

Imai, K.: Effect of doxapram hydrochloride administered to pregnant rats on pre– and post–natal development of their offsprings. Oyo Yakuri 8:237–243, 1974b.

892 Doxazosin CAS 74191–85–8

Horimoto and Ohtsuki (**1990**) gave this antihypertensive to rats and rabbits in doses of up to 120 and 100 mg per kg respectively during organogenesis and found no teratogenicity. No behavioral changes were found in the offspring of rats given 50 mg per kg. Fertility studies in rats at 100 mg found no effect.

Horimoto, M. and Ohtsuki, I.: Reproductive and developmental toxicity studies with doxazosin in rats and rabbits. Oyo Yakuri 39(1):29–38, 1990.

893 Doxepin HCl *N,N–Dimethyldibenz[b,e]oxepin–δ–11 (6H) γ–propylamine HCl* CAS 1229–29–4

Owaki et al. (1971A,B) treated rabbits and rats during pregnancy with this psychotherapeutic agent. In rabbits doses up to 100 mg per kg during organogenesis produced some increase in neonatal death but no increase in gross malformations. The rats were treated with maximal doses of 270 mg per kg per day. At the highest levels postnatal survival was decreased but no congenital defects were found.

Owaki, Y.; Momiyama, H. and Onodera, N.: Effects of doxepin hydrochloride administered to pregnant rabbits upon the fetuses. Oyo Yakuri 5:905–912, 1971A.

Owaki, Y.; Momiyama, H. and Onodera, N.: Effects of doxepin hydrochloride administered to pregnant rats upon the fetuses and their postnatal development. Oyo Yakuri 5:913–924, 1971B.

894 Doxifluridine *5'–Deoxy–5–fluorouridine*

Shimizu (1985) studied this antineoplastic in pregnant male and female mice orally for 14 days before mating and for 7 days of gestation. With the highest doses (50 and 100 mg per kg), resorptions and reduced fetal weight occurred. Eye defects were increased at the 100 mg dose level.

Shimizu, M.: Toxicity study of doxifluridine (8): Reproduction segment I study in rats. Yakuri to Chiryo 13(2):469–479, 1985.

895 Doxycycline *Vibramycin* CAS 564–25–0

Cahen and Fave (1972) found no teratogenicity in rats, rabbits and mice in a preliminary report. They used dosages which were more than 100 times those used clinically. Delahunt et al. (1967) found non–teratogenicity in rats, rabbits and monkeys.

Cahen, R.L. and Fave, A.: Absence of teratogenic effect of 6–α–deoxy–5–oxytetracycline. Fed. Proc. 31:238 only, 1972.

Delahunt, C.S.; Jacobs, R.T.; Stebbins, R.B. and Rieser, N.: Toxicology of vibramycin. Toxicol. Appl. Pharmacol. 10:402–408, 1967.

896 Doxylamine Succinate *Decapryn*^TM CAS 562–10–7

Gibson et al. (1968), using 100 mg per kg per day in the rabbit and 100 mg per kg per day in the rat during periods of organogenesis, produced no increase in congenital defects as compared to controls. At the highest doses which were toxic to the mother, a reduction in fetal weight occurred.

Gibson, J.P.; Staples, R.E.; Larson, E.J.; Kuhn, W.L.; Holtkamp, D.E. and Newberne, J.W.: Teratology and reproduction studies with an antinauseant. Toxicol. Appl. Pharmacol. 13:439–447, 1968.

897 Duazomycin CAS 1403–47–0

Thiersch (1971) gave rats 4 to 10 mg per kg on day 7 of gestation and found increased resorptions and fetal stunting with minor unspecified malformations.

Thiersch, J.B.: Investigations into the differential effect of compounds on rat litter and mother. In: Malformations Congenitales Des Mammiferes. H. Tuchmann–Duplessis (ed.), Paris: Masson et Cie, 95–113, 1971.

898 DV–7572 *MRI Contrast Media*

Harada and coworkers (**1993** A,B,C,D) studied this MRI contrast media in rats using the intravenous mode of doses of up to 1.0 mmol per kg. Fertilization was decreased mainly due to effects on the testes. No teratogenic effects were found at doses up to 2.5 mmol per kg. Perinatal and postnatal studies were also negative. In the rabbit Harada (**1993**

E) found an increase in skeletal defects at 0.5 and 1.0 mmol per kg given daily through organogenesis.

Harada, S. and Nemec, M.D.: Intravenous fertility and general reproductive performance study of DV–7572 injection in rats. Yakuri to Chiryo 21:s799–s820, 1993A.

Harada, S.; Much, J.D. and Margitich, D.J.: Intravenous teratology study of DV–7572 injection in rats. Yakuri to Chiryo 21:s821–s826, 1993B.

Harada, S. and Itabashi, M.: Intravenous teratology study of DV–7572 injection in rats: Additional investigation. Yakuri to Chiryo 21:s827–s846, 1993C.

Harada, S. and Nemec, M.D.: Intravenous perinatal and postnatal study of DV–7572 injection in rats. Yakuri to Chiryo 21:s855–s872, 1993D.

Harada, S.; Much, J.D. and Margitich, D.J.: Intravenous teratology study of DV–7572 injection in rabbits. Yakuri to Chiryo 21:s847–s854, 1993E.

899 DWA2114R *2–Aminomethylpyrrolidine (1,1–cyclobutanedicarboxylato platinum monohydrate)*

Igarashi et al. (**1992**) gave rats this anticancer drug before fertilization and during the first week of gestation. At 0.6 and 3.0 mg per kg intravenously a decrease in implantation and fetal body weight was found. Sugiyama et al. (**1992 A**) using a similar treatment during organogenesis found increased fetal mortality, dwarfs and skeletal defects. Perinatal studies by Sugiyama et al. (**1992 B**) produced pups with decreased body weights. No treatment–related effects of 4.5 mg per day during organogenesis was found in rabbits (Hara et al. **1992**).

Hara, H.; Nakagawa, T.; Yahashi, H.; Sugiyama, O. and Deki, T.: Teratological study of DWA2114R in rabbits. Yakuri to Chiryo 20:s891–s896, 1992.

Igarashi, S.; Watanabe, S.; Watanabe, K. and Sugiyama, O.: Fertility study of DWA2114R in rats. Yakuri to Chiryo 20:s865–s874, 1992.

Sugiyama, O.; Igarashi, S.; Watanabe, K.; Watanabe, S. and Satoh, T.: Teratological study of DWA2114R in rats. Yakuri to Chiryo 20:s875–s890, 1992A.

Sugiyama, O.; Igarashi, S.; Takeuchi, H. Tanaka, N.; Aoki, A. and Hara, H.: Perinatal and postnatal study of DWA2114R in rats. Yakuri to Chiryo 20:s897–s908, 1992B.

900 E–3810

Shimizu et al. (**1993 A&B**) gave this anti–ulcer medication to rats intravenously before and during early pregnancy and in the perinatal and postnatal period. No effects on fertility were noted but the perinatal studies found reduced weight in the fetuses. No teratogenicity was found.

Shimizu, M.; Uto, K.; Kobayashi, Y.; Yamashita, Y.; Kato, M. and Shinoda, A.: Fertility study of E–3810 in rats by intravenous administration. G.I. Research 1:601–613, 1993.

901 E–643
4–Amino–2–(4–butanoylhexahydrahydro–1H– 1, 4–diazepin–1–yl)6, 7–dimethoxyquinazoline HCL

Okada et al. (1983) gave this antihypertensive to pregnant rats and rabbits. Oral doses of 100–400 mg per kg during days 7–17 were associated with growth retardation but no malformations. Fertility was decreased at 25 and 100 mg per kg. In perinatal studies at 100 mg, neonatal mortality was increased and postweaning, there was growth retardation. Maternal toxicity (hepato–renal) occurred at 25 mg per kg and above. In rabbits treated during organogenesis, no ill effects were seen with oral doses of up to 200 mg per kg.

Okada, F.; Nishimura, O.; Kondo, F.; Suzuki, Y.; Mikami, T.; Takamura, N.; Ohsumi, I.; Ogura, A.; Mizuno, T.; Goto, M.; Matsubara, Y.; Hiroe, K.; Tagaya, O. and Hanari, H.: Reproductive studies of 4–amino–2–(4–butanoylhexahydra–1H–1,4–diazepin–1–yl)6, 7–dimethoxy–quinazoline HCl (E–643) in rats and rabbits. Clin. Report. 17:907–939, 1983.

902 E–6010 *Plasminogen Activator Monteplase*

Shimzu et al. (**1994**) gave this plasminogen activator intravenously (10 mg per kg) to rats before fertilization and during the first week of gestation and found no adverse effects. Ogura et al. (**1994**) gave similar treatment during organogenesis without encountering teratogenicity. Perinatal and postnatal studies were performed at the same dose and no behavioral effects were observed. In the rabbit Niwa et al. (**1994**) used up to 3 mg per kg daily during organogenesis and found no teratogenicity.

Matsubara, Y.; Yamanaka, H.; Motooka, S.; Okada, F.; Kawaguchi, T.; Sagami, F. and Yamatsu, K.: Perinatal and postnatal study in rats treated intravenously with E6010. Yakuri to Chiryo 22:s299–s313, 1994.

Niwa, N.; Yamanaka, H.; Nakanowatari, J.; Okada, F.; Matsubara, Y.; Sagami, F. and Yamatsu, K.: Teratogenicity study in rabbits treated intravenously with E6010. Yakuri to Chiryo 22:s291–s298, 1994.

Ogura, K.; Kawaguchi, T.; Nakanowatari, J.; Tagaya, O.; Sagami, F. and Yamatsu, K.: Teratogenicity study in rats treated intravenously with E6010. Yakuri to Chiryo 22:s273–s289, 1994.

Shimizu, M.; Uto, K.; Kato, M.; Shinoda, A.; Kimura, H.; Matsubara, Y.; Sagami, F. and Yamatsu, K.: Fertility study in rats treated intravenously with E6010. Yakuri to Chiryo 22:s255–s272, 1994.

903 Ebastine *4'–tert–Butyl–4–[4–(dephenylmethoxy) piperidino] butyrophene*

Aoki et al. (**1994 A,B,C**) gave this antihistamine to rats before pregnancy and in the first week of gestation, during orgagnogenesis and in the peri and postnatal period. Oral doses of up to 300 mg per kg during organogenesis produced no adverse fetal effects except for weight reduction in the highest dose group. No teratogenicity or abnormal postnatal behavior was found. In the perinatal studies a dose level of 200 mg per kg was associated with decreased weight gain.

Aoki, Y.; Terada, Y.; Nakamura, H.; Uchiyama, H. and Morita, H.: Reproductive and developmental toxicity studies of ebastine (1): Fertility study in rats: Yakuri to Chiryo 22:1179–1192, 1994A.

Aoki, Y.; Funabashi, H.; Terada, Y.; Nakamura, H. and Morrrita, H.: Reproductive and developmental toxicity studies of ebastine (2): Teratogenicity study in rats. Yakuri to Chiryo 22:1193–1215, 1994B.

Aoki, Y.; Terada, Y.; Shigematsu, K.; Imurar, Y.; Inaoka, T. Tateishi, Y. and Nakamura, H.: Reproductive and developmental toxicity studies of ebastine (3): Perinatal and postnatal study in rats. Yakuri to Chiryo 22:1217–1239, 1994C.

904 Ebimar™ CAS 9013–42–7

Saito et al. (1970) found no teratogenic effects in rats or mice after oral doses of up to 4,500 mg per kg.

Saito, S.; Takagi, Y.; Iijima, Y.; Tokunaga, Y.; Kawashima, K.; Yamamoto, T. and Maeda, N.: Teratologic studies of Ebimar in mice and rats. Nichidai Igaku Zasshi 30:7–16, 1970.

905 Ecabet 12–Sulfodehydroabietic Acid Monosodium

Nakagawa et al. (**1991**) gave this peptic ulcer medication to rats before and in early pregnancy and during organogenesis in amounts of up to 2,000 mg per kg and found only a retardation in vertebral ossification. Perinatal studies with up to 600 mg per kg were associated with a decrease in fertility in the offspring at the 3,000 mg per kg dose level. No ill effects resulted after rabbits were treated during organogenesis with up to 600 mg per kg.

Nakagawa, H.; Shibano, T.; Yamamoto, M.; Sasaki, M.; Imado, N. and Ariyuki, F.: Reproductive and developmental toxicity studies of 12–sulfodehydroabietic acid monosodium talt (TA–2711) in rats and rabbits. The Clinical Report 25:941–951, 1991.

906 Econazole CAS 27220–47–9

Goorsman et al. (1985) found no malformations among 107 infants born to mothers using this topical vaginal medication.

This antimycotic agent was given to mice and rabbits subcutaneously in maximum daily doses of 100 and 75 mg per kg, respectively. Although no teratologic changes occurred, the mice at 100 mg per kg had prolonged gestation and increased fetal deaths. The rabbits had maternal deaths at 75 mg and some increase in fetal deaths at 25 mg per kg (Maruoka et al., 1978).

Goorsman, E.; Beck, J.M.; Declercq, J.A.; Loendersloot, E.W.; Roelofs, H.J.M. and Van Zanton, A.: Efficacy of econazole in vaginal candidosis during pregnancy. Current Med. Res. Opin. 9:371–372, 1985.

Maruoka, H.; Kadota, Y.; Ueshima, M.; Uesako, T.; Takemoto, Y. and Sato, H.: Toxicological studies on econazole nitrate. 4. Teratological studies in mice and rabbits. Iyakuhin Kenkyu 9:955–970, 1978.

907 E.D.T.A. Ethylenediaminetetracetic Acid CAS 60–00–4

Tuchmann–Duplessis and Mercier–Parot (1956) reported that 20 to 40 mg E.D.T.A. injected during embryogenesis caused tail defects and polydactyly in rat fetuses. Swenerton and Hurley (1971) produced cleft palate, brain and eye defects and skeletal anomalies in rat fetuses exposed to 2 or 3 percent E.D.T.A. in the diet after day 6 of gestation. By adding 1000 PPM of zinc to their experimental diet they were able to prevent the defects. Schardein et al. (1981) gave up to 1000 mg per kg by gavage pm days 7 through 14 of the rat gestation. Neither EDTA nor its sodium and calcium salts produced adverse fetal changes. Congenital defects in the quail were produced with EDTA by Craig et al. (1968).

Craig, R.M.; Kratzer, F.H. and Vohra, P.: Growth and reproductive performance of coturnix fed several levels of E.D.T.A. (abs) Poultr. Sci. 47:1664–1665, 1968.

Swenerton, H. and Hurley, L.S.: Teratogenic effects of a chelating agent and their prevention by zinc. Science 173:62–64, 1971.

Schardein, J.L.; Sakowski, R.; Petreve, J. and Humphrey, R.R.: Teratogenesis studies with EDTA and its salts in rats. Toxicol. Appl. Pharmacol. 61:423–428, 1981.

Tuchmann–Duplessis, H. and Mercier–Parot, L.: Influence d'un corps de chelation, l'acide ethlenediaminetetraacetique sur la gestation et le developppment foetal du rat. C. R. Acad. Sci. (Paris) 243:1064–1066, 1956.

908 Edatrexate

Epstein et al. (**1992**) gave this analogue of Methotrexate to rats and rabbits intravenously in doses of up to 1.0 and 2.0 mg per kg respectively. No defects were produced by administration on single days during organogenesis and they did not produce embryo–fetotoxicity. Maternal toxicity was seen in rabbits at about 6 mg per kg.

Epstein, D.L.; Raab, D.M.; Melando, A.R.; Hazelette, J.R.; Yau, E.T. and Traina, V.M.: Edatrexate: intravenous teratology (segment II) studies in rats and rabbits (A). Teratology 45:471, 1992.

909 Egg White Hydrolysate

Yanagimoto et al. (1983) fed rats up to 27.5 gm per 200 gm of diet during organogenesis and found no adverse effects. The fetal weight was significantly increased.

Yanagimoto, Y.; Niii, N.; Fujii, Y.; Murakami, H.; Matsuda, M.; Yamamoto, E.; Sano, Y.; Yamazaki, K.; Takao, S.; Hara, H. and Watanabe, M.: Effects of egg white hydrolysate on reproduction. Clin. Report 17:3894–3903, 1983.

910 Electric Blankets

Dlugosz et al. (**1992**) studied electric blanket and heated water bed exposures in 542 pregnancies ending with facial clefts or neural tube defects. There was no increased risk as compared to randomize and matched controls. Milunsky et al. (**1992**) found no increase in risk of neural tube defects among 2,876 users. /eb

Dlugosz, L.; Vena, J.; Buyers, T.; Sever, L.; Bracken, M. and Marshall, E.: Congenital defects and electric bed heating in New York State: A register–based case–control study. American Journal of Epidemiology 135:1000–1011, 1992.

Milunsky, A.; Ulcickas, M.; Rothman, K.J.; Willett, W.; Jick, S.S. and Jick, H.: Material heat exposure and neural tube defects. JAMA 268:882–885, 1992.

911 Electroconvulsive Therapy

Impastato et al. (**1964**) studied 318 women treated with electric shock therapy for psychiatric disease. No increase in defects or pregnancy loss was found. The convulsions were modified in most cases by succinylcholine.

Impastato, D.J.; Gabriel, A.R. and Lardaro, H.H.: Electric and insulin shock therapy during pregnancy. Dis Nerv Syst 25:542–546, 1964.

912 Electromagnetic Field

This subject is discussed also under Video Display terminals. Brent et al. (**1993**) have comprehensively reviewed the literature on video display terminals and find that the results are generally negative and suggest that a lower priority to these exposures be given. For electromagnetic exposures consistent findings indicate even at high exposures that there is no increase in reproductive failures.

Kowalczuk et al. (**1994**) exposed mice to 50 Hz magnetic field at 200 mT throughout gestation and found no evidence of ill effects on the fetuses. Postnatal studies on the same animals did not find any gross impairments in development and behavior (Sienkiewicz et al., **1994**). Mevissen et al. (**1994**) exposed pregnant rats to static or flux density 50 Hz (30 mT) magnetic fields for 20 days and found no serious malformations. In both conditions the fetal skeletal ossification was enhanced. Resorptions were 13.9% in the static treated group and 3.8% in the control. Postnatal behavior was not affected.

Robert (**1993**) studied 1688 malformed infants in a case control fashion and found no excess of specific malformations among those whose mothers resided in municipalities with potentially high exposure to electromagnetic fields from overhead power lines.

Landesman and Douglas (**1990**) studied regenertion of amputated newt limbs exposed to a Bi–Osteogen system 204 for the first 30 days of the regeneration process. In the control 2% were abnormal while in the treated 28% were abnormal with reduced patterns.

Zusman et al. (**1990**) found inhibition of hatching of mouse blastocysts after exposure for 72 hours to 20 or 50 Hz of pulsed electromagnetic waves. Embryos of 10–12 days were grown in culture and growth retardation with abnormolities of the brain, forelimbs and yolk sac were increased at 50 and 70 Hz but not at 20 Hz.

Konermann and Monig (**1986**) exposed mice to 1T on days 7, 10, or 13 and found no adverse fetal effects. Brain weight and other CNS measures were not different from controls postnatally.

Meyer et al. (**1989**) reviewed animal work with electromagnetic fields and found the few positive effects difficult to interpret or reproduce. Chernoff et al. (**1992**) reviewed the experimental and epidemiologic work dealing with electric and magnetic fields and concluded there was no conclusive data to support that they adversely effect reproduction.

Wiley et al. (**1992**) exposed large numbers of mice to 20–kHz sawtooth magnetic currents (similar to video display terminals) on days 1–18 of pregnancy. The field strengths were 3.6 to 200 microT. No adverse reproductive or fetal effects were found.

Brent, R.L.; Gordon, W.E.; Bennett, W.R. and Beckman, D.A.: Reproductive and teratologic effects of electromagnetic fields. Reproductive Toxicology 7:535–580, 1993.

Chernoff, N.; Rogers, J.M. and Kavet, R.: A review of the literature on potential reproductive and developmental toxicity of electric and magnetic fields. Toxicology 74:91–126, 1992.

Konermann G. and Monig, H.: Effect of static magnetic fields on the prenatal development of the mouse. Radiologe 26:490–497, 1986.

Kowalczuk, C.I.; Robbins, L.; Thomas, J.M.; Butland, B.K. and Saunders, R.D.: Effects of prenatal exposure to 50 hz magnetic fields on development in mice: I. Implantation rate and fetal development. Bioelectromagnetics 15:349–361, 1994.

Landesman, R.H. and Douglas, W.S.: Abnormal limb regeneration in adult newts exposed to a pulsed electromagnetic field. Teratology 42:137–145, 1990.

Mevissen, M.; Buntenkotter, S. and Loscher, W.: Effects of static and time–varying (50–Hz) magnetic fields on reproduction and fetal development in rats. Teratology 50:229–237, 1994.

Meyer, R.E.; Aldrich, T.E. and Easterly, C.E.: Effects of noise and electromagnetic fields on reproductive outcomes, Environmental Health Perspectives 81, 193–200, 1989.

Robert E.: Birth defects and high voltage power lines: an exploratory study based on registry data. Reproductive Toxicology 7:283–287, 1993.

Sienkiewicz, Z.J.; Robbins, L.; Haylock, R.G.E. and Saunders, R.D.: Effects of prenatal exposure to 50 hz magnetic fields on development in mice: II. Postnatal development and behavior. Bioelectromagnetics 15:363–375, 1994

Wiley, M.J.; Kavet, C.R.; Charry, J.; Harvey, S.; Agnew, D. and Walsh, M.: The effects of continuous exoisure to 20–kHz sawtooth magnetic fields on the litters of cd–1 mice. Teratology 46:391–398, 1992.

Zusman, I.; Yaffe, P.; Pinus, H. and Ornoy, A.: Effects of pulsing electromagnetic fields on the prenatal and postnatal development in mice and rats: in vivo and in vitro studies. Teratology 42:157–170, 1990.

913 Ellagic Acid

Frank et al. (**1992**) studied this naturally occurring plant phenol in whole rat embryo culture. The compound which is anticarcinogenic and antimutagenic protected the embryos from exposure to N–methyl–N–nitrosurea. The concentration used was 50 micromolar. No teratogenicity was found at this concentration.

This naturally-occurring plant phenol was studied in whole embryo culture by Frank et al. (**1993**). Its effect on N–methyl–N–nitrosourea teratogenicity was studied by distribution methods and by measuring adducts. O^6–methylguanine DNA adduct was decreased but not N^7–methylguanine.

Frank, A.A.; Collier, J.M.; Forsyth, C.S.; Heur, Y. and Stoner, G.D.: Ellagic acid protects rat embryos in culture from the embryotoxic effects n–methyl–n–nitrosourea. Teratology 46:109–115, 1992.

Frank, A.A.; Collier, J.M.; Forsyth, C.S.; Zeng, W. and Stoner, G.D.: Ellagic Acid embryoprotection in vitro: distribution and effects on dna adduct formation. Teratology 47:275–280, 1993.

914   Elymoclavine CAS 548–43–6

Witters et al. (1975) injected 3, 30 or 40 mg per kg on the 10th day and produced rib and vertebral defects in the offspring mouse fetuses.

Witters, W.L.; Wilms, R.A. and Hood, R.D.: Prenatal effects of elymoclavine administration and temperature stress. J. Animal Science 41:1700–1705, 1975.

915   EM$_{12}$ 2–(2,6–Dioxopiperiden–3–yl)phthalimidine CAS 26581–81–7

This analogue of thalidomide is teratogenic in monkeys and rabbits (Schumacher et al., 1972). In the rat, oral doses of 250 mg per kg or intravenous doses of 10 mg per kg for three days failed to produce typical thalidomide malformations. Jackson and Schumacher (1979) administered 150 mg per kg intraperitoneally to rats maintained on a low zinc diet and produced phocomelia in over 50 percent of the fetuses. EM$_{12}$ treatment was on days 8–10 of pregnancy. Merker et al. (1988) using 12 mg per kg, produced typical phocomelia in marmosets. Treatment was given on days 51–57 post ovulation.

Jackson, A.J. and Schumacher, H.J.: The teratogenic activity of thalidomide analogue EM$_{12}$ in rats on a low–zinc diet. Teratology 19:341–344, 1979.

Merker, H–J.; Wolfgang, H.; Sames, K.; Sturje, H. and Neubert, D.: Embryotoxic effects of thalidomide-derivatives in the non–human primate Callithrix jacchus. I. Effects of 3–(1, 3–dihydro–1–oxo–2H–isoindol–2–yl)–2,6–dioxopiperidine (EM$_{12}$) on skeletal development. Arch. Toxicol. 61:165–179, 1988.

Schumacher, H.J.; Terapane, J.J.; Jordan, R.L. and Wilson, J.G.: The teratogenic activity of a thalidomide analogue, EM$_{12}$ in rabbits, rats and monkeys. Teratology 5:233–240, 1972.

916   EM–240

This metabolite of CG 3033 a structural analog of thalidomide was given to monkeys at 15 mg per kg from day 19 or 22, for 25 days and no teratogenicity was found (Scott et al., 1980).

Scott, W.J.; Wilson, J.G. and Helm, F.CH.: A metabolite of a structural analog of thalidomide lacks teratogenic effect in pregnant rhesus monkeys. Teratology 22:183–185, 1980.

917   Empenthrim Vaporthrin^{TM}

Kaneko et al. (1992 gave this pyrethinoid insecticide to rats and rabbits during organogenesis in doses of up to 500 and 1000 mg per kg respectively and no adverse reproductive changes occurred. Postnatal development and reproduction were normal.

Kaneko, H.; Kawaguchi, S.; Misaki, Y.; Koyama, Y.; Nakayama, A.; Kawasaki, H.; Hirohashi, A.; Yoshitake, A. and Yamada, H.: Mammalian toxicity of empenthrin (vamprthrin^P, S–2852F). J. Tox. Sc. 17:313–334, 1992.

918   Emitefur 3–[3–(6–benzoyloxy–3–cyano–2–pyridyloxycarbonyl)benzoyl]–1–ethoxymethyl–5–fluoro–1,2,3,4–tetrahydro–2–4–pyrimidinedione

Oi et al. (1994) treated rats before and during the first week of pregnancy and found weight reduction of the dam and fetus at 20 mg per kg orally but no adverse reproductive effects. Schardein et al. (1994) found no teratogenicity using a similar dose on days 7–17. Ishizuka et al. (1994) found no adverse effects of 20 mg per kg given during the perinatal and postnatal period. Development and reproduction of the pups was not affected. In rabbits, Oi et al. (1994) at a dose of 5 mg per kg during organogenesis there was an increase in rib and vertebral variations. At 1.5 mg per kg there were no increased defects.

Ishizuka, Y.; Miyahara, T.; Nagao, T.; Mizutani, M. and Oi, A.: Reproductive and developmental toxicity studies of emitefur, a new antieoplastic agent (IV): Perinatal and postnatal study in rats with oral administration. Yakuri to CHiryo 22:233–249, 1994.

Oi, A.; Iriguchi, K.; Hamabuchi, T.; Nishioeda, R. and Tamagawa, M.: Reproductive and developmental toxicity studies of emitefur, a new antieoplastic agent(I): Fertility study in rats with oral administration. Yakuri to Chiryo 22:191–200, 1994.

Oi, A.; Nishioeda, R. and Tamagawa, M.: Reproductive and developmental toxicity studies of emitefur, a new antieoplastic agent (III): Teratology study in rabbits with oral administration. Yakuri to Chiryo 22:223–232, 1994.

Schardein, J.L.; West, A.; York, R.G. and Oi, A.: Reproductive and developmental toxicity studies of emitefur, a new antineoplastic agent (II): Teratology study in rats with oral administration. Yakuri to Chiryo 22:201–222, 1994.

919   Emedastine Difumarate

Kanemoto et al. (1990,A) gave rats orally up to 140 mg per kg during organogenesis at 40 mg and 140 mg levels the maternal and fetal weights were reduced. No teratogenicity was found. Similar treatments before fertilization in both sexes and for 7 days into pregnancy in the female revealed no adverse effects. (Kanemoto et al. 1990, B)

Kanemoto, I.; Yokoi, Y.; Mitsumori, T.; Fukunishi, K.; Nagano, M. and Nurimoto. S.: Reproduction study of emedastine difumarate (kg–2413): teratological study in rats. Oyo Yakuri 39(3):329–342, 1990.

Kanemoto, I.; Yokkoi, Y.; Mitsumori, T.; Fukunishi, K.; Nagano, M.; Maekawa, T. and Nurimoto, S.: Reproduction study of emedastine difumarate (kg–2413): fertility study in rats. Oyo Yakuri 39(3):319–328, 1990.

920   Emonapride YM–09151

Cozens et al. (1989) gave rats orally up to 30 mg per kg on days 6 through 17. No teratogenicity was found and postnatal development was normal. Shabita and Uchida (1989) gave 3, 10 or 30 mg per kg daily during organogenesis and found fetal weight reduction at 10 and 30 mg per kg but no teratogenicity. Kawakami and Shibata (**1989**) gave males and females up to 100 mg per kg before pregnancy and to females in the first 7 days of gestation and found no ill effects.

Cozens, D.D.; Clar, R.; James, P.; Smith J.A. and Offer, J.M.: Pre– and post natal development of the rat following oral administration of emonapride (ym–09151) during organogenesis. Kiso to Rinsho 23:4847–4856, 1989.

Kawaami, A. and Shibata, M.: Effect of orally administered emonapride (ym–09151) on male fertility in rats. Oyo Yakuri 38:109–112, 1989.

Shibata, M. and Uchida, T.: Teratology of orally administered emonapride (ym–09151) in rabbits. Oyo Yakuri 38:121–124, 1989.

921 Emorfazone
4–Ethoxy–2–methyl–5–morpholino–3(2H)–pyridazinone
CAS 38957–41–4

This anti–inflammatory agent was given to rabbits orally in doses of up to 300 mg per day on days 6–18 of gestation. Tanigawa et al. (1979A,B) gave rats up to 600 mg per kg orally before and during early pregnancy or in the late gestational period and observed no ill effects in the offspring.

Tanigawa, H.; Obori, R.; Tanaka, H.; Yoshida, J. and Kosazuma, E.: Reproductive study of 4–ethoxy–2–methyl 5–morpholino–3(2H)–pyridazinone (M 73101) on rabbits administration of M73101 during the period of major organogenesis. J. Toxicol. Sci. 4:162–174, 1979A.

Tanigawa, H.; Yoshida, J.; Tanaka, H.; Obori, R. and Kosazuma, E.: Reproductive study of 4 ethoxy–2–methyl–5–morpholino–3(2H)–pyridazone (M73101) in rats. Oyo Yakuri 18:417–447, 1979B.

922 Emotional Stress *Stress, Psychological*

Scialli (1988) reviewed the subject in experimental animals and humans. Bensen et al. (1987) measured maternal and fetal heart rates in mothers who rated themselves anxious or non–anxious. The fetal rates were significantly lower in the anxious mothers, but the fetal responses to auditory stimuli given by head phones did not differ significantly. Stott (1973) made a study of 153 pregnancies and found no increase in physical defects in 14 where personal tension existed during gestation. Murata and Takigawa (1989) exposed mice to noise on days 7–12 for 6 hours daily and found 2.7% defects with intermittant noise and 2.8% with continuous wide–band 100db was used. The control was 2.1%.

Rosenzweig (1966) reviewed the experimental literature on psychologic stress and production of cleft palates. A number of stress techniques have caused an increase in cleft palate in the naturally susceptible AJAX strain of mice. Warkany and Kalter (1962) exposed two strains of mice to audiogenic stimuli and observed no increase in cleft palate in the offspring. Besides handling, restraint and avoidance, an effect of air transportation was shown (Brown et al., 1972).

Hartel and Hartel (1960) subjected pregnant rats to six hour periods of intermittent ringing of bells and flashing of light on the 9th through the 12th days and found no defects in 13 litters. Ishii and Yokobori (1960) exposed mice to noise for six hours daily on days 11 through 14 of gestation and produced seven defects among 130 survivors. The fetal weight was reduced and the stillbirth rate increased by the treatment. Kimmel et al. (1976) in a carefully controlled exposure of mice and rats to noise stress (100 DB), found no teratogenicity but in mice maternal weight was reduced and also resorption rates were increased.

Nawrot et al. (1980) studied various types of noise stimuli in mice (126 dBA for 4 hours daily) and found that the extremely high frequency (jet type) had an embryo lethal effect on days 1–6 and that very high frequency (110 dBA) had the same effect on days 6–15. No malformations were produced. Meyer et al (1989) have reviewed briefly the studies in animals and humans.Maternal corticosterone levels were not elevated. Hamburgh et al. (1974) produced a 9.5 percent incidence of malformations in the offspring of mice which were stressed by overcrowding.

Meyer et al. (**1989**) reviewed the animal data on noise exposure in experimental and human studies. In general there was little evidence for teratogenicity but several studies showed embryo–feto toxicity. Studies in the human population especially near airports yielded some evidence for reproductive toxicity but the import of confounding factors and differing exposure assessment made analysis difficult.

Bensen, P.; Little, B.C.; Talbert, D.G.; Dewhurst, J. and Priest, R.G.: Foetal heart rate and maternal emotional state. Br. J. Med. Psychol. 60:151–154, 1987.

Brown, K.S.; Johnston, M.C. and Niswander, J.D.: Isolated cleft palate in mice after transportation during gestation. Teratology 5:119–124, 1972.

Hamburgh, M.; Mendoza, L.A.; Rader, M.; Lang, A.; Silverstein, H. and Hoffman, K.: Malformations induced in offspring of crowded and parabiotically stressed mice. Teratology 10:31–38, 1974.

Hartel, A. and Hartel, G.: Experimental study of teratogenic effect of emotional stress in rats. Science 132:1483–1484, 1960.

Ishii, H. and Yokobori, K.: Experimental studies on teratogenic activity of noise stimulation. Gunma J. Med. Sci. 9:153–167, 1960.

Kimmel, C.A.; Cook, R.O. and Staples, R.E.: Teratogenic potential of noise in mice and rats. Toxicol. Appl. Pharmacol. 36:239–245, 1976.

Meyer, R.E.; Aldrich, T.E. and Easterly, C.E.: Effects of noise and electromagnetic fields on reproductive outcomes. Environmental Health Perspectives 81:193–200, 1989.

Murata, M. and Takigawa, H.: Teratogenic effects of noise in mice. J. Sound and Vibration 132:11-18, 1989.

Nawrot, P.S.; Cook, R.O. and Staples, R.E.: Embryotoxicity of various noise stimuli in the mouse. Teratology 22:279–289, 1980.

Rosenzweig, S.: Psychological stress in cleft palate etiology. J. Dent. Res. 45:1585–1593, 1966.

Scialli, A.R.: Is stress a developmental toxin? Reproductive Toxicology 1(3):163–172, 1988.

Stott, D.H.: Follow–up study from birth of the effects of prenatal stress. Develop. Med. Child Neurol. 15:770–787, 1973.

Warkany, J. and Kalter, H.: Maternal impressions and congenital malformation. Plast. Reconstr. Surg. 30:628–637, 1962.

923 Encainide

Jones et al. (**1994**) reported two offspring whose mothers were exposed to this antiarrhythmic. One had a submucous cleft and the other congenital heart disease with dextrocardia and aspleenism.

Jones, K.L.; Braddock, S.; Curry, C. and Benirschke, K.: Possible teratogenesis of encainide. (abs) Teratology 49:412, 1994.

924 Enalapril *MK–421*

See discussion under captopril Barr (**1994**) has reviewed the subjects and recommends the ACE inhibitors not be used in pregnant women. The syndrome of fetal oligohydramnios and/or neonatal anuria associated with hypoplasia of the skull bones has been associated with this antihypertensive agent (Brent and Beckman, **1991**). Kreft–Jais et al. (1988) reported nine pregnancies where hypertension was treated. Two spontaneous abortions, one miscarriage and no congenital defects occurred.

Fujii et al. (1985a,b) studied this antihypertensive in rats on days 6–17 giving orally up to 1200 mg per kg daily. Maternal and fetal weight were decreased at the highest dose but no other adverse findings occurred at day 20 or postnatally. Minsker et al. (1990) found nephrotoxicity in the offspring of rabbits receiving 1 mg per kg or more on days 14–27. Fetal wastage was also observed at doses as low as 3 mg per kg. Fetal hypotension was thought to be the cause and supplementation with NaCl protected the fetuses.

Barr Jr., M.: Teratogen update: angiotensin–converting enzyme inhibitors. Teratology 50:399–409, 1994.

Brent, R.L. and Beckman, D.A.: Angiotensin–converting enzyme inhibitors, an embryopathic class of drugs with unique properties: Information for clinical teratology counselors. Teratology 43:543–546, 1991.

Fujii, T.; Nakatsuka, T.; Hanada, S.; Komatsu, T. and Okiyama, H.: Enalapril (MK–421) oral teratogenicity study in the rat. Yakuri to Chiryo 13:519–528, 1985a.

Fujii, T. and Nakatsuka, T.: Enalapril (MK–421) oral teratogenicity study in the rat. Yakuri to Chiryo 13:529–548, 1985b.

Kreft–Jais, C.; Plouin, P–F.; Tchobroutsky, C. and Boutroy, M–J.: Angiotensin–converting enzyme inhibitors during pregnancy: A survey of 22 patients given captopril and nine given enalapril. Br. J. Obstet. Gynecol. 95:420–422, 1988.

Minsker, D.H.; Bagdon, W.J.; MacDonald, J.; Robertson, R.T. and Bokelman, D.L.: Maternotoxicity and Fetotoxicity of an angiotensin-converting enzyme inhibitor, enalapril, in rabbits. Fundam Appl Toxicol. 14:461–470, 1990.

925 Endobenzyline Bromide

Back et al. (1961) fed this cholinergic blocking agent in amounts up to 0.2 percent of the diet. The newborns had an increase in mortality rate. No defects were mentioned.

Back, K.C.; Newberne, J.W. and Weave, L.C.: A toxicopathologic study of endobenzyline bromide, a new cholinergic blocking agent. Toxicol. Appl. Pharmacol. 3:422–430, 1961.

926 3–(1,4–Endomethylene–cyclohexane–2, 3–endo–cis–dicarboximido)–piperidine–2,6–dione

This hypnotic drug structurally similar to thalidomide was tested in the pregnant rabbit and found to be non–teratogenic (Stockinger and Koch, 1969). It was given orally from day 7 through 12 in amounts of 150 mg per kg. The authors suggest that their findings give evidence that aromatic phthalidimide moiety of thalidomide is responsible for the embryotoxic activity.

Stockinger, L. and Koch, H.: Teratologische Untersuchung einer neuen, dem Thalidomid Stuckturell Nahestehenden Sedativ–hypnotisch Wirksamen Verbindung (K–2004). Arzneim. Forsch. 19:167–169, 1969.

927 Endosulfan CAS 115–29–7

This insecticide was given orally on days 6 through 14 in doses of 5 and 10 mg per kg to rats and no soft tissue defects were found. Some delayed ossification occurred with treatment (Gupta et al., 1978).

Gupta, P.K.; Chandra, S.V. and Saxena, D.K.: Teratogenic and embryotoxic effects of endosulfan on rats. Acta Pharmacol. Toxicol. 42:150–152, 1978.

928 Endotoxin CAS 11034–88–1

Ornoy and Altshuler (1976) treated pregnant rats with E. coli endotoxin intraperitoneally. When the treatment was given on days 10, 13 and 16, or on days 11, 14 and 17, hydrocephalus and or neuronal necrosis was increased in the fetus. Kanoh and Ema (1980) gave up to 1000 microgm intravenously to rats on day 7, 12 or 17 and produced resorptions and dead fetuses but no congenital defect increase.

Hilbelink et al. (1986) gave guinea pigs 1, 10, 20 or 200 microgm per kg of endotoxin intravenously on the morning of day 8. At 10 and 20 microgm levels, malformations were increased and included encephalocele, renal agenesis and skeletal defects. Resorptions were increased at doses of over 10 microgm. Abdominal temperature elevations were between 38 and 39 degrees C.

Hilbelink, D.R.; Chen, L.T. and Bryant, M.: Endotoxin–induced hyperthermia in golden hamsters. Teratogenesis Carcinog. Mutagen. 6:209–217, 1986.

Kanoh, S. and Ema, M.: Effects of bacterial endotoxin on pregnant rats and their offspring. Cong. Anom. 20:151–155, 1980.

Ornoy, A. and Altshuler, G.: Maternal endotoxemia, fetal anomalies, and central nervous system damage: A rat model of a human problem. Am. J. Obstet. Gynecol. 124:196–204, 1976.

929 3–(1, 4–Endoxo–cyclohexane–2–exo–3–exo–di-carboximido)–piperidine–2–6–dione

This sedative chemically related to thalidomide was non–teratogenic in the offspring of rabbits receiving 150 mg per kg orally from days 7 through 13 (Koch and Stockinger, 1971).

Koch, H. and Stockinger, L.: Teratologische Untersuchung des 3–(1,4–endoxocyclohexan–2–exo–3–dicarboximido)–piperidin–2,6–dion N, einer Weiteren dem Thalidomide strukturell Nahestehenden, Sedativhypnotisch Wirksamen Verbindung. Arzneim. Forsch. 21:2022–2025, 1971.

930 Endrin CAS 72–20–8

Noda et al. (1972) gave rat and mice 0.58 mg per kg four times weekly for a month and then after a week or more without treatment, the animals were allowed to become pregnant. A reduced survival rate was found in both species. Nine mouse fetuses with club foot were found in the treated group of 177 and only one in the control group of 303. For further information, see Cyclodiene Pesticides.

Noda, K.; Hirabayashi, M.; Yonemura, I.; Maruyama, M. and Endo, I.: Influence of pesticides on embryos. 2. On the influence of organochloric pesticides. (Japanese) Oyo Yakuri 6:673–679, 1972.

931 Enflurane 2–Chloro–1,1,2–trifluoro Ethyl Diethy Methyl Ether CAS 13838–16–9

Saito et al. (1974) exposed pregnant rats and mice to one hour of this anesthetic daily during major organogenesis. Concentrations for the rat were 1.25 percent and for the mouse 0.75 percent. No teratogenicity was found. No postnatal changes were found in a six week study.

Wharton et al. (1979) confirmed this work using four hour daily exposures. Mazze et al. (1986) found no adverse reproductive effects in rats exposed for six hours during three day periods of organogenesis to 1.65 percent.

Mazze, R.I.: Fujinaga; M.; Rice, S.; Harris, S.B. and Baden, J.M.: Reproductive and teratogenic effects of nitrous oxide, halothane, isoflurane, and enflurane in Sprague–Dawley rats. Anesthesiology 64:339–344, 1986.

Saito, N.; Urakawa, M. and Ito, R.: Influence of enflurane on fetus and growth after birth in mice and rats. (Japanese) Oyo Yakuri 8:1269–1276, 1974.

Wharton, R.S.; Wilson, A.I.; Rice, S.A. and Mazze, R.I.: Teratogenicity of enflurane in Swiss–ICR mice. (abs) Teratology 19:53A only, 1979.

932 Enovid™ CAS 8015–30–3

This oral progestin which contains 9.85 mg 17—α–ethinyl–17–OH–5(10) estren–3–one and 0.15 mg ethinylestradiol 3 methyl ether per 10 mg tablet has been given in 10 mg daily doses to a mother who produced a girl with labio–scrotal fusion and clitoromegaly (Grumbach et al., 1959).

Grumbach, M.M.; Ducharme, J.R. and Moloshok, R.E.: On the fetal masculinizing action of certain oral progestins. J. Clin. Endocrinol. Metab. 19:1369–1380, 1959.

933 Enoxacin AT–2266 CAS 74011–58–8

This antibiotic was given by gavage to male and female rats before and to female rats during the first 7 days of gestation and during days 7–17 in doses up to 1000 mg per kg. No adverse fetal effects occurred but decreased fertility in the male occurred at 1000 mg per kg (Terada et al., 1983). Peri–and postnatal studies by Nishimura et al., (1983) were negative in the fetus but at 1000 mg per kg the dams had soft stools. Nishimura et al. (1983) found 60 mg per kg to be non–teratogenic in dogs.

Nishimura, K.; Mukumoto, K.; Terada, Y.; Takenaka, H. and Yoshida, K.: Reproduction studies of AT–2266 teratogenicity study in dogs. Chemotherapy 32:327–333, 1983.

Terada, Y.; Nishimura, K.; Komurasaki, M.; Yoshioka, M. and Yoshida, K.: Reproduction studies of AT–2266 fertility and teratological studies in rats. Chemotherapy 32:279–292, 1983.

934 Enteroviruses

Brown and Karunas (1972) collected sera from 22,935 women early in pregnancy and again at delivery. Among 82 mothers who gave birth to children with urogenital malformation a significantly higher number (37) were found to have had infections with one of the 5 coxsackie B viruses (B2 and B4 mainly). Among a group of 77 infants with cleft lip or palate, pyloric stenosis or other gastrointestinal tract anomalies, there was an increase in the number of women with serologic evidence of enterovirus A–9. A significant increase in the number of maternal titers against coxsackie B3 and B4 was found in pregnancies associated with congenital heart conditions.

Brown, G.C. and Karunas, R.S.: Relationship of congenital anomalies and maternal infection with selected enteroviruses. Am. J. Epidemiol. 95:207–217, 1972.

935 Ephedrine CAS 299–42–3

Discussed under Epinephrine

936 E–Phtaloyl–imido–caproic Acid

Bussi et al. (1992) studied this bleaching chemical in rabbits. At 300 mg per kg fetal body weight was reduced.

Bussi, R.; Bianchi, U.; Giavini, E.; Kreiling, R. and Malinverno, G.: Rabbit teratology study of e–phtsaloyl–imido–caproic acid (Abst). Teratology 46:20A, 1992.

937 Epichlorohydrin 1–Chloro–2,3–epoxypropane CAS 106–89–8

Marks et al. (1982) gavaged mice and rats with up to 160 mg per kg daily and found no adverse fetal effects. John et al. (1983) exposed rats and rabbits to vapors of 2.5 and 25 ppm for seven hours daily during active organogenesis. Maternal toxicity occurred at 25 ppm but no adverse fetal effects were found.

John, J.A.; Gushow, T.S.; Ayres, J.A.; Hanley, T.R., Jr.; Quast, J.F. and Rao, K.S.: Teratogenic evaluation of inhaled epichlorohydrin and allyl chloride in rats and rabbits. Fund. Appl. Toxicol. 3:437–442, 1983.

Marks, T.A.; Gerling, F.S. and Staples, R.E.: Teratogenic evaluation of epichlorohydrin in the mouse and rat and glycidol in the mouse. J. Toxicol. Environ. Health 9:87–96, 1982.

938 Epidermal Growth Factor CAS 62229-50-9

Bedrick and Ladda (1978) injected pregnant mice on days 11–15 with 40 microgm per gm of animal and produced no defects. Epidermal growth factor did cause an increase in cortisone–induced cleft palates. Saitou et al. (1992 A,B,C) gave human urinary epidermal growth factor (MG–III) to mice before pregnancy and during the first week of gestation, during organogenesis and in the peri and postnatal period. Intraperitonal doses of up to 250 thousand units per kg were given except for the organogenesis study where a dose of up to 100 thousand units was employed. No adverse reproductive effects or teratogenicity were found. In the postnatal period the fetal livers were increased in weight. Mizutani et al. (1992) gave rabbits up to 250 thousand units per kg intravenously during organogenesis and found reduced fetal and placental weight at the highest dose but no teratogenicity.

Bedrick, A.D. and Ladda, R.L.: Epidermal growth factor potentiates cortisone–induced cleft palate in the mouse. Teratology 17:13–18, 1978.

Mizutani, H.; Matuda, H.; Hirano, K. and Yamazaki, Y.: Teratological study of epidermal growth factor MG111 extracted from human urine in rabbits. Yakuri to Chiryo 20:s2977–s2987, 1992.

Saitou, N.; Watanabe, Y.; Azuma, R.; Hirano, K.; Ohashi, T.; Noguchi, K.; Yamazaki, Y.; Irie, D. and Takeuchi, M.: Fertility study of human urinary epidermal growth factor MG111 in mice intraperitoneal administration. Yakuri to Chiryo 20:s2951–s2960, 1992.

Saitou, N.; Watanabe, Y.; Azuma, R.; Hirano, K.; Ohashi, T.; Noguchi, K.; Yamazaki, Y.; Irie, D. and Takeuchi, M.: Perinatal and postnatal study of human urinary epidermal growth factor MG111 in mice by intraperitoneal administration. Yakuri to Chiryo 20:s2989–s3002, 1992.

Saitou, N.; Watanabe, Y.; Azuma, R.; Hirano, K.; Ohashi, T.; Noguchi, K.; Yamazaki, Y.; Irie, D.; Mizutani, M. and Takeuchi, M.: Teratological study of human urinary epidermal growth factor MG111 in mice by intraperitoneal administration. Yakuri to Chiryo 20:s2961–s2976, 1992.

939 Epidihydrocholestrin

This steroid precursor was given subcutaneously in doses of up to 200 mg per kg to mice and no fetal alterations were detected. Treatment was on days 6 through 15 (Matumoto et al., 1978).

Matumoto, H.; Tujitani, N.; Ouchi, S.; Jida, T.; Tomizawa, S. and Kamata, J.: Teratogenicity test of epidihydrocholesterine in mice. (Japanese) Clinical Report 12:479–492, 1978.

940 Epilepsy also see *Individual medications*

The subject of epilepsy and its treatment was reviewed comprehensively by Kaneko (1991) who gives the results of retrospective studies on IQ and malformations. Small but significant reductions in IQ were associated and increased with age. The average percentage of malformations in 14 studies was about doubled (5.2 vs 2.9 percent). In 12 prospective studies the treated epileptic offspring had 11.1% malformations while the untreated (11 studies) had 5.7% (controls 4.8%). Four studies of the malformation rate of offspring from epileptic fathers average 8.4% (control 6.8%). Polytherapy including valproic acid or carbamazepine may represent a higher risk for malformation (meningomyeloceles). Kallen (1986) discussed the difficulties in interpretation of the studies on antiepilepsy drugs. He felt that monotherapy was probably associated with a lower risk for malformations than was polytherapy. Only marginal differences in malformation rates among therapy groups existed but valproic acid seemed to have a stronger association with spina bifida than other anticonvulsants. Kallen et al. (1989) analyzed 318 malformed infants born to mothers with epilepsy. A significant association between valproic acid and spina bifida was found but an increased non–significant association was found with carbamazepine. The odds ratio for facial clefts was 6.0 with phenobarbitone and 1.7 for diphenylhydantoin.

Gaily and Granstrom (1992) summarized six prospective studies on the presence of minor anomalies among the offspring of women with epilepsy. They concluded that except for distal digital hypoplasia (Dilantin) and possibly hypertelorism none were related to drug therapy. Czeizel et al. (1992) in comparing 10,698 malformed infants to 21,546 controls found 144 women who took anticonvulsants during pregnancy. The rate of anticonvulsant use was 2.9 times higher among the malformed group. The odds ratio was 10.1 (Conf. int. 3.0–24.1) for sultiame a carbonic anhydrase inhibitor used to enhance other antiepileptic drugs. For single drugs taken, phenytoin had an odds ration of 1.8 (Conf. int. 0.9–3.6) and carbamazepine was 2.4 (Conf. int. 0.8–7.6).

Nakane et al. (1980) studied 657 women receiving antiepileptic drugs and 162 epileptic patients who were untreated. Eighteen of 43 women on trimethadione and 23 of 133 women on primidone had offspring with malformations but only 39 of 325 of the phenytoin users had malformed offspring. One malformation occurred among 19 treated with phenobarbital alone. In the treated series, 63 (9.9 percent) were malformed which was five times higher than in the group which was not treated. Cleft lip and palate and heart defects were the most common defects. Seizures during pregnancy were associated with an increased defect rate.

Kaneko et al. (1988) studied 172 pregnancies in which epileptic drugs were used. For the 31 in which only one drug was used, there was 6.5 percent malformations while among the remaining 141 patients which were treated with more than one drug, the rate was 15.6 percent. Certain minor defects such as pilonidal sinus, inguinal hernia, and strabismus were included. By Wilcoxon rank–sum tests, valproate use was significantly associated with a defect increase and carbamazepine was almost significantly associated (p = 0.053). Among the polydrug users who used carbamazepine, there were 11 malformations among 105. When primidone was used in polydrug use, there were 8 malformations among 80 cases. In this study of Japanese children, 4 facial clefts and 3 heart defects occurred among 172. Yerby et al (1992) reported a significant increase (5.05 vs 3.65) p:0.0001 in minor congenital anomalies in the offspring of mothers treated with antiepileptics.

The interaction if polydrug use and genetic predisposition has been discussed by Robert. She proposed that genetic susceptibility to teratogenic action of the various drugs used in treatment might account for the increase in malformations.

Phillips and Lockard (1986) studied 9 monkey offspring of epileptic mothers on phenytoin. The blood level during the third trimester correlated with infant motor developmental delay but the seizure activity was unrelated.

Czeizel, A.E.; Bod, M. and Halasz, P.: Evaluation of anticonvulsant drugs during pregnancy in a population–based Hungarian study. Eur. J. Eipdemiol. 8:122–127, 1992.

Gaily, A.E.; Bod, M. and Halasz, P.: Evaluation of anticonvulsant drugs during pregnancy in a population–based Hungarian study. Eur. J. Epidemiol 8:122–127, 1992.

Gaily, E. and Granstrom, M.L.: Minor anomalies in children of mothers with epilepsy. Neurology 42:128–131, 1992.

Kallen, B.: Maternal epilepsy, antiepileptic drugs and birth defects. Pathologica 78:757–768, 1986.

Kallen, B.; Robert, E.; Mastroiacovo, P.; Martinez–Frias, M.L.; Castilla, E.E. and Cocchi, G.: Anticonvulsant drugs and malformations is there a drug specificity? Eur. J. Epidemiol. 5(1):31–36, 1989.

Kaneko, S.: Antiepileptic drug therapy and reproductive consequences: functional and morphologic effects. Reproductive Toxicology 5:179–198, 1991.

Kaneko, S.; Otani, K.; Fukushima, Y.; Ogawa, Y.; Nomura, Y.; Ono, T.; Nakane, Y.; Teranishi, T. and Goto, M.: Teratogenicity of antiepileptic drugs: Analysis of possible risk factors. Epilepsia 29(4):459–467, 1988.

Nakane, Y.; Okuma, T.; Takahashi, R.; Sato, Y.; Wada, T.; Sato, T.; Fukushima, Y.; Kumashiro, H.; Omo, T.; Takahashi, T.; Aoki, Y.; Kazamatsuri, H.; Inami, M.; Komai, S.; Seino, M.; Miyakoshi, M.; Tanimura, T.; Hazama, H.; Kawahara, R.; Otsuki, S.; Hosokawa, K.; Inanaga, K.; Nakazawa, Y. and Yamamoto, K.: Multi–institutional study on the teratogenicity and fetal toxicity of antiepileptic drugs. A report of a collaborative study group in Japan. Epilepsia 21:663–680, 1980.

Phillips, N.K. and Lockard, J.S.: A gestational monkey model: Effects of phenytoin versus seizures on neonatal outcome. Epilepsia 26:697–703, 1986.

Robert, E.: Risques teratogenes de l'epilepsie et des antiepileptiques. Pediatrie 46:579–583, 1991.

Yerby, M.S.; Leavitt, A.; Erickson, D.M.; McCormick, K.B.; Loewenson, R.B.; Sells, C.J. and Benedetti, T.J.: Antiepileptics and the development of congenital anomalies. Neurology 42:132–140, 1992.

941 Epinastine

Niggeschulze and Kast (1991) studied this antihistamine in rats before mating, during embryogenesis and prenatally. Oral doses of up to 200 mg per kg produced maternal toxicity but no fetal ill effects when given during embryogenesis. At 120 mg per kg the estrus cycle was disturbed giving fewer conceptions. Studies in the rabbit at 75 mg per kg caused resorptions but no teratogenic effect. Post natal studies in rats were performed without significant findings.

Niggeschulze, A. and Kast, A.: Oral reproduction toxicology of epinastine. Oyo Yakuri 41:355–369, 1991.

942 Epinephrine *Adrenalin* CAS 51–43–4

Heinonen et al. (1977) studied 189 offspring of women exposed to epinephrine in the first 4 lunar months. The hospital standardized relative risk was 1.57. Combining all sympathomimetics a statistically significant increase in minor and major defects was found. The relative risk for eye and ear malformations was 1.77 and for musculo–skeletal defects 1.04. Jost et al. (1969) summarized their experiments with adrenalin–induced limb defects in the rabbit. They injected 5 to 50 microgms of adrenalin directly into rabbit fetuses at 18 to 22 days of gestational age and noted hemorrhages, edema and necrosis of the distal extremities. A recessively inherited syndrome (br) in the rabbit produces similar types of defects. They showed that the concentration of adrenalin is elevated in the adrenal medulla of these affected fetuses and this raised the possibility of excess endogenous adrenalin leading to the genetic limb defect.

Jost, in his original work (1953), produced limb defects in rat fetuses injected directly with 1 to 50 microgm of adrenalin on the 17th day.

Gatling (1962) dropped 20 to 200 microgm of this material on the chorio–allantoic membrane of the chick once on the 10th, 11th or 12th day and produced hemorrhages of the head, skin and extremities. Similar findings occurred when norepinephrine and Neosynephrine™ were used. Ephedrine and serotonin gave negative results. Hodach et al. (1974) administered 5 microgm of epinephrine to chick embryos at intervals between 24 and 190 hours of incubation. Treatment between 96 and 124 hours produced over 50 percent aortic arch anomalies in the surviving embryos.

Heinonen et al. (1977) studied the offspring of 373 women exposed to ephedrine in the first 4 lunar months. Seventeen defects were found for a relative risk of 0.98.

Chernoff and Grabowski (1971) found that injection of adrenalin intraperitoneally in rats on days 15–21 caused a decrease in fetal heart rate. Fetal hyperkalemia occurred. Weir (1965) injected epinephrine intravenously into rabbits on day 22. The fetal weight was reduced by using 0.159 ml per kg of a 1:10,000 dilution.

Chernoff, N. and Grabowski, C.T.: Responses of the rat foetus to maternal injections of adrenaline and vasopressin. Br. J. Pharmac. 43:270–278, 1971.

Gatling, R.R.: The effect of sympathomimetic agents on the chick embryo. Am. J. Pathol. 40:113–127, 1962.

Heinonen, O.P.; Slone, D. and Shapiro, S.: Birth Defects and Drugs in Pregnancy. John Wright Publishing Sciences Group, Inc., Littleton, Mass. 1977.

Hodach, R.J.; Gilbert, E.F. and Fallon, J.F.: Aortic arch anomalies associated with administration of epinephrine in chick embryos. Teratology 9:203–210, 1974.

Jost, A.: Degenerescence des extremites du foetus de rat provoquee par l'adrenaline. C. R. Acad. Sci. (Paris) 236:1510–1512, 1953.

Jost, A.; Roffi, J. and Cowitat, M.: Congenital amputations determined by the br gene and those induced by adrenalin injection in the rabbit fetus. In: Limb Development and Deformity: Problems of Evaluation and Rehabilitation. C.A. Swinyard (ed.), Springfield: C.C. Thomas, 187–199, 1969.

Wier, K.: Effect on the weights of fetuses and fetal lymphoid organs of adrenalin given to rabbits at a critical period of pregnancy: Observations on spontaneous and induced runting. Anat. Rec. 153:373–376, 1965.

943 2–α, 3–α–Epithio–5– α–androstan–17–β–01 CAS 2363–58–8

This long–acting antiestrogenic compound was given to mice and rats during pregnancy and at doses of 10 and 40 mg per kg per day delayed parturition and maternal death occurred. Administration late in pregnancy of 2.5 mg per kg caused no effect on the reproductive ability of the male offspring but none of the female offspring showed reproductive ability. Deformities of the female mice sex organs were seen with doses of 10 mg per kg per day (Mineshita et al., 1973).

Mineshita et al. (1972) found total embryo lethality at 10 and 40 mg per kg intramuscularly in rat offspring. Limb defects amd masculinization of female fetuses were found after treatment of mice with 2.5 mg per kg during mid–pregnancy.

Mineshita, T.; Hasegawa, Y.; Yoshida, T.; Kozen, T. and Sakaguchi, I.: Teratology study on epithio steroids in rats and mice. I. Effects on fetuses of 2–α, 3–α–epithio–5–α–androstan–17–β–01 (10275–S) given to mothers during mid–pregnancy. (abs) Teratology 6:113, 1972.

Mineshita, T.; Hasegawa, Y.; Yoshida, T.; Kozen, T.; Sakaguchi, I.; Okamoto, A. and Ohara, T.: Teratological and reproductive studies on 2–α, 3–α–epithio–5 α–androstan–17–β–01 in mice and rats. (Japanese) Oyo Yakuri 7:723–752, 1973.

944 β–(3,4–Epoxycyclohexyl)ethyl trimethoxysilane CAS 3388–04–3

Tyl et al. (1988) studied pregnant rats and rabbits on days 6–15 with up to 2.5 ml or 0.75 ml per kg, respectively. The material found in some epoxy resins was gavaged. No embryotoxicity or teratogenicity was found.

Tyl, R.W.; Ballantyne, B.; Fisher, L.C. and France, K.A.: Evaluation of the developmental toxicity of β–(3,4–epoxycyclohexyl)ethyl trimethoxysilane in Fisher 344 rats and New Zealand white rabbits. Fund. Appl. Toxicol. 10:439–452, 1988.

945 EPN

Courtney et al. (1985) administered this pesticide to mice intragastrically on days 6–16 in doses of up to 12.0 mg per kg. No adverse reproductive effects were found.

Courtney, K.D.; Andrews, J.E.; Springer, J. and Dalley, L.: Teratogenic evaluation of the pesticides baygon, carbofuran, dimethoate and EPN. J. Environ. Sci. Health B20(4):373–406, 1985.

946 Epstein–Barr Virus *Infectious Mononucleosis*

Epstein–Barr infection of the mother has been associated with at least three newborns with congenital defects (Brown and Stenchever, 1978). No pattern of defects has been identified: Cleft lip and palate, microphthalmus, cardiovascular defects and cataracts were included in the case reports. Fleisher and Bologonese (1984) followed 4063 pregnancies and found seroconversion in only 3. None of the three had intrauterine infection with the virus. They reviewed other reports and failed to find conclusive evidence of transplacental infection.

Icart and Didier (1981) reported a higher rate (21 percent) of pathologic pregnancies among 91 women with persistent antibody to E–B virus. In the controls, 7 percent had a pathologic pregnancy. Two case reports with evidence of Epstein–Barr infection in utero appeared (Goldberg et al., 1981; Joncas et al., 1981). Both had platelet reduction and severe nervous system involvement. Celery stalking of the femurs occurred in one.

Brown, Z.A. and Stenchever, M.A.: Infectious mononucleosis and congenital malformations. Am. J. Obstet. Gynecol. 131:108–109, 1978.

Fleisher, G. and Bologonese, R.: Epstein–Barr virus infections in pregnancy: A prospective study. J. Pediatr. 104:374–379, 1984.

Goldberg, G.N.; Fulginiti, V.A.; Ray, G.; Ferry, P.; Jones, J.F.; Cross, H. and Minnich, L.: In utero Epstein–Barr virus (infectious mononucleosis) infection. JAMA 246:1579–1581, 1981.

Icart, J. and Didier, J.: Infections due to Epstein–Barr virus during pregnancy. J. Infect. Dis. 143:499 only, 1981.

Joncas, J.H.; Alfieri, C.; Leyritz–Wills, M.; Brochu, P.; Jasmin, G.; Boldogh, I. and Huang, E.S.: Simultaneous congenital infection with Epstein–Barr virus and cytomegalovirus. Medical Intelligence 304:1399–1403, 1981.

947 Eptazocine *L–1,4–Dimethyl–10–hydroxy–2,3,4,5,6,7 hexahydro–1,6–methano–1H–4–benzazonine hydrobromide* CAS 725–22–13–5

Matsuda et al. (1980) in a series of papers, reported reproductive studies in the mouse and rabbit. They used up to 125 mg per kg subcutaneously during organogenesis and 100 mg per kg subcutaneously before mating and the first 7 days of gestation or during days 15 through 21. They observed no ill effects except for a decrease in fetal weight and delayed ossification at the higher doses.

Transient behavioral alterations were observed. In rabbits, 100 mg per kg were given on days 6 through 18 and no adverse fetal effects were found.

Matsuda, M.; Minami, Y.; Kawakami, T.; Nogami, M.; Ikeda, M.; Kumada, S.; Urata, M. and Kihara, T.: Reproductive studies of l–1,4–dimethyl–10–hydroxy–2, 3, 4, 5, 6, 7–hexahydro–1, 6–methano–1H–4–benzazonine hydrobromide (Eptazocine HBr, 1–ST–2121). Oyo Yakuri 20:501–526, 703–714, 803–811, 1980.

948 Ergocornine CAS 564–36–3

Carpent and Desclin (1968) found increased eye and heart defects in fetuses of rats injected with 1.0 mg on day 8 of gestation. Doses of 0.18 to .30 mg on the 5th or 12th days had little effect on the gestation.

Carpent, G. and Desclin, L.: Effects of ergocornine on the mechanism of gestation and on fetal morphology in the rat. Endocrinol. 84:315–324, 1968.

949 Ergotamine *Ergonovine* CAS 113–15–5

David (1972) reported that among ten newborns with Poland syndrome, five were adopted and of the remaining five, three mothers had attempted abortion. One of these mothers took ergonovine. Wainscott et al. (1978) studied the outcome of women treated during pregnancy for migraine and reported no adverse effects. Graham et al. (1983) reported jejunal atresia in a mother who took cafergot. Among 9,460 children with congenital defects Czeizel (**1989**) reported 3 neural tube defects whose mothers were exposed to the drug in the first 3 months of pregnancy.

Grauwiler and Schon (1973) studied the effect of this drug on pregnant mice, rats and rabbits and found no specific defects but compression effects on the fetus (reduced fetal weight and skeletal retardation) were found at oral doses of 100 mg per kg in mice and 10 mg per kg in rats. The rodents were treated on days 6–15 and the rabbits on days 6–18.

Czeizel, A.: Teratogenicity of ergotamine. Journal of Medical Genetics 26:69–71, 1989.

David, T.J.: Nature and etiology of Poland anomaly. N. Eng. J. Med. 287:487–489, 1972.

Graham, J.M.; Marin–Padilla, M. and Hoefnagel, D.: Jejunal atresia associated with cafergot ingestion during pregnancy. Clinical Pediatrics 22:226–228, 1983.

Grauwiler, J. and Schon, H.: Teratological experiments with ergotamine in mice, rats and rabbits. Teratology 7:227–236, 1973.

Wainscott, G.; Sullivan, F.M.; Volans, G.N. and Wilkinson, M.: The outcome of pregnancy in women suffering from migraine. Postgraduate Med. J. 54:98, 1978.

950 Ergotoxine CAS 8006–25–5

Sommer and Buchanan (1955) injected rats intraperitoneally with 0.5 mg twice daily from the 12th day until the 21st day after vaginal sperm were found. Defects were not studied but the newborn fetuses were reduced in weight.

Sommer, A.F. and Buchanan, A.R.: Effects of ergot alkaloids on pregnancy and lactation in the albino rat. Am. J. Physiol. 180:296–300, 1955.

951 Erythromycin CAS 114–07–8

Takaya (1965) gave 10 to 25 mg per kg subcutaneously to rats on days 6–10 and found no increase in defects or decrease in fetal weight.

Takaya, M.: Teratogenic effects of antibiotics. J. Osaka City Med. Cen. 14:107–115, 1965.

952 Erythropoietin *EPOCH KRN 5702*

Igarashi et al. (**1990**) treated rats intravenously with up to 100 microgm daily during organogenesis and found decrease fetal weight at 1.0 microgm but no teratogenicity. All the dams had increased hematocrits. In perinatal studies Sugiyama et al. (**1990**) found decreased weight gain in the pup of dams treated with 1.0 microgm or more. The decrease was associated with a decline in lactation. In rabbits Hara et al. (**1990**) found decreased weight in fetuses exposed during organogenesis to 10 or 100 microgm per kg.

Furuhashi et al. (**1988** A) gave this recombinant human erythropoietin intravenously to rats intravenously on days 7–17. At the highest dose 500 iu per kg fetal weight and ossification were reduced. No increase in defects was found. In rabbits Furuhashi et al. (**1988** B) treated similarly on days 6–18 no adverse fetal effects were found.

Furuhashi, T.; Ushida, K.; Ishizaka, Y.; Izumi, H. and Amano, K.: Teratology study of krn5702 in rats. Kiso to Rinsho 22:5451–5468, 1988.

Furuhashi, T.; Itagaki, Y,; Kurihara, H.; Murase, E.; Izumi, H. and Amano, K.: Teratology study of krn5702 by intravenous administration in rabbits. Kiso to Rinsho 22:5469–5476, 1988.

Hara, H,; Yahasi, H.; Nakagawa, T.; Kiyosawa, K.; Sugiyama, O. and Deki, T.: Teratological study of epoch in rabbits. Rinsko Iyaku 6(2):450–455, 1990.

Igarashi, S.; Watanabe, K.; Sugiyama, O.; Watanabe, S.; Ebihara, Y. and Usami, M.: Teratological study of epoch in rats. Rinsho Iyaku (Journal of Clinical Therapeutics and Medicines) 6(2):429–455, 1990.

Igarashi, S.; Sugiyama, O.; Watanabe, S.; Watanabe, K.; Ebihara, Y.; Sato, T. and Usami, M.: Fertility study of epoch in rats. Rinsho Iyaku 6(2):411–428, 1990.

953 Estradiol CAS 50–28–2

Using 0.8 to 35 mg estradiol during the 12th to 19th gestational days of the rat, Greene et al. (1939, 1940) produced complete feminization of the external genitalia of male fetuses. Testes were retained in the female position with reduction or absence of epididymis, vas, seminal vesicles and prostate. A rudimentary vagina was present in the males. In the female fetus there was enlargement of the uterus and a paradoxical retention of the wolffian bodies. In both sexes the number of nipples was increased. At high dose levels, fetal loss was common. Burns (1955) produced ovotestes by treating premature opossum embryos with topical estradiol diprorionate (0.1 to 5 microgm per day).

Nishihara (1958) produced 14 percent cleft palates in the mouse fetus by injecting the mother with 1 mg on the days 11–16. Closure of the fetal mouse eye lids was also delayed by the use of estradiol on the 17th gestational day (Raynaud, 1942).

Yasuda et al. (1985) gave 0.02, 0.2 or 2.0 mg per kg to Jc1:ICR mice on days 11–17 and produced ovotestes and intraabdominal testes with persistent Mullerian ducts and Wolffian ducts in male fetuses. Ovarian hypoplasia was produced in female fetuses. They suggested that the presence of the ethinyl estradiol affected Sertoli cell differentiation which resulted in suppression of Mullerian inhibiting factor. In the developing ovary the estradiol affected the contact of follicular cells with oocytes. Yasuda et al. (1987) gave ethinyl estradiol 0.2 mg per kg to mice on days 11–17 and observed medullary hyperplasia of the fetal ovaries with follicular cell hypoplasia. Yasuda et al. (1988) studied the long term effects on male reproduction and found atrophy of the seminiferous tubules, Leydig cell hyperplasia and in one case precancerous changes in the epididymis.

Burns, R.K.: Experimental reversal of sex in the gonads of the opossum Didelphis virginiana. Proc. Nat. Acad. Sci. USA 41:669–676, 1955.

Greene, R.R.; Burrill, M.W. and Ivy, A.C.: Experimental intersexuality: The paradoxicol effects of estrogens on the sexual development of the female rat. Anat. Rec. 74:429–438, 1939.

Greene, R.R.; Burrill, M.W. and Ivy, A.C.: Experimental intersexuality: The effects of estrogens on the antenatal sexual development of the rat. Am. J. Anat. 67:305–345, 1940.

Nishihara, G.: Influences of female sex hormones in experimental teratogenesis. Proc. Soc. Exp. Biol. Med. 97:809–812, 1958.

Raynaud, A.: Inhibition de l'allongement et de la soudre des paupieres des embryons de souris. C. R. Soc. Biol. 136:337–338, 1942.

Yasuda, Y.; Kihara, T.; Tanimura, T. and Nishimura, H.: Gonadal dysgenesis induced by prenatal exposure to ethyinyl estradiol in mice. Teratology 32:219–227, 1985.

Yasuda, Y.; Konish, H. and Tanimura, T.: Ovarian follicular cell hyperplasia in fetal mice treated transplacentally with ethinyl estradiol. Teratology 36:35–43, 1987.

Yasuda, Y.; Ohara, I.; Konishi, H. and Tanimura, T.: Long–term effects on male reproductive organs of prenatal exposure to ethinyl estradiol. Am. J. Obstet. G. 159(5):1246–1250, 1988.

954 Estramustine Phosphate Disodium CAS 52205–73–9

This chemical combination of estradiol and nitrogen mustard was tested in rats and rabbits by Nomura et al. (1980 and 1981). Rats received orally up to 4.0 mg per kg before mating on days 7–17 or on days 17–21. Dosing prenatally and during the first 7 days of gestation was not associated with any adverse effects. Treatment with 4.0 mg per kg during organogenesis was followed by an increase in ectopic kidneys and skeletal defects. Late treatment with 4.0 mg per kg was associated with increased abnormalities of the external genitalia and decreased fertility in the female offspring. Both males and females had increased exploratory behavior at 55–58 postnatal days. Rabbits treated on days

6–18 had interruption of pregnancy at 0.2 mg per kg and at 0.1 mg per kg resorptions were increased. At 0.1 mg per kg no increase in defects was found.

Nomura, A.; Watanabe, M.; Ninomiya, H. and Enomoto, H.: Reproductive studies of estramustine phosphate disodium (EMP). Oyo Yakuri 20:1211–1236, 21:41–65, 1980 and 1981.

955 Estrone CAS 53–16–7

Nishihara (1958) produced 12.4 percent cleft palate in mouse fetuses whose mothers were injected with 1 mg estrone on the gestational days 11–16. Among injection–control fetuses, the incidence of cleft palate was 0.7 percent. Garmasheva et al. (1971) injected rabbits with 25 IU of folliculin, an estrone like hormone, on the 9th day and on the 12th day found increased wet weight and decreased dry weight of the embryos.

Garmasheva, N.L.; Konstantinova, N.N.; Zhakhova, Z.N. and Bakkal, T.P.: The effect of folliculin and sigetin on the development of the embryo and the placenta in rabbits with normal and impaired sympathetic innervation of the uterus. Am. J. Obstet. Gynecol. 111:1083–1091, 1971.

Nishihara, G.: Influence of female sex hormones in experimental teratogenesis. Proc. Soc. Exp. Biol. Med. 97:809–812, 1958.

956 Ethambutol CAS 74–55–5

Bobrowitz (1974) studied 42 pregnancies in which 15 to 25 mg per kg was given. There was neither intrauterine growth retardation nor increase in prematurity. Although there were some minor malformations observed, there was no increase in the rate. Lewit et al. (1974) could find no abnormalities in aborted embryos exposed to the drug. Snider et al. (1980) reviewed 638 pregnancies treated with this and other drugs and found no increase in congenital anomalies. Three hundred and twenty were treated in the first 3 months.

Bobrowitz, I.D.: Ethambutol in pregnancy. Chest 66:20–24, 1974.

Lewit, T.; Nebel, L.; Terracina, S. and Karman, S.: Ethambutol in pregnancy: Observations on embryogenesis. Chest 66:25–26, 1974.

Snider, D.E.; Layde, P.M. and Johnson, M.W.: Treatment of tuberculosis during pregnancy. Am Rev Respir Dis 122:65–79, 1980.

957 Ethamoxytriphetol MER-25
1-(p–2–Diethylaminoethoxyphenyl)–1–phenyl–2–p–methoxyphenyl Ethanol

Heller and Jones (1964) produced ovarian dysgenesis in rat fetuses by administering this anti–estrogenic compound in oral doses of 25 mg per kg on day 13 through 19. The fetal ovaries were necrotic and hemorrhagic in about half of the female fetuses. Although the number of fetuses per litter and fetal weight were reduced in the treated group, no other effects were noted. The fetal testes were histologically normal but increased in weight.

Heller, R.H. and Jones, H.W.: The production of ovarian dysgenesis in the rat by ethamoxytriphetol (MER–25). Am. J. Obstet. Gynecol. 90:264–270, 1964.

958 Ethidine

2,6–Dimethyl–3,5–dicarbethoxy–1,4–dihydropirine

Maganova and Zaitsev (1978) studied the effect of this compound in pregnant rats (100 and 500 mg per kg) and mice (1000 mg per kg) and found no mutagenic, embryotoxic or any adverse postnatal effect.

Maganova, N.B. and Zaitsev, A.N.: A study of mutagenic and embryotoxic action produced by ethidine. Vopr. Pitaniia (USSR) 4:70–74, 1978.

959 Ethinylestradiol CAS 57–63–6

Discussed under Pregnancy Test Tablets and Oral Contraceptives

960 17–Ethinyl–7α–methyl–19–nortestosterone

Pharriss (1970) found that implantation sites were smaller than normal or resorbing when rats were given 100 microgm per day on days 4–8.

Pharriss, B.B.: Biological properties of 17–ethinyl–7α–methyl–19–nortestosterone. Contraception 1:87–100, 1970.

961 17–α–Ethinyl–19–nor–testosterone *Norlutin*™ CAS 68–22–4

This oral progestin can produce masculinization of the external genitalia in the female fetus. Details appear under 17–α–Ethinyl–testosterone.

962 17–α–Ethinyl–testosterone *Pranone*™ *Lutocylol*™ *Pregneninolone Ethisterone* CAS 434–03–7

Although 17-hydroxyprogesterone is not teratogenic (Johnston and Franklin, 1964; Suchowsky and Junkmann, 1961), certain derivatives can cause virilizing effects in animals and man. Wilkins et al. (1958) and Wilkins (1960) reported that 17–α–ethinyl–testosterone in daily doses of 20–200 mg used during the first trimester of pregnancy caused masculinization of the external genitalia of female offspring. Jacobson (1962) with daily doses of 10–20 mg of 17–α–ethinyl–19–nor–testosterone reported 15 percent of female newborns to be masculinized. Grumbach et al. (1959) pointed out that labio–scrotal fusion was produced only in fetuses exposed prior to the 13th gestational week, but clitoromegaly could still be produced after this period. Serment and Ruf (1968) gave a good general summary of the types of progestins that cause masculinization. They cite four incidences where normethandrone was implicated.

There is a fairly extensive experimental literature on the effect of progesterones on masculinization of the female fetus. Courrier and Jost (1942) first warned that this compound could masculinize the fetus. Johnstone and Franklin (1964) used the mouse, Revesz et al. (1960) and Suchowsky et al. (1961) the rat and Jost (1946) the rabbit. Some of these fetuses also had blind vaginas or absence of portions of the uterus.

Courrier, R. and Jost, A.: Intersexualite foetale provoqvee par pregneninolone au cours de la grossesse. C. R. Soc. Biol. (Paris) 136:395–396, 1942.

Grumbach, M.M.; Ducharme, J.R. and Moloshok, R.E.: On the fetal masculinizing action of certain oral progestins. J. Clin. Endocrinol. Metab. 19:1369–1380, 1959.

Jacobson, B.D.: Hazards of norethindrone therapy during pregnancy. Am. J. Obstet. Gynecol. 84:962–968, 1962.

Johnstone, E.E. and Franklin, R.R.: Assay of progestins for fetal virilizing properties using the mouse. Obstet. Gynecol. 23:359–362, 1964.

Jost, A.: Recherches sur la differenciation sexuelle de l'embryon de lapin. Action des androgenes de synthese sur l'histogenese genitale. Arch. Anat. Microsc. Morphol. Exp. 36:242–270, 1946.

Revesz, C.; Chappel, C.I. and Gaudry, R.: Masculinization of the female fetuses in the rat by progestational compounds. Endocrinology 66:140–144, 1960.

Serment, H. and Ruf, H.: Therapeutiques hormonales. Bull. Fed. Soc. Gynecol. Obstet. Lang. Fr. 20:69–76, 1968.

Suchowsky, G.K. and Junkmann, K.: A study of the virilizing effect of progestogens on the female rat fetus. Endocrinol. 68:341–349, 1961.

Wilkins, L.: Masculinization due to orally given progestins. JAMA 172: 1028–1032, 1960.

Wilkins, L.; Jones, H.W.; Holman, G.H. and Stempfel, R.S.: Masculinization of the female fetus associated with administration of oral and intramuscular progestins during gestation: Non–adrenal female pseudohermaphrodism. J. Clin. Endocrinol. Metab. 18:559–585, 1958.

963 Ethionamide CAS 536–33–4

Zierski (1966) studied 38 newborns whose mothers took 15 mg per kg daily during pregnancy and all were healthy. Potworowski et al. (1966) found seven deformities among 23 exposed infants but two of these were Down's syndrome. No pattern of malformation was apparent and they included congenital heart, spina bifida and atresia of the gastrointestinal tract.

Fujimori et al. (1965) gave rats by mouth, 200 mg per kg from gestational days 6–14 and found a low incidence of omphalocele (5.2 percent), exencephaly (2.7 percent) and cleft palate (1.4 percent). Takekoshi (1965) gave mice 75 mg orally during days 6 through 13 of gestation and found no defects. When hydroxymethylpyrimidine (0.75 mg) was added to 7.5 mg of ethionamide, 23 percent defects occurred (exencephaly and facial clefts).

Fujimori, H.; Yamada, F.; Shibukawa, N.; Goda, S. and Itani, I.: The effect of tuberculostatics on the fetus: An experimental production of congenital anomaly in rats by ethionamide. (abs) Proceedings of the Congenital Anomalies Research Association of Japan 5:34–35, 1965.

Potworowski, M.; Sianozecka, E. and Szufladowicz, R.: Ephionamide treatment and pregnancy. Pol. Med. J. 5:1152–1158, 1966.

Takekoshi, S.: Effects of hydroxymethylpyrimidine on isoniazid–and ethionamide–induced teratosis. Gunma J. Med. Sci. 14:233–244, 1965.

Zierski, M.: Influence of ethionamide on development of human fetus. Grazlica I Choroby Plac (Warsaw) 34:349–352, 1966.

964   Ethionine CAS 13073–35–3

Feldman and Waddington (1955) injected approximately 1 mg of this substance into chick eggs before incubation and found after 42 hours incubation that neurulation was defective. Methionine given concommitantly resulted in normal development. Proffit and Edwards (1962) injected 200 mg of D–L ethionine during the second week of the rats' gestation and produced no congenital defects. Landauer and Salam (1974) produced beak abnormalities, muscle hypoplasia and malformations of the cervical vertebrae in chicks by injecting 4.5 mg at 96 hours of incubation.

Feldman, M. and Waddington, C.H.: The uptake of methionine ^{35}S by the chick embryo and its inhibition by ethionine. J. Embryol. Exp. Morphol. 3:44–58, 1955.

Landauer, W. and Salam, N.: Experimental production in chicken embryos of muscular hypoplasia and associated defects of beak and cervical vertebrae. Acta Embryol. Experimentalis 1:51–66, 1974.

Proffit, W.R. and Edwards, L.E.: Effects of ethionine administration during pregnancy in the rat. J. Exp. Zool. 150:135–142, 1962.

965   Ethoheptazine

Heinonen et al. (1977) found no increase in defects among the offspring of 60 mothers treated during the first trimester.

Heinonen, O.P; Slone, D. and Shapiro, S.: Birth Defects and Drugs in Pregnancy. John Wright Publishing Sciences Group, Inc., Littleton, Mass. 1977.

966   Ethotoin

Brown et al. (1982) gave mice intraperitoneally up to 2.70 mmol per kg on days 8,9, and 10 and found up to 12% malformations per litter. Skeletal and cardiovascular anomalies, exencephaly and cleft palate were found.

Brown, N.A.; Shull, G.; Kao, J.; Goulding, E.H. and Fabro, S.: Teratogenicity and lethality of hydantoin derivatives in the mouse: Structure–toxicity relationships. Toxicology and Applied Pharmacology 64:271–288, 1982.

967   Ethoxyquin

In a group of 57 women exposed to this antielepticx there were two with malformations (metatarsus varus and clubfoot) (Lindhout et al., 1992). DeSesso and Goeringer (1990) gave rabbits 950 mg per kg subcutaneously on day 12 and found no increase in defects. This antioxidant was associated with a significant increase in fetal body weight.

DeSesso, J.M. and Goeringer, G.C.: Ethoxyquin and nordihyroguaiaretic acid reduce hydroxyurea developmental toxicity. Reproductive Toxicology 4:267–275, 1990.

Lindhout, D.; Meinardi, H.; Meijer, W.J.A. and Nau, H.: Antiepileptic drugs and teratogenesis in two consecutive cohorts: Changes in prescription policy paralleled by changes

in pattern of malformations. Neurology 42(Suppl 5):94–110, 1992.

968   Ethosuximide $Zarotin^{TM}$ CAS 77–67–8

In a group of 57 women exposed to this antiepileptic there were two with malformations (metatarsus varus and clubfoot) (Lindhout et al. 1992). The suxinimides (ethosuximide, methsuximide and phensuximide) appear to have much less teratogenic potential than the oxazolidine–2,4–diones (trimethadione and parmethadione) (Fabro and Brown, 1979). These authors summarized a large number of case histories and found only 6 percent (5 per 89) malformations in the suxinimide group while the other group had 36 percent (5 per 14). Kao et al. (1979) have shown in mice that the suxinimides are less teratogenic than the oxazolidine–2,4–diones.

Fabro, S. and Brown, N.A.: The teratogenic potential of anticonvulsants. N. Eng. J. Med. 300:1280–1281, 1979.

Kao, J.; Brown, N.A.; Shull, G. and Fabro, S.: Chemical structure and teratogenicity of anticonvulsants. (abs) Fed. Proc. 38:438 only, 1979.

Lindhout, D.; Meinardi, H.; Meijer, S.J.A. and Nau, H.: Antiepileptic drugs and teratogenesis in two consecutive cohorts: Changes in prescription policy paralleled by changes in pattern of malformations. Neurology 42(Suppl 5):94–110, 1992.

969   Ethoxyacetic Acid

Rawlings et al. (1985) exposed explanted day 9.5 rat embryos and observed growth retardation and structural malformations.

Rawlings, S.J.; Shuker, D.E.G.; Webb, M. and Brown, N.: The teratogenic potential of alkoxy acids in post-implantation rat embryo culture: Structure–activity relationships. Toxicol. Lett. 1472:49–58, 1985.

970   o–Ethoxybenzoylhydrazone of Pyruvic Acid

Pisanti and Volterra (1970) treated rats and rabbits with up to 60 mg per kg during organogenesis and found no teratogenicity.

Pisanti, N. and Volterra, G.: Un nuovo antiflogistico-antalgico: L'o–etossibenzoilidrazone dell'acido piruvico. Farmaco Ed. Prat. 25:105–121, 1970.

971   2–Ethoxyethyl Acetate

This glycol ether was studied in rats by Nelson et al. (1985) who used 130, 390 and 600 ppm for 7 hours on days 7–15. Total resorption occurred at 600 ppm. At 390 ppm there were 9 heart defects among 116 fetuses and at 130 ppm one was found among 142 studied. Resorptions and decreased fetal weight was found at 390 ppm.

Nelson, B.K.; Setzer, J.V.; Brightwell, W.S.; Mathinos, P.R; Kuczuk, M.H.; Weaver, T.E. and Goad, P.T.: Comparative inhalation teratogenicity of four glycol ether solvents and an amino derivative in rats. Environmental Health Perspectives 57:261–271, 1984.

972 Ethoxzolamide CAS 452–35–7

Wilson et al. (1968) fed this carbonic anhydrase inhibitor to rats as 0.3 or 0.6 percent of their diet throughout pregnancy. The acetazolamide syndrome consisting of localized post–axial forelimb defects was observed in about 10 percent of the fetuses. Occasionally other organ systems were involved also.

Wilson, J.G.; Maren, T.H.; Takano, K. and Ellison, A.C.: Teratogenic action of carbonic anhydrase inhibitors in the rat. Teratology 1:51–60, 1968.

973 1–Ethyl–2–acetylhydrazine

von Kreybig et al. (1970) gave 100 mg per kg to two rats on day 13 of gestation and found no adverse fetal effects.

von Kreybig, T.; Preussmann, R. and von Kreybig, I.: Chemische Konstitution und Teratogene Wirkung bei der Ratte. Arzneim–Forsch. 20:363–367, 1970.

974 Ethyl Acrylate CAS 140–88–5

Murray et al. (1981) exposed rats for 6 hours daily on days 6–15 to 50 or 150 ppm. Maternal toxicity occurred at the highest dose but no adverse fetal effects were found.

Murray, J.S.; Miller, R.R.; Deacon, M.M.; Hanley, T.R., Jr.; Hayes, W.C.; Rao, K.S. and John, J.A.: Teratological evaluation of inhaled ethyl acrylate in rats. Toxicol. Appl. Pharmacol. 60:106–111, 1981.

975 2–(Ethylamino–2–thienyl)cyclohexanone Hydrochloride

Four cats anesthetized with 20 mg per kg intramuscularly between 10 to 50 days of gestation gave birth to normal offspring (Bennett, 1969).

Bennett, R.R.: The clinical use of 2–(ethylamino)–2–thienyl)cyclohexanone HCl (CI–634) as an anesthetic in the cat. Am. J. Vet. Res. 30:1469–1470, 1969.

976 2–Ethylamino–1,3,4 Thiadiazole also see *Acetazolamide;* also see *Thiadiazole* CAS 13275–68–8

Murphy et al. (1957) using one–quarter to one–half the maternal LD–50 (200 mg per kg) produced skeletal defects in rat fetuses with treatment on the 11th day. On the 12th day, cleft lips and palates and encephaloceles were seen. But the most distinctive feature was absence of the tail. Maren and Ellison (1972) confirmed the work of Murphy et al. (1957).

Maren, T.H. and Ellison, A.C.: The teratological effect of certain thiadiazoles related to acetazolamide with a note on sulfanilamide and thiazide diuretics. Johns Hopkins Med. J. 130:95–104, 1972.

Murphy, M.L.; Dagg, C.P. and Karnofsky, D.A.: Comparison of teratogenic chemicals in the rat and chick embryos. Pediatrics 19:701–714, 1957.

977 Ethyl Apovincaminate *Cavinton™*

Cholnoky and Domok (1976) reported teratogenicity and fertility studies in rats and rabbits. Rats (male and female) were given up to 50 mg orally before mating and no effect on fertility was found. Oral doses to 135 mg per kg and

intravenous doses to 12.5 mg per kg were given rats during organogenesis and no teratogenicity was found. At the highest oral dose fetal growth retardation and maternal death occurred. Rabbits received up to 18 mg per kg without adverse fetal effects but growth retardation was found in the fetuses. Perinatal studies with up to 45 mg per kg were done without adverse fetal effects.

Cholnoky, E. and Domok, L.I.: Summary of safety tests of ethylapovincaminate. Arzneim–Forsch. 26:1938–1944, 1976.

978 Ethylbenzene CAS 100–41–4

Ungvary and Tatrai (1985) exposed rats, mice and rabbits to up to 2400 mg per square meter during organogenesis. Extra ribs and urinary tract defects were found to be increased in the rodent offspring.

Ungvary, G. and Tatrai, E.: On the embryotoxic effects of benzene and its alkyl derivatives in mice, rats and rabbits. Arch. Toxicol. Suppl. 8:425–430, 1985.

979 Ethylene Chlorhydrin *2–Chlorethanol* CAS 107–07–3

This breakdown product of ethylene oxide was given by gavage to mice during organogenesis (Courtney et al., 1982). Doses of up to 227 mg per kg in the water or 100 mg per kg by gavage produced no malformations. With the gavage dosing, both maternal and fetal weights were reduced.

Courtney, K.D.; Andrews, J.E. and Grady, M.: Teratogenic evaluation of ethylene chlorhydrin. J. Environ. Sci. Health B17:381–391, 1982.

980 Ethyl Chloride

Hanley et al. (**1987**) studied the effect of this gas on mice exposed 6 hours daily to up to 5000 ppm during organogenesis and found no adverse reproductive effects.

Hanley, Jr., T.R.; Scortichini, B.H.; Johnson, K.A. and Momany–Pfruender, J.J.: Effects of inhaled ethyl chloride on fetal development in CF–1 mice. Toxicologist 7:189, 1987.

981 Ethylene Dibromide CAS 106–93–4

Haar (1980) reviewed the existing human epidemiologic data and concluded there was no reproductive adverse effect among workers.

Hardin et al. (1981) injected 55 mg per kg into rats on days 1–15. Maternal toxicity occurred but no teratogenicity. Mitra et al. (**1992**) have shown that ethylene dibromide can be bioactivated by rat liver glutathione S–transferases.

Haar, T.G.: An investigation of possible sterility and health effects from exposure to ethylene dibromide. In: Banbury Report No. 5, Ethylene Dichloride. B. Ames, P. Infante and R. Reitz (eds.), Cold Spring Harbor Laboratory, 149–161, 1980.

Hardin, B.D.; Bond, G.P.; Sikov, M.R.; Andrew, F.D.; Beliles, R.P. and Niemeier, R.W.: Testing of selected workplace chemicals for teratogenic potential. Scand. J. Work Environ. Health 7(4):66–75, 1981.

Mitra, A.; Hilbelink, D.R.; Dwornik, J.J. and Kulkarni, A.: Rat hepatic glutathione s–transferase–mediated embryotoxic bioactivation of ethylene dibromide. Teratology 46:439–446, 1992.

982 Ethylene Fluorohydrin

Mankes et al. (1992) gavaged rats with 0.06, 0.36 or 0.6 mg per kg on days 6–15. Hydronephrosis was increased at the two highest doses. Growth retardation and skeletal changes occurred at all doses. No decrease in maternal weight occurred.

Mankes, R.F.; LeFevre, R. and Calvano, C.J.: Teratogenic effects of the halogenated ethanol 2–fluoro ethanol in long evans rats (A). Teratology 45:463, 1992.

983 Ethylene Glycol CAS 107–21–1

Carney (1994) has reviewed the literature on toxicity in humans and prenatal animals. Extreme exposure measures would be necessary to produce dermal toxicity in humans and the low vapor pressure makes inhalation toxicity unlikely. Nearly continuous daily inhalation exposure has resulted in little or no increase in blood levels and no toxicity.

Rosica (1994) has pointed out that the higher molecular weight ethylene glycol ethers such as ethylene glycol butyl ether are not associated with developmental toxicity.

Lamb et al. (1985) fed up to 1 percent of ethylene glycol in the drinking water of mice for two generations. Slight but statistically significant decreases in number of litters and number of offspring were found in the group treated with the highest dose. Price et al. (1985) gavaged mice and rats on days 6 through 15 with doses of up to 3,000 and 5,000 mg per kg, respectively. At dose levels of 750 mg in the mouse and 1250 mg per kg in rats, increased defects were found. They consisted of cleft palates, facial defects, neural tube closure defects and other visceral and skeletal anomalies.

Wistar rats received (by gavage) from 1 to 15 days of pregnancy ethanol, ethylene glycol or glycerol in doses less or close to LD–50 (Bariljak et al. 1988). In low doses (for ethanol–1/12 of LD–50, for ethylene glycol–1/500 of LD–50, for glycerol–1/50 of LD–50) all three substances did not express embryotoxic activity. Ethylene glycol (140 and 1400 mg/kg) reduced heavy malformations of many organs but glycerol at the same doses possessed no teratogenic activity. All studied agents in lowest doses disturbed the structure and function of embryonic liver and induced changes in the activity of alcohol dehydrogenase and aspartate aminotransferase. Tyl et al. (1989) exposed rats and rabbits to near lethal gas concentrations of ethylene glycol monohexyl ether during gestation and produced no teratogenicity or developmental toxicity.

Bariljak, I.R.; Korach, V.I.; Spitovskaya, L.D. and Kalinovskaya, L.A.: Effects of alcohols on the rat embryos liver. Dokl. Aad. Nauk Ukr. SSR 6(5):67–70, 1988.

Carney, E.W.: An integrated perspective on the developmental toxicity of ethylene glycol. Reporductive Toxicology 8:99–113, 1994.

Lamb, J.C.; Maronpot, R.R.; Gulati, D.K.; Russell, V.S.; Hommel–Barnes, L. and Sabharwal, P.S.: Reproductive and developmental toxicity of ethylene glycol in the mouse. Toxicol. Appl. Pharmacol. 81:100–112, 1985.

Price, C.J.; Kimmel, C.A.; Tyl, R.W. and Marr, M.C.: The developmental toxicity of ethylene glycol in rats and mice. Toxicol. Appl. Pharmacol. 81:113–127, 1985.

Rosica, K.A.: Letters: Teratology 50:321, 1994.

Tyl, R.W.; Ballantyne, B.; France, K.A.; Fisher, L.C.; Klonne, D.R. and Pritts, I.M.: Evaluation of the developmental toxicity of ethylene glycol monohexyl ether vapor in Fischer 344 rats and New Zealand white rabbits. Fundamental and Applied Toxicology 12:269–280, 1989.

984 Ethylene Glycol Monoethyl Ether CAS 111–15–9
Cellosolve™ Glycol Ethers

Cook et al. (1982) studied exposed male workers and could not demonstrate decreased fertility.

Stenger et al. (1971) tested this material by mouth and injection in pregnant mice, rats and rabbits. Some fetal skeletal defects were found in rats with doses over 100 microliters per kg per day but no other defects were described in the three species. The type of skeletal defects was not mentioned. Hardin et al. (1982) applied 0.25 ml two times daily to the skin of pregnant rats on days 6–15. Maternal and fetal weight were reduced as were fetal survivors. Five of 35 survivors had cardiovascular defects and 13 had enlarged cerebral ventricles. Wickramaratne (1986) applied ethylene glycol monomethyl ether to rats dermally on days 6–17. Three percent solution applied as 10 ml per kg of body weight produced no adverse effects but 10 percent produced small litter size. Postnatal behavior effects in rats were reported for 2–ethoxyethanol (Nelson et al., 1981) but no changes were found with 2–methoxyethanol (Nelson et al., 1984).

Hardin et al. (1981) exposed rats and rabbits to 200–765 or 160–615 PPM for 7 hours daily during most of gestation. In rabbits, the higher concentration was lethal to the embryos but at the lower concentration 10 of 167 had cardiovascular defects and some increase in vertebral defects was found. Cardiovascular and skeletal defects were increased in the rat fetuses exposed to the lower dose at which maternal toxicity did not occur. Nagano et al. (1981) gavaged mice during organogenesis and found skeletal defects at 31 mg per kg and soft tissue defects (exencephaly) at higher levels. Nagano et al. (1979) reported testicular atrophy in mice. Hardin (1983) reviewed the subject of glycol ether reproductive toxicity.

Plasterer et al. (1985) studied ethylene glycol dimethyl ether, diethylene glycol dimethyl ether, triethylene glycol dimethyl ether and diethylene glycol diethyl ether in pregnant mice gavaged on days 7–14 at the highest respective doses of 3600, 5360, 7120 and 9000 mg per kg. Although there were no malformation increases in the neonates these sublethal doses reduced the fetal survivors exposed to ethyl glycol dimethyl ether, diethyl glycol dimethyl ether and triethylene glycol dimethyl ether. Hardin and Eisenmann (1987) compared equimolar concentrations of ethylene glycol di and monomethyl ether, triethylene glycol dimethyl ether and diethylene glycol dimethyl ether in mice treated on day 11. Only triethylene glycol dimethyl ether failed to produce paw defects. They concluded that the defects might be due to in vivo conversion to methoxyacetic acid.

Nolen et al. (1985) gave rats diethylene glycol monobutyl ether orally in doses of up to 1000 mg per kg before and during gestation and found only a small reduction in fetal birth weight. No paternal effect was noted. Studies in the rabbit with similar doses on days 7–18 were not associated with adverse fetal effects. Ema et al. (1988) using dietary amounts up to one percent during the entire rat gestation, found no adverse effects on the fetus or postnatal development.

Cook, R.R.: A cross–sectional study of ethylene glycol monomethyl ether process employees. Arch. Environ. Health 37:346–351, 1982.

Ema, M.; Itami, T. and Kawasaki, H.: Teratology study of diethylene glycol mono–n–butyl ether in rats. Drug Chem. Toxicol. 11(2):97–111, 1988.

Hardin, B.D.: Reproductive toxicity of glycol ethers. Toxicology 27:91–102, 1983.

Hardin, B.D.; Bond, G.P.; Sikov, M.R.; Andrew, F.D.; Beliles, R.P. and Niemeier, R.W.: Testing of selected workplace chemicals for teratogenic potential. Scand. J. Work Environ. Health 7(4):66–75, 1981.

Hardin, B.D. and Eisenmann, C.J.: Relative potency of four ethylene glycol ethers for induction of paw malformations in the CD–1 mouse. Teratology 35:321–328, 1987.

Hardin, B.D.; Niemier, R.W.; Smith, R.J.; Kuczuk, M.H.; Mathinos, P.R. and Weaver, T.E.: Teratogenicity of 2 ethoxyethanol by dermal application. Drug Chem. Toxicol. 5:277, 1982.

Nagano, K.; Nakayama, E.; Koyano, M.; Oobayashi, H.; Adachi, H. and Yamada, T.: Testicular atrophy in mice induced by ethylene glycol mono alkyl ethers. Jpn. J. Ind. Health 21:29–35, 1979.

Nagano, K.; Nakayama, E.; Oobayashi, H.; Yamada, T.; Adachi, T.; Ozawa, H.; Nakaichi, M.; Okuda, H.; Minami, K. and Yamazaki, K.: Embryotoxic effects of ethylene glycol monomethyl ether in mice. Toxicology 20:335–343, 1981.

Nelson, B.K.; Brightwell, W.S.; Burg, J.R. and Massari, V.J.: Behavioral and neurochemical alterations in the offspring of rats after maternal or paternal inhalation exposure to the industrial solvent 2–methoxyethanol. Pharmacol. Biochem. Behav. 20:269–279, 1984.

Nelson, B.K.; Brightwell, W.S.; Setzer, J.V.; Taylor, B.J.; Hornung, R.W. and O'Donohue, T.L.: Ethoxyethanol behavioral teratology in rats. Neurotoxicol. 2:231–249, 1981.

Nolen, G.A.; Gibson, W.B.; Benedict, J.H.; Briggs, D.W. and Schardein, J.L.: Fertility and teratogenic studies of diethylene glycol monobutyl ether in rats and rabbits. Fund Appl. Toxicol. 5:1137–1143, 1985. Piasterer, M.R.; Bradshaw, W.S.; Booth, G.M.; Carter, M.W.; Schuler, R.L. and Hardin, B.D.: Developmental toxicity of nine selected compounds following prenatal exposure in the mouse: Naphthalene, p–nitrophenyl, sodium selenite, dimethyl phthalate, ethylene–thiourea and four glycol ether derivatives. J. Toxicol. Environ. Health 15:25–38, 1985.

Stenger, E.G.; Aeppli, L.; Muller, D.; Peheim, E.

and Thomann, P.E.: Zur Toxikologie des Athylenglykol–Monoathylathers. Arzneim. Forsch. 21:880–885, 1971.

Wickramaratne, G.A.: The teratogenic potential and dose response of dermally administered ethylene glycol monomethyl ether (EGME) estimated in rats with the Chernoff–Kavlock assay. J. Appl. Toxicol. 6:165–166, 1986.

985 Ethylene Glycol Monohexyl Ether

Tyl et al. (1988) exposed rats and rabbits to 300 ppm for 6 hours daily during organogenesis. Maternal toxicity was found at 100–300 ppm fetal toxicity and malformations were increased. In the rat, facial, skeletal and cardiovascular defects were found. In the rabbit, renal, cardiac and brain defects predominated.

Tyl, R.W.; Pritts, I.M.; France, K.A.; Fisher, L.C. and Tyler, T.R.: Developmental toxicity evaluation of inhaled 2–ethoxyethanol acetate in Fischer 344 rats and Nex Zealand white rabbits. Fund. Appl. Toxicol. 10:20–39, 1988.

986 Ethylene Glycol Monopropyl Ether CAS 2807–30–9

Krasavage and Katz (1985) exposed rats on days 6–15 to up to 400 ppm for 6 hours daily. No adverse reproductive or fetal effects were found. Some variations in ossification were

Krasavage, W.J. and Katz, G.V.: Developmental toxicity of ethylene glycol monopropyl ether in the rat. Teratology 32:93–102, 1985.

987 Ethylene Oxide CAS 75–21–8

Laborde and Kimmel (1980) injected this material intravenously on several days during organogenesis in the mouse. Skeletal malformations occurred in fetuses whose mothers received 150 mg per kg which produced maternal toxicity. Doses of 75 mg per kg caused no defects. Snellings et al. (1982) exposed rats on days 6–15 of gestation for 6 hours daily to 10 to 100 PPM. At the highest dose fetal growth retardation occurred but there was no increase in congenital defects. Kimmel et al. (1984) reviewed the subject. Rutledge and Generoso (1989) exposed female mice to 1 hour of inhalation (1,200 ppm) at 1 hour or 6 hours after mating and found fetal death and increases in hydrops and eye defects. Treatment at 9 or 25 hours after mating was associated with only marginal effects. A few cardiac and skeletal defects were also found.

Hemminki et al. (1982) studied the abortion rate in 1443 hospital workers exposed during pregnancy to sterilizing procedures by ethylene oxide. The frequency was 16.7 as compared to appropriate hospital controls with 5.6 percent. Based on 38 pregnancies in another study, Hemminki et al. (1985) found the odds ratio for malformations in the exposed offspring to be 4.7. Gordon and Meinhardt (1983) raised several questions about the interpretation of this work.

Generoso et al. (1987) exposed mice by inhalation to 1200 ppm for one hour before mating, during early pronuclear stages, later zygotic stages or during the early two–cell stage. Only the second period (early pronuclear stage) was associated with embryonic and fetal loss and congenital defects. Cleft palates, skeletal, abdominal wall defects and hydrops occurred with a total rate of 31 percent. Smith (1988) presented a pharmacokinetic model for extrapolating risks in the exposed male from mouse studies.

Generoso, W.M.; Rutledge, J.C.; Cain, K.T.; Hughes, L.A. and Braden, P.W.: Exposure of female mice to ethylene oxide within hours after mating leads to fetal malformation and death. Mutation Res. 176:269–274, 1987.

Gordon, J.E. and Meinhardt, T.J.: Spontaneous abortions in hospital sterilizing staff. Br. Med. J. 286:976 only, 1983.

Hemminki, K.; Kyyronen, P. and Lindbohm, M–L.: Spontaneous abortions and malformations in the offspring of nurses exposed to anaesthetic gases, cytostatic drugs and other potential hazards in hospitals, based on registered information of outcome. J. Epidemiol. Comm. Health 39:141–147, 1985.

Hemminki, K.; Mutanen, P.; Saloniemi, I.; Niemi, M–L. and Vainio, H.: Spontaneous abortions in hospital staff engaged in sterilizing instruments with chemical agents. Br. Med. J. 285:1461–1463, 1982.

Kimmel, C.A.; LaBorde, J.B. and Hardin, B.D.: Reproductive and developmental toxicology of selected epoxides. In: Toxicology and the Newborn, S. Kacew and M.J. Reasor (eds.), Elsevier Sc. Publ. Amsterdam, Chapter 13, 270–287, 1984.

Laborde, J.B. and Kimmel, C.A.: The teratogenicity of ethylene oxide administered intravenously to mice. Toxicol. Appl. Pharmacol. 56:16–22, 1980.

Rutledge, J.C. and Generoso, W.M.: Fetal pathology produced by ethylene oxide treatment of the murine zygote. Teratology 39:563–572, 1989.

Smith, T.J.: Extrapolation of laboratory findings to risks from environmental exposures: Male reproductive effects of ethylene oxide. Chapter 5, Genetic Risk Assessment, A.D. Bloom and P.K.F. Poskitt (eds.), Environmental Health Institute. 79–100, 1988.

Snellings, W.M.; Maronpot, R.R.; Zdenak, J.P. and Laffoon, C.P.: Teratology study in Fischer 344 rats exposed to ethylene oxide by inhalation. Toxicol. Appl. Pharmacol. 64:478–481, 1982.

988 Ethylenebisisothiocyanate CAS 3688–08–2

Discussed under Ethylenethiourea

989 Ethylenediamine Dihydrochloride

DePass et al. (1987) treated rats on days 6–15 with up to 1.0 gm per kg per day and found no evidence for teratogenicity.

DePass, L.R.; Yang, R.S.H. and Woodside, M.D.: Evaluation of the teratogenicity of ethylenediamine dihydrochloride in Fischer 344 rats by conventional and pair–feeding studies. Fund. Appl. Toxicol. 9:687–697, 1987.

990 Ethylenethiourea and Degradation Products CAS 96–45–7

Ruddick and Khera (1975) studied this fungicide degradation product in rats during several periods of organogenesis at 10 to 80 mg per kg per day. Above 10 mg per kg, neu-

ral tube closure defects, hydrocephalus and other malformations of the brain were found along with kinky tails and limb defects. In the rabbit at 80 mg per kg, decreased brain weight was found. Ruddick et al. (1976) studied the distribution and metabolism of the compound in pregnant rats. Lu and Staples (1978) produced defects in the rat even following thyroparathyroidectomy and suggested the mode of action was not via a maternal thyroid effect. Khera (1989) studied the pathogenesis of hydrocephalus in in vitro culture and in vivo rat embryos. At doses above 45 mg per kg on day 12, hydrocephalus was found in vivo.

Khera and Iverson (1981) gavaged (200 mg per kg) or gave intraperitoneally (400 mg per kg) of a similar antithyroid N–methyl–2–thioimidazole and found no adverse effects in rat fetuses. Teramoto et al. (1981) gave 250 per kg orally to rats on day 12 and 2000 mg per kg to mice on day 10 and found no increase in the defect rate or resorptions in the mice. A wide spectrum of defects was found in the rat fetuses.

Ruddick et al. (1976) studied the correlation of teratogenicity with molecular structure of 16 related chemicals. The only chemical which was teratogenic was 4–methyl-ethylene–thiourea at a dose of 240 mg per kg on day 12 or 13 in the rat. The following chemicals were not teratogenic (although minor skeletal changes were present): Allylthiourea (240), allylisothiocyanate (60), ethylene–bis–isothiocyanate (25), ethylenethiuram monosulfide (60), ethyleneurea (240), N,N–dimethyl–thiourea (240), imidazole (240), 3–(2–imidazoline–2–yl)–2–imidazolidinethione (Jaffe's base, 200), 2–mercaptobenzimidazole (120), 2–mercapto–1–methyl–imidazole (240), 2–mercaptothiazoline (240), N–methylethylenethiourea (240), 3,4,5,6–tetrahydro–2–pyrimidinethiol (240), 1,1,3,3–tetramethyl–2–thiourea (240). The figure in parenthesis following the drug is the dosage used in mg per kg on day 12 or 13. A sulfur in the 2 position and the imidazolidine ring appeared to be necessary in producing the teratogenic activity of ethylenethiourea.

Stula and Krauss (1977) exposed rats cutaneously to 50 mg per kg (one forty–fifth the adult lethal dose) on days 17 and 13 and found a high incidence of limb reduction, encephalocele and other skeletal defects. Khera (1987) reviewed the risk and animal studies.

Khera, K.S.: Ethylenethiourea–induced hydrocephalus in vivo and in vitro with a note on the use of a constant gaseous atmosphere for rat embryo cultures. Teratology 39:277–286, 1989.

Khera, K.S.: Ethylenethiourea: A review of teratogenicity and distribution studies and an assessment of reproduction risk. CRC Critical Reviews in Toxicology 18:129–139, 1987.

Khera, K.S. and Iverson, F.: Effects of pretreatment with SKF–525A, N–methyl–2–thioimidazole sodium phenobarbital or methyl cholanthrene on ethylenethiourea–induced teratogenicity in rats. Teratology 24:131–137, 1981.

Lu, M.H. and Staples, R.E.: Teratogenicity of ethylenethiourea and thyroid function in the rat. Teratology 17:171–178, 1978.

Ruddick, J.A. and Khera, K.S.: Pattern of anomalies following single oral doses of ethylene–thiourea to pregnant rats. Teratology 12:277–282, 1975.

Ruddick, J.A.; Newsome, W.H. and Nash, L.: Correlation of teratogenicity and molecular structure. Ethylenethiourea and related compounds. Teratology 13:263–266, 1976.

Ruddick, J.A.; Williams, D.T.; Hierlihy, L. and Khera, K.S.: [14–C]Ethylenethiourea: Distribution, excretion and metabolism in pregnant rats. Teratology 13:35–39, 1976.

Stula, E.F. and Krauss, W.C.: Embryotoxicity in rats and rabbits from cutaneous application of amide–type solvents and substituted ureas. Toxicol. Appl. Pharmacol. 41:35, 1977.

Teramoto, S.; Kaneda, M.; Aoyama, H. and Shirasu, Y.: Correlation between the molecular structure of N–alkylureas and N–alkylthioureas and their teratogenic properties. Teratology 23:335–342, 1981.

991   Ethylenethiuram Monosulfide CAS 33813–20–6

Discussed under Ethylenethiourea

992   Ethyleneurea CAS 120–93–4

Discussed under Ethylenethiourea

993   Ethylestrenol

Kawashima et al. (/77) found that this nortestosterone–like agent caused masculinization of female rat offspring after exposure to 1 mg per kg.

Kawashima, K.; Nakaura, S.; Nagao, S.; Tanaka, S.; Kuwamura, T. and Omori, Y.: Virilizing activities of various steroids in female rat fetuses. Endocrinol. Japon. 24:77–81, 1977.

994   Ethyl Ether

Schwetz (1970) anesthetized rats and mice for one hour periods over 3 day periods of organogenesis and found an increase in fetal resorptions and skeletal anomalies but no increase in soft tissue defects. See anesthesia for a more complete discussion.

Schwetz, B.A.: Teratogenicity of maternally administered volatile anesthetics in mice and rats. Pharmacology 31:3599, 1970.

995   2–Ethyl–1–hexanol

Nelson et al. (1988) exposed rats to vapor for seven hours daily during gestation. At the highest concentration, 14,000 mg per meter cubed, no adverse reproductive effects were found. Limited maternal toxicity was seen.

Nelson, B.K.; Brightwell, W.S.; Khan, A.; Hoberman, A.M. and Krieg, E.F.: Teratological evaluation of 1–pentanol, 1–hexanol, and 2–ethyl–1–hexanol administered by inhalation to rats. Teratology 37:479–480, 1988.

996   Ethyl Icosapentate

Saito et al. (1989) gave rats up to 3.0g orally on days 7–17. No adverse fetal effects were seen and postnatal development was not altered. The dam had reduced weight increase at 1.0 and 3.0g per kg.

Saito, M.; Narama, I. and Obata, M.: Toxicity studies of 5, 8, 11, 14, 17–eicosapentaenoic acid ethyl ester (epa–e) (v) teratogenicity study in rats. Iyakuhin Kenyu 20:853–866, 1989.

997   Ethyl Methanesulphonate EMS Alkane Sulphonates

Hemsworth (1968) administered 200 mg per kg intraperitoneally on day 13 and produced cleft palate and limb defects in rat offspring.

Hemsworth, B.N.: Embryopathies in the rat due to alkane sulphonates. J. Reprod. Fertil. 17:325–334, 1968.

998   4'–Ethyl–2–methyl–3–pyrvolidinopropiophenone Hydrochloride HY–770

James et al. (1992) studied this muscle relaxant in pregnant rats during organogenesis in oral doses up to 200 mg per kg and found no adverse fetal effects. Behavior was uneffected. In rabbits Tateda et al. (1992) found no adverse fetal effects from a similar type study.

James, P.; James, R.W.; Smith, J.A.; Hughes, E.W.; John, D.M.; Aoki, Y. and Inomata, N.: Reproductive studies of HY–770, a new muscle relaxant, in rats–II teratology study. Yakuri to Chiryo 20:1097–1115, 1992.

Tateda, C.; Ichikawa, K.: Kiwaki, S,; Ono, C.; Yamamae, H.; Nakai, N.; Oketani, Y. and Inomata, N.: Teratological study of HY–770 in rabbits by oral administration. Yakuri to Chiryo 20:1117–1126, 1992.

999   N–Ethylnicotinamide CAS 4314–66–3

Landauer and Wakasugi (1967) injected 2.5 to 10 mg of this compound into the yolk sac at 24 or 96 hours of incubation and observed shortening of the legs, abnormal upper beaks and a low incidence of rumplessness in the 17–day survivors. Supplementation with ADP tended to protect the embryos.

Landauer, W. and Wakasugi, N.: Problems of acetazolamide and N–ethylnicotinamide as teratogens. J. Exp. Zool. 164:499–516, 1967.

1000   Ethylnitrosourea CAS 759–73–9

Druckrey et al. (1966) injected 80 mg per kg on the 15th day and found abnormalities of the paws of the rat offspring. Three of five surviving fetuses subsequently developed brain tumors. Subsequent work (Ivankovic and Druckrey, 1968) indicated that a single dose of only 5 mg per kg on the 15th day was sufficient to produce postnatal tumors. Druckrey et al. (1972) found that ethylnitrosabiuret produced postnatal neurogenic tumors. Druckrey (1973) and Druckrey et al. (1972) summarized the role of this compound and other indirect alkylating compounds.

Alexandrov and Janisch (1971) fed pregnant rats a combination of ethyl urea and nitrite on the 9th and 10th gestational days and produced hydrocephalus, anencephalus and absence of the eye. These defects were felt to be due to absorption of ethyl nitrosourea which was formed in the gastrointestinal tract. The dosage used was 0.3 to 1 percent nitrite in the feed and 0.5 percent ethyl urea in the water with tube feeding of ethyl urea 100 or 200 mg per kg per day. Alexandrov (1976) reviewed the extensive Russian experimental work on transplacental carcinogenesis of the compound and related chemicals. He reported that when 15 mg per kg was given intravenously on the 21st day, 78 percent of the rat offspring had postnatal tumors of the central nervous system. Dreosti et al. (1983) gavaged rats on days 7–16 and at doses of 10 mg per kg or higher found defects.

Lovell et al. (1985) gave 250 mg per kg to male mice during the spermatogonial stage and mated them. They were unable to find any evidence for skeletal changes in the offspring.

Alexandrov, V.A.: Some results and prospects of transplacental carcinogenesis studies. Neoplasia 23:285–299, 1976.

Alexandrov, V.A. and Janisch, W.: Die Teratogene Wirkung von Athylharnstoff und Nitrit bei Ratten. Experienta 27:538–539, 1971.

Dreosti, I.E.; McMichael, A.J. and Bridle, T.M.: Teratogenic effect of administration of N–nitrosoethylurea and ethylurea–nitrate in rats. Res. Commun. Chem. Pathol. Pharmacol. 41:265–281, 1983.

Druckrey, H.: Specific carcinogenic and teratogenic effects of indirect alkylating methyl and ethyl compounds, and their dependency on stages of ontogenic developments. Xenobiotica 3:271–303, 1973.

Druckrey, H.; Ivankovic, S. and Preussmann, R.: Teratogenic and carcinogenic effects in the offspring after single injection of ethylnitrosourea to pregnant rats. Nature 210:1378–1379, 1966.

Druckrey, H.; Ivankovic, S.; Preussmann, R.; Zulch, K. and Mennel, H.D.: Selective induction of malignant tumors of the nervous system by resorptive carcinogens. In: The Experimental Biology of Brain Tumors. W.M. Kirsch, E.G. Paoletti and P. Paoletti (eds.), Springfield, C.C. Thomas, 107–108, 1972.

Ivankovic, S. and Druckrey, H.: Transplacentare Erzeugnung Maligner Tumoren des Nervensystems: Athyl–Nitroso–Harnstoff (ANH) an BD IX–Ratten. Z. Krebsforsch. 71:320–360, 1968.

Lovell, D.P.; Willis, D.B. and Johnson, F.M.: Lack of evidence for skeletal abnormalities in offspring of mice exposed to ethylnitrosourea. Proc. Natl. Acad. Sci. USA 82:2852–2856, 1985.

1001 Ethylparathion CAS 56–38–2

Noda et al. (1972) applied this pesticide percutaneously once a day to rats for 7 days starting on day zero or the 7th day of gestation. No effects were seen in the group treated starting on the 7th day but a decrease in implantations was found in the group started on day zero.

Noda, K.; Numata, H.; Hirabayashi, M. and Endo, I.: Influence of pesticides on embryos. 1. On influence of organophosphoric pesticides. Oyo Yakuri 6:667–672, 1972.

1002 Ethylpyrimidine

Discussed under 2,4–Diamino–5–(p–chlorophenyl–6––ethylpyrimidine)

1003 Ethylthiourea

Teramoto et al. (1981) gave 2000 mg per kg orally to rats on day 12 and to mice on day 10 and found no increases in resorptions or defects in the mice. In rats treated on day 12 or 14, they found hydrocephalus, short tails, craniomenigoceles and other defects at a dose of 500 and 1000 mg per kg.

Teramoto, S.; Kaneda, M.; Aoyama, H. and Shirasu, Y.: Correlation between the molecular structure of N–alkylureas and N–alkylthioureas and their teratogenic properties. Teratology 23:335–342, 1981.

1004 Ethylurea

Teramoto et al. (1981) gave 2000 mg per kg orally to rats on day 12 and to mice on day 10 and found no increase in the defect rates or resorptions.

Teramoto, S.; Kaneda, M.; Aoyama, H. and Shirasu, Y.: Correlation between the molecular structure of N–alkylureas and N–alkylthioureas and their teratogenic properties. Teratology 23:335–341, 1981.

1005 Ethynodiol Diacetate $Ovulen^{TM}$ $Mutrulen^{TM}$ CAS 1231–93–2

The effect of oral contraceptives on pregnant women is discussed under Oral Contraceptives.

Spira et al. studied 143 infants with malformations and found no increase in the rate of use of this progesterone like drug.

Saunders and Elton (1967) treated rats and rabbits buccally and subcutaneously with the compound and found neither defects nor infertility in the offspring. The treatment periods were for various periods of gestation as well as the entire period. The highest dose during the entire period was 2 mg per kg per day for rats and 0.05 mg per kg per day for rabbits. In mice, Andrew et al. (1972) found 30 percent malformations when 1.0 mg per kg was given on days 7 through 9.

Andrew, F.D.; Williams, T.L.; Gidley, J.T. and Wall, M.E.: Teratogenicity of contraceptive steroids in mice. (abs) Teratology 5:249 only, 1972.

Saunders, F.J. and Elton, R.L.: Effects of ethynodiol diacetate and mestranol in rats and rabbits, on conception, on the outcome of pregnancy and on the offspring. Toxicol. Appl. Pharmacol. 11:229–244, 1967.

Spira, N.; Goujard, J.; Huel, G. and Rumeau–Ronquette, C.: Investigation into the teratogenic action of sex hormones: First results of an epidemiologic survey involving 20,000 women. Rev. Med. 41:2683–2694, 1972.

1006 Etidronate Disodium $Ethane–1–hydroxy$ $1,1–diphosphonate$ CAS 7414–83–7

Eguchi et al. (1982) administered this calcification inhibitor subcutaneously to mice on days 11–17 of gestation in doses of 200 mg per kg daily. Decrease in mineralization and angulations of fetal long bones were produced. Sakiyama et al. (1988) using intraperitoneal injection of mice on day 10 found selective damage to odontoblasts. Nolen and Buehler

(1971) tested this material in pregnant rats and rabbits and found no evidence of teratogenicity. The rats were given 500 mg per kg per day.

This drug with osteogenic effects was given by Hirohashi et al (**1989**) orally to rats on days 7–17 in doses of up to 1500 mg per kg. At 1000 mg and over curved scapulae, humeri and radii were found as well as wavy ribs. These changes were corrected spontaneously postnatally.

Eguchi, M.; Yamaguchi, R.; Shiota, E. and Handa, S.: Fault of ossification and calcification and angular deformities of long bones in the mouse fetuses caused by high doses of ethane–1–hydroxy–1, 1–diphosphonate (EHDP) during pregnancy. Cong. Anomal. 22:47–52, 1982.

Hirohashi, A.; Kannan, N.; Matsumot, Y.; Katoh, T. and Yamada, H.: Reproduction studies of sm–5600 in rats. Kiso to Rinsho 23:1317–1335, 1989.

Nolen, G.A. and Buehler, E.U.: The effects of disodium etidronate on the reproductive functions and embryology of albino rats and New Zealand rabbits. Toxicol. Appl. Pharmacol. 18:548–561, 1971.

Sakiyama, Y.; Yamamoto, H.; Soeda, Y.; Iwasa, K. and Ikeo, T.: Effect of ethane–1–hxdrosy–l, 1–diphosphonate (EHDP) on fetal mice during pregnancy. Skika Igaku 5:997–1003, 1988.

1007 Etizolam

6–(o–Chlorophenyl)8–ethyl–1–methyl–4H–S–triazolo (3,4-C)thieno(2,3–e)(1,4)–diazepine CAS 40054–69–1

Hamada et al. (1979) studied this anti–anxiety drug in pregnant rats, mice and rabbits at maximum oral doses of 100, 500 and 25 mg per kg, respectively. In rats and rabbits there were no increases in congenital defects. Some decrease in fetal weight gain occurred in all species at the highest doses. In mice at 500 mg per kg, six exencephalic fetuses were found from 167 exposed while only one occurred in the control group of 229. No adverse effects on reproduction in rats was found when 25 mg per kg was given 63 days before mating and for 7 days after fertilization (Hamada and Imanishi, 1979). No adverse perinatal or postnatal effects were observed in rats after 100 mg per kg from days 17–21 (Hamada and Imanishi, 1979).

Hamada, Y. and Imanishi, M.: Fertility study of etizolam in rats. Oyo Yakuri 17:781–785, 1979.

Hamada, Y. and Imanishi, M.: Perinatal and postnatal study of etizolam in rats. Oyo Yakuri 17:787–797, 1979.

Hamada, Y.; Imanishi, M.; Onishi, K. and Hashiguchi, M.: Teratogenicity study of etizolam (P–INN) in mice, rats and rabbits. Oyo Yakuri 17:763–779, 1979.

1008 Etlornithine Hydrochloride *DFMO*

This irreversible inhibitor of L–ornithine decarboxylase was given in the drinking water of rats and rabbits (O'Toole et al., 1989). At one percent in rats (1,270 mg per kg) and three percent (915 mg per kg) in rabbits, maternal weight gain was decreased and all implantations were aborted or resorbed. The drug was given during organogenesis or on separate days during organogenesis at 200–600 mg per kg.

The fetuses were reduced in number and weight. No ill effects were seen at 60 mg per kg and lower. No teratogenicity was found.

O'Toole, B.A.; Huffman, K.W. and Gibson, J.P.: Effects of eflornithine hydrochloride (DFMO) on fetal development in rats and rabbits. Teratology 39:103–114, 1989.

1009 Etofenamate CAS 30544–47–9

Terada et al. (1982) studied the effect of this non–steroidal anti–inflammatory agent on reproduction in the rat. Using doses of up to 80 mg per kg daily in males and females, no loss of reproductive capacity was found. The pregnant females were treated on days 7–17 subcutaneously with the same dose and no adverse fetal effects were found. At 80 mg per kg a significant number of the mothers had intestinal ulcers.

Terada, Y.; Nishimura, K.; Shigematsu, K.; Imura, Y.; Sasaki, H.; Yoshioka, M.; Tasumi, H.; Hashimoto, M. and Yoshida, K.: Reproductive study of etofenamate: Fertility and teratogenicity studies in the rat. Iyakuhin Kenkyu 13:886–909, 1982.

1010 Etofylline

Ujhazy et al. (1989) gave mice up to 585 mg per kg orally on days 7–16 of gestation. Fetal weight was reduced but no other adverse effects were found.

Ujhazy, E.; Onderova, E.; Horaova, M.; Bencova, E.; Durisova, R.; Nosal, R.; Balonova, T. and Zeljenova, D.: Teratological study of the hypolipidaemic druge etofylline clofibrate (VULM) and fenofibrate in swiss mice. Pharmacology and Toxicology 64:286–290, 1989.

1011 Etoperidone

2–3–(4(M–Chlorophenyl–1–piperazinyl)–opyl)4,5–diethyl–2,4 dihydro–3H 1,2,4–triazol–3–one monohydrochloride CAS 52942–31–1

Barcellona et al. (1977) gave oral doses of up to 300 mg per kg in the rat and 50 mg per kg in the rabbit during organogenesis and found no increase in fetal malformations.

Barcellona, P.S.; Fanelli, O. and Campana, A.: Teratological study of etoperidone in the rat and rabbit. Toxicology 8:87–94, 1977.

1012 Etoposide *VP–16–213 Epipodophyllotoxin* CAS 33419–42–0

This podophyllotoxin derivative was injected intraperitoneally into mice in amounts of 0.5 and 1.0 mg per kg on day 6, 7 or 8. Numerous malformations were found but dextrocardia, exencephaly and skeletal defects were most common (Sieber et al., 1978). Mirkes and Zwelling (1990) studied this antineoplastic in whole embryo rat culture. At 2.0 micromolar concentration growth retardation and abnormolities occurred.

Mirkes, P.E. and Zwelling, L.A.: Embryotoxicity of the intercalating agents in m–AMSA and o–AMSA and the epipodophyllotoxin vp–16 in postimplantation rat embryos in vitro. Teratology 41:679–688, 1990.

Sieber, S.M.; Whang–Peng, J.; Botkin, C. and Knutsen, J.: Teratogenic and cytogenic effects of some plant–derived antitumor agents (vincristine, colchicine, maytansine, VP–16–213 and VM–16) in mice. Teratology 18:31–48, 1978.

1013 Etretin

This chemical is the free acid analog of etretinate. Kistler and Hummler (1985) treated pregnant mice, rats and rabbits. The lowest teratogenic doses were 3, 15 and 0.6 mg per kg daily in mice, rats and rabbits, respectively. Skeletal and soft tissue malformations were found. Reiners et al. (1988) found limb defects and cleft palate after oral doses of 100 mg per kg to the mouse.

Kistler, A. and Hummler, H.: Teratogenesis and reproductive safety evaluation of the retinoid etretin (RO 10–1670). Arch. Toxicol. 58:50–56, 1985.

Reiners, J.; Lofberg, B.; Creech Kraft, J.; Kochhar, D.M. and Nau, H.: Transplacental pharmacokinetics of teratogenic doses of etretinate and other aromatic retinoids in mice. Reproductive Toxicology 2:19–30, 1988.

1014 Etretinate

(all–E)–9(4–Methoxy–2,3,6–trimethyl–phenyl)–3, 7–dimethyl–2,4, 6,8–nonatetracnoate Tigason™ CAS 54350–48–0

Grote et al. (1985) reported a fetus with limb reductions in a woman who conceived four months after discontinuing the medication. Lammer (1988) reported an infant with the clinical picture of retinoic acid embryopathy and microcephaly from a woman who had discontinued the medication 51 weeks before conception. The maternal serum contained etretinate by HPLC three months after birth. Verloes (**1990**) reported on a fetal autopsy performed on a women who took etretinate 7 to 8 months before becoming pregnant. Small normally formed ears, intraatrial septal defect and horseshoe kidneys were found. Bonnivert, et al. (**1990**) report fetal autopsy findings of this case. Geiger et al. (**1994**) have summarized the outcome of 75 women exposed during pregnancy. Nine out of 29 exposed during pregnancy had malformations. Of 11 cases including examined fetuses 7 had malformation of the digits. Of the 70 women exposed before pregnancy 18 had malformations.

Aikawa et al. (1982) gave this vitamin A analog to rats orally in doses of 2–8 mg per kg daily on days 7–17. The defect rate was higher after 4 and 8 mg dosage schedules. Exencephaly, craniofacial defects, cleft palate and skeletal defects were found. Using 40 mg per kg on a single day on days 8–13, they found increases in defects in each study group. Williams et al. (1984) studied the teratogenicity of this compound in the hamster. Agnish et al. (1990) studied the effects of oral administration on the pregnant rat. Plasma levels of the compound and its metaboliter were reported. Verloes et al. (**1990**) autopsied a 23 week fetus with very small external ears, atrial septal defect and horseshoe kidney. The mother had discontinued treatment 7 to 8 months before conception.

Takizawa et al. (**1994**) studied the effect of prenatal etretinate on growth inhibition postnatally and found increased levels of growth hormone and IFG–I in pups on the 21st day. Six mg per kg was given prenatally and cross-fostering performed.

Aikawa, M.; Sato, M.; Noda, A. and Udaka, K.: Toxicity study of etretinate. III. Reproductive segment. 2. Study in rats. Yakuri to Chiryo 9:5095–5108, 5117–5143, 1982.

Agnish, N.D.; Vane, F.M., Rusin, G.; DiNardo, B. and Dashman, T.: Teratogenicity of etretinate during early pregnancy in the rat and its correlation with maternal plasma concentrations of the drug. Teratology 42:25–33, 1990.

Bonnivert, J.; Lamgotte, R. and Verloes, A.: Etretinate embryotoxicity 7 months after discontinuation of treatment. American Journal of Medical Genetics 37:437–438, 1990.

Geiger, J.M.; Baudin, M. and Saurat, J.H.: Teratogenic risk with etretinate and acitretin treatment. Dermatology 189:109–116, 1994.

Grote, W.; Harms, D.; Janig, U.; Keitzmann, H.; Ravens, V. and Schwarze, I.: Malformations of a fetus conceived 4 months after termination of maternal etretinate treatment. Lancet 2:1276 only, 1985.

Lammer, E.J.: Embryopathy in infant conceived one year after termination of maternal etretinate. Lancet 2:1080–1081, 1988.

Takizawa, S.; Fukatsu, N.; Horii, I. and Fujii, T.: Effect of etretinate (aromatic retinoid) treatment during gestation and lactation periods on viability and somatic growth in f1 rats. J. Tox. Sc. 19:107–117, 1994.

Verloes, A.: Etretinate embryotoxicity 7 months after discontinuation of treatment. American Journal of Medical Genetics 37:437–438, 1990.

Williams, K.J.; Ferm, V.H. and Willhite, C.C.: Teratogenic dose–response relationships to etretinate in the golden hamster. Fund. Appl. Toxicol. 4:977–982, 1984.

1015 Evans Blue CAS 314–13–6

Wilson (1955) reported that injection of 10 mg on days 7, 8 and 9 produced brain defects and also occasional eye and heart anomalies in the rat fetus. This azo dye probably affects yolk sac nutrition in a way similar to the action of trypan blue. A review on the subject of teratogenicity of azo dyes was written by Beck and Lloyd (1966).

Beck, F. and Lloyd, J.B.: Teratogenic effects of azo dyes. In: Advances in Teratology, D.H.M. Woollam (ed.), London: Logos Press, 1:131–193, 1966.

Wilson, J.G.: Teratogenic activity of several azo dyes chemically related to trypan blue. Anat. Rec. 123:313–334, 1955.

1016 Exaprolol

Ujhazy et al. (**1981**) gave rats up to 20 mg per kg from the 4th through the 16th day without producing embryotoxicity.

Ujhazy, E.; Balonova, T.; Rippa, S.; Buran, L. and Babulova, A.: Zhodnotenie ucinku exaprololu (vulm 111) na prenatalny vyvoj mysi a potkanov. Bratisl. lek. Listy 76:641–768, 1981.

1017 Exell™

Murray et al. (**1994**) gave rats and rabbits up to 50 or 100 mg per kg orally during organogenesis respectively. Above 20 mg per kg the rat dams showed maternal toxicity. In both species no adverse reproductive findings occurred.

Murray, S.M.; Martin, T.; Hoberman, A.M.; Huritt, M.E. and Staples, R.E.: Developmental toxicity of Exell™ in rats and rabbits. (abs) Teratology 49:402, 1994.

1018 Exercise

Boehnke et al. (1987) exercised mice heavily and moderately on a tread–mill and observed no ill effects in the offspring.

Boehnke, W.H.; Chernoff, G.F. and Finnell, R.H.: Investigation of the teratogenic effects of exercise on pregnancy outcome in mice. Teratogenesis, Carcinogenesis and Mutagenesis 7:391–397, 1987.

1019 Famotidine CAS 76824–35–6

Shibata et al. (1983) studied this arthritis drug in pregnant rats and rabbits in oral doses of up to 2000 mg per kg and no adverse effect on reproduction, organogenesis or postnatal behavior in the rat were found. Intravenous dosing on days 7–17 using 200 mg per kg produced no teratologic change in rats and similar results were seen when rabbits were given up to 100 mg per kg intravenously. Burek et al. (1985) gave 2,000 mg per kg orally or 200 mg per kg intravenously on days 7–17 in the rat and no teratogenicity was found. Rabbits received up to 500 mg per kg orally on days 6–18 without evidence of teratogenicity.

Burek, J.D.; Majka, J.A. and Bokelman, D.L.: Famotidine: Summary of preclinical safety assessment. Digestion 32:7–14, 1985.

Shibata, M.; Kawano, K.; Shiobara, Y.; Yoshinaga, T.; Fujiwara, M.; Uchida, T. and Odani, Y.: Reproductive studies of famotidine (YM 11170) in rats and rabbits. Oyo Yakuri 26:489–497, 543–578, 831–840, 1983.

1020 Fasting *Starvation*

The effect of starvation in the human was carefully reported in a book by Stein et al. (1975). During the war famine of 1944–45 in the Netherlands the head circumference of newborns was decreased but no measurable intellectual deficit was found in adult survivors. For infants exposed to famine during early gestation there was a significant increase in the incidence of hydrocephalus and meningomyelocele.

Evidence that nutritional deprivation in early childhood may lead to congenital nervous system defects in the subsequent offspring of those deprived was summarized by Emanuel and Sever (1973).

The mouse appears to be the animal most susceptible to fasting teratogenesis (see Kalter and Warkany, 1959, for a concise review). A 24 to 30 hour fast in the mouse during the 7th to 10th gestational day produced vertebral and rib defects and occasionally exencephaly (Miller, 1962). Fasting on the 8th or 9th day produced the highest incidence

of defects. Interruption of the fast with glucose or various aminoacids prevented the defects (Runner and Miller, 1956). Kalter (1950) reported that cortisone has a synergetic effect with fasting on teratogenesis in the mouse.

Miller (1973) made a study of the effect of fasting in the mouse and found that cleft palate incidence was increased especially when fasting occurred on day 13. Fasting on days 11 through 13 produced 58 percent clefts. Administration of succinate or glucose reduced the number of clefts.

A comprehensive paper on protein malnutrition and the growth of the rat central nervous system is available (Morgane et al., 1978).

Emanuel, I. and Sever, L.E.: Questions concerning the possible association of potatoes and neural–tube defects, and an alternative hypothesis relating to maternal growth and development. Teratology 8:325–332, 1973.

Kalter, H.: Teratogenic action of a hypocaloric diet and small doses of cortisone. Proc. Soc. Exp. Biol. Med. 104:518–520, 1950.

Kalter, H. and Warkany, J.: Experimental production of congenital malformations in mammals by metabolic procedure. Physiol. Rev. 39:69–115, 1959.

Miller, J.R.: A strain difference in response to the teratogenic effect of maternal fasting in the mouse. Can. J. Genet. Cytol. 4:69–78, 1962.

Miller, T.J.: Cleft palate formation: The effects of fasting and iodoacetic acid on mice. Teratology 7:177–182, 1973.

Morgane, P.J.; Miller, M.; Kemper, T.; Stern, W.; Forbes, W.; Hall, R.; Bronzino, J.; Kissane, J.; Hawrylewicz, E. and Resnick, O.: Effects of protein malnutrition on the developing central nervous system in the rat. Neurosci. Behav. Rev. 2:137–230, 1978.

Runner, M.N. and Miller, J.R.: Congenital deformity in the mouse as a consequence of fasting. (abs) Anat. Rec. 124:437–438, 1956.

Stein, Z.A.; Susser, M.; Saenger, G. and Marolla, F.: Famine and Human Development, the Dutch Hunger Winter of 1944–45. New York: Oxford University Press, 1975.

1021 Fasudil *HA–177 HA–1077 AT–877*

Sasaki et al. (**1994**) gave rats up to 25 mg per kg before fertilization and in the first week of pregnancy. Intravenous dosage of this vasodilator did not decrease reproduction but the animals ate less and had fewer implantation sites. Perinatal and postnatal studies at up to 40 mg per kg caused no abnormalities in behavior (Takahashi, **1994**). In rabbits treated with up to 60 mg per kg during organogenesis there was toxicity in the dams but no adverse fetal effects (Nakazima et al., **1994**).

Kobayashi et al. (**1992**) gave this diazepine during organogenesis to rats intravenously in doses of up to 40 mg per kg. Maternal and decreased fetal weight were found at the highest dose.

Kobayashi, Y.; Sasaki, M.; Takahashi, H.; Iida, T.; Murofushi, K.; Nakazima, M.; Yano, J.; Ichijyo, K. and Shibuya, C.: Reproductive and developmental toxicity study of fasudil hydrochloride (HA–1077, AT–877) (2)–study on intravenous administration during the period of organogenesis in rats. Yukuri to Chiryo 20:s1469–s1490, 1992.

Nakazima, M.; Sasaki, M.; Iida, T.; Murofushi, K.; Matsuda, K.; Takahashi, H.; Yano, J.; Ichijyo, K.; Kobayashi, Y. and Shibuya, C.: Reproductive and developmental toxicity study of fasudil hudrochloride (HA–1077, AT-877) (III): Study on intravenous administration during the period of organogenesis in rabbits. Yakuri to Chiryo 22:51–59, 1994.

Sasaki, M.; Takahashi, H.; Iida, T.; Murofushi, K.; Nakazima, M.; Yano, J.; Ichijyo, K.; Kobayashi, Y. and Shibuya, C.: Reproductive and developmental toxicity study of fasudil hydrochloride (HA–1077, AT–877) (i): Study on intravenous administration prior to and in the early stages of pregnancy in rats. Yakuri to Chiryo 22:41–50, 1994.

Takahashi, H.; Sasaki, M.; Nakazima, M.; Iida, T.; Murrofushi, K.; Yano, J.; Ichijyo, K.; Kobayashi, Y. and Shibuya, C.: Reproductive and developmental toxicity study of fasudil hudrochloride (HA–1077, AT–877) (IV): Study on intravenous administration during the perinatal and lactation periods in rats. Yakuri to Chiryl 22:61–80, 1994.

1022 Fatty Acids

Martinet (1952) reported that rats maintained on a bread, sucrose, caseine and oat diet with added vitamins A and D produced fetal hemorrhages in 20 percent of the offspring. By supplementing this diet with various dietary substances, he concluded that even small amounts of cottonseed oil, rich in linoleic acid, prevented the hemorrhages. No other defects were described.

Martinet, M.: Hemorragies embryonnaires par deficience en acid linoleique. Ann. Med. Interne 53:286–333, 1952.

1023 Feline Panleucopenia Virus

This virus can cross the placenta and by interfering with proliferating cerebellar cells result in ataxia in offspring of the cat (Johnson et al., 1967). Kalter (1968) summarized this subject.

Johnson, R.H.; Margolis, G. and Kilham, L.: Identity of feline ataxia virus with feline panleucopenia virus. Nature 214:175–177, 1967.

Kalter, H.: Teratology of the Central Nervous System. Chicago: University of Chicago Press, 263 only, 1968.

1024 Fazadinium Bromide

Blogg et al. (1973) studied placental transfer of this muscle relaxant and reported that the cord levels at term were about one–tenth the mothers levels.

Blogg, C.E.; Simpson, B.R.; Tyers, M.B.; Martin, L.E.; Bell, J.A.; Arthur, A.; Jackson, M.R. and Mills. J.: Placental transfer of ah8165. Brit. J. Anaesth. 45:638–639, 1973.

1025 Fenbufen y–Oxo[1,1,biphenyl]–4 Butanoic Acid CAS 36330–85–5

Jackson et al. (1980) studied this non–steroidal anti–inflammatory in pregnant rats, rabbits and mice during organogenesis. Doses of up to 40 mg per kg had no adverse

fetal effects. Parturition was delayed and prolonged in rats. No ill effects on fertility in male and female rats was found.

Jackson, B.A.; Tonelli, G.; Chiesa, F. and Alvarez, L.: Reproductive toxicology of fenbufen. Arzneim–Forsch. 30:725–727, 1980.

1026 Fenfluramine CAS 458–24–2

Gilbert et al. (1971) reported negative teratologic testing in rats, rabbits and mice. Doses of up to 45 mg per kg were given orally to mice on days 5 through 14 of gestation. Postnatal studies of rats whose mothers received 20 mg per kg daily during most of gestation were reported to be different from controls (Vorhees et al., 1979). Locomotor tests (pivoting) were the most altered. Brain weight, but not DNA, was significantly reduced in the pups at 70 days of postnatal life.

Gilbert, D.L.; Franko, B.V.; Ward, J.W.; Woodard, G. and Courtney, K.D.: Toxicologic studies of fenfluramine. Toxicol. Appl. Pharmacol. 19:707–711, 1971.

Vorhees, R.C.; Brunner, R.L. and Butcher, R.E.: Psychotropic drugs as behavioral teratogens. Science 205:1220–1225, 1979.

1027 Fenofibrate

Ujhazy et al. (1989) gave mice up to 585 mg per kg orally during days 7–16 of gestation. At the highest dose some fetal weight loss was found but no adverse fetal effects.

Ujhazy, E.; Onderova, E.; Horaova, M.; Bencova, E.; Durisova, M.; Nosal, R.; Balonova, T. and Zeljenova, D.: Teratological study of the hypolipidaemic drugs etofylline clofibrate (VULM) and fenofibrate in swiss mice. Pharmacology and Toxicology 64:286–290, 1989.

1028 Fenoprofen B–L–2–(3–Phenoxy–phenyl)proprionic Acid CAS 31879–05–7

No teratologic effects were found in the rat or rabbit given up to 100 mg per kg orally (Emmerson et al., 1973). Powell and Cochrane (1978) administered 50 mg per kg to the rat orally starting on the 18th day and produced premature closure of the ductus arteriosus in 4 to 10 percent of the offspring treated for 3.5 to 5 days.

Emmerson, J.L.; Gibson, W.R.; Pierce, E.C. and Todd, G.C.: Preclinical toxicology of fenoprofen. (abs) Toxicol. Appl. Pharmacol. 25:44 only, 1973.

Powell, J.G. and Cochrane, R.L.: The effects of the administration of fenoprofen or indomethacin to rat dams during late pregnancy with special reference to the ductus arteriosus of the fetuses and neonates. Toxicol. Appl. Pharmacol. 45:783–796, 1978.

1029 Fenoterol 1–(3,5–Dihydroxyphenyl)–2–[1–(4–hydroxy–benzyl) ethyl]–amino–ethanol hydrobromide CAS 13392–18–2

Nishimura et al. (1981) tested this β–adrenoreceptor stimulant orally in rats and rabbits during organogenesis at maximum doses of 25 (rats) and 100 mg per kg (rabbits) per day. No teratogenic effects were found. Early embryogenesis and postnatal function was also studied. No ill effects were found except for some delay in partuition.

Nishimura, M.; Kast, A. and Tsunenari, Y.: Reproduction studies of fenoterol (Th 1165a) in rats and rabbits. (Japanese) Iyakuhin Kenkyu 12:742–761, 1981.

1030 Fenpentadiol
2–(p–Chlor–phenyl)–4–methyl–pentan–2,4–diol

Kriegel (1971) treated rats and mice with up to 100 mg per kg during all of gestation and found no adverse fetal effects.

Kriegel, V.H.: Untersuchungen zur Frage einer teratogenen Wirkung von Fenpentadiol bei Maus und Ratte. Arzneim–Forsch. 21:13–15, 1971.

1031 Fenretinide

Kenel et al. (1987) treated rats and rabbits orally during organogenesis with this amide of all–trans retinoic acid. In the rat at 125 and 800 mg per kg, hydrocephaly and eye defects were found in both species.

Kenel, M.F.; Krayer, J.H.; Merz, E.A. and Pritchard, J.F.: Teratogencity of fenretinide in rats and rabbits. Teratology 56A, 1987.

1032 Fentanyl
N–(1–Phenethyl–4–piperidyl)propionanilide Sublimaze^TM

Fujinaga et al. (1986) found no adverse reproductive effects of this narcotic given in doses of up to 500 microgm per day from two weeks before conception and until the end of pregnancy in the rat.

Fujinaga, M.; Stevenson, J.B. and Mazze, R.I.: Reproductive and teratogenic effects of fentanyl in Sprague–Dawley rats. Teratology 34:51–57, 1986.

1033 Fenthion™ CAS 55–38–9

This organophosphorus insecticide was given to mice intraperitoneally on single or in 3 day periods during organogenesis in doses up to 80 mg per kg. Some reduction in fetal weight occurred but no defects were found (Budreau and Singh, 1973).

Budreau, C.H. and Singh, R.P.: Teratogenicity and embryotoxicity of demeton and fenthion in CF No. 1 mouse embryos. Toxicol. Appl. Pharmacol. 24:324–332, 1973.

1034 Fentiazac
4–(p–Chlorophenyl)–2–phenyl–5–thiazoleacetic Acid CAS 18046–21–4

Shimazu et al. (1979) gave this non–steroidal anti–inflammatory agent to rats and rabbits by mouth. They gave up to 50 mg in rats and 100 mg per kg in rabbits. An increase in stillbirths was found at 50 mg per kg in the rat. No adverse fetal effects were found in either species.

Shimazu, H.; Ichibana, T.; Matsuura, M. and Kojima, N.: Reproductive studies on 4(p–chlorophenyl)–2–phenyl–5–thiazoleacetic acid. Kiso to Rinsho 13:1929–1945, 1979.

1035 Feprazone
4–Prenyl–1,2–diphenyl–3,5–pyrazolidinedione CAS 30748–29–9

Kato et al. (1979) in a series of papers studied reproduction in the rat and rabbit using oral doses of up to 480 mg and 240 mg per kg, respectively. At the highest dose in the rat, preimplantation losses were increased and in the rabbits at doses above 120 mg per kg fetal lethality was increased. In neither species was any increase in defects found. Behavioral studies in the rat offspring exposed up to 120 mg per day were not different from controls.

Kato, M.; Matsuzawa, K.; Enjo, H.; Makita, T. and Hashimoto, Y.: Reproductive studies of feprazone 1–3. Iyakuhin Kenkyu 10:142–175, 1979.

1036 Fern

Yasuda et al. (1974) fed a diet containing 33 percent dried bracken fern to pregnant mice. The offspring had an increase in cervical and lumbar ribs and a delay in ossification as well as growth retardation. The treated group did not have an increase in other types of malformations.

Yasuda, Y.; Kihara, T. and Nishimura, H.: Embryotoxic effects of feeding bracken fern (Pteridium aquilinum) to pregnant mice. Toxicol. Appl. Pharmacol. 28:264–268, 1974.

1037 Fertilysin WIN–18446
N–N–Bis–(dichloracetyl)–1,8–octamethylenediamine CAS 1477–57–2

Oster et al. (1974) studied this male antifertility agent in pregnant rats and found the drug to be highly embryocidal. When they used 200 mg per kg per day on days 11 and 12 a high incidence of fetal edema, cleft lip and microcephaly was found. In a later publication, these authors also detailed septal heart defects, diaphragmatic hernias and small irregular thymuses (Taleporos et al., 1978). More detailed studies were published by Kilburn et al. (1982). Hayakawa et al. (1983) studied the congenital heart defects found in rats after administration of this compound.

Hayakawa, K.; Okishima, T.; Ohdo, S. and Okamoto, N.: Experimental study on pathogenesis and morphogenesis of congenital heart defects: Conotruncal abnormality produced by bis(dichloracetyl)diamine. Cong. Anom. 23:267–277, 1983.
Kilburn, K.H.; Hess, R.A.; Lesser, M. and Oster, G.: Perinatal death and respiratory apparatus dysgenesis due to bis(dichloroacetyl)diamine. Teratology 26:155–162, 1982.
Oster, G.; Salgo, M.P. and Taleporos, P.: Embryocidal action of bis(dichloracetyl)diamine: An oral abortifacient in rats. Am. J. Obstet. Gynecol. 119:583–588, 1974.
Taleporos, P.; Salgo, M.P. and Oster, G.: Teratogenic action of a bis(dichloroacetyl)diamine on rats. Patterns of malformations produced in high incidence at time–limited periods of development. Teratology 18:5–16, 1978.

1038 α–Fetoprotein Antibodies

In a preliminary report, Smith (1972) produced abnormalities including neural tube closure defects in chick and rat embryos given species–specific anti–α–fetoprotein antiserum.

Smith, J.A.: α–Fetoprotein: A possible factor necessary for normal development of the embryo. Lancet 1:851 only, 1972.

1039 Filariasis

Steel et al. (**1994**) studied the immune response of children born to mothers infected with Wuchereria bancrofti and found them hyporesponsive to microfilaraemic antigens. The eleven exposed newborns did not differ from those without the infection.

Steel, C.; Guines, A.; McCarthy, J.S. and Ottesen, E.A.: Long–term effect of prenatal exposure to maternal mirofilaraemia on immune responsiveness to filarial parasite antigens. The Lancet 343:890–893, 1994.

1040 Finasteride *Proscar™*

Clark et al. (1990) fed rats this 5–α–reductase inhibitor during pregnancy. Hypospadias and decreased anogenital distance was increased in the male offspring at 1mg per kg and above. Anderson and Clark (**1990**) studied this 5–α–reductase inhibitor in developing rats. Although there was no effect on the female a hypospadias like condition was produced in the males. Doses over 30 mg per kg daily on days 6–20 were associated with the defect. Reduction in the anorectal distance in males was found down to a dose of 0.03 mg per kg.

Anderson, C.A. and Clark, R.L.: External genitalia of the rat: normal development and the histogenesis of 5–α reductase inhibitor–induced abnormalities. Teratology 42:483–496, 1990.

Clark, R.L.; Antonello, J.M.; Grossman, S.J.; Wise, L.D.; Anderson, C.; Bagdon, W.J.; Prahalada, S.; MacDonald, J.S. and Robertson, R.T.: External genitalia abnormalities in male rats exposed in utero to finasteride, a 5α–reductase inhibitor. Teratology 42:91–100, 1990.

1041 Fish Oil

Based on increased length of pregnancy among fish–eating women from the Faroe Islands, Olsen et at (**1992**) randomly assigned women in their third trimester to groups ingesting fish oil (4 gm), olive oil or no supplement. The group with fish oil averaged 4 days longer gestation (95% CI 1.5–6.4) and had newborns 102 gm heavier than those taking olive oil. The gestational length and birth weight of the non–supplemented group were intermediate between the two treated groups.

Olsen, S.F.; Sorensen, J.D.; Secher, N.J.; Hedegaard, M.; Henriksen, T.B.; Hansen, H.S. and Grant, A.: Randomised controlled trial of effect of fish–oil supplementation on pregnancy duration. The Lancet 339:1003–1007, 1992.

1042 Flecainide

Nishimura et al. (**1989**) treated rats orally on days 7–17 with up to 80 mg per kg. At the highest doses increased fetal death occurred and fetal weight ws reduced. A small but significant increase in defects occurred at 30 mg per kg and above. Seletal variations were also increased. Thymic reminants in the neck also occurred.

Nishimura, O.; Okada, F.; Ohsumi, I.; Gotoh, M.; Hosokawa, S. and Matsubara, Y.: Reproduction study of flecainide teratological study in rats with oral administration. Kiso to Rinsho 23:1797-1814, 1989.

1043 Fletazepam *SCH–15698*

Arthaud et al. (**1981**) gave male rats 630 to 2500 mg per kg and found a decreased pregnancy rate and lower weights of the testes.

Arthaud, L.E.; Davis, G.J.; Black, H.E. and Heller, J.: Effects of a single oral dose of SCH 15698, a 1,4 benzodiazepine, on testicular weight and morphology and on reproductive performance in rats. Toxicologist 1:105, 1981.

1044 Floctafenine

Glomot et al. (**1976**) treated rats, mice and rabbits with this analgesic and found no teratogenic effect. Gavage doses were up to 320 mg per kg in the mice and up to 160 mg per kg in the rats and rabbits. Fetal loss was increased in the mice at the highest doses.

Glomot, R.; Chevalier, B. and Vannier, B.: Toxicological studies on floctafenine. Toxicology and Applied Pharmacology 36:173–185, 1976.

1045 Floxuridine *2–Deoxy–5–fluorouridine* CAS 50–91–9

This pyrimidine analogue used in cancer chemotherapy is teratogenic in chicks, mice and rats (see review by Chaube and Murphy, 1968). In the rat the fetal LD–100 is 8 times greater than the maternal LD–100. Given on the 11th or 12th day, 75 to 150 mg per kg causes defects of the central nervous system, palate and skeleton. Bro–Rasmussen et al. (1971) described embryonic hemorrhages which they believe contribute to production of the defects. Ferguson (1977, 1978) presented data that the mechanism producing cleft palate is related to lack of water binding by mucopolysaccharides and elevation of the shelves. By injecting into the yolk sac of the alligator embryo, Ferguson (1981) produced facial clefts.

Bro–Rasmussen, F.; Jensen, B.; Hansen, O.M. and Ostergaard, A.H.: Fluorodeoxyuridine–induced malformations in mice. Studies of early embryogenesis. Acta Pathol. Microbiol. Scand. (a) 79:55–60, 1971.

Chaube, S. and Murphy, M.L.: The teratogenic effects of the recent drugs active in cancer chemotherapy. In: Advances in Teratology, D.H.M. Woollam (ed.), New York: Academic Press, 3:181–237, 1968.

Ferguson, M.W.J.: The mechanism of palatal shelf elevation and the pathogenesis of cleft palate. Virchows Arch. A. Path. Anat. Histol. 375:97–113, 1977.

Ferguson, M.W.J.: The teratogenic effects of 5–fluoro–2–desoxyuridine (FUDR) on the Wistar rat fetus, with particular reference to cleft palate. J. Anat. 126:37–49, 1978.

Ferguson, M.W.J.: Review: The value of the American alligator as a model for research in craniofacial development. J. Craniofac. Genet. Dev. Biol. 1:123–144, 1981.

1046 Flubendazole *Flumoxal™* CAS 31430–15–6

This benzimidazole derivative is related chemically to mebendazole. Yoshimura (1987) gavaged mice daily on days 8–15 with 2.5 to 160 mg per kg. Doses of 40 mg and above were teratogenic producing hydrocephaly, anophthalmia and microphthalmia as well as omphalocele, ectrodactyly, anal atresia and skeletal defects. At the 10 mg per kg level no adverse effects were found.

Yoshimura, H.: Teratogenicity of flubendazole in rats. Toxicology 43:133–138, 1987.

1047 Fluocinolone

This fluorinated cortiocosteroid used topically in humans was given to rats (Casilli et al., **1977**) and rabbits (Kihlstrom and Lundberg **1987**) and the offspring at a dose of 50 microgm per kg had reduced weights. In the rabbits there was fetal resorption. No terata were reported.

Casilli, L.; Sauro, M.; Biraschi, L.M.; Tarantino, P. and Michelazzi, L.: Toxicological studies on halpredone acetate. Arzneim–Forsch 27:2102–2108. 1977.

Kihlstrom, I. and Lundberg, C.: Teratogenicity study of the new flucocorticosteroid budesonide in rabbits. Arzneim–Forsch 31(1):43–46, 1987.

1048 Flumazenil

8–Fluoro–5,6–dihydro–5–methyl–6–oxo–4H–imidazo [1,5–a] benzodiazepine–3–carboxylate; Anexate™

This benzodiazepine antagonist was studied in rats and rabbits by Schlappi et al. (1988). Both species received up to 150 mg per kg by gavage during organogenesis. No increase in defects occurred. Fertility and postnatal studies were also carried out and no adverse effects found.

Schlappi, B.; Bonetti, E.P.; Burgin, H. and Strobel, R.: Toxicological investigations with the benzodiazepine antagonist flumazenil. Arzneim–Forsch. Drug Res. 38:247–250, 1988.

1049 Flunarizine *1–[Bis(4–fluorophenyl)methyl] 4–(3–phenyl–2–propenyl) piperizine Dihydrochloride* CAS 52468–60–7

This drug increased cerebral blood flow. Miyazaki et al. (1982) gave it orally to pregnant rats and rabbits in amounts of up to 30 or 36 mg per kg, respectively. Some growth retardation and increase in embryo lethality was found in the rat. In neither species was there an increase in congenital defects. After postnatal studies in rats there were decreased nursing rates among the offspring.

Miyazaki, E.; Haro, T.; Nishikawa, S. and Oguro, T.: Toxicologic studies of K–W–3149. Kiso to Rinsho 16:1832–1839, 1840–1859, 1860–1871, 1982.

1050 Flunisolide CAS 3385–03–3

This anti–inflammatory steroidal agent was studied by Itabashi et al. (1982) in pregnant rats and mice. Oral administration of 50 microgm per kg before mating and for the first seven days produced a decrease in viability of embryos and fetuses. At 100 microgm per kg during organogenesis, umbilical hernias, anasarca, cleft palate and other defects were increased. Perinatal treatment was associated with growth retardation up to 10 weeks and the offspring had decreased implantation and viability of their fetuses. Mice were treated during organogenesis and at 0.2 mg per kg and above cleft palate was increased. At 0.04 mg no fetal defects were found.

Itabashi, M.; Yomazaki, M.; Watanabe, H.; Takehara, K.; Inoue, T.; Yokata, M. and Tajima, M.: Reproductive studies of flunisolide in rats and mice. Oyo Yakuri 24:631–672, 741–750, 1982.

1051 Flunitrazepam *5–(2–Fluorophenyl)–1,3–dihydro– 1–methyl–7–nitro–2H–1,4–benzodiazepin–2–one*

Suzuki et al. (1983) studied this hypnotic in rat and rabbit pregnancy. Oral doses of up to 100 mg per kg before and during the first 7 days, and up to 50 mg per day on days 7–17 were used. At the 50 mg dose during organogenesis enlarged cerebral ventricles, ventricular septal defects and remnants of thymus in the neck were increased. At 100 mg per kg during the first seven days visceral malformation also resulted. In the offspring whose mothers received 25 mg and above daily on days 7–17, effects on learning and motor activity were found.

McClain and Hoar (1980) treated rats orally with up to 25 mg per kg on days 7 through 15. There was no increase in fetal defects. Fetal survival was decreased in part due to maternal depression.

McClain, R.M. and Hoar, R.M.: The effect of flunitrazepam on reproduction in the rat. The use of cross–fostering in the evaluation of postnatal parameters in rat reproduction studies. Toxicol. Appl. Pharmacol. 53:92–100, 1980.

Suzuki, Y.; Mikami, T.; Gotoh, M.; Nishimura, O.; Seto, T. and Chiba, T.: Teratological and reproduction studies with a new hypnotic, 5–(2–fluorophenyl)–1,3–dihydro–1–methyl–7–nitro–2H–1, 4–diazepine–2–one (flunitrazepam). Clin. Reports 17:2585–2593, 1983.

1052 Fluphenazine Decanoate

Kawakami et al. (**1990**) gave rats up to 25 mg per kg intramuscularly on day 7 of gestation. At 5 and 25 mg the dams were sedated and had other toxic signs. There were no ill effects to the fetuses. Nomura and Shimomura (**1990**) gave rabbits up to 6.4 mg per kg on day 6 and observed no adverse fetal effects. In rats treated before fertilization and up to the 7th day of gestation no fetal effects occurred but diestrus to proestrus was delayed (Nomura and Hatekeyama, **1990**).

Kawakami, Y.; Hatakeyama, Y. and Hironaka, N.: Reproduction study by intramuscular administration of fluphenazine decanoate during organogenesis period in rat. Preclin. Rep. Cent. Inst. Exp. Anim. 16(1):11–41, 1990.

Nomura, G. and Hatakeyama, Y.: Reproduction study by intramuscular administration of fluphenazine decanoate during the pre–and early gestation periods in rats. Preclin. Rep. Cent. Inst. Exp. Anim. 16(1):1–19, 1990.

Nomura, G. and Shimomura, K.: Reproduction study by intramuscular administration of fluphenazine decanoate during the fetal organogenesis periods in rabbits. Preclin. Rep. Cent. Inst. Exp. Anim. 16(1):63–70, 1990.

1053 Fluocortolone CAS 152–97–6

Ezumi et al. (1976, 1977) treated mice and rabbits with this steroid. At 20 mg per kg injected subcutaneously or given orally on day 13, the mouse fetuses had increased number of cleft palates. Rabbit dams injected with 2.5 mg per kg produced fetuses with cleft palates and exencephaly.

Ezumi, Y.; Tomoyama, J. and Kodama, N.: Teratogenicity (cleft palate formation) in mouse embryos of fluocortolone by a single injection. (Japanese) Yakabutsu Ryoho 9:1623–1632, 1976.

Ezumi, Y.; Tomoyama, J.; Kodama, N. and Tanaka, M.: Effects of subcutaneous injection of flucortolone to rabbit embryos. Yakabutsu Ryoho 10:151–156, 1977.

1054 Fluometuron

Fluometuron is a herbicide. Khamidov et al. (1986) studied tissue and subcellular distribution of tritium–labeled fluometuron (cotoran) administered in pregnant rats by stomach tube on days 19 through 21. The distribution was not random and fluometuron accumulated in the intestine.

Khamidov, D.Kh.; Vdovina, S.K.; Sagatova, G.A.; Muchnik, S.E. and Mirakhmedov, A.K.: The transfer of fluometuron (cotoran) through the placenta in the late stages of pregnancy. Uzbekskiy Biologicheskiy Jurnal (USSR) 2:57–58, 1986.

1055 Fluoracizine 10–Diethylaminopropionyl–3–trifluoromethyl–phenothiazine Hydrochloride

Smolnikova et al. (1973) gavaged mice, rats and rabbits with maximum doses of 100, 50 and 25 mg per kg respectively. No evidence of teratogenicity was found but increased embryonic lethality was found in mice and rabbits at the highest doses.

Smolnikova, N.M.; Strekalova, S.N. and Garibova, T.L.: Different species–specifric sensitivity of animal embryos to fluoracizine. (Russian) Farmakol. Toksikol. 36:399–402, 1973.

1056 Fluoranthene

This isolate of diesel soot was studied in vitro in a rat embryo culture and by intraperitoneal dosing in mice. By both routes embryotoxicity was found but the doses were not reported in the abstract (Irvin and Martin, 1987).

Irvin, T.R. and Martin, J.E.: In vitro and in vivo embryotoxicity of fluoranthene, a major prenatal toxic component of diesel soot. (abs) Teratology 35:65A, 1987.

1057 N–2–Fluorenylacetamide CAS 53–96–3

Izumi (1966) injected a single dose of 0.1 mg per gm into mice (D–D strain) on the 8th to the 15th gestational days and found mainly skeletal defects, but some cleft lips and palates and cerebral hernias occurred.

Izumi, T.: Developmental anomalies in offspring of mice induced by administration of 2–acetylaminofluorene during pregnancy. Acta Anat. Nippon 37:239–249, 1966.

1058 Fluorine CAS 7782–41–4

Mottled dental enamel was found in the deciduous teeth of children whose mothers used well water containing 12 to 18 parts of fluorine per million (Smith and Smith, 1935). This fluorine level is over 20 times higher than usually found. Glenn (1981) reviewed strong evidence that prenatal sodium fluoride reduces dental caries. In her group he found 95 percent to be caries–free at age 5 when the mother was supplemented with a 2.2 mg tablet of NaF.

The lack of association between Down syndrome and fluoridation was well studied and discussed (Needleman et al., 1974). Erickson (1980) added further evidence for lack of association between fluoridation and Down syndrome.

In explanted chick embryos both Spratt (1950) and Duffy and Ebert (1957) reported a differential inhibition of cardiac development. Concentrations over 5×10^{-3}M had less effect on heart development; pyruvate in concentrations of 2×10^{-2}M prevented the effect of fluoride. The rat was also used for experimental study of this subject (Schour and Smith, 1935). Flemming and Greenfield (1954) found decalcification and histologic changes in the teeth of mouse fetuses when the mothers received 600 microgm of CaF_2 or 1,000 microgm of NaF.

Messer et al. (1973) maintained mice on a low fluoride diet (0.1–0.3ppm) and observed progressive adevelopment of infertility and anemia in two successive generations.

Duffy, L.M. and Ebert, J.D.: Metabolic characteristics of the heart–forming areas of the early chick embryo. J. Embryol. Exp. Morphol. 5:324–339, 1957.

Erickson, J.D.: Down syndrome, water fluoridation and maternal age. Teratology 21:177–180, 1980.

Flemming, H.S. and Greenfield, V.S.: Changes in the teeth and jaws of neonatal Webster mice after administration of NaF and CaF_2 to the female parent during gestation. J. Dent. Res. 33:780–788, 1954.

Glenn, F.B.: The rationale for administration of a NaF tablet supplement during pregnancy and postnatally in a private practice setting. March–April 118–122, J. Dentist. Child., 1981.

Messer, H.H.; Armstrong, W.D. and Singer, L.: Influence of fluoride intake on reproduction in mice. J. Nutr. 103:1319–1326, 1973.

Needleman, H.L.; Pueschel, S.M. and Rothman, K.J.: Fluoridation and the occurrence of Down's syndrome. N. Eng. J. Med. 291:821–823, 1974.

Schour, I. and Smith, M.C.: Mottled teeth: An experimental and histologic analysis. J. Am. Dent. Assoc. 22:796–813, 1935.

Smith, C.M. and Smith, H.V.: The occurrence of mottled enamel on the temporary teeth. J. Am. Dent. Assoc. 22:814–817, 1935.

Spratt, N.T.: Nutritional requirements of the early chick embryo. III. The metabolic basis of morphogenesis and differentiation as revealed by the use of inhibitors. Biol. Bull. 98:120–135, 1950.

1059   Fluoroacetate CAS 513–62–2

This inhibitor of citrate oxidation via the Krebs cycle was given to rats on day 9, 10 or 11 in a dose of 1 mg per kg and no malformations detected (Spielmann et al., 1973). These same authors noted a reduction in the Q O2 of rat embryos studied in the presence of fluoroacetate.

Spielmann, H.; Meyer–Wendecker, R. and Spielmann, F.: Influence of 2–deoxy–D–glucose and sodium fluoroacetate on respiratory metabolism of rat embryos during organogenesis. Teratology 7:127–134, 1973.

1060   5–Fluorocytosine CAS 2022–85–7

Chaube and Murphy (1969) produced defects in the offspring of rats receiving 700 to 1000 mg per kg on day 11 or 12 of gestation. The defects included cleft lip and palate, micrognathia and other skeletal defects. They comment that this compound was considerably less effective as a teratogen than 5–fluorodeoxycytidine.

Takeuchi et al. (1976 A,B) studied the agent in rats and mice and found teratogenicity at doses over 10 mg per kg. Vertebral fusions were found at 40 mg per kg given orally. At 400 mg per kg, the mouse fetuses were reduced in size and a small number had cleft palate.

Chaube, S. and Murphy, M.L.: The teratogenic effects of 5–fluorocytosine in the rat. Cancer Res. 29:554–557, 1969.

Takeuchi, I.; Takagaki, T.; Yoshino, T.; Noda, A.; Shimizu, M. and Udaka, K.: Toxicological studies on flucytosine (5-FC) 4. Teratology study in rats. (Japanese) Basic Pharmacol. Ther. 4:59–80, 1976A.

Takeuchi, I.; Shimizu, M.; Tamitani, S.; Noda, A. and Udaka, K.: Toxicologic studies on flucytosine (5–FC). 5. Embryotoxicity study in mice. (Japanese) Basic Pharmacol. Ther. 4:101–123, 1976B.

1061   5–Fluoro–2–deoxycytidine CAS 10356–76–0

Chaube and Murphy (1968) in their general review of fluoropyrimidines, reported that this compound is 200 times more lethal in the rat fetus than in the mother. A dose of 0.15 to 2.5 mg per kg on the day 11 or 12 of gestation causes defects of the central nervous system, palate and skeleton. In the rat, this is by weight the most potent teratogen of the fluoropyrimidine group. Chaube and Murphy (1968) were unable to prevent teratogenicity by thymidine administration. Degenhardt et al. (1968) published detailed studies of the effect of timed dosages on incidence and type of congenital defect produced.

Rogers et al. (1994) studied cell cycle changes and cell death in mouse embryos. Cell death occurred in areas of greatest proliferation.

Chaube, S. and Murphy, M.L.: The teratogenic effects of the recent drugs active in cancer chemotherapy. In: Advances in Teratology, D.H.M. Woollam (ed.), London: Logos Press, 3:181–237, 1968.

Degenhardt, K.H.; Franz, J. and Yamamura, H.: A model in comparative teratogenesis: Dose response to 5–fluoro–2–deoxycytidine (FCDR, RO5–1090) in organogenesis of mice strains C57Bl–6JHANFFM and C57Bl–10JFM. Teratology 1:311–334, 1968.

Rogers, J.M.; Francis, B.M.; Sulik, K.K.; Alles, A.J.; Massaro, E.J.; Zucker, R.M.; Elstein, K.H.; Rosen, M.B. and Chernoff, N.: Cell death and cell cycle perturbartion in the developmental toxicity of the demethylating agent, 5–aza–2'–deoxycytidine. Teratology 50:332–339, 1994.

1062
9–Fluoro–6,7–dihydro–8–(4–hydroxy–1–piperidyl)–5–methyl–1–oxo–1H–benzo[i,j]quinolizine–2–carboxylic acid. OPC–7251

Matsuzawa et al. (1991 studied this antibiotic in rats giving up to 75 mg per kg on days 17–21. No adverse changes in behavior or reproduction of the offspring were found.

Matsuzawa, A.; Yoshida, M.; Hamabuchi, T.; Kajiyoshi, K.; Okamoto, R.; Nishioeda, R. and Tamagawa, M.: Reproductive and developmental toxicity studies of 9–fluoro–6,7–dihydro–8–(4–hydroxy–1–piperidyl)–5–methyl–1–oxo–1H,5H–benzo[i,j] quinolizine–2–carboylic acid (OPC–7251), a synthetic antibacterial agent (IV): perinatal and postnatal study in rats with subcutaneous administration. Iyakuhin Kenkyu 22:61–76, 1991.

1063   M–Fluorodimethylaminoazobenzene

Discussed under Aminoazobenzene

1064   5–Fluoro–n–(4)–methyl–2–deoxycytidine CAS 2248–73–9

This fluoropyrimidine is teratogenic in the mouse at doses of 5 to 300 mg per kg. Defects of the palate, central nervous system and skeleton were reported (Chaube and Murphy, 1968).

Chaube, S. and Murphy, M.L.: The teratogenic effects of the recent drugs active in cancer chemotherapy. In: Advances in Teratology, D.H.M. Woollam (ed.), New York: Academic Press, 3:181–237, 1968.

1065   5–Fluoroorotic Acid CAS 703–95–7

Chaube and Murphy (1968) reported that 150 to 200 mg per kg on the 11th or 12th gestational day produced deformities of the skeleton in the rat.

Chaube, S. and Murphy, M.L.: The teratogenic effects of the recent drugs active in cancer chemotherapy. In: Advances in Teratology, D.H.M. Woollam (ed.), New York: Academic Press, 3:181–237, 1968.

1066   p–Fluoro–phenylalanine CAS 60–17–3

Waddington and Perry (1958) produced growth inhibition with large blisters on each side of the embryonic axis in chick embryos explanted to media containing 0.2 to 0.4 mg per ml. Equimolar concentrations of phenylalanine prevented the effect.

Waddington, C.H. and Perry, M.M.: Effects of some amino–acid and purine antagonists on chick embryos. J. Embryol. Exp. Morphol. 6:365–372, 1958.

1067 5–Fluorouracil CAS 51–21–8

Stephens et al. (1980) reported a human pregnancy during which the mother received 600 mg intravenously five times weekly at 11 to 12 fetal weeks. Radial aplasia, imperforate anus, esophageal aplasia and hypoplasia of the duodenum, lung and aorta were present. Kopelman and Miyazawa (**1990**) report three women who applied a 5% cream vaginally during early pregnancy. No defects were found in the 3 newborns. The problem of inadvertent exposurre to the topical medication has been discussed by Robert and Scialli (**1994**) who do not recommend termination of the pregnancy.

This pyrimidine analogue given on the 11th or 12th day in doses of 12 to 37 mg per kg produces defects of the central nervous system, palate and skeleton in the rat (Chaube and Murphy, 1968). Defects in the mouse fetus were studied by Dagg (1960). The mechanism of action for the 5–fluoropyrimidines is through interference with RNA and DNA synthesis. They inhibit thymidylate synthetase producing thymidine deficiency and a syndrome termed thymidineless death. The general subject was reviewed by Chaube and Murphy (1968). Puchkov (1967) studied the effect of this compound on 4 to 23 somite chick embryos. Naya et al. (1990) found that two glutathione depletors, phorone and buthionine sulfoximine, enhanced the teratogenicity of 5–fluorouracil. Shuey et al. (**1994**) found fetal anemia in rat embryos exposed to 20–40 mg per kg on day 14.

Chaube, S. and Murphy, M.L.: The teratogenic effects of the recent drugs active in cancer chemotherapy. In: Advances in Teratology, D.H.M. Woollam (ed.), New York: Academic Press, 3:181–237, 1968.

Dagg, C.P.: Sensitive stages for the production of developmental abnormalities in mice with 5–fluorouracil. Am. J. Anat. 106:89–96, 1960.

Kopelman, J.N. and Miyazawa, K.: Inadvertent 5–fluorouracil treatment in early pregnancy: A report of three cases. Reproductive Toxicology 4:233–235, 1990.

Naya, M.; Mataki, Y.; Takahira, H.; Deguchi, T. and Yasuda, M.: Effects of phorone and/or buthionine sulfoximine on teratogenicity of 5–fluorouracil in mice. Teratology 41:275–280, 1990.

Puchkov, V.F.: Teratogenic action of aminopterin and 5–fluorouracil on 4–23 somite chick embryos after application in ovo. Bull. Exp. Biol. (Russian) 7:99–102, 1967.

Robert, E. and Scialli, A.R.: Topical medications during pregnancy. Reproductive Toxicology 8:197–202, 1994.

Shuey, D.L.; Zucker, R.M.; Elstein, K.H. and Rogers, J.M.: Fetal anemia following maternal exposure to 5–fluorouracil in the rat. Teratology 49:311–319, 1994.

Stephens, T.D.; Globus, M.; Miller, J.R.; Wilber, R.R. and Epstein, C.J.: Multiple congenital anomalies in a fetus exposed to 5–fluorouracil during the first trimester. Am. J. Obstet. Gynecol. 137:747–749, 1980.

1068 5–Fluorouridine CAS 316–46–1

This fluoropyrimidine produces defects of the central nervous system, palate and skeleton in rat fetuses when the mother rats are treated with 25 to 50 mg per kg on the 11th or 12th gestational day (Chaube and Murphy, 1968). The maternal LD–50 is 400 to 800 mg per kg.

Chaube, S. and Murphy, M.L.: The teratogenic effects of the recent drugs active in cancer chemotherapy. In: Advances in Teratology, D.H.M. Woollam (ed.), New York: Academic Press, 3:181–237, 1968.

1069 Fluoxetine *Prozac*^(TM)

Pastuszak et al. (**1993**) studied 98 women who took the medication during pregnancy and found no increase in defects. Women taking tricylic antidepressants or fluoxetine had higher rates of miscarriage. Byrd et al. (1989) studied this serotonin–reuptake inhibitor in rats and rabbits. Rats were given up to 12.5 mg and rabbits 15 mg per kg daily during organogenesis. Maternal toxicity was seen in both species but no decreased fetal weight or mortality occurred.

Byrd, R.A.; Brophy, G.T. and Markham, J.K.: Developmental toxicology studies of fluoxetine hydrochloride (I) administered orally to rats and rabbits. Teratology 39:P67, 1989.

Pastuszak, A.; Schick–Boschetto, B.; Zuber, C.; Feldkamp, M.; Pinelli, M.; Donnenfeld, A.; McCormack, M.; Leen–Mitchell, M.; Woodland, C.; Gardner, A.; Hom, M. and Koren, G.: Pregnancy outcome following first trimester exposure to fluoxetine (Prozac). JAMA 269:2246–2248, 1993.

1070 Fluoxymesterone

Jost (**1960**) found that 20 mg per kg injected after ovarectomy on the 16th day of gestation maintained pregnancy in the rat.

Jost, A.: Action de divers steroides sexuels et voisins sur la croissance et la differenciation sexuelle des foetus. First International Congress of Endocrinology, 7:119–123, 1960.

1071 Fluphenazine *Prolixin*^(TM) *Apazone* *3–Dimethylamino–7–methyl–1–2, [N–propylmalonyl]–1,2–dihydro–1,2,4 benzotriazine; Azapropazone CAS 13539–59–8 CAS 69–23–8*

This medication belongs to the phenothiazine group of drugs which are relatively safe during pregnancy. Mild withdrawal extrapyramidal signs were reported several weeks after birth (Cleary, 1977). Merlob et al. (**1994**) have reviewed the reports of diverse congenital defects in several offspring exposed to the drug.

Jahn and Adrian (1969) treated rats during organogenesis with up to 100 mg per kg orally and found no adverse fetal effects. Adrian (1973) found no fetal changes in the offspring of rats treated with 100 mg per kg per day during the entire period of pregnancy.

Adrian, R.W.: Reproduktionstoxikologische Untersuchungen mit Azapropazon an Ratten. Arzneim Forsch 23:658–660, 1973.

Cleary, M.F.: Fluphenazine deconate during pregnancy. Am. J. Psychiatry 134:815–816, 1977.

Jahn, V. and Adrian, R.W.: Pharmacologische und toxikologische Prufung des neuen Antiphlogisticums Azapropazon = 3–Dimethylamino–7–methyl–1,2–(n–propylmalonyl–1,2–dihydro–1,2, 4–benzotriazin. Arzneim–Forsch 19:36–52, 1969.

Merlob, P.; Stahl, B. and Maltz, E.: Is fluphenazine a teratogen? American Journal of Medical Genetics 52:231–232, 1994.

1072 Fluproquazone 4–(p-Fluorophenyl-1-isopropyl-7-methyl-2-(1H)quinazolinone) CAS 40507–23–1

Ruttimann et al. (1981) studied reproduction in rats and rabbits. Oral doses of up to 50 mg per kg in rats and 50 mg in rabbits did not adversely affect organogenesis.

Ruttimann, G.; Schon, H.; Madorin, M.; VanRyzin, R.J.; Richardson, B.P. and Matter, B.E.: Toxicological evaluation of fluproquazone. Arzneim Forsch. 31:882–892, 1981.

1073 Flurazepam Dalmane™ CAS 17617–23–1

In doses up to 80 mg per kg during various periods of gestation in the rat, neither teratogenic nor postnatal changes were produced in the offspring. In the rabbit, doses of adverse fetal effects (Hoffmann–LaRoche Company, 1979).

Hoffmann–LaRoche Co.: Personal communication to the author, 1979.

1074 Flutazolam CAS 27060–91–9

Sato et al. (1978) administered this benzodiazepin derivative orally to rats on days 8 through 15. The maximum dose of 1000 mg per kg per day did not alter fetal development. Ishimura et al. (1978) gave this substance orally to rabbits in doses of up to 1000 mg per kg from the 6th to 18th day. Viable fetuses were reduced at 1000 mg per kg and maternal toxicity was present. No teratogenic effect was seen.

Ishimura, K.; Honda, Y.; Neda, K.; Kawaguchi, Y.; Hayashi, T. and Maebara, K.: Teratological studies on flutazolam (MS–4101) teratogenicity study in rabbits. Oyo Yakuri 16:709–714, 1978. Sato, R.; Sato, H. and Kashima, M.: Studies on the possible teratogenicity of flutazolam. 3. Teratogenicity test in rats. (Japanese) Basic Pharmacol. Ther. 6:1692–1738, 1978.

1075 Fluticasone Propionate

Shimpo et al. (1992 A) gave this corticosteroid subcutaneously to rats before and in early pregnancy in doses of up to 50 microgm per day. Embryonic mortality was increased and fetal weight suppressed at 15 mg per kg but no decrease in fertility occurred. Treatment during organogenesis in doses of 100 mg per kg produced an increase in omphaloceles and skeletal variations (Shimpo et al., 1992 B). Perinatal treatment at 4 mg per kg was associated with

decreased weights of the pups (Rawlings et al., 1992). In rabbits treated during organogenesis variation in the pulmonary and live lobulation was increased in the fetuses. A low incidence of various other defects were found at the highest dose (4 mg per kg) (Ezaki et al., 1992).

Ezaki, H.; Yokoyama, S.; Takahashi, N.; Takamatsu, M.; Utsumi, K.; Suzuki, R.; Masuoka, M.; Tokado, H. and Takeda, K.: Teratogenicity study on fluticasone propionate in rabbits by subcutaneous administration. Jpn. Pharmacol. Ther. 20:1643–1656, 1992.

Rawlings, S.J.; Adams, M.J.; Parkinson, M.M.; Sibley, P.R. and Fluck, P.A.: Peri–and post–natal study on fluticasone propionate in rats by subcutaneous route. Jpn. Pharmacol. Ther. 20:1633–1641, 1992.

Shimpo, K.; Kondo, K.; Kobayashi, H.; Tsunemi, K.; Yahata, A.; Kamada, S.; Ezaki, H.; Takamatsu, M. and Tokado, H.: Fertility study on fluticasone propionate administered subcutaneously in rats. Jpn. Pharmacol. Ther. 20:1573–1595, 1992A.

Shimpo, K.; Tsunemi, K.; Kobayashi, H.; Yahata, A.; Kamada, S.; Ezaki, H.; Takamatsu, M. and Tokado, H.: Teratogenicity study on fluticasone propionate administered subcutaneously to rats. Jpn. Pharmacol. Ther. 20:1597–1632, 1992B.

1076 Flutoprazepam 7–Chloro–1–cyclopropylmethyl–1,3–dihydro–5–(2H–1,4–benzodiazepin–2–one CAS 25967–29–7

Yokoi et al. (1981) gave this tranquilizer to rats before mating (700 mg per kg), during organogenesis (1000 mg per kg) and on days 17–21 (700 mg per kg). No adverse effects were noted following the first two treatments but perinatal test doses of 100 mg per kg or more were associated with increased death of the offspring. Learning ability and reproduction were not altered. Fukunishi et al. (1982) gave rabbits orally up to 50 mg per kg during organogenesis and produced no adverse fetal effects.

Fukunishi, K.; Yoshida, H.; Hirano, K.; Yokoi, Y.; Nagano, M.; Mitsumori, T.; Terasaki, M. and Nose, T.: Teratological studies on flutoprazepam in rabbits. Kiso to Rinsho 16:658–666, 1982.

Yokoi, Y.; Yoshida, H.; Nagano, M.; Sagara, J.; Mitsumori, T.; Hirano, K.; Tsunawaki, M. and Nose, T.: Reproductive studies of 7–chloro–1–cyclopropylmethyl–1,3–dihydro–5 (2–fluorphenyl) 2H–1,4–benzodiazepin–2–one (KB–509). Oyo Yakuri 21:1–40, 1981.

1077 FM-100

Nemec et al. (1992) exposed rats and rabbits by inhalation for 6 hours daily during organogenesis. At concentrations of 4,000 ppm in the rat and 1.000 ppm in the rabit there was maternal toxicity. In the rabbit decreased fetal weight occurred at 1,000 ppm and above. In the rat exposed to 10,000 ppm osseous retardation and microphthamlmia occurred.

Nemec, M.; Holson, J.; Naas, D.; Knapp, J.; Lamb, I. and McAllister, D.: The developmental toxicity of fm-100 in the rat and rabbit following repeated exposure by inhalation (A). Teratology 45:475, 1992.

1078 Folic Acid Deficiency *Pteroylglutamic Acid Deficiency* CAS 59–30–3

Pritchard et al. (1970) reported that among women with folate deficiency during pregnancy, no increase in malformations of their offspring was found. Pritchard et al. (1971) studied women with anticonvulsant–induced lowered serum folate and found no increase in fetal complications. Dansky et al. (1987), in a prospective study of 49 pregnancies in women treated for epilepsy, found 8 with serum folate levels less than 4 nanogm per ml. Three had spontaneous abortions and two had malformations. The abnormal outcome was compared to 19 pregnancies with normal folates where one abortion and two malformations occurred.

Evidence was presented that neural tube closure defects in the human may be related to folic acid deficiency (or some other nutritional supplement). Smithells et al. (1981) supplemented the diet of 202 mothers who had given birth to infants with neural tube defects. Only two recurrences of the defect were found as compared to a group of 198 controls with ten recurrences. The controls consisted of women who entered the study after becoming pregnant. The supplement used included vitamins, calcium phosphate and iron. The folic acid content was 0.36 mg. Smithells et al. (1983) extended this study and observed a 0.7 percent recurrence among 454 supplemented and a 4.7 percent recurrence among 519 unsupplemented. More details of this study are given by Seller and Nevin (1984). Laurence et al. (1981) gave a daily supplement of 4 mg of folic acid to 44 mothers who had previous pregnancies with neural tube defects and had no recurrences. There were six recurrences among the 77 controls or non–compliers.

Smithells et al. (1989) recruited mothers who had previous neural tube defects by notifying their family doctor and recommending a multiple vitamin preparation containing folic acid (Pregnavite Forte F). Of 187 women who were fully or partially supplemented before their next pregnancy only one recurrence was found while in the unsupplemented 16 recurrences were found among 320. During the 7 year period there was a drop in the recurrence rate of unsupplemented from 6.0% to 4.7% while essentially no change was found in the supplemented (0.6%). Pregnancy terminations were included but the proportion in each group was not given. Bias factors in the family doctor's recruitment of the families may have been present.

Oakley et al. (1983) pointed out that with the marked drop in recurrence rate of neural tube defects since the mid–1970's, a randomized control of vitamin supplementation is still urgently needed. Wald and Polani (1984) discussed the existing data on vitamins and neural tube defects and called for a randomized clinical trial. Molloy et al. (1985) compared early pregnancy serum folate and B_{12} concentrations for 32 mothers who had neural tube defects and compared them to 395 randomly selected controls. No differences were noted.

Bower and Stanley (1989) in a case control study found that dietary folate in the first 6 weeks protected against neural tube defects in western Australia. The odds ratios for the risk of having a neural tube defect increased gradually with reduction of folic and total folate intake. No association with postpartum serum or red cell folate levels was found. Reduced fiber, calcium, vitamin C and carotene were also associated with increase risk.

Milunsky et al. (1989) identified 49 women who had neural tube defects among 22,776 pregnancies tested for α–fetoprotein. By telephone interview the folic acid intake during the first weeks was determined. The interviews were done at 15–20 weeks of pregnancy. Those taking folic acid had a rate of 0.9 per 1000 and those not taking it had 3.5 per 1000. Women who took multivitamins without folic acid or those who took vitamins after the first 6 weeks had a prevalence of 1.0 per 1000.

Mills et al. (1989) studied 571 mothers who had previous children with neural tube defects, 546 women bearing children with another defect and 573 normals. Interviews carried out by workers who did not know the outcome of the pregnancy were performed to determine whether periconceptional vitamins was taken and what the folic acid intake was around the time of neural tube closure. The pregnancies occurred between 1985 and 1987 in Illinois or California. There was no significant difference found (odds ratio for NTD = 0.95). The authors raise the question that geographical or temporal effects may have prevented coroboration of earlier findings on the subject. The percent taking folic acid was 5.8% in the neural tube group, 3.3% in the other abnormality group and 6.9% in controls. Sixteen (2.8%) women of the 571 had recurrence of a neural tube defect. Mills et al. (**1992**) measured serum folic acid, retinol and B_{12} in 89 mothers who gave birth to neural tube defects in a low prevalence and found no significant differences from controls. Most of the specimens were drawn within 8 weeks of neural tube closure.

Adams et al. (**1993**) reported in abstract that women with maternal alpha–feto protein elevations and neural tube defects had methyl malonic acid serum levels that were significantly elevated. Mills et al. (**1995**) studied homocysteine metabolism in 81 pregnancies complicated by neural tube defects and found elevated homocysteine suggesting that methionine synthase was deficient. Folic acid and B_{12} are both required cofactors for this enzyme.

A recent double–blind prevention trial conducted at 33 centers strongly supports the role of folic acid in prevention of neural tube defects (Wald et al., **1991**). Mothers of previous offspring with anencephaly, meningomyelocele or encephocele were randomly divided into four preconceptual treatment groups: 1. folic acid (4 mg) 2. Multiple vitamins without folic acid 3. Both folic acid and multiple vitamins and 4. Neither folic acid or multiple vitamins. Of 1195 who completed a subsequent pregnancy 27 had neural tube defects. Since 21 were in the non–folic acid groups and only 6 were in the folic acid groups the study was terminated and a firm recommendation for folic acid supplementation was made. The group that took multiple vitamins without folic acid had essentially the same high recurrence as those that took no vitamins. The rates of other types of congenital malformations did not differ significantly between the groups. Shaw et al. (**1994**) found a reduced risk for conotruncal congenital heart defects in women ingesting folic acid periconceptually. Odds 0.65 (95 CI 0.44–0.96).

The occurrance of neural tube defects and possibly hydroancephaly has been reported to be decreased in a case–control study by (Martinez–Frias and Rodriguez–Pinilla **1992**). Based on a total of 9,337 newborns the odds ratio for neural tube defect among those taking 0.4 mg folic acid was 0.69 (CI 0.51–0.94, p 0.01). Czeizel (**1993**) in a prospective preconception study of 2,420 women found that malformation rates were significantly less than in 2,333 women who took trace elements. The main difference was due to decreased neural tube defects. In another report of the same data Czeizel and Dudas (**1992**) reported 6 neural tube defects in the control group and none in the vitamin treated group (0.8 mg folic acid). There was no change in facial clefts with multivitamin treatment.

Deficiency of this substance was used widely to experimentally produce a wide variety of congenital defects. The general literature was reviewed by Kalter and Warkany (1959) and by Giroud and Tuchmann–Duplessis (1962). By altering the timing and period of the deficiency regimen, the incidence of particular anomalies can be altered (Nelson et al., 1955). During days 7 through 8 before major organogenesis, the deficiency state had no effect. In most animal models the deficient diet is supplemented by a folic acid antagonist such as X–methyl–pteroylglutamic acid or its more purified component 9–methyl–pteroylglutamic acid. Aminopterin, another antimetabolite, produces defects in animals and man and is covered under its separate heading.

Although the mouse can be used, the rat has been the most common test subject. Cardiovascular defects were produced by Baird et al. (1954) in 57 percent of rat fetuses exposed to a deficient diet al.ong with X–methyl––pteroylglutamic acid and succinylsulfathiazole on days 7 through 9 of gestation. No cardiovascular defects were produced when the diet was started on day 11. Genitourinary tract defects produced by a deficiency diet on days 10 through 13 were described by Monie et al. (1957) and included renal hypoplasia, renal ectopia, absence of kidney, hydronephrosis and hydroureter. Asling et al. (1955) made detailed studies of the skeletal defects. Heid et al. (**1992**) fed mice folic acid deficient synthetic diets and found a high rate of resorption at levels below 680 nmol per kg of diet. At a level of 453 nmol per kg reduced fetal weight occurred but no increase in congenital defects was found among day 12 embryos.

Johnson (1964), by careful studies of early deficient embryos, suggested that mitotic arrest at metaphase might play a role in the mechanism of action of this teratogen. Further studies by Chepenik et al. (1970) showed that the embryonic deficiency is associated with decreased levels of adenosine triphosphate. In a model produced by ethanol or heat in the hamster, Graham and Ferm (1985) found no protection by folate. Miller et al. (1989) studied the teratogenic effect of folic–acid–deficient serum in whole rat embryo culture. Conversion to dorsiflexion was impaired and anemia

was observed. DeSesso and Goeringer (**1991**) found that leucovorin ameliorated methotrexate teratogenicity in the rat.

Adams, M.J.; Khoury, M.J.; Stevenson, R.; Haddow, J.; Knight, G. and Allen, R.H.: Midtrimester serum methylmalonic acid as a risk marker for neural tube defects. (abs) Teratology 47:384, 1993.

Asling, C.W.; Nelson, M.M.; Wright, H.V. and Evans, H.M.: Congenital skeletal abnormalities in fetal rats resulting from maternal pteroylglutamic acid deficiency during gestation. Anat. Rec. 121:775–800, 1955.

Baird, C.D.C.; Nelson, M.M.; Monie, I.W. and Evans, H.M.: Congenital cardiovascular anomalies induced by pteroylglutamic acid deficiency during gestation in the rat. Circ. Res. 2:544–554, 1954.

Bower, C. and Stanley F.J.: Dietary folate as a risk factor for neural–tube defects–evidence from a case–control study in Western Australia. The Med. J. of Australia 150:613—619, 1989.

Chepenik, K.; Johnson, E.M. and Kaplan, S.: Effects of transitory maternal pteroylglutamic acid (PGA) deficiency on levels of adenosine phosphates in developing rat embryos. Teratology 3:229–236, 1970.

Czeizel, A.E.: Controlled studies of multivitamin supplementation on pregnancy outcomes. Maternal Nutrition and Pregnancy Outcome ed by C.L. Keen, A. Benedich and C.C. Willhite. New York Acad. of Sciences vol 678:266–275, 1993.

Czeizel, A.E. and Dudas, I.: Prevention of the first occurrence of neural–tube defects by periconceptual vitamin supplementation. N. Eng. J. Med. 327:1832–1835, 1992.

Dansky, L.V.; Andermann, E.; Rosenblatt, D.; Sherwin, A.L. and Andermann, F.: Anticonvulsants, folate levels, and pregnancy outcome: A prospective study. Annals of Neurology 21:176–182, 1987.

DeSesso, J.M. and Goeringer, G.C.: Amelioration by leucovorin of methotrexate developmental toxicity in rabbits. Teratology 43:201–215, 1991.

Giroud, A. and Tuchmann–Duplessis, H.: Malformations congenitales role des facteurs exogenes. Pathol. Biol. 10:119–151, 1962.

Graham, J.M. and Ferm, V.H.: Heat–and alcohol–induced neural tube defects: Interactions with folate in a golden hamster odel. Pediatr. Res. 19:247–251, 1985.

Heid, M.K.; Bills, N.D.; Hinrichs, S.H. and Clifford A.J.: Folate deficiency. J Nutri 122:888–894, 1992.

Johnson, E.M.: Effects of maternal folic acid deficiency on cytologic phenomena in the rat embryo. Anat. Rec. 149:49–56, 1964.

Kalter, H. and Warkany, J.: Experimental production of congenital malformations in mammals by metabolic procedure. Physiol. Rev. 39:69–115, 1959.

Laurence, K.M.; James, N.; Miller, M.H.; Tennant, G.B. and Campbell, H.: Double–blind randomized controlled trial of folate treatment before conception to prevent recurrence of neural tube defects. Br. Med. J. 282:1509–1511, 1981.

Martinez–Frias, M.L. and Rodriguez–Pinilla, E.: Folic aid supplementation and neural tube defects. The Lancet 340:620, 1992.

Miller, P.N.; Pratten, M.K. and Beck, F.: Growth of 9.5 day rat embryos in folic–acid–deficient serum. Teratology 39:375–386, 1989.

Mills, J.L.; Rhoads, G.G.; Simpson, J.L.; Cunningham, G.C.; Conley, M.R.; Lassman, M.R.; Walden, M.E.; Depp, O.R.; Hoffman, H.J. and The National Institute of Child Health and Human Development Neural Tube Defects Study Group: The absence of a relation between the periconceptional use of viatmins and neural–tube defects. The New England Journal of Medicine 321: 430–435, 1989.

Mills, J.L.; Tuomilehto, J.; Yu, K.F.; Colman, N.; Blaner, W.S.; Koskela, P.; Rundle, W.E.; Forman, M.; Taivanen, L. and Rhoads, G.G.: Maternal vitamin levels during pregnancies producing infants with neural tube defects. The Journal of Pediatrics 120:863–871,1992.

Mills, J.L.; McPartlin, J.M.; Kirke, P.N.; Lee, Y.J.; Conley, M.R. and Weir, D.G.: Homocysteine metabolism in pregnanies complicated by neural–tube defects. The Lancet 345:149–151, 1995.

Milunsky, M.; Jick, H.; Jick, S.S.; Bruell, C.L.; MacLaughlin, D.S.; Rothman, K.J. and Willett, W.: Multivitamin /folic acid supplementation in early pregnancy reduces the prevalence of neural tube defects. JAMA 262:2847–2851, 1989.

Molloy, A.M.; Kirke, P.; Hillary, I.; Weir, D.G. and Scott, J.M.: Maternal serum folate and vitamin B_{12} concentrations in pregnancies associated with neural tube defects. Arch. Dis. Child. 60:660–665, 1985.

Monie, I.W.; Nelson, M.M. and Evans, H.M.: Abnormalities of the urinary system of rat embryos resulting from transitory deficiency of pteroylglutamic acid during gestation. Anat. Rec. 127:711–724, 1957.

Nelson, M.M.; Wright, H.V.; Asling, C.W. and Evans, H.M.: Multiple congenital abnormalities resulting from transitory deficiency of pteroylglutamic acid during gestation in the rat. J. Nutr. 56:349–370, 1955.

Oakley, G.P.; Adams, M.J. and Levy, L.M.: Vitamins and neural tube defects. Lancet 2:798–799, 1983.

Pritchard, J.A.; Scott, D.E. and Whalley, P.J.: Maternal folate deficiency and pregnancy wastage. IV. Effects of folic acid supplements, anticonvulsants and oral contraceptives. Am. J. Obstet. Gynecol. 109:341–346, 1971.

Pritchard, J.A.; Scott, D.E.; Whalley, P.J. and Haling, R.F.: Infants of mothers with megaloblastic anemia due to folate deficiency. JAMA 211:1982–1984, 1970.

Seller, M.J. and Nevin, N.C.: Periconceptual vitamin supplementation and the prevention of neural tube defects in south–east England and Northern Ireland. J. Med. Genetics 21:325–330, 1984.

Shaw, G.M.; Wasserman, C.R. and O'Malley, C.D.: Periconceptional vitamin use and reduced risk for conotruncal and limb defects in California. (abs) Teratology 49:372, 1994.

Smithells, R.W.; Seller, M.J.; Nevin, N.C.; Sheppard, S.; Harris, R.; Read, A.P.; Fielding, D.W.; Walker, S.; Schorah, C.J. and Wild, J.: Further experience of vitamin supplementation for prevention of neural tube defect recurrences. Lancet 1:1027–1031, 1983.

Smithells, R.W.; Sheppard, S.; Schorah, C.J.; Seller, M.J.; Nevin, N.C.; Harris, R.; Read, A.P.; Fielding, D.W. and Walker, S.: Vitamin supplementation and neural tube defects. Lancet 1:425 only, 1981.

Smithells, R.W.; Sheppard, S.; Wild, J. and Schorah, C.J.: Prevention of neural tube defect recurrences in Yorkshire: final report. The Lancet 498–499, 1989.

Wald, N.J. and Polani, P.E.: Neural–tube defects and vitamins: The need for a randomized clinical trial. Br. J. Obstet. Gynecol. 91:516–523, 1984.

Wald, N.; Snedden, J.; Densem, J.; Frosi, C.; Stone, R. and MRC Vitamin Study Research Group: Prevention of neural tube defects: Results of the medical research council vitamin study. The Lancet 338:131–137, 1991.

1079 **Folpet** *N–Trichloromethylthiophthalamide* CAS 133–07–3

Verrett et al. (1969) injected the chick egg via the air cell or yolk sac with 3–20 mg before incubation and found 8.2 percent defects. Abnormalities of the eye and central nervous system with skeletal defects were present. Amelia and phocomelia were the most noteworthy. Doses of 75 mg per kg in rabbits and 500 mg per kg in rats given at critical stages of organogenesis were non–teratogenic (Kennedy et al., 1968).

Fabro et al. (1966) fed 80 mg per kg daily to rabbits on days 7–12 and found no defects in the offspring.

Fabro, S.; Smith, R.L. and Williams, R.T.: Embryotoxic activity of some pesticides and drugs related to phthalimide. Food Cosmet. Toxicol. 3:587–590, 1966.

Kennedy, G.L.; Fancher, O.E. and Calandra, J.C.: An investigation of the teratogenic potential of captan, folpet, and difolatan. Toxicol. Appl. Pharmacol. 13:420–430, 1968.

Verrett, M.J.; Mutchler, M.K.; Scott, W.F.; Reynaldo, E.F. and McLaughlin, J.: Teratogenic effects of captan and related compounds in the developing chicken embryo. Ann. N.Y. Acad. Sci. 160:334–343, 1969.

1080 **Fominoben** *PB 89Cl* CAS 18053–31–1

This antitussive was studied by Iida et al. (1978) in rats. No teratogenicity was found at oral doses of 1,500 mg per kg during organogenesis. Maternal weight was reduced at 1,000 and 1,500 mg levels, and sucklings had reduced weight gain at 1,000 mg per kg given to the mother. Postnatal reproduction was not modified.

Iida, H.; Matsuo, A.; Kast, A. and Tsunenari, Y.: Reproductive studies with fominoben hydrochloride (PB 89Cl) in rats. Iyakuhin Kenkyu 9:724–735, 1978.

1081 **Fonazine** *2-Dimethylsulfamido-10(2-dimethyl-aminopropyl)phenothiazine* CAS 7456–24–8

Tanaka and Matsuura (1970) gave this antihistamine to pregnant rabbits from day 8 through 16 of gestation in doses up to 60 mg per kg per day and observed no teratogenicity.

Tanaka, O. and Matsuura, M.: Effect of 8599RP administered to pregnant rabbits on pre– and postnatal development of their offspring. Oyo Yakuri 4:373–379, 1970.

1082 Food Dye and Coloring Red No. 40 *Allura Red*™

Collins et al. (1989) gavaged up to 1000 mg per kg on days 0–19 in the rat and found no adverse effects in the dams or fetuses.

Collins, T.F.X.; Black, T.N.; Welsh, J.J. and Brown, L.H.: Study of the teratogenic potential of FD and C red no. 40 when given by intubation. Teratology 39:P64, 1989.

1083 Food Dye and Coloring Red No. 102

This dye was fed in amounts of one percent of the diet to pregnant rats and no teratogenic or postnatal alterations were found (Kihara et al., 1977).

Kihara, T.; Yasuda, Y. and Tanimura, T.: Effects on pre– and postnatal offspring of pregnant rats fed food red no. 102. (abs) Teratology 16:111–112, 1977.

1084 Food Dye and Coloring Red No. 104 *Phloxine B*
2,4,5,7-Tetrabromo-12,13,14,15-tetrachlorofluorescein CAS 6441-77-6

Uchida and Enomoto (1971) studied mice in a three generation dietary experiment using 0.5 or 1.2 percent in the diet. The weight increase was decreased with long term treatment. Alopecia and malformations of the limbs (hindlimbs) was observed in each generation. It was not clear whether this was a developmental defect or postnatal toxicity. Nakaura et al. (1975) found no adverse effects of up to 3 percent in the rat diet.

This chemical dye present in the diet of a few countries was fed to mice on days 6 through 16. At levels of 3 and 5 percent, which were maternally toxic, cleft palate and exencephaly were increased. At the 1 percent level and higher, splitting of the cervical vertebrae was found (Seno et al., 1984).

Nakaura, S.; Kawashima, K.; Nagao, S.; Tanaka, S.; Takanaka, A.; Kuwamura, T. and Omori, Y.: Studies on the teratogenicity of food additives (4). Effects of food dye red no. 104 (phloxine) on the pre– and postnatal development in rats in relation to fetal distribution. J. Food Hyg. Soc. Jpn. 16:34–40, 1975.

Seno, M.; Fukuda, S. and Umisa, H.: A teratogenicity study of phloxine B in ICR mice. Food Chem. Toxicol. 22:55–60, 1984.

Uchida, Y. and Enomoto, N.: Action of red dye R104 on mice and E. Coli and their DNA. Proceedings of Annual Meeting. Jpn. Assoc. Agric. Chem. 46:1F–09, 1971.

1085 Food Dye and Coloring Red No. 105

Kanoh and Hori (1982) fed a diet of 2.5 percent dye to rats on days 8–21 of gestation. Slight fetal weight decrease occurred and 4 out of 55 fetuses had dilated lateral cerebral ventricles.

Kanoh, S. and Hori, Y.: Fetal toxicity of food red no. 105. Oyo Yakuri 24:391–397, 1982.

1086 Food Dye and Coloring Yellow No. 4

Kanoh et al. (1982) fed rats up to 5 percent dye in the diet on days 7–14 and observed no increase in defects but the neonatal death rate was increased.

Kanoh, S.; Ema, M. and Kawasaki, H.: Fetal toxicity of food yellow no. 4. Oyo Yakuri 24:399–404, 1982.

1087 Food Dye and Coloring Yellow No. 5 *Tartrazine*

Collins et al. (1992) fed rats in their drinking water up to 1,064 mg per kg during gestation and found no adverse fetal effects.

Collins, T.F.X.; Black, T.N.; O'Donnell Jr., M.W. and Bulhack, P.: Study of the teratogenic potential of fd&c yellow no. 5 when given in drinking–water. Fd. Chem. Toxic. 30:263–268, 1992.

1088 Formaldehyde CAS 50-00-0

Ma and Harris (1988) reviewed the lack of evidence for teratogenicity in experimental animals and comment that there is no evidence for embryo toxicity in the human. No significant association with maternal occupational exposure was found among 164 nurses who had spontaneous abortions or 34 nurses whose infants had anomalies (Hemminki et al., 1985).

Pushkina et al. (1968) exposed rats continuously during pregnancy to formaldehyde vapors (1 mg per cubic meter) and found no visible fetal malformations. The ascorbic acid content of the treated fetuses was lower than controls but the body weight was increased. The fetal DNA content was decreased and the RNA content was increased. Gofmekler (1968) found a 14–15 percent increase in the duration of rat gestation with exposure of 1 mg per cu. meter. Hurni and Ohder (1973) fed beagles 125 and 375 PPM in the diet from 4 through 56 days of gestation and found no reproductive or fetal changes.

Marks et al. (1981) gave mice formaldehyde or potentiated glutaraldehyde (Sonacide™) by gavage on days 6–15 of gestation. Maximum doses of formaldehyde (185 mg per kg) and glutaraldehyde (5.0 ml of a 2 percent solution per kg) did not produce increased malformations except in the group receiving the highest dose of glutaraldehyde which was lethal to a number of pregnant mice. Overman (1985) applied formaldehyde topically to the hamster on days 8, 9, 10 and 11 and found increased resorption rates in the hamster but no weight reduction or increased malformations in their fetuses. Sheveleva (1971) exposed rats for 4 hours per day to up to 4 ppm and made postnatal functional studies. Some decrease in activity of the one month old offspring was found.

Gofmekler, V.A.: Effect on embryonic development of benzene and formaldehyde in inhalation experiments. Gig. Sanit. 33:12–16, 1968.

Hemminki, K.; Kyyronen, P. and Lindbohm, M–L.: Spontaneous abortions and malformations in the offspring of nurses exposed to anaesthetic gases, cytostatic drugs and other potential hazards in hospitals, based on registered information of outcome. J. Epidemiol. Comm. Health 39:141–147, 1985.

Hurni, H. and Ohder, H.: Reproductive study of formaldehyde and hexamethylene tetramine in beagle dogs. Food Cosmet. Toxicol. 11:459–462, 1973.

Ma, T–H. and Harris, M.M.: Review of the genotoxicity of formaldehyde. Mutation Research 196:37–59, 1988.

Marks, T.A.; Worthy, W.C. and Staples, R.E.: Influence of formaldehyde and Sonacide™ (potentiated acid glutaraldehyde) on embryo and fetal development in mice. Teratology 22:51–58, 1981.

Overman, D.O.: Absence of embryotoxic effects of formaldehyde after percutaneous exposure in hamsters. Toxicol. Lett. 24:107–110, 1985.

Pushkina, N.N.; Gofmekler, V.A. and K'ertsova, G.N.: Changes in content of ascorbic acid and nucleic acids produced by benzene and formaldehyde. Bull. Exp. Biol. Med. (Russian) 66:868–869, 1968.

Sheveleva, G.A.: Specific action of formaldehyde on the embryogeny and progeny of white rats. (Russian) Toksikol. Nov. Prom. Khim. Veschestv. 12:78–86, 1971.

1089 Formamide CAS 75–12–7

Thiersch (1971) gave one ml intraperitoneally to rats on gestational days 11 through 16 and found 36 percent resorption. In addition, 46 percent of the survivors were stunted. Defects of the palate and extremities also occurred. Stula and Krauss (1977) exposed rats cutaneously to 600 mg per kg on days 12 and 13 and found 4 of 60 fetuses with subcutaneous hemorrhages. Exposure on day 7 did not produce a defect increase. Kennedy (1986) summarized the developmental toxicity of this chemical and its related congeners.

Kennedy, G.L.: Biological effects of acetamide, formamide and their monomethyl and dimethyl derivatives. CRC Critical Reviews in Toxicology 17(2): 129–182, 1986.

Stula, E.F. and Krauss, W.C.: Embryotoxicity in rats and rabbits from cutaneous application of amide–type solvents and substituted ureas. Toxicol. Appl. Pharmacol. 41:35–55, 1977.

Thiersch, J.B.: Investigations into the differential effect of compounds on rat litter and mother. In: Malformations Congenitales Des Mammiferes. H. Tuchmann–Duplessis (ed.), Paris: Masson, 95–113, 1971.

1090 Formhydroxamic Acid

Oral doses of 500–550 mg per kg to rats on day 13 produced cleft lip and palate (Pfeifer and von Kreybig, 1973).

Pfeifer, G. and von Kreybig, T.: Uber die Atiologie der Lippen–Kieferspaltformen und Gaumenspalten beim Menschen und im Tier–experiment. Experentia 29:225–228, 1973.

1091 Formic Acid

Ebron–McCoy et al. (1994) studied the toxicity of formic acid and formate in whole embryo cultures of rats and mice. Some growth reduction was seen at 2 and 3 mg per ml.

Ebron–McCoy, M.T.; Nichols, H.P. and Andrews, J.E.: Effects of altered pH of culture medium on embryogenesis in WEC. (abs) Teratology 49:393, 1994.

1092 Formoterol Fumarate CAS 43229–80–7

Sato et al. (1984) studied this bronchodilator in rats and rabbits using oral doses of 60 and 600 mg, respectively.

No adverse fertility, teratogenicity or postnatal functional changes were noted in the rats. In rabbits the 500 mg dose level on days 6–18 caused some reduction in fetal viability but no malformations.

Sato, T.; Kaneko, Y. and Saegusa, T.: Reproductive studies with formoterol fumarate in rats and rabbits. Oyo Yakuri 27:239–265 and 375–385, 1984.

1093 Formylglycin

Pfeifer and von Kreybig (1973) did not find teratogenicity in rat offspring exposed to 3000 mg per kg on day 13.

Pfeifer, G. and von Kreybig, T.: Uber die Atiologie der Lippen–Kieferspaltformen und Gaumenspalten beim Menschen und im Tier–experiment. Experentia 29:225–228, 1973.

1094 N–Formylpiperidine

Nair et al. (**1992**) gavaged rats with this solvent using 110, 220, or 440 mg per kg on days 6–20. No significant increase in defects occurred but fetal death and growth retardation were present after the two highest doses which also produced maternal toxicity.

Nair, R.S.; Alvarez, L. and Johannsen, F.R.: Evaluation of teratogenic potential of n–formylpiperidine in rats. Fundamental and Applied Toxicology 18:96–101, 1992.

1095 Forskolin

This potent cardio–activating agent did not produce cardiac defects in chick embryos but potentiated the cardiac teratogenicity of ephedrine in this system. (Nishikawa et al., **1991**)

Nishikawa, T.; Kasajima, T. and Kanai, T.: Potentiating effects of forskolin on the cardiovascular teratogenicity of ephedrine in chick embryos. Toxicology Letters 56:145–150, 1991.

1096 Fosfomycin CAS 23155–02–4

Koeda and Moriguchi (1979, 1980) gave this antibiotic intraperitoneally during organogenesis to rats and rabbits at maximum daily doses of up to 150 mg and 800 mg per kg, respectively. Maternal and fetal toxicity was seen at the highest doses but no increase in defects or decrease in postnatal performance was found. Male and female rats treated before mating and during the first 7 days of gestation with up to 1500 mg per kg were found to have normal fertility. Perinatal studies were also done and no differences from controls were found. Similar studies were done using fosfomycin–calcium orally. The rats received up to 14,000 mg per kg and the rabbits 420 mg per kg daily during organogenesis. No adverse fetal effects were found. Fertility and perinatal and postnatal studies were carried out without finding any adverse effects except for a reduction of fetal weight and survival at the 2,800 mg per kg dose level.

Koeda, T. and Moriguchi, M.: Effects of fosfomycin–Na on reproduction of rats and rabbits. Jpn. J. Antibiot. 32:155–163, 164–170, 171–179, 1979.

Koeda, K. and Moriguchi, M.: Effect of fosfomycin–calcium on reproduction of the rat and rabbit. Jpn. J. Antibiot. 32:546–554, 1979; 33:613–617, 733–737, 478–486, 1980.

1097 FRC–8653 *2–Methoxyethyl(E)–3–phenyl–2 propen–1–yl (+–)–1, r–dihydro–2, 6 dimethyl–4–(3–nitrophenyl) pyridine–3, 5–dicarboxylate*

Oguihari et al. (**1992**) gave this calcium antagonist to rats before they reproduced and in the early stages of gestation. Oral doses of up to 200 mg per kg were used. At 25 mg per kg there was increased fetal loss. Tateda et al. (**1992 A,B**) found fetotoxicity at doses of 50–200 mg per kg during organogenesis but no fetal ill effects were found at 25 mg per kg. Peri and postnatal studies using up to 24 mg per kg produced no abnormal behavior (Wada et al. **1992**). Shibano et al. (**1992**) gave rabbits up to 400 mg per kg during organogenesis and found no ill effects in the dams or fetuses.

Oguihara, S.; Yokoi, O. and Shyioya, S.: Study on oral administration of FRC–8653 to rats prior to and in the early stages of pregnancy. Yakuri to Chiryo 20:s1905–s1924, 1992.

Shibano, T.; Sakai, Y.; Yamamoto, M.; Tsubuku, S. and Shioya, S.: Study on oral administration of FRC–8653 to rabbits during the period of organogenesis. Yakuri to Chiryo 20:s1963–s1973, 1992.

Tateda, C.; Kiwaki, S.; Iwadate, K.; Ohno, S.; Yamamae, H. and Hayashi, Y.: Study on oral administration of FRC–8653 to rats during the period of organogenesis. Yakuri tp Chiryo 20:s1925–s1943, 1992A.

Tateda, C.; Yamashita, Y.; Sakai, K.; Yamamae, H.; Kiwaki, S. and Hayashi, Y.: Study on oral administration of FRC–8653 to rats during the period of organogenesis (additional study). Yakuri to Chiryo 20:s1945–s1961, 1992B.

Wada, S.; Akamatsu, H.; Taniguchi, N.; Shikuma, H. and Fukuyama, M.: Study on oral administration of FRC–8653 to rats during the period of perinatal and lactation. Yakuri to Chiryo 20:s1975–1988, 1992.

1098 Freezing

Kola et al. (1988) studied the effect of freezing, vitrification and storage of mouse oocytes on survival after transfer to the uterus. Only 5 percent of the vitrified oocytes formed viable fetuses by day 15 as compared to the controls which had a 47 percent survival rate. Chromosomal aneuploidy was increased threefold in both the group vitrified as well as those that were DMSO slow–frozen. Six to ten percent of the vitrified embryos had morphologic defects including neural tube defects while no defects were found among 46 embryos after slow–cooling.

Kola, I.; Kirby, C.; Shaw, J.; Davey, A. and Trounson, A.: Vitrification of mouse oocytes results in aneuploid zygotes and malformed fetuses. Teratology 38:467–474, 1988.

1099 Fropenem Sodium *SY5555*

Okamoto et al. (**1994**) gave this antibiotic to rabbits intravenously to rabbits in doses up to 200 mg per kg during organogenesis. At 100 and 200 mg per kg some dams died with severe diarrhea. No ill effects other than reduced fetal weight were found in the offspring. Muller et al. (**1994**) tested this antibiotic in rats before fertilization, during organogenesis and perinatally. No ill effects on reproduction were found at 1,620 or 2,000 mg per kg doses.

Muller, W.; Osterburg, I and Korte, R.: SY555. Chemotherapy 42:161–173, 1994.

Okamoto, M.; Nakanishi, Y. and Otaka, T.: Intravenous teratogenicity study of fropenem sodium (SY5555) in rabbits. Oyo Yakuri 47:139–145, 1994.

1100 Fumagillin CAS 23110–15–8

This antibiotic produced complete litter destruction in the rat when given on the 7th gestational day in amounts of 25 mg per kg (Thiersch, 1971). Treatment on the 11th day caused 68 percent of the survivors to be stunted.

Thiersch, J.B.: Investigations into the differential effect of compounds on rat litter and mother. In: Malformations Congenitales Des Mammiferes. H. Tuchmann–Duplessis (ed.), Paris: Masson, 95–113, 1971.

1101 Furazolidone

Heinonen et al. (1977) found no increase in defects among the offspring of 132 women treated in the first four lunar months.

Jackson and Robson (1957) administered 750 mg per kg to mice on the 7th day and interrupted pregnancy. Doses of up to 1,250 mg per kg during days 10–11 did not increase malformations but fetal loss was increased. Greenaway et al. (1986) using in vitro culture of rat embryos found axial asymmetry and growth retardation at concentrations of 0.01 millimoles.

Greenaway, J.C.; Fantel, A.G. and Juchau, M.R.: On the capacity of nitroheterocyclic compounds to elicit an unusual axial asymmetry in cultured rat embryos. Toxicol. Appl. Pharmacol. 82:307–315, 1986.

Heinonen, O.P.; Slone, D. and Shapiro, S.: Birth Defects and Drugs in Pregnancy. Publishing Sciences Group Inc., Littleton, Mass., 1977.

Jackson, D. and Robson, J.M.: The action of furazolidone on pregnancy. J. Endocrin. 15:355–359, 1957.

1102 Furbiprofen *2–(2–Fluoro–4–biphenylyl)proprionic Acid* CAS 51543–40–9

Yoshinaka et al. (1976) studied the teratogenicity of this compound in rats and rabbits giving oral doses on days 9 through 14 (rat) and 8 through 17 (rabbit). The maximum daily dose per kg was 10 mg in the rat and 25.5 mg in the rabbit. The only positive finding was an increase in skeletal variation (hypoplasia of sternebrae) in the rat fetuses.

Yoshinaka, I.; Saito, K.; Hikida, S.; Komori, S.; Okuda, T.; Matubara, T.; Moriji, H. and Saito, H.: Studies on toxicity of FP–70 2. Teratogenicity test in rats and rabbits. (Japanese) Clinical Report 10:1890–1915, 1976.

1103 Furosemide CAS 54–31–9

Christianson and Page (1976) analyzed 4,035 pregnancies where diuretics were used and found increased uterine inertia, birth weight and meconium staining. Whether the complications were due to the edema condition or treatment could not be determined. Only 8 percent took the diuretic before the 26th week. Briggs et al. (1983) in a review of the pharmacology of furosemide during pregnancy, found no reports of fetal or newborn adverse effects.

This diuretic was given twice daily in amounts of 37.5 to 300 mg per kg to rats on days 6–17 of gestation (Robertson et al., 1981). Decreased fetal weight was seen at 150 and 300 mg levels. Wavy ribs and some skeletal defects were found at these levels. These defects were not seen when KCl was given concurrently. Nakatsuka, et al. (**1993**) found that sodium bicarbonate in the drinking water increased wavy ribs while ammonium chloride reduced it.

Briggs, G.G.; Bodendorfer, T.W.; Freeman, R.K. and Yaffe, S.J.: Drugs in Pregnancy and Lactation. Williams and Wilkins, Baltimore, 1983.

Christianson, R. and Page, E.W.: Diuretic drugs and pregnancy. Obstet. Gynecol. 48:647–652, 1976.

Nakatsuka, T.; Fujikake, N.; Hasebe, M. and Ikeda, H.: Effects of sodium bicarbonate and ammonium chloride on the incidence of furosemide–induced fetal skeletal anomaly, wavy rib, in rats. Teratology 48:139–147, 1993.

Robertson, R.T.; Minsker, D.H.; Bokelman, D.L.; Durand, G. and Conquet, P.: Potassium loss as a causative factor for skeletal malformations in rats produced by indacrinone: A new investigational loop diuretic. Toxicol. Appl. Pharmacol. 60:142–150, 1981.

1104 Fursultiamine *Thiamine Tetrahydrofurfuryl Disulfide TTFD* CAS 804–30–8

Mizutani et al. (1972) fed this compound to pregnant monkeys and rabbits in maximum doses of 300 and 500 mg per kg, respectively, during organogenesis and found no increase in fetal malformations.

Mizutani, M.; Ihara, T. and Kaziwara, K.: Effects of orally administered thiamine tetrahydrofurfuryl disulfide on foetal development of rabbits and monkeys. Jpn. J. Pharmacol. 22:115–124, 1972.

1105 Furterene *Triamino–2,4,7–Furyl–2)–6 pteridine*

Nouvel and David (**1966**) studied this diuretic in pregnant mice, rats, rabbits and dogs during most of organogenesis. The animals were given oral doses of 200 mg (mouse & rat), 100 mg (rabbit) and 33 mg per kg (dogs). No malformations were observed.

Nouvel, P.G. and David, J.: Un nouveau diuretique de la serie des aminopteridines: Le Furterene. Therapie 21:1317–1326, 1966.

1106 Furylfuramide
2–(2–Furyl)–3–(5–nitro–2–furyl)acrylamide CAS 3688–53–7

Miyaji (1971) fed pregnant mice a diet containing 0.2 percent of this compound after the 7th day of mating and observed no increase in malformations. The fertility of males was diminished in a three–generation test using a diet with 0.0125 percent furylfuramide. Tanaka et al. (1977) found fetal growth retardation at their highest dietary exposure (0.6 percent) but no teratogenicity.

Miyaji, T.: Effect of furylfuramide on reproduction and malformation. Tohoku J. Exp. Med. 103:381–388, 1971.

Tanaka, S.; Onoda, K–I.; Kawashima, K.; Nakaura, S.; Nagao, S.; Kuwamura, T. and Omori, Y.: Studies on the teratogenicity of furylfuramide in fetuses and offspring of rats in relation to fetal distribution. J. Toxicol. Sci. 2:149–159, 1977.

1107 Fusarenone–X

Ito et al. (1980) treated mice with oral and subcutaneous routes. At 5, 10 and 20 ppm in the diet inhibition of implantation was found. Using seven–day periods during embryogenesis they found resorptions at all dose levels. At 0.6 to 1.6 subcutaneously during organogenesis, body weight reduction of the fetus occurred but no teratogenicity.

Ito, Y.; Ohtsubo, K. and Saito, M.: Effects of Fusarenon–X, a trichothecene produced by fusarium nivale, on pregnant mice and their fetuses. Jpn. J. Exp. Med. 50:167–172, 1980.

1108 Fusaric Acid–Ca CAS 21813–99–0

Matsuzaki et al. (1976, 1977) did teratologic studies in mice, rats and rabbits. In the mouse and rat, a maximum daily dose of 125 mg per kg orally caused no defects but delay in ossification and growth of the mouse fetuses was found. At the same dose, no effects were found in the rabbit.

Matsuzaki, A.; Akutsu, S.; Mukaikawa, H. and Shimamura, T.: Studies on the possible teratogenicity of fusaric acid–Ca. 3. Effects on the mouse and rat embryo. Jpn. J. Antibiot. 5:543–551, 1976.

Matsuzaki, A.; Akutsu, S.; Shimamura, T. and Nakatani, H.: Studies on the possible teratogenicity of fusaric acid–Ca. 3. Effects on the embryos of oral administration to pregnant rabbits. Jpn. J. Antibiot. 30:321–333, 1977.

1109 Gabexate *Ethyl p–(6–guanidinohexanoyloxy) Benzoate Methanesulfonate* CAS 39492–01–8; CAS 56974–61–9

Fujita et al. (1975) studied this chemical in pregnant mice and rats giving up to 100 mg per kg to mice and 30 mg per kg to rats during organogenesis. No teratogenic activity was detected.

Fujita, T.; Suzuki, Y.; Yamamoto, Y.; Yokohama, H.; Yonezawa, H.; Ozeki, Y.; Mori, T. and Matsuoka, Y.: Toxicities and teratogenicity of ethyl p–(6–guanidinohexanyloxy) benzoate methanesulfate (FOY). (Japanese) Oyo Yakuri 9:743–760, 1975.

1110 Galactoflavin also see *Riboflavin Deficiency* CAS 5735–19–3

This analogue of riboflavin was used to augment the riboflavin–deficient state in experimental animal models (Nelson et al., 1956). The usual dose of galactoflavin is 60 mg per kg of riboflavin–deficient diet, and this combination given during the entire gestation produces in the rat fetus a syndrome characterized by growth retardation, hypoplasia of mandible and reduction defects of the extremities. Hydronephrosis (26 percent), cleft palate (34 percent) and subcutaneous edema (30 percent) are frequently observed (Shepard et al., 1968). The exact mechanism by which galactoflavin and riboflavin deficiency interfere with the development of the terminal electron transport system is not known (Aksu et al., 1968), however, if 600 mg of riboflavin is added to the deficient diet with galactoflavin, the anomalies are prevented.

Aksu, O.; Mackler, B.; Shepard, T.H. and Lemire, R.J.: Studies of the development of congenital anomalies in embryos of riboflavin–deficient galactoflavin fed rats. II. Role of the terminal electron transport systems. Teratology 1:93–102, 1968.

Nelson, M.M.; Baird, C.D.C.; Wright, H.V. and Evans, H.M.: Multiple congenital abnormalities in the rat resulting from riboflavin deficiency induced by the antimetabolite galactoflavin. J. Nutr. 58:125–134, 1956.

Shepard, T.H.; Lemire, R.J.; Aksu, O. and Mackler, B.: Studies of the development of congenital anomalies in embryos of riboflavin–deficient galactoflavin fed rats. I. Growth and embryonic pathology. Teratology 1:75–92, 1968.

1111 Galactose CAS 59–23–4

Chen et al. (**1981**) fed a diet with 50% galactose to pregnant rats and found oocyte reduction in the offspring. The premiotic phase was the most sensitive time for exposure. Premature ovarian failure has been observed in galactosemics even when dietary treatment is started early.

Evidence that galactose toxicity reduces the weight and the content of protein and DNA of the fetal brain was given by Haworth and Ford (1973).

Bannon et al. (1945) fed a 25 percent galactose diet to rats and found cataracts starting on day 15 of gestation. Demeyer (1959) fed rats 55 percent galactose diets and produced anophthalmia and lens abnormalities in the fetuses. The most sensitive period was the 8th, 9th and 10th days of gestation. Segal and Bernstein (1963) demonstrated that galactose crosses the placenta of the rat.

Bannon, S.L.; Higginbottom, R.M.; McConnell, J.M. and Kaan, H.W.: Development of the galactose cataract in the albino rat. Arch. Ophthalmol. 33:224–228, 1945.

Chen, Y.; Mattison, D.R.; Feigenbaum, L.; Fukui, H. and Schulman, J.D.: Reduction in oocyte number following prenatal exposure to a diet high in galactose. Science 214:1145–1147, 1981.

Demeyer, R.: Action teratogene du galactose administre a la rate gravide. Ann. Endocrinol. (Paris) 20:203–211, 1959.

Haworth, J.C. and Ford, J.D.: Effect of galactose toxicity on incorporation of tritiated thymidine into fetal brain. Brain Res. 63:470–473, 1973.

Segal, S. and Bernstein, H.: Observations on cataract formation in the newborn offspring of rats fed a high galactose diet. J. Pediatr. 62:363–370, 1963.

1112 Gallamine Triethiodide CAS 65–29–2

Jacobs (1971) using 8–64 mg per kg on day 13.5 of gestation, produced flaccid paralysis in the rat fetus but no cleft palates resulted.

Jacobs, R.M.: Failure of muscle relaxants to produce cleft palate in mice. Teratology 4:25–30, 1971.

1113 Gallium CAS 7440–55–3

Ferm and Carpenter (1970) injected 40 mg of sodium sulfate per kg intravenously into hamsters on day 8 of gestation and found no increase in malformation rate in 237 fetuses. Mast et al. (**1991**) exposed mice and rats to gallium arsenide by inhilation for 6 hours on days 4–17. Decreased fetal rat weight and maternal toxicity were found at 37 mg per m^3 but no teratogenicity was found in either species. In mice fetal weight decreases were found at 10, 37 and 75 mg per square meters.

Ferm, V.H. and Carpenter, S.J.: Teratogenic and embryopathic effects of indium, gallium and germanium. Toxicol. Appl. Pharmacol. 16:166–170, 1970.

Mast, T.J.; Dill, J.A.; Greenspan, B.J.; Evanoff, J.J.; Morrissey, R.E. and Schwetz, B.A.: The developmental toxicity of inhaled gallium arsenide in rodents. (Abs) Teratology 43:455–456, 1991.

1114 Gallium Nitrate

Gomez et al. (**1992**) treated mice intraperitoneally with up to 100 mg per kg on days 6,8,10,12 and 14. At 50 mg per kg and below no adverse effects were seen in the fetus but at 100 mg per kg cleft palate and renal hypoplasia were increased. Stunting was increased at 50 mg per kg.

Gomez, M.; Sanchez, D.J.; Domingo, J.L. and Corbella, J.: Developmental toxicity evaluation of gallium nitrate in mice. Arch. Toxicol. 66:188–192, 1992.

1115 Gamma Irradiation of Diet

Takeshita et al. (**1993**) fed rats diets that were sterilized by 50 kGy electron radiation or gamma irradiation. No reproductive or teratogenic effects were found in either group.

Takeshita, M.; Shibatani, M.; Kobayashi, T.; Imahie, H.; Susami, M. and Ariyuki, F.: Comparison of electron irradiation diets and gamma irradiation diets for reproductive effects on rats. Exp. Anim. 42:405–410, 1993.

1116 Gas Fuel

Discussed under Methane

1117 Gasoline

Two children with profound mental retardation and neurological abnormalities were reported following pregnancies during which gasoline was inhaled recreationally (Hunter et al., 1979).

Schreiner (1983) reported "negative" results after exposing rats to 400 or 1600 ppm of unleaded gasoline. The exposures were from day 6 through 15 and for 6 hours daily.

Hunter, A.G.W.; Thompson, D. and Evans, J.A.: Is there a fetal gasoline syndrome? Teratology 20:75–80, 1979.

Schreiner, C.A.: Petroleum and petroleum products: A brief review of studies to evaluate reproductive effects. Environmental and Health Sciences Laboratory, Mobil Oil Corporation. In: Assessment of Reproductive and Teratogenic Hazards, volume III. M.S. Christian, W.M. Galbraith, P. Voytek, M.A. Mehlman (eds), Princeton Scientific Publishers Inc., 29–46, 1983.

1118 Gastric Bypass

Haddow et al. (1986) reported three pregnancies with neural tube defects from women with gastrojejunostomy bypasses. All occurred in women living in Maine where an estimated 261 women underwent the procedure during a four year period. All three women had decreased vitamin B_{12} levels at various times post–surgically. Martin et al. (1988) identified 908 women who had gastric by–pass surgery for obesity. Eighty–seven had a pregnancy after the operation and two infants had neural tube defects (a twelve–fold increased expectancy). Knudsen and Kallen (1986) studied 77 women who had gastric bypass operations prior to their pregnancies. Two defects occurred: One with limb reduction and the other with cleft lip and palate. Short gestation, growth retardation and increased low birth weight was found.

Haddow, J.E.; Hill, L.E.; Kloza, E.M. and Thanhauser, D.: Neural tube defects after gastric bypass. Lancet 1:1330, 1986.

Knudsen, L.B. and Kallen, B.: Intestinal bypass operation and pregnancy outcome. Acta. Obstet. Gynecol. Scand. 65:831–834, 1986.

Martin, L.; Chavez, G.F.; Adams, M.J.; Mason, E.E.; Hanson, J.W.; Haddow, J.E. and Currier, R.W.: Gastric bypass surgery as maternal risk factor for neural tube defects. Lancet 1:640–641, 1988.

1119 Gemcadiol 2,2,9,9–Tetramethyl–1,10–decanediol

Fitzgerald et al. (1986) gave rats up to 200 mg per kg on days 6–15 and rabbits up to 150 mg per kg on days 6–18 and found no teratogenicity.

Fitzgerald, J.E.; Petrere, J.A.; McGuire, E.J. and De La Iglesia, F.A.: Preclinical toxicology studies with the lipid–regulating agent gemcadiol. Fund. Appl. Toxicol. 6:520–531, 1986.

1120 Gemcitabine

This antimetabolite was studied in mice given 0.05, 0.25 or 1.5 mg per kg intravenously on days 6–15 (Eudaly, et al. 1993). At the highest dose runting and congenital defects (mostly cleft palates) occurred. Post natal studies of behavior and reproduction were not altered but startle amplitudes in males and ovarian weights were reduced.

Eudaly, J.A.; Tizzano, J.P.; Higdon, G.L. and Todd, G.C.: Developmental toxicity of gemcitabine, an antimetabolite oncolytic, administered during gestation to CD–1 mice. Teratology 48:365–381, 1993.

1121 Gemfibrozil CAS 25812–30–0

5–(2,5–Dimethylphenoxy)–2,2–dimethylpentanoic Acid

Rosa (1994) reported only one defect among the offspring of 15 women taking this medication.

Fitzgerald et al. (1987) administered up to 200 mg per kg in the diet of rats or rabbits before fertilization during organogenesis or in the perinatal period. In the perinatal studies, some reduction in weight gain occurred but no adverse outcomes were found.

Kurtz et al. (1976) gave 81 and 28 mg per kg to rats on day 6–15 and except for some reduction in implantation no teratologic findings were detected. Rabbits treated with 200 mg per kg orally on days 6–18 had normal fetuses.

Fitzgerald, J.E.; Petrere, J.A. and De La Iglesia, F.A.: Kurtz, S.M.; Fitzgerald, J.E.; Fisken, R.A.; Schardein, J.L.; Reutner, T.H. and Lucas, J.A.: Toxicological studies on Gemfibrozil. Proc. roy. Soc. Med. 69:15–23, 1976.

Rosa, F.: Anti–cholesterol agent pregnancy exposure outcomes. (abs.) Reproductive Toxicology 8:445–446, 1994.

1122 Gentamicin CAS 1403–66–3; CAS 1405–41–0

Chahoud et al. (1986) treated rats subcutaneously with gentamicin on either days 10–15 or days 15–20. In the female offspring of those treated during both periods there was a significant increase in blood pressure after one year. Mallie et al. (1986) gave 75 mg per kg to rats on days 7–11 and found cortical thinning and reduced glomeruli in the neonates. Lelievre–Pegorier et al. (1987) made detailed histologic exams of the kidneys of intrauterine exposed guinea pigs. Gilbert et al. (1987) gave rats 75 mg per kg from day 10 to term and found that the number of nephrons at birth were decreased by 20 percent. Focal tubular changes were found. Birth weight was also reduced in the fetuses. The defect rate was not mentioned.

Nishio et al. (1987) gave mice 1, 10 and 100 mg per kg on days 6–10. Fetal loss was higher in the 10 and 100 mg level groups. Some defect increase was found at 10 mg per kg but it was not significant statistically. The renal weights were not reduced.

Chahoud, I.; Stahlmann, R. and Neubert, D.: Blood pressure elevation in one–year–old female rats after prenatal exposure to gentamicin. (abs) Teratology 33:13A, 1986.

Gilbert, T.; Lelievre–Pegorier, M.; Malienou, R.; Meulemans, A.; Merlet–Benichou, C.M.: Effects of prenatal and postnatal exposure to gentamicin on renal differentiation in the rat. Toxicology 43:301–313, 1987.

Lelievre–Pegorier, M.; Gilbert, T.; Sakly, R.; Meulemans, A. and Merlet–Benichou, C.M.: Effect of fetal exposure to gentamicin on kidneys of young guinea pigs. Antimicrobial Agents and Chemotherapy. 31:33–92, 1987.

Mallie, J.P.; Gerard, H. and Gerard, A.: In–utero gentamicin–induced nephrotoxicity in rats. Pediatr. Pharmacol. 5:229–239, 1986.

Nishio, A.; Ryo, S. and Miyao, N.: Effects of gentamicin, exotoxin and their combination on pregnant mice. The Bulletin of the Faculty of Agriculture, Kagoshima University 37:129–136, 1987.

1123 Gentian Violet

Heinonen et al. (1977) found no increase in congenital defects among the offspring of 40 women treated topically during the first four lunar months.

No teratogenicity in the offspring of rats was found when they were given 5.0 mg per kg. Renal and skeletal defects were increased at the maternal toxic day of 10 mg per kg (Kimmel et al., 1986). Vaginal tampons used by women contain only 5 mg of the dye.

Heinonen, O.P.; Slone, D. and Shapiro, S.: Birth Defects and Drugs in Pregnancy. Publishing Sciences Group Inc., Littleton, Mass., 1977.

Kimmel, C.A.; Price, C.J.; Tyl, R.W.; Ledoux, T.A.; Reel, J.R. and Marr, M.C.: Developmental toxicity of gentian violet (GV) in rats and rabbits. Teratology 33:P48, 1986.

1124 Germanium Trioxide CAS 7440–56–4

Ferm and Carpenter (1970) injected hamsters intravenously on day 8 of pregnancy with 40 or 100 mg per kg and reported no increase in congenital defects.

Ferm, V.H. and Carpenter, S.J.: Teratogenic and embryopathic effects of indium, gallium, and germanium. Toxicol. Appl. Pharmacol. 16:166–170, 1970.

1125 Gliclazide *Sulfonylurea Gliclazide* CAS 21187–98–4

Kawanishi et al. (1981) administered this new antidiabetic to rats and rabbits orally in maximum doses of 800 and 120 mg per kg, respectively. No adverse fetal effects were found except for fetal weight reduction in the rat at 400 and 800 mg per kg. Maternal weight was also reduced. Learning ability and reproductive function of rat offspring was not affected after dosing on days 17 of gestation through day 21 postnatally.

Kawanishi, H.; Takeshima, T.; Igarashi, N. and Tauchi, K.: Reproductive studies of gliclazide a new sulfonylurea antidiabetic agent. Yakuri to Chiryo 9:3551–3571, 1981.

1126 Glimepiride

Baeder (1993) gave this antidiabetic to rats orally in amounts up to 2,500 mg per kg during pregnancy and organogenesis and found no maternal or fetal drug–related ill effects. In rabbits hypoglycemia occurred at 0.0067 mg per kg and above. At 0.32 mg per kg there was a small increase in lenticular aplasia and skeletal effects.

Baeder, C.: Embryotoxicilogical/teratological investigation, including effects on postnatal development, of the new oral antidiabetic glimepiride after oral administration to rats and rabbits. The Clinical Report 27:1477–1492, 1993.

1127 Gliquidone *ARDF 26 SE* CAS 33342–05–1

Iida et al. (1976) treated pregnant rats and rabbits orally with this oral sulfonylurea antidiabetic medication. Doses up to 2500 mg and 250 mg per kg were used in the rat and rabbit, respectively. No adverse effects were found in rats receiving gavaged feedings on days 9–14. Resorptions

were increased in the rabbit at 10 mg per kg and above. No significant increase in defects was reported.

Iida, H.; Kast, A. and Tsunenari, Y.: Studies on the teratogenicity of a new sulfonylurea derivative (ARDF 26 SE) on rats and rabbits. (Pharmacometrics) Oyo Yakuri 11:119–131, 1976.

1128 Glisoxepid *1–(Hexahydro–1–H–azepin–1–yl)–3 p–(2(5–methyl–isoxazol–3–carbox-amido)ethyl)phenyl-sulfonyl) Acid* CAS 25046–79–1

Tettenborn (1974) studied this antidiabetic compound in pregnant mice, rats and rabbits during organogenesis. Maximum daily doses of 400, 100 and 125 mg per kg in the mouse, rat and rabbit, respectively, had no adverse effect on the offspring.

Tettenborn, D.: Zur Toxikologie von Glisoxepid, einem neuen oralen Antidiabetikum. Arzneim–Forsch. 24:409–419, 1974.

1129 γ–Globulin, Human, treated with Polyethylene Glycol

Saito et al. (1984) did studies on this solution using up to 500 mg per kg in the rat and 200 mg per kg in the rabbit. No adverse reproductive events were noted when it was administered before fertilization to males and females, during organogenesis or in the perinatal period of rats. The rabbits were treated during organogenesis and no adverse effects were noted. In all studies, the intravenous route was used.

Saito, M.; Narama, I.; Satoh, T.; Kaneko, Y. and Naito, Y.: Reproduction studies in rats and rabbits of polyethylene glycol treated immunoglobulin. Oyo Yakuri 27:63–72, 173–198, 1984.

1130 Glucagon CAS 9007–92–5

Scaglione (1960) reported production of fetal cataracts after daily administration of 20 or 200 microgm of glucagon to pregnant rats. Tuchmann–Duplessis and Mercier–Parot (1962) produced glaucoma in fetal rats by injecting 300 microgm on the 7th, 8th and 9th days. When the dosage was increased to 400 or 500 microgm, microphthalmia and defects of the skeleton were found.

Scaglione, S.: Recherches sur l'action du cortisone de l'acth, du glucagone et de l'iode (131) sur les embryons des rates pleines. Folia Hered. Pathol. 9:143–150, 1960.

Tuchmann–Duplessis, H. and Mercier–Parot, L.: Production de malformations congenitales chez le rat traite par le glucagon. C. R. Acad. Sci. (Paris) 254:2655–2657, 1962.

1131 Glucose CAS 50–99–7

Discussed under Hyperglycemia

1132 β–L–Glutamyl–amino Propionitrile

This compound is an active principle of lathyrus odoratus seeds (Schilling and Strong, 1954) and produces the lathyrus syndrome in mammalian fetuses (see Lathyrism).

Schilling, E.D. and Strong, F.M.: Isolation, structure and synthesis of a lathyrus factor from l. Odoratus. J. Am. Chem. Soc. 76:2848 only, 1954.

1133 Glufonsinate

Ebert et al. (**1990**) studied this herbicide in rats and rabbits at oral doses of up to 250 mg or 20 mg per kg daily during organogenesis. At the highest doses there was maternal and fetal toxicity but no increase in malformations.

Ebert, W.; Leist, K.H. and Mayer, D.: Summary of safety evaluation toxicity studies of glufosinate ammonium. Fd. Chem. Toxic. 28:339–349, 1990.

1134 Glutaraldehyde CAS 111–30–8

Marks et al. (1980) using Sonacide™ , a potentiated acid glutaraldehyde, gave mice the equivalent of 100 mg per kg on days 6–15. Although this dose was lethal in 19 of 35 dams, no evidence of teratogenicity was detected.

Marks, T.A.; Worthy, W.C. and Staples, R.E.: Influence of formaldehyde and Sonacide™ (potentiated acid glutaraldehyde) on embryo and fetal development of mice. Teratology 22:51–58, 1980.

1135 Glutethimide CAS 77–21–4

McColl et al. (1963) reported that a diet containing 0.4 percent glutethimide fed during pregnancy failed to produce defects in the rat fetus. Double vertebral centra which occurred in 4 of 17 fetuses represent probably only a delay in ossification. Tuchmann–Duplessis and Mercier–Parot (1963) produced resorptions but no congenital defects in the rabbit, mouse and rat.

Kotin and Ignatyeva (1982) injected pregnant rats with 200–400 mg per kg on days 14 and 15 and observed some behavioral abnormalities in the first progeny of animals under test.

Kotin, A.M. and Ignatyeva, T.V.: Variation in rat behavior after exposure to glutethimide during antenatal neurogenesis. Farmakol. Toksikol. (USSR) 4:73–78, 1982.

McColl, J.D.; Globus, M. and Robinson, S.: Drug induced skeletal malformations in the rat. Experientia 19:183–184, 1963.

Tuchmann–Duplessis, H. and Mercier–Parot, L.: Repercussion d'un somnifere, le glutethimide, sur la gestation et le developpement foetal du rat, de la souris et du lapin. C. R. Acad. Sci. (Paris) 256:1841–1843, 1963.

1136 Glyburide N–[4–(2–(5–Chloro–2–methoxybenzamido)–ethylphenylsulfonyl] N–Cyclohyexylurea; HB 419 CAS 10238–21–8

Miyamoto et al. (1970) fed this compound to pregnant rats in maximal doses of 2,000 mg per kg per day and found no fetotoxicity or teratogenicity. The rats were fed from day 6–15 and the mice from day 4–13 of gestation.

This oral antidiabetic drug was tested in mice, rats and rabbits with negative results (Baeder and Sakaguchi, 1969). Doses up to 350 mg per kg per day were used.

Baeder, C. and Sakaguchi, T.: Teratologische Untersuchungen mit HB 419. Arzneim. Forsch. 19:1419–1420, 1969.

Miyamoto, M.; Ohtsu, M.; Kumai, M.; Takayama, K. and Sakaguchi, T.: Influence of N–[4–(2–(5–chloro–2–methoxybenzamido)–ethyl)phenylsufonl] N–cyclohexylurea

(HB–419) on embryos. (Japanese) Oyo Yakuri 4:271–283, 1970.

1137 Glyceraldehyde

Slott and Hales (1985) injected 10, 100 or 250 microgm directly into day 13 amniotic sacs and found embryolethality at the highest dose.

Slott, V.L. and Hales, B.F.: Teratogenicity and embryolethality of acrolein and structurally related compounds in rats. Teratology 32:65–72, 1985.

1138 Glycerol Formal

Aliverti et al. (1980) gave rats 0.25, 0.50 or 1.00 ml per kg intramuscularly, subcutaneously or orally on days 6 through 15. Malformations (septal defects and wavy ribs) were found at the .50 and 1.0 ml levels. Fetal death and growth retardation also occurred.

Aliverti, V.; Bonanomi, L.; Giavini, E.; Leone, V.G. and Mariani, L.: Effects of glycerol formal on embryonic development in the rat. Toxicol. Appl. Pharmacol. 56:93–100, 1980.

1139 Glycidol

Slott and Hales (1985) injected 10, 100 or 1000 microgm directly into day 13 rat amniotic sacs and found an increase in defects at 1000 microgm. The defects were of the limbs and low set ears.

Slott, V.L. and Hales, B.F.: Teratogenicity and embryolethality of acrolein and structurally related compounds in rats. Teratology 32:65–72, 1985.

1140 γ–Glycidoxypropyltrimethoxysilane

This chemical was given by gavage to rats on days 6–15 and no adverse reproductive effects occurred at up to 1,000 mg per kg daily (Siddiquai and Hobbs, 1984).

Siddiquai, W.H. and Hobbs, E.J.: Teratological evaluation of γ–glycidopropyltrimethoxysilane in rats. Toxicology 31:1–8, 1984.

1141 Glycodiazine

2–Benzenesulfonamido–5(β–methyloxy–ethoxy)pyrimidine

Kramer et al. (1964) found no teratogenic action in rats given 200 mg per kg in various six–day periods during gestation.

Kramer, V.M.; Hecht, G.; Gunzel, P.; Harwart, A.; Richier, K.D. and Gloxhuber, Ch.: Vertraglichkeit von 2–Benzolsulfonamido–5(β–methoxy–athoxy)–pyrimidin (Glycodiazin) bei langdauernder Verabreichung im Tierversuch. Arzneim–Forsch. 14:389–396, 1964.

1142 Glycopyrrolate CAS 596–51–0

Kagiwada et al. (1973) gave this anticholinergic agent orally to pregnant mice and rats during active organogenesis. The maximum dose was 150 mg per kg per day in the rat and 100 mg in the mouse. No increase in defects was found and postnatal studies for three weeks were negative.

Kagiwada, K.; Ishizaki, O. and Saito, G.: Effects of glycopyrrolate on pre– and post–natal developments of the off-springs in pregnant mice and rats. (Japanese) Oyo Yakuri 7:617–626, 1973.

1143 Glycyrrhizinate–Disodium

Itami et al. (1985) fed this sweetner in amounts of up to 2.0 percent in the diet of pregnant rats on days 0–20 and found no adverse effects in the fetuses.

Itami, T.; Ema, M. and Kanoh, S.: Effect of disodium glycyrrhizinate on pregnant rats and their offspring. J. Food Hyg. Soc. Jpn. 26(5):460–464, 1985.

1144 GMK–527 rt–PA

Tanaka et al. (1988) gave rats this tissue-type plasminogen activator intravenously on days 7–17 and found no evidence of teratogenicity. Postnatally the F_1 males exposed to the highest dose (10 mg per kg) ate less and gained weight more slowly. Kojima et al. (1988) treated rabbits similarly on days 6–18 and found no teratogenicity but the male fetuses were slightly lighter.

Kojima, N.; Naya, M.; Imoto, H.; Hara, T.; Deguchi, T. and Takahira, H.: Reproduction studies of gmk–527 (rt–pa)–(iii) teratogenicity study in rabbits treated intravenously with gmk–527. Yakuri to Chiryo 16:1129–1142, 1988.

Tanaka, E.; Mizuno, F.; Ohtsuka, T.; Komatsu, K.; Umeshita, C.; Mizusawa, R. and Toshida, J.: Reproduction studies of gmk–527 (rt–pa)–(ii) Teratogenicity study in rats treated intravsnously with gmk–527. Yakuri to Chiryo 16:1129–1142, 1988.

1145 L–Goitrin CAS 500–12–9

This thioamide present as its glycoside in vegetables such as turnips and cabbage can inhibit the metabolism of the thyroid gland in the developing chick. A total dose of 1.0 mg given during the 7th to 15th days of incubation caused thyroid hypertrophy and reduction of radioiodine uptake in the chick embryo on the 17th day (Shepard, 1960).

Shepard, T.H.: Unpublished data, 1960.

1146 Gold CAS 7440–57–5

Transfer of gold across the placenta of a 20 week human pregnancy was documented (Rocker and Henderson, 1976) but Hollander (1972) documented a series where several women received gold without untoward effects on their fetuses. Miyamoto et al. (1974) in a survey, found 26 patients who received gold during pregnancy and all of these offspring were normal. Another 93 patients took gold during only the first part of their pregnancy and among these offspring there was one with dislocated hip and one with a flattened acetabulum.

Ridgway and Karnofsky (1952) using $AuCl_3$, determined that the LD–50 for chick embryos was less than 20 mg per egg. No congenital defects were noted.

Kidston et al. (1971) reported a 25 percent malformation rate in rats exposed to somewhat higher than human serum levels of gold (aurothiomalate). These malformations included hydronephrosis, hydrocephalus with some

eye, heart and palate defects. Gold accumulated in the yolk sac lysosomes. Szabo et al. (1978A,B) studied two oral forms and a thiomalate subcutaneous compound in rats and rabbits treated during days 6–15 and 6–18, respectively. At the higher dose levels of 6–44 mg gold per kg per day, they found a small increase in defects. Hydrocephalus, eye defects and rib fusion were the most frequent defects in rats while gastroschisis and umbilical hernia were most common in rabbits.

Hollander, J.L.: Arthritis and Allied Conditions. J.C. Hollander and D.J. McCarthy (eds.), Philadelphia, 479, 1972.

Kidston, M.E.; Beck, F. and Lloyd, J.B.: Effects of myocrisin injection in rats. J. Anat. 108:590–591, 1971.

Miyamoto, T.; Miyaji, S. and Horiuchi, Y.: Gold therapy in bronchial asthma: Special emphasis upon blood level of gold and its teratogenicity. J. Jpn. Soc. Int. Med. 63:1190–1197, 1974.

Ridgway, L.P. and Karnofsky, D.A.: The effects of metals on the chick embryo: Toxicity and production of abnormalities in development. Ann. N.Y. Acad. Sci. 55:203–215, 1952.

Rocker, I. and Henderson, W.J.: Transfer of gold from mother to fetus. Lancet 2:1246 only, 1976.

Szabo, K.T.; DiFebbo, M.E. and Phelan, D.G.: The effects of gold–containing compounds on pregnant rabbits and their fetuses. Vet. Path. 15 (5):97–102, 1978A.

Szabo, K.T.; Guerriero, F.J. and Kang, Y.J.: Effects of gold–containing compounds on pregnant rats and their fetuses. Vet. Path. 15(5):89–96, 1978B.

1147 Gold Sodium Thiomalate Myochrysine CAS 12244–57–4

This drug used in rheumatoid arthritis therapy was injected into pregnant rats between 7.5 and 11.5 days (Kidston et al., 1971). About one–half of the fetuses had malformations including hydronephrosis, hydrocephalus, cleft palate, ventricular septal defects and eye defects. The mechanism of action was via lysosomal inhibition in the yolk sac. Unfortunately the dose was not given.

Kidston, M.E.; Beck, F. and Lloyd, J.B.: The teratogenicity of myochrysine injection in rats. (abs) J. Anat. 108:590–591, 1971.

1148 Gonadorelin Luteinizing Hormone–Releasing Hormone LH–RH CAS 52699–48–6

Tanabe et al. (1974) injected this synthetic neuroendocrine compound intraperitoneally into mice in doses of 0.04–4 mg per kg on days 6 through 15 and produced delayed parturition but there was no growth inhibition or increases in congenital defects. Postnatal studies to 8 weeks revealed no differences. Ishihara et al. (1980) gave 0.002–2.0 mg per kg per day intraperitoneally to male and female rats for 63 days before mating and during the first seven days of gestation. Mating performance in both sexes was decreased at all dose levels. No teratogenic effects were noted.

Ishihara, H.; Asano, Y.; Nito, S. and Higaki, K.: Fertility study of rats intraperitoneally treated with synthetic luteinizing hormone–releasing hormone. Oyo Yakuri 20:149–161, 1980.

Tanabe, Y.; Ariyuki, F. and Higaki, K.: Effects of synthetic LH–RH administration to pregnant mice on pre- and post–natal development of their offsprings. Oyo Yakuri 8:685–695, 1974.

1149 Gonadotropin, Human Chorionic CAS 9002–61–3

Caspi et al. (1976) studied 110 pregnancies following induction of ovulation by gonadotropins. The multiple pregnancy rate was 27 percent at 20 weeks and the incidence of malformations was 7 percent, which was not considered to be elevated. The abortion rate was 54 percent and an increase in female newborns was found (83 vs. 69).

Hultquist and Engfeldt (1949) gave human chorionic gonadotropin injections in the rat and produced some increase in weight of the newborn fetuses. Whether this size increase was due to delay in parturition or reduction in number of fetuses per litter was not evident. No defects were found.

Sakai and Endo (1987) treated mice intraperitoneally with 2.5, 5 or 10 IU of human chorionic gonadotropin. The number of fetuses per litter and the percent resorbed was increased but the fetal weights were reduced. Increases in open eyelids and cleft palates were found. Nishimura and Shikata (1958) injected mare's gonadotropic hormone subcutaneously into mice before fertilization. Increased litter numbers occurred. Various deformations of the skeleton were found. These included curved vertebral column, club foot, flexed wrist joints and a few cleft palates. The dosage given was 5 r.u. (100 IU).

Caspi, E.; Ronen, J.; Schreyer, P. and Goldberg, M.D.: The outcome of pregnancy after gonadotrophin therapy. Br. J. Obstet. Gynecol. 83:967–973, 1976.

Hultquist, G.T. and Engfeldt, B.: Growth of rat fetuses produced experimentally by means of administration of hormones to the mother during pregnancy. Acta Endocrinol. (KBH) 3:365–376, 1949.

Nishimura, H. and Shikata, A.: The maldevelopment of the fetuses of mice treated with gonadotropic hormone before the conception. Okajimas Folia Anat. Jpn. 31:195–203, 1958.

Sakai, N. and Endo, A.: Potential teratogenicity of gonadotropin treatment for ovulation induction in the mouse offspring. Teratology 36:229–233, 1987.

1150 Gossypol CAS 303–45–7

This product of the cotton plant used for birth control in China reduced the weight and hatchability of cotton leaf worm larvae (El–Sabae et al., 1981). Beaudoin (1985) gavaged male rats before mating or pregnant females up to 20 mg per kg and found no adverse reproductive changes. The females were treated on day 8, 10, 12, 14 or 16. Beaudoin (1988) found no adverse behavioral changes in the offspring of rats treated with up to 20 mg per kg. Sein and Phil (1986) gave 50 to 75 mg per kg intragastrically to mice on days 1–15. Immune depression was found in the offspring at both

dose levels. Li et al. (1989) studied the compound in chicks and mice. Some embryotoxicity was found in the mouse at 120 mg per kg. Exencephaly and micromelia occurred with increased frequency in the chicks. Randel et al. (1992) have reviewed the reproductive toxicity.

Beaudoin, A.R.: The embryotoxicity of gossypol. Teratology 32:251–257, 1985.

Beaudoin, A.R.: A developmental toxicity evaluation of gossypol. Contraception 37(2):197–219, 1988.

El–Sabae, A.H.; Sherby, S.I. and Manscrit, N.A.: Gossypol as inducer and inhibitor in spodoptera littoralis larvae. J. Environ. Sci. Health B. 16:167–178, 1981.

Li, Y.F.; Booth, G.M. and Seegmiller, R.E.: Evidence for embryotoxicity of gossypol in mice and chicks with no evidence of mutagenic activity in the ames test. Reprod. Toxicol. 3:59–62, 1989.

Randel, R.D.; Chase, C.C. and Wyse, S.J.: Effects of gossypol and cottonseed products on reproduction of mammals. J. Animal Sci. 70:1628–1638, 1992.

Sein, G.M. and Phil, M.: Embryotoxic and immunodepressive effects of gossypol. Am. J. Chin. Med. XIV (3 and 4) 110–115, 1986.

1151 Gramicidin CAS 113–73–5

Heinonen et al. (1977) report 61 women exposed during the first four lunar months. There were seven malformed children with a standardized risk of 1.85 which was not an increase statistically.

Heinonen, O.P.; Slone, D. and Shapiro, S.: Birth Defects and Drugs in Pregnancy. Publishing Sciences Group Inc., Littleton, Mass., 1977.

1152 Granisetron CAS 109889–09–0

Baldwin et al. (1990,A) gave this antinauseant intravenously to rats and rabbits during organogenesis at maximum daily doses of 9 and 3 mg per kg respectively. No adverse fetal effects were found. Baldwin et al. (1990,B) gave rats 6 mg per kg subcutaneously before fertilization or in the perinatal and postnatal period without changes in reproduction or postnatal behavior. Baldwin et al. (1993) studied rats prenatally and during the first part of pergnancy, during organogenesis and peri and postnatally. No ill effects in the fetuses were found at oral doses of up to 100 mg per kg for the organogenesis studies. Two dams of 22 died at the highest dose (100 mg per kg) in the perinatal studies but no drug related effects on behavior were found.

Baldwin, J.A.; Davidson, E.J.; Goodwin, J.; Pritchard, A.L. and Ridings, J.E.: Intravenous administration study during organogenesis in rats and rabbits. Kiso to Rinsho 24:5043–5053, 1990.

Baldwin, J.A.; Davidson, E.J.; Goodwin, J.; Pritchard, A.L. and Ridings, J.E.: Fertility and general reproductive performance and peri- and post–natal toxicity studies with subcutaneous administration in rats. Kiso to Rinsho 24:5055–5069, 1990.

Baldwin, J.A.; Caton, F.D.; Davidson, E.J.; Goodwin, J.; Pritchard, A.L.; Ohta, M.; Yasuda, E. and Nishioka, Y.: Toxicity study of granisetron hydrochloride: Reprroduction studies in rats by oral administration. Yakuri to Chiryo 21:1753–1769, 1993.

1153 Granulocyte Colony Stimulating Factor *G–CSF*

Sugiyama et al. (**1990**) gave rats intravenously doses of up to 100 microgm on days 7–17. No adverse reproductive effects were found and there was no effect on reproduction in the exposed fetuses (F–2). Similar studies were done in rabbits and a high fetal mortality was found at 10 and 100 microgram doses. No teratogenic changes were found. (Hara et al. **1990**) Tatsuhiro et al (**1991**) gave rats up to 1000 microgm per kg during organogenesis without ill effects. Ito et al. (**1991**) gave rabbits up to 200 microgm per kg during organogenesis and found fetotoxicity at the highest dose but no teratogenicity. Maternal spleen weights were increased.

Ito, T.; Takenaka, C.; Sakuma, T.; Kato, Y.; Hara, T. and Deguchi, T.: Teratogenicity study of marograstim (KW–2228) in rats. The Clinical Report 25:3757–3774, 1991.

Ito, T.; Imoto, H.; Kato, Y.; Hara, T. and Deguchi, T.: Teratogenicity study of marograstim (KW–2228) in rabbits. The Clinical Report 25:3775–2787, 1991.

Hara, H.; Kiyosawa, K.; Tanaka, N.; Igarashi, S.; Takeuchi, H.; Aoki, A. and Sugiyama, O.: Teratological study of recombinant human g–csf(rg csf) in rabbits. Yakuri to Chiryo 18:s2371–s2399, 1990.

Sugiyama, O.; Watanabe, S.; Masuda, K.; Igarashi, S.; Watanabe, K.; Ebihara, Y. and Sato, T.: Teratology study of recombinant human g–csf(rg csf) in rats. Yakuri to Chiryo 18:s2355–s2369, 1990.

1154 Granulocyte–Macrophage Colony Stimulating Factor *GM–CSF*

Hoon et al. (**1993** A) gave up to 1000 microgm per day subcutaneously to rats during organogenesis and did not observe differences from the control fetuses, parturition or postnatal growth and reproduction. Similar studies perinatally did not affect the offspring (Hoon et al.,**1993** B).

Kim, S.; Chung, M.; Han, S. and Roh, J.: Teratogenicity study (segment II) of recombinant granylocyte–macrophage colony stimulating factors (LBD–005) in rats. J. Tox. Sc. 18:1–17, 1993.

Kim, S.; Chung, M.; Han, S. and Roh, J.: Peri–and postnatal study (segment III) of recombinant granulocyte–macrophage colony stimulating factors (LBD–005) in rats. J. Tox. Sc. 18:19–35, 1993.

1155 Gravity

Duke (1983) exposed mouse limb bud explants to 2.6 G force and found retardation in limbs removed and grown from mice on gestational days 12 through 13.5.

Duke, J.C.: Suppression of morphogenesis in embryonic mouse limbs exposed to excess gravity. Teratology 27:427–436, 1983.

1156 Grayanotoxin 1

Kobayashi et al. (**1990**) gave this plant toxin to mice in amounts of 1.5 mg per kg on three consequitive days in

organogenesis and found no fetal ill effects. There was some lethality among the dams. In chicks no teratogenicity was found following intraamniotic injections of 0.1–1.o microgm on day 3 or 4.

Kobayashi, T.; Yasuda, M. and Seyama I.: Developmental toxicity potential of grayanotoxin 1 in mice and chics. The Journal of Toxicological Sciences 15:227–234, 1990.

1157 Green S Dye *Wool Green BS EEC E 142 CI (1971) No. 44090*

Clode (1987) gavaged rats on days 0–19 of pregnancy with up to 1000 mg per kg per day of this food dye. No adverse effects resulted in dams or fetuses. The skeletal development in the treated fetuses was slightly advanced. More tracheal mucous was found in the treated fetuses.

Clode, S.A.: Teratogenicity and embryotoxicity study of Green S in rats. Food Chem. Toxicol. 25(12):995–997, 1987.

1158 Griseofulvin CAS 126–07–8

Rosa et al. (1987) reported two sets of infants with the condition of thoracopagus conjoined twinning. Both mothers received the drug during early pregnancy. These authors cite a study of over 4000 mothers with spontaneous abortion and in that report, 7 women had exposure with a relative risk of 2.5. Knudsen (1987) drawing on data from the International Clearing House for Birth Defects Monitoring Systems, reported 47 conjoined twins, none of whom were exposed to griseofulvin.

Slonitskaya (1969) injected 50 to 500 mg per kg into pregnant rats on the 11th through the 14th day. With the largest dose, 22 percent of the fetuses were abnormal. A wide range of defects was seen including the eye, skeleton, urogenital tract and central nervous system. Injection during the first four days of pregnancy did not cause defects or embryonic death. Chronic injection on each day of pregnancy was less toxic to the embryo than acute administration during the critical periods of organogenesis.

Klein and Beall (1972) using doses of 1250 to 1500 mg per kg daily from the 6th through the 15th day of gestation in the rat, produced an 8 to 10 percent incidence of tail defects in the offspring. Occasionally exencephaly was found. These dosages are 60 to 75 times those used therapeutically in man.

Klein, M.F. and Beall, J.R.: Griseofulvin: A teratogenic study. Science 175:1483–1484, 1972.

Knudsen, L.B.: No association between griseofulvin and conjoined twinning. Lancet 2:1097, 1987.

Rosa, F.W.; Hernandez, C. and Carlo, W.A.: Griseofulvin teratology including two thoracopagus conjoined twins. Lancet 1:171, 1987.

Slonitskaya, N.N.: Teratogenic effect of grisofulvinforte on rat fetus. (Russian) Antibiotiki 14:44–48, 1969.

1159 GYKI 13504

This tamoxifen analogue was studied during organogenesis in rabbits at doses of 1,4 and 15 mg per kg. A dose–dependent increase in intrauterine mortality was found as well as fetal growth retardation.

Kovacs, E.; Meggyesy, K. and Druga, A.: teratology study of the compound GYKI 13504 in New Zealand white rabbits (Abst). Teratology 46:24A, 1992.

1160 Guanabenz CAS 5051–62–7

This antihypertensive agent was studied in rats and rabbits by Akatsuka et al. (1982). At oral doses of 15 and 30 mg per kg, the dams were sedated and there was a decrease in corpora lutea and implantation sites. At the same dose levels given after the 7th day, fetal growth retardation occurred but malformations were not significantly increased. Perinatal mortality was increased at doses that sedated the mother (15 and 30 mg per kg). In rabbits receiving up to 10 mg per kg on the 6–18th days there was no increase in fetal malformations.

Akatsuka, K.; Hashimoto, T.; Takeuchi, K.; Yanagisawa, Y. and Kogure, M.: Reproduction studies of guanabenz in the rat and rabbit. J. Toxicol. Sci. 11:93–151, 1982.

1161 Guanazodine CAS 32059–15–7

Sawano et al. (1978) administered the anti–hypertensive orally up to 400 mg per kg to mice on the 7th to 12th day. No defect increase or postnatal development changes were found.

Sawano, J.; Yamamura, H.; Oyama, K.; Hada, R. and Kobayashi, Y.: Effects of Guanazodine administered to pregnant mice on pre– and postnatal development of the offspring. Oyo Yakuri 15:333–340, 1978.

1162 Guanethidine CAS 55–65–2

West (1962) gave daily subcutaneous injections to pregnant rats using 10 mg per kg and found no increase in fetal death.

West, G.B.: Drugs and rat pregnancy. Letter to the Editor. J. Pharm. Pharmacol. 14:828–830, 1962.

1163 Guanfacine N–Amidino–2(2,6–dichlorophenyl) acetamide hydrochloride CAS 29110–47–2

Esaki and Nakayama (1979) gave 0.5 to 2 mg per kg daily to pregnant rabbits from the 6th to 18th days of gestation. Even though the weight gain of the mothers was reduced at the 1 mg level and above, no increase in defects was found. In another series of papers using similar dosing in the mouse, no evidence of teratogenicity or reproductive dysfunction was observed (Esaki and Hirayama, 1979). Postnatal studies were also negative although some neonatal mortality increase was found at 1.0 mg per kg (Esaki et al., 1980).

Esaki, K. and Hirayama, M.: Effect of oral administration of BS–100–141 on reproduction in the mouse. Preclin. Rept. Cent. Exp. Animal 5:107–128, 1979.

Esaki, K. and Nakayama, T.: Effects of oral administration of BS–100–141 on the rabbit fetus. Preclin. Rept. Cent. Exp. Animal 5:129–136, 1979.

Esaki, K.; Oshio, K. and Yamaguchi, K.: Effects of oral administration of BS–100–141 on reproduction in mice: Observations on behavior. Preclin. Rept. Cent. Exp. Animal 6:117–122, 1980.

1164 Guanine also see 2–Deoxyguanosine CAS 73–40–5

This substance was injected by Karnofsky and Lacon (1961) into 4–day chick eggs and no abnormalities were found. The LD–50 for the embryo on the day 4 was 8 mg per egg.

Karnofsky, D.A. and Lacon, C.R.: Effects of physiological purines on the development of the chick embryo. Biochem. Pharmacol. 7:154–158, 1961.

1165 Guanine Monophosphate

Kaziwara et al. (1971) gavaged rats with 100 mg per kg during organogenesis and found no adverse effects.

Kaziwara, K.; Mizutani, M. and Ihara, T.: On the fetotoxicity of Disodium 5'–Ribonucleotide in the mouse, rat and monkey. Journal of the Takeda Research Laboratories 30(2):314–321, 1971.

1166 Guanosine also see 2–Deoxyguanosine CAS 118–00–3

Karnofsky and Lacon (1961) found that the LD–50 for 4–day chick embryos was 8 mg per egg. Using 2 to 12 mg, they reported that 12 percent of the embryos were abnormal. The type of defects was not specified.

Karnofsky, D.A. and Lacon, C.R.: Effects of physiological purines on the development of the chick embryo. Biochem. Pharmacol. 7:154–158, 1961.

1167 5–Guanylate, Disodium

Kojima (1974) fed rabbits 200 mg per kg on days 6–18 and found no adverse effects. Negative findings were seen with 200 mg per kg given to pregnant rats on days 9–15.

Kojima, K.: Safety evaluation of disodium 5'–inosinate, disodium 5'–guanylate and disodium 5'–ribonucleotide. Toxicology 2:185–206, 1974.

1168 Guar Gum

Collins et al. (1987) fed male and female rats up to 15 percent in the diet before gestation started and up to 11.8 gm per kg during the pregnancy. No malformations or delays in fetal development were observed.

Collins, T.F.X.; Welsh, J.J.; Black, T.N.; Graham, S.L. and O'Donnell, M.W., Jr.: Study of the teratogenic potential of guar gum. Food Chem. Toxicol. 25(11):807–814, 1987.

1169 Gum Arabic

Collins et al. (1987) fed up to 15 percent in the diet during gestation of rats and found no adverse fetal or maternal effects.

Collins, T.F.X.; Welsh, J.J.; Black, T.N.; Graham, S.L. and Brown, L.H.: Study of the teratogenic potential of gum arabic. Food Chem. Toxicol. 25(11): 815-821, 1987.

1170 Guthion

Short et al. (1980) injected mice and rats for 10 days starting on day 6 with 5.0 mg per kg daily and found no adverse fetal effects.

Short, R.D.; Minor, J.L.; Lee, C.C.; Chernoff, N. and Baron, R.L.: Developmental toxicity of guthion in rats and mice. Arch. Toxicol. 43:177–186, 1980.

1171 Hadacidin *N–Formyl–N–hydroglycine* CAS 689-13-4

Chaube and Murphy (1963) reported that rats receiving 1 to 5 mg per kg on days 9 through 12 or single injections on day 10, 11 or 12 produced fetuses with exencephaly, cleft palate or other skeletal defects. Two analogues of hadacidin, N–acetyl hydroxyaminoacetic acid and hydroxyaminoacetic acid at the same dose ranges were toxic to the fetuses but did not produce malformations. Lejour–Jeanty (1966) studied the pathoembryogenesis of the cleft lip deformity and found that cell death played a prominent role in preventing the lateral nasal process to fuse with the medial nasal and maxillary processes. Milaire (1971) made a study of the pathogenesis of the syndactyly which occurs and found an associated reduction in proliferation of the limb blastema along with some hyperplasia of the apical ectodermal ridge.

Chaube, S. and Murphy, M.L.: Teratogenic effect of hadacidin (a new growth inhibitory chemical) on the rat fetus. J. Exp. Zool. 152:67–73, 1963.

Lejour–Jeanty, M.: Becs–de–lievre provoques chez le rat par un derive de penicilline, l'hadacidine. J. Embryol. Exp. Morphol. 15:193–211, 1966.

Milaire, J.: Etude morphogenetique de la syndactylie postaxiale provoquee chez le rat par l'hadacidine: 2 les bourgeous de membres chez les embryons de 12 a 14 jours. Arch. Biol. (Liege) 82:253–322, 1971.

1172 Hair Dyes

A composite of a series of commercially available semipermanent hair colorings (mostly phenylendiamines) was added to the diet of rats in the maximal amount of 7800 ppm from days 6 through 15 and no fetal effects were found. Rabbits were also tested by gavage on days 6–18 of gestation with doses up to 97.5 mg per kg per day and no teratogenicity was found (Wernick et al., 1975).

Marks et al. (1981) studied the effects of 4–nitro–p–phenylenediamine (2–NPPD), 4–nitro–o–phenyldiamine (4–NOPD) and 2,5–toluenediamine sulfate (2,5TDS) by subcutaneous injection on days 6–15 in the pregnant mouse. At doses which were toxic to the mother, defects were produced with 2–NPPD and 4–NOPD (160 mg and 256 mg per kg per day, respectively). Cleft palate was the most common defect but mineralization of the myocardium also was observed. 2,5–TDS was not teratogenic at 64 mg per kg. DiNardo et al. (1985) administered five oxidative hair dyes by gavage to rats on days 6 through 15 and observed no adverse fetal effects. The dyes and maximum daily dose per kg were the following: 4,4 Diaminodi–phenylenesulfate (50 mg), N–(2,–hydroxyethyl)–4–nitro–o–p–phenylenediamine (200 mg), 2,3–dihydroxynaphthalene (450 mg), N,N–dimethyl–p–phenylenediamine (150 mg) and resorcinol (500 mg). Heinonen et al. (1977) reported no increase in defects among 118 women who used this dermatologic agent. Only 18 used it in the first 4 lunar months.

DiNardo et al. (1985) gavaged rats with up to 500 mg per kg daily on days 6–15. No teratogenic effect was seen.

Marks et al. (1981) gave mice up to 1024 mg per kg on days 6–15. At 64 mg per kg subcutaneously there was an increase in maternal mortality but no malformation increase. Some alizarin red staining material was found at higher doses in the fetal myocardium.

DiNardo, J.C.; Picciano, J.C.; Schnetzinger, R.W.; Morris, W.E. and Wolf, B.A.: Teratological assessment of five oxidative hair dyes in the rat. Toxicol. Appl. Pharmacol. 78:163–166, 1985.

Heinonen, O.P.; Slone, D. and Shapiro, S.: Birth Defects and Drugs in Pregnancy. Littleton, Massachusetts: John Wright–PSG, Inc.,; pp 410–411, 444, 499, 1977.

Marks, T.A.; Gupta, B.N.; Ledoux, T.A. and Staples, R.E.: Teratogenic evaluation of 4–nitro–p–phenylenediamine, 4–nitro–o–phenylenediamine and 2,5–toluenediamine sulfate in the mouse. Teratology 24:253–265, 1981.

Wernick, T.; Lanman, B.M. and Fraux, J.L.: Chronic toxicity, teratologic and reproductive studies with hair dyes. Toxicol. Appl. Pharmacol. 32:450–460, 1975.

1173 Halazepam 7
Chloro–1–(2,2,2–trifluoroethyl)–5–phenyl 1,3–dihydro–2H,1,4 benzodiazepin–2–one

Beall (1972) reported that this diazepam–like drug was not teratogenic in rats gavaged with 2, 50 or 200 mg per kg on days 6–15.

Beall, J.R.: Study of teratogenic potential of diazepam and SCH–12041. C.M.A. Jouurnal 106:1061, only, 1972.

1174 Halofantrine

This antimalarial has been tested in pregnant rats and rabbits at maximum oral doses of 45 and 120 mg per kg respectively. No increase in malformations was found but at maternal toxic doses some fetal weight reduction was found (Schuster and Canfield, 1989).

Schuster, B.G. and Canfield, C.J.: Halofantrine in the treatment of multidrug–resistant malaria. Cambridge, UK: Elsevier Publications, 1989, pp3–4.

1175 Haloperidol CAS 52–86–8

Van Waes and Van De Velde (1969) studied 100 pregnancies in which 0.6 mg was administered twice daily for varying periods. Not a single malformation was found in 94 newborns examined and no adverse effects on the pregnancies were observed.

Tuchmann–Duplessis and Mercier–Parot (1971) showed that this tranquilizer has a remarkable delaying effect on time of implantation in the mouse and rat. In the rat treated immediately after mating with 2 to 10 mg per kg daily, about 50 percent of the fetuses on day 20 were the size of 14–day embryos. If delivery were allowed to occur, normal fetuses appeared 2 to 8 days later than the controls. No defects were associated with this delay in growth. Yamamura et al. (1982) found a delay in the first three cleavages of the mouse morula when 3.5 mg per kg was administered subcutaneously.

Jurand and Martin (1990) injected mice intraperitoneally on the 9th day with 22 mg per kg or more. Exencephaly, hydrocephalus and occassionally ectopia of the neural tube was observed. Maternal lethality was observed at 34 mg per kg. Holson et al. (**1994**) gave rats two 5 mg per kg doses subcutaneously on days 12–16 or 16–20 and found significant reductions in whole brain weight in the adult. Most parts of the brain were affected. Pair–fed rat controls were used. Whole embryo cultured embryos were delayed in development at 5 micromolar or greater concentrations. The authors interpreted their work as evidence that compounds which interfer with neurotransmission may inhibit brain growth.

Holson, R.R.; Webb, P.J.; Grafton, T.F. and Hansen, D.K.: Prenatal neuroleptic exposure and growth stunting in the rat: An in vivo and in vitro examination of sensitive periods and possible mechanisms. Teratology 50:125–126, 1994.

Jurand, A. and Martin, L.V.H.: Teratogenic potential of two neurotropic drugs, haloperidol and dextromoramide, tested on mouse embryos. Teratology 42:45–54, 1990.

Tuchmann–Duplessis, H. and Mercier–Parot, L.: Influence of neuroleptics on prenatal development in mammals. In: Malformations, Tumors and Mental Defects, Pathogenetic Correlations. H. Tuchmann–Duplessis, G. Fanconi and G.R. Burgio (eds.), Milan, Carlo Erba Foundation, 1971.

Van Waes, A. and Van De Velde, W.: Safety evaluation of haloperidol in the treatment of hyperemesis gravidarum. J. Clin. Pharmacy 9:224–227, 1969.

Yamamura, H.; Kukui, K.; Fukui, Y. and Inamoto, M.: Effects of haloperidol, an antipsychotic agent, on preimplantation development in the mouse. Cong. Anom. 22:145–160, 1982.

1176 Halopredone CAS 57781–14–3

Imoto et al. (1985) administered this steroid subcutaneously to rats on days 7–17 in doses of up to 12.5 mg per kg and found no adverse fetal effects on day 20 or in postnatal development. Maternal body weight was reduced by the higher doses.

Imoto, S.; Kamada, S.; Yahata, A.; Kosaka, M.; Takeuchi, M.; Shinpo, K.; Sudo, J. and Tanabe, T.: Teratogenicity study on halopredone acetate in rats. J. Toxicol. Sci. 10:83–103, 1985.

1177 Halothane CAS 151–67–7

Basford and Fink (1968) reported that when rats were exposed to 0.8 percent halothane for 24 hours on day 9, a significant increase in defects of the ribs and vertebrae were found in the offspring as compared to controls exposed to air and also starved during day 9. Koeter and Rodier (1986) exposed mice to 4 or 6 hours on day 14 or 2 days postnatally, and found a delay in the appearance of developmental landmarks as well as delay in appearance of righting and locomotion. Halothane was given in the amount of 0.5 percent. Mazze et al. (1986) exposed rats to 0.8 percent halothane for

6 hours daily during two–day periods of organogenesis and found no adverse findings in the fetuses.

Basford, A.B. and Fink, B.R.: The teratogenicity of halothane in the rat. Anesthesiology 29:1167–1173, 1968.

Koeter, H.B. and Rodier, P.M.: Behavioral effects in mice exposed to nitrous oxide or halothane: Prenatal vs. postnatal exposure. Neurobehav. Toxicol. 8:189–194, 1986.

Mazze, R.I.; Fujinaga, M.; Rice, S.A.; Harris, S.B. and Baden, J.M.: Reproductive and teratogenic effects of nitrous oxide, halothane, and enflurane in Sprague–Dawley rats. Anesthesiology 64:339–344, 1986.

1178 Harmaline Hydrochloride

Poulson and Robson (1963) gave 0.4 mg per day to pregnant mice on days 1–6, 4–7 and 6–11 and produced no decrease in the number of live litters.

Poulson, E. and Robson, J.M.: The effect of amine oxidase inhibitors on pregnancy. J. Endocrin. 27:147–152, 1963.

1179 Hazardous Waste Sites

Geschwind et al. (1992) studied 9,313 newborns of mothers living within a mile of hazardous waste sites. The odds ratio for congenital defects was slightly higher (1.12, 95% confidence interval 1.06–1.18). A careful attempt was made to break down contributions of the type of exposure and type of general defect. The increased rate was based on defects of the nervous, musculoskeletal and integument systems. Odds ratios for the occurrance of general type of defects with general chemicals were significantly increased for pesticides and musculoskeletal, metals and nervous system and solvents and nervous system. The authors state that further research is needed to corroborate these findings.

Geschwind, S.A.; Stolwijk, J.A.J.; Bracken, M.; Fitzgerald, E.; Stark, A.; Olsen, C. and Melius, J.: Risk of congenital malformations associated with proximity to hazardous waste sites. American Journal of Epidemiology 135:1197–1207, 1992

1180 α–Hederin

This saponin isolated from ivy reduces hepatic metallothionein which causes redistribution of Zn. Daston et al. (**1994**) treated rats on day 11 with 30 or 300 micromol per kg and produced resorption increases. At 300 micromoles per kg encephalocele, umbilical hernia, undescended testes and renal defects were found. Whole embryo cultures were done and showed that the treated rat serum after 18 hours did not support growth, a finding reversed by addition of Zn.

Daston, G.P.; Overmann, G.J.; Baines, D.; Taubeneck, M.W.; Lehman-McKeeman, L.D.; Rogers, J.M. and Keen, C.L.: Altered Zn status by α–hederin in the pregnant rat and its relationship to adverse developmental outcome. Reproductive Toxicology 8:15–24, 1994.

1181 Heliotrine Dehydroheliotridine CAS 303–33–3

This pyrrolizidine alkaloid occurring in several plant families can cause liver disease in animals and man. Green and Christie (1961) injected 150 or 200 mg per kg into the rat. Vertebral and rib defects occurred when the rats were treated after the 11th day. Mandibular hypoplasia and cleft palate were seen when the higher dose was used. Peterson and Jago (1980) confirmed the above work and showed that dehydro–heliotridine was similarly teratogenic.

Green, C.R. and Christie, G.S.: Malformations in foetal rats induced by the pyrrolizidine alkaloid heliotrine. Br. J. Exp. Pathol. 42:369–378, 1961.

Peterson, J.E. and Jago, M.V.: Comparison of the toxic effects of dehydroheliotridine and heliotrine in pregnant rats and their embryos. Pathology 131:339–355, 1980.

1182 Hemangiomas by Genetic Insertion of Middle T Oncogene of Polyoma Virus

Williams et al. (1988) inserted this gene for middle T oncogene of polyoma virus into embryonic stem cells. They found that mouse blastocysts receiving these cells died at midgestation when multiple hemangiomas disrupted blood vessel formation. The hemangiomas were located in the yolk sac.

Williams, R.L.; Courtneidge, S.A. and Wagner, E.F.: Embryonic lethalities and endothelial tumors in chimeric mice expressing polyoma virus middle T oncogene. Cell 52:121–131, 1988.

1183 Hemlock γ–Coniceine

Edmonds et al. (1972) reported an outbreak of congenital defects in the offspring of sows who ingested poison hemlock (Conium maculatum). Seven of 55 newborn piglets had defects of the hind legs and 34 had classic signs of hemlock poisoning including trembling and ataxia. Sows from the same farm pastured in an area where poison hemlock did not grow gave birth to normal offspring. Panter et al. (1985) determined that treatment between days 43–53 and 51–61 produced the effects. They postulated that γ–coniceine is the responsible alkaloid. Cleft palates were observed in the offspring.

Edmonds, L.D.; Selby, L.A. and Case, A.A.: Poisoning and congenital malformations associated with consumption of poison hemlock by sows. J. Am. Vet. Med. Ass. 160:1319–1324, 1972.

Panter, K.E.; Keeler, R.F. and Buck, W.B.: Congenital skeletal malformations induced by maternal ingestion of Conium maculatum (poison hemlock). Am. J. Vet. Res. 46:2064–2066, 1985.

1184 Heparin CAS 9005–49–6

With a molecular weight of 20,000, heparin apparently does not cross the placenta. The author has found no studies of this drug in pregnant animals. Complications of pregnancy are common and may involve the fetal central nervous system (Pauli and Hall, 1980).

Ginsberg and Hirsh (1989) identified 355 cases treated during pregnancy with heparin. Although 21.7% of the pregnancies had adverse outcomes, they were able to separate out 317 women who did not have existing maternal risk fac-

tors such as toxemia glomerulonephritis or recurrent spontaneous abortions; in this group there was no increase in prematurity (6.8%) or fetal neonatal death (2.5%). There is controversy regarding the number of complications which occur when pregnant women with prosthetic valves are treated with heparin or warfarin (Oakley, 1994).

Sbarouni and Oakley (1994) reported in a large series a higher rate of thromboembolic complications with continuous use of heparin (15 vs 5 for warfarin). Ginsberg and Barron (1994) recommend use of heparin between the 6th and 12th week of gestation.

Bertoli and Borelli (1986) administered low molecular weight heparin in dosages of up to 10 mg per kg daily during organogenesis in the rat and produced no adverse fetal effects.

Low molecular weight heparin was given intravenously to rats on days 7–17 in doses of up to 10,000 mg per kg. At the highest dose fetal mortality was increased but no other adverse effects were found. (Shimazu et al., 1989) Postnatal development was normal. No skeletal changes were seen in 4 day old pups. Rabbits were given up to 4,000 mg per kg intravenously on days 6–18. Fetal weight was reduced at 2,500 mg per kg and increased fetal mortality occurred at 40,000 per kg but no teratogenicity was found. Kesby (1992) found damage in whole rat embryo cultures containing 25 microgm per ml or more of heparin.

Bertoli, D. and Borelli, G.: Peri–and postnatal, teratology and reproductive studies of a low molecular weight heparin in rats. Arzneim–Forsch. 36:1260–1263, 1986.

Ginsberg, J.S. and Barron, W.M.: Pregnancy and prosthetic heart valves. The Lancet 344:1170–1171, 1994.

Ginsberg, J. and Hirsh, J.: Anticoagulants during pregnancy. Annu Rev Med 40:79–86, 1989.

Kesby, G.J.: In vitro development of rat embryos undergoing organogenesis in heparin–plasma. Teratology 45:293–301, 1992.

Oakley, C.M.: Pregnancy and prosthetic heart valves. The Lancet 344:1643–1644, 1994.

Pauli, R.M. and Hall, J.G.: Warfarin embryopathy. Am. J. Med. 68:122–144, 1980.

Sbarouni, E. and Oakley, C.M.: Outcome of pregnancy in women with valve protheses. Br. Heart J. 71:196–201, 1994.

Shimazu, H.; Ishida, S.; Shiota, Y.; Tamura, K.; Isaji, M. Ohba, M. and Naitoh, J.: Reproduction studies of low molecular weight heparin sodium (fr–860) (2nd report)– teratological study in rats. Kiso to Rinsho 23:6433–6456, 1989.

1185 Heparin, Low Molecular Weight

Itahashi et al. (1992 A,B,C) studied this compound which had an average molecular weight of 4,000–6,000 in rats before fertilization and in the first week, during organogenesis and in the perinatal period. At intravenous doses of 360 mg per kg they found increased placental weights and delayed ossification. No teratogenicity was found but in the postnatal study a decrease in grooming and open field activity occurred in the highest group. Studies in rabbits using up to 360 mg per kg intravenously during organogenesis found increased hydrocephalus, microphthalmia and rib fusions in fetuses exposed to the highest dose. (Kudow et al., **1992**)

Itahashi, M.; Yamazako, M.; Amano, Y.; Suzuki, H.; Aihara, H.; Sannai, S.; Ogata, Y. and Mori, A.: Study on intravenous administration of low molecular weight heparin (LHG) prior to and in the early stages of pregnancy in rats. Yakuri to Chiryo 20:s281–s294, 1992A.

Itabashi, M.; Inoue, T.; Nakajima, K.; Aihara, H.; Sannai, of S.; Ogata, Y. and Mori, A.: Study on intravenous administration of low molecular weight heparin (LHG) during the period of organogenesis in rats. Yakuri to Chiryo 20:s295–s328, 1992B.

Itabashi, M.; Yamashita, K.; Ikura, K.; Aihara, H.; Sannai, S. and Mori, A.: Study on intravenous administration of low molecular weight heparin (LYG) during the perinatal and lactation periods in rats. Yakuri to Chiryo 20:s339–s358, 1992C.

Kudow, S.; Sadako, S.; Suzuki, K.; Yoshida, K.; Aihara, H. and Mori, A.: Study of intravenous administration of low molecular weight heparin (LHG) during the period of organogenesis in rabbits. Jpn. Pharmacol. Ther. 20:329–337, 1992.

1186 Heparitinase

Tuckett and Morriss–Kay (1989) produced open neural folds in rat embryos receiving a intraamniotic enzyme or when the enzyme was added to the culture medicine.

Tuckett, F. and Morriss–Kay, M.: Heparitinase treatment of rat embryos during cranial neurulation. Anat. Embryol. 180:393–400, 1989.

1187 Hepatitis

Siegel and Fuerst (1966) in a prospective study of the offspring of 60 mothers with hepatitis, found no increase in congenital defect rate. Low birth rate was found in 37 percent of mothers having the infection after 20 weeks of gestation. Adams and Combes (1965) found no fetal wastage in 34 patients.

Schweitzer et al. (1973) followed 31 infants whose mothers had hepatitis B during or shortly after pregnancy. Neonatal infection occurred in only 1 out of 10 women infected in the first 2 trimesters but 16 out of 21 infants exposed in the third trimester. Of the infected babies, 35 percent were less than 2,500 gm at birth. Heiber et al. (1977) in a study of 50 pregnancies confirmed these findings by observing no congenital defects. They found a higher incidence of low birth weight and some asymptomatic carriers among infants whose mothers were infected in the last trimester. Drew et al. (1978) found that when either parent was HBsAg positive, there was a very high sex ratio in their offspring (60 males and 24 females). Beasley et al. (1981) significantly reduced the carrier rate in offspring of HBsAg carrier mothers by immunizing from birth with hepatitis B immune globulin. Zhaomeng (**1988**) studied 96 pregnant women with positive HBsAg titers and found no increase in malformations as compared to a matched control.

Adams, R.H. and Combes, B.: Viral hepatitis during pregnancy. JAMA 192:195–198, 1965.

Beasley, R.P.; Lin, C–C.; Wang, K–Y.; Hsieh, F–S.; Hwang, L–Y.; Stevens, C.E.; Sun, T.S. and Szmuness, W.: Hepatitis B immune globulin (HBIG) efficiency in the interruption of perinatal transmission of Hepatitis B virus carrier state. Lancet 2:388–393, 1981.

Drew, J.S.; London, W.T.; Lustbader, E.D.; Hesser, J.E. and Blumberg, B.S.: Hepatitis B virus and sex ratio of offspring. Science 201:687–692, 1978.

Heiber, J.P.; Dalton, D.; Shorey, J. and Combes, B.: Hepatitis in pregnancy. J. Pediatr. 91:545–549, 1977.

Schweitzer, I.L.; Dunn, A.E.G.; Peters, R.L. and Spears, R.L.: Viral hepatitis B in neonates and infants. Am. J. Med. 55:762–771, 1973.

Siegel, M. and Fuerst, H.T.: Low birth weight and maternal virus diseases. A prospective study of rubella, measles, mumps, chicken pox and hepatitis. JAMA 197:680–684, 1966.

Zhaomeng, H.: The relationship between congenital malformation of newborn and hepatitis b virus infection in pregnant women. Chin. J. Epidemiol. 9(6):360–363, 1988.

1188 Hepatitis B Vaccine

Ayoola and Johnson (**1987**) immunized 72 women with two intramuscular injections in the 3rd trimester. No abnormal birth defects were observed.

Ayoola, E.A. and Johnson, A.O.K.: Hepatitis B vaccine in pregnancy: Immunogenicity, safety and transfer of antibodies to infants. Int. J. Gynaecol. Obstet. 25:297–301, 1987.

1189 Hepatitis C

Zanetti et al. (**1995**) reported that of 94 mothers infected with hepatitis C alone none transmitted the virus to their newborns. Of 15 mothers coinfected with HIV, eight transmitted to their babies.

Zanetti, A.R.; Tanzi, E.; Paccagnini, S.; Principi, N.; Pizzocolo, G.; Caccamo, M.L.; D'Amico, E.; Cambie, G.; Vecchi, L and Lombardy Study Group on Vertical HCV Transmission: Mother–to–infant transmission of hepatitis C virus. The Lancet 345:289–291, 1995.

1190 Heptachlor

Yamaguchi et al. (1987) fed up to 20 mg per kg to rats on days 7–17 and found no adverse fetal effects. At the highest dose, there was maternal toxicity. Le Marchand et al. (1986) found no increase in defect rates on the island of Oahu following a 27–29 month contamination of the milk with heptachlor. The range of concentrations in cow's milk was 1.20 to 5.0 ppm.

Le Marchand, L.; Kolonel, L.N.; Dendle, W.H. and Siegel, B.Z.: Trends in birth defects for a Hawaiian population exposed to heptachlor and for the United States. Arch. Environ. Health 41(3):145–148, 1986.

Yamaguchi, M.; Tanaka, S.; Kawashima, K.; Nakaura, S. and Takanaka, A.: Effects of heptachlor on fetal development of rats. Bull. Natl. Inst. Hyg. Sci. 105:33–36, 1987.

1191 Heptafluoropropane *FM–200™*

Nemec et al. (**1994**) exposed rats by inhalation to doses up to 105,000 ppm for 6 hours daily during organogenesis. The dams required extra oxygen at the highest dose. No adverse fetal effects were reported.

Nemec, M.J.; Holson, J.: Naas, D.; Knapp, J.; Lamb, I. and Biesemeier, J.: An assessment of the potential developmental toxicity of heptafluoropropane (FM–200™) by inhalation in the rat. (abs) Teratology 49:420, 1994.

1192 2–Hepta–5–methylbenzimidazole

Discussed under Benzimidazole

1193 Heptanoic Acid

Dawson (**1991**) using a frog embryo teratogenesis assay found that at 51 mg per liter 50% of the survivors were malformed. Microcephaly, abnormal gut coiling and edema were the most common types of abnormalities.

Dawson, D.: Additive incidence of developmen;tal malformation for xenopus embryos exposed to a mixture of ten aliphatic carboxylic acids. Teratology 44:531–546, 1991.

1194 Heptylhydrazine

Poulson and Robson (1963) gave 0.133 mg per day intraperitoneally to pregnant mice on days 1–6, 6–11 and 11–16 and produced a decrease in the number of live litters in the first two groups but not in those treated on the 11–16th days.

Poulson, E. and Robson, J.M.: The effect of amine oxidase inhibitors on pregnancy. J. Endocrinol. 27:147–152, 1963.

1195 Herbicides

Discussed under Occupation; 2,3,7,8-Tetrachlorodibenzo-p-dioxin; Trichlorophenoxyacetic Acid and 2,4–Dichlorophenoxy Acetic Acid

1196 Hercules 14503

Robens (1970) gave hamsters 1 ml per 100 gm of body weight orally on days 6–10 or 7 or 8. Curved tails and limb defects were found at 250 mg and 500 mg per kg levels.

Robens, J.F.: Teratogenic activity of several phthalimide derivatives in the golden hamster. Toxicol. Appl. Pharmacol. 16:24–34, 1970.

1197 Heroin *Diacetylmorphine* CAS 561–27–3

Naeye et al. (1973) reported that newborns from addicted mothers have diminished numbers of cells in most of their tissues. Except for a higher rate of infection, no other disorders were increased in these 39 infants. Heinonen et al. (1977) found no increase in defects among 11 exposed infants. The perinatal complications are discussed by Briggs et al. (1983). Withdrawal in the newborn is commonly seen. One study purporting an increase in congenital defects gives a rate of 2.4% but the control group had a rate of only 0.5% which would cause one to be skeptical. (Ostrea and Chavez, **1979**)

Taeusch et al. (1973) injected 3 to 12 mg of uncut heroin intravenously to the rabbit on days 24 through 26 of gestation. The fetal body weight was reduced but lung maturation was accelerated.

Briggs, G.G.; Bodendorfer, T.W.; Freeman, R.K. and Yaffe, S.J.: Drugs in Pregnancy and Lactation. Williams and Wilkins, Baltimore, 1983.

Heinonen, O.P.; Slone, D. and Shapiro, S.: Birth Defects and Drugs in Pregnancy. Publishing Sciences Group Inc., Littleton, Mass., 1977.

Naeye, R.L.; Blanc, W.; LeBlanc, W. and Khatamee, M.A.: Fetal complications of maternal heroin addiction. J. Pediatr. 83:1055–1061, 1973.

Ostrea, E.M. and Chavez, C.J.: Perinatal problems excluding neonatal withdrawal in maternal addiction: a study of 830 cases. J. Pediat. 94:292–295, 1979.

Taeusch, H.W.; Carson, S.H.; Wang, N.S. and Avery, M.E.: Heroin induction of lung maturation and growth retardation in fetal rabbits. J. Pediatr. 82:869–875, 1973.

1198 Herpes Simplex Virus I and II

South et al. (1969) reported a newborn infant with microcephaly, intracranial calcifications, eye defects and vesicular skin lesions. Type II herpes virus was recovered from the child and the mother reported genital blisters and vaginal discharge in the early weeks of pregnancy. Schaffer (1965) reported a similar patient with the herpes virus isolated. Florman et al. (1973) reported an infant infected with herpes I who had microcephaly, intracranial calcifications and owl eye inclusion bodies in the urine. They summarized five reported cases of herpes II infection. Eye and retinal disease are associated with the brain damage. All of the five cases reported were suspected because of a vesicular rash in the newborn period.

Hutto et al. (1985) reported that about 8 percent of neonatal herpes cases are acquired before onset of labor. Of 8 infants, all were premature. Manifestations included skin lesions with or without scars in 5 of 8, choreoretinitis in 2 of 4, microphthalmia in 3 of 7, microcephaly in 5 of 8, and other CNS lesions in 6 of 8. Four of the 8 mothers had primary HSV infection between 6–16 weeks.

Brown et al. (1987) followed 29 patients who had HSV type II during pregnancy. Fifteen had a primary first infection. Birth weight was less in those having the infection in the third trimester and the morbidity was 40 percent (intrauterine growth retardation and prematurity). In the group with non–primary infections, there were no adverse fetal effects. Baldwin and Whitely (1989) reviewed and discussed 71 cases of intrauterine acquired herpes. Herpes II accounted for 61 percent while HSV–I was found in seven percent. Thirty-two percent were not reported. Chorioretinitis, nervous system lesions (hydranencephaly) and cutaneous lesions within the first 48 hours of delivery were the main findings.

In a multicentered study of 210 babies, Whitely et al. (**1991**) found no difference in outcome of viralogic positive infants treated with vidarabine or acyclovir. About 60% of the parents were asmptomatic.

Heath et al. (1956) produced microcephaly and flexion deformities in chick embryos receiving herpes simplex at 48 hours of incubation.

Baldwin, S. and Whitley, R.J.: Intrauterine herpes simplex virus infection. Teratology 39:1–10, 1989.

Brown, Z.A.; Vontver, L.A.; Benedetti, J.; Critchlow, C.W.; Sells, C.J.; Berry, S. and Corey, L.: Effects on infants of a first episode of genital herpes during pregnancy. N. Eng. J. Med. 317:1246–1250, 1987.

Florman, A.L.; Gershon, A.A.; Blackett, P.R. and Nahmias, A.J.: Intrauterine infection with herpes simplex virus. Resultant congenital anomalies. JAMA 225:129–132, 1973.

Heath, H.D.; Shear, H.H.; Imagawa, D.T.; Jones, M.H. and Adams, J.M.: Teratogenic effects of herpes simplex vaccinia, influenza–A (NWS) and distemper virus infections on early chick embryos. Proc. Soc. Exp. Biol. Med. 92:675–682, 1956.

Hutto, C.; Willett, L.; Yeager, A. and Whitely, R.: Congenital herpes simplex virus (HSV) infection. Early vs late gestational acquisition. Pediatr. Res. 19:296A, 1985.

Schaffer, A.J.: Diseases of the Newborn. Philadelphia: W.B. Saunders Co. (2nd ed.), 733–734, 1965.

South, M.A.; Thompkins, W.A.F.; Morris, C.R. and Rawls, W.E.: Congenital malformation of the central nervous system associated with genital type (type 2) herpes virus. J. Pediatr. 75:13–18, 1969.

Whitley, R.; Arvin, A.; Prober, C.; Burchett, S.; Corey, L.; Powell, D.; Plotkin, S.; Starr, S.; Soong, S.J. and National Institue of Allergy and Infectious Diseases Collaborative Antiviral Study Group: A controlled trial comparing vidarabine with acyclovir in neonatal herpes simplex virus infection. The New England Journal of Medicine 324:444–449, 1991.

1199 Hexabrominated Naphthalenes

Miller and Birnbaum (1986) gavaged pregnant mice on days 6 to 15 with 0.5–10 mg per kg. Fetal survival was decreased at 5.0 mg per kg and above. Fetal hydrocephalus was increased at 0.5 mg per kg. Reduced weight of the thymus and spleen and subcutaneous edema was also reported. Skeletal anomalies and cleft palate were also increased.

Miller, C.P. and Birnbaum, L.S.: Teratologic evaluation of hexabrominated naphthalenes in C57BL6N mice. Fund. Appl. Toxicol. 7:398–405, 1986.

1200 Hexabromobenzene

Khera and Villeneuve (1975) gavage fed rats up to 200 mg per kg on days 6–15 of gestation. No adverse effects were found.

Khera, K.S. and Villeneuve, D.C.: Teratogenicity studies on halogenated benzenes (pentachloro–, pentachloronitro–, and hexabromo–) in rats. Toxicol. Appl. Pharmacol. 33:125, 1975.

1201 Hexachlorobenzene CAS 118–74–1

Khera (1974) gave single oral doses to rats during several periods of organogenesis and until day 21 of gestation. The doses varied from 10 to 120 mg per kg per day. At the higher doses, extra 14th ribs were found but this was associated in the maternal toxicity. The fetal weights and survival rates did not differ from controls and no other defects were found. Dominant lethal studies were negative. Courtney et al. (1976) used 100 mg per kg in the mouse and found no significant fetal changes. Andrews and Courtney (1986) treated rats and mice on days 15–20 and 6–16, respectively, and found hydronephrosis postnatally. The dose in both species was 10 mg per kg.

In rat offspring of dams given 10 or 100 mg per kg before breeding Goldey and Taylor (**1992**) found no physical differences from controls at birth but negative geotoxic reflex, olfactory discrimination and exploratory activities were altered.

The transplacental passage of the chemical was found in the rat and mouse (Svendsgaard et al., 1979); swine, (Hansen et al., 1979); rat (Villeneuve and Hierlihy, 1975); rabbit, (Villeneuve et al., 1974) and human (Astolfi et al., 1974).

Andrews, J.E. and Courtney, K.D.: Hexachlorobenzene-induced renal maldevelopment in CD–1 mice and CD rats. Hexachlorobenzene: Proceedings of an international symposium, C.R. Morris and J.R.P. Cabral (eds.), International Agency for Research on Cancer Scientific Publications 77:381–391, 1986.

Astolfi, E.; Alonso, A.H.; Mendizabal, A. and Zubizarreta, E.: Pesticides chlores de l'accorichee et du cordin ombilical des nouveau–nes. J. European Toxicol. 1:330–338, 1974.

Courtney, K.D.; Copeland, M.F. and Robbins, A.: The effects of pentachloronitrobenzene, hexachlorobenzene and related compounds on fetal development. Toxicol. Appl. Pharmacol. 35:239–256, 1976.

Goldey, E.S. and Taylor, D.H.: Developmental neurotoxicity following premating maternal exposure to hexachlorobenzene in rats. Neurotoxicology and Teratology 14:15–21, 1992.

Hansen, L.G.; Simon, J.; Dorn, S.B. and Teske, R.H.: Hexachlorobenzene distribution in tissues of swine. Toxicol. Appl. Pharmacol. 51:1–7, 1979.

Khera, K.S.: Teratogenicity and dominant lethal studies of hexachlorobenzene in rats. Food Cosmet. Toxicol. 12:471–477, 1974.

Svendsgaard, D.J.; Courtney, K.D. and Andrews, J.E.: Hexachlorobenzene (HCB) deposition in maternal and fetal tissues of rat and mouse. Environ. Res. 20:267–281, 1979.

Villeneuve, D.C. and Hierlihy, S.L.: Placental transfer of hexachlorobenzene in the rat. Bull. Environ. Contamin. Toxicol. 13:489–491, 1975.

Villeneuve, D.C.; Panopio, L.G. and Grant, D.L.: Placental transfer of hexachlorobenzene in the rabbit. Environ. Physiol. Biochem. 4:112–115, 1974.

1202 3,3',4,4',5,5'–Hexachlorobiphenyl CAS 32774–16–6

Marks et al. (1981) gavaged mice on days 6–15 of gestation with 0.1 to 16 mg per kg per day. At 8 mg per kg the dams had a decrease in weight gain. Cleft palate and hydronephrosis were found to be increased beginning at 2 and 4 mg dose levels, respectively.

Marks, T.A.; Kimmel, G.L. and Staples, R.E.: Influence of symmetrical polychlorinated biphenyl isomers on embryo and fetal development in mice. Toxicol. Appl. Pharmacol. 61:269–276, 1981.

1203 Hexachlorobutadiene

Hardin et al. (1981) injected 10 mg per kg intraperitoneally into rats on days 1–15. Maternal and fetal toxicity occurred but no teratogenicity.

Hardin, B.D.; Bond, G.P.; Sikov, M.R.; Andrew, F.W.; Beliles, R.P. and Niemeier, R.W.: Testing of selected workplace chemicals for teratogenic potential. Scand. J. Work Environ. Health 7(4):66–75, 1981.

1204 Hexachlorocyclohexane

Das et al. (1990) studied immune function in the offspring of mice treated with 19 or 100 mg per kg of this insecticide. Treatment was from day 0 to 18. The lower dose was found to be associated with an increase in many immune functions studied but the higher dose was not. At the highest dose about 25% of the offspring died. Although the newborn weights after 10 mg per kg were not significantly reduced the spleen weights were increased.

Das, S.N.; Paul, B.N.; Saxena, A.K. and Ray, P.K.: Effect of in utero exposure to hexachlorocyclohexane on the developing immune system of mice. Immunopharmacology and Immunotoxicology 12(2):293–310, 1990.

1205 Hexachlorocyclopentadiene

John et al. (1979) gavaged pregnant mice and rabbits with up to 75 mg per kg per day during active organogenesis and observed no teratogenic effect.

John, J.A.; Murray, F.J.; Murray, J.S.; Schwetz, B.A. and Staples, R.E.: Evaluation of environmental contaminants tetrachloroacetone, hexachlorocyclopentadiene and sulfuric acid aerosol for teratogenic potential in mice and rabbits. (abs) Teratology 19:32A–33A, 1979.

1206 Hexachlorophene CAS 70–30–4

Halling (1977) reported that among 65 women exposed to hexachlorophene by hand washing, there were 5 with serious defects: (2) Anal atresia, (1) kidney malformation, (1) esophageal atresia and (1) cleft lip and palate. Baltzar et al. (1979) studied 3007 pregnancies of hospital workers who used hexachlorophene extensively and compared them to 1653 pregnancies where hospital workers did not use it. No increase in defects or perinatal death was found.

Levels of 100 to 500 PPM fed to rats have not produced gross changes in the fetal brain; but at the highest levels which were toxic to the mothers, 3 of 44 fetuses had cleft palate (Oakley and Shepard, 1972). Kimbrough and Gaines (1971) reported cystic changes in the white matter of the brain of rats fed 500 PPM (25 mg per kg per day) hexachlorophene for two weeks. Kimmel et al. (1972) administered hexachlorophene to rats by intravaginal application

on days 7, 8, 9 and 10. The approximate dosage was 300 mg per kg daily. Microphthalmia, hydrocephalus and wavy ribs were produced in the fetuses. Only two cleft palates were produced out of 82 surviving fetuses. Kennedy et al. (1975) using 6 mg per kg on days 6 through 18 in the pregnant rabbit produced no soft tissue anomalies and only 3 out of 175 had rib malformations. Transplacental studies of the material in mice were reported (Brandt et al., 1979).

Baltzar, B.; Ericson, A. and Kallen, B.: Pregnancy outcome among women working in Swedish hospitals. N. Eng. J. Med. 300:627–628, 1979.

Brandt, I.; Dencker, L. and Larsson, Y.: Transplacental passage and embryonic–fetal accumulation of hexachlorophene in mice. Toxicol. Appl. Pharmacol. 49:393–401, 1979.

Halling, H.: Suspected link between exposure to hexachlorophene and birth of malformed infants. (Swedish) Lakartidningen 74:542–546, 1977.

Kennedy, G.L.; Smith, S.H.; Keplinger, M.L. and Calandra, J.C.: Evaluation of the teratological potential of hexachlorophene in rabbits and rats. Teratology 12:83–88, 1975.

Kimbrough, R.D. and Gaines, T.B.: Hexachlorophene effects on the rat brain. Arch. Environ. Health 23:114–118, 1971.

Kimmel, C.A.; Moore, W. and Stara, J.F.: Hexachlorophene teratogenicity in rats. Lancet 2:765 only, 1972.

Oakley, G.P. and Shepard, T.H.: Possible teratogenicity of hexachlorophene in rats. (abs) Teratology 4:264 only, 1972.

1207 Hexacol

Watanabe et al. (1965) gave this antitussive to mice and rats during organogenesis in amounts up to 100 mg per kg and found no adverse effect on the fetuses.

Watanabe, N.; Iwanami, K. and Nakahara, N.: Hexacol toxicology. Yakugaku Kenkyu 3:95–110, 1965.

1208 Hexafluoroacetone CAS 648–16–2

Brittelli et al. (1979) applied this organic solvent to the skin of pregnant rats on days 6 through 16 and at absorbed doses of 5 and 25 mg per kg, found hematomas, hydronephrosis, hydrocephalus and other defects. Some hydronephrosis increase was found at a dose of 1 mg per kg.

Brittelli, M.R.; Culik, R.; Dascheill, O.L. and Fayerweather, W.E.: Skin absorption of hexafluoroacetone: Teratogenic and lethal effects in the rat fetus. Toxicol. Appl. Pharmacol. 47:35–39, 1979.

1209 2–(β–Hexamethyleniminoethyl)cyclohexanone–2–carboxylic Acid–benzylester–hydrochloride
Amicibone

Kraushaar et al. (1964) gavaged rats during gestation with 45 mg per kg and found no adverse effects on the fetuses.

Kraushaar, V.A.E.; Schunk, R.W. and Thym, H.F.: Ein neues antitussivum: 2–(β–hexamethyleniminoathyl)–cyclohexanon–2–carbonsaurebenzylester–HCl. Arzneim. Forsch 14:986–995, 1964.

1210 2(β–Hexamethyleniminoethyl)cyclohexanone–2–carboxylic acid-nenzylester HCl)

Kraushaar et al. (1964) studied this antitussive agent in pregnant rats using up to 100 mg per kg by gavage daily during gestation. The only adverse effect was a slight increase in resorptions at the highest dose.

Kraushaar, A.E.; Schunk, R.W. and Thym, H.F.: Ein neues Antitussivum: 2–(β–Hexamethyleniminoethyl)–cyclo-hexanon–2–carbonsaurebenzyl ester–HCl. Arzneim Forsch. 14:986–995, 1964.

1211 Hexamethylphosphoramide Triamide CAS 680–31–9

Rats and rabbits were treated before and during pregnancy with 2 or 10 mg per kg orally. No teratologic effects were found (Shott et al., 1971).

Shott, L.D.; Borhovec, A.B. and Knapp, W.A., Jr.: Toxicology of hexamethylphosphoric triamide in rats and rabbits. Toxicol. Appl. Pharmacol. 18:499–506, 1971.

1212 N–Hexane and Metabolites CAS 110–54–3

N–hexane can be metabolically activated to methyl n–butyl ketone (MBK) and 2,5–hexanedione (2,5–HD). Pregnant rats were exposed 6 hours per day to 1000 ppm N–hexane on days 8–12, 12–16, or 8–16 of gestation (Bus et al., 1979). No increase in resorptions, fetal deaths or malformations were observed in the fetus. Concentrations in the fetuses of N–hexane, MBK and 2,5–HD were approximately equal to that in the maternal blood.

Bus, J.J.; White, E.L.; Tyl, R.W. and Barrow, C.S.: Perinatal toxicity and metabolism of n–hexane in Fischer 344 rats after inhalation exposure during gestation. Toxicol. Appl. Pharmacol. 51:295–302, 1979.

1213 Hexanoic Acid

Dawson (1991) using a frog embryo teratogenesis assay found that at a concentration of 108 mg per liter, 50% of the offspring were malformed. Microcephaly, gut and skeletal defects were the main defects found.

Dawson, D.A.: Additive incidence of developmental malformation for xenopus embryos exposed to a mixture of ten aliphatic carboxylic acids. Teratology 44:531–546, 1991.

1214 1–Hexanol

Nelson et al. (1988) exposed rats to vapor for seven hours daily during gestation. At the highest concentration, 14000 mg per meter cubed, no adverse reproductive effects were found. Limited maternal toxicity was seen.

Nelson, B.K.; Brightwell, W.S.; Khan, A.; Hoberman, A.M. and Krieg, E.F.: Teratological evaluation of 1–pentanol, 1–hexanol and 2–ethyl–1–hexanol administered by inhalation to rats. Teratology 37:479–480, 1988.

1215 Hexazinone CAS 51235–04–2

Kennedy and Kaplan (1984) fed this chemical to pregnant rabbits and rats during organogenesis. No adverse effects were found in the rabbit (125 mg per kg) or in the rat (5,000 mg per kg).

Kennedy, G.L. and Kaplan, A.M.: Chronic toxicity, reproductive and teratogenic studies of hexazinone. Fund. Appl. Toxicol. 4:960–971, 1984.

1216 Hexobarbital also see *Barbituric Acid* CAS 56–29–1

Persaud (1965) treated 10 pregnant rats with 50 to 100 mg injections on the 4th, 8th and 18th days and observed no defective offspring.

Persaud, T.V.N.: Tierexperimentelle Untersuchungen zur Frage der Teratogenen Wirkung von Barbituraten. Acta Biol. Med. Ger. 14:89–90, 1965.

1217 Hexoprenaline

Pinder et al. (**1977**) gave rats 5 mg per kg orally between days 6–15 and produced no evidence of teratogenicity although there was a slight decrease in fertility index. Rabbits receiving 500 microgm per kg during days 16–18 of pregnancy had an increase in stillborn but no dysmorphogenic effects.

Pinder, R.M.; Brogden, R.N.; Speight, T.M. and Avery, G.S.: Hexoprenaline: A review of its pharmacological properties and therapeutic efficacy with particular reference to asthma. Drugs 14:1–28, 1977.

1218 1–Hexylcarbamoyl–5–fluorouracil

Sato et al. (1980) gave this antimetabolite orally to pregnant rats and rabbits. No ill effects on fertility were found using up to 50 mg per kg in both rat sexes. Administration of the same dose during organogenesis was associated with a few skeletal variations in the fetuses. At a dose of 100 mg per kg, there was increased fetal death. No adverse postnatal findings were detected. In the rabbit they used 50 mg per kg and no adverse fetal effects occurred.

Sato, T.; Nagaoka, T.; Kaneko, Y.; Osuga, F.; Naramo, I. and Sejima, Y.: Reproductive studies of 1–hexylcarbamoyl–5–fluorouracil. Kiso to Rinsho 14:1373–1402, 1980.

1219 Hippurate

Rodwell et al. (1980) fed rats 5 percent sodium hippurate in their diet during pregnancy and found an increased number of fetuses with anencephaly and hydrocephalus.

Rodwell, D.E.; Schoenig, G.P.; Goldenthal, E.I. and Wazeter, F.X.: The teratogenic effect of sodium hippurate administered in the diet. (abs) Teratology 21:65A, 1980.

1220 Histamine CAS 51–45–6

Gatling (1962) dropped 50 microgm of histamine on the chorio–allantoic membrane of the 11–day chick embryo and found no defects. Kameswaran et al. (1962) studied the physiologic role of histamine in rat pregnancy and concluded that it controls the blood flow through the placenta.

Gatling, R.R.: The effect of sympathomimetic agents on the chick embryo. Am. J. Pathol. 40:113–127, 1962.

Kameswaran, L.; Pennefeather, J.N. and West, G.B.: Possible role of histamine in rat pregnancy. J. Physiol. 164:138–149, 1962.

1221 HLA Antigens *Major Histocompatibility Antigens*

Gill (1987) has summarized data indicating that couples or mice sharing major histocompatibility loci have higher abortion rates. In the mouse the MHC genes are in close association with the t–gene which is associated with embryonic lethality and congenital defects. He hypothesizes that complementation of lethal genes from the two parents produces lethality. Other genetic mechanisms are suggested such as a multiple chromosome hypothesis with two recessive lethals.

Gill, T.J.: Genetic factors in fetal losses. American Journal of Reproductive Immunology 15:133–137, 1987.

1222 Hog Cholera Vaccine

Sautter et al. (1953) injected this vaccine on the 14th to 16th gestational days and produced a syndrome in pigs consisting of generalized edema, ascites, mottling of the liver and skeletal defects. Small lungs and pitted kidneys were found in other effected fetuses (Young et al., 1955).

Cerebellar hypoplasia with congenital tremors and hypomyelinogenesis was associated with vaccination against hog cholera when given from the 20th to 97th gestational days (Emerson and Delez, 1965). Macrocephaly, contraction of tendons, edema and hemorrhages were also found in the pig fetuses.

Emerson, J.L. and Delez, A.L.: Cerebellar hypoplasia, hypomyelinogenesis and congenital tremors of pigs, associated with prenatal hog cholera vaccination of sows. J. Am. Vet. Med. Assoc. 147:47–54, 1965.

Sautter, J.H.; Young, G.A.; Luedke, A.J. and Kitchell, R.L.: The experimental production of malformations and other abnormalities in fetal pigs by means of attenuated hog cholera virus. J. Am. Vet. Med. Assoc. 90:146–150, 1953.

Young, G.A.; Kitchell, R.L.; Luedke, A.J. and Sautter, J.H.: The effect of viral and other infections of the dam on fetal development in swine. I. Modified live hog cholera viruses, immunological, virological and gross pathologic studies. J. Am. Vet. Med. Assoc. 126:165–171, 1955.

1223 Homatropine CAS 87–00–3

Heinonen et al. (1977) found no increase in defects among the offspring of 26 women exposed in the first trimester.

Heinonen, O.P.; Slone, D. and Shapiro, S.: Birth Defects and Drugs in Pregnancy. Publishing Sciences Group Inc., Littleton, Mass., 1977.

1224 Homocysteinuria *Homocysteine*

Steegers–Theunissen et al. (**1992**) found 6 of 24 women with two or more successive fetal losses had homocysteine levels greater than normal after a loading test. Folate and B^{12} levels were studied and found lower than the controls. Burke et al (**1992**) could not identify elevated homocysteine levels in women who gave birth to infants with intrauterine growth failure.

Vanaerts et al. (**1994**) found in whole rat embryo culture that concentrations of 6 and 12 mM were toxic but at 1–2 mM there was growth enhancement. Homocystine at 2mM also enhanced growth. A number of other experiments using combinations of folic acid and B$_{12}$ were carried out with these studies.

Burke, G.; Robinson, K.; Refsum, H.; Stuart, B.; Drumm, J. and Graham, I.: Intrauterine growth retardation, perinatal death, and maternal homocysteine levels. The New England Journal of Medicine 326:69–70, 1992.

Steegers–Theunissen, R.P.M.; Boers, G.H.; Blom, H.J.; Trijbels, F.J.M. and Eskes, T.K.A.B.: Hyperhomocysteinaemia and recurrent spontaneous abortion or abruptio placentae. The Lancet 339:1122–1123, 1992.

Vanaerts, L.A.G.J.M.; Blom, H.J.; Deabreu, R.A.; Trijbels, F.J.M.; Eskes, T.K.A.B.; Peereboom–Stegeman, J.H.J.C and Noordhoek, J.: Prevention of neural tube defects by and toxicity of l–homocysteine in cultured postimplantation rat embryos. Teratology 50:348–360, 1994.

1225 Hopantenic Acid *Homopantothenic Acid Butanoic Acid HOPA*
4–((2,4–Dihydroxy–3,3–dimethyl–1–oxobutyl)–amino) CAS 18679–90–8

Asano et al. (1980) gave rats up to 60 mg per kg on days 17–21 of gestation. An increase in stillbirths and a decrease in postnatal survival was found at the highest dose. Reproduction by the offspring was not adversely affected. Nishizawa et al. (1969) found no evidence of teratogenicity in the rat and mouse. They gave 2 gm per kg orally to both species on days 7–14, during active organogenesis.

Asano, Y.; Ariyuki, F. and Higaki, K.: Pre– and post-natal studies of calcium D(+)–4–(2,4–dihydroxy–3,3–dimethylbutramide) butyrate hemidrate (HOPA) in rats. Oyo Yakuri 19:1011–1017, 1980.

Nishizawa, Y.; Kodama, T.; Noguchi, Y.; Nakayama, Y.; Hori, M. and Kowa, Y.: Chronic toxicity and teratogenic effect of homopantothenic acid. J. Vitaminol. 15:26–32, 1969.

1226 HP 1325 *2–2(p–Phenylenedioxy)–di–(ethyl hydrazinium)*

Poulson and Robson (1963) gave 2 mg per day and produced a decrease in the number of live litters. No decrease was found when the administrations were on days 11–16.

Poulson, E. and Robson, J.M.: The effect of amine oxidase inhibitors on pregnancy. J. Endocrin. 27:147–152, 1963.

1227 HVJ Virus *Hemagglutinating Virus of Japan*

Ohba (1959) administered chorio–allantoic fluid containing this virus intranasally or intravenously to mice on the 8th day of pregnancy and some abnormalities of the central nervous system were observed in the embryos. The author found 20.9 percent abnormal embryos in the treated group and only 4.4 percent in the controls, but he was cautious in ascribing the cause to the viral agent itself.

Ohba, N.: Formation of embryonic abnormalities of the mouse by a viral injection of mother animals. Acta Pathol. Jpn. 9:149–157, 1959.

1228 Hyaluronate, Sodium

Furuhashi et al. (1985) studied reproduction in rats and rabbits. At maximum doses of 60 mg per kg subcutaneously in rats and intraperitoneally in rabbits no adverse reproductive effects were found. Treatments in the rat were before fertilization and during the first 7 days, during organogenesis and in the perinatal period. The rabbits were treated on days 6–18. Ono et al. (1992 A&B) gave up to 50 mg per kg to rats before fertilization and in the early stages of pregnancy and found no adverse fetal effects. Similar studies in the perinatal and postnatal period had no adverse effects on the behavior or reproduction of the offspring (Ono et al., 1993 C). Rabbits given up to 50 mg per kg did not evidence teratogenicity. In both species the treated dams had increased weight due to subcutaneous retention of hyaluronate. Similar studies by Tanaka et al. (1991) and Wada et al. (1991) confirmed the lack of teratogenicity in rats and rabbits.

Furuhashi, T.; Nakayoshi, H.; Nakazawa, M.; Uehara, M. and Honda, T.: Reproduction studies of sodium hyaluronate in rats and rabbits. Oyo Yakuri 29:95–153, 1985.

Ono, C.; Fujiwara, Y.; Koura, S.; Tsuchida, H. and Nakamura, T.: Reproductive and developmental toxicity study on sodium hyaluronate (SH)–(2) study on subcutaneous administration to rats prior to and in the early stages of pregnancy. Yakuri to Chiryo 20:729–737, 1992A.

Ono, C.; Iwama, A.; Nakajima, Y.; Kitsuya, A. and Nakamura, T.: Reproductive and developmental toxicity study on sodium hyaluronate (SH)–(1) Study on subcutaneous administration to rats during the period of organogenesis. Yakuri to Chiryo 20:713–728, 1992B.

Ono, C.; Ishitobi, H.; Kuzuoka, K.; Konagai, S. and Nakamura, T.: Reproductive and developmental toxicity study on sodium hyaluronate (SH)–(3) Study on subcutaneous administration to rats during the perinatal and lactation period. Yakuri to Chiryo 20:739–752, 1992C.

Tanaka, C.; Sasa, H.; Hirama, S.; Inaba, T.; Tokunaga, S.; Eiro, H. and Kuramoto, M.: Reproductive and developmental toxicity studies of sodium hyaluronate (SL–1010) (I): Fertility study in rats. Yakuri to Chiryo 19: s81–s92, 1991.

Tateda, C.; Nagaoka, S.; Nagai, T. and Nakamura, T.: Reproductive and developmental toxicity study on sodium hyaluronate (SH)–(4) Study on subcutaneous administration to rabbits during the period of organogenesis. Yakuri to Chiryo 20:753–760, 1992.

Wada, K.; Hashimoto, Y.; Mizutani, M. and Tanaka, C.: Reproductive and developmental toxicity studies of sodium hyaluronate (SL001919) (III): Teratogenicity study in rabbits. Yakuri to Chiryo 19:s111–s119, 1991.

1229 Hyaluronate, Sodium High Molecular Weight
NRD101

Matsuura et al. (1994 A,B,C) gave this intraperitoneally to rats in doses of up to 64 mg per kg and found no adverse effects on reproduction, fetal development or behavior of the pups. In the rabbit, Matsuura et al. (1994) found no teratogenic effects of doses of up to 40 mg per kg during organogenesis.

Matsuura, T.; Nakajima, H.; Maeda, H.; Ozaki, K.; Kurio, W.; Uechi, S.; Hiramatsu, Y.; Ogawa, Y.; Ishihara, R. and Miyoshi, T.: Fertility study of high molecular weight sodium hyaluronate (NRD101) in rats. Yakuri to Chiryo 22:s565–s576, 1994A.

Matsuura, T.; Nakajima, H.; Maeda, H.; Ozaki, K.; Kurio, W.; Uechi, S.; Hiramatsu, Y.; Ogawa, Y.; Ishihara, R. and Miyoshi, T.: Teratological study of high molecular weight sodkum hyaluronate (NRD101) in rats. Yakuri to Chiryo 22:s577–s595, 1994B.

Matsuura, T.; Nakajima, H.; Maeda, H.; Ozaki, K.; Kurio, W.; Uechi, S.; Hiramatsu, Y.; Ogawa, Y.; Ishihara, R. and Miyoshi, T.: Perinatal and postnatal study of high molecular weight sodium hyaluronate (NRD101) in rats. Yakuri to Chiryo 22:s607–s625, 1994C.

Matsuura, T.; Nakajima, H.; Maeda, H.; Ozaki, K.; Kurio, W.; Uechi, S.; Hiramatsu, Y.; Ogawa, Y.; Ishihara, R. and Miyoshi, T.: Teratological study of high molecular weight sodium hyaluronate (NRD101) in rabbits. Yakuri to Chiryo 22:s597–s605, 1994D.

1230 Hycanthone Methane Sulfate CAS 23255–93–8

Moore (1972) administered intramuscularly, either 35 or 50 mg per kg to mice on the 7th gestational day. There was a high incidence of congenital defects which were mainly exencephaly, hydrocephaly, microphthalmia or skeletal. At a dose level of 10 mg per kg there was no increase in fetal defects. Human therapeutic doses are 3 to 4 mg per kg.

Moore, J.A.: Teratogenicity of hycanthone in mice. Nature 239:107–109, 1972.

1231 Hydergine

Sommer and Buchanan (1955) injected rats intraperitoneally with 0.5 mg twice daily from the 12th day until the 21st after vaginal sperm were found. Defects were not studied but the newborn fetuses were not reduced in weight.

Sommer, A.F. and Buchanan, A.R.: Effects of ergot alkaloids on pregnancy and lactation in the albino rat. Am. J. Physiol. 180:296–300, 1955.

1232 Hydralazine *1–Hydrazinephthalazine* CAS 86–54–4

Briggs et al. (1983) summarized 196 treated pregnancies in pre–eclamptic and eclamptic women and found no increase in congenital malformations.

Neonatal thrombocytopenia and bleeding were found (Widerlov et al., 1980). This compound which inhibits hydroxylation steps in collagen synthesis was used to produce skeletal defects which are similar to those produced by manganese deficiency (Rapaka et al., 1977). Matsuo et al. (1986) fed rats 80 ppm in their water during their entire pregnancy or during the last two weeks. Retardation of ossification occurred but no decrease in fetal weight.

Danielsson et al. (1989) gavaged rabbits with 763 micromoles per kg on day 16 and found reduction of the 4th hind paw digit on day 29.

Briggs, G.G.; Bodendorfer, T.W.; Freeman, R.K. and Yaffe, S.J.: Drugs in Pregnancy and Lactation. Williams and Wilkins, Baltimore, 1983.

Danielsson, B.R.G.; Reukabdm S,; Rundqvist, E. and Danielson, M.: Digital defects induced by vasodilating agents: relationship to reduction in uteroplacental blood flow. Teratology 40:351–358, 1989.

Matsuo, T.; Sakamoto, M.; Kihara, T. and Tanimura, T.: Effects of hydralazine hydrochloride on the skeletal development of spontaneously hypertensive rats (SHR). Med. J. Kinki Univ. 11:1–6, 1986.

Rapaka, R.S.; Parr, R.W.; Lin, T–Z. and Bhatnagar, R.S.: Biochemical basis of skeletal defects induced by hydralazine. Teratology 15:185–194, 1977.

Widerlov, E.; Karlman, I. and Stoisater, J.: Hydralazine–induced neonatal thrombocytopenia. N. Eng. J. Med. 303:1235–1238, 1980.

1233   Hydrazine CAS 302–01–2

Stoll et al. (1967) using doses close to the LD–50, produced a low incidence of skeletal defects in chicks receiving 30–200 microgm on the third day of incubation.

Several derivatives of hydrazine have been found to be embryotoxic, but only in the early embryo (Poulson and Robson, 1963). Lee and Aleyassine (1970) injected rats with 8 mg per kg subcutaneously on the last 10 days of gestation. This dose caused maternal weight loss and the fetuses were small, pale and edematous but without defects.

Lee, S.H. and Aleyassine, H.: Hydrazine toxicity in pregnant rats. Arch. Environ. Health 21:615–619, 1970.

Poulson, E. and Robson, J.M.: The effect of amine oxidase inhibitors on pregnancy. J. Endocrinol. 27:147–152, 1963.

Stoll, R.; Bodit, F. and Maraud, R.: Sur l'action teratogene de hydrazine et de substances voisines chez l'embryon de poulet. C. R. Soc. Biol. (Paris) 161:1680–1684, 1967.

1234   p–Hydrazinobenzoic Acid CAS 619–67–0

Discussed under Semicarbazide

1235   Hydrochlorothiazide CAS 58–93–5

Heinonen et al. (1977) reported no significant increase in defect rate among 107 women exposed during the first four months.

Maren and Ellison (1972) using 250 mg per kg on gestational days 9, 10, 11 and 12 produced no defects in the rat fetus.

Heinonen, O.P.; Slone, D. and Shapiro, S.: Birth Defects and Drugs in Pregnancy. Publishing Sciences Group Inc., Littleton, Mass., 1977.

Maren, T.H. and Ellison, A.C.: The teratological effect of certain thiadiazoles related to acetazolamide, with a note on sulfanilamide and thiazide diuretics. Johns Hopkins Med. J. 130:95–104, 1972.

1236   Hydrocortisone *Cortisol* also see *Cortisone* CAS 50–23–7

Pinsky and DiGeorge (1965) compared the cleft palate producing effect of this compound with several other corticoids.

Kalter and Fraser (1952) observed that when hydrocortisone was given to mice in doses of 2.5 mg on the 10th or 11th day many offspring had cleft palates. Using the AJAX mouse they found that injections of 4 mg on days 11 through 18 produced 18 percent cleft palate. Aoyama et al. (1974) studied the effect of hydrocortisone–17–α butyrate in pregnant mice and rats. Doses were given subcutaneously up to 1.0 mg per kg per day in mice and 9 mg per kg per day in rats. No increase in congenital malformations was found.

Hydrocortisone 17–butyrate 21–propionate was studied by subcutaneous route in rats and rabbits by Yamada et al. (1981A). In the male and female rat, doses of up to 0.4 mg per kg per day subcutaneously did not interfere with reproduction. During organogenesis doses of 10 and 50 mg per kg increased congenital defects (omphaloceles, cleft palate and edema). Doses of one mg perinatally had no adverse effect on postnatal behavior.

In the rabbit, at 0.5 mg per kg during organogenesis cleft palate and open neural tube defects were increased over controls. Using single daily doses subcutaneously of up to 200 mg per kg, Yamada et al. (1981B) found resorption, fetal growth retardation and an increase in umbilical hernias occurred at the highest dose. Some fetal weight decrease and lethality was found with percutaneous doses of 0.5 percent cream or ointment (Yamada et al., 1981C).

Aoyama, T.; Furuoka, R.; Hasegawa, N. and Terabayashi, M.: Teratological studies on hydrocortisone–17–α–butyrate (H–17B) in mice and rats. Oyo Yakuri 8:1035–1047, 1974.

Kalter, H. and Fraser, F.C.: Production of congenital defects in the offspring of pregnant mice treated with compound f. Nature 169:665 only, 1952.

Pinsky, L. and DiGeorge, A.M.: Cleft palate in the mouse: A teratogenic index of glucocorticoid potency. Science 147:402–403, 1965.

Yamada, T.; Nogariya, T.; Ichikawa, A.; Nakane, S.; Sasajima, M. and Ohzeki, M.: Teratogenicity study in rats by a single injection. Yakuri to Chiryo 9:3083–3104, 1981B.

Yamada, T.; Nogariya, R.; Ichikawa, A.; Nakane, S.; Sasajima, M. and Ohzeki, M.: Hydrocortisone 17–butyrate 21–propionate teratogenicity study in rats by percutaneous administration. Yakuri to Chiryo 9:3045–3082, 1981C.

Yamada, T.; Suzuki, H.; Matsumoto, S.; Nakane, S.; Sasajima, M. and Ohzeki, M.: Reproductive studies of hydrocortisone 17–butyrate 21–propionate in rats and rabbits. Oyo Yakuri 21:427–482, 1981A.

1237   Hydrodextran Sulfate Potassium

Pares et al. (1969) gave 20 mg per kg in the drinking water of rats and mice during the entire pregnancy and found no adverse effects in the fetuses. Guinea pigs were treated similarly with 100 mg per kg and no adverse effects were observed in the fetuses.

Pares, P.J.; Drobnic, M.M.; Taxonera, F. and Sabater–Tobella, J.: Proprietes physico–chimiques et etude pharmacologique et toxicologique de l'hydrodextrane–sulfate potassique. Therapie XXIV:1071–1087, 1969.

1238 Hydrogen Peroxide CAS 7722–84–1

Moriyama et al. (1982) fed pregnant rats a diet containing up to 10 percent hydrogen peroxide. Maternal and fetal weights were reduced but no significant malformations were reported.

Moriyama, I.; Fuyita, M.; Hiraoka, K.; Ichija, M. and Kanoh, S.: Effects of food additive hydrogen peroxide on fetal development. (abs) Teratology 26:28A, 1982.

1239 Hydrogen Sulfide

Saillenfait et al. (**1989**) exposed rats to 100 ppm of hydrogen sulfide for 6 hours daily on days 6–20. No adverse fetal or maternal effects were found. The same concentration did enhance fetal toxicity when combined with carbon disulfide.

Saillenfait, A.M.; Bonnet, P. and de Ceaurriz, J.: Effects of inhalation exposure to carbon disulfide and its combination with hydrogen sulfide on embryonal and fetal development in rats. Toxicology Letters 48:57–66, 1989.

1240 Hydromorphone *Dilaudid*™

Behm et al. (1985) injected mice with up to 5 mg per kg using a minipump on single days from the the 7th through the 12th day. Minor skeletal defects were found to be slightly increased. A dose response was not consistently found. Controls with saline injection were used but pair feeding was not.

Behm, M.C.; Stout–Caputi, M.V.; Mahalik, M.P. and Gautieri, R.F.: Evaluation of the teratogenic potential of hydromorphone administered via a miniature implantable pump in mice. Research Communications in Substances of Abuse. 6(3):165–177, 1985.

1241 Hydroquinone

Krasavage et al. (**1992**) gavaged rats on days 6–15 with up to 300 mg per kg. Slight maternal weight loss occurred. No adverse reproductive effects were found except for increased variations in the vertebrae.

Krasavage, W.J.; Blacker, A.M.; English, J.C. and Murphy, S.J.: Hydroquinone: a developmental toxicity study in rats. Fundamental and Applied Toxicology 18:370–375, 1992.

1242 Hydroxybenzoic Acids

Koshakji and Schulert (**1973**) injected rats subcutaneously with 380 mg per kg on days 9–16 and produced no increase in resorptions or defects in the fetuses. Various dihydro and trihydrobenzoic acids were used. Larsson and Bostrom (**1965**) gave 10 mg of p–hgydroxybenzoic acid to rats on the 9th or 12th gestational day and did not produce an increase in defects.

Koshakji, R.P. and Schulert, A.R.: Biochemical mechanisms of salicylate teratology in the rat. Biochemical Pharmacology 22:407–416, 1973.

Larsson, K.S. and Bostrom, H.: Teratogenic action of salicylates related to the inhibition of mucopolysaccharide synthesis. Acta Paediatrica Scandinavica 54:43–38, 1965.

1243 β–Hydroxybutyrate

Horton et al. (1985) studied the effect of this ketone on development and metabolism of the explanted mouse embryo and fetus. During early embryonic stages (day 9) addition of the ketone produced no significant alteration in CO_2 production. On days 11 and 17, an effect was seen and evidence for oxidation of hydroxybutyrate was obtained. Zusman et al. (1987) found that the development of mouse ova was inhibited by concentrations of 6 mg per ml or higher. Hunter and Sadler (1987) studied the effect of the naturally occurring D–isomer of hydroxybutyrate in whole mouse culture. Malformations were increased in a dose dependent manner. Glucose utilization by the pentose cycle and Krebs cycle was decreased. Shum and Sadler (**1990**) studied the effect of this ketone on invitro grown mouse embryos and on their pentose shunt pathway.

Horton, W.E.; Sadler, T.W. and Hunter, E.S.: Effects of hyperketonemia on mouse embryonic and fetal glucose metabolism in vitro. Teratology 31:227–233, 1985.

Hunter, E.S. and Sadler, T.W.: D–(–)–β–Hydroxybutyrate induced effects on mouse embryos. Teratology 36:259–264, 1987.

Shum, L. and Sadler, T.W.: Biochemical basis for d,l,—beta–hydroxybutyrate–induced teratogenesis. Teratology, 42:553–563, 1990.

Zusman, I.; Yaffe, P. and Ornoy, A.: Effects of metabolic factors in the diabetic state on the in vitro development of preimplantation mouse embryos. Teratology 35:77–85, 1987.

1244 7–α–Hydroxycholesterol

Chan et al. (**1994**) studied the effect of this bishemisuccinate on rat whole embryo culture and found that at 100 microgm per ml embryotoxicity occurred. No changes in growth were found at 50 microgm per ml.

Chan, W.Y.; Ng, T.B.; Rong, S.H.; Yeung, H.W. and Mok, S.C.: Antiproliferative and teratogenic activities of the bishemisuccinates of 7–α–hydroxycholesterol and 7–β–hydroxycholesterol. Gen. Pharmac. 25:767–772. 1994.

1245 7–β–Hydroxycholesterol

Chan et al. (**1994**) studied the effect of this bishemisuccinate on rat whole embryo culture and found that at 100 microgm per ml embryotoxicity occurred. No changes in growth were found at 50 microgm per ml.

Chan, W.Y.; Ng, T.B.; Rong, S.H.; Yeung, H.W. and Mok, S.C.: Antiproliferative and teratogenic activities of the bishemisuccinates of 7–α–hydroxycholesterol and 7–β–hydroxycholesterol. Gen. Pharmac. 25:767–772. 1994.

1246 6–Hydroxydopamine

MacDonald and Airaksinen (1974) gave 100 mg per kg weekly to rats and found increased stillbirths and resorptions.

MacDonald, E.J. and Airaksinen, M.M.: The effect of 6–hydroxydopamine on the oestrus cycle and fertility of rats. J. Pharm. Pharmacol. 16:518–521, 1974.

1247 N–Hydroxyethylpromethazine Chloride *Aprobit*™
CAS 2090–54–2

West (1962) reported that this antihistamine which accumulates in the placenta without crossing was toxic to the rat fetus at doses of 0.25 mg per kg. Other antihistamines, cyproheptadine and promethazine produced fetal death at a much higher dose (25 mg per kg). Details including the presence or absence of fetal defects were not included in this report.

West, G.B.: Drugs and rat pregnancy. J. Pharm. Pharmacol. 14:828–830, 1962.

1248 Hydroxyethyl Starch *Hetastarch Hespander*™

Irikura et al. (1972) injected this plasma substitute intravenously into mice and rabbits during organogenesis. Doses of up to 60 and 75 ml per kg did not produce teratogenic effects. Ivankovic and Bulow (1975) found no teratogenicity from intraperitoneal doses of up to 50 g per kg daily in mice and rats. Fetal death was increased at maternally toxic doses.

Irikura, T.; Hosomi, J.; Ishiyama, N. and Suzuki, H.: Studies on hydroxyethyl starch solution (Hespander™) as a plasma subsitute (IX). Teratological studies in mice and rabbits. Oyo Yakuri 6:1119–1128, 1972.

Ivankovic, S. and Bulow, I.: On the lacking teratogenic effect of the plasmaexpander hydroxyethyl starch in the rat and mouse. Anaesthesist 24:244–245, 1975.

1249 N–(2, Hydroxyethyl)4–nitro–o–p–phenylenediamine

Discussed under Hair Dyes

1250 4–(2–Hydroxy–3–isopropylaminopropoxy)–indole
LB–46

This β adrenergic inhibitor was studied in the pregnant mouse and rabbit during organogenesis using oral doses of up to 56 and 32 mg per kg, respectively. No adverse outcome was seen (Maruyama et al., 1970).

Maruyama, S.; Kawai, Y.; Katano, Y.; Takahashi, K. and Matsuda, K.: Studies on acute, subacute, and chronic toxicities and teratogenic effect of 4–(2–hydroxy–3–isopropylaminopropoxy)–indole (LB–46) in animals. Niigata Igakhai Zasshi 84:438–456, 1970.

1251
1L–[1(OH)2,4,5/3]–5–[2–hydroxy–1(hydroxymethyl)ethyl] amino–1–C–hydroxymethyl01,2,3,4–cyclohexanetetrol

Morseth and Nakatsu (1991 A,B,C,D) studied this antidiabetic in pregnant rats and rabbits. In rats exposed to 100 mg per kg the survival rate at weaning was decreased but no teratogenic effects were found. No effect on fertility was found at 100 mg per kg. Postnatal death was increased at 200 mg per kg which affected lactation. In the rabbit treated during organogenesis resorption was increased at 100, 100 and 1000 mg per kg and maternal weight gain decreased. No teratoginic effect was found.

Morseth, S.L. and Nakatsu, T.: Reproduction study in rats with AO–128, Jpn. Pharm. Therap. 19:4325–4340, 1991A.

Morseth, S.L. and Nakatsu, T.: Teratological study in rats with AO–128. Jpn. Pharm. Therap. 19:4341–4363, 1991B.

Morseth, S.L. and Nakatsu, T.: Perinatal and postnatal study in rats with AO–128. Jpn. Pharm. Therap. 19:4375–4396, 1991C.

Morseth, S.L. and Nakatsu, T.: Teratological study in rabbits with AO–128. Jpn. Pharm. Therap. 19:4365–4374, 1991D.

1252 Hydroxylamine CAS 7803–49–8

Chaube and Murphy (1966) injected rats intraperitoneally on single days 9–12 of gestation with up to 125 mg per kg and did not produce an increase in defects. Increased fetal loss was found at doses of 59 mg per kg or above. Intravenous administration to rabbits (Zimmerman and Gottschewski, 1966) on day 3 or 8 with 10 mg per kg was associated with an increase in defects of the head regions of embryos. DeSesso (1980) found craniofacial defects in rabbits after administering up to 200 microgm into the intracolomic sites. Winter and Hokanson (1964) fed heifers the hydrochloride form and found no defects in four calves.

Chaube, S. and Murphy, M.L.: The effects of hydroxyurea and related compounds on the rat fetus. Cancer Res. 26:1448–1457, 1966.

DeSesso, J.M.: Demonstration of the embryotoxic effects of hydroxylamine in the New Zealand white rabbit. Anat. Rec. 196:45A–46A, 1980.

Winter, A.J. and Hokanson, J.F.: Effect of long–term feeding of nitrate, nitrite or hydroxylamine on pregnant dairy heifers. Am. J. Vet. Res. 25:353–361, 1964.

Zimmermann, W. and Gottschewski, G.H.M.: Die Wirkung bestimmer Substanzen auf die DNS–und Protein-synthese in der fruhen Embryonalentwicklung des Kaninchens. Bull. Schweiz. Akad. Med. Wiss. 22:166–183, 1966.

1253 6–Hydroxylaminopurine CAS 5667–20–9

Chaube and Murphy (1969) injected this adenine antagonist into pregnant rats on day 11 or 12 of gestation and produced malformations with doses of 200 to 900 mg per kg. The defects included cleft palate, micrognathia and deformed appendages. Inosine provided complete protection against this teratogen.

Chaube, S. and Murphy, M.L.: Teratogenic effects of 6–hydroxylaminopurine in the rat: Protection by inosine. Biochem. Pharmacol. 18:1147–1156, 1969.

1254
7–Hydroxymethyl–12–methylbenz–(a)–anthracene
7–OHM–12–MBA CAS 568–75–2

Bird et al. (1970) injected intravenously 25 mg per kg on day 17 of gestation and produced adrenal necrosis in rat fetuses.

Bird, C.C.; Crawford, A.M. and Currie, A.R.: Foetal adrenal necrosis induced by 7–hydroxymethyl–12–methyl-benz(a)–anthracene and its prevention. Nature 228:72–73, 1970.

1255
17–Hydroxymethyl–12–methylbenz–(a)–anthracene CAS 568–75–2

Discussed under 7–12–Dimethylbenz–(a)–anthracene

1256 Hydroxy–methylmorphinan

O'Callaghan and Holtzman (**1977**) found that administration of 1.0 mg increasing to 1.75 mg per kg subcutaneously on days 5–12 of gestation produced tolerance to morphine injected into the 9 week offspring.

O'Callaghan, J.P. and Holtzman, S.G.: Prenatal administration of levorphanol or dextrorphan to the rat: analgesic effect of morphine in the offspring. The Journal of Pharmacology and Experimental Therapeutics 200:255–262, 1977.

1257 Hydroxymethylpyrimidine

Discussed under Ethionamide

1258 δ–Hydroxy–γ–oxo–L–norvaline

This asparagine analogue was studied by Mizutani and Ihara (1973) in mice, rats, hamsters and rabbits. The minimal dose by mouth or intraperitoneally that produced defects in rats and rabbits was 500 mg per kg and 1000 mg per kg in mice and hamsters. In all species digital defects were produced. In addition, neural tube closure defects, short tails and cleft palate were produced in mice and rats, hydrocephalus in hamsters and cleft palate in rabbits.

Mizutani, M. and Ihara, T.: Teratogenicity of δ–hydroxy–γ–oxo–L–norvaline. 1. Studies in mice, rats, hamsters and rabbits. (abs) Teratology 8:99 only, 1973.

1259 p–Hydroxyphenyl Lactic Acid CAS 306–23–0

Zharova et al. (1979) injected CC–57BR and C–57BL mice with 5 mg of this substance either on days 1 through 10 or during the last week of pregnancy and found no teratogenic effect. Malignant and benign neoplasms in the progeny of CC–57BR and C–57BL mouse strains occurred in 88 percent and 78 percent of cases, respectively.

Zharova, E.I.; Sergeeva, T.I.; Malakhova, N.V.; Romanenko, V.I.; Chitoridi, N.G. and Raushenbakh, M.O.: Transplacental blastomogenic action of p–hydroxyphenyl lactic acid. Byull. Eksper. Biol. Med. (USSR) 87:46–48, 1979.

1260 3–α–Hydroxy 5–α–pregnane–II, 20–dione with Cremophor *CT–1341*

Esaki et al. (1975) gave this anesthetic combination subcutaneously to rats and mice during organogenesis. Some reduction in bone maturation of the fetuses was found but no teratogenicity.

Esaki, K.; Tsukada, M.; Izumiyama, K. and Oshio, K.: Influence of CT 1341 on the fetuses of the mouse and rat. Jicchuken Zenrinsho Kenkyusho 1:165–172, 1975.

1261 17–Hydroxyprogesterone CAS 68–96–2

Suchowsky and Junkmann (1961) found that this compound had no masculinizing effects on the female rat fetus. Seegmiller et al. (1983) gave up to 200 times the human therapeutic dose to mice during organogenesis and found no increase in defects or androgenic effects.

The rare occurrence of masculinization of the female human fetus was summarized by Serment and Ruf (1968) who stated that in more than 1,500 treated cases only two cases of clitoral hypertrophy were found. Michaelis et al. (1983) prospectively studied 648 patients given hydroxyprogesterone and found no increase in major defects. Ressequie et al. (1985) reported that among 988 infants exposed to progesterone or 17–α–hydroxyprogesterone caproate during pregnancy, there was no increase in congenital defects including hypospadias and limb reduction.

Michaelis, J.; Michaelis, H.; Gluck, E. and Koller, S.: Prospective study of suspected associations between certain drugs administered during early pregnancy and congenital malformations. Teratology 27:57–64, 1983.

Ressequie, L.J.; Hick, J.F.; Bruen, J.A.; Noller, K.L.; O'Fallon, W.M. and Kurland, L.T.: Congenital malformations among offspring exposed in utero to progestins, Olmstead County, Minnesota 1936–1974. Fertil. Steril. 43:514–519, 1985.

Seegmiller, R.E.; Nelson, G.W. and Johnson, C.K.: Evaluation of the teratogenic potential of Delalutin (17 α–hydroxy–progesterone caproate) in mice. Teratology 28:201–208, 1983.

Serment, H. and Ruf, H.: Therapeutiques hormonales. Bull. Fed. Soc. Gynecol. Obstet. Lang. Fr. 20:69–76, 1968.

Suchowsky, G.K. and Junkmann, K.: A study of the virilizing effect of the progestogens on the female rat fetus. Endocrinology 68:341–349, 1961.

1262 Hydroxypropylcellulose of Low–substitution

Kitagawa et al. (1978A, B) fed up to 5.0 g per kg on days 7–17 to rats and rabbits and found no adverse effects in the fetuses.

Kitagawa, T.S.; Satoh, T.; Saito, H.; Katoh, M.; Makita, T. and Hashimoto, Y.: Teratological study of hydroxypropylcellulose of low substitution (L–HPC) in rats. Oyo Yakuri 16:271–298, 1978A.

Kitagawa, T.S.; Satoh, T.; Saito, H.; Katoh, M.; Makita, T. and Hashimoto, Y.: Teratological study of hydroxypropylcellulose of low substitution (L–HPC) in rabbits. Oyo Yakuri 16:259–269, 1978B.

1263 Hydroxypropyl Methanethiolsulfonate

Adam et al. (1990) gavaged rats and rabbits at up to 75 mg and 7.5 mg per kg daily during organogenesis. No teratogenicity was found but maternal and fetal toxicity were found in rats at 30 and 75 mg per kg.

Adam, G.A.; Drake, K.D.; Helmhout, S.L.; Mckenzie, M. and McKenzie, J.J.: Developmental toxicity study of 2–hydroxypropyl methanthiolsulfonate (HPMTS) in rats and rabbits. (abst)Teratology 41:535, 1990.

1264 Hydroxypropylmethylcellulose Acetate Succinate

Hoshi et al. (1985a) fed rats up to 2.5 g per kg on days 7–17 and found no adverse fetal effects. No behavioral changes were found after treatment on days 17–21 (Hoshi et al., 1985b). A rabbit study at the same dose during organogenesis produced no evidence of teratogenicity (Hoshi et al., 1985c).

Hoshi, N.; Ueno, K.; Igarashi, T.; Kitagawa, H.; Fujita, T.; Ichikawa, N.; Kondo, Y. and Isoda, M.: Teratological studies of hydroxypropylmethylcellulose acetate succinate in rats. J. Toxicol. Sci. 10(II):203–226, 1985a.

Hoshi, N.; Ueno, K.; Igarashi, T.; Kitagawa, H.; Fujita, T.; Ichikawa, N.; Kondo, Y. and Isoda, M.: Effects on offspring induced by oral administration of hydroxypropylmethylcellulose acetate succinate to the female rats in peri– and postnatal periods. J. Toxicol. Sci. 10(II):235–255, 1985b.

Hoshi, N.; Ueno, K.; Igarashi, T.; Kitagawa, H.; Fujita, T.; Ichikawa, N.; Kondo, Y. and Isoda, M.: Teratological studies of hydroxypropylmethylcellulose acetate succinate in rabbits. J. Toxicol. Sci. 10(II):227–234, 1985c.

1265 Hydroxystreptomycin

Discussed under Streptomycin

1266 5–Hydroxytryptamine

Discussed under Serotonin

1267 Hydroxyurea CAS 127–07–1

Chaube and Murphy (1966) injected pregnant rats with 185 to 1000 mg per kg on gestational day 9, 10, 11 or 12 and produced defects of the central nervous system, palate and skeleton. Large doses of hydroxylamine or urethan did not cause fetal abnormalities. Scott et al. (1971) showed that DNA synthesis is depressed by the presence of hydroxyurea in the embryo. Extensive cell death in the limb buds and central nervous system was noted 5 hours after maternal administration of 750 mg per kg. Soukup et al. (1967) found no chromosome changes in embryos after the rat was treated with 750 mg per kg on the 13th day. Murphy and Chaube (1964) showed that this substance produced a low incidence of beak defects in the chick embryo. Ferm (1966) caused neural tube defects and heart defects in hamsters given 50 mg intravenously on day 9, 10 or 11. Khera (1979) found teratogenicity in the cat given 50 or 100 mg per kg daily orally during organogenesis. Wilson et al. (1975) studied the teratogenicity in monkeys. DeSesso et al. (**1994**) found that a free radical scavenger, d–mannitol reduced the agent's toxicity.

Adlard and Dobbing (1975) studied the postnatal maze learning after administering 1 or 2 gm per kg to the mother rat on the 14th day. A decrease in performance was found along with body and brain growth reduction. Pigmentation was also less in the treated offspring. In another study using 150 mg per kg on days 6, 9, 12 15 and 18 of rat gestation, Brunner et al. (1978) were not able to show alterations in postnatal behavioral tests. The authors point out that in this case behavioral changes were less sensitive than morphological measures.

Adlard, B.P.F. and Dobbing, J.: Maze learning by adult rats after inhibition of neuronal multiplication in utero. Pediatr. Res. 9:139–142, 1975.

Brunner, R.L.; McLean, M.; Vorhees, C.V. and Butcher, R.E.: A comparison of behavioral and anatomic measures of hydroxyurea induced abnormalities. Teratology 18:379–384, 1978.

Chaube, S. and Murphy, M.L.: The effects of hydroxyurea and related compounds on the rat fetus. Cancer Res. 26:1448–1457, 1966.

DeSesso, J.M.; Scialli, A.R. and Goeringer, G.C.: D–Mannitol, a specific hydroxyl free radical scavenger, reduces the developmental toxicity of hydroxyurea in rabbits. Teratology 49:248–259, 1994.

Ferm, V.H.: Severe developmental malformations. Arch. Path. 81:174–177, 1966.

Khera, K.S.: Teratogenicity study on hydroxyurea and diphenylhydantoin in cats. Teratology 20:447–452, 1979.

Murphy, M.L. and Chaube, S.: Hydroxyurea (NSC–32065) as a teratogen. Cancer Chemother. Rep. 40:1–7, 1964.

Scott, W.J.; Ritter, E.J. and Wilson, J.G.: DNA synthesis inhibition and cell death associated with hydroxyurea teratogenesis in rat embryos. Dev. Biol. 26:306–315, 1971.

Soukup, S.; Takas, E. and Warkany, J.: Chromosome changes in embryos treated with various teratogens. J. Embryol. Exp. Morphol. 18:215–226, 1967.

Wilson, J.G.; Scott, W.J.; Ritter, E.J. and Fradkin, R.: Comparative distribution and embryotoxicity of hydroxyurea in pregnant rats and rhesus monkeys. Teratology 11:169–178, 1975.

1268 Hydroxyurethane CAS 589–41–3

Chaube and Murphy (1966) injected 225 to 700 mg per kg into rats on the 9th, 10th, 11th or 12th day of gestation and produced defects of the central nervous system, palate or skeleton. Tail defects were especially common after treatment on the 11th or 12th day.

Chaube, S. and Murphy, M.L.: The effects of hydroxyurea and related compounds on the rat fetus. Cancer Res. 26:1448–1457, 1966.

1269 Hydroxyzine *Atarax*™ CAS 68–88–2

Heinonen et al. (1977) reported five malformed children from 50 mothers who took the medication in the first four lunar months of gestation. This increase included four defects which were classified as non–uniform, that is of such variable severity that different hospitals classified them differently. Heinonen et al. (1977) reported that among 187 women who took this drug any time in pregnancy, six malformed infants occurred. This relative risk of 1.35 was not significantly increased. Erez et al. (1971) studied 100 women treated in the first trimester with 50 mg daily and found no increase in defects in the offspring. Fifty women took the drug in the first four lunar months and only one malformation was found.

This tranquilizer belongs to the benzhydrylpiperazine series of compounds. It has been used as a teratogen in the rat. The details are discussed under the heading, Meclizine.

Steffek et al. (1968) gave 5–12 mg per kg to five rhesus monkeys during organogenesis and observed three abortions and two normal offspring. King et al. (1966) gave the rat 60–200 mg per kg during organogenesis and found an increase in cleft palates, other orofacial defects and micromelia.

Erez, S.; Schifrin, B.S. and Dirim, O.: Double–blind evaluation of hydroxyzine as an antiemetic in pregnancy. J. Reprod. Med. 7:57–59, 1971.

Heinonen, O.P.; Slone, D. and Shapiro, S.: Birth Defects and Drugs in Pregnancy. Publishing Sciences Group Inc., Littleton, Mass., 1977.

King, C.T.G. and Howell, J.: Teratogenic effect of buclizine and hydroxizine in the rat and chlorcyclizine in the mouse. Am. J. Obstet. Gynecol. 95:109–111, 1966.

Steffek, A.J.; King, C.T.G. and Wilk, A.L.: Abortive effects and comparative metabolism of chlorcyclizine in various mammalian species. Teratology 1:399–406, 1968.

1270 Hyoscyamine CAS 101–31–5

Heinonen et al. (1977) reported 322 pregnancies in which exposure occurred in the first 4 lunar months and they found no increase in defect rates.

Heinonen, O.P.; Slone, D. and Shapiro, S.: Birth Defects and Drugs in Pregnancy. Publishing Sciences Group Inc., Littleton, Mass., 1977.

1271 Hyperbaric Air

Bolton and Alamo (1981) exposed gravid rats to 6 atmospheres of air during several periods of organogenesis and found no adverse changes in their fetuses.

Bolton, M.E. and Alamo, A.L.: Lack of teratogenic effects of air at high ambient pressure in rats. Teratology 24:181–185, 1981.

1272 Hypercalcemia *Hypervitaminosis D* also see *Cholecalciferol*

The association of hypercalcemia with congenital heart disease (supravalvular aortic stenosis) was suggested (Black and Bonham–Carter, 1963). The subject is covered more fully under Cholecalciferol.

Black, J.A. and Bonham–Carter, R.E.: Association between aortic stenosis and facies of severe infantile hypercalcemia. Lancet 2:745–749, 1963.

1273 Hyperemesis Gravidarum

Kallen (1987) studied the outcome of 3068 pregnancies with hyperemesis gravidarum. This group accounted

for about 3 per 1,000 pregnancies. Congenital malformations were present slightly more often than expected (undescended testes, hip dysplasia and Down syndrome). The rate was 5.8 percent as compared to 5.0 percent in the general population of Sweden. There were 9 infants with Down syndrome, 60 with hip dysplasia and 21 with undescended testicles. The rate of hyperemesis in twinning was increased 2.2 times. Klebanoff and Mills (1986) studied the effect of emesis among 16,398 women. After adjusting for confounding variables they found no relationship between vomiting and malformations.

Kallen, B.: Hyperemesis during pregnancy and delivery outcome: A registry study. Eur. J. Obstet. Gynecol. Reprod. Biol. 26:291–302, 1987.

Klebanoff, M. and Mills, J.L.: Is vomiting during pregnancy teratogenic? Br. Med. J. 292:724–726, 1986.

1274 Hyperglycemia *Glucose*

Cockroft and Coppola (1977) studied the effect of 12 and 15 mg of D–glucose per ml on explanted rat embryos and found fusion of the anterior to the posterior neural folds in many of the specimens. L–glucose retarded growth and did not produce abnormalities. Sadler (1980) exposed mouse embryos to 5 and 8 times the normal blood glucose level in vitro and produced a high frequency of exencephaly.

Cockroft, D.L. and Coppola, P.T.: Teratogenic effects of excess glucose on head–fold rat embryos in culture. Teratology 16:141–146, 1977.

Sadler, T.W.: Effects of maternal diabetes on early rat embryogenesis. 2. Hyperglycemia–induced exencephaly. Teratology 21:349–356, 1980.

1275 Hyperoxia

Ferm (1964) produced umbilical hernias, exencephaly, spina bifida and limb defects in a small but significant number of hamster fetuses by exposing the mother to 3.0 to 4.0 atmospheres of oxygen for periods of 2 to 3 hours on the 6th, 7th or 8th day of gestation. Grote (1965) exposed rabbits to 70 to 96 percent oxygen and produced no vertebral or other fetal defects.

The effect of oxygen concentration on the metabolism of salicylates, cyclophosphamide, niridazole and phosphoramide mustard during rat embryo culture was reported (Greenaway et al., 1985).

Ferm, V.H.: Teratogenic effects of hyperbaric oxygen. Proc. Soc. Exp. Biol. Med. 116: 975–976, 1964.

Greenaway, J.C.; Mirkes, P.E.; Walker, E.A.; Juchau, M.R.; Shepard, T.H. and Fantel, A.G.: The effect of oxygen concentration on the teratogenicity of salicylate, niridazole, cyclophosphamide and phosphoramide mustard in rat embryos in vitro. Teratology 32:287–295, 1985.

Grote, W.: Stoerung der Embryonalentwicklung bei Erhoehtem CO_2 und O_2–Partialdruck und bei Unterdruck. Z. Morphol. Anthropol. 56:165–194, 1965.

1276 Hypertension *Preeclampsia Eclampsia*

Brazy et al. (1982) studied the offspring of 28 women with severe hypertension before the 36th week of pregnancy. All the mothers had diastolic pressures above 110 mm Hg and four had hypertension prior to pregnancy. Most were treated with intravenous magnesium sulfate and other antihypertensive agents. Growth retardation, microcephaly, thrombocytopenia, leukopenia, patent ductus arteriosus and hypotonia of the skeletal and gut musculature were significantly increased over controls with similar gestational ages. Eight of the infants had head circumferences below the tenth percentile.

Preeclampsia and eclampsia are often associated with placental insufficiency which can lead to intrauterine growth retardation. The evidence for this was summarized by Dancis (1975). Naeye and Peters (1987) in an analysis of factors contributing to low IQ at age 7, reported that hypoxia, chronic hypertension, maternal hemoglobin less than 9 gm per DL, smoking and chronic hypotension were associated with low IQ.

Brazy, J.E.; Grimm, J.K. and Little, V.A.: Neonatal manifestations of severe maternal hypertension occurring before the thirty–sixth week of pregnancy. J. Pediatr. 100:265–271, 1982.

Dancis, J.: Fetomaternal interaction in Avery, G.B. Neonatology, J.B. Lippincott, Philadelphia, 44, 1975.

Naeye, R.L. and Peters, E.C.: Antenatal hypoxia and low IQ values. AJDC 141:50–54, 1987.

1277 Hyperthermia *Sauna Bathing*

The evidence for human teratogenicity from very high fever is continuing to accumulate but sauna bathing does not seem to be associated with adverse effects (see Sauna Bathing).

Maternal Fever and Congenital Defects

Miller et al. (1978) studied 63 mothers who gave birth to anencephalics and found 5 with temperatures from 38.9 degrees C to 40.0 degrees C and two who had possible hyperthermia related to sauna bathing. Of special note, all the mothers had their hyperthermia during the period the anterior neural tube closes (14–28 days).

Fraser and Skelton (1978) randomly reviewed records of children with congenital defects to determine whether or not any particular type of defect was associated with maternal fever. In a group of 55, they found six with microphthalmia and this number was significantly higher than in controls. Chance and Smith (1978) studied the pregnancy history of 43 women who gave birth to infants with meningomyeloceles, and in three, fevers of over 102 degrees F (38.9 degrees C) occurred between the 25th to 28th day of gestation. None of the 63 controls had fever at this time. Layde et al. (1980) found a significant increase in fever among mothers who delivered infants with spina bifida. Christo et al. (1987) studied the maternal records of 30 infants born with major CNS malformations. Four of the mothers had significant fevers in the first trimester. In a prospective study of 165 women with first trimester fever from 55,000 pregnancies, no significant differences in defect rate or intelligence of the offspring were identified as compared to matched controls (Clarren et al., 1979). Most of these fevers were for brief episodes. Ivarsson and Hen-

riksson (1984) reported a patient with arthrogryposis and mental retardation after the mother had hyperthermia and septic shock during part of the 20th week of gestation.

Fisher and Smith (1981) interviewed 17 women who had infants with encephaloceles, and found 4 who had fever elevations of 1.5C above normal during the gestational period 20–28 days. Hunter (1984) asked 264 women giving birth to infants with neural tube defects whether they had fever during pregnancy. In total, 229 recalled no fever, 3 were unsure and 32 (12 percent) reported fever. Thirteen of those with fever had the fever during the first 4 weeks. Nine of 13 fevers occurring during the first 4 weeks were associated with anencephaly. Little et al. (**1991**) studied 54 women who had fever over 101 degrees F in the first trimester and found 5 offspring with abdominal wall defects (i.e. diastasis recti or malformed umbilicus). The overall defect rate was not increased as compared to controls.

In a prospective follow–up study of 23,491 women who were screened by serum alpha–fetoprotein or amniocentesis, Milunsky et al. (**1992**) found that the neural tube relative risk was 2.8 (95% confidence 1.2–6.5) for hot tub use but for sauna exposure it was not significantly elevated, (1.8 confidence limit 0.4–7.9). These women reported the use of hot tubs or sauna bathing in the first 2 months. For fever during the first 3 months the relative risk was 1.8 (95% confidence, 0.8–4.1). They concluded that hot tub, sauna or fever in early pregnancy were associated with increased risk for neural tube defects. Electric blanket exposure was not associated with a risk for neural tube defects.

Kleinbrecht et al. (1979) found no significant increase in neural tube or eye defects among the offspring of women having fever in the first 12 weeks. Smith et al. (1978) in a retrospective study of 13 infants exposed to hyperthermia during the first trimester, reported that all had nervous system complications with microcephaly in three and hypotonia in nine. Microphthalmia occurred in five.

Shiota (1982) made a study of the histories from 113 women who donated embryos with neural tube defects to the Kyoto University embryo collection of Professor Nishimura. The histories were supplied at the time the therapeutically aborted specimen was sent to the collection. The obstetrician did not know the condition of the embryo. Maternal fever was mentioned in 18 percent of the 50 with anterior neural tube defects, significantly higher than the matched controls of 4.9 percent. Among the 63 embryos with myeloschisis, 11.1 percent had a history of fever.

Graham et al. (1988) reported three children with Moebius syndrome and in all three there was exposure to hyperthermia in the late first trimester or early second trimester. Kline et al. (1985) found that among women who had euploid spontaneous abortions, 18 percent had fever of 100 degrees F or more while among those who had aneuploid abortuses, only 7.1 percent had febrile episodes. Lipson (1988) interviewed the mothers of 40 infants with Hirschsprung disease and found a significant number (11) who had a history of increased body temperature during gestation. Among a control group of mothers whose children had congenital defects, only one case had fever.

Lipson et al. (1989) reported that two mothers of 15 infants with Moebius syndrome had "high fevers" at 7-8 weeks gestation.

Pleet et al. (1981) reviewed the clinical syndromes associated with hyperthermia. Warkany (1986) summarized the clinical and experimental findings. Miller and Ziskin (**1989**) have reviewed the biologic effects of ultrasound and constructed a time versus temperature plot on which to estimate damage. Adverse effects were not found in animals exposed to temperatures below 39 degrees C.

Animal Studies

A moderately large literature exists on hyperthermia as an experimental agent for producing congenital defects. Edwards and Wanner (1977) reviewed the animal teratogenicity of hyperthermia. Edwards (1981) reviewed his extensive work on the effect of hyperthermia on the guinea pig brain. Edwards (1986) reviewed the subject of teratogenicity in chickens, mice, rats, rabbits, sheep, pigs, guinea pigs and monkeys.

In the rabbit, Brinsmade and Rubsaamen (1957) produced fever by injecting milk on the 7th and 8th gestational days and found microcephaly or encephalocele in 3 of 65 embryos. Skreb and Frank (1963) immersed one uterine horn in water of 40 to 41 degrees C for 40 to 60 minutes on the 8th to 16th day of pregnancy of the rat and produced closure defects or severe histological effects of the central nervous system. Defects of the eye, extremity or palate occurred also. A high resorption rate was found on each day. Diathermy has been used to produce defects of the central nervous system in rat fetuses (Hofmann and Dietzel, 1966). Exencephaly has been produced by hyperthermia in the hamster (Kilham and Ferm, 1976) and in the mouse (Webster and Edwards, 1984). Harding and Edwards (**1991**) studied rats exposed to 43.5 degrees C for 5 minutes on the 13th or 14th day of gestation and found that although the brain weights were not reduced at birth a significant weight reduction was found after 3 weeks.

Lecyk (1966) subjected mice to 40 to 41 degrees C for 20–hour periods on gestational days 7 through 12 and observed a 25 percent incidence of rib or vertebral defects in the survivors. About one–third of the mothers died, and strict controls to exclude the teratogenic effect of fasting were not performed. Edwards (1969) studied the effect of a one hour daily exposure of guinea pigs to 43 degree external temperature. When exposed from the 18th to the 25th day, 86 percent of the fetuses had multiple anomalies which included microcephaly, reduction defects of the extremities, exomphalos and renal agenesis. An unusual form of amyoplasia was found also. Fetal death and resorptions were common when treatment was carried out before the 18th day.

German et al. (1985) found that in rats immersed in water, an elevation of temperature for one hour at 2.5 degrees C was the threshold combination for teratogenesis. Postnatal studies of this model indicated persistent impairment of learning (Jonson et al., 1976). Shiota (1988) studied mouse embryos exposed by maternal immersion in hot water and found a brief period of decreased mitoses followed by a burst of activity in the central nervous system. With a 42 degree

C exposure at 10 minutes, no increase in defects were seen, but at 15 minutes, 31 percent were malformed.

Shiota and Kayamura (1989) studied behavior and learning in mice exposed 10 minutes once or twice daily to 42 or 43 degree water on days 12 through 15. Open field activity was decreased in those exposed once per day to 43 degrees C or twice daily to 42 degrees. Water maze testing was significantly reduced in offspring exposed twice daily to 42 degrees C. Brain weights were below normal in those exposed to 43 degrees C once daily or 42 degrees twice daily.Harding and Edwards (**1993**) found microcephaly in rats exposed twice to 44 degrees C for 10 minutes on days 13 and 14.

Cockroft and New (1978) isolated the rat embryo from the mother by embryo culture from the egg cylinder stage to the 25 somite stage. Temperature exposure of 40.5 C retarded whole embryo growth: at 40 degrees C both the size and protein content of the brain were reduced. Mirkes (1985) has determined the temperature and duration of hyperthermia necessary to produce cell necrosis in the nervous system in the explanted day 10 rat embryo. Necrosis was found after 60 minutes at 42 degrees C and after 2–4 hours at 41 degrees C. Rao et al. (1990) found that nerve tissue from rat embryos exposed to hyperthermia had significantly lower acetylcholine levels when grown in tissue culture.

Hartley et al. (1974) exposed ewes to increased temperature for nine hours daily during the last one–third or two–thirds of pregnancy and produced a high incidence of brain cavitation and microcephaly in a large proportion of the lambs. The ewes body temperatures ranged from 40 to 41 degrees centigrade after treatment. Poswillo et al. (1974) in a preliminary study exposed marmosets to an incubator temperature of 42 degrees C for one hour daily from days 25 through 50 of gestation and the offspring were small and had a higher incidence of skeletal changes resembling rickets than did the controls.

Brinsmade, A.B. and Rubsaamen, H.: Zur Teratogenetischen Wirkung von Unspezifischem Fieber auf den sich Entwickelnden Kaninchenembryo. Beitr. Pathol. Anat. 117:154–164, 1957.

Chance, P.F. and Smith, D.W.: Hyperthermia and meningomyelocele and anencephaly. Lancet 1:769–770, 1978.

Christo, G.G.; Urala, M.S.; Duvvi, H.V. and Venkatesh, A.: Maternal hyperthermia as a teratogenic agent. Ind. Pediatr. 24:597–600, 1987.

Clarren, S.K.; Smith, D.W.; Harvey, M.A.S.; Ward, R.H. and Myrianthopoulos, N.C.: Hyperthermia: A prospective evaluation of a possible teratogenic agent in man. J. Pediatr. 95:81–83, 1979.

Cockroft, D.L. and New, D.A.T.: Abnormalities induced in rat embryos by hyperthermia. Teratology 17:277–284, 1978.

Edwards, M.J.: Hyperthermia as a teratogen: A review of experimental studies and their clinical significance. Teratogenesis, Carcinogenesis and Mutagenesis 6:563–582, 1986.

Edwards, M.J.: Congenital defects in guinea pigs: Fetal

resorptions, abortions and malformations following induced hyperthermia during early gestation. Teratology 2:313–328, 1969.

Edwards, M.J.: Clinical disorders of fetal brain development: effects due to hyperthermia in fetal brain disorders. In: Recent Approaches to Mental Deficiency, B.S. Hetzel and R.M. Smith (eds.), Elsevier North Holland, Amsterdam, Chapter 14, 335–363, 1981.

Edwards, M.J. and Wanner, R.A.: Extremes of temperature. In: Handbook of Teratology, J.G. Wilson and F.C. Fraser (eds.), Plenum Press, Chicago, 1:421–444, 1977.

Fisher, N.L. and Smith, D.W.: Occipital encephalocele and early gestational hyperthermia. Pediatrics 68:480–483, 1981.

Fraser, F.C. and Skelton, J.: Possible teratogenicity of maternal fever. Lancet 2:634 only, 1978.

German, M.A.; Webster, W.S. and Edwards, M.J.: Hyperthermia as a teratogen: Parameters determining hyperthermia–induced head defects in the rat. Teratology 31:265–272, 1985.

Graham, J.M.; Edwards, M.J.; Lipson, A.H. and Webster, W.S.: Gestational hyperthermia as a cause for Moebius syndrome. Teratology 37(5):461–462, 1988.

Harding, A.J. and Edwards, M.J.: As a result of prenatal hyperthermia three week old rats display microcephaly absent in the newborn. Teratology 44:477–478, 1991.

Harding, A.J. and Edwards, M.J.: Micrencephaly in rats caused by maternal hyperthermia on days 13 and 14 of pregnancy. Cong. Anom. 33:203–209, 1993.

Hartley, W.J.; Alexander, G. and Edwards, M.J.: Brain cavitation and microcephaly in lambs exposed to prenatal hyperthermia. Teratology 9:299–304, 1974.

Hofmann, D. and Dietzel, F.: Abort und Missbildungen nach Kurzwellendurchflutung in der Schwangerschaft. Geburts. Frauenheilk 26:378–390, 1966.

Hunter, A.G.W.: Neural tube defects in eastern Ontario and western Quebec: Demography and family data. Am. J. Med. Genet. 19:45–63, 1984.

Ivarsson, S.A. and Henrikkson, P.: Septic shock and hyperthermia as possible teratogenic factors. Acta Paediatr. 73:875–876, 1984.

Jonson, K.M.; Lyle, J.G.; Edwards, M.J. and Penny, R.H.C.: Effect of prenatal heat stress on brain growth and serial discrimination reversal learning in the guinea pig. Brain Research Bulletin 1:133–150, 1976.

Kalter, H.: Teratology of Central Nervous System. Chicago: University of Chicago Press, 180–181, 1968.

Kilham, L. and Ferm, V.H.: Exencephaly in fetal hamsters exposed to hyperthermia. Teratology 14:323–326, 1976.

Kline, J.; Stein, Z.A.; Susser, M. and Warburton, D.: Fever during pregnancy and spontaneous abortion. Am. J. Epidemiol. 121:832–842, 1985.

Kleinbrecht, J.; Michaelis, H.; Michaelis, J. and Koller, S.: Fever in pregnancy and congenital defects. Lancet 1:1403, 1979.

Layde, P.M.; Edmonds, L.D. and Erickson, J.D.: Maternal fever and neural tube defects. Teratology 21:105–108, 1980.

Lecyk, M.: The effect of hyperthermia applied in the given stages of pregnancy on the number and form of vertebrae in the offspring of white mice. Experientia 22:254–255, 1966.

Lipson, A.H.: Hirschsprung disease in the offspring of mothers exposed to hyperthermia during pregnancy. Am. J. Med. Genet. 29:117–124, 1988.

Lipson, A.H.; Webster, W.S.; Brown-Woodman, P.D.C. and Osborn, R.A.: Moebius syndrome: animal model—human correlations and evidence for a brainstem vascular etiology. Teratology 40:339–350, 1989.

Little, B.B.; Ghali, F.E.; Snell, L.M.; Knoll, K.A.; Johnston, W. and Gilstrap III, L.C.: Is hyperthermia teratogenic in the human. American Journal of Perinatology 8:185–189, 1991.

Miller, P.; Smith, D.W. and Shepard, T.H.: Hyperthermia as one possible etiology of anencephaly. Lancet 1:519–521, 1978.

Miller, M.W. and Ziskin, M.C.: Biological consequences of hyperthermia. Ultrasound in Med. and Biol. 15:707–722, 1989.

Milunsky, A.; Ulcickas, M.; Rothman, K.J.; Willett, W.; Jick, S.S. and Jick, H.: Maternal heat exposure and neural tube defects. JAMA 268:882–885, 1992.

Mirkes, P.E.: Effects of acute exposures to elevated temperatures on rat embryo growth and development in vitro. Teratology 32:259–266, 1985.

Pleet, H.B.; Graham, J.M. and Smith, D.W.: Central nervous system and facial defects associated with maternal hyperthermia at four to 14 weeks gestation. Pediatrics 67:785–795, 1981.

Poswillo, D.E.; Nunnerly, H.; Sopher, D. and Keith, J.: Hyperthermia as a teratogenic agent. Ann. Roy. College of Surgery of England 55:171–174, 1974.

Rao, G.S.; Abraham, V.; Fink, B.A.; Margulies, N. and Ziskin, M.C.: Biochemical changes in the developing rat central nervous system due to hyperthermia. Teratology 41:327–332, 1990.

Shiota, K.: Induction of neural tube defects and skeletal malformations in mice following brief hyperthermia in utero. Biology of the Neonate 53:86–97, 1988.

Shiota, K.: Neural tube defects and maternal hyperthermia in early pregnancy: Epidemiology in a human embryo population. Am. J. Med. Genet. 12:281–288, 1982.

Shiota, K.; Kayamura, T.: Effects of prenatal heat stress on postnatal growth, behavior and learning capacity in mice. Biology of the Neonate 56:6–14, 1989.

Skreb, N. and Frank, Z.: Developmental abnormalities in the rat induced by heat shock. J. Embryol. Exp. Morphol. 11:445–457, 1963.

Smith, D.W.; Clarren, S.K. and Harvey, M.A.S.: Hyperthermia as a possible teratogenic agent. J. Pediatr. 92:877–883, 1978.

Warkany, J.: Teratogen update: Hyperthermia. Teratology 33:365–371, 1986.

Webster, W.S. and Edwards, M.J.: Hyperthermia and the induction of neural tube defects in mice. Teratology

29:417–425, 1984.

1278 Hyperthyroidism

Mujtaba and Burrow (1975) reported 26 pregnancies complicated by hyperthyroidism. All mothers were treated with propylthiouracil or methimazole. Four infants had goiters at birth and three percent of these had thyrotoxicosis. Four children had congenital defects which included aortic atresia, hypospadias, imperforate anus and scalp skin defects (2). The two with scalp defects were both exposed to methimazole which was previously associated with congenital scalp defects (see Methimazole). One child had developmental retardation but had developed hypothyroidism at four months while on antithyroid therapy.

Mujtaba, Q. and Burrow, G.N.: Treatment of hyperthyroidism in pregnancy with propylthiouracil and methimazole. Obstet. Gynecol. 46:282–286, 1975.

1279 Hypervitaminosis A *Vitamin A Excess* also see *Retinoic Acid*

Bernhardt and Dorsey (1974) reported that the offspring of a mother taking 25,000 IU daily during the first three months of gestation was found to have an aberrant ureter which was dilated and entered the vagina. Fantel et al. (1977) cited another case with urogenital anomaly in a mother who ingested large amounts of vitamin A. A third report (Mounoud et al., 1975) gave a description of an infant with Goldenhar's syndrome following inadvertent ingestion of the vitamin. Martinez–Frias and Salvador (**1990**) performed a case control study of over 12,000 newborns with congenital defects and found 16 where the mother was exposed to 10,000 to 100,000 i.u. Diverse malformations were found but included only one ear defect (preauricular pit) and nothing suggesting a retinoic acid syndrome. Five patients had hip instability or club feet as the only defect. The odds ratio was 5.4 for those cases exposed in the 1st or 2nd month.

Some concern has been raised about avid consumption of liver as a source of vitamin A–excess. Unless the domestic animal is overdosed with vitamin A, the liver content is usually less than 15,000 iu per 100 gm. One unit is equivalent to 0.30 microgm of vitamin A.

Von Lennep et al. (1985) reported a fetus with partial sirenomelia and single cystic kidney; the mother had taken 150,000 IU daily for two weeks before and three weeks after conception. They reviewed five other reported cases. Rosa et al. (1986) reported briefly on 18 pregnancies exposed to excess vitamin A. In 12, the malformations were similar to the findings in animals and humans exposed to retinoic acid. Nine had involvement of the external ear. Zuber et al. (1987) followed prospectively 27 pregnancies where the mother took high doses of vitamin A and found no congenital defects. Most of the daily dosage levels were 25,000 IU but one was 50,000 and another 150,000 IU. Martinez–Frias and Salvador (1988) interviewed 12,000 women who had congenital defects in their offspring and compared their intake of vitamin A. The odds ratio for offspring of mothers receiving 40,000 IU or more was 2.7 which was of borderline statistical significance. There were 10 exposures to vitamin

A alone (average 66,700 IU) in the group with defects and only one among the normal controls.

Since the discovery by Cohlan (1954) that large doses of vitamin A given by mouth could produce congenital defects, this experimental model has been used widely (see summary by Kalter, 1968). Fantel et al. (1977) reported cleft palate, craniofacial, skeletal and urogenital anomalies in pigtail monkeys treated with 10 mg of retinoic acid per kg from day 20 through 44. Cohlan tube–fed rats 35,000 IU of vitamin A from the 2nd, 3rd or 4th to the 16th day of gestation and found 52 percent of the offspring to be abnormal. All the defective fetuses had exencephaly which was associated with an increased amount of bloody amniotic fluid. Thirty–eight percent had cleft palate and eye defects also were found. Giroud and Martinet (1956) studied the effect of time of administration of vitamin A to the rat and type of malformation. The peak effect on neural tube closure was seen when treatment was started on day 8 and the highest incidence of cleft palate and skeletal defects was noted when treatment commenced on day 11.

Giroud et al. (1956) demonstrated a small increase in vitamin A content of the treated fetuses. Giroud and Martinet (1957) showed that the exencephaly was associated with failure of the anterior portion of the neural tube to close and externalization of the choroid plexuses occurred. Kalter and Warkany (1961) produced defects in the mouse. A number of the fetuses with genitourinary abnormalities had absence of the umbilical artery and a compensatory retention of the vitelline (superior mesenteric) artery. Marin–Padilla and Ferm (1965) and Marin–Padilla (1966) studied the syndrome in hamsters and noted that cell necrosis in the somites and notochord was an early effect of treatment. A comprehensive review of the animal studies on vitamin A teratogenesis was made by Geelen (1979).

Postnatal studies of rats receiving 60,000 units of vitamin A on days 14 and 15 or 17 and 18 revealed learning and fine motor changes, respectively (Hutchings et al., 1973; Hutchings and Gaston, 1973). Vorhees et al. (1979) confirmed this. Newman et al. (1982) found that vitamin A administration on days 15–19 was associated with decreased oxygen consumption in the neonates.

Bernhardt, I.B. and Dorsey, D.J.: Hypervitaminosis A and congenital renal anomalies in a human infant. Obstet. Gynecol. 43:750–755, 1974.

Cohlan, S.Q.: Congenital anomalies in the rat produced by the excessive intake of vitamin A during pregnancy. Pediatrics 13:556–569, 1954.

Fantel, A.G.; Shepard, T.H.; Newell–Morris, L.L. and Moffett, B.C.: Teratogenic effects of retinoic acid in pigtail monkeys (Macaca nemistrina). Teratology 15:65–72, 1977.

Geelen, J.A.G.: Hypervitaminosis A induced teratogenesis. CRC Critical Reviews in Toxicology. 6:351–376, 1979.

Giroud, A.; Gounelle, H. and Martinet, M.: Concentration de la vitamine A chez la mere et le foetus au cours de la teratogenese par hypervitaminose A. C. R. Soc. Biol. 150:2064–2065, 1956.

Giroud, A. and Martinet, M.: Teratogenese par hautes doses de vitamine A en fonction des stades du developpement. Arch. Anat. Micr. Morph. Exp. 45:77–98, 1956.

Giroud, A. and Martinet, M.: Morphogenese de l'anencephalie. Arch. Anat. Micr. Morph. Exp. 46:247–264, 1957.

Hutchings, D.E.; Gibbon, J. and Kaufman, M.A.: Maternal vitamin A excess during the early fetal period: Effects on learning and development in the offspring. Dev. Psychol. 6:445–457, 1973.

Hutchings, D.E. and Gaston, J.: The effects of vitamin A excess administered during mid–fetal period on learning and development in rat offspring. Dev. Psychol. 7:225–233, 1973.

Kalter, H.: Teratology of the Central Nervous System. Chicago: University of Chicago Press, 45–56, 1968.

Kalter, H. and Warkany, J.: Experimental production of congenital malformations in strains of inbred mice by maternal treatment with hypervitaminosis A. Am. J. Pathol. 38:1–20, 1961.

Marin–Padilla, M.: Mesodermal alterations induced by hypervitaminosis A. J. Embryol. Exp. Morphol. 15:261–269, 1966.

Marin–Padilla, M. and Ferm, V.H.: Somite necrosis and developmental malformations induced by vitamin A in the golden hamster. J. Embryol. Exp. Morphol. 13:1–8, 1965.

Martinez–Frias, M.L. and Salvador, J.: Megadose vitamin A and teratogenicity. Lancet 1:236 only, 1988.

Martinez–Frias, M.L. and Salvador, J.: Epidemiological aspects of prenatal exposure to high doses of vitamin A in Spain. European Journal of Eipdemiology 6:118–123, 1990.

Mounoud, R.L.; Klein, D. and Weber, F.: A propos d'un case de syndrome de Goldenhar intoxication aique a la vitamin A chez la mere pendent la grossesse. J. Genet Hum. 23:135–154, 1975.

Newman, L.M.; Johnson, E.M. and Cadogan, A.S.A.: Predictability of postnatal survival in prenatally hypervitamin A exposed rat pups based on 0–2 consumption values on day of delivery. (abs) Teratology 25:64A, 1982.

Rosa, F.W.; Wilk, A.L. and Kelsey, F.O.: Teratogen update: Vitamin A congeners. Teratology 33:355–364, 1986.

Sanders, T.A.B.: Vitamin a and pregnancy (Letter). Lancet 336:1375, 1990.

Von Lennep, E.; Khazen, N.E.; De Pierreux, G.; Amey, J.J.; Rodesch, F. and Regemorter, N.V.: A case of partial sirenomelia and possible vitamin A teratogenesis. Prenatal Diag. 5:35–40, 1985.

Vorhees, C.V.; Brunner, R.L. and Butcher, R.E.: Psychotropic drugs as behavioral teratogens. Science 205:1220–1225, 1979.

Zuber, C.; Librizzi, R.J. and Vogt, B.L.: Outcome of pregnancies exposed to high dose vitamin A. (abs) Teratology 42A, 1987.

1280 Hypoglycemia also see *Fasting*

Although hypoglycemia was not studied specifically, Runner and Miller (1956) prevented defects in fasted mice by administration of glucose (also see the headings: Insulin, Carbutamide and Tolbutamide). Buchanan et al. (1986) infused pregnant rats with insulin and produced hypoglycemia for 1 hour. Controls were treated similarly but their blood glucose was maintained. Treatment on days 9.5 or 9.75 was associated with significant reductions in somite numbers, crown–rump lengths, protein and DNA in embryos examined 2 days later. Hunter et al. (1987) found that mouse embryos grown at a glucose concentration of 80 mg per deciliter from 3–6 hours developed exencephaly and craniofacial defects in about 20 percent of the embryos. At 40 mg per deciliter the malformation rate was 100 percent. Smoak and Sadler (**1990**) using mouse embryos in vitro found that at 60 mg per dl for 2 hours there was a significant increase in malformations. A decrease in protein content of the embryos was found at 60 mg per dl for 24 hours or 40 mg per dl for 2 hours or more. In day 9 1/2 rat embryos cultured for 17 hours, concentrations of 30 mg per dl or more supported normal growth (Shepard and Park, **1994**).

Buchanan, T.A.; Schemmer, J.K. and Freinkel, N.: Embryotoxic effects of brief maternal insulin–hypoglycemia during organogenesis in the rat. J. Clin. Invest. 78:643–649, 1986.

Hunter, E.S.; Smoak, I. and Sadler, T.W.: Hypoglycemia induced biochemical alterations in neurulating mouse conceptuses in vitro. Teratology 76A, 1987.

Runner, M.N. and Miller, J.R.: Congenital deformity in the mouse as a consequence of fasting. (abs) Anat. Rec. 124:437–438, 1956.

Shepard, T.H. and Park, H.W.: Neural plate microvillus lengthening in rat embryos grown in various concentrations of glucose and further studies of the mechanism. Teratology 50:340–347, 1994.

Smoak, I.W. and Sadler, T.W.: Embryopathic effects of short term exposure to hypoglycemia in mouse embryos in vitro. Am. J. Obstet and Gynecol. 163:619–624, 1990.

1281 Hypoglycin A CAS 156–56–9

This analogue of leucine present in the fruit Blighia sapida was administered by Persaud and Kaplan (1970) to rats intraperitoneally on the 1st through the 6th gestational days in amounts of 30 mg per kg. Ninety–two percent of the offspring were defective; gastroschisis, stunting, encephalocele and syndactyly were found. Chick embryos were not effected by the substance, and this result was thought to be related to the large leucine pools present in the egg. Persaud (1972) reviewed the subject.

Persaud, T.V.N.: Teratogenic effect of hypoglycin–A. In: Advances in Teratology, D.H.M. Woollam (ed.), New York, Academic Press, 5:78–95, 1972.

Persaud, T.V.N. and Kaplan, S.: The effects of hypoglycin–A, a leucine analogue, on the development of the rat and chick embryos. Life Sci. 9:1305–1313, 1970.

1282 Hypothermia

Kalter (1968) reviewed this subject. Generally, studies at low temperature have resulted in reduction of litter size. Smith (1957) maintained pregnant hamsters for 30 minutes in baths at minus 5 degrees C. The pregnancies were unaffected except for uterine hemorrhage followed by resorption when treatment was on days 9, 10 or 11. When cooling was extended to 45 minutes on days 6, 7 and 8, defects of the neural tube, palate and extremities occurred. Lecyk (1965) subjected mice to 24 hours at 20 degrees C on day 7.5 or 8.5 and produced defects of the vertebral column in over one–half of the offspring. The control animals also receiving chlorpromazine were placed in an incubator and produced only 9 skeletal defects among 175 offspring. Unfortunately, this author did not perform control experiments to exclude the effect of fasting which can produce similar defects in the mouse fetus. Smoak and Sadler (1991) studied rat embryos in vitro exposed to 35 or 32 degrees for 4 or 24 hours. Little change was found at 4 hours but with the 24 hour exposure growth retardation and changes in neuralation occurred.

Kalter, H.: Teratology of the Central Nervous System. Chicago: University of Chicago Press, 181–182, 1968.

Lecyk, M.: The effect of hypothermia applied in the given stages of pregnancy on the number and form of vertebrae in the offspring of white mice. Experientia 21:452–453, 1965.

Smoak, I.W. and Sadler, T.W.: Hypothermia: Teratogenic and protective effects on the development of mouse embryos in vitro. Teratology 43:635–641, 1991.

Smith, A.U.: The effects on fetal development of freezing pregnant hamsters (Mesocricetus auratus). J. Embryol. Exp. Morphol. 5:311–323, 1957.

1283 Hypothyroidism also see *Thyroidectomy*

Lu and Staples (1978) parathyroidectomized rats and 20 days later mated them. Decrease in implantations and a 10 percent malformation rate was found. Hydrocephalus and convoluted retinae were the most common type of defect.

Lu, M.H. and Staples, R.E.: Teratogenicity of ethylenethiourea and thyroid function in the rat. Teratology 17:171–178, 1978.

1284 Hypoxanthine also see *2–Deoxyinosine* CAS 68–94–0

Fujii and Nishimura (1972) gave up to 1000 mg per kg intraperitoneally on day 10 to the rat. A reduction in fetal survival and malformations occurred. Skeletal defects including reduction types, facial clefts and other types of defects occurred. At 800 mg per kg defects were not increased.

Karnofsky and Lacon (1961) injected 2 to 8 mg into the chick yolk sac on the 4th day and produced no defects. Inosine and 2–deoxyinosine were teratogenic under these conditions.

Fujii, T. and Nishimura, H.: Comparison of teratogenic action of substances related to purine metabolism in mouse embryos. Jpn. J. Pharmacol. 22:201–206, 1972.

Karnofsky, D.A. and Lacon, C.R.: Effects of physiological purines on the development of the chick embryo. Biochem. Pharmacol. 7:154–158, 1961.

1285 Hypoxia

Naeye and Peters (1987) in an analysis of factors contributing to low IQ at age 7, reported that chronic hypoxia,

hypertension, maternal hemoglobin less than 9 gm per DL, smoking and chronic hypotension were associated with low IQ.

In a study of children born at high altitude, Lichty et al. (1957) found their birth weight reduced but no increase in the rate of congenital defects. Penaloza et al. (1964) summarized data suggesting that the frequency of patent ductus is increased at high altitudes and rises to 1 percent of newborns at altitudes over 4500 meters. Warkany (1971) discussed critically the reports dealing with hypoxia and congenital heart disease.

This experimental tool was used extensively by teratologists and was reviewed critically by Kalter and Warkany (1959). In general, lowered atmospheric pressure was used. Hermans et al. (1992) exposed rats to 4 hours of 10.5% oxygen on days 15 to 20 and cross-fostered the offspring which were tested. Some behavioral changes occurred but disappeared with maturation. The authors reviewed the results of other animal studies of hypoxia.

In the mouse, exposed for 5 hours to 260 to 280 mm Hg on the 9th, 10th or 15th gestational day, Ingalls et al. (1952) and Ingalls and Curley (1957) produced hemivertebral fused ribs, cleft palate and cranioschisis. Curley and Ingalls (1957), using a mixture of 6 percent oxygen and nitrogen, produced similar defects in mice. The rat and hamster (Ferm, 1964) appeared to be resistant to the teratogenic effect of hypoxia, but Degenhardt and Knoche (1959) produced vertebral and rib defects in the rabbit and observed that the most effective day of treatment was the 9th. Murakami and Kameyama (1963) described in detail the vertebral defects produced in the mouse.

Petter et al. (1971) produced a syndrome of generalized edema followed by hemorrhages in the extremities of rat fetuses after maternal exposure to a variety of hypoxia–producing conditions at 16.5 days of gestation. These conditions included vascular occlusion of maternal blood supply, reduced respiratory oxygen, fuel gas and potassium ferric cyanide.

De Grauw et al. (1986) subjected rats to 11.6, 10.7 and 9.0 percent oxygen from day 11 until day 21. Placental weights were increased while body weights were reduced. In the pair–fed controls the placenta was reduced in weight. The brains were reduced in size and weight but no gross or microscopic abnormality was found.

Pregnant mice (C57 BL x CBA) were exposed by Udalova (1978) to low atmosphere pressure (230 mm Hg) for three hours on the 7–10th days of pregnancy. No damage to embryonic development or effect on sex ratio of embryos was found in the treated animals. Khokhlova et al. (1979) reported that rats subjected on the 15th day of pregnancy to two–hour hypoxia corresponding to 8,000 m altitude delivered offspring with delay of differentiation and maturation of the brain cortex neurons. Fantel et al. (1988) presented evidence that the in vitro teratogenicity was associated with oxygen depletion secondary to redox cycling.

Curley, F.J. and Ingalls, T.H.: Hypoxia at normal atmospheric pressure as a cause of congenital malformations in mice. Proc. Soc. Exp. Biol. Med. 94:87–88, 1957.

Degenhardt, K.H. and Knoche, E.: Analysis of intrauterine malformations of vertebral column induced by oxygen deficiency. Can. Med. Assoc. J. 80:441–445, 1959.

De Grauw, T.; Myers, R.E. and Scott, W.J.: Fetal growth retardation in rats from different levels of hypoxia. Biol. Neonate 49:85–89, 1986.

Fantel, A.G.; Juchau, M.R.; Burroughs, C.J. and Person, R.E.: Studies of embryotoxic mechanisms of niridazole: Evidence that oxygen depletion plays a role in dysmorphogenicity. Teratology 39:243–251, 1989.

Ferm, V.H.: Teratogenic effects of hyperbaric oxygen. Proc. Soc. Exp. Biol. Med. 116:975–976, 1964.

Ingalls, T.H. and Curley, F.J.: Principles governing the genesis of congenital malformations induced in mice by hypoxia. N. Eng. J. Med. 257:1121–1127, 1957.

Hermans, R.H.M.; Hunter, D.E.; McGivern, R.F.; Cain, C.D. and Longo, L.D.: Behavioral sequelae in young rats of acute intermittent antenatal hypoxia. Neurotoxicology and Teratology 14:119–129, 1992.

Ingalls, T.H.; Curley, F.J. and Prindle, R.A.: Experimental production of congenital abnormalities: Timing and degree of anoxia as factors causing fetal deaths and congenital abnormalities in the mouse. N. Eng. J. Med. 247:758–768, 1952.

Kalter, H. and Warkany, J.: Experimental production of congenital malformations in mammals by metabolic procedure. Physiol. Rev. 39:69–115, 1959.

Khokhlova, V.A. and Kazakova, P.B.: Effect of maternal hypoxia on neurogenesis of the brain cortex in rat progeny (autoradiographic study). Byull, Eksper. Biol. Med. (USSR) 87(5):485–487, 1979.

Lichty, J.A.; Ting, R.Y.; Bruns, P.D. and Dyar, E.: Studies of babies born at high altitude. I. Relation of altitude to birth weight. A.M.A. J. Dis. Child. 93:666–669, 1957.

Murakami, U. and Kameyama, Y.: Vertebral malformations in the mouse foetus caused by maternal hypoxia during early stages of pregnancy. J. Embryol. Exp. Morphol. 11:107–118, 1963.

Naeye, R.L. and Peters, E.C.: Antenatal hypoxia and low IQ values. AJDC 141:50–54, 1987.

Penaloza, D.; Arias–Stella, J.; Sime, F.; Recavarren, S. and Marticorena, E.: The heart and pulmonary circulation in children at high altitudes: Physiological, anatomical and clinical observations. Pediatrics 34:568–582, 1964.

Petter, C.; Bourbon, J.; Maltier, J. and Jost, A.: Production d'hemorragies des extremites chez le foetus de rat soumis a une hypoxie in utero. C. R. Acad. Sci. (D) (Paris) 272:2488–2490, 1971.

Udalova, L.D.: The influence of acute hypoxia on adult and embryonic mortality and sex ratio of the embryos. Byull. Eksper. Biol. Med. (USSR) 86(7):88–89, 1978.

Warkany, J.: Congenital Malformations: Notes and Comments. Chicago: Year Book Medical Publishers, 465 only, 1971.

1286 Ibopamine Hydrochloride CAS 66195–31–1

Taniguchi et al. (**1990** A) fed rats up to 600 mg per kg on days 7–17. At the highest dose the dams were sleepy and gained less weight. No teratogenicity or decrease in fetal size was seen. Taniguchi et al. (**1990** B) in studies on days 17–21 with up to 400 mg per kg found normal behavior in the offspring.

Taniguchi, H.; Himeno, Y.; Chono, M.; Nakamura, M. and Haruguchi, T.: Teratogenicity study of ibopamine hydrochloride in rats. Oyo Yakuri 40(4):409–428, 1990.

Taniguchi, H.; Himeno, Y.; Chono, M.; Araki, E.; Nakamura, M. and Haruguchi, T.: Perinatal and postnatal study of ibopamine hydrochloride in rats. Oyo Yakuri 40(4)429–445, 1990.

1287 Ibuprofen 2–(4–Isobutylphenyl)–proprionic Acid CAS 15687–27–1

Aselton et al. (1985) found one child with a congenital defect among 51 exposed during the first trimester.

The anti–inflammatory antipyretic was tested in rabbits and rats by Adams et al. (1969) and no evidence for teratogenicity was found. The rabbits received oral doses of up to 60 mg per kg, and the rats received up to 180 mg per kg. The treatments were given throughout pregnancy.

Ono et al. (1982) administered up to 100 mg per kg rectally before and for 7 days following fertilization and found a reduction in the number of implants. At 200 mg per kg rectally during organogenesis no adverse fetal effects were found. Administration of 100 mg rectally on days 17–21 produced no significant postnatal changes in the offspring. Ono et al. (1984) administered this drug rectally to pregnant rats and rabbits in doses of up to 200 mg per kg and found a decrease in fetal rat weights.

Adams, S.S.; Bough, R.G.; Cliffe, E.E.; Lessel, B. and Mills, R.F.N.: Absorption, distribution and toxicity of ibuprofen. Toxicol. Appl. Pharmacol. 15:310–330, 1969.

Aselton, P.A.; Jick, H.; Milunsky, A.; Hunter, J.R. and Stergachis, A.: First–trimester drug use and congenital disorders. Obstet. Gynecol. 65:451–455, 1985.

Ono, M.; Ogihara, K.; Nagase, M.; Asami, K.; Kodama, M. and Sakakibara, Y.: Reproduction studies of ibuprofen by rectum in rats and rabbits. Clin. Report. 18:537–556, 1984.

Ono, M.; Ogawa, Y.; Ogihara, K.; Nagase, M. and Asimi, K.: Reproductive studies of ibuprofen by rectal administration. Oyo Yakuri 24:467–473, 539–547, 1982.

1288 Ibutilide Fumerate

Marks and Terry (**1994**) gavaged rats with 20, 40, or 80 mg per kg from days 6–15. At the highest doses resorption was high. The survivors at all dose levels had increased scoleosis, cleft palate and adactyly. At 10 mg per kg some scoleosis was found.

Marks, T.A. and Terry, R.D.: Developmental toxicity iof ibutilide fumerate in rats. (abs) Terratology 49:406, 1994.

1289 ICI 199,456

This thromboxane receptor antagonist was studied by Freeman et al. (**1991**). Oral doses of up to 350 mg per kg were given on days 6–15. Maternal and fetal toxicity occurred at 100 mg per kg. At 350 mg per kg diaphragmatic hernias were increased.

Freeman, S.J.; Evans, S.J.; Martin, V.J. and Siddall, R.A.: Teratogenic effects in rat of ICI 199, 454, a thromboxane receptor antagonist. (Abs) Teratology 43:473, 1991.

1290 Idarubicin Hydrochloride

Yamashita et al. (1992) gave this anthracycline antineoplastic agent intravenously to rats in amounts of up to 0.2 mg per kg. At the highest dose given before and in early pregnancy there was early fetal loss and decreased ossification. Fetuses exposed to 0.2 mg per kg during organogenesis were small and had skeletal defects such as fused ribs (Ono et al., 1992). The second generation may have had reduced fertility. Similar treatment perinatally was not associated in differences from the controls.

Ono, C.; Iwama, A.; Fujiwara, Y.; Kitsuya, A. and Nakamura, T.: Teratology study of idarubicin hydrochloride in rats in intravenous injection. Shinyaku to Rinsho 41:2958–2987, 1992.

Tateda, C.; Kajiwara, A. and Nakamura, T.: Perinatal study of idarubicin hydrochloride in rats by intravenous injection. Shinyaku to Rinsho 41:2989–3007, 1992.

Yamashita, Y.; Tateda, C.; Yamamae, H.; Kuzuoka, K.; Ishitobi, H. and Nakamura, T.: Fertility study of idarubicin hydrochloride in rats by intravenous injection. Shinyaku to Rinsho 41:29488–2957, 1992.

1291 Idebenone 6–(10–Hydroxydecyl)–2,3–dimethoxy–5–methyl–1,4–benzoquinone

Ihara et al. (1985) gave up to 500 mg on days 6–17 in the rat. Gavage feedings did not cause adverse effects to the fetuses.

Ihara, T.; Ooshima, Y. and Yoshida, T.: Teratogenic effects of Ibedenone (CV–2619) in the rat. Yakuri to Chiryo 13(7):4033–4046, 1985.

1292 Idoxuridine IUDR 5–Iododeoxyuridine CAS 54–42–2

Skalko and Packard (1973) administered this compound on single days intraperitoneally to mice from day 7 through 11. Doses of 300 and 500 mg per kg were highly embryolethal and teratogenic while at 100 mg per kg fewer defects and resorptions were found. Exencephaly, polydactyly, skeletal reduction defects and cleft palate were found.

Percy (1975) treating pregnant rats and mice for 3 days toward the end of gestation produced cerebellar dysplasia, microcystic kidneys and rare retinal changes. The changes could be produced by doses of 50 mg per kg. Itoi et al. (1975) applied a 0.1 percent solution to rabbit eyes four times daily for 12 days during organogenesis and found exophthalmos and clubbing of the forefeet in the fetuses. Trifluorothymidine (1 percent) did not produce abnormalities when applied in a similar manner.

Itoi, M.; Gefter, J.W.; Kaneko, N.; Ishii, Y.; Ramer, R.M. and Gasset, A.R.: Teratogenicities of ophthalmic drugs. 2. Antiviral ophthalmic drugs. Arch. Ophth. 93:46–49, 1975.

Percy, D.H.: Teratogenicity effects of pyrimidine analogues 5–iododeoxyuridine and cytosine arabinoside in late fetal mice and rats. Teratology 11:103–108, 1975.

Skalko, R.G. and Packard, D.S.: The teratogenic response of the mouse embryo to 5–iododeoxyuridine. Experientia 29:198–200, 1973.

1293 Ifenprodil CAS 23210–56–2

This α receptor blocker was given intraperitoneally to the rat and mouse in maximum daily doses of 40 mg per kg during organogenesis. Maternal weight gain in the rat was found at 40 mg per kg. No effects were observed in the fetus of either species.

Kihara, J.; Sugisawa, A.; Kumura, I.; Fukui, Y. and Sakaokibara, E.: The effect of ifenprodil (FX–505) on rat and mice embryos. Yakubutsu Ryoho 9:45–65, 1979.

1294 Ifosfamide CAS 3778–73–2

This structural analogue of cyclophosphamide when injected intraperitoneally into pregnant mice on day 11 at a dose of 20 mg per kg produced skeletal and renal malformations and hydrocephalus in the offspring (Bus and Gibson, 1973).

Ifosfamide was tested intravenously by Nagaoka et al. (1982) in rats and rabbits. Treating rats of both sexes before mating and during the first 7 days of gestation with 2.5 or 5.0 mg per kg they found decreased viability at term. In perinatal studies using up to 10 mg per kg an increase in stillbirths and hydrocephalus was found. Studies during organogenesis were followed by an increase in defects mostly involving the central nervous system when 5.0 mg per kg was used. In rabbits, ectrodactylia was produced when 20 mg per kg was given during organogenesis.

Bus, J.S. and Gibson, J.E.: Teratogenicity and neonatal toxicity of ifosfamide in mice. Proc. Soc. Exp. Biol. Med. 143:965–970, 1973.

Nagaoka, T.; Oishi, M. and Narama, I.: Reproductive studies of ifosfamide. Kiso to Rinsho 16:508–516, 517–541, 542–552, 553–568, 1982.

1295 Ildamen™
L–3–Methoxy–ω(1–hydroxy–1,phenyl–isopropyl–amino) propiopheneone HCl

Habersang et al. (1967) studied this compound by oral route in mice, rats and rabbits during organogenesis. The rabbits and mice received up to 50 mg per kg and the rats 563 mg per kg. No adverse effects on the fetuses were found.

Habersang, S.; Leuschner, F. and Schlichtegroll, A.: Toxikologische Untersuchungen uber eine neue Myocard- und Coronarwirksame Verbindung aus der Reihe der β–Aminoketone. Arzneim–Forsch. 17:1478–1491, 1967.

1296 Imidapril Hydrochloride

Asano et al. (1992) gave orally up to 1,500 mg per kg on days 7–17 of the rat gestation. At the highest dose some dams died and the fetal weights were reduced. No adverse effects were found at 150 mg per kg.

Asano et al. (1992) gave this ACE inhibitor to rats before and in early pregnancy and during organogenesis. Oral doses of up to 1,500 mg per kg caused no increase in defects but there was reduced weight in the term female fetuses. Rabbits received up to 1.0 mg per kg during organogenesis and no adverse fetal effects ere seen. Behavior and reproduction of rats treated perinatally were normal. The studies in rats treated before fertilization did not reveal ill effects.

Asano, Y.; Nito, S.; Ariyuki, F. and Shimizu, M.: Reproductive and developmental toxicity study of imidapril hydrochloride (TA–6366). Oral administration to rats during the period of fetal organogenesis (seg II). Oyo Yikuri/Pharmacometrics 43:469–487, 1992.

Asano, Y.; Mito, S.; Ariyuki, F.; Shimizu, M.: Reproductive and developmental toxicity studies of imidapril hydrochloride in rats and rabbits. The Clinical Report 26:4669–4676, 1992.

1297 3–(2–Imidazoline–2–yl)–2–imidazolidinethione

Discussed under Ethylenethiourea

1298 Imidazopyridine *Zolpidem™ Tartrate*

Sasaki et al. (**1993**) gave this hypnotic orally on days 7–17 in the pregnant rat and found no increase in congenital defects. Fetal body weight was decreased at 5, 25 and 125 mg per kg. Wavy ribs were increased at 125 mg per kg.

Saski, M.; Nakajima, I.; Kawakami, T.; Kinoshita, K.; Shimizu, T.; Koyama, A.; Katsuki, S.; Saegusa, T. and Noguchi, H.: Study on administration of zolpidem tartrate during the period of organogenesis in rats. The Clinical Report 27:137–145, 1993.

1299 β, β–Iminodipropionitrile CAS 111–94–4

Miike and Chou (1981) gave approximately 40–70 mg per kg daily during gestation to rats. Hind limb paresis was found in 7 of 137 offspring and about 50 percent of the newborns died. The paresis appeared 3–6 weeks after birth and was associated with subluxation and A–D angulation of the thoracic vertebrae. The effects were considered to be due to the monomer of the chemical β–aminopropionitrile.

Miike, T. and Chou, S.M.: Lordosis of thoracic vertebrae column and following paraparesis of hind–limbs induced by maternal administration of β, β–iminodipropionitrile (IDPN) during pregnancy in rats. Cong. Anom. 21:407–413, 1981.

1300 Imipramine *5–(3–Dimethylaminopropyl)–10, 11–dihydro–5–H–dibenz(b,f)–azepine hydrochloride* CAS 50–49–7

In a Finnish study of 2,784 congenital malformations, three mothers took this drug as compared to one in a matched control (Idanpaan–Heikkila and Saxen, 1973). Although a possible association between this drug and human defects was reported from Australia (McBride, 1972), surveillance groups in the United States and Canada reviewed histories of hundreds of mothers who gave birth to children with limb reduction defects and found no supporting evidence of an association (Rachelefsky et al., 1972; Banister et al., 1972).

Robson and Sullivan (1963) injected rabbits with about 15 mg per kg starting on the first day and until the 13th to 20th day of gestation. Out of 12 mothers, they observed one fetus with encephalocele and another with incomplete palatal development and spina bifida occulta. Some of the other fetuses had hemorrhages of the cranium and abnormal limb flexion. Larsen (1963), using another strain of rabbit and conditions quite similar to the above, found only one abnormal fetus (single kidney) out of 53 examined. Harper et al. (1965) using rabbits found no defective fetuses among 46 whose mothers were given by injection 15 mg per kg from days 6–16. Administration by the injection route was associated with some increased fetal loss. At a dosage range of 30 mg per kg which was toxic to the mother, 4 of 29 surviving fetuses were abnormal. Harper et al. (1965) produced no fetal defects in the mouse or rat exposed to 150 and 15 mg per kg respectively. The offspring of rats given 5 and 10 subcutaneously on days 8–20 were studied postnatally.

These negative results with the rabbit and rat were further supported by Stenger et al. (1965) and Hendrickx (1975) who found no teratogenicity in monkeys given up to 1000 mg per kg per day during twenty–day periods of early gestation.

Ali, S.F.; Buelke–Sam, J.B.; Newport, G.D. and Slikker Jr., W.: Early neurobehavioral and neurochemical alterations in rats prenatally exposed to imipramine. Neurotoxicol. 7:365–380, 1986.

Banister, P.; DaFoe, C.; Smith, E.S.O. and Miller, J.R.: Possible teratogenicity of tricyclic antidepressants. Lancet 1:838–839, 1972.

Harper, K.H.; Palmer, A.K. and Davies, R.E.: Effect of imipramine upon the pregnancy of laboratory animals. Arzneim. Forsch. 15:1218–1221, 1965.

Hendrickx, A.G.: Teratologic evaluation of imipramine hydrochloride in bonnet monkeys Macaca mulatta. Teratology 11:219–222, 1975.

Idanpaan–Heikkila, J. and Saxen, L.: Possible teratogenicity of imipraminechloropyramine. Lancet 2:281–283, 1973.

Larsen, V.: Teratogenic effects of thalidomide, imipramine HCl and imipramine N–oxide HCl on white Danish rabbits. Acta Pharmacol. Toxicol. 20:186–200, 1963.

McBride, W.G.: Limb deformities associated with iminodibenzyl hydrochloride. Med. J. Aust. 1:492 only, 1972.

Rachelefsky, G.S.; Flynt, J.W.; Ebbin, A.J. and Wilson, M.G.: Possible teratogenicity of tricyclic antidepressants. Lancet 1:838 only, 1972.

Robson, J.M. and Sullivan, F.M.: The production of foetal abnormalities in rabbits with imipramine. Lancet 1:638–639, 1963.

Stenger, E.G.; Aeppli, L. and Fratta, I.: Zur Frage der Keimschaedigenden Wirkung von N–(γ–dimethylaminopropyl)–iminodibenzyl–HCl am Tier. Arznei–Mittelforschung 15:1222–1224, 1965.

1301 Immobilization

Immobilization of the mouse fetus by muscle relaxants at the time of palate closure failed to produce cleft palate (Jacobs, 1971). Drachman and Coulombre (1962) produced arthrogryposis and clubfoot in chick embryos immobilized by D–tubocurarine.

Hartel and Hartel (1960) produced 36.7 percent cleft palates in rat fetuses when the vitamin–A–treated mother was immobilized for 3 to 4 hours on the 9th through the 12th days of gestation. The vitamin A alone (15,000 IU daily on the 8th through 12th days) produced only 4.5 percent cleft palates and immobilization alone produced no defects.

Drachman, D.B. and Coulombre, A.J.: Experimental clubfoot and arthrogryposis multiplex congenita. Lancet 2:523–526, 1962.

Hartel, A. and Hartel, G.: Experimental study of teratogenic effect of emotional stress on rats. Science 132:1483–1484, 1960.

Jacobs, R.M.: Failure of muscle relaxants to produce cleft palate in mice. Teratology 4:25–30, 1971.

1302 Incidal
3–N–Methyl–9–benzyl–tetrahydro–y–gamma–carbolin

Hamada (**1970**) gave this antihistamine orally to rats on days 7–14 and found no fetal changes. The dose was 50 and 150 mg per kg.

Hamada, H.: The effect of incidal, administered to pregnant wistar strain rats, upon the fetuses. Bull. Osaka Med. Sch. 16:100–108, 1970.

1303 Indacrinone CAS 57296–63–6

Robertson et al. (1981), using this diuretic at 40–120 mg per kg on days 6–17 of rat gestation, produced wavy ribs and skeletal abnormalities in the fetus. This finding was reversed by coadministration of extra potassium chloride.

Robertson, R.T.; Minsker, D.H.; Bokelman, D.L.; Durand, G. and Conquet, P.: Potassium loss as a causative factor for skeletal malformations in rats produced by indacrinone: A new investigational loop diuretic. Toxicol. Appl. Pharmacol. 60:142–150, 1981.

1304 Indanazoline
N–(2–Imidazolin–2–yl)–N–(4–indanyl)amine Monohydrochloride CAS 40507–78–6

Worstmann et al. (1980) treated rats with up to 90 mg orally on days 0–6, 0–15, or 0–19 and found no teratogenicity although at the maternal toxic doses the fetal weight was reduced. The rabbits were treated with up to 30 mg per kg and no fetal changes were noted.

Worstmann, W.; Leuschner, F.; Neumann, W. and Kretzschmar, R.: Untersuchungen zur Toxizitat von N–(2–Imidazolin–2–yl)–N–(4–indanyl) amin–monohydrochlorid (Indanazolin), einem neuen Rhinologikum. Arzneim–Forsch. 30:1760–1771, 1980.

1305 1,3–Indandion

Kohler et al. (1975) gave mice single intraperitoneal doses of 100 mg per kg on day 9 and produced no abnormalities. 2–1, 3–Methyl indandion, bindon and bindonethylester were similarly negative.

Kohler, F.; Fickentscher, K.; Halfmann, U. and Koch, H.: Embryotoxicitat und Teratogenitat von Derivaten des 1,3–Indandion. Arch. Toxicol. 33:191–197, 1975.

1306 Indapamide fcas 26807–65–8

Seki et al. (1982) administered this antihypertensive orally to rats and rabbits in maximum doses of 1000 and 80 mg per kg daily. Fertility was unaffected in the rat. Studies during organogenesis and on days 17–21 were done and no adverse effects were found in the rat fetuses or in the behavior of the offspring. Some growth retardation was seen in rat fetuses at the highest dose schedule. No adverse fetal effects were found in the rabbit.

Seki, T.; Fujitani, M.; Osumi, S.; Yamamoto, T.; Eguchi, K.; Inoue, N.; Sakka, N. and Suzuki, M.R.: Reproductive studies of indapamide. Yakuri to Chiryo 10:1325–1335, 1337–1353, 1355–1362, 1363–1374, 1982.

1307 India Ink

Wilson et al. (1959) injected one ml of India Ink (50 percent on days 7, 8 and 9) and produced no malformations. The maternal death rate was high.

Wilson, J.G.; Beaudoin, A.R. and Free, H.J.: Studies on the mechanism of the teratogenic action of trypan blue. Anat. Rec. 133:115–128, 1959.

1308 Indiogofera Extract

The extract of Indiogofera spicata, a hardy pasture plant, has been fed to the rat and cleft palate was produced in about 60 percent of the offspring (Pearn, 1967).

Pearn, J.H.: Report of a new site–specific cleft palate teratogen. Nature 215:980–981, 1967.

1309 Indium CAS 7440–74–6

Ferm and Carpenter (1970) injected indium nitrate (0.5 to 1.0 mg per kg intravenously) on day 8 of gestation of the hamster and observed digital defects in over one–half of the surviving fetuses. Dosage over 1 mg per kg caused intrauterine death.

Ferm, V.H. and Carpenter, S.J.: Teratogenic and embryopathic effects of indium, gallium and germanium. Toxicol. Appl. Pharmacol. 16:166–170, 1970.

1310 Indoleacetic Acid *Indole–3–acetic Acid* CAS 87–51–4

John et al. (1979) gavaged mice and rats on days 7–15 of gestation with 50 to 500 mg per kg per day. At the highest dose, fetuses from both species had increased numbers of cleft palate. The mouse fetuses also had increased rates of exencephaly, dilated cerebral ventricles and crooked tails.

Ruddick et al. (1974) gave pregnant rats up to 100 mg per kg by gavage on days 6 through 15 and found no fetal changes.

John, J.A.; Blogg, C.D.; Murray, F.J.; Schwetz, B.A. and Gehring, P.J.: Teratogenic effects of the plant hormone indole–3–acetic acid in mice and rats. Teratology 19:321–326, 1979.

Ruddick, J.A.; Harwig, J. and Scott, P.M.: Nonteratogenicity in rats of blighted potatoes and compounds contained in them. Teratology 9:165–168, 1974.

1311 Indol–yl–ethyl–oxa–oxo–diazaspiro–undecane

Andrew et al. (1985) studied this antihypertensive agent in rats by gavage on days 8–17. A dose related increase in maternal and fetal deaths and orofacial defects was found.

Andrew, F.; Terrell, T.; Thacker, G.; Onizuka, N. and Zustak, C.: Reproductive and developmental toxicity of an alpha 1–adrenergic blocker in rats. Toxicologist 5:188, 1985.

1312 Indometacin Farnesil

Kondoh et al. (1989) gave rats up to 400 mg per kg orally on daya 7–17 and found no fetal ill effects.

Kondah, S.; Okada, F.; Gotoh, M.; Nishimura, O.; Ohsumi, I.; Aoki, T. and Matsubara, Y.: Reproduction study of indometacin farnesil (II)–teratogenicity study in rats by oral administration. Yakuri to Chiryo 17:63–85, 1989.

1313 Indomethacin CAS 53–86–1

Aselton et al. (1985) reported only one congenital defect among 50 offspring of mothers who took the medication. Levin et al. (1978) reported an infant with pulmonary artery changes following short term prenatal exposure (25 mg daily) to the mother. The developmental toxicity in animals and humans was reviewed by Lione and Scialli (1995). The drug is used commonly for stopping preterm labor with few adverse effects. They also review the drug's action on constriction of the ductus arteriosus and decreased fetal urine flow.

Kalter (1973) administered this antipyretic agent to mice on days 9–15 by oral, subcutaneous and intramuscular route. Dosage up to the maternal lethal dose (7.5 mg per kg) did not produce teratogenic activity. Kusanagi et al. (1977), using 7.5 mg per kg orally on days 7–15, produced fused ribs, vertebral abnormalities and other skeletal defects in mouse fetuses. Similar negative findings in the rat using 4 mg per kg during organogenesis were reported by Klein et al. (1981). Powell and Cochrane (1978) used 1.6 mg per kg orally starting on day 18 in the rat and caused premature closure of the ductus arteriosus in 1 to 3 percent of the offspring whose mothers were treated for 3.5 to 5 days. Morozova (1989) treated rats on days 18–21 with 1 or 2.5 mg per kg and found decreased lymphocytes and structural changes in the joints.

Aselton, P.A.; Jick, H.; Milunsky, A.; Hunter, J.R. and Stergachis, A.: First–trimester drug use and congenital disorders. Obstet. Gynecol. 65:451–455, 1985.

Kalter, H.: Nonteratogenicity of indomethacin in mice. (abs) Teratology 7:A–19, 1973.

Klein, K.L.; Scott, W.J.; Clark, K.E. and Wilson, J.G.: Indomethacin–placental transfer, cytotoxicity and teratology in the rat. Am. J. Obstet. Gynecol. 141:448–452, 1981.

Kusanagi, T.; Ihara, T. and Mizutani, M.: Teratogenic effects of non–steroid anti–inflammatory agents in mice.

(Japanese) Cong. Anom. 17:177–185, 1977.

Levin, D.L.; Fixler, D.E.; Morriss, F.C. and Tyson, J.: Morphologic analysis of pulmonary vascular bed in infants exposed in utero to prostaglandin synthetase inhibitors. J. Pediatr. 92:478–483, 1978.

Lione, A. and Scialli, A.R.: The developmental toxicity of indomethacin and sulindac. Reproductive Toxicology 9:7–20, 1995.

Morozova, E.V.: Structure of the mesenteric lymph nodes in fetuses and offspring of white rats under indomethacin treatment. Arh. Anat., Gistol. Embriol. 96(3):48–55, 1989.

Powell, J.G. and Cochrane, R.L.: The effects of the administration of fenoprofen or indomethacin to rat dams during late pregnancy with special reference to the ductus arteriosus of the fetuses and neonates. Toxicol. Appl. Pharmacol. 45:783–796, 1978.

1314 Indospicine

Keeler (1983) reviewed the studies on this chemical extracted from Indigofera spicata. Cleft palate and fetal death in the rat were produced by administration on day 13. It may compete with arginine for incorporation into protein.

Keeler, R.F.: Naturally occurring teratogens from plants. In: Handbook of Natural Toxins, R.F. Keeler and A.T. Tu (eds). Marcel Dekker Inc., New York, 186, 1983.

1315 Influenza Vaccine

Heinonen et al. (1973, 1977) studied 3,051 pregnancies where influenza immunization was given and found no increase in defect rate or postnatal malignancies. Sarnat et al. (1979) reported one infant with cerebral malformation after the mother received swine influenza vaccine six weeks post–conception.

Heinonen, O.P.; Shapiro, S.; Monson, R.R.; Hartz, S.C.; Rosenberg, L. and Slone, D.: Immunization during pregnancy against poliomyelitis and influenza in relation to childhood malignancy. Int. J. Epidemiol. 2:229–235, 1973.

Heinonen, O.P.; Slone, D. and Shapiro, S.: Birth Defects and Drugs in Pregnancy. Publishing Sciences Group, Inc., Littleton, Mass. 1977.

Sarnat, H.B.; Rybak, G.; Kotagal, S. and Blair, J.D.: Cerebral embryopathy in late first trimester: Possible association with swine influenza vaccine. Teratology 20:93–100, 1979.

1316 Influenza Viruses

There is a major body of literature on the role that influenza viruses might play in production of congenital defects (see review by Brown, 1966). Some epidemiologic studies reported a small non–specific increase in defects (Hardy et al., 1961; Saxen et al., 1960; Coffey and Jessop, 1959) an almost equal number of studies reported none (Wilson et al., 1959; Ingalls, 1960; Walker and McKee, 1959; Doll et al., 1960; Korones et al., 1970). Hardy et al. (1961) reported an increase in fetal wastage associated with an influenza outbreak. Leck (1963) noted that the incidence of esophageal atresia was especially high; and Doll et al. (1960), Coffey and Jessop (1959) and Hakosalo and Saxen (1971) found anencephaly to be more frequent following epidemics of Asian influenza.

The data from many of these reports are difficult to assess because of the following factors: Lack of serologic proof of the infection, unknown attack rates in the control population, restricted numbers of congenital defects or inadequate controls. A number of investigators have suggested that the small increase in attack rates during epidemics may be caused by fever, medications or other non–viral associated factors. Saxen (1975) analyzed the effect of non–viral components of influenza attacks.

Studies of the effect of influenza virus on chicks (Adams et al., 1956) and mice (Siem et al., 1960) induced some decrease in number of offspring but no clear–cut congenital defects were produced. In the study by Siem et al. (1960), the virus did not cross the mouse placenta until late in the gestation period.

Adams, J.M.; Heath, H.D.; Imagawa, D.T.; Jones, M.H. and Shear, H.H.: Viral infections in the embryo. Am. J. Dis. Child. 92:109–114, 1956.

Brown, G.C.: Recent advances in the viral aetiology of congenital anomalies. In: Advances in Teratology, D.H.M. Woollam (ed.), London: Logos Press, 1:55–80, 1966.

Coffey, V.P. and Jessop, W.J.E.: Maternal influenza and congenital deformities: A follow–up study. Lancet 2:935, 1959.

Doll, R.; Hill, A.B. and Sukula, J.: Asian influenza in pregnancy and congenital defects. Br. J. Prev. Soc. Med. 14:167–172, 1960.

Hakosalo, J. and Saxen, L.: Influenza epidemic and congenital defects. Lancet 2:1346–1347, 1971.

Hardy, J.B.; Azarowicz, E.N.; Mannini, A.; Medearis, D.N. and Cooke, R.E.: The effect of Asian influenza on the outcome of pregnancy, Baltimore, 1957–1958. Am. J. Public Health 51:1182–1188, 1961.

Ingalls, T.H.: Prenatal human ecology. Am. J. Public Health 50:50–54, 1960.

Korones, S.B.; Todaro, J.; Roane, J.A. and Sever, J.L.: Maternal virus infection after the first trimester of pregnancy and status of offspring to 4 years of age in a predominantly Negro population. J. Pediatr. 77:245–251, 1970.

Leck, I.: Incidence of malformations following influenza epidemics. Br. J. Prev. Soc. Med. 17:70–80, 1963.

Saxen, L.: Newborn monitoring. In: Methods for Detection of Environmental Agents Which Produce Congenital Defects, T.H. Shepard, J.R. Miller and M. Marois (eds.), Amsterdam: North–Holland American Elsevier, 205–216, 1975.

Saxen, L.; Hjelt, L.; Sjostedt, J.E.; Hakosalo, J. and Hakosalo, H.: Asian influenza during pregnancy and congenital malformations. Acta Pathol. Microbiol. Scand. 49:114–126, 1960.

Siem, R.A.; Ly, H.; Imagawa, D.T. and Adams, J.M.: Influenza virus infections in pregnant mice. J. Neuropathol. Exp. Neurol. 19:125–129, 1960.

Walker, W.M. and McKee, A.P.: Asian influenza in pregnancy relationship to fetal anomalies. Obstet. Gynecol.

13:394–398, 1959.

Wilson, M.G.; Heins, H.L.; Imagawa, D.T. and Adams, J.M.: Teratogenic effects of Asian influenza. JAMA 171:116–119, 1959.

1317 5–Inosinate, Disodium

Kojima (1974) gave orally 100 mg per kg on days 9–15 of pregnancy and found no teratogenicity. No reproductive ill effects were found in three–generation studies of rats maintained on 2.0 percent of the chemical in the diet. At a dose of 200 mg per kg, there was no teratogenicity found in rabbits.

Kojima, K.: Safety evaluation of disodium 5'–inosinate, disodium 5–guanylate and disodium 5'–ribonucleotide. Toxicology 2:185–206, 1974.

1318 Inosine also see 2–Deoxyinosine CAS 58–63–9

Karnofsky and Lacon (1961) injected 2–8 mg into four–day–old chick yolk sacs and found a 9 percent incidence of unspecified defects. The LD–50 under these conditions was 4–8 mg.

Mercier–Parot and Tuchmann–Duplessis (1973) gave 500 mg per kg on the 12th day of rat gestation intraperitoneally and found no abnormal fetuses.

Karnofsky, D.A. and Lacon, C.R.: Effects of physiological purines on the development of the chick embryo. Biochem. Pharmacol. 7:154–158, 1961.

Mercier–Parot, L. and Tuchmann–Duplessis, H.: Malformations squelettiques chez le rat produites par la 6–hydroxylaminopurine. Essais de Prevention. Compt. R. S. Soc. Biol. 167:5–10, 1973.

1319 Inositol

Cockroft et al. (**1992**) found cranial neural tube defects in mouse embryos grown in deficient culture medium. Curly tail ct/ct embryos were especially sensitive to inositol deficiency.

Cockroft, D.L.; Brook, F.A. and Cop, A.J.: Inositol deficiency increases the susceptibility to neural tube defects of genetically predisposed (curly tail) mouse embryos in vitro. Teratology 45:223–232, 1992.

1320 Insulin Insulin Shock CAS 9004–10–8

Sobel (1960) identified 17 incidences where the mother was treated for insulin shock. Four fetal deaths and two fetuses with multiple congenital anomalies were reported. Impastato et al. (1964) reported that among 19 women given insulin shock therapy two had macerated fetuses at term, two had spontaneous abortions and two had developmentally retarded offspring.

Kalter (1968) and Kalter and Warkany (1959) reviewed this subject. A number of negative reports using the rat appeared (see Kalter, 1968). Ferrill (1943), using 20 to 40 units per kg from weaning, found no defects in the rat offspring that were produced during five generations. Love et al. (1964), giving 0.5 units every 12 hours during the last two weeks of pregnancy, produced no defects in the rat.

The mouse was used by Smithberg et al. (1956) and Smithberg and Runner (1963) who injected 0.1 units of protamine zinc insulin 8.5 days post coitum and found a high incidence of exencephaly, rib and vertebral defects. The teratogenic effects of insulin or tolbutamide alone or in conjunction with nicotinamide were compared by Smithberg and Runner (1963) in three strains of mice. The primary role of insulin is hard to judge when the mouse is sensitive also to fasting teratogenesis.

In the rabbit, Chomette (1955) and Brinsmade et al. (1956) produced microcephaly and other central nervous system defects. Landauer (1947) produced rumplessness, micromelia and other skeletal defects in chicks by injecting 2 units of insulin into the yolk sac at 24 or 120 hours of incubation. Riboflavin given with the insulin caused an increase in the number of deformities (Landauer, 1952) but nicotinamide reduced the incidence (Landauer and Rhodes, 1952). Landauer (1972) reviewed the evidence that insulin alone is teratogenic. Travers et al. (1989) grew rat embryos in insulin depleted serum and found growth retardation which was restored by adding back physiologic amounts of insulin.

Brinsmade, A.B.; Buchner, F. and Rubsaamen, H.: Missbildungen am Kaninchenembryo durch Insulininjektion beim Muttertier. Naturwissenschaften 43:259 only, 1956.

Chomette, G.: Entwicklungsstoerungen nach Insulinschock beim Traechtigen Kaninchen. Beitr. Pathol. Anat. 115:439–451, 1955.

Ferrill, H.W.: Effect of chronic insulin injections on reproduction in white rats. Endocrinology 32:449–450, 1943.

Impastato, D.J.; Gabriel, A.R. and Lardaro, H.H.: Electric and insulin shock therapy during pregnancy. Diseases of the Nervous System 25:542–546, 1964.

Kalter, H.: Teratology of the Central Nervous System. Chicago: University of Chicago Press, 1968.

Kalter, H. and Warkany, J.: Experimental production of congenital malformations in mammals by metabolic procedure. Physiol. Rev. 39:69–115, 1959.

Landauer, W.: Insulin–induced abnormalities of the beak, extremities and eyes in chickens. J. Exp. Zool. 105:145–172, 1947.

Landauer, W.: Malformations of chicken embryos produced by boric acid and the probable role of riboflavin in their origin. J. Exp. Zool. 120:469–508, 1952.

Landauer, W.: Is insulin a teratogen? Teratology 5:129–135, 1972.

Landauer, W. and Rhodes, M.B.: Further observations on the teratogenic nature of insulin and its modification by supplementary treatment. J. Exp. Zool. 119:221–261, 1952.

Love, E.J.; Kinch, R.A.H. and Stevenson, J.A.F.: Effect of protamine. Zinc insulin on outcome of pregnancy in the normal rat. Diabetes 13:44–48, 1964.

Smithberg, M. and Runner, M.N.: Teratogenic effects of hypoglycemic treatment in inbred strains of mice. Am. J. Anat. 113:479–489, 1963.

Smithberg, M.; Sanchez, H.W. and Runner, M.N.: Congenital deformity in the mouse induced by insulin. (abs) Anat. Rec. 124:441 only, 1956.

Sobel, D.E.: Fetal damage due to ECT, insulin coma, chlorpromazine, or reserpine. A.M.A. Gen. Psych. 2:606–611, 1960.

Travers, J.P.; Pratten, M.K. and Beck, F.: Effects of low insulin levels on rat embryonic growth and development. Diabetes 38:773–778, 1989.

1321   α Interferon

No teratogenicity was found in rats and rabbits given up to 10,000 IU per kg during organogenesis. The rabbits received this human protein intravenously. The rats received it by the intramuscular or intraperitoneal mode (Matsumoto et al., 1986A,B,C and Shibutani et al., 1987). Perinatal administration to rats produced no adverse developmental changes in the offspring (Matsumoto et al., 1986C.)

Hasegawa et al. (**1993**) gave this intramuscularly to rats before and during the first week of gestation in doses of up to 12 million IV per kg and found no adverse reproductive or fetal effects. Similar treatment during organogenesis and perinatally was not associated with adverse fetal or behavioral effects (Hasegawa et al., **1993** B,C). In rabbits Faruhashi et al. (**1993**) used a similar treatment during organogenesis and found reduced fetal weight and a delay in ossification.

Furuhashi, T.; Kato, M.; Takahashi, M.; Nishiwaki, M.; Matsui, E.; Hayashi, H. and Araki, H.: Teratogenicity study of FPI–31 by intramuscular administration in rabbits. Yakuri to Chiryo 21:s1377–s1386, 1993.

Hasegawa, M.; Kihara, K.; Kakigi, K.; Yasugi, D.; Kuwahara, Y.; Isaka, K.; Sakai, Y.; Habuchi, M.; Kobayashi, J.; Ishigami, M.; Tsuji, K.; Hagino, T.; Usmura, H. and Araki, H.: Study on the intramuscular injection of FPI–31 prior to and in the early stages of pregnancy in mice. Yakuri to Chiryo 21:s1337–s1356, 1993.

Hasegawa, M.; Kihara, K.; Shimada, C.; Yasugi, D.; Kakigi, K.; Kuwahara, T.; Isaki, K.; Sakai, Y.; Habuchi, M.; Kobayashi, J.; Ishigami, M.; Tsuji, K.; Hagino, T.; Uemura, H. and Araki, H.: Study on the intramuscular injection of FPI–31 during the period of organogenesis in mice. Yakuri to CHiryo 21:s1357–s1375, 1993.

Hasegawa, M.; Kiharea, K.; Shimada, C.; Yasugi, D.; Kuwahara, y.; Kakigi, K.; Isaka, K.; Habuchi, M.; Kobayashi, J.; Ishigami, M.; Tsuji, K.; Hagino, T.; Ito, C.; Uemura, H.; Nishida, M. and Araki, H.: Study on the intramuscular injection of FPI–31 during the perinatal and lactation periods in mice. Yakuri to Chiryo 21:s1387–s1406, 1993.

Matsumoto, T.; Nakamura, K.; Imai, M.; Aoki, H.; Okugi, M.; Shimoi, H. and Hagita, K.: Reproduction studies of human interferon α (interferol alpha). (I) Teratological study in rabbits. Iyakuhin Kenkyu 17:397–404, 1986A.

Matsumoto, T.; Nakamura, K.; Imai, M.; Aoki, H.; Okugi, M.; Shimoi, H. and Hagita, K.: Reproduction studies of human interferon α (interferol alpha). (III) Teratological study in rats. Iyakuhin Kenkyu 17:417–438, 1986B.

Matsumoto, T.; Nakamura, K.; Imai, M.; Aoki, H.; Okugi, M.; Shimoi, H. and Hagita, K.: Perinatal studies of human interferon α (interferol alpha). (IV) Teratological study in rats. Iyakuhin Kenkyu 17:439–457, 1986C.

Shibutani, Y.; Hamada, Y.; Kurokawa, M.; Inoue, K.

and Shichi, S.: Toxicity studies of human lymphoblastoid interferon α. Teratogenicity study in rats. Iyakuhin Kenkyu 18:60–78, 1987.

1322 Interferon β *GKT–β*

Naya et al. (**1988**) gave this human recombinant intravenously to rats on days 7–17. At the highest dose 10 MU no adverse fetal effects were seen.

Naya, M.; Fujita, T.; Takahashi, H.; Hara, T. and Takahira, H.: Toxicology study of gkt–β–teratogenicity study in rats administered intravenously. Kiso to Rinsho 22:137–145, 1988.

1323 Interferon-γ Human *SUN 4800 OH-6000*

Ohnishi et al. (**1992** A) gave up to 1.0 mg intravenously to rats before fertilization and during the first 7 days of pregnancy. Although males had an increase in liver and kidney weight no other reproductive or fetal effects were found. Ohnishi et al. (**1992** B) used a similar treatment during organogenesis and found no alteration in development or postnatal behavior. Perinatal studies were also performed without adverse fetal effects (Ohnishi et al., **1992** C). Rabbit studies at the same dose during organogenesis did not produce any teratogenicity (Ohnishi et al., **1992** D).

Hatagkeyama et al. (**1993** A,B,C) gave this compound to rats intramuscularly with up to $2x 10^6$ units IV per kg before pregnancy and during the first 7 days of gestation and found no ill effects on reproduction. Similar studies during organogenesis and perinatal and postnatal periods produced no ill effects on the fetuses. Behavior and reproduction of the exposed fetuses were within normal limits. Morita et al. (**1993**) performed similar studies during organogenesis on the rabbit without identifying any fetal ill effects.

Hatakeyama, Y.; Iizuka, T.; Nishioeda, R, and Kotosai, K.: A reproductive toxicity study by intramuscular administration of OH–6000 before and during the early stages of gestation in rats. Yakuri to Chiryo 21:3579–3589, 1993A.

Hatakeyama, Y.; Ando, K.; Iizuka, T.; Nishioeda, R. and Kotosai, K.: A reproductive toxicity study by intramuscular administration of OH–6000 during the fetal organogenesis period in rats. Yakuri to Chiryo 21:2591–3616, 1993B.

Hatakeyama, Y.; Kawakami, Y.; Shimomura, K.; Hironaka, N.; Nishioeda, R. and Kotosai, K.: A reproductive toxicity study by intramuscular administration of OH–6000 during the peripartum and nursing periods in rats. Yakuri to Chiryo 21:3629–3649, 1993C.

Morita, H.; Naito, Y.; Uchiyama, H.; and Oi, A.: A reproductive toxicity study by intramuscular administration of OH-6000 during the fetal organogenesis period in rabbits. Yakuri to Chiryo 21:3617–3627, 1993.

Ohnishi, S.; Yamamori, K.; Satoh, T. and Narama, I.: Intravenous fertility study of human interferon–γ (SUN 4800) in rats. Rinsho Iyaku 8, suppl 2:165–172, 1992A.

Ohnishi, S.; Yamamori, K.; Satoh, T. and Narama, I.: Intravenous teratogenicity study of human interferon-γ (SUN 4800) in rats. Rinsho Iyaku 8 suppl 2:179–193, 1992B.

Ohnishi, S.; Yamamori, K.; Satoh, T. and Narama, I.: Intravenous perinatal and postnatal study of human

interferon-γ (SUN 4800) in rats. Rinsho Iyaku 8 suppl 2:207–218, 1992C.

Ohnishi, S.; Yamamori, K.; Saegusa, T. and Narama, I.: Intravenous Teratogenicity study of human interferon–γ (SUN 4800) in rabbits. Rinsho Iyaku 8 Suppl 2:195–205, 1992D.

1324 Interleukin–2, Recombinant Human *S–6820 r–IL–2*

Furuhashi et al. (**1989**) gave rats intravenously up to 100 microgm per kg on days 6–17 and found no adverse fetal effects. Ihara et al. (**1989**) gave rabbits up to 60 microgm per kg on days 6–18. About one–half of the dams aborted on days 22–27. Cleft palate and seletal defects were increased. Eventration was also found. Cozens et al. (**1989**) gave rats 1.5 or 25 microgm per kg intravenously before mating and until day 7 of gestation or during all of pregnancy and part of weaning. No teratologic or reproductive effects were found. Nishioeda et al. (**1994** A,B) studied interleukin–1–β in rats and rabbits during organogenesis giving subcutaneous doses of up to 30 and 3 microgram per day respectively. In the rat fetuses at the highest dose there was an increase in skeletal variations, flexed tails and fetal death. In the rabbits an increase in fetal death and a decrease in survival was found. Behavior and reproduction of the rat offspring was normal. Perinatal and postnatal studies in the rat were performed by Uchiyama et al. (**1994**) who found decreased weight gain at 3 mg per kg given to the dam. Reproductive studies in the rat by Nishioeda et al. (**1994** C) indicated a decreased weight gain and fewer implantations at 1, 10 and 100 mg per day.

Cozens, D.D.; Jones, K.; John, D.M.; Hughes, E.W. and Gibson, W.A.: Reproductive studies on S–6280 (r–IL–2) (1)–Effect of S–6820 on fertility and general reproductive performance of the rat. Kiso to Rinsho S:3001–3003, 1989.

Furuhashi, T.; Suda, H.; Ishizaka, Y. and Ihara, T.: Teratogenic effects of recombinant human interleukin–2 (tgp–3) in rats. Kiso to Rinsho 23:4249–4262, 1989.

Ihara, T.; Ohshima, Y.; Nakamura, H. and Sugimoto, T.: Teratogenic effects of recombinant human interleukin–2 (tgp–3) in rabbits. Kiso to Rinsho 23:4857–4864, 1989.

Nishioeda, R.; Tao, K. and Tamagawa, M.: Reproductive and developmental toxicity studies of OCT–43 (1): Fertility study in rats by subcutaneous administration. Yakuri to Chiryo 22:2571–2581, 1994A.

Nishioeda, R.; Tao, K. and Tamagawa, M.: Reproductive and developmental toxicity studies of C'OCT–43 (3): Teratogenicity study in rabbits by cubcutaneous administration. Yakuri to Chiryo 22:2603–2611, 1994.

Uchiyama, H.; Kitagaki, T.; Sekiya, K. and Nishioeda, R.: Reproductive and developmental toxicity studies of OCT–43 (2): Teratogenicity study in rats with subcutaneous administration. Yakuri to Chiryo 22:2583, 2401, 1994.

Uchiyama, H.; Kitagaki, T.; Sekiya, K. and Nishioeda, R.: Reproductive and developmental toxicity studies of OCT–43 (4): Perinatal and postnatal study in rats with subcutaneous administration. Yakuri to Chiryo 22:2613–2624, 1994.

1325 In Vitro Fertilization

Lancaster (1987) studied the records of 16 in–vitro fertilization clinics and 1,694 live births and stillbirths. Thirty–six or 2.2 percent had major congenital malformations. Six had spina bifida and four had transposition of the great vessels. Data from 1510 pregnancies in Australia and New Zealand gives the live birth rate at 57.5 percent and ectopic pregnancy at 5.2 percent (Australian In–Vitro Fertilization Collaborative Group, 1988). Spontaneous abortion occurred in 24.3 percent. In another study Morin et al. 1989 studied 83 in vitro fertilized offspring and compared them with 93 matched controls. No increase in defects or decrease in developmental scores was found.

Australian In–Vitro Fertilization Collaborative Group: In–vitro fertilization pregnancies in Australia and New Zealand, 1979–1985. Med. J. Australia 148:429–436, 1988.

Lancaster, P.A.L.: Congenital malformations after in-vitro fertilizations. Lancet 2:1392 only, 1987.

Morin, N.C.; Wirth, F.H.; Johnson, D.H.; Frank, L.M.; Presburg, H.J.; Van de Water, V.L.; Chee, E.M. and Mills, J.L.: Congenital malformations and psychosocial development in children conceived by in vitro fertilization. The Journal of Pediatrics 115:222–227, 1989.

1326 Iodide

Protracted ingestion of iodide containing medications (expectorants) by the mother may result in fetal thyroid enlargement which can lead, in some cases, to tracheal compression and choking in the newborn (Parmelee et al., 1940; Galina et al., 1962). These goiters are due to fetal thyroid inhibition with secondary compensatory hypertrophy. Klevit (1969) reviewed this subject. Pharoah et al. (1971) prevented endemic cretinism and the associated mental retardation by injection of iodized oil before conception.

Galina, M.P.; Avnet, N.L. and Einhorn, A.: Iodides during pregnancy. An apparent cause of death. N. Eng. J. Med. 267:1124–1127, 1962.

Klevit, H.D.: Iatrogenic thyroid disease. In: Endocrine and Genetic Diseases of Childhood. L.I. Gardner (ed), Philadelphia, W.B. Saunders, 246–247, 1969.

Parmelee, A.H.; Allen, E.; Stein, I.F. and Buxbaum, H.: Three cases of congenital goiter. Am. J. Obstet. Gynecol. 40:145–147, 1940.

Pharoah, P.O.D.; Buttfield, I.H. and Hetzel, B.S.: Neurological damage to the fetus from severe iodine deficiency during pregnancy. Lancet 1:308–310, 1971.

1327 Iodine[131] *Radioiodine* CAS 10043–66–0

This isotope of iodine, if given in millicurie doses, can damage or ablate the developing thyroid of the human fetus. Hypothyroidism, either congenital or of late onset, was reported in at least five children whose mothers were treated with I[131] during pregnancy (Russell et al., 1957; Hamill et al., 1961; Fisher et al., 1963; Green et al., 1971). The risk of fetal radiochemical thyroidectomy increases with the onset of the iodide concentrating ability of the fetal thyroid at around the 74th gestational day (Shepard, 1967).

Stoffer and Hamburger (1976) surveyed 517 physicians and obtained information on 182 pregnancies. The general complication rate was not increased. However, six infants had hypothyroidism and of these, four were mentally retarded.

Speert et al. (1951) first studied the thyroid damaging effect of I[131] in the mouse fetus whose mother was given 200 microcuries after the 15th gestational day.

Fisher, W.D.; Voorhess, M.L. and Gardner, L.I.: Congenital hypothyroidism in infant following maternal I[131] therapy. J. Pediatr. 62:132–146, 1963.

Green, H.G.; Gareis, F.J.; Shepard, T.H. and Kelley, V.C.: Cretinism associated with maternal sodium iodide I[131] therapy during pregnancy. Am. J. Dis. Child. 122:247–249, 1971.

Hammill, G.C.; Jarman, J.A. and Wynne, M.D.: Fetal effects of radioactive iodine therapy in a pregnant woman with thyroid cancer. Am. J. Obstet. Gynecol. 81:1018–1023, 1961.

Russel, K.P.; Rose, H. and Starr, P.: The effects of radioactive iodide on maternal and fetal thyroid function during pregnancy. Surg. Gynecol. Obstet. 104:560–564, 1957.

Shepard, T.H.: Onset of function in the human fetal thyroid: Biochemical and radioautographic studies from organ culture. J. Clin. Endocrinol. Metab. 27:945–958, 1967.

Speert, H.; Quimby, E.H. and Werner, S.C.: Radioiodine uptake by the fetal mouse thyroid and resultant effects in later life. Surg. Gynecol. Obstet. 93:230–242, 1951.

Stoffer, S.S. and Hamburger, J.I.: Inadvertent I[131] therapy for hyperthyroidism in the first trimester of pregnancy. J. Nucl. Med. 17:146–149, 1976.

1328 Iodine Deficiency

Warkany (1971) comprehensively reviewed the role that iodine deficiency played in production of endemic and congenital goiters. Before the introduction of modern transportation of iodinated foods isolated communities drawing their food and water supply from glaciated terrain were the source of most cases of endemic goiter.

Warkany, J.: Congenital Malformations: Notes and Comments. Chicago: Year Book Medical Publishers, 106–107, 1971.

1329 Iodoacetate CAS 64–69–7

Runner and Dagg (1960) injected 1 mg of iodoacetate intraperitoneally into mice on day 8 of gestation and found deformed ribs and vertebrae in 62 percent of the offspring. Miller (1973) injected 0.5 mg intramuscularly on days 12 through 14 and produced 14 percent cleft palates in the mouse and was able to lower the incidence by giving succinate concurrently.

Miller, T.J.: Cleft palate formation: The effects of fasting and iodoacetic acid on mice. Teratology 7:177–182, 1973.

Runner, M.N. and Dagg, C.P.: Metabolic mechanisms of teratogenic agents during morphogenesis. Natl. Cancer Inst. Monogr. 2:41–54, 1960.

1330 Iodothiouracil

Peterson (1953) gave guinea pigs 0.15 percent in their drinking water beginning on day 38 of gestation and found increased thyroid weights in the newborns.

Peterson, R.R.: Comparison of the effects of placental transmission of propyl–and iodothiouracil in the guinea pig. Anat. Record 115:359–360, 1953.

1331 Ionizing Radiation

Discussed under Radiation, Ionizing

1332 Ioxilan *(+-)–N–(2,3–Dihydroxypropyl)–5–[N–(2,3–dihyroxypropyl)–2, 4,6–triiodoiso–phthalamide*

Studies in pregnant rats using this chemical intravenously at up to 4 g of iodide per kg were done by Hara et al. (1993), Shibano et al. (1993) and Koyama et al. (1993). No adverse reproductive teratogenic or behavioral adverse effects were found. In the rabbit Igarashi et al. (1993) found no adverse fetal effects of intravenously administration of up to 2 g per kg.

Hara, H.; Kiyosawa, K.; Igarashi, S.; Tanaka, N.; Takeuchi, H.; Aoki, A. and Sugiyama, O.: A fertility study of ioxilan in rats. Yakuri to Chiryo 21:s1513–s1520, 1993.

Igarashi, S.; Takeuchi, H.; Aoki, A.; Tanaka, N.; Hara, H.; Kiyosawa, K. and Sugiyama, O.: A teratological study of ioxilan in rabbits. Yakuri to Chiryo 21:s1543–s1549, 1993.

Koyama, A.; Kawakami, T.; Kinoshita, K. and Igarashi, S.: A perinatal and postnatal study of ioxilan in rats. Yakuri to Chiryo 21:s1551–s1570, 1993.

Shibano, T.; Kawakami, T.; Kinoshita, K. and Igarashi, S.: A teratological study of ioxilan in rats. Yakuri to Chiryo 21:s1521–s1542, 1993.

1333 Ioxynil Octanate *3,5–Diiodo–4–octanoyloxybenzonitrile*

Kobayashi et al. (1976) tested this herbicide in three generations of mice at 2.5 and 100 ppm in the diet. The growth of the newborns was slightly retarded at the highest dose. No other adverse effects were found.

Kobayashi, F.; Ando, M.: Ito, M.; Shigemura, M.; Hara, K.; Sasaki, K. and Muranaka, R.: Reproductive study on 3,5–diiodo–4–octanoyloxybenzonitrile in mice. Oyo Yakuri 11:881–894, 1976.

1334 Ipecac CAS 8012–96–2

Heinonen et al. (1977) found no increase in defects among the offspring of 68 women who took the medication in the first four lunar months. Similar findings were present among 379 women exposed anytime in pregnancy.

Heinonen, O.P.; Slone, D. and Shapiro, S.: Birth Defects and Drugs in Pregnancy. Publishing Sciences Group Inc., Littleton, Mass., 1977.

1335 Ipratropium Bromide *SCH 1000* CAS 22254–24–6

Nishimura et al. (1978) gavage fed rats and rabbits with maximum daily doses of 500 and 125 mg per kg during organogenesis. The only positive finding was a slight reduction in weight of the rat fetuses at the highest dose.

Nishimura, M.; Kast, A. and Tsunenari, Y.: Reproduction studies of ipratropium bromide (SCH 1000) on rats and rabbits. Iyakuhin Kenkyu 9:393–416, 1978.

1336 Ipriflavone *7–Isopropoxy–3–phenyl–4H–1–benzopyran–4–one*

Mitzutani et al. (1985a) gave up to 3000 mg per kg by gavage daily to rabbits on days 6–18. No teratogenicity or embryolethality was found. In rats, similar treatment on days 6–17 produced no adverse fetal effects (Mitzutami et al., 1985b). After doses of 3000 mg per kg on days 17–21 in the rat, reduced fetal body weight occurred but no other significant behavioral changes resulted (Mitzutami et al., 1985c).

Mitzutani, M.; Izutsu, M.; Matsuda, H. and Hashimoto, Y.: Teratogenic effects of Ipriflavone (TC–80) in the rabbit. Yakuri to Chiryo 13(9):4987–4995, 1985a.

Mitzutani, M.; Izutsu, M.; Shirota, M.; Hashimoto, Y. and Nagao, T.: Teratogenic effects of Ipriflavone (TC–80) in the rat. Yakuri to Chiryo 13(9):4967–4985, 1985b.

Mitzutani, M.; Izutsu, M.; Hashimoto, Y.; Nagao, T. and Shirota, M.: Effects of Ipriflavone (TC–80) on peri–and postnatal development in the rat. Yakuri to Chiryo 13(9):4997–5019, 1985c.

1337 Iproniazid CAS 54–92–2

This compound which inhibits the breakdown of 5–hydroxytryptamine was used by Poulson et al. (1960) to produce hemorrhages in rat placentas. They gave 5 to 10 mg per day and reported a reduction in fertility when it was administered during the first half of pregnancy. Teratogenic results were not reported.

Werboff et al. (1961) gave up to 8.0 mg per kg intraperitoneally to rats on days 8–11, 11–14 or 17–20. A neonatal (to postnatal day 30) mortality rate of 35 to 43 percent was found for each treatment period. Total mortality in the "control" group receiving water was increased when treatment was on days 17–20.

Poulson, E.; Botros, M. and Robson, J.M.: Effect of 5–hydroxytryptamine and iproniazid on pregnancy. Science 131:1101–1102, 1960.

Werboff, J.; Gottlieb, J.S.; Dembicki, E.L. and Havlena, J.: Postnatal effect of antidepressant drugs administered during gestation. Experimental Neurology 3:542–555, 1961.

1338 Iron

Tadokoro et al. (1979) fed a slow–releasing iron to rats in doses up to 1200 mg per kg during during 6 days of organogenesis and found no adverse effects on the offspring. Linkheimer (1964) studied uptake of iron in pregnant rats using ferrous sulfate or iron–fructose chelate. The fructose chelate enhanced selective deposit of iron in the fetus. Heinonen et al. (1977) found no increase in defects among the offspring of 66 women treated with parenteral iron during the first four lunar months or among 1,864 women treated anytime during pregnancy.

Heinonen, O.P.; Slone, D. and Shapiro, S.: Birth Defects and Drugs in Pregnancy. Publishing Sciences Group Inc., Littleton, Mass., 1977.

Linkheimer, W.H.: The placental transfer of orally administered iron. Toxicology and Applied Pharmacology 6:669–675, 1964.

Tadokoro, T.; Miyaji, T. and Okumura, M.: Teratogenicity studies of slow–Fe in mice and rats. Oyo Yakuri 17:483–495, 1979.

1339 Iron Deficiency

Discussed under Anemia, Hemorrhagic

1340 Iron Dextran

The distribution of radioactive iron from iron dextran was studied by Cotes et al. (1966) in the pregnant monkey. Between 0.25 to 4.5 percent of the dose was transferred to the fetus at 133–148 days of gestation after an intravenous dose.

Cotes, P.M.; Moss, G.F.; Muir, A.R. and Scheuer, P.J.: Distribution of iron in maternal and foetal tissues from pregnant rhesus monkeys treated with a single intravenous infusion of (Fe^{59}) iron dextran. Br. J. Pharmacol. 26:633–648, 1966.

1341 Irradiation *X–ray*

Discussed under Radiation

1342 Isepamicin *HAPA–B* CAS 58152–03–7

Sasaki et al. (1986) studied this aminoglycoside antibiotic in pregnant rats using up to 200 mg per kg intramuscularly on days 7–17. No adverse fetal effects were found and postnatal development was not altered.

Sasaki, M.; Kawaguchi, K. and Yamada, H.: Effect of Isepamicin (HAPA–B) on reproduction. Jpn. J. Antibiot. 39:3291–3310, 1986.

1343 Isoamyl

5,6–Dihydro–7,8–dimethyl–4,5–dioxo–4H–pyrano(3,2–c) quinoline–2–carboxylate; MY–5116

Ikeda et al. (1986a,b,c) studied this drug in rats and rabbits at maximum doses of 1600 mg and 400 mg per kg orally during organogenesis. No evidence of teratogenicity was found. Peri and postnatal studies in the rat did not reveal any adverse fetal effects.

Ikeda, Y.; Sukegawa, J. and Fujii, O.: Reproduction studies on isoamyl 5,6-dihydro-7,8-dimethyl-4,5-dioxo-4H-pyrano(3,2-c) quinoline-2-carboxylate (MY-5116) I. Teratogenicity study in rabbits. Iyakuhin Kenkyu 17(1):51–56, 1986a.

Ikeda, Y.; Sukegawa, J. and Iwase, T.: Reproduction studies on isoamyl 5,6-dihydro–7,8-dimethyl–4,5-dioxo–4H-pyrano(3,2-c) quinoline-2-carboxylate (MY-5116) I. Teratogenicity study in rats. Iyakuhin Kenkyu 17(1):57–68, 1986b.

Ikeda, Y.; Sukegawa, J.; Iwase, T. and Hiraga, Y.: Reproduction studies on isoamyl 5,6–dihydro–7, 8–dimethyl–4,5–dioxo–4H–pyrano(3,2–c)quinoline–2–carboxylate (MY–

5116) I. Perinatal and postnatal study in rats. Iyakuhin Kenkyu 17(5):998–1010, 1986c.

1344 Isobenzan *Telodrin* CAS 297–78–9

Ware and Good (1967) fed this polychlorinated insecticide in the diet to mice. A dietary level of one ppm had no effect on fertility or litter size.

Ware, G.W. and Good, E.E.: Effects of insecticides on reproduction in the laboratory mouse. Toxicol. Appl. Pharmacol. 10:54–61, 1967.

1345 Isobromindione *Uridion*™ *5–Br–2–phenyl–indan 1,3–dione*

This uricosuric drug was tested in pregnant rats and rabbits with oral doses of 15 and 40 mg per kg, respectively, during active organogenesis. Resorptions and fetal growth were observed but no congenital defects (Fanelli et al., 1974).

Fanelli, O.; Mazzoncini, V. and Ferri, S.: Toxicological and teratological study of 5–Br–2–phenyl–indan–1,3 dione, an uricosuric drug. Arzneim–Forsch. 24:1609–1613, 1974.

1346 2,2'–Isobutylidend–bis(4,6–dimethylphenol)

Ishii et al. (**1991**) gavaged rats with up to 45 mg per kg during organogenesis and although there was adult toxicity no adverse fetal effects were found.

Ishii, H.; Usami, M.; Kawashima, K. and Takanaka, A.: Studies on the teratogenic potential of 2,2'–isobutylidene–bis (4,6–dimethylphenol) in rats. Bull. Nat. Hygienic Sciences 109:37–42, 1991.

1347 Isocarboxazid

Werboff et al. (1961) gave up to 4.0 mg per kg intraperitoneally to rats on days 5–8, 11–14 or 17–20. The mortality rate until 30 days postnatal was increased to 46 percent in those receiving treatment on days 5–8. Sixteen percent were born dead in this group.

Werboff, J.; Gottlieb, J.S.; Dembicki, E.L. and Havlena, J.: Postnatal effect of antidepressant drugs administered during gestation. Experimental Neurology 3:542–555, 1961.

1348 Isoconazole Nitrate

Iida et al. (1981) studied this antimycotic drug in rats and rabbits. The drug was given subcutaneously at doses of up to 125 mg per kg in rats and 30 mg in rabbits. Fertility, teratogenicity and postnatal development was studied and no adverse effects were found. The higher doses were slightly toxic to the dams.

Iida, H.; Kast, A. and Tsunenari, Y.: Reproduction studies with isoconazole nitrate in rats and rabbits. Iyakuhin Kenkyu 12:762–783, 1981.

1349 Isofenphos *Oftanol*™ CAS 25311–71–1

Mast et al. (1985) exposed rats to up to 10 mg per kg during gestation and found no reproductive ill effects.

Mast, T.J.; Bracco, C.A.; Rowland, J.R. and Hendrickx, A.G.: Oftanol exposure during organogenesis in rats: Serum cholinesterase depression and pregnancy outcome. (abs) Teratology 31:47A, 1985.

1350 Isoflurane CAS 26675–46–7

Mazze et al. (1985) exposed mice to 0.006 percent isoflurane for 4 hours daily on days 6 through 15 and found no adverse fetal effects. Similar exposure to 0.06 percent gas was asssociated with decreased fetal weight, decreased ossification and minor renal changes. Cleft palate was found in 12 percent of the offspring. Pair feeding was not done but similar degrees of anethesia with halothane and enflurane produced only 1.2 and 1.9 percent cleft palate, respectively. Mazze et al. (1986) exposed rats to 1.05 percent concentrations during periods of organogenesis and found no evidence of teratogenicity.

Mazze, R.I.; Fujinaga, M.; Rice, S.A.; Harris, S.B. and Baden, J.M.: Reproductive and teratogenic effects of nitrous oxide, halothane, isoflurane, and enflurane in Sprague–Dawley rats. Anesthesiology 64:339–344, 1986.

Mazze, R.I.; Wilson, A.I.; Rice, S.A. and Baden, J.M.: Fetal development in mice exposed to isoflurane. Teratology 32:339–345, 1985.

1351 Isoflurophate *Diisopropyl Phosphorofluoridate DFP* CAS 55–91–4

Fish (1966) administered this material intraperitoneally to rats in amounts of 1 to 4 mg per kg on days 7, 8, 9, 10 or 12 and found no defects. Treatment did not increase the resorption rate.

Fish, S.A.: Organophosphorus cholinesterase inhibitors and fetal development. Am. J. Obstet. Gynecol. 96:1148–1154, 1966.

1352 Isomalic Acid

Cima (1966) fed this in the diet in amounts up to 5 percent and rats gave birth to normal fetuses.

Cima, L.: Acute and subacute toxicity of isomalic acid. Toxicol. Appl. Pharmacol. 9:274–278, 1966.

1353 Isoniazid *Isonicotinic Acid Hydrazide* CAS 54–85–3

There are a number of reports on the effect of treating pregnant women with antituberculous drugs including isoniazid. Lowe (1964) examined 74 exposed infants and Marcus (1967), 19 children and found no increase in congenital defects. Varpela (1964) reported 12 anomalies in 123 children exposed in utero to various antituberculous drugs. Heinonen et al. (1977) reported 10 malformed children from 85 mothers exposed during the first 4 months of pregnancy. The types of defects were not given. Monnet (1967) studied five children with severe encephalopathies after their mothers were treated with INH. Warkany (1979) published a complete review of the effects of antituberculosis drugs in human pregnancy.

Heinonen, O.P.; Slone, D. and Shapiro, S.: Birth Defects and Drugs in Pregnancy. Publishing Sciences Group Inc., Littleton, Mass., 1977.

Lowe, C.R.: Congenital defects among children born to women under supervision or treatment for pulmonary tuberculosis. Br. J. Prev. Soc. Med. 18:14–16, 1964.

Marcus, J.C.: Non–teratogenicity of antituberculous drugs. S. Afr. Med. J. 41:758–759, 1967.

Monnet, P.; Kalb, J.C. and Pujol, M.: Doit–on craindre une influence teratogene eventuelle de l'isoniazide. Rev. Tubercul. (Paris) 31:845–848, 1967.

Varpela, E.: On the effect exerted by first–line tuberculosis medicines on the fetus. Acta Tuberculosea et Pneumologica Scandinavica 35:53–69, 1964.

Warkany, J.: Antituberculous drugs. Teratology 20:133–138, 1979.

1354 Isonicotinic Acid–2–isopropylhydrazide CAS 55–22–1

Discussed under 1–Methyl–formylhydrazine

1355 p–Isooctylpolyoxyethylphenol Polymer

Discussed under Triton W–R 1339

1356 Isoprene

Mast et al. (**1990**) exposed rats and mice to doses of up to 7,000 ppm 6 hours a day during organogenesis. At the highest dose maternal toxicity and fetal skeletal variations occurred. No significant teratogenic activity was found.

Mast, T.J.; Rommereim, R.L.; Weigel, R.J.; Stoney, K.H.; Schwetz, B.A. and Morrissey, R.E.: Inhalation developmental toxicity of isoprene in mice and rats. Toxicologist 10:42, 1990.

1357 Isopropamide CAS 71–81–8

Heinonen et al. (1977) found no increase in defects among the offspring of 180 women exposed during the first four lunar months or among 1,071 women exposed any time in pregnancy.

Heinonen, O.P.; Slone, D. and Shapiro, S.: Birth Defects and Drugs in Pregnancy. Publishing Sciences Group, Inc.; Littleton, Mass., 1977.

1358 Isopropyl Alcohol CAS 67–63–0

Antonova and Salmina (1978) reported that this compound induced death in 31 percent of embryos and malformations in 14 percent of embryos. The substance was given to pregnant rats at 0.018–1008 mg per kg. Behaviour abnormalities in progeny of treated females were also found.

Nelson et al. (**1988**) exposed rats for 7 hours daily on day 1–19 to 10,000, 7,000 or 3,500 ppm. At 3,500 ppm no adverse fetal effects were found while at the higher doses malformations, resorptions and fetal deaths were increased. Skeletal defects were the main defects. In the work place levels from 3–263 ppm have been measured.

Antonova, V.I. and Salmina, Z.A.: The maximum permissible concentration of isopropyl alcohol in water bodied with due regard for its action on the gonads and the progeny. Gigiena i Sanitariya (USSR) 1:8–11, 1978.

Nelson, B.K.; Brightwell, W.S.; MacKenzie–Taylor, D.R.; Khan, A.; Burg, J.R. and Weigel, W.W.: Teratogenicity of n–propanol and isopropanol administered at high inhalation concentrations to rats. Fd. Chem. Toxic. 26:247–254, 1988.

1359 Isopropyl Methanesulphonate *IMS Alkane Sulphonates*

Hemsworth (1968) administered 50 or 75 mg per kg intraperitoneally on day 13 and produced limb defects and temporary reduction in seminiferous tubule cells of the offspring.

Hemsworth, B.N.: Embryopathies in the rat due to alkane sulphonates. J. Reprod. Fertil. 17:325–334, 1968.

1360 Isoproterenol CAS 7683–59–2

Heinonen et al. (1977) found no increase in defects among the offspring of 31 women treated during pregnancy.

Vogin et al. (1970) administered isoproterenol to rats and rabbits by aerosol spray and produced no congenital defects. The rats received 150 or 450 microgm per kg per day from day 5 through day 16. The rabbits received 36.5 microgm per kg daily from day 6 through day 16. Hodach et al. (1975) administered 0.4 micromoles in 5 microliters directly to the chick embryo and produced cardiovascular anomalies. Increased heart rates in explanted rat embryos have been found at exposure doses of 500 nanograms per ml of medium (Robkin et al., 1976).

Heinonen, O.P.; Slone, D. and Shapiro, S.: Birth Defects and Drugs in Pregnancy. Publishing Sciences Group Inc., Littleton, Mass., 1977.

Hodach, R.J.; Hodach, A.E.; Fallon, J.F.; Folts, J.D.; Bruyere, H.J. and Gilbert, E.F.: The role of β–adrenergic activiy in the production of cardiac and aortic arch anomalies in chick embryos. Teratology 12:33–49, 1975.

Robkin, M.A.; Shepard, T.H. and Dyer, D.C.: Autonomic receptors of the early rat embryo heart. Proc. Soc. Exp. Biol. Med. 151:799–803, 1976.

Vogin, E.E.; Goldhamer, R.E.; Scheimberg, J.; Carson, S. and Boxill, G.C.: Teratology studies in rats and rabbits exposed to an isoproterenol aerosol. Toxicol. Appl. Pharmacol. 16:374–381, 1970.

1361 Isoprothiolane CAS 50512–35–1

This fungicide was given orally to mice on days 6 through 12 in doses to 600 mg per kg and only reduced fetal weight and ossification centers were found (Sukurai and Kasai, 1976).

Sukurai, K. and Kasai, T.: Teratological studies of isoprothiolane in mice. (abs) Teratology 14:251 only, 1976.

1362 Isorubijervine

Keeler and Binns (1968) fed 0.2–1.3 gm to 8 ewes on the 14th day of gestation and found no cyclopia in the offspring.

Keeler, R.F. and Binns, W.: Teratogenic compounds of veratrum californicum (Durand). Teratology 1:5–10, 1968.

1363 Isouracil CAS 4874–29–7

Kosmachevskaya and Tichodeeva (1968) produced malformations in 18 percent of chick embryos injected with 0.5 to 4 mg of isouracil at 24 hours of incubation. Microphthalmia, rumplessness, beak and axial skeletal defects and sirenomelia were found. Embryonic mortality was insignificant.

Kosmachevskaya, E.A. and Tichodeeva, I.I.: Relation between embryotoxic activity of some pyrimidine derivatives and their chemical structure. Chick Embryo Test. Pharmacol. Toxicol. (Russian) 5:618–620, 1968.

1364 Isoxsuprine

Heinonen et al. (1977) studied the offspring of 54 women taking this vasodilator in the first four lunar months and 858 women taking it anytime in pregnancy and found no increase in congenital defects.

Heinonen, O.P.; Slone, D. and Shapiro, S.: Birth Defects and Drugs in Pregnancy. Publishing Sciences Group Inc., Littleton, Mass., 1977.

1365 Ivermectin

Pacque et al. (**1991**) studied 203 women who received this drug as propholaxis for onchocerciasis. No adverse effects were found including developmental status.

Pacque, M.; Munoz, B.; Poetsche, G.; Foose, J.; Greene, B.M. and Taylor, H.R.: Pregnancy outcome after inadvertent invermectin treatment during community–based distribution. The Lancet 2:1486–1489, 1990.

1366 JKMS 201

Tsunemi et al. (**1990**) gave this oral contraceptive consisting of norethisterone and ethinylestradiol (20:1) orally to rats on days 7–17 and found no adverse fetal effects. The dose was 0.004 to 0.5 mg per kg.

Tsunemi, K.; Kaneko, T.; Kobayashi, H.; Kamada, S.; Shinpo, K.; Koyama, K. and Koshugo, I.: Teratogenicity study of jkms 201 administered orally to rats. Kiso to fbnsko 24:4737–4756, 1990.

1367 Janus Green B CAS 2869–83–2

Braun (1954) inoculated chicken eggs at the beginning of incubation or once during the first three days of incubation with 65 to 300 microgm of Janus green. Over 90 percent of the embryos receiving the dosage schedule of 100 microgm were defective. The defects included hydrocephalus, microphthalmia, cerebral dysplasia, phocomelia and hemorrhages in various places. The author postulated that the dye reduced the oxygen consumption of the cells which led to the defective development.

Braun, S.: Janus green B teratological action in embryonated hen's eggs and embryogenetic and carcinogenic bearings of its mechanism of action. Acta Morphologica 4:61–79, 1954.

1368 Japanese B Encephalitis Virus

Burns (1950) isolated Japanese B encephalitis virus from newborn pigs with encephalomalacia or internal hydrocephalus. At the time there was a large epidemic of this infection in both man and pigs, and the fetal morbidity in swine was 60 to 70 percent Shimizu et al. (1954) injected the virus intravenously into susceptible pigs and produced hydrocephalus in some, but virus recovery was successful from the fetuses only during early gestation.

Burns, K.F.: Congenital Japanese encephalitis infection of swine. Proc. Soc. Exp. Biol. Med. 75:621–625, 1950.

Shimizu, T.; Kawakami, Y.; Fukuhara, S. and Matumoto, M.: Experimental stillbirth in pregnant swine infected with Japanese encephalitis virus. Jpn. J. Exp. Med. 24:363–375, 1954.

1369   Japanese Equine Encephalitis

A high incidence of hydrocephalus in newborn calves was reported by Tabuchi et al. (1953) during an epidemic of Japanese equine encephalitis. Isolation of the virus from the fetuses was unsuccessful.

Tabuchi, A.; Narita, R.; Ebi, Y. and Hosoda, T.: Studies on hydrocephalus of newborn calves in Aomori, Akita and Iwate prefectures. Experimental Report of the Government Experimental Station of Animal Hygiene (Tokyo) 26:21–26, 1953.

1370   Jervine also see Cyclopamine CAS 469–59–0

The alkaloid prepared from Veratrum californicum was administered at 0.9 to 2.4 gm to ewes on the 14th gestational day and produced cyclopian fetuses (Keeler and Binns, 1968). Omnell et al. (1990) studied the interstrain differances in the mouse and suggested that the precusor chondrocytes might account for the defects of the palate.

Keeler, R.F. and Binns, W.: Teratogenic compounds of Veratrum californicum (Durand): Five comparisons of cyclopian effects of steroidal alkaloids from plant and structurally related compounds from other sources. Teratology 1:5–10, 1968.
Omnell, M.L.; Sim, F.R.P.; Keeler, R.F.; Harne, L.C. and Brown, K.S.: Expression of veratrum alkaloid teratogenicity in the mouse. Teratology 42:105–119, 1990.

1371   Jet Fuel

Schreiner (1983) reported "negative" results after exposing rats to jet fuel A for 6 hours daily on days 6 through 15 for either 100 or 400 ppm.

Schreiner, C.A.: Petroleum and petroleum products: A brief review of studies to evaluate reproductive effects. Environmental and Health Sciences Laboratory, Mobil Oil Corporation. In: Assessment of Reproductive and Teratogenic Hazards, volume III. M.S. Christian, W.M. Galbraith, P. Voytek, M.A. Mehlman (eds), Princeton Scientific Publishers Inc., 29–46, 1983.

1372   Jimsonweed

Leipold et al. (1973) reported congenital arthrogryposis in 25 pig offspring among 8 litters. The dams were exposed to a dense vegetation of Jimsonweed. In subsequent unexposed pregnancies, the offspring were normal.

Leipold, H.W.; Oehme, F.W. and Cook, J.E.: Congenital arthrogryposis associated with ingestion of Jimsonweed by pregnant sows. J.A.V.M.A. 162:1059–1060, 1973.

1373   Josamycin Proprionate CAS 40922–77–8

Oshima and Iwadare (1973) administered this compound orally to pregnant mice and rats in doses up to 2000 mg per kg during 7 days of their organogenesis and found no significant fetal effects.

Oshima, T. and Iwadare, M.: Studies of josamycin proprionate. Jpn. J. Antibiot. 26:148–153, 1973.

1374   KJK–945

3–(2–(4–(p–Fluorobenzoyl)–l–piperidinyl)ethyl)–2, 4(1H,3H) quinazolinedione L–Tartrate Ketansarin Tartrate

Naya et al. (1988 A&B) gave this antihypertensive orally to rats and rabbits in maximum daily doses of up to 50 and 72.5 mg per kg respectively. Treatment during organogenesis had no adverse fetal effects in either species.

Naya, M.; Sakuma, T.; Fujita, T.; Hara, T. and Takahira, H.: Reproduction study of kjk–945 (ketanserin tartrate) (2)–teratogenicity study in rats. Kiso to Rinsho 22:1335–1348, 1988.
Naya, M.; Sakuma, T.; Fujita, T.; Hara, T. and Takahira, H.: Reproduction study of kjk–945 (Ketanserin tartrate) (3)–teratogenicity study in rabbits. Kiso to Rinsho 22:1349–1356, 1988.

1375   KW–1062

Hara et al. (1977) gave this aminoglycoside to rats, mice and rabbits during organogenesis and except for slight inhibition of body weight increase in the rats at 225 mg per kg, no adverse effects were found. The rats were also treated before reproduction and in the perinatal period without significant effects.

Hara, T.; Imamura, S.; Miyazaki, H. and Ohguro, Y.: Safety evaluation of KW–1062. Jpn. J. Antibiot. 6:432–449, 1977.

1376   KW–1070

Nishikawa et al. (1981) studied this aminoglycoside in rats and rabbits. Intramuscular doses to 500 mg per kg in rats and 300 mg per kg in rabbits produced no teratogenic effects when given during organogenesis. The fertilization and perinatal studies in the rat revealed no adverse effects.

Nishikawa, S.; Hara, T.; Miyazaki, H. and Ohguro, Y.: Safety evaluation of KW–1070, III. Chemotherapy 29:167–175, 1981.

1377   KW–2307

Sadanaga et al. (1993) treated rats orally during organogenesis with up to 0.50 mg per kg. Axial skeletal defects were increased at the highest dose but not at 0.22 mg per kg. Postnatal development was normal. No other defects were increased.

Sadanaga, O.; Arima, J.; Sakuma, T.; Naya, M. and Deguchi, T.: Reproductive and developmental toxicity of KW–2307–study with administration during the fetal organogenesis period in rats. The Clinical Report 27:1375–1399, 1993.

1378 Kainic Acid

Hata (**1994**) injected rats on days 11–14 or 17–19 with 3 mg per kg and studied postnatal development. Those injected on days 17–19 had increased body weight and behavioral changes. The changes were 1) a reduction of the latent time to the first line crossing at 15 days 2) a decrease in ambulation and 3) early appearance of rearing behavior at 15 days of age.

Hata, M.: Effects of maternal exposure to a single dose of kainic acid on the functional development of the brain in the rat. Nihon Yakuri Gakkai Zasshi 104:7–18, 1994.

1379 Kallidinogenase, Human Urinary

Komai et al. (**1993**,A,B,C) gave this compound intravenously to mice before fertilization and in the first 7 days, during organogenesis and in the perinatal period. The dose was up to 2.0 PNAU per kg for the latter two . No adverse developmental changes were found. Komai (**1993** D) found no ill effects in rabbits exposed to 5 PNAU during organogenesis.

Komai, Y.; Hattori, M.; Fukuda, T.; Ishimura, K.; Hayasaka, I. and Koide, M.: Reproductive and developmental toxicity study of human urinary kallidinogenase (SK–827): Fertility study in mice given the substance intravenously. Oyo Yakuri 45(2):75–82, 1993.

Komai, Y.; Hattori, M.; Fukuda, T.; Ishimura, K.; Hayasaka, I. and Koide, M.: Reproductive and developmental toxicity study of human urinary kallidinogease (SK–827): Teratogenicity study in mice given the substance intravenously. Oyo Yakuri 45(2):83–97, 1993.

Komai, Y.; Hattori, M.; Inoue, S.; Fukuda, T.; Ishimura, K.; Hayasaka, I. and Koide, M.: Reproductive and developmental toxicity study of human urinary kallidinogenase (SK–827): Peri–and postnatal study in mice given the substance intravenously. Oyo Yakuri 45(2):99–112, 1993.

Komai, Y.; Fukuda, T.; Hattori, M.; Ishimura, K.; Hayasaka, I and Koide, M.: Reproductive and developmental toxicity study of human urinary kallidinogenase (SK–827): Teratogenicity study in rabbits given the substance intravenously. Oyo Yakuri 45(2):113–118, 1993.

1380 Kanamycin also see *Streptomycin* CAS 59–01–8

Jones (1973) reported a woman who, at 28 weeks, had renal failure treated with kanamycin and other drugs and who developed total deafness. Her infant was permanently deaf.

Akiyoshi et al. (1977) studied the effect of 100–200 mg per kg given intramuscularly on the guinea pig fetuses from day 7 through day 56. At both doses cochlear hair cell damage was found and hearing impairment using the pinna reflex test was reduced at higher frequencies in some offspring. Tests with amikacin at the same dose showed no hearing impairment but some hair cell damage occurred. Bevelander and Cohlan (1962) injected rats with 100 mg per kg from the 8th through 16th days and produced no fetal

changes. No cochlear damage was found in rat offspring exposed to 400 mg per kg on days 10 through 20 of gestation (Onejeume and Khan, 1984). On days 8 through 16, postnatally, they did find an effect on hair cells.

Akiyoshi, M.; Yano, S.; Tajima, T.; Matsuzaki, M.; Akutsu, S.; Nishimoto, K. and Maeda, M.: Ototoxic effect of BB–K–8 administered to pregnant guinea pigs on development of inner ear of intrauterine litters. Jpn. J. Antibiot. 196:53–64, 1977.

Bevelander, G. and Cohlan, S.Q.: The effect on the rat of transplacentally acquired tetracycline. Biol. Neonate 4:365–370, 1962.

Jones, H.C.: Intrauterine toxicity: A case report and review of literature. J. Natl. Med. Assoc. 65:201–203, 1973.

Onejeume, A.U. and Khan, K.M.: Morphologic study of effects of kanomycin on the developing cochlea of the rat. Teratology 29:57–71, 1984.

1381 Kanechlor also see *Chlorobiphenyls*

This polychlorinated biphenyl was administered orally to pregnant rats from days 8 to 14 or from days 15 through 21 in amounts of 20 or 100 mg per kg per day (Shiota et al., 1976A). At the highest dose some fetal and maternal toxicity occurred but no congenital defect increase was found. Postnatal function was measured and maze learning was slower in the offspring exposed in utero. At dietary levels of 500 PPM Kanechlor 300 and Kanechlor 500 were associated with fetal weight decrease but there was no increase in defects (Shiota, 1976B).

Shiota, K.: Embryotoxic effects of polychlorinated biphenyls (Kanechlors 300 and 500) in rats. Okijimas Fol. Anat. 53:93–104, 1976A.

Shiota, K.: Postnatal behavioral effects of prenatal treatment with PCBs (Polychlorinated biphenyls) in rats. Okajimas Fol. Anat. 53:105–114, 1976B.

1382 Kerosene

Schreiner (1983) reported "negative" results after exposing rats to 100 or 365 ppm four 6 hours daily on days 6 through 15.

Schreiner, C.A.: Petroleum and petroleum products: A brief review of studies to evaluate reproductive effects. Environmental and Health Sciences Laboratory, Mobil Oil Corporation. In: Assessment of Reproductive and Teratogenic Hazards, volume III. M.S. Christian, W.M. Galbraith, P. Voytek, M.A. Mehlman (eds), Princeton Scientific Publishers Inc., 29–46, 1983.

1383 Ketamine CAS 6740–88–1

This general anesthetic agent related to phencyclidine was tested in pregnant rats by El–Karim and Benny (1976). They gave 120 mg per kg intramuscularly from the 9th through the 13th day. No adverse fetal effects were noted.

El–Karim, A.H.B. and Benny, R.: Embryotoxic and teratogenic action of ketamine hydrochloride in rats. Ain Shavis Med. J. 27:459–463, 1976.

1384 Ketoconazole™ *KW 1414*

Nishikawa et al. (1984) studied this antimycotic in pregnant rats and rabbits. Rats received up to 80 mg per kg orally before fertilization and during the first 7 days, during days 7–17 or during the perinatal period. Fertility was decreased in the female at 20 mg per kg and above. At 40 mg per kg and above, malformations were increased and included cleft palate and patent incisive foramena. The neonatal death rate in the perinatally treated fetuses was increased at 40 and 80 mg per kg. In the rabbits embryolethality was found at 80 mg per kg but there was no malformation increase.

Buttar et al. (1989) found malocclusion in the offspring of rats given 25 mg per kg. Intrauterine growth retardation occurred also. Doses of up to 40 mg per kg in the mouse on days 6–18 had lttle effect on the fetus.

Buttar, H.S.; Moffatt, J.H. and Bura, C.: Pregnancy outcome in ketoconazole–treated rats and mice. Teratology 39:P117, 1989.

Nishikawa, S.; Hara, T.; Miyazaki, H. and Ohguro, Y.: Reproduction studies of KW–1414 in rats and rabbits. Clin. Report 18:1433–1488, 1984.

1385 4–Ketocyclophosphamide CAS 27046–19–1

Discussed under Cyclophosphamide

1386 Ketoprofen *M–Benzoylhyratropic Acid* CAS 22071–15–4

This anti–inflammatory analgesic has been tested in mice, rats and monkeys (Esaki et al., 1975A,B; Tanioka and Koizumi, 1977; Tanioka et al., 1975). Mice and rats were treated on days 7 through 13 and 9 through 15, respectively. The maximum daily subcutaneous doses were 100 mg and 5 mg and orally 10 and 3 mg, respectively, in the mouse and rat. There were no ill effects to the fetuses, but in several groups in both species the treated fetuses were heavier than controls. In monkeys, intramuscular (30 mg) and oral doses (150 mg) were given from day 23 through 35 without detectable effect on the fetuses at day 60.

Esaki, K.; Tsukada, M.; Izumiyama, K. and Oshio, K.: Teratogenicity of sodium ketoprofen (19583RP–Na) tested by subcutaneous administration in mice and rats. CIEA Preclinical Report 1:101–109, 1975A.

Esaki, K.; Tsukada, M.; Izumiyama, K. and Oshio, K.: Teratogenicity of ketoprofen (19583RP) tested by oral administration in mice and rats. (Japanese) CIEA Preclinical Report 1:91–100, 1975B.

Tanioka, Y. and Koizumi, H.: Teratogenicity test by intramuscular administration of ketoprofen–Na (19583 RP–Na) in rhesus monkeys. CIEA Preclinical Report 3:87–96, 1977.

Tanioka, Y.; Koizumi, H.; Ogata, T. and Esaki, K.: Teratogenicity of ketoprofen (19583RP) in the rhesus monkey. CIEA Preclinical Report 1:67–73, 1975.

1387 Ketotifen CAS 34580–13–7

Nakajima et al. (1979) studied this antiallergic agent in pregnant rats using 30 mg per kg daily by mouth. Dosing was in both sexes before mating and during the first 7 days, during days 7–17 or on days 17–21. No adverse fertility effects were found and the fetuses were similar to controls. Some decrease in postnatal survival and weight was found.

Nakajima, T.; Ishizaka, K.; Hamada, M. and Matsuda, M.: Reproductive studies of HC–20–511 in rats. Kiso to Rinsho 13:4096–4114, 1979.

1388 Krypton[85]

Sikov et al. (1984) found no increase in incidence of malformations in rats exposed to [85]krypton during organogenesis. The atmospheres of gas were calculated to deliver approximately 50R to the maternal fetal unit.

Sikov, M.R.; Ballou, J.E.; Willard, D.H. and Andrew, F.D.: Disposition and effects of [85]Kr in pregnant rats. Health Physics 47:417–427, 1984.

1389 Labetalol Hydrochloride CAS 32780–64–6

In a randomized double–blind prospective study of 152 women with hypertension there were no congenital defects or deaths in either the treated or control group. Some reduction in perinatal complications occurred in the treated group (100–200 mg three times daily) (Pickles et al., 1989). Jorge et al. (**1982**) in a study of 28 exposed infants found no abnormalities.

Nagaoka et al. (1981) studied this β blocker in rats and rabbits using oral routes of up to 300 mg and 200 mg per kg, respectively. The copulation rate was decreased in rats at the 300 mg level. No defects were found after treatment. In the perinatal studies (days 17–21) decreased viability and size was found with the 300 mg and 150 mg doses. Only a slight decrease in survival was found in rabbit fetuses at the 200 mg per kg level.

Jorge, C.S.; Fernandes, L. and Cunha, S.: The investigation of labetalol in the management of hypertension in pregnancy. Amsterdam: Excerpta Medica, 1982, pp 123–130.

Nagaoka, T.; Shigemura, T. and Narama, I.: Reproductive studies on labetalol hydrochloride. Yakuri to Chiryo 9:839–850, 851–867, 869–877, 879–893, 1981.

Pickles, C.J.; Symonds, E.M. and Broughton, P.F.: The fetal outcome in a randomized double–blind controlled trial of labetalol versus placebo in pregnancy–induced hypertension. Br. J. Obstet. Gynaecol. 98:38–43, 1989.

1390 Lacidipine

Wada et al. (**1994**) gave this calcium blocker orally before and during the first week of pregnancy. Embryonic mortality increase was found at the highest dose (15 mg per kg) and placental weights were increased at 2.5 mg per kg or more. Ogawa et al. (**1994**) gave up to 2 mg per kg peri and postnatally and at the highest dose body weight of the pups was reduced. Postnatal development and reproduction was normal. Wada et al. (**1994** B) gave up to 18 mg per kg during rabbit organogenesis and found fetal weight reduction but no teratogenicity.

Ogawa, M.; Wada, A.; Sekino, S.; Katoh, H.; Nagao, T. and Mizutani, M.: Reproductive study on oral administration of lacidipine during the perinatal and lactation periods (seg III). Yakuri to Chiryo 22:401–413, 1994.

Wada, A.; Sekino, S.; Katoh.; Nagao, T. and Mizutani, M.: Reproductive study on oral administration of lacidipine prior to and in the early stages of pregnancy in rats (seg I). Yakuri to Chiryo 22:377–388, 1994.

Wada, K.; Nagao, T. and Mizutani, M.: Reproductive study on oral administration of lacidipine during the period of organogenesis in rabbits (seg II). Yakuri to Chiryo 22:389–397, 1994B.

1391 Lactitol

Koeter and Bar (**1992**) fed rats up to 10% of this polyol sugar substitute preceeding and during pregnancy. No ill effects were observed in fertility or the offspring.

Sinkeldam et al. (**1992**) fed rats up to 10% in their diet for three generations and found no adverse effect on fertility or viability on day one postnatally. Some weight gain decrease and increased mortality was found during the first 3 weeks postnatally.

Koeter, H.B.W.M. and Bar, A.: Embryotoxicity and teratogenicity studies with lactitol in rats. Journal of the American College of Toxicology 11:249–257, 1992.

Sinkeldam, E.J.; Hollanders, V.M.H.; Woutersen, R.A.; Koeter, H.B.W.M. and Bar, A.: Multigeneration reproduction study of lactitol in rats. Journal of the American College of Toxicology, 11:233–248, 1992.

1392 Lactose

Beltrame and Cantone (1973) administered 400–4000 mg per kg to rats, mice and rabbits. High maternal mortality occurred in rabbits. External and visceral defects were increased in rats and mice but the type was not given.

Beltrame, D. and Cantone, A.: Maternal and fetal toxicity induced by lactose. (abs) Teratology 8:215, 1973.

1393 Laetrile *D,L–Amygdalin* CAS 1332–94–1

Oral D,L–amygdalin, the main constituent of laetrile administered orally in amounts of 250 mg per kg or more to hamsters was associated with neural tube defects (Willhite, 1982). Intravenous administration was not teratogenic. D–Prunasin was also associated with neural tube defects at oral doses of 177 mg per kg. Since thiocyanide blocked the teratogenicity of amygdalin it was suggested that cyanide was the teratogenic metabolite.

Willhite, C.C.: Congenital malformations induced by laetrile. Science 215:1513–1515, 1982.

1394 Laminin Antibodies

Weeks et al. (1989) immunized 4 monkeys to laminin and found that the maternal serum became toxic to whole rat embryo cultures. Two of the monkeys previously fertile were found to be infertile for 2 years after treatment. Further preliminary studies supported their hypothesis that laminin antibodies might interfere with normal embryonic uptake of aminoacids. Robbins et al. (**1991**) extended this

work by adding sera from patients with Chaga's disease (containing anti–laminin antibodies) to whole embryo culture. All 20 embryos were abnormal. By preabsorption of the serum with purified laminin toxicity was reduced in 6 of 7 samples. Purified antibody added to normal serum was associated with toxicity.

Robbins, B.; Klein, N.W. and Cavalcanti, H.: Toxicity of sera from individuals with chagas' disease to cultured rat embryos: role of antibodies to laminin. Teratology 44:561–570, 1991.

Weeks, B.S.; Klein, N.W.; Kleinman, H.; Frederickson, T. and Sackett G.P.: Laminin immunized monkeys develop sera toxic to cultured rat embryos and fail to reproduce. Teratology 40:47–57, 1989.

1395 Lanolin

No teratogenicity was found when rabbits were fed up to maternally toxic doses (Anonymous, 1982). Treatment was on days 2–16.

Anonymous: Final report on the safety assessment of laneth–10 acetate group. J. Am. Coll. Toxicol. 1:1–23, 1982.

1396 Lansoprazole

Schardein et al. (**1990**) carried out studies on reproduction, organogenesis and postnatal development in rats and rabbits. In the rat the highest doses for reproductive teratogenesis and postnatal studies were 50,300 and 150 mg per kg respectively. At the highest doses some decrease in fetal and pup weight occurred but no teratogenicity or other adverse effects were found. In rabbits up to 30 mg per kg was given during organogenesis and no increawe in defects was found.

Schardein, J.L.; Furuhashi, T. and Ooshima, Y.: Reproductive and developmental toxicity studies of lansoprazole (ag–1749) in rats and rabbits. Yakuri to Rinsho 18:S2773–S2783, 1990.

1397 Lanthanum Chloride CAS 1099–58–8

Abarmczuk (1985) injected 44 mg of metal intraperitoneally on days 4 and 6 of the mouse pregnancy and found the litter size was reduced but no malformations were observed. Direct addition to the ova in culture had the peculiar result of increasing the percent that developed into blastocysts from the one cell stage.

Abarmczuk, J.W.: The effects of lanthanum chloride on pregnancy in mice and on preimplantation mouse embryos in vitro. Toxicology 34:315–320, 1985.

1398 Lapachol *2–Hydroxy–3–(3–methyl–2–butenyl)–1, 4–naphtoquinone*

Rodrigues de Almeida et al. (**1988**) treated pregnant rats orally with 100 or 500mg per kg of this anticancer drug. Treatment early in pregnancy caused a high resorption rate. When treatment was given on the 7th through the 14th day exophthalmia, leporine lip were increased. At the highest dose gastroschisis occurred in 87% of the survivors.

Rodrigues de Almeida, E.; de Mello, A.C.; de Santa, C.F.; da Silva Filho, A.A. and dos Santos, E.R.: The action of 2–hydroxy–3–(3–methyl–2–butenyl)–1,4–naphtoquinone (lapachol) in pregnant rats. Revista Portuguesa de Farmacia 38:21–23, 1988.

1399 Laporotomy

Johnson (1971) laporotomized rabbits on the 7th, 8th, 9th and 10th post–coital days. Pentobarbital (40 mg per kg) which was used as an anesthetic did not cause congenital defects, but on the 9th and 10th day, the resorption rate was increased. Laporotomy with uterine exposure to the atmosphere caused an increase in resorptions on all days, but no congenital defects were produced. Brent and Franklin (1960) laporotomized rats on the 9th day and in the non–traumatized horns found significant increases in resorption rates.

Brent, R.L. and Franklin, J.B.: Uterine vascular clamping: New procedure for the study of congenital malformations. Science 132:89–91, 1960.

Johnson, W.E.: Fetal loss from anesthesia and surgical trauma in the rabbit. Toxicol. Appl. Pharmacol. 18:773–779, 1971.

1400 Lathyrism

Stamler (1955) fed a diet containing 50 percent ground lathyrus odoratus peas and found that after the 17th day, the lathyrism syndrome could be produced in the rat fetus. This syndrome consisted of poorly developed muscles and connective tissue with scanty collagen formation, dissecting aneurisms of the aorta and severe spinal deformities.

Beta–aminopropionitrile, bis–(β–cyanoethyl)amine and aminoacetonitrile produced similar findings when fed in the diet at 0.01 to 0.05 percent levels. These compounds had little effect on fetuses when fed to the rat during the first 16 days of gestation. Tris(β–cyanoethyl)amine and β–dimethylaminopropionitrile had no effect when given after the 17th day. Steffek et al. (1972) produced cleft palates in rat fetuses by feeding the mothers sweet pea seeds or giving single 500 mg doses of β–aminopropionitrile. The most effective time of administration was on day 15. Rosenberg (1957) produced the syndrome in chicks. Barrow et al. (1974) reviewed the subject.

Barrow, M.V.; Simpson, C.F. and Miller, E.J.: Lathyrism: A review. Quart. Rev. Biol. 49:101–128, 1974.

Rosenberg, E.E.: Teratogenic effects of β–aminopropionitrile in the chick embryo. Nature 180:706–707, 1957.

Stamler, F.W.: Reproduction in rats fed lathyrus peas or aminonitriles. Proc. Soc. Exp. Biol. Med. 90:294–298, 1955.

Steffek, A.J.; Verrusio, A.C. and Watkins, C.A.: Cleft palate in rodents after maternal treatment with various lathyrogenic agents. Teratology 5:33–40, 1972.

1401 LC 9018

Hashimoto et al. (**1989**) gave this anticancer drug to rats on days 6–17 intrapleurally in doses of up to 10 mg per kg. Maternal food consumption was decreased but no reproductive effects occurred. Similar treatment in the pregnant rabbit caused no reproductive changes. (Wada et al., **1989**)

Hashimoto, Y.; Kawaguchi, H.; Miyahara, T. and Mizutani, M.: Reproduction and developmental toxicity study of lc 9018–teratogenicity study in rats with intrapleural treatment of lc 9018. Yakuri to Chiryo 17:2089–2106, 1989.

Wada, K.; Hashimoto, Y. and Mizutani, M.: Reproduction and developmental toxicity study of lc 9018–teratogenicity study in rabbits with intrapleural tratment of lc 9018. Yakuri to Chiryo 17:2121–2129, 1989.

1402 Lead CAS 7439–92–1

Congenital Defects

The evidence that lead affects the development of the central nervous system of the human fetus is not complete. Needleman et al. (1984) measured lead levels in 5,183 umbilical cord bloods and found a small but significant increase in minor malformations among those with high levels. Neither a pattern of malformations nor an increase in major malformations was found. Factors associated with lower socioeconomic class were associated with the higher levels. Needleman (**1990**) has discussed the strong association betwen lead and I.Q. in children and presented a meta analysis of 13 studies. Bellinger (**1994**) in a review of the literature on prenatal lead discusses the problems of studies on postnatal development. Although significantly slower mental development in infancy has been associated with lead levels, the follow-up studies indicate that the effect is transient and not detectable by school age.

Borella et al. (1986) measured lead content in electively aborted material and could not show a correlation with the mothers' concentration or evidence that lead accumulated in the embryo–fetus during the first trimester. Angle and McIntyre (1964) and Cantarow and Trumper (1944) summarized the evidence that lead poisoning increases the incidence of abortion and stillbirths in humans. Rabinowitz (1988) summarized the effects of lead on pregnancies and especially its effects on maternal blood pressure and the sources of environmental lead.

Mental Retardation

Beattie et al. (1975) suggested that lead from water pipes might produce mental retardation. Bellinger et al. (1984) studied in detail 249 six–month old infants with known cord blood levels. The low level group averaged 1.8 microgm per dl and the mid and high were 6.5 and 14.6 microgm per dl. Although none of the levels were over 30 microgm, multiple regression analyses indicated that high cord blood levels were associated with lower covariance–adjusted scores on the Bayley Mental Development Index. The psychomotor index scores were not significantly related to cord blood levels. Moore et al. (1977) found some increase in blood lead among newborns who subsequently were found to be mentally retarded.

Bellinger et al. (1987), in a subsequent follow–up report, found that a prenatal value above 10 microgm per deciliter was associated with an IQ drop of 4.8 points on overall performance after adjusting for potential confounders. Barltrop

(1969) measured lead in human fetal tissues and cord blood. The cord blood fetal levels were about two–thirds of the maternal levels. Lead was measured in the fetal brain after 14 weeks.

Winneke et al. (1982), in a preliminary study of lead in deciduous teeth, compared two groups of 26 children. One group had lead levels averaging 2.4 ppm and the other had an average of 9.2 ppm in the teeth. Winneke's group found significant reduction in perceptual–motor integration in double–blind studies. A near significant decrease of IQ by 5–7 points was found. Increased perinatal risk factors in the higher group lead the authors to caution in the interpretation. Cooney et al. (**1989**) prospectively followed 207 children with lead levels in the range of 12–18 micrograms per dl. They could find no correlation between the levels and multiple intelligence test carried out until age 4 yrs. Dietrich et al. (**1993**) studied 253 school children after factoring out other variables such as maternal I.Q. they felt that blood levels in excess of 20 microgm per dL were associated with deficits in performance I.Q. of about 7 points as compared to children with levels of 20 microgm. The Center for Disease Control has reduced the upper limit of normal lead to 25 micrograms per dl.

Animal Studies

Ridgway and Karnofsky (1952) reported brain hemorrhage and damage followed by hydrocephalus in chick embryos exposed to 0.10 mg of lead nitrate on the 4th day. Murakami et al. (1954) gave examples of nervous system defects in mice receiving unspecified amounts of lead carbonate on the 7th and 8th days of gestation. In hamsters, Ferm and Carpenter (1967) injected 50 mg per kg of various lead salts on day 8 and found a high incidence of tail and sacral defects in the offspring. Later, they studied the genesis of these caudal defects and found that edematous blebs and hemorrhage led to the production of the malformations (Carpenter and Ferm, 1977).

McClain and Becker (1975) administered lead nitrate intravenously to rats and on the 9th day using 35 to 70 mg per kg and a teratogenic effect was observed. The malformation consisted of urorectalcaudal syndrome which included a decrease in vertebral bodies, often associated with sirenomelia, absence of the rectum or external genitalia and tail defects. Treatment on the 16th day resulted in hydrocephalus and brain hemorrhage. Gerber and Maes (1978), using diets containing 0.25, 0.5 and 1.0 percent lead acetate, showed reduced iron uptake and weight in fetal mice. Carson et al. (1974) found learning defects in lambs exposed in utero to maternal blood levels of 34 microgm per 100 ml.

Rabe et al. (1985) fed pregnant rats for two weeks before and during gestation 0.5 percent lead nitrate in the drinking water and then cross fostered the offspring with healthy dams. In spite of high lead levels in the pups (99 microgm per dl), they showed no delay, impairment or any other change in functional measures. Minsker et al. (1979) exposed rat fetuses during late gestation and=or lactation to 5 or 25 mg per kg of lead nitrate intravenously (maternal) or by mouth and found no behavioral or neurohistologic alterations. Hackett et al. (1979) studied the kinetics of lead

exposure in the rat and reported only small amounts in the fetus.

Studies in mice by Nagymajtenyi et al. (1984) found embryotoxicity as evidenced by skeletal retardation but no teratogenicity. Exposures were 0.25 or 2.35 mg per meter sq for four hours on the 9th through 12th days. Increased resorptions were found at the highest dose and the fetal cells had increased chromosomal aneuploidy and deletions. Nayak et al. (1989) found chromosomal deletions to be increased in the mouse fetus whose mother received 100, 150 or 200 mg per g of lead nitrate on day 9 of pregnancy.

Wide (1978) found that intravenously–administered lead did not interfere with blastocyst attachment or growth in the rat but in–vitro exposure did cause changes. The primordial germ cells of the mouse embryo were found to be reduced in number and in alkaline phosphatase staining after exposure to 20 microgm of lead chloride on day 8 (Wide and D'Argy, 1986). Wide (1985) found smaller litters and increased fetal death in mouse fetuses exposed to intravenous lead on day 8 of gestation. Similarly treated fetuses were raised and mated and an increase in fetal loss and small litters was observed. A decrease in primordial follicles in the ovaries of the exposed mice was observed.

Summaries

Kimmel (1984) reviewed the developmental effects. Swinyard et al. (1983) reviewed the animal work and epidemiologic studies. Gerber et al. (1980) summarized the data from animal experiments and humans. Rom (1976) reviewed the ancient and modern literature on reproduction in lead exposed women. Davis et al (1990) summarized the literature on neurotoxicity of lead in humans and animals. Ernhart (**1992**) has critically reviewed the effect of low level lead on pregnancy. She indicated that there was no evidence that low lead changed fertility or survival of the embryo and fetus. Gross human defects also were not associated. Ernhart (**1992**) concluded that much work was needed before clear conclusions could be reached.

Angle, C.R. and McIntyre, M.S.: Lead poisoning during pregnancy: Fetal tolerance of calcium disodium edetate. Am. J. Dis. Child. 108:436–439, 1964.

Barltrop, D.: Transfer of lead in the human fetus. In: Mineral Metabolism, Pediatrics, D. Barltrop (ed.), Glaxo Symposium, Blackwell, Oxford, Chapter 9, 135–151, 1969.

Beattie, A.D.; Moore, M.R.; Goldberg, A.; Finlayson, M.J.W.; Mackie, E.M.; Graham, J.F.; Main, J.C.; McLaren, D.A.; Murdock, R.M. and Stewart, G.T.: Role of chronic low-level lead exposure in the aetiology of mental retardation. Lancet 1:589–592, 1975.

Bellinger, D.C.; Needleman, H.L.; Leviton, L.; Waternaux, C.; Rabinowitz, M.B. and Nichols, M.I.: Early sensory–motor development and prenatal exposure to lead. Neurobehav. Toxicol. Teratol. 6:387–402, 1984.

Bellinger, D.C.; Leviton, A.; Waternaux, C.; Needleman, H.L. and Rabinowitz, M.: Longitudinal analyses of prenatal and postnatal lead exposure and early cognitive development. N. Eng. J. Med. 316:1037–1043, 1987.

Bellinger, D.: Teratogen update: lead. Teratology 50:367–373, 1994.

Borella, P.; Picco, P. and Masellis, G.: Lead content in abortion material from urban women in early pregnancy. Int. Arch. Occup. Environ. Health 57:93–99, 1986.

Cantarow, A. and Trumper, M.: Lead Poisoning. Williams and Wilkins, Baltimore, 84–86, 142–144, 1944.

Carpenter, S.J. and Ferm, V.H.: Embryopathic effects of lead in the hamster. Lab. Invest. 37:369–385, 1977.

Carson, T.L.; Vangelder, G.A.; Karas, G.G. and Buck, W.B.: Development of behavioral tests for the assessment of neurologic effects of lead in sheep. Environ. Health Perspect., May, 233–237, 1974.

Cooney, G.H.; Bell, A.; McBride, W. and Carter, C.: Low-level exposures to lead: the Sidney lead study. Developmental Medicine and Child Neurology 31:640–649, 1989.

Davis, J.M.; Otto, D.A.; Weil, D.E. and Grant, L.D.: The comparative developmental neurotoxicity of lead in humans and animals. Neurotoxicology and Teratology 12:215–229, 1990.

Dietrich, K.N.; Berger, O.G.; Succop, P.A.; Hammond, P.B. and Bornschein, R.L.: The developmental consequences of low to moderate prenatal and postnatal lead exposure: intellectual attainment in the Cincinnati lead study cohort following school entry. Neurotoxicology and Teratology 15:37–44, 1993.

Ernhart, C.B.: A critical review of low–level prenatal lead exposure in the human: 1. effects on the fetus and newborn. Reproductive Toxicology 6:9–19, 1992.

Ferm, V.H. and Carpenter, S.J.: Developmental malformations resulting from administration of lead salts. Exp. Mol. Pathol. 7:208–213, 1967.

Gerber, G.B.; Leonard, A. and Jacquet, P.: Toxicity, mutagenicity and teratogenicity of lead. Mutation Research 76:115–141, 1980.

Gerber, G.B. and Maes, J.: Heme synthesis in the lead–intoxicated mouse embryo. Toxicology 9:173–179, 1978.

Hackett, P.L.; Hess, J.O. and Sikov, M.R.: Cross–placental transfer and distribution of inhaled or ingested lead nitrate in rats. (abs) Teratology 19:28A, 1979.

Kimmel, C.A.: Critical periods of exposure and developmental effects of lead. In: Toxicology and the Newborn, S. Kacew and M.J. Reasor (eds.), Elsevier Sci. Publ., Amsterdam 218–235, 1984.

McClain, R.M. and Becker, B.A.: Teratogenicity, fetal toxicity, and placental transfer of lead nitrate in rats. Toxicol. Appl. Pharmacol. 31:72–82, 1975.

Minsker, D.H.; Moskalski, N.; Peter, C.P.; Robertson, R.T. and Bokelman, D.L.: Effects of lead exposure in utero or postpartum on brain histomorphology and behavior in rat offspring. (abs) Teratology 19:40A only, 1979.

Moore, M.R.; Meredith, P.A. and Goldberg, A.: A retrospective analysis of blood–lead in mentally retarded children. Lancet 1:717–719, 1977.

Murakami, U.; Kameyama, Y. and Kato, T.: Basic processes seen in disturbance of early development of the central nervous system. Nagoya J. Med. Sci. 17:74–88, 1954.

Nagymajtenyi, L.; Selypes, A. and Berencsi, G.: Study of the mutagenic and teratogenic effect of aerogenic lead exposure in the mouse. Toxicovigilance Industrielle Actes 10th Congres du Medichem No. 125, E. Fournier and M.L. Efthmiou (eds), Masson, Paris, 1984.

Nayak, B.N.; Ray, M.; Persaud, T.V.N. and Nigli, M.: Relationship of embryotoxicity to genotoxicity of lead nitrate in mice. Exp. Pathol. 36:65–73, 1989.

Nedleman, H.L.: What can the study of lead teach us about other toxicants. Environmental Health Perspectives 86:183–189, 1990.

Needleman, H.L.; Rabinowitz, M.; Leviton, A.; Linn, S. and Schoenbaum, S.: The relationship between prenatal exposure to lead and congenital anomalies. JAMA 251:2956–2959, 1984.

Rabe, A.; French, J.H.; Sinha, B. and Fersko, R.: Functional consequences of prenatal exposure to lead in immature rats. Neurotoxicol. 6:43–54, 1985.

Rabinowitz, M.: Lead and pregnancy. Birth 15(4):236–241, 1988.

Ridgway, L.P. and Karnofsky, D.A.: The effects of metal on the chick embryo: Toxicity and production of abnormalities in development. Ann. N.Y. Acad. Sci. 55:203–215, 1952.

Rom, W.N.: Effects of lead on the female and reproduction: A review. Mount Sinai J. Med. 43:542–552, 1976.

Swinyard, C.A.; Sutton, D.B. and Saloum, L.M.: Lead in the environment: Experimental studies of lead toxicity in animals and their relevance to marginal lead toxicity in children. Cong. Anom. 23:29–60, 1983.

Winneke, G.; Hrdina, K–G. and Brockhaus, A.: Neuropsychological studies in children with elevated tooth–lead concentrations. Int. Arch. Occup. Environ. Health 51:169–183, 1982.

Wide, M.: Effect of inorganic lead on the mouse blastocyst in vitro. Teratology 17:165–170, 1978.

Wide, M.: Lead exposure on critical days of fetal life affects fertility in the female mouse. Teratology 32:375–380, 1985.

Wide, M. and d'Argy, R.: Effect of inorganic lead on the primordial germ cells in the mouse embryo. Teratology 34:207–212, 1986.

1403 Lebaycide

Fytizas–Danielidou (**1971**) gave rats 5 mg per kg on days 1–13 or 6–10. This insecticide was given by gavage and produced an increase in resorptions but no teratogenicity.

Fytizas–Danielidou, R.: Effets des pesticides sur la reproduction des rats blancs, i. lebaycide. Meded. Fac. Landbouwwet. Rijksuniv. Gent. 36:1146–1150, 1971.

1404 Lectins

Discussed under concanavolin A.

1405 Lenacil 3–Cyclohexyl–5,6–trimethyleneuracil

Worden et al. (1974) fed 500 ppm in the diet over three generations and found no reproductive changes.

Worden, A.N.; Noel, P.R.B.; Mawdesley–Thomas, L.E.; Palmer, A.K. and Fletcher, M.A.: Feeding studies on lenacil in the rat and dog. Toxicol. Appl. Pharmacol. 27:215–224, 1974.

1406 Letimide

Barriga–Arceo et al. (**1991**) gavaged mice with up to 200 mg per kg on days 6–15. This salicylate derivative did not produce teratogenicity or fetal toxicity.

Barriga–Arceo, S.D.; Madrigal–Bujaidar, E.; Salazar, M. and Chamorro, G.: Cytogenetic and teratogenic evaluation of letimide. Toxocilogy Letters 56:99–107, 1991.

1407 Leupeptin N–Ethylmaleimide

Lysozymal proteinase activity of the yolk sac is important in embryonic nutrition. Leupeptin and E–64 both inhibitors of this activity produced growth retardation and increased defects when 5 micrograms per ml were added to embryo culture. N–ethylmaleimide was embryolethal at 1.0 microgm per ml.

Miyata et al. (**1991**) extended studies on this inhibitor of lysosomal proteolytic enzymes by administering 30 or 50 mg per kg intreaperitoneally on day 9. Heart, brain, eye and renal anamalies were increased at both treatment levels. The mechanism of action via disruption of yolk sac nutrition has been detailed by Freeman and Lloyd 1983.

Daston, G.P.; Baines, D.; Yonker, J.E. and Lehman–McKeeman, L.D.: Effects of lysosomal proteinase inhibitor on the development of the ratembryo in vitro. Teratology 43:253–261, 1991.

Freeman, S.J. and Lloyd, J.B.: Inhibition of proteolysis in rat yolk sac as a cause of teratogenesis. Effects of leupeptin in vitro and in vivo. J. Embryol. Exp. Morphol. 78:183–193, 1983.

Miyata, K.; Kodama, A.; Chen, S.; Tachikura, T.; Oku, S.; Yamasaki, T. and Nakamura, M.: Induction of cardiovascular malformations by leupeptin in the rat. Cong. Anom. 31:41–45, 1991.

1408 LFP 83 Flurbiprofen

Imai et al. (**1988** A&B) gave this non–steroidal anti inflammatory to rats and rabbits intravenously during organogenesis in doses of up to 10 and 80 mg respectively. In the rat some decrease in postnatal growth was found. In rabbits the dams had digestive disorders which caused some lethality. In neither species was there teratogenicity found.

Imai, M.; Ishii, S.; Ohkochi, M.; Shibata, H.; Abe, S.; Takahashi, J. and Kagitani, Y.: Reproduction studies of lfp 83. Yakuri to Chiryo 16:3689–3912, 1988.

Imai, M.; Ohkochi, M.; Shibata, H.; Ishi, S.; Abe, S.; Takahashi, J. and Kagitani, Y.: Reproduction studies of lfp 83. Yakuri to Chiryo 16:3731–3741, 1988.

1409 Lenampicillin Hydrochloride KBT–1585

Tauchi et al. (1984) gave oral doses of up to 2,500 mg per kg during organogenesis to rats and found no adverse effects on development. Toyohara et al. (1985) studied this antibiotic in pregnant rats and rabbits. In fertility studies, 2,000 mg per kg was given orally without adverse effects. In the perinatal studies a dose of 2,500 mg per kg was associated with decreased fetal weight. Oral administration during organogenesis to rabbits produced only some retardation of ossification at the maximum dose (25 mg per kg).

Tauchi, K.; Igarashi, N.; Takeshima, T. and Kou, K.: Teratogenicity study in rats by oral administration of Lenampicillin hydrochloride (KBT–1585). Chemotherapy 32:130–145, 1984.

Toyohara, S.; Tauchi, K.; Takeshima, T.; Imai, S.; Huang, K–J.; Sudo, T.; Aoyama, T.; Nose, T. and Kashima, M.: Reproduction studies of lenampicillin in the rat and rabbit. Clin. Report 19:857–890, 1985.

1410 Lens Ablation by Transgenic Insertion Cataract

Breitman et al. (1987) produced microphthalmia with lens reduction in mice after introducing α–2–crystallin promoter fused to the coding region of diptheria A toxin gene into the pronucleus of the ova. The pathogenesis of the lens lesions in the fetus were studied (Klein et al., 1989).

Breitman, M.L.; Clapoff, S.; Rossant, J.; Tsui, L–C.; Glode, L.M.; Maxwell, I.H. and Bernstein, A.: Genetic ablation: Targeted expression of a toxin gene causes microphthalmia in transgenic mice. Science 238:1563–1565, 1987.

Klein, K.L.; Klintworth, G.K.; Breitman, M.L. and Bernstein, A.: Embryologic development of microphthalmia in transgenic mice. Teratology 39:463, 1989.

1411 Leptophos O–Methyl–O–(4–bromo–2, 5–dichlorophenyl) Phenyl Phiophosphonate; Phosvel[TM] CAS 21609–90–5

Kanoh et al. (1981) gave this insecticide to rats in the diet from days 8–20 at levels of 12.5, 50 and 125 ppm. Some fetal growth retardation was found at 12.5 and 125 mg dose levels. There was some delay in ossification and 5 out of 150 fetuses had dilated renal pelves. Two fetuses with abnormal nasal cavities were found at the 50 ppm exposure level.

Kanoh, S.; Ema, M. and Hori, Y.: Studies on the toxicity of insecticides and food additives in the pregnant rats. (1) Fetal toxicity of O–methyl–O–(4–bromo–2, 5–dichlorophenyl) phenyl thiophosphonate. Oyo Yakuri 22:373–380, 1981.

1412 Leucine CAS 61–90–5

Persaud (1969) injected pregnant rats intraperitoneally with 15 mg per kg on days 1 through 6 or days 6 through 9. In each group, about 45 percent of the fetuses were defective. The defects included microphthalmia, anophthalmia, encephalocele, skeletal defects and eventration of the abdominal wall. Bergstrom et al. (1967) reported inward flexion of the toes in chicks treated with leucine on the 9th day.

Bergstrom, R.M.; Erila, T. and Pirskanen, R.: Teratogenic effects of the amino acid leucine in the chicken. Experientia 23:767–768, 1967.

Persaud, T.V.N.: Developmental abnormalities in the rat induced by the amino acid leucine. Naturwissenschaften 56:37–39, 1969.

1413 Leuprolide Acetate TAP–144–SR

No congenital anomalies were found among 15 liveborn offspring of women treated for ovulation induction. (Wilshir et al, **1993**; Young et al., **1993**)

Ooshima et al. (**1990** A) administered this LH–RH analogue subcutaneously on day 6 of rat pregnancy in doses of 0.0024, 0.008 and 0.024 mg per kg. At the highest dose fetal mortality was high. At the middle and lowest dose fetal weight and skeletal ossification was retarded but no teratogenic findings were seen. Post natal development was normal. Rabbits were similarly treated at 0.00024, 0.0024 and 0.024 mg per kg and although fetal weight was reduced at the highest and middle doses no teratogenicity was found. (Ooshima et al., **1990** B).

Ooshima, Y.; Negishi, R.; Yoshida, T.; Kanamori, H.; Sugitani, T. and Ihara. T.: Teratological study of Tap–144–sr in rats. Yakuri to Chiryo 18:s609–s623, 1990.

Ooshima, Y.; Nakamura, H.; Negishi, R.; Sugimoto, T. and Ihara, T.: Teratological study of Tap–144–sr in rabbits. Yakuri to Chiryo 18:s633–s639, 1990.

Wilshire, G.B.; Emmi, A.M.; Gagliardi, C.C. and Weiss, G.: Gonadotropin–releasing hormone agonist administration in early human pregnancy is associated with normal outcomes. Fertil. Steril. 60(6):980–983, 1993.

Young, D.C.; Snabes, M.C. and Poindexter, A.N. III: GnRH agonist exposure during the first trimester of pregnancy. Obstet. Gynecol. 81:587–589, 1993.

1414 Levodropropizine

Bestetti et al. (1988) studied reproduction in the rabbit and the rat after gavaging with up to 100 and 150 mg per kg, respectively. At the highest doses, maternal toxicity was found and in the rat, some reduced litter size or postnatal growth was reported.

Bestetti, A.; Giuliani, P.; Nunziata, A.; Melillo, G. and Tonon, G.C.: Safety and toxicological profile of the new antitussive levodropropizine. Arzneim–Forsch. Drug Res. 38:1150–1155, 1988.

1415 Levomepate CAS 428–07–09

Bianchi et al. (1967) gave rats up to 200 mg per kg orally on days 2–21 and observed no adverse effects in the fetuses. Subcutaneous injections of 10 mg per kg were given on days 2–21 to rabbits and no adverse effects were seen.

Bianchi, G.; Dezulan, M.V.; Kramer, M.; Maffii, G.; Quinton, R.M. and Serralunga, M.G.: Spasmolytic and cholinergic blocking properties and toxicity of levomepate, l–tropine α–methyltropate. Toxicol. Appl. Pharmacol. 10:424–443, 1967.

1416 Levonordefrin

Heinonen et al. (1977) found no increase in congenital defects among the offspring of twenty-six women who were given this vasoconstrictor locally primarily during dental procedures.

Heinonen, O.P.; Slone, D. and Shapiro, S.: Birth Defects and Drugs in Pregnancy. Publishing Sciences Group Inc., Littleton, Mass., 1977.

1417 Lidocaine CAS 137–58–6

Heinonen et al. (1977) recorded 947 women exposed to this during during pregnancy and the relative risk for

malformation was not increased (0.99). Two hundred and ninety–three women were exposed in the first four lunar months and the relative risk was only 0.54.

Ramazzotto et al. (1985) injected intraperitoneally 56 mg per kg into rats on days 5–7, 9–11, 12–14 and 15–17 and observed no adverse fetal effects. Fujinaga and Mazze (1986) used up to 500 mg per kg daily by infusion pump in rats and found only a reduction in fetal weight at the highest dosage. Smith et al. (1986) found that 6 mg per kg on day 11 produced significant postnatal effects in the offspring.

Fujinaga, M. and Mazze, R.I.: Reproductive and teratogenic effects of lidocaine in Sprague–Dawley rats. Anesthesiology 65:626–632, 1986.

Heinonen, O.P.; Slone, D. and Shapiro, S.: Birth Defects and Drugs in Pregnancy. Publishing Sciences Group Inc., Littleton, Mass., 1977.

Ramazzotto, L.J.; Curro, F.A.; Paterson, J.A.; Tanner, P. and Coleman, M.: Toxicological assessment of lidocaine in the pregnant rat. J. Dent. Res. l64:1214–1217, 1985.

Smith, R.F.; Wharton, G.G.; Kurtz, S.L.; Mattran, K.M. and Hollenbeck, A.R.: Behavioral effects of mid–pregnancy administration of lidocaine and mepivacaine in the rat. Neurobehav. Toxicol. Teratol. 8:61–68, 1986.

1418 Limb Defect from Transgenic Insertion

Woychik et al. (1985) introduced a gene segment (mouse mammary tumor virus–myc) into the pronucleus of mouse ova and found a line of animals with syndactyly and oligodactyly. The defect was inherited as an autosomal recessive. The new mutant was bred with two known mutations, 1d–J and 1d–OR and similar malformations were produced by the mating of heterozygotes, suggesting that the mutations might be allelic.

McNeish et al. (1988) described a syndrome in mice expressed in the homozygous autosomal state. The DNA construct was made of a heat shock gene and a herpes thymidine kinase gene. The absence of lower extremities was associated with preaxial defects of the upper extremities, facial clefts and midline defects in the prosencephalon. The pattern of cell necrosis was increased in the effected embryos.

McNeish, J.D.; Scott, W.J., Jr. and Potter, S.S.: Legless, a novel mutation found in PHT1-1 transgenic mice. Science 241:837–839, 1988.

Woychik, R.P.; Stewart, T.A.; Davis, L.G.; D'Eustachio, P. and Leder, P.: An inherited limb deformity created by insertional mutagenesis in a transgenic mouse. Nature 318:36–40,1985.

1419 D–Limonene *D–p–Mentha–1,8–diene* CAS 138–86–3

The gallstone solubilizer was given to pregnant rats in doses of up to 2,869 mg per kg during organogenesis. Although maternal toxicity and fetal growth reduction occurred no teratogenicity was found (Tsuji et al., 1975).

Tsuji, M.; Fujisaki, Y.; Okubo, A.; Arikawa, Y.; Noda, K.; Hiroyuki, I. and Ikeda, T.: Studies on D–limonene as a gallstone solubilizer. 5. Effects on development of rat fetuses and offsprings. Oyo Yakuri 10:179–186, 1975.

1420 Linamarin

Discussed under Cassava Root

1421 Lincomycin CAS 154–21–2

Gray et al. (1966) injected rats subcutaneously once with 300 mg per kg on individual days 7 through 16. No teratogenicity was found.

Gray, J.E.; Purmalis, A. and Mulvihill, W.J.: Further toxicologic studies of lincomycin. Toxicol. Appl. Pharmacol. 9:445–454, 1966.

1422 Lindane *Hexachlorocylohexane Kwell*™ CAS 58–89–9

Palmer et al. (1978A) studied three generations of rats exposed to up to 100 ppm and found no adverse effects on the reproduction or fetuses. Administration of up to 15 mg per kg in the rat and rabbit during organogenesis produced no teratogenic effect (Palmer et al., 1978B).

The International Agency for Research on Cancer (Anonymous, 1979) reviewed the embryotoxicity and teratogenicity. Neither the author nor Briggs et al. (1983) found reports of adverse effects associated with the use of lindane. Saxena et al. (1981) measured this insecticide in maternal and cord blood and found the cord level to be about two–thirds of the mother's.

McNutt and Harris (**1994**) found that whole rat embryos in culture were damaged by concentrations of 100 micromolar or greater. Embryonic glutathione levels were reduced by treatment.

Anonymous: Hexachlorocyclohexane (technical HCH and Lindane). IARC Monographs an Evaluation of Carcinogenic Risk of Chemicals to Humans. 20:217–218, 1979.

Briggs, G.G.; Bodendorfer, T.W.; Freeman, R.K. and Yaffe, S.J.: Drugs in Pregnancy and Lactation. Williams and Wilkins, 2nd ed., Baltimore, 1983.

McNutt, T.L. and Harris, C.: Lindane embryotoxicity and differential alteration of cysteine and glutathione levels in rat embryos and visceral yolk sacs. Reproductive Toxicology 8:251–362, 1994.

Palmer, A.K.; Bottomley, A.M.; Worden, A.N.; Frohberg, H. and Bauer, A.: Effect of lindane on pregnancy in the rabbit and rat. Toxicology 9:239–247, 1978B.

Palmer, A.K.; Cozens, D.D.; Spicer, E.J.F. and Worden, A.N.: Effects of lindane upon reproduction function in a 3–generation study of rats. Toxicology 11:45–54, 1978A.

Saxena, M.C.; Siddiqui, M.K.J.; Bhargava, A.K.; Murti, C.R.K. and Kutty, D.: Placenta transfer of pesticides in humans. Arch. Toxicol. 48:127–134, 1981.

1423 Linear Alkylbenzene Mixture *Alkylate 215*

This mixture of decyl and tridecyl benzenes was studied by Robinson and Schroeder (**1992**) in pregnant rats. The dams were gavaged on days 6–15 with up to 2,000 mg per kg daily. No adverse fetal effects were found. Two generation studies were also performed.

Robinson, E.C. and Schroeder, R.E.: Reproductive and developmental toxicity studies of a linear alkylbenzene mixture in rats. Fundamental and Applied Toxicology 18:549–556, 1992.

1424 Linoleic Acid CAS 60–33–3

Cutler and Schneider (1973) fed rats and mice a diet containing 10 percent oxidized linoleic acid. Although there was no change in survival rate, a significant increase in urogenital malformations was found only in the rat fetuses. These defects included agenesis of uterine horn and ovary, pelvic kidneys and cysts of the kidneys.

Cutler, M.G. and Schneider, R.: Malformations produced in mice and rats by oxidized lineolate. Food Cosmet. Toxicol. 11:935–942, 1973.

1425 Linuron CAS 330–55–2

Khera et al. (1978) gavage fed pregnant rats on days 6 through 15 of gestation with 100 or 200 mg per kg. Formulation from one supplier was teratogenic at 200 mg per kg. Wavy ribs were the most common defect.

Khera, K.S.; Whalen, C. and Trivett, G.: Teratogenicity studies on linuron, malathion, and methoxychlor in rats. Toxicol. Appl. Pharmacol. 45:435–444, 1978.

1426 Lipase AP CAS 9001–62–1

This fat solubility agent extracted from mold was given orally to rats and mice on days 7–13 of gestation by Tsutsumi et al. (1981). Maximum doses in both species of 4,500 mg per kg produced no adverse fetal effects.

Tsutsumi, S.; Kawaguchi, M.; Yoshida, H.; Simomura, H. and Sakuma, N.: Teratological study of lipase AP in mice and rats. Kiso to Rinsho 15:2577–2584, 1981.

1427 Lipase Pseudomonas *GA–56*

Tamura et al. (1974) fed rats and mice up to 2.0 gm per kg and found no adverse fetal effects slight increase in fetal deaths was found at 8.0 gm per kg.

Tamura, S.; Tsutsumi, S. and Nozaki, S.: Pharmacological studies of a lipase GA–56 produced by pseudomonas sp. V. teratogenic effect in mice and rats of GA–56. Folia Pharmacol. Japon. 70:107–118, 1974.

1428 Liranaftate *M–732*

Ishihara et al. (**1993** A,B,C,D) studied this antifungal drug in rats using the subcutaneous route and in rabbits intravenously. Prenatal studies using 100 mg per kg showed normal reproduction and early stages of pregnancy. No significant behavioral or structural changes occurred after 200 mp either during organogenesis or perinatally. In rabbits doses of up to 500 mg per kg. Fetal weight reduction was found at 100 mg per kg and at 500 mg per kg fetal deaths were increased.

Ishihara, M.; Higashikawa, K.; Yamada, H.; Ochiai, T.; Andoh, K.; Tatek Y.; Masaki, F.; Ishikawa, H. and

Nakazawa, M.: Reproductive and developmental toxicity studies of liranaftate (I) fertility study in rats. Iyakuhin Kenkyu 23:363–375, 1993A.

Ishihara, M.; Fujioka, S.; Andoh, K.; Ishiyama, Y.; Sakaguchi, Y.; Ichikawa, A.; Masaki, F. and Nakazawa, M.: Reproductive and developmental toxicity studies of liranaftate (II) teratological study in rats. Iyakuhin Kenkyu 24:376–401, 1993B.

Ishihara, M.; Yanada, H.; Sakaguchi, Y.; Arai, C.; Ichikawa, A.; Mori, S.; Hieno, Y.; Masaki, F.; Shikuma, H. and Nakazawa, M.: Reproductive and developmental toxicity studies of liranaftate (III) perinatal and postnatal study in rats. Iyakuhin Kenkyu 24:402–421, 1993C.

Isshihara, M.; Yamada, H.; Sakaguchi, M.; Sakaguchi, Y.; Arai, C.; Ichikawa, A.; Masaki, F.; Funahashi, N. and Nakazawa, M.: Reproductive and developmental toxicity stydies of liranaftate (IV) teratological study in rabbits. Iyakuhin Kenkyu 24:422–432, 1993D.

1429   Lisinopril *1-[N²-[(S)-1-carboxy-3-phenylpropyl] lysyl] proline Dihydrate*

Bagdon et al. (**1993**) treated rats before and in early pregnancy, during organogenesis and perinatally and postnatally with up to 300 mg per kg of this ACE inhibitor. No adverse effects on the fetuses or behavior of the offspring were found. Lower fetal body weights in the highest dosage group were found. Mice and rabbits receiving up to 1000 mg per kg during organogenesis had normal fetuses but saline supplementation was necessary to prevent resorption increase (Bagdon et al., **1993** B).

Bagdon, W.J.; Clark, R.L.; Minsker, D.H. and Robertson, Jr, R.T.: Reproductive and developmental toxicity studies of lisinopril in rats. Yakuri to Chiryo 21:2093–2111, 1993.

Bagdon, W.J.; Clark, R.L.; Minsker, D.H. and Robertson, Jr., R.T.: Developmental toxicity studies of lisinopril in mice and rabbits. Yakuri to Chiryo 21:2113–2120, 1993B.

1430   Listeriosis

L. monocytogenes is a gram positive motile rod which can localize in the placenta and fetus and produce abortion. The maternal infection which may be flu–like with high fever is often followed closely by abortion. Granulomas of the placenta are found. Seeliger and Finger (1983) reviewed the subject.

Seeliger, H–P. and Finger, H.: Listeriosis, Infectious Diseases of the Fetus and Newborn. Remington, J.S. and Klein, J.O. (eds), 2nd Ed., Saunders, Philadelphia, Chapter 6, 264–289, 1983.

1431   Lisuride Hydrogen Maleate CAS 19875–60–6

Kodama et al. (1981) fed mice up to 30 mg per kg daily before mating and during the first 6 days of gestation and found no changes in fertility. Similar doses did not increase malformations and postnatal function was not altered. Rabbits were given up to 10 mg per kg orally during organogenesis without producing an increase in congenital defects.

Kodama, N.; Tsubota, K. and Ezumi, Y.: Reproductive studies of lisuride hydrogen maleate. Kiso to Rinsho 15:2299–2310, 2311–2337, 2338–2345, 2346–2371, 1981.

1432   Lithium CAS 7439–93–2

Based on early studies of case reports an increased risk for congenital heart disease and especially Ebstein's anomaly has been reported. With the appearance of several prospective studies reported below, the risk has been reduced considerably. If congenital heart disease is increased the risk is probably only doubled. Ebsteins seems to be more frequent than expected.

Schou and Amidsen (1971) found three malformations in 60 children born of mothers receiving lithium. They point out that although the number studied was small, the incidence of anomalies was not increased over that of the general population. Nora et al. (1974) first observed an increase in Ebstein's anomaly among the offspring of women taking lithium. Weinstein (1979) found 17 cardiovascular defects among 212 offspring exposed in utero to lithium therapy. Six of these children had a rare cardiac defect, Ebstein's anomaly. A registry of lithium treated pregnancies is maintained in the Department of Psychiatry at the University of Wisconsin.

Kallen and Tandberg (1983) studied pregnancy outcome among 287 women with manic depression. Among the 228 women not on lithium, there were 11 with congenital defects. Among 59 on lithium there were 4 with congenital heart disease but none with Ebstein's anomaly. Schou (1976) followed up 60 exposed children after they were school age and found no increase in physical or mental anomalies as compared to their unexposed siblings. Cunniff et al. (1989) identified 50 women who were taking lithium during pregnancy and examined the offspring. One myelomeningocele and one unilateral hernia was identified but none had heart disease. In a preliminary abstract Jacobson et al. (**1992**) reported that of 148 pregnancies where the mother was identified during gestation only one case of Ebstein's anomaly was identified and the major malformation incidence was not increased as compared to similarly collected controls.

Warkany (1988) reviewed the subject of congenital heart disease and lithium treatment. Warkany referred to an unpublished analysis of 16 cases of Ebstein's anomaly where Shepard and Van Allen identified one mother who took lithium. In an accompanying comment, Kallen (1988) reported from a joint European study that among 40 Ebstein's anomalies and 44 tricuspid atresias, no maternal ingestion of lithium was found. Jacobson et al. (**1992**) prespectively followed 138 women on lithium and identified one with a defect (Ebstein's anomaly).

Robert and Francannet (**1990**) have cited reports with lingual thyroid and hypothyroidism or goiter in the offspring of lithium treated mothers.

Wright et al. (1971) gave rats 50 mg LiCl intraperitoneally on the 1st, 4th, 7th and 9th days followed by 20 mg per day until the 17th day and produced defects of the palate, eye and external ear. Szabo (1970) produced cleft palates in mouse fetuses whose mothers were gavaged with 300 to 465 mg of lithium carbonate per kg daily from day 6 to day 15 of gestation. Smithberg and Dixit (1982) found defects in mice receiving intraperitoneal doses of 200 mg per kg. The recommended human therapeutic dose is 90 to 1,800 mg per day. Johansen (1971), using a different strain of rat, found only one defect in 42 animals injected with 212 mg $LiCl_2$ per kg on day 4, 7 or 9 and followed until day 19 by a daily dose of 85 mg per kg. Gralla and McIlhenny (1972) found no teratogenicity testing rats, rabbits and monkeys during organogenesis. The rats received 27 mg, the rabbits 7 mg, and the monkeys 4 mg per kg per day.

Fritz (1988), using 100 mg per kg in rats on days 16–20, reported dilated renal pelves and increased perinatal death. Wilby et al. (1987) found increased heart and pericardium defects in rat embryos grown at 100, 150 and 200 mg of lithium carbonate per ml of medium. Some hearts were dilated and others malrotated. No changes occurred in the fetuses exposed in vivo to a maternal dose of 100 mg per kg. Hansen et al. (1990) have studied rat and mouse embryos in vitro and found the mouse to have increased open neural tubes (5.0 meq/L). Growth decrease occurred in both species at concentrations above 1.8 meq/L.

Cunniff, C.M.; Sahn, D.J.; Reed, K.L.; Chambers, C.C.; Johnson, K.A. and Jones, K.L.: Pregnancy outcome in women treated with lithium. Teratology 39:146, 1989.

Fritz, H.: Lithium and the developing rat kidney in transplacental target organ toxicity. Arzneim–Forsch Drug Research 38(1):50–54, 1988.

Gralla, E.J. and McIlhenny, H.M.: Studies in pregnant rats, rabbits and monkeys with lithium carbonate. Toxicol. Appl. Pharmacol. 21:428–433, 1972.

Hansen, D.K.; Walker, R.C.; and Grafton, T.F.: Effect of lithium carbonate on mouse and rat embryos in vitro. Teratology 41:155–160, 1990.

Jacobson, S.J.; Johnson, K.; Ceolin, L.; Kaur, P.; Sahn, D.; Donnenfeld, A.E.; Rieder, M.; Santelli, R.; Pastuszak, A.; Koren, G.; The Motherisk Program, division of Clinical Pharmacology, The Hospital for SIck Children, Toronto: A perspective multicenter study of pregnancy outcome following lithium exposure during the first trimester of pregnancy. Pediat. Res. 31:69A, 1992.

Jacobson, S.J.; Jones, K.; Johnson, K.; Ceolin, L.; Kaur, P.; Sahn, D.; Donnenfeld, A.E.; Rieder, M.; Santelli, R.; Smythe, J.; Pastuszak, A.; Einarson, T. and Koren, G.: Prospective multicentre study of pregnancy outcome after lithium exposure during first trimester. The Lancet 339:530–533, 1992.

Johansen, K.T.: Lithium teratogenicity. Lancet 1:1026–1027, 1971.

Kallen, B.: Comments on teratogen update: Lithium. Teratology 38:597, 1988.

Kallen, B. and Tandberg, A.: Lithium and pregnancy: A cohort on manic–depressive women. Acta Psychiatr. Scand. 68:134–139, 1983.

Nora, J.J.; Nora, A.H. and Toews, W.H.: Lithium, Ebstein's anomaly and other congenital heart defects. Lancet 2:594–595, 1974.

Robert, E. and Francannet, C.: Comments on teratogen update on lithium by J. Warkany. Teratology 42:205 only, 1990.

Schou, M.: What happened later to lithium babies? A follow–up study of children born without malformations. Acta Psychiat. Scand. 54:193–197, 1976.

Schou, M. and Amidsen, A.: Lithium teratogenicity. Lancet 1:1132 only, 1971.

Smithberg, M. and Dixit, P.K.: Teratogenic effects of lithium in mice. Teratology 26:239–246, 1982.

Szabo, K.T.: Teratogenic effect of lithium carbonate in the foetal mouse. Nature 225:73–75, 1970.

Warkany, J.: Teratogen update: Lithium. Teratology 38:593–596, 1988.

Weinstein, M.R.: Lithium teratogenesis. In: Lithium, Controversies and Unresolved Issues, T.B. Cooper, S. Gershon, N.S. Kline and M. Schou (eds.), Excerpta Medica, Amsterdam, 432–446, 1979.

Wilby, O.K.; Tesh, S.A.; Ross, F.W. and Tesh, J.M.: Effects of lithium on development in vitro and in vivo in the rat. (abs) Teratology 35:P69, 1987.

Wright, T.L.; Hoffman, L.H. and Davies, J.: Teratogenic effects of lithium in rats. Teratology 4:151–156, 1971.

1433 Lithium Hydroxybaterate

Smolnikova et al. (**1984**) gavaged rats with 240 mg per kg daily during gestation. Postnatally no changes in physicl growth or learning occurred but increased motor activity and a shorter latency of stress avoidance was found.

Smolnikova, N.M.; Allakhverdiev, V.D. and Lyubimov, B.I.: Development of progeny after prenatal exposure to lithium hydroxybutyrate. Farmakol. Toksikol. (Mosc.) 48:73–76, 1984.

1434 Lividomycin CAS 11111–23–2

Mori et al. (1972) gave pregnant mice up to 400 mg per kg per day from days 7–14 and found no defects. The treated fetuses on the 18th day had a delay in ossification but no postnatal differences were observed in groups allowed to suckle. Studies in the pregnant rabbit showed no teratogenicity using 100 mg per kg intramuscularly during organogenesis (Mori et al., 1973).

Mori, H.; Kakishita, T. and Kato, Y.: The safety test of lividomycin. 2. The effect of lividomycin on development of fetuses and newborns of mouse. (Japanese) Oyo Yakuri 6:813–820, 1972.

Mori, H.; Saito, N. and Kato, Y.: The safety test of lividomycin. 5. Effect of lividomycin on development of fetuses and newborn rabbits. Oyo Yakuri 7:1241–1250, 1973.

1435 Locoweed

Locoweed, Astragalus pubentissimus, was fed to pregnant sheep at various periods during the first 120 days of gestation, and many of the lambs exhibited weakness and contractures of the joints of the legs (James et al., 1969). The sheep ate 400 to 680 gm of the dried material daily. The authors postulated a lathyrogenic mechanism of teratogenesis.

James, L.F.; Keeler, R.F. and Binns, W.: Sequence in the

abortive and teratogenic effects of locoweed fed to sheep. Am. J. Vet. Res. 30:377–380, 1969.

1436 Lofepramine CAS 23047–25–8 *N–Methyl–N–(4– chlorophenacyl)–3–[10,11–dihydro–5H–dibenz (b,f)–azepin–5–yl]–propylamine Hydrochloride*

Suzuki et al. (1976) treated mice and rats orally with up to 200 and 100 mg per kg, respectively, during active organogenesis. At the higher doses, maternal weight was reduced in both species and fetal weight in the rat fetus.

Suzuki, K.; Watanabe, T.; Oura, K.; Matuhashi, K.; Kouchi, T.; Morita, T. and Akimoto, K.: Effects of a new antidepressant lofepramine on the reproduction of small laboratory animals. Clinical Report 10:2186–2205, 1976.

1437 Lomefloxacin

Tesh et al. (1988 A,B,C) gavaged rats with this antibiotic in doses of up to 300 mg per kg on days 7–17. Except for decreasd maternal and fetal weights at 150 and 300 mg per kg no adverse effects were found. Slight decrease in the weight of the offspring occurred but no decrease in development or reproduction was seen. Reproduction studies at a dose level of 300 mg per kg relealed no adverse effects.

Test, J.M.; McAnulty, P.A.; Willoughby, C.R.: Tesh, S.A. and Wilby, O.K.: Reproductive studies of ny–198 in rats. II. Teratology study. The Japanese Journal of Antibiotics 12:1352–1369, 1988.

Tesh, J.M.; McAnulty, P.A.; Willoughby, C.R.; Higgins, C.; Tesh, S.A. and Wilby, O.K.: Reproductive studies of ny–198 in rats. III. Perinatal and postnatal study. The Japanese Journal of Antibiotics 12:1370–1384, 1988.

Tesh, J.M.; McAnulty, P.A.; Willoughby, C.R.; Higgins, C.; Tesh, S.A. and Wilby, O.K.: Rproductive studies of ny–198 in rats. I. Fertility study. The Japanese Journal of Antibiotics 12:1341–1351, 1988.

1438 Loperamide *4–(p–Chlorophenyl)–4–hydroxy–N,N– dimethyl–a–a–diphenyl–1–pipeributyramide HCl* CAS 53179–11–6

Marsboom et al. (1974) gave up to 40 mg per kg in the diet of rats and found no adverse reproductive effects or teratogenicity at 2.5 and 10 mg given in the diet daily. Reproduction did not occur at 10 and 40 mg per kg. No adverse functional effects were noted in those fetuses exposed in the latter part of pregnancy.

Marsboom, R.; Herin, V.; Verstraeten, A.; Vandesteene, R. and Fransen, J.: Loperamide (R 18 553): A novel type of antidiarrheal agent. Arzneim–Forsch. Drug. Res. 24:1645–1649, 1974.

1439 Loprinone

Fumihiro et al. (**1994**) gave rats 0.1, 1.0 or 10 mg per kg during organogenesis and found no teratogenic effect but fetal weight was reduced at the two highest doses. Postnatal behavior and reproduction of the offspring was not altered. In rabbits Yoshio et al. (**1994**) used a similar treatment and produced increased resorptions and fetal deaths at the two highest doses. There was no increase in fetal defects.

Fumihiro, Ok. Yoshio, M.; Takashi, K.; Satoru, K.; Osamu, N.; Isamu, O.; Osamu, T. and Kiyomi, Y.: Teratogenicity study in rats treated intravenously with loprinone hydrochloride. The Clinical Report 28:2569–2584, 1994.

Yoshio, M.; Isamu, O.; Takashi, K.; Osamu, T.; Satoru, K. and Kiyomi, Y.: Teratogenicity study in rabbits treated intravenously with loprinone hydrochloride. The Clinical Report 28:2585–2592, 1994.

1440 Lormetazepam

Kodama et al. (1985) gave rats up to 100 mg per kg on days 7–17 or 17–21 and found no evidence of teratogenicity at 100 but not 10 mg per kg. Perinatal deaths and decreased fetal body weights were found in those exposed during organogenesis. At 10 and 100 mg the neonatal mortality was increased in those exposed on days 17–21.

Kodama, N.; Tsubota, K.; Kato, K.; Ishida, K.; Urabe, K.; Ezumi, K. and Nakao, H.: Reproduction studies of Lormetazepam in rats. Yakuri to Chiryo 13(3):605–647, 1985.

1441 Lorazepam CAS 846–49–1

Whitelaw et al. (**1981**) studied 51 mothers given this drug for 3rd trimester hypertension. There was a high incidence of low Apgar scores, need for ventilation, hypothermia and poor suckling in the preterm neonate.

Esaki et al. (1975) gave up to 4.0 mg per kg daily during organogenesis to the mouse and rat. Some reduction in fetal weight was observed in both species. No malformation increase occurred.

Esaki, K.; Tanioka, Y.; Tsukada, M. and Izumiyama, K.: Teratogenicity of lorazepam (WY–4036) in mice and rats. CIEA Preclinical Reports 1:25–34, 1975.

Whitelaw, A.G.L.; Cummings, A.J. and McFadyen, I.R.: Effect of maternal lorazepam on the neonate. British Medical Journal 282:1106–1108, 1981.

1442 Losulazine CAS 81435–67–8

Poppe et al. (1987) studied rats orally using 0, 4 or 8 mg per kg perinatally and found no adverse effect including fertility. The pups exposed only via maternal milk had reduced growth, development and reproductive capacity.

Poppe, S.M.; Marks, T.A.; Mesfin, G.M.; Soule, D.L.; Shaw, C.I.; Morris, D.F. and Black, D.L.: Reproductive and developmental effects on rats after prenatal, postnatal, or pre–and postnatal exposure to the hypotensive agent losulazine. Teratology 36:171–180, 1987.

1443 Lovastatin *Mevacor™*

Rosa (**1994**) reported 114 pregnancies where this medication was used. Two congenital defects were reported, one of which was cardiovascular.

Rosa, F.: Anti–cholesterol agent pregnancy exposure outcomes. (abs) Reproductive Toxicology 8:445–446, 1994.

1444 Loxapine

2–Chloro–11(4–methyl–1–piperazinyl)dibenzo [b,f] [1,4] oxazepine CAS 1977–10–2

Mineshita et al. (1970) administered this neuroleptic orally to pregnant rats and mice during organogenesis. At the dose level of 12 mg per kg per day fetal survival was decreased and a low incidence of exencephaly was found. All the abnormal fetuses occurred in one litter of the 20 treated at that level. No defects were found in the rat offspring. The thiazepine derivative of this drug was similarly studied and no teratogenicity was found.

Mineshita, T.; Hasewaga, Y.; Inoue, Y.; Kozen, T. and Yamamoto, A.: Teratological studies on fetuses and suckling young mice and rats of S–805. Oyo Yakuri 4:305–316, 1970.

1445 Luprostiol

This estrus regulator for livestock was studied in rats at doses of 0.12, 1.2, 12 and 1500 microgm per kg daily on days 6–15. No effects were seen at 12 but at 2500 microgm increased fetal death and resorptions occurred (Lochry et al., 1987).

Lochry, E.A.; Theodorides, V.J.; Roberman, A.M. and Christian, M.S.: Embryofetal toxicity and teratogenic potential study of luprostiol in pregnant rats. (abs) Teratology 35:P71, 1987.

1446 Luteinizing Hormone–Releasing Hormone
LH–RH CAS 9034–40–6

Discussed under Gonadorelin

1447 LY–117018

Henry and Miller (1986) failed to find that this anti–estrogen blocked estrogen–induced teratogenesis. Doses of 1–50 microgm injected directly into the day 19 rat fetus produced oviduct malformations and cleft phallus in the offspring.

Henry, E.C. and Miller, R.K.: The antiestrogen LY 117018 is estrogenic in the fetal rat. Teratology 34:59–63, 1986.

1448 LY 170053

Hagopian et al. (1987) studied this benzodiazepine antipsychotic in rats and rabbits using up to 18 and 30 mg per kg during organogenesis. In the rat decreased fetal weights were found at 4 mg per kg and at higher doses runting was found. Maternal toxicity occurred at 18 mg per kg. In the rabbit at the highest dose maternal and fetal toxicity occurred.

Hagopian, G.S.; Meyers, D.B. and Markham, J.K.: Teratology studies of LY 170053 in rats and rabbits (a). Teratology 35:60A, 1987.

1449 LY 171883

1–[2–Hydroxy–3–propyl–4–[4–(1H–tetrazol–5–yl) butoxy] phenyl] ethanone

This leukotriene LTD–4 antagonist was studied by Hagopian et al. (1988) in pregnant rats and rabbits. Mater-

nal toxicity occurred but no fetal toxicity or increase in defects was observed at doses of 425 and 200 mg per kg during organogenesis in the rat and rabbit, respectively. In the rabbit, twenty percent fetal death occurred, probably secondarily to maternal toxicity.

Hagopian, G.S.; Hoover, D.M. and Markham, J.K.: Teratology studies of compound LY 171883 administered orally to rats and rabbits. Fund. Appl. Toxicol. 10:672–681, 1988.

1450 LY 275585

This proinsulin human analog was given to rats during gestation doses of up to 20 units per kg and fetal runts were increased at the highest doses (Buelke–Sam et al., 1994).

Buelke–Sam, J.; Byrd, R.A.; Hoyt, H.A. and Zimmermann, J.L.: Implementing the ICH guidelines: a combined segment I/II/III study in CD rats of LY275585, [Lys(B28), Pro(B29)]–human insulin analog. (abs) Teratology 49:400, 1994.

1451 Lyme Disease

Markowitz et al. (1986) reported studies of 19 women infected with this spirochete during pregnancy. Thirteen were treated. In one of the treated pregnancies, the infant was developmentally delayed and had cortical blindness. Schlesinger et al. (1985) reported one newborn who died of congenital heart disease after the mother had untreated Lyme disease in the first trimester. Spirochetes were found in the infants' spleen, kidneys and bone marrow.

Markowitz, L.E.; Steere, A.C.; Benach, J.L.; Slade, J.D. and Broome, C.V.: Lyme disease during pregnancy. JAMA 255:3394–3396, 1986.

Schlesinger, P.A.; Duray, P.H.; Burke, B.A.; Steere, A.C. and Stillman, M.T.: Maternal-fetal transmission of the Lyme disease spirochete, Borrelia burgdorferi. Annals Int. Med. 103:67–68, 1985.

1452 Lymphyocytic Choriomeningitis Virus

Ackermann et al. (1974) reported two newborns with internal hydrocephalus and choreoretinitis. Both mothers were exposed to golden hamsters with lymphocytic choreomenigitis and both had falling titers of complement binding serum antibodies against the virus.

Kreschover and Hancock (1956) studied the effect of this virus in mice and found that fetal infection and resorption was much more common following inoculation during the first seven days of gestation. Dental abnormalities which were relatively infrequent in the young were characterized by disturbed amelogenesis.

Ackermann, R.; Korver, G.; Turss, R.; Wonne, R. and Hochgesand, P.: Pranatale Infektion mit dem Virus der Lymphozytaeren Choriomeningitis. Dtsch. Med. Wschr. 99:629–632, 1974.

Kreschover, S.J. and Hancock, J.A.: Effect of lymphocytic choriomeningitis on pregnancy and dental tissues in mice. J. Dent. Res. 35:467–478, 1956.

1453 Lynestrenol 17
β–Hydroxy–17–α–ethynyl–4–estrene

Kawashima et al. (1977) administered orally one mg on days 17–20 to rats and found reduction of the length of the urovaginal septum in female fetuses.

Kawashima, K.; Nakaura, S.; Nagao, S.; Tanaka, S.; Kuwamura, T. and Omori, Y.: Virilizing activities of various steroids in female rat fetuses. Endocrinol. Japon. 24:77–81, 1977.

1454 Lysergide Lysergic Acid Diethylamide LSD CAS 50–37–3

A good deal of controversy exists about whether or not lysergic acid diethylamide is teratogenic in either animals or man. Although sporadic case reports of defective infants born to mothers taking LSD have appeared, no specific pattern or clear evidence was produced (Zellweger et al., 1967; Aase et al., 1970; Assemany et al., 1970). McGlothlin et al. (1970) carried out a study of 121 pregnancies and found no increase in defects but a possible increase in spontaneous abortions in the mothers who took LSD as compared to pregnancies where only the father took it.

A critical analysis of the experimental teratologic work is beyond the scope of this catalog. Accordingly, the references showing teratogenicity will be mentioned and in a following paragraph the published negative experiments will be given.

Auerbach and Rugowski (1967) reported central nervous system abnormalities in day 11 mouse embryos after injecting 0.05 to 1.0 microgm on day 7. No dose response effect was seen. They used inbred mouse lines. Hanaway (1969) injected Swiss–Webster mice on days 6, 7, 8 or 9 of pregnancy with 5 microgm of LSD–25 and found a high incidence of histologic abnormalities of the lens. No mention of gross defects was made. In the Wistar–O'Grady rat, Alexander et al. (1970) produced fetal loss by administering LSD by mouth or subcutaneously (20 microgm and 5 microgm, respectively) between the 1st and 4th day of gestation. No specific congenital defects were reported. Geber (1967), using 0.08 to 410 microgm per kg in the hamster on the 8th day, found a small increase in defects of the central nervous system in 12–day embryos. No dose response effect was found. Dipaolo et al. (1968) could not produce defects in the hamster, but in one of two mouse strains (A–CUM), an increase in congenital defects was found when 30 microgm was injected during organogenesis.

Warkany and Takacs (1968) gave total doses of 1.5 to 300 microgm to Wistar rats on the 7th through 12th gestational days and found no increases in congenital defects. They also gave 1 to 100 microgm on the 4th or 5th day and did not confirm the fetal loss effect found by Alexander et al. (1970). Roux et al. (1970) injected rats with 5 to 100 microgm on the 4th through the 7th days or the 7th through the 13th days and found no increase in fetal death or deformity. They injected Swiss mice with 5 to 500 microgm per kg between the 4th and 14th days of gestation and found no increased fetal mortality or defects. In their experiments with hamsters, they injected 50 to 500 microgm per kg during the 7th to 13th days and found no fetal changes. Including all three species, a total of 1,723 fetuses were examined. Fabro and Sieber (1968) reported no teratogenicity in rabbits injected with 100 microgm per kg on the 7th, 8th and 9th days of pregnancy. Long (1972) critically reviewed the literature.

Aase, J.M.; Laestadius, N. and Smith, D.W.: Children of mothers who took L.S.D. in pregnancy. Lancet 1:100–101, 1970.

Alexander, G.J.; Gold, G.M.; Miles, B.E. and Alexander, R.B.: Lysergic acid diethylamide intake in pregnancy: Fetal damage in rats. J. Pharmacol. Exp. Ther. 173:48–59, 1970.

Assemany, S.R.; Neu, R.L. and Gardner, L.I.: Deformities in a child whose mother took L.S.D. Lancet 1:1290 only, 1970.

Auerbach, R. and Rugowski, J.A.: Lysergic acid diethylamide: Effects on embryos. Science 157:1325–1326, 1967.

Dipaolo, J.A.; Givelber, H.M. and Erwin, H.: Evaluation of teratogenicity of lysergic acid diethylamide. Nature 220:490–491, 1968.

Fabro, S. and Sieber, S.M.: Is lysergic acid a teratogen? Lancet 1:639 only, 1968.

Geber, W.F.: Congenital malformations induced by mescaline, lysergic acid and bromolysergic acid in the hamster. Science 158:265–267, 1967.

Hanaway, J.K.: Lysergic acid diethylamide: Effects on the developing mouse lens. Science 164:574–575, 1969.

Long, S.Y.: Does LSD induce chromosomal damage and malformation? A review of the literature. Teratology 6:75–90, 1972.

McGlothlin, W.H.; Sparkes, R.S. and Arnold, D.O.: Effect of LSD on human pregnancy. JAMA 212:1483–1487, 1970.

Roux, C.; Dupuis, R. and Aubry, M.: LSD: No teratogenic action in rats, mice and hamsters. Science 169:588–589, 1970.

Warkany, J. and Takacs, E.: Lysergic acid diethylamide (LSD): No teratogenicity in rats. Science 159:731–732, 1968.

Zellweger, H.; McDonald, J.S. and Abbo, G.: Is lysergic-acid diethylamide a teratogen? Lancet 2:1066–1068, 1967.

1455 Lysine CAS 56–87–1

Bergstrom et al. (1970) injected a 1 percent solution of lysine into the chick amniotic sac at 7 to 9 days of incubation and found anomalies of the legs with muscle spasticity and weakness. The total amount of lysine was not stated.

Bergstrom, R.M.; Erila, T. and Pirskanen, R.: Teratogenic effects of lysine in the chicken. Naturwissenschaften 57:134 only, 1970.

1456 M&B 30227 R,S[4–(2–hydroxyethyl)piperazin–1–yl]–(4–isopropylphenyl)–(isothiazol–5–yl) methane dihydrochloride

Steele et al. (1983) reported malformations in rats given 100 mg per kg on days 5–18. Anal agenesis microstomia and other defects were found in 6 of 124 exposed fetuses. They demonstrated that the chemical had a direct effect on embryos cultured in 15 microgm per ml.

Steele, C.E.; New, D.A.J.; Ashford, A. and Copping, G.P.: Teratogenic action of hypolipidemic agents: an in vitro study with postimplantation rat embryos. Teratology 28:229–236, 1983.

1457 M&B 31426

Steele et al. (**1983**) demonstrated that this hypolipemic agent produced craniofacial defects including microstomia genitourinary tract and other defects in rat fetusres exposed to 100 mg per kg on days 5–18 of gestation. They demonstrated that the chemical had a direct effect by observing toxicity in cultured embryos in 15 micrograms per ml.

Steele, C.E.; New, D.A.J.; Ashford, A. and Copping, G.P.: Teratogenic action of hypolipidemic agents: an in vitro study only. Teratology 28:229–236, 1983.

1458 MBL–A *Mecobalamin*

Oada et al. (**1988** A,B) gave this vitamin B $_{12}$ preparation to rats and rabbits intravenously during organogenesis in doses of up to 50 mg per kg and found no ill effects on the fetuses.

Okada, K.; Suzuki, T.; Hiramatsu, Y.; Nakagawa, K.; Kondo, S.; Matsubara, Y.; Sugiyama, K. and Ohgoh, T.: Teratological study of mecobalamin (MBL–A) in rats by intravenous administration. The Clinical Report 22:3899–3916, 1988.

Okada, K.; Suzuki, T.; Hiramatsu, Y.; Nakagawa, KK.; Kondo, S.; Matsubara, Y.; Sugiyama, K. and Ohgoh, T.: Teratological study of mecobalamin (MBL–A) in rabbits by intravenous administration. The Clinical Report 22:3931–3938, 1988.

1459 Mabuterol *4–Amino–3–chloro–trifluoromethyl–α– [(tertbutylamino)methyl)benzylalcohol HCl]*

This β stimulator was injected into rats on days 7–17 in amounts of up to 40 mg per kg. Although at the highest dose some fetal weight reduction occurred, no functional or physical defects were found (Hoberman et al., 1985)

Hoberman, A.M.; Weatherholtz, W.M. and Durloo, R.S.: A teratologic evaluation and postnatal behavioral screen of KF 868 in rats. J. Am. Coll. Toxicol. 4:91–110, 1985.

1460 Mafenide Acetate CAS 13009–99–9

This antibacterial agent was given by Tokunaga et al. (1973) subcutaneously to mice and rats during active organogenesis. At the maximal dose (1.0 gm per kg per day), some decrease in implants was found but congenital defects were not increased.

Tokunaga, Y.; Kawada, K.; Nagano, A.; Kunimatu, H.; Miyakubo, H. and Miyagawa, E.: Influence of mafenide acetate on the offspring of rats and mice. Nichidai Igaku Zasshi 32:973–995, 1973.

1461 Magnesium Deficiency

Gunther et al. (1973) produced a very high resorption rate in rats placed on magnesium deficient diets for periods of their gestation. No significant increase in congenital defects was found.

Hurley (1971) reported studies in pregnant rats on a deficient diet (0.2 mg per 100 gm of diet.) When the diet was administered from days 6 through 12, 14 percent of the fetuses had defects and many others were resorbed. The defects included skeletal, cleft lip, hydrocephalus, heart, lung and urogenital anomalies. The author reviewed the incomplete data on magnesium and human pregnancy.

Gunther, T.; Dorn, F. and Merker, H–J.: Embryo–toxic effects produced by magnesium deficiency in rats. Z. Klin. Chem. Klin. Biochem. 11:87–92, 1973.

Hurley, L.S.: Magnesium deficiency in pregnancy and its effects on the offspring. In: First Symposium International Sur Le Deficit Magnesique en Pathologie Humaine. J. Durlach (ed.), 1 Volume des rapports (S.G.E.M.V.; ed), Vittel, 481–492, 1971.

1462 Magnesium Sulfate CAS 7487–88–9

Heinonen et al. (1977) found no increase in defects among the offspring of 141 women treated in pregnancy but only six were treated in the first four lunar months.

Heinonen, O.P.; Slone, D. and Shapiro, S.: Birth Defects and Drugs in Pregnancy. Publishing Sciences Group Inc., Littleton, Mass., 1977.

1463 Magnetic Resonance Imaging

In a preliminary study, Tyndall (1989) found an increase in eye malformations following exposure of mice to "clinically realistic conditions".

Heinrichs et al. (**1988**) exposed mice on day 8.75 for 16 hours to MRI (0.35T, 15 MHz). A small but significant decrease in crown–ramp length was seen but no resorption, weight decrease or increase in congenital defect rate was found.

Heinrichs, W.L.; Fong, P.; Flannery, M.; Heinrichs, S.C.; Crooks, L.E.; Spindle, A. and Pedersen, R.A.: Midgestational exposure of pregnant balb/c mice to magnetic resonance imaging conditions. Magnetic Resonance Imaging 6:305–313, 1988.

Tyndall, D.A.: The effects of magnetic resonance imaging and x–irradiation on eye development in the C57BL mouse. Teratology 39:487, 1989.

1464 Malaria

Watkinson and Rushton (1983) found that 27 placentas of 65 studied had malaria pigment histologically. The patients were from a region where falciparum malaria was endemic. Birth weight was 2580 gm in the affected pregnancies and 3150 in those not affected. The rate of preterm delivery was not increased.

Watkinson, M. and Rushton, D.I.: Plasmodial pigmentation of placenta and outcome of pregnancy in West African mothers. Br. Med. J. (Clin. Res.) 287:251–254, 1983.

1465 Malathion CAS 121–75–5

Grether et al. (**1987**) found no increase in defects after low dose aerial spraying in three counties of northern California. The cohort for exposed was 22,465 and an earlier year control of 54,904.

This compound was given to rats in amounts of 240 mg per kg and no teratogenic activity was detected (Kalow and Marton, 1961). There was some increase in the mortality rate of the newborns from the treated mothers. Khera et al. (1978) gavage fed pregnant rats with 300 mg per kg on days 6 through 15 and found no teratogenicity. Machin and McBride (**1989**) treated rabbits with 100 mg per kg on days 7–12 of gestation and found no adverse fetal effects.

Grether, J.K.; Harris, J.A.; Neutra, R. and Kizer, K.W.: Exposure to aerial malathion application and the occurrence of congenital anomalies and low birthweight. Am. J. Public Health 77:1009–1010, 1987.

Kalow, W. and Marton, A.: Second–generation toxicity of malathion in rats. Nature 192:464–465, 1961.

Khera, K.S.; Whalen, C. and Trivett, G.: Teratogenicity studies on linuron, malathion, and methoxychlor in rats. Toxicol. Appl. Pharmacol. 45:435–444, 1978.

Machin, M.G.A. and McBride, W.G.: Teratological study of malathion in the rabbit. Journal of Toxicology and Environmental Health 26:249–253, 1989.

1466 Maleic Anhydride CAS 108–31–6

Short et al. (1986) gave up to 140 mg per kg to rats on days 6–15 and observed no treatment–related fetal effects. No ill effects were found in a two–generation study using 55 mg per kg.

Short, R.D.; Johannsen, F.R.; Levinskas, G.J.; Rodwell, D.E. and Schardein, J.L.: Teratology and multigeneration reproduction studies with maleic anhydride in rats. Fund. Appl. Toxicol. 7:359–366, 1986.

1467 Malonate, Sodium

Spratt (1950) used $10^{-3}M$ sodium malonate in the medium for explanted chick embryos and found that after 20 hours the central nervous system degenerated, while the heart continued to develop and function. Addition of succinate partially prevented the effect of malonate.

Spratt, N.T.: Nutritional requirements of the early chick embryo. III. Metabolic basis of morphogenesis and differentiation as revealed by the use of inhibitors. Biol. Bull. 98:120–135, 1950.

1468 Malonic Dialdehyde

Pregnant rats were given daily 4mg malonic dialdehyde (MDA) in 2ml of 25mM tris–HCl–buffer, pH 7,4 (with 0,175 M Cl diluted with physiological solution 1:1) for 5–14 days before and 1–2 weeks after birth. Decrease brain weight and retarded physiologic maturation were found.

Banova, V.V.; Simutenko, L.V. and Barsegyan, G.G.: Influence of malonic dialdehyde in the ration of pregnant and lactating rats on lipid peroxidation and physical development of newborn rats. Vapr. Pitan. 4:57–61, 1989.

1469 Maltol *Methyl Maltol*

Gralla et al. (1969) found no adverse reproductive effects when male and female rats were given up to 200 mg

per kg during gestation.

Gralla, E.J.; Stebbins, R.B.; Coleman, G.L. and Delahunt, C.S.: Toxicity studies with ethyl maltol. Toxicol. Appl. Pharmacol. 15:604–613, 1969.

1470 Maltose CAS 69–79–4

Maruoka and Kume (1973) gave this compound to rabbits intravenously in maximum doses of 10 gm per kg per day and found no teratogenicity. This sugar was administered to rats and mice intravenously daily during organogenesis and no teratogenic effects were found. Up to 30 gm per kg were given to the rats and 10 gm per kg per day to the mice (Maruoka et al., 1972).

Maruoka, H.; Kume, M. and Horie, K.: Toxicological studies of maltose. 3. Teratological study. 1. Influence of maltose on fetuses and suckling youngs of mouse and rat. (Japanese) Oyo Yakuri 6:751–768, 1972.

Maruoka, H. and Kume, M.: Toxicological studies of maltose. Teratological study. 2. Effect of maltose on growth and differentiation of fetuses and suckling youngs of rabbits. Oyo Yakuri 7:1359–1369, 1973.

1471 Mancozeb *Manganese Ethylenebis(diethylcarbamate)*

Lu and Kennedy (1986) exposed rats on days 6–15 to up to 890 mg per square meter. At 55 mg per meter, decreased maternal body weight was found and the fetuses had increased numbers of wavy ribs and external hemorrhage. An increase in total litter loss also occurred.

Lu, M.H. and Kennedy, G.L.: Teratogenic evaluation of mancozeb in the rat following inhalation exposure. Toxicol. Appl. Pharmacol. 84:355–368, 1986.

1472 Manganese Deficiency

Hurley et al. (1958, 1960) demonstrated that rats maintained on a manganese–deficient diet gave birth to offspring with an ataxic condition secondary to defective morphogenesis of the vestibular portion of the inner ear. In a subsequent publication, Erway et al. (1966) showed that in the mouse, the defective morphogenesis was characterized by absence of the vestibular otoliths. Their diets were 1 and 3 PPM of manganese and the defect rate increased from 4 to 100 percent during the course of three consequent litters. A genetic condition in the mouse (Pallid) also exhibits defective otolith formation (Lyon, 1955). Erway et al. (1966), in an interesting demonstration, treated the genetically defective (Pallid) mice with 1000 PPM of manganese and were able to completely inhibit the genetic development of the otolithic defect.

Erway, L.C.; Hurley, L.S. and Fraser, A.: Neurological defect: Manganese in phenocopy and prevention of a genetic abnormality of the ear. Science 152:1766–1767, 1966.

Hurley, L.S.; Everson, G.J. and Geiger, J.F.: Manganese deficiency in rats: Congenital nature of ataxia. J. Nutr. 66:309–319, 1958.

Hurley, L.S.; Wooten, E.; Everson, G.J. and Asling, C.W.: Anomalous development in the inner ear of offspring of manganese–deficient rats. J. Nutr. 71:15–19, 1960.

Lyon, M.F.: The developmental origin of hereditary absence of otoliths in mice. J. Embryol. Exp. Morphol. 3:230–241, 1955.

1473 Manidipine HCl

Morseth and Ihara (**1989**) fed rats 3 to 30 mg per kg on days 6–17. Some fetal growth retardation and maternal toxicity (hair loss and decreased intake) were noted at the highest doses but no teratogenicity was found. Morseth and Ihara (**1989** A,B,C) studied this drug in rats and rabbits by gavaging up to 30 and 100 mg per kg respectively daily during organogenesis. No tertogenicity was found. In the rat parturition was delayed.

Morseth, S.L. and Ihara, T.: Teratology study in rats with manidipine hydochloride. Yakuri to Chiryo (Suppl 4) 17:1119–1139, 1989.

Morseth, S.L. and Ihara, T.: Teratology study in rabbits with manidipine hydrochloride [dv–4093(2hcl)]. Yakuri to Chiryo 17:114–1149, 1989.

Morseth, S.L. and Ihara, T.: Perinatal and postnatal study in rats with manidipine hydrochloride [(cv–4093(shcl)]. Yakuri to Chiryo 17:1151–1173, 1989.

1474 Mannitol

Petter (1967) injected hypertonic mannitol intravenously into rabbits on day 16 1/2 and found reduced weight with occasional hemorrhages of the limbs in the fetuses.

Petter, C.: Lesions of the extremities provoked in the fetus of the rat by intravenous injection of hypertonic mannitol into the mother. C. R. Soc. Biol. 161(5):1010–1014, 1967.

1475 Maprotiline CAS 10262–69–8

Esaki et al. (1976) administered orally up to 30 mg per kg to mice and rats during active organogeneis and observed no fetal changes other than slight weight reduction at the highest dose. Postnatal changes were not detected.

Esaki, K.; Tanioka, Y.; Tsukada, M. and Izumiyama, K.: Teratogenicity of maprotiline tested by oral administration to mice and rats. (Japanese) CIEA Preclinical Report 2:69–77, 1976.

1476 Marijuana *Cannabis δ–9–Tetrahydrocannabinol* CAS 8063–14–7

Linn et al. (1983) studied a group of 12,424 women of whom 1,246 used marijuana. The odds ratio for major malformation after controlling simultaneously for other characteristics, such as alcohol use, age, smoking and previous miscarriage, was 1.36 which was not statistically significant. The odds ratio for low birth weight was only 1.07. Only sporadic case reports appeared on humans and no pattern of malformation emerged (Hecht et al., 1968; Neu et al., 1969). Hingson et al. (1982) observed, after controlling for other variables, that there was a 105 gm decrease in birth weight of offspring of marijuana smokers. In addition there

was a five–fold increase in infants with features compatible with the fetal alcohol syndrome. O'Connell and Fried (1984) found a 0.8 week reduction in length of pregnancy in heavy marijuana users. This was found even after correcting for 6 contributing factors. Hatch and Braken (1986) found that women using marijuana regularly had an increased risk for low birth weight and small fetuses for gestational age. O'Connell and Fried (1984) studied the offspring of 25 cannabis users and found no increase in minor or major defects as compared to matched controls. Witter and Niebyl (1990) studied 8,350 mothers and found 417 had used marijuana. No increase in premature rates orcongenital defects was found. Fried (1989) studied over 50 offspring whose mothers were exposed to marijuana and could detect no motor, mental or language changes. O'Connell and Fried (**1991**) followed 28 children into school age after controlling for maternal age, mothers personality and home environment could not detect a difference from controls. Day et al. (**1994**) reported significant negative effects of exposure on the performance of 3–year–old children as assessed by Stanford–Binet test. The results remained significant after controlling for social, economic and psychological factors. Astley et al. (**1992**) made detailed morphometric measurements of the offspring of 40 mothers who used marijuana frequently during gestation. No consistent pattern was found.

In hamsters using up to 500 mg per kg of D–9 tetrahydrocannabinol, Joneja (1977) could not produce malformations. Intragastric doses were given from days 7 through 13 of gestation. Joneja (1976) treated pregnant mice intravenously, subcutaneously and by gavage with the same material. Most dose levels gave no malformations but 200 mg per kg by mouth on day 8, 9 or 10 produced a low increase in malformation rate over controls. Umbilical hernias with occasional club foot, cleft palate and exencephaly were found. The intravenous and subcutaneous doses did not produce defects.

Geber and Schramm (1969) injected crude extract of marijuana into rabbits and hamsters on gestational days 7 through 10 and 6 through 8, respectively. At doses over 150 mg per kg neural tube closure defects, phocomelia and other defects were reported. Persaud and Ellington (1968) gave resin (4.2 mg per kg) to pregnant rats and found 57 percent of the offspring had defects which included stunting, syndactyly, encephalocele and limb reductions. Banerjee et al. (1975) injected up to 100 mg per kg of tetrahydrocannabinol in rats on days 7 through 16 of gestation and found no significant defects in the offspring. Sofia et al. (1979), using synthetic δ–9–hydrocannabinol in rabbits subcutaneously at doses of up to 60 mg per kg per day on days 7–19, produced no teratogenicity.

Gianutsos and Abbatiello (1972) injected 250 mg per kg of an extract of cannabis resin into rats on days 8 through 11 and then observed significant loss of maze learning in the offspring. No congenital defects were found in the fetuses after delivery. Borgen et al. (1973) studied the offspring of rats injected subcutaneously with 10 mg per kg of tetrahydrocannabinol on days 10, 11 and 12 of gestation. Significant increases in open field activity and decreases in growth rate were found until weaning. Cross fostering techniques helped to exclude maternal effects on the postnatal activity. In the mouse, Fleischman et al. (1975) gave up to 150 mg of tetrahydrocannabinol per kg per day on days 6 through 15 of gestation and produced no fetal effects. Abel (1984), using a dose of 5 to 50 mg of tetrahydrocannabinol on days 0 through 5 and then either 50 or 150 mg per kg for the remainder of the pregnancy, found a decrease in fetal body weight. No effects on postnatal behavior were found. Grilly et al. (1974) exposed male and female chimpanzees to long term tetrahydrocannabinol (1.4 to 2.0 mg per kg per day) and then after discontinuing the drug, mated the animals and found 7 normal offspring. Rosenkrantz et al. (1986) exposed rabbits to marijuana smoke on days 6–18 and found a slight increase in early and late resorptions but no congenital defect increase. Plasma THC levels were 45 nanograms per ml in the highest dose group. Fleischman et al. (1975) reviewed the animal literature.

Morgan et al. (**1988**) studied DNA, RNA and protein levels in embryo–fetal rat brains exposed to 15 or 50mg per kg on days 2 through 22. A transient decrease in brain protein was found on days 7 and 14 but not at day 21. DNA and RNA were not different from pair–fed controls.

Abel, E.L.: Effects of δ–9–THC on pregnancy and offspring in rats. Neurobehav. Toxicol. Teratol. 6:29–32, 1984.

Astley, S.J.; Clarren, S.K.; Little, R.E.; Sampson, P.D. and Daling, J.R.: Analysis of facial shape in children gestationally exposed to marijuana, alcohol, and/or cocaine. Pediatrics 89:67–77, 1992.

Banerjee, B.N.; Galbreath, C. and Sofia, R.D.: Teratological evaluation of synthetic δ–9–tetrahydrocannabinol. Teratology 11:99–102, 1975.

Borgen, L.A.; Davis, W.M. and Pace, H.B.: Effects of prenatal δ-9-tetrahydrocannabinol on development of rat offspring. Pharmacol Biochem Behav 1:203–206, 1973.

Day, N.L.; Richardson, G.A.; Goldschmidt, L.; Robles, N.; Taylor, P.M.; Stoffer, D.S.; Cornelius, M.D. and Geva, D.: Effect of prenatal marijuana exposure on the cognitive development of offspring at age three. Neurotoxicology and Teratology 16: 169–175, 1994.

Fleischman, R.W.; Hayden, D.W.; Rosenkrantz, H. and Braude, M.C.: Teratologic evaluation of δ–9–tetrahydrocannabinol in mice, including a review of the literature. Teratology 12:47–50, 1975.

Fried, P.A.: Cigarettes and marijuana: are there measurable long–term neurobehavioral teratogenic effects. NeuroToxicology 10:577–584, 1989.

Geber, W.F. and Schramm, L.C.: Effect of marijuana extract on fetal hamsters and rabbits. Toxicol. Appl. Pharmacol. 14:276–282, 1969.

Gianutsos, G. and Abbatiello, E.R.: The effect of prenatal cannabis sativa on maze learning ability in the rat. Psychopharmacologia (Berl) 27:117–122, 1972.

Grilly, D.M.; Ferraro, D.P. and Braude, M.C.: Observations on the reproductive activity of chimpanzees following long–term exposure to marihuana. Pharmacology 11:304–307, 1974.

Hatch, E.E. and Bracken, M.B.: Effect of marijuana use in pregnancy on fetal growth. Am. J. Epidemiol. 124:986–993, 1986.

Hecht, F.; Beals, R.K.; Lees, M.H.; Jolly, H. and Roberts, P.: Lysergic–acid–diethylamide and cannabis as possible teratogens in man. Lancet 2:1087 only, 1968.

Hingson, R.; Alpert, J.J.; Day, N.; Dooling, E.; Kayne, H.; Morelock, S.; Oppenheimer, A. and Zuckerman, B.: Effects of maternal drinking and marijuana. Pediatrics 70:539–545, 1982.

Joneja, M.G.: A study of teratological effects of intravenous, subcutaneous and intragastric administration of δ–9–tetrahydrocannabinol in mice. Toxicol. Appl. Pharmacol. 36:151–162, 1976.

Joneja, M.G.: Effects of δ–9–tetrahydrocannabinol on hamster fetuses. J. Toxicol. Environ. Health 2:1031–1040, 1977.

Linn, S.; Schoenbaum, S.C.; Monson, R.R.; Rosner, R.; Stubblefield, P.C. and Ryan, K.J.: The association of marijuana use with outcome of pregnancy. AJPH 73:1161–1164, 1983.

Morgan, B.; Brake, S.C.; Hutchings, D.E.; Miller, N. and Gamagari, Z.: Delta–9–tetrahydrocannabinol during pregnancy in the rat: effects on development of RNA, DNA, and protein in offspring brain. Pharm Bio B. 31:365–369, 1988.

Neu, R.L.; Powers, H.; Kings, S. and Gardner, L.I.: Cannabis and chromosomes. Lancet 1:675 only, 1969.

O'Connell, C.M. and Fried, P.A.: An investigation of prenatal cannabis exposure and minor physical anomalies in a low risk population. Neurobehav. Toxicol. Teratol. 6:345–350, 1984.

O'Connell, C.M. and Fried, P.A.: Prenatal exposure to cannabis: a preliminary report of postnatal consequences in school–age children. Neurotoxicology and Teratology 13:631–639, 1991.

Persaud, T.V.N. and Ellington, A.C.: Teratogenic activity of cannabis resin. Lancet 2:406–407, 1968.

Rosenkrantz, H.; Grant, R.J.; Fleischman, R.W. and Baker, J.R.: Marihuana–induced embryotoxicity in the rabbit. Fund. Appl. Toxicol. 7:236–243, 1986.

Sofia, R.D.; Strasbaugh, J.E. and Banerjee, B.N.: Teratologic evaluation of synthetic δ–9–tetrahydrocannabinol. Teratology 19:361–366, 1979.

Witter, F.R. and Niebyl, J.R.: Marijuana use in pregnancy and pregnancy outcome. American Journal of Perinatology 7:36–38, 1990.

1477 Mathamidophos

Wang and Huang (1987) studied postnatal behavior in mice treated on days 16–21 with up to 4.0 mg per kg per day. Most of the functional tests were decreased as compared to controls, as was the neuron density.

Wang, S. and Huang, X–S.: Behavioral toxicity in the offspring of mice following maternal exposure to mathamidophos. (Chinese) J. Pharmacol. Toxicol. 1(5):375–376, 1987.

1478 Maytansine CAS 35846–53–8

Sieber et al. (1978) studied this plant extract using 0.1 to 0.25 mg per kg intraperitoneally on day 6, 7 or 8 and found numerous malformations including dextrocardia, axial skeleton defects and hydrocephalus.

Sieber, S.M.; Whang–Peng, J.; Botkin, C. and Knutsen, T.: Teratogenic and cytogenic effects of some plant–derived antitumor agents (vincristine, colchicine, maytansine, VP–16–213 and VM–26) in mice. Teratology 18:31–48, 1978.

1479
1-(1(o-(M-Chorobenzyloxy)phenyl)vinyl)1H-imidazole HCl 710674-S

Kobayashi et al. (1984) studied this antimycotic in pregnant rats and rabbits. In studies before fertilization and during the first seven days, using up to 50 mg per kg subcutaneously, some decrease in live fetuses was found. The neonatal mortality was increased at the same dose when administered perinatally. Studies during organogenesis at 250 mg per kg found some generalized fetal edema in six of the 274 fetuses. In rabbits, dermal doses of up to 20 mg per kg during organogenesis caused no fetal adverse effects.

Kobayashi, F.; Hara, K.; Hasegawa, Y.; Takegawa, Y.; Yoshida, T.; Yamagata, H.; Fukiishi, Y.; Ando, M. and Ito, M.: Reproduction studies of the antimycotic 710674–S in rats and rabbits. Clin. Report 18:4917–5010, 1984.

1480 Mazaticol 6,6,9-Trimethyl-9-azabicylo (3,3,1) non–3 β–yl–α,α–di(2–thienyl) Glycolate HCl Monohydrate CAS 32891–29–5

Yamaguchi et al. (1974) studied this compound in rat and mouse pregnancy and except for fetal weight reduction no fetal alterations were found. Dosage orally was up to 100 mg per kg in mice and 200 mg per kg in rats, and it was given during major organogenesis.

Yamaguchi, K.; Ishihara, H.; Ariyuki, F.; Noguchi, Y. and Kowa, Y.: Teratological study of 6,6,9–trimethyl-9–azabicyclo(3,3,1) non–3–β–yl–α,α–di(2–thienyl) glycolate hydrochloride monohydrate. Oyo Yakuri 8:1213–1218, 1974.

1481 ME 1207

This agent is an oral cephem antibiotic. Hata et al. (1992) found no effect of up to 1,000 mg per kg on rat reproduction. Hata et al. (1992) found only decrease in sacrococcygeal ossification centers after giving 1000 mg per kg to rats during organogenesis. Perinatal studies by Hattori et al. (1992) were done using up to 750 mg per kg and no adverse effect on development or behavior were found.

Hata, T.; Asaoka, H.; Ito, M.; Okano, K.; Izawa, M.; Kosugi, I. and Fujita, M.: Reproductive and developmental toxicity study of me1207 in rats by oral administration prior to and in the early stages of pregnancy. Chemotherapy 40S-2):247–255, 1992.

Hata, T.; Asaoka, H.; Ito, M.; Okano, K.; Izawa, M.; Shindo, Y.; Kosugi, I. and Fujita, M.: Reproductive and developmental toxicity study of me1207 in rats by oral administration during the period of organogenesis. Chemotherapy 40(S-2):256–172, 1992.

Hattori, M.; Inoue, S.; Katano, T.; Tachibana, I.; Isowa, K.; Komai, Y.; Ito, M.; Kosugi, I. and Fujita, M.: Reproductive and developmental toxicity study of me1207 of rats by oral administration during the perinatal and lactation period. Chemotherapy 40(s-2):272–283, 1992.

1482 Mebendazole Vermox™ Methyl 5–benzoylbenzimidazole–2–carbamate CAS 31431–39–7

In 112 human pregnancies with exposure to the drug, only one malformation is known to have occurred and this was a digital reduction of one hand (Sargent, 1979).

In doses up to 40 mg per kg on days 7 through 10 of rat gestation, no significant increases in defect rates were found. Studies in the rabbit at the same dose were negative (Sargent, 1979).

Sargent, E.C.: Ortho Pharmaceutical Corporation, personal communication to the author, 1979.

1483 Mechlorethamine Mustargen™ Nitromin™ HN2 Methylamine CAS 51–75–2

Murphy et al. (1958) found that this alkylating agent given on the 12th day to the rat (about 0.7 mg per kg) caused cleft palate and defects of the central nervous and skeletal systems. They also produced defects in the chick embryo. Okano et al. (1958) found further defects including skeletal exophthalmus as well as renal, ovarian and adrenal histologic changes.

Murphy, M.L.; Moro, A.D. and Lacon, C.R.: The comparative effects of five polyfunctional alkylating agents, with additional notes on the chick embryo. Ann. N.Y. Acad. Sci. 68:762–781, 1958.

Okano, K.; Esumi, K.; Ito, S.; Kashiyama, S.; Fujita, H.; Toba, T. and Ito, H.: Influences of nitromin on rat embryo. (abs) Acta. Pathol. Jpn. 8:561 only, 1958

1484 Meclizine 1–(p–Chloro–α phenyl benzyl)–4–(M–methyl benzyl)–piperazine Antivert™ CAS 569–65–3

In the wake of the thalidomide disaster, some of these compounds were suspected to be teratogens in the human (Watson, 1962), but on further analysis no association could be found (Carter and Wilson, 1962; David and Goodspeed, 1963). Lenz (1966) summarized the reports on 3,333 infants exposed to meclizine during pregnancy and believed that the finding of 12 patients with cleft lip or palate in this group does not completely clear the drug of teratogenicity. Yerushalmy and Milkovich (1965) studied prospectively 8,090 births and did not observe an increased defect rate in the 330 fetuses exposed to meclizine early in pregnancy.

The same series was enlarged to 613 offspring exposed during the first 84 days of pregnancy and no teratogenicity was detected (Milkovich and Van den Berg, 1976). Michaelis et al. (1983) studied prospectively 472 matched pairs of newborns and found no increase in frequency of major defects.

Since Tuchmann–Duplessis and Mercier–Parot's original report (1963), the benzhydrylpiperazine series of compounds were studied in the rat extensively (King et al.,

1965). The members of this series of drugs are buclizine, cyclizine, chlorcyclizine, hydroxyzine, meclizine and norchlorcyclizine. The syndrome in the rat fetus consists of cleft palate, micrognathia, microstomia and glossopalatine fusion (Posner and Darr, 1970). The drug (20 to 80 mg per kg) is usually given on the 13th through 16th days. Congenital defects were found in the ferret (Steffek et al., 1968) and mouse (King and Howell, 1966). The major tissue metabolite of these compounds, norchlorcyclizine, is teratogenic and produced cleft palate even when applied directly over the amniotic sac (Wilk, 1969). A number of antihistamines lacking the ethylamine as a ring structure were non–teratogenic (King et al., 1965). Massive edema of the embryonic thorax may play a role in the mechanism which produces these defects (Posner and Darr, 1970). Chlorcyclizine, hydroxyzine and cyclizine produce abortions in the monkey (Steffek et al., 1968). Tuchmann–Duplessis and Mercier–Parot (1963) described eye defects resulting from treatment of the rat, rabbit and mouse with cyclizine.

Carter, M.P. and Wilson, F.W.: 'Ancoloxin' and fetal abnormalities. Br. Med. J. 2:1609 only, 1962.

David, A. and Goodspeed, A.H.: 'Ancoloxin' and fetal abnormalities. Br. Med. J. 1:121 only, 1963.

King, C.T.G.: Teratogenic effects of meclizine hydrochloride on the rat. Science 141:353–355, 1963.

King, C.T.G. and Howell, J.: Teratogenic effect of buclizine and hydroxyzine on the rat and chlorcyclizine in the mouse. Am. J. Obstet. Gynecol. 95:109–111, 1966.

King, C.T.G.; Weaver, S.A. and Narrod, S.A.: Antihistamines and teratogenicity in the rat. J. Pharmacol. Exp. Ther. 147:391–398, 1965.

Lenz, W.: Malformations caused by drugs in pregnancy. Am. J. Dis. Child. 112:99–106, 1966.

Michaelis, J.; Michaelis, H.; Gluck, E. and Koller, S.: Prospective study of suspected associations between certain drugs administered during early pregnancy and congenital defects. Teratology 27:57–64, 1983.

Milkovich, L. and Van den Berg, B.J.: An evaluation of the teratogenicity of certain antinauseant drugs. Am. J. Obstet. Gynecol. 125:244–248, 1976.

Posner, H.S. and Darr, A.: Fetal edema from benzhydrolpiperazines as a possible cause of oral–facial malformations in the rat. Toxicol. Appl. Pharmacol. 17:67–75, 1970.

Steffek, A.J.; King, C.T.G. and Wilk, A.L.: Abortive effects and comparative metabolism of chlorcyclizine in various mammalian species. Teratology 1:399–406, 1968.

Tuchmann–Duplessis, H. and Mercier–Parot, L.: Action du chlorhydrate de cyclizine sur la gestation et le developpement embryonnaire du rat, de la souris et du lapin. C. R. Acad. Sci. (Paris) 256:3359–3362, 1963.

Watson, G.I.: Meclizine ('Ancoloxin') and foetal abnormalities. Br. Med. J. 2:1446 only, 1962.

Wilk, A.L.: Production of fetal rat malformations by norchlorcyclizine and chlorcyclizine after intrauterine application. Teratology 2:55–65, 1969.

Yerushalmy, J. and Milkovich, L.: Evaluation of the teratogenic effect of meclizine in man. Am. J. Obstet. Gynecol. 93:553–562, 1965.

1485 Meclofenamic Acid CAS 644–62–2

Schardein et al. (1969) reported negative teratologic studies of this anti–inflammatory drug in rats and rabbits. Up to 3.5 mg per kg per day was given daily during the period of active organogenesis. Petrere et al. (1985) dosed rats orally with 6 and 9 mg per kg and found a postimplantation loss but at maternally toxic doses (20 mg per kg) no adverse effects were found. F_1 and F_2 generations were studied.

Petrere, J.A.; Humphrey, R.R.; Anderson, J.A.; Fitzgerald, J.E. and De La Iglesia, F.A.: Studies on reproduction in rats with meclofenamate sodium, a nonsteroidal anti–inflammatory agent. Fund. Appl. Toxicol. 5:665–671, 1985.

Schardein, J.L.; Blatz, A.T.; Woosley, E.T. and Kaup, D.H.: Reproductive studies on sodium meclofenamate in comparison to aspirin and phenylbutazone. Toxicol. Appl. Pharmacol. 15:46–55, 1969.

1486 Medazepam

Banerjee (**1975**) studied CAR–learning in the offspring of rats given 0.02% in their drinking water during pregnancy and found a significant decrease at 6 weeks of age.

Banerjee, U.: Conditioned learning in young rats born of drug–addicted parents annd raised on addictive drugs. Psychopharmacologia (Berl.)42:113–116, 1975.

1487 Medroxyprogesterone Provera™ CAS 520–85–4

Burstein and Wasserman (1964) treated 172 women before the 12th menstrual week with 5 to 50 mg per day and observed transient clitoral hypertrophy in only one newborn. The mother had received 25 mg daily for 7 days during the 6th menstrual week. Katz et al. (1985) compared 1,608 newborns exposed to progestagen (1,274 received 20 mg daily of medroxyprogesterone) to controls and found no significant increase in any form of malformation. Ressequie et al. (1985) found no increase in congenital malformations among 244 offspring exposed to progesterone-like compounds. Pardthaisong et al. (1988) studied 1,229 women where the drug had been injected. Polysyndactyly was increased (10 cases) but half of them were cases where the mother received the drug more than 9 months before conception. Other types of defects were not increased significantly and the authors felt that there was no causal association with polysyndactyly.

Aarskog (1979) reported that hypospadias could be associated with synthetic progesterones. Mau (1981) who studied 3,602 male newborns, of whom 33 had hypospadias, could not confirm it. Of these, eight had been exposed, giving a 1.4 percent rate which was not significantly different from the controls at 0.8 percent The type of hypospadias was correlated with the time of ingestion in the study by Aarskog but not in that of Mau.

Suchowsky and Junkmann (1961) produced masculinization of the female rat fetus with this compound. Andrew and Staples (1977) produced no malformations in rats and mice using up to 3000 mg per kg during various days of organogenesis, but in the rabbit with 0.3 to 30 mg per kg, cleft palates were found. Eibs et al. (1982A), using 30 mg per kg subcutaneously early in the mouse gestation (days 1–12), produced defects of palate, urinary and respiratory tract. They pointed out that the drug has a long half–life. Eibs et al. (1982B) studied the toxicity of the compound on preimplantation stages of the rat in vitro.

Prahalada and Hendrickx (1982) gave monkeys 300 mg per day from days 23 to day 41 of gestation. The fetuses removed at around day 100 had hypoplastic adrenals, thymus and thyroids and the females had masculinized external genitalia. The male monkey fetuses were hypospadic with micropenis. Prahalada et al. (1985A) gave single doses of 25 or 100 mg per kg to monkeys on day 27 and found masculinization of female fetuses. The same workers (Prahalada et al., 1985B) found that a single injection at contraceptive dose level into baboons on day 27 was not teratogenic. Dickmann (1973) found that this compound led to destruction of fertilized eggs when given on day 1 (12.5 mg) followed on day 3 by oestrone. Carbone et al. (1990) studied endochondral bone development in mouse fetus. At a dose 2,500–fold higher than the equivalent human dose no effect was seen and no non–genital bone defects occurred.

Aarskog, D.: Maternal progestins as a possible cause of hypospadias. N. Eng. J. Med. 300:75–78, 1979.

Andrew, F.D. and Staples, R.E.: Prenatal toxicity of medroxyprogesterone acetate in rabbits, rats and mice. Teratology 15:25–32, 1977.

Burstein, R. and Wasserman, H.C.: The effect of provera on the fetus. Obstet. Gynecol. 23:931–934, 1964.

Carbone, J.P.; Figurska, K.; Buck, S. and Brent, R.L Effect of gestational sex steriod exposure on limb development and endochondral ossification in the pregnant c57bi/6j mouse: i medroxyprogesterone acetate. Teratology 42:121–130, 1990.

Dickmann, Z.: Postcoital contraceptive effects of medroxyprogesterone acetate and oestrone in rats. J. Reprod. Fertil. 32:65–69, 1973.

Eibs, H.G.; Spielmann, H. and Hagele, M.: Teratogenic effects of cyproterone acetate and medroxyprogesterone treatment during the pre– and postimplantation period of the mouse. Teratology 25:27–36, 1982A.

Eibs, H.G.; Spielmann, H.; Jacob–Muller, U. and Klose, J.: Cyproterone acetate and medroxyprogesterone acetate treatment before implantation in vivo and in vitro. Teratology 25:291–299, 1982B.

Katz, Z.; Lancet, M.; Skornik, J.; Chemke, J.; Mogilner, B.M. and Klinberg, M.: Teratogenicity of progestagens given during the first trimester of pregnancy. Obstet. Gynecol. 65:775–780, 1985.

Mau, G.: Progestins during pregnancy and hypospadias. Teratology 24: 285–287, 1981.

Pardthaisong, T.; Gray, R.H.; McDaniel, E.B. and Chandacham, A.: Steroid contraceptive use and pregnancy outcome. Teratology 38:51–58, 1988.

Prahalada, S. and Hendrickx, A.G.: Teratogenicity of medroxyprogesterone acetate (MPA) in cynomolgus monkeys. (abs) Teratology 25:67A–68A, 1982.

Prahalada, S.; Carroad, M.; Cukierski, M. and Hendrickx, A.G.: Embryotoxicity of a single dose of medroxypro-

gesterone acetate (MPA) and maternal serum MPA concentrations in Cynomolgus monkey (Macaca fascicularis). Teratology 32:421–432, 1985A.

Prahalada, S.; Carroad, E. and Hendrickx, A.G.: Embryotoxicity and maternal serum concentrations of medroxyprogesterone acetate in baboons (Papio cynocephalus). Contraception 32:497–515, 1985b.

Ressequie, L.J.; Hick, J.E.; Bruen, J.A.; Noller, K.L.; O'Fallon, W.M. and Kurland, L.: Congenital malformations among offspring exposed in utero to progestins, Olmstead County Minnesota 1936–1974. Fertil. Steril. 43:514–519, 1985.

Suchowsky, G.K. and Junkmann, K.: A study of the virilizing effect of the progestogens on the female rat fetus. Endocrinology 68:341–349, 1961.

1488 Medrysone

This steroid was given to rats in doses of 1.14 or 3.42 mg per kg daily. The fetuses were small with fused ribs and retarded skeletal development. No internal abnormalities were found in rabbit offspring. In neither species was the time of exposure reported (Anonymous, 1971).

Anonymous.: Medrysone: A review. Drugs 2:5–19, 1971.

1489 Megestrol Acetate 17 α–Acetoxy–6–methyl–4, 6–pregnadiene–3, 20–dione

Kawashima et al. (1977) administered 5 mg orally to rats on days 17–20 and produced a reduction in urovaginal septum length in the female fetuses.

Kawashima, K.; Nakaura, S.; Nagao, S.; Tanaka, S.; Kuwamura, T. and Omori, Y.: Virilizing activities of various steroids in female rat fetuses. Endocinol. Japon. 24:77–81, 1977.

1490 Meglutol CAS 503–49–1 3–Hydroxy–3–methylglutaric Acid

Savoie and Lupien (1975) treated rats with up to 500 mg per kg orally or intraperitoneally on days 7–11 and found no adverse reproductive effects. Mice were treated with 3.7g per kg orally or 1.6g per kg intraperitoneally with similar negative results.

Savoie, L.L. and Lupien, P.J.: Preliminary toxicological investigations of 3–hydroxy–3–methylglutarid acid (HMG). Arzneim. Forsch (Drug Res) 25:1284–1286, 1975.

1491 Mefloquine

Balocco and Bonati (1992) reported that there were no malformations among 11 offspring who were exposed to this malarial prophylactic. In another study Nosten et al. (1990) reported normal pregnancies in 20 women.

Minor et al. (1976) gave rats and mice 10 or 100 mg per kg respectively on days 6–15 and found no adverse effects. Fetal weight reduction was found at higher doses but maternal toxicity was present.

Balocco, R. and Bonati, M.: Mefloquine prophylaxis against malaria for female travfellers of childbearing age. Lancet 340:309–310, 1992.

Minor, J.L.; Short, R.D.; Heiffer, M.H.; Lee, C.C.: Reproductive effects of mefloquine HCL (MFQ) in rats and mice. Pharmacologist 18:171, 1976.

Nosten, F.; Karbwang, J.; White, N.J.; Honeymoon; Bangchang, K.N.; Bunnag, D. and Harinasuta, T.: Mefloquine antimalarial prophylaxis in pregnancy: dose finding and pharmacokinetic study. Br. J. Clin. Pharmac. 30:79–85, 1990.

1492 Meglumine CAS 6284–40–8

Kodama et al. (1980, 1981) administered the contrast medium to rats and rabbits in doses of up to 1800 mg per kg daily. For the reproductive studies in rats an intraperitoneal route was used. For studies during organogenesis in both species intravenous routes were used. The perinatal studies were done by the intravenous route followed after birth by intraperitoneal administration to the dam. The only adverse effect was some postnatal growth retardation in rats exposed during organogenesis to 1800 and 360 mg per kg.

Kodama, N.; Tsubota, K. and Ezumi, Y.: Effects of meglumine on reproduction in the rat and rabbit. Nichi Doku Iho (Japanisch–Deutsche Medizinische Berichte). 25:398–405, 1980.

Kodama, N.; Tsubota, K. and Ezumi, Y.: Effects of meglumine on reproduction in the rat and rabbit. Nichi-Doku Iho (Japanisch–Deutsche Medizinische Berichte). 26:110–118, 119–135, 1981.

1493 Melamine 2,4,6–Triamino–s–Triazine CAS 108–78–1

Thiersch (1957) gave 70 mg per kg intraperitoneally to rats on days 4,5 or 7,8 or 11,12. A resorption rate of 91 to 97 percent was found but no defects were seen.

Thiersch, J.B.: Effect of 2,4,6 Triamino–"S"–Triazine (TR), 2,4,6 "Tris" (Ethyleneimino)–"S"–Triazine (TEM) and N,N',N"–Triethylenephosphoramide (TEPA) on Rat Litter in Utero. Proc. Soc. Exp. Biol. Med. 94:36–40, 1957.

1494 Meperidine Demerol CAS 57–42–1

Heinonen et al. (1977) reported that among 268 women taking this drug in the first four lunar months there was no increase in defect rate.

Brackbill (1986) studied behavioral development of rats exposed postnatally via mothers' milk. The dams were given 0.44 mg per kg intraperitoneally. Three of four behavioral measures were reduced (surface righting, negative geotaxis and auditory startle).

Brackbill, Y.: Teratogenicity for rat pups of meperidine administered postnatally, via breast milk. NeuroToxicol. 7:349–364, 1986.

Heinonen, O.P.; Slone, D. and Shapiro, S.: Birth Defects and Drugs in Pregnancy. Publishing Sciences Group Inc., Littleton, Mass., 1977.

1495 Menichlopholan
3,3–Dichloro–5,5–dinitro–O,O–biphenol Distolon

This compound used to treat liver flukes in sheep was given by Juszkiewicz et al. (1971) to hamsters by mouth on the 8th day of gestation. With 20 mg per kg a 4 percent incidence of skeletal retardation was found. One meningomyelocele occurred among 77 examined fetuses.

Juszkiewicz, T.; Rakalska, Z. and Dzierzawski, A.: Effet embryopathique du 3,3–di–chloro–5,5–dinitro–O,O–biphenyl (Bayer 9015) chez le hamster dore. Eur. J. Toxicol. 4:523–528, 1971.

1496 Mephobarbital Mebaral™ CAS 115–38–8

Heinonen et al. (1977) reported that among eight women treated during the first four months of pregnancy no congenital defects occurred in their offspring.

Heinonen, O.P.; Slone, D. and Shapiro, S.: Birth Defects and Drugs in Pregnancy. Publishing Sciences Group Inc., Littleton, Mass., 1977.

1497 Mepirodipine Hydrochloride

Ohata and Shibata (1990,A) gave rats orally up to 20 mg per kg during organogenesis. At a dose of 10 mg per kg that was maternal toxicity and fetal growth retardation but no teratogenic or postnatal behavioral changes resulted. Ohata and Shibata (1990,B) treated rabbits with up to 80 mg per kg during days 6–18 and at this dose found some increase in fetal loss. No teratogenicity was found. Fujiwara et al. (1991) found no adverse effects of perinatal treatment on parturition, behavior or reproductive ability.

Fujiwara, M.; Itou, S. and Shibata, M.: Perinatal and postnatal studies in rats given administered mepirodipine hydrochloride (YM730) orally. Oyo Yakurri 42:283–289, 1991.

Ohata, T, and Shibata, M.: Oral teratology study of mepirodipine hydrochloride (ym730) in rabbits. Kiso to Rinsho 24;4341–4355.

Ohata, T. and Shibata, M.: Oral teratology study of mepirodipine hydrochloride (ym730) in rats. Kiso to Rinsho 24:4335–4340, 1990.

1498 Mepivacaine CAS 96–88–8

Heinonen et al. (1977) observed 82 offspring of women exposed in the first four lunar months and found eight with malformations. This was 1.96 times the expected. The type of defect was not given. This local anesthetic has been shown to cross the placenta and produce fetal bradycardia (Gordon, 1968).

Smith et al. (1986) found that six mg per kg on day eleven in the rat produced significant postnatal effects in the offspring.

Gordon, H.R.: Fetal bradycardia after paracervical block: Correlation with fetal and maternal blood levels of local anesthetic (mepivacaine). N. Eng. J. Med. 279:910–914, 1968.

Heinonen, O.P.; Slone, D. and Shapiro, S.: Birth Defects and Drugs in Pregnancy. Publishing Sciences Group Inc., Littleton, Mass., 1977.

Smith, R.F.; Wharton, G.G.; Kurtz, S.L.; Mattran, K.M. and Hollenbeck, A.R.: Behavioral effects of mid–pregnancy administration of lidocaine and mepivacaine in the rat. Neurobehav. Toxicol. Teratol. 8:61–68, 1986.

1499 Meprobamate *Miltown™ Equanil™* CAS 57–53–4

Milkovich and Van den Berg (1974) in a study of 19,044 births, used a computer system of record linkage to study the association between prescriptions filled for women and serious congenital defects in their offspring. A four–fold increase in defect rate was found when mothers took meprobamate or chlordiazepoxide during the first 42 days of gestation. No pattern of malformations was identified but 5 infants from the 66 meprobamate takers had congenital heart disease. A group of women taking phenobarbital did not have an increased malformation rate in their offspring. Hartz et al. (1975), in a follow–up of 50,282 pregnancies, could not associate an increased defect rate with maternal ingestion of meprobamate or chlordiazepoxide. They had 1345 mothers in the meprobamate exposure group. Crombie et al. (1975) reported four malformations among a group of 67 mothers who took meprobamate.

Werboff and Kesher (1963) gave pregnant rats 60 mg per kg on days 5–8, 11–14 or 17–20 and then carried out several learning tests postnatally. Although they reported decreased scores in all treated groups, Kletzkin and Berger (1971) using similar conditions were unable to confirm their work. These later workers also reviewed the clinical literature and reported absence of teratogenic activity. Caldwell and Spille (1964) using up to 128 mg per kg in the rat's diet during the entire gestation period, could not demonstrate learning impairment in the offspring. Nishikawa (1963) using near lethal doses (750 mg per kg) in the mouse produced some skeletal defects, but lower doses did not produce defects (Clavert, 1963; Brar, 1969).

Brar, B.S.: The effect of meprobamate on fertility gestation and offspring viability and development of mice. Arch. Int. Pharmacodyn. 177:416–422, 1969.

Caldwell, M.B. and Spille, D.F.: Effect on rat progeny of daily administration of meprobamate during pregnancy and lactation. Nature 202:832–833, 1964.

Clavert, J.: Etude de l'action du meprobamate sur la formation de l'embryon. C. R. Soc. Biol. (Paris) 157:1481–1482, 1963.

Crombie, D.L.; Pinsent, R.J.; Fleming, D.M.; Rumeau–Rouquette, C.; Goujard, J. and Huel, G.: Fetal effects of tranquilizers in pregnancy. N. Eng. J. Med. 293:198–199, 1975.

Hartz, S.C.; Heinonen, O.P.; Shapiro, S.; Siskind, V. and Slone, D.: Antenatal exposure to meprobamate and chlordiazepoxide in relation to malformations, mental development and childhood mortality. N. Eng. J. Med. 292:726–728, 1975.

Kletzkin, M. and Berger, F.M.: Influence of mebrobamate on the fetus, fertility and post–natal development. In: Malformations Congenitales des Mammiferes. H. Tuchmann–Duplessis (ed.), Paris: Masson, 255–272, 1971.

Milkovich, L. and Van den Berg, B.J.: Effects of prenatal meprobamate and chlordiazepoxide hydrochloride on human embryonic and fetal development. N. Eng. J. Med. 291:1268–1271, 1974.

Nishikawa, M.: Effect of meprobamate on the development of the fetus on pregnant mice. Acta Anat. Nippon 38:258, 1963.

Werboff, J. and Kesner, B.: Learning deficits of offspring after administration of tranquilizing drugs to the mothers. Nature 197:106–107, 1963.

1500 Mequitazine

10–(3–Quinuclidinylmethyl)phenothiazine CAS 29216–28–2

Maeda et al. (1982) gave oral doses of 1.25, 5 and 20 mg per kg to rats before pregnancy and during the entire gestation or on days 7–17. At the highest dose the maternal weight was decreased but no adverse fetal effects were found. Behavioral and fertility studies of the exposed offspring did not show significant differences from controls. Rabbits were given orally up to 125 mg per kg and no increase in malformations was found in the offspring.

Maeda, H.; Yoshifune, S. and Shimizu, Y.: Reproductive studies of 10–(3–quinclidinylmethyl)phenothiazine. Oyo Yakuri 21:855–898, 1982.

1501 n–Butyl Mercaptan CAS 109–79–5 *n–Butanethiol N–BM*

Thomas et al. (1987) exposed rats and mice by inhalation for 6 hours daily during organogenesis. No defects occurred in the rat fetuses exposed to 152 ppm. Maternal toxicity including death occurred in the mice at 68 and 152 ppm. Malformations of cleft palate and hydrocephalus were increased at 68 ppm.

Thomas, W.C.; Seckar, J.A.; Johnson, T.; Ulrich, C.E.; Klonne, D.R.; Schardein, J.L. and Kirwin, C.J.: Inhalation teratology studies of n–butyl mercaptan in rats and mice. Fund Appl. Toxicol. 8:170–178, 1987.

1502 Mercaptobenzimidazole

Discussed under Ethylenethiourea

1503 Mercaptobenzothiazole

Hardin et al. (1981) injected 200 mg per kg into rats intraperitoneally on days 1–15. Neither teratogenicity nor maternal or fetal toxicity were found.

Hardin, B.D.; Bond, G.P.; Sikov, M.R.; Andrew, F.D.; Beliles, R.P. and Niemeier, R.W.: Testing of selected workplace chemicals for teratogenic potential. Scand. J. Work. Environ. Health 7(4):66–75, 1981.

1504 Mercaptobenzothiazole Disulfide

Histological study of placentas from women working with acrolein, methylmercaptopropionaldehyde and rubber accelerators, including the above chemical, showed collagenization of the stroma of the villi and walls of the blood vessels and an increase in the number of Hofbauer cells in the second half of pregnancy (Ferste•, 1970).

Mirkova (1980) found that 100 mg per kg on days 1–21 in the rat produced hydrocephalus, intracerebral hematomas and disturbances in cranial ossification. Ten mg per kg had no effect. Microphthalmia and hydrocephalus resulted in 5 percent of mouse fetuses at 100 mg per kg per day.

Ferster, L.N.: Morphological changes in the chorion and placenta in women under the effect of some toxic products of organic synthesis. SB Nauchin. Rab. Volgogr. Med. Inst. 23:169–171, 1970.

Mirkova, E.: Study of the hazard of embryotoxic and teratogenic action of the vulcanization accelerator. Altax. Probl. Khig. (5):83–91, 1980.

1505 2–Mercaptoethane Sulfonate

Slott and Hales (1986) gave up to 30 mg per kg by intraamniotic injection on day 13 and did not cause fetal adverse effects. This chemical protected against cyclophosphamide teratogenicity.

Slott, V.L. and Hales, B.F.: Sodium 2–mercaptoethane sulfonate protection against cyclophosphamide–induced teratogenicity in rats. Toxicol. Appl. Pharmacol. 82:80–86, 1986.

1506 Mercapto–1–methyl–imidazole

Discussed under Ethylenethiourea

1507 N–(2–Mercapto–2–methylpropionyl)–L–cysteine
SA96

Yamamoto et al. (1985a,b) studied the teratogenicity of this anti–rheumatoid in mice and rabbits. In both species, at maternal doses that were toxic, some decrease in fetal growth was found but no teratogenicity. The mice were orally treated on days 6–15 with up to 480 mg per kg and the rabbits on days 6–18 with up to 300 mg per kg.

Yamamoto, Y.; Horie, S. and Iso, T.: Reproduction study of N–(2–mercapto–2–methylpropionyl)–L–cysteine (SA96). II. Teratogenicity study in mice. Iyakuhin Kenkyu 16(4):626-653, 1985a.

Yamamoto, Y.; Horie, S.; Fujimura, K–I.; Nakayama, T. and Iso, T.: Reproduction study of N–(2–mercapto–2–methylpropionyl)–L–cysteine (SA96). III. Teratogenicity study in rabbits. Iyakuhin Kenkyu 16(4):654–664, 1985b.

1508 6–Mercaptopurine and Related Derivatives CAS 50–44–2

Experience with 6–mercaptopurine during human pregnancy was reviewed by Sokal and Lessmann (1960). Five pregnant women treated during pregnancy produced five offspring without defects, but two were prematurely born. Moloney reported that 5 of 21 pregnancies treated mainly with 6–mercaptopurine ended with stillbirth or neonatal death associated with immaturity. No congenital malformations were found.

Degeneration of the embryonic discs appeared during the blastocyst stage and no conceptus survived after the mother received 75 mg per kg (Chaube and Murphy, 1968).

This SH–containing purine analogue and related derivatives are teratogenic in the chick, mouse, rabbit and rat. Chaube and Murphy (1968) reported results

on 13 purine analogues and found the following teratogenic: 6–Mercaptopurine, 6–mercaptopurine riboside, 6–mercaptopurine–3–N–oxide, 9–butyl–6–mercaptopurine, 9–ethyl–6–mercaptopurine, 6–chloropurine, 6–thioguanine, 6–thioguanosine and 6–hydroxylaminopurine. 6–Mercaptopurine in doses of 31 to 125 mg per kg on the 11th or 12th day produced defects of the extremities and tail in rat fetuses. 6–Mercaptopurine riboside and 6–mercaptopurine–3–N–oxide under similar conditions produced essentially the same syndrome except for inclusion of cleft palate. 9–Butyl–6–mercaptopurine and 6–hydroxylaminopurine (in the mouse) and 9–ethyl–6–mercaptopurine and 6–chloropurine (in the rat) produced skeletal defects when the dosages were close to the LD–50 for the mother (400 to 900 mg per kg). 6–Thioguanine and 6–thioguanosine produced skeletal defects in rats at dosage ranges of 12 to 50 mg per kg. Chaube and Murphy (1968) reported that 9–ethyl–6–mercaptopurine and 6–methyl–mercaptopurine riboside were non–teratogenic.

In the chick, Karnofsky (1960) used 400 microgm on the 4th day and produced some facial defects. Adams et al. (1961) injected 25 to 250 mg per kg into rabbits during ovulation and cleavage stages and found no morphologic changes in the embryos.

Adams, C.E.; Hay, M.F. and Lutwak–Mann, C.: The action of various agents upon the rabbit embryo. J. Embryol. Exp. Morphol. 9:468–491, 1961.

Chaube, S. and Murphy, M.L.: The teratogenic effects of the recent drugs active in cancer chemotherapy. In: Advances in Teratology. D.H.M. Woollam (ed.), New York: Logos and Academic Press, 3:181–237, 1968.

Karnofsky, D.A.: Influences of antimetabolites inhibiting nucleic acid metabolism on embryonic development. Trans. Assoc. Am. Physicians 73:334–347, 1960.

Moloney, W.C.: Management of leukemia in pregnancy. Ann. N.Y. Acad. Sci. 114:857–867, 1964.

Sokal, J.E. and Lessmann, E.M.: Effects of cancer chemotherapeutic agents on the human fetus. JAMA 172:1765–1771, 1960.

1509 2–Mercaptothiazoline CAS 96–53–7

Discussed under Ethylenethiourea

1510 Mercury CAS 7439–97–6

An epidemic of cerebral palsy with microcephaly occurred in Minamata, Japan, and the cause was felt to be maternal ingestion of fish contaminated with methyl mercury. Two autopsy reports with mercury analyses and detailed brain examination are given by Matsumoto et al. (1965). Murakami (1972) reviewed the 25 cases of fetal Minamata disease and reported that cerebral palsy was the primary feature, but in seven an association with microcephaly occurred. Subtle dental changes were observed also, but with the exception of one case of auricular deformity other congenital defects did not occur. Harada and Noda (1988) summarized the history and clinical findings of the epidemic.

Women exposed to inorganic levels less than 0.01 mg per m^3 in a mercury vapor lamp factory did not have increases in low birth weight or malformations. The 153 pregnancies were compared to those in another factory (DeRosis et al., **1985**). Inorganic (or elemental) mercury is not thought to be converted in tissues to organic mercury which is the toxic form.

Snyder (1971) reported severe central nervous system damage in an infant whose mother ate meat from a pig contaminated by a mercury–containing grain diet. The ingestion occurred during the third gestational month.

Larsson and Sagulin (**1990**) reviewed studies reporting ratios of maternal to fetal mercury and concluded that the fetal level was usually less for exprimental animals and humans. Since the daily uptake of mercury from dental amalgam is low (2–5 microgm) restriction of amalgam therapy in pregnant women was unwarrented according to their findings. Vimy et al. **1990** reported accumulation of mercury in fetal pituitary liver and other tissues after the ewes received fillings with dental amalgam containing [203]Hg. They asked the question about whether dental amalgam containing Hg should be used during pregnancy. Most of the Hg remained in the elemental or inorganic form. Similar findings in the subhuman primate have been reported (Hahn et al. **1990**).

Wannag and Skjaerasen (1975) studied mothers exposed to elemental mercury through their dental work place and found significantly increased mercury content in their babies' placentae and membranes. Brodsky et al. (1985) analyzed over 40,000 questionaires to dental workers exposed to mercury. No increase in spontaneous abortions or defects were found in the offspring of women exposed to the highest concentrations of mercury. Koos and Longo (1976) summarized human experience with mercury poisoning giving exposure limits for women of childbearing age and levels at which toxicity might be expected. For the fetus and newborn, the toxic level was given as 3 microgm Hg per gm.

Amin–Zaki et al. (1974) reported studies on 15 infant–mother pairs poisoned in Iraq due to ingestion of home–made bread prepared from wheat treated with a methyl–mercury fungicide. In all cases but one, the infants blood mercury level was higher than that of the mother. Six of the infants were severely impaired in their motor and mental development. Follow–up neurological examinations on 32 prenatally exposed children are given by Amin–Zaki et al. (1979). Cerebral palsy occurred even when exposure was during the third trimester. Milder cases with developmental retardation in addition to exaggerated tendon reflexes and pathological extensor plantar reflexes were described.

Murakami (1972) reviewed the work in experimental animals. Specific defects following inorganic mercury are uncommon. Inouye et al. (1972) injected methyl mercury chloride (30 mg per kg) into mice from days 6 to 13 of pregnancy. Treatment after day 7 was associated with a high incidence of cleft palate and hydrocephalus. Spyker and Smithberg (1972), using methyl mercuric dicyandiamide and Khera and Nera (1971), using one mg of methyl mercuric chloride in mice found cleft palate and histopathologic and functional changes in neuronal development. Mottet (1974) studied the offspring of rats exposed to chronic low dose methylmercury hydroxide (total about 20 mg per kg)

and could find no microscopic malformations or changes in behavior. Murakami et al. (1953) used a vaginal tablet containing 0.1 mg of phenylmercuric acetate on day 7 of the rat gestation and found tail and neural tube abnormalities. Inouye et al. (1972) studied the rat fetus also. Harris et al. (1972) studied methyl mercuric chloride teratogenicity in the hamster.

Shtenberg and Safronova (1979) reported that in experiments on Wistar rats, oral methyl mercuric iodide at doses of 0.85, 0.64, 0.42 and 0.21 mg per kg daily exerted no teratogenic or embryotoxic action. Administration at doses of 0.85 and 0.64 mg per kg produced in females a significant decrease in the SH–group content in renal and cerebral tissues. Also, a decrease in the activity of acetylcholine esterase in the cerebellum by 18–48 percent and a rise in the activity of glucose–6–phosphatase in the kidneys by 48–70 percent was produced. The doses of 0.42 and 0.21 mg per kg administered at various periods of pregnancy did not exert any effect on the body of females under experimental conditions. The biochemical parameters of these animals did not differ from those of the control.

Ignatyev (1980) treated male rats with inhalation of metallic mercury during 115 days. Alterations of spermatogenesis and DNA and RNA synthesis in the testis were found. Offspring of treated males (mated with intact females) had decreased weight and vitality. Exposure of rat females during the entire gestation with 6 and 1 mg per cubic millimeter produced slight embryotoxic and gonadotoxic effect.

Using 12.5 ppm of methyl mercury during gestation and the suckling period, Howard and Mottet (1986) showed impaired cell proliferation in the cerebellum. Burbacher et al. (1988) studied monkeys with intakes of 50, 70 or 90 microgm of methyl mercury per kg per day and found no change in menstrual cycle or pregnancy rate but maternal blood mercury levels above 1.5 ppm reduced the number of viable deliveries. There was no maternal toxicity at these levels. Burbacher et al. (1990) found decreased social behavior in infants of monkeys receiving 50 microgm per kg of methylmercury daily. Danielsson et al. (**1993**) in a further attempt to study metallic mercury in rats exposed females to 1.8 mg per cubic meter for 1 or 3 hours daily on days 11–14 and 17–20. Most of the behavioral tests were decreased at 3 months of age but not at 6 months. At 6 months hyperactivity was found. Spatial learning was decreased at 4 months.

Curle et al. (1987) studied chromosomes in embryonic mice and found increased clumping at metaphase. They also found an increase in sister chromatid exchange with increasing dosage. Ramel and Magnusson (1969) presented some evidence that organic mercury may produce meiotic nondisjunction in drosophila. The disposition of organic mercury in the maternal–fetal system of the rat indicates a preferential concentration in the fetal brain (Yang et al., 1972; King et al., 1976).

Burbacher et al. (1990) reviewed the neurobehavioral effects and hemopathology in humans and animals exposed to methyl mercury. Data on brain doses of 12–20 ppm indicate severe functional and anatomic defects. At doses of

3–11 ppm decreased brain mass and cortical damage was present in small mammals.

Amin–Zaki, L.; Elhassani, S.B.; Majeed, M.A.; Clarkson, T.W.; Doherty, R.A. and Greenwood, M.R.: Intrauterine methylmercury poisoning in Iraq. Pediatrics 54:587–595, 1974.

Amin–Zaki, L.; Majeed, M.A.; Elhassani, S.B.; Clarkson, T.W.; Greenwood, M.R. and Doherty, R.A.: Prenatal mercury poisoning, clinical observations over five years. A.M.A. Dis. Child 133:172–177, 1979.

Brodsky, J.B.; Cohen, E.N.; Whitcher, C.; Brown, B.W. and Wu, M.L.: Occupational exposure to mercury in dentistry and pregnancy outcome. J. Am. Dent. Assoc. 111:779–780, 1985.

Burbacher, T.M.; Mohamed, M.K. and Mottet, N.K.: Methylmercury effects on reproduction and offspring size at birth. Reproductive Toxicology 1:267–278, 1988.

Burbacher, T.M.; Rodier, P.M. and Weiss, B.: Methylmercury developmental neurotoxicity: a comparison of effects in humans and animals. Neurotoxicology and Teratology 12:191–202, 1990.

Burbacher, T.M.; Sackett, G.P.; and Mottet, N.K.: Methylmercury effects on the social behavior of macaca fascicularis infants. Neurotoxicology and Teratology 12:65–71, 1990.

Curle, D.C.; Ray, M. and Persaud, T.V.N.: In vivo evaluation of teratogenesis and cytogenetic changes following methylmercuric chloride treatment. The Anatomical Record 219:286–295, 1987.

Danielsson, B.R.G.; Fredriksson, A.; Dahlgren, L.; Gardlund, A.T.; Olsson, L.; Dencker, L. and Archer, T.: Behavioural effects of prenatal metallic mercury inhalation exposure in rats. Neurotoxicology and Teratology 15:391–396, 1993.

DeRosis, F.; Anastasio, S.P.; Selvaggi, L.; Beltrame, A. and Moriani, G.: Female reproductive health in two lamp factories. Effects of exposure to inorganic mercury vapour and stress factors. Brit. J. Indust. Med 42:488–494, 1985.

Hahn, L.J.; Kloiber, R.; Leininger, R.W.; Vimy, M.J. and Lorscheider, F.L.: Whole–body imaging of the distribution of mercury released from dental fillings into monkey tissues. FASEB J. 4:3256–3260, 1990.

Harada, Y. and Noda, K.: How it came about the finding of methyl mercury poisoning in Minimata district. Cong. Anom. 28:S59–S69, 1988.

Harris, S.B.; Wilson, J.G. and Printz, R.H.: Embryotoxicity of methyl mercuric chloride in golden hamsters. Teratology 6:139–142, 1972.

Howard, J.D. and Mottet, N.K.: Effects of methylmercury on the morphogenesis of the rat cerebellum. Teratology 34:89–95, 1986.

Ignatyev, V.M.: Gonadotoxic and embryotoxic effects of metallic mercury. Gigiena Sanit. 3:72–73, 1980.

Inouye, M.; Hoshino, K. and Murakami, U.: Effect of methyl mercuric chloride on embryonic and fetal development in rats and mice. Ann. Report Res. Inst. Environ. Med. Nagoya Univ. 19:69–74, 1972.

Khera, K.S. and Nera, E.A.: Maternal exposure to methyl mercury and postnatal cerebellar development in mice. (abs) Teratology 4:233 only, 1971.

King, R.B.; Robkin, M.A. and Shepard, T.H.: Distribution of ^{203}Hg in the maternal and fetal rat. Teratology 13:275–290, 1976.

Koos, B.J. and Longo, L.D.: Mercury toxicity in the pregnant woman, fetus, and newborn infant. Am. J. Obstet. Gynecol. 126:390–409, 1976.

Larsson, K.S. and Sagulin, G.B.: Placental transfer of mercury from amalgam. The Lancet 2:1251, 1990.

Matsumoto, H.; Koya, G. and Takeuchi, T.: Fetal Minimata disease: A neuropathological study of two cases of intrauterine intoxication by a methyl mercury compound. J. Neuropathol. Exp. Neurol. 24:563–574, 1965.

Mottet, N.K.: Effects of chronic low–dose exposure of rat fetuses to methylmercury hydroxide. Teratology 10:173–190, 1974.

Murakami, U.: Organic mercury problem affecting intrauterine life. Proceedings of the International Symposium on the Effect of Prolonged Drug Usage on Fetal Development. In: Advances in Experimental Biology and Medicine, M.A. Klingberg (ed.), New York: Plenum Publishing Corp. 27:301–336, 1972.

Murakami, U.; Kameyama, Y.; Kato, T.; Tsuji, S.; Imai, M. and Furakawa, E.: An experiment by mercury compounds (preliminary report). Influences of the contraceptive agent upon embryo and mother animal. Ann. Report Res. Inst. Environ. Med. Nagoya Univ. 5:167–168, 1953.

Ramel, C. and Magnusson, J.: Genetic effects of organic mercury compounds. II. Chromosome segregation in drosophila melanogaster. Hereditas 61:231–254, 1969.

Shtenberg, A.I. and Safronova, A.M.: Effect of minor quantities of methyl mercuric iodide on embryogenesis and some biochemical parameters. Vopr. Pitaniya (USSR) 5:53–57, 1979.

Snyder, R.D.: Congenital mercury poisoning. N. Eng. J. Med. 284: 1014–1016, 1971.

Spyker, J.M. and Smithberg, M.: Effects of methyl mercury on prenatal development in mice. Teratology 5:181–190, 1972.

Vimy, M.J.; Takahashi, Y. and Lorscheider, F.L.: Maternal–fetal distribution of mercury (^{203}Hg) released from dental amalgam fillings. Am. U. Physiol. 258:R939–R945, 1990.

Wannag, A. and Skjaerasen, J.: Mercury accumulation in placenta and foetal membranes. A study of dental workers and their babies. Environ. Physiol. Biochem. 5:348–352, 1975.

Yang, M.G.; Krawford, K.S.; Garcia, J.D.; Wang, J.H.C. and Lei, K.Y.: Deposition of mercury in fetal and maternal brain. Proc. Soc. Biol. Med. 141:1004–1007, 1972.

1511 Meropenem

Russel et al. (**1992**) studied rats during preconception, organogenesis and perinatal periods. Doses of up to 750 mg per kg were used. The adults had soft stools with this intravenous mode of therapy. The F–1 females were lighter at birth after treatment during organogenesis at 240 mg per kg but heavier in the 1000 mg per kg dosage group.

Russel, A.W.; Ferrman, S.J. and Siddall, R.A.: Reproduction and developmental study of meropenem in rats. Chemotherapy 40(S-1):238–250, 1992.

1512 Mervan *4–Allyloxy–3–chlorophenylacetic Acid*

Lambelin et al. (1970) studied this drug during organogenesis in the rat using up to 150 mg per kg orally and up to 100 mg per kg intraperitoneally. No adverse fetal effects were noted.

Lambelin, G.; Roba, J.; Gillet, C.; Gautier, M. and Buu–Hoi, N.P.: Toxicity studies of 4–allyloxy–3–chlorphenylacetic acid, a new analgesic, antipyretic and anti–inflammatory agent. Arzneim Forsch. 20:618–630, 1970.

1513 Mesalazine

Ota et al. (**1994**) treated rats orally with this medication developed to treat colitis and Crohn's disease. Doses up to 400 mg per kg were given from day 6 through the 21st day of lactation and no treatment-related effects were found in the newborns or their behavior including reproduction.

Ota, T.; Fujimura, T.; Hongyo, T. and Kawase, S.: Prenatal and postnatal study of mesalazine in rats. Oyo Yakuri 47:513–522, 1994.

1514 Mescaline *3,4,5–Trimethoxyphenethylamine* CAS 54–04–6

Geber (1967) reported production of congenital defects in guinea pigs given a single intravenous dose of 0.45 to 3.25 mg per kg on the 8th day. The defects involved the central nervous system mainly. There was no correlation between dose and rate of congenital defects, but resorptions and runts increased with dose.

Geber, W.F.: Congenital malformations induced by mescaline, lysergic acid diethylamide and bromolysergic acid in the hamster. Science 158:265–266, 1967.

1515 Mesna

Komai et al. (**1990** A) studied this agent in rats giving it intravenously on days 7–17 in doses of up to 800 mg per kg. Fetal weight was decreased at the highest dose and lumbar ribs were seen at 400 mg per kg. The pups had increased open field activity when exposed in utero to 400 mg or above. Similarly treated rabbits (Komai et al., **1990** B) had decreased fetal weight and lumbar ribs at levels of 600 mg and above.

Komai, Y.; Itoh, I.; Iriyam, K.; Ishimura, K.; Fuchigami, K. and Kobayashi, F.: Reproduction study of mesna–teratogenicity study in rats by intravenous administration. Kiso to Rinsho 24:6653–6594, 1990.

Komai, Y.; Itoh, I.; Ishimura, K.; Fuchigami, K. and Kobayashi, F.: Reproduction study of mesna–teratogenicity study in rabbits by intravenous administration. Kiso to Rinsho 24:6596–6602, 1990.

1516 Mesoinositol Pentanicotinate

Linari (1970) gave six times the single therapeutic dose per kg to rats for a four month period and did not detect any effects on fertility or reproduction.

Linari, G.: Observations on the synthesis, chemical hydrolysis in vitro and toxicity of a new ester of nicotinic acid: The mesoinositol pentanicotinate. Arzneim–Forsch. 20:723–724, 1970.

1517 Mesoridazine CAS 5588–33–0

Van Ryzin et al. (1971) gave 10 or 40 mg per kg to female rats for two successive litters and found that only the higher dose reduced reproductive performance. Teratologic studies in rats and rabbits at 100 mg per kg gave no evidence of teratogenesis.

Van Ryzin, R.J.; Carson, S.E.; Hartman, H.A. and Trapold, J.H.: Animal safety evaluation studies on the antipsychotic phenothiazine mesoridazine. (abs) Toxicol. Appl. Pharmacol. 19:363, 1971.

1518 Mestranol CAS 72–33–3

The effect of oral contraceptives on women is discussed under Oral Contraceptives.

Saunders and Elton (1967) reported negative teratogenic studies in the rat and rabbit. There was no fertility problem in fetuses raised after intrauterine exposure. The rabbits received up to 0.1 mg per kg buccally or 0.25 mg per kg subcutaneously and the rats up to 0.1 mg per kg orally. Anorectal distances in the rat fetuses were not changed.

Saunders, F.J. and Elton, R.L.: Effects of ethynodiol diacetate and mestranol in rats and rabbits on conception, on the outcome of pregnancy and on the offspring. Toxicol. Appl. Pharmacol. 11:229–244, 1967.

1519 Metabisulfite–Potassium

Ema et al. (1985) fed rats up to 10 percent of this food additive on days 7–14 of gestation. Maternal and fetal weights were reduced at the highest dose but no increase in malformations occurred.

Ema, M.; Itami, T. and Kanoh, S.: Effect of metabisulfite on pregnant rats and their offspring. J. Food Hyg. Soc. Jpn. 26(5):454–459, 1985.

1520 Metahexamide CAS 565–33–3

Bariljak (1968) gave 2000 mg per kg to rats on the 9th and 10th days of gestation and produced a high incidence of congenital defects which included hydrocephalus, microcephaly, hydronephrosis and heart malformation.

Bariljak, I.R.: Comparison of antithyroidal and teratogenic activity of some hypoglycemic sulphanylamides. Probl. Endocrinol. (Russian) 14(6):89–94, 1968.

1521 Metaproterenol *Alupent™ Orciprenaline Sulfate 1–(3,5–Dihydroxyphenyl)–1–hydroxy–2–isopropylaminoethane* CAS 586–06–1

Iida et al. (1988) gavaged mice with up to 500 mg per kg on days 6–15, 11–13 or day 12 of gestation. At doses of

50 and 500 mg per kg, there was a small but significant increase in defects, mostly cleft palate. The cleft palate was associated with a threefold increase in maternal corticosterone. Matsuo et al. (1982) gave up to 50 mg per kg per day orally to rabbits on days 6 through 18 and found no adverse fetal effects. Hollingsworth et al. (1969) gave 50 mg per kg in the diet daily on days 7–16 of the rabbit and found no evidence of teratogenicity.

Hollingsworth, R.L.; Scott, W.J.; Woodard, M.W. and Woodard, G.: Fetal rabbit ductus arteriosus assessed in a teratologic study on isoproterenol and metaproterenol. (abs) Toxicol. Appl. Pharmacol. 14:641, 1969.

Iida, H.; Kast, A.; Tsunenari, Y. and Asakura, M.: Corticosterone induction of cleft palate in mice dosed with orciprenaline sulfate. Teratology 38:15–27, 1988.

Matsuo, A.; Kast, A. and Tsunenari, Y.: Teratology study with orciprenaline sulfate in rabbits. Arzneim. Forsch. Drug Research 32:808–810, 1982.

1522 Metasystox ™ S[2–(ethylsulfinyl) ethyl] 0,0–dimethylphophorothioate

Lenselink et al. (1992) administered 0.1 mg to chick embryos and produced ventricular septal defects and amelia.

Lenselik, D.R.; Midtling, J.E.; Lawrence, J.M. and Kolesari, G.L.: Teratogenic effects of s–[2–(ethylsulfinyl(ethyl] 0,0–dimethylphophorothioate (metasystox–r) in the stage 12 chick embryo (A). Teratology 45:500, 1992.

1523 Metavanadate

Gomez et al. (1992) gave mice up to 8 mg per kg per day intraperitoneally on day 6–15 of gestation and found reduced fetal weight at 4 and 8 mg levels. Cleft palate without other defects was increased at 8 mg per kg.

Gomez, M.; Sanchez, D.J. and Domingo, J.L.: Embryotoxic and teratogenic effects of intraperitoneally administered metavanadate in mice. Journal of Toxicology and Environmental Health 37:47–56, 1992.

1524 Metepa

1,1',1"-Phosphinylidynetris(2–methyl–aziridine) CAS 57–39–6

Gaines and Kimbrough (1966) injected rats with 30 mg per kg on the 12th day of pregnancy and found a high resorption rate and malformations in the offspring. Ectrodactyly occurred in 100 percent with some kinky tails and meningoceles.

Gaines, T.B. and Kimbrough, R.D.: The sterilizing, carcinogenic and teratogenic effects of metepa on rats. Bull. W.H.O. 34:317–320, 1966.

1525 Metformin *1,1–Dimethylbiguanide* CAS 657–24–9

Tuchmann–Duplessis and Mercier–Parot (1961) administered 500 to 1000 mg per kg to rats by tube and found anophthalmia and anencephaly in a few fetuses. Major malformations occurred in less than 0.5 percent of the fetuses suggesting that the material was not strongly teratogenic. Denno and Sadler (1994) exposed mouse embryos to 0.15 to 1,8 mg per ml of media and produced no malformations or growth alterations.

Denno, K.M. and Sadler, T.W.: Effects of the biguanide class of oral hypoglycemic agents on mouse embryogenesis. Teratology 49:260–266, 1994.

Tuchmann–Duplessis, H. and Mercier–Parot, L.: Repercussions sur la gestation et le developpement foetal du rat d'un hypoglyciemant, le chlorhydrate de N,N–dimethylbiguanide. C. R. Acad. Sci. (Paris) 253:321–323, 1961.

1526 Methacrylate Esters and Methacrylic Acid *Acrylic Acid*

Singh et al. (1972) administered six types of methacrylate esters to pregnant rats on days 5, 10 and 15 of gestation in doses up to one–third the acute intraperitoneal LD–50. The maximum doses used were 0.44, 0.40, 0.78, 0.46 and 0.82 ml per kg for the methyl, ethyl, n–butyl, isobutyl, isodecyl methacrylate respectively. Hemangiomas were increased at the highest doses as were resorptions. The fetal weight was reduced by treatment. Acrylic acid was injected in volumes of up to 0.0075 ml per kg and this was associated with resorptions and hemangiomas. Fetal mortality and an incidence of up to 16 percent malformations were reported but only in an abstract.

Merkle and Klimisch (1983) exposed rats to 25, 135 and 250 ppm of n–butyl acrylate for 6 hours daily on the 6th through 15th days of gestation. The two highest concentrations caused maternal toxicity and increased embryo lethality but no defects were found in surviving fetuses. McLaughlin et al. (1978) found no adverse effects in mice after exposure of methylmethacrylate vapor (1330 ppm) for two hours twice daily on days 6–15. Rogers et al. (1986) found concentrations of 1.5 microM of methacrylic acid to cause malformations and growth retardation in in–vitro rat embryo cultures. Murray et al. (1981) found no statistical increase in malformations when rats were exposed to ethyl acrylate 150 ppm for 6 hours daily on days 6 through 15. Edwards (1976) fed rats acrylamide in the diet, 400 ppm during pregnancy or l00 mg per kg intravenously on day 9. At 400 ppm the fetal weight was reduced but defects were not increased. No adverse effect was found after intravenous dosing. The concentration in the fetal blood was the same as in adult males treated with the same doses.

Solomon et al. (1993) exposed rate to methyl methacrylate at doses up to 2,028 ppm for 6 hours on days 6–15 and found no adverse reproductive effects even at maternally toxic doses.

Edwards, P.M.: The insensitivity of the developing rat fetus to the toxic effects of acrylamide. Chem. Biol. Interactions 12:13–18, 1976.

McLaughlin, R.E.; Reger, S.I.; Barkalow, B.S.; Allen, M.S. and Difazio, C.A.: Methylmethacrylate: A study of teratogenicity and fetal toxicity of the vapor in the mouse. J. Bone Joint Surg. 60:355–358, 1978.

Merkle, J. and Klimisch, H–J.: N–butyl acrylate: Prenatal inhalation toxicity in the rat. Fund. Appl. Toxicol. 3:443–447, 1983.

Murray, J.; Miller, R.; Deacon, M.; Hanley, T.R., Jr.; Hayes, W.; Rao, K. and John, J.: Teratological evaluation of inhaled ethyl acrylate in rats. Toxicol. Appl. Pharmacol. 60:106–111, 1981.

Rogers, J.G.; Greenaway, J.C.; Mirkes, P.E. and Shepard, T.H.: Methacrylic acid as a teratogen in rat embryo culture. Teratology 33:113–117, 1986.

Singh, A.R.; Lawrence, W.H. and Autian, J.: Embryo-fetal toxicity and teratogenic effects of a group of methacrylate esters in rats. J. Dent. Res. 51:1632–1638, 1972.

Solomon, H.M.; McLaughlin, J.E.; Swenson, R.E.; Hagan, J.V.; Wanner, F.J.; O'Hara, G.P. and Krivanek, N.D.: Methyl methacrylate: inhalation developmental toxicity study in rats. Teratology 48:115–125, 1993.

1527 Methadone

D–1–6–Dimethylamino–4,4–diphenyl–3–heptanone Hydrochloride CAS 76–99–3

Briggs et al. (1983) stated that the offspring of methadone users do not have an increased congenital defect rate. They discussed the withdrawal and low birth weight problems.

Markham et al. (1971) gave pregnant rats and rabbits up to 40 mg per kg during days 6 through 15 and 6 through 18, respectively and detected no drug–related defects in the offspring. Geber and Schramm (1969) injected guinea pigs on day 8 of gestation and found some defects in the embryos examined on the 12th day. Jurand (1973) injected subcutaneously 22 to 24 mg per kg into pregnant mice on the 9th day and produced exencephaly in 11 percent of the surviving 13 day embryos.

Geber and Schramm (1975) produced CNS defects in hamsters by giving 67 or more mg per kg subcutaneously on day 8. Nalorphine and other antagonists blocked the teratogenic action. Lichtblau and Sparber (1984) critically analyzed the results of animal experiments with this agent. Hutchings et al. (**1993**) administered 10 mg per kg by minipump and found increased startle reflex in the offspring who were cross-fostered to untreated mothers.

Briggs, G.G.; Bodendorfer, T.W.; Freeman, R.K. and Yaffe, S.J.: Drugs in Pregnancy and Lactation. Williams and Wilkins, Baltimore, 1983.

Geber, W.F. and Schramm, L.C.: Congenital malformations of the central nervous system produced by narcotic analgesics in the hamster. Am. J. Obstet. Gynecol. 123:705–713, 1975.

Geber, W.F. and Schramm, L.C.: Comparative teratogenicity of morphine, heroin and methadone in the hamster. (abs) Pharmacologist 11:248 only, 1969.

Hutchings, D.E.; Zmitrovich, A.C.; Brake, S.C.; Church, S.H. and Malowany, D.: Prenatal administration of methadone in the rat inccreases offspring acoustic startle amplitude at age 3 weeks. Neurotoxicology and Teratology 15:157–164, 1993.

Jurand, A.: Teratogenic activity of methadone hydrochloride in mouse and chick embryos. J. Embryol. Exp. Morph. 30:449–458, 1973.

Lichtblau, L. and Sparber, S.B.: Opioids and development: A perspective on experimental models and methods. Neurobehav. Toxicol. Teratol. 6:3–8, 1984.

Markham, J.K.; Emmerson, J.L. and Owen, N.V.: Teratogenicity studies of methadone HCl in rats and rabbits. Nature 233:342–343, 1971.

1528 Methallibure

1–α–Methyl–allyl–6–methyldithiourea CAS 926–93–2

King (1969) fed sows 100 mg daily for 20 days starting on the 29th or 30th day of gestation and found nearly all of the piglets to have contractures of the distal extremities with distorted mandibles and cranial bones. Treatment after the 49th day had no effect. Low (1972) extensively reviewed the pharmacology of this compound including its teratogenicity. Schafer et al. (1973) found that this pituitary inhibitor, when given to pigs between the 30th to 50th day of gestation in doses of 100 mg per day, produced alopecia, cranial bone thickening, dysplasia of the renal cortex and some musculo–skeletal defects in the offspring.

King, G.J.: Deformities in piglets following administration of methallibure during specific stages of gestation. J. Reprod. Fertil. 20:551–553, 1969.

Low, O.: Chemie, Pharmakologie und Anwendung des 1–methylthiocarbamoyl–2–(1–methylallyl) thiocarbamoylhydrazines. Arch. Exp. Veterinaermed. 58: 883–938, 1972.

Schafer, J.H.; Christensen, R.K.; Teaque, H.S. and Grifo, A.P.: Effects of methallibure on early pregnancy in the swine. J. Anim. Sci. 36:722–725, 1973.

1529 Methandriol *Androstenediol CAS 21526–76–5*

Jost (1953) found masculinization of female rat fetuses when the diprorionate of methyl androstenediol was given during the last days of gestation. The diprorionate of androstenediol had less effect.

Jost, A.: Intersexualite foetale provoquee par le methyl-androstenediol chez le rat. C.R. Soc. Biol. (Paris) 147:1920–1933, 1953.

1530 Methane *Natural Gas CAS 74–82–8*

Kato (1958) exposed pregnant mice on the 8th day for one hour to 5 to 8 percent concentration of fuel–gas. In addition to 85 percent methane, most natural gases contain small amounts of ethane, propane and butane. Abnormalities of the fetal brains were found to result in brain hernia and hydrocephalus.

Kato, T.: Embryonic abnormalities of the central nervous system caused by fuel–gas inhalation of the mother animal. Folia Psychiatr. Neurol. Jpn. 11:301–307, 1958.

1531 Methanesulphonate Diesters *Alkane Sulphonates*

Hemsworth (1968) administered various amounts (15–100 mg per kg) intraperitoneally to rats on day 13. Limb defects were found when the side chain had one or four carbons.

Hemsworth, B.N.: Embryopathies in the rat due to alkane sulphonates. J. Reprod. Fertil. 17:325–334, 1968.

1532 1–Methane–sulphonyl–3–(1–methyl–5–nitro–1H–imidazole–2,yl)–2–imidazolidinone

This new amoebecide and trichomonicide was studied in rats by Rao and Bhat (1983). Doses up to 600 mg per kg orally on days 6–15 produced no adverse fetal effects.

Rao, R. and Bhat, N.: Evaluation of the teratogenic potential in Ciba–Geigy GO 10213 a new nitroimidazole derivative, 1–methane–sulphonyl–3–(1–methyl–5–nitro–1H–imidazole–2–yl) 2–imidazolidinone, an amoebicide trichomonicide and giardicide, in rats. Toxicology 29:157–161, 1983.

1533 Methanol CAS 67–56–1

Nelson et al. (1985) exposed rats during their entire gestation for seven hours daily to 5,000, 10,000 and 20,000 ppm. Significant increases in defects occurred at the highest level. The defects involved the skeletal, cardiac and urinary system. Infurna and Weiss (1986) added 2 percent methanol to the drinking water of rats on days 15–17 or 17–19. The average intake was 2.5 gm per kg. Time of initiation of suckling and time to locate nesting material was increased in the exposed offspring. Rogers et al. (**1993**) exposed mice by inhalation to up to 10,000 ppm for 7 hours daily on days 6–15 of gestation. Increases in exencephaly and cleft palate were found at the 5,000 ppm level. No fetal effect was found at 1,000 ppm. Bolon et al. (**1994**) exposed mice to a high concentration of methanol (15,000 ppm) for 6 hours daily on days 7–9 and found neurol tube defects and morphologic changes in the cortex of fetuses. Youssef et al. (**1991**) gavaged rats on day 10 with 1.3; 2.6 or 5.2 ml per kg and found anomalies at all levels (undescended testes and eye defects mainly). Andrews et al. (**1991**) in an invitro culture system of rat embryos showed embryotoxicity started at 8,000 microgm per ml of media.

Andrews, J.E.; Ebron–McCoy, M. and Rogers, J.M.: Embryotoxic effects of methanol in whole embryo culture. (Abs) Teratology 43:461, 1991.

Bolon, B.; Welsch, F. and Morgan, K.T.: Methanol induced neural tube defects in mice: Pathogenesis during neuralation. Teratology 49:497–517, 1994.

Infurna, R. and Weiss, B.: Neonatal behavioral toxicity in rats following prenatal exposure to methanol. Teratology 33:259–265, 1986.

Nelson, B.K.; Brightwell, W.S.; Mackensie, D.R.; Khan, A.; Burg, J.R.; Weigel, W.W. and Goad, P.T.: Teratological assessment of methanol and ethanol at high inhalation levels in rats.

Rogers, J.M.; Mole, L.M.; Chernoff, N.; Barbee, B.D.; Turner, C.I.; Logsdon, T.R. and Kavlock, R.J.: The developmental toxicity of inhaled methanol in the cd–1 mouse, with quantitative dose–response modeling for estimation of benchmark doses. Teratology 47:175–186, 1993.

Youssef, A.F.; Baggs, R.B.; Weiss, B. and Miller, R.K.: Methanol teratogenicity in pregnant Long Evans rats. (Abs) Teratology 43:467, 1991.

1534 Methaqualone CAS 72–44–6

McColl et al. (1963) fed rats a diet containing 0.8 percent methaqualone (a sedative) and found double vertebral centra and extra lumbar ribs in some of the offspring. Bough et al. (1963) found no teratogenicity in rabbits given 200 mg per kg orally from days 1 through 29 and in rats given 100 mg per kg from day 1 through day 20.

Bough, R.G.; Gurd, M.R.; Hall, J.E. and Lessel, B.: Effect of methaqualone hydrochloride in pregnant rabbits and rats. Nature 200:656–657, 1963.

McColl, J.D.; Globus, M. and Robinson, S.: Drug induced skeletal malformations in the rat. Experientia 19:183–184, 1963.

1535 Methazolamide CAS 554–57–4

Discussed under Sulphonamides

1536 Methenamine CAS 100–97–0

This chemotherapeutic agent is used for urinary tract infections. Heinonen et al. (1976) found no increase in defects among the offspring of 49 women treated in the first four lunar months. A small but significant increase in anomalies was found among 299 women treated at anytime in pregnancy. There were twelve congenital defects but the types were not given.

Hurni and Ohder (1972) treated beagle dogs with up to 1250 ppm in the diet on days 4–56 and found no increase in malformations.

Heinonen, O.P.; Slone, D. and Shapiro, S.: Birth Defects and Drugs in Pregnancy. Publishing Sciences Group Inc., Littleton, Mass., 1977.

Hurni, H. and Ohder, H.: Reproduction study with formaldehyde and hexamethylenetetramine in beagle dogs. Food Cosmet. Toxicol. 11:459–462, 1973.

1537 Methergine

Sommer and Buchanan (1955) injected rats intraperitoneally with 0.5 mg twice daily from the 12th day until the 21st, after vaginal sperm were found. Defects were not studied but the newborn fetuses were not reduced in weight.

Sommer, A.F. and Buchanan, A.R.: Effects of ergot alkaloids on pregnancy and lactation in the albino rat. Am. J. Physiol. 180:296–300, 1955.

1538 Methimazole *Tapazole*™ CAS 60–56–0

Milham and Elledge (1972) reported that ulcerlike midline defects of the scalp occurred in the offspring of eleven mothers and that three of these mothers were under treatment for hyperthyroidism with methimazole. Mujtaba and Burrow (1975) reported scalp defects in two of five infants exposed to methimazole. Twenty infants exposed to propylthiouracil did not have scalp defects. Milham (1985) added several other cases where the association was reported to him privately. Japanese workers (Momotani et al., 1984) reported finding no scalp lesions among 177 women treated with methimazole during pregnancy. Van Dijke et al. (1987) also raised questions about the association and report 25 exposed pregnancies with no scalp defects recorded in the offspring.

Martinez–Frias et al. (**1992**) reported an increase in congenital aplasia cutis in Spain and suggested that it was associated with methimazole addition to animal feed. The methimazole was added along with clenbuterol a beta–agonist to increase the animals' weight. Their new cases were predominantly from regions where poisoning from clenbuterol was reported.

Mujtaba and Burrow (**1975**) followed 26 pregnancies treated with methimazole or propylthiouracil. Four had goiters and of these 3 were thryotoxic. Burrow (**1985**) has reviewed the management of hyperthyroidism during pregnancy. Fetal goiter and hypothyroidism may occur from the antithyroid effect of high doses of thioamides. Greenberg (**1987**) reported a case of choanal atresia and athelia in an exposed infant.

Rice et al. (1987) gave 0.1 mg per ml of drinking water to the mouse dam on day 16 through day 10 postpartum and produced developmental delays in the offspring. Brain weights on day 120 were not reduced as compared to controls. Stanisstreet et al. (1990) added the agent to in vitro rat cultures and found abnormal development at levels higher than those attained with human therapy.

Burrow, G.N.: The management of thyrotoxicosis in pregnancy. N. Engl. J. Med. 313:562–565, 1985.

Greenberg, F.: Brief clinical report: choanal artresia and athelia: methimazole teratogenicity or a new syndrome? Am. J. Med. Genet. 28:931–934, 1987.

Martinez–Frias, M.L.; Cereijo, A.; Rodriguez–Pinilla, E. and Urioste, M.: Methimazole in animal feed and congenital aplasia cutis. The Lancet 339:742–743, 1992.

Milham, S.: Scalp defects in infants of mothers treated for hyperthyroidism with methimazole or carbimazole during pregnancy. Teratology 32:321 only, 1985.

Milham, S. and Elledge, W.: Maternal methimazole and congenital defects in children (letter). Teratology 5:125 only, 1972.

Momotani, N.; Ito, K.; Hamada, N.; Ban, Y.; Nishikawa, Y. and Mimura, T.: Maternal hyperthyroidism and congenital malformations in the offspring. Clinical Endocrinology 20:695–700, 1984.

Mujtaba, Q. and Burrow, G.N.: Treatment of hyperthyroidism in pregnancy with propylthiouracil and methimazole. Obstet. Gynecol. 46:282–286, 1975.

Mujtaba, Q. and Burrow, G.N.: Treatment of hyperthyroidism in pregnancy with propylthiouracil and methimazole. Obstet. Gunecol. 46:282–286, 1975.

Rice, S.A.; Millan, D.P. and West, J.A.: The behavioral effects of perinatal methimazole administration in Swiss Webster mice. Fund. Appl. Toxicol. 8:531–540, 1987.

Stanisstreet, M.; Herbert, L.C. and Pharoah, P.O.D.: Effects of thyroid antagonists on rat embryos cultured in vitro. Teratology 41:721–729, 1990.

Van Dijke, C.P.; Heydendael, R.J. and De Kleine, M.J.: Methimazole, carbimazole and congenital skin defects. Annals of Internal Medicine 106(1):60-61, 1987.

1539 Methionine

Viau and Leathem (1973) fed pregnant rats four percent

of their diet as this amino acid and found no adverse fetal effects as compared to pair–fed controls.

Fujinaga and Badaen (**1994**) found in whole embryo culture that methionine prevented the maldevelopment produced by nitrous oxide.

Fujinaga, M. and Baden, J.M.: Methionine prevents nitrous oxide–induced teratogenicity in rat embryos grown in culture. Anesthesiology 81:184–189, 1994.

Viau, A.T. and Leathem, J.H.: Excess dietary methionine and pregnancy in the rat. J. Reprod. Fertil. 33:109–111, 1973.

1540 Methionine Deficiency

Coelho and Klein (**1990**) studied neural tube closure in rat embryos grown in methionine–deficient media. Failure of closure was associated with lack of elevation and rounding of the neural folds while the surface topography of the neuroepithelium was unchanged.

Ferrari et al. (**1986**) using sera from 6 recurrent aborting women and two women with infertility cultured rat embryos and found exencephaly and anopthalmia after 48 hours growth. The sera could be corrected by supplementing directly or via the mothers diet with amino acids including methionine. Fecal free amino acids were increased in the subjects. Flynn et al. (**1991**) in a double blind study of 11 women with reproductive failure found only one that supported normal in vitro embryonic growth. In the 10 specimens which did not support growth seven were improved by addition of amino acids and vitamins. A low incidence of neural tube defect was found when embryos were grown in sera from healthy laboratory workers. Essein (**1992**) showed that injection of methionine (70 mg/kg) to Axd mice reduced the expected neural tube defects from 29% to 17%.

Scialli et al. (**1993**) studied sera from infertile, habitual aborters or normal women by growing rat embryos in with or without added vitamins and aminoacids (Methionine 50 mg per liter). Four of 10 controls had embryotoxic sera while 13 of 27 infertile and 5 of 15 habitual aborters had serum which did not maintain normal rat embryo growth. The authors cite two other studies of normal women of whom 27 and 30% had embryotoxic serum.

Coelho, C.N.D. and Klein, N.W.: Methionine and neural tube closure in cultured rat embryos: morphological and biochemical analyses. Teratology 42:437–451, 1990.

Essien, F.B.: Maternal methionine supplementation promotes the remediation of axial defects in axd mouse neural tube mutants. Teratology 45:205–212, 1992.

Ferrari, D.; Klein, N.; Fogelsonger, R.; Lammi–Keefe, C.; Marella, L.; Hillman, R.; Maier, O.D.; Olsen, P. and Gross, P.: Use of rat embryo cultures to identify amino acid absorption defects in association with human reproductive failures. (Abstract) Teratology 33:84C, 1986.

Flynn, T.J.; Scialli, A.R.; and Gibson, R.R.: Cultured organogenesis–staged rat embryos as biomarkers for nutritional factors in human reproductive failure. (Abstract) Teratology 43:468, 1991.

Scialli, A.R.; Flynn, T.J. and Gibson, R.R.: Rat embryo culture to detect nutritional deficiency in women with poor reproductive histories. Reproductive Toxicology 7:581–587, 1993.

1541 Methocarbamol

This muscle relaxant was used to treat 22 women in the first four lunar months and one child with inguinal hernia was reported (Heinonen et al., 1977). Hall and Reed (1982) reported an infant with arthrogryposis. The mother took 750 mg two or three times daily. Frontal bossing and a midline hemangioma were present.

Hall, J.G. and Reed, S.D.: Teratogens associated with congenital contractures in humans and in animals. Teratology 25:173–191, 1982.

Heinonen, O.P.; Slone, D. and Shapiro, S.: Birth Defects and Drugs in Pregnancy. Publishing Sciences Group Inc., Littleton, Mass., 1977.

1542 Methomyl *Methyl N–[[(methylamino)carbonyl] oxy] ethanimidothioate* CAS 16752–77–5

Kaplan and Sherman (1977) fed rabbits 0, 50 and 100 ppm in the diet during days 8 through 16 of pregnancy and found no adverse effects in the fetuses on days 29 and 30.

Kaplan, M.A. and Sherman, H.: Toxicity studies with methyl N–[[methylamino)carbonyl]oxy]–ethanimidothioate. Toxicol. Appl. Pharmacol. 40:1–17, 1977.

1543 Methotrexate *Amethopterin Methylaminopterin* CAS 59–05–2

This methyl derivative of aminopterin is a folic acid antagonist. Milunsky et al. (1968) reported defects in a child whose mother ingested 2.5 mg daily for 5 days between the 8th to 10th weeks. The defects included absence of the frontal bones, premature craniosynostosis, rib defects and absence of digits (see Aminopterin and Folic Acid Deficiency).

Powell and Ekert (1971) reported a similar child from the pregnancy of a mother who received 5 mg daily during the first two months for treatment of psoriasis. Sieber and Adamson (1975), summarizing the human experience, reported three women treated after the first 60 days gave birth to newborns without defects. Feldkamp and Carey (1993) reviewed 20 first trimester exposures to methotrexate. Three of the 17 live births had the aminopterin syndrome and all were exposed during the 6–8 week period after conception. They state that a dose of over 10 mg per week is probably necessary in order to produce the syndrome.

Adams et al. (1961) reported that 6.5 mg per kg in early pregnancy caused no visible effects on 6.5–day rabbit embryos. Berry (1971) used the compound in pregnant rats to study DNA inhibition in the fetus and embryo. Skalko and Gold (1974) reported on the teratogenicity in mice and found no defects at 10 mg per kg but at 25 and 50 mg per kg exencephaly, omphalocele, ectrodactyly and cleft palate were found. Giving 30 mg per kg intravenously on days 29 through 32 in the monkey caused transitory embryonic growth retardation but was not teratogenic (Wilson

et al., 1979). DeSesso and Goeringer (1992) described intracellular swelling and nuclear changes in mesenchyme of rabbit embryonic limb buds and these changes were reduced by concurrent treatment with 1–(p–tosyl)–3,4,4–trimethylimidazolidine (tti) a reestablisher of one carbon metabolism. TTI itself was not teratogenic.

Adams, C.E.; Hay, M.F. and Lutwak–Mann, C.: The action of various agents on the rabbit embryo. J. Embryol. Exp. Morphol. 9:468–491, 1961.

Berry, C.L.: Transient inhibition of DNA synthesis by methotrexate in the rat embryo and fetus. J. Embryol. Exp. Morphol. 26:469–474, 1971.

DeSesso, J.M. and Goeringer, G.C.: Methotrexate–induced developmental toxicity in rabbits is ameliorated by 1–(p–tosyl)–3,4,4–trimethylimidazolidine, a functional analog for tetrahydrofolate–mediated one–carbon transfer. Teratology 45:271–283, 1992.

Feldkamp, M. and Carey, J.C.: Clinical teratology counseling and consultation case report: low dose methotrexate exposure in the early weeks of pregnancy. Teratology 47:533–539, 1993.

Milunsky, A.; Graef, J.W. and Gaynor, M.F.: Methotrexate–induced congenital malformations with a review of the literature. J. Pediatr. 72:790–795, 1968.

Powell, H.R. and Ekert, H.: Methotrexate–induced congenital malformations. Med. J. Aust. 2:1076–1077, 1971.

Sieber, S.M. and Adamson, R.H.: Toxicity of antineoplastic agents. In: Advances in Cancer Research, G. Klein and S. Weinhouse (eds.), Academic Press, New York, 22:57–155, 1975.

Skalko, R.G. and Gold, M.P.: Teratogenicity of methotrexate in mice. Teratology 9:159–164, 1974.

Wilson, J.G.; Scott, W.J.; Ritter, E.J. and Fradkin, R.: Comparative distribution and embryotoxicity of methotrexate in pregnant rats and rhesus monkeys. Teratology 19:71–80, 1979.

1544 Methotrimeprazine CAS 60–99–1

Discussed under Phenothiazines

1545 Methoxyacetic Acid CAS 625–45–6

Brown et al. (1984) gave rats 0.1 to 2.5 milliM per kg intraperitoneally on day 8, 10, 12 or 14. Skeletal defects and hydrocephalus were found after all doses. Rawlings et al. (1985) studied the effect of direct exposure of 2 and 5 milliM concentrations of day 9.5 rat embryos in vitro and found irregularlity of the neural suture line and growth retardation.

Brown, N.A.; Holt, D. and Webb, M.: The teratogenicity of methoxyacetic acid in the rat. Toxicology Letters 22:93–100, 1984.

Rawlings, S.J.; Shuker, D.E.G.; Webb, M. and Brown, N.A.: The teratogenic potential of alkoxy acids in postimplantation rat embryo culture: Structure–activity relationships. Toxicology Letters 28:49–58, 1985.

1546 Methoxsalen *Psoralen*

Stern et al. (1991) reported 86 pregnancies where the mother or father used methoxsalen plus ultraviolet therapy around the time of fertilization and 326 pregnancies during which the treatment was used. No increase in adverse pregnancy outcome or malformations was found.

Stern, R.S. and Lange, R.: Outcomes of pregnancies among women and partners of men with a history of exposure to methoxsalen photochemotherapy (PUVA) for the treatment of psoriasis. Arch. Dermatol. 127:347–350, 1991.

1547 Methoxychlor CAS 72–43–5

Khera et al. (1978) gavage fed pregnant rats with 50 to 400 mg per kg on days 6 through 15. At 200 and 400 mg per kg, maternal and fetal toxicity occurred and many rat fetuses had wavy ribs.

Khera, K.S.; Whalen, C. and Trivett, G.: Teratogenicity studies on linuron, malathion, and methoxychlor in rats. Toxicol. Appl. Pharmacol. 45:435–444, 1978.

1548 Methoxyflurane

Wharton et al. (1980) exposed mice for 4 hours daily to 2, 60 or 2,000 ppm on days 6–15 of pregnancy. No adverse effects were seen except at 2,000 ppm. At this highest dose, fetal weight and ossification were reduced and a delay in renal maturation was found, but no increase in malformations.

Wharton, R.S.; Sievenpiper, T.S. and Mazze, R.I.: Developmental toxicity of methoxyflurane in mice. Anesth. Analg. 59:421–425, 1980.

1549 3–(o-Methoxyphenoxy)-2-hydroxypropyl Nicotinate CAS 20026–81–7

Discussed under Niacin Acid

1550 1-[2-[p–[α–(p–Methoxyphenyl)–β-nitrostyryl] phenoxy]ethyl]pyrrolidine Monocitrate *CI628* CAS 5863–35–4

Schardein et al. (1973) studied this estrogen antagonist in dogs treated on days 1–15 of gestation with 0.125 or 0.25 mg per kg. At the higher dose, 3 of 5 pups had major defects including cleft palate, thoracogastroschisis, microphthalmia, tail defects, persistent cloaca, diaphragmatic hernia and other skeletal defects. Three pups were stillborn. Doses of 0.5 to 5 mg prevented pregnancy.

Schardein, J.L.; Reutner, T.F.; Fitzgerald, J.E. and Kurtz, S.M.: Canine teratogenesis with an estrogen antagonist. Teratology 7:199–204, 1973.

1551 2–Methoxypropylacetate–1

Merkle et al. (1987) exposed rats and rabbits by inhalation to up to 14.9 mg and 3.0 mg per liter, respectively, six hours daily during organogenesis. Some vertebral anomalies were increased at the highest dose in the rats. In the rabbit studies, skeletal defects, cleft palate, heart and kidney defects were increased at 550 ppm (3 mg per liter).

Merkle, J.; Klimisch, H.J. and Jackh, R.: Prenatal toxicity of 2–methoxypropylacetate–1 in rats and rabbits. Fund. Appl. Toxicol. 8:71–79, 1987.

1552 5–Methoxypsoralen CAS 484–20–8

Herold et al. (1981) fed up to 560 mg per kg to rats on days 6–15 and found no increase in fetal anomalies. In rabbits receiving 70 or 560 mg per kg on days 7–18, malformations were increased but the types were not given.

Herold, H.; Berbey, B.; Angignare, D. and LeDuc, R.: Toxicological study of the compound 5–methoxy–psoralen (5–MOP). In: Psoralens in Cosmetics and Dermatology, J. Cahn, P. Forlot, C. Grupper, A. Maybeck and F. Urbach (eds.), New York, Pergamon Press, 303–309, 1981.

1553 Methyclothiazide CAS 135–07–9

Heinonen et al. (1977) found five infants with persistent ductus arteriosus among 942 women who took the medication sometime during pregnancy. This was 6.2 times the expected rate. Only three of the 942 women took the agent in the first four lunar months.

Heinonen, O.P.; Slone, D. and Shapiro, S.: Birth Defects and Drugs in Pregnancy. Publishing Sciences Group Inc., Littleton, Mass., 1977.

1554 M–Methyl Acetamide

Discussed under Acetamide

1555 1–Methyl–Acetylhydrazine CAS 3530–13–0

Discussed under 1–Methyl–formylhydrazine

1556 1–Methyl–2–p–allophanoyl–benzyl–hydrazine

Mercier-Parot and Tuchmann-Duplessis (1968) gave the bromhydrate form of this compound orally to rats at various periods from the 8th through the 14th days. Daily doses of as low as 5 mg per kg resulted in fetal defects localized to the eye and extremities. Earlier treatment was associated with microphthalmia and anophthalmia while treatment after the 12th day produced limb defects. The same authors (1969) extended this study by showing that the compound was teratogenic in mice and rabbits.

Mercier–Parot, L. and Tuchmann–Duplessis, H.: Action d'une methyl–hydrazine, le bromhydrate de 1–methyl–2–p–allophanoyl–benzyl–hydrazine sur la morphogenese du rat. C. R. Acad. Sci. (d) (Paris) 267:444–447, 1968.

Mercier–Parot, L. and Tuchmann–Duplessis, H.: Mise en evidence chez deux autres rongeurs: La souris et le lapin, de l'action teratogene du bromhydrate de 1–methyl–2–p–allophanoyl–benzyl–hydrazine. C. R. Acad. Sci. (d) (Paris) 268:1088–1091, 1969.

1557 Methylamine Hydrochloride

Miller (1972) administered 20 ml per kg by intracardiac injection to rats on day 13 and found no gross malformations.

Miller, L.R.: Teratogenicity of degradation products of 1–methyl–1–nitrosourea. Anat. Record 169:379–380, 1972.

1558 Methylamphetamine CAS 29088–49–1

Discussed under Amphetamine

1559 Methyl Arsenate

Ancel (1946) used 1.0 mg of the disodium salt dropped onto the developing 26 hour chick to produce spina bifida.

Ancel, P.: Reserche experimentale sur le spina bifida. Arch. Anat. Microscop. Morphol. Exp. 36:45–68, 1946.

1560 Methylazoxymethanol *Cycasin* CAS 590–96–5

This chemical is the glycone of cycasin which occurs in the seeds of the tropical plants Cycas circinalis and C. revoluta. Spatz et al. (1967) injected 25 mg per kg into hamsters on the 8th gestational day and produced fetal defects including exencephaly, spina bifida, craniorachischisis and oligodactyly. This compound is also mutagenic and carcinogenic. Spatz (1969) found neoplasms of the jejunum and brain in the offspring of mothers fed three percent cycasin. Haddad et al. (1975) reported postnatal studies of rats which were treated with this compound. Rodier et al. (**1991**) presented data which suggested the effect of this neurotoxic agent was via hypopituitary axis change. Yamamoto and Tanimura (**1989**) found decreased motor behavior in rats exposed in utero to up to 30 mg per kg on day 13 of gestation.

Haddad, R.; Rabe, A. and Dumas, R.: Functional consequences of chemically induced cerebellar dysplasia in the rat. (abs) Teratology 11:20A only, 1975.

Rodier, P.M.; Kates, B.; White, W.A. and Muhs, A.: Effects of prenatal exposure to methylazoxymethanol (MAM) on brain weight, hypothalamic cell number, pituitary structure, and postnatal growth in the rat. Teratology 43:241–251, 1991.

Spatz, M.: Toxic and carcinogenic alkylating agents from cycads. Ann. N.Y. Acad. Sci. 163:848–855, 1969.

Spatz, M.; Dougherty, W.J. and Smith, D.W.E.: Teratogenic effects of methylazoxymethanol. Proc. Soc. Exp. Biol. Med. 124:467–478, 1967.

Yamamoto, Y. and Tanimura, T.: Effect of prenatal methylazoxymethanol acetate exposure on the motor behavior of the rat offspring. Cong. Anom. 29:51–58, 1989.

1561 Methyl Benzimidazole Carbamate

Discussed under Benomyl

1562 1–Methyl–2–benzyl–hydrazine CAS 10309–79–2

Discussed under 1,2–Diethylhydrazine

1563 Methyl Bromide CAS 74–83–9

Hardin et al. (1981) exposed rats and rabbits to 20 or 70 ppm for six hours daily on days 1–15 or 1–19, respectively. No teratogenicity was found but 96 percent of the mother rabbits died at the higher doses.

Hardin, B.D.; Bond, G.P.; Sikov, M.R.; Andrew, F.D.; Beliles, R.P. and Niemeier, R.W.: Testing of selected workplace chemicals for teratogenic potential. Scand. J. Work Environ. Health 7(4):66–75, 1981.

1564 Methyl t–Butyl Ether

Conaway et al. (1985) exposed mice and rats to up to 2,500 ppm during periods of organogenesis. No adverse effects in the fetuses were found.

Conaway, C.C.; Schroeder, R.E. and Snyder, N.K.: Teratology evaluation of methyl tertiary butyl ether in rats and mice. J. Toxicol. Environ. Health 16:797–809, 1985.

1565 3–Methylbutyric Acid

Dawson (**1991**) using a frog embryo teratogenesis assay found that 50% of the survivors were malformed when exposure was to 575 mg per liter. Microcephaly and abnormal gut coiling were the most common types of defects found.

Dawson, D.A.: Additive incidence of developmental malformation for xenopus embryos exposed to a mixture of ten aliphatic carboxylic acids. Teratology 44:531–546, 1991.

1566 Methyl Cellosolve *2–Methoxyethanol* CAS 109–86–4

Nelson et al. (1984) exposed rats to 50 or 100 ppm for six hours daily on days 7–15 of gestation. An increase in skeletal and cardiac malformations was reported. No embryotoxicity was found with 2(2–ethoxy ethoxy) ethanol at 100 ppm. 2–Butoxyethanol was toxic to the dams at 200 ppm but no increase in fetal defects was found. Welsch et al. (1987) postulated from metabolic pathway studies that abnormal macromolecules generated from anabolic reactions disrupt normal paw development. The tetrahydrofolic acid pathway may be central to this activity. Sleet et al. (**1988**) studied the pharmacokinetics in the mouse and found teratogenesis depended mainly on methoxyacetic acid production. Khenta. ra (**1993**)reported the pathologic changes seen in the mouse placenta. Terry et al. (**1994**) studied the timing and pharmacokinetics of the chemical in mice.

Scott et al. (1989) gavaged monkeys with up to 0.47 mM per kg (36 mg per kg) on days 20–45. All 8 pregnancies at the highest dose and 4 of 11 at 0.32 mM per kg aborted. One of the fetuses at the highest dose had a missing digit on each forelimb. Concentration of the material in the yolk sac was noted. Sleet et al. (**1987**) determined the critical periods of teratogenic exposure in the mouse. Feuston et al. (**1990**) applied it dermally to rats and found growth retardation resorptions and congenital heart and skeletal defects. Malformations occurred at 500 mg per kg and above applied on single days.

Feuston, M.H.; Kerstetter, S.L. and Wilson, P.D.: Teratogenicity of 2–methoxyethanol applied as a single dermal dose to rats. Fundamental and Applied Toxicology 15:448–456, 1990.

Khera, K.S.: Mouse placenta: hemodynamics in the main maternal vessel and histopathologic changes induced by 2–methoxyethanol and 2–methoxyacetic acid following maternal dosing. Teratology 47:299–310, 1993.

Nelson, B.K.; Setzer, J.V.; Brightwell, W.S.; Mathinos, P.R.; Kuczuk, M.H.; Weaver, T.E. and Goad, P.T.: Comparative inhalation teratogenicity of four glycol ether solvents and an amino derivative in rats. Environmental Health Perspectives 57: 261–271, 1984.

Scott, W.J.; Fradkin, R.; Wittfoht, W. and Nau, H.: Teratologic potential of 2–methoxyacetic acid, in non–human primates. Teratology 39:363–376, 1989.

Sleet, R.B.; Greene, J.A. and Welsch, F.: Tertogenicity and disposition of the glycol ether 2–methoxyethanol and their relationship in cd–1 mice. Approaches to Elucidate Mechanisms in Teratogenesis. Ed. Welsch, F. Hemisphere Publishing Corp., Washington, D.C. 1987. 99 33–56.

Sleet, R.B.; Greene, J.A. and Welsch, F.: The relationship of embryotoxicity to disposition of 2–methoxyethanol in mice. Toxicology and Appliedd Pharmacology

Terry, K.K.; Elswick, B.A.; Stedman, D.B. and Welsch, F.: Developental phase alters dosimetry–teratogenicity relationship–2–methoxyethanol in CD–1 mice. Teratology 49:218–227, 1994.

Welsch, F.; Sleet, R.B. and Greene, J.A.: Attenuation of 2-methoxyethanol and methoxyacetic acid–induced digit malformations in mice by simple physiological compounds: Implications for the role of further metabolism of methoxyacetic acid in developmental toxicity. J. Biochem. Toxicol. 2:225–240, 1987.

1567 Methyl Chloride *Chloromethane* CAS 74–87–3

Wolkowski–Tyl et al. (1983a,b) exposed rats and mice to 100, 500 and 1500 ppm during organogenesis using a six–hour exposure period each day. At 500 ppm a small but statistically significant increase in congenital right–sided heart defects was found in the mouse. No teratogenicity was found in the rat nor with 100 ppm in the mouse.

Hardin et al. (1980) exposed rate to 4500 ppm during the first seven days of gestation. No increase ind efects was found but frowth retardation was reported.

Hardin, B.D. and Manson, M.J.: Absence of dichloromethane teratogenicity with inhalation exposure in rats. Toxicol. Appl. Pharmacol. 52:22–28, 1980.

Wolkowski–Tyl, R.; Phelps, M. and Davis, J.K.: Structural teratogenicity evaluation of methyl chloride in rats and mice after inhalation exposure. Teratology 27:181–195, 1983a.

Wolkowski–Tyl, R.; Lawton, A.D.; Phelps, M. and Hamm, T.E.: Evaluation of heart malformations in $B_6C_3F_1$ mouse fetuses induced by in utero exposure to methyl chloride. Teratology 27:197–206, 1983b.

1568 1–Methyl–5–chloroindoline Methylbromide CAS 32179–45–6

Irikura et al. (1973) studied this parasympathomimetic drug in mice and rabbits using the oral and subcutaneous routes during active organogenesis. No teratogenic effects were seen with oral maximum doses of 30 mg per kg per day and subcutaneous maximal doses of 10 mg.

Irikura, T.; Suzuki, H. and Sugimoto, T.: Teratological study of 1–methyl–5–chloroindole methylbromide. (Japanese) Oyo Yakuri 7:1171–1180, 1973.

1569 Methylcholanthrene CAS 56–49–5

Savkur et al. (1961) injected 2.5 microgm into each embryonic site of the pregnant mouse on the 10th day and found a high incidence of tail defects and subcutaneous hemorrhages. The limited extent and description of the defects casts some doubt about the teratogenicity of this compound.

The incidence of the defects did not increase with dose and a persistence of defects in subsequent untreated litters was noted.

Tomatis et al. (1971) administered orally 8.4 mg of 3–methylcholanthrene to mice during their last week of pregnancy. They found that the treatment group had a three–fold increase in the incidence of tumors. The most common type was lymphoma and lung tumor. Khera and Iverson (1981) injected intraperitoneally 20 mg per kg on days 11, 12 and 13 of the gravid rat and produced no adverse effects in the fetuses.

Khera, K.S. and Iverson, F.: Effects of pretreatment with SKF–525A, N–methyl–2–thioimidazole, sodium phenobarbital, or methyl cholanthrene on ethylenethiourea-induced teratogenicity in rats. Teratology 24:131–137, 1981.

Savkur, L.D.; Batra, B.K. and Sridharan, B.N.: Effect of 20–methylcholanthrene on mouse embryos. II Strain C3–H (JAX). J. Reprod. Fertil. 2:374–380, 1961.

Tomatis, L.; Turusov, V.; Guibbert, D.; Duperray, B.; Malaveille, C. and Pacheco, H.: Transplacental carcinogenic effect of 3–methylcholanthrene in mice and its quantitation in fetal tissues. J. Natl. Cancer Inst. 47:645–651, 1971.

1570 Methyldigoxin

Discussed under Digoxin

1571 Methyldihydrotesterone

Schultz and Wilson (1974) administered 4 mg subcutaneously to rats on days 14 through 21 and found virilization of the female Wolffian ducts.

Schultz, F.M. and Wilson, J.D.: Virilization of the wolffian duct in the rat fetus by various androgens. Endocrinology 94:979–986, 1974.

1572 Methyldopa *Aldomet*™ CAS 555–30–6

Redman et al. (1976) carried out a controlled study in which 122 women with hypertension were treated with this drug. Only one malformation occurred (absent kidney with 2 umbilical vessels). A better pregnancy outcome was associated with the treatment. Sleet et al. (1987) studied pregnant rats and mice by gavaging up to 750 or 500 mg per kg respectively during organogenesis. At the highest doses there was maternal toxicity and a marginal increase in congenital defects.

Peck et al. (1985) studied multiple generation of mice and rats at 1000 and 100 mg per kg respectively and found no increase in congenital defects. Studies of the rabbit during organogenesis did not reveal an increase in congenital defects.

Peck, H.M.; Mattiss, P.A.; and Zawoiski, E.J.: The evaluation of drugs for their effects on reproduction and fetal development. Drug–Induced Diseases 2nd Symposium. Excerpta Medica Foundation, New York, pp 19–29, 1965.

Redman, C.W.G.; Beilin, L.J.; Bonnar, J. and Ounsted, M.K.: Fetal outcome in trial of antihypertensive treatment in. Lancet 2:753–756, 1976.

Sleet, R.B.; George, J.D. Price, C.J.; Marr, M.C.; Kimmel, C.A.; and Schwetz, B.A.: Conceptus development in mice and rats exposed to α–methyldopa. Toxicologist 7:1, 1987.

1573   3–O–Methyldopa

Kitchin and DiStefano (**1976**) gave 100 mg per kg orally to rats on days 1–21 and found hemorrhage in the brown fat of the neonates.

Kitchin, K.T. and DiStefano, V.: L–dopa and brown fat hemorrhage in the rat pup. Toxicology and Applied Pharmacology 38:251–263, 1976.

1574   α–Methyldopamine

Kitchin and DiStefano (**1976**) administered 100 mg per kg on gentational days 0–7 to rats and found hemorrhage into the brown fat in the newborn.

Kitchin, K.T. and DiStefano, V.: L–dopa and brown fat hemorrhage in the rat pup. Toxicology and Applied Pharmacology 38:251–263, 1976.

1575   2–2'–Methylenebis(4–ethyl–6–tert–buthylphenol)

This antioxidant was given orally by Tanaka et al. (**1989**) to rats on days 7–17 at doses of up to 750 mg per kg. At the highest dose the lethality rate in the dams and fetuses was slightly increased; otherwise there was no sign of fetal ill effects. In perinatal studies Tanaka et al. (**1989**, B) used similar treatment and found no malformations in spite of diarrhea which occurred in the dams at 187 and 375 mg per kg levels.

Tanaka, S.; Kawashima, K.; Nakaura, S.; Djajalaksana, S.; Huang, M. and Takanaka, A.: Studies on the teratogenic potential of 2,2'–methylenebis (4–ethyl–6–tert–buthylphenol) in rats. Eiseishikenjo Hokoku (Bulletin of National Institute of Hygenic Sciences) 107:51–55, 1989.

Tanaka, S.; Kawashima, K.; Usami, M.; Nakaura, S.; Kodama, Y. and Takanaka, A.: Studies on the teratogenic potential of 2,2'–methylenebis (4–methyl–6–tert–butlphenol) in rats. Eisenshikenjo Hokoku (Bulletin of National Institute of Hygenic Sciences) 108:52–57, 1990.

1576   Methylenebisacrylamide

Rutledge and Generoso (1989) studied semisterile male mice recovered from the offspring of male mice given five daily doses of 90 mg per kg before mating. All stocks had an increase in resorption moles. Mid and late gestation deaths were also increased. Three of the 24 stocks had increased defects including palatal clefts, abnormal eyes and exencephaly.

Rutledge, J.C. and Generoso, W.M.: Evaluating the risk of congenital anomalies arising in the offspring of mutagen induced balanced chromosomal translocations. Teratology 39:P60, 1989.

1577   N,N–Methylene–bis(2–amino–1,3,4–thiadiazole)
MATDA

Li et al. (1986) studied reproductive outcomes of 6,173 pregnancies where the pesticide had been applied to the ingested rice. The population was compared to 3,326 women who had pregnancies prior to the introduction of the pesticide in the same area. No significant differences were found for fetal loss, birth defects, sex ratio or birth weight.

In rats, one mg per kg by gavage was reported to be teratogenic (Wang and Zhong, 1981). It was also studied in the hamster by Mizutani et al. (1974).

Li, D–K.; Zhou, Q–D.; Qin, X–B.; Sun, R–M.; Zhu, X–L.; Cheng, H–J.; Wang, C–S.; He, J–P.; Qian, C.; Xue, S–Z. and Gu, X–Q.: An epidemiological study on the effect of N,N'–methylene–bis–(2–amino–1,3,4–thiadiazole) (MATDA) on outcomes of pregnancy. Teratology 33:289–297, 1986.

Mizutani, M.; Ihara, T. and Sugitani, T.: Protective effects of nicotinamide and tryptophan against the teratogenicity of N,N–methylene–bis–(2–amino–1,3,4–thiadiazole) in the hamster. (abs) Teratology 9:A28–A29, 1974.

Wang, F.Y. and Zhong, Y.: A primary study on the teratogenesis of thiadiazole compounds. Acta Pharmaceuticasinica 16:654–660, 1981.

1578   Methylene Blue CAS 61–73–4

Heinonen et al. (**1977**) reported on 46 women who were treated during pregnancy. There was no increase in malformations but only nine werre treated during the first trimester. Gillman et al. (1951) reported that this dye was not teratogenic in the rat.

Nicolini and Monni (**1990**) have reported that intraamniotic injection of methylene blue (10–30 mg) at 15–17 weeks was complicated in 7 twins with multiple intestinal atresia. In all, the atresia was discordant and in 3 the twin receiving the dye was shown to be effected. Two other series have found intestinal atresia in twins receiving intra–amniotic methylene blue (van der Pol et al., **1992** and Lancaster et al., **1992**).

Gillman, J.; Gilbert, C.; Spence, I. and Gillman, T.: A further report on congenital anomalies in the rat produced by trypan blue. S. Afr. J. Med. Sci. 16:125–135, 1951.

Heinonen, O.P.; Slone, D. and Shapiro, S.: Birth Defects and Drugs in Pregnancy. John Wrright Publishing Sciences Group, Inc., Littleton, Mass. 1977.

Lancaster, P.A.L.; Pedisich, E.L.; Fisher, C.C. and Robertson, R.D.: Intra–amniotic methylene blue and intestinal atresia in twins. J. Perinat. Med. 20(Suppl 1):262, 1992.

Nicolini, U. and Monni, G.: Intestinal obstruction in babies exposed in utero to methylene blue. Lancet 336:1258–1259, 1990.

van der Pol, J.G.; Wolf, H.; Boer, K.; Treffers, P.E. Leschot, N.J.: Hey, H.A. and Vos, A.: Jejunal atresia related to the use of methylene blue in genetic amniocentesis in twons. Br. J. Obstet. Gynaecol. 99:141–143, 1992.

1579   Methylene Chloride *Dichloromethane* CAS 75–09–2

Taskinen et al. (1986) found the odds ratio for abortion was 2.3 (p = 0.06) among pharmaceutical workers exposed to this solvent.

Hardin and Manson (1980) exposed rats to 4500 ppm for 6 hours per day before and during the first 17 days of gestation. Some fetal weight reduction occurred but no malformation increase was found. In the same treatment group as above behavioral studies were done (Bornschein et al., 1980). Postnatal growth, activity and avoidance learning were not impaired but behavioral habituation was more rapid in the exposed group.

Schwetz et al. (1975) exposed pregnant mice and rats to this vapor in concentrations which were twice the maximal allowable limit for human industrial exposure (1225 ppm). Both species were exposed for seven hour daily periods on days 6 through 15 of gestation. No fetal toxicity or teratogenicity was found.

Bornschein, R.L.; Hastings, L. and Manson, J.M.: Behavioral toxicity in the offspring of rats following maternal exposure to dichloromethane. Toxicol. Appl. Pharmacol. 52:29–37, 1980.

Hardin, B.D. and Manson, J.M.: Absence of dichloromethane teratogenicity with inhalation exposure in rats. Toxicol. Appl. Pharmacol. 52:22–28, 1980.

Schwetz, B.A.; Leong, B.K.J. and Gehring, P.J.: The effect of maternally inhaled trichoroethylene, perchloroethylene, methyl chloroform and methylene chloride on embryonal and fetal development in mice and rats. Toxicol. Appl. Pharmacol. 32:84–96, 1975.

Taskinen, H.; Lindbohm, M–L. and Hemminki, K.: Spontaneous abortions among women working in the pharmaceutical industry. Br. J. Ind. Med. 43:199–205, 1986.

1580 Methylenedioxymethamphetamine *MDMA*

Omer et al. (**1991**) gavaged rats with 2.5 or 10 mg per kg on alternate days from day 6–18. Postnatal behavioral changes were not found except for negaive geotaxis which was delayed in the female pups. Serotonin metabolism in the fesuses was not changed.

Omer, V.E.V.; Ali, S.F.; Holson, R.R.; Duhart, H.M.; Scalz, F.M. and Slikker, W.: Behavioral and neurochemical effects of prenatal methylenedioxymethamphetamine (MDMA) exposure in rats. Neurotoxicology and Teratology 13:13–20, 1991.

1581 4–Methyl–N–ethylamino Phenol Sulfate

Picciano et al. (1984) treated rats by gavage with up to 1200 mg per kg on days 6–15. No adverse effects were found. This chemical is used as a hair coloring product.

Picciano, J.C.; Morris, W.E. and Wolf, B.A.: Evaluation of the teratogenic potential of the oxidative dyes 6–chloro–4–nitro–2–aminophenol and o–chlor–p–phenylenediamine. Food Chem. Toxicol. 22:147–149, 1984.

1582 Methylethylenethiourea

Discussed under Ethylenethiourea

1583 Methyl Ethyl Ketone CAS 78–93–3

Schwetz et al. (1974) exposed rats for 7 hours a day on days 6 through 15 of gestation to 1000 and 3000 ppm. At the highest dose mandibular hypoplasia and tail defects were increased among the fetuses. Schwetz et al. (**1991**) exposed mice for 7 hours daily on days 6–15 to up to 3,000 ppm. Maternal liver weights were increased and male fetal weights were reduced at the highest dose but no teratogenicity was found.

Schwetz, B.A.; Leong, B.K.J. and Gehring, P.J.: Embryo–and fetotoxicity of inhaled carbon tetrachloride, 1,1–dichoroethane and methyl ethyl ketone in rats. Toxicol. Appl. Pharmacol. 28:452–464, 1974.

Schwetz, B.A.; Mast, T.J.; Weigel, R.J.; Dill, J.A. and Morrissey, R.E.: Developmental toxicity of inhaled methyl ketone in swiss mice. Fundamental and Applied Toxicology 16:742–748, 1991.

1584 Methylformamide

Pfeifer and von Kreybig (1973) found 500 mg per kg on day 13 caused cleft lip and brain and skeletal defects in rats.

Pfeifer, G. and von Kreybig, T.: Uber die Atiologie der Lippen–Kieferspaltformen und Gaumenspalten beim Menschen und im Tier–experiment. Experentia 29:225–228, 1973.

1585 N–Methylformamide

Stula and Krauss (1977) exposed rats by the cutaneous route to 600 mg per kg (one–eighteenth of the lethal dose) on day 7, 8 or 10 and produced a high incidence of encephalocele in the offspring.

Stula, E.F. and Krauss, W.C.: Embryotoxicity in rats and rabbits from cutaneous application of amide–type solvents and substituted ureas. Toxicol. Appl. Pharmacol. 41:35–55, 1977.

1586 Methylformhydroxamic Acid

Pfeifer and von Kreybig (1973) found brain and skeletal defects in fetuses whose mothers were given 700 mg per kg orally on day 13.

Pfeifer, G. and von Kreybig, T.: Uber die Atiologie der Lippen–Kieferspaltformen und Gaumenspalten beim Menschen und im Tier–experiment. Experentia 29:225–228, 1973.

1587 1–Methyl–formylhydrazine CAS 758–17–8

von Kreybig et al. (1970) reported that this compound given at 100 mg per kg on day 13 to rats produced congenital malformations. They found misshaped skulls with hypoplasia of the telencephalon and skeletal defects of mandible and extremities. 1–Methyl–acetylhydrazine and 1–ethyl–2–acetyl hydrazine were not teratogenic. Podophyllinic acid ethylhydrazide and isonicotinic acid 2–isopropylhydrazide and isonicotinic acid 2–isopropylhydrazide phosphate (Iproniazid™) were toxic at high levels but not teratogenic.

von Kreybig, T.; Preussmann, R. and von Kreybig, I.: Chemische Konstitution und Teratogene Wirkung bei der Ratte. 3. N–Alkylcarbonhydrazide weitere Hydrazinderivate. Arzneim–Forsch. 20:363–367, 1970.

1588 Methyl Hesperidin

Kawashima et al. (**1986**) gavaged rats with up to 8 g per kg on days 7–17 and found no ill effects.

Kawashima, K.; Nakaura, S.; Usami, M.; Yamaguchi, M.; Tanaka, S.; Takanaka, A. and Omori, Y.: Effect of methyl hesperidin on prenatal developments of rats. Eiseishikenjo Hokoku (Bulletin of National Institute of Hygienic Sciences) 104:64–68, 1986.

1589 Methylhydrazine CAS 60–34–4

Mercier–Parot and Tuchmann–Duplessis (1969) extended their initial observations on rats to the mouse and rabbit. With oral doses of 20 mg per kg when administered on days 8 through 12 of the mouse pregnancy, they found 36 percent of the offspring malformed. A dose of 200 mg per kg in the rabbit on the 14th day produced an equal number of defective offspring. The anomalies consisted of anencephaly, eye defects and complex facial bone deformities.

Mercier–Parot, L. and Tuchmann–Duplessis, H.: Action embryotoxique et teratogene d'une methylhydrazine chez la souris et le lapin. C. R. Soc. Biol. (Paris) 163:16–20, 1969.

1590 N–Methyl–N–(7–propoxynaphthalene–2–ethyl) Hydroxylamine QAB

N–methyl–n–(7–propoxynaphthalene–2–ethyl) hydroxylamine a model compound related to cocaine and acetaminophen was studied by Terlouw et al. (**1990**) in whole rat embryo culture.. At 122 microM growth was decreased. N–desoxy–QAB was thought to be the active metabolite.

Terlouw, G.D.C.; Namkung, M.J.; Juchau, M.R. and Bechter, R.: In vitro embryotoxicity of n–methyl–n(7–propoxynaphthalene–2–ethyl) hydroxylamine (QAB): evidence for n–dehydroxylated metabolite as a proximate dysmorphogen. Teratology 48:431–439, 1993.

1591 Methyl o-(4-hydroxy-3-methoxycinnamoly)reserpate CAS 35440–49–4 Rescimetol

Shimazu et al. (1979) gave this Rauwolfia alkaloid to pregnant rats and rabbits in maximum oral doses of 200 and 30 mg per kg respectively. Delayed implantation was found in rats receiving 100 mg per kg. At 200 mg per kg during organogenesis an increase in dead fetuses and resorptions was found. Perinatal administration of 100 mg per kg was associated with decreased suckling and weight gain. General behavior was not altered. No effect was seen following treatment of rabbits during organogenesis.

Shimazu, H.; Ikka, T.; Matsuura, M.; Tamada, T. and Fujimoto, Y.: Teratological and reproductive studies of methyl o-(4-hydroxy-3-methoxycinnamoyl) reserpate in rats and rabbits. Oyo Yakuri 18:105–124, 1979.

1592 Methyl Isobutyl Ketone CAS 108–10–1

Tyl et al. (1987) exposed rats and mice to up to 3000 ppm during organogenesis and found at the higher levels some maternal toxicity and a reduction in fetal weight and retardation of ossification.

Tyl, R.W.; France, K.A.; Fisher, L.C.; Pritts, I.M.; Tyler, T.R.; Phillips, R.D. and Moran, E.J.: Developmental toxicity evaluation of inhaled methyl isobutyl ketone in Fischer 344 rats and CD–1 Mice. Fund. Appl. Toxicol. 8:310–327, 1987.

1593 Methyl Isocyanate Bhopal Disaster

In the wake of the Bhopal disaster, questions were raised about the teratogenicity of this vapor. Varma et al. (1987) exposed mice for 3 hours on day 8 of gestation to 2,6,9 and 15 ppm. Resorptions were increased at 2 ppm and above. Total resorption was common at 9 ppm and maternal weight was about one–half the controls. At 9 and 15 ppm there was an increase in defects including cleft palate, hydronephrosis, diaphragmatic hernia, myocardial enlargement and meningomyelocele. Guest et al. (**1992**) found that 0.1–0.2 mM concentrations of S–(N–methylcarbamoyl) GSH, the labile conjugate of methyl isocyanote, produced decreased growth and development in explanted mouse embryos.

Guest, I.; Baille, T.A. and Varma, D.R.: Toxicity of the methyl isocyanate metabolite s–(n–methylcarbamoyl)gsh on mouse embryos in culture. Teratology 46:61–67, 1992.

Varma, D.R.: Reproductive toxicity of methyl isocyanate in mice. J. Toxicol. Environ. Health 21:265–275, 1987.

1594 Methyl–ISP

Wu et al. (**1989**) studied this organophorphorons insecticide in mice and rats given 0.625 to 2.5 mg per g on days 7–16. No teratogenicity was observed.

Wu, H.; Zhiwei, L.; Xu, H.; Ruikun, S.; Ma, T.; Shi, N.; Siu, R. and Liu. Y.: Toxicological studies on the organophosphorous insecticide methyl–isp. J. Tongji. Med. Univ. 9(1):58–64, 1989.

1595 1–Methyllysergic Acid Butanolamide

West (1962) gave daily subcutaneous injections to pregnant rats using 4 mg per kg and found no increase in fetal death.

West, G.B.: Drugs and rat pregnancy. Letter to the Editor. J. Pharm. Pharmacol. 14:828–830, 1962.

1596 Methyl Methanesulfonate MMS Alkane Sulphonates CAS 66–27–3

Fabro et al. (1984) exposed mouse ova at the 16 cell stage for one hour to 1.0 milliM concentrations of this mutagen. They were then transplanted into the ova duct of mice and the conceptuses were shown to be growth retarded but without malformations. The implantation rate was not changed by treatment. Hemsworth (1968) administered 100 mg per kg intraperitoneally on day 13 and produced cleft palate and limb defects in the offspring.

Fabro, S.; McLachlan, J.A. and Dames, N.M.: Chemical exposure of embryos during preimplantation stages of pregnancy: Mortality rate and intrauterine development. Am. J. Obstet. Gynecol. 148:929–936, 1984.

Hemsworth, B.N.: Embryopathies in the rat due to alkane sulphonates. J. Reprod. Fertil. 17:325–334, 1968.

1597 N–Methyl–N–(1 naphthyl)–fluoroacetamide CAS 5903–13–9

This pesticide was given to mice orally from day 1 through day 12 of gestation in maximum doses of 20 mg per kg. Some growth retardation occurred with the highest dose but no malformation increase was found (Makita et al., 1970).

Makita, T.; Hashimoto, Y. and Noguchi, T.: Teratological studies of N–methyl–N–(1–naphthyl)–fluoroacetamide in mice. Oyo Yakuri 4:463–468, 1970.

1598 3–Methyl–4–nitrophenol

Nehez et al. (1985) reported no adverse reproductive effects from 25 mg per kg orally on the 7th, 9th and 11th days of the mouse pregnancy.

Nehez, M.; Mazzay, E.; Huszta, E. and Bevencs, Gy.: The teratogenic embryotoxic and prenatal mutagenic effect of 3–methyl–4–nitrophenol in the mouse. Ecotoxicol. Environ. Safety 9:230–232, 1985.

1599

2-Methyl-4-nitro-1-(4–nitrophenyl)imidazole–Imidazole Derivative CAS 21721–92–6

Bauer et al. (1972) reported that this chemical given orally to rats, mice and rabbits was non–teratogenic. The multiple oral doses given were 200 mg per kg for rats, 100 mg per kg for mice and 50 mg per kg for rabbits.

Bauer, A.; Frohberg, H.; Jochmann, G. and Schilling, B.V.: Reproduction and mutagenicity trials of 2–methyl--4–nitro–1–(4–nitrophenyl)imidiazole. (abs) Naunyn Schmiedebergs Arch. Pharmakol. Suppl. 274:R15 only, 1972.

1600 N–Methyl–N–nitro–N–nitrosoguanidine CAS 70–25–7

Inouye and Murakami (1978) studied this potent mutagen and carcinogen in pregnant mice at dose levels of 40 to 80 mg per kg given as a single intraperitoneal injection on day 7, 8, 9, 10, 11 or 12. Hydrocephalus, cleft palate, micrognathia and reduction defects of the extremities were found in the fetus. The highest incidence of hydrocephalus was found in the group treated on day 10 with 60 mg per kg. At 80 mg per kg approximately one–third of the mothers died.

Inouye, M. and Murakami, U.: Teratogenic effect of N–methyl–N–nitro–N–nitrosoguanidine in mice. Teratology 18:263–266, 1978.

1601 Methylnitrosoaniline *Nitrosomethylaniline* CAS 614–00–6

Alexandrov (1968, 1973) reported that this carcinogen given to rats intraperitoneally (140 mg per kg) produced an increase in congenital defects in the offspring. The most sensitive days were the 9th and the 13th and the malformations included microphthalmos hydrocephaly, exencephaly and reduction defects of the forelimbs. Nitrosoethylaniline was also teratogenic at 180 mg per kg given on the 9th day.

Alexandrov, V.A.: Effect of N–nitroso–N–methyl–aniline and N–nitroso–N–ethylaniline on the rat embryo. Vop. Onkol. 14:37–38, 1968.

Alexandrov, V.A.: Embryotoxic and teratogenic effects of chemical carcinogens. International Agency for Research in Cancer, Lyons, 112–126, 1973.

1602 Methylnitrosonitrosoguanidine *MNNG*

Faustman et al. (1989) compared this chemical to other direct acting alkylating agents in a whole embryo in vitro system and found it to cause malformations and lethality at micromolar levels of 2 to 4. Defects of the CNS and eyes were noted. This agent was compared to other alkylaters and found to be active at the lowest concentration.

Faustman, E.M.; Zamyat, K.; Gage, D. and Varnum, M.: In vitro developmental toxicity of five direct–acting alkylating agents in rodent embryos: structure–activity patterns. Teratology 40:199–210, 1989.

1603 Methylnitrosourea CAS 684–93–5
N–Methyl–N–nitrosourea MNU

Napalkov and Alexandrov (1968) found a 80 percent malformation rate in rat fetuses whose mothers received injections of 20 mg per kg on the 9th day. Koyama et al. (1970), using 10 mg of methylnitrosourea per kg in the rat on single gestational days from the 8th through the 15th day, produced hydrocephalus, exencephaly, hypoplasia of the pallium or microcephaly. The exencephaly occurred in the earlier treated group with hydrocephalus on the 9th and 10th days and microcephalus after the 11th day. Alexandrov (1976) reviewed the extensive Russian experimental work on transplacental carcinogenesis of this and related compounds. Platze et al. (1988) found increased skeletal and palatal defects when they injected mice intraperitoneally with 3 to 10 mg per kg. The 5th digit was perferentially effected.

Alexandrov and Schreiber (1978) and Alexandrov (1979) studied brain blastomogenesis against the background of the developmental deformities induced by the combined transplacental effect of methylnitrosourea (MNU) and ethylnitrosourea (ESU). To induce brain defects such as microcephaly, MNU was injected on the 15th day. To induce cerebellar defects, MNU was injected on the 21st day of embryogenesis. Moreover, at the 13th or 17th day, ENU was additionally injected and found to be highly effective for inducing brain tumors. It was found that MNU exposure at the 15th day until ENU exposure at the 17th day of embryogenesis, no reliable decrease in brain tumor occurrence was noted compared to when only ENU was employed. In the reverse sequence, that is, first the exposure to ENU on the 13th day and then to MNU on the 15th day, the occurrence of tumors located in cerebral hemispheres was three times less. It is assumed that cytotoxic effect of NMU leading to microcephaly is likely to cause the death of a considerable amount of the cell population previously transformed. Dimant and Beniashvili (1978) reported the results of exposure with a number of carcinogenic agents in rabbits. They remarked that NEU had the greatest carcinogenic and neurotropic effects.

Nagao et al. (1986) administered this agent in doses of 5–25 mg per kg to male mice before breeding to untreated females. A dose response and time specific response was found in the defect rate of fetuses on day 18. Bossert et al. (**1990**) exposed preimplanted embryos to 5 or 10 micrograms per liter and then transfered the embryos to pseudopregnant mice. Altered 2–dimensional protein electrophoresis was found in midgestaion embryo extracts. This treatment produces a decrease in fetal weight and defects of the CNS and skeletal system as well as nasopharyngeal dysgenesis. Platze et al. (**1988**) studied in detail the skeletal defects observed in mice.

Faustman et al. (1989) have compared this agent with other alkalates using a whole embryo culture system.

Alexandrov, V.A.: Some results and prospects of transplacental carcinogenesis studies. Neoplasia 23:285–299, 1976.

Alexandrov, V.A.: The pattern of N–nitrosoethylurea action in rats during embryogenesis. Vopr. Onkol. (USSR) 25.6:60–65, 1979.

Alexandrov, V.A. and Schreiber, D.: Combined transplacental carcinogenic action of N–nitrosomethylurea (NMU) and N–nitrosoethylurea in rats. Vopr. Onkol. (USSR) 24.4:38–43, 1978.

Bossert, N.L.; Hitselberger, M.H. and Iannaccone, P.M.: Protein alterations associated with n–methyl–n–nitrosourea exposure of preimplantation mouse embryos transferred to surrogate mothers. Teratology 41:147–156, 1990.

Dimant, I.N. and Beniashvili, D.Sh.: Some aspects of the transplacental blastomogenesis in rabbits. Byull. Eksper. Biol. Med. (USSR) 85.3:343, 1978.

Faustman, E.M.; Zamyat, K.; Gage, D. and Varnum, M.: In vitro developmental toxicity of five direct–acting alkylating agents in rodent embryos:structure–activity patterns. Teratology 40:199–210, 1989.

Oyama, T.; Handa, J.; Handa, H. and Matsumoto, S.: Methylnitrosourea–induced malformations of brain in SD–JCL rat. Arch. Neurol. 22:342–347, 1970.

Nagao, T.; Ishizuka, Y. and Mizutani, M.: Induction of congenital malformations in mice after treatment of spermatogonial stem cells with methyl nitrosourea. (abs) Toxicol. Lett. 31:68, 1986.

Napalkov, N.P. and Alexandrov, V.A.: On the effects of blastomogenic substances on the organism during embryogenesis. Z. Krebsforsch. 71:32–50, 1968.

Platzek, T.; Bochert, G.; Pauli, B.; Meister, R. and Neubert, D.: Embryotoxicity induced by alkylating agents:5. dose–response relationships of teratogenic effects of methylnitrosourea in mice. Arch. Toxicol. 62:411–423, 1988.

1604   o–Methyl Pantothenic Acid

Discussed under Pantothenic Acid Deficiency

1605   X–Methyl Pantothenic Acid

Discussed under Pantothenic Acid Deficiency

1606   Methyl Parathion *Dimethyl–O–4–nitrophenyl Phosphorothioate* CAS 298–00–0

This cholinesterase inhibitor was given to rats in amounts of 24 mg per kg on day 9 or 15 of gestation, and no defects were seen in the offspring (Fish, 1966). Some perinatal mortality increase was seen and the growth rate after birth was reduced. Cerebral cortical cholinesterase was reduced in the fetuses.

Tanimura et al. (1967) found no teratogenicity with doses of 15 mg per kg on day 12 in the rat. In the mouse, cleft palate occurred at 60 mg per kg on day 10. The material was given intraperitoneally.

Fish, S.A.: Organophosphorus cholinesterase inhibitor and fetal development. Am. J. Obstet. Gynecol. 96:1148–1154, 1966.

Tanimura, T.; Katsuya, T. and Nishimura, H.: Embryotoxicity of acute exposure to methyl parathion in rats and mice. Arch. Environ. Health 15:609–613, 1967.

1607   2–Methylpentanoic Acid

Dawson (**1991**) using a frog embryo teratogenesis assay found that 50% of the survivors were malformed when exposure was to 172 mg per liter. Gut coiling, edema and microcephaly were the most common types of defects.

Dawson, D.A.: Additivew incidence of developmental malformation for xenopus embryos exposed to a mixture of ten aliphatic carboxylic acids. Teratology 44:531–546, 1991.

1608   Methylperone Hydrochloride *γ(4–Methylpiperidine)-p-(fluorobutyrophenone HCl)*

Heywood and Palmer (1974) studied this neuroleptic in rats and rabbits giving up to 60 mg per kg during organogenesis. No adverse effects were found, but the rabbit pups in the treated groups had lower weights.

Heywood, R. and Palmer, A.K.: Prolonged toxicological studies in the rat and beagle dog with methylperone hydrochloride, with embryo toxicity studies in the rabbit and rat. Farmaco. Ed. Pr. 29:586–593, 1974.

1609   Methylphenidate *Ritalin*™ CAS 113–45–1

Heinonen et al. (1977) included 11 mothers who took this drug in the first 4 lunar months of gestation among a group of 96 in which there was no significant increase in defect rate among the offspring.

Heinonen, O.P.; Slone, D. and Shapiro, S.: Birth Defects and Drugs in Pregnancy. Publishing Sciences Group Inc., Littleton, Mass., 1977.

1610   1–Methyl–4–phenyl–1,2,3,6–tetrahydropyridine *MPTP*

Schmahl and Usler (**1991**) treated mice with 20, 40 or 60 mg per kg intraperitoneally on day 15. Significant reduction in placental weight occurred at the 20 and 40 mg levels and histologic alterations were seen. At 40 mg per kg tail abnormalities severe growth retardation and edema occurred. This drug produces Parkinsonism in animals.

Schmahl, W. and Usler, B.: Placental toxicity of 1–methyl–4–penyl–1,2,3,6–tetrahydropyridine (MPTP) in mice. Toxicology 67:63–74, 1991.

1611 Methylprednisolone CAS 83–43–2

Walker (1971) produced cleft palate in the mouse (AJAX) using 0.5 mg daily on days 11 through 14. Doses of up to 8.0 mg daily in the rat did not produce cleft palates.

Walker, B.: Induction of cleft palate with anti–inflammatory drugs. Teratology 4:39–42, 1971.

1612 Methylprednisolone Aceponate

Kageyama et al. (1991 A,B,C) treated rats subcutaneously before during early pregnancy with up to 0.1 mg per kg and during organogenesis and perinatally with up to 1 mg per kg. No increase in abnormalities of the fetuses occurred and postnatal development was uneffected.

Kageyama, A.; Kato, K.; Urabe, K.; Kawakita, Y.; Sanada, M.; Kodama, N. and Nakagawa, H.: Toxicity study of methylprednisolone aceponate (ZK 91 588) (IV) fertility study in rats. Jpn. Pharm. Therap. 19:3063–3071, 1991A.

Kageyama, A.; Katt(o, K.; Urabe, K.; Sanada, M.; Kodama, N. and Nakagawa, H.: Toxicity study of methylprednisolone aceponate (ZK 91 588) (V) teratogenicity study in rats. Jpn. Pharm. Therap. 19:3073–3088, 1991B.

Kageyama, A.; Kato, K.; Urabe, K.; Kawakita, Y.; Sanada, M.; Kodama, N. and Nakagawa, H.: Toxicity study of methylprrednisolone aceponate (ZK 91 588) (VI) perinatal and postnatal study in rats. Jpn. Pharm. Therap. 19:3089–3099, 1991C.

1613 9–Methyl–pteroylglutamic Acid also see *Folic Acid Deficiency* CAS 2179–16–0

This analogue of folic acid is used to potentiate the folic acid deficiency state in experimental models.

1614 X–Methyl–pteroylglutamic Acid also see *Folic Acid Deficiency*

This is an unpurified substance which was used to potentiate the folic acid deficiency state in experimental models. It contains 9–methyl–pteroylglutamic acid.

1615 N–Methyl–2–pyrrolidone CAS 872–50–4 *NMP*

Lee et al. (1987) exposed rats to 0.1 and 0.36 mg per liter by inhalation for 6 hours daily on days 6–15. No significant fetal changes occurred. Previous teratology studies at high levels were reviewed. At 129 mg per kg intraperitoneally teratogenicity was found in the mouse (Schmidt, 1976). Hass et al.(1994) to rats to 150 ppm for 6 hours on days 7–20. The offspring were lighter and had normal motor behavior. However, reversal procedure in water maze and operant delayed spatial alternation were impaired.

Hass, U.; Lund, S.P. and Elsner, J.: Effects of prenatal exposure to N–methylpyrrolidone on postnatal development and behavior in rats. Neurotoxicology and Teratology 16:241–249, 1994.

Lee, K.P.; Chromey, N.C.; Culik, R.; Barnes, J.R. and Schneider, P.W.: Toxicity of N–methyl–2–pyrrolidone (NMP): Teratogenic, subchronic, and two–year inhalation studies. Fund. Appl. Toxicol. 9:222–235, 1987.

Schmidt, R.: Tierexperimentelle Untersuchungen zur Embryotoxischen und Teratogenen Wirkung von N–Methyl–Pyrrolidon (NMP). Biol. Rund. 14:38–41, 1976.

1616 2–Methylresorcinol

Re et al. (1986) fed rats up to 1.5 percent in the diet during gestation and found no teratogenic effects. The dams did not gain weight as well as the controls.

Re, T.A.; Loehr, R.F.; Rodriguez, S.C.; Rodwell, D.E. and Burnett, C.M.: A subchronic, teratologic and dominant lethal study of 2–methylresorcinol in rats. Fund. Appl. Toxicol. 7:293–298, 1986.

1617 Methylsalicylate

Discussed under Salicylate

1618 Methyl Tertiary–butyl Ether

Conaway (1985) exposed pregnant rats in amounts of up to 2500 ppm on days 6–15 of gestation. No adverse effects were seen on the fetuses.

Conaway, C.C.; Schroeder, R.E. and Snyder, N.K.: Teratology evaluation of methyl tertiary butyl ether in rats and mice. J. Toxicol. Environ. Health 16:797–809, 1985.

1619 Methyl Styrene

Hardin et al. (1981) injected rats with 250 mg per kg intraperitoneally on days 1–15. No teratogenicity or maternal toxicity was found. Some fetal toxicity occurred.

Hardin, B.D.; Bond, G.P.; Sikov, M.R.; Andrew, F.D.; Beliles, R.P. and Niemeier, R.W.: Testing of selected workplace chemicals for teratogenic potential. Scand. J. Work Environ. Health 7(4):66–75, 1981.

1620 Methyltestosterone CAS 58–18–4

This androgen is capable of masculinizing the human female fetus (Grumbach and Ducharme, 1960) as well as female fetuses of experimental animals (Jost, 1955). See Testosterone for more complete coverage.

Golubeva et al. (1978) reported that this drug applied directly to the skin of pregnant rats (0.01–0.05 mg per kg from day 1–15) induced in the offspring some behavior abnormalities and slight disturbances in the function of cardio–vascular system.

Golubeva, M.I.; Shashkina, L.F.: Starkov, M.V. and Fedorova, Z.A.: Development of the progeny of rats after application of androgens to the skin throughout the entire pregnancy. Gig. Tr. Prof. Zabol. (USSR) 6:25–28, 1978.

Grumbach, M.M. and Ducharme, J.R.: The effects of androgens on fetal sexual development androgen–induced female pseudohermaphroditism. Fertil. Steril. 11:157–180, 1960.

Jost, A.: Biologie des androgenes chez l'embryon. Reunion Des Endocrinologists De Langue Francaise (third), Paris: Masson, 160–180, 1955.

1621 **N–Methyl–2–thioimidazole** CAS 3247–70–9

Discussed under Ethylenethiourea

1622 **Methylthiouracil** CAS 56–04–2

Freiesleben and Kjerulf–Jensen (1947) reported a five–month old fetus with thyroid hypertrophy following maternal treatment during pregnancy.

In pregnant rats fed 0.25 mg per 10 gm diet, they demonstrated fetal thyroid changes histologically. Toriumi (1959) fed rabbits 50 mg per kg and observed histologic changes in the fetal thyroid after the 18th day. By the 21st day, grossly enlarged fetal thyroids occurred. Klosovskii (1963) gave chinchillas 300 mg per kg and produced cretinism, hypoplastic brains and some dextrocardia and transposition of the great vessels.

Freiesleben, E. and Kjerulf–Jensen, K.: The effect of thiouracil derivatives on fetuses and infants. J. Clin. Endocrinol. 7:47–51, 1947.

Klosovskii, B.N.: The Development of the Brain and its Disturbance by Harmful Factors. Translated from Russian, B. Haigh (ed), New York: MacMillian, 161–167, 1963.

Toriumi, K.: Embryological studies on the experimental congenital goiter due to methylthiouracil in rabbits. J. Osaka City Med. Cen. 8:1281–1293, 1959.

1623 **Methylthiourea**

Teramoto et al. (1981) gave 2000 mg per kg orally to mice on day 10 and found no increase in the defect rates or resorptions. In the rat, 250 mg given on day 12 or 14, tail defects, hydrocephalus and cranial meningoceles were significantly increased.

Teramoto, S.; Kaneda, M.; Aoyama, H. and Shirasu, Y.: Correlation between the molecular structure of N–alkylureas and N–alkylthioureas and their teratogenic properties. Teratology 23:335–342, 1981.

1624 **α–Methyltyrosine** CAS 305–96–4

Kvist and Rubin (1975) reported studies in the chick egg with this tyrosine analogue. Twenty mg were injected at 20–22 hours of incubation and 30 percent of the survivors had defects including anencephaly and spina bifida.

Kvist, T.N. and Rubin, C.: The role of catecholamines in neural tube closure and head flexure formation in the chick embryo. (abs) Teratology 11:26A–27A, 1975.

1625 **4–Methylumbelliferyl–β–D–xyloside** CAS 6734–33–4

Discussed under β–D–Xyloside

1626 **4–Methyl Uracil** CAS 626–48–2

Kosmachevskaya and Tichodeeva (1968) injected chick eggs with four mg of this substance and produced abnormalities of the brain or eye in 24 percent. No defects were produced in rat fetuses when the mother received 3000 mg per kg of body weight on the 9th and 10th days of gestation (Kosmachevskaya and Chebotar, 1968).

Kosmachevskaya, E.A. and Chebotar, N.A.: The damaging effect of 4–methyl uracil on rat embryogenesis in conditions of maternal stress. Bull. Exp. Biol. (Russian) 12:89–91, 1968.

Kosmachevskaya, E.A. and Tichodeeva, I.I.: Relation between embryotoxic activity of some pyrimidine derivatives and their chemical structure. Chick Embryo Test. Pharmacol. Toxicol. (Russian) 5:618–620, 1968.

1627 **Methylurea** CAS 598–50–5

von Kreybig et al. (1969) showed that the dimethyl, trimethyl and tetramethyl form of this compound was teratogenic in rats. Reduction defects of the skeletal system were produced by administration of 500 to 1000 mg per kg on the 13th or 14th day of gestation. Cros et al. (1972) gave detailed studies of the teratogenicity of tetramethylurea in mice. They found exencephaly and reduction defects of the extremities.

Teramoto et al. (1981) gave 2000 mg per kg orally to rats on day 12 and 1000 mg to mice on day 10 and found no increase in the defect rates or resorptions.

Cros, S.B.; Moisand, C. and Tollon, Y.: Influence de la tetramethyluree sur le developpement embryonnaire de la souris. Ann. Pharm. Franc. 9:585–593, 1972.

Teramoto, S.; Kaneda, M.; Aoyama, H. and Shirasu, Y.: Correlation between the molecular structure of N–alkylureas and N–alkylthioureas and their teratogenic properties. Teratology 23:335–342, 1981.

von Kreybig, T.; Preussmann, R. and von Kreybig, I.: Chemische Konstitution und Teratogene Wirkung bei der Ratte. II. N–Alkylharnstoffe, N–Alkylsulfonamide, N,N–Dialkylacetamide, N–Methylthioacetamid, Chloracetamid. Arzneim. Forsch. 19:1073–1076, 1969.

1628 **Metiapine** CAS 5800–19–1

Gibson and Newberne (1973) gave up to 30 mg per kg orally to pregnant rats and rabbits during organogenesis and produced no defects. At the highest doses, postnatal survival was reduced in the rats.

Gibson, J.P. and Newberne, J.W.: Teratology and reproductive studies with metiapine. Toxicol. Appl. Pharmacol. 25:212–219, 1973.

1629 **Metiazinic Acid** *10–Methyl–2–phenothiazinyl Acetic Acid* CAS 13993–65–2

Julou et al. (1969) found no teratogenicity in mice, rats and rabbits when they used up to 60 mg per kg per day during active organogenesis. This anti–inflammatory agent was tested orally in pregnant rats by Nakamura et al., (1974). Doses up to 80 mg per kg per day were given from the 8th through the 14th days of gestation. No teratogenicity or postnatal effects of treatment were found.

Julou, L.; Ducrot, R.; Fournel, J.; Ganter, P.; Populaire, P.; Durel, J.; Myon, J.; Pascal, S. and Pasquet, J.: Etude toxicologique de l'acide metiazinique (16091 R.P.). Arzneim. Forsch. 19:1207–1214, 1969.

Nakamura, E.; Kimura, M.; Kato, R.; Honma, K.; Tsuruta, M.; Uchida, S.; Kaneka, K. and Sato, H.: Teratogenic studies on metiazinic acid. Oyo Yakuri 8:1587–1631, 1974.

1630 Meticrane CAS 1084–65–7

Nomura et al. (1974) gave up to 10,000 mg per kg orally to mice on days 7–12 and rats on days 9–14. No adverse effects occurred.

Nomura, A.; Watanabe, M.; Yamagata, H. and Ohata, K.: Influence of meticrane (6–methyl–7–sulfamido–thiochroman–1,1–dioxide) on pregnant mice and rats and their fetuses. Oyo Yakuri 8:217–227, 1974.

1631 Metoclopramide *Reglin* CAS 364–62–5

Pinder et al. (1976) reviewed the toxicology of this antiemetic and indicated that in one study of 120 women treated, there were no offspring with congenital defects.

They also report a study where rats, mice and rabbits were treated with 10 mg per kg orally, subcutaneously and intravenously and no adverse effects were found.

Pinder, R.M.; Brogden, R.N.; Sawyer, P.R.; Speight, T.M. and Avery, G.S.: Metoclopramide, a review of its pharmacological properties. Drugs. 12:81–131, 1976.

1632 Metofenazate also see *Chlorpromazine, Prochlorperazine and Phenothiazine* CAS 388–51–2

Horvath and Druga (1975) administered this tranquilizer to pregnant rats by mouth. With doses of 10 mg per kg per day from the 7th through the 14th days malformations were found which included micrognathia and micromelia of the hind limbs. With doses of 100 mg per kg given on the 8th, 9th, 10th or 11th day, a high fetal mortality was found. Single large doses on the 14th day produced cleft palate and hydronephrosis was common. These authors believed that the teratogenic action of phenothiazines is increased when the length of the N–alkyl side chain is lengthened.

Horvath, C. and Druga, A.: Action of the phenothiazine derivative methophenazine on prenatal development in rats. Teratology 11:325–330, 1975.

1633 Metolazone 7-*Chloro 1,2,3,4 tetrahydro–2–methyl–4–oxo–3–o–tolyl–6–quinazolinesulfonamide* CAS 17560–51–9

Nakajima et al. (1978A,B) studied the effect of this diuretic on pregnant rats and rabbits. The dose range was 2–250 mg per kg daily orally during active organogenesis. Fetal hydronephrosis was increased in the rat fetuses exposed to 2 mg per kg but not at higher doses. Ureteric dilatation was found in fetuses exposed to 2, 10 and 50 mg per kg. No limb changes were detected. In the rabbit the only fetal change was a weight reduction at 10 mg per kg.

Nakajima, T.; Ishisaka, K.; Taylor, P. and Matuda, S.: Effects of metolazone on the reproduction function of rats. 2. Teratogenicity test. (Japanese) Clinical Report 12:3394–3406, 1978A.
Nakajima, T.; Ishisaka, K.; Taylor, P. and Matuda, S.: Effects of metolazone on reproduction of rabbits. Teratogenicity test. (Japanese) Clinical Report 12:3417–3421, 1978B.

1634 Metoprolol Tartrate CAS 56392–17–7

Sandstrom (1982) used this medication in combination with other drugs for treating hypertension in the third trimester among 184 women. Two groups were treated with metoprolol and either bendroflumethiazide or hydralazine and the third with bendroflumethiazide and hydralazine. The maternal and fetal outcomes in the three groups were not statistically different but there were five intrauterine deaths in the third group as compared to one each in the first two groups. Kolesari and Shields (1989) found that this beta–receptor antagonist prevented cardiac defects from dopamine in the chick embryo.

Fukuhara et al. (1979) studied the adrenergic blocker in rats and rabbits. The rabbits received 64 mg and the rats 500 mg per kg. In the rat implantation was inhibited at 500 mg per kg. No adverse fetal effects were noted in either species except for slight growth retardation in the rat. Neonatal mortality was increased among fetuses exposed on days 17–21 of gestation. Some increase in embryolethality was noted in the rabbit fetuses exposed to 64 mg per kg.

Fukuhara, Y.; Fujii, T.; Emi, Y.; Kado, Y. and Watanabe, N.: Reproductive studies of metoprolol tartrate. Kiso to Rinsho 13:3216–3224, 1979.
Kolesari, G.L. and Shields, H.E.: The effect of metoprolol on the absorption and distribution of a teratogenic dose of dopamine in the stage 24 chick embryo. Teratology 39:463, 1989.
Sandstrom, B.: Adrenergic β–receptor blockers in hypertension of pregnancy. Clin. Exp. Hypertens. B. (1), 127–141, 1982.

1635 Metoxadiazone

This insecticide was given subcutaneously to rats on days 7–17 in doses of up to 50 mg per kg by Saito et al. (1987). No adverse fetal effects were found and at the highest dose, there was maternal toxicity.

Saito, M.; Kumagai, Y. and Narama, I.: A teratological evaluation following subcutaneous administration of metoxadiazone (S-21074) to rats. Pharmacometrics (Oyo Yakuri) 34:147–162, 1987.

1636 Metrizamide CAS 31112–62–6

Kodama et al. (1979) administered up to 1800 mg per kg of this contrast medium to rats before mating and during the first 7 days, on days 7–17 or on days 17–21. The fertility studies were done by intraperitoneal route and the teratological and perinatal by intravenous route. Implantations were reduced in the treated group and a borderline decrease in viability was reported. Other adverse fetal effects were not reported.

Kodama, N.; Tsubota, K. and Ezumi, Y.: Effects of metrizamide on rat reproduction. Nichi–doku Iho Japanisch–Deutsche Medizinische Berichte 24:277–285, 287–302, 303–318, 1979.

1637 Metronidazole *Flagyl*ᵀᴹ CAS 443–48–1

Monitoring of the offspring of treated pregnant women has been carried out for 20 years, and several large studies are summarized by Berget and Weber (1972). Their review included 1,469 pregnant women, of whom 206 were treated in the first trimester. No increase in the incidence of malformation, abortion, or stillbirth was found. Postnatal followup was done in some of the patients. Royer (1983) reviewed publications involving 2,139 pregnant exposed women and found no increase in malformations as compared to a group of non–treated women with trichomonas.

In a record linkage study, Rosa et al. (1987) found the spontaneous abortion relative risk to be 1.37 among 1020 women exposed to the drug 300–180 days before delivery. Among 50,381 women who had legal abortions, there were 2,148 (4.26 percent) exposed to metronidazole while among 4,264 who had spontaneous abortions 135 (3.16 percent) were exposed. Greenberg (1985) cited two reports of clefting and added another case in women who were treated during organogenesis.

Legator et al. (1975) found evidence of mutagenic activity using a salmonella typhimurium test when urine from patients taking metronidazole was assayed. The importance of this test is still unresolved. We have determined that about half of the chemicals found to be mutagenic in these bacterial tests are teratogenic in animal tests. Positive animal teratogenicity tests do not by any means imply that the drug is embryo–or fetotoxic in the human.

Ganter et al. (1960) used 100 mg per kg during the entire rat pregnancy and observed neither change in number of viable offspring nor any malformations.

Berget, A. and Weber, T.: Metronidazole and pregnancy. Ugeskr. Laeger 134:2085–2089, 1972.

Ganter, P.; Julou, L. and Cosar, C.: Study of the action of metronidazole (No. 8823RP.) on the genital system of the rat. Obstet. Gynecol. 59:609–620, 1960.

Greenberg, F.: Letter to the editor: possible metronidazole teratogenicity and clefting. Am J of Med Genetics 22:825, 1985.

Legator, M.S.; Conner, T.H. and Stoeckel, M.: Detection of mutagenic activity of metronidazole and niridazole in body fluids of humans and mice. Science 188:1118–1119, 1975.

Rosa, F.W.; Baum, C. and Shaw, M.: Pregnancy outcomes after first–trimester vaginitis drug therapy. Obstet. Gynecol. 69:751–755, 1987.

Royer, M.E.: Innocuite du metronidazole (Flagyl) prescrit pendant la grossesse. Medecine et Malades Infectieuses 13:727–729, 1983.

1638 Mevinolinic Acid *Lovostatin* CAS 77550–67–5

Robertson et al. (1981) fed 800 mg mevinolin per kg to the gravid rat on days 6–17 and produced fetal malformations of the vertebrae and ribs. Ghidini et al. (**1992**) reported a single infant with VATER syndrome. The mother had taken lovestatin 10 mg daily and dextroamphetamine 10 mg daily during the first trimester.

Ghidini, A.; Sicherer, S. and Willner, J.: Congenital abnormalities (VATER) in baby born to mother using lovastatin. The Lancet 339:1416–1417, 1992.

Robertson, R.T.; Minsker, D.H.; MacDonald, J.S.; Bokelman, D.L. and Christian, M.S.: Mevalonic acid antagonism of the teratogenic effects of mevinolic acid, a potent inhibitor of hydroxymethylglutaryl–coenzyme A reductase. (abs) Teratology 23:58A, 1981.

1639 Mexiletine HCl CAS 5370–01–4

Matsuo et al. (1983) reported no teratogenic effect of this anti–arrhythmia medication in pregnant rats and rabbits. Oral doses up to and including maternal toxicity were used and consisted of 150 mg per kg in the rat and 75 mg in the rabbit. Five mg and 2.5 mg per kg given intravenously to rats and rabbits, respectively, had no ill effects on reproduction (Nishimura et al., 1983).

Matsuo, A.; Kast, A. and Tsunenari, Y.: Reproduction studies of mexiletine hydrochloride by oral administration. Iyakuhin Kenkyu 14:527–549, 1983.

Nishimura, M.; Kast, A. and Tsunenari, Y.: Reproduction studies of mexiletine hydrochloride by intravenous administration. Ikakuhin Kenkyu 14:550–570, 1983.

1640 Mezlocillin CAS 51481–65–3

Hamada and Imanishi (1978) gave intravenously up to 1000 mg per kg on days 7 through 17 and found no fetal changes in rats. Tanioka and Koizumi (1978) found no evidence of teratogenicity in monkeys given up to 100 mg per kg from days 23 through 47 of gestation.

Hamada, Y. and Imanishi, M.: Reproduction study of mezlocillin in rats. 2. Teratogenicity study. (Japanese) Iyakuhin Kenkyu 9:986–996, 1978.

Tanioka, Y. and Koizumi, H.: Influence of sodium mezlocillin on fetuses of rhesus monkeys. (Japanese) CIEA Preclin. Report 4:11–22, 1978.

1641 Mibolerone *17–β–Hydroxy–7–α 17–dimethylestr–4–en–3–one* CAS 3704–09–4

Sokolowski and Kasson (1978) administered 20 to 60 microgm daily to beagle dogs by oral route during gestation. At both levels the female external genitalia were masculinized.

Sokolowski, H.G. and Kasson, C.W.: Effects of mibolerone on conception, pregnancy, parturition, and offspring in the beagle. Am. J. Vet. Res. 39:837–840, 1978.

1642 Miconazole CAS 22916–47–8

Jick et al. (1981) found no increase in congenital defect rates among 360 pregnancies treated in the first trimester. Rosa et al. (1987) found a relative risk of spontaneous abortion to be 1.36 in a linkage study but they warn that since many associations were being tested in the study no firm conclusions could be drawn.

Ito et al. (1976A,B) tested this antimycotic in rats and rabbits during active organogenesis. At the maximum dose of 100 mg per kg, both species had an increase in fetal mortality. In the rat, difficult labor occurred. No malformation rate increase was found.

Ito, C.; Shibutani, Y.; Inoue, K.; Nakano, K. and Ohnishi, H.: Toxicological studies of miconazole. 2. Teratological studies of miconazole in rats. Iyakuhin Kenkyu 7:367–376, 1976A.

Ito, C.; Shibutani, Y.; Taya, K. and Ohnishi, H.: Toxicological studies of miconazole. 3. Teratological studies of miconazole in rabbits. Iyakuhin Kenkyu 7:377–381, 1976B.

Jick, H.; Holmes, L.B.; Hunter, J.R.; Madsen, S. and Stergachis, A.: First–trimester drug use and congenital disorders. JAMA 246:343–346, 1981.

Rosa, F.W.; Baum, C. and Shaw, M.: Pregnancy outcomes after first–trimester vaginitis drug therapy. Obstet. Gynecol. 69:751–755, 1987.

1643 Miconomicin CAS 52093–21–7

Hara et al. (1983) studied this antibiotic in rats and rabbits. Giving up to 75 mg per kg intravenously before pregnancy, during organogenesis or in the later part of pregnancy produced no adverse fetal outcomes in rats. No adverse effects were found in rabbits given 70 mg per kg intravenously during organogenesis.

Hara, T.; Nishikawa, S.; Miyazaki, H. and Ohguro, Y.: Safety evaluation of miconomicin. Teratogenicity, fertility and perinatal and postnatal study in rats and teratologic studies in rabbits. Jpn. J. Antibiot. 36:177–181, 315–318, 319–329, 1983.

1644 Microwave Radiations *Radio Frequency Waves*

The effects of microwave and shortwave radiation were studied extensively and reviews were published by Michaelson (1969) and by Brent (1977). Michaelson gave a discussion of why the energy from these sources is too small regardless of dose to produce the type of excitation necessary for ionization. Consequently, there seems to be agreement that any damage from these sources would be related to hyperthermia which may be teratogenic under certain conditions (see Hyperthermia). Radiation at frequencies below 1000 MHz causes heat primarily in deep tissues. With increasing frequencies and especially over 3,000, proportionally more surface heating occurs. Most microwave ovens and diathermy generate approximately 2,450 MHz and according to Brent (1977), even at the maximum permissible level of 1 to 10 mW, no hazard to the human embryo would be expected. Saito et al. (1991) exposed chick eggs to radio frequency radiation at 428 MHz at 5.5 mW/cm^2 for 20 days. They found in the treated that 87% had a functional defect in creeping. No increase in temperature was demonstrable in gelatin replaced eggs.

A more detailed account of the large literature on ultrasound is given under the heading, Ultrasound. Logue (1985) studied the 4,732 offspring of 5,300 male members of the American Physical Therapy Association and found a defect rate of 3.7 percent without increases in any special category.

Rubin and Erdman (1959) reported four case histories of pregnant women inadvertently treated with microwave for chronic pelvic infection. The dose was 2,450 MHz with a 100–W machine. One miscarried after 10 days treatment, the others delivered normal infants. The women who mis-carried became pregnant again, treatment was continued and she gave birth to a normal child. Fetal heart detectors used in human pregnancy monitoring produce 5 to 20 mW per cm squared (Mannor et al., 1972). Lary and Conover (1987) reviewed the effects of radio frequency (300 MHz to 3000 GH3). Fetal damage was associated with hyperthermia.

Hofmann and Dietzel (1966) reported that diathermy treatment of rats on the 13th or 14th gestational day produced defects of the tail and extremities. They used a 70 or 100 watt intensity at 27 MHz for 10 minutes. Pregnancy in the rabbit was interrupted especially by treatment on the 10th day. Umeda (1941) applied diathermy to rabbits for 20 minutes at various times during pregnancy and produced fetal death and some histological changes in the viscera of surviving fetuses. Mannor et al. (1972) used levels of 164 to 1050 mW per cm squared at 2.28 MHz for up to 60 minutes in the mouse for varying periods of gestation. Four hundred and ninety mW did not cause critical temperature rises and higher intensity produced tissue defects identical to those from overheating. No defects were produced and postnatal fertility and chromosomal findings in the offspring were normal. Shoji et al. (1975) treated mice for five hours during day 8 with 2.25 MHz and power of 40 mW per cm squared. They found a low but significant increase in severe brain and facial defects in one of two strains.

Tachibana (1977) exposed two strains of mice to up to ten minutes of 200 mW per cm squared at a frequency of two MHz from the 7th to the 13th days of gestation. Although there was no increase in fetal loss or significant increase in malformations, a few exencephalies and umbilical hernias occurred in the treated group. Nawrot et al. (1981) exposed mice on multiple days to eight–hour daily periods of 2.45–GH2 CW at 5–30 mW per cm squared. The higher energy levels raised body temperature. At the highest energy, 3.2 percent of the fetuses were malformed. This was significantly ($p < 0.05$) more than in the hyperthermic control group which had 1.7 percent defects. Decreased implantation was found when treatment was given on days 1–6. Lary et al. (1982) exposed pregnant rats to 27.12 MHz radiations at 300 volts per meter and found malformations of the central nervous system, skeleton and palate when treatment was on day 9, 11, 13, or 15. The defects were related to measured hyperthermia to 43.0 degrees C. Preimplantation exposures were followed by some increase in defects.

Inouye et al. (1983) made detailed studies of the brains of rat offspring exposed to 2,450 MHz microwave radiation at 10 mW per cm squared for 3 hours from day 4 through 40 postpartum and failed to demonstrate any significant effect. In similar type experiments in mice, Berman et al. (1984) found some brain weight decrease. Brown–Woodman et al. (1988) found a close association between the amount of heating and defects produced in rats at nine days by 27.12 MHz. Brown–Woodman and Hadley (1988) used pulsed shortwave radiation (27.12 MHz) for periods up to 60 minutes on day 9. These conditions did not raise temperature in a phantom model. There was an increase in resorptions but no increase in defects or lowering of fetal body weight.

Berman, E.; Carter, H.B. and House, D.: Growth and development of mice offspring after irradiation in utero with 2,450–MHz microwaves. Teratology 30:393–402, 1984.

Brent, R.L.: Radiations and other physical agents. In: Handbook of Teratology, J.G. Wilson and F.C. Fraser (eds.), Plenum Press, 1:153–223, 1977.

Brown–Woodman, P.D.C. and Hadley, J.A.: Studies of the teratogenic potential of exposure of rats to 27.12 MHz pulsed shortwave radiation. J. Bioelectric. 7(1):57–67, 1988.

Brown–Woodman, P.D.C.; Hadley, J.A.; Waterhouse, J. and Webster, W.S.: Teratogenic effects of exposure to radiofrequency radiation (27.12 MHz) from a shortwave diathermy unit. Ind. Health 26:1–10, 1988.

Hofmann, D. and Dietzel, F.: Aborte und Missbildungen nach Kurwellendurchflut–ung in der Schwangerschast. Geburts. Frauen Heilk. 26:378–390, 1966.

Inouye, M.; Galvin, M.J. and McRee, D.I.: Effect of 2,450 MHz microwave radiation on the development of the rat brain. Teratology 28:413–419, 1983.

Lary, J.M.; Conover, D.L.; Fole, E.D. and Hanser, P.L.: Teratogenic effects of 27.12 MHz radiofrequency radiation in rats. Teratology 26:299–309, 1982.

Lary, J.M. and Conover, D.L.: Teratogenic effects of radio frequency radiation. IEEE Eng., Medicine and Biology Magazine, 42–26, March, 1987.

Logue, J.N.; Hamburger, S.; Silverman, P.M. and Chiacochierino, R.P.: Congenital anomalies and paternal occupational exposure to shortwave, microwave, infrared and acoustic radiation. J. Occup. Med. 27:451–452, 1985.

Mannor, S.M.; Serr, D.M.; Tamari, I.; Meshorer, A. and Frei, E.H.: The safety of ultrasound in fetal monitoring. Am. J. Obstet. Gynecol. 113:653–661, 1972.

Michaelson, S.M.: Biological effects of microwave exposure. In: Biological Effects and Implications of Microwave Radiation. Symposium Proceedings, Richmond, Virginia, Sept. 17–19, S.F. Cleary (ed). U.S. Public Health Service, Richmond, Virginia, 35–58, 1969.

Nawrot, P.S.; McRee, D.I. and Staples, R.E.: Effects of 2.4 GHz CW microwave radiation on embryofetal development in mice. Teratology 24:303–314, 1981.

Rubin, A. and Erdman, W.J.: Microwave exposure of the human female pelvis during early pregnancy and prior to conception. Case rep. Am. J. Phys. Med. 38:219:220, 1959.

Saito, K.; Suzuki, K. and Motoyoshi, S.: Lethal teratogenic effects of long–term low–intensity radio frequency radiation at 428 mhz on developing chick embryo. Teratology 43:609–614, 1991.

Shoji, R.; Murakami, U. and Shimizu, T.: Influence of low–intensity ultrasonic irradiation on prenatal development of two in–bred mouse strains. Teratology 12:227–232, 1975.

Tachibana, M.: Effects of irradiation of high–energy continuous–wave ultrasounds on the fetuses of mice (dd–I and CH3HE strains). (Japanese) Acta Obstet. Gynaec. Jpn. 29:1097–1105, 1977.

Umeda, S.: Supplementary information on the biological effects of short waves. The effects on the course of pregnancy and development of fetuses. Sanka Fujinka Kiyo 24:265–346, 1941.

1645 Midaflur

4–Amino–2,2,5,5–tetrakis(trifluoromethyl)–3–imidazoline

Clark et al. (1971) tested this drug in rats and rabbits orally using 2 mg per kg during organogenesis. No adverse fetal effects were found but there was some delay in ossification in the rabbit fetuses.

Clark, R.; Lynes, T.E.; Price, W.A.; Smith, D.H.; Woodward, J.K.; Marvel, J.P. and Vernier, V.G.: The pharmacology and toxicology of midaflur. Toxicol. Appl. Pharmacol. 18:917–943, 1971.

1646 Mifepristone *RU486*

Preliminary case reports suggest lack of human teratogenicity but Pons et al. (**1991**) have reported an exposed fetus with cleft palate and lip and sirenomelia. Jost (1986) gave rabbits about 250 micro gm per kg for 1 to 5 days starting on day 11. About one–half aborted but among the 9 surviving fetuses they had malformations which included growth retardation and cranial rachischisis. Hardy and New (**1991**) using a whole embryo culture of rat embryos at 9.5 days gestation found inhibition of development when the concentrations were 7–10 times those in human serum.

Hardy, R.P. and New, D.A.T.: Effects of the anti–progestin RU 38486 on rat embryos growing in culture. Fd. Chem. Toxic. 29:361–362, 1991.

Jost, A.: Animal reproduction. C R Acad Sci Ser III 303:281–284, 1986.

Pons, J.C.; Imbert, M.C.; Elefant, E.; Roux, C.; Herschkorn, P. and Papiernik, E.: Development after exposure to mifepristone in early pregnancy. The Lancet 338:763, 1991.

1647 Milacemide

Garny et al. (1986) gave up to 1000 mg per kg to rats on days and produced no increase in congenital defects.

Garny, V.; DeBoelpaep, C. and Roba, J.: Teratology study of milacemide and sodium valproate in the OF$_1$ mouse. (abs) Teratology 33:18A, 1986.

1648 Miloxacin CAS 37065–29–5

Yamada et al. (1980) studied this antibiotic in the rat before mating and during organogenesis in doses of up to 300 mg per kg daily. Fetal weight was reduced at doses above 37.5 mg per kg but no defects were found. At the highest dose some decrease in viability and fetal weight occurred when treatment was given on days 17–21.

Yamada, T.; Tarumoto, Y.; Hosoda, K.; Koike, M.; Furasawa, S.; Sasajima, M. and Ohzeki, M.: Reproductive studies of miloxacin in rat fertility, teratogenicity and peri and postnatal studies. Oyo Yakuri 19:651–662, 815–831, 833–844, 1980.

1649 Milrinone *YM018*

Ono et al. (**1993**) gave 5 mg per kg intravenously to rabbits before and in early pregnancy and found no adverse effects on reproduction. In rats intravenous doses of 5 mg before pregnancy and during early gestation and 10 mg per kg during organogenesis were given without ill effects to the fetuses. Perinatal dosing was followed by normal behavior and reproduction in the next generation. In both species the highest doses caused flushing of the ears and limbs.

Ono, C.; Ishitobi, H.; Iwama, A.; Fujiwara, M. and Shibata, M.: Reproductive and developmental toxicity studies in rats andrabbits given milrinone (YM018) intravenously. Oyo Yakuri 46:305–316, 1993.

1650 Mimosine CAS 500–44–7

Dewreede and Wayman (1970) fed rats diets containing this amino acid extracted from Leucaena leucocephala. With diets containing 0.7 percent the resorption rate was increased, and 3.5 percent of the fetuses were deformed. The deformities were associated with uterine perforations which caused constriction of the protruding fetal parts.

Dewreede, S. and Wayman, O.: Effect of mimosine on the rat fetus. Teratology 3:21–28, 1970.

1651 Minocycline CAS 10118–90–8

Jackson et al. (1975) administered 8.7–17.4 mg per kg orally to rhesus monkeys during embryogenesis and the period of fetal skeletal formation. No adverse effects were observed.

Jackson, B.A.; Rodwell, D.E.; Kanegis, L.A. and Noble, J.F.: Effect of maternally administered minocycline on embryonic and fetal development in the rhesus monkey. (abs) Toxicol. Appl. Pharmacol. 33:156 only, 1975.

1652 Minoxidil CAS 38304–91–5

Kaler et al. (1987) reported an infant born after the mother took 10 mg daily during gestation. Hypertrichosis was prominent and suggests a role of Minoxidil which can produce hypertrichosis in adult patients. Captopril (50 mg) and propanolol (160 mg) were also taken during gestation. The infant had a depressed nasal bridge, micrognathia, undescended testes, fifth finger clinodactyly, mid–phallic constriction and small omphalocele. A large ventriculoseptal defect was present. Transient hypotension occurred postnatally. Rosa et al. (1987) summarized this case and reported one newborn who had hypertrichosis without other defects.

Carlson and Feenstra (1977) studied the effect of this vasodilator in pregnant rats and rabbits. Oral doses during organogenesis did not produce adverse effects at 10 mg per kg. Changes in the retinal folds of the rat fetuses were probably the result of poor fixation. Yamada et al. (1992) gave up to 80 mg per kg to rats before and during the early days of gestation. No abnormalities were found but the number of corpora lutea and implants was reduced. At 120 mg per kg during organogenesis fetal weight was reduced. Eleven percent of the fetuses had visceral anomolies (Yamada et al.,1992). Skeletal defects also occurred. No increase in defects was found at 11 mg per kg. Rats receiving up to 80 mg per kg preinatally had decreased body weight gain and increased body weight gain and increased stillbirths. The

pups were small but developed normally and reproduction was normal (Yamada et al., **1992**).

Carlson, R.G. and Feenstra, E.S.: Toxicologic studies with the hypotensive agent minoxidil. Toxicol. Appl. Pharmacol. 39:1–11, 1977.

Kaler, S.G.; Patrinos, M.E.; Lambert, G.H.; Myers, T.F.; Karlman, R. and Anderson, C.L.: Hypertrichosis and congenital anomalies associated with maternal use of minoxidil. Pediatrics 79:434–436, 1987.

Rosa, F.W.; Idanpaan–Heikkila, J. and Asanti, R.: Fetal minoxidil exposure. Pediatrics 80:120 only, 1987.

Yamada, T.; Nishiyama, T.; Ohno, H. and Nakane, S.: Reproductive and developmental toxicity studies of minoxidil (I) fertility study in rats. Iyakuhin Kenkyu 23:748–755, 1992.]bib Yamada, T.; Uchida, H.; Ohno, H.; Ohsawa, K. and Nakane, S.: Reproductive and developmental toxicity studies of minoxidil (II) teratogenicity study in rats. Iyakuhin Kenkyu 23:756–773, 1992.

Yamada, T.; Inoue, T. and Nakane, S.: Reproductive and developmental toxicity studies of minoxidil (III) perinatal and postnatal study in rats. Iyakuhin Kenkyu 23:774–785, 1992.

1653 Miporamicin

Sasaki et al. (**1989**) gave this antibiotic orally to rats in doses of up to 1000 per kg on days 7–17. No adverse fetal or postnatal effects were seen except for birth weight reduction in male fetuses at the 200 and 1000 mg per kg level. In rabbits gavaged with up to 200 mg per kg on days 6–18 the only adverse effect was weight reduction in the dams treated with the highest dose. (Hazelden el al, **1989**)

Hazelden, K.P.; Wilson, J.A.; Sasaki, M.; Takahashi, H. and Yamamoto, H.: Miporamicin teratogenicity study in rabbits. The Japanese Journal of Antibiotics 42:2488–2499, 1989.

Sasaki, M. Nakajima, M. and Yamamoto, H.: Oral dosage study of miporamicin administered during the period of fetal organogenesis in rats. The Japanese Journal of Antibiotics 42:2472–2487, 1989.

1654 Miracil–D CAS 548–57–2

Karnofsky and Lacon (1962) briefly reported that at the LD–50 of 2 to 4 mg per egg a slight feather inhibition was noted in the chick embryo.

Karnofsky, D.A. and Lacon, C.R.: Survey of cancer chemotherapy service center compounds for teratogenic effect in the chick embryo. Cancer Res. 22:84–86, 1962.

1655 Mirex™ CAS 2385–85–5

This polychlorinated insecticide was given in the diet at 5 PPM to mice by Ware and Good (1967). They found that this low dosage produced a significant reduction in litter size and the number of offspring produced. Congenital defects were not studied.

Khera et al. (1976) studied this compound in rats at levels of six and 12.5 mg per kg which were toxic to the mother. Gavaging the mothers on days 6 through 15, they found decreased fetal survival with an increased incidence of cleft palate, subcutaneous edema along with several other types of defects. Grabowski and Payne (1983) gave rats 6 mg per kg on days 8.5–15.5. At term, 18 percent had heart block and many of these fetuses were edematous.

Grabowski, C.T. and Payne, D.B.: The causes of perinatal death induced by prenatal exposure of rats to the pesticide, Mirex. Part 1: Preparturition observations of the cardiovascular system. Teratology 27:7–11, 1983.

Khera, K.S.; Villeneuve, D.C.; Terry, G.; Panopio, L.; Nash, L. and Trivett, G.: Mirex: A teratogenicity, dominant lethal and tissue distribution study in rats. Food Cosmet. Toxicol. 14:25–29, 1976.

Ware, G.W. and Good, E.E.: Effects of insecticides on reproduction in the laboratory mouse. Toxicol. Appl. Pharmacol. 10:54–61, 1967.

1656 Miroprofen 2–(p–Imidazo(1,2–a)pyridyl)phenyl Proprionic Acid CAS 55843–86–2

Hamada and Imanishi (1981) studied this anti–inflammatory agent in rats and rabbits in a series of papers. Doses of 25 mg per kg before mating had no adverse effect in the rat. During organogenesis both rabbits and rats received up to 100 mg per kg and no increase in major defects was found. These higher doses were particularly toxic in the rabbit where fetal death and lumbar ribs were more common. At 5 and 25 mg per kg given on day 17 through day 21 postnatally, there was prolongation of gestation, parturition and excessive maternal vaginal bleeding. Neonates and suckling pups had a higher mortality rate. Studies done after weaning showed no changes in behavioral tests.

Hamada, Y. and Imanishi, M.: Fertility, teratogenicity and perinatal and postnatal studies of miroprofen in rats. Iyakuhin Kenkyu 12:802–841, 1981.

1657 Misoprostol

Schuler et al. (1992) followed 29 women in Brazil who were exposed to this PGF, analog. In three spontaneous abortions occurred and no major defects were found among 17 newborns. Abortion does not follow administration in 6–10% of women. (Costa and Vessey, 1993) Brent (1993) has described the difficulty in interpreting case reports from an exposed population that could number 5 million.

Brent, R.L.: Congenital malformation case reports: the editor's and reviewers dilemma. Am. J. Med. Genetics 47:872–874, 1993.

Costa, S.H. and Vessey, M.P.: Misoprosol and illegal abortion in Rio de Janeiro, Brazil. Lancet 341:1258–1281, 1993.

Schuler, L.; Ashton, P.W.; Sanseverino, M.T.: Teratogenicity of misoprostol. The Lancet 339:437, 1992.

1658 Mitomycin C CAS 50–07–7

This growth inhibitor isolated from streptomyces caespitosus produces defects in the mouse when given 5 to 10 mg per kg on single gestational days 7 through 13 (Tanimura, 1968). Skeletal defects were most common with some defects of the palate and brain. Chaube and Murphy (1968)

found that in the rat, both the maternal and fetal LD–50 were 2 to 2.5 mg per kg and no teratogenicity was found. Snow (1983) summarized his laboratory's work on repair of the cell necrosis associated with this agent. Spielmann et al. (1986) injected 5, 10 or 20 mg per kg intraperitoneally on day 2 in the mouse and found increased lethality on day 17.

Tam (1988) found that mouse embryos treated at the morula stage had reduced inner cell masses. Those with reductions to 70 percent or above were able to regenerate and grow normally to days 12–14.

Chaube, S. and Murphy, M.L.: The teratogenic effects of the recent drugs active in cancer chemotherapy. In: Advances in Teratology, D.H.M. Woollam (ed.), New York: Logos and Academic Press, 3:181–237, 1968.

Snow, M.H.L.: Restorative growth in mammalian embryos. In: Issues and Reviews in Teratology, H. Kalter (ed.), Plenum Press, New York, 1(9):251–281, 1983.

Spielmann, H.; Kruger, C.; Tenschert, B. and Vogel, R.: Studies on the embryotoxic risk of drug treatment during the preimplantation period in the mouse. Arzneim–Forsch. 36:219–223, 1986.

Tam, P.P.L.: Postimplantation development of mitomycin C-treated mouse blastocysts. Teratology 37:205–212, 1988.

Tanimura, T.: Effects of mitomycin C administered at various stages of pregnancy upon mouse fetuses. Okajimas Folia Anat. Jpn. 44:337–355, 1968.

1659 Mizoribine Bredinin™

4–Carbamoyl–1–β–D–ribofuranosylimidazolium–5–olate CAS 50924–49–7

This immunosuppressive agent was given intraperitoneally to rats on day 8, 9, 10 or 11 in amounts of 5 mg per kg and produced anophthalmos, hydrocephaly and visceral defects as well as some skeletal defects. Day 9 was teratogenically the most susceptible (Okamoto et al., 1978).

Kubota et al. (1983) and Sasaki et al. (1983) studied this immunosuppressant in pregnant rats and rabbits. In rats treated on days 7–17 of gestation with one mg per kg, the fetuses had an increase in congenital defects which included ventricular septal defects, diaphragmatic hernias and hydrocephaly. At the 0.5 mg dose, there was no increase in defects. No postnatal behavioral changes were detected. No reproductive changes occurred in females treated with 3 mg per kg. An increase in hydronephrosis and decreased renal weight was found in rats treated perinatally with 2 mg per kg. In rabbits treated on days 6–18 with maximum doses of 4 mg per kg, an increase in resorption and dead fetuses was found.

Kubota, H.; Sasaki, M.; Suda, M. and Kobayashi, Y.: Reproduction studies on mizoribine (Bredinin™) in rats. Oyo Yakuri 26:377–407, 1983.

Okamoto, K.; Kobayashi, Y.; Yoshida, K.; Nozaki, Y.; Kawai, Y.; Kawano, H.; Mayumi, T. and Hama, T.: Teratogenic effects of Bredinin™ , a new immunosuppressive agent, in rats. Cong. Anom. 18:227–233, 1978.

Sasaki, M.; Kubota, H.; Suda, M.; Kobayashi, Y. and Hayano, K.: Reproduction studies of mizoribine (Bredinin™

) in rabbits. Oyo Yakuri 26:409–414, 1983.

1660 MMR–701 *Plasminogen Activator*

Sato et al. (1993 A,B,C) gave this human tissue plasminogen activator to rats before pregnancy, during organogenesis and perinatally in intravenous doses of up to 10 mg per kg and found no adverse effects on reproduction, fetal development or behavior. Similar studies using up to 6 mg per kg in rabbits were done without finding teratogenicity(Kumada et al., **1993**).

Kumada, J.; Komai, Y.; Ogura, H.; Isowa, K.; Hattori, M.; Furutaki, K. and Hayakawa, M.: Study on intravenous administration of MMR701 during the period of organogenesis in rabbits. Yakuri to Chiryo 21:s2007–s2016, 1993.

Sato, S.; Ushimaru, T.; Kawakami, T.; Kinoshita, K.; Amada, Y. and Yokoyama, Y.: Study on intravenous administration of MMR701 prior to and in the early stages of pregnancy in rats. Yakuri to Chiryo 21:s1973–s1987, 1993A.

Sato, S.; Suyama, S.; Ushimaru T.; Kinoshita, K.; Yamamoto, M. and Hayakawa, M.: Study on intravenous administration of MMR701 during the period of organogenesis in rats. Yakuri to Chiryo 21:s1989–s2006, 1993B.

Sato, S.; Kinoshita, K. and Hayakawa, M.: Study on intravenous administration of MMR701 during the perinatal and lactation periods in rats. Yakuri to Chiryo 21:s2017–s2031, 1993C.

1661 MO–911 *N–Benzyl–N–methyl–2–propynylamine HCl*

Poulson and Robson (1963) gave 2 mg per day orally to pregnant mice on days 1–6 and produced a decrease in the number of live litters. Treatment on days 11–16 was not associated with a decrease in live litters.

Poulson, E. and Robson, J.M.: The effect of amine oxidase inhibitors on pregnancy. J. Endocrinol. 27:147–152, 1963.

1662 Mofebutazone *Monazan™ Monophenylbutazone*

Larsen and Bredahl (**1966**) gave rabbits 60 or 150 mg per kg subcutaneously during the first 20 days of gestation. No increase in defects was found.

Larsen, V. and Bredahl, E.: The embryotoxic effect of rabbits of monophenylbutazone (monazan™) compared with phenylbutazone and thalidomide. Acta Pharmacol. et Toxicol. 24:443–455, 1966.

1663 Mofezolac

Toteno et al. (**1990**) gave rats oral doses up to 150 mg per kg of this non–steroiddal antiinflammatory on days 7–17. At the highest dose fetal weight was reduced but no teratogenicity or interference with development was found. Fuchigami et al. (**1990**) treated rabbits with up to 200 mg per kg daily during days 6–18 and found no fetal ill effects. Perinatal studies in rats were done and at 50 and 100 mg per kg the dams had poor weight gain and did not care for their pups well. Behavior and reproduction of the exposed pups was normal.

Fuchigami, K.; Otsuka, T.; Sameshima, K.; Matsunaga, K.; Kodama, R. and Yamakita, O.: Reproductive and developmental toxicity study of mofezolac (n–22) (3)–teratogenicity studdy in rabbits by oral administration. The Journal of Toxicological Sciences 15:209–218, 1990.

Toteno, I.; Haguro, S.; Furukawa, S.; Morinaga, T.; Morino, K.; Fujii, S. and Yamakita, O.: Reproduction and developmental toxicity study of mofezolac (n–22) (2)–study by oral administration of n–22 during the period of fetal organogenesis in rats. The Journal of Toxicological Sciences 15:165–208, 1990.

1664 Molybdenum CAS 7439–98–7

Wide (1984) found decreased fetal weight in the offspring of mice receiving 100 millimoles intravenously on day 8. Ridgway and Karnofsky (1952) found that the LD–50 for the four–day chick embryo was 0.8 mg. No defects were found. Schroeder and Mitchener (1971) fed mice a diet containing 0.45 PPM and found some runting in the offspring of the third generation.

Ridgway, L.P. and Karnofsky, D.A.: The effects of metals on the chick embryo: Toxicity and production of abnormalities in development. Ann. N.Y. Acad. Sci. 55:203–215, 1952.

Schroeder, H.A. and Mitchener, M.: Toxic effects of trace elements on the reproduction of mice and rats. Arch. Environ. Health 23:102–106, 1971.

Wide, M.: Effect of short–term exposure to five industrial metals on the embryonic and fetal development of the mouse. Environ. Res. 33:47–53, 1984.

1665 Mometasone Furoate

Morita et al. (**1990**) gave this steroid subcutaneously to rats on days 7–17. The drug is designed for topical use. Fetal growth, birth rate and ossification was decreased at 0.006 mg per kg and above. Postnatal growth was decreased at 0.03 mg per kg. No significant increase in malformations was seen in doses up to 0.03 mg per kg. In rabbits Wada et al. (**1990**) applied up to 1.0 mg per kg dermally on days 6–18. At doses of 0.2 mg per kg external malformations were seen including cleft lip, omphalocele and brain hernias. Other types of defects were seen at 1.0 mg per kg.

Morita, Y.; Ohta, R.; Watanabe, C.; Mizutani, M. and Kobayashi, F.: Teratogenicity study of mometasone furoate in rats. Kiso to Rinsho 24:2517–2543, 1990.

Wada, K.; Hashimoto, Y.; Mizutani, M. and Kobayashi, F.: Teratogenicity study of mometasone furoate in rabbits. Kiso to Rinsho 24:2545–2555, 1990.

1666 Momorcharines

Chan et al. (1986) studied the effect of α and β momorcharins on mouse embryos grown in vitro at 50 and 100 microgm per ml of medium. There was lethality and some defects such as absence of forelimbs and unclosed cranial folds were found.

The compounds are from the Chinese herb Kaguazi and have been described as abortifacient in experimental animals.

Chan, W.Y.; Tam, P.P.L.; Choi, H.L.; Ng, T.B. and Yeung, H.W.: Effects of momorcharins on the mouse embryo at the early organogenesis stage. Contraception 34:537–544, 1986.

1667 MON 18500

Schroeder et al. (**1994**) fed this herbicide to up to 1000 ppm in the diet for two generations. No adverse reproductive effects were seen at 200 ppm. At 1000 ppm growth retardation and enlarged internal organs were found in the fetuses.

Schroeder, R.E. and Farmer, D.R.: A two–generation reproduction–fertility study in rats with MON 185000. (abs) Teratology 49:389–390, 1994.

1668 Monensin

Fetal growth retardation was found by Atef et al. (1986) in rats but no teratogenicity was observed at oral doses of 1.75 on days 9–17. No fetal survivors were found at dosages of 3.5 mg per kg per day.

Atef, M.; Shalaby, M.A.; El–Sayed, M.G.A.; El–Din, S.; Youssef, A.H. and El–Sayed, M.A.I.: Influence of monensin on fertility in rats. Clin. Exper. Pharm. Physiol. 13:113–121, 1986.

1669 Monocrotaline

Sriraman et al. (**1988**) gavaged rats on the 10th through 14th day of gestation with 18mg of this plant alkaloid. Postnatally lesions of the pulmonary artery and liver were found.

Sriraman, P.K.; Naidu, N.R.G. and Roa, P.R.: Effect of monocrotaline, a pyrrolizidine alkaloid on the progeny of rats. Indian Journal of Animal Sciences 58(11):1292–1295, 1988.

1670 Monocrotophos *Azodrin*
3–(Dimethoxyphosphinyloxyl)–N–methyl–cis–crotonamide CAS 6923–22–4

Schom and Abbott (1977) produced teratogenicity in avian embryos with exposure levels of as little as 0.4 mg per kg.

Schom, C.B. and Abbott, U.K.: Temporal, morphological and genetic responses of avian embryos to azodrin, an organophosphate insecticide. Teratology 15:81–88, 1977.

1671 Monomethylaminobenzene

Discussed under Aminoazobenzene

1672 Monomethylformamide *N–Methylformamide* CAS 123–39–7

Thiersch (1971) demonstrated that this compound was highly toxic to rat fetuses even when it was administered by means of tail painting. Tail painting on days 7 through 14 produced fetal death in 87 percent, and all of the survivors were malformed. The malformations consisted of hydronephrosis, hydrocephalus, along with hydramnion. An oral dose of 1.0 cc per kg produced similar results in the rat. Oettel and Frohberg (1964) injected 0.1 ml per kg on the

11th and 12th days and produced a 50 percent mortality rate in the rat fetus. Tuchmann–Duplessis and Mercier–Parot (1965) demonstrated by skin application that the most sensitive period for teratogenicity was between the 10 and 12th days of gestation.

Oettel, H. and Frohberg, H.: Teratogene Wirkung einfacher Saureamide im Tierversuch. Naunyn–Schmiedebergs. Arch. Pharmakol. Exp. Pathol. 247:363–364, 1964.

Thiersch, J.B.: Investigations into the differential effect of compounds on rat litter and mother. In: Malformations Congenitales Des Mammiferes, H. Tuchmann–Duplessis (ed.), Paris: Masson, 95–113, 1971.

Tuchmann–Duplessis, H. and Mercier–Parot, L.: Production chez le rat, d'anomalies apres applications cutanees d'un solvent industriel: La mono–methyl–formamide. C. R. Acad. Sci. (Paris) 261:241–243, 1965.

1673 Monophenylbutazone

Larsen and Bredahl (1966) gave up to 150 mg of this chemical to rabbits subcutaneously during the first 20 days of gestation. No adverse effects were found.

Larsen, V. and Bredahl, E.: The embryotoxic effect on rabbits of monophenyl butazone (Monazan™) compared with phenylbutazone and thalidomide. Acta Pharmacol. Toxicol. 24:443–455, 1966.

1674 Monosodium Glutamate *Glutamate* CAS 142–47–2

Murakami and Inouye (1971) injected mice on the 17th or 18th day of gestation with 5 mg per kg. Nuclear pyknosis was found in the cells of the arcuate and ventromedial nuclei of the fetuses after three hours. Examination of treated fetuses after a 24–hour period did not show any abnormal lesion. Olney (1969) reported that immature mice injected with substantial doses of monosodium glutamate developed brain lesions.

Murakami, U. and Inouye, M.: Brain lesions in the mouse fetus caused by maternal administration of monosodium glutamate (preliminary report). Congenital Anomalies 11:171–177, 1971.

Olney, J.W.: Brain lesions, obesity and other disturbances in mice treated with monosodium glutamate. Science 164:719–721,

1675 Cis–1[4–(p–Monthane–8–yloxy)phenyl] Piperdine *YM9429*

Shibata (**1993**) found cleft palates in rat fetuses exposed to 400 to 800 mg per kg orally during various periods of organogenesis. Skeletal defects were also found.

Shibata, M.: A new potent teratogen in cd rats inducing cleft palate. The Journal of Toxicological Sciences 18:171–178, 1993. 1969.

1676 Moquizone *1-Morpholino–acetyl–3–phenyl–2,3–di-hydro-4(1H)-quinazolinone* CAS 19395–58–5

Setnikar and Magistretti (1970) gave mice, rabbits and

rats up to 60 mg per kg daily by mouth during active organogenesis and found no increase in congenital defects.

Setnikar, I. and Magistretti, M.J.: Maternal and fetal toxicity of moquizone. Arzneim. Forsch. 20:1559–1561, 1970.

1677 Morphine CAS 57–27–2

Heinonen et al. (1977) found no significant increase in congenital defects among the offspring of 448 mothers who took morphine any time in pregnancy.

In a series of papers on the effect of morphine on rat spinal cord size (summary by Kirby, 1983), evidence was presented that both the drug and the associated nutritional effects contributed to reduced spinal cord volume. Twenty mg per kg was given daily from day 12 through the second day postpartum. Friedler and Cochin (1972) pretreated rats with morphine, and then after a five–day non–treatment interval, pregnancy was initiated. Although there were no reported changes in the litters, the fetuses after a three–to four–week period exhibited significant but transient growth retardation. The effect was not eliminated by cross–foster feeding, excluding a long–term effect through maternal nutrition. In subsequent work with mice treated pre–gestationally, postnatal behavioral effects were found in the offspring (Friedler, 1978). Friedler and Wheeling (1979) also treated males with opioids prior to mating and found behavioral effects in the offspring.

Iuliucci and Gautieri (1971) gave 200 to 400 mg of morphine per kg on gestational day 8 or 9 of the mouse and produced a few exencephalic and axial skeletal defects. The authors point out that the hypoxic effect alone of such large doses could account for the defects. Johannesson and Becker (1972) were unable to produce fetal changes in rat fetuses exposed to maternal doses of 20 mg per kg for periods before and during organogenesis. Fujinaga and Mazze (1988) gave rats up to 70 mg per kg daily from day 5 through day 20 using miniperfusion pumps. No teratogenicity was found at doses up to 70 mg per kg but a higher postnatal mortality occurred even in cross–fostered animals.

Geber and Schramm (1975) injected 35 to 322 mg per kg subcutaneously into hamsters on day 8 of gestation and produced with the higher doses, 20–30 percent congenital defects in the day 12 embryos. Cranioschisis was the predominant type of defect. Morphine antagonists nalorphine, naloxone and cyclazocine blocked the teratogenic activity when given concurrently. Eleven other narcotic analgesics including codeine, heroin and meperidine were studied and found to produce similar teratogenicity. O'Callaghan and Holtzman (1977) injected rats subcutaneously with 1.0 mg per kg of levorphanol or dextrorphan (a cogener and inactive isomer of morphine, respectively). Treatment on days 5–12 did not reduce the fetal weight or survival. Teratogenicity was not studied.

Friedler, G.: Pregestational administration of morphine sulfate to fetal mice: Long term effects on development of subsequent progeny. J. Pharmacol. Exp. Therap. 205:33–39, 1978.

Friedler, G. and Cochin, J.: Growth retardation in offspring of female rats treated with morphine prior to conception. Science 175:654–655, 1972.

Friedler, G. and Wheeling, H.S.: Behavioral effects on offspring of males injected with opioids prior to mating. Pharmacol. Biochem. 11:23–28, 1979.

Fujinaga, M. and Mazze, R.I.: Teratogenic and postnatal developmental studies of morphine in Sprague–Dawley rats. Teratology 38:401–410, 1988.

Geber, W.F. and Schramm, L.C.: Congenital malformations of the central nervous system produced by narcotic analgesics in the hamster. Am. J. Obstet. Gynecol. 123:705–713, 1975.

Harpel, H.S. and Gautieri, R.F.: Morphine–induced fetal malformations: Exencephaly and skeletal fusions. J. Pharm. Sci. 57:1590–1597, 1968.

Heinonen, O.P.; Slone, D. and Shapiro, S.: Birth Defects and Drugs in Pregnancy. Publishing Sciences Group Inc., Littleton, Mass., 1977.

Iuliucci, J.D. and Gautieri, R.F.: Morphine–induced fetal malformations II. Influence of histamine and diphenhydramine. J. Pharm. Sci. 60:420–424, 1971.

Johannesson, T. and Becker, B.A.: The effects of maternally–administered morphine on rat foetal development and resultant tolerance to the analgesic effect of morphine. Acta Pharm. Toxicol. 31:305–313, 1972.

Kirby, M.L.: Recovery of spinal cord volume in postnatal rats following prenatal exposure to morphine. Dev. Brain Res. 6:211–217, 1983.

O'Callaghan, J.P. and Holtzman, S.G.: Prenatal administration of levorphanol or dextrorphan to the rat: Analgesic effect of morphine on the offspring. J. Pharmacol. Exp. Ther. 200:255–262, 1977.

1678 Morphine–N–oxide

This intermediate product of morphine was studied by Fennessy and Fearn (1969) in rats on the 3rd, 4th and 5th days of gestation. Subcutaneous doses of 50 mg per kg had no ill effect on the fetuses at term.

Fennessy, M.R. and Fearn, H.J.: Some observations on the toxicology of morphine–n–oxide. J. Pharm. Pharmac. 21:668–673, 1969.

1679 Mosapride Citrate 4–amino–5–chlloro–ethoxy–N–[[4–(4–fluorobenzyl)–2–morpholinyl methyl]–Benzamide Citrate Dihydrate

Funabashi et al. (1993 A,B,C) gavaged rats before fertilization and during the first 7 days of pregnancy, during organogenesis and peri and postnatally with up to 300 mg per kg. No adverse fetal effects were found but at the highest dose there was maternal toxicity and reduced fetal weight. Postnatal weight and survival were decreased at the highest dose but behavioral development was normal. Funabashi et al. (1993 D) gave rabbits up to 125 mg per kg without adverse fetal effects.

Funabashi, H.; Terada, Y.; Nakamura, H.; Ishida, S. and Ikeya, M.: Reproductive and developmental toxicity studies of mosapride citrate (1): Fertility study in rats. Yakuri to Chiryo 21:3411–3422, 1993A.

Funabashi, H.; Nishimura, K.; Terada, Y.; Shigematsu, K.; Tateishi, Y. and Nakamura, H.: Reproductive and developmental toxicity studies of mosapride citrrate (2): Teratogenicity study in rats. Yakuri to Chiryo 21:3423–3445, 1993B.

Funabashi, H.; Terada, Y.; Nakamura, H.; Shimazu, H. and Sugisawa, K.: Reproductive and developmental toxicity studies of mosapride citrate (3): Perinatal and postnatal study in rats. Yakuri to Chiryo 21:3447–3468, 1993C.

Funabashi, H.; Terada, Y.; Nakamura, H.; Matsuoka, T. and Takigami, H.: Reproductive and developmental toxicity studies of mosapride citrate (4): Teratogenicity study in rabbits. Yakuri to Chiryo 21:3469–3479, 1993D.

1680   Motretinide

This aromatic retinoid was given to mice on day 11 orally at a dose of 100 mg per kg and skeletal reduction defects and cleft palates were found in the fetuses (Reiners et al., 1988). They presented evidence that the drug was metabolized to etretin.

Reiners, J.; Lofberg, B.; Creech Kraft, J.; Kochhar, D.M. and Nau, H.: Transplacental pharmacokinetics of teratogenic doses of etretinate and other aromatic retinoids in mice. Reproductive Toxicology 2:19–30, 1988.

1681   Moveltipril Calcium *MC–838*

Sugiyama et al. (**1990**) gave the ACE inhibitor to rats on days 7–17 at doses of up to 250 mg per kg. No adverse fetal effects were found and behavioral and reproductive functions of the offspring were not effected.

Sugiyama, O.; Igarashi, S.; Watanabe, K.; Toya, M.; Watanabe, S. and Kouichiro, T.: Teratological Study of MC–838 (moveltipril calcium) in rats. Yakuri to Chiryo 18:3311–3324, 1990.

1682   Moxalactam CAS 64952–97–2

This oxacephalosporin was given intraperitoneally by Kobayashi and Hara (1980) to rats of both sexes before pregnancy and during the first 7 days of gestation. Up to 200 mg per kg per day had no adverse effect on reproduction. Using the intravenous route and the same dose, rats were treated on days 7–17 or 17–21 and no adverse effects were found in the fetuses or in postnatal development of the offspring (Kobayashi and Ando, 1980; Hasegawa and Yoshida, 1980).

Hasegawa, Y. and Yoshida, T.: Teratology study of 6059–S on rats. Chemotherapy 28:1119–1141, 1980.

Kobayashi, F. and Ando, M.: Perinatal–postnatal study on 6059–S in rats. Chemotherapy 28:1141–1157, 1980.

Kobayashi, F. and Hara, K.: Fertility study of 6059–S in rats. Chemotherapy 28:1108–1118, 1980.

1683   Moxisylyte *Thymoxamine Hydrochloride* CAS 964–52–3

Shoji et al. (1982) studied this adrenergic blocker in rats and rabbits using 120 mg per kg per day orally. No effects on fertility, fetal development or postnatal function was reported. Rabbit fetuses did not have demonstrable adverse effects.

Shoji, S.; Kida, M.; Wada, S.; Kurimoto, T.; Shikuma, N.; Okuma, Y.; Harada, H.; Machida, N. and Hirayama, Y.: Reproductive studies of thymoxamine hydrochloride. Yakuri to Chiryo 10:589–602, 603–615, 617–627, 629–640, 1982.

1684   Mumps Virus

A number of studies of the offspring of mothers having mumps during gestation demonstrated no increase in congenital defects (Hill et al., 1958; Siegel et al., 1966; Korones et al., 1970).

An intriguing hypothesis that intrauterine mumps virus might cause congenital endocardial firbroelastosis was tested by St. Geme et al. (1971) who produced persistent virus infection in the chick embryo with associated myocarditis. St. Geme et al. (1974) extended their testing to the monkey but could not detect the persistence of virus in the fetuses. Delayed hypersensitivity without neutralizing antibody was demonstrated in this monkey fetal model. They reported that during gestation the risk of endocardial fibroelastosis is less than two percent of the exposed fetuses. Direct inoculation of the virus into fetal monkey brains did not produce any adverse effects (Moreland et al., 1979).

Shone et al. (1966) observed a very high incidence of positive mumps skin tests in patients with either fibroelastosis or congenital mitral stenosis but could not show a rise in mumps serum antibodies in 23 patients with fibroelastosis. Gersony et al. (1966), using a more rigorous criterion for skin test interpretation, could not show an increased incidence in this condition.

Gersony, W.M.; Katz, S.L. and Nadas, A.S.: Endocardial fibroelastosis and the mumps virus. Pediatrics 37:430–434, 1966.

Hill, B.; Doll, R.; Galloway, T.M. and Hughes, J.P.W.: Virus diseases in pregnancy and congenital defects. Br. J. Prev. Soc. Med. 12:1–7, 1958.

Korones, S.B.; Todaro, J.; Roane, J.A. and Sever, J.L.: Maternal virus infection after the first trimester of pregnancy and status of offspring to 4 years of age in a predominantly Negro population. J. Pediatr. 77:245–251, 1970.

Moreland, A.F.; Gaskin, J.M.; Schimpff, R.D.; Woodard, J.C.; Olson, G.A.: Effects of influenza, mumps and western Equine viruses on fetal rhesus monkeys (Macaca mulatta). Teratology 20:53–64, 1979.

Shone, J.; Armas, S.M.; Manning, J.A. and Keith, J.D.: The mumps antigen skin test in endocardial fibroelastosis. Pediatrics 37:423–429, 1966.

Siegel, M. and Fuerst, H.T.: Low birth weight and maternal virus diseases: A prospective study of rubella, measles, mumps, chickenpox, and hepatitis. JAMA 197:680–684, 1966.

St. Geme, J.W.; Davis, C.W.C. and Noren, G.R.: An overview of primary endocardial fibroelastosis and chronic viral cardiomyopathy. Perspectives in Biology and Medicine. Summer: 495–505, 1974.

St. Geme, J.W.; Peralta, H.; Farias, E.; Davis, C.W.C.

and Noren, G.R.: Experimental gestational mumps virus infection and endocardial fibroelastosis. Pediatrics 48:821–826, 1971.

1685 Muzolimine

Hoffmann and Luckhaus (**1983**) gave this antidiuretic to pregnant rats and rabbits at oral doses of up to 30 mg per kg. Studies during organogenesis did not alter reproduction except at the highest doses fetal weight was decreased. Perinatal studies in the rat did not change development.

Hoffmann, K. and Luckhaus, G.: Toxicological investigations of muzolimine. Clinical Nephrology 19:S-20–S-25, 1983.

1686 MY–1

Hattori et al. (**1990**) and Sato et al. (**1990**) gave this antitumor agent to rats and rabbits subcutaneously during organogenesis in doses of up to 24 mg per kg daily. No adverse fetal effects occurred in either species. Similar treatment before fertilization in the rat did not effect fertility (Takahashi et al **1990**).

Hattori, M.; Katano, T.; Tachibana, I.; Ogura, H.; Komai, Y. and Yokoyama, Y.: Reproduction study of my–1 - teratogenicity study in rats by subcutaneous administration. Yakuri to Chiryi 18:1405–1432, 1990.

Sato, K.; Hayakawa, M.; Furukawa, M.; Genra, Y.; Gomita, S. and Yokoyama, Y.: Reproduction study of my–1 - teratogenicity study in rabbits by subcutaneous administration. Yakuri to Chiryi 18:1433–1441, 1990.

Takahashi, M.; Hiraide, Y.; Karasawa, N.; Miyazaki, N.; Sakurai, T. and Yokoyama, Y.: Reproduction study of my–1 - fertility study in rats by subcutaneous administration. Yakuritochiryi 18:1387–1403, 1990.

1687 Mycoheptin

Slonitskaya and Mikhailets (**1975**) gave this fungicide to rats in doses of up to 1000 mg per kg on day 9 or on days 7–10 at 100 mg per kg. No congenital malformation increase was found but those treated on days 7–10 had a higher post implantation loss.

Slonitskaya, N.N. and Mikhailets, G.A.: Effect of nystatin and mycpheptin on the intrauterine development of rat fetus (Russian). Antibiotiki 20:45–47, 1975.

1688 Mycoplasma Pneumonia

There is increasing interest in the role of this organism in spontaneous abortion. Bray and Hackett (1976) reported an infant whose mother had mycoplasma pneumonia at two months of gestation. The infant had unexplained hydrocephalus with evidence of intrauterine infection. In addition, pedunculated skin tags of the eyelids, alopecia and punctate skin defects of the scalp were present.

Bray, P.F. and Hackett, T.N.: Multiple birth defects in a newborn exposed to mycoplasma pneumoniae in utero. Am. J. Dis. Child. 130:312–314, 1976.

1689 Myasthenia gravis

Holmes et al. (1980) reported the case of a newborn with congenital contractures. The mother had myasthema gravis during her entire pregnancy. Several similar cases are cited.

Holmes, L.B.; Driscoll, S.G. and Bradley, W.F.: Contractures in a newborn infant of a mother with myasthema gravis. J. Pediatr. 96:1067–1069, 1980.

1690 Nabilone

Markham et al. (1979) gave up to 12 mg per kg to rats on days 6–15 or 14–20 and up to 3.3 mg per kg to rabbits on days 6–18. No defect increase was found, but liveborn rat fetuses were decreased in number.

Markham, J.K.; Hanasono, G.K.; Adams, E.R. and Owen, N.V.: Reproduction studies on nabilone, a synthetic 9–keto–cannabinoid. (abs) Toxicol. Appl. Pharmacol. 48:A119, 1979.

1691 Nabumetone

Toshiyaki et al. (**1988**) gave rats up to 160 mg per kg during organogenesis and found no fetal ill effects. At 60 mg per kg and above the postnatal weight gain was reduced. In rabbits given up to 320 mg per kg during organogenesis no defects increase was found, but at the highest dose fetal mortality was increased.

Toshiyaki, F. Kadoh, Y.; Fujimoto, Y.; Tenshio, A,; Fuchigami, K. and Ohtsuka, T.: Toxicity study of nabumetone (iii)–teratological studies. Kiso to Rinsho 22:2975–2985, 1988.

1692 Nadifloxacin

Nagao et al. (**1990**) gave this antibiotic to rats subcutaneously in doses of up to 75 mg per kg on days 7–17. At the highest dose some decrease in fetal weight and bone maturation was found but no teratogenicity. Studies of reproduction in rats using up to 15 mg per kg showed no adverse effects. (Matsuzawa et al., **1990** A) Rabbits received up to 15 mg per kg during days 6–18 and no adverse fetal effects were seen. (Matsuzawa et al., **1990** B)

Nagao, T.; Wada, A.; Hashimoto, Y.; Mizutani, M. and Matsuzawa, A.: Reproductive and developmental toxicity studies o (+)(-)–9–fluoro–6,7–dihyrdo–8–(4–hydroxy–1–piperidyl)- 5–methyl–l–oxo–1h,5h–benzo[i,j]quinolizine–2–carboxylic acid (opc-7251), a synthetic antibacterial agent (ii) teratogenicity study in rats with subcutaneous administration. Iyakuhin Kenkyu 21:647–662, 1990.

Matsuzawa, A.; Kajiyoshi, K.; Azuma, K.; Tamagawa, M. and Nishioeda, R.: Reproductive and developmental toxicity studies o (+)(-)–9–fluoro–6,7–dihydro–8–(4–hydroxy–1–piperdyl)–5–methyl–1–oxo–1h,5h–benzo[i,j]quinolizine–2–carboxylic acid (opc-7251), a synthetic antibacterial agent (i) study by subcutaneous administration prior to and in the early stages of pregnancy in rats. Iyakuhin Kenkyu 21:636–646, 1990.

Matsuzawa, A.; Yoshida, M.; Tamagawa, M. and Nishioeda, R.: Reproductive and developmental toxicity studies of (+)(-)–9–fluoro–6,7–dihydro–8–(4hydroxy–1–piperidyl)–5–methyl–1–oxo–1h,5h–benzo[i,j]quinolizine–2–carboxylic acid (opc–7251), a synthetic antibacterial agent (iii) teratogenicit study in rabbits with subcutaneous administration. Iyakuhin Kenkyu 21:663–670, 1990.

1693   Nadolol CAS 42200–33–9

Saegusa et al. (1983) studied this β–adrenergic blocking agent in rats treating orally before pregnancy on days 7–17 and days 17 until 21 days after birth. For the first two studies a maximum dose of 500 mg per kg was given and for the perinatal study 250 mg per kg was used. No adverse fetal effects were found although at 500 mg per kg the dams did not gain weight normally.

Saegusa, T.; Suzuki, T. and Narama, I.: Reproduction studies of nadolol a new β–adrenergic blocking agent. Yakuri to Chiryo 11:5119–5138, 1983.

1694   Nafamstat Mesilate *FUT–175*

Tauchi et al. (1984) studied this drug in pregnant rats and rabbits. No adverse fetal effects were found in rabbit fetuses after the dams received 4 mg per kg intravenously on days 6–18. In the rat intraperitoneal doses of 12 and 16 mg per kg in fertility and organogenesis studies caused neither adverse effects or decrease in fertility. The perinatal studies with up to 12 mg per kg did not produce adverse fetal effects.

Tauchi, K.; Igarashi, N.; Kawanishi, H.; Koh, K.; Hasegawa, N.; Terabayashi, M.; Kuramoto, S.; Suzuki, S.; Minakawa, A. and Shimamura, K.: Reproduction studies of FUT–175 (nafamstat mesilate) in rats and rabbits. Clin. Report 18:3901–3942, 1984.

1695   Nafcillin CAS 147–52–4

Schardein (1976) cites Mizutani et al. (1970) who studied this antibiotic in mice and rats and found no teratogenicity.

Mizutani, M.; Ihara, T.; Kanamori, H.; Takatani, O. and Kaziwara, K.: Influence of sodium nafcillin upon the development of fetuses of mice and rats. Takeda Kenkyusho Ho 29:283–296, 1970.

Schardein, J.L.: Drugs as Teratogens, CRC Press, Cleveland, 157, 1976.

1696   Nafronyl

Fontaine et al. (**1979**) studied this vasodilator during organogenesis in the mouse, rat and rabbit at maximum doses of 360, 480 and 5 mg per kg respectively. No increase in malformation was found but a decrease in fetal weight was found in the mouse.

Fontaine, L.; Chabert, J.; Grand, M.; Depin, J.C. and Szarvasi, E.: Etude toxicologique et teratologique du naftidrofuryl. J. Eur. Toxicol. 11:40–50, 1979.

1697   Naftalofos *Phthalophose* CAS 732–11–6

Kagan et al. (1978) observed teratogenic and embryotoxic action of this pesticide in rats treated during all the gestation (0.3–15 mg per kg). These doses were non–toxic for the maternal organism. Abbasov et al. (1980) obtained similar results.

Abbasov, T.G.; Karavaeva, G.H. and Makhno, P.M.: Embryotoxic effect of phthalophose. Veterinariya. (USSR) 1:62–63, 1980.

Kagan, Y.S.; Voronina, V.M. and Akkerman, G.A.: Effect of phthalophose on the embryogenesis and its metabolism in the body of albino rats and their embryos. Gig. Sanit. (USSR) 9:28–31, 1978.

1698   Naftidrofuryl Oxalate *LS–121*

Umemura et al. (1985a,b) studied this drug in rabbits using the intravenous route (2 mg per kg) or oral route (400 mg per kg). No adverse fetal effects were found when treatment was given during organogenesis. In rats exposed to 600 mg per kg during organogenesis only, a reduction in fetal weight was found (Umemura et al., 1986a). Perinatal and prenatal studies had no significant effect in the rat (1986b,c). Intravenous studies (Umemura, 1986d,e) were also done in rats. Before, during and after organogenesis, auditory responses were not affected (Umemura, 1986f).

Umemura, T.; Esaki, K.; Sasa, H. and Yanagita, T.: Teratogenicity of intravenous administration of naftidrofuryl oxalate (LS–121) in rabbits. Preclin. Rep. Cent. Inst. Exp. Anim. 11(1):111–119, 1985a.

Umemura, T.; Esaki, K.; Sasa, H. and Yanagita, T.: Teratogenicity of intragastric administration of naftidrofuryl oxalate (LS–121) in rabbits. Preclin. Rep. Cent. Inst. Exp. Anim. 11(1):91–102, 1985b.

Umemura, T.; Yamaguchi, K.; Ando, K.; Esaki, K. and Yanagita, T.: Effects of oral administration of naftidrofuryl oxalate (LS–121) on reproduction in rats; III. Experiment on drug administration during organogenesis period. Preclin. Rep. Cent. Inst. Exp. Anim. 12(2):89–113, 1986a.

Umemura, T.; Yamaguchi, K.; Ando, K.; Esaki, K. and Yanagita, T.: Effects of oral administration of naftidrofuryl oxalate (LS–121) on reproduction in rats; IV. Experiment on drug administration during peripartum and nursing periods. Preclin. Rep. Cent. Inst. Exp. Anim. 12(2):115–130, 1986b.

Umemura, T.; Yamaguchi, K.; Esaki, K. and Yanagita, T.: Effects of oral administration of naftidrofuryl oxalate (LS–121) on reproduction in rats. II. Experiment on drug administration in the pre– and early gestation periods. Preclin. Rep. Cent. Inst. Exp. Anim. 12(2):79–88, 1986c.

Umemura, T.; Yamaguchi, K.; Sasa, H.; Esaki, K. and Yanagita, T.: Effects of intravenous administration of naftidrofuryl oxalate (LS–121) on reproduction in rats. II. Experiment on drug administration in the pre– and early gestation periods. Preclin. Rep. Cent. Inst. Exp. Anim. 12(3):139–147, 1986d.

Umemura, T.; Yamaguchi, K.; Sasa, H.; Ando, K.; Esaki, K. and Yanagita, T.: Effects of intravenous administration of naftidrofuryl oxalate (LS–121) on reproduction in rats. III. Experiment on drug administration during organogenesis period. Preclin. Rep. Cent. Inst. Exp. Anim. 12(3):149–172, 1986e.

Umemura, T.; Yamaguchi, K.; Sasa, H.; Ando, K.; Esaki, K. and Yanagita, T.: Effects of intravenous administration of naftidrofuryl oxalate (LS–121) on reproduction in rats.

IV. Experiment on drug administration during peripartum and nursing periods. Preclin. Rep. Cent. Inst. Exp. Anim. 12(3):173–188, 1986f.

1699 Naftopidil *1-[-4-(2-methoxyphenyl)piperazinyl] 3-1-naphthyloxy) propan-2-01 KT-611*

Barton et al. (**1994** and Leuschner et al. (**1994**) treated pregnant rats orally with this α–1 receptor blocking agent. Doses of 160 mg per kg from two weeks before pregnancy until term or through lactation. At 40 and 160 mg per kg maternal toxicity was present and fetal retardation in growth and increase death among the neonates. No teratogenicity was found and postnatal behavior was normal. Bode et al. (**1994**) gave up to 100 mg per kg to rabbits and found maternal toxicity and increased fetal loss but no increase in defects.

This alpha–1 receptor block was studied in the rat pregnancy during organogenesis (Ihara et al., **1992**). No teratogenicity was found but at the highest dose (200 mg per kg) the newborn weights were reduced.

Barton, S.J.; Bode, G.; Sterz, H.G.; Fukunishi, K. and Kobayashi, Y.: Reproductive and developmental toxicity study: Effect of naftopidil on fertility and general reproductive performance in rats. Oyo Yakuri 48(1):17–30, 1994.

Bode, G.; Vierling, T.; Sterz, H.G.; Fukunishi, K. and Kobayashi, Y.: Reproductive and developmental toxicity study in rabbits given naftopidil orally during the period of organogenesis. Oyo Yakuri 48(1):1–6, 1994.

Ihara, T.; Oneda, S.; Magata, K.; Ohta, K.; Kobayashi, Y. and Nurimoto, S.: Teratogenic effect of naftopidil (KT–611) in the rat. The Clinical Report 26:75–93, 1992.

Leuschner, F.; Bode, G.; Fukunishi, K. and Kobayashi, Y.: Reproductive and developmental toxicity study in rats given naftopidil orally during the perinatal and lactation periods. Oyo Yakuri 48(1):7–15, 1994.

1700 Nalidixic Acid

Murray (1981) traced 63 women treated during pregnancy and found one offspring with congenital heart disease. Other types of defects and hypertension were not identified.

Sato et al. (1980, A&B) studied the teratogenicity in rats and rabbits. In the rat fetus growth retardation without an increase in defects was found at doses of 200 and 300 mg per kg given on days 7–17. At 800 mg per kg during organogenesis in the rabbit no adverse fetal effects were found.

Murray, EDS: Nalidixic acid in pregnancy. Br Med J. 282:224, 1981.

Sato, T.; Kaneko, Y. and Saegusa, T.: Reproduction studies of cinoxacin in rats. Chemotherapy 28(Suppl 4):484–507, 1980.

Sato, T.; Kobayashi, F.: Teratological study on cinoxacin in rabbits. Chemotherapy 28(Suppl 4):508–515, 1980.

1701 Naloxone

Shepanek et al. (**1989**) studied behavior and neuroanatomical sequelae to 1 or 5 mg per kg given on days 4–19 to rats. The high dose accelerated development of neg-

ative geotaxis and righting while the lower dose tended to slow development.

Shepanek, N.A.; Smith, R.F.; Tyer, Z.E.; Royall, G.D. and Allen, K.S.: Behavioral and neuroanatomical sequelae of prenatal naloxone administration in the rat. Neurotoxicology and Teratology 11:441–446, 1989.

1702 Naltrexone CAS 16590–41–3

Kennedy et al. (1975) studied this drug in pregnant rats and rabbits using up to 200 mg per day orally during active organogenesis. No fetal changes or teratogenicity occurred.

Christian (1984) reviewed the reproductive effects of this compound on adult animals and reported some postnatal effects in rat offspring. A slight but statistically significant delay in the acoustic startle response in the group exposed to 100 mg per kg daily was found.

Christian, M.S.: Reproductive toxicity and teratology evaluations of naltrexone. J. Clin. Psychiatr. 45:7–10, 1984.

Kennedy, G.L.; Smith, S.; Keplinger, M.L. and Calandra, J.C.: Reproductive and teratogenic studies with naltrexone in rats and rabbits. (abs) Toxicol. Appl. Pharmacol. 33:173–174, 1975.

1703 Nandrolone Phenpropionate *17 β–Hydroxy–19–nor–4–androsten–3–one*

Kawashima et al. (1977) administered orally up to 10 mg to rats on days 17–20 and produced no reduction of the length of the urovaginal septum in female fetuses.

Kawashima, K.; Nakaura, S.; Nagao, S.; Tanaka, S.; Kuwamura, T. and Omori, Y.: Virilizing activities of various steroids in female rat fetuses. Endocrinol. Japon. 24(1):77–81, 1977.

1704 Naphthalene CAS 91–20–3

Van der Hoeve (1913) administered a metabolite of naphthalene, 2–naphthol to pregnant rabbits and found cataracts and retinal damage in the offspring. He gavaged the dams with one gm per kg on days 20, 22 and 24 of gestation. Plasterer et al. (1985) gavage fed mice on days 7–14 with a dose considered to be just below adult lethality (300 mg per kg). No gross congenital defects were detected in neonates. There was a slight reduction in number of neonate survivors.

Plasterer, M.R.; Bradshaw, W.S.; Booth, G.M.; Carter, M.W.; Schuler, R.L. and Hardin, B.D.: Developmental toxicity of nine selected compounds following prenatal exposure in the mouse: Naphthalene, p–nitrophenol, sodium selenite, dimethylphthalate, ethylene–thiourea and four glycol ether derivatives. J. Toxicol. Environ. Health 15:25–38, 1985.

Van der Hoeve, J.: Wirkung von Naphthol auf die und auf foetale Augen. Graefes Arch. Ophthal. 85:305–315, 1913.

1705 Naphtha Solvents *Petroleum Solvents* CAS 64741–66–8; CAS 64742–88–7

This light alkylate and medium aliphatic forms have been given by inhalation for 6 hour daily periods to rats on days 6–15 of gestation. The light alkylate (Isopar C) was given in concentrations of up to 1200 ppm and the medium aliphatic (Varsol–40) was given in doses of up to 300 ppm. No adverse reproductive or fetal changes were reported (Exxon Company, 1984).

Exxon Company, U.S.A.: FYI–OTS–06 84–0313 Supplement sequence H. Review of toxicity information on Isopar C and Varsol 40 (Varsol 1). Washington, DC: Office of Toxic Substances, U.S. Environmental Protection Agency, 1984.

1706 Naproxen *D–2(6–Methoxy–2–naphthyl) Proprionic Acid* CAS 22204–53–1

Wilkinson et al. (1979) reported persistent pulmonary hypertension in three premature infants born to mothers receiving this prostaglandin synthetase inhibitor.

Wilkinson, A.R.; Aynsley–Green, A. and Mitchell, M.D.: Persistent pulmonary hypertension and abnormal prostaglandin E levels: In preterm infants after maternal treatment with naproxen. Arch. Dis. Childhood 54:942–945, 1979.

1707 Narasin CAS 55134–13–9

This antifungal was given to rats on the 4th day in amounts equal to one–tenth the LD50 caused slight pre and post implantation loss. On day 13 the same dose did not cause any ill effects (Chakurov and Luu, 1988).

Chakurov, R. and Luu, N.D.: Embryotoxic and teratogenic effect of narasin. Vet. Sb 86(7): 52–53, 1988.

1708 Nebularine *Xanthosine 9–β–D–Ribofuranosyl xanthine* CAS 146–80–5

This compound isolated from mushrooms inhibits tumor cell growth. It was administered to mice in doses of greater than 800 mg per kg (the maternal LD–50) and did not produce fetal defects (Chaube and Murphy, 1968).

Chaube, S. and Murphy, M.L.: The teratogenic effects of the recent drugs active in cancer chemotherapy. In: Advances in Teratology, D.H.M. Woollam (ed.), New York: Logos and Academic Press, Volume 3, 181–237, 1968.

1709 Nebracetam

Nishimura et al. (**1990**) gave rats up to 800 mg per kg orally before pregnancy, during organogenesis and perinatally and found no teratogenicity. At the highest dose the fetuses and pups had reduced weight. In rabbits 400 mg per kg was given withouut teratogenicity.

Nishimura, M.; Tsunenari, Y. and Kast, A.: Oral reproductive and developmental toxicology of nebracetam, a central cholinergic agent. Iyakuhin Kenyu 21(6):1308–1327, 1990.

1710 Nefopam HCL CAS 23327–57–3

The analgesic was given orally to pregnant mice and rabbits in maximum doses of 75 or 80 mg per kg per day of active organogenesis and no teratogenic activity was observed (Case et al., 1975).

Case, M.T.; Smith, J.K. and Nelson, R.A.: Reproductive, acute and subacute toxicity studies with nefopam in laboratory animals. Toxicol. Appl. Pharmacol. 33:46–51, 1975.

1711 Neocarzinostatin *Zinostatin* CAS 9014–02–2

Tomizawa et al. (1976) gave intraperitoneal doses of 0.1–0.5 mg per kg during organogenesis to rats and mice. Growth retardation and increased lethality was found in the fetuses from each species. Hydronephrosis and defects of the heart and gastrointestinal tract were found in the rat. Tracheal abnormalities occurred at 0.5 mg per kg. Ono et al. (**1991**) gave up to 0.05 mg per kg intravenously on days 7–17 of the rat and rabbit and found no adverse fetal effects. The postnatal development and reproduction in the rats was not effected.

Ono, C.; Kiwaki, S.; Oketani, Y. and Imamura, K.: Teratogenicity study of intravenously administered zinostatin stimalamer in rats and rabbits. Iyakuhin Kenkyu 22:399–415, 1991.

Tomizawa, S.; Kamada, K. and Segawa, M.: Influences of neocarzinostatin on fetuses of mice and rats. Oyo Yakuri 3:329–339, 1976.

1712 Neomycin CAS 1404–04–2

Heinonen et al. (1977) found only 2 defects among 30 women using this antibiotic during the first 4 lunar months.

Skalko and Gold (1974) found no increase in defects of offspring of mice treated with drinking water containing 4 gm per liter on days 1–10.

Heinonen, O.P.; Slone, D. and Shapiro, S.: Birth Defects and Drugs in Pregnancy. Publishing Sciences Group Inc., Littleton, Mass., 1977.

Skalko, R.G. and Gold, M.P.: Teratogenicity of methotrexate in mice. Teratology 9:159–164, 1974.

1713 Neopyrithiamine CAS 534–64–5

This analogue of thiamine was injected into chick eggs in amounts of 0.25 to 2.0 mg by Naber et al. (1954). Doses above 0.5 mg were lethal to the embryos if given at the start or at 5 days of incubation. Injection on the 10th or 15th day produced ataxia, paralysis and polyneuritis in the hatched chicks. Kosterlitz (1960) administered an unstated amount to pregnant rats and found fetal weight reduction but no malformations.

Kosterlitz, H.W.: Ciba Foundation Symposium on Congenital Malformations. G.W.E. Wolstenholme and C.M. O'Connor (eds.), Boston: Little Brown, 275 only, 1960.

Naber, E.C.; Cravens, W.W.; Baumann, C.A. and Bird, H.R.: The effect of thiamine analogues on embryonic development and growth of the chick. J. Nutr. 54:579–591, 1954.

1714 Nerve Growth Factor

Purified salivary nerve growth factor was injected in microgm amounts into 7 to 10 day chick embryo yolk sacs and 3 to 4 days later, the embryos were shown to have hypertrophic and hyperplastic sensory and sympathetic nerve ganglia. The viscera and skin were flooded with nerve fibers. The same results were found in rat and embryo–fetuses after maternal injection. Explanted ganglia of human fetuses also responded to the material with a dense growth of nerve fibers. These experiments are summarized by Levi–Montalcini and Cohen (1960).

Levi–Montalcini, R. and Cohen, S.: Effects of the extract of the mouse submaxillary salivary glands on the sympathetic system of mammals. Ann. N.Y. Acad. Sci. 85:324–341, 1960.

1715 Nervenruh Forte

Shoji (**1968**) treated mice orally or intraperitoneally with doses of up to 400 mg per kg on the 9th day of pregnancy. Exencephaly, open eye, and exophthalmos were increased in the intraperitoneally treated group.

Shoji, R.: Aspects on the effects of nervenruh forte, a psychotropic drug, on developing mouse embryos. Zoological Magazine 77:220–225, 1968.

1716 Neticonazole

Ito et al. (**1990**) gave this antimycotic to rats and rabbits subcutaneously during organogenesis in doses of up to 50 mg per kg. No adverse effects were found except for lumbar ribs in rat fetuses exposed to the highest dose.

Ito, R. and Tanihata, T.: Teratological studies with an antifungal agent ss717 in rats and rabbits. Kiso to Rinsho 24:6753–6772, 1990.

1717 Netilmicin CAS 56391–56–1

This aminoglycoside antibiotic was given by Nomura et al. (1982) and Furuhashi et al. (1982) to rats and rabbits intramuscularly in amounts of up to 100 mg per kg daily. Fertility, teratogenicity and postnatal studies were done in rats and the only significant finding was reduced fetal weights when 50 or 100 mg per kg was given during organogenesis. No changes were produced in the rabbit except for reduced weight of the fetuses at 35 and 100 mg per kg.

Furuhashi, T.; Nomura, A. and Nakayoshi, H.: Reproductive study on netilmicin in rabbits. Jpn. J. Antibiot. 35:659–666, 1982.
Nomura, A.; Furuhashi, T.; Komuro, E.; Uehara, M.; Miyoshi, K. and Nakayoshi, H.: Reproductive study on netilmicin in rats. Jpn. J. Antibiot. 35:614–629, 630–642, 1982.

1718 Newcastle Disease Virus

Robertson et al. (1955) inoculated chicken eggs with live virus at 1.5 to 3.5 days of incubation and found that the neural tube, lens, auditory vesicles, visceral arches and limb buds were affected. Cytoplasmic degeneration followed by nuclear disintegration was seen in the ectodermal tissues of these structures. Williamson et al. (1965) studied the pathogenesis of this virus in the chick model.

Maroun et al. (1986) injected the virus into rabbits 8–10 hours after ovulation and found a four–fold increase in midterm embryonic mortality. Aneuploidy of a small acrocentric chromosome was observed in three of the live midterm embryos recovered from the treated does.

Maroun, L.E.; Degner, M.; Mourey, M.E.; Rowan, D.F. and Amankwah, K.: Increased mortality and aneuploidy in embryos from rabbits injected with Newcastle disease virus at the time of ovulation and conception. Teratology 34:201–206, 1986.
Robertson, G.G.; Williamson, A.P. and Blattner, R.J.: A study of abnormalities in early chick embryos inoculated with Newcastle disease virus. J. Exp. Zool. 129:5–43, 1955.
Williamson, A.P.; Blattner, R.J. and Robertson, G.G.: The relationship of viral antigen to virus–induced defects in chick embryos. Newcastle disease virus. Dev. Biol. 12:498–519, 1965.

1719 Niacin *Nicotinic Acid* CAS 59–67–6

Rosa (**1994**) reported a total of 87 women who took niacin to lower cholesterol. Only two congenital defects were found and they were both in the group of 25 who took the medication in the first trimester. Hansborough (1947) replaced 2 ml of egg white of chick eggs with a solution containing 20 mg of nicotinic acid at 2, 3 or 4 days of incubation. The embryos were found to have a high incidence of neural tube closure defects, abnormal neural development and abnormalities of the cardiovascular system.

Takaori et al. (1973) gave 3–(o–methoxyphenoxy)–2–hydroxypropyl nicotinate, a nicotinic acid derivative, orally to pregnant rats and rabbits in doses of 100 or 1,000 mg per kg during the days of active organogenesis. No gross or skeletal defects were found.

Hansborough, L.A.: Effect of increased nicotinic acid in the egg on the development of the chick embryo. Growth 11:177–184, 1947.
Rosa, F.: Anti–cholesterol agent pregnancy exposure outcomes. (abs) Reproductive Toxicology 8:445–446, 1994.
Takaori, S.; Usui, H. and Kondo, M.: Studies of a new nicotinic acid derivative, 3–(o–methoxyphenoxy)–2–hydroxypropyl nicotinate (H–1) teratogenicity in rats and rabbits. (Japanese) Oyo Yakuri 7:441–447, 1973.

1720 Niagara Blue CAS 2602–46–2

Beck and Lloyd (1966) reviewed the chemistry and experimental work done with this series of azo dyes. Niagra blue 2B was not teratogenic in experiments on rats by Beaudoin (1962) but Beck and Lloyd (1966) injecting 150 mg per kg on the 8.5th day found 25 percent abnormalities. Doses of 200 mg per kg caused 100 percent resorptions. Wilson (1955) found that Niagra Blue 4B and 6B were slightly teratogenic. The syndrome of malformation is similar to that seen with Trypan Blue. Kernis and Marshall (1969) showed an effect of Niagara 2B on ionic absorption by the yolk sac.

Beaudoin, A.R.: Interference of Niagara 2B with teratogenic action of trypan blue. Proc. Soc. Exp. Biol. Med. 109:709–711, 1962.
Beck, F. and Lloyd, J.B.: The teratogenic effects of azo dyes. In: Advances in Teratology, D.H.M. Woollam (ed.), New York: Logos and Academic Press, 1:131–193, 1966.
Kernis, M.M. and Johnson, E.M.: Effects of Trypan Blue and Niagra Blue 2B on the in vitro absorption of ions by the

rat visceral yolk sac. J. Embryol. Exp. Morphol. 22:115–125, 1969.

Wilson, J.G.: Teratogenic activity of several azo dyes chemically related to Trypan Blue. Anat. Rec. 123:313–326, 1955.

1721 Nialamide Isonicotinic Acid
2-[2-(Benzylcarbamoyl)-ethyl]-hydrazide CAS 51-12-7

Tuchmann–Duplessis and Mercier–Parot (1963) maintained female rats on 10 mg per kg per day for 6 to 10 months before gestation. During gestation, the same dose of this monoamine oxidase inhibitor was maintained. After weaning, the newborns were given 5 mg per kg per day. Although the newborns and young rats showed no external changes, their fertility was markedly reduced and the females had an abnormal tendency to mount each other while refusing to accept males. Reserpine did not produce this effect.

Tuchmann–Duplessis, H. and Mercier–Parot, L.: Modifications du comportement sexual chez des descendants de rats traites par un inhibiteur des monoamine–oxydases. C. R. Acad. Sci. (Paris) 256:2235–2237, 1963.

1722 Nicardipine CAS 54527-84-3
2-(N-Benzyl-N-methylamino)ethyl methyl
2,6-dimethyl-4(M-nitrophenyl) 1, 4-dihydropyridine-3,
5-dicarboxylate Hydrochloride; YC-93 CAS 55985-32-5

Sejima and Sado (1979) gave this vasodilator orally to rats in maximum doses of 100 mg per kg on days 7–17 of gestation and found no adverse fetal effects. Sato et al. (1979) gave the same dose before mating to male and female rats and during the first 7 days of gestation and observed no fertility decrease. Treatment on days 17–21 did not adversely affect postnatal function and fertility of the offspring. Rabbits given 150 mg per kg during organogenesis had fetuses which were not different from controls.

Sato, T.; Nagaoka, T.; Fuchigami, K.; Ohsuga, F. and Hatano, M.: Reproductive studies of 2–(N–benzyl–N–methylamino) ethyl methyl 2,6–dimethyl–4–m–nitrophenyl)–1,4–dihydropyridine–3,5–dicarboxylate hydrochloride (YC–93) in rats and rabbits. Kiso to Rinsho 13:1160–1176, 1979.

Sejima, Y. and Sado, T.: Teratological study of 2–(N–benzyl–N–methylamino) ethyl methyl 2,6–dimethyl–4–m–nitrophenyl)–1,4–dihydropyridine–3, 5–dicarboxylate hydrochloride (YC–93) in rats. Kiso to Rinsho 13:1149–1159, 1979.

1723 Nickel CAS 7440-02-0

Ridgway and Karnofsky (1952) found no defects in chick embryos using the hydrated chloride salt at the estimated LD–50 dose of 0.2 mg on the 4th day of incubation. In a three–generation study of rats fed 5 ppm in their water, an increase in number of newborn runts was found by Schroeder and Mitchener (1971). Sunderman et al. (1979) produced anophthalmia and microphthalmia in rat fetuses exposed to 0.08 to 0.3 mg of nickel carbonyl per liter of air for 15 minute periods on either day 7 or day 8 of gestation.

Very few extraocular anomalies occurred. Lu et al. (1974) found teratogenic results when they injected pregnant mice with nickel chloride (1–6.9 mg per kg) on individual days from the 7th through the 11th days of gestation. The distribution of [63]N in pregnant rats was studied by Mas et al. (1985). Nadeenko et al. (1979) gave nickel in drinking water to rats for 7 months before pregnancy and during pregnancy and some increase of preimplantation mortality was found. Some cases of malformed fetuses were noted.

Lu, C.C.; Matsumoto, N. and Iijime, S.: Teratogenic effects of nickel chloride on embryonic mice and its transfer to embryonic mice. Teratology 19:137–142, 1979.

Mas, A.; Holt, D. and Webb, M.: The acute toxicity and teratogenicity of nickel in pregnant rats. Toxicology 35:47–57, 1985.

Nadeenko, V.G.; Lenchenko, B.T.; Arkhipenko, G.A. and Saichenko, S.P.: Embryotoxic effect of nickel getting by organism with drinking water. Gig. Sanit. (USSR) 6:86–88, 1979.

Ridgway, L.P. and Karnofsky, D.A.: The effects of metals on the chick embryo: Toxicity and production of abnormalities in development. Ann. N.Y. Acad. Sci. 55:203–215, 1952.

Schroeder, H.A. and Mitchener, M.: Toxic effects of trace elements on the reproduction of mice and rats. Arch. Environ. Health 23:102–106, 1971.

Sunderman, F.W.; Allpass, P.R.; Mitchell, J.M.; Basett, R.C. and Albert, D.M.: Eye malformations in rats: Induction by prenatal exposure to nickel carbonyl. Science 203:550–553, 1979.

1724 Nicorandil

Kawanishi et al. (1991 A,B,C) gave rats this angina pectoralis drug intravenously before and during early pregnancy, during organogenesis and peri and postnatally. No adverse fetal effects or changes in behavior were found at doses up to 20 mg per kg. Adult toxicity was found at 10 and 20 mg per kg. Kawanishi et al. (1991 D) gave rabbits up to 10 mg per kg intravenously during organogenesis and found no teratogenicity.

Kawanishi, H.; Shiraishi, M.; Igarashi, Y.; Takeshima, T.; Toyohara, S.; Imai, S. and Sugiyama, O.: Fertility study of nicorandil (SG–75) in rats by intravenous injection. Jpn. Pharm. Therap. 7:2625–2633, 1991A.

Kawaanishi, H.; Shiraishi, M.; Igarashi, Y.; Takeshima, T.; Toyohara, S.; Imai, S. and Sugiyama, O.: Teratological study of nicorandil (SG–75) in rats by intravenous injection. Jpn. Pharm. Therap. 7:2635–2649, 1991B.

Kawanishi, H.; Shiraishi, M.; Igarashi, Y.; Takeshima, T.; Toyohara, S.; Imai, S. and Sugiyama, O.: Perinatal and postnatal study of nicorandil (SG–75) in rats by intravenous injection. Jpn. Pharm. Therap. 7:2651–2662, 1991C.

Kawanishi, H.; Shiraishi, M.; Igarashi, Y.; Takeshima, T.; Toyohara, S.; Imai, S. and Sugiyama, O.: Teratological study of nicorandil (SG–75) in rabbits by intravenous injection. Jpn. Pharm. Therap. 7:2663–2670, 1991D.

1725 Nicotinamide Deficiency also see

6–Aminonicotinamide Niacin Deficiency

Although 6–aminonicotinamide, a nicotinamide antagonist, showed to be teratogenic, the effect on fetal development of maternal diets deficient in nicotinamide does not appear to have been reported. Fratta et al. (1964) fed rats a diet deficient in nicotinamide and its precursors, nicotinic acid and tryptophan, from day 2 through day 13 of gestation and found no viable fetuses. Chlorpromazine or imipramine protected the fetuses exposed to this regime. No defects were described.

Fratta, I.; Zak, S.B.; Greengard, P. and Sigg, E.B.: Fetal death from nicotinamide–deficient diet and its prevention by chlorpromazine and imipramine. Science 145:1429–1430, 1964.

1726 Nicotinamidomethylaminopyrazolone

Tubaro et al. (1970) gave up to 200 mg per kg subcutaneously to pregnant rats from the first through the 18th day and found no adverse effects. Oral doses of 500 mg per kg to rats did not reduce litter size or fetal weights. Five dogs received up to 250 mg per kg without adverse effects on their litters.

Tubaro, E.; Bulgini, M.J.; DelGrande, P. and Monai, A.: Some toxicologic aspects of nicotinamidomethylaminopyrazolone (Ra 101). Arzneim–Forsch. 20:1024–1029, 1970.

1727 Nicotine also see *Cigarette Smoking* CAS 54–11–5

Nicotine via chewing of gum or patch application deliver nicotine to the embryo and fetus. Studies f nicotine levels indicate that the maternal venus level is generally lower than with continued smoking. In addition, it is unlikely that the very high peaks of nicotine in the arterial blood of smokers will be found with the patch or gum delivery systems. The manufacturers recommend complete cessation of tobacco use and a behavioral support system.

Essenberg et al. (1940) reduced the size of rat offspring by exposing the mothers to nicotine or smoke. Schoeneck (1941) using rabbits, found an increased stillbirth rate and reduced fetal size. Nishimura and Nakai (1958) produced skeletal defects and occasional cleft palates in the mouse fetus when the mother was injected with 25 mg per kg of nicotine on the 9th, 10th and 11th days. The fetal size was reduced but not statistically. In the rat, Mosier and Armstrong (1964) used 0.05 mg per ml of drinking water and found a reduced size in the newborn. Menges et al. (1970) reported an epidemic of limb deformities in the offspring of swine which fed on tobacco stalks containing 1058 PPM of nicotine and 115 PPM of maleic hydrazide.

Mosier and Jansons (1972), using tritiated–labelled nicotine in rats showed that the substance and its breakdown product, cotinine, equilibrated at a higher concentration in fetal than in maternal plasma during one period after administration. Sieber and Fabro (1971) observed a fourfold increase in concentration of nicotine in the rabbit blastocyst. Nash and Persaud (1988) gave rats 25 mg of nicotine by mini–osmotic pump over the period of days 6–12. A significant (p < 0.05) increase in hydrocephalus was found in the treated group, but when cotreatment with caffeine was studied, no increase in hydrocephalus was found. Luck et at (1985) demonstrted that the 1st trimester placenta, amniotic fluiid and term fetal serum levels of nicotine were significantly higher than materna serum. Continine levels were about the same as maternal in the fetus and amniotic fluid.

Zahalka et al. (**1992**) found that [³H]hemicholinium–3 binding which labels the presynaptic high affinity cholinergic transporter was deficient in both the striatum and hippocampus of rat fetuses exposed on days 4 through 20 to 2 or 6 mg per kg per day. Brain weights and choline acetyltransterase were not altered.

Ignatieva et al. (1987) reported adverse effects on germ cells produced by nicotine, administered to rats from the first through 21 days of pregnancy. Sheveleva (1987) administered nicotine in a total dose 0.4–1.5 or 5 mg per kg in rats in different periods of pregnancy. Some of the rats exposed to the cigarette smoke inhalation showed an increased concentration of nicotine in the blood (0.4 mg per kg). The result was retardation of placental development and alteration of its permeability. There were many mases of embryonic death, the cranio–caudal size and the body mass of fetuses decreased, fetal hemorrhage and developmental disturbances in the cardiovascular system were observed. Mardanova et al. (1986) studied the ultrastructure of oocytes in the ovary of 21–day–old rat fetuses after the action of nicotine (0.4, 1.5 and 5 mg per kg) in the antenatal period. Nicotine was shown to have toxic action on oocytes which was directly dependent on its concentration.

Baldwin and Racowsky (1988) studied the effect of nicotine and cotinine on the two–cell mouse embryo and found no effect at the level expected in the average smoker (0.4 to 4 micromolar). Concentration of nicotine at or below 0.5 millimolar and concentrations of cotinine at or below 0.008 millimolar did not adversely affect development.

Baldwin, K.V. and Racowsky, C.: Nicotine and cotinine effects on development of two–cell mouse embryos in vitro. Reproductive Toxicology 1(3):173–178, 1988.

Essenberg, J.M.; Schwind, J.V. and Patras, A.R.: The effects of nicotine and cigarette smoke on pregnant female albino rats and their offspring. J. Lab. Clin. Med. 25:708–716, 1940.

Ignatieva, E.L.; Kurilo, L.F.; Mardanova, G.L. and Sheveleva, G.A.: Adverse effects of nicotine on sexual cells of female fetus in rats. Tsitologiya i Genetika (USSR) 2:91–94, 1987.

Luck, W.; Nau, H.; Hansen, R. and Steldinger, R.: Extent of nicotine and cotinine transfer to the human fetus, placenta and amniotic fluid of smoking mothers. Dev. Pharmacol. Ther. 8:384–395, 1985.

Mardanova, G.V.; Kurilo, L.F.; Sheveleva, G.A. and Ignatieva, E.L.: The ultrastructure of oocytes of rat fetuses under the action of nicotine. Proceedings of USSR Acad. Med. Sci. (USSR) 1:15–22, 1986.

Menges, R.W.; Selby, L.A.; Marienfeld, C.J.; Ave, W.A. and Greer, D.L.: A tobacco related epidemic of congenital limb deformities in swine. Environ. Res. 3:285–302, 1970.

Mosier, H.D. and Armstrong, M.K.: Effects of maternal

intake of nicotine on fetal newborn rats. Proc. Soc. Exp. Biol. Med. 116:956–958, 1964.

Mosier, H.D. and Jansons, R.A.: Distribution and fate of nicotine in the rat fetus. Teratology 6:303–312, 1972.

Nishimura, H. and Nakai, K.: Developmental anomalies in offspring of pregnant mice treated with nicotine. Science 127:877–878, 1958.

Nash, J.E. and Persaud, T.V.N.: Influence of nicotine and caffeine on rat embryonic development. Histol. Histopath. 3:377–388, 1988.

Schoeneck, F.J.: Cigarette smoking in pregnancy. N.Y. State J. Med. 41:1945–1948, 1941.

Sheveleva, G.A.: The influence of smoking and nicotine on reproductive function. Akusherstvo i Ginekologiya (USSR) 12:46–51, 1987.

Sieber, S.M. and Fabro, S.: Identification of drugs in the preimplantation blastocyst and in the plasma, uterine secretion and urine of the pregnant rabbit. J. Pharmacol. Exp. Therap. 176:65–75, 1971.

Zahalka, E.A.; Seidler, F.J.; Lappi, S.E.; McCook. E.C.; Yanai, J. and Slotkin, T.A.: Deficits in development of central cholinergic pathways caused by fetal nicotine exposure: differential effects on choline acetyltransferase activity and [^{3}]hemicholinium–3 binding. Neurotoxicology and Teratology 14:375–382, 1992.

1728 Nicotinyl Alcohol *3–Pyridinemethanol*

Cekanova et al. (**1974**) gavaged mice during organogenesis with up to 1000 mg per kg and found no adverse fetal effects.

Cekanova, E.; Larsson, K.S.; Morck, E. and Aberg, G.: Interactions between salicylic acid and pyridyl–3–methanol: Anti–inflammatory and teratogenic effects. Acta Pharmacol. et Toxicol. 35:107–118, 1974.

1729 Nidroxyzone *5–Nitro–2–furaldehyd [2–(2–hydroxyethyl)–semicarbazone]; Furadroxyl* CAS 405–22–1

Nelson and Steinberger (1953) reported that 1.5 gm per kg of diet fed to pregnant rats caused termination of pregnancy in about 70 percent. No comment was made about associated congenital defects.

Nelson, W.O. and Steinberger, E.: Failure of pregnancy in rats treated with furadroxyl. (abs) Anat. Rec. 115:352–353, 1953.

1730 Nifedipine *Vasodilators Hydialazine Nitroendipine Felodipine*

Magee et al. (**1994**) found no significant increase in defects among 28 offspring of women identified prospectively during their first trimester. Danielsson et al. (1989) gavaged rabbits with 400–100 micromoles per kg a day 16 and found a significant increase in reduced 4th digit of the hindpaw. Fetal weight was not reduced. Other vasodilators hydralazine, nitrendipine and felodipine produced the same defect. They postulated that the mechanisms involved maternal hypotension.

Momma and Takao (**1989**) studied the effect on near–term rat fetal hearts. A slight dilatation of the ventricles was found with 0.1mg per kg and the effect increased with dosage. Akaike et al. (**1992**) gave up to 11.5 mg per kg perinatally and found no adverse behavioral changes in the pups.

Akaike, M.; Ohno, H.; Omosu, M. and Kobayashi, T.: Reproductive and developmental toxicity study of orally administered felodipine–Perinatal and postnatal study in rats. Jpn. Pharmacol. Ther. 20:3025–3036, 1992.

Danielsson, B.R.G.; Reiland, S.; Rundqvist, E. and Danielson, M.: Digital defects induced by vasodilating agents: relationship to reduction in uteroplacental blood flow. Teratology 40: 352–358, 1989.

Magee, C.A.; Lurie, I.; Kuehl, K.; Ferencz, C.; Loffredo, C. and Baltimore–Washington Infant Study Group: Looping anomalies of the heart and great vessels; associated risk factors. (abs) Teratology 49:372, 1994.

Momma, K. and Takao, A.: Fetal Cardiovascular effects of nifedipine in rats. Pediatric Research 26(5):442–447, 1989.

1731 Nifuroxime

Greenaway et al. (1986) using in vitro culture of rat embryos found rat axial asymmetry and growth retardation at concentrations of 0.02 millimoles.

Greenaway, J.C.; Fantel, A.G. and Juchau, M.R.: On the capacity of nitroheterocyclic compounds to elicit an unusual axial asymmetry in cultured rat embryos. Toxicol. Appl. Pharmacol. 82:307–314, 1986.

1732 Nifurtimox *Lampit^{TM} 3–Methyl–4–(5–nitrofurfurylideneamino)–tetrahydro–4H–1,4–thiazine 1,1–dioxide* CAS 23256–30–6

Lorke (1972) tested this drug used in chagas disease in mice and rats and found no evidence of teratogenicity although at the highest doses fetal weight was reduced. Doses up to 125 mg per kg were given orally during organogenesis. 600 PPM in the diet of male rats impaired their fertility.

Lorke, D.: Embryotoxicity studies of fertility and general reproductive performance. Arzneim–Forsch 22:1603–1607, 1972.

1733 Nigericin CAS 28380–24–7

Vedel–Macrander and Hood (1986) gave this carboxylic ionophore to pregnant mice intraperitoneally at doses of 2.5, 5.0 and 7.0 mg per kg. Maternal toxicity occurred at all dose levels. Treatment on day 9 with 5.0 mg per kg produced 26 percent malformations which included exencephaly, eye and other facial defects. Median cleft palates were also seen. Treatment on days 10 through 12 produced defects also but at lower rates.

Vedel–Macrander, G.C. and Hood, R.D.: Teratogenic effects of nigericin, a carboxylic ionophore. Teratology 33:47–51, 1986.

1734 Niludipine CAS 22609–73–0

Hamada et al. (1981) gave this vasodilator orally to

mice, rats and rabbits during their respective organogenesis periods. Maternal toxicity occurred at 250 mg per kg in the rats and rabbits. Some increase in fetal deaths occurred in the rat at 250 mg per kg. No evidence of teratogenic activity was found in any species. Postnatal studies in the rat were done and no differences from control were found.

This vasodilator was studied in pregnant mice, rats and rabbits by Hamada et al. (**1981**) at oral doses of up to 250 mg per kg. No teratogenicity was found but fetal deaths were increased in the rats at the highest dose. The rabbit dams were killed at 250 mg but no fetotoxicity was found at 100 mg per kg. /eb

Hamada, Y.; Imanishi, M. and Hashiguchi, M.: Reproductive study of niludipine 1 and 2. Ikakuhin Kenkyu 12:1082–1109, 1981.

Hamada, Y.; Imanishi, M. and Hashiguchi, M.: Reproduction study of niludipine I. Teratogenicity studies in mice, rats and rabbits. Iyakuhin Kenkyu 12:1082–1099, 1981.

1735 Nimetazepam CAS 2011–67–8

Saito et al. (1984) gavaged rats with 100 mg per kg on days 8–12 and found no defects in the offspring. Resorptions were increased.

Saito, H.; Kobayashi, H.; Takeno, S. and Sakai, T.: Fetal toxicity of benzodiazepine in rats. Res. Commun. Chem. Pathol. Pharmacol. 46:437–447, 1984.

1736 Niprofazone 1–phenyl–2,3–dimethyl–4(N–nicotinamidomethyl–N–isopopyl) amino–5–pyrazolone

Tubaro et al. (**1970**) treated mice subcutaneously with up to 200 mg per kg for various periods during organogenesis or most of gestation and produced no significant increase in fetal defects. Rats received up to 250 mg per kg either orally or intraperitoneally and produced no abnormalities among the offspring. Dogs received up to 240 mg per kg with normal offspring.

Tubaro, E.; Bulgini, M.J.; Del Grande, P. and Monai, A.: Some toxicological aspects of a nicotinamidomethylaminopyrazolone (Ra 101). Arzneim. Forsch 20:1024–1039, 1970.

1737 Niridazole CAS 61–57–4

Fantel et al. (1986) found that this antischistosomal agent added directly to an in–vitro culture of rat embryos produced right–sided hypoplasia and eye defects at concentrations of 25 and 50 micrograms per ml of medium. Oral administration of up to 250 mg per kg. on days 11 through 14 produced no adverse effects in rat fetuses. Greenaway et al. (1986) compared the minimal teratogenic dose in in–vitro rat culture for a group of nitroheterocyclic compounds and found that the single–electron redox potential was directly related to teratogenic potential.

In subsequent studies, Fantel et al. (1988) gave evidence that 1–thiocarbamoyl–2–imidazolidinone was the active breakdown component and acts by producing localized hypoxia as a result of redox cycling. Fantel et al. (1988) presented evidence that the in vitro teratogenicity was associated with oxygen depletion secondary to redox cycling.

Fantel, A.G.; Greenaway, J.C.; Walker, E.A. and Juchau, M.R.: The toxicity of niridazole in rat embryos in vitro. Teratology 33:105–117, 1986.

Fantel, A.G.; Juchau, M.R.; Burroughs, C.J. and Person, R.E.: Studies of embryotoxic mechanisms of niridazole: Evidence that oxygen depletion plays a role in dysmorphogenicity. Teratology 39:243–251, 1989.

Fantel, A.G.; Person, R.E.; Tracy, J.W. and Juchau, M.R.: Niridazole metabolism by rat embryos in vitro. Teratology 37:213–221, 1988.

Greenaway, J.C.; Fantel, A.G. and Juchau, M.R.: On the capacity of nitroheterocyclic compounds to elicit an unusual axial asymmetry in cultured rat embryos. Toxicol. Appl. Pharmacol. 82:307–315, 1986.

1738 Nitrapyrin CAS 1929–82–4
2–Chloro–6–(trichloromethyl) Pyridine

This nitrogen stabilizer was studied in rats and rabbits during organogenesis. Maximum doses of 50 (rat) and 30 (rabbit) mg per kg were given orally without evidence of teratogenicity. Crooked hyoid bones were found in the rabbit fetuses at the highest dose.

Berdasco, N.M.; Lomax, L.G.; Zimmer, M.A. and Hanley Jr, T.R.: Teratologic evaluation of orally administered nitrapyrin in rats and rabbits. Fundamental and Applied Toxicology 11:464–471, 1988.

1739 Nitrates in Drinking Water

Dorsch et al. (1984) studied 218 case control pairs for the association between congenital defects and nitrates in the water. The controls drank rain water during pregnancy. The relative rate was 2.8 with increases in the central nervous and musculoskeletal systems. The increase was threefold for women taking 5–15 ppm and fourfold for those drinking over 15 ppm. They postulated that the nitrates were converted to nitrites and then to amides including N–nitorsamines.

Dorsch, M.M.; Scragg, R.K.R.; McMichael, A.J.; Baghurst, P.A. and Dyer, K.F.: Congenital malformations and maternal drinking water supply in rural South Australia: A case–control study. Am. J. Epidemiol. 119:473–485, 1984.

1740 Nitrazepam CAS 146–22–5

Czeizel (1988) studied prenatal exposures to nitrazepam in among 1201 offspring with facial clefts or multiple congenital anomalies. There were six exposed in the index patients and five among the matched controls.

Saito et al. (1984) gavaged rats on days 8 through 14 with 50, 75 and 100 mg per kg. At 50, 75 and 100 mg levels 15.7, 20 and 61.3 percent defects were found. The defects consisted mainly of ectrodactyly but other types including omphalocele and kinky tail were found. It was more teratogenic in the rat than five other benzodiazepines studied. Other types of defects were reported to include cleft palate and exencphaly (Takeno et al., **1990**). The teratogenic dose in animals is approximately 5,000 times that of the human.

Czeizel, A.: Lack of evidence of teratogenicity of benzodiazepine drugs in Hungary. Reproductive Toxicology 1(3):183–188, 1988.

Saito, H.; Kobayashi, H.; Takemo, S. and Sakai, T.: Fetal toxicity of benzodiazepines in rats. Res. Commun. Chemical Pathol. Pharmacol. 46:437–447, 1984.

Takeno, S.; Nakagawa, M. and Sakai, T.: Teratogenic effects of nitrazepam in rats. Research Communications in Chemical Pathology and Pharmacology 69:59–70, 1990.

1741 Nitrendipine CAS 39562–70–4

Shimizu et al. (**1988**) treated rats orally with up to 1000 mg per kg on days 7–17. Maternal toxicity was found at 100 mg or above and at 1000 mg per kg the fetal weight was reduced. No statistical increase in defects was found. Postnatal studies of growth neurodevelopment and reproduction were normal.

Shimizu, Y.; Mariko, S.; Kamiya, Y.; Watanabe, C.; Masahiro, M. and Masanori, I.: Teratogenicity study on nitrendipine in rats. Oyo Yakuri 36(2):145–158, 1988.

1742 Nitric Oxide

Pierce et al. (**1994**) used a nitric oxide synthase inhibitor N^G–nitro–L–arginine methyl ester in the drinking water of rats at 0.3 and 1 mg per ml on days 13–19 and found growth retardation and hind limb reduction in the offspring. L–arginine addition reversed the limb defect findings. Diket et al. (**1994**) published a full account of this work. Nitroprusside coadministration prevented the limb defects. Maternal blood pressure was not altered by treatment.

Diket, A.L.; Pierce, M.R.; Munshi, U.K.; Voelker, C.A.; Eloby–Childress, S.; Greenberg, S.S.; Zhaug, X.J.; Clark, D.A. and Miller, M.J.S.: Nitric oxide inhibition causes intrauterine growth retardation and hind–limb disruptions in rats. Am. J. Obstet. Gynecol. 171:1243–1250, 1994.

Pierce, M.R.; Diket, A.L.; Volker, C.A.; Clark, D.A. and Miller, M.J.S.: Constitutive nitric oxide (NO) inhibition in rats causes intrauterine growth retardation and hindlimb disruptions: effects of supplemental L–aginine. Pediatric Res. 35:92A, 1994.

1743 Nitrilotriacetic Acid *NTA* CAS 139–13–9

This amino acid chelating agent has been recommended as a partial substitute for phosphates as a detergent–building agent. Nolen et al. (1971) tested the material in rabbits and rats and found no evidence for teratogenicity. The pregnant rats received up to 0.5 percent of NTA in their diet, and the rabbits were intubated with up to 250 mg per kg per day. Negative findings were observed in the mouse (Tjalve, 1972).

Nolen, G.A.; Klusman, L.W.; Black, D.L. and Buehler, E.U.: Reproduction and teratology of trisodium nitrilotriacetate in rats and rabbits. Food Cosmet. Toxicol. 99:509–518, 1971.

Tjalve, H.: A study of the distribution and teratogenicity of nitrilotriacetic acid (NTA) in mice. Toxicol. Appl. Pharmacol. 23:216–221, 1972.

1744 Nitrite *Sodium Nitrate* CAS 7632–00–0

Sleight et al. (1972) produced severe toxicosis in pregnant sows with 21 to 35 mg of sodium nitrite per kg subcutaneously. The treatment performed on various single days during the first 100 days of gestation did not produce any fetal defects. Fetal methemoglobin remained at a very much lower level than that in the mother. Globus and Samuel (1978) administered 0.5 mg of sodium nitrite to mice during pregnancy and found no adverse effects. Fetal erythropoiesis was stimulated. Shimada (1989) administered either 100 or 1000 mg per liter of drinking water to mice on days 7–18 and observed no teratogenicity or mutagenic effects gaps and breaks of chromosomes were studied in fetal liver.

Globus, M. and Samuel, D.: Effect of maternally administered sodium nitrite on hepatic erythropoiesis in fetal CD–1 mice. Teratology 18:367–378, 1978.

Shimada, T.: Lack of teratogenic and mutagenic effects of nitrite on mouse fetuses. Arch. Environ. Health 44:59–63, 1989.

Sleight, S.D.; Sinha, D.P. and Uzoukwu, M.: Effect of sodium nitrite on reproductive performance of pregnant sows. J.A.V.M.A. 161:819–823, 1972.

1745 Nitrite, Potassium

Sleight and Atallah (**1968**) fed guinea pigs up to 352 mg per kg. At 119 mg per kg and above there was increased fetal loss. This dose was 4,000 ppm in the drinking water.

Sleight, S. and Atallah, O.A.: Reproduction in the guinea pig as affected by chronic administration of potassium nitrate and potassium nitrite.

1746 Nitrobenzene

Dodd et al. (1987) exposed rats to up to 40 ppm and found reduced fertility in the males at the highest dose. Exposure was for six hours daily, five days a week. No adverse fetal effects were found when pregnant females were exposed. Tyl et al. (1987) exposed rats on days 6–15 to similar concentrations and found no evidence for teratogenicity.

Dodd, D.E.; Fowler, E.H.; Snellings, W.M.; Pritts, I.M.; Tyl, R.W.; Lyon, J.P.; O'Neal, F.O. and Kimmerle, G.: Reproduction and fertility evaluation in CD rats following nitrobenzene inhalation. Fund. Appl. Toxicol. 8:493–505, 1987.

Tyl, R.W.; France, K.A.; Fisher, L.C.; Dodd, D.E.; Pritts, I.M.; Lyon, J.P.; O'Neal, F.O. and Kimmerle, G.: Developmental toxicity evaluation of inhaled nitrobenzene in CD rats. Fund. Appl. Toxicol. 8:482–492, 1987.

1747 Nitrofen *2,4–Dichlorophenyl–4–nitrophenyl Ether Diphenyl Ethers* CAS 1836–75–5

Francis and Metcalf (1982) administered 18 mg per kg transcutaneously to mice and rats from implantation through organogenesis and produced a high incidence of eye defects. No increase in fetal mortality was found. Kavlock and Gray (1982) gave rats 75 mg per kg orally per day on day 11 and found hydronephrosis and reduced renal function in the offspring. Gray et al. (1983) found cleft palate, diaphragmatic hernias and defects in the reproductive tracts of rat fetuses where the mothers were given 100 mg or more per kg on days 7–17. Costlow et al. (1983) using dermal applications of 12 mg per kg during organogenesis in rats produced diaphragmatic hernias and absent Harderian glands in the offspring. No gross behavioral effects were noted. Lau et al. (1988) found that rat fetuses exposed on days 10–13 to 20 or 40 mg per kg daily had hypoplastic lungs which led to neonatal death. Wickman et al. (**1993**) gavaged mice on days 8 and 9 with 200 or 500 mg per kg and produced diaphragmatic defects cleft palate, renal agenesis, excephaly or the DiGeorge sequence.

Francis (1986) applied analogs of nitrofen to the backs of mice on day 8 and then determined the number of surviving pups. 2,4,5 Trichlorophenyl 4–nitrophenyl ether was significantly more toxic than nitrofen but 2,4,6–trichlorophenyl 4–nitrophenyl ether was not toxic at 175 mg per kg. Neither of the 2 monochlorinated 4–nitrophenyl ethers nor the unchlorinated 4–nitrophenyl ethers were fetotoxic at 1750 mg per kg. Francis (1990) gavaged mice with the unchlorinated, monochlorinated and dichlorinated–phenyl ethers of nitrofen. Maternal toxicity postnatal survived and small or absent Harderian glands were studied and none of the diphenyl ethers was found to be as active as nitrofen.

Costlow, R.D.; Hirsekorn, J.M.; Stiratelli, R.G.; O'Hara, G.P.; Black, D.L.; Kane, W.W.; Burke, S.S.; Smith, J.M. and Hayes, A.W.: Effects on rat pups when nitrofen 4(2,4–dichlorophenoxy)nitrobenzene was applied dermally to the dam during organogenesis. Toxicology 28:37–50, 1983.

Francis, B.M.: Role of structure in diphenyl ether teratogenesis. Toxicology 40:297–309, 1986.

Francis, B.M.: Relative teratogenicity of nitrofen analogs in mice: unchlorinated,monochlorinated, and dichlorinated–phenyl ethers. Teralology 41:443–451, 1990.

Francis, B.M. and Metcalf, R.L.: Percutaneous teratogenicity of nitrofen. (abs) Teratology 25:41A only, 1982.

Gray, L.E., Jr.; Kavlock, R.J.; Chernoff, N.; Ostby, J. and Ferrell, J.: Postnatal developmental alteration following prenatal exposure to the herbicide 2,4–dichlorophenyl–p–nitrophenyl ether a dose response evaluation in the mouse. Toxicol. Appl. Pharmacol. 67:1–14, 1983.

Kavlock, R.J. and Gray, L.E., Jr.: Postnatal evaluation of the mophological and functional effects of prenatal nitrofen exposure. (abs) Teratology 25:53A only, 1982.

Lau, C.; Cameron, A.M.; Irsula, O.; Antolick, L.L.; Langston, C. and Kavlock, R.J.: Teratogenic effects of nitrofen on cellular and functional maturation of the rat lung. Toxicol. Appl. Pharmacol. 95:412–422, 1988.

Wickman, D.S.; Siebert, J.R. and Benjamin, D.R.: Nitrofen–induced congenital diaphragmatic defects in cd1 mice. Teratology 47:119–125, 1993.

1748 Nitrofurantoin CAS 67–20–9

Heinonen et al. (1977) listed 83 pregnant women who took this medication in the first four lunar months and they found five malformations which was not a significant increase. Hailey et al. (1983) followed 91 pregnancies with urinary infection and nitrofurantoin treatment and did not identify any adverse effect.

Prytherch et al. (1984) administered oral doses of up to 30 mg per kg to pregnant rats and rabbits before and during gestation and no adverse fetal effects were identified.

Hailey, F.J.; Furt, H.; Williams, J.C. and Hammers, B.: Foetal safety of nitrofurantoin macrocrystals therapy during pregnancy: A retrospective analysis. J. Int. Med. Res. 11:364–369, 1983.

Heinonen, O.P.; Slone, D. and Shapiro, S.: Birth Defects and Drugs in Pregnancy. Publishing Sciences Group Inc., Littleton, Mass., 1977.

Prytherch, J.P.; Sutton, M.L. and Denine, E.P.: General reproduction, perinatal–postnatal and teratology studies of nitrofurantoin macrocrystals in rats and rabbits. J. Toxicol. Environ. Health 13:811–823, 1984.

1749 Nitrofurazone

Greenaway et al. (1986) using in vitro culture of rat embryos, found axial asymmetry and growth retardation at concentrations of 0.05 millimoles. Nomura et al. (1975) gave pregnant mice 300 microgm per gm of body weight on day 10 of gestation. An increase in limb reduction defects was found.

Greenaway, J.C.; Fantel, A.G. and Juchau, M.R.: On the capacity of nitroheterocyclic compounds to elicit an unusual axial asymmetry in cultured rat embryos. Toxicol. Appl. Pharmacol. 82:307–315, 1986.

Nomura, T.; Kimura, S.; Isa, Y.; Tanaka, M. and Sakamoto, Y.: Teratogenic and carcinogenic effects of nitrofurazone in the mouse embryo and newborn. (abs) Teratology 12:206–207, 1975.

1750 Nitrogen Dioxide

Nikiforov and Tabacova (1985) exposed pregnant rats to 1 or 10 mg per square meter during all of pregnancy. Hexabarbitol sleeping time was increased in the offspring suggesting a deficiency in liver drug metabolism.

Nikiforov, B. and Tabacova, S.: Drug metabolism in NO_2–treated female rats and their generation. (abs) Teratology 32:30A, 1985.

1751 Nitrogen Mustard
2,2–Dichloro–N–methyldiethylamine N–oxide
Hydrochloride
Methyl–bis(β–chloroethyl)aminehydrochloride

This alkylating agent was used as a teratogen in rats and mice (see reviews by Chaube and Murphy, 1968; Kalter, 1968). Skeletal defects, cleft palate, exencephaly and encephalocele were reported. Kalter emphasized that many of the neural closure type of defects occurred following treatment given relatively late in pregnancy, at times when the neural tube is apparently already normally closed. The dose given in the rat is 0.5 to 0.7 mg per kg on the 11th or 12th gestational day. The maternal rat LD–50 is 2.0 mg per kg.

Jurand (1961) using electron microscopy studied histologic changes which occur in the central nervous system.

Muller (1966) produced defects by direct intrauterine injection. Salzgeber (1969) studied in detail the action of nitrogen mustard on the production of limb defects in the chick embryo. Sanyal et al. (1981) exposed rat embryos in vitro to 1–5 micrograms per ml and retarded growth severely. Tabakova and Balabaeva (1986) exposed rats to 1 and 10 mg per square meter for six days throughout gestation. Late fetal lethality, growth retardation and hydrocephalus were found.

Chaube, S. and Murphy, M.L.: The teratogenic effects of the recent drugs active in cancer chemotherapy. In: Advances in Teratology, D.H.M. Woollam (ed.), New York: Logos and Academic Press, 3:181–237, 1968.

Jurand, A.: Further investigations on the cytotoxic and morphogenetic effects of some nitrogen mustard derivatives. J. Embryol. Exp. Morphol. 9:492–506, 1961.

Kalter, H.: Teratology of the Central Nervous System. Chicago: University of Chicago Press, 139–140, 1968.

Muller, M.: Does nitrogen mustard affect the foetus directly or secondarily by its effect on the mother? Experientia 22:247 only, 1966.

Salzgeber, B.: Etude comparative des effets de l'yperite azotee sur les constituants, mesodermique et ectodermique, des bourgeons de membres de l'embryon de poulet. J. Embryol. Exp. Morphol. 22:373–394, 1969.

Sanyal, M.K.; Kitchin, K.T. and Dixon, R.L.: Rat conceptus development in vitro. Comparative effects of alkalating agents. Toxicol. Appl. Pharmacol. 57:14–19, 1981.

Tabakova, S. and Balabaeva, L.: Nitrogen dioxide embryotoxicity and lipid peroxidation. (abs) Teratology 33:58A, 1986.

1752 Nitroglycerin CAS 55–63–0

Oketani et al. (1981) administered up to 4 mg daily on days 6–18 to rabbits and found no adverse fetal effects. Rats were given up to 20 mg per kg intraperitoneally before mating, during organogenesis or during late gestation. No adverse effects were reported.

Sato et al. (1984) studied the dermal application of this chemical in rats and rabbits. In rats, in amounts of up to 7 gm per kg on days 7–17, days 17–21 postnatally, before fertilization and for 7 days after. In the organogenesis study of rats, abortion occurred at 1.4 and 7.0 gm per kg. No increased defects or postnatal changes were noted in either species.

Oketani, Y.; Mitsuzona, T.; Ichikawa, K.; Itono, Y.; Gojo, T.; Gofuku, M. and Konoha, N.: Teratological studies in rabbits and reproductive and teratological studies in rats. Oyo Yakuri 22:633–648, 737–763, 1981.

Sato, K.; Taniguchi, H.; Ohtsuka, T.; Himeno, Y.; Uchiyama, K.; Koide, M. and Hoshino, K.: Reproductive studies of nitroglycerin applied dermally to pregnant rats and rabbits. Clin. Report 18:3511–3586, 1984.

1753 Nitroimidazole

Greenaway et al. (1986) using in vitro culture of rat embryos found axial asymmetry and growth retardation at concentrations of 0.1 millimoles.

Greenaway, J.C.; Fantel, A.G. and Juchau, M.R.: On the capacity of nitroheterocyclic compounds to elicit an unusual axial asymmetry in cultured rat embryos. Toxicol. Appl. Pharmacol. 82:307–315, 1986.

1754 p–Nitrophenol CAS 100–02–7

Plasterer et al. (1985) gavage fed mice on days 7–14 with a dose considered to be just below adult lethality (400 mg per kg). No gross congenital defects were detected in neonates and the neonatal death rate was not increased.

Plasterer, M.R.; Bradshaw, W.S.; Booth, G.M.; Carter, M.W.; Schuler, R.L. and Hardin, B.D.: Developmental toxicity of nine selected compounds following prenatal exposure in the mouse. Naphthalene, p–nitrophenol, sodium selenite, dimethyl phthalate, ethylene–thiourea and four glycol ether derivatives. J. Toxicol. Environ. Health. 15:25–38, 1985.

1755 N–Nitro–p–phenylenediamine

Marks et al. (1981) studied the effect of this hair dye in pregnant mice and found at 160 mg per kg daily subcutaneously from days 6 through 15, an increase in cleft palate, resorptions and intrauterine growth retardation. The no effect level was 64 mg per kg per day. Some alizarin red staining spots were found in the heart ventricles of the fetuses.

Marks, T.A.; Gupta, B.N.; Ledoux, T.A. and Staples, R.E.: Teratogenic evaluation of N–nitro–p–phenylenediamine, 4–nitro–o–phenylenediamine and 2,5–toluenediamine sulfate in the mouse. Teratology 24:253–265, 1981.

1756 4–Nitro–o–phenylenediamine

This hair dye was given subcutaneously to pregnant mice from days 6–15 at levels of 256 mg per kg per day and an increase in cleft palate and major blood vessel anomalies were found. Maternal and fetal weights were reduced. At 128 mg per kg no fetal effects occurred (Marks et al., 1981).

Marks, T.A.; Gupta, B.N.; Ledoux, T.A. and Staples, R.E.: Teratogenic evaluation of N–nitro–p–phenylenediamine, 4–nitro–o–phenylenediamine and 2,5–toluenediamine sulfate in the mouse. Teratology 24:253–265, 1981.

1757 p–Nitrophenyl–β–D–xyloside

Discussed under β–D–Xylosides

1758 2–Nitro–p–phenylenediamine

Marks et al. (1981) gave mice up to 256 mg per kg subcutaneously on days 6–15. At doses over 160 mg per kg there was maternal toxicity and fetotoxicity. Some malformation increase also occurred at the higher doses. Alizarin red spots were found in the myocardium in the higher dose group.

Marks, T.A.; Gupta, B.N.; Ledoux, T.A. and Staples, R.E.: Teratogenic evaluation of 2–nitro–p–phenylenediamine, 4–nitro–o–phenylenediamine, and 2,5–toluenediamine sulfate in the mouse. Teratology 24:253–265, 1981.

1759 p–Nitrophenyl–p'–quanidino Benzoate

This acrosin inhibitor was studied by Beyler and Zaneveld (**1980**) in mice. The concentration which inhibited 50% in vitgro fertilization was 1.3 x 10 [8]M. Subcutaneous administration reduced fertility when administered to the female by minipump.

Beyler, S.A. and Zaneveld, L.J.D.: Antifertility effect of acrosin inhibitors in vitro and in vivo. Fed. Proc. 39:624, 1980.

1760 Nitropropane

Hardin et al. (1981) injected rats intraperitoneally with 170 mg per kg on days 1–15. No teratogenicity or maternal toxicity was reported.

Hardin, B.D.; Bond, G.P.; Sikov, M.R.; Andrew, F.W.; Beliles, R.P. and Niemeier, R.W.: Testing of selected workplace chemicals for teratogenic potential. Scand. J. Work Environ.

1761 Nitroprusside, Sodium

Ivankovic (**1979**) treated rats and rabbits intervenously with up to 7.5 and 1.25 mg per kg respectively and found no teratogenicity. Lewis et al. (**1977**) found that the ewe fetus develops higher levels of cyanide than the dam after administration of nitripruside. Lee and Juchau (**1994**) injected sodium nitroprusside directly into the amniotic cavity of d. 10.5 rats and produced whitened zones of dead cells in the mesencephalon. This chemical is converted to nitric oxide. By blocking nitric oxide synthetase with N^G–monomethyl–L–anine they suggested that nitric oxide plays a role in embryogenesis. The inhibitor itself was associated with open neural tube production.

Kawai et al. (**1994** A,B,C) gave rats up to 7.4 mg per kg intravenously in a reproduction, teratology and perinatal study and found no adverse fetal effects. Postnatal development was unaffected. Ishikara et al. (**1994**) gave a similar treatment at dose levels of up to 1.6 mg per kg to rabbits during organogenesis and found no adverse fetal effects although the dams had flushing and increased respiratory rates for a few hours after each treatment.

Ishihara, M.; Fukushima, T.; Masaki, F.; Matsumura, H.; Kawai, Y.; Tamura, T. and Sato, N.: Study of sodium nitroprusside (SNP) by intravenous administration on the period of fetal organogenesis in rabbits. Yakuri to Chiryo 22:s2017–s2024, 1994.

Ivankovic, V.S.: Fehlende teratogene wirkung von Arzneim. Forsch./Drug Res. 29:1092–1094, 1979. kaninchen.

Kawai, Y.; Shibata, K.; Watanabe, Y.; Nakamura, Y.; Okumura, N.; Tamura, T. and Sato, N.: Study of sodium nitroprusside (SNP) by intravenous administration on prior to and in the early stages of pregnancy in rats. Yakuri to Chiryo 22:s1987–s1997, 1994A.

Kawai, Y.; Inoue, T.; Shibata, K.; Nakamura, Y.; Tanigami, J.; Tsujimoto, H.; Heike, C.; Tamura, T. and Sato, N.: Study of sodium nitroprusside (SNP) by intravenous administration on the period of fetuses organogenesis in rats. Yakuri to Chiryo 22:s1999–s2016, 1994B.

Kawai, Y.; Inoue, T.; Shibata, K.; Watanabe, Y.; Nakamura, Y.; Tanigami, J.; Tsujimoto, H.; Heike, C.; Ueda, T.; Tamura, T. and Sato, N.: Study of sodium nitroprusside (SNP) by intravenous administration during the perinatal and lactation periods in rats. Yakuri to Chiryo 22:s2025–s2039, 1994C.

Lee, Q.P. and Juchau, M.R.: Dysmorphogenic effects of nitric oxide (NO) and NO–synthase inhibition: studies with intra–amniotic injections of sodium nitroprusside and N^Γ–Monomethyl L–arginine. Teratology 49:452–464, 1994.

Lewis, P.E.; Cefalo, R.C.; Naulty, J.S. and Rodkey, F.L.: Placental transfer and fetal toxicity of sodium nitroprusside. Gyn. Invest. 8:46, 1977.

1762 4–Nitroquinoline 1–oxide CAS 56–57–5

Nomura (1977) injected pregnant mice subcutaneously on day 9, 10 or 11 with 15 micrograms per gm of body weight. At this maximum tolerated dose, no increased malformations or fetal deaths occurred. When one microgram was injected directly into the amniotic cavity on day 11 cleft palate, tail and leg defects were found.

Nomura, T.: Similarity of the mechanism of chemical carcinogen–initiated teratogenesis and carcinogenesis in mice. Cancer Research 37:969–973, 1977.

1763 Nitrosodiethylamine *Diethylnitrosamine* CAS 55–18–5

Druckrey (1973) summarized his work with indirect alkylating compounds. Giving 70 mg per kg intravenously on the 15th day to the rat produced no postnatal tumors but a few were found when the dose was given on the 22nd day. After oral administration a very high incidence of liver carcinomas, nephroblastomas and neurogenic tumors were found postnatally. Dimethylnitrosamine given on the last day of pregnancy produced only a few uncharacteristic tumors. Dimethylnitrosamine was not teratogenic in rats when 30 mg was given orally in single days during gestation (Napalkov and Alexandrov, 1968).

Druckrey, H.: Specific carcinogenic and teratogenic effects of indirect alkylating methyl and ethyl compounds, and their dependency on stages of ontogenic developments. Xenobiotica 3:271–303, 1973.

Napalkov, N.P. and Alexandrov, V.A.: On the effects of blastomogenic substances on the organism during organogenesis. Z. Krebsforsch 71:32–50, 1968.

1764 Nitrosoethylenethiourea

Khera and Iverson (1980) gave single oral doses of 120 to 240 mg per kg on day 13 of rat gestation. At the higher doses there was maternal lethality but at 120 and 160 mg per kg hydrocephalus and other defects were found.

Khera, K.S. and Iverson, F.: Hydrocephalus induced by N-nitrosoethylene-thiourea in the progeny of rats treated during gestation. Teratology 21:367–370, 1980.

1765 Nitrosourea CAS 13010–20–3

Discussed under Ethylnitrosourea

1766 **Nitrous Oxide** also see *Anesthetics* CAS 10024–97–2

Cohen et al. (1980) analyzed questionnaires from 30,650 dentists and 30,547 chairside assistants. The chairside assistants reported a rate of 19 abortions among heavy users while the non–users had a rate of 8. In the same group there were 7.7 percent defects compared to 3.6 percent in the controls. These defects were mainly of the nervous and musculoskeletal systems. Mazze and Lecky (1985) outlined the flaws in the epidemiological evidence which suggested a small risk ratio (1.3) for increased spontaneous abortions among operating room workers. He also pointed out that exposure levels to these workers was reduced tenfold. In a survey of 405 dental assistants, the group of 19 exposed for more than 5 hours per week to unscavenged nitrous oxide had a significant reduction in fecundability (adjusted 0.41, confidence interval 0.23–0.74). No reduction was found in groups of women who worked in places that used a scavenger system or who were exposed to less than 5 hours per week (Rowland, **1992**).

Fink et al. (1967), using 50 percent nitrous oxide exposure for 2, 4 or 6 days starting on gestational day 8, produced rib and vertebral defects in nearly all surviving rat fetuses. A low incidence of hydrocephalus, cardiac and renal defects was observed. In later work 24–hour exposure on day 9 to 70 percent nitrous oxide was shown to produce the greatest number of defects (Shepard and Fink, 1968). Gofmekler et al. (1977), using an exposure of 0.34–0.8 mg per cubic meter during the entire gestation, produced some developmental anomalies and increased number of non viable rat fetuses. Rector and Eastwood (1964) and Smith et al. (1965) showed that nitrous oxide causes lethality in chick eggs.

Mazze et al. (1982) found no reproductive ill effects in male mice treated for long periods. Mazze et al. (1986) found increased resorptions but no defects in rats exposed to 75 percent nitrous oxide for 6 hours daily during three–day periods of organogenesis. Using larger numbers, Fujinaga et al. (1987) found increased defects in the rat treated with 50 percent nitrous oxide on day 8. They also observed that the concurrent use of isoflurane with nitrous oxide led to significant reductions in resorptions and defects. Fujinaga et al. (1990) observed altered body laterality in rats exposed in vivo on day 8 to 75% nitrous oxide a steep increase in embryonic mortality occurred between day 13 and 14. Lane et al. (1980) exposed rats on day 9 to 70–75 percent nitrous oxide and found increased malformations. Xenon, another anesthetic, did not produce defects. Koeter and Rodier (1986) exposed mice to 4 or 6 hours on day 14 or 2 days postnatally and found a delay in the appearance of righting reflex and locomotion. Nitrous oxide was administered as 75 percent.

Fujinaga and Baden (**1994**) found that methionine but not folinic acid or prazosin reversed the maldevelopment of cuture of rat embryos exposed to nitrous oxide.

Cohen, E.N.; Brown, B.W.; Wu, M.L.; Whitcher, C.E.; Brodsky, J.B.; Gift, H.C.; Greenfield, W.; Jones, T.W. and Driscoll, E.J.: Occupational disease in dentistry and chronic exposure to trace anesthetic gases. J. Am. Dent. Assoc. 101:21–31, 1980

Fink, B.R.; Shepard, T.H. and Blandau, R.J.: Teratogenic activity of nitrous oxide. Nature 214:146–148, 1967.

Fujinaga, M.; Baden, J.M.; Yhap, E.O. and Mazze, R.I.: Reproductive and teratogenic effects of nitrous oxide, isoflurane, and their combinations in Sprague–Dawley rats. Anesthesiology 67:960–964, 1987.

Fujinaga, M.; Baden, J.M.; Shepard, T.H. and Mazze, R.I.: Nitrous oxide alters body laterality in rats. Teratology 41:131–135, 1990.

Fujinaga, M. and Baden, J.M.: Methionine prevents mitrous oxide–induced teratogenicity in rat embryos grown in culture. Anesthesiology 81:184–189, 1994.

Gofmekler, V.A.; Brekhman, I.I.; Golotin, V.G.; Sheparev, A.A.; Kpivelevich, E.B.; Kamynina, L.N.; Dobryakova, A.I. and Gonenko, V.A.: Embryotoxic effect of nitrous dioxide and atmosphere pollution. Gigiena i Sanitariya (USSR) 12:22–25, 1977.

Koeter, H.B.M. and Rodier, P.M.: Behavioral effects in mice exposed to nitrous oxide or halothane: Prenatal vs. postnatal exposure. Neurobehav. Toxicol. Teratol. 8:189–194, 1986.

Lane, G.A.; Nahrwold, M.L.; Taylor–Busch, M. and Cohen, P.J.: Anesthetics as teratogens: Nitrous oxide is fetotoxic, xenon is not. Science 210:899–901, 1980.

Mazze, R.I.; Fujinaga, M.; Rice, S.A.; Harris, S.B. and Baden, J.M.: Reproductive and teratogenic effects of nitrous oxide, halothane, isoflurane, and enflurane in Sprague–Dawley rats. Anesthesiology 64:339–344, 1986.

Mazze, R.I. and Lecky, J.H.: The health of operating room personnel. Anesthesiology 62:226–228, 1985.

Mazze, R.I.; Wilson, A.I.; Rice, S.A. and Baden, J.M.: Reproduction and fetal development in mice chronically exposed to nitrous oxide. Teratology 26:11–16, 1982.

Rector, G.H.M. and Eastwood, D.W.: The effects of an atmosphere of nitrous oxide and oxygen on the incubating chick. Anesthesiology 25:109 only, 1964.

Rowland, A.S.; Baird, D.D.; Weinberg, C.R.; Shore, D.L.; Shy, C.M. and Wilcox, A.J.: Reduced fertility among women employed as dental assistants exposed to high levels of nitrous oxide. N. Eng. J. Med. 327:993–997, 1992.

Shepard, T.H. and Fink, B.R.: Teratogenic activity of nitrous oxide in rats. In: Toxicity of Anesthetics, B.R. Fink (ed.), Baltimore: Williams and Wilkins, 308–323, 1968.

Smith, B.E.; Gaub, M.L. and Moya, F.: Teratogenic effects of anesthetic agents: nitrous oxide. Anesth. Analg. (Cleve) 44:726–732, 1965.

1767 **Nivalenol** *3,4,7,15–tetrahydroxy–12, 13–epoxytrichothec–9–en–8–one*

Ito et al. (**1986**) gave mice up to 1.5 mg per kg intraperitoneally and at 0.5 and 1.5 mg per kg embryolethality was found but no malformations. The treatment was on days 7–15 of gestation.

Ito, Y.; Ohtsubo, K.; Ishii, K. and Ueno, Y.: Effects of nivalenol on pregnancy and fetal development of mice. Mycotoxin Research 2:71–77, 1986.

1768 Nizatidine

Buelke–Sam et al. (**1989** A) treated rats orally with up to 1500 mg per kg on days 7–17 and found no teratogenicity. This histamine H_2–receptor antagonist at the highest dose increased fetal resorption, decrease the percentage of male vs female fetuses and reduced female fetal weight. Similar treatment on days 17–20 did not alter postnatal development and reproduction. Some reduction in weight was found at 6 weeks of age in all the treated groups. (Buelke–Sam et al., **1989** B)

Buelke–Sam, J.B.; Hagopian, G.S.; Probst, K.S. and Fisher, L.F.: Nizatidine: teratogenicity study in rats. Yakuri to Chiryo 17:547–570, 1989.

Buelke–Sam, J.; Tizzano, J.P.; Probst, K.S. and Fisher, L.F.: Nizatidine: perinatal and postnatal study in rats. Yakuri to Chiryo 17:571–589, 1989.

1769 Nizofenone Fumarate

Imanishi et al. (1985) gave rats up to 20 mg per kg intravenously on days 7–17. The food consumption of the dams was decreased at the highest dose but no malformations or functional deficits were found.

Imanishi, M.; Yoneyama, M.; Takeuchi, M. and Kato, Y.: Teratogenicity study of nizofenone fumarate in rats. Ikakuhin Kenkyu 16:1–19, 1985.

1770 NKT–01 *1–Amino–19–quadidino–11–hydroxy–4,9,12–triazanonadcane–10, 13–dione Trihydrochloride Gusperimus Hydrochloride*

Shimizu et al. (**1991**,A,B,C) gave this immunosuppressant to rats during organogenesis. At the highest intravenous dose the adults did not gain weight normally but the fetuses were normal. Similar findings were reported after perinatal and prenatal studies. No teratogenicity was found. Rabbits were given up to 0.9 mg per kg and no teratogenicity was found although at the highest dose an increase in fetal mortality occurred (Handa et al, **1991**).

Handa, J.; Aoki, A.; Norikoshi, M.; Irie, Y.; Hayashi, M.; Tubosaki, M. and Nakamori, K.: Toxicological studies on NKT–01 (VI) Intravenous administration during the period of fetal organogenesis in rabbits.

Hiramatsu, Y.; Shimizu, M.; Suzuki, T.; Uto, K.; Kato, M.; Kobayashi, Y.; Iwasaki, S.; Kaike, T. and Suzuki, R.: Toxocological studies on NKT–01 (V) Intravenous administration during the period of fetal organogenesis in rats. The Clinical Report 25:3005–3024, 1991B.

Shimizu, M.; Ota, T.; Kato, M.; Uto, K.; Kobayashi, Y.; Yamashita, Y.; Handa, J.; Tsubosaki, M. and Nakamori, K.: Toxicological studies on NKT–01 (IV) intravenous administration prior to and in the early stages of pregnancy in rats. The Clinical Report 25:2993–3004, 1991A.

Shimizu, M.; Ota, T.; Kato, M.; Kobayashi, Y.; Yamashita, Y.; Asano, M.; Handa, J.; Tsubosaki, M. and Nakamori, K.: Toxicological studies on NKT–01 (VII) intravenous administration during the perinatal and lactation periods in rats. The Clinical Report 25:3035–3049, 1991C.

1771 Nocodazole

Bishop et al. (1989) used this inhibitor in mice one hour after mating and found developmental anomalies which were associated with numerical chromosome changes.

Bishop, J.B.; Rutledge, J.C. and Generoso, W.M.: Effects of nocodazole (NOC) and ethylene glycol monomethyl ether (EGMME) on mouse zygotes. Teratology 39:442, 1989.

1772 Noise

Discussed under Emotional Stress

1773 Nomifensine Maleate CAS 32795–47–4

Sugisaki et al. (1984) gave pregnant mice orally up to 60 mg per kg before, during and after pregnancy. No adverse effects were found in growth or development or behavior of the offspring.

Sugisaki, T.; Takagi, S.; Seshimo, M.; Takayama, K. and Miyamoto, M.: Reproduction studies of nomifensine maleate in mice. Oyo Yakuri 28:171–185, 1984.

1774 Nonachlazine CAS 49780–10–1

Smolnikova and Strekalova (1979) reported that nonachlazine given orally to rats from days 1 through day 17 of pregnancy in a dose of 35 mg per kg (one–eighth LD–50) produced no adverse effect on embryogenesis.

Smolnikova, N.M. and Strekalova, S.N.: Results of studies on embryotoxic and teratogenic action of nonachlazine. Farmakol. Toksikol. (USSR) 42.3:302–303, 1979.

1775 Nonoxynol–9 *Vaginal Spermicides* CAS 26027–38–3

Scholl et al. (1983) observed an increase in spontaneous abortion but Harlap et al. (1980) found no increase based on a large cohort study. If there is an increase, the cause may be indirect and in some way related to other factors associated with the form of birth control.

Although an increase in defect rate was suggested from the studies of Jick et al. (1981), subsequently, more detailed studies of larger groups by Shapiro et al. (1982) and Cordero and Layde (1983) failed to find any increased risk. Bracken (1985) critically reviewed the multiple studies and calculated that there is no asssociation with birth defects. Simpson (1985) summarized studies of congenital defects in the offspring of mothers using spermicides and other contraceptives. He concluded that a "clear scientific consensus exists" that there is no association with increased malformations. Warburton et al. (1987) did not find increased spermacide exposure rates by 154 women with trisomic fetuses.

Louik et al. (1987) reported a case control study of vaginal spermicide exposure during periconceptual, the first trimester or at any time in life. All cases had deformity. No increase in risk was found among specific groups such as Down's Syndrome, hypospadias, limb reduction or neural tube defects.

Jick et al. (1982) observed a higher rate for early spontaneous abortion among vaginal spermicide users. Polednak et al. (1982) compared the offspring of 302 women using spermicides to those of 715 women using no contraceptives and found no increased relative risk in malformations. Strobino et al. (1986) studied spermicide use among 7,184 women who had spontaneous abortions. About two–thirds of the abortuses were karyotyped. No correlation was found between spermicide use and those with normal karyotypes. Among the group of private patients there was one category (trisomy) where a borderline significant increase was found (probability 0.05). The exposed female newborns were lighter and a non–significant increase in hypospadias and limb reduction defects occurred in the user's group. In another study, Strobino et al. (1988) studied congenital anomalies among 2,712 pregnancies. In 149, the spermicide was taken periconceptually. There were 661 women who took spermicides preconceptionally. No adverse effects were noted but a slight reduction in the weight of female newborns was recorded. The authors thought that this was a "chance occurence."

This non–ionic surfactant commonly used in vaginal spermicidal formulations was studied in the rat by intravaginal application of 4 and 40 mg per kg on days 6 through 15 of gestation (Abrutyn et al., 1982). No maternal or fetal changes were observed. In similar experiments up to 400 mg per kg produced no fetal effects (Shiota and Shepard, unpublished data). Buttar (1982) showed that 44 percent of the intravaginal dose is absorbed. Buttar et al. (1985) found that mouse eggs studied in vitro did not cleave properly at media dose levels of 0.25 microgram per ml. Tryphonas and Buttar (1986) found inflammatory changes in the genital tract of rats given vaginal doses of 5 mg per 100 gm body weight after implantation. Associated with these changes, there was a significant reduction in viable fetuses, but no data was given on the surviving fetuses.

Abrutyn, D.; McKenzie, B.E. and Nadaskay, N.: Teratology study of intravaginally administered nonoxynol–9 containing contraceptive cream in rats. Fertil. Steril. 37:113–117, 1982.

Bracken, M.B.: Spermicidal contraceptives and poor reproductive outcomes: The epidemiologic evidence against an association. Am. J. Obstet. Gynecol. 151:552–556, 1985.

Buttar, H.S.: Transvaginal absorption and disposition of nonoxynol–9 in gravid rats. Toxicol. Letters 13:211–216, 1982.

Buttar, H.S.; Moffatt, J.H. and Bura, C.: Effects of vaginal spermicides, nonoxynol–9 and octoxynol–9 on the development of mouse embryos in vitro. (abs) Teratology 31:51A, 1985.

Cordero, J.F. and Layde, P.M.: Vaginal spermicides, chromosomal abnormalities and limb reduction defects. Perspectives 15:16–18, 1983.

Harlap, S.; Shiono, P.H. and Ramcharan, S.: Spontaneous foetal losses in women using different contraceptives around the time of conception. Int. J. Epidemiol. 9:49–56, 1980.

Jick, H.; Shiota, K.; Shepard, T.H.; Hunter, J.R.; Stergachis, A.; Madsen, S. and Porter, J.B.: Vaginal spermicides and miscarriage seen primarily in the emergency room. Teratogenesis, Mutagenesis and Carcinogenesis. 2:205–210, 1982.

Jick, H.; Walker, A.M.; Rothman, K.J.; Hunter, J.R.; Holmes, L.B.; Watkins, R.N.; Dewart, D.C.; Danford, A. and Madsen, S.: Vaginal spermicides and congenital disorders. JAMA 1329–1332, 1981.

Louik, C.; Mitchell, A.A.; Werler, M.M.; Hanson, J.W. and Shapiro, S.: Maternal exposure to spermicides in relation to certain birth defects. New England Journal of Medicine 317:474–478, 1987.

Polednak, A.P.; Janerich, D.T. and Glebatis, D.M.: Birthweight and birth defects in relation to maternal spermicide use. Teratology 26:27–38, 1982.

Scholl, T.O.; Sobel, E.; Tanfer, K.; Soefer, E.F. and Saidman, B.: Effects of vaginal spermicides in pregnancy outcome. Fam. Plann. Persp. 15:244–249, 1983.

Shapiro, S.; Slone, D.; Heinonen, O.P.; Kaufman, D.W.; Rosenberg, L.; Mitchell, A.A. and Heinrich, S.P.: Birth defects and vaginal spermicides. J. Am. Med. Assoc. 247:2381–2384, 1982.

Simpson, J.L.: Relationship between congenital anomalies and contraception. Adv. Contracept. 1:3–30, 1985.

Strobino, B.; Kline, J.; Lai, A.; Stein, Z.A.; Susser, M. and Warburton, D.: Vaginal spermicides and spontaneous abortion of known karyotype. Am. J. Epidemiol. 123:431–443, 1986.

Strobino, B.; Kline, J. and Warburton, D.: Spermicide use and pregnancy outcome. Am. J. Public Health 78:260–263, 1988.

Tryphonas, L. and Buttar, H.S.: Effects of the spermicide nonoxynol–9 on the pregnant uterus and the conceptus of rat. Toxicology 39:177–186, 1986.

Warburton, D.; Neugut, R.H.; Lustenberger, A.; Nicholas, A.G. and Kline, J.: Lack of association between spermicide use and trisomy. The New England Journal of Medicine 317:478–482, 1987.

1776 Nonoxynol–30

Meyer et al. (1988) gave up to 1000 mg orally to rats on days 6–15 or 1–20 and found no adverse effects.

Meyer, O.; Andersen, P.H.; Hansen, E.V. and Larsen, J.C.: Teratogenicity and in vitro mutagenicity studies on nonoxynol–9 and nonoxynol–30. Pharmacol. Toxicol. 62:236–238, 1988.

1777 Norchlorcyclizine also see *Meclizine* CAS 303-26-4

This antihistamine belongs to the benzhydrylpiperazine series of compounds. It is the teratogenically active breakdown product of several other compounds, and the experimental studies in rats are described more fully under the heading, Meclizine. Wilk (1969) applied this substance directly over the amniotic sac of the rat and produced cleft palates.

Wilk, A.L.: Production of fetal rat malformations by norchlorcyclizine and chlorcyclizine after intrauterine application. Teratology 2:55–66, 1969.

1778 Nordihydroguaiaretic Acid

DeSesso and Goeringer (**1990**) gave rabbits 950 mg per kg subcutaneously on day 12 and found no increase in defects. This antitoxidant was associated with a significant increase in body weight.

DeSesso, J.M. and Goeringer, G.C.: Ethoxyquin and nordihydroguaiaretic acid reduce hyroxyurea developmental toxicity. Reproductive Toxicology 4:267–275, 1990.

1779 Norea 1–[5–(3a,4,5,6,7,7a–Hexahydro–4, 7–methanoindanyl]–3,3–dimethylurea CAS 18530–56–8

This organophosphate cholinesterase inhibitor was tested by Robens (1969) in hamsters using 2000 mg per kg by mouth on days 6 through 8 and no adverse effect on reproduction was found.

Robens, J.F.: Teratologic studies of carbaryl, diazinon, norea, disulfiram and thiram in small laboratory animals. Toxicol. Appl. Pharmacol. 15:152–163, 1969.

1780 Norepinephrine *Noradrenalin* CAS 51–41–2

Gatling (1962) dropped norepinephrine on the chorio–allantoic membranes of the chick once on the 10th, 11th or 12th day of incubation and produced cephalic, skin and extremity hemorrhages. Pitel and Lerman (1962) injected 25 microgm of noradrenalin directly into rat fetuses at day 16 or 17 of gestation and produced a 50 percent incidence of cataract. The mechanism of action was postulated to be by spasm of the hyaloid arteries. Similar results were obtained using adrenalin.

Fujinaga et al. (**1992**) found an increase in situs inversus among persomite embryos grown for 2 days in culture. Levels of 10 and 100 micromoles produced the defect but 1 micromole did not. The L–form was active but not the D–form. An Alpha–2 adrenergic antagonist (Prazosin) inhibited the effect but a β adrenergic agonist (isoproterenol) did not cause situs inversus.

Fujinaga, M.; Maze, M.; Hoffman, B.B. and Baden, J.M.: Activation of α–1 adrenergic receptors modulates the control of left/right sidedness in rat embryos. Developmental Bioloby 150:419–421, 1992.

Gatling, R.R.: The effect of sympathomimetic agents on the chick embryo. Am. J. Pathol. 40:113–127, 1962.

Pitel, M. and Lerman, S.: Studies on the fetal rat lens. Effects of intrauterine adrenalin and noradrenalin. Invest. Ophthalmol. 1:406–412, 1962.

1781 Norethandrolone *17 β–Hydroxy–17 α–ethyl–19–nor–4–androsten–3–one*

Kawashima et al. (1977) administered orally one mg on days 17–20 to rats and found reduction of the length of the urovaginal septum in female fetuses.

Kawashima, K.; Nakaura, S.; Nagao, S.; Tanaka, S.; Kuwamura, T. and Omori, Y.: Virilizing activities of various steroids in female rat fetuses. Endocrinol. Japon. 24(1):77–81, 1977.

1782 Norethindrone also see *Oral Contraceptives 17–α–Ethinyl–19–nor–testosterone* CAS 68–22–4

Prahalada and Hendrickx (1983) gave rhesus monkeys Norlestrin (2.5 mg norethindrone and 0.05 mg ethinyl estradiol per tablet) in amounts of 5 to 50 mg per kg per day. The overall prenatal mortality was 38.5 percent as compared to the control colony which had a 21 percent loss. No malformations were observed in the fetuses. At the dose level of 25 mg per kg, there were three prenatal deaths and three normal term infants. All three fetuses exposed to 50 mg per kg were sacrificed at day 50 and were normal.

Prahalada, S. and Hendrickx, A.G.: Embryotoxicity of Norlestrin, a combined synthetic oral contraceptive, in rhesus macaques (Macaca mulatta). Teratology 27:215–222, 1983.

1783 Norethisterone Acetate CAS 51–98–9

Discussed under Pregnancy Test Tablets and Oral Contraceptives

1784 Norethynodrel CAS 68–23–5

Gidley et al. (1970) gave this steroid orally to mice on days 8–10 or 11–13 in amounts of up to 0.6 mg per kg. Incomplete development of the parietal bones and exencephaly (5 percent) occurred. A crytorchid–like lesion occurred when two diol metabolites of Norethynodrel were used on days 11–13.

Gidley, J.T.; Christensen, H.D.; Hall, I–H.; Palmer, K.H. and Wall, M.E.: Teratogenic and other side effects produced in mice by norethynodrel and its 3–hydroxy–metabolites. Teratology 3:339–344, 1970.

1785 Norfloxacin *Ciprofloxacin Quiniolones* CAS 70458–96–7

Berkovitch et al. (**1994**) followed 28 mothers treated with norfloxacin and 10 treated with ciprofloxacin and found no congenital defects. Most of the treatments were during the first trimester. The Denver development scale was normal and no signs of arthritis were found after the offspring walked.

Cukierski et al. (1989) gave monkeys this antibiotic at up to 200 or 300 mg per day during a 29 day period of organogenesis and found no teratogenicity. Some increase of embryolethality was reported at the highest doses which were also toxic to the mother. The plasma concentrations at the highest doses were three times those observed in humans at therapeutic doses (5.7 mg per kg).

Irikura et al. (1981) gave this chemotherapeutic agent orally to rats and rabbits during organogenesis. Maximum doses of 500 in rats and 100 mg per kg in rabbits failed to produce any increase in defects. At 100 mg per kg in the rabbit there was embryolethality. Irikura et al. (1981) gave up to 500 mg per kg daily before mating and during the first 6 days of gestation, on days 6–15 and on day 15 through the 21st postnatal day in the mouse. No adverse effects were found on fertility, the fetuses or on the postnatal function of the offspring.

Berkovitch, M.; Pastoszak, A.; Gazarian, M.; Lewis, M. and Koren, G.: Safety of quinolones in pregnancy. (Abst) Pediat. Res. 35:91A, 1994.

Cukierski, M.A.; Prahalada, S.; Zacchei, A.G.; Peter, C.P.; Rodgers, J.D.; Hess, D.L.; Cukierski, M.J.; Tarantal, A.F.; Nyland, T.; Robertson, R.T. and Hendrickx, A.G.: Embryotoxicity studies of norfloxacin in cynomolgus monkeys: I. Teratology studies and norfloxacin plasma concentration in pregnant and nonpregnant monkeys. Teratology 39:39–52, 1989.

Irikura, T.; Imada, O.; Suzuki, H. and Abe, Y.: Teratological study of 1–ethyl–6–fluoro–1, 4–dihydro–4–oxo–7–(1–piperazinyl)–3–quinolinecarboxilic acid (AM–715). Kiso to Rinsho 15:5251–5263, 1981.

Irikura, T.; Suzuki, H. and Sugimoto, T.: Reproductive studies of AM–715. Chemotherapy 29:886–894, 895–914, 915–931, 1981.

1786 d–Norgestrel

Dasgupta et al. (1973) gave oral doses of 2 microgm per day to rats on days 1–7 and found decreased implantation.

Dasgupta, P.R.; Srivastava, K. and Kar, A.B.: Effect of d–Norgestrel on early pregnancy in rats. Ind. J. Exp. Biol. 11:321–322, 1973.

1787 Normethandrone CAS 514–61–4

Discussed under 17 α–Ethinyl–testosterone

1788 Nortriptyline

One case report by Bourke (1974) reported an infant with a dermoid cyst and reduction defects of toes and bones of the lower leg. The woman received 10 mg three times daily starting two weeks after her last menstrual period and for two weeks.

A chemically similar drug amitriptyline, was studied in 87 women who took it in the first trimester and no teratogenicity was found.

Bourke, G.M.: Antidepressant teratogenicity? Lancet 1:98, 1974.

1789 NW–138 *Norshin White*

Ishikawa et al. (1988) gave rats orally up to 300 mg per kg on days 6–15. At the highest dose maternal weight was decreased. No adverse fetal effects were found.

Ishikawa, N.; Ogawa, S.; Arakawa, E.; Murofushi, A.; Yamada, H.; Imai, K. and Ohmura, C.: Reproduction study of nw–138: teratogenicity study in mice. Kiso to Rinsho 22:3009–3021, 1988.

1790 Nystatin CAS 1400–61–9

Heinonen et al. (1977) found 142 women who were exposed to nystatin during the first four lunar months. Ten malformations occurred. Jick et al. (1981) found 5 congenital disorders among the offspring of women exposed during the first trimester. In Jick's group 2.7 malformations would have been expected for this number. In a subsequent study the same group (Aselton et al., 1985) found among another group of 176 women, only three infants with congenital defects.

Aselton, P.A.; Jick, H.; Milunsky, A.; Hunter, J.R. and Stergachis, A.: First–trimester drug use and congenital disorders. Obstet. Gynecol. 65:451–455, 1985.

Heinonen, O.P.; Slone, D. and Shapiro, S.: Birth Defects and Drugs in Pregnancy. Publishing Sciences Group Inc., Littleton, Mass., 1977.

Jick, H.; Holmes, L.B.; Hunter, J.R.; Madsen, S. and Stergachis, A.: First–trimester drug use and congenital disorders. JAMA 246:343–346, 1981.

1791 NZ–105 *(+-)–2–[Benzyl(phenyl)amino]–ethyl 1,4–dihydro 2,6–dimethyl–5–(5,5–dimethyl–2 oxo 1,3,2 dioxaphosphorinan–2–yl)–4–(3–nitrophenyl)–3–pyridin carboxylate Hydrochloride Ethanol*

Nishi et al. (**1991**) studied this antihypertensive in rabbits before and during early gestation and found weight reduction in the dams and fetuses at oral doses of 200 mg per kg. Ishikawa et al. (**1991** A) found reduced fetal weight and viability at 30 and 100 mg per kg given during organogenesis. Similar findings by Ishikawa et al. (**1991** B) were found in peri and postnatal studies. Behavior and sexual function of the offspring was not affected. Ishikawa et al. (**1993** C) found no rabbit teratogenicity at doses of up to 900 mg per kg.

Nishi, N.; Matsuda, K.; Sakashita, M. and Furuhashi, T.: Fertility study of NZ–105 in rats. Yakuri to Chiryo 19:s1465–s1493, 1991.

Ishikawa, H.; Higashikawa, K.; Ishihara, M.; Yamada, H.; Fujioka, S.; Shinoda, K.; Masaki, F.; Nakazawa, M.; Sakashita, M. and Nishi, N.: Teratogenicity study of NZ–105 in rats. Yakuri to Chiryo 19:s1495–s1516, 1991A.

Ishikawa, H.; Higashikawa, K.; Ishihara, M.; Fugioka, S.; Ichikawa, A.; Masaki, F.; Kitazawa, Y.; Nakazawa, M.; Sakashita, M. and Nishi, N.: Perinatal and postnatal study of NZ–105 in rats. Yakuri to Chiryo 19:s1527–s1544, 1991B.

Ishikawa, H.; Ishihara, M.; Yamada, H.; Masaki, F.; Funahashi, N.; Nakazawa, M.; Sakashita, M. and Nishi, N.: Teratogenicity study of NZ–105 in rabbits. Yakuri to Chiryo 19:s1517–s1525, 1991C.

1792 Occupation *Anesthesiologists Arsenic Smelter Workers Art Workers Dental Workers Farmworkers Firefighters Hairdressers Insecticide Workers Laborotory Workers Leather Workers Nurses Painters and Printers Petroleum Industry Pharamaceutical Workers Physiotherapists Plastics Industry Pulp and Paper Industry Rubber Industry Workplace Hazards Workplace Solvents*

The teratogenesis of occupational chemicals has been reviewed (Hemminki et al., 1980; Hemminki and Vineis, 1985; Hemminki, 1980; Hunt, 1979). Books on the subject are available: Hunt (1979) and Barlow and Sullivan (1982).

General Work Conditions and Multiple Exposures

Peters et al. (1981) reported an increase in the rate of brain tumors among the offspring of mothers exposed to chemicals, fathers exposed to solvents and fathers employed in the aircraft industry. The 92 cases were carefully matched but not to children with congenital health problems. Maternal occupation among 93,000 Swedish workers was studied by Ericson et al. (1987) and congenital malformations and perinatal deaths were not significantly increased according to type of work. There were large groups of nurses, childcare providers and teachers included and 3,181 cleaners. The health of 257 pregnant women who worked at the car-building factory (directly on production line) was inspected comparing to 167 women–clerks in the same factory. In the first group of women there were more complications of pregnancy and higher rate of newborn mortality. (Chirova et al. **1988**)

Gibson et al. (1983) interviewed prospectively 6,267 patients and found a 3.38 relative risk of malformations in the group of 109 exposed to chemicals around the time of conception. This risk factor was within the 90 percent expected confidence limits. The exposures consisted of workplace solvents and pesticides. No pattern of defects was identified. Armstrong et al. (**1989**) analyzed the effect of work conditions on the rate of preterm deliveries and low birth weight. Long working hours, and fatigue were associated with preterm delivery while heavy lifting and shift work were more related to low birth weight.

Cavedon and Figa–Talamanca (1987) studied early fetal death among workers in the pharmaceutical and the rubber and plastics industry in Italy. Among 28 factories, 4,121 pregnancies were identified. Age, pregnancy order, educational level and smoking were significantly associated with increased fetal death but the type of industrial work was not. Lindbohm et al. (**1984**) studied spontaneous abortion rates among Finish women (294, 309) and found no significant association with general work exposure categories. Vaughn et al. (**1984**) give fetal deaths among 12 maternal occupations in the State of Washington for 130,000 birth records but many important risk factors were not ascertained. Saurel–Lubizolles and Kaminski (**1987**) studied 2387 French women and found that heavy physical work was associated with increased preterm delivery and low birth weight. Assembly line work was particularly associated with preterm delivery.

Naeye and Tafari (1983) presented data from the Collaborative Perinatal Project that suggests birth weights were lower in women who continued working outside their homes after 28 weeks of gestation. This weight retardation was greatest where the mother was either underweight or hypertensive. Work which involved standing was much more likely to be associated with newborns who were underweight. Head circumferences and birth lengths were not reduced in those women working after 28 weeks (Naeye and Peters, 1982). Shift work during pregnancy was found to be related to a higher miscarriage rate by Axelsson et al. (1984). Mamelle et al (**1984**) used a fatigue index to detect strenuous work conditions. Among 1928 women workers the index was related to the prematurity rate which started at 2.3% and rose to 11.1%.

Bonashevakaya et al. (1987) made a study of 88 healthy women with their first pregnancies at the age of 18–30.

One group served as control (44), the other (44) consisted of women from regions with a high level of pollution (carbon monoxide, ammonia, nitric oxides). Air pollution was found to be responsible for a shift in indices indicating the exchange of nucleic acids in bioenergetics, organospecific liver enzymes of the fetus and hormonal homeostasis found to be present in the placenta and amniotic fluid in the blood serum of pregnant women. Tikkanen et al. (1988) studied occupational exposures of the mothers of 160 infants with congenital heart disease. Neither occupation nor maternal working were found to be associated, but the classifications contained small numbers. Shaw and Gold (1988) discussed the problems faced by epidemiologists in analysis of work exposures. In their review they list in table form studies done where the exposure was agricultural chemicals, industrial chemicals and anesthetics. Savitz et al. (1989) reported data on the National Natality and Fetal Mortality Survey for 1980 in the United States. The odds ratios for preterm death were increased (1.8) for workers in the plastic rubber and synthetics industries. Preterm birth was increased (2.3) for those with lead exposure. Some of the paternal occupations were also associated with increases in preterm delivery (They were for those who worked in the glass, clay, textile and mining industries).

Tabacova et al. (**1994**) studied pregnant women residing in the vicinity of a copper smelter and found a three–fold increase in toxemia. Their measurements supported the hypothesis that multiple heavy metal exposure enhanced the complications by increasing lipid peroxidation via depletion of reduced glutathione reserves.

Anesthesiologists (also see Anesthesia)

Kallen and Moritz (1982) in Sweden carried out analyses of reproductive hazards of the workplace in their country. Nearly all Swedish women having offspring were monitored. Among over 500 women working in operating rooms no adverse fetal effects were found (Ericson and Kallen, 1979). These workers expanded their studies of infants born from 1973–1978 to 1,323 (Ericson and Kallen, 1985). The malformation rate was lower in infants of anesthesiology operating room nurses than in the control group of nurses working on medical floors or the nationwide average. Hemminki et al. (1985) studied abortion rates among 169 nurses exposed to anesthetics daily and found no significant increase as compared to other nurses. Johnson et al. (1987) studied reproductive outcome among veterinary personnel exposed to anesthetics and found no adverse effects in a large questionaire survey.

Arsenic Smelter Workers

Nordstrom et al. (1978) studied spontaneous abortions, birth weight of offspring and congenital defects in Swedish workers in and around a smelter. Significant increases in malformation rates of the female workers as compared to women living in the region were found. Thirteen malformations among 253 women who worked in the plant was compared to 694 among 24,018 in the region (p < 0.05). Birth weights were reduced and spontaneous abortions increased.

Art Workers

Art workers may face special reproductive hazards such

as lead, cadmium, mercury, copper, carbon monoxide, dyes and organic solvents. McCann (1978) discussed precautions that may be taken.

Dental Workers

Brodsky et al. (1985) analyzed over 40,000 questionaires to dental workers exposed to mercury. No increase in spontaneous abortion or defects were found in the offspring of women exposed to the highest concentration of mercury. Ericson and Kallen (1989) studied 8,157 offspring of dentists, dental assistants and dental technicians and found no increase in malformations or spontaneous abortions.

Farmworkers

Nurminen et al. (**1995**) studied 1,206 malformation cases in Finland. Of the 158 women who were farm workers the odds ratio for orofacial defects was slightly elevated (1.4, CI=0.9–2.0). Skeletal and central nervous system defect odds ratios were o.8 ans 1.2 respectively.

Firefighters

Evanoff and Rosenstock (1986) discussed the potential reproductive hazards of female firefighters.

Hairdressers

McDonald et al. (**1988**) studied the outcome of 714 pregnancies among hairdressers and found an odds ratio of 0.94.

Insecticide Workers

Roan et al. (1984) found no adverse reproductive effects associated with pilots who spray insecticides.

A major study of the reproduction of workers in flower growing companies in Bogota, Colombia was reported by Restrepo et al (**1990,A&B**). All workers had been engaged in work in the industry for 6 months and included 2951 men and 5916 women. Pesticides numbered 125 types but Captan™ was the most common exposure. Pregnancies were divided into those occurring before and those after employment. The odds ratios for female workers for spontaneous abortion (1.79), prematurity (2.75) and malformed offspring (1.53) were all increased as was the odds ratio for induced abortion (3.63). Recall bias was suspected as an explanation since the risks decreased in the more recent periods. When 403 of the reportedly deformed children were examined confirmation of a defect occurred in only 38%. There was no increase in defects among the exposed children except for hemaniomas (relative risk: 6.6) (Restrepo et al. 90,B).

Laboratory Workers

Among the offspring of laboratory workers (total 1,161), there was a significant increase in malformed or dead infants (3.45 percent vs 2.55 percent). No specific type of exposure was identified (Ericson et al., 1984). Gastrointestinal atresia appeared to be increased in this group. Seven (4.3 percent) of 164 mothers giving birth to infants with atresia was compared to five (1.4 percent) among 367 controls (Ericson, et al., 1982).

Heidam (1983) studied the spontaneous abortion rate in 1571 industrial and hospital laboratory workers and found no significant increase as compared to office workers and physical and occupational therapists. The laboratory worker exposures were broken down into organic solvents (209), lead (39), cadmium (18), mercury (114), benzi-

dine (112), cyanide compounds (203), azo compounds (83) and other smaller groups. None of the groups had elevated abortion rates. (The numbers in parentheses are the number exposed).

Olsen (1983) studied Danish laboratory workers, painters and printers and found no significant association between defect rate and occupation.

Axelsson et al. (1984) studied pregnancy outcome in 782 Swedish laboratory workers. Among those exposed to organic solvents there was a slight but not statistically significant increase in miscarriage rate but no differences in perinatal or malformation rate were found. Saurel–Cubizolles et al. (1985) studied 621 hospital workers with special attention to standup and heavy load carrying and found a sifnigicant association between arduous work conditions and preterm delivery. Among the non–nurse subjects there was 3 to 5 fold increase in preterm delivery when heavy work conditions were present.

Leather Workers

Clarke and Mason (1985) observed a doubling of perinatal death among pregnant leather workers. Congenital malformations and macerated fetuses accounted for the major part of the increase as compared to other workers of the same social class. This was not explained by any of the demographic differences. Three cases of trisomy 18 occurred in about 1200 births.

Nurses

Among 1500 women employed in medical occupations there was no increase in congenital or perinatal problems but vacuum extraction and caeserian section were more frequent (Baltzar et al., 1979). Nurses exposed to anesthetics, sterilizing gases, soaps or X–rays were studied by Hemminki et al. (1985) in Finland. There were 217 with spontaneous abortions and 46 with malformed offspring: Neither number was significantly greater than nurses matched for age and hospital employment. The odds ratio based on eight malformed cases out of 38 exposed to cytostatic drugs was 4.7 (p < 0.02). No specific pattern of defects was identified.

Silverman et al. (1985) studied abortus chromosomes from 1,252 women from various occupations and found among the lower social economic class an increased odds ratio of 3.11 and 1.86 for chromosomal aneuploidy for women who worked during pregnancy or both before and during pregnancy. They felt that their data on factory and hospital nursing workers were of significant number to exclude a two–fold or greater increase in abortion rate.

Selevan et al. (1985) in a study drawing on 17 Finnish hospitals, identified 124 nurses who were occupationally exposed to antineoplastic drugs. The fetal loss odds ratio was 2.30 (p < 0.01) but details of the type of defect or loss were not given. The nurses were exposed mainly to doxorubicin, vincristine and cyclophophamide. The odds ratio for nurses exposed to anesthetics was 0.96. Stellman and Zoloth (1986) reviewed the literature on effects of chemotherapeutic agents on health workers and gave guidelines for safety.

Painters and Printers

A preliminary report showed some increase in defects of

the abdominal wall of infants born to mothers in the printing industry (Erickson et al., 1978). Olsen (1983) found no association between printers and the defect rate in their offspring.

Fathers who were painters had an increased risk (4.9) for offspring with malformations of the central nervous system but this was not statistically significant (Olsen, 1983). Hooisma et al. (**1993**) studied 47 young and 45 old painters comparing them to controls in the building trade and found no differences in behavior tests.

Scialli (1989) has reviewed the subject of paint exposure during pregnancy and concluded that the available studies do not indicate a risk for abortion or malformation. The studies of cancer risk in the offspring were not deemed adequate for a conclusion.

Petroleum Industry

Gainullina (1985) studied the influence on pregnant women of saturated, unsaturated and aromatic hydrocarbons, SH–4, SO–2 or CO which did not exceed the maximum allowable concentration (MAC) and was group I. The influence of more complex compounds, including phenol, aromatic hydrocarbons, fatty acids, alcohols at concentrations exceeding MAC was group II. Fetal hypoxia was reported in 53 percent (group I) and 16.9 percent (group II).

Pharmaceutical Workers

Taskinen et al. (1986) studied pregnancy outcome among 1795 women working in pharmaceutical industries and found that the rate during time of working was 10.9 and 10.6 percent after employment. In a corresponding hospital district the rate was 8.5 percent. The odds ratio for miscarriage for women doing heavy lifting was 5.7.

Photographic Workers

Lappe (1983) reviewed the animal data on chemicals found in photographic materials. Hydroquinone in developers increased fetal resorptions in the hamster at daily doses of 0.5 gm. Methylaminophenol, a replenisher was associated with some fetotoxicity after a two–hour skin exposure in the hamster.

Physiotherapists

Among physiotherapists (2,018) who were studied for their non–ionizing radiation exposure, there was an increase in frequent exposure to shortwaves among those who had malformed infants. There were 11 malformed or perinatally dead infants in this group and 9 in the control groups which totaled twice the study group (Kallen and Moritz, 1982).

Plastics Industry

Abortions among workers in the plastics industry of Finland were reported by Lindbohm et al. (1985). The odds ratio was 1.9 for women exposed to polyurethane and 3.0 for all 16 workers in the polyurethane plant. The later ratio was significantly increased to the 0.02 level. Workers exposed to polymerized plastics, heated plastics, polyvinyl chloride and styrene did not have increased abortion rates. Ahlborg et al. (1987) studied 1,685 female workers in plastic plants of Sweden and Norway. Adverse outcome (low birth weight, stillbirth or congenital defects) among workers processing polyvinyl chloride was increased (odds ratio 2.2; p < 0.05).

The authors cautioned against interpreting the data as a risk factor since the number of workers studied was small (24).

Tabacova (1986) summarized studies of workers exposed to toluene, xylene and styrene. Ryabko (1986) studied conditions of the production of epoxy resin E-41 and the mixture of epoxy E-40 and phenol–formaldehyde (SF-302 IC) resins (enamels EP-525 and FL-62, respectively). He reported an occasional example of the reproductive malfunction of men, complications during their wives' pregnancy, increase in postnatal mortality and some diseased newborns.

Pulp and Paper Industry

Among 890 women in the pulp and paper industry, no increase in defects or perinatal loss was recorded (Blomqvist et al., 1981).

Rubber Industry

Shirinyan et al. (1986) made a study of women workers and wives of men engaged in the chloroprene rubber production and found a high rate of spontaneous abortion.

Welders

Hemminki and Lindbohn (**1986**) reviewed animal and human data on metal exposure of welders. In their own studies of 317 male workers and 28 female workers they found a slight increase in spontaneous abortions among the exposed women (9.5% vs 8.2% in other industries).

Ahlborg, G., Jr.; Bjerkedal, T. and Egenaes, J.: Delivery outcome among women employed in the plastics industry in Sweden and Norway. Am. J. Ind. Med. 12:507–517, 1987.

Armstrong, B.G.; Nolin, A.D. and McDonald, A.D.: Work in pregnancy and birth weight for gestational age. British Journal of Industrial Medicine 46:196–199, 1989.

Axelsson, G.; Lutz, C. and Rylander, R.: Exposure to solvents and outcome of pregnancy in university laboratory employees. Br. J. Ind. Med. 41:305–312, 1984.

Baltzar, B.; Ericson, A. and Kallen, B.: Delivery outcome in women employed in medical occupations in Sweden. J. Occup. Med. 8:543–548, 1979.

Barlow, S.M. and Sullivan, F.M.: Reproductive Hazards of Industrial Chemicals: An Evaluation of Animal and Human Data. Academic Press Inc. (London), 1982.

Blomqvist, V.; Ericson, A.; Kallen, B. and Westerholm, P.: Delivery outcome for women working in the pulp and paper industry. Scand. Work Environ. Health 7:114–118, 1981.

Bonashevskaya, T.I.; Lamentova, T.G. and Tsapok, P.I.: Changes in the fetoplacenta system due to atmospheric conditions: The morphological aspect. Gigiena i Sanitariya (USSR) 9:4–7, 1987.

Brodsky, J.B.; Cohen, E.N.; Whitcher, C.E.; Brown, B.W. and Wu, M.L.: Occupational exposure to mercury in dentistry and pregnancy outcome. J. Am. Dent. Assoc. 111:779–780, 1985.

Cavedon, G. and Figa–Talamanca, I.: Correlates of early fetal death among women working in industry. Am. J. Ind. Med. 11:497–504, 1987.

Chirkova, A.V.; Oshchepksv, V.I.; Kornyaeva, Z.S.; Khamitov, R.L.; Zhdanova, V.I. and Petrushkova, N.L.: Course of pregnancy and delivery in female workers of a machine–building plant. Kazan Med ZH 69(3):194–195,

1988.

Clarke, M. and Mason, E.S.: Leatherwork: A possible hazard to reproduction. Br. Med. J. 290:1235–1237, 1985.

Erickson, J.D.; Cochran, W.M. and Anderson, C.E.: Birth defects and printing. Lancet 1:385 only, 1978.

Ericson, A.; Eriksson, M.; Kallen, B.; and Zetterstrom, R.: Maternal occupation and delivery outcome: A study using central registry data. Acta Paediatr. Scand. 76:512–518, 1987.

Ericson, A. and Kallen, B.: Survey of infants born in 1973 to 1975 to Swedish women working in operating rooms during their pregnancies. Anesth. Analg. 58:302–305, 1979.

Ericson, A.; Kallen, B.; Meirik, O. and Westerholm, P.: Gastrointestinal atresia and maternal occupation during pregnancy. J. Occup. Med. 24:515–518, 1982.

Ericson, A.; Kallen, B.; Zetterstrom, R.; Eriksson, M. and Westerholm, P.: Delivery outcome of women working in laboratories during pregnancy. Arch. Environ. Health. 39:5–10, 1984.

Ericson, A. and Kallen, B.: Pregnancy outcome in women working as dentists, dental assistants or dental technicians. Int. Arch. Occup. Environ. Health, in press (657), 1989.

Ericson, A. and Kallen, B.: Hospitalization for miscarriage and delivery outcome among Swedish nurses working in operating rooms 1973–1978. Anesth. Analg. 64:981–988, 1985.

Evanoff, B.A. and Rosenstock, L.: Reproductive hazards in the workplace: A case study of women firefighters. Am. J. Ind. Med. 9:503–515, 1986.

Gainullina, M.K.: Functional state of the feto–placenta complex in women working in the petroleum industry. Proceedings of Moscow Research Institute of Hygiene. (USSR) 15:32–38, 1985.

Gibson, G.T.; Colley, D.P. and Baghurst, P.A.: Maternal exposure to environmental chemicals and aetiology of teratogenesis. Aust. N.Z. Obstet. Gynecol. 23:170–174, 1983.

Heidam, L.Z.: Spontaneous abortions among laboratory workers: A follow-up study. J. Epidemiol. Commun. Health 38:36–41, 1983.

Hemminki, K.: Occupational chemicals tested for teratogenicity. Int. Arch. Occup. Environ. Health 47:191–207, 1980.

Hemminki, K.; Kyrronen, P. and Lindbohm, M–L.: Spontaneous abortions and malformations in the offspring of nurses exposed to anaesthetic gases, cytostatic drugs, and other potential hazards in hospitals, based on registered information of outcome. J. Epidemiol. Commun. Health 39:141–147, 1985.

Hemminki, K.; Mutanen, P.; Luoma, K. and Saloniemi, I.: Congenital malformations by parental occupation in Finland. Int. Arch. Occup. Environ. Health 46:93–98, 1980.

Hemminki, K. and Vineis, P.: Extrapolation of the evidence on teratogenicity of chemicals between humans and experimental animals. Teratogenesis, Carcinogenesis, Mutagenesis 5:251–318, 1985.

Hemminki, K. and Lindbohm, M.L.: Reproductive effects of welding fumes: Experimental studies with special reference to chromium and nickel compounds in Health Hazards and Biological Effects of Welding Fumes and Gases, ed by R.M. Stern, A. Berlin, A.C. Fletcher and J. Jarvisalo:Excerpta Medica International Congress Services 767, Amsterdam, 1986, pp. 291–309.

Hooisma, J.; Hanninen, H.; Emmen, H.H. and Kulig, B.M.: Behavioral effects of exposure to organic solvents in dutch painters. Neurotoxicology and Teratology 15:397–406, 1993.

Hunt, V.R.: Work and the Health of Women, CRC Press, Boca Raton, Florida, 1979.

Hunt, V.R.; Smith, M.K. and Worth, D.: Environmental factors in human growth and development. Banbury report, Vol. 2, Cold Spring Harbor Laboratory, 1982.

Johnson, J.A.; Buchan, R.M. and Reif, J.S.: Effect of waste anesthetic gas and vapor exposure on reproductive outcome in veterinary personnel. Am. Ind. Hyg. Assoc. J. 48:62–66, 1987.

Kallen, B. and Moritz, V.: Delivery outcome among physiotherapists in Sweden. Is non–ionizing radiation a fetal hazard? Arch. Environ. Health 37:81–84, 1982.

Lappe, M.: Risks from maternal exposure to photographic chemicals in pregnancy. Birth 10:173–177, 1983.

Lindbohm, M–L.; Hemminki, K. and Kyrronen, P.: Spontaneous abortions among women employed in the plastics industry. Am. J. Ind. Med. 8:579–586, 1985.

Lindbohm, M.; Hemminki, K. and Kyrronen, P.: Parental occupational exposure and spontaneous abortions in Finland. American Journal of Epidemiology 120:370–378, 1984.

Mamelle, N.; Laumon, B. and Lazar, P.: Prematurity and occupational activity during pregnancy. American Journal of Epidemiology 119:309–322, 1984.

McCann, M.: The impact of hazards in art on female workers. Preventive Medicine 7:338–348, 1978.

McDonald, A.D.; McDonald, J.C.; Armstrong, B.; Cherry, N.M.; Cote, R.; Lavoie, J.; Nolin, A.D. and Robert, D.: Congenital defects and work in pregnancy. British Journal of Industrial Medicine 45:581–588, 1988.

Naeye, R.L. and Peters, E.C.: Working during pregnancy, effects on the fetus. Pediatrics 69:724–727, 1982.

Naeye, R.L. and Tafari, N.: Risk Factors in Pregnancy and Diseases of the Fetus and Newborn. Williams and Wilkins, Baltimore, 57–61, 1983.

Nordstrom, S.; Beckman, L. and Nordenson, I.: Occupational and environmental risks in and around a smelter in northern Sweden. VI. Congenital Malformations. Hereditas 90:297–307, 1978.

Nurminen, T.; Rantala, K.; Kurppa, K. and Holmberg, P.C.: Agricultural work during pregnancy and selected structural malformations in Finland. Epidemiology 6:23–30, 1995.

Olsen, J.: Risk of exposure to teratogens amongst laboratory staff and painters. Dan. Med. Bull. 30:24–28, 1983.

Peters, J.M.; Preston–Martin, S.; Yu, M.C.: Brain tumors in children and occupational exposure of parents. Science 213:235–237, 1981.

Restrpo, M.; Munoz, N.; Day, N.E.; Parra, J.E.; de

Romero, L. and Nguyen–Dinh, X.: Prevalence of adverse reproductive outcomes in a population occupationally exposed to pesticides in Colombia. Scand. J. Wor . Environ. Health 16:232–238, 1990.

Restrepo, M.; Munoz, N.; Day, N.; Parra, J.E.; Hernandez, C.; Blettner, M. and Giraldo, A.: Birth defects among children born to a population occupationally exposed to pesticides in Colombia. Scand. J. Work Environ Health 16:239–246, 1990.

Roan, C.C.; Matanoski, G.E.; McIlnay, C.Q.; Olds, K.L.; Pylant, F.; Trout, J.R.; Wheeler, P. and Morgan, D.P.: Spontaneous abortions, stillbirths and birth defects in families of agricultural pilots. Arch. Environ. Health 39:46–60, 1984.

Ryabko, T.P.: The influence of polymer finishing materials on the reproductive function in men. Theses: Protection of people's health and the environment from harmful chemical factors. I. All–Union Congress of Toxicologists. (USSR) Rostov–on–Don, 244–245, 1986.

Saurel–Cubizolles, M.J.; Aminsi, M. and Llado–Arkhipoff, J.: Pregnancy and its outcome among hospital personnel according to occupation and working conditions. J. Epidemiol. Commun. Health 39:129–134, 1985.

Saurel–Cubizolles, M.J. and Kaminski, M.: Pregnant women's working conditions and their changes during pregnancy: a national study in France. British Journal of Industrial Medicine 44:236–243, 1987.

Savitz, D.A.; Whelan, E.A. and Kleckner, R.C.: Effect of parents occupational exposures on risk of stillbirth, preterm delivery, and small–for–gestational–age infants. American Journal of Epidemiology 129:1201–1218, 1989.

Scialli, A.R.: Who should paint the nursery. Reproductive Toxicology 3:159–164, 1989.

Selevan, S.G.; Lindbohm, M–L.; Hornung, R.W. and Hemminki, K.: A study of occupational exposure to antineoplastic drugs and fetal loss in nurses. N. Eng. J. Med. 313:1173–1178, 1985.

Shaw, G.M. and Gold, E.B.: Methodological considerations in the study of parental occupational exposures and congenital malformations in offspring. Scandd. J. Work Environ. Health 14:344–355, 1988.

Shirinyan, G.S.; Tumanyan, E.R. and Arutyunyan, R.R.: Spontaneous abortion in women engaged in chemical industry. Biologichesky Jurnal Armeniy (USSR) 9:8–12, 1986.

Silverman, J.; Kline, J.; Hutzler, M.; Stein, Z.A. and Warburton, D.: Maternal employment and the chromosomal characteristics of spontaneously aborted conceptions. J. Occup. Med. 27:427–438, 1985.

Stellman, J.M. and Zoloth, S.R.: Cancer chemotherapeutic agents as occupational hazards: A literature review. Cancer Investigation 4(2):127–135, 1986.

Tabacova, S.: Maternal exposure to environmental chemicals. NeuroToxicol. 7(2):421–440, 1986

Tabacova, S.; Little, R.E.; Balabaeva, L.; Pavlova, S. and Petrov, I.: Complications of pregnancy in relation to maternal lipid peroxides,glutathione, and exposure to metals. Reproductive Toxicology 8:217–224, 1994.

Taskinen, H.; Lindbohm, M–L. and Hemminki, K.:

Spontaneous abortions among women working in the pharmaceutical industry. Br. J. Ind. Med. 43:199–205, 1986.

Tikkanen, J.; Kurppa, K.; Timonen, H.; Holmberg, P.C.; Kuosma, E. and Rantala, K.: Cardiovascular malformations, work attendance, and occupational exposures during pregnancy in Finland. Am. J. Ind. Med. 14:197–204, 1988.

Vaughn, T.L.; Daling, J.R. and Starzyk, P.M.: Fetal death and maternal occupation an analysis of birth records in the State of Washington. Journal of Occupational Medicine 23:676–678, 1984.

1793 Ochratoxin A CAS 303–47–9

This mycotoxin was studied in pregnant mice by Hayes et al. (1974). Intraperitoneal injection with 5 mg per kg on one of gestation days 7 through 12 resulted in fetal growth retardation and malformations. Exencephaly and anomalies of the eyes, face, digits and tail were found most commonly. Hood et al. (1976) found teratogenicity in hamsters using 5 to 20 mg per kg intraperitoneally on day 7, 8 or 9. Hoshino et al. (1988) injected mice with either 2 or 3 mg per kg on day 10 and made detailed studies of the significant brain changes 6 weeks after birth.

Hayes, A.W.; Hood, R.D. and Lee, H.L.: Teratogenic effects of ochratoxin A in mice. Teratology 9:93–98, 1974.

Hood, R.D.; Naughton, M.J. and Hayes, A.W.: Perinatal effects of ochratoxin A in hamsters. Teratology 13:1–14, 1976.

Hoshino, K.; Fukui, Y.; Hayasaka, I. and Kameyama, Y.: Developmental disturbance of the cerebral cortex of mouse offspring from dams treated with ochratoxin A during pregnancy. Cong. Anom. 28:287–294, 1988.

1794 Octanoic Acid

Dawson (1991) using a frog embryo teratogenesis assay found that at a concentration of 28 mg per liter 50% of the ofspring were malformed. Defects of the gut, edema and microcephaly were among the defects found.

Dawson, D.A.: Additive incidence of developmental malformation for xenopus embryos exposed to a mixture of ten aliphatic carboxylic acids. Teratology 44:531–546, 1991.

1795 Octoxynol–9 CAS 9002–93–1

This spermicidal agent was administered intravaginally to rats in doses of 0.5 and 5 mg per kg on days 6 through 15 (Saad et al., 1984). No adverse fetal effects were found.

Saad, D.J.C.; Kirsch, R.M.; Kaplan, L.L. and Rodwell, D.E.: Teratology of intravaginally administered contraceptive jelly containing octoxynol–9 in rats. Teratology 30:25–30, 1984.

1796 Octyl acetate CAS 108419–32–5

Doughtrey et al. (1989) gavaged this solvent on days 6–15 to rats in doses of 0.1, 0.5, and 1.09 per kg. Maternal toxicity was found at the two highest doses. Some increase in skeletal and dilated brain ventricles was observed in fetuses at the highest dose.

Doughtrey, W.C.; Wier, P.J.; Traul, K.A.; Biles, R.W. and Egan, G.F.: Evaluation of the teratogenic potential of octyl acetate in rats. Fundamental and Applied Toxicology 13:303–309, 1989.

1797 Ofloxacin Tarivid™

Takayama et al. (1986) treated rats on days 7–17 with up to 810 mg per kg and rabbits on days 6–18 with up to 160 mg per kg and found no evidence of teratogenicity. At the highest doses the fetuses were reduced in weight.

Takayama, S.; Watanabe, T.; Akiyama, Y.; Ohura, K.; Harada, S.; Matsuhashi, K.; Mochida, K. and Yamashita, N.: Reproductive toxicity of ofloxacin. Arzneim Forsch. 36:1244–1248, 1986.

1798 Oleandomycin CAS 3922–90–5

Oleandomycin was injected intramuscularly (20 mg/kg) into Wistar rats from 8 to 14 or from 15 to 20 days of pregnancy. (Dolgova andd Savitskaja 1989) Tracheobroncheal and mesenteric lymph nodes were investigated in fetuses and 5–, 14–, 30–day old rats using anatomical, histological and electron microscopical methods. This drug induced in lymph nodes morphological changes as well as changes in cell composition. These observations support the assumption that therapeutic doses of antibiotic can disturb the development and immunological function of lymph nodes.

Dolgova, M.A. and Savitskaja, T.N.: Changes in anatomical structure and cell composition of lymph nodes after prenatal action of oleandomycin phosphate. Arh. Anat. Gistol. Embriol. 91(11):53–60, 1989.

1799 α–Olefin Sulfonate

Discussed under Surfactants

1800 Oleylamine ODA

Mercieca et al. (1990) gavaged this 18–carbon fatty amine during organogenesis in the rat and rabbit. The rats received up to 80 mg per day and the rabbits up to 30 mg per day. At the higher doses the does showed toxicity but no adverse changes occurred in fetuses from either species.

Mercieca, M.D.; Bisinger, E.C.; Gerhart, J.M.; Rodwell, D.E. and Merriman, T.N.: Developmental toxicity study of oleylamine in two species. (abst)Teratology 41:577, 1990.

1801 Omeprazole CAS 7390–58–6

Shimazu et al. (1988) gave up to 320 mg orally to rats before fertilization, during the first week of gestation, during organogenesis or in the perinatal period. No teratogenic effects were found and although there was growth retardation in the highest dose given perinatally behavior was not changed.

Shimazu, H.; Ishida, S.; Ikeya, M.; Yamazai, E.; Fujii, T. and Takeuchi, M.: Reproduction studies of omeprazole in rats. Oyo Yakuri 36(3):189–208, 1988.

1802 Ondansetron Hydrochloride

This 5–hydroxytryptamine type 3 receptor antagonist was studied by Kaneko et al. (1992) in rats giving up to 8.0 mg per kg daily intravenously before pregnancy and during the first week of gestation and no ill effects were observed. Sutherland et al. (1992) gave it orally from before pregnancy until weaning at a dose of 15 mg per kg without fetal ill effects. Shimizu et al. (1992 A) gave intravenous doses of up to 10 mg per kg and orally (Shimizu et al.,1992 B) 40 mg per kg during organogenesis and found no teratogenicity. Secker et al. (1992 A,B) gave up to 15 mg per kg orally and 4 mg per kg intravenously in the perinatal and postnatal period and found some inhibition in body weight growth at the highest oral dose. Ezaki et al. (1992) gave rabbits up to 10 mg per kg during organogenesis and found delayed growth and ossification at the highest dose.

Ezaki, H.; Yokoyama, S.; Takahashi, N.; Takamatsu, M.; Tokado, H. and Takeda, K.: Reproduction study (seg II) of ondansetron hydrochloride in rabbits by oral route. Yakuri to Chiryo 20:s1187–s1196, 1992.

Kaneko, Y.; Kuwagata, M.; Hashimoto,Y.; Mizutani, M.; Ezaki, H.; Yokoyama, S. and Tokado, H.: Reproduction study (seg I) of ondansetron hydrochloride in rats by intravenous route. Yakuri to Chiryo 20:s1119–s1130, 1992.

Secker, R.C.; Parkinson, M.M.; Kelly, J.A.; Sparrow, S.J.; Libretto, S.E.; Ezaki, H.; Yokoyama, S. and Tokado, T.: Reproduction study (seg III) of ondansetron hydrochloride in rats by intravenous route. Yakuri to Chiryo 20:21197–s1209, 1992A.

Secker, R.C.; Parkinson, M.M.; Kelly, J.A.; Sparrow, S.J.; Libretto, S.E.; Ezaki, H.; Yokoyama, S. and Tokado, H.: Reproduction study (seg III) of ondansetron hydrochloride in rats by oral route. Yakuri to Chiryo 20:s1211–s1226, 1992B.

Shimizu, M.; Ohta, T.; Kato, M.; Kobayashi, Y.; Yamashita, Y.; Asano, M.; Mori, K.; Koike, T.; Ezaki, H.; Yokoyama, S. and Takeda, K.: Reproduction study (seg II) of ondansetron hydrochloride in rats by intravenous route. Yakuri to Chiryo 20:s1145–s1164, 1992A.

Shimizu, M.; Ohta, T.; Kato, M.; Kobayashi, Y.; Yamashita, Y.; Asano, M.; Mori, K.; Fujimura, T.; Ezaki, H.; Yokoyama, S. and Takeda, K.: Reproduction study (seg II) of ondansetron hydrochloride in rats by oral route. Yakuri to Chiryo 21:s1165–s1185, 1992B.

Sutherland, M.F.; Adams, M.J.; Parkinson, M.M.; Sparrow, S.J.; Fluck, P.A.; Ezaki, H.; Yokoyama, S. and Tokado, H.: Reproduction study (seg I) of ondansetron hydrochloride in rats by oral route. Yakuri to Chiryo 20:s1131–s1141, 1992.

1803 Oral Contraceptives

Although there was some initial concern that oral contraceptives might be teratogenic in the human, the preponderance of data now supports their safety (Wilson and Brent, 1981). No increase in congenital defect rate has been shown in the offspring of 1250 women having used oral contraceptives (Robinson, 1971). Similar negative findings were reported in 5,530 pregnancies reported to the Royal College of General Practitioners (1976). Janerich et al. (1974) found a higher incidence of oral contraceptive users among mothers of children with limb defects and all these children were males. Jaffe et al. (1975) did not observe an increase in inadvertent oral contraception continuance in seven cases where limb reduction was found. Kallen (1989) in a study of 277 Swedish infants with limb reduction, could not find an association with oral contraceptives or intrauterine devices.

Nora and Nora (1974) suggested that congenital heart disease was more common in pregnancies during which birth control pills were continued. Heinonen et al. (1976) studied 1042 women who received female hormones during early pregnancy and found a heart defect rate of 18 per 1000 which was compared to a control of 7.8 per 1000.

Wiseman and Dodds–Smith (1984) evaluated the case histories tabulated by Heinonen et al. (1977) and found that only 8 instead of 17 were exposed during the critical period of cardiac organogenesis. This number was not significantly different from the controls. Savolainen et al. (1981) reported no significant differences in malformation rates among 800 users as compared to matched controls. Harlap et al. (1975) studied prospectively 11,468 pregnancies in Israel and found 5 with heart disease when the expected rate was 2.6. Neither limb reduction nor esophageal atresia occurred in the studied group. Goujard and Rumeau–Rouquette (1977) reported on a prospective study of 830 women receiving mostly hormonal pregnancy tests. General malformation rates, limb and heart defects were not increased but a microcephaly rate of 3.4 per 1000 as compared to a control of 0.6 was found. Ferencz et al. (1980) studied mothers of 110 children with heart disease and found no association with maternal hormone therapy. Reports which did not find significantly associated teratogenicity are available (Mulvihill et al., 1974; Yasuda and Miller, 1975; Rothman et al., 1979). Lammer et al. (1986), in a study of 34 infants with VACTERL, compared to 1,024 infants with malformations reported no increase in prenatal hormone exposure. A well balanced discussion of this important problem was published (Anonymous, 1974; Yerushalmy, 1972).

Janerich et al. (1980) listed 18 papers reporting the rate of malformation in the offspring of women exposed to oral contraceptives. Approximately one–half found no increase. Wilson and Brent (1981) reviewed the literature and concluded there was no proof that exogenous sex hormones produce nongenital malformations. Schardein (1980) reviewed the clinical data on birth defects and hormones during pregnancy and concluded there is little hazard.

Carr (1967) reported six abnormal karyotypes in eight specimens from mothers previously taking oral contraceptives; the overall incidence and especially that of triploidy was significantly higher than that found in abortuses of mothers not taking contraceptives. Boue (1970) reporting on studies from 333 unselected spontaneous abortions, found a 61.6 percent incidence of abnormal chromosomes with no increase in abortuses from mothers who took oral contraceptives. Boue observed a much higher incidence of abnormal karyotypes in the younger specimens and pointed out that the apparent increase in Carr's series may have been related to the presence of older specimens in the control group. Nelson et al. (1971) and Dhadial et al. (1970) also failed to observe an increase in abnormal karyotypes from abortuses of mothers taking contraceptives. Boue et al. (1975) reported no increase in chromosome aberrations among 520 spontaneous abortions following the use of birth control pills.

Harlap et al. (1985) traced contraceptive use in 45 de novo aneuploid conceptuses among 33,551 pregnancies and found no increased risk for those who used oral contraceptives. Janerich et al. (1976) found no increase in oral contraception use among 103 mothers giving birth to children with Down's syndrome. Jagiello and Lin (1974) studied the effect of a large number of oral contraceptives on the oocyte of several species. In over 175 human ova from oral–contraceptive–taking women about one–third divided in vitro, a result similar to the control group. Poland and Ash (1973) found a significantly higher incidence of 'disorganized' abortuses from women who took oral contraceptives as compared to women using other forms of contraception.

Tuchmann–Duplessis and Mercier–Parot (1972) using 1 mg per kg per day of both norethyndrel and mestranol in the rat either before or during gestation observed no congenital defects. Treatment during two generations did not cause any functional or histologic abnormalities of the fetal genital tracts.

Kwarta et al. (**1991** A,B,C) studied the effect of levonorgestrel and ethinyl estradiol (50:30 ratio) on rats before fertilization and during organogenesis. At gavage doses of 16 and 80 microgm per kg the estrous cycles of some of the dams were interrupted and some postnatal retardation in reflexes was found. Reproduction in the offspring was normal. No teratogenicity was found after treatment during days 7–17. A significant reduction in the weight of the dams occurerd at both treatment levels. The no observable level of exposure was 3.2 microgm per kg. In rabbits Kwarta et al. (**1991** D) found no adverse effects of up to 40 microgm per kg given on days 6–11.

Anonymous: Synthetic sex hormones and infants. Br. Med. J. 4:485–486, 1974.

Boue, J.: Etude chromosomique des avortements spontanes apres inhibition physiologique or therapeutique de l'ovulation. In: L'Inhibition de L'Ovulation, A. Netter (ed.), Paris: Masson, 349–356, 1970.

Boue, J.; Boue, A. and Lazar, P.: Retrospective and prospective epidemiological studies of 1500 karyotyped spontaneous human abortions. Teratology 12:11–26, 1975.

Carr, D.H.: Chromosomes after oral contraceptives. Lancet 2:830–831, 1967.

Dhadial, R.K.; Machin, A.M. and Tait, S.M.: Chromosomal anomalies in spontaneously aborted human fetuses. Lancet 2:20–21, 1970.

Ferencz, C.; Matanoski, G.M.; Wilson, P.D.; Rubin, J.D.; O'Neill, C.A. and Gutberlet, R.: Maternal hormone therapy and congenital heart disease. Teratology 21:225–239, 1980.

Goujard, J. and Rumeau–Rouquette, C.: First trimester exposure to progestagenoestrogen and congenital abnormalities. Lancet 1:482–483, 1977.

Harlap, S.; Prywes, R. and Davies, A.M.: Birth defects and oestrogens and progesterones in pregnancy. Lancet 1:682–683, 1975.

Harlap, S.; Shiono, P.H.; Ramcharan, S.; Globus, M.; Bachman, R.; Mann, J. and Lewis, J.P.: Chromosomal abnormalities in the Kaiser–Permanente birth defects study, with special reference to use around time of conception. Teratology 31:381–387, 1985.

Heinonen, O.P.; Slone, D.; Monson, R.R.; Hook, E.B. and Shapiro, S.: Cardiovascular birth defects in antenatal exposure to female sex hormones. N. Eng. J. Med. 296:67–70, 1976.

Heinonen, O.P.; Slone, D. and Shapiro, S.: Birth Defects and Drugs in Pregnancy. Publishing Sciences Group Inc., Littleton, Mass., 1977.

Jaffe, P.; Liberman, M.M.; McFadyen, I. and Valman, H.B.: Incidence of congenital limb–reduction deformities. Lancet 1:526–527, 1975.

Jagiello, G. and Lin, J.S.: Oral contraceptive compounds and mammalian oocyte meiosis. Am. J. Obstet. Gynec. 120:390–406, 1974.

Janerich, D.T.; Flink, E.M. and Keogh, M.D.: Down's syndrome and oral contraceptive usage. Br. J. Obstet. Gynecol. 8:617–620, 1976.

Janerich, D.T.; Piper, E.M. and Glebatis, D.M.: Oral contraceptives and congenital limb reduction defects. N. Eng. J. Med. 291: 697–700, 1974.

Janerich, D.T.; Piper, E.M.; Glebatis, D.M.: Oral contraceptives and birth defects. Am. J. Epidemiol. 112:73–79, 1980.

Kallen, B.: A prospective study of some aetiological factors in limb reduction defects in Sweden. J. Epidemiol. Comm. Health 43:00–00, 1989.

Kwarta Jr., R.F.; Hemm, R.D.; Pollock, J.J.; Christian, M.S.; Usui, T.; Suzuki, M. and Yago, Nagataka. Levonorgestrel/ethinyl estradiol study of fertility and reproductive performance with behavioral and reproductive assessment of the offspring (Seg I–phase I–female rat). Oyo Yakuri 42:313–326, 1991A.

Kwarta Jr., R.F.; Hemm, R.D.; Pollock, J.J.; Christian, M.S.; Usui, T.; Suzuki, M. and Yago, N.: Levonorgestrel/ethinyl estradiol preimplantation reproduction study with behavioral and reproductive assessment of the offspring (Seg I–phase II–female rat). Oyo Yakuri 42:351–363, 1991B.

Kwarta Jr., R.F.; Hemm, R.D.; Pollock, J.J.; Christian, M.S.; Usui, T.; Suzuki, M. and Yago, N.: Levonorgestrel/ethinyl estradiol developmental toxicity study with behavioral and reproductive assessment of the offspring (Seg II–rat). Oyo Yakuri 42:327–340, 1991C.

Kwarta Jr., R.F.; Hemm, R.D.; Pollock, J.J.; Christian, M.S.; Usui, T.; Suzuki, M. and Yago, N.: Levonorgestrel/ethinyl estradiol study of developmental toxicity in rabbits. Oyo Yakuri 42:341–349, 1991D.

Lammer, E.J.; Cordero, J.F. and Khoury, M.J.: Exogenous sex hormone exposure and the risk for VACTERL Association. Teratology 34:165–169, 1986.

Levy, E.P.; Cohen, A. and Fraser, F.C.: Hormone treatment during pregnancy and congenital heart disease. Lancet 1:611 only, 1973.

Mulvihill, J.J.; Mulvihill, C.G. and Neill, C.A.: Congeni-

tal heart defects and prenatal sex hormones. Lancet 1:1168, 1974.

Nelson, T.; Oakley, G.P. and Shepard, T.H.: A centralized laboratory for collection of human embryos and fetuses: Seven years experience. II. Classification and tabulation of conceptual wastage with observations on type of malformations, sex ratio and chromosome studies. In: Monitoring birth defects and environment, the problem of surveillance. E.B. Hook, D.T. Janerich and I.H. Porter (eds.), New York: Academic Press, 45–81, 1971.

Nora, J.J. and Nora, A.H.: Can the pill cause birth defects? N. Eng. J. Med. 291:731–732, 1974.

Poland, B.J. and Ash, K.A.: The influence of recent use of an oral contraceptive on early intrauterine development. Am. J. Obstet. Gynecol. 116:1138–1142, 1973.

Robinson, S.C.: Pregnancy outcome following oral contraceptives. Am. J. Obstet. Gynecol. 109:354–358, 1971.

Rothman, K.J.; Fyler, D.C.; Goldblatt, A. and Kreidberg, M.B.: Exogenous hormones and other drug exposures of children with congenital heart disease. Am. J. Epidemiol. 109:433–439, 1979.

Royal College of General Practitioners: The outcome of pregnancy in former contraceptive users. Br. J. Obstet. Gynecol. 83:608–616, 1976.

Savolainen, E.; Saksela, E. and Saxen, L.: Teratogenic hazards of oral contraceptives analyzed in the National Register. Am. J. Obstet. Gynecol. 140:521–524, 1981.

Schardein, J.L.: Congenital abnormalities and hormones during pregnancy: A clinical review. Teratology 22:251–270, 1980.

Shapiro, S. and Slone, D.: Effects of exogenous female hormones on the fetus. Epidemiol. Rev. 1:110–123, 1979.

Tuchmann–Duplessis, H. and Mercier–Parot, L.: Action d'un steroide anticonceptionnel sur la descendance. J. Gynecol. Obstet. Biol. Reprod. 1:141–159, 1972.

Wilson, J.G. and Brent, R.L.: Are female sex hormones teratogenic? Am. J. Obstet. Gynecol. 141:567–580, 1981.

Wiseman, R.A. and Dodds–Smith, I.C.: Cardiovascular birth defects and antenatal exposure to female sex hormones: A reevaluation of some base data. Teratology 30:359–370, 1984.

Yasuda, M. and Miller, J.R.: Prenatal exposure to oral contraceptives and transposition of the great vessels in man. Teratology 12:239–244, 1975.

Yerushalmy, J.: Methodologic problems encountered in investigating the possible teratogenic effects of drugs. In: Drugs and Fetal Development, M.A. Klingberg, A. Abramovici and J. Chemke (eds.), Plenum Press, New York, 427–440, 1972.

1804 Orbiviruses

Whistler and Swanepoel (**1991**) studied the Palyam serogroup of orbiviruses by injecting them into 4 day chick eggs. Death, growth retardation and congenital defects were produced. Arthrogryposis and reduced feathering were the most common defects. Many of these viruses are abortigenic in cattle.

Whistler, T. and Swanepoel, R.: Teratogenicity of the

palyam serogroup orbiviruses in the embryonated chicken egg model. Epidemiol. Infect. 106:179–188, 1991.

1805 ORF 3858
2–Methyl–3–ethyl–4–phenyl–4–cyclohexane Carboxylic Acid

Van Petten and King (1971) treated pregnant rats with unspecified amounts and found no increase in visceral or skeletal anomalies. Perinatal studies revealed a decrease in neonatal birth weight.

Van Petten, L.E. and King, T.O.: Toxicologic and teratologic studies with ORF 3858 (2–methyl–3–ethyl–4–phenyl–4–cyclohexane carboxylic acid). (abs) Toxicol. Appl. Pharmacol 19:412 only, 1971.

1806 Org OD 14

This steroid studied by Van Julsingha (1973) reported that the hepatotoxicity of the compound in rabbits was associated with fetal death and malformations. The dose and type of malformation was not given in the abstract. No embryotoxicity was found in rats.

Van Julsingha, E.B.: Prediction of outcome of pregnancy on the basis of SGOT and SGPT activities on day 19 of pregnancy of rabbits dosed with Org OD 14. Teratology 8:224, 1973.

1807 Orgotein CAS 9016–01–7

Carson et al. (1973) gave up to 50 times the clinical dose to rats and rabbits and found no adverse effects on reproduction. This metalloprotein was given intramuscularly in amounts of 2.0 mg per kg during organogenesis in the rat and rabbit and on days 15 through 21 in the rat.

Carson, S.; Vogin, E.E.; Huber, W. and Schulte, T.L.: Safety tests of orgotein, an anti–inflammatory protein. Toxicol. Appl. Pharmacol. 26:184–202, 1973.

1808 Orpanoxin

Sutton and Denine (**1985**) gave this anti–inflammatory to rats before and during pregnancy and in the perinatal period. Oral doses of up to 100 mg per kg did not produce drug related changes.

Sutton, M.L. and Denine, E.P.: Teratology studies of orpanoxin in Sprague–Dawley rats and New Zealand white rabbits. Toxicologist 5:184, 1985.

1809 Orphenadrine

Beall (1972) gave 3.0, 15 or 30 mg orally to rats on days 6–15. At the two highest doses there were 8 of 159 fetuses with enlarged bladders which contained blood. Other anomalies were not found.

Beall, J.R.: A teratogenic study of chlorpromazine, orphenadrine, perphenazine, and LSD–25 in rats. Toxicol. Appl. Pharmacol. 21:230–236, 1972.

1810 Oryzalin *3,5–Dinitro–N–4, N–4–di(N–propyl) sulfanilamide)*

This is the active ingredient of the herbicide SURFLANTM. Byrd et al. (1990) gavaged rats and rabbits with up to 1000 mg and 125 mg per kg daily during organogenesis. Decreased fetal weight gain in the rat fetus was seen at 125 and 1000 mg per kg. No teratogenicity was found in either species even at maternally toxic doses.

Byrd, R.A.; Jordan, W.H. and Markham, J.K.: Developmental toxicity of dinitroanilines. III. Oryzalin. (abs)Teratology 41:542, 1990.

1811 Osaterone Acetate CAS 105149–00–6
17α–acetoxy–6–chloro–2–oxa–4,6–pregnadiene–3, 20–dione TZP–4238

Usui et al. (**1994** A) gave this antiandrogen to rats before pregnancy and at 3 and 30 mg per kg the females did not allow copulation. Usui et al. (**1994** B) gave up to 30 mg orally during day 7–17 and found no decreased growth or fetal death but the anogenital distance was reduced in the males and increased in the females. Behavior was normal but in the F_2 generation there was an increased neonatal death rate. Rabbits were treated by Shimpo et al. (**1994**) with doses up to 15 mg per kg on days 6–18 of gestation. Decreased fetal body weight and decreased ossification were found in fetuses exposed to 2.5 or 15 mg per kg. No effect was seen at 0.42 mg pe kg.

Shimpo, K.; Kamada, S.; Yahata, A.; Nasu, Y.; Kobayashi, K.; Usui, T.; and Suzuki, M.R.: Teratogenicity study of osaterone acetate (TSP–4238) administered orally to rabbits. Oyo Yakuri 47:289–295, 1994.

Usui, T.; Eguchi, K.; Ogawa, C.; Sone, H.; Ogawa, T.; Horiuchi, T. and Suzuki, M.R.: Effect of oral administration of a new antiandrogen, osaterone acetate (TSP–4238), prior to and in the early stages of pregnancy in rats. Oyo Yakuri 48:83–95, 1994A.

Usui, T.; Eguchi, K.; Ogawa, C.; Sone, H.; Ogawa, T.; Yamamoto, T.; Makino, M. and Suzuki, M.R.: Effect of oral administration of a new antiandrogen, osaterone acetate (TZP–4238), during the period of fetal organogenesis in rats. Oyo Yakuri 48:97–108, 1994B.

1812 Ouabaine

Petter (**1969**) studied the effect of ouabaine on the amniotic fluid of the rat. Intra amniotic injections were made of 0.4 mg on day 16, 18, 20 or 21. Hydramnios was produced with increased sodium content of the fluid but no effect on the fetuses.

Petter, C.: Production experimentale d'hydramnios par administration locale d'ouabaine au niveau des membranes foetales de rat. C.R. Soc. Biol. (Paris) 163:1023–1027, 1969.

1813 Oxacephem *Flomoxef™*

Hasegawa et al. (1987) gave this antibiotic intravenously to rats on days 7–18 in doses of up to 3,200 mg per kg. No teratogenicity or fetal loss was found but growth retardation was found at 200 mg or above. Postnatal studies did not show any adverse developmental changes.

Hasegawa et al. (1987A,B) studied this antibiotic during organogenesis in mice and rabbits. In subcutaneous doses of up to 3,200 mg per day in mice and intravenous doses of up to 25 mg per kg in rabbits, no malformation increases were found. In the rabbit, doses of 6.25 mg per kg or more decreased viability and weight gain of fetuses.

Hasegawa, Y.; Fukiishi, Y.; Hara, K.; Andou, M.; Yoshida, T.; Itoh, M.; Kishi, K.; Muranaka, R.; Takegawa, Y.; Kanamori, S.; Miyago, M.; Hirashiba, M.; Uchida, H. and Matsuo, E.: A teratology study of flomoxef (6315–S) in rabbits. Pharmacometrics (Oyo Yakuri) 34(6):475–491, 1987A.

Hasegawa, Y.; Hara, K.; Andou, M.; Yoshida, T.; Itoh, M.; Kishi, K.; Muranaka, R.; Fukiishi, Y.; Takegawa, Y.; Kanamori, S.; Miyago, M.; Hirashiba, M.; Uchida, H. and Matsuo, E.: A teratology study of flomoxef (6315–S) on mice. Pharamacometrics (Oyo Yakuri) 34(4):335–349, 1987B.

Hasegawa, Y.; Takegawa, Y. and Yoshida, T.: Reproduction of rats under 6315–S (Flomoxef). 2. Intravenous administration during fetal organogenesis. Chemotherapy 35:370–403, 1987.

1814 Oxacillin CAS 66–79–5

Korzhova et al. (1981) gave rats 40,000 units orally on the 4th through 13th days of gestation. Growth retardation, mortality and hemorrhages of the brain and kidneys were increased in the treated fetuses. No disturbances of central nervous system function were found in the living young.

Korzhova, V.V.; Lisitsyna, N.T. and Mikhailova, E.G.: Effect of ampicillin and oxacillin on fetal and neonatal development. Bull Exp. Biol. Med. (USSR) 91:169–171, 1981.

1815 Oxalic Acid

Sheikh–Omar and Schiefer (1980) fed 35 or 45 mg per kg to rats during pregnancy and found no reproductive ill effects. The dams did have oxalate crystals in their kidneys.

Sheikh–Omar, A.R. and Schiefer, H.B.: Effects of feeding oxalic acid to pregnant rats. Pertanika 3:25–31, 1980.

1816 Oxamyl *Vydate–L* CAS 23135–22–0

Kennedy (1986) found no evidence of teratogenicity in rats or rabbits. At doses of 100 ppm or greater the dams had reduced weight gain.

Kennedy, G.L.: Chronic toxicity, reproductive, and teratogenic studies with oxamyl. Fund. Appl. Toxicol. 7:106–118, 1986.

1817 Oxapium Iodide *N–Methyl–N–(2–cyclohexyl–2– phenyl–1,3–dioxolan–4–yl–methyl)piperidinium Iodide* CAS 6577–41–9

Takai and Nakada (1970) reported no teratogenicity or fetal changes in mice and rats treated orally during organogenesis. The maximal dose in the rat was 50 mg per kg and in the mouse 100 mg per kg per day.

Takai, A. and Nakada, H.: Teratological studies on N–methyl–N–(2–cyclohexyl–2–phenyl–1, 3–dioxolan–4–yl–methyl) piperidinium iodide (SH–100). Oyo Yakuri 4:109–112, 1970.

1818 Oxaprozin *4,5–Diphenyl–2–oxazolepropionic Acid* CAS 21256–18–8

Yamada et al. (1984a) gave this anti–inflammatory agent orally to pregnant rats and rabbits. In the rat treated

before fertilization with the highest dose (300 mg per kg) a decrease in corpora lutea occurred. At doses of 500 mg per kg during days 7–17 no adverse effects on reproduction were found. Postnatal studies were also found to be unchanged. Rabbits were given up to 30 mg per kg on days 6–18 and the only abnormality was decreased growth of the fetus at the 30 mg level.

Yamada et al. (1984b) studied this non–steroidal anti–inflammatory drug in pregnant rats giving up to 500 mg per kg on days 7–17 and up to 400 mg per kg on day 17 until weaning. No adverse effects were found. They also measured the diameter of the ductus arteriosus 4 hours after administering it to day 21 fetuses and found significant reduction at 10 mg per kg and above. The reduction was dose dependent and the results were similar to those they found with indomethacin, ibuprofen and aspirin.

Yamada, T.; Nishiyama, T.; Sasagima, M. and Nakane, S.: Reproduction studies of oxaprozin in the rat and rabbit. Iyakuhin Kenkyu 15:207–292, 1984a.

Yamada, T.; Inoue, T.; Hara, M.; Ohba, Y.; Nakame, S. and Uchida, H.: Reproductive studies of oxaprozin and studies on the fetal ductus arteriosus. Clin. Report 18:514–525, 528–536, 1984b.

1819 Oxazepam CAS 604–75–1

Saito et al. (1984) gavaged 100 mg per kg to rats on days 8–14 and found no adverse reproductive effects. Alleva and Bignami (1986) gave mice up to 100 mg per kg daily on days 12–16 by the oral route. Maternal lethality was increased at the highest dose. At the 30 mg per kg level postnatal studies were done. Transient growth and neurobehavioral development was found. A reduced hyperactivity response to amphetamine was found and selective impairment of adult active avoidance was present. A number of the findings were normal.

Alleva, E. and Bignami, G.: Prenatal benzodiazepine effects in mice. Postnatal behavioral development, response to drug challenges, and adult discrimination learning. Neuro-Toxicol. 7:303–318, 1986.

Saito, H.; Kobayashi, H.; Takeno, S. and Sakai, T.: Fetal toxicity of benzodiazepine in rats. Res. Commun. Chemical Pathol Pharmacol. 46:437–447, 1984.

1820 Oxazolam *Oxazolazepam* CAS 24143–17–7

Tanase et al. (1969) treated rats and mice orally with up to 600 and 1000 mg per kg, respectively, during organogenesis. Weight reduction was present in the fetuses at the highest doses but no teratogenic effect was found.

Tanase, H.; Hirose, K.; Shimada, K.; Aoki, K.; Suzuki, H. and Suzuki, Y.: The safety test of oxazolazepam. II. Effect of oxazolazepam on embryos during pregnancy and post–natal offsprings of experimental animals. Ann. Sankyo Res. Lab. 21:107–119, 1969.

1821 Oxazolidinethione *LC–11,634*

Webster et al. (1967) reported that 2.5 mg administered to rats orally or subcutaneously inhibited pregnancy. The inhibition was not associated with its antithyroid activity. A dose of 5 mg per day on days 0–6 also prevented viable implants.

Webster, H.D.; Johnston, R.L. and Duncan, G.W.: Toxicologic and interrelated studies with an oxazolidinethione contraceptive. Toxicology and Applied Pharmacology 10(2):322–333, 1967.

1822 Oxepinac
6,11–Dihydro–11–oxodibenz(b,e)oxepin–3–acetic Acid

Arauchi et al. (1978) gave this medication orally during organogenesis to mice and rabbits in maximum doses of 90 and 20 mg per kg, respectively. No teratogenic effects were found. Postnatal studies in mice were carried out without significant postnatal effects.

Arauchi, T.; Watanabe, K.; Nakashima, K.; Matsuchashi, K.; Morita, H. and Akimoto, T.: Teratogenicity study of oxepinac in mice and rabbits. Arzneim–Forsch. 28:451–455, 1978.

1823 Oxiracetam

Morimoto et al. (1991) gave this drug orally to rats during organogenesis in doses of up to 5000 and found only fetal weight reduction and doses which were toxic to the dams. Similar findings were reported for rabbits treated at 3000 mg per kg during organogenesis (Sakai et al., 1991)

Morimoto, K.; Kusumoto, S.; Tadokoro, T. and Miyamoto, M.: A study of reproductive and developmental toxicity in rats administered oxiracetam during fetal organogenesis. Jpn. Pharm. Therap. 19:3545–3563, 1991.
Sakai, Y.; Kawakami, T.; Kinoshita, K.; Wada, H. and Miyamoto, M.: A study of reproductive and developmental toxicity in rabbits administered oxiracetam during fetal organogenesis. Jpn. Pharm. Therap. 19:3565–3576, 1991.

1824 Oxisuran

Hornyak et al. (1973) gave rats up to 400 mg per kg by gastric gavage on days 6 through 15. No reproductive or adverse fetal effects were found.

Hornyak, E.P.; Jones, S.M.; Williams, K.W.; Cerniski, A. and Schwartz, E.: Teratogenic studies with oxisuran (a differential immunosuppressive) and azathioprine in the rat. Toxicol. Appl. Pharmacol. 25:462, 1973.

1825 Oxitropium Bromide

Savary and Glomot (1989) studied the oral effect of this scopolamine derivative in rats and rabbits. In the rabbit doses up to 200 mg were severely toxic to the dams but no malformations occurred. Treatment during organogenesis in the rat produced severe fetotoxicity at 600 mg per kg and four cranio–facial defects were found among approximately 300 fetuses. Wavy ribs also occurred. Fertility studies in the rat indicated some decrease in estrus at toxic doses. Postnatal development in the rat was not significantly changed.

Savary, M.H. and Glomot, R.: Oral reproduction toxicology of oxitropium bromide. Oyo Yakuri 37(5):413–428, 1989.

1826 4–Oxo–8–[4–)4–phenylbutoxy)benzoylamino]–2–(tetrazol–5–yl)–4H–1–benzopyran hemiydrate
ONO–1078

Ozeki et al. (1992) found no ill effects of up to 1000 mg per kg of this leukotreine antagonist given orally before pregnancy and in early gestation. In similar studies done during organogenesis Komai et al. (1992) found no increase in fetal ill effects. Wada et al. (1992) found no ill effects after perinatal dosing. Wada et al. (1992) gave rabbits up to 500 mg per kg during organogenesis and found no increase in defects.

Komai, Y.; Ito, I.; Hibino, H.; Iriyama, K.; Isokazu, K.; Matsuoka, Y. and Fujita, T.: Reproductive and developmental toxicity study of ONO–1078 (2) teratological study in rats. Iyakuhin Kenkyu 23:832–845, 1992.
Ozeki, Y.; Wada, M.; Tamagawa, N.; Mor, H.; Nishimura, T.; Takada, S.; Oku, H.; Komeno, M.; Sugai, S.; Shinomiya, K. and Matsuoka, Y.: Reproductive and developmental toxicity study of ONO–1078 (1) fertility study in rats. Iyakuhin Kenkyu 23:819–831, 1992.
Wada, M.; Shinomiya, K.; Narita, M.; Nishimura, T.; Takada, S.; Oku, H.; Chihara, N.; Sawaragi, H.; Yanagizawa, Y.; Narita, M.; Shimouchi, K.; Itagaki, I.; Tanaka, M.; Suzuki, Y.; Ozeki, Y. and Matsuoka, Y.: Reproductive and developmental toxiicity study of ONO–1078 (3) peri–and postnatal study in rats. Iyakuhin Kenkyu 23:846–872, 1992.
Wada, M.; Narita, M.; Shinomiya, K.; Nishimura, T.; Takada, S.; Oku, H.; Tanaka, M.; Itagaki, I.; Shichino, Y.; Yanagizawa, Y.; Ozeki, Y. and Matsuoka, Y.: Reproductive and developmental toxicity study of ONO–1078 (4) teratological study in rabbits. Iyakuhin Kenkyu 23:873–888, 1992.

1827 Oxoisotretinoin

Webster (1985) cultured rat embryos in this major metabolite of 13–cis–retinoic acid. At concentrations of above 250 nanograms per ml of medium the branchial arches were small and malformed. No effect was seen at 125 ng per ml.

Webster, W.S.: Isotretinoin embryopathy: The effects of isotretinoin and 4–oxo–isotretinoin on postimplantation rat embryos in vitro. (abs) Teratology 31:59A, 1985.

1828 Oxolamine
5–(2–Diethyl–aminoethyl)–3–phenyl–1,2,4–oxadiazole Citrate CAS 959–14–8

This anti-inflammatory drug was tested in pregnant mice by Nilsson (1967). No significant gross defects occurred with 2 mg daily injections although some increase in number of ribs and abnormal centers in the sternum occurred in the treatment and sham injected groups.

Nilsson, L.: Teratogenic studies on mice with an antiphlogistic substance, 5–(2–diethyl–aminoethyl)–3–phenyl–1,2,4–oxadiazole citrate. Arzneim–Forsch. 17:781–782, 1967.

1829 Oxomemazine CAS 3689–50–7

Discussed under Phenothiazines

1830 **Oxonic Acid** *Uric Acid Elevation*

Gralla and Crelin (1976) blocked uric acid cataboles in mice by adding 3 percent potassium oxonate to their diet. On days 8–10 fetal resorptions occurred. No changes occurred with treatment on days 10–13 unless sodium urate was injected. With combined therapy resorptions were increased to 47 percent. A 3.6 percent increase of cleft palate occurred with this latter treatment.

Gralla, E.J. and Crelin, E.S.: Oxonic acid and fetal development. I. Embryotoxicity in mice. Toxicology 6:289–297, 1976.

1831 **Oxophenarsine Hydrochloride** *Mapharsen*™ CAS 538–03–4

Mosher (1938) administered ten one mg doses in nine hours to a pregnant guinea pig. Hemorrhages in the fetal scala vestibuli and vestibule were found.

Mosher, H.P.: Does animal experimentation show similar changes in the ear of the mother and fetus after the ingestion of quinine by the mother? Laryngoscope 48:361–395, 1938.

1832 **11–Oxo–11H–pyrido(2,1–b) quinazoline–2–carboxylic Acid** *Sm 857 SE*

Nishimura et al. (1988) studied this antiallergic medication in rats and rabbits at 400 mg per kg on days 7–17. Microphthalmia and vertebral–costal defects occurred in the rat fetuses. Maternal toxicity occurred at 400 mg per kg at dose levels of 20, 90 and 200 mg per kg. No significant malformation increase occurred. Rabbits exposed to 400 mg had much lower serum levels than rats and no adverse fetal effects.

Nishimura, M.; Kast, A.; Tsunenari, Y. and Kobayashi, S.: Teratogenicity of the antiallergic Sm 857 SE in rats versus rabbits. Teratology 38:351–367, 1988.

1833 **Oxycodone** *Percodan*™

Heinonen et al. (1977) included 8 exposed pregnancies among a group of 46 other opium alkaloids in which there was no increase in congenital defects.

Heinonen, O.P.; Slone, D. and Shapiro, S.: Birth Defects and Drugs in Pregnancy. Publishing Sciences Group Inc., Littleton, Mass., 1977.

1834 **Oxydemeton–Methyl**

This pesticide has been tested in chicks and rats. Clemens et al. (1990) gave rats up to 4.5 mg per kg orally on days 6–15 of gestation and found no fetal ill effects. The dams did not gain weight normally. Lenselink et al. (1993) produced defects in chicks starting at 0.06 mg per egg applied to the vitalline membrane.

Clemens, G.R.; Hartnagel, R.E.; Bare, J.J. and Thyssen, J.H.: Teratological, neurochemical and postnatal neurobehavioral assessment of Metasystox™, an organophosphate pesticide in the rat. Fund. Appl. Pharm. 14:131–143, 1990.

Lenselink, D.R.; Midtling, J.E. and Kolesari, G.L.: Teratogenesis associated with oxydemeton–methyl in the stage 12 chick embryo. Teratology 48:207–211, 1993.

1835 **Oxyfedrine** CAS 15687–41–9

This compound was tested in mice, rabbits and rats and no evidence of teratogenicity was found (Habersang et al., 1967). The rats were given up to 600 mg per kg daily, and the mice and rabbits received 50 mg per kg.

Habersang, S.; Leuschner, F. and Schlichtegroll, A.: Toxikologische Untersuchungen uber eine neue myocard– und coronarwirksame Verbindung aus der Reihe der β–Aminoketone. Arzneim. Forsch. 17:1478–1491, 1967.

1836 **Oxygen** CAS 7782–44–7

Discussed under Hyperoxia and Hypoxia

1837 **Oxymetholone** *17 β–Hydroxy–2–hydroxy-methylene–17α-methyl–5α-androstane–3–one*

Kawashima et al. (1977) administered 5 mg orally to rats on days 17–20 and produced a reduction in urovaginal septum length in the female fetuses.

Kawashima, K.; Nakaura, S.; Nagao, S.; Tanaka, S.; Kuwamura, T. and Omori, Y.: Virilizing activities of various steroids in female rat fetuses. Endocrinol. Japon. 24:77–81, 1977.

1838 **5–Oxymethyl Uracil**

Kosmachevskaya and Tichodeeva (1968) injected 4 mg of this substance into chick eggs at 24 hours of incubation and found no lethal effect but 15 percent of the embryos had malformations which included microphthalmia, rumplessness, beak and skeletal defects and celosomia.

Kosmachevskaya, E.A. and Tichodeeva, I.I.: Relation between embryotoxic activity of some pyrimidine derivatives and their chemical structure. Chick embryo test. Pharmacol. Toxicol. (Russian) 5:618–620, 1968.

1839 **p–Oxyphenyl**

Zharova et al. (1979) studied this metabolite of tyrosine in pregnant mice giving 5 mg daily subcutaneously during the first 10 days of gestation. Increase malignant and benign neoplasms occurred in two strains of mice and at an earlier age than in the controls.

Zharova, E.I.; Sergeeva, T.I. and Malakhova,N.V.: Transplacental blastomogenic action of p–hydroxyphenyl lactic acid (Russian). Byull. Eksp. Biol. Med. 37:39–41, 1979.

1840 **Oxyquinoline**

Heinonen et al. (1977) found no increase in defects among 21 women treated in the first four lunar months.

Heinonen, O.P.; Slone, D. and Shapiro, S.: Birth Defects and Drugs in Pregnancy. Publishing Sciences Group Inc., Littleton, Mass., 1977.

1841 **Oxytetracycline** *Terramycin*™ CAS 79–57–2

This antibiotic is capable of crossing the placenta and causing staining of the deciduous teeth. It stains to a lesser degree than tetracycline (Baden, 1970).

Morrissey et al. (1986) treated mice and rats orally on days 6–15 with up to 1,500 and 2100 mg per kg, respectively. They found no increase in congenital defects.

Baden, E.: Environmental pathology of the teeth. In: Oral Pathology, R.J. Gorlin and H.M. Goldman (eds.), St. Louis, C.V. Mosby, 1:190 only, 1970.

Morrissey, R.E.; Tyl, R.W.; Price, C.J.; Ledoux, T.A.; Reel, J.R.; Paschke, L.L.; Marr, M.C. and Kimmel, C.A.: The developmental toxicity of orally administered oxytetracycline in rats and mice. Fund. Appl. Toxicol. 7:434–443, 1986.

1842 Oxythiamine CAS 582–36–5

This analogue of thiamine was used by Naber et al. (1954) in the chick. With 0.25 to 2.0 mg injected at the start or after 5 days of incubation a high mortality was found. Hemorrhage, edema and abdominal hernias were found in the dying embryos.

Naber, E.C.; Cravens, W.W.; Baumann, C.A. and Bird, H.R.: The effect of thiamine analogues on embryonic development and growth of the chick. J. Nutr. 54:579–591, 1954.

1843 Oxytocin

Kumaresan (1974) produced a decrease in oxytocin in rats by giving a rabbit antibody raised in rabbits against oxytocin coupled with bovine albumin. A small but significant reduction in litter size was found in rats receiving the antibody before mating. Boer (1993) implanted oxytocin into rats on day 17 and then followed their development postnatally. A decrease in weight and reduced cell counts in the brain occurred. When cross fostered at birth no effects were seen.

Boer, G.J.: Chronic oxytocin treatment during late gestation and lactation impairs development of rat offspring. Neurotoxicology and Teratology 15:383–389, 1993.

Kumaresan, P.: The effect of oxytocin antibodies on the litter size in rats. Am. J. Obstet. Gynecol. 118:68–72, 1974.

1844 Ozagrel Hydrochloride

Imai et al. (1990) gave this antiasthmatic orally to rats on days 7–17 in amounts up to 2,000 mg per kg. At 2,000 mg per kg fetal weight and ossification was decreased. No teratogenicity was found.

Imai, K.; Nakajoh, M.; Ohba, M.; Ozawa, S.; Naitoh, J. and Matsuoka, Y.: Reproduction studies of oky–046 hcl (2nd report)–teratological study in rats. Kiso to Rinsho 24:3703–3719, 1990.

1845 Ozone CAS 10028–15–6

Kavlock et al. (1979) exposed rats to up to 1.97 ppm during parts or all of organogenesis and produced no defects in the offspring. Resorption rates were increased. Veninga (1967) reported blepharophimosis and jaw anomalies in mouse fetuses exposed in utero to 0.2 ppm for 7 hours five days a week.

Kavlock, R.J.; Daston, G. and Grabowski, C.T.: Studies on the developmental toxicity of ozone. 1. Prenatal effects. Toxicol. Appl. Pharmacol. 48:19–28, 1979.

Veninga T.S.: Toxicity of ozone in comparison with ionizing radiation. Strahlentherapie 134:469–477, 1967.

1846 Paclitaxel

Kai et al. (1994 A) gave this antineoplastic to rats before pregnancy and during the first week of gestation and found a decrease in implantations and live fetuses at the highest dose of 1.0 mg per kg given intravenously. Studies during organogenesis at doses of up to 0.6 mg per day were not associated with significant teratogenic activity or change in behavior including reproduction of the offspring (Kai et al., 1994 B). Peri and postnatal studies (Kai et al., 1994 C) were also unremarkable.

Kai, S.; Kohmura, H.; Hiraiwa, E.; Koizumi, S.; Ishikawa, K.; Kawano, S.; Kuroyanagi, K.; Hattori, N.; Chikazawa, H.; Dondoh, H.; Sakakura, K.; Kadota, T. and Takahashi, N.: Reproductive and developmental toxicity studies of paclitaxel (I)–intravenous administration to rats prior to and in the early stages of pregnancy. J. Tox. Sc. 19:s57–s67, 1994A.

Kai, S.; Kohmura, H.; Hiraiwa, E.; Koizumi, S.; Kshikawa, K.; Kawano, S.; Kuroyanagi, K.; Hattori, N.; Chikazawa, H.; Kondoh, H.; Sakakura, K.; Kadota, T. and Takahashi, N.: Reproductive and developmental toxicity studies of paclitaxel (II)–intravenous administration to rats during the fetal organogenesis J. Tox. Sc. 19:s69–s91, 1994B.

Kai, S.; Kohmura, H.; Hiraiwa, E.; Koizumi, S.; Ishikawa, K.; Kawano, S.; Kuroyanagi, K.; Hattori, N.; Chikazawa, H.; Kondoh, H.; Sakakura, K.; Hisada, S.; Kadota, T. and Takahashi, N.: Reproductive and developmental toxicity studies of paclitaxel (III)–intravenous administration to rats during the perinatal and lactation periods. J. Tox. Sc. 19:s93–s111, 1994C.

1847 Palladium

Ridgway and Karnofsky (1952) tested the chloride salt in chick eggs and found it non–teratogenic. The LD–50 was greater than 20 mg per egg on the 4th day of incubation.

Ridgway, L.P. and Karnofsky, D.A.: The effects of metals on the chick embryo: Toxicity and production of abnormalities in development. Ann. N.Y. Acad. Sci. 55:203–215, 1952.

1848 Palm Oil CAS 8002–75–3

Singh (1980) gave this food substance which contains carotene (32–48 mg per 100 ml) to rats in amounts of 1–3 ml daily on days 5 through 15 of pregnancy. Exencephaly was found at 1,2 and 3 mg levels. At 3 mg levels, eye defects and cleft palate occurred.

Singh, J.D.: Palm oil induced congenital anomalies in rats. Cong. Anom. 139–142, 1980.

1849 Palmotoxin BO *Palmotoxin GO* CAS 39450–10–7; CAS 39450–11–8

Bassir and Adekunle (1970) injected chick eggs with 0.2 to 0.7 microgm of the B(O) form and 2.0 to 6.3 microgm of the

G(O) form and observed an increased death rate and twisting of the lower extremities, crossed beak and roughness of plumage.

Bassir, O. and Adekunle, A.: Teratogenic action of aflatoxin B$_1$, palmotoxin B(O) and palmotoxin G(O) on the chick embryo. J. Pathol. 102:49–51, 1970.

1850 Palonidipine Hydrochloride

3–(Benzlmethylamino)–2,2–dimethylpropyl methyl 4–(2–fluoro–5–nitriphenyl)–1,4–dihydro–2, 6–dimethyl–3, 5–pyridinedicarboxylate Hydrochloride TC–81

This calcium antagonist was given by Matsuzawa et al. (**1992** A&B) to rats orally before and during early pregnancy, during organogenesis and perinatally. At 2.5 mg per kg there was fetal toxicity (creeping, piloerection and lacrimation). No teratogenic effect was found and postnatal development and reproduction were unchanged. Perinatal and postnatal studies found decreased weight gain in the 2.5 mg per kg group. (Sugawara et al., **1992**). In rabbits treated with up to 50 mg per kg during organogenesis there were no adverse fetal effects (Matsuzawa et al, **1992** C).

Matsuzawa, K.; Enjyo, H.; Nishizawa, S.; Izawa, Y. and Makita, T.: Toxicity studies of palonidipine hydrochloride (TC–81)–Fertility study in rats. The Clinical Report 26:2993–3005, 1992A.

Matsuzawa, K.; Nishizawa, S.; Takano, H.; Izawa, Y. and Makita, T.: Toxicity studies of palonidipine hydrochloride (TC–81)–Teratogenicity study in rats. The Clinical Report 26:3007–3025, 1992B.

Matsuzawa, K.; Ikegawa, S.; Nishiazwa, S.; Hisada, T.; Izawa, Y. and Makita, T.: Toxicity studies of palonipipide hydrochloride (TC–81)–Teratogenicity study in rabbits. The Clinical Report 26:3027–2026, 1992C.

Sugawara, S.; Oguri, M.; Izawa, Y. and Makita, T.: Toxicity studies of palonipidine hydrochloride (TC–81)–Perinatal and postnatal study in rats. The Clinical Report 26:3037–3057, 1992.

1851 Pancreatic Ablation by Transgenic Insertion

Palmiter et al. (1987) injected a hybrid gene promoter from elastase I gene fused with diphtheria A toxin gene construct into mouse pronuclei and produced transgenic mice which were deficient or absent in pancreatic tissue. Of 24 transgenic mice produced, seven had macroscopic absence of the pancreas.

Palmiter, R.D.; Behringer, R.R.; Quaife, C.J.; Maxwell, F.; Maxwell, I.H. and Brinster, R.L.: Cell lineage ablation in transgenic mice by cell–specific expression of a toxin gene. Cell 50:435–443, 1987.

1852 Pantothenic Acid Deficiency

Kalter and Warkany (1959) reviewed and summarized the literature about this experimental model in rodents. Boisselot (1948) was the first to feed a pantothenic acid–deficient diet and observe defective rat offspring. Nearly all organ systems can be affected, and a deficiency lasting 36 hours during active embryogenesis may produce specific de-

fects (Nelson et al., 1957). A dietary level in the rat below 10 microgm per day produces complete litter resorption, while a level of 20 to 25 microgm produces a large number of defective offspring; the minimal amount required to prevent fetal defects in the rat is 50 microgm (Giroud et al., 1954; Lefebvres, 1954). Three chemical antagonists have been used to augment the deficiency state (X–methyl-pantothenic, sodium ω methyl pantothenic and pantoyltaurine).

Boisselot, J.: Malformations congenitales provoquees chez le rat par une insuffisance en acide pantothenique du regime maternel. C. R. Soc. Biol. (Paris) 142:928–929, 1948.

Giroud, A.; Levy, G. and Lefebvres, J.: Recherches sur le taux de l'acide pantothenique chez les meres et les foetus normaux et chez les meres carencees. Int. Z. Vitaminforschung 25:148–153, 1954.

Kalter, H. and Warkany, J.: Experimental production of congenital malformations in mammals by metabolic procedure. Physiol. Rev. 39:69–115, 1959.

Lefebvres, J.: Influence d'une deficience pantothenique legere sur les resultats de la gestation chez la ratte. C. R. Acad. Sci. (Paris) 238:2123–2125, 1954.

Nelson, M.M.; Wright, H.V.; Baird, C.D.C. and Evans, H.M.: Teratogenic effects of pantothenic acid deficiency in the rat. J. Nutr. 62:395–406, 1957.

1853 Pantoyltaurine CAS 2545–84–8

This pantothenic acid antagonist was used by Zunin and Borrone (1954) to produce defects in the offspring of treated rats. They used 0.5 to 1.5 mg by injection daily during various periods of gestation. The syndrome of defects resembled that seen with pantothenic acid deficiency. Generalized edema, hemorrhage and neural tube closure defects were seen.

Zunin, C. and Borrone, C.: Embriopatie da carenza di acido pantotenico. Acta Vitaminol. Enzymol. (Milano) 8:263–268, 1954.

1854 Papaverine CAS 58–74–2

Neural tube closure defects were found in the chick after culture in medium containing 50 micrograms per ml of medium (Lee and Nagele, 1979). Jurand (1980) injected subcutaneously 140 mg per kg into mice on the 9th day. This is approximately five times higher than the human dose. Kinking of the tail or dilation of the 3rd ventricle was found in 5% of the 13 day embryos examined.

Jurand, A.: Malformations of the central nervous system induced by neurotropic drugs in mouse embryos. Dev Growth Differ 22:61–78, 1980.

Lee, H. and Nagele, R.G.: Neural tube closure defects caused by papaverine in explanted early chick embryos. Teratology 20:321–332, 1979.

1855 Paradimethylaminobenzene

Pizzarello and Ford (1968) injected chick eggs at 48 hours of incubation with 6 mg of this carcinogen and produced shortening of the legs and feather defects in the survivors.

Pizzarello, D.J. and Ford, R.V.: Effects of paradimethyl-aminoazobenzene and the antioxidant N,N–diphenyl–p–phenylene diamine in developing chicks. Experientia 24:621–622, 1968.

1856 Paraffins, Chlorinated

Serrone et al. (**1987**) gavaged rats and rabbits with up to 5,000 mg per kg daily on days 6–19 or 6–27 respectively and found no teratogenicity.

Serrone, D.M.; Birtley, R.D.N.; Weigand, W. and Millischer, R.: Summaries of toxicological data toxicology of chlorinated paraffins. Fd. Chem. Toxic. 25:553–562, 1987.

1857 Parahydroxypropiophenone

Morris et al. (1967) studied the effect in rabbits giving 25–50 mg per kg on days 1–8 and 60 mg on days 13–15. No malformation increase was found.

Morris, J.M.; Van Wagenen, G.; McCann, T. and Jacob, D.: Compounds interfering with ovum implantation and development. Fertil. Steril. 18:18–34, 1967.

1858 Paramethadione CAS 115–67–3

Discussed under Trimethadione

1859 Paraquat *1,1–Dimethyl–4,4–dipyridilium Dichloride* CAS 4685–14–7

Talbot et al. (1988) reported nine cases where the mothers deliberately ingested 24 percent paraquat. Levels in the fetus and amniotic cavity were higher than in maternal serum. All fetuses died and only two of the mothers survived. The amount ingested by the mothers was a mouthful or more.

This herbicide was given to hens in their drinking water at a concentration of 40 ppm and about 0.1 ppm was found in their eggs (Fletcher, 1967). A small but significant increase in the number of abnormal eggs was found in the treated group. The type of defect was not described. Khera et al. (1970) reported a small increase in incidence of costal cartilage defects in the offspring of rats injected with 0.5 mg per kg per day. Ahmed et al. (**1988**) gavaged 7.8 or 15.6 mg per kg on days 8–15 in the rat. Some fetal weight reduction occurred but no skeletal abnormalities were found.

Ahmed, A.A.; Soliman, M.M.; Khalifa, B.A.A.; El Sadek, S.E. and Nounou, A.H.: Embryocidal and teratogenic effects of paraquat on chick embryos and white rats. Arch. Exp. Veterinaermed 42(6):848–853, 1988.
Fletcher, K.: Production and viability of eggs from hens treated with paraquat. Nature 215:1407–1408, 1967.
Khera, K.S.; Whitta, L.L. and Clegg, D.J.: Embryopathic effects of diquat and paraquat. In: Pesticides Symposia, Interamerican Congress on Toxicology and Occupational Medicine. W.B. Deichmann, J.L. Radomski and R.A. Penalver (eds.), Miami, Halos and Assoc. Inc., 257–261, 1970.
Talbot, A.R.; Fu, C.C. and Hsieh, M.F.: Paraquat intoxication during pregnancy: A report of 9 cases. Vet. Hum. Toxicol. 30:12–17, 1988.

1860 Parathion *Phosphorothioic Acid O,O–Diethyl–O–(4–nitrophenyl) Ester* CAS 56–38–2

Fish (1966) gave rats 3.8 mg per kg on day 8, 9, 15 or 16 of gestation and found no abnormalities in the offspring. The perinatal death rate was increased and some subcutaneous hemorrhages were found. The weight gain post-delivery was slower in the treated group. Fetal cerebral cortical cholinesterase was decreased. Postnatal studies in the mouse exposed prenatally to 3 mg per kg were negative (Al–Hachim and Fink, 1968).

Al–Hachim, G.M. and Fink, G.B.: Effect of DDT or parathion on condition avoidance response from DDT or parathion treated mothers. Psychopharmacologia 12:424–427, 1968.
Fish, S.A.: Organophosphorus cholinesterase inhibitor and fetal development. Am. J. Obstet. Gynecol. 96:1148–1154, 1966.

1861 Paratoluic Methylate

Krotov and Chebotar (**1972**) exposed rats to 750 mg per cubic meter on days 7–14 and found no teratogenicity or increase in embryonic lethality. Intraperitoneal doses of 750 mg per kg were toxic to the dams but no fetal effects were seen.

Krotov, J.A. and Chebotar, N.A.: Embryotoxic and teratogenic action of some industrial substances formed during production of dimethyl terephtalate. Gig. Tr. Prof. Zabol. 16:40–43, 1972.

1862 Paraxanthine *1,7–Dimethylxanthine*

York et al. (1986) studied this major metabolite of caffeine in pregnant mice giving up to 300 mg per kg intraperitoneally on days 11 and 12. A dose related increase in cleft palate and limb defects was found. The pattern was similar to that seen after caffeine administration but it was somewhat less toxic than caffeine.

York, R.G.; Randall, J.L. and Scott, W.J.: Teratogenicity of paraxanthine (1,7–dimethylxanthine) in C57BL6J mice. Teratology 34:279–282, 1986.

1863 Para–xylene

Krotov and Chebotar (**1972**) exposed rats to 500 mg per cubic meter during the entire gestation and found embryotoxicity but no evidence of teratogenicity.

Krotov, J.A. and Chebotar, N.A.: Embryotoxic and teratogenic action of some industrial substanced formed during production of dimethyl terephtalate. Gig. Tr. Prof. Zabol. 16:40–43, 1972.

1864 Parbendazole CAS 14255–87–9

This animal anthelmintic is teratogenic in sheep (30 mg per kg), rats and mice but not in rabbits (Lapras et al., 1974). Abnormalities of the limbs and vertebral column were found.

Wildgoose and Steele (1987) cultured rat embryos in serum from rats dosed with 12.5 mg or 25 mg per kg and found embryo lethality. In serum from animals given 11 mg per kg, growth retardation of the embryos occurred.

Lapras, M.; Lourge, G.; Gastellu, J.; Regnier, B.; Delatour, P.; Deschanel, J.P. and Lombard, M.: Parbendazole and congenital abnormalities. Cornell Vet. 457, 1974.

Wildgoose, C.L. and Steele, C.E.: In vitro teratogenicity of parbendazole. Arch. Toxicol. 11:172–174, 1987.

1865   Pargyline

Koren et al. (1965) found that intraamniotic injection of 2.8–4.6 mg per kg killed rat fetuses at every stage of pregnancy. Poulson and Robson (1963) injected mice on days 1–6 or 11–16 with 2 mg per day and found in the first group no signs of implantation and in the second group, no adverse effects.

Koren, Z.; Pfeifer, Y. and Sulman, F.G.: Deleterious effect of the monoamine oxidase inhibitor pargyline on pregnant rats. Fertil. Steril. 16:393–400, 1965.

Poulson, E. and Robson J.M.: The effect of amine oxidase inhibitors on pregnancy. J. Endocrinol. 27:147–152, 1963.

1866   Paroxetine BRL 29060A

Baldwin et al. (1989) gave rabbits and rats up to 5.1 or 43.0 mg per kg respectively. Fetal toxicity was found at maternally toxic doses but no teratogenicity was found. Pregnancy wastage and fertility was noted in rats given 13 mg or more per kg.

Baldwin, J.A.; Davidson, E.J.; Pritchard, A.L. and Ridings, J.E.: The reproductive toxicology of paroxetine. Acta Psychiatr. Scand. 80:37–39, 1989.

1867   Paroxypropione p–Hydroxypropiophenone

Morris (1970) gave 10 to 130 mg per kg to rabbits on days 1 through 15 and found only one defect among 91 fetuses.

Morris, J.M.: Postcoital antifertility agents and their teratogenic effect. Contraception 2:85–97, 1970.

1868   Parvovirus B19 Erythema Infectiosum

Jordan and Sever (1994) have comprehensively reviewed the effects of this viral infection in humans and animals. The greatest risk appears to be during the second trimester. The inclusions are found in erythocyte progenitors as well as myocardial cells. Maternal IgM may be undetectable after 60 days while IgG persists and is found in about 60% of adults. The absence of IgM in the newborn does not rule out intrauterine infection. Maternal alpha fetoprotein may be increased.

Van Elsacker–Niele et al. (1989) reported on 10 cases diagnosed prenatally. Five were normal at birth; two had malformations including microphthalmia. Nuclear inclusions were seen in various tissues including the placenta. Brown et al. (1984) reported a male fetus with generalized edema and effusions in the serous cavities. There was a marked leucoerythroblastic response in the liver, spleen and kidneys. There was specific IgM and IgG evidence of maternal infection with parvovirus. Dot hybridization studies with virion DNA and cloned DNA probes revealed parvovirus DNA in the placenta and other fetal organs. Knott and Welply (1984) published another case where a stillborn had evidence of infection. Mortimer et al. (1985) studied sera from 253 abnormal infants and found neither specific IgM nor HPV antigen.

Bond et al. (1986) obtained definite evidence of transplacental human parvovirus from a stillbirth. Placenta, adrenal and liver were positive by DNA hybridization, antigen radioimmunoassay and immune electron microscopy. Weiland et al. (1987) described a fetus with microphthalmia following infection with parvovirus B19 in the 6th week of her pregnancy. Hartwig et al. (1988) described an embryo with parvovirus infection. Endothelial damage and perivascular mononuclear infiltrations were found in the embryo and placental vessels. Eye abnormalities were also found.

Carrington et al. (1987) reported two pregnancies in which fetal hydrops occurred with evidence of overwhelming parvovirus (B19) infection. Maternal α–fetoprotein was elevated and in one case, fetal aplastic anemia was present. Burton and Caul (1988) detailed the type of inclusion in the precursor red cells and commented that they may be seen with light microscopy. Schwarz et al. (1988) reported 42 women with specific IgM against B19 and found that 36 percent were clinically symptom–free. Ten or 26 percent had fetuses with hydrops. Three were successfully treated by "intrauterine exchange transfusion" and the others died. Three of those with hydrops were women working in kindergartens. Anand et al. (1987) reported details of two autopsies of infected fetuses. Increased hematopoiesis and ballooning of the hepatocyte nuclei were noted. Mead (1989) has reviewed evidence suggesting that the virus does not produce defects but only fetal death. He also summarizes the clinical management which involves testing for B–19 IgM and IgG and if positive monitoring maternal alpha feto protein. Morey et al. (1991) report a case of intrauterine infection withCongenital anemia has been reported after intrauterine infection (Brown et al. 1994). occurred.

Anand, A.; Gray, E.S.; Brown, T.; Clewley, J.P. and Cohen, B.J.: Human parvovirus infection in pregnancy and hydrops fetalis. N. Eng. J. Med. 316:183–186, 1987.

Bond, P.R.; Caul, E.O.; Usher, J.; Cohen, B.J.; Clewley, J.P. and Field, A.M.: Intrauterine infection with human parvovirus. Lancet 1:448–449, 1986.

Brown, T.; Anand, A.; Ritchie, L.D.; Clewley, J.P. and Reid, T.M.S.: Intrauterine parvovirus infection associated with hydrops fetalis. Lancet 2:1033–1034, 1984.

Brown, K.E.; Green, S.W.; de Mayolo, J.A.; Bellanti, J.A.; Smith, S.D.; Smith, T.J. and Young, N.S.: Congenital anaemia after transplacental B19 parvovirus infection. The Lancet 343:895–896, 1994.

Burton, P.A. and Caul, E.O.: Fetal cell tropism of human parvovirus B19. Lancet 1:767 only, 1988.

Carrington, D.; Whittle, M.J.; Gibson, A.A.M.; Brown, T.; Field, A.M.; Cohen, B.J.; Gilmore, D.H.; Aitken, D.; Patrick, W.J.A.; Huaux, J.P.; Caul, E.O. and Clewley, J.P.:

Maternal serum α–fetoprotein: A marker of fetal aplastic crisis during intrauterine human parvovirus infection. Lancet 1:433–435, 1987.

Hartwig, N.G.; Vermeij–Keers, C.; Van Elsacker–Niele, A.M.W. and Fleuren, G.J.: Embryonic malformations in a case of intrauterine

Jordan, E.K. and Sever, J.L.: Fetal damage caused by parvoviral infections. Reproductive Tox. 8:161–189, 1994.

Knott, P.D.; Welply, G.A.C. and Anderson, M.J.: Serologically proved intrauterine infection. Br. Med. J. 289:1660 only, 1984.

Mead, P.B.: Parvovirus B19 infection and pregnancy. Contemporary Obst. and Gyn. 34:56–70, 1989.

Morey, A.L.; Nicolini, U.; Welch, C.R.; Economides, D.; Chamberlain, P.F. and Cohen, B.J.: Parvovirus b19 infection and transient fetal hydrops. The Lancet 337:496 only, 1991.

Mortimer, P.P.; Cohen, B.J.; Buckley, M.M.; Cradock–Watson, J.E.; Ridehalgh, M.K.S.; Burkhardt, F. and Schilt, U.: Human parvovirus and the fetus. Lancet 2:1012 only, 1985.

Schwarz, T.F.; Roggendorf, M.; Hottentrager, B.; Deinhardt, F.; Enders, G.; Gloning, K.P.; Schramm, T. and Hansmann, M.: Human parvovirus B19 infection in pregnancy. Lancet 2:566–567, 1988.

Van Elsacker–Niele, A.M.W.; Salimans, M.M.M.; Weiland, H.T.; Vermey–Keers, C.; Anderson, M.J. and Versteeg, J.: Fetal pathology in human parvovirus b19 infection. Br. J. Obstet. Gynaecol. 96(7):768–775, 1989.

Weiland, H.T.; Vermey–Keers, C.; Salimans, M.M.; Fleuren, G.J. and Verwey, R.A.: Parvovirus B19 associated with fetal abnormality. Lancet March 1:682–683, 1987.

1869 Passiflora incarnata Extract

Hirakawa et al. (1981) fed up to 400 mg per kg to rats on days 7–17 and found no adverse fetal effects.

Hirakawa, T.; Suzuki, T.; Sano, Y.; Kamata, T. and Nakamura, M.: Reproductive studies of Passiflora incarnata extract. Teratological study. Kiso to Rinsho 15:3431–3451, 1981.

1870 Patulin CAS 149–29–1

This mycotoxin was given in doses of up to 2.0 mg per kg on days 6 through 17 to mice and although fetal weight reduction occurred, no teratogenicity was detected (Reddy et al., 1978).

Reddy, C.S.; Chan, P.K. and Hayes, A.W.: Teratogenicity and dominant lethal studies of patulin in mice. Toxicology 11:219–223, 1978.

1871 PCB

Discussed under Kanechlor

1872 Penbutolol Sulfate CAS 38363–32–5

This β adrenergic blocker was administered orally to mice at doses of 6 to 60 mg per day. Dosing before mating, during organogenesis or on the last portion of gestat'on had no adverse effect on the fetuses. Behavioral studies were also negative (Sugisaki et al., 1981).

Sugisaki, T.; Takagi, S.; Seshimo, M.; Hayashi, S. and Miyamoto, M.: Reproductive studies of penbutolol sulfate given orally to mice. Oyo Yakuri 22:289–305, 1981.

1873 Penfluridol 4–(4–chloro,α,α,α–trifluoro–m–tolyl)–1–(4,4,bis(p–fluorophenyl) butyl)–4–piperidinol CAS 26864–56–2

Asano et al. (1979) gave orally 0.5 to 8 mg per kg to rats on days 15 through 20 of gestation. At doses of 2 mg per kg and above an increase in stillbirths and a decrease in body weight and neonatal survival were found.

Asano, Y.; Ariyuki, F. and Higaki, K.: Peri and postnatal studies of penfluridol (TLP–607) in rats. Oyo Yakuri 17:849–857, 1979.

1874 D–Penicillamine CAS 52–67–5

A single case report of a newborn with signs of Ehlers–Danlos syndrome was recorded following administration of large doses of penicillamine (2,000 mg per day) to a mother with cystinuria (Mjolnerod et al., 1971). Solomon et al. (1977) reported that a woman with rheumatoid arthritis receiving 900 mg per day until the 16th week of gestation gave birth to a 2,000 gm infant with lax skin, inguinal hernias and flexion contractures of the knees and hips. The infant died of an undiagnosed abdominal obstruction. Rosa (1986) has summarized the 5 case reports of cutis laxa associated with animal and human teratogenicity. Roubenoff et al. (**1988**) have also summarized animal and human teratology.

Scheinberg and Sternlieb (1975) summarized 29 pregnancies in women being treated for Wilson's disease with approximately 1000 mg of penicillamine per day and they reported no adverse fetal effects. The possibility was raised that in Wilson's disease the effective penicillamine exposure is reduced due to its loss in the urine.

Merker et al. (1975) observed skeletal alterations in rat fetuses whose mothers received 200 mg per animal intraperitoneally on days 15 through 19. Rib fusions, incomplete mineralization and swelling of collagen fibrils were found. Kilburn and Hess (1982) gavaged rats on days 9–14 of gestation with 250 mg per kg and found tracheobronchomegaly in 40 percent of the offspring. Yamada et al. (1982) found no increase in defects after giving 500 mg per kg orally to rats.

Kilburn, K.H. and Hess, R.A.: Neonatal deaths and pulmonary dysplasia due to d–penicillamine in the rat. Teratology 26:1–9, 1982.

Merker, H–J.; Franke, L. and Gunther, T.: The effect of D–penicillamine on skeletal development of rat foetuses. Naunyn Schmiedbergs Arch. Pharm. 287:359–376, 1975.

Mjolnerod, O.K.; Rassmussen, K.; Dommerud, S.A. and Gjeruldsen, S.T.: Congenital connective–tissue defect probably due to penicillamine treatment in pregnancy. Lancet 1:673–675, 1971.

Rosa, F.W.: Teratogen update: penicillamine. Teratology 33:127–131, 1986.

Roubenoff, R.; Hoyt, J.; Petri, M.; Hochberg, M.C. and Hellmann, D.B.: Effects of antiinflammatory and immuno-suppressive drugs on pregnancy and fertility. Seminars in Arthritis and Rheumatism 18:88–110, 1988.

Scheinberg, I.H. and Sternlieb, I.: Pregnancy in penicillamine–treated patients with Wilson's disease. N. Eng. J. Med. 293:1300–1302, 1975.

Solomon, L.; Abrams, G.; Dinner, M. and Berman, L.: Neonatal abnormalities associated with D–penicillamine treatment during pregnancy. N. Eng. J. Med. 296:54–55, 1977.

Yamada, T.; Otome, S.; Tanaka, Y.; Sasajima, M. and Ohzeki, M.: Reproductive studies of d–penicillamine in rats. Fertility study. Pharmacometrics 18:553–560, 1982.

1875 Penicillic Acid CAS 90–65–3

Hayes and Hood (1978) injected mice intraperitoneally with 30 or 50 mg per kg on single days between day 7–10. No adverse effects were noted. At 90 mg per kg, most of the dams died.

Hayes, A.W. and Hood, R.D.: Effects of prenatal administration of penicillic acid and penitrem A to mice. Toxicon. 16:92–96, 1978.

1876 Penicillin CAS 1406–05–9

Heinonen et al. (1977) found no increase in defect rate among 3,546 pregnancies where there was treatment with penicillin derivatives during the first 4 lunar months. Kullander and Kallen (1976) found no increase in penicillin exposure among 194 infants with malformations. Stoll et al. (1982) gave intramuscular penicillin to 453 women with gonorrhea and detected no adverse fetal effects.

Boucher and Delost (1964) failed to produce defects in the mouse fetus after giving the mother 50 or 500 units per gm on the 14th gestational day. They found that the treated fetuses grew more rapidly in early postnatal life. Brown et al. (1968) maintained pregnant rabbits on penicillin G or V (100 mg per kg per day) and found no abortions or evidence of teratogenic action.

Boucher, D. and Delost, P.: Developpement post–natal des descendents issus de meres traitees par la penicilline au cours de la gestion chez la souris. C. R. Soc. Biol. (Paris) 158:528–532, 1964.

Brown, D.M.; Harper, K.H.; Palmer, A.K. and Tesh, S.A.: Effects of antibiotics upon pregnancy in the rabbit. (abs) Toxicol. Appl. Pharmacol. 12:295 only, 1968.

Heinonen, O.P.; Slone, D. and Shapiro, S.: Birth Defects and Drugs in Pregnancy, Publishing Sciences Group Inc., Littleton, Mass., 1977.

Kullander, S. and Kallen, B.: Prospective study of Drugs and Pregnancy. 4. Miscellaneous drugs. Acta Obstet. Gynecol. Scand. 55:287–295, 1976.

Stoll, B.J.; Kanto, Jr., W.P.; Glass, R.I. and Pushkin, Joan: Treated maternal gonorrhea without adverse effect on outcome of pregnancy. Southern Medical Journal 75:1236–1238, 1982.

1877 Penitrem A CAS 12627–35–9

Hayes and Hood (1978) injected mice intraperitoneally on single days on days 7–10 of gestation. At doses of 2 or 3 mg per kg some increase in fused ribs or malformed verte-brae occurred.

Hayes, A.W. and Hood, R.D.: Effects of prenatal administration of penicillic acid and penitrem A to mice. Toxicon. 16:92–96, 1978.

1878 Pentabromotoluene

Ruddick et al. (1984) gave up to 600 mg per kg orally to rats during organogenesis and found no adverse fetal effects.

Ruddick, J.A.; Black, W.D.; Villeneuve, D.C. and Valli, V.E.: A teratological evaluation of pentachloro–and pentabromotoluene following oral treatment in the rat. (abs) Teratology 29:56A only, 1984.

1879 Pentacaine

Ujhazy et al. (1989) gavaged rabbits on days 6–20 with doses up to 10 mg per kg daily and found no adverse fetal or embryonic effects.

Ujhazy, E.; Zeljenkova, D.; Balonova, T.; Nosal, R.; Chalupa, I.; Blasko, M. and Siracky, J.: teratologicka a cy-togeneticka studia loalneho anestetika penainu na ralikoch. Cesk. Fysiol. 38:278–284, 1989.

1880 Pentachloroaniline

Courtney et al. (1976) gave up to 200 mg per kg orally on days 7–18 to mice and found some fetal mortality increase but no defective fetuses. The dams gained only 2–3 gm during the treatment.

Courtney, K.D.; Copeland, M.F. and Robbins, A.: The effects of pentachloronitrobenzene, hexachlorobenzene, and related compounds on fetal development. Toxicol. Appl. Pharmacol. 35:239–256, 1976.

1881 Pentachloroanisole

Welsh et al. (1987) exposed rats during mating and pregnancy to 4, 12 and 41 mg in the diet. At the highest dose embryolethality was found and the number of corpora lutea was decreased. No teratogenicity was found.

Welsh, J.J.; Collins, T.F.X.; Black, T.N.; Graham, S.L. and O'Donnell, M.W., Jr.: Teratogenic potential of purified pentachlorophenol and pentachloroanisole in subchron-ically exposed Sprague–Dawley rats. Food Chem. Toxicol. 25:163–172, 1987.

1882 Pentachloronitrobenzene *Quintozene* CAS 82–68–8

Jordan et al. (1975) administered this compound orally to pregnant rats during the active period of organogenesis in doses up to 125 mg per kg per day and found no fetal tox-icity or teratogenicity. Khera and Villeneuve (1975) fed rats up to 200 mg per kg on days 6–15 and observed at the high-est dose reduced fetal weight. A dose related increase in mi-nor skeletal anomalies occurred.

Jordan, R.L.; Sperling, F.; Klein, H.H. and Borzelleca, J.F.: A study of the potential teratogenic effects of pentachloronitrobenzene in rats. Toxicol. Appl. Pharmacol. 33:222–230, 1975.

Khera, K.S. and Villeneuve, D.C.: Teratogenicity studies on halogenated benzenes (pentachloro–, pentachloronitro–, and hexabromo–) in rats. Toxicol. Appl. Pharmacol. 33:125, 1975.

1883 Pentachlorophenol CAS 87–86–5

Schwetz et al. (1974) administered orally 5, 15, 30 and 50 mg per kg to rats on days 6–15. No effect was seen at 5 mg but embryolethality and toxicity occurred at 15 mg and higher. Hinkle (1973) found resorptions and embryotoxicity at oral levels of 1.25 to 20 mg per kg in the hamster. Small amounts have been shown to cross the placenta (Larsen et al., 1975). Welsh et al. (1987) exposed rats during mating and pregnancy to 4, 13 and 43 mg per kg in the diet. At 43 mg per kg the maternal lethality occurred and at lower doses fetal weight reduction was seen as well as skeletal variations.

Hinkle, D.K: Fetotoxic effects of pentachlorophenol in the golden Syrian hamster. Toxicol. Appl. Pharmacol. 25:455 only, 1973.

Larsen, R.V.; Boin, G.S.; Kessler, W.V.; Shaw, S.M. and Von Sickle, D.C.: Placenta transfer and teratology of pentachlorophenyl in rats. Environ. Lett. 10:121–128, 1975.

Schwetz, B.A.; Keeler, P.A. and Gehring, P.J.: The effect of purified and commercial grade pentachlorophenol on rat embryonal and fetal development. Toxicol. Appl. Pharmacol. 28:151–161, 1974.

Welsh, J.J.; Collins, T.F.X.; Black, T.N.; Graham, S.L. and O'Donnell, M.W., Jr.: Teratogenic potential of purified pentachlorophenol and pentachloroanisole in subchronically exposed Sprague–Dawley rats. Food Chem. Toxicol. 25:163–172, 1987.

1884 Pentachlorotoluene

Ruddick et al. (1984) gave up to 300 mg per kg orally to rats during organogenesis and found no adverse fetal effects.

Ruddick, J.A.; Black, W.D.; Villeneuve, D.C. and Valli, V.E.: A teratological evaluation of pentachloro–and pentabromotoluene following oral treatment in the rat. (abs) Teratology 29:56A, 1984.

1885 Pentaerythritol Tetranicotinate

Sugawara et al. (1977) gave up to 1000 mg orally to rabbits during organogenesis and found no adverse fetal effects.

Sugawara, T.; Uchiyama, K.; Asano, K. and Asano, O.: Toxicological studies on pentaerythritol tetranicotinate (SK–1). VII. Studies on teratogenicity test of SK–1 in rabbits. Oyo Yakuri 14:903–911, 1977.

1886 Pentamidine

Harstad et al. (**1990**) gave intravenous doses of this anti–pneumocystis drug to rats for various periods during gestation. On days 6–11 at a dose of 4 mg per kg resorptions were increased. No malformation increase was found with

doses up to 20 mg per kg. Lower weight gain in the dam was found at 20 mg per kg.

Harstad, T.W.; Little, B.B.; Bawdon, R.E.; Knoll, K.; Roe, D. and Gilstrap III, L.C.: Embryofetal effects of pentamidine isethionate administered to pregnant Sprague–Dawley rats. Am. J. Obstet. Gynecol. 163:912–916, 1990.

1887 1–Pentanol

Nelson et al. (1988) exposed rats to vapor for seven hours daily during gestation. At the highest concentration, 14000 mg per meter cubed, no adverse reproductive effects were found. Limited maternal toxicity was seen.

Nelson, B.K.; Brightwell, W.S.; Khan, A.; Hoberman, A.M. and Krieg, E.F.: Teratological evaluation of 1–pentanol, 1–hexanol and 2–ethyl–1–hexanol administered by inhalation to rats. Teratology 37:479–480, 1988.

1888 Pentazocine CAS 359–83–1

Geber and Schramm (1975) gave hamsters a single subcutaneous dose of this anelgesic on day 8. At 98 and 400 mg per kg the rate of malformations was not significantly increased. However at 196 and 570 mg per kg, a significant increase in exencephaly and other defects was found.

Geber, W.F. and Schramm, L.C.: Congenital malformations of the central nervous system produced by narcotic analgesics in the hamster. Am. J. Obstet. Gynecol. 123:705–713, 1975.

1889 Pentobarbital CAS 76–74–4

Discussed under Barbituric Acid

1890 Pentoic Acid

Dawson (**1991**) using a frog embryo teratogenesis assay found that 50% of the survivors were malformed at 227 mg per liter. The main type of malformation was microcephaly and envolvement of the gut but edema occurred.

Dawson, D.A.: Additive incidence of developmental malformation for xenopus embryos exposed to a mixture of ten aliphatic carboxylic acids. Teratology 44:531–546, 1991.

1891 Pentoxifylline *1–(5–Oxohexyl)theobromine* CAS 6493–05–6

Sugisaki et al. (1981) administered this xanthine derivative intravenously in daily doses of up to 25 mg per kg on days 6 through 18 in the rabbit. No adverse fetal effects were found. Sugisaki et al. (1981) gave this xanthine vasodilator intravenously to mice before mating and during the first 7 days or on days 6–15 of gestation. Maximum doses of 50 mg failed to produce changes in fertility or in the fetuses.

Sugisaki, T.; Hayashi, S. and Miyamoto, M.: Teratological study of pentoxifylline in rabbits by the intravenous route. Oyo Yakuri 22:451–458, 1981.

Sugisaki, T.; Kitatani, T.; Seshimo, M.; Takagi, S.; Hayashi, S. and Miyamoto, M.: Reproductive studies of pentoxifylline: intravenous administration in mice. Kiso to Rinsho 15:1895–1916, 1981.

1892 Perathiepine

Jelinek et al. (**1967**) gave rats orally 20 mg per kg on days 10–17 and found no increase in defective fetuses.

Jelinek, P.V.; Zikmund, E. and Reichlova, R.: L'influence de quelques medicaments psychotropes sur le developpement du foetus chez le rat. Therapie 23:1429–1433, 1967.

1893 Pergolide Mesylate

Buelke–Sam et al. (**1991**,A) gavaged mice with up to 60 mg per kg on days 6–15. This long acting dopamine agonist is used in Parkinsonism and hyperprolactinemia. At the highest dose fetal weight was reduced but no teratogenicity was found. Startle amplitudes were decreased in postnatal males exposed to 60 mg per kg. Perinatal administration did not interfer with postnatal growth or fertility (Buelke–Sam et al., **1991** B). Wada et al. (**1994**) found no teratogenicity in rabbits given up to 0.5 mg per kg during organogenesis.

Buelke–Sam, J.; Byrd, R.A.; Johnson, J.A.; Tizzano, J.P. and Owen, N.V.: Developmental toxicity of the dopamine agonist pergolide mesylate in cd–1 mice. I: Gestational exposure. Neurotoxicology and Teratology 13:283–295, 1991.

Buelke–Sam, J.; Cohen, I.R.; Tizzano, J.P. and Owen, N.V.: Developmental toxicity of the dopamine agonist pergolide mesylate in cd–1 mice. II: Perinatal and postnatal exposure. Neurotoxicology and Teratology 13:297–306, 1991.

Wada, K.; Nagao, T.; Mizutani, M. and Kawai, M.: Teratogenicity study of perfolide mesylate by oral administration in rabbits. Iyakuhin Kenkyu 25:147–155, 1994.

1894 Perflavon CAS 1715–55–5

This drug used in humans for angina was given by Ito et al. (1972) to pregnant mice and rats in amounts up to 250 mg per kg per day orally. At the highest dose the mice had delayed parturition with dead fetuses. No teratogenicity was found.

Ito, R.; Kawamura, H.; Tokoro, Y.; Tosaka, K.; Nakagawa, S.; Toida, S.; Matsuura, S.; Ozaki, M. and Hiyama, T.: 1,3–dimethylxanthine–7–acetic acid–7(β–dimethyl) (-amino–ethoxy) flavon (perflavon). J. Med. Soc. Toho Japan 19:116–125, 1972.

1895 Perfluorodecanoic Acid

Harris and Birnbaum (**1989**) gavaged mice on days 10–13 or 6–15 with up to 32 or 12.8 mg per kg. At the highest doses maternal toxicity and reduced viability and body weight of the fetuses was found. No congenital defect increase was seen.

Harris, M.W. and Birnbaum, L.S.: Developmental toxicity of perfluorodecanoic acid in c57bl/6n mice. Fundamental and Applied Toxicology 13:723–726, 1989.

1896 Pericyazine

2–Cyanno–10–[3–(4–hydroxy–piperidine)propyl] pheothiazine Neuleptil

No significant abnormalities were observed in rat offspring given up to 5 mg per kg orally during gestation.

Anonymous: Neurolpetil. Rx Bull 1:5–8, 1970.

1897 Perindopril

Harada et al. (**1994**) gave this ACE inhibitor to rats before pregnancy and during the first week, during organogenesis and in the peri and postnatal period. Oral doses of up to 16 mg per kg in rats and up to 10 mg per kg in rabbits. No teratogenicity was reported but at the highest dosage in rats impaired weight gain and decreased locomotion of the pups was found. Viability in the rat pups was only 43 percent at the time of weaning.

Harada, S.; Tawara, K.; Autissier, C.; Osterburg, I. and Korte, R.: Reproductive toxicity studies of perindopril in rats and rabits. Yakuri to Chiryo 22:1729–1734, 1994.

1898 Perisoxal

3–(1–hydroxy–2–piperidinoethyl)–5–phenylisoxazole Citrate

Hasegawa et al. (**1972**) studied this analgesic in pregnant rats and mice using an oral dose of up to 150 mg per kg during organogenesis. No teratogenic findings occurred but fetal growth retardation was found in mice exposed to the highest doses.

Hasegawa, Y.; Yoshida, T.; Kozen, T.; Ohara, T.; Sakaguchi, I.; Okamoto, A.; Matsuyama, T. and Minesita, T.: Studies on 5–aminoalkyl–and 3–aminoalkyliosoxazoles and related derivatives. Annual Report of Shionogi Research Laboratory 22:22–120, 1972.

1899 Permethrin *Pyrethrin*

This ester of pyrethrin was studied in rats by Spencer and Berhane (**1982**) who dosed with up to 4,000 ppm and found no decrease in implantation sites or live fetuses.

Spencer, F. and Berhane, Z.: Uterine and fetal characteristics in rats following a post–implantational exposure to permethrin. Bull. Environm. Contam. Toxicol. 29:84–88, 1982. Toxicology 12:442–448, 1989.

1900 Perphenazine CAS 58–39–9

Heinonen et al. (**1977**) did not find an increase in congenital defect rate among 63 women who took the drug in the first four lunar months.

This phenothiazine neuroleptic was administered by gavage to pregnant rats during days 7 through 14 in doses of 20 to 150 mg per kg (Druga, 1976). Significant increases in cleft palate, retrognathia and micromelia occurred at all dose levels. Single doses of 90 mg per kg on days 9–12 also produced defects, compared to a chemical derivative, methophenazine, the drug was more teratogenic in the rat.

Druga, A.: The effect of perphenazine treatment during the organogenesis in rats. Acta Biol. Acad. Sci. Hung. 27:15–23, 1976.

Heinonen, O.P.; Slone, D. and Shapiro, S.: Birth Defects and Drugs in Pregnancy. Publishing Sciences Group Inc., Littleton, Mass., 1977.

1901 Pertussis Vaccine *Typhoid Vaccine*

When subteratogenic doses of cytochalasin D were given with whole cell Pertussis vaccine to mice during early organogenesis, exencephaly was produced (Au–Jensen and Heron, 1987). Experiments with purified vaccines indicated that lipopolysaccharides in the pertussis and typhoid vaccine were the factors causing the teratogenicity. The authors did not believe that hyperthermia was associated.

Au–Jensen, M. and Heron, I.: Synergistic teratogenic effect produced in mice by whole cell pertussis vaccine. Vaccine 5:215–218, 1987.

1902   Phenacetin CAS 62–44–2

Heinonen et al. (1977) found no increase in the defect rate among 5,546 exposed pregnancies.

Heinonen, O.P.; Slone, D. and Shapiro, S.: Birth Defects and Drugs in Pregnancy. Publishing Sciences Group Inc., Littleton, Mass., 1977.

1903   Phenazepam 7–Brom–t(o–chlorphenyl)–1,2–dehydro–3H–1,4–benziazepin–2–OH CAS 51753–57–2

Smolnikova and Strekalova (1980) showed that this drug (100 mg per kg taken daily during the entire pregnancy) did not damage fetal development but caused some behavioral abnormalities in the offspring. Different doses were also tested in pregnant mini–pigs, dogs and guinea pigs but no congenital malformations were observed (Lyubimov et al., 1979a,b).

Lyubimov, B.I.; Smolnikova, N.M.; Strekalova, S.N.; Boiko, S.S.; Yavorsky, A.N.; Dushkin, V.A. and Poznakhirev, P.R.: Study of the embryotropic action of phenazepam in mini pigs, a new species of experimental animals. Byull. Eksp. Biol. Med. (USSR) 88(11):557–560, 1979a.

Lyubimov, B.I.; Smolnikova, N.M.; Strekalova, S.N.; Kurochkin, I.G.; Mitrofanov, V.S.; Porfirieva, R.P.; Markin, V.A. and Sharow, P.A.: Preclinical trials of the new tranquilizer phenazepam safety. Farmakol. Toksikol. (USSR) 42,5:464–467, 1979b.

Smolnikova, N.M. and Strekalova, S.N.: Development of the progeny in antenatal exposure to phenazepam. Farmakol. Toksikol. (USSR) 43(3):293–302, 1980.

1904   Phenazopyridine

Jick et al. (1981) and Heinonen el al (1977) reported that among over 500 women who took this medication during pregnancy there were no increase in congenital disorders.

Heinonen, O.P.; Slone, D.; Shapiro, S.: Birth Defects and Drugs in Pregnancy. Littleton, Massachusetts: Publishing Sciences Group, Inc., pp 308, 486, 1977.

Jick, H.; Holmes, L.B.; Hunter, J.R.: First–trimester drug use and congenital disorders. JAMA 246:343–346, 1981.

1905   Phencyclidine PCP CAS 77–10–1

Wachsman et al. (1989) followed 57 infants exposed in utero. Sixty–five percent had withdrawal symptoms. Many were exposed to narcotics also. Two infants were referred to the genetic department because of possible dysmorphy but

no consistent clinical pattern was found. The follow up rate was only 57% by the 4th month. At one year 24% were less than the 10th percentile for weight and 29% were less than the 10 percent for head circumference.

Golden et al. (1984) found evidence of phenycyclidine use in 0.8 percent of pregnancies but no fetal outcome was reported. Chasnoff et al. (1983) observed that the offspring of seven women who abused the drug had outbursts of agitation and rapid changes in consciousness. No significant changes in the Bayley Scales were found at three months of age.

Jordan et al. (1978) injected 25 to 30 mg per kg into rats of gestational days 6 through 15 and produced skeletal dysplasias and cleft palates in the offspring. Marks et al. (1980) gavaged mice on days 6–15 with 60 to 120 mg per kg. At the highest level some maternal mortality occurred and a 6 percent rate of defective offspring was found. Nickola and Schreiber (1983) found growth retardation and delay in reflexes in the offspring of mice exposed to 5 to 20 mg per kg.

Chasnoff, I.J.; Burns, W.J.; Hatcher, R.P. and Burns, K.A.: Phencyclidine: Effects on fetus and neonate. Dev. Pharmacol. Ther. 6:404–408, 1983.

Golden, N.L.; Kahnert, R.R.; Sokoll, R.J.; Martier, S. and Bagby, B.S.: Phencyclidine use during pregnancy. Am. J. Obstet. Gynecol. 148:254–259, 1984.

Jordan, R.L.; Young, T.R. and Harry, G.J.: Teratology of phencyclidine in rats: Preliminary results. (abs) Teratology 17:40A only, 1978.

Marks, T.A.; Worthy, W.C. and Staples, R.E.: Teratogenic potential of phencyclidine in the mouse. Teratology 21:241–246, 1980.

Nickola, J.M. and Schreiber, E.C.: Phencyclidine exposure and the developing mouse. Behavioral teratological implications. Teratology 28:319–326, 1983.

Tonge, S.R.: Neurochemical teratology: 5–Hydroxyindole concentrations in discrete areas of the rat brain pre– and neonatal administration of phencyclidine and imipramine. Life Sci. 12:481–486, 1972.

Wachsman, L.; Schuetz, S.; Chan L.S. and Wingert, W.A.: What happens to babies exposed to phencyclidine (PCP) in utero? Am. J. Drug Alcohol Abuse 5(1):31–39, 1989.

1906   Phenelzine β–Phenylethylhydrazine Hydrogen Sulfate CAS 51–71–8

Heinonen et al. (1977) reported three defects among 21 women taking monoamine oxidase inhibitors. Three of these women took phenelzine in the first four lunar months.

This monoamine oxidase inhibitor decreases implantation in mice when given during the first six days of gestation in amounts of 25 mg per kg per day (Poulson and Robson, 1964). Robson et al. (1971) studied a large number of phenelzine derivatives (designated by WL, LON and a number) in mice subcutaneously. Many of the derivatives delayed implantation but teratologic effects were not found.

Heinonen, O.P.; Slone, D. and Shapiro, S.: Birth Defects and Drugs in Pregnancy. Publishing Sciences Group Inc., Littleton, Mass., 1977.

Poulson, E. and Robson, J.M.: Effect of phenelzine and some related compounds on pregnancy and on sexual development. J. Endocrinol. 30:205–215, 1964.

Robson, J.M.; Sullivan, F.M. and Wilson, C.: The maintenance of pregnancy during the preimplantation period in mice treated with phenelzine derivatives. J. Endocrinol. 49:635–648, 1971.

1907 Phenformin

This drug is used in Europe to treat non–insulin dependent diabetes. Denno and Sadler (**1994**) exposed mouse embryos in culture to 2.5 x 10^{-5} Nitric O8[3] to 0.3 mg per ml and produced a concentration dependent increase in neural tube defects and craniofacial defects. The blood levels in humans range from 2.8 x 10^{-5} to 1.14 x 10^{-4} mg per ml.

Denno, K.M. and Sadler, T.W.: Effects of three biguanide class of oral hypoglycemic agents on mouse embryogenesis. Teratology 49:260–266, 1994.

1908 Phenglutarimide $Aturban^{TM}$
2–(2–Diethylaminoethyl)2–phenylglutarimide CAS 1156–05–4

Tuchmann–Duplessis and Mercier–Parot (1964) administered 50 to 60 mg per kg to rats, mice and rabbits during organogenesis and produced no defective fetuses.

Tuchmann–Duplessis, H. and Mercier–Parot, L.: Action sur la gestation et le developpement foetal d'un derive glutarimique, l aturbane. C. R. Acad. Sci. (Paris) 258:2666–2669, 1964.

1909 Phenmetrazine $Preludin^{TM}$ CAS 134–49–6

Milkovich and Van den Berg (1977) did a prospective study of 406 children exposed in utero to phenmetrazine and found no increase in serious defects. Heinonen et al. (1977) reported no increase in congenital defect rate among 58 women who took the drug during the first four months.

Heinonen, O.P.; Slone, D. and Shapiro, S.: Birth Defects and Drugs in Pregnancy. Publishing Sciences Group Inc., Littleton, Mass., 1977.

Milkovich, L. and Van den Berg, B.J.: Effects of antenatal exposure to anorectic drugs. Am. J. Obstet. Gynecol. 129:637–642, 1977 also see *Barbituric Acid* CAS

1910 Phenobarbital also see *Barbituric Acid* CAS 50–06–6

Robert et al. (1986) identified 40 pregnancies during which the drug was used alone and found one offspring with ventricular septal defect and another with hypospadias. Other studies are detailed under Barbituric Acid and Derivatives.

McColl et al. (1963) found double vertebral centra in the offspring of mice fed 0.16 percent phenobarbital. This skeletal finding could be due to a nutritional effect causing delayed ossification rather than representing a true congenital defect. McColl et al. (1967) reported skeletal and aortic arch defects in the offspring of rabbits treated with 50 mg per kg from days 8 through 16. Finnell et al. (1987) treated three strains of mice with 60, 120 or 240 mg per kg orally from

20 days before pregnancy and during the entire gestation. A dose response increase in fetal defects was found. The defects were dilated cerebral ventricles, failure of the testes to descend or other malformations such as cleft palate and renal defects.

Gupta and Yaffe (1981) administered 40 mg per kg subcutaneously to pregnant rats during the last several days of gestation. The female offspring had delay in onset of puberty, disorders of the estrus cycle and a 50 percent infertility rate.

Finnell, R.H.; Shields, H.E.; Taylor, S.M. and Chernoff, G.F.: Strain differences in phenobarbital–induced teratogenesis in mice. Teratology 35:177–185, 1987.

Gupta, C. and Yaffe, S.J.: Reproductive dysfunction in female offspring after prenatal exposure to phenobarbital: Critical period of action. Pediatr. Res. 15:1488–1491, 1981.

McColl, J.D.; Globus, M. and Robinson, S.: Drug induced skeletal malformations in the rat. Experientia 19:183–184, 1963.

McColl, J.D.; Robinson, S. and Globus, M.: Effect of some therapeutic agents on the rabbit fetus. Toxicol. Appl. Pharmacol. 10:244–252, 1967.

Robert, E.; Lojkvist, E.; Mauguiere, F. and Robert, J.M.: Evaluation of drug therapy and teratogenic risk in a rhone–alpes district population of pregnant epileptic women. Eur. Neurol. 25:436–443, 1986.

1911 Phenol CAS 108–95–2

Minor and Becker (1971) injected rats intraperitoneally on days 8–10 or 11–13 with up to 200 mg per kg. No adverse fetal effects were found. Chapman et al. (**1994**) found that at a concentration of 100 micromoles cultured rat embryos had reduced protein content and at 10 and 50 micromolar concentrations the prosencephalic measures were significantly reduced. The effect was seen only when rat hepatic microsomes were used in conjunction with the phenol.

Gray and Kavloc (1990) fed $C14^{v-1}$ labelled phenol to rats and determined that the levels in placenta and embryo were equivalent to maternal serum.

Chapman, D.E.; Namkung, M.J. and Juchau, M.R.: Benzene and benzene metabolites as embryotoxic agents: Effects on cultured rat embryos. Toxicology and Applied Pharmacology 128:129–137, 1994.

Gray, J.A. and Kavclock, R.J.: A pharmacokinetic analysis of phenol in the pregnant rat: deposition in the embryo and maternal tissues. (abst)Teratology 41:561, 1990.

Minor, J.L. and Becker, B.A.: A comparison of the teratogenic properties of sodium salicylate, sodium benzoate and phenol. (abs) Toxicol. Appl. Pharmacol. 19:373 only, 1971.

1912 Phenolphthalein

Heinonen et al. (1977) reported no increase in defects among offspring of 236 women who took the laxative during the first four lunar months. Similar findings were reported for 806 women who took it anytime during pregnancy.

Heinonen, O.P.; Slone, D. and Shapiro, S.: Birth Defects and Drugs in Pregnancy. Publishing Sciences Group Inc., Littleton, Mass. 1977.

1913 Phenols, Para–Substituted

Oglesby et ak (**1992**) carried out invitro rat embryo cultures of 12 para–substituted phenols. These included p–amino, p–bromo, p–chlore, p–cyano, p–fluoro, p–n–heptyloxy, p–hydroxy, p–iode, p–methoxy, p–methyl, p–nitro and p–n–pentyloxy substitutes. They compared their results with and without hepatocyte activation to in vivo animal studies. In subsequest work Fisher et al. (**1993**) studied the uptake kinetics and structure activity relations of these chemicals in whole embryo culture.

Fisher, H.L.; Sumler, M.R.; Shrivastava, S.P.; Edwards, B.; Oglesby, L.A.; Ebron-McCoy, M.T.; Copeland, F.; Kavlock, R.J. and Hall, L.L.: Toxicokinetics and structure–activity relationships of nine para–substitited phenols in rat embryos in vitro. Teratology 48:285–297, 1993.

Oglesby, L.A.; Ebron–McCoy, M.T.; Logsdon, T.R.; Copeland, F.; Beyer, P.E. and Kavlock, R.J.: In vitro embryotoxicity of a series of para–substituted phenols: structure, activity, and correlation with in vivo data. Teratology 45:11–33, 1992.

1914 Phenothiazines *Acetylpromazine Prochlorpromazine Trimeprazine Thioridazine*

This group of antipsychotic and antinauseant drugs consists of over two dozen structurally related chemicals. Rumeau–Rouquette et al. (1977) studied prospectively 12,764 pregnancies and observed eleven malformed children from 304 women who took phenothiazines during the first 3 months. The rate in this group was 3.5 percent which was significantly higher than the 1.6 percent found in mothers not taking the drugs. The three–carbon side chained phenothiazines (chlorpromazine, methotrimeprazine, trimeprazine and oxomemazine) were taken by 133 mothers and 8 malformed infants occurred in the offspring. Other forms with two–carbon, piperazine or piperidine side chains were taken by fewer women but were not associated with significant increases in defects.

In another larger group of pregnancies in which phenothiazines were taken during the first four lunar months, 66 malformed infants were found among 1,309 exposed women (Heinonen et al., 1977). Of this large group, 877 took prochlorperizine. When the exposed group was matched carefully for parity, age and social class, no significant difference from the controls was found. It is of interest that this study found approximately a 5.0 percent defect rate in both control and treated groups while the study reported above found 3.5 percent in the exposed and 1.6 percent in the control group. This study was also reported by Slone et al. (1977). Milkovich and Van den Berg (1976) studied 543 pregnancies exposed to phenothiazines and 433 exposed to prochlorthiazine and found no significant increases in serious congenital malformations in the offspring at birth, one year or at 5 years. Mellin (1975), in another study, did not identify an increased defect rate in 74 women who took prochlorperazine. The general subject was reviewed by Nahas and Goujard (1979).

McElhatton (**1992**) comprehensively reviewed the teratogenicity studies of phenothiazines as well as data on breast milk exposure. Based on literature reporting some 600 pregnancies in psychotic women no increased tendency for malformations or decreased IQ was found. Seven cases treated with high doses throughout pregnancy and at the time of delivery had extrapyramidal signs which persisted during the early post-natal months.

Animal studies and some additional clinical data on these agents are reported under their individual headings (Chloromazine, Prochlorperazine, Promazine, Methophenazine and Trifluoperazine). The role of the piperazine ring in teratogenicity in the rat was studied by Druga et al. (1980).

McElhatton (**1992**) reviewed the reproductive effects of several phenothiazines and did not identify any significant and reproduceable ill effects. She points out that the risks to the fetus of mothers who are not treated for psychotic episodes may be higher.

No congenital anomalies were found among 23 infants born to women treated with thioridazine in the first trimester (Scanlan, **1972**).

Druga, A.; Nyitra, M. and Szaszovszky, E.: Experimental teratogenicity of structurally similar compounds with or without piperazine–ring: A preliminary report. Pol. J. Pharmacol. Pharm. 32:199–204, 1980.

Heinonen, O.P.; Slone, D. and Shapiro, S.: Birth Defects and Drugs in Pregnancy. Publishing Sciences Group Inc., Littleton, Mass., 1977.

McElhatton, P.R.: The use of phenothiazines during pregnancy and lactation. Reproductive Toxicology 6:745–490, 1992.

McElhatton, P.R.: The use of phenothiazines durring pregnancy and lactation. Reproductive Toxicology 6:475–490, 1992.

Mellin, G.W.: Report of prochlorperazine utilization during pregnancy from fetal life study data bank. (abs) Teratology 11:28A, 1975.

Milkovich, L. and Van den Berg, B.J.: An evaluation of the teratogenicity of certain antinauseant drugs. Am. J. Obstet. Gynecol. 125:244–248, 1976.

Nahas, G. and Goujard, J.: Phenothiazines, benzodiazepines, and the fetus. In: Reviews in Perinatal Medicine, E.M. Scarpelli and E.V. Cosmi (eds.), Raven Press, New York, 243–280, 1979.

Rumeau–Rouquette, C.; Goujard, J. and Huel, G.: Possible teratogenic effect of phenothiazines in human beings. Teratology 15:57–64, 1977.

Scanlan, F.J.: The use of thioridazine (Melleril) during the first trimester. Med. J. Aust 1:1271–1271, 1972.

Slone, D.; Siskind, V.; Heinonen, O.P.; Monson, R.R.; Kaufman, D.W. and Shapiro, S.: Antenatal exposure to phenothiazines in relation to congenital malformations, perinatal mortality rate, birth weight, and intelligence quotient scores. Am. J. Obstet. Gynecol. 128:486–488, 1977.

1915 Phenoxyacetamide *B1420*

Lear et al. (1964) injected dogs with 40 mg per kg intravenously on alternate days among the first 30 days of gestation. One litter contained fetuses with phocomelia (2) and small bowel atresia (2). A third fetus had a urachal cyst. Another litter gave birth to 2 pups which developed severe anemia.

Lear, E.; Tangoren, G.; Chiron, A.E.; Pallin, I.M. and Allen, A.: New phenoxyacetamide systemic anesthetic, toxicity and clinical studies. New York State J. Med. 2177–2184, 1964.

1916 Phenoxyacetic Acid CAS 122–59–8

Hood et al. (1979) gavaged pregnant mice with 800–900 mg per kg on single days 8 through 15 or with 250–300 mg per kg on multiple days during organogenesis and observed no adverse fetal effects.

Hood, R.D.; Patterson, B.L.; Thacker, G.T.; Sloan, G.L. and Szczech, G.M.: Prenatal effect of 2,4,5–T, 2,4,5–trichlorophenol and phenoxyacetic acid in mice. J. Environ. Sci. Health C13:189–204, 1979.

1917 Phenoxybenzamine CAS 59–96–1

Hornblad et al. (1970) found that 1 mg intraperitoneal doses to full term guinea pigs produced wider diameters of their ductus arteriosus.

Hornsblad, P.Y.; Boreus, L.O. and Larsson, K.S.: Studies in closure of the ductus arteriosus 8. Reduced closure rate in guinea pigs treated with phenoxybenzamine. Cardiology 55:237–241, 1970.

1918 2–Phenoxyethanol *Ethylene Glycol Monophenyl Ether*

Scortichini et al. (1987) applied this chemical to the shaved skin of rabbits on days 6–18 at doses of up to 1000 mg per kg daily. Maternal toxicity and death occurred but no evidence of teratogenicity was found.

Scortichini, B.H.; Quast, J.F. and Rao, K.S.: Teratologic evaluation of 2–phenoxyethanol in New Zealand white rabbits following dermal exposure. Fund. Appl. Toxicol. 8:272–279, 1987.

1919 Phenoxy–isobutyric Acid Ethyl Ester CAS 18672–04–3

Amels et al. (1974) gave this anticholesterol compound to rats and mice at various periods during gestation. An oral dose for two days during early embryogenesis produced some fetuses with edema and hemorrhages in both species. The dose was 33 mg per kg, close to the presumptive human therapeutic dose.

Amels, D.; Fazekas–Todea, I. and Sandor, S.: The prenatal noxious effect of a blood cholesterin level lowering compound. Rev. Roum. Morphol. Embryol. 19:37–43, 1974.

1920 Phenylalanine CAS 63–91–2 *PKU Phenylketonuria*

The offspring of mothers with phenylketonuria showed a high frequency of mental retardation with microcephaly and intrauterine growth retardation (reviewed by Hsia, 1970; Frankenburg et al., 1968). Those mothers with hyperphenylalaninemia, but with serum phenylalanine levels below 15 mg per 100 ml during pregnancy, produced

11 normal infants out of a total of 12 (Hsia, 1970). Lenke and Levy (1980) reviewed 524 pregnancies in mothers with phenylketonuria. Among 34 women who were treated, there was a tendency for higher IQ and normal head circumference with treatment begun earlier in gestation. Lenke and Levy (1982) summarized 34 pregnancies of mothers with PKU. Treatment during the first trimester of 11 mothers did not prevent congenital heart disease which occurred in four. The offspring of fathers with PKU have not had increased malformations. (Fisch et al., **1991**) One normal pregnancy was reported in a women with phenylketonuria but without elevated serum phenylalanine (Dorland et al., **1993**).

Optimal results were achieved when treatment was started before conception (Rohr et al., 1987). Other anomalies associated with these children were congenital heart disease, dislocation of the hips and strabismus (Stevenson and Huntley, 1967). In a report of two families, they found eight out of ten children had congenital heart disease including coarctation of the aorta, patent ductus arteriosus and other types. Of 26 documented pregnancies, 16 (62 percent) terminated in abortion. Montenegro and Castro (1965) and Fisch et al. (1969) reported similar types of defects in the offspring of phenylketonuric mothers. Gandier et al. (1972) summarized the literature and reported 3 new cases. Lipson et al. (1984) reported on 34 children. They observed mental retardation was much more common (91 percent) in the offspring of mothers with serum phenylalanine over 1.2 mmol per L. Six of the 34 had congenital heart disease. Most were below the 10th percentile for height. A decrease in development of the face, especially the philtrum and simple ears was reported. Drogari et al. (1987) presented data supporting the necessity for dietary restriction before conception. All of their 17 pregnancies treated before conception had normal outcomes but of twenty–nine women on relaxed diets at conception (and then treated), the risk for microcephaly or defect was the same as those who were not treated. Waisbren and Levy (1990) reported that the maternal plasma level at birth was significantly related to the IQ of the offspring.

Aspartame sweetener does not significantly increase phenylalanine plasma levels in phenylketonuria heterozygotes (see Aspartame).

Maternal non–PKU mild hyperphenylalaminemia is believed to be benign. Levy et al. (**1994**) collected data on 86 mothers with phenylalanine levels of 167–715 micromoles per liter and found no increase in defects among the offspring of 219 untreated pregnancies. I.Q. averaged 100 with those who had higher levels and 108 for those with lower levels (less than 400 micromoles per liter). They concluded that maternal phenylalamine levels of less than 400 micromoles per liter did not warrant intervention.

Kerr et al. (1968) fed phenylalanine to pregnant rhesus monkeys and found mental retardation, but this was not associated with microcephaly or other congenital defects. They observed higher phenylalanine levels in the cord blood than in the maternal blood. Loo et al. (1983–1984) reported that either phenylacetate or p–chlorophenylalanine with L–phenylalanine given by subcutaneous infusion to maternal rats produced the PKU syndrome with retarded brain growth in offspring. Kronick et al. (1987) treated guinea pigs with p–chlorophenylalanine and phenylalanine during organogenesis and produced a fetal syndrome consistent with that in humans with untreated PKU. Suyama et al. (1989) studied a rat model system in which phenylalanine administration was associated with increased fetal heart defects. Vorhees and Berry (1989) presented evidence that the addition of large neutral fatty acids (valine, isoleucine and leucine) blocked the damaging effect of phenylalanine on the rat fetus. Denno and Sadler (**1990**) studied the effects of phenylalanine and its metabolites on whole mouse embryos in vitro. At 10 mM phenylalanine inhibited neural tube closure whereas at 0.1 mM phenylethylamine was inhibitory. Other metabolites (phenylactic acid, phenylpyruvic acid, phenyl acetic acid, 20 h–phenylacetic acid) were toxic at higher concentrations.

Denno, K.M. and Sadler, T.W.: Phenylalanine and its metabolites induce embryopathies in mouse embryos in culture. Teratology, 42:565–570, 1990.

Dorland, L.; Poll–The, B.T.; Duran, M.; Smeitink, J.A.M. and Berger, R.: Phenylpyruvate, fetal damage, and maternal phenylketonuria syndrome. The Lancet 341:1351–1352, 1993.

Drogari, E.; Beasley, M.; Smith, I. and Lloyd, J.K.: Timing of strict diet in relation to fetal damage in maternal phenylketonuria. Lancet 2:927–934, 1987.

Fisch, R.O.; Doeden, D.; Lansky, L.L. and Anderson, J.A.: Maternal phenylketonuria. Detrimental effects on embryogenesis and fetal development. Am. J. Dis. Child. 118:847–858, 1969.

Fisch, R.O.; Matalon, R.; Weisberg, S. and Michals, K.: Children of fathers with phenylketonuria: an international survey. The Journal of Pediatrics 118:739–741, 1991.

Frankenburg, W.K.; Duncan, B.R.; Coffelt, R.W.; Koch, R.; Coldwell, J.G. and Son, C.D.: Maternal phenylketonuria: Implications for growth and development. J. Pediatr. 73:560–570, 1968.

Gandier, B.; Ponte, C.; Duquennoy, G.; Callens, M. and Ballester, L.: Retard de croissance intra–uterin avec microcéphalie chez trois enfants nes de mere hyperphenylalaninemique. Ann. Pediatr. (Paris) 19:269–276, 1972.

Hsia, D.Y.: Phenylketonuria and its variants. In: Progress fn Medical Genetics. A.G. Steinberg and E.G. Bearn (eds.), New York: Grune and Stratton, 7:53–68, 1970.

Kerr, G.R.; Chamove, A.S.; Harlow, H.F. and Waisman, H.A.: 'Fetal PKU' the effect of maternal hyperphenylalaninemia during pregnancy in the rhesus monkey (Macaca mulatta). Pediatrics 42:27–36, 1968.

Kronick, J.B.; Whelan, D.T. and McCallion, D.J.: Experimental hyperphenylalanemia in the pregnant guinea pig: Possible phenylalanine teratogenesis and p–chlorophenylalanine embryotoxicity. Teratology 36:245–258, 1987.

Lenke, R.R. and Levy, H.L.: Maternal phenylketonuria and hyperphenylalaninemia. N. Eng. J. Med. 303:1202–1208, 1980.

Lenke, R.R. and Levy, H.L.: Maternal phenylketonuria: Results of dietary therapy. Am. J. Obstet. Gynecol. 142:548–553, 1982.

Levy, H.L.; Waisbren, S.E.; Lobbregt, D.; Allred, E.; Schuler, A.; Trefz, F.K.; Schweitzer, S.M.; Sardharwalla, I.B.; Walter, J.H.; Barwell, B.E.; Berlin, Jr., C.M. and Leviton, A.: Maternal mild hyperphenylalaninaemia: an international survey of offspring outcome. The Lancet 344:1589–1594, 1994.

Lipson, A.H.; Beuhler, B.; Bartley, J.; Walsh, D.A.; Yu, J.; O'Halloran, M. and Webster, W.S.: Maternal hyperphenylalaninemia fetal effects. J. Pediatr. 104:216–224, 1984.

Loo, Y.H.; Rabe, A.; Potempska, A.; Wang, P.; Fersko, R. and Wisniewski, H.M.: Experimental maternal phenylketonuria. An examination of the animal model. Dev. Neurosci. 6:227–234, 1983–1984.

Montenegro, J.E. and Castro, G.L.: Fenilcetonuria materna: Anomalieas en la descendencia. Acta Med. Venezolana 12:233–236, 1965.

Rohr, F.J.; Doherty, L.B.; Waisbren, S.E.; Bailey, I.V.; Ampola, M.G.; Benacerraf, B. and Levy, H.: New England maternal PKU project: Prospective study of untreated and treated pregnancies and their outcomes. J. Pediatr. 110:391–398, 1987.

Stevenson, R.E. and Huntley, C.C.: Congenital malformations in offspring of phenylketonuric mothers. Pediatr. 40:33–45, 1967.

Suyama, I.; Tani, M.; Matsumura, M.; Isshiki, G.; Okano, Y.; Oura, T. and Nishimura, K.: Fetal heart malformations in experimental hyperphenylalaninemia in pregnant rats. Cong. Anom. 29:15–30, 1989

Vorhees, C.V. and Berry, H.K.: Branched chain amino acids improve complex maze learning in rat offspring prenatally exposed to hyperphenylalaninemia: Implications for maternal phenylketonuria. Pediatr. Res. 25:568–572, 1989.

Waisbren, S.E. and Levy, H.L.: Effects of untreated maternal hyperphenylalaninemia on the fetus: further study of families identified by routine cord blood screening. Journal of Pediatrics 116:926–929, 1990.

1921 2–Phenyl–5–benzothiazole Acetic Acid CAS 36774–74–0

This non–steroidal anti–inflammatory agent was given orally to rabbits in doses of up to 200 mg per kg on days 8 through 16 of gestation (Ito et al., 1977). At the highest dose level which was toxic to the mother, fetal death was increased significantly. No increase in malformations was found although two dead fetuses at the highest dose level had exencephaly. There were 86 live normal fetuses in this group. They cite their previous work stating that no teratogenicity was found in mice but cleft palate occurred in exposed rat embryos.

Ito, T.; Yamamoto, M. and Kamimura, K.: Teratogenicity of a non–steroidal anti–inflammatory agent in rabbits. Acta Medica et Biologica 24:173–178, 1977.

1922 Phenylbutazone CAS 50–33–9

Kullander and Kallen (1976) reported that 18 women taking the drug in the first trimester had one miscarriage, 6 minor and one major malformation.

Schardein et al. (1969) found no teratogenic action in rats and rabbits using 42 and 50 mg per kg, respectively, daily during organogenesis. Larsen and Bredahl (1966) fed rabbits up to 60 mg for the first 20 days of gestation and found no adverse fetal affects.

Kullander, B. and Kallen, B.: A prospective study of drugs in pregnancy. Acta Obstet. Gynecol. Scand. 55:289–295, 1976.

Larsen, V. and Bredahl, E.: The embryotoxic effects on rabbits of monophenylbutazone (Monazen™) compared with phenylbutazone and thalidomide. Acta Pharmacol. Toxicol. 24:453–455, 1966.

Schardein, J.L.; Blatz, A.T.; Woosley, E.T. and Kaup, D.H.: Reproductive studies on sodium meclofenamate in comparison to aspirin and phenylbutazone. Toxicol. Appl. Pharmacol. 15:46–55, 1969.

1923 1–Phenyl–3,3–dimethyl–triazene CAS 7227–91–0

Druckrey (1973) and Druckrey et al. (1972) reviewed work with the triazenes. In the rat, 1–phenyl–3,3–dimethyl–triazene produced postnatal brain tumors when given in single parenteral doses of up to 110 mg per kg during the last part of gestation or at birth. The 3–dimethyl–aryltriazenes were teratogenic after the 10th day when skeletal defects and cleft palates were produced. Pyridyl–dimethyltriazene given in a dose of 5 mg per kg on the 14th day, produced microcephaly.

Druckrey, H.: Specific carcinogenic and teratogenic effects of indirect alkylating methyl and ethyl compounds and their dependency on stages of ontogenic developments. Xenobiotica 3:271–303, 1973.

Druckrey, H.; Ivankovic, S.; Preussmann, R.; Zulch, K. and Mennel, H.D.: Selective induction of malignant tumors of the nervous system by resorptive carcinogens. In: The Experimental Biology of Brain Tumors. W.M. Kirsch, E.G. Paoletti and P. Paoletti (eds.), Springfield: C.C. Thomas, 111–112, 1972.

1924 o–Phenylenediamine CAS 95–54–5

Karnofsky and Lacon (1962) reported 0.5 ml of this material injected into the yolk sac of the four–day old chick produced facial coloboma, cleft palates and skeletal defects.

Karnofsky, D.A. and Lacon, C.R.: Survey of cancer chemotherapy service center compounds for teratogenic effect in the chick embryo. Cancer Res. 22:84–85, 1962.

1925 Phenylephrine Neosynephrine CAS 59–42–7

Heinonen et al. (1977) found no increase in defect rate among 1,249 exposed pregnancies. There were 8 eye and ear malformations as compared to an expected number of 2.9. The difference was not significantly increased.

Gatling (1962) dropped neosynephrine on the chorio–allantoic membrane of the chick once on the 10th, 11th or 12th day of incubation and produced hemorrhages of the head, skin and extremities. Fujinaga and Baden (**1991**) added 0.5 mM to whole embryo rat cultures on day 9 and produced situs inversus in about 50 percent.

Fujinaga, M. and Baden, J.M.: Critical period of rat development when sidedness of asymmetric body structures is determined. Teratology 44:453–462, 1991.

Gatling, R.R.: The effect of sympathomimetic agents on the chick embryo. Am. J. Pathol. 40:113–127, 1962.

Heinonen, O.P.; Slone, D. and Shapiro, S.: Birth Defects and Drugs in Pregnancy. Publishing Sciences Group Inc., Littleton, Mass., 1977.

1926 Phenylethanol CAS 60–12–8

Mankes et al. (1983) gavaged rats on days 6–15 with 4.3, 43 and 432 mg per kg. At all levels, defects were increased in the offspring. The anomalies included the eye, neural tube and limb. Hydronephrosis was also found. Maganova and Zaitsev (1973) did not find defects in rats given 508 mg per kg in sunflower oil on the 4th or 10–12th day.

Maganova, N.B. and Zaitsev, A.N.: Effect of food additives on fetal development in the rat. (Phenylacetic acid, phenylethanol, and cinnamic alcohol). Vopr Pitan 4:50–54, 1973.

Mankes, R.F.; LeFevre, R.; Bates, H. and Abraham, R.: Effects of various exposure levels of 2–phenylethanol on fetal development and survival in Long–Evans rats. J. Toxicol. Environ. Health 12:235–244, 1983.

1927 Phenylethylacetic Acid Diethylaminoethoxyethanol Ester Citrate

Hiyama and Nakajima (1970) gave this antitussive orally during organogenesis to mice and rats. No teratogenic effect was seen after doses of up to 100 mg per kg per day.

Hiyama, T. and Nakajima, S.: Phenyl acetic acid diethylaminoethoxyethanol ester citrate (HH–197), a new remedy for antitussive (3). J. Med. Soc. Toho, Japan 17:524–530, 1970.

1928 Phenylglycidyl Ether CAS 122–60–1

Terrill et al. (1982) exposed rats to 2 to 11 ppm (six hours daily, five days a week) in a two–generation study and during organogenesis. No adverse fetal or reproductive changes were found.

Terrell, J.B.; Lee, K.P.; Culik, R. and Kennedy, G.L.: Inhalation toxicity of phenylglycidyl ether: Reproductive, mutagenic, teratogenic and cytogenic studies. Toxicol. Appl. Pharmacol. 64:204–212, 1982.

1929 Phenylhydrazine CAS 100–63–0

Tamaki et al. (1974) studied postnatal function of rat fetuses made icteric by administering the mother 10 mg per kg intraperitoneally on the 17th, 18th and 19th days of gestation. Conditioned avoidance learning was found to be significantly retarded. Studies of the brains of these animals did not show kernicterus (Yamamura et al., 1973).

Yamamura, H.; Semba, H.; Keino, H.; Ohta, K. and Murakami, U.: Experimental studies on developmental disorder due to icterus gravis neonatorum: Perinatal hemolytic jaundice and its effect on postnatal development. (abs) Teratology 8:110 only, 1973.

Tamaki, Y.; Ito, M.; Semba, R.; Yamamura, H. and Kiyono, S.: Functional disturbances in adult rats suffered from icterus gravis neonatorum due to maternal application of phenylhydrazine hydrochloride. Cong. Anom. 14:95–103, 1974.

1930 Phenylketophosphamide and Derivatives

Hales et al. (1989) found teratogenicity at 10 and 25 microM for phenylketophosphamide, and phenylketoisophosphoramide, respectively, in rat embryo culture. Phenyl vinyl ketone was not teratogenic at concentrations up to 100 microM. Metabolic activation did not enhance the observed teratogenicity.

Hales, B.F.; Ludeman, S.M. and Boyd, V.L.: Embryotoxicity of phenyl ketone analogs of cyclophosphamide. Teratology 39:31–37, 1989.

1931 Phenylmercuric Acetate CAS 62–38–4

Discussed under Mercury

1932 Phenylmethylcyclosiloxane

Palazzolo et al. (1972) gave doses of up to 1000 mg per kg dermally or subcutaneously to rabbits on days 6–18 and observed two neural tube defects and some club feet. The increases were not statistically significant.

Palazzolo, R.J.; McHard, J.A. and Hobbs, E.J.: Investigation of the toxicologic properties of a phenylmethylcyclosiloxane. Toxicol. Appl. Pharmacol. 21:15–28, 1972.

1933 Phenyl–β–Naphtilamine *Neozon–D*

This antioxidant was given intragastrically to SHK strain of mice during all of gestation and postnatally (9 mg per animal). Malignant tumors in the offspring were observed. Salnikova et al. (1979) did not detect any embryotoxic action of this substance.

Salnikova, L.C.; Vorontsov, R.S.; Pavlenko, G.I. and Kotosova, L.D.: Mutagenic, embryotropic and blastomogenic effect of neozon–D (phenyl–β–naphtilamine). Gigiena Tr. Prof. Zabol. (USSR) 9:57, 1979.

1934 Phenylpropanolamine CAS 14838–15–4

Heinonen et al. (1977) studied the offspring of 726 women exposed to this drug in the first four lunar months and reported 71 defects which included malformations of the eye, ear and hypospadias. The rate was 1.40 times higher than the control and this was a statistically significant increase (p < 0.01). Three of the exposed fetuses had cataracts. Aselton et al. (1985) found 2 infants with congenital defects among 82 women exposed to the drug and chlorpheniramine (Ornade™) during the first trimester.

Aselton, P.A.; Jick, H.; Milunsky, A.; Hunter, J.R. and Stergachis, A.: First–trimester drug use and congenital disorders. Obstet. Gynecol. 65:451–455, 1985.

Heinonen, O.P.; Slone, D. and Shapiro, S.: Birth Defects and Drugs in Pregnancy. Publishing Sciences Group, Inc., Littleton, Mass., 1977.

1935 1–Phenylsemicarbazide CAS 103–03–7

Discussed under Semithiocarbazide

1936 Phenyltoloxamine CAS 92–12–6 CAS 1176–0805

Heinonen et al. (1977) reported no increase in congenital defects among 45 women using this antihistamine during pregnancy.

Heinonen, O.P.; Slone, D. and Shapiro, S.: Birth Defects and Drugs in Pregnancy. Publishing Sciences Group Inc., Littleton, Mass., 1977.

1937 Phleomycin CAS 11006–33–0

In explanted chick embryos, Lee et al. (1972) found that concentrations above 0.05 microgram per ml inhibited mitosis and neural tube closure.

Lee, H–Y.; Cortes, J.L. and Levin, M.A.: Teratogenic effects of phleomycin in early chick embryos. Teratology 6:201–206, 1972.

1938 Phosalone CAS 2310–17–0

Khera et al. (1979) administered this cholinesterase-inhibiting insecticide to rats by gavage in amounts of up to 50 mg per kg from day 6–15 and found no teratogenicity.

Khera, K.S.; Whalen, C.; Angers, G. and Trivett, G.: Assessment of the teratogenic potential of piperonyl butoxide, biphenyl, and phosalone in the rat. Toxicol. Appl. Pharmacol. 47:353–358, 1979.

1939 Phosmet *Imidan*

O,O–Dimethyl–S–phthalidomethyl Phosphorodithioate
CAS 732–11–6

Fabro et al. (1966) fed 35 mg per kg daily to rabbits on gestational days 7 through 12 and found no congenital defects in the offspring. Martson and Voronina (1976) found hydrocephalus in rat fetuses whose mothers received 30 mg per kg on day 14 of gestation. Staples et al. (1976) gavage fed rats up to 30 mg per kg from days 6 through 15 and produced only an increase in minor skeletal fetal anomalies.

Fabro, S.; Smith, R.L. and Williams, R.T.: Embryotoxic activity of some pesticides and drugs related to phthalimide. Food Cosmet. Toxicol. 3:587–590, 1966.

Martson, L.V. and Voronina, V.M.: Experimental study of the effect of a series of phosphoroorganic pesticides (dipterex and imidan) on embryogenesis. Environ. Health Perspect. 13:121–125, 1976.

Staples, R.E.; Kellam, R.G. and Haseman, J.K.: Developmental toxicity in the rat after ingestion or gavage of organophosphate pesticides (dipterex, imidan) during pregnancy. Environ. Health Persp. 113:133–140, 1976.

1940 Phosphamidon

Bhatnagar and Soni (1988) treated mice orally on days 6, 7, 10, or 13 with up to 1.0 mg per 100 g of body weight. Some stunting of the fetuses occurred. Treatment on days 6 through 15 also produced low fetal weight and a reduction in litter size. Maternal body weight gain was reduced and 12.5 percent of the dams died with the multiple day treatment. No external defects were found.

Bhatnagar, P. and Soni, I.: Evaluation of teratogenic potential of phosphamidon in mice by gavage. (abs) Toxicol. Lett. 42:101–107, 1988.

1941 Phosphonacetyl–L–aspartic Acid CAS 51321–79–0

This antitumor agent was studied by Sieber et al. (1980) in mice at doses of 0.75 to 6.25 mg per kg given intraperitoneally on days 7–11 of gestation or on single days. The four–day regime at 1.5 mg per kg was associated with 53 percent fetal lethality. Increased malformation rates were found after treatment on days 7 and 8. Skeletal, heart and renal defects were most common.

Sieber, S.M.; Botkin, C.; Soong, P.; Lee, E.C. and Whang–Peng, J.: Embryotoxicity in mice of phosphonacetyl–L–aspartic acid (PALA), a new antitumor agent. 1. Embryolethal, teratogenic and cytogenetic effects. Teratology 22:311–319, 1980.

1942 Phosphoramide Mustard CAS 10159–53–2

Discussed under Cyclophosphamide

1943 Phosphorus[32]

Burstone (1951) injected mice for 5 days before parturition subcutaneously with carrier–free radioactive phosphorous. At doses of 5 microcurie per gm of body weight no adverse effects were seen but at dosages of 10 to 17 microcurie, the offspring were stunted in growth.

Burstone, M.S.: The effect of radioactive phosphorus upon the development of the embryonic tooth bud and supporting structures. Am. J. Pathol. 27:21–26, 1951.

1944 Photodieldrin CAS 13366–73–9

This photo–degradation product of dieldrin was given by intubation to pregnant rats and mice on days 7 through 16 in doses up to 0.6 mg per kg and neither fetal effects nor teratogenic action were found (Chernoff et al., 1975).

Chernoff, N.; Kavlock, R.J.; Katherein, J.R.; Dunn, J.M. and Haseman, J.K.: Prenatal effects of dieldrin and photodieldrin in mice and rats. Toxicol. Appl. Pharmacol. 31:302–308, 1975.

1945 Photomirex CAS 39801–14–4

This photodegradation product of mirex was fed in the diet to rats in amounts of 5–40 ppm. At 40 and 20 ppm the survival indices of pups were decreased (Chu et al., 1981).

Chu, I.; Villeneuve, D.C.; Secours, V.F.; Valli, V.E. and Becking, G.C.: Effects of photomirex and mirex on reproduction in the rat. Toxicol. Appl. Pharmacol. 60:549–556, 1981.

1946 Phthalate Esters Di(2-ethylhexyl) Phthalate Di-n-butyl Phthalate Monoethylhexyl Phthalate Di(N–octyl)tin S,S'–bis (isooctyl) Mercaptoacetate Dibenzltin S,bis(isooctyl) Mercaptoacetate 2–Ethylhexanol 2–Ethylhexanoic Acid

Singh et al. (1972) studied eight phthalate esters using the rat. Injections were made intraperitoneally on gestational days 5, 10 and 15. Doses were generally from about 0.2 ml to 1.0 ml per kg. Few or no defects occurred in the groups receiving the following esters: Dimethyl, diethyl, bibutyl, diisobutyl, butyl carbobutoxy methyl and di–2–ethylhexyl. The dimethoxyethyl and dioctyl forms were associated with congenital defects which included absence of the tail, anophthalmia, twisted hind legs and hematomas. Ritter et al. (1987) presented evidence that the diethyl form acts via hydrolysis to 2–ethylhexanol which in turn is metabolized to 2–ethylhexanoic acid, the proximate teratogen. Bower et al. (1970) studied the effect of various phthalic esters on the development of the chick embryo. Dibutoxyethyl (0.05 to 0.1) injected into the yolk at 2.5 to 3 days of incubation was associated with crania bifida and anophthalmia. Parkhie et al. (1982) studied the possible role of Zn in the mechanism of teratogenicity for dimethyoxyethyl phthalate in the rat. Teratogenic results were found with the intraperitoneal injection of 0.6 ml per kg on individual days 10–14.

Shiota et al. (1980) gave mice diets containing 0.05–1.0 percent di–2–ethylhexyl (DEHP) or di–n–butyl phthalate (DBP) throughout gestation. In the group treated with 0.4 and 1.0 percent DEHP, all the implanted ova died. At 0.2 percent, there was an increase in defects which was of borderline significance (4 out of 77 implants). With 1.0 percent DBP, two of the 181 implants had exencephaly. They estimated that the maximum no effect level was 70 mg per kg per day–far higher than current estimated human intake. Merkle et al. (1988) exposed rats for six hours daily on days 6–15 for up to 0.3 mg per liter. No adverse effects were found on day 20 nor in postnatal studies.

Shiota and Mima (1985) found that di(2–ethylhexyl)phthalate was teratogenic in mice if given orally but not if administered intraperitoneally. Doses of 250 to 2,000 mg per kg were given on days 7–9 and neural tube defects and occasional other malformations were found at 500 mg per kg. The mono(2–ethylhexyl)phthalate was not teratogenic given by mouth or intraperitoneally. Tyl et al. (1988) fed rats and mice up to 2.0 percent and 0.15 percent, respectively, during most of gestation and found no teratogenicity in the rat but increased defects in the mice exposed to 0.05 percent or above. Eye defects, exencephaly and skeletal defects were found. Thomas et al. (1984) reviewed the teratogenicity and effects in the neonate.

Hendrickx et al. (**1993**) gavaged rats and rabbits during organogenesis with closer of up to 500 or 250 mg per kg respectively with 2–ethylhexanoic acid. At 500 mg per kg the rat fetuses had increased rates of hydrocephalus but the rabbits had no increase in malformations at 250 mg per kg.

Thomas et al. (1979) gave rabbits intravenously up to 11 mg per kg on the 6th through 30th day. No adverse fetal effects were found. Nikonorow et al. (1973) gave up to 0.04 and 0.09 gm per day of di–(N–octyl)tin S,S'–bis(isooctyl) mercaptoacetate or dibenzltin S,S bis(isooctyl)mercaptoacetate, respectivley, to rats by gavage for 21 days during gestation. No defects were found but resorptions were increased.

Ema et al. (**1993**) gavaged rats on days 7–15 with up to 1.0 gm per kg of di–n–butyl phthalate. At levels of 0.63 gm per kg maternal toxicity increased litter loss and cleft palate occurred. Ema et al. (**1994**) provided further details about the teratogenicity in rats.

Bower, R.K.; Haberman, S. and Minton, P.D.: Teratogenic effects in the chick embryo caused by esters of phthalic acid. J. Pharmacol. Exp. Ther. 171:314–324, 1970.

Ema, M.; Amano, H.; Itami, T. and Kawasaki, H.: Teratogenic evaluation of di–n–butyl phthalate in rats. Toxicology Letters 69:197–203, 1993.

Ema, M.; Amano, H. and Ogawa, Y.: Characterization of the developmental toxicity of di–n–butyl phthalate in rats. Toxicology 86:163–174, 1994.

Hendrickx, A.G.; Peterson, P.E.; Tyl, R.W.; Fisher, L.C.; Fosnight, L.J.; Kubena, M.F.; Vrbanic, M.A. and Katz, G.V.: Assesement of the developmental toxicity of 2–ethylhexanoic acid in rats and rabbits. Fundamental and Applied Toxicology 20:199–209, 1993.

Merkle, J.; Klimisch, H–J. and Jackh, R.: Developmental toxicity in rats after inhalation exposure of di–2–ethylhexylphthalate (DEHP). Toxicology Letters 42:215–223, 1988.

Nikonorow, M.; Mazur, H. and Piekacz, H.: Effect of orally administered plasticizers and polyvinyl chloride stabilizers in the rat. Toxicol. Appl. Pharmacol. 26:253–259, 1973.

Parkhie, M.R.; Webb, M. and Norcross, M.A.: Dimethoxyethylphthalate: Embryopathy, teratogenicity and fetal metabolism and the role of zinc in the rat. Environ. Health Perspect. 45:89–97, 1982.

Ritter, E.J.; Scott, W.J.; Randall, J.L. and Ritter, J.M.: Teratogenicity of di(2–ethylhexyl)phthalate, 2–ethylhexanol, 2–ethylhexanoic acid, and valproic acid, and potentiation by caffeine. Teratology 35:41–46, 1987.

Shiota, K.; Chou, M.J. and Nishimura, H.: Embryotoxic effects of di–2–ethylhexyl phthalate (DEHP) and di–n–butyl phthalate (DBP) in mice. Environ. Res. 22:245–255, 1980.

Shiota, K. and Mima, S.: Assessment of the teratogenicity of di(2–ethylhexyl)phthalate and mono(2–ethylhexyl)phthalate in mice. Arch. Toxicol. 56:263–266, 1985.

Singh, A.R.; Lawrence, W.H. and Autian, J.: Teratogenicity of phthalic esters in rats. J. Pharm. Sci. 61:51–55, 1972.

Thomas, J.A.; Felice, P.R.; Schein, L.G.; Gupta, P.K. and McCafferty, R.E.: Effects of monoethylhexyl phthalate (MEHP) on pregnant rabbits and their offspring. (abs) Toxicol. Appl. Pharmacol. 48:A33 only, 1979.

Thomas, J.A.; Wierda, D. and Thomas, H.J.: Phthalate acid esters: Teratogenicity, fetotoxicity and effects on neonates. In: Toxicology and the Newborn, Chapter 11, S. Kacew and M.J. Reasor (eds). Elsevier Sci. Publ. Amsterdam, 238–249, 1984.

Tyl, R.W.; Price, C.J.; Marr, M.C. and Kimmel, C.A.: Developmental toxicity evaluation of dietary di(2–ethylhexyl) phthalate in Fischer 344 rats and CD–1 mice. Fund. Appl. Toxicol. 10:395–412, 1988.

1947 Phthalazinol

7–Ethoxycarbonyl–4–hydroxymethyl–6,8–dimethyl–1–1

(2H)–phthalazinone CAS 85–73–4

This analgesic was given orally by Matsuzaki et al. (1982) to rats and rabbits in maximum doses of up to 800 and 400 mg, respectively. No effects on fertility or postnatal function were found in the rat. During organogenesis, doses of 800 mg per kg were associated with increased malformations which included hematomas, edema and skeletal abnormalities. Heart defects were also found. The rabbit fetuses had an increase in heart and skeletal defects after exposure during organogenesis to 200 mg per kg.

Matsuzaki, M.; Akutsu, S.; Karwana, K.; Kato, M.; Shimamura, T. and Nagami, K.: Reproductive studies of phthalazinol in rats and rabbits. Kiso to Rinsho 16:6357–6364, 6365–6380, 6381–6388, 6389–6396, 1982.

1948 Phthalic Acid CAS 88–99–3

Verrett et al. (1969) reported a 4 percent incidence of congenital defects in chicks receiving 3 to 20 mg of this material via the yolk sac or air cell before incubation. Tetrahydrophthalimide, phthalimide and phthalamide were similarly teratogenic. Defects of the head and eyes and phocomelia or amelia were noted. The compound, which is related to thalidomide, was not teratogenic in rabbits when used in amounts of 150 mg per kg on days 7 through 12 of pregnancy (Smith et al., 1965).

Smith, R.L.; Fabro, S.; Schumacher, H.J. and Williams, R.T.: Studies on the relationship between chemical structure and embryotoxic activity of thalidomide and related compounds. In: Embryopathic Activity of Drugs. J.M. Robson, F.M. Sullivan and R.L. Smith (eds.), Boston: Little Brown, 194–209, 1965.

Verrett, M.J.; Mutchler, M.K.; Scott, W.F.; Reynaldo, E.F. and McLaughlin, J.: Teratogenic effects of captan and related compounds in the developing chicken embryo. Ann. N.Y. Acad. Sci. 160:334–343, 1969.

1949 Phthalimide CAS 85–41–6

This chemical structurally related to thalidomide was tested in rabbits and found not to be teratogenic. Smith et al. (1965) gave 150 mg per kg on days 7 through 12 of pregnancy.

Smith, R.L.; Fabro, S.; Schumacher, H.J. and Williams, R.T.: Studies on the relationship between chemical structure and embryotoxic activity of thalidomide and related compounds. In: Embryopathic Activity of Drugs. J.M. Robson, F.M. Sullivan and R.L. Smith (eds.), Boston: Little Brown, 194–209, 1965.

1950 4–Phthalimidobutyric Acid CAS 3130–75–4

Kohler et al. (1973) found fetal skeletal defects when this chemical was given to mice in doses of 800 mg per kg intraperitoneally on day 9 of gestation.

Kohler, F.; Ockenfels, H. and Meise, W.: Teratogene Aktivitat von N–Phthalylglycin und 4–Phthalimidobuttersaure. Pharmazie 28:680–681, 1973.

1951 Phthalimidoethanesulphon–N–isopropylamide *MY–117*

This taurine derivative proposed as an anticonvulsant was tested in mice orally from the 6th through the 12th day of gestation in doses of up to 1040 mg per kg. No adverse fetal effects were found (Lankinen et al., 1982).

Lankinen, S.; Linden, I–B. and Gothoni, G.: Teratological studies on a new anticonvulsive taurine derivative in mice. (abs) Teratology 26:19A, 1982.

1952 N–Phthalylglycine CAS 4702–13–0

Kohler et al. (1973) found fetal skeletal defects when this chemical was given to mice in doses of 200 mg per kg intraperitoneally on day 9 of gestation.

Kohler, F.; Ockenfels, H. and Meise, W.: Teratogene Aktivitat von n–Phthalylglycin und 4–Phthalimidobuttersaure. Pharmazie 28:680–681, 1973.

1953 Phthorotanum

Anisimova (1981) treated by inhalation (1024 plus or minus 10.4 mg per cubic meter) pregnant rats during the entire gestation and observed gonadotoxic effects. Embryotoxic and teratogenic effects were not found.

Anisimova, I.G.: Gonadotoxic and embryotoxic effect of phthorotanum. Gig. Sanit. (USSR) 4:21–24, 1981.

1954 Physostigmine *Eserine* CAS 57–47–6

Landauer (1954) injected this material into the yolk sac of chick embryos and produced micromelia, parrot beaks, syndactylism and clubbed down. Bueker and Platner (1956) injected 0.05 to 15.0 mg into the yolk sac during the first 12 days of incubation and produced severe vertebral defects and micromelia in chicks. Ancel (1946) used 0.05 mg at 26 hours to produce brachymelia and spina bifida in the chick.

Ancel, P.: Reserche experimentale sur le spina bifida. Arch. Anat. Microscop. 36:45–68, 1946.
Bueker, E.D. and Platner, W.S.: Effect of cholinergic drugs on development of chick embryo. Proc. Soc. Exp. Biol. Med. 91:539–543, 1956.
Landauer, W.: On the chemical production of developmental abnormalities and of phenocopies in chicken embryos. Cell Comp. Physiol. 43:261–305, 1954.

1955 Picenadol

This drug is a mixed opioid agonist–antagonist. Tizzano et al gave up to 210 mg per kg orally before mating to males and females and to females during pregnancy. There were no adverse effects on fertility. At 70 mg per kg fetal weight was reduced and at higher doses the fetal mortality was increased. No behavior changes were found in the offspring from 70 mg per kg treatment schedule.

Tizzano, J.P.; Hoyt, J.A.; Hanasono, G.K.; Helton, D.R. and Buelke–Sam, J.: Developmental toxicology studies of picenadol administered in the diet to rats. (abst)Teratology 41:620, 1990.

1956 Picloram *4–Amino–3,5,6–trichloropicolinic Acid* CAS 1918–02–1

Thompson et al. (1972) found no teratogenic action of this compound in rats fed up to 100 mg per kg on gestational days 6 through 15. No neonatal adverse effects were noted. John–Greene et al. (1985) found no adverse effects in rabbit offspring whose mothers received up to 400 mg per kg orally during organogenesis.

John–Greene, J.A.; Ouellette, J.H.; Jeffries, T.K.; Johnson, K.A. and Rao, K.S.: Teratological evaluation of picloram potassium salt in rabbits. Food Chem. Toxicol. 23:753–756, 1985.
Thompson, D.J.; Emerson, J.L.; Strebing, R.H.; Gerbig, C.G. and Robinson, V.B.: Teratology and postnatal studies on 4–amino–3,5,6–trichloropicolinic acid (picloram) in the rat. Food Cosmet. Toxicol. 10:797–803, 1972.

1957 Picosulfate, Sodium CAS 10040–45–6

This laxative was given by gastric intubation to rats and rabbits during organogenesis. At 100 mg per kg no increase in malformations occurred, but in the rabbit, early resorptions were increased (Nishimura, 1977). At 10,000 mg per kg in the rat and 1,000 mg per kg in the rabbit, there was no increase in malformations. In both species, maternal treatment postnatally was associated with increased death in the offspring.

Nishimura, M.; Kast, A. and Tsunenari, Y.: Reproduction studies of sodium picosulfate (DA–1773, Laxoberon) on rats and rabbits. (Japanese) Iyakuhin Kenkyu 8:366–396, 1977.

1958 Pilocarpine CAS 92–13–7

Landauer (1953) used this parasympathomimetic drug to produce defects in chick embryos. He injected 3 to 12 mg into the yolk at 24 to 96 hours of incubation. Rumplessness occurred in some of the early injected embryos; injection at a later period was associated with tarsometatarsus, beak and other skeletal defects. Nicotinamide given concurrently protected against the teratogenic effects.

Landauer, W.: On teratogenic effects of pilocarpine in chick development. J. Exp. Zool. 122:469–483, 1953.

1959 Pilsicainide Hydrochloride *SUN 1165*

Yamamori et al. (1991 A) fed this anti–arrythmic drug to rats before pregnancy and during early gestation, during organogenesis and in the perinatal period. For the highest doses 160 and 225 mg per kg the dams salivated and drank more water. No adverse reproductive effects were found and postnatal development was normal. Yamamori et al. (1991 B) treated rabbits with up to 20 mg per kg and found no treatment–related abnormalities.

Yamamori, K.; Ohnishi, S.; Tesh, J.M. and McAnulty, P.A.: Reproductive and developmental toxicity studies in rats given pilsicainide hydrochloride (SUN 1165) orally. Oyo Yakuri 42:507–518, 1991A.
Yamamori, K.; Ohnishi, S.; Tesh, J.M. and Ross, F.W.: Teratogenicity study in rabbits given pilsicainide hydrochloride (SUN 1165) orally. Oyo Yakuri 42:519–527, 1991B.

1960 Pimefylline

Ciaceri and Attaguile (**1973**) found no adverse fetal effects after exposing rats to 300 mg per kg.

Ciaceri, G. and Attaguile, G.: First results from a pharmacological study of a new theophylline derivative: 7-[2-(pyridyl)-methylamino–ethyl]–theophylline nicotinate. NoteI–Tolerance and foetal toxicity. Gaz. Med. It. 132:36–40, 1973.

1961 Pimeprofen 2–*Pyridylmethyl* 2–(p–(2–methylpropyl)phenyl) proprionate CAS 64622–45–3

Fuchigami et al. (1982) gave rats up to 75 mg per kg subcutaneously on days 7–17 of gestation and observed no adverse fetal effects. Postnatal studies with doses to 150 mg per kg produced no significant changes. Using rabbits and a subcutaneous dose of up to 40 mg per kg during organogenesis they observed no adverse fetal effects.

Fuchigami, K.; Hatano, M.; Shimamura, K.; Iwaki, M.; Aoyama, T.; Tsuji, M. and Noda, K.: Reproductive studies on 2–pyridylmethyl 2–(p–(2–methylpropyl)phenyl) proprionate (pimeprofen). Oyo Yakuri 23:883–893, 24:1–47, 1982.

1962 Pimobendan

Matsuo et al. (**1992**) studied this cardiotonic drug before and during early gestation, during organogenesis and in the perinatal period. The maximum oral dosages were 300, 200 and 100 mg per kg for the prenatal, organogenesis and perinatal studies. At the highest dose prenatally and in early gestation there was an increase in resorptions. No adverse fetal effects were seen except a decrease in weight which was associated with decreased maternal weight. No adverse fetal effects were found in rabbits treated with 100 mg per kg during organogenesis.

Matsuo, A.; Honma, M. and Kohei, H.: Oral reproductive and developmental toxicology or pimobendan. Oyo Yakuri 43:415–430, 1992.

1963 Pimozide CAS 2062–78–4

Fukuhara et al. (1980) gave this diphenylbutyl piperidine drug to rats orally in amounts of up to 3.2 mg per kg daily from 3 weeks of age through pregnancy and found no ill effects on fertility or the offspring.

Fukuhara, Y.; Fujii, T.; Kado, Y. and Watanabe, N.: Reproductive studies of pimozide administered from young age in rats. Kiso to Rinsho 14:2163–2170, 1980.

1964 Pinacidil 2–*Cyano–1–(4–pyridyl)–3–(1,2,2–trimethylpropyl) guanadine Monohydrate S–1230*

Komai et al. (**1991** A,B,C) gave rats orally up to 50 mg per kg before and during the first 7 days of gestation or during organogenesis. At the highest dose fetal weight was reduced and an increase in fetal deaths occurred. In the perinatal and postnatal period the exposed pups (25 mg per kg) gained weight more slowly but their behavior was not altered. At 4 mg per kg there was no effect.

Komai, Y.; Ito, I.; Hibino, H.; Ishimura, K.; Fuchigami, K. and Kobayashi, H.: Reproduction study of S–1230: Fertility study in rats by oral administration. Jpn. Pharm. Therap. 19:2491–2501, 1991A.

Komai, Y.; Ito, I.; Ishimura, K.; Fuchigami, K. and Kobayashi, H.: Reproduction study of S–1230: Teratogenicity study in rats by oral administration. Jpn. Pharm. Therap. 19:2503–2529, 1991B.

Komai, Y.; Hibino, H.; Inoue, S.; Iriyama, K.; Ito, I.; Ishimura, K.; Fuchigami, K. and Kobayashi, H.: Reproduction study of S–1230: Peri–and postnatal study in rats by oral administration. Jpn. Pharm. Therap. 19:2531–2555, 1991C.

1965 Pinazepam 7–*Chloro–1–propargyl–5–phenyl–3 H–1, 4–benzodiazepin–2–one* CAS 52463–83–9

Scrollini et al. (1975) gave up to 200 mg per kg orally to rats and up to 25 mg per kg to rabbits during organogenesis without evidence of teratogenicity. At the highest dose in the rabbit resorptions were increased.

Scrollini, F.; Caliari, S.; Romano, A. and Torchio, P.: Toxicological and pharmacological investigations of pinazepam (7–chloro–1–propargyl–5–phenyl–3H–1,4, benzodiazepin–2–one): A new psychotherapeutic agent. Arzneim Forsch. 25:934–940, 1975.

1966 Pindolol *Nipradilol*

Dubois et al. (1982) treated 38 hypertensive women with this drug in late pregnancy. One newborn had a cleft palate and 3 were less than 2500 gr at birth. Ellenbogen et al. (1986) treated 16 women and found no side effects in the newborn. Koga et al. treated rats and rabbits with doses which were maternally toxic and found growth retardation and osseous retardation. The rats received up to 400 mg per kg and the rabbits 10 or 20 mg per kg per day.

Dubois, D.; Petitcolas, J.; Temperville, B. and Klepper, A.: Treatment of hypertension in pregnancy with beta–adrenoceptor antagonists. Br J Clin Pharmac 13:375S–378S, 1982.

Ellenbogen, A.; Jaschevatzy, O.; Davidson, A. and Anderman, S.: Management of pregnancy–induced hypertension with pindolol—comparative study with methyldopa. Int J Gynaecol Obstet 24:3–7, 1986.

Koga, T.; Ohta, T.; Aoki, Y. and Sugasawa, M.: Teratological study of nipradilol (K–351) in rats and rabbits. Oral administration during the period of fetal organogenesis. Oyo Yakuri 29:747–759, 1985.

1967 Pine Needles

Chow et al. (1972) summarized the work in cattle and mice on the effect of pine needle ingestion on pregnancy. In cows, abortions or weak non–viable calves followed pine needle consumption. Mice fed after the 4th, 10th or 15th day of pregnancy had reduction in litter–size as well as fetal weight. No congenital defects were reported. An aqueous fraction of the pine needles was the most active. Subsequent work by this group indicated that the active agent was heat labile product of fungi associated with the pine needles (Chow et al., 1974).

Chow, F.C.; Hamar, D.W. and Udall, R.H.: Myotoxic effect on fetal development: Pine needle abortion in mice. J. Reprod. Fertil. 40:203–204, 1974.

Chow, F.C.; Hanson, K.J.; Hamar, D.W. and Udall, R.H.: Reproductive failure of mice caused by pine needle ingestion. J. Reprod. Fertil. 30:169–172, 1972.

1968 Pipamazine CAS 84–04–8

Discussed under Phenothiazines

1969 Pipecolinomethylhydroxyindane

Boris et al. (1974) gave this chemical orally to rats during their entire gestation at maximum doses of 50 mg per kg daily and did not observe any adverse fetal effects.

Boris, A.; Ng, C. and Hurley, J.F.: Antitesticular and antifertility activity of a pipecolinomethylhydroxyindane in rats. J. Reprod. Fertil. 38:387–394, 1974.

1970 Pipemidic Acid CAS 51940–44–4

Nishimura et al. (1976) gave this antibacterial agent to pregnant rats by gavage in doses of up to 3,200 mg per kg on days 8 through 14 of gestation. At the highest dose, a slight decrease in fetal weight and delay in ossification occurred. Dilation of the ureters and renal pelves were significantly more common in the fetuses exposed to 800 and 3,200 mg per kg. Postnatally 8 of 132 pups in the 3,200 mg per kg group showed dilation of the renal pelves.

Nishimura, K.; Nanto, T.; Mukumoto, K.; Yasuba, J.; Sasaki, H.; Terada, Y.; Fukagawa, S.; Shigematsu, K. and Tatsumi, H.: Reproduction studies of pipemidic acid in rats. 2. Teratogenicity study. (Japanese) Iyakuhin Kenkyu 7:321–329, 1976.

1971 Piperacetazine

Rats and rabbits were treated by gavage during organogenesis with up to 30 mg and 75 mg per kg respectively and no teratogenicity was found. Postnatal development in the rat offspring was normal. Prenatal treatment reduced libido. (Anonymous, 1972)

Anonymous: Piperacetazine: Rx Bull 3:122–125, 1972.

1972 Piperacillin Sodium T–1220 CAS 59703–84–3

This new penicillin analog was given intravenously in amounts of up to 2,000 mg per kg daily to mice on days 6 through 15 and no teratologic or postnatal growth changes occurred (Takai et al., 1977A). The same authors (1977B) found no teratogenicity in the rat given 1,000 mg per kg daily on days 7 through 17 of gestation.

Takai, A.; Yoneda, T.; Nakada, H.; Nakamura, S. and Inaba, J.: Toxicity tests of T–1220. 6. Reproduction study in mice. (Japanese) Chemotherapy 25:915–927, 1977A.

Takai, A.; Yoneda, T.; Nakada, H.; Nakamura, S. and Inaba, J.: Toxicity tests of T–1220. 7. Teratology tests in rats. (Japanese) Chemotherapy 25:928–933, 1977B.

1973 Piperanitrozole

Nohara (1966) treated mice with 60 or 120 mg per kg on days 8–14 and produced no adverse fetal effects.

Nohara, S–I.: Experimental studies on the effect of oral anti–trichomonas drugs on mice fetus. J. Antibiot., Ser. B XIX–3:163–173, 1966.

1974 Piperazine

Wilk (1969) reported in abstract that injection of 50 microgm into rat amniotic fluid on day 13 produced no increase in cleft palates.

Wilk, A.L.: Relation between teratogenic activity and cartilage–binding affinity of norchlorcyclizine analogues (A). cleft palate. J. Tox. Sc. 18:171–178, 1993.

1975 Piperilate PB–106 CAS 4546–39–8 CAS 23182–46–9

This anticholinergic and ganglion–blocking agent was tested in pregnant rats and mice by Ohata and Nomura (1970). Treatment was given during active organogenesis using the oral, intraperitoneal or subcutaneous route. Oral doses of up to 1000 mg per kg were used. No teratogenicity or fetal effects were observed.

Ohata, K. and Nomura, A.: Influence of 2–(1–piperidino)–ethyl benzilate ethylbromide (PB–106) on pregnant mice and rats, and on their fetuses. (Japanese) Oyo Yakuri 4:59–68, 1970.

1976 Piperonyl Butoxide CAS 51–03–6

Khera et al. (1979) administered this compound to rats by gavage at doses up to 500 mg per kg on days 6–15 of gestation and found no evidence of teratogenicity. Kennedy et al. (1977) also had negative findings at a dose of 3000 mg per kg. Tanaka et al (1994) found oligodachyly at 1385 and 1800 mg per kg in fetal mice treated by gavage on day 9.

Kennedy, G.L.; Smith, S.H.; Kinoshita, F.K. and Keplinger, M.L.: Teratogenic evaluation of piperonyl butoxide in the rat. Food Cosmet. Toxicol. 15:337–339, 1977.

Khera, K.S.; Whalen, C.; Angers, G. and Trivett, G.: Assessment of the teratogenic potential of piperonyl butoxide, biphenyl, and phosalone in the rat. Toxicol. Appl. Pharmacol. 47:353–358, 1979.

Tanaka, T.; Fujitani, T.; Takahashi, O. and Oishi, S.: Developmental toxicity evaluation of piperonyl butoxide in CD–1 mice. Toxicology Letters 71:123–129. 1994.

1977 Pipethiadene Tartrate

4–(1–Methyl–4–piperidinylidene–4,9–dihydrothienol (2,3,c) (2)benzothiepine

This antihistamine was studied in mice using orally 0.24, 0.6 or 1.2 mg per kg on days 4–16 (Ujhazy et al., 1988). No teratogenicity was found but fetal weight reduction was found at all dose levels. The mitotic indices of maternal bone marrow were also reduced at all levels.

Ujhazy, E.; Nosal, R.; Zeljenkova, D.; Balonova, T.; Chalupa, I.; Siracky, J.; Blasko, M. and Metys, J.: Teratological and cytogenetical evaluation of two antihistamines (pipethiadene and pizotifen maleate) in mice. Agents and Actions 23:376–378, 1988.

1978 Pipobroman *1,4–Bis(3–bromopropionyl)piperazine* CAS 54–91–1

Nagai (1972) injected this anti–neoplastic agent into pregnant mice. At doses of 30 mg per kg given on days 11 through 14 of gestation, brachygnathia and small noses occurred.

Nagai, H.: Effects of transplacentally injected alkalating agents upon development of embryos. Bull. Tokyo Dent. Coll. 13:103–119, 1972.

1979 Piquizium

Tsuruzaki et al. (1981) gave this anticholinergic to rats daily in amounts of up to 250 mg per kg before mating and during the first 7 days and found no decrease in reproduction. Up to 500 mg per kg was given during organogenesis and no adverse fetal effects were observed. Postnatal function was unaffected.

Tsuruzaki, T.; Inui, H.; Kato, H. and Yamamoto, M.: Effects of 3–(di–2–thienyl–methylene)–5–methyl–trans– quinolizidinium bromide on reproduction. Kiso to Rinsho 15:6183–6193, 6194–6244, 6251–6253, 1981.

1980 Pirbuterol CAS 38677–81–5

Sakai et al. (1980) gave rats up to 300 mg per kg orally before mating and for 7 days of gestation, on days 7–17 or on days 17–21. The only ill effect was growth retardation when 300 mg per kg was given during organogenesis. Postnatal effects were not seen. The rabbits received up to 300 mg per kg orally on days 6–18. At the highest dose abortion occurred in 4 of 13 litters. Body weight of the fetuses was reduced at 30, 100, and 300 mg per kg.

Sakai, T.; Owaki, Y. and Noguchi, Y.: Reproduction studies of pirbuterol hydrochloride. Yakuri to Chiro 8:731–743, 1980.

1981 Pirenzepine *Gastrozepin™* CAS 28797–61–7

Iida et al. (1980) gave this anti–ulcer medication to rats and rabbits preconceptually, during organogenesis and perinatally and found no adverse effects on reproduction. Iida et al. (1986) gave this anti–ulcer drug intravenously to pregnant rats and rabbits before and during implantation, during organogenesis and from late pregnancy through weaning. No adverse fetal effects were found at doses 3 to 60 mg, but maternal toxicity was found at doses above 30 mg per kg. Fertility and postnatal activity were not significantly affected.

Iida, H.; Kast, A. and Tsunenari, Y.: Reproduction toxicology of intravenous pirenzepine dihydrochloride. Iyakuhin Kenkyu 17:859–881, 1986.

Iida, H.; Matsuo, A.; Kast, A. and Tsunenari, Y.: Reproductive studies: Pirmenrenzepine (LS519Cl(2)) in rats and rabbits. Iyakuhin Kenkyu 11:424–436, 1980.

1982 Piretanide
4–Phenoxy–3–(1–pyrrolidynyl)–5–sulfamoylbenzoic Acid

CAS 55837–27–9

Kitantani et al. (1980) gave this diuretic drug to rats and rabbits by both oral and intravenous routes. With oral doses of up to 500 mg per kg in the rat, no adverse effects were found except that the pregnancy rate was reduced when treatment was on the first 7 days of gestation. The intravenous daily dose (60 mg per kg) did not alter postnatal growth or produce increased defects in the rat. Doses of 1.25 mg or 1.0 mg per kg were given intravenously or by mouth, respectively, during organogenesis and rabbit fetuses were similar to controls.

Kitantani, T.; Sugisaki, T.; Takagi, S.; Hayashi, S. and Miyamoto, M.: Reproductive studies of piretanide. Kiso to Rinsho 14:4330–4347, 4348–4363, 4364–4366, 4367–4373, 1980.

1983 Pirmenol Hydrochloride

Schardein et al. (1980) tested this antiarrhythmic agent in gravid rats and rabbits. Oral doses of up to 150 mg per kg in rats and 50 mg per kg in rabbits did not alter organogenesis. At 150 mg per kg, the rat fetuses increased death and decreased weight. Six maternal deaths occurred among 20 treated at this level. Anderson et al. (1986) studied this antiarrhythmic agent in rats giving up to 100 mg per kg orally before, during and after gestation. No adverse effects were found in the fetuses or newborns.

Anderson, J.A.; Petrere, J.A.; Fitzgerald, J.E. and De La Iglesia, F.A.: Studies on reproduction in rats with pirmenol, an antiarrhythmic agent. Fund. Appl. Toxicol. 7:221–227, 1986.

Schardein, J.L.; Fitzgerald, J.E.; Sanyer, J.L.; McGuire, E.J. and De La Iglesia, F.A.: Preclinical toxicology studies with a new antiarrhythmic agent pirmenol hydrochloride (Cl–845). Toxicol. Appl. Pharmacol. 56:294–301, 1980.

1984 Piroheptine *3(10,11–Dihydro–5H–dibenzo(a,d)– cyclohepten–5–ylidene)–1–ethyl–2–methylpyrrolidine*

Hitomi et al. (1972) fed up to 50 mg per kg to mice and rats during pregnancy and found no adverse reproductive or fetal changes.

Hitomi, M.; Watanabe, N.; Kumadaki, N. and Kumada, S.: Pharmacological study of piroheptine, a new antiparkinson drug. II. Anticholinergic, antihistaminic and psychopharmacological actions and toxicity. Arzneim Forsch 22(6):961–966, 1972.

1985 Piroxicam CAS 36322–90–4

Sakai et al. (1980) studied this non–steroidal anti–inflammatory in rabbits and rats using maximum oral doses of up to 10 and 70 mg per kg, respectively. At 2.5 mg per day on days 7–17 there was decreased fetal growth. At 10 mg per kg, 11 of 39 dams died. No fetal malformation increase was found. After perinatal studies, growth retardation was found at 10 mg per kg. Studies of rabbits during organogenesis were negative except that most of the dams died at 10 mg per kg.

Sakai, T.; Ofsuki, I. and Noguchi, F.: Reproduction studies on piroxicam. Yakuri to Chiryo 8:4655–4671, 1980.

1986 Pirozadil

Grau and Balasch (**1979**) carried out a study in rats on days 6–16 using 500 mg per kg with no teratogenic effects.

Grau, M. and Balasch, J.: Teratogenic study of prirozadil in rats. Cien. & Ind. Farm. 11:67–70, 1979.

1987 Pirprofen CAS 31793–07–4

Hirooka et al. (1984) gave this non–steroidal anti–inflammatory orally to pregnant rats on days 7–17 in amounts of 15 mg per kg and found no adverse effects on reproduction.

Hirooka, T.; Takahashi, S. and Kitagawa, S.: Oral teratogenicity study of pirprofen in rats. Clin. Report. 18:5651–5673, 1984.

1988 Pituitary Ablation by Transgenic Insertion

Behringer et al. (1988) inserted a human growth hormone gene along with a diptheria toxin producing gene into the nucleus of mouse ova and produced dwarf mice with absent pituitary somatotropes.

Behringer, R.R.; Mathews, L.S.; Palmiter, R.D. and Brinster, R.L.: Dwarf mice produced by genetic ablation of growth hormone–expressing cells. Genes Dev. 2:453–461, 1988.

1989 Pizotifen 9, 10–Dihydro–4–(1–methyl–4–piperidinyl–idene)–4H–benzo(4,5)cyclohepta(1,2–b) thiophene

This antihistamine was studied in mice using orally 0.24, 0.6 or 1.2 mg per kg on days 4–16 (Ujhazy et al., 1988). No teratogenicity was found but fetal weight reduction was found at all dose levels. The mitotic indices of maternal bone marrow were also reduced at all levels.

Ujhazy, E.; Nosal, R.; Zeljenkova, D.; Balonova, T.; Chalupa, I.; Siracky, J.; Blasko, M. and Metys, J.: Teratological and cytogenetical evaluation of two antihistamines (pipethiadene and pizotifen maleate) in mice. Agents and Actions 23:376–378, 1988.

1990 Plafibride N–2–(p–Chlorophenoxy)isobutyryl–N–morpholinmethylurea CAS 1407–93–8

Sanfeliu et al. (1981) gave rats up to 500 mg per kg and rabbits 100 mg per kg during organogenesis. At 250 mg per kg, hydrocephalus in the rat fetuses was increased. In the rabbit, at 25 mg per kg, vertebral agenesis, fused kidneys, gonadal hypoplasia, malformation of the genito–anal region and other osseous defects were found to be increased.

Sanfeliu, C.; Zapatero, J. and Bruseghini, L.: Toxicological studies of plafibride. Arzneim Forsch. 31:1831–1834, 1981.

1991 Plasminogen Activator, Tissue–Type SM–9527

Shibano et al. (**1991** A,B,C) gave this material to rats intravenously in amounts up to 40 mg per kg daily before and during early pregnancy, during organogenesis and peri– and postnatally. No adverse effects were found. Treatment of rabbits during organogenesis in doses to 10 mg per kg and found increased fetal death and a slight increase in vertebral variations at the highest dose.

Kawamura, S.; Higuchi, H.; Hirohashi, A.; Kato, T. and Yamada, H.: Teratogenicity study of SM–9527 in rabbits. The Clinical Report 25:117–125, 1991.

Shibano, T.; Satoh, S.; Yoneyama, S. and Kinoshita, K.: Study on intravenous administration of SM–9527 prior to and in the early stages of pregnancy in rats. The Clinical Report 25:887–894, 1991A.

Shibano, T.; Satoh, S.; Yoneyama, S. and Kinoshita, K.: Study on intravenous administration of SM–9527 during the period of organogenesis in rats. The Clinical Report 25:895–908, 1991B.

Shibano, T.; Kamijima, M.; Yoneyama, S. and Kinoshita, K.: Study of intravenous administration of SM–09527 during the perinatal and lactation periods in rats. The Clinical Report 25:989–921, 1991C.

1992 Platonin

Kimoto et al. (1977) injected this dye in amounts of 10 or 200 γ per kg intraperitoneally in mice on days 7–13. No adverse effects were found.

Kimoto, T.; Nishitani, K. and Nishioka, Y.: Study on the toxicity of a photosensitizing dye, platonin. Report 2. Effects of platonin on fetus of pregnant mice. Kanko Shikiso 86:25–33, 1977.

1993 Plicamycin Mithramycin CAS 18378–89–7

Chaube and Murphy (1968) reported that this antibiotic which inhibits nucleic acid and protein synthesis was not teratogenic in the rat. Single doses of less than the fetal LD–100 were given on the 5th through the 12th days.

Chaube, S. and Murphy, M.L.: The teratogenic effects of the recent drugs active in cancer chemotherapy. In: Advances in Teratology, D.H.M. Woollam (ed.), New York: Logos and Academic Press, 3:181–237, 1968.

1994 Plutonium[239]

Kelman et al. (1982) administered this isotope intravenously to rabbits and on days 9–28. Using 40 microcuries per kg fetal growth retardation was found. No defect increase was found.

Kelman, B.J.; Sikov, M.R. and Hackett, P.L.: Effects of monomeric [239]Pu on the fetal rabbit. Health Physics 43:80–83, 1982.

1995 Podophyllinic Acid Ethylhydrazide

von Kreybig et al. (1970) gave 75 or 100 mg per kg to rats on day 10 or 13 and found no malformations. Fetotoxicity was found.

von Kreybig, T.; Preussmann, R. and von Kreybig, I.: Chemische Konstitution und Teratogene Wirkung bei der Ratte. Arzneim–Forsch. 20:363–367, 1970.

1996 Podophyllotoxin CAS 518–28–5

This chemical derived from the American May–apple is an antitumor agent. Chaube and Murphy (1968) reported that single injections of 0.3 to 1.0 mg per kg into pregnant rats once from the 9th to the 12th days produced no fetal changes. Litter resorption occurs if the drug is used earlier than the 9th day in the rat (Thiersch, 1963) or mouse (Wiesner and Yudkin, 1955). Thiersch (1963) reported no congenital defects in the rat fetuses. Robert and Scialli (**1994**) have summarized the literature on this compound and state that inadvetent use is not an indication for pregnancy interruption.

Chaube, S. and Murphy, M.L.: The teratogenic effects of the recent drugs active in cancer chemotherapy. In: Advances in Teratology, D.H.M. Woollam (ed.), New York: Logos and Academic Press, 3:181–237, 1968.

Robert, E. and Scialli, A.R.: Topical medications during pregnancy. Reproductive Toxicology 8:197–202, 1994.

Thiersch, J.B.: Effect of podophyllin (P) and podophyllotoxin (PT) on the rat litter in utero. Proc. Soc. Exp. Biol. Med. 113:124–127, 1963.

Wiesner, B.P. and Yudkin, J.: Control of fertility by antimitotic agents. Nature (London) 176:249–250, 1955.

1997 Poliovirus Vaccine

No adverse effects were reported by Heinonen et al. (1977) among 1628 women receiving the live vaccine in the first 4 lunar months. Some increase in central nervous system tumors has been reported associated with contamination with SV–40 virus. (see SV–40 virus) The use of inactivated vaccine has not been associated with adverse effects but published reports are not available.

Ornoy et al. (1990) studied a large population in Israel after mass vaccination with trivalent oral poliovirus vaccine and found no increase in spontaneous abortions among over 20,000 pregnancies. In addition no increase in malformation rate was found (Ornoy and Ishai, **1993**).

Heinonen, O.P.: Slone, D. and Shapiro,S.: Birth Defects and Drugs in Pregnancy. John Wright Publishing Sciences Group Inc., Littleton, Mass. 1977. pp 376–7, 317–18, 423,445, 473, 486.

Ornoy, A.; Arnon, J.; Feingold, M. and Ishai, P.B.: Spontaneous abortions following oral poliovirus in first trimester. The Lancet 1:800, 1990.

Ornoy, A. and Ishai, P.B.: Congenital anomalies after oral poliovirus vaccination during pregnancy. The Lancet 341:1162, 1993.

1998 Polybrominated Biphenyls *Firemaster BP–6*

Weil et al. (1981) found no physical or psychological differences between a group of 20 in utero–exposed children and a control group.

Corbett et al. (1978) fed mice and rats up to 1,000 parts per million during gestation and produced no significant increase in defects in the offspring. Fisher (1980) administered the compound to rats 24 hours before explanting their embryos. He found chemical growth retardation and malformations in these embryos when they were grown in vitro.

Henck et al. (**1994**) fed up to 2 mg per kg orally to maternal rats starting on day 6 through the 24th postpartum day. At the highest dose various neurobehavioral behaviors were delayed.

Corbett, T.H.; Simmons, J.L. and Endres, J.: Teratogenicity and tissue distribution studies of polybromated biphenyls (Firemaster BP–6) in rodents. (abs) Teratology 17:37A only, 1978.

Fisher, D.L.: Effects of polybrominated biphenyls on the accumulation of DNA, RNA, and protein in cultured rat embryos following administration. Environ. Res. 23:334–340, 1980.

Henck, J.W.; Mattsson, J.L.; Rezabek, D.H.; Carlson, C.L. and Rech, R.H.: Developmental neurotoxicity of polybrominated biphenyls. Neurotoxicology and Teratology 16:391–399, 1994.

Weil, W.B.; Spencer, M.; Benjamin, D. and Seagull, E.: The effect of polybrominated biphenyl on infants and children. J. Pediatr. 98:47–58, 1981.

1999 Polybromodiphenyl Oxide

This mixture proposed as a flame retardant was studied in rabbits by Breslin et al. (1989). Slight fetal toxicity was found with oral doses of 15 mg per kg given on days 7–19. No teratogenicity occurred at these maternally toxic doses.

Breslin, W.J.; Kirk, H.D. and Zimmer, M.A.: Teratogenic evaluation of a polybromodiphenyl oxide mixture in New Zealand white rabbits following oral exposure. Fund. Appl. Toxicol. 12:151–157, 1989.

2000 Polychlorobiphenyls

Discussed under Chlorobiphenyls and Kanechlor

2001 Polydimethylsiloxane

Bates et al. (1985) injected this polymer used in mammary prosthetic devices into rats at up to doses of 20 mg per kg and found no teratogenicity.

Bates, H.; Filler, R. and Kimmel, C.A.: Developmental toxicity study of polydimethyl–siloxane injection in the rat. (abs) Teratology 31:50A, 1985.

2002 Polyriboinosinic *Polyribocytidylic Acid Poly I–poly–C* CAS 24939–03–5

This synthetic double stranded RNA was injected into rabbits in amounts of 1 or 2 mg per kg on days 8 and 9 or 11 and 12 (Adamson and Fabro, 1969). A very high resorption rate occurred but the number of defects in the survivors was not greater than in the control fetuses.

Adamson, R.H. and Fabro, S.: Embryotoxic effect of poly 1–poly–C. Nature 223:718 only, 1969.

2003 Polythiazide

Heinenon etal (1977) reported 505 women treated with this diuretic and no increase in defects was seen. Only 10 were treated in early pregnancy.

Heinonen, O.P.; Slone, D. and Shapiro, S.: Birth Defects and Drugs in Pregnancy. Littleton, Massachusetts: John Wright–PSG, Inc.; pp 372, 436, 441, 1977.

2004 Ponceau 4R

Larsson (1975) gavaged mice on days 0–7 or 6–18 with up to 100 mg per kg. No adverse reproductive findings were reported. Brantom et al. (1988) found no increase in cancer among offspring of rats treated with up to 1250 mg per kg for 60 days before mating.

Brantom, P.G.; Stevenson, B.I. and Wright, M.G.: Long–term toxicity study of Ponceau 4R in rats using animals exposed in utero. Food Chem. Toxicol. 25:955–962, 1988.

Larsson, K.S.: A teratological study with the dyes amaranth and ponceau 4R in mice. Toxicology 4:75–82, 1975.

2005 Potassium CAS 7440–09–7

Crocker and Vernier (1970) showed in organ culture of fetal mouse kidney that lowered concentrations of potassium produced abnormal branching of tubules and occasional cystic dilatations of the ureteral buds.

Wilson et al. (1968) showed that potassium deficiency does not account for the limb defects associated with acetazolamide administration.

Crocker, J.F.S. and Vernier, R.L.: Fetal kidney in organ culture: Abnormalities of development induced by decreased amounts of potassium. Science 169:485–487, 1970.

Wilson, J.G.; Maren, T.H.; Takano, K. and Ellison, A.C.: Teratogenic action of carbonic anhydrase inhibitors in the rat. Teratology 1:51–60, 1968.

2006 Potassium Ferrocyanide

Besedina (1987) studied the embryotoxic action of potassium ferrocyanide on white rats exposed to inhalation during pregnancy. The nontoxic concentration was found to be the dose 0.036 mg per m³.

Besedina, E.I.: Embryotoxic action of potassium ferrocyanides. Gigiena Naselennykh Mest (USSR), Kiev 26:72–74, 1987.

2007 Potassium Nitrate

Ema and Kanoh (1983) fed up to 2.5 percent in the diet to rats on days 7–14 of gestation. In the first–generation offspring there were no malformations or changes in behavior. The offspring were reared and bred and given similar diets. At the 2.5 percent diet level only, there were malformations, namely 6 per 133 fetuses. The malformations consisted of exencephaly, facial clefts and generalized edema.

Ema, M. and Kanoh, S.: Studies on the pharmacological bases of fetal toxicity of drugs. Fetal toxicity of potassium nitrate in two generations of rats. Folia Pharmacol. Jpn. 81:469–480, 1983.

2008 Potato Blight

Renwick (1972) published a hypothesis that anencephaly and spina bifida were associated with maternal exposure to an unknown substance in blighted potatoes. A geographical and temporal correlation between the severity of the late–blight of potatoes and the incidence of neural tube closure defects was made. An explanation for the higher incidence of these defects in the lower socioeconomic groups

could have been explained by this hypothesis. This work stimulated a large number of studies which were not supportive (see below). Renwick (1974) modified his approach to the hypothesis suggesting long term storage in the body of some component from potatoes. Other types of moldy foods perhaps containing cytochalasins might be involved (Shepard, 1973). Tea drinking (Fedrick, 1974) and corned beef consumption (Knox, 1974) also were shown to be more commonly found in the diets of mothers producing anencephalics.

A comprehensive review of potato ingestion and anencephaly can be found in Lemire et al. (1977).

The following paragraph lists the negative evidence against Renwick's hypothesis. Macmahan et al. (1973) in Maine, Elwood (1973) in eastern Canada, Field and Kerr (1973) in Australia, Smith et al. (1973) and Kinlen and Hewitt (1973) in Scotland were not able to associate blighted potato epidemics with neural tube closure defects. Clark et al. (1973) and Roberts et al. (1973) studied the dietary intake and storage methods of potatoes used by mothers giving birth to children with neural tube closure defects and neither could support the hypothesis that blighted potatoes was a causative factor. Emanuel (1972) pointed out that in Taiwan where potato eating is more common in the economically richer groups, a higher rate of neural defects persists in the poorer population.

Although Poswillo et al. (1972) reported midline skull defects in marmosets after maternal feeding of blighted potato, these workers were unable to duplicate their experiment (Poswillo et al., 1973). Allen et al. (1977) fed phytophthora blighted potatoes to pregnant rhesus monkeys and marmosets and produced no neural tube defects. Two of 32 rhesus fetuses had hydrocephalus.

The feeding of blighted potato to rats produced no congenital defects (Ruddick et al., 1974; Swinyard and Chaube, 1973). Keeler et al. (1975) found no teratogenicity from blighted or aged potatoes fed to pregnant rabbits, hamsters, rats and mice. A few neural defects were reported in the pig and rabbit by Sharma et al. (1978).

Allen, J.R.; Marlar, R.J.; Chesney, C.F.; Helgeson, J.P.; Kelman, A.; Weckel, K.G.; Traisman, E. and White, J.W.: Teratogenicity studies on late blighted potatoes in non–human primates (Macaca mulatta and Saguinus labiatus). Teratology 15:17–24, 1977.

Clark, C.A.; McKendrick, D.M. and Sheppard, P.M.: Spina bifida and potatoes. Br. Med. J. 3:251–254, 1973.

Elwood, J.M.: Anencephaly and potato blight in eastern Canada. Lancet 1:769 only, 1973.

Emanuel, I.: Non–tuberous neural tube defects. Lancet 2:879 only, 1972.

Fedrick, J.: Anencephalus and maternal tea drinking: evidence for a possible association. Proc. Roy. Soc. Med. 67:356–359, 1974.

Field, B. and Kerr, C.: Potato blight and neural–tube defects. Lancet 2:507–508, 1973.

Keeler, R.F.; Douglas, D.R. and Stallknecht, G.F.: The testing of blighted, aged, and control Russett Burbank potato tuber preparations for ability to produce spina bifida and anencephaly in rats, rabbits, hamsters and mice. Am. Potato J. 52:125–132, 1975.

Kinlen, L. and Hewitt, A.: Potato blight and anencephalus in Scotland. Brit. J. Prev. Soc. Med. 27:208–213, 1973.

Knox, E.G.: Anencephalus and dietary intakes. Proc. Roy. Soc. Med. 67:355–356, 1974.

Lemire, R.J.; Beckwith, J.B. and Warkany, J.: Anencephaly (chapter 2). Incidences, Etiology and Epidemiology. Raven Press, New York, 12–47, 1977.

MacMahon, B.; Yen, S. and Rothman, K.J.: Potato blight and neural–tube defects. Lancet 1:598–599, 1973.

Poswillo, D.E.; Sopher, D. and Mitchell, S.: Experimental induction of foetal malformation with blighted potato: A preliminary report. Nature 239:462–464, 1972.

Poswillo, D.E.; Sopher, D.; Mitchell, S.J.; Coxon, D.T.; Curtis, R.F. and Price, K.R.: Further investigations into the teratogenic potential of imperfect potatoes. Nature 244:367–368, 1973.

Renwick, J.H.: Hypothesis. Anencephaly and spina bifida are usually preventable by avoidance of a specific but unidentified substance present in certain potato tubers. Br. J. Prev. Soc. Med. 26:67–88, 1972.

Renwick, J.H.; Possamai, A.M. and Munday, M.R.: Potatoes and spina bifida. Proc. Roy. Soc. Med. 67:360–364, 1974.

Roberts, C.J.; Revington, C.J. and Lloyd, S.: Potato cultivation and storage in South Wales and its relation to neural tube malformation prevalence. Brit. J. Prev. Soc. Med. 27:214–216, 1973.

Ruddick, J.A.; Warwig, J. and Scott, P.M.: Nonteratogenicity in rats of blighted potatoes and compounds contained in them. Teratology 9:165–168, 1974.

Sharma, R.P.; Willhite, C.C.; Wu, M.T. and Salunkhe, D.K.: Teratogenic potential of blighted potato concentrate in rabbits, hamsters and miniature swine. Teratology 18:55–62, 1978.

Shepard, T.H.: Anencephaly and potatoes. Lancet 1:79 only, 1973.

Smith, C.; Watt, M.; Boyd, A.E.W. and Holmes, J.C.: Anencephaly, spina bifida and potato blight in the Edinburgh area. Lancet 1:269 only, 1973.

Swinyard, C.A. and Chaube, S.: Are potatoes teratogenic for experimental animals? Teratology 8:349–358, 1973.

2009 Povidone *Polyvinylpyrrolidone* CAS 9003–39–8

Claussen and Breuer (1975) studied this suspending agent by injecting it into rabbit yolk sacs on day 9. No increase in defects occurred.

Mahillon et al. (**1989**) showed that douching with povidone–I caused significant increase in iodine in the amniotic fluid. Siegemund and Weyers (**1987**) injected 0.5 ml per kg during organogenesis of rabbits and found reduced fetal weight but no increase in defects.

Claussen, U. and Breuer, H–W.: The teratogenic effects in rabbits of doxycycline, dissolved in polyvinylpyrrolidone, injected into the yolk sac. Teratology 12:297–302, 1975.

Mahillon, I.; Peers, W.; Bourdoux, P.; Ermans, A.M. and Delange, F.: Effect of vaginal douching with povidone–iodine

during early pregnancy on the iodine supply to mother and fetus. Biol. Neonate 56:210–217, 1989.

Siegemund, V.B. and Weyers, W.: Teratologische untersuchungen eines niedermolekularen polyvinylpyrrolidon–jod–komplesex an kaninchen. Arzneim–Forsch./ Drug Res. 37:340–341, 1987.

2010 Practolol

4–(2–Hydroxy–3–isopropylaminopropoxy)acetanilide CAS 6673–35–4

Ito et al. (1974) studied this β blocker in pregnant rats and mice and found no teratogenicity. Oral doses in the rat during organogenesis was associated with fetal weight reduction at 50, 250 and 750 mg per kg. The mice given 2,000 mg per kg had fetuses with reduced weight.

Ito, R.; Toida, S.; Kato, T. and Nakamoto, N.: Teratological safety of a new selective β blocker, practolol in rats and mice. Toho Igaku Zasshi 21:567–574, 1974.

2011 Prajmaline *N–Propylajmaliniumhydrogentartrate*

This antiarrhythmmic was studied by von Philipsborn and Stalder (**1972**) in the pregnant rat and mouse at the highest doses of 12 and 5 mg per kg respectively. At the highest dose in the rat fetal weight was reduced and fetal deaths increased.

von Philipsborn, V.G. and Stalder, B.: Vertraglichkeitsprufung von N–Propylajmaliniumhydrogentartrat in tierversuchen. Arzneim–Forsch. (Drug Res.)22:2085–2090, 1972.

2012 d–d–T80–Prallethrin *S–4068SF*

Saegusa et al. (1987) studied the effect of this insecticide on rabbits giving up to 10 mg per kg subcutaneously on days 6–18. No toxicity or teratogenicity was found.

Saegusa, T.; Naito, Y. and Narama, I.: Teratogenicity study of subcutaneously administered d–d–T80–prallethrin (S–4068SF) in rabbits. Oyo Yakuri 34:319–325, 1987.

2013 Pranidipine

Nishioeda et al. (**1993** A) gave up to 100 mg per kg to rats before and during the first week of pregnancy and found no fetal ill effects. This dihydropyridine antihypertensive was given by Takeuchi et al. (**1993**) during organogenesis in amounts up to 30 mg per kg and found no ill effects. The body and placental weights of the fetuses were increased. Peri and postnatal studies at doses up to 3.0 mg were done and showed no adverse effects on the behavior or reproduction of the offspring (Furuhashi et al., **1993**). A higher mortality in the pups of dams receiving 3 mg per kg was found. Rabbits exposed to 100 mg per kg had fetuses with reduced weight and increased death rates but no teratogenicity (Takeuchi et al., **1993**).

Furuhashi, T.; Kato, M.; Shinoda, A.; Yamashita, Y.; Kobayashi, Y.; Kimura, H. and Takeuchi, R.: Reproductive and developmental toxicity studies of OPC–13340 (4): Perinatal and postnatal study in rats by oral administration. Yakuri to Chiryo 21:1415–1430, 1993.

Nishioeda, R.; Matsuzawa, A.; Takeuchi, R.; Yoshida, M. and Tamagawa, M.: Reproductive and developmental toxicity studies of OPC–13340 (1): Fertility study in rats by oral administration. Yakuri to Chiryo 21:1375–1387, 1993A.

Takeuchi, R.; Kajiyoshi, K.; Azuma, K. and Tamagawa, M.: Reproductive and developmental toxicity studies of OPC–13340 (2): Teratogenicity study in rats by oral administration. Yakuri to Chiryo 21:1389–1403, 1993.

Takeuchi, R.; Azuma, K. and Tamagawa, M.: Reproductive and developmental toxicity studies of OPC–13340 (3): Teratogenicity study in rabbits by oral administration. Yakuri to Chiryo 21:1405–1413, 1993.

2014 Pranoprofen *Y–8004 2–(5H–)Benzopyrano [2–3–b] pyridin–7yl) proprionic Acid*

Hamada and Imamura (1976) studied this anti–inflammatory agent in pregnant mice and rats. At doses of up to 25 mg daily in the rat and 5 mg in the mouse during active organogenesis, no adverse fetal effects occurred. Postnatal studies were also negative.

Hamada, Y. and Imamura, H.: Teratological studies of 2–(5H–[1]Benzopyrano[2,3–b]pyridin–7yl)proprionic acid (Y–8004) in mice and rats. Iyakuhin Kenkyu 7:301–311, 1976.

2015 Pravadoline

This analgesic was given to male and female rats in amounts up to 300 mg per kg orally for a period before mating and to females for the 1st 19 days of gestation. An increase in wavy and bulbous ribs was found. Increased duration of pregnancy and increased neonatal death occurred.

Dennis, M.M.; Bradford, J.C.; Brown, G.L. and Blazak, W.F.: A three–generation study in rats with pravadoline administered orally. (abst)Teratology 41:548–549, 1990.

2016 Prazepam CAS 2955–38–6

Kuriyama et al. (1978) gave this tranquilizer orally to rats on days 8–17 in doses of up to 2,000 mg per kg. The pregnant dams had increased mortality at doses of 1,000 and 2,000 mg per kg. At these doses short tails and edema of the fetuses were found. At 250 mg per kg no effects were seen.

Kuriyama, T.; Nishigaki, K.; Ota, T.; Koga, T.; Okubo, M. and Otani, G.: Safety studies of prazepam (K–373) (VI). Teratological study in rats. Oyo Yakuri 15:797–811, 1978.

2017 Praziquantel

Muermann et al. (1976) found no teratogenicity in rats and rabbits given doses of up to 300 mg per kg.

Muermann, P.; Von Eberstein, M. and Frohberg, H.: Notes on the tolerance of Droncit. Summary of trial results. Vet Med Rev 2:142–165, 1976.

2018 Prazosin Hydrochloride CAS 19237–84–4

Noguchi and Ohwaki (1979) gave this hypotensive drug orally up to 300 mg per kg to pregnant rats on days 9–14 and observed no adverse effects in the fetuses. In rabbits on days 8–16 at 200 mg per kg, some fetal weight reduction and kinky tails was found. No teratogenic effects were found at this dose which was toxic to the dams. Late gestational treatment in the rat was associated with slight reduction in fetal weight and survival.

Noguchi, Y. and Ohwaki, Y.: Reproductive and teratologic studies with prazosin hydrochloride in rats and rabbits. Oyo Yakuri 17:57–62, 1979.

2019 Preeclampsia

Discussed under Hypertension

2020 Prednisolone also see *Cortisone* CAS 50–24–8

Pinsky and DiGeorge (1965) gave 0.5 mg daily to A–JAX mice during mid–pregnancy and produced a 77 percent incidence of cleft palate in the offspring. In clinically equivalent doses, these authors felt that hydrocortisone had less ability to produce cleft palate than prednisolone. Balika and Kartasheva (1979) gave 30 mg daily to rats on days 10–15 of gestation. Clinically equivalent doses caused a stimulation of erythropoiesis, but retardation of myelopoiesis in the offspring. The disturbance observed was considered to be a result of drug transplacental passage.

Hasegawa et al. (1974) studied betamethasone, 17, 21–dipropionate, a long–acting derivative of prednisolone in pregnant rats and mice. Doses up to 2.5 mg per kg were given subcutaneously during midpregnancy. Adrenal hypertrophy and hemorrhage along with decreased viability occurred in the rat fetus. Also cleft palates and umbilical hernias were increased in the rat fetus. In the mouse fetus at the 2.5 mg dose level, 96 percent of the fetuses had cleft palates compared to a 11 percent incidence in those treated with 20.0 mg prednisolone.

Koga et al. (1980A,B) studied the 17 valerate 21–acetate form of prednisolone in the pregnant rat and rabbit. Using up to 10 mg per kg subcutaneously on days 7 through 17 in the rat, they found growth retardation and omphaloceles in the fetuses. In the rabbit using 0.1 and 0.25 mg per kg subcutaneously on days 6 through 18, they found increased resorptions and cleft palates in the fetuses. Fertility studies in the rat using 1.0 mg per kg were done and no adverse effects were found (Koga et al., 1980C).

Balika, Yu.D. and Kartasheva, V.E.: Transplacental action of prednisolone on fetuses blood–forming system. Akush. Ginekol. (USSR) 9:29–30, 1979.

Hasegawa, Y.; Yoshida, T.; Kozen, T.; Ohara, T.; Okamoto, A.; Sakaguchi, I. and Kozen, T.: Teratology studies on βmethasone 17,21–diproprionate, prednisolone and βmethasone 21–disodium phosphate in mice and rats. (Japanese) Oyo Yakuri 8:705–720, 1974.

Koga, T.; Ota, T.; Aoki, Y.; Nishigaki, K. and Suganuma, Y.: Reproductive studies of prednisolone 17–valerate 21–acetate teratologic studies in the rats. Oyo Yakuri 20:67–86, 1980A.

Koga, T.; Ota, T.; Aoki, Y. and Suganuma, Y.: Reproductive studies of prednisolone 17–valerate 21–acetate teratologic studies in rabbits. Oyo Yakuri 20:87–98, 1980B.

Koga, T.; Ota, T.; Nishigaki, K.; Aoki, Y. and Suganuma, Y.: Fertility study in rats of prednisolone 17–valerate 21 acetate. Yakuri to Chiryo 8:2169–2181, 1980C.

Pinsky, L. and DiGeorge, A.M.: Cleft palate in the mouse: A teratogenic index of glucocorticoid potency. Science 147:402–403, 1965.

2021 Prednisolone Farnesylate

Taniguchi et al. (**1992** A) gave adult rats up to 1 mg per kg subcutaneously before pregnancy and during the 1st 7 days and found no reproductive ill effects. Taniguchi et al. (**1992** B) gave up to 25 mg per kg during rat organogenesis and found no teratogenicity. Taniguchi et al. (**1992** C) gave up to 5 mg per kg to rats in the perinatal period and found decreased fetal weight but no changes in behavioral development. In rabbits, Aso et al. ok92) gave up to 0.25 mg per kg during organogenesis and found fetal weight decrease but no increase in defects at the highest dose.

Aso, S.; Morita, S.; Kajiwara, Y.; Horiwaki, S.; Tanaka, H. and Shinomiya, M.: Reproductive and developmental toxicity study of prednisolone farnesylate (PNF)–teratogenicity study in rabbits by subcutaneous administration. The Journal of Toxicological Sciences 17:241–250, 1992.

Taniguhi, q.; Yoshitomi, H.; Miyazaki, K.; Koga, N.; Himeno, Y.; Hara, Y.; Nakamura, M.; Tsuji, M.; Tanaka, H. and Shinomiya, M.: Reproductive and developmental toxicity study of prednisolone farnesylate (PNF)–study by subcutaneous administration of PNF prior to and in the early stages of pregnancy in rats. The Journal of Toxicological Sciences 17:201–215, 1992A.

Taniguchi, H.; Himeno, Y.; Chono, M.; Araki, E.; Nakamura, M.; Tsuji, M.; Tanaka, H. and Shinomiya, M.: Reproductive and developmental toxicity study of prednisolone farnesylate (PNF)–study by subcutaneous administration of PNF during the period of fetal organogenesis in rats. The Journal of Toxicological Sciences 17:217–239, 1992B.

Taniguchi, H.; Araki, E.; Nagai, M.; Masaki, K.; Mizoguchi, S.; Nakamura, M.; Tsuji, M.; Tanaka, H. and Shinomaya, M.: Reproductive and developmental toxicity study of prednisolone farnesylate (PNF)–study of subcutaneous administration of PNF during the perinatal and lactation periods in rats. The Journal of Toxicological Sciences 17:251–267, 1992C.

2022 Prednisone also see *Cortisone* CAS 53–03–2

Renisch et al. (1978) gave mice 100 or 400 micrograms subcutaneously from day 13 until day 18 and observed significant weight reduction in the offspring. They also reported reduced weight in the newborns of 119 women treated with 10 mg daily for infertility problems. The newborn infants weighed approximately 300 gm less than the controls. Unfortunately, in neither the mouse nor the human study was the important effect of maternal weight gain reported or discussed.

Renisch, J.M.; Simon, J.N.; Karow, W.G. and Gandelman, R.: Prenatal exposure to prednisone in humans and animals retards intrauterine growth. Science 202:436–438, 1978.

2023 Pregnancy Test Tablets

Gal (1972) summarized the evidence that oral hormone pregnancy test tablets (Primodos™ or Amenorone Forte™) are associated with neural tube malformations. From a group of 100 mothers giving birth to defective children 19 had been given the hormone pregnancy test. In a matched control of 100 only four women were tested. Maternal age and the number of infections in the test group were also increased significantly but were shown not to account for the increased malformation rate. Smithells (1965) found only three congenital defects (cardiac) among 189 infants whose mothers were pregnancy tested with tablets. Oakley et al. (1973) reported a survey which gave no definite evidence for teratogenicity of this treatment. Dubowitz (1962) reported a single case of a virilized female with a cloaca following pregnancy testing sometime after the 35th gestational day.

Neither Smithells (1965) nor Kullander and Kallen (1976) found hypospadias in the male offspring of 253 exposed mothers. Lammer and Cordero (1986) in a case control study of malformed children, found that the exposure odds ratio for esophageal atresia tripled among the 72 with the condition. Twelve of the 72 had exposure to hormonal pregnancy tests while in the other malformation categories, 140 of 2,110 were exposed.

Pulkkinen et al. (1984) carried out a double–blind randomized trial of 25 women undergoing therapeutic abortion. After a single dose of norethisterone acetate (20 mg) and ethinylestradiol (0.4 mg), the curettage material was studied after 96 hours for evidence of hemorrhage. No increase in hemorrhage into the endometrium or any change in progesterone, estradiol, hydroxyprogesterone or follicle stimulating hormone levels were found as compared to the 25 controls.

Dubowitz, V.: Virilisation and malformation of a female infant. Lancet 2:405–406, 1962.

Gal, I.: Hormonal imbalance in human reproduction. In: Advances in Teratology, D.H.M. Woollam (ed.), New York: Academic Press, 5:161–173, 1972.

Kullander, S. and Kallen, B.: A prospective study of drugs and pregnancy. Acta Obstet. Gynecol. Scand. 55:221–224, 1976.

Lammer, E.J. and Cordero, J.F.: Exogenous sex hormone exposure and the risk for major malformations. JAMA 255:3128–3132, 1986.

Oakley, G.P.; Flynt, J.W. and Falek, A.: Hormone pregnancy tests and congenital anomalies. Lancet 2:256–257, 1973.

Pulkkinen, M.O.; Dusterberg, B.; Hasan, H.; Kivikoski, A. and Laajoki, V.: Norethisterone acetate and ethinylestradiol in early human pregnancy. Teratology 29:241–249, 1984.

Smithells, R.W.: The problem of teratogenicity. Practitioner 194:104–110, 1965.

2024 Prenalterol

L–1–[4–Hydroxyphenoxy]–3–isopropyl–amino–2–propanol

CAS 57526–81–5

Bruyere et al. (1983) produced heart defects in chick embryos exposed topically to this cardiac β–receptor stimulant. At stages 17 through 25, 30 mM injections produced arch and ventricular septal defects.

Bruyere, H.J.; Matsuoka, R.; Carlsson, E.; Cheung, M.O.; Dean, R. and Gilbert, E.F.: Cardiovascular malformations associated with administration of prenalterol to young chick embryos. Teratology 28:75–82, 1983.

2025 Prifinium Bromide 1,1

Diethyl–2–methyl–3–diphenylmethylene–pyrrolidinium bromide

Kumada et al. (1972) administered this atropine–like drug to pregnant rats, mice, and rabbits and found no adverse fetal effects. The maximum doses were 50 and 20 mg pr kg to mice by the oral or subcutaneous route, in rats 100 mg per kg orally and in rabbits 1.0 mg per kg. The treatment was carried out during organogenesis.

Kumada, S.; Watanabe, N. and Nakai, T.: Toxicological and teratological studies of 1,1-Diethyl-2–methyl–3–diphenylmethylene–pyrrolidinium bromide (prifinium bromide), a new atropine–like drug. Arzneim–Forsch. 22: 706–710, 1972.

2026 Primidone *Mysoline™ Primaclone* CAS 125–33–7

This anticonvulsant has a major metabolite which is a barbiturate. Rating et al. (1982) followed 14 mothers treated with only primidone or in combination with other drugs. One offspring had a ventricular septal defect and five of the thirteen newborns had head circumferences at or less than the third percentile. Two were slightly retarded and one was severely retarded in psychomotor development. Myhre and Williams (1981) reported two children exposed solely to the drug during gestation. Both had low nasal bridges and ocular hypertelorism. One had pulmonic stenosis and the other developmental delay. Rudd and Freedom (1979) reported another case. More clinical data is needed to assess the associated risk.

Nakane et al. (1980) studied 133 women who took primidone usually with other seizure medications. Eighteen malformations were found and they included menigocele (1), cleft lip (4), cleft lip and palate (6), heart disease (7) and other types of defects. The increase in cleft lip and palate was significant at the p < 0.05 level.

Kaneko et al. (1988) studied 172 pregnancies in which epileptic drugs were used. For the 31 in which only one drug was used, there was 6.5 percent malformations while among the remaining 141 patients which were treated with more than one drug, the rate was 15.6 percent. By Wilcoxon rank–sum tests, valproate use was significantly associated with a defect increase and carbamazepine was almost significantly associated (p = 0.053). Among the polydrug users who used carbamazepine, there were 11 malformations among 105. When primidone was used in polydrug use, there were 8 malformations among 80 cases. In this study of Japanese children, 4 facial clefts and 3 heart defects occurred among 172.

Kaneko, S.; Otani, K.; Fukushima, Y.; Ogawa, Y.; Nomura, Y.; Ono, T.; Nakane, Y.; Teranishi, T. and Goto, M.: Teratogenicity of antiepileptic drugs: Analysis of possible risk factors. Epilepsia 29:459–467, 1988.

Myhre, S.A. and Williams, R.: Teratogenic effects associated with maternal primidone therapy. J. Pediatr. 99:160–162, 1981.

Nakane, Y.; Okuma, T.; Takahashi, R.; Sato, Y.; Wada, T.; Sato, T.; Fukushima, Y.; Kumashiro, H.; Ono, T.; Takahashi, T.; Aoki, Y.; Kazamatsuri, H.; Inami, M.; Komai, S.; Seino, M.; Miyakoshi, M.; Tanimura, T.; Hazama, H.; Kawahara, R.; Otsuki, S.; Hosokawa, K.; Inanaga, K.; Nakazawa, Y. and Yamamoto, K.: Multi–institutional study on the teratogenicity and fetal toxicity of antiepileptic drugs: A report of a collaborative study group in Japan. Epilepsia 21:663–680, 1980.

Rating, D.; Nau, H.; Jager–Roman, E.; Gopfert–Geyer, I.; Koch, S.; Beck–Mannagetta, G.; Schmidt, D. and Helge, H.: Teratogenic and pharmacokinetic studies of primidone during pregnancy and in the offspring of epileptic women. Acta Paediatr. Scand. 71:301–311, 1982.

Rudd, N.L. and Freedom, R.M.: A possible primidone embryopathy. J. Pediatr. 94:835–837, 1979.

2027 Probucol

4,4-(Isopropylidenedithio)bis(2,6d-t-butyl-phenol) CAS 23288–49–5

Rosa (**1994**) reported only one defect among the offspring of 15 women taking this medication. Molello et al. (1979) gave up to 1000 mg per kg to pregnant rats during organogenesis and found no adverse effects. Rabbits were also treated with unspecified amounts and no adverse effects seen.

Molello, J.A.; Thompson, D.J. and LeBeau, J.E.: Eight year toxicity study in monkeys and reproduction studies in rats and rabbits with a new hypocholesterolemic agent, probucol. (abs) Toxicol. Appl. Pharmacol. 648:A98, 1979.

Rosa, F.: Anti–cholesterol agent pregnancy exposure outcomes. (abs) Reproductive Toxicology 8:445–446, 1994.

2028 Procaine

Heinonen et al. (1977) reported no increase in defects among 1,340 women treated with procaine in the first four lunar months. Mellin (1964) conducted a case control study of 266 infants with defects and found that the frequency of maternal exposure was not increased over controls.

Heinonen, O.P.; Slone, D. and Shapiro, S.: Birth Defects and Drugs in Pregnancy. Publishing Sciences Group Inc., Littleton, Mass., 1977.

Mellin, G.W.: Drugs in the first trimester of pregnancy and the fetal life of Homo sapiens. Am. J. Obstet. Gynecol. 90:1169–1180, 1964.

2029 Procarbazine

N–Isopropyl–α–(2–methylhydrazino)–p–toluamide CAS 671–16–9

Mennuti et al. (1975) reported a woman treated with 100 mg daily of this drug from the 28th through the 35th gestational days was found to have a fetus with small pelvic kidneys. She also received as treatment of her Hodgkin's disease, a single intravenous dose of nitrogen mustard and vincristine on the 28th day. Wells et al. (1968) reported a normal infant from a woman treated in early pregnancy. Schapira and Chudley (1984) reported the delivery of a normal 2.3 kg newborn after treatment with procarbazine and BCNU during the first two trimesters. From this review of the literature, two of six offspring exposed in the first trimester were malformed.

Tuchmann–Duplessis and Mercier–Parot (1967) administered this hydrazine orally to rats on the 8th to 14th days of gestation in a dose of 5 to 10 mg per kg. Treatment before the 12th day produced almost exclusively eye defects whereas after the 12th day they found defects of the limbs. Chaube and Murphy (1968) did teratologic studies in the rat and found that five 1–methyl–2–benzyl hydrazines derivatives were similarly teratogenic. Ivankovic (1972) produced neurogenic tumors in the offspring of rats given 125 mg per kg intravenously on day 22 of gestation. Johnson et al. (1985) found a dose–related reduction in fetal brain weight when 1–10 mg per kg was given gravid rats on days 12–15. No obvious behavioral changes occurred.

Chaube, S. and Murphy, M.L.: The teratogenic effects of the recent drugs active in cancer chemotherapy. In: Advances in Teratology, D.H.M. Woollam (ed.), New York: Academic Press, 3:181–237, 1968.

Ivankovic, S.: Erzeugung von Malignomen bei Ratten nach Transplazentarer Einwirkung von N–isopropyl–α––2–(methyl–hydrazino)–p–tolvamid HCl. Arzneim–Forsch 22:905–907, 1972.

Johnson, J.M.; Thompson, D.J.; Haggerty, G.C.; Dyke, I.L. and Lower, C.E.: The effect of prenatal procarbazine treatment on brain development in the rat. Teratology 32:203–212, 1985.

Mennuti, M.T.; Shepard, T.H. and Mellman, W.J.: Fetal renal malformation following treatment of Hodgkin's disease during pregnancy. Obstet. Gynecol. 46:194–196, 1975.

Schapira, D.V. and Chudley, A.E.: Successful pregnancy following continuous treatment with combination chemotherapy before conception and throughout pregnancy. Cancer 54:800–803, 1984.

Tuchmann–Duplessis, H. and Mercier–Parot, L.: Production chez le rat de malformations oculaires et squelettiques par administration d'une methyl–hydrazine. C. R. Soc. Biol. (Paris) 161:2127–2131, 1967.

Wells, J.H.; Marshall, J.R. and Carbone, P.P.: Procarbazine therapy for Hodgkin's disease in early pregnancy. JAMA 205:935–937, 1968.

2030 **Procaterol** *5–(1–Hydroxy–2–isopropylaminobutyl)–8–Hydroxycarbostyril Hydrochloride Hemihydrate* CAS 60443–17–6

This β adrenergic stimulant was administered orally to rats on days 7 through 17 in doses as high as 250 mg per kg daily (Minami et al., 1979). Retarded ossification occurred in the treated fetuses. At 250 mg per kg, there was an increase in number of enlarged renal pelves in the fetuses. This oc-

curred in 52 percent of the treated and 28.7 percent of the controls. In the rabbit, doses of up to 500 mg per kg on days 6 through 18 were given orally (Tamagawa et al., 1979). The fetal weight was reduced and fetal deaths increased at the 150 mg per kg level. No increase in defects was found.

Minami, J.; Hatori, M. and Tanaka, N.: Reproduction studies of procaterol. 2. Teratogenicity study in rats. (Japanese). Iyakuhin Kenkyu 10:102–111, 1979.

Tamagawa, M.; Kita, K.; Okabe, M. and Tanaka, N.: Reproduction studies of procaterol. 3. Teratogenicity study in rabbits. Iyakuhin Kenkyu 10:80–101, 1979.

2031 **Prochlorperazine** *Prochlorpemazine Compazine™* CAS 58–38–8

In a group of 4,295 pregnant women, Mellin (1975) identified 74 who took Compazine™ during their pregnancy. No increase in malformation rate or pattern of malformations was found. Among the 543 offspring of mothers taking this medication in the first 84 days of gestation, Milkovich and Van den Berg (1976) found no increase in malformations. Heinonen et al. (1977) found 877 mothers who took the drug during the first four lunar months and no increased defect rate was found.

Vorhees et al. (1979) gave 20 mg per kg orally to rats from days 7 to 20 and found a significant postnatal weight decrease and increase in fetal mortality. Only minor behavioral changes occurred. Roux (1959) administered this tranquilizer to pregnant mice and rats and found an increased incidence of cleft palate. A few anencephalic defects and one double monster were observed. The rats received 2.5 to 10 mg per day parenterally or 10 to 20 mg per day orally.

Heinonen, O.P.; Slone, D. and Shapiro, S.: Birth Defects and Drugs During Pregnancy. Publishing Sciences Group Inc., Littleton, Mass., 1977.

Mellin, G.W.: Report of prochlorperazine utilization during pregnancy from fetal life study data bank. (abs) Teratology 11:28a, 1975.

Milkovich, L. and Van den Berg, B.J.: An evaluation of the teratogenicity of certain antinauseant drugs. Am. J. Obstet. Gynecol. 125:244–248, 1976.

Roux, C.: Action teratogene de la prochlorpemazine. Arch. Fr. Pediatr. 16:968–971, 1959.

Vorhees, C.V.; Brunner, R.L. and Butcher, R.E.: Psychotropic drugs as behavioral teratogens. Science 205:1220–1225, 1979.

2032 **Progesterone** *Progestins* CAS 57–83–0

Although there are many reports in the human of masculinization by the synthetic progestins (see 17–α–ethinyl–testosterone), the use of progesterone itself has been infrequently reported. Hayles and Nolan (1957) reported two masculinized female infants whose mothers received progesterone (10 mg by injection on 3 days and in the second case, up to 60 mg by mouth).

Aarskog (1979) reported evidence that progesterone at doses of 5 to 250 mg per day may be associated occasionally with hypospadias. Among 130 patients with hypospadias, there were 11 whose mothers were exposed during pregnancy to progestins (medroxyprogesterone, hydroxyprogesterone, nor–ethisterone). In the control group of mothers giving birth to infants with facial clefts, there were only 2 of 111 with similar exposures. He summarized two similar studies of infants with hypospadias, both of which found a slight increased number of exposed mothers (Sweet et al., 1974; Kupperman, 1961). Mau (1981) studied 33 males with hypospadias and found that eight were exposed to progestins during early pregnancy. This was not a significant increase over the 11 percent of control mothers exposed to progestins. The degree of hypospadias was not related to the period in pregnancy during which the medication was taken.

Ressequie et al. (1985) found no increase in defects among 244 offspring exposed to progestins. Rock et al. (1985) followed 42 pregnancies treated with progesterone suppositories (average total dose 2236) and observed 28 percent spontaneous abortions but no malformations. They also followed 45 pregnancies treated with intramuscular progesterone (average total dose, 1009 mg) and found 5.8 percent spontaneous abortions and two malformations (a unilateral undescended testes and a meningomyelocele). Kallen et al. (**1991**) studied fertility in 846 women giving birth to children with hypospadias and found no significant differences.

Yerushalmy (1972), in an analysis of hormones used for bleeding, concluded that the bleeding was associated with an increase in malformation rate but the hormone treatment without bleeding caused no increase. Simpson (1985) summarized reports which indicate a lack of evidence for human teratogenicity. This includes many papers on hypospadias and defects of limb, heart and neural tube.

Neither progesterone nor 17–hydroxyprogesterone were shown to masculinize the external genitalia of female fetuses from experimental animals. Johnstone and Franklin (1964) injected 0.25 mg from day 16 to day 19 in mice with negative results. Revesz et al. (1960), using up to 200 mg in the rat, reported no virilization of the female fetuses. Suchowsky and Junkmann (1961) found no virilizing effects in the rat fetus exposed to progesterone or 17–hydroxyprogesterone.

Kawashima et al. (**1977**) studied female rat offspring exposed to up to 200 mg per kg daily and found no change in urogenital differentiation.

Schardein (**1993**) has comprehensively reviewed the human and animal data on progesterone and progestins.

Aarskog, D.: Maternal progestins as a possible cause of hypospadias. N. Eng. J. Med. 300:75–78, 1979.

Hayles, A.B. and Nolan, R.B.: Masculinization of the female fetus, possibly related to administration of progesterone during pregnancy. Proceedings of Staff Meeting, Mayo Clinic 32:200–203, 1957.

Johnstone, E.E. and Franklin, R.R.: Assay of progestins for fetal virilizing properties in the mouse. Obstet. Gynecol. 23:359–362, 1964.

Kallen, B.; Castilla, E.E.; Kringelbach, M.; Lancaster, P.A.L.; Martinez-Frias, M.L.; Mastroiacovo, P.; Mutchinick, O. and Robert, E.: Parental fertility and infant hypospadias: an international case–control study. Teratology 44:629–634, 1991.

Kawashima, K.; Nakaura, S.; Nagao, S.; Tanaka, S.; Kuwamura, T. and Omori, Y.: Virilizing activities of various steroids in female rat fetuses. Endocrinol. Japon. 24:77–81, 1977.

Kupperman, H.S.: Progesterone and Related Steroids in the Management of Abortion. Brook Lodge Symposium, Barnes, A.K. (ed). Brook Lodge Press, Augusta, 105–107, 1961.

Mau, G.: Progestins during pregnancy and hypospadias. Teratology 24: 285–287, 1981.

Ressequie, L.J.; Hick, J.F.; Bruen, J.A.; Noller, K.L.; O'Fallon, W.M. and Kurland, L.T.: Congenital malformations among offspring exposed in utero to progestins, Olmstead County, Minnesota, 1936–1974. Fertil. Steril. 43:514–519, 1985.

Revesz, C.; Chappel, C.I. and Gaudry, R.: Masculinization of female fetuses in the rat by progestational compounds. Endocrinol. 66:140–144, 1960.

Rock, J.A.; Wentz, A.C.; Kimball, A.W.; Zacur, H.A.; Early, S.A. and Jones, G.S.: Fetal malformations following progesterone therapy during pregnancy: A preliminary report. Fertil. Steril. 44:17–20, 1985.

Schardein, J.L.: Chemically Induced Birth Defects 2nd Edition. Marcel Dekker, New York, 1993, pp384–301.

Simpson, J.L.: Relationship between congenital anomalies and contraception. Adv. Contracept. 1:3–30, 1985.

Suchowsky, G.K. and Junkmann, K.: Study of the virilizing effect of progestogens on the female rat fetus. Endocrinol. 68:341–349, 1961.

Sweet, R.A.; Schrott, H.G.; Kurland, R. and Culp, A.S.: Study of the incidence of hypospadias in Rochester, Minnesota, 1940–1970, and a case–control comparison of possible etiologic factors. Mayo Clin. Proc. 49:52–58, 1974.

Yerushalmy, J.: Methodologic problems encountered in investigating the possible teratogenic effects of drugs. In: Drugs and Fetal Development, M.A. Klingberg, A. Abramovici and J. Chemke (ed.), Plenum Press, New York, 427–440, 1972.

2033 Proglumide

D–L–4–Benzamido–N–N–dipropylglutaramic Acid CAS 6620–60–6

This compound used in peptic ulcer therapy was given orally to pregnant rats and mice during major organogenesis. At the highest doses (rats 3,350 mg per kg, mouse 225 mg per kg), no gross defects or postnatal changes were found, but an increase in extra ribs and abnormal vertebrae was observed (Ishizaki et al., 1971).

Ishizaki, O.; Saito, G. and Kagiwada, K.: The pharmacological study on proglumide 4. Effects of proglumide (KXM) on pre– and post–natal developments of the offsprings. Oyo Yakuri 5:225–237, 1971.

2034 Proguanil CAS 500–92–5

Discussed under Cycloguanil

2035 ProHance ™

This MRI agent was given intravenously to rats during

gestation. At doses of 6 mmol per kg per day some of the animals died but not fetal effects were found (Lochry et al., **1993**).

Lochry, E.A.; Myhre, J.L. and Martin, P.M.: Fertility and general reproduction evaluation of ProHance ™ in rats. (abs) Teratology 47:426, 1993.

2036 Prolinomethyltetracycline

Fujita et al. (1972a,b) injected this compound in amounts up to 300 mg per kg intraperitoneally into mice and rats during major organogenesis. Pyrrolidin–methyl tetracycline was also tested in the mouse and both drugs produced some maternal toxicity and death at doses of 150 mg or more per kg as well as an increase in the incidence of polydactyly in the mice. Polydactyly of the hindlimbs was also found in tetracycline–treated mice (150 mg per kg). Postnatal studies were done in both species without major findings.

Fujita, M.; Moriguchi, M. and Koeda, T.: Teratological studies on prolinomethyltetracycline (PM–TC) in mice. Iyakuhin Kenkyu 3:75–81, 1972a.

Fujita, M.; Moriguchi, M. and Koeda, T.: Teratological studies on prolinomethyltetracycline (PM–TC) in rats. Iyakuhin Kenkyu 3:69–74, 1972b.

2037 L–prolyl–l–leucyl–glycinamide *MIF*
Melanocyte–stimulating horomone releasing factor

This melanocyte stimulating hormone inhibitor was studied in chick eggs with injection on days 1.5, 2, 3 or 4. when amounts of as little as 2 times $10-5^{v-1}$ moles was injected. After 1.5 days malformations were increased (Kosar and Vanzura, 1988). Body wall and facial defects predominated.

Kosar, K. and Vanzura, J.: Embryotoxicity of l–prolyl–l–leucyl–glycinamide, cyclo(1–amino–cyclopentanecarbonyl–alanyl) and cyclo(glycyl–leucyl), new potential neuropeptides in chick embryos. Pharmazie 43:715–716, 1988.

2038 Promazine CAS 58–40–2

The Collaborative Perinatal Project monitored 50 women exposed in the first trimester and found no adverse effects including intelligence testing at age four years (Heinonen et al., 1977).

Murphree et al. (1962) observed an increased postnatal mortality in rats when the mother was given 5 mg per kg during 18 days of gestation.

Heinonen, O.P.; Slone, D. and Shapiro, S.: Birth Defects and Drugs in Pregnancy. Publishing Sciences Group Inc., Littleton, Mass., 1977.

Murphree, O.D.; Monroe, B.L. and Seager, L.D.: Survival of offspring of rats administered phenothiazines during pregnancy. J. Neuropsych. 3:295–297, 1962.

2039 Promethazine *Phenergan™* CAS 60–87–7

This phenothiazine antihistamine was given to 114 women daily during the first four lunar months and 746 anytime during pregnancy and no significant increase

in malformation risk was found (Heinonen et al., 1977). Wheatley (1964) reported on the offspring of 165 women who took the drug in the first trimester and found no increase in malformations or other adverse effects. Respiratory depression in the perinate was reported but others have not confirmed this (Briggs et al., 1983).

Briggs, G.G.; Bodendorfer, T.W.; Freeman, R.K. and Yaffe, S.J.: Drugs in Pregnancy and Lactation. Williams and Wilkins, Baltimore, 1983.

Heinonen, O.P.; Slone, D. and Shapiro, S.: Birth Defects and Drugs in Pregnancy. Publishing Sciences Group Inc., Littleton, Mass., 1977.

Wheatley, D.: Drugs and the embryo. Br. Med. J. 1:863 only, 1964.

2040 Prometryn

Lyapkalo et al. (1986) found that herbicides of cimtriazine type (cimazine, cimerone and prometrine) possess embryotoxic properties.

Lyapkalo, A.A. and Zdolnik, T.G.: Susceptible periods of embryogenesis as revealed by cim–triazines. Proceedings of Ryazan Medical Institute (USSR) 89:118–121, 1986.

2041 Propanidid

This short acting non–barbiturate used in anesthesia was studied by Giovanelli et al. (1979) in rats anesthetized briefly with the agent. No teratogenic effects were found.

Giovanelli, L.; Brandolin, P.; Zanozi, A.; Fregnan, L. and Sarorelli, M.: Fetal abnormolities induced by anesthetic drugs in early stages of pregnancy. Riv. Ital. Ginecol 53:770–777, 1969.

2042 Propanediol–1,3 CAS 504–63–2

Gebhardt (1968) tested a number of glycols in the developing chick and found that propanediol–1,3 caused micromelia if given by injection into the air or yolk sac on the 4th day. He used 0.05 ml of the compound. It also produced micromelia if injected at the beginning of incubation. Similar amounts of propylene glycol (propanediol–1,2), butanediols, glycerol and diethyl and ethyl glycol were not teratogenic. Glycerol, propylene glycol and propanediol–1,3 were all moderately lethal when injected into the air cell on the 4th day. Studies in rats and rabbits exposed to 500 ppm and summarized by Kimmel et al. (1984). Teratogenicity was not found. Schumacher et al. (1968) administered 20 mg of propylene glycol intravenously to rabbits on days 8–12 and found no increase in defective fetuses but growth retardation and resorptions increased.

Gebhardt, D.O.E.: The teratogenic action of propylene glycol (propanediol–1,2) and propanediol–1,3 in the chick embryo. Teratology 1:153–162, 1968.

Kimmel, C.A.; LaBorde, J.B. and Hardin, B.D.: Reproductive and developmental toxicology of selected epoxides. In: Toxicology and the Newborn, Chapter 13, S. Kacew and M.J. Reasor (eds.), Elsevier Sci. Publish., Amsterdam, 270–287, 1984.

Schumacher, H.J.; Blake, D.A.; Gurian, J.M. and Gillette, J.R.: A comparison of the teratogenic activity of thalidomide in rabbits and rats. J. Pharmacol. Exp. Therap. 160:189–200, 1968.

2043 N–Propanol

Nelson et al. (**1988**) exposed rats to 3,500, 7,000 ir 10,000 ppm for 7 hours daily on days 1–19. At the two highest doses skeletal and heart defects wre increased. At 3,500 ppm fetal weight was reduced.

Nelson, B.K.; Brightwell, W.S.; MacKenzie–Taylor, D.R.; Khan, A.; Burg, J.R. and Weigel, W.W.: Teratogenicity of n–propanol and isopropanol administered at high inhalation concentrations to rats. Fd. Chem. Toxic. 26:247–254, 1988.

2044 Propentofylline

Kitatani et al. (1986) gave up to 500 mg per kg on days 6–15 in the mouse and observed no increase in defects or behavioral changes in the offspring. Akaike et al. (1986) found no teratogenicity in rabbits given up to 150 mg per kg daily during days 6–18.

Akaike, M.; Sugisaki, T. and Miyamoto, M.: Propentofylline: Teratological study of propentofylline given orally in rabbits. Pharmacometrics 31(2):397–407, 1986.

Kitatani, T.; Akaike, M.; Sugisaki, T.; Takagi, S. and Miyamoto, M.: Teratological study of propentofylline given orally in mice. Oyo Yakuri 31(2):373–385, 1986.

2045 Propionaldehyde

Slott and Hales (1985) injected 10, 100 or 1000 micrograms directly into the day 13 rat amnion and found a dose dependent increase in embryolethality but no significant increase in defects.

Slott, V.L. and Hales, B.F.: Teratogenicity and embryolethality of acrolein and structurally related compounds in rats. Teratology 32:65–72, 1985.

2046 Propionic Acid

Dawson (**1991**) using a frog embryo teratogenesis assay found that 50% of the survivors were malformed at a concentration of 1,514 mg per liter. Microcephaly, abnormal gut coiling and edema were the most common defects.

Dawson, D.A.: Additive incidence of developmental malformation for xenopus embryos exposed to a mixture of ten aliphatic carboxylic acids. Teratology 44:531–546, 1991.

2047 Propionitrile CAS 107–12–0

This compound was injected intraperitoneally into hamsters on day 8 and at 30 mg per kg, exencephaly and encephaloceles were increased among the fetuses. Sodium thiosulfate, a cyanide antagonist, had a protective action. (Willhite et al. **1981**) Johannsen et al. (1986) gavaged up to 80 mg per kg to rats on days 6–19 and found no teratogenicity but at high levels, maternal and fetal toxicity were found.

Johannsen, F.R.; Levinskas, G.J.; Berteau, P.E. and Rodwell, D.E.: Evaluation of the teratogenic potential of three aliphatic nitriles in the rat. Fund. Appl. Toxicol. 7:33–40, 1986.

Willhite, C.C.; Ferm, V.H. and Smith, R.P.: Teratogenic effects of aliphatic nitriles. Teratology 23(3):317–324, 1981.

2048 Propionyllencomycin

This antimicrobial agent was given orally to rats in dosages of up to 1000 mg per kg during the first 7 days, 7–17 and 17–21 days of gestation. No adverse effects occurred (Sasaki et al., 1984, Kobayashi et al., 1984A,B).

Kobayashi, Y.; Suda, M.; Sasaki, M.; Kubota, H. and Hayano, K.: Reproductive study of TMS–19–Q. Chemotherapy 32:209–213, 1984A.

Kobayashi, Y.; Suda, M.; Sasaki, M.; Kubota, H. and Hayano, K.: Reproduction study of TMS-19-Q. Perinatal and postnatal studies in rats. Chemotherapy 32:214–221, 1984B.

Sasaki, M.; Suda, M.; Kubota, H.; Kobayashi, Y. and Hayano, K.: Reproduction study of TMS-19-Q. Teratological study in rats. Chemotherapy 32:200–208, 1984.

2049 Propiramfumarat

N–(1–Methyl–2–piperidino–athyl–N–(2–pyridyl) propionamid–fumarat)

Tettenborn (1974) gave rats and rabbits oral doses of up to 450 mg per kg and 300 mg per kg, respectively. Treatment during the period of organogenesis did not produce adverse fetal effects.

Tettenborn, D.: Toxikologische Untersuchungen mit Propiramfumarat. Arzneim–Forsch. 24:624–631, 1974.

2050 Propiverine Hydrochloride

Saito et al. (**1989**,A) gave rats up to 50 mg per kg orally on days 7–17. Maternal toxicity occurred at 10 mg per kg. Except for decreased fetal weight no adverse fetal effects were found. Postnatal development and reproduction was uneffected. Some decrease in viability of perinatally treated pups was found. (Saito et al., **1989**, B) Fertility was not changed in rats treated before pregnancy. (Saito et al., **1989**, C) Saito et al. (**1989**,D) found no fetal ill effects in rabbits treated with up to 60 mg per kg daily during organogenesis.

Saito, M.; Narama, I. and Yoshida, R.: Reproduction study of propiverine hydrochloride (2) - teratological study in rats by oral administration. The Journal of Toxicological Sciences 14:179–205, 1989.

Saito, M.; Ogawa, H.; Narama, I. and Yoshida, R.: Reproduction study of propiverine hydrochloride (1)–fertility study in rats by oral administration. The Journal of Toxicological Sciences 14:161–177, 1989.

Saito, M.; Ogawa, H.; Narama, I. and Yoshida, R.: Reproducrion study of propiverine hydrochloride (4)–perinatal and postnatal study in rats by oral administration. The Journal of Toxicological Sciences 14:221–247, 1989.

Saito, M.; Suzuki, T.; Narama, I. and Yoshida, R.: Reproduction study of propiverine hydrochloride (3) - teratological study in rabbits by oral administration. The Journal of Toxicological Sciences 14:207–219, 1989.

2051 n–Propoxyacetic Acid

Rawlings et al. (1985) exposed explanted day 9.5 rat embryos to 5 ml and found only minimal embryotoxicity.

Rawlings, S.; Shuker, D.E.G.; Webb, M. and Brown, N.A.: The teratogenic potential of alkoxy acids in post–implantation rat embryo culture: Structure–activity relationships. Toxicol. Lett. 1472:49–58, 1985.

2052 Propoxyphene Napsylate *Darvon*™ CAS 26570–10–5

Barrow and Souder (1971) reported a single case of a newborn with arthrogryposis and Pierre Robin syndrome following maternal ingestion of the drug. Heinonen et al. (1977) reported that in over 600 exposed women there was no increase in malformation rate.

Emmerson et al. (1971) studied the effect of the analgesic compound in the pregnant rat and rabbit and found no teratogenic action. The rats receiving 200 and 400 mg per kg per day had some reduced fertility, fetal deaths and stunting of fetuses, but these doses produced about a 20 percent maternal death rate. The rabbits were given up to 80 mg per kg per day. Mineshita et al. (1970) administered the drug by gavage to pregnant mice and rats during active organogenesis and no teratogenic or postnatal effects were found. The mice received up to 600 mg per kg and the rats, 100 mg per kg per day. Vorhees et al. (1979) reported behavioral changes in rats whose mothers received 75 mg per kg daily during most of gestation.

Barrow, M.V. and Souder, D.E.: Propoxyphene and congenital malformations. JAMA. 217:1551–1552, 1971.

Emmerson, J.L.; Owen, N.V.; Koenig, G.R.; Markham, J.K. and Anderson, R.C.: Reproduction and teratology studies on propoxyphene napsylate. Toxicol. Appl. Pharmacol. 19:471–479, 1971.

Heinonen, O.P.; Slone, D. and Shapiro, S.: Birth Defects and Drugs in Pregnancy. Publishing Sciences Group, Inc., Littleton, Mass., 1977.

Mineshita, T.; Hasegawa, Y.; Yoshida, T.; Kozen, T.; Maeda, T.; Sakaguchi, I. and Yamamoto, A.: Teratological effects of dextropropoxyphene napsylate on foetuses and suckling young of mice and rats. (Japanese) Oyo Yakuri 4:1031–1038, 1970.

Vorhees, C.V.; Brunner, R.L. and Butcher, R.E.: Psychotropic drugs as behavioral teratogens. Science 205:1220–1225, 1979.

2053 Propranolol *Inderal*™ CAS 525–66–6

Although this drug has been shown to cross the placenta and produce respiratory depression, hypoglycemia and bradycardia in the human neonate (Gladstone et al., 1975), there have been no reports of the production of congenital defects.

Propranolol has been shown to be pharmacologically active during the early somite stages of the explanted rat embryo (Robkin et al., 1974). There is evidence that propranolol is associated with decreased fetal growth in the rat (Redmond, 1981) and suggestive evidence in humans (Redmond, 1982). Injection of 10 mg per kg intravenously on

day 13 of the mouse produced no fetal defects (Fujii and Nishimura, 1974). Harmon et al. (1986) gave rats up to 50 mg per kg subcutaneously on days 8–20. The fetal body weight and heart weight were reduced. Protein in the heart was reduced but not DNA content, which suggested that cell size was reduced but not cell number.

Fujii, T. and Nishimura, H.: Reduction in frequency of fetopathic effects of caffeine in mice by pretreatment with propranolol. Teratology 10:149–151, 1974.

Gladstone, G.R.; Hardof, A. and Gersony, W.M.: Propranolol administration during pregnancy: Effects on the fetus. J. Pediatr. 86:962–964, 1975.

Harmon, J.R.; Delongchamp, R.R.; Kimmel, G.L. and Webb, P.J.: Effect of prenatal propranolol exposure on development of the postnatal rat heart. Teratogenesis, Carcinogenesis and Mutagenesis 6:139:150, 1986.

Redmond, G.P.: Propranolol inhibits brain and somatic growth in the rat. Pediatr. Res. 15:645 only, 1981.

Redmond, G.P.: Propranolol and fetal growth retardation. Seminars in Perinatalogy. 6:142–147, 1982.

Robkin, M.A.; Shepard, T.H. and Baum, D.: Autonomic drug effects on the heart rate of early rat embryos. Teratology 9:35–44, 1974.

2054 Propyl Gallate

Tanaka et al. (1979) fed up to 2.5 percent during the gestation of rats and found no adverse effects in the newborns.

Tanaka, S.; Kawashima, K.; Nakaura, S.; Nagao, S. and Omori, Y.: Effects of dietary administration of propyl gallate during pregnancy on the prenatal and postnatal developments of rats. J. Food Hyg. Soc. Jpn. 20:378–384, 1979.

2055 N–Propylajmaliniumhydrogentartrate *Neo–Gilurytmal*™

von Philipsborn and Stalder (1972) gave rats and rabbits up to 12 mg per kg or 5 mg per kg, respectively. No adverse effects were found following administration during organogenesis.

von Philipsborn, G. and Stalder, B.: Vertraglichkeit–sprufung von N–Propylajmaliniumhydrogentartrate in Tierversuchen. Arzneim–Forsch. 22:2085–2090, 1972.

2056 Propylene Glycol CAS 57–55–6

El–Shabrawy and Arbid (**1988**) gave 0.2 ml of 10% propylene glycol to rats during the first 10 days of pregnancy. No adverse fetal effects were found.

El–Shabrawy, O.A. and Arbid, M.: Evaluation of some drug solvents for teratological investigations in rats. Egypt. J. Vet. Sci. 24:143–152, 1988.

2057 Propylene Oxide *1,2–epoxypropane*

Hayes et al. (**1988**) exposed male and female rats to concentrations of up to 300 ppm for 6 hours daily before conception and then postnatally. No adverse reproductive effects were found. Harris et al. (**1989**) exposed rats to concentrations of up to 500 ppm for 6 hours daily during organogenesis. The weight gain of the dams was decreased and extra cervical ribs were found in the fetuses exposed to the highest concentration.

Harris, S.B.; Schardein, J.L.; Ulrich, C.E. and Ridlon, S.A.: Inhalation developmental toxicity study of propylene oxide in Fischer 344 rats. Fundamental and Applied Toxicology 13:323–331, 1989.

Hayes, W.C.; Kirk, H.D.; Gushow, T.S. and Young, J.T.: Effect of inhaled propylene oxide on reproductive parameters in Fischer 344 rats. Fundamental and Applied Toxicology 10:82–88, 1988.

2058 Propylthiouracil *PTU* also see *Hyperthyroidism*
CAS 51–52–5

Although methimazole has been associated with some scalp defects in the exposed offspring, propylthiouracil has not. Klevit (1969) reviewed the goitrogenic action of this drug on the human fetus.

Freiesleben and Kjerulf–Jensen (1947) reviewed some clinical cases of goiter and reported that feeding 2000 mg per kg of diet to maternal rats produced fetuses which were then fed to untreated rats and produced thyroid hypertrophy.

Freiesleben, E. and Kjerulf–Jensen, K.: The effect of thiouracil derivatives on fetuses and infants. J. Clin. Endocrinol. 7:47–51, 1947.

Klevit, H.D.: Iatrogenic thyroid disease. In: Endocrine and Genetic Diseases of Childhood. L.I. Gardner (ed.), Philadelphia: W.B. Saunders, 243–252, 1969.

2059 Proquazone *Isopropyl–7 methyl–4–phenyl–2(1H)–quinazolinone*

This anti–inflammatory drug was studied in pregnant rats rabbits by Van Ryzin and Trapold (1980). No teratogenicity was found with oral doses of up to 50 mg per kg in the rat and up to 80 mg per kg in the rabbits. Perinatal studies in the rat at doses of up to 100 mg per kg were done with no significant differences found.

Van Ryzin, R.J. and Trapold, J.H.: The toxicology profile of the anti–inflammatory drug proquazone in animals. Drug and Chemical Toxicology 3(4):361–379, 1980.

2060 Prostaglandin A$_1$ CAS 14152–28–4

Jackson and Persaud (1976) treated rats on days 9 through 12 or 12 through 15 with 200 micrograms subcutaneously and found no growth retardation or increase in resorptions or malformations. Using intrauterine injections on day 17, a significant increase in resorptions was found.

Jackson, C.W. and Persaud, T.V.N.: Pregnancy and progeny in rats treated with prostaglandin A$_1$. Acta Anat. 95:40–49, 1976.

2061 Prostaglandin E$_1$ *Alprostadil ONO–802*

Marks et al. (1987) studied rats using up to 2.0 mg or 6.0 mg per kg by either subcutaneous or intravenous routes, respectively. Treatment was given during organogenesis. Malformations occurred at 2.0 mg per kg, subcutaneously,

and consisted of hydrocephaly, anophthalmia and micro-ophthalmia. Maternal weight gain was reduced at doses of 0.5 mg per kg and above.

Ichikawa et al. (1982) administered this compound intravaginally to the rat on days 0–7 and found decreased implantations, but when pregnancy was established there was no adverse fetal effect. The dose was 1.0 mg per kg. Second pregnancies were not abnormal. They also gave up to 0.5 mg per kg intraperitoneally on days 7–21, and for 20 days postnatally to dams. Labor was complicated and fetal viability was reduced.

Petrere et al. (**1984**) found pre and post implantation loss in rabbits receiving 250 or 12.5 micrograms per kg intravaginally on days 6–18. No teratogenic effect was found.

Ichikawa, Y.; Ozeki, K.; Yamamoto, Y.; Suzuki, Y. and Toh, A.: Studies of the administration of 16, 16–dimethyl–trans–δ–2–prostaglandin E$_1$ in the pregnant rat. Gendai Iryo 14:593–618, 809–819, 1982.

Marks, T.A.; Morris, D.F. and Weeks, J.R.: Developmental toxicity of alprostadil in rats after subcutaneous administration or intravenous infusion. Toxicol. Appl. Pharmacol. 91:341–357, 1987.

Petrere, J.A.; Humphrey, R.R.; Sakowski, R.; Fitzgerald, J.E. and de la Iglesia, F.A.: Teratology study with the synthetic prostaglandin ono–802 given intravaginally to rabbits. Teratogenesis, Carcinogenesis, and Mutagenesis 4:225–231, 1984.

Petrere, J.A.; Humphrey, R.R.; Sakowski, R.; Fitzgerald, J.E. and de la Iglesia, F.A: Two–phase teratology study with the synthetic prostaglandin ono–802 given intravaginally to rats. Teratogenesis, Carcinogenesis, and Mutagenesis 4:233–243, 1984.

2062 Prostaglandin E$_2$ *Methylhesperidin Complex Arbaprostil^{TM}*

Fujita et al. (1973) gave this compound orally to mice and rats during active organogenesis in the maximum dose of 240 mg per kg daily and found an increase in resorptions but no malformation increase. Mercier–Parot and Tuchmann–Duplessis (1977) confirmed these findings in the rat but at intraperitoneal doses of 12.5 or more per kg on days 6 to 10, malformations occurred. The defects included anophthalmia, anencephaly, cleft lip and shortening of the mandible.

Daidohji et al. (1981) gave oral doses of up to 2000 mg per kg to rats and mice during organogenesis and found no significant fetal effects. Postnatal function and reproduction in the rat offspring were normal. Administration on days 17 through weaning in the rat was associated with a decreased survival rate. Sugiyama et al. (1988) gave rats up to 2.0 mg per kg of 15(R)-15-methyl-prostaglandin E$_2$ orally and found a significant decrease in the live birth index but no teratogenicity at the highest dose.

Daidohji, S.; Ishizaki, O.; Horiguchi, T.; Kimura, K.; Shibuya, K. and Ohmori, Y.: Reproductive studies of prostaglandin E$_2$ methylhesperidin complex (KPE). Yakuri to Chiryo 9:1369–1394, 1981.

Fujita, T.; Suzuki, Y.; Yokohama, H.; Yonezawa, H.;

Ozeki, Y.; Ichikawa, Y.; Yamamoto, Y. and Matsuoka, Y.: Toxicity and teratogenicity of prostaglandin E$_2$. Oyo Yakuri 8:787–796, 1973.

Mercier–Parot, L. and Tuchmann–Duplessis, H.: Action of prostaglandin E$_2$ on pregnancy and embryonic development of the rat. Toxicology Letters 1:3–7, 1977.

Sugiyama, O.; Tanaka, K.; Igarashi, S.; Watanabe, K.; Toya, M.; Watanabe, S. and Tsuji, K.: 15(R)–15–methylprostaglandin E$_2$ (Arbaprostil) in rats. Jpn. Pharmacol. Ther. 16(4):957–972, 1988.

2063 Prostaglandin F$_1$-α

Persaud (1980) reported that doses of 50, 100 and 200 microgram given subcutaneously to rats on days 12–15 produced no fetal defects or lethality but growth retardation and placental necrosis were found.

Persaud, T.V.N.: Pregnancy and progeny in rats treated with prostaglandin F$_1$-α. Prostaglandins and Medicine 4:101–106, 1980.

2064 Prostaglandin F$_2$ *Dinoprost*

Collins and Mahoney (1983) reported an infant born with hydrocephalus and attenuated distal phalanges after exposure to 15-methyl F$_2$ α prostaglandin intravaginally five weeks after conception. The medication was given at the mother's request.

Matsuoka et al. (1971) gave this compound to pregnant mice intraperitoneally and intravenously to rats during active organogenesis. No significant changes were found in the mouse fetuses. The maximum daily dose in mice was 0.25 mg per kg and in rats, 2.0 mg per kg. At the highest dose in the rat, an increase in short tails was found.

Chang and Hunt (1972) studied the effect of this prostaglandin on the uterus, fallopian tube and embryo of the rabbit. Five mg per kg soon after ovulation caused the disappearance of eggs and administration during the next 5 days disturbed development of the corpus luteum and embryo. Abortion was produced with 2 to 5 mg per kg given subcutaneously on day 21. The effective dose was close to the lethal dose in the rabbit.

Chang, M.C. and Hunt, D.M.: Effect of prostaglandin F$_2$-α on the early pregnancy of rabbits. Nature 236:120–121, 1972.

Collins, F.S. and Mahoney, M.J.: Hydrocephalus and abnormal digits after failed first–trimester prostaglandin abortion attempt. J. Pediatr. 102:620–621, 1983.

Matsuoka, Y.; Fujita, T.; Nozato, T.; Yokohama, H.; Onishi, Y. and Ohta, K.: Toxicity and teratogenicity of prostaglandin F$_2$ α. Iyakuhin Kenkyu 2:403–413, 1971.

2065 Prostaglandin ICI 74205

Labhsetwar (1972) treated rats on days 4–6 with 0.05–0.1 mg per kg subcutaneously or 1.0 mg per kg orally and observed decreased fertility.

Labhsetwar, A.P.: New antifertility agent: An orally active prostaglandin, ICI 74205. Nature 238:400–401, 1972.

2066 Prothrin *5–Propargyl–2–furylmethyl dl–cis, trans–chrysanthemate*

Yamamoto et al. (1970) injected this insecticide intraperitoneally for 6 days during organogenesis of rats and mice in doses of up to 100 mg per kg. No adverse fetal effects were reported.

Yamamoto, H.; Kuchii, M. and Hayano, T.: Teratological studies on prothrin in mice and rats. Oyo Yakuri 5:779–787, 1970.

2067 Protizinic Acid

Ito et al. (**1975**) gave rats and mice orally up to 100 and 200 mg respectively. No changes occurred in the mouse offspring but at the highest dose in the rat there was maternal toxicity with edema and growth retardation in the fetuses. No teratogenic activity was reported.

Ito, C.; Hayashi, Y.; Fujii, M.; Kubota, H.; Ohnishi, H. and Ogawa, N.: Toxicological studies of protizinic acid. Teratological Studies of protizinic acid in mice and rats. Iyakuhin Kenkyu 6:77–84, 1975.

2068 Protverine *Desatrine*

Keeler and Binns (1968) gave 0.3 to 0.8 gm to ewes and found in four instances, normal lambs.

Keller, R.F. and Binns, W.: Teratogenic compounds of Veratrum californicum (Durand): v. Comparison of cyclopian effects of steroidal alkaloids from plant and structurally related compounds from other sources. Teratology 1:5–10, 1968.

2069 Proxibarbal

This hypnotic was studied in pregnant dogs by Sanz et al. (**1970**) and no adverse reproduction was observed at 100 mg per kg.

Sanz, F.; Jurado, R.; Tarazona, J.M.; Frias, J.; Illera, M. and Perez, M.: Method for the study of embryopathic and teratogenic effects in dogs. Arch. Inst. Farmacol Exp. (Madr.) 22:7–11, 1970.

2070 Proxigermanium

Hayasaka et al. (**1990**, A) gave this antiviral compound orally to rats on days 7–17 in amounts of up to 4,000 mg per kg and found no adverse fetal effects. A slight decrease in open field activity ws found at 6 wks. In rabbits given up to 300 mg per kg on days 6–18 no adverse fetal effects occurred. (Hayasaka et al., **1990** B).

Hayasaka, I.; Murakami, K.; Katoh, Z.; Tamakki, F.; Shibata, T. and Koide, M.: Teratogenicity study of proxigermanium (sk–818) in rats. Kiso to Rinsho 24:227–252, 1990.

Hayasaka, I.; Murakami, K.; Katoh, Z.; Tamaki, F.; Shibata, T.; Niino, K.; Katoh, T. and Koide, M.: Teratogenicity study of proxigermanium (sk–818) in rabbits. Kiso to Rinsho 24:253–261, 1990.

2071 Prozyme CAS 9074–07–1

Tsutsumi et al. (1978) fed this protease to pregnant rats and mice on days 7 through 13 in amounts up to 2,500 mg per kg per day. No adverse fetal effects were found.

Tsutsumi, S.; Yamamoto, R.; Tamura, A.; Sakuma, N. and Fuikiage, S.: Investigations on the possible teratogenicity of prozyme in mice and rats. (Japanese) Clin. Report 12:767–774, 1978.

2072   D–Prunasin CAS 99–18–3

Discussed under Laetrile

2073   Pseudoephedrine CAS 90–82–4

Heinonen et al. (1977) reported that in 39 pregnancies where exposure was during the first four lunar months, there was only one malformed offspsring. Werler et al. (1992) reported significantly elevated relative risk (3.2) for gastroschisis among users of pseudoephedrine. They compared 76 cases with 2,142 other types of malformations. Other decongestants and analgesics (and antipyretics) were not associated with increased relative risks. They considered their association to be tentative and requiring confirmation by other studies. Two studies (Jick et al., 1981 and Aselton et al., 1985) of 902 women taking the medication during the first 3 months found no increase in defects.

Aselton, P.; Jick, H.; Milunsky, A.; Hunter, J.R. and Stergachis, A.: First–trimester drug use and congenital disorders. Obstet Gynecol 65:451–455, 1985.
Jick, H.; Holmes, L.B.; Hunter, J.R.; Madsen, S. and Stergachis, A.: First–trimester drug use and congenital disorders. JAMA 246:343–346, 1981.
Heinonen, O.P.; Slone, D. and Shapiro, S.: Birth Defects and Drugs in Pregnancy. Publishing Sciences Group Inc., Littleton, Mass., 1977.
Werler, M.M.; Mitchell, A.A. and Shapiro, S.: First trimester maternal medication use in relation to gastroschisis. Teratology 45:361–367, 1992.

2074   Pseudojervine

Keeler and Binns (1968) fed 0.25 gm to 1 ewe on the 14th day of gestation and found no cyclopia in the offspring.

Keeler, R.F. and Binns, W.: Teratogenic compounds of veratrum californicum (Durand). Teratology 1:5–10, 1968.

2075   Pseudothymine

Rottkay (1987) studied the effect of gavaging 350 mg per kg in mice. Treatment during preimplantation or embryogenesis did not produce malformation increases or weight reduction and postnatal development was normal.

Rottkay, F.V.: Prenatal toxicity of ambazone. Studia Biophysica 117:193–198, 1987.

2076   Putrescine CAS 110–60–1

Manen et al. (1983) administered putrescine and two of its analogs intraperitoneally to mice on days 10–14. Growth retardation was produced but no gross malformations.

Manen, C–A.; Hood, R.D. and Farina, J.: Ornithine decarboxylase inhibitors and fetal growth retardation in mice. Teratology 28:237–242, 1983.

2077   PVP–1 Povidone iodine powder

Kurebe et al. (1988) gave rats up to 400 mg subcutaneously on days 7–17. No adverse fetal effects were found

but at the highest dose postnatal growth was inhibited at 4–14 days after birth. Fetal thyroid weights wre normal. Similar treatment in rabbits on days 6–18 with doses of up to 80 mg per kg did not adversely effect the fetuses.

Kurebe, M.; Kawamura, K.; Moriguchi, M.; Ishimura, K.; and Komai, Y.: Teratogenicity of pvp–1 in rats by subcutaneous administrations. Kiso to Rinsho 22:4633–4651, 1988.

2078   Pyrabital

Nomura et al. (1977) studied this barbiturate in combination with aminopyrine (1:3) in mice giving it subcutaneously on days 9, 10 and 11. A dose of 0.2 mg per gm of body weight was associated with ruptured ompholoceles.

Nomura, T.; Isa, H.; Tanaka, H.; Kanzaki, T.; Kimura, S. and Sakamoto, Y.: Teratogenicity of aminopyrine and its molecular compound with barbital. Teratology 16:118 only, 1977.

2079   Pyrantel Pamoate

trans–1,4,5,6–tetrahydro–1–methyl–2–[–2(2–thienyl) vinyl] pyrimidine Pamic Acid Salt CAS 22204–24–6

Owaki et al. (1971a) fed this anthelminthic to pregnant rats on days 9 through 14 of gestation in the highest dose of 3,000 mg per kg. No teratogenic or postnatal effects were seen. The same authors (1971b) found no changes in fetuses from treated rabbits (1,000 mg per kg).

Owaki, Y.; Sakai, T. and Momiyama, H.: Teratological studies on pyrantel pamoate in rats. Oyo Yakuri 5:41–50, 1971a.
Owaki, Y.; Sakai, T. and Momiyama, H.: Teratological studies on pyrantel pamoate in rabbits. Oyo Yakuri 5:33–39, 1971b.

2080   Pyrazole and Derivatives CAS 288–13–1

Pyrazole (6 mg per kg), 4–methylpyrazole (6 mg per kg) or decyclopyrazole (6 mg per kg) were injected intraperitoneally on days 8, 10, 12 and 14 of the mouse gestation (Giknis and Damjanov, 1982). Malformations did not increase but each of the two derivatives increased embryo lethality when combined with ethanol. Ukita et al. (1993) gave 100 mg per kg intraperitoneally on day 7 and found no adverse fetal effects in the mouse. When combined with ethanol (4 gm per kg) this alcohol dehydrogenase inhibitor increased the malformations suggesting that the ethanol was mainly responsible for the defects rather than a metabolite.

Giknis, M.L.A. and Damjanov, I.: The effects of pyrazole and its derivatives on the transplacental embryotoxicity of ethanol. (abs) Teratology 25:43A–44A, 1982.
Ukita, K.; Fukui, Y. and Shiota, K.: Effects of prenatal alcohol exposure in mice: incluence of an adh inhibitor and a chronic inhalation study. Reproductive Toxicology 7:273–281, 1993.

2081 Pyrethrum Extract

Lutz–Ostertag and Lutz (1970) applied an extract containing pyrethrum and piperonyl butoxide to the chorio–allantoic membrane and produced damaged testes with absence of gonadocytes in the surviving chicken embryo. Khera et al. (1982) gavaged rats on days 6–15 with 50, 100 and 150 mg per kg and found increased resorptions at the 100 and 150 mg levels No significant increase in defect rate was found.

Khera, K.S.; Whalen, C. and Angers, G.: Teratogenicity study on pyrethrum and rotenone (natural origin) in pregnant rats. J. Toxicol. Environ. Health 10:111–119, 1982.

Lutz–Ostertag, Y. and Lutz, H.: Action teratogene et sterilisante des pyrethrines synergisees sur l'embryon de poulet. C. R. Soc. Biol. 164:777–779, 1970.

2082 Pyridine Derivatives

(β–Diethyl—aminoethyl)–pyridine dichlorohydrate
(β–Propyl—aminoethyl)–pyridine dichlorohydrate
(β–Dipropyl–aminoethyl)–pyridine dichlorohydrate
(β–Morpholinoethyl)–pyridine dichlorohydrate
(β–Morpholinomethyl)–pyridine dichlorohydrate
(β–Piperidinoethyl)–pyridine dichlorohydrate
(β–Piperidinomethyl)–pyridine dichlorohydrate
(β–Dimethylaminoethyl)–pyridine dichlorohydrate
(β–Dimorpholinoaminoethyl)–pyridine dichlorohydrate
(β–Dipiperidylaminoethyl)–pyridine dichlorohydrate
(β–Diethylaminomethyl)–pyridine dichlorohydrate

Bariljak and Tarakhovskii (1973) gave mice about 1/3 of the LD–50 of 12 pyridine derivatives intraperitoneally on days 2, 10, or 14. All compounds were associated with increased defects of the genitourinary systems and skeleton.

Bariljak, I.R. and Tarakhovskii, M.L.: Embryotoxic effect and toxicity of a series of pyridine derivatives (Russian). Farmakol. Toksikol. 8:99—102, 1973.

2083 Pyridine, Substituted also see
6–Aminonicotinamide CAS 110–86–1

Landauer and Salam (1974) injected pyridine (20 mg per egg) into the developing chick egg at 96 hours and found muscular hypoplasia of the legs. Landauer and Salam (1973) summarized their studies on chick teratogenicity of 12 substituted pyridines. 3–Hydroxypridine and 3–hydroxy–6–methyl pyridine are highly teratogenic, producing micromelia and beak defects at doses of 2.5 mg injected at 96 hours of incubation. 2–Hydroxypyridine caused the same type of malformations but at a lower frequency and higher dose. 4–Hydroxypyridine was non–teratogenic.

2–Amino–3hydroxypyridine was highly teratogenic and produced acromelia. 2,3–Dihydroxypyridine was teratogenic at 1.0 mg levels and produced microphthalmia and rumplessness. 2,6–Dihydroxypyridine was non–teratogenic.

The teratogenic effects were felt due to interference with pyridine nucleotide utilization with substitutions at the 3–position playing a dominant role.

Landauer, W. and Salam, N.: The experimental production in chicken embryos of muscular hypoplasia and associ-

ated defects of beak and cervical vertebrae. Acta Embryol. Experimentalis 1:51–66, 1974.

Landauer, W. and Salam, N.: Quantitative and qualitative distinctions in developmental interference produced by various substituted pyridines. Molecular shape and teratogenicity as studied on chicken embryos. Acta Embryol. Experimentalis 1:179–197, 1973.

2084 Pyridine–2–thiol–1–oxide

The zinc salt of this compound was applied to pregnant pigs daily from days 8 through 32 of gestation and no teratogenic effects were found. The concentration of the ointment was as high as 400 mg per kg (Wedig et al., 1975).

Wedig, J.H.; Kennedy, G.L.; Jenkins, D.H.; Henderson, R. and Keplinger, M.L.: Teratologic evaluation of zinc omadine when applied dermally on Yorkshire pigs. (abs) Toxicol. Appl. Pharmacol. 33:123 only, 1975.

2085 N–3–Pyridoyltryptamine Tryptamide

Dluzniewski et al. (1987) gave mice and rats up to 500 mg per kg on days 6–14 and found no teratogenicity. In mice the highest dose caused reduced survival and fetal weight.

Dluzniewski, A.; Gastol–Lewinska, L.; Buczynska, B. and Moniczewski, A.: Influence of n–3–pyridoyltryptamine (tryptamide) on fetal development in rats and mice. Pol. J. Pharmacol. Pharm. 39:779–786, 1987.

2086 Pyridostigmine

Levine and Parker (1991) gave rats up to 30 mg per kg before fertilization on days 6–15 of gestation or perinatally. Maternal and fetal toxicity (weight decrease) were found at the highest dose but no increase in defects was found.

Levine, B.S. and Parker, R.M.: Reproductive and developmental toxicity studies of pyridostigmine bromide in rats. Toxicology 69:291–300, 1991.

2087 Pyridoxine Deficiency Vitamin B₆ Deficiency

Davis et al. (1970) produced defects in the offspring of rats maintained during pregnancy on a pyridoxine deficient diet and with 4–deoxypyridoxine added to the drinking water (0.1 mg per ml). Reduction defects of the digits, cleft palate, omphalocele and exencephaly were found. The treated fetuses had significant reductions in the weight of their spleen and thymus but not their kidney. The effects of the experimental regime were prevented by adding 1.0 mg pyridoxine to each ml of drinking water.

Kirksey et al. (1990) have summarized their extensive work on the effects of pyridoxine deficiency on development of the nervous system of the rat and human.

Davis (1974) produced a postnatal immune deficiency by prenatal deficiency in the rat. The deficiency was terminated on day 21 of gestation. Runting syndrome occurred in some of the offspring. The non–runted offspring at six weeks of age were immunized with mycobacterium tuberculosis and when skin–tested had significantly less response than the controls.

Davis, S.D.: Immunodeficiency and runting syndrome in rats from congenital pyridoxine deficiency. Nature 251:548–550, 1974.

Davis, S.D.; Nelson, T. and Shepard, T.H.: Teratogenicity of vitamin B_6 deficiency: Omphalocele, skeletal and neural defects, and spleenic hypoplasia. Science 169:1329–1330, 1970.

Kirksey, A.; Morre, D.M. and Wasynczuk, A.Z.: Neuronal development in vitamin B_6 deficiency. Annals of the New York Academy of Sciences 585:202–218, 1990.

2088 1–Pyridyl–3,3–diethyltriazine

Discussed under 1–Phenyl–3,3–dimethyl–triazine

2089 3–Pyridymethlsalicylate

Cekanova et al. (**1974**) found skeletal defects when they gave 2,000 mg per kg orally to mice on day 9. Gross defects were not increased.

Cekanova, E.; Larsson, K.S.; Morck, E. and Aberg, G.: Interactions between salicylic acid and pyridyl–e–methanol: Anti–inflammatory and teratogenic effects. Acta Pharmacol. et Toxicol. 35:107–118, 1974.

2090 Pyrilamine CAS 59–33–6 CAS 91–84–9

Heinonen et al. (1977) found no increase in malformations among the offspring of 121 women taking the antihistamine in the first four lunar months or 392 women taking it any time in pregnancy.

Heinonen, O.P.; Slone, D. and Shapiro, S.: Birth Defects and Drugs in Pregnancy. Publishing Sciences Group Inc., Littleton, Mass., 1977.

2091 Pyrimethamine *Ethylpyrimidine Primethamine Daraprim™ Chloridin™*
2,4–Diamino–5–(p–chlorophenyl)–6–ethylpyrimidine CAS 58–14–0

This antimalarial drug which is also a folic acid antagonist was used by Dyban et al. (1965) to produce defects in the rat. By using 3 to 10 mg per animal on gestational day 9, 10 or 13, a high incidence of cleft palate, mandibular hypoplasia and limb defects was found. Neural tube closure defects were also recorded. When 0.6 mg was given by mouth on days 9, 10 and 11, a defect rate of 36 percent was produced. Anderson and Morse (1966) produced similar defects in the rat. Sullivan and Takacs (1971) reported a low incidence of congenital defects in hamsters after 20 mg on day 9. These authors point out that the teratogenic dose of 0.5 mg per day in the rat does not differ greatly from the 25 mg daily dose being used in humans for treatment of toxoplasmosis. The dose used weekly for prevention of malaria is less than 0.5 mg per kg. A few single case reports of defects in exposed human pregnancies appeared (Harpey et al., 1983) but no systematic studies are available.

Thiersch (1954) used the methyl pyrimidine form and found defects in the rat with 0.5 mg per kg. Kotb (1972) found that the methyl form in amounts of 75 mg per kg on the first day of gestation caused mortality in nearly all embryos. Folinic acid was shown to antagonize the

action of ethyl pyrimidine (Sullivan and Takacs, 1971). Stanzheuskaya (1966) showed that the compound is teratogenic in the chick. Dyban and Udalova (1967) induced various chromosomal aberrations in five–day rat embryos when the mother received the drug 8 hours after insemination.

Pyrimethamine was given by gavage (2 or 5 mg per kg) to the adult female rat just before ovulation. A significant dose–dependent decrease in the rate of cleavage was observed. In a dose of 2 mg per kg, no chromosomal aberrations were observed in five–day rat embryos. Five mg per kg induced chromosomal aberrations in 14 percent of all five–day rat embryos. In the control group only 6 percent of the embryos had chromosomal aberrations. Pyrimethamine induced mostly trisomy or chromosomal mosaicism in rat embryos (Baranov and Chebotar, 1980).

Male rats were treated with pyrimethamine (10 mg per kg once every day by gavage during 1–3 weeks) before mating with normal females. In these experiments no increased embryonic mortality was found (Udalova, 1976). This drug was given to rats on 9 or 10 days of gestation and severe damage of decidual tissue was observed. It was shown that this effect resulted from the direct inhibition of dihydrofolate reductase not only in embryonic but also in the decidual cells. Different sensitivity of mesometral and antimesometral parts of the decidua was observed. Under the effect of pyrimethamine, differences in DNA synthesis between the mesometral and antimesometral regions were manifested which were connected with differences in thymidine monophosphate synthesis in the two populations of decidual cells (Dyban et al., 1976a).

Pyrimethamine given intragastrically (50 mg per kg) to female rats on the 13th day of pregnancy induced typical severe malformations in all embryos. It was shown that pyrimethamine inhibited DNA and RNA synthesis within 15 to 20 minutes following the administration of the drug. The rate of 14C–thymidine incorporation in DNA of abnormal embryos did not change within 4 hours following the administration of pyrimethamine. The data presented agree with the earlier results which suggested that the mechanism of teratogenic action of pyrimethamine on rat embryonic development was due to the primary inactivation of dihydrofolate reductase and the inhibition of folate cycle in the embryonic cells (Kotin and Repina, 1973).

Pyrimethamine (50 mg per kg given by gavage on the first day of pregnancy) did not induce any disturbances in preimplantation development of CBA mouse embryos while a single dose of 25 mg per kg on the first day of pregnancy caused death of all cleaving rat embryos. The cleavage of mouse embryos cultured in medium containing blood serum from rats treated with pyrimethamine was abnormal and significant numbers of embryos died. It was concluded that mouse and rat embryos had the same sensitivity to pyrimethamine in culture in vitro but not during the development in the maternal organism because of a barrier function of the oviduct to this drug in mice (Dyban et al., 1976b).

Female rats were treated on the 10th day of pregnancy with pyrimethamine (25 mg per kg intragastrically) and after different time intervals (from 15 minutes to 24 hours), animals were sacrificed, embryos removed from the uterus and cultured in vitro in blood serum from control (non–treated) rats. It was shown that teratogenic action of pyrimethamine became irreversible 6 hours after exposure of pregnant rats to this drug. In the other experiments, embryos from intact control females were cultured (1–4 hours) in the serum from rats treated with the same dose of pyrimethamine. In these conditions the irreversible teratogenic action of pyrimethamine was observed after culturing rat embryos 3 hours in this serum. These data supported the conclusion that to produce embryotoxic and teratogenic action, the presence in blood of pyrimethamine and=or its metabolites for at least 5–6 hours is needed (Popov, 1982). There is some evidence in rats that pyrimethamine embryotoxicity is increased by coadministration of folic acid but not folinic acid (Chung et al., **1993**).

The teratogenic and embryotoxic effects of the following 6 derivatives of 2,4–diamino–5–phenylpyrimidine with different length of the alkyl radical in the 6 position of pyrimidine ring was studied in rats (2,4–diamino–5–phenylpyrimidine; 2,4–diamino–5–phenyl–6–methylpyrimidine; 2,4–5–phenyl–6–ethylpyrimidine; 2,4–diamino–5–phenyl–6–propylpyrimidine and 2,4–diamino–5–phenyl–6–butylpyrimidine). It was found that among these analogues of pyrimidine the highest teratogenic and embryotoxic activity was the substance with an ethyl group and the lowest was the H–radical (Dyban et al., 1976d). The mitotic activity of liver cells in rat embryos from treated animals and corneal epithelium in adult rats obtained after administration of these substances was studied. A definite correlation was established between antimitotic, embryotoxic and teratogenic activities of previously mentioned pyrimethamine analogues (Dyban et al., 1976c).

Anderson, I. and Morse, L.M.: The inluence of solvent on the teratogenic effect of folic acid antagonist in the rat. Exp. Mol. Pathol. 5:134–145, 1966.

Baranov, V.S. and Chebotar, N.A.: Mutagenic effect of pyrimethamine injection before ovulation on preimplantational rat embryos. Tsitol. Genet. (USSR) 14.6:20–24, 1980.

Chung, M.; Han, S. and Roh, J.: Synergistic embryotoxicity of combination pyrimethamine and folic acid in rats. Reproductive Toxicology 7:463–468, 1993.

Dyban, A.P.; Akimova, I.M. and Svetlova, V.A.: Effect of 2–4–diamino–5–chloro–phenyl–6–ethylpyrimidine on the embryonic development of rats. Akad. Nauk. U.S.S.R. 163:455–458, 1965.

Dyban, A.P.; Bariljak, I.R. and Tichodeeva, I.I.: The teratogenic and embryotoxic activity of some 2,4–diaminopyrimidine derivatives. Arkh. Anat. Gistol. Embryol. (USSR) 7:29–35,1976d.

Dyban, A.P.; Samoshkina, N.A.; Weisman, B.L. and Golinsky, G.F.: Pattern of DNA synthesis in antimesometral and mesometral regions of the decidual tissue in rats. Ontogenez (USSR) 7.4:323–332, 1976a.

Dyban, A.P.; Sekirina, G.G. and Golinsky, G.F.: Sensitivity of the early mouse embryos to antifolic preparation chloridine (pyrimethamine). Byull. Eksper. Biol. Med. (USSR) 82(10):1247–1250, 1976b.

Dyban, A.P. and Udalova, L.D.: Study of chromosome aberrations at the early stages of mammalian embryogenesis. I. The experiments with the effect of 2,4–di–amino,5–p–chlorophenyl, 6–ethyl pyrimidine (pyrimethamine) on rats. Genetika (Russian) 4:52–65, 1967.

Dyban, A.P.; Bariljak, I.R.; Tichodeeva, I.I. and Chebotar, N.A.: On correlation of teratogenic and antimitotic activity of some derivatives of 2,4–diaminopyrimidine. Ontogenez. (USSR) 7.1:58–63, 1976c.

Harpey, J–P.; Darbois, Y. and Lefebvre, G.: Teratogenicity of pyrimethamine. Lancet 2:399 only, 1983.

Kotb, M.M.: Peculiarities of damage effect of the structural analogue of antimalarial drug chloridine (lem–687) on different stages of embryogenesis in the rat. Arch. Anat. (Russian) 63(8):88–96, 1972.

Kotin, A.M. and Repina, V.S.: Effect of pyrimethamine on nucleic acid metabolism in white rat embryos. Ontogenez. (USSR) 4.2:128–137, 1973.

Popov, V.B.: Study into the time of realizing embryotoxic action of cyclophosphamide and chloridine in vivo and in vitro. Farmakol. Toksikol. (USSR) 45.1:79–83, 1982.

Stanzheuskaya, T.L.: Effect of chloridin on chick embryogenesis. Bull. Exp. Biol. Med. 61:427–429, 1966.

Sullivan, G.E. and Takacs, E.: Comparative teratogenicity of pyrimethamine in rats and hamsters. Teratology 4:205–210, 1971.

Thiersch, J.B.: The effect of substituted 2,4–diamino–pyrimidines on the rat fetus in utero. Proceedings of the International Congress on Chemotherapy 3:367–372, 1954.

Udalova, L.D.: Postimplantation development of rat embryos after the action of some chemical agents on male gametes. Arkh. Anat. Gistol. Embriol. (USSR) 1:46–50, 1976.

2092 Pyrithione CAS 1121–30–8

The zinc salt of this anti–dandruff agent was applied to the skin of pregnant rabbits from days 7 through 18 and no embryotoxicity or teratogenicity was observed (Nolen et al., 1975).

Nolen, G.A.; Patrick, L.F. and Dierckman, T.A.: A percutaneous teratology study of zinc pyrithione in rabbits. Toxicol. Appl. Pharmacol. 31: 430–433, 1975.

2093 Pyronaridine

Shao et al. (1985) studied this antimalarial in pregnant rats using 10 or 20 mg per kg orally from 60 days before fertilization through pregnancy. No external or skeletal defects were found in the fetuses.

Shao, B–R.; Zhan, C–Q. and Ha, S–H.: Influence of pyronaridine phosphate on three–generation reproduction in rats. Acta. Pharmacol. Sinica 6(2):131–134, 1985.

2094 Pyrrolazote

Shelesnyak and Davies (1955) suppressed implantation with subcutaneous injections of 15 mg in rats. After implantation the same treatment caused resorptions.

Shelesnyak, M.C. and Davies, A.M.: Disturbance of pregnancy in mouse and rat by systemic antihistaminic treatment. Proc. Soc. Exp. Biol. Med. 89:629–632, 1955.

2095 QA 208–199 (N–Hydroxy–N–methyl–7–propoxy–2–naphtalinethanamine)

This lipoxygenase inhibitor has been studied in in vitro cultures of rat embryos. Embryotoxicity was lower with addition of S–9 from rats treated with 3–methylcholanthrene or phenobarbital. (Bechter and Terlouw, 1990)

Becher, R. and Terlous, G.D.C.: Embryotoxicity, metabolism and tissue accumulation of QA 208–199 in vitro can be modified by the use of the cytochrome–P450 enzyme inducers aroclor 1254, phenobarbital and 3–methylcholanthrene. (abst)Teratology 41:537, 1990.

2096 Quazepam

Black et al. (1987) treated mice and rabbits during organogenesis up to 120 and 40 mg per kg respectively and found no teratogenicity. Some reduction in fetal growth was found at the highest doses.

Black, H.E.; Szot, R.J.; Arthaud, L.E.; Massa, T.; Mylecraine, L.; Klein, M.; Lake, A.; Fabry, A.; Kaminska, G.Z.; Sinha, D.P. and Schwartz, E.: Preclinical safety evaluation of the benzodiazepine quazepam. Arzneimittelforschung 37(8):906–913, 1987.

2097 Quercetin 3,3',4',5,7–Pentahydroxyflavone CAS 117–39–5

This flavinol found in edible plants was fed to rats on days 6–15 in doses up to 2000 mg per kg (Willhite, 1982). No adverse effects were found although some fetal weight reduction was found at the higher doses.

Willhite, C.C.: Teratogenic potential of quercetin in the rat. Food Chem. Toxicol. 20:75–79, 1982.

2098 Quinacillin 3–Carboxy–2–quinoxalinylpenicillin CAS 985–32–6

Bough et al. (1971) injected pregnant rats and rabbits with 250 and 100 mg per kg respectively and found no fetal changes. The rats were injected from day 1 through day 20 and the rabbits from day 1 through day 16 of gestation. They demonstrated that the antibiotic reached the fetal serum and amniotic fluid.

Bough, R.G.; Everest, R.P.; Hale, L.J.; Lessel, B.; Mason, C.G. and Spooner, D.F.: Chemotherapeutic and toxicological properties of quinacillin. Chemotherapy 16:183–195, 1971.

2099 Quinacrine Atabrine CAS 83–89–6

Rothschild and Levy (1950) injected 120 mg per kg of quinacrine subcutaneously into rats on the 13th through the 19th gestational days. Fetal death was increased but no defective fetuses were found. The level of quinacrine in the fetal liver was only 9 microgm as compared to the maternal liver level of 549 microgm. No systematic studies in humans have been reported.

Rothschild, B. and Levy, G.: Action de la quinacrine sur la gestation chez le rat. C. R. Soc. Biol. 144:1350–1352, 1950.

2100 Quinapril Hydrochloride

Imanishi (1993) gave this ACE inhibitor to rats before and in the early time of gestation and during organogenesis. At the highest dose (100 mg per kg) the dams failed to gain weight normally but no adverse effects were found in the fetuses or their behavior or reproduction.

Imanishi, M.; Shimohata, R.; Mori S. and Takeuchi, M.: Fertility and general reproductive performance in rats given quinapril hydrochloride. Oyo Yakuri 46:67–74, 1993.

2101 Quinestrol

Chang et al. (1971) found that subcutaneous implantation of two mg of this steroid in rabbits caused degeneration of the fertilized eggs for up to 25 days later.

Chang, M.C.; Casas, J.H. and Hunt, D.M.: Suppression of pregnancy in the rabbit by subcutaneous implantation of silastic tubes containing various estrogenic compounds. Fertil. Steril. 22:383–388, 1971.

2102 Quinidine CAS 56–54–2

No reports of adverse fetal effects of quinidine were identified. Hill and Malkasian (1979) reported one patient with a normal pregnancy outcome.

Hill, L.M. and Malkasian, G.D.: The use of quinidine sulfate throughout pregnancy. Obstet. Gynecol. 54:366–368, 1979.

2103 Quinine CAS 130–95–0

Dannenberg et al. (1983) reviewed 70 pregnancies complicated by quinine suicide and among the 18 infants studied 10 had deafness and 2 mental retardation. Robinson et al. (1963) reported quinine ingestion during early pregnancy in two out of 200 mothers giving birth to congenitally deaf children. Tanimura (1972) reviewed the human literature and reported that from 21 attempted abortions, 10 central nervous system defects (6 hydrocephalics), 8 limb, 6 face, 5 digestive and 3 urogenital malformations resulted. The same author summarized results in experimental animals and reported the absence of defects or abortions in pigtail monkeys receiving up to 200 mg per kg for 3 days.

Mosher (1938) using guinea pigs, gave quinine at different times during pregnancy. The average total dose was 1.8 gm. Hemorrhages, practically always localized to the scala tympani of the fetal inner ear, were found. Covell (1936) reported histologic changes in the cochlea of fetal guinea pigs after maternal treatment with 200 mg per kg.

Covell, W.P.: A cytologic study of the effects of drugs on the cochlea. Arch. Otolaryngol. 23:633–641, 1936.
Dannenberg, A.L.; Dorfman, S.F. and Johnson, J.: Use of quinine for self–induced abortion. Southern Medical Journal 76:846–849, 1983.
Mosher, H.P.: Does animal experimentation show similar changes in the ear of mother and fetus after the ingestion of quinine by the mother? Laryngoscope 48:361–395, 1938.

Robinson, G.C.; Brommitt, J.R. and Miller, J.R.: Hearing loss in infants and preschool children. II. Etiological Considerations. Pediatrics 32:115–124, 1963.

Tanimura, T.: Effects on macaque embryos of drugs reported or suspected to be teratogenic to humans. In: The Use of Non–human Primates in Research on Human Reproduction. E. Diczfalusy and C.C. Standley (eds.), Stockholm: WHO Research and Training Centre on Human Reproduction, 293–308, 1972.

2104 2–Quinoline Thioacetamide CAS 31293–15–9

In doses as high as 400 mg per kg on days 8 to 14, no teratogenic activity was found in rats. However, when the drug was given in doses of 200 mg per kg after organogenesis, a single dose produced a high incidence of digital abnormalities. This phenomenon was associated with protracted fetal vascular spasm (Sugitani et al., 1976).

Sugitani, T.; Ihara, T. and Mizutani, M.: Teratologic study of 2–quinoline thioacetamide in the rat. (abs) Teratology 14:254–255, 1976.

2105 R2323 13β–ethyl–17α–ethinyl–17–hydroxy–gona–4, 9,11–triene–3–one

Hiramatsu et al.l 1988 gave this antigonadotropin to mice on days 5–15 in doses of up to 5 mg per kg daily. No adverse fetal effects were found.

Hiramatsu, Y.; Suzuki, T.; Shimizu, M.; Udo, K.; Yamashita, Y.; Koike, T.; Katoh, M. and Wada, H.: Reproduction study by oral administration of r2323 during the period of fetal organogenesis in mice. Yakuri to Chiryo 16:713–736, 1988.

2106 Rabbit Serum Protein

Adachi (1979) injected rabbit serum intraperitoneally into mice in 1 and 2 ml amounts on day 10, 12 or 14. Treatment on day 12 was followed by increases in cleft palate and skeletal defects. The average number of live fetuses was also reduced in the treatment groups.

Adachi, K.: Congenital malformations induced by heterologous protein. Cong. Anom. 19:57–64, 1979.

2107 Radar

Sigler et al. (1965) studied the radiation exposures of parents of children with Down's syndrome. In the study group 8.7 percent of the fathers had been intimately exposed to radar. This was significantly greater than exposures in a matched healthy control group (3.3 percent).

Sigler, A.T.; Lilienfeld, A.M.; Cohen, B.H. and Westlake, J.E.: Radiation exposures in parents of children with mongolism (Down's syndrome). Johns Hopkins Hosp. Bull. 117:374–399, 1965.

2108 Radiation X–irradiation Ionizing Radiation

A large body of information is available on the adverse effects of irradiation on the human and animal embryo and fetus. The effects of radiation on the human conceptus has centered on (1) damage to the fetal central nervous system, (2) early embryonic death with sex ratio changes and (3) long–term effects on carcinogenesis. The effects of accidental X–radiation on the developing fetus are well documented. Several reviews are available: Hicks and D'Amato (1966), Yamazaki (1966), Kalter (1968) and Brent (1971,1977,1980) gave general summaries, and Jacobsen (1970) reviewed data related to low dose irradiation. Sikov and Mahlum (1969) edited a large symposium on radiation biology of the fetal mammal. The main concentration was on gene mutation and development of the central nervous system. Schull et al. (1990) reviewed data on the effects on the nervous system of experimental animals and humans. They conclude that sensitivity of all mammalian species when developmental periods are compared are markedly similar. Hawkins and Smith (1989) studied the reproductive outcome of women treated with abdominal radiation for tumors mostly Wilm's tumor. For firstborn children there was a significant (300 gm) reduction in birth weight as compared to non-radiated women with tumors. The exposed women had higher nulliparity and spontaneous abortions. Otake et al. (1990) have reanalyzed reproduction data from atomic bomb survivors and found a non–statistical increase in major malformations, stillbirths and/or dying in the first 14 days of life.

Brent et al. (1987) discussed the human risk based on animal studies. At the most sensitive period (18–36 days), the minimal dose which could produce gross malformations was estimated to be 20cGy (20 rads or about 20 rem). This dose level is considerably higher than can be received during diagnostic radiation.

Plummer (1952) observed that microcephaly was a common complication of intrauterine radiation after the atomic bomb explosion at Hiroshima, and that the degree of microcephaly was directly related to the distance the mother was from the epicenter. Blot and Miller (1973) found mental retardation after 50 rad doses in Hiroshima but 200 rad doses in Nagasaki and suggested the lower dose effect may have been due to a higher neutron exposure in Hiroshima. Miller (1956) and Neel and Schull (1956) did not find further health problems at that time in survivors or significant increases in defects in the subsequent offspring of parents exposed to the Japanese atomic bomb explosions. Driscoll et al. (1963) reported on histologic changes occurring in human fetuses following dosages of about 500 rads.

Otake and Schull (1984) reported the prevalance of mental retardation among 1600 offspring exposed to the Japanese atomic bombs. They found that the critical period was between 8 to 15 weeks after fertilization (six of nine children were retarded). This period corresponds to the time when major neuronal proliferation is occurring. A dose response was found. For all gestation periods the percentage of retardation was 1.4, 2.4, 17.6 and 36.8 for the 1–9, 10–49, 50–99, and over 100 absorbed fetal rads. The control was 0.8 percent. Mole (1987) analyzed the findings of Otake and Schull (1984) and used the data to discuss mechanisms which could produce malformations by ionizing radiation. Yamazki and Schull (1990) have summarized the neurological anormalities found in the offspring of exposed mothers at Nagasaki and Hiroshima. If there is a threshold it would exist between 0.1 to 0.2 Gy of fetal dose. The few autopsies that have been performed support the hypothesis that errors in neuronal migration are important in the mechanism of damage.

Schull et al. (**1990**) summarized the literature on the animal and human findings of central nervous system function after radiation. For animals, in general, positive findings were found at doses of 0.25 to .75 Gy (25 to 75R) but the time of exposure and type of test were important variables. For humans, exposed between 8–15 weeks a dose of 1.0 Gy was estimated to reduce IQ by 21–33 points.

Miller (1956) did find a subsequent leukemia incidence of 1 in 1000 in children who were under 10 years of age and were within 1500 meters of the epicenter of the atomic bomb explosion. Yoshimoto et al. (1988) found that among 1630 individuals exposed in utero to the atomic bombs, there were two cases of childhood cancer in the first 14 years of follow-up. The relative risk for adult type cancer 40 years after exposure was 3.9 in the 0.30+ Gy exposure group. The crude cancer rate per 100,000 was 23, 32, 72 and 91 for the uterus organ doses of 0, 0.01–0.29, 0.30–0.59 and 0.60+ Gy, respectively. There were no differences in risks that could be associated with exposure in different gestational periods.

Long term effect of maternal radiation on the incidence of malignancies in the offspring was reported by MacMahon (1962). The extensive data of the Oxford Survey of Childhood Cancers (Bithell and Stewart, 1975) indicated a relative risk estimate of 1.47 for mothers with prenatal radiation exposure. Translated into numbers of childhood cancers per 10,000, the increase would be from 10 to 15 cases. The risk was dependent on the number of films taken and could be described as a linear relationship. Exposure in the earlier months of pregnancy appeared to carry a much higher risk. Other factors leading to maternal radiation are hard to separate from the radiation effect. A special committee of the United Nations (1972) carefully assessed the reports dealing with this subject. Diamond et al. (1973) studied 20,000 children exposed to radiation during gestation and found a tripling of the leukemia death rate in the treated white group but none was observed in an equal–sized group of Black children similarly exposed.

Sever et al. (1988) studied malformations among 672 malformed offspring of workers around the Hanford plant in the state of Washington. The cumulative whole–body radiation dose of each of the 195 workers was known. Some increase in congenital dislocation of the hip and tracheo esophageal fistula was found among the twelve malformation types studied but no association with dose was found. Given the number of statistical tests, the authors felt some or all of these correlations could represent false findings. Macht and Lawrence (1955) surveyed the offspring of radiologists and could detect no increase in congenital defects. Wagner and Hayman (1982) summarized the relative safety of pregnancy in female radiologists.

Schull et al. (1981) compared gonadal doses from atomic bomb exposed parents with life expectancy, chromosomal aneuploidy, and electrophoretic mutants of their offspring.

Their pregnancy outcomes were also studied. Although all four indicators were found to be changed as expected, there was no statistical significance. The average genetic doubling dose for the four indicators was 156 rems.

An association between Down's syndrome and maternal X–ray exposure was suggested by three retrospective studies (Uchida and Curtis, 1961; Sigler et al., 1965; Alberman et al., 1972a), but Carter et al. (1961) could not show a connection. Uchida (1977) recently summarized 11 studies of which 9 showed an increase in radiation exposure of the mothers giving birth to Down's infants. Although Neel and Schull (1956) did not report significant sex ratio changes in the offspring of irradiated parents, Scholte and Sobels (1964) offered some evidence for a change in sex ratio after parents were given radiation therapy. Boue et al. (1975) reported an increase in chromosomally abnormal abortuses from fathers who were occupationally exposed to X–ray.

An increase in spontaneous abortion has been associated with gonadal radiation (Alberman et al., 1972b). They reported that matched controls received 180 mR while all forms of spontaneously aborting women received 245 mR. Among the group with abnormal karyotype, the average exposure was 331. Mothers of triploid embryos averaged 735 mR and these authors point out that most of the increased risk is expressed by non–viable conceptuses. Strobino et al. (**1978**) have reviewed the effect of radiation on human reproduction.

Among commonly used animal models the mouse was studied extensively by Russell (1950). She reported that preimplantation irradiation tended to be lethal or to have no effect. Exposure on days 6.5 through 13.5 produced little or no prenatal death but a high incidence of growth retardation and abnormalities which in general were related to dose and time of administration. Eye defects (microophthalmia and coloboma) were most common after treatment on days 7.5 through 9.5, while renal changes were associated with treatment at day 9.5 and skeletal changes appeared after exposure during days 9.5 to 12.5. After treatment on day 14.5 abnormalities were uncommon, but cataracts, hydrocephalus and skin defects did develop in later life. Dr. Lillian Russell used radiation doses of 100 to 400 rads. Kuno et al. (**1994**) confirmed Russel's studies of eye defects in mice and detailed the type of eye defect seen in mice exposed on days 8 or 9 to 4.6 Gy. Devi et al. (**1994**) treated mice at 11.5 days postcoitus and found a concentration response decrease in brain growth starting at 0.05 Gy and extending through 0.5 Gy.

Hicks and D'Amato (1966) concentrated their studies in rats and mice on the central nervous system effects occurring when treatment is given in the late embryonic and fetal periods. Jensh et al. (1987) radiated mice with 0.4 or 0.6 Gy on day 9 or 17. Most postnatal studies were not different from the sham controls but those exposed to the highest dose on day 17 exhibited higher conditioned avoidance. Norton and Kimler (1987) administered 1.0 Gy on day 11 or day 178. The behavioral effects were correlated with the thickness of 5 cortical layers. Significant association of behavior was found with layers 5 and 6. Jensh and Brent (1988A) gave

0.75 or 1.5 Gy to rats from the 14th to the 18th day. Weight reduction at birth and delay in acquisition of several reflexes was found. After weaning, the body weight to organ weight ratios were reduced for the brain, kidneys, testes, ovary but not liver. Jensh and Brent (1988B) radiated day 18 rats with 1.5 Gy and found testicular hypoplasia and some decreased mating performance and positive inseminations.

Wilson et al. (1953) reported the effect of timed radiation on type of defect including those of the cardiovascular system. The effect of radiation on the skeletal system of mouse fetuses was detailed by Degenhardt and Franz (1969) and Murakami and Kameyama (1964). Rugh et al. (1964a,b) carried out studies of the association of X–rays with cataract formation and skeletal retardation. The cataracts developing in mice were most common after exposure immediately after fertilization and were interpreted as overall damage rather than direct effects on the organ primordia. Okamoto et al. (1968) studied the effect of fast neutron irradiation on the 7th through the 11th days in the rat fetus and found a dose–related increase in congenital malformations. Cardiovascular anomalies were the most frequent and an increased mortality of female fetuses was observed.

Kalter (1968) reviewed the work that was done in the rabbit and hamster. The teratogenic action of different isotopes such as tritium, strontium, and ^{131}I are listed under their separate headings.

Alberman, E.; Polani, P.E.; Fraser–Roberts, J.A.; Spicer, C.C.; Elliott, M. and Armstrong, E.: Parental exposure to x–irradiation and Down's syndrome. Ann. Hum. Genet., London 36:195–208, 1972a.

Alberman, E.; Polani, P.E.; Fraser–Roberts, J.A.; Spicer, C.C.; Elliott, M.; Armstrong, E. and Dhadial, R.K.: Parental x–irradiation and chromosome constitution in their spontaneously aborted foetuses. Ann. Hum. Genet., London 36:185–194, 1972b.

Bithell, J.F. and Stewart, A.M.: Prenatal irradiation and childhood malignancy: A review of the British data from the Oxford survey. Br. J. Cancer 31:271–287,1975.

Blot, W.J. and Miller, R.W.: Mental retardation following in utero exposure to the atomic bombs of Hiroshima and Nagasaki. Radiology 106:617–619, 1973.

Boue, J.; Boue, A. and Lazar, P.: Retrospective and prospective epidemiological studies of 1500 karyotyped spontaneous abortions. Teratology 12:11–26, 1975.

Brent, R.L.: Irradiation in pregnancy. In: Gynecology and Obstetrics, J.J. Lovinsky (ed.), Vol. 2, Chap. 32, Hagerstown, Maryland: Harper and Row, 1–30, 1971.

Brent, R.L.: Radiation and other physical agents. In: Handbook of Teratology, F.C. Fraser and J.G. Wilson (eds.), Plenum Press, New York, 153–201, 1977.

Brent, R.L.: Radiation teratogenesis. Teratology 21:281–298, 1980.

Brent, R.L.; Beckman, D.A. and Jenson, R.P.: Relative radiosensitivity of fetal tissues. Adv. Radiat. Biol. 12:239–256, 1987.

Carter, C.O.; Evans, K.A. and Stewart, A.M.: Maternal radiation and Down's syndrome (mongolism). Lancet 2:1042 only, 1961.

Degenhardt, K.H. and Franz, J.: Models in comparative teratogenesis. Arch. Biol. (Liege) 80:257–298, 1969.

Devi, P.U.; Baskar, R. and Hande, M.P.: Effect of exposure to low–dose gamma radiation during late organogenesis in the mouse fetus. Radiation Research 138:133–138, 1994.

Diamond, E.L.; Schmerler, H. and Lilienfeld, A.M.: The relationship of intra–uterine radiation to subsequent mortality and development of leukemia in children: A prospective study. Am. J. Epidemiol. 97:283–313, 1973.

Driscoll, S.G.; Hicks, S.P.; Copenhaver, E.H. and Easterday, C.L.: Acute radiation injury to two human fetuses. Arch. Pathol. 76:113–119, 1963.

Hawkins, M.M. and Smith, R.A.: Pregnancy outcomes in childhood cancer survivors: probable effects of abdominal irradiation. Int. J. Cancer 43:399–402, 1989.

Hicks, S.P. and D'Amato, C.J.: Effects of ionizing radiations on mammalian development. In: Advances in Teratology, D.H.M. Woollam (ed.), New York: Academic Press, 1:195–250, 1966.

Jacobsen, L.: Radiation induced foetal damage. A quantitative analysis of seasonal influence and possible threshold effect following low dose x–irradiation. In: Advances in Teratology, D.H.M. Woollam (ed.), New York: Academic Press, 4:95–124, 1970.

Jensh, R.P.; Brent, R.L. and Vogel, W.H.: Studies of the effect of 0.4–Gy and 0.6–Gy prenatal x–irradiation on postnatal adult behavior in the Wistar rat. Teratology 35:53–61, 1987.

Jensh, R.P. and Brent, R.L.: Effects of prenatal x–irradiation on the 14th–18th days of gestation on postnatal growth and development in the rat. Teratology 38:431–442, 1988A.

Jensh, R.P. and Brent, R.L.: Effects of prenatal x–irradiation on the postnatal testicular development and function in the Wistar rat: Development/Teratology/Behavior/Radiation. Teratology 38:443–451, 1988B.

Kalter, H.: Teratology of the Central Nervous System. Chicago: University of Chicago Press, 90–138, 1968.

Kuno, H.; Kemi, M. and Matsumoto, H.: Critical period for induction of ocular anomalies produced by soft x–ray irradiation (4.6Gy) in the rat F[1] offspring. Exp. Anim. 43:115–119, 1994.

Macht, S.H. and Lawrence, P.S.: National survey of congenital malformations resulting from exposure to roentgen radiation. Am. J. Roentgenol. Radium Ther. Nucl. Med. 73:442–466, 1955.

MacMahon, B.: Prenatal x–ray exposure and childhood cancer. J. Natl. Cancer Inst. 28:1173–1191, 1962.

Miller, R.W.: Delayed effects occurring within the first decade after exposure of young individuals to the Hiroshima atomic bomb. Pediatrics 18:1–18, 1956.

Mole, R.H.: Irradiation of the embryo and fetus. Br. J. Radiol. 60:17–31, 1987.

Murakami, U. and Kameyama, Y.: Vertebral malformation in the mouse foetus caused by x–radiation of the mother during pregnancy. J. Embryol. Exp. Morphol. 12:841–850,

1964.

Neel, J.V. and Schull, W.J.: The effect of exposure to the atomic bombs on pregnancy termination in Hiroshima and Nagasaki. National Academy of Science, N.R. Council Publication 461, Washington, 1956.

Norton, S. and Kimler, B.F.: Correlation of behavior with brain damage after in utero exposure to toxic agents. Neurotoxicol. Teratol. 9:145–150, 1987.

Okamoto, N.; Ikeda, T.; Satow, Y.; Sawasaki, M. and Inoue, A.: Effects of fast neutron irradiation on the developing rat embryo. Hiroshima J. Med. Sci. 17:169–190, 1968.

Otake, M. and Schull, W.J.: In utero exposure to A–bomb radiation and mental retardation: A reassessment. Br. J. Radiol. 57:409–414, 1984.

Otake, M.; Schull, W.J. and Neel, J.V.: Congenital malformations, stillbirths, and early mortality among the children of atomic bomb survivors: A reanalysis. Radiation Research 122:1–11, 1990.

Plummer, G.: Anomalies occurring in children exposed in utero to the atomic bomb in Hiroshima. Pediatrics 10:687–693, 1952.

Rugh, R.; Duhamel, L.; Chandler, A. and Varma, A.: Cataract development after embryonic and fetal x–irradiation. Radiat. Res. 22:519–534, 1964a.

Rugh, R.; Duhamel, L.; Osborne, A.W. and Varma, A.: Persistent stunting following x–irradiation of the fetus. Am. J. Anat. 115:185–198, 1964b.

Russell, L.B.: X–ray induced developmental abnormalities in the mouse and their use in the analysis of embryological patterns. J. Exp. Zool. 114:545–602, 1950.

Scholte, P.J.L. and Sobels, F.H.: Sex ratio shifts among progeny from patients having received therapeutic x–radiation. Am. J. Human Genetics 16:26–37, 1964.

Schull, W.J.; Norton, S. and Jensh, R.P.: Ionizing radiation and the developing brain. Neurotoxicology and Teratology 12:249–260, 1990.

Schull, W.J.; Otake, M. and Neal, J.V.: Genetic effect of the atomic bombs: A reappraisal. Science 213:1220–1227, 1981.

Sever, L.E.; Gilbert, E.S.; Hessol, N.A. and McIntyre, J.M.: A case–control study of congenital malformations and occupational exposure to low–level ionizing radiation. Am. J. Epidemiol. 127:226–242, 1988

Sigler, A.T.; Lilienfeld, A.M.; Cohen, B.H. and Westlake, J.E.: Radiation exposure in parents of children with mongolism (Down's syndrome). Bulletin of Johns Hopkins Hospital 117:374–399, 1965.

Sikov, M.R. and Mahlum, D.D.: Radiation Biology of the Fetal and Juvenile Mammal. Proceedings of the 9th Annual Hanford Biology Symposium at Richland, Washington, May 5–8, 1969. U.S. Atom Energy Commission, Oakridge, Tenn., Division of Technical Information, 1969.

Strobino, B.R.; Kline, J. and Stein, A.: Chemical and physical exposures of parents. Effects on human reproduction and offspring. Early Human Dev. 1:371–399, 1978.

Uchida, I.: Maternal radiation and trisomy 21. In: Population Cytogenetics, Studies in Humans. E.B. Hook and I.H. Porter (eds.), Academic Press, New York, 285–299, 1977.

Uchida, I. and Curtis, E.J.: A possible association between maternal radiation and mongolism. Lancet 2:848–850, 1961.

United Nations Committee: Ionizing radiation: Levels and effects. United Nations, New York, II:427–428, 1972.

Wagner, L.K. and Hayman, L.A.: Pregnancy and women radiologists. Radiology 145:559–562, 1982.

Wilson, J.G.; Brent, R.L. and Jordan, H.C.: Differentiation as a determinant of the reaction of rat embryos to x–irradiation. Proc. Soc. Exp. Biol. Med. 82:67–70, 1953.

Yamazaki, J.N.: A review of the literature on the radiation dosage required to cause manifest central nervous system disturbances from in utero and postnatal exposure. Pediatr. 37:877–903, 1966.

Yamazaki, J.N. and Schull, W.J.: Perinatal loss and neurological abnormalities among children of the atomic bomb. JAMA 264:605–609, 1990.

Yoshimoto, Y.; Kato, H. and Schull, W.J.: Risk of cancer among children exposed in utero to A–bomb radiations, 1950–84. Lancet 2:665–669, 1988.

2109 Radium

Gudernatsch and Bagg (1920) gave 5 millicuries intravenously or subcutaneously late in the rat pregnancy and caused fetal death associated with hemorrhage mainly in the head and dorsal areas.

Gudernatsch, J.F. and Bagg, H.J.: Disturbances in the development of mammalian embryos caused by radium emanation. Proc. Soc. Exp. Biol. Med. 183–187, 1920.

2110 Ralitoline

Dostal et al. (**1992**) gave this anticonvulsant orally to rats on days 6–15. Maternal toxicity occurred at 60 mg per kg and maternal death at 180 mg per kg. Above 120 mg per kg aortic arch anomalies were found in the fetuses.

Dostal, L.A.; Bleck, J. and Anderson, J.A.: Developmental tosicity of the anticonvulsant drug candidate, ralitoline, in rats (A). Teratology 45:471, 1992.

2111 Ramosetron *YM060*

Tabata et al. (**1994**) studied this serotonin (5 HT)$_3$–receptor on pregnant rats and rabbits using an intravenous dosage. In rabbits dosed with up to 20 mg per kg some decrease in maternal and fetal weights was found but no teratogenicity. The rats were treated before and in early pregnancy, during organogenesis and in the perinatal period and no ill effects were found in the fetuses or their behavior including reproduction.

Tabata, H.; Matsuzawa, T.; Kamada, S.; Ono, C. and Barrow, P.C.: Intravenous reproductive and developmental toxicity of ramosetrorn (YM060), a new serotonin (5HT)$_3$–receptor antagonist, in rats and rrabbits. Oyo Yakuri/Pharmacometrics 47:199–209, 1994.

2112 Ranitidine Hydrochloride *Zantac*^{TM} CAS 71130–06–08

This histamine H–2–receptor antagonist was tested in pregnant rats and rabbits using maximum oral doses of

800 and 400 mg per kg (Higashida et al., 1983, 1984). No changes were observed in either species. Fertility, organogenesis and perinatal behavioral studies were done in the rat and studies of organogenesis were done in the rabbit.

Higashida, N.; Kamada, S.; Sakanove, M.; Takeuchi, M.; Simpo, K. and Tanabe, T.: Teratogenicity studies in rats and rabbits. J. Toxicol. Sci. 8:101–150, 1983; 9:53–72, 1984.

2113 Rapeseed Oil

Beare et al. (1961) fed rats 20 percent rapeseed oil in their diet during pregnancy and reported a reduction in weight and number of weanlings.

Beare, J.L.; Gregory, E.R.W.; Smith, D.M. and Campbell, J.A.: The effect of rapeseed oil on reproduction and on the composition of rat milk fat. Can. J. Biochem. Physiol. 39:195–201, 1961.

2114 Rat Virus *H–1 Strain*

Ferm and Kilham (1965) injected guinea pigs intravenously with the H–1 strain of rat virus on days 6, 7 or 8 of pregnancy. The virus was cultured from the fetuses' tissues which had widespread intranuclear inclusions. Exencephaly, microcephaly and spina bifida were found in the embryos examined. Some enlarged livers and hearts were also reported.

Ferm, V.H. and Kilham, L.: Histopathologic basis of the teratogenic effects of H–1 virus on hamster embryos. J. Embryol. Exp. Morphol. 13:151–158, 1965.

2115 Rauwolfia Alkaloids

Discussed under Deserpidine and Reserpine

2116 Rebamipide
(+-)-2–(4–Chlorobenzoylamino)–3–[2(1H)–quinolinon–4–yl] propionic acid

Saito and Kotosai (**1989**) gave this antiulcer agent orally to rats during organogenesis in doses of up to 1000 mg per kg. At all doses there was no teratogenic effect and postnatal development was normal. It was without effect when given to males and females before mating or to females in the first week. (Oi et al., **1989**).

Oi, A.; Takeuchi, R.; Kajiyoshi, K.; Kotosai, K. and Nishioeda, R.: Reproduction studies on the anti–ulcer agent (+-)–2–(4–chlorobenzoylamino)–3–[2(1h)–quinolinon–4–yl] propionic acid (opc–12759) Fertility study in rats with oral administration. Iyakuhin Kenkyu 20:436–447, 1989.

Saito, M. and Kotosai, K.: Reproduction studies on the anti–ulcer agent (+-)–2–(4–chlorobenzoylamino)–3–[2(1h)–quinolinon–4–yl propionic acid (opc–12759) (ii) Teratological study in rats with oral administration. Iyakuhin Kenkyu 20:448–469, 1989.

2117 Rentiapril

Cozens et al. (1987) gave up to 500 mg per kg by gastric tube. At the highest dose, the rat dams had reduced weight gain. Administration to rats on days 7–17 did not adversely affect fetal development. At doses of 100 mg per kg on days 17 through 21, some reduction in developmental landmarks was observed. No adverse effects were found in the rabbit fetuses treated during organogenesis.

Cozens, D.D.; Barton, S.J.; Clark, R.; Hughes, E.W.; Offer, J.M. and Yamamoto, Y.: Reproductive toxicity studies of rentiapril. Arzneim–Forsch. Drug Res. 37(1):164–169, 1987.

2118 Reovirus

Kilham and Margolis (1974) reported the transplacental passage of reovirus type 3. In the hamster, inoculation during the first 5 days of gestation caused fetal death. On days 9 through 11, the fetuses became infected but survived and developed normally. The pathogenesis and fetal recovery after this virus infection in the rat has been described by Margolis and Kilham (1973).

Kilham, L. and Margolis, G.: Congenital infections due to reovirus type 3 in hamsters. Teratology 9:51–64, 1974.

Margolis, G. and Kilham, L.: Pathogenesis of intrauterine infections in rats due to reovirus type 3. Pathologic and fluorescent antibody studies. Lab Inv. 28:605–613, 1973.

2119 Reproterol *7–3–(2(3,5–Dihydroxyphenyl–2–hydroxy–ethylamino)propyl) theophylline*

Habersang et al. (1977) tested this bronchodilator in rats and rabbits. In rats, oral doses of 320 mg and intravenous doses of 120 mg per kg did not produce adverse effects. In the rabbit, oral doses of 180 mg per kg had no effect on the fetuses, but intravenous doses of 30 mg which were toxic to the dams, produced fetal death.

Habersang, S.; Leuschner, F. and Schlichtegroll, A.: Toxikologische Prufung von Reproterol. Arzneim Forsch. 27:45–52, 1977.

2120 Reserpine CAS 50–55–5

Sobel (1960) reported pregnancy outcome from 15 women treated with reserpine. One stillborn and one pair of twins with congenital lung cysts resulted. Heinonen et al. (1977) reported 4 malformations among the offspring of 48 women taking reserpine or other rauwolfia alkaloids during the first 4 lunar months. Czeizel (1988) studied a group of 60 pregnancies exposed to reserpine and found no increase in defects as compared to 9,892 controls. Budnick et al. (1955) found nasal congestion with cyanosis, costal retraction and lethargy in newborns whose mothers were treated close to the time of parturition. Pauli and Pettersen (1986) reported a single case of an infant with craniofacial, abdominal and central nervous system malformations.

Kehl et al. (1956) reported that 0.75 to 4.5 mg administered during 6 to 10 day periods early in pregnancy prevented implantation in the rabbit. They found no effect on the fetus with treatment later in pregnancy. Tuchmann–Duplessis et al. (1957) studied the rat and found increased abortions. Goldman and Yakovac (1965) giving 1.5 mg per kg on the 9th or 10th day, produced anophthalmia and other defects in slightly over 20 percent of the surviving rat fetuses.

Kalter (1968) reviewed the teratologic work on rauwolfia alkaloids. Buelke–Sam et al. (1984) studied rat offspring exposed to up to 1.0 mg per kg prenatally. At higher doses maternal and fetal toxicity were observed and it is believed that the functional and neurotransmitter changes were caused by the toxicity at high doses. Holson et al. (1994) gave rats subcutaneously .01 mg per kg on days 12–16 or 16–20 and found that the adult brain (and most regions) was reduced in weight after the day 12–16 treatment. The controls were pair–fed. The in vitro growth of rat embryos was inhibited by 100 micromolar concentrations in the medicine.

Budnick, I.J.; Leiken, S. and Hoeck, L.E.: Effect in the newborn infant of reserpine administered ante partum. A.M.A. J. Dis. Child. 90:286–289, 1955.

Buelke–Sam, J.B.; Kimmel, G.L.; Webb, P.O.J.; Slikker, W.; Newport, G.D.; Nelson, C.J. and Kimmel, C.A.: Postnatal toxicity following prenatal reserpine exposures in rats: Effects of dose and dosing schedule. Fund. Appl. Toxicol. 4:983–991, 1984.

Czeizel, A.: Reserpine is not a human teratogen (letter). J. Med. Gen. 25:787, 1988.

Goldman, A.S. and Yakovac, W.C.: Teratogenic action in rats of reserpine alone and in combination with salicylate and immobilization. Proc. Soc. Exp. Biol. Med. 118:857–862, 1965.

Heinonen, O.P.; Slone, D. and Shapiro, S.: Birth Defects and Drugs in Pregnancy. Publishing Sciences Group Inc., Littleton, Mass., 1977.

Holson, R.R.; Webb, P.J.; Grafton, T.F. and Hansen, D.K.: Prenatal neuroleptic exposure and growth stunting in the rat: An in vivo and in vitro examination of sensitive periods and possible mechanisms. Teratology 50:125–136, 1994.

Kalter, H.: Teratology of the Central Nervous System. Chicago: University of Chicago Press, 147–148, 1968.

Kehl, R.; Audibert, A.; Gage, C. and Amarger, J.: Action de la reserpine a differentes periodes de la gestation chez la lapine. C. R. Soc. Biol. (Paris) 150:2196–2199, 1956.

Pauli, R.M. and Pettersen, B.J.: Is reserpine a human teratogen? J. Med. Genet. 23:267–268, 1986.

Sobel, D.E.: Fetal damage due to ECT, insulin coma, chlorpromazine or reserpine. A.M.A. Gen. Psychiatry 2:606–611, 1960.

Tuchmann–Duplessis, H.; Gershon, R. and Mercier–Parot, L.: Troubles de la gestation chez la ratte, provoques par la reserpine et essais d'hormontherapie compensatrice. J. Physiol. 49:1007–1019, 1957.

2121 Resorcinol CAS 108–46–3

Discussed under Hair Dyes

2122 Retinoic Acid *Tretinoin Isotretinoin Accutane*™ CAS 302–79–4

The all–trans retinoic acid is Vitamin A acid or tretinoin. Isotretinoin is 13-cis-retinoic acid. A discussion and comparison of retinoid teratogenicity is given in a position paper of the Teratology Society (Anonymous, 1987). The Public Affairs Committee of the Teratology Society (Anonymous, 1991) has summarized the animal and human teratogenicity in a position paper. In prospective studies 23% of women treated in the first trimester had major malformations while 52% of 5 year olds had intellectual defects. Jick et al. (1993) identified 215 women using topical tretinoin during pregnancy and found no increase in major defects. Buchan (1993) and Nau (1993) have discussed the low exposure from cutaneously administered retinoids. They estimate that a maximal daily dose would be about 0.05 mg per kilogram.

Benke (1984) reported two infants exposed to 20 mg or 50 mg of the drug in utero during the first month of gestation. Both had hydrocephalus and ear defects, one had tetralogy of Fallot and the other cleft palate. Case reports by Fernhoff and Lammer (1984) and Lott et al. (1984) described craniofacial, cardiac and nervous system effects of this drug. Rosa (1983) found of 12 cases reported to FDA, 2 had hydrocephalus, 2 had hydrocephalus with microtia and one had microtia. Details of the neuropathologic lesion from one case were reported by Hansen and Pearl (1985). Lammer et al. (1985) collected over 154 pregnancies during which treatment was inadvertently given. There were 26 normal at birth and 21 malformed. The characteristic pattern included craniofacial (21) cardiac (8) and central nervous system involvement (18). Thymic aplasia occurred in seven. Rosa (1987) reported that 34 of 36 affected infants were exposed to 0.8 mg per kg or more. Lammer et al. (1988) reported that the average highest daily dose taken by 11 mothers with malformed offspring was 0.96 mg per kg as compared to a dose of 0.86 mg per kg for 34 mothers who had infants without malformations. There is no evidence that maternal use which terminates before conception, increases the risk for a malformed child. The mean serum half–life is 16–20 hours. Dai et al. (1989) prospectively collected 88 cases where the mother discontinued treatment before pregnancy. No evidence of teratogenicity was found. In 10 cases the mother discontinued the medication within 5 days of conception.

Lynberg et al. (1990) reported the defects in 61 exposed offspring. Seventy percent had defects of the external ear or canal, 49% had CNS envolvement and 33% had cardiovascular defects. Rizzo et al. (1991) reported two cases with limb defects after exposure. One had missing shoulder bones and the other had absense of the thumb with abnormal radius and ulna.

This form of vitamin A is biologically more active than retinol or retinylesters but fails to protect a vitamin A–deficient animal from blindness. Most of the animal studies used the all–trans form of retinoic acid. The medication used for skin conditions was cis–retinoic acid or isotretinoin (Accutane™).

Kochhar (1967) using pregnant rats and mice, produced the same defects as with excess retinyl acetate. A maternal oral dose of 50 mg per kg on day 9 or 10 produced over 40 percent malformations in the fetal mice. Ehlers et al. (1992) studied the morphology of caudal neural defects in mice and reported the greatest sensitivity to occur on day 8. Shenefelt (1972) used the material as a teratogen in the hamster. Wilson (1971) reported a rhesus monkey with malformed face and ears and hydrocephalus following administration of 40 mg per kg on days 23, 24 and 25 of gestation. Fantel et al. (1977) reported oral–facial, limb and urogenital anomalies in a series of pigtail monkeys treated with 10 mg per kg per day from days 20 through 44. Kamm (1982) reviewed the small animal work. Hendrickx and Hummler (**1992**) found defects in cynomologus monkeys treated at 10 mg per kg but not 5 mg per kg. Seegmiller et al. (1990) applied all-trans retinoic acid in DMSO to rats on days 11–14. At the doses of 25–250 mg per kg there was marked maternal toxicity (20% died). Tail defects and neural tube closure malformations were encountered. Frierson et al. (**1990**) studied a wide number of retinoids in two assays (hamster and limb bud spot culture) and found that the hydrophobic region of the molecules had the greatest effect on potencies. A single case report of an infant with an ear defect identical to those seen following oral retinoic acid was reported after a woman used the cream (0.05%) during the first 11 weeks of pregnancy (Camera and Pregliasco, **1992**).

Limb and lower–body duplications have been induced in mice using 37.5 to 100 mg per kg intraperitoneally on days 4.5, 5 or 5 1/2. (Rutledge et al., **1994**) Other defects such as exencephaly, eye defects and abdominal wall defects were found. Similar treatment on day 4 or 6 did not result in these malformations. Therefore a very early window of susceptibility existed.

Gunning et al. (**1993**) studied all–trans retinoic acid, all–trans retinoyl β–glucose and all-trans retinoyl β–glucuronide in rats and found the first two chemicals teratogenic but not the third which failed to accumulate in the embryo.

Eichele (**1993**) has reviewed the role of retinoids in embryofnic development.

Nolen (1986) reported behavioral changes in the off-spring of rats given a subteratogenic dose (4–6 mg per kg on various days during organogenesis). Webster et al. (1986) produced evidence from in vitro and in vivo mouse models that the agent acts preferentially against neural crest cells. Because the degradation product 4–oxo–isotretinoin was approximately three times higher in human serum than isotretinoin, they suggest that it plays a major role in the pathogenesis. Kochhar and Penner (1987) gave 4–oxo-isotretinoin, a metabolite of isotretinoin, orally to mice on day 11 in amounts of 100 mg per kg and found cleft palates and limb defects in the offspring. Reiners et al. (1988) compared the teratogenicity and metabolism of tretinoin with that of etretinate, etretin and motretinide. Burk and Willhite (**1992**) described inner ear abnormalities in a hamster model and reviewed those seen in humans exposed to retinoic acid.

Satre et al. (1989) found that 4–oxo–all–trans–retinoic acid, a metabolic product of oxidation, was teratogenic in mice and had much longer half–life in the fetus than in the mother. The spectrum of malformations was similar to that of retinoic acid. Goulding et al. (1988) studied a number of benzoic acid derivatives of retinoic acid in whole mouse embryo culture. RO–13–7410, Ch55, 13–cis retinoic acid, RO–14–3899, Ch80, SRI–4529–19 and Am580; all produced

visceral arch abnormalities but at widely different concentrations. All were teratogenic at concentrations less than that for all–trans–retinoic acid. RO–15–0778 (containing a terminal benzene ring) was not teratogenic. Kochhar et al. (**1992**) presented evidence that hydrolysis of the amide group on N–(Retinyl) glycine produced levels of retinoic acid sufficient to be teratogenic.

Kraft et al. (1989) compared the teratogenic dose of trans and cis forms of retinoic acid and found that the cis form and its 4–oxo derivative were much less teratogenic. They suggested that the teratogenicity of the cis–form could be its partial conversion to the trans–form. Klug et al. (1989) found the all-trans form to be more active than the 13–cis form in whole rat embryo culture and suggested based on analysis of the two that the tertogenicity of 13–cis was due to isomerisation of the all–trans. Kocchar et al. (1988) offered evidence that conversion of retinol to retinoic acid could account for the teratogenic action of retinol. Eckhoff et al. (**1994**) detailed the teratogenicity and placental pharmacokinetics of 13–cis–retinoic acid in the rabbit. Low amounts of 13–cis–retinoic acid were found while all–trans–4–oxoretinoic acid was transfered at a higher rate.

Jiang and Kochhar (**1992**) found that apoptosis in treated mouse embryo limbs was associated with a rise in transglutaminase levels.

Chachoud et al. (1989) applied 0.05% tretinoin to the shaved back of rats on day 11 and found mild tertogenicity at this dose which was about 100 fold that expected in the human. Howard et al. (**1989**) reported the pharmacokinetics and non–teratogenicity of N–ethyl–all–trans–retinamide and its 13 cis congener in the pregnant hamster. Willhite et al. (**1990**) applied radiolabelled all–trans retinoic acid to hamster skin and studied its distribution. Up to 100 mg per kg was applied and no teratogenicity was found.

Anonymous: Teratology Society position paper: Recommendations for vitamin A use during pregnancy. Teratology 35:269–275, 1987.

Anonymous: Teratology Society position paper: Recommendations for isotretinoin use in women of childbearing potential. Teratology 44:1–6, 1991.

Benke, P.J.: The isotretinoin teratogen syndrome. JAMA 251:3267–3269, 1984.

Buchan, P.: Evaluation of the teratogenic risk of cutaneously administered retinoids. Skin Pharmacol 6:45–52, 1993.

Burk, D.T. and Willhite, C.C.: Inner ear malformations induced by isotretinoin in hamster fetuses. Teratology 46:147–157, 1992.

Camera, G. and Pregliasco, P.: Ear malformation in baby born to mother using tretinoin cream. The Lancet 339:687, 1992.

Chahoud, I.B.; Loefberg, B.; Mittmann, B. and Nau, H.: Teratogenicity and pharmacokinetics of vitamin a acid (tretinoin, all–trans retinoic acid) after dermal application in the rat. Naunyn–Schmiedeberg's Arch Pharmacol 339(Suppl):R30, 1989.

Dai, W.S.; Hsu, M. and Itri, L.M.: Safety of pregnancy after discontinuation of isotretinoin. Arch. Dermatol.

125:362–365, 1989.

Eckhoff, C.; Chari, S.; Kromka, M.; Staudher, H.; Juhasz, L.; Rudiger, H. and Agnish, N.: Teratogenicity and transplacental pharmacokinetics of 13–cis–retinoic acid in rabbits. Toxicol. Appl. Pharm. 125:34–41, 1994.

Ehlers, K.; Sturje, H. Merker. H. and Nau, H.: Spina bifida aperta induced by valproic acid and by all–trans–retinoic acid in the mouse: distinct differences in morphology and periods of sensitivity. Teratology 46:117–130,

Eichele, G.: Retinoids in embryonic development. Maternal Nutrition and Pregnancy Outcome. C.L. Keen, A. Bendish and C.C. Willhite, eds. N.Y. Acad Sciences. 678:22–36, 1993.

Fantel, A.G.; Shepard, T.H.; Newell–Morris, L.L. and Moffett, B.C.: Teratogenic effects of retinoic acid in pigtail monkeys (Macaca nemestrina). Teratology 15:65–72, 1977.

Fernhoff, P.M. and Lammer, E.J.: Craniofacial features of isotretinoin embryopathy. J. Pediatr. 105:595–597, 1984.

Frierson, M.R.; Mielach, F.A. and Kochhar, D.M.: Computer–automated structure evaluation (case) of retinoids in teratogenesis bioassays. Fundamental and Applied Toxicology 14:408–428, 1990.

Goulding, E.H.; Jetten, A.M.; Abbott, B.D. and Pratt, R.M.: Teratogenicity of benzoic acid derivatives of retinoic acid in cultured mouse embryos. Reprod. Toxicol. 2:91–98, 1988.

Gunning, D.B.; Barua, A.B. and Olson, J.A.: Comparative teratogenicity and metabolism of all–trans retinoic acid, all–trans β–glucose, and all–trans retinoyl β–glucuronide in pregnant Sprague–Dawley rats. Teratology 47:29–36, 1993.

Hansen, L.A. and Pearl, G.S.: Isotretinoin teratogenicity: A case report with neuropathologic findings. Acta Neuropathol. (Berl.) 65:335–337, 1985.

Hendrickx, A.G. and Hummler, H.: Teratogenicity of all–trans retinoic acid during early embryonic development in the cynomolgus monkey (macaca fascicularis). Teratology 45:65–74, 1992.

Howard, W.B.; Willhite, C.C.; Omaye, S.T. and Sharma, R.P.: Pharmacoinetics, tissue distribution, and placental permeability of all–trans–and 13–cis–n–ethyl retinamides in pregnant hamsters. Fundamental and Applied Toxicology 12:621–627, 1989.

Jiang, H. and Kochhar, D.M.: Induction of tissue transglutaminase and apoptosis by retinoic acid in the limb bud. Teratology 46:333–340, 1992.

Jick, S.S.; Terris, B.Z. and Jick, H.: First trimester topical tretinoin and congenital disorders. The Lancet 341:1181-1182, 1993.

Kamm, J.J.: Toxicology, carcinogenicity and teratogenicity of some orally administered retinoids. J. Am. Acad. Dermatol. 6:652–659, 1982.

Klug, S.; Kraft, J.C.; Wildi, E.; Merker, H.J.; Persaud, T.V.N.; Nau, H. and Neubert, D.: Influence of 13–cis and all–trans retinoic acid on rat embryonic development in vitro: correlation with isomerisation and drug transfer to the embryo. Arch. Toxicol. 63:185–192, 1989.

Kochhar, D.M. and Penner, J.D.: Developmental effects of isotretinoin and 4–oxo–isotretinoin: The role of metabolism in teratogenicity. Teratology 36:67–75, 1987.

Kochhar, D.M.; Penner, J.D. and Satre, M.A.: Derivation of retinoic acid and metabolites from a teratogenic dose of retinol (vitamin a) in mice. Toxicology and Applied Pharmacology 96:429–441, 1988.

Kochhar, D.M.: Teratogenic activity of retinoic acid. Acta Pathol. Microbiol. Scand. 70:398–404, 1967.

Kochhar, D.M.; Shealy, Y.F.; Penner, J.D. and Jiang, H.: Retinamides: hydrolytic conversion of retinoylglycine to retinoic acid in pregnant mice contributes to teratogenicity. Teratology 45:175–185, 1992.

Kraft, J.C.; Chahoud, L.I.; Bochert, G. and Nau, H.: Teratogenicity and -placental transfer of all–trans–,13–cis–,4–oxo–all–trans–and 4–oxo–13–cis–retinoic acid after administration of a low oral dose during organogenesis in mice. Toxicology and Applied Pharmacology 100:162–176, 1989.

Lammer, E.J.; Chen, D.T.; Hoar, R.M.; Agnish, N.D.; Benke, P.J.; Braun, J.T.; Curry, C.J.; Fernhoff, P.M.; Grix, A.W.; Lott, I.T.; Richard, J.M. and Sun, C.C.: Retinoic acid embryopathy a new human teratogen and mechanistic hypothesis. N. Eng. J. Med. 313:837–841, 1985.

Lammer, E.J.; Schunior, A.; Hayes, A.M. and Holmes, L.B.: Isotretinoin dose and teratogenicity. Lancet 2:503–504, 1988.

Lott, I.T.; Bocian, M.; Pribram, H.W. and Leitner, M.: Fetal hydrocephalus and ear anomalies associated with maternal use of isotretinoin. J. Pediatr. 105:597–600, 1984.

Lynberg, M.C.; Khoury, M.J.; Lammer, E.J.; Waller, K.O.; Cordero, J.F. and Erickson, J.D.: Sensitivity, specificity, and positive predictive value of multiple malformations in isotretinoin embryopathy surveillance. Teratology, 42:513–519, 1990.

Nau, H.: Embryotoxicity and teratogenicity of topical retinoic acid. Skin Pharmacol. 6:35–44, 1993.

Nolen, G.A.: The effects of prenatal retinoic acid on the viability and behavior of the offspring. Neurobehav. Toxicol. Teratol. 8:643–654, 1986.

Reiners, J.; Lofberg, B.; Kraft, J.C.; Kochhar, D.M. and Nau, H.: Transplacental pharmacokinetics of teratogenic doses of etretinate and other aromatic retinoids in mice. Reproductive Toxicology 2:19–30, 1988.

Rizzo, R.; Lammer, E.J.; Parano, E.; Pavone, L. and Argyle, J.C.: Limb reduction defects in humans associated with prenatal isotretinoin exposure. Teratology 44:599–604, 1991.

Rosa, F.W.: Teratogenicity of isotretinoin. Lancet 2:513, 1983.

Rosa, F.W.: Isotretinoin dose and teratogenicity. Lancet endogenous retinoic acid metabolite. Teratology 39:341–348, 1989.

Rutledge, J.C.; Shourbaji, A.G.; Hughes, L.A.; Polifka, J.E.; Cruz, Y.P.; Bishop, J.B. and Generoso, W.M.: Limb and lower–body duplications induced by retinoic acid in mice. Proc. Natl. Acad. Sci. 91:5436–5440, 1994.

Seegmiller, R.E.; Carter, M.W.; Ford, W.H. and White, R.D.: Induction of maternal toxicity in the rat by dermal application of retinoic acid and its effect on fetal outcome. Reproductive Toxicology 4:277–281, 1990.

Shenefelt, R.E.: Morphogenesis of malformations in hamsters caused by retinoic acid. Relation to dose and stage of treatment. Teratology 5:103–118, 1972.

Webster, W.S.; Johnston, M.C.; Lammer, E.J. and Sulik, K.K.: Isotretinoin embryopathy and the cranial neural crest: An in vivo and in vitro study. J. Craniofac. Genet. Dev. Biol. 6:211–222, 1986.

Willhite, C.C.; Sharma, R.P.; Allen, P.V. and Berry, D.L.: Percutaneous retinoid absorption and embryotoxicity. The Journal of Investigative Dermatology 95:523–529, 1990.

Wilson, J.G.: Use of primates in teratological research and testing. In: Malformations Congenitales Des Mammiferes. H. Tuchmann–Duplessis (ed.), Paris: Masson, 277–280, 1971.

2123 Retinylidene Methyl Nitrone

This form of retinoic acid has been modified by addition of a methyl nitrone function. Willhite and Balogh–Nair (1984) gave single oral doses of 50–100 mg per kg to hamsters of the all trans form and produced the same syndrome as observed with all–trans–retinoic acid. Treatment was at 10:00 am on the 8th day.

Willhite, C.C. and Balogh–Nair, V.: Developmental toxicity of retinylidene methyl nitrone in the golden hamster. Toxicology 33:331–340, 1984.

2124 Retroprogesterone *9 β, 10 α–4–Pregnene–3, 20–dione*

Kawashima et al. (1977) administered up to 50 mg orally to rats on days 17–20 and produced no decrease in the urovaginal septum length in female fetuses.

Kawashima, K.; Nakaura, S.; Nagao, S.; Tanaka, S.; Kuwamura, T. and Omori, Y.: Virilizing activities of various steroids in female rat fetuses. Endocrinol. Japon. 24(1):77–81, 1977.

2125 Rh Immune Globulin *Rhogam*

Crane et al. (1984) followed 147 women receiving Rh immune globulin after second trimester amniocentesis. No adverse maternal or fetal effects were attributed to the treatment.

Crane, J.P.; Rohland, B. and Larson, D.: Rh immune globulin after genetic amniocentesis: impact on pregnancy outcome. American Journal of Medical Genetics 19:763–768, 1984.

2126 Rheumatic Disease of Mother *Lupus Erythematosis*

Of 22 children with congenital heart block, 14 were born to mothers with rheumatic disease, primarily systemic lupus erythematosis (McCue et al., 1977). Among 58 pregnancies in mothers with lupus erythematosis there were three spontaneous abortions and 14 with premature labor or induction. Eighteen had term delivery but no comment was made about congenital defects (Fine et al., 1981) Nicholas (**1988**) has reviewed the general problem of rheumatic disease in pregnancy. Brucato et al. (**1991**) discussing the frequency of congenital heart block cite a study of Lockshin et al. (Arthritis Rheum 31:697, 1988) in which 91 pregnancies were studied prospectively among patients with lupus erythematosis and no congenital heart blocks of infants were found.

Bierman et al. (1988) reported successful treatment of fetal heart block by treating the mother who had increased antinuclear antibody titers with betamethasone. Ho et al. (1986) reported the presence of fibrous and adipose tissue in the area of infants with congenital heart block from mothers who had anti–Ro antibodies.

Bierman, F.Z.; Baxi, L.; Jaffe, I. and Driscoll, J.: Fetal hydrops and congenital complete heart block: Response to maternal steroid therapy. J. Pediatr. 112:646–648, 1988.

Brucato, A.; Ferraro, G. and Gasparini, M.: Congenital heart block and maternal SLE. The Lancet 338:892, 1991.

Fine, L.G.: Systemic lupus erythematosis in pregnancy. Annals Int. Med. 94:667, 1981.

Ho, S.Y.; Esscher, E.; Anderson, R.H. and Michaelsson, M.: Anatomy of congenital heart block and relation to maternal anti–Ro antibodies. Am. J. Cardiol. 58:291–294, 1986.

McCue, C.M.; Mantakas, M.E.; Tingelstad, J.B. and Ruddy, S.: Congenital heart block in newborns of mothers with connective tissue disease. Circulation 56:82–90, 1977.

Nicholas, N.S.: Rheumatic diseases in pregnancy. Br. J. Hosp. M. 39(1):50–53, 1988.

2127 Rheumatoid Synovium Agent

Warren et al. (1970) injected pregnant mice with a raw slurry of synovial fluid from patients with rheumatoid arthritis and produced a redness and swelling in the joints of over one–half the offspring. The condition which resembled rheumatoid arthritis was transmitted to four generations without further injections.

Warren, S.L.; Marmor, L.; Liebes, D.M. and Hollins, R.L.: Congenital deformities of mice transmitted by a human rheumatoid synovium agent. In: Clinical Orthopedics and Related Research, No. 70., M. Urist (ed.), Philadelphia, J.B. Lippincott. 216–219, 1970.

2128 Rhodamines CAS 81–88–9

No fetal abnormalities were found after oral administration to dogs and rats during organogenesis. The highest no effect dose from two–year studies was used (Burnett et al., 1974). Ranganathan and Hood (1988) found that Rh 6G and Rh 123 uncoupled mitochondria ATPase activity in day 12 mice. Hood et al. (1989) compared cationic (Rh 123 and Rh 6G) with neutral rhodamines (Rh 116 and Rh B) and found the neutral forms did not effect prenatal survival or growth in mice. The use of 2–deoxyglucose enhanced the effect of the cationic forms.

Burnett, C.M.; Agersborg, H.P.K., Jr.; Borzelleca, J.F.; Eagle, E.; Ebert, A.G.; Pierce, E.C.; Kirschman, J.C. and Scala, R.A.: Teratogenic studies with certified colors in rats and rabbits. Toxicol. Appl. Pharmacol. 29:121, 1974.

Hood, R.D.; Jones, C.L. and Ranganathan, S.: Comparative developmental toxicity of cationic and neutral rhodamines in mice. Teratology 40:143–150, 1989.

Ranganathan, S. and Hood, R.D.: Inhibition of oxidative phosphorylation by cationic rhodamines as a possible teratogenicity mechanism. Teratology 37:484, 1988.

2129 Rhodium CAS 7440–16–6

Ridgway and Karnofsky (1952) exposed chick embryos at eight days of incubation to Rh chloride (4.3 microatoms) and caused stunting, mild micromelia and inhibition of feather growth.

Ridgway, L.P. and Karnofsky, D.A.: The effects of metals on the chick embryo: Toxicity and production of abnormalities in development. Ann. N.Y. Acad. Sci. 55:203–215, 1952.

2130 Ribavirin *1–β–D–Ribofuranosyl–1,2,4 triazole–3–carboximide* CAS 36791–04–5

Studies reported by the Center for Disease Control detected 0.44 microgm per ml in the RBC sample from only one sample from ten hospital workers exposed to aerosols (Anonymous, 1988). The estimated absorbed dose exceeded one–hundredth of the short–term daily dose levels teratogenic in animals. Health–care workers should be advised of the potential risk. No epidemiologic studies or case reports were found so far.

Kilham and Ferm (1977) gave single intraperitoneal doses of 1.25 to 6.15 mg per kg to hamsters on day 8 and found a high incidence of defects which included the limbs, ribs, eyes, and central nervous system. Anophthalmia and exencephaly were found. In mice, Kochhar et al. (1980) produced craniofacial bone defects when doses were 25 mg per kg daily intraperitoneally on 3 days during organogenesis. At doses above 25 mg per kg on single days during organogenesis Kochhar (**1990**) reported skeletal defects spina bifida and cleft palate. Ferm et al. (1978) found head defects in rats exposed intraperitoneally or orally to 37.5 mg per kg on day 9 of gestation. Johnson (**1990**) has reviewed the animal work with this drug.

Using doses of up to 620 mg per meter square for six-hour periods for 10 to 30 days postnatally in ferrets, Hoffmann et al. (1987) were unable to find changes in lung function or find permanent lung histologic alterations.

Anonymous: Morbidity and Mortality Weekly Report, Center for Disease Control, Atlanta, GA. 560–563, Sept. 16, 1988.

Ferm, V.H.; Willhite, C.C. and Kilham, L.: Teratogenic effects of ribavirin on hamster and rat embryos. Teratology 17:93–102, 1978.

Hoffmann, S.H.; Staffa, J.A. and Smith, R.A.: Inhalation toxicity of ribavirin in suckling ferrets. J. Appl. Toxicol. 7:343–351, 1987.

Johnson, E.M.: The effects of ribarvirin on development and reproduction: A critical review of published and unpublished studies in experimental animals. Journal of the American College of Toxicology 9:551–561, 1990.

Kilham, L. and Ferm, V.H.: Congenital anomalies induced in hamster embryos with ribavirin. Science 195:413–414, 1977.

Kochhar, D.M.: Effects of exposure to high concentrations of ribavirin in developing embryos. Pediatr. Infect. Dis.

J. 9:s88–s90, 1990.

Kochhar, D.M.; Penner, J.D. and Knudsen, T.B.: Embryotoxic, teratogenic and metabolic effects of ribavirin in mice. Toxicol. Appl. Pharmacol. 52:99–112, 1980.

2131 Riboflavin Deficiency also see *Galactoflavin*

Experiments by Warkany and Nelson (1940) established for the first time that a syndrome of skeletal malformations could be induced in mammals by withholding a single dietary factor. A high number of offspring from deficient rats have short mandibles, cleft palate, syndactylism and reduction defects of the extremities. Hydronephrosis occurs also but defects of the central nervous system and eye are uncommon.

The standard method for producing this experimental model is at the beginning of pregnancy to place the rat on a riboflavin deficient diet containing galactoflavin (60 mg per kg of diet) a riboflavin analog. The mechanism of teratogenesis is associated with a lack of the high energy generating source, the terminal electron transport system (Aksu et al., 1968). Warkany and Kalter (1959) reviewed the general subject. Kalter and Warkany (1957) produced congenital hydrocephalus in the mouse fetus made riboflavin deficient. Romanoff and Bauernfeind (1957) produced micromelia and mandibular hypoplasia and increased mortality in chicks from hens maintained for three weeks on a riboflavin–deficient diet.

There is no compelling evidence that riboflavin deficiency is a cause of congenital defects in the human fetus.

Aksu, O.; Mackler, B.; Shepard, T.H. and Lemire, R.J.: Studies of the development of congenital anomalies in embryos of riboflavin–deficient, galactoflavin fed rats. II. Role of the terminal electron transport systems. Teratology 1:93–102, 1968.

Kalter, H. and Warkany, J.: Congenital malformations in inbred strains of mice induced by riboflavin–deficient galactoflavin–containing diets. J. Exp. Zool. 136:531–566, 1957.

Romanoff, A.L. and Bauerfeind, J.C.: Influence of riboflavin–deficiency in eggs on embryonic deficient galactoflavin–containing diets. J. Exp. Zool. 136:531–566, 1957.

Warkany, J. and Kalter, H.: Experimental production of congenital malformations in mammals by metabolic procedure. Physiol. Rev. 39:69–115, 1959.

Warkany, J. and Nelson, R.C.: Appearance of skeletal abnormalities in the offspring of rats reared on a deficient diet. Science 92:383–384, 1940.

2132 Ribonuclease

Matousek et al. (1973) injected purified fractions of bull seminal fluid containing ribonuclease activity into pregnant guinea pigs and rabbits. An early postimplantation loss of embryos was found in both species. Teratology studies were not performed.

Matousek, J.; Fulka, M.J. and Pavlok, A.: Effect of ribonuclease fractions isolated from bull seminal vesicle fluid on embryonic mortality in guinea pigs, rabbits and pigs. Int. J. Fertil. 18:13–16, 1973.

2133 5–Ribonucleotide, Disodium

Kaziwara et al. (1971) fed pregnant mice, rats and monkeys 2000, 500 and 1000 mg per kg, respectively. No adverse embryonic effects were found.

Kaziwara, K.; Mizutani, M. and Ihara, T.: On the fetoxicity of disodium 5'–Ribonucleotide in the mouse, rat and monkey. Journal of the Takeda Research Laboratories 30(2):314–321, 1971.

2134 Rifampin *Rifamycin Methyl–4–piperazinyl–1 iminomethyl–3 rifamycine S.V. Rifampicin* CAS 6998–60–3

Warkany (1979) reviewed this subject and 82 exposed pregnancies where no increase in malformation rate occurred. Steen and Stainton–Ellis (1977) reported nine malformations among 202 exposed newborns. The malformations included (1) anencephaly, (2) hydrocephalus, (2) genitourinary anomalies, (1) dislocated hip and (1) skeletal reduction anomalies. The presence of three skeletal reduction defects in such a small group is unusual. The method for selecting the treated women was not given in the report and without this knowledge about the total exposed population, evaluation is difficult. In a personal communication from Dr. Steen, the writer was informed that the total number of treated pregnancies was not known. Snider et al. (1980) reviewed 15 studies of women exposed during pregnancy to rifampin and among 446 pregnancies, the abortion rate was 1.67 percent and the malformation rate was 3.35 percent.

Tuchmann–Duplessis and Mercier–Parot (1969) gave this antibiotic by mouth to mice, rats and rabbits during the active period of organogenesis. In the mice and rats doses above 150 mg per kg produced spina bifida in both and cleft palates in the mouse fetuses. Similar treatment of pregnant rabbits had no effect on the fetuses. Anufrieva et al. (1980A,B) exposed by inhalation (in doses 6.1 and 0.81 mg per cubic kilometer) Wistar rats during the entire gestation and this drug produced no congenital malformations. Some functional disturbances of the offspring's organs were found. The authors thought that the drug did not affect the structure or function of the placenta.

Greenaway et al. (1981) produced open neural tubes in rat embryos grown in vitro at concentrations of 12.5 to 50 micrograms per ml of medium. Bioactivation by a liver monooxygenation was necessary to produce the defects but not the reduction in growth measures. Spielmann et al. (1986) gave 1000 mg per kg intraperitoneally to mice on day 2 and at this dose (two–thirds lethal), found some increase in fetal lethality on day 17.

Anufrieva, R.G.; Zeltser, I.Z.; Balabanova, E.L.; Lapchinskaya, A.V.; Baru, R.V. and Svinogeeva, T.P.: Experimental study of rifampicin effect on albino rat embryogenesis. Antibiotiki (USSR) 25,4:280–284, 1980A.

Anufrieva, R.G.; Zeltser, I.Z. and Svinogeeva, T.P.: Placenta permeability by rifampicin. Antibiotiki (USSR) 25,3:199–201, 1980B.

Greenaway, J.C.; Fantel, A.G. and Shepard, T.H.: In vitro metabolic activation of rifampicin teratogenicity. (abs) Teratology 23:37A, 1981.

Snider, D.E.; Layde, P.M.; Johnson, M.W. and Lyle, M.A.: Treatment of tuberculosis during pregnancy. Amer. Rev. Resp. Dis. 122:65–79, 1980.

Spielmann, H.; Kruger, C.; Tenschert, B. and Vogel, R.: Studies on the embryotoxic risk of drug treatment during the preimplantation period in the mouse. Arzneim-Forsch. 36:219–223, 1986.

Steen, J.S.M. and Stainton–Ellis, D.M.: Rifampicin in pregnancy. Lancet 2:604–605, 1977.

Tuchmann–Duplessis, H. and Mercier–Parot, L.: Influence d'un antibiotique, la rifampicine, sur le developpement prenatal des ronguers. C. R. Acad. Sci. (d) (Paris) 269:2147–2149, 1969.

Warkany, J.: Antituberculous drugs. Teratology 20:133–138, 1979.

2135 Rimantadine

α–Methyl–1–adamanthylmethylaminohydrochloride CAS 13392–28–4

Alexandrov et al. (1982) gave this drug intragastrically to pregnant rats (500 mg per kg) and mice (300 mg per kg) on the 8th, 9th or 13th day, respectively, and found no fetal damage. However, when the drug was given to rats on the 8th day of gestation, embryonic mortality increased.

Alexandrov, V.A.; Pozharsky, K.M.; Likhachev, A.Y.; Anisimov, V.N.: Okulov, V.B. and Ivanov, M.N.: The result of testing rimantadine for carcinogenicity, teratogenicity and embryotoxicity. Rimantadine and other viruses inhibitors. Riga (USSR) 154–165, 1982.

2136 Rioprostil

Clemens et al. (1992) treated rabbits with up to 1500 microgm per kg of this E1 prostoglanin. Neither the days of treatment nor mode of administration were given in the abstract. A syndrome of gastroschisis, umbilical hernia, spina bifida and caudal vertebral defect was found increased at 300 microgm per kg and above.

Clemens, G.R.; Hartnagel, R.E.; Hilbish, K.G.; Abrutyn, D. and Schlueter, G.: Developmental toxicity of rioprostil, an E1 class prostaglandin in the rabbit (A). Teratology 45:470, 1992.

2137 Risperidone

This antipsychotic drug was given to rats orally in doses up to 2.5 mg per kg before and during early pregnancy and up to 10 mg per kg during organogenesis. At 10 mg per kg some fetal decrease in weight was found but no teratogenic or fertility effects were encountered.

Van Cauteren, H.; Coussement, W.; Dirkx, P.; Lampo, A. and Usui, T.: Reproductive and developmental toxicity studies in rats with risperidone. The Clinical Report 27:3023–3024, 1993.

2138 Ritodrine Hydrochloride

Erythro–1–(4–hydroxyphenyl)–2–

[2–4–hydroxyphenyl)ethylamino)–1–propanol HCL CAS 23239–51–2

Nuchpuckdee et al. (1986) studied the thickness of ventricular septum in infants exposed prenatally to this inhibitor of preterm labor. The increase was correlated with duration of treatment. The echocardiogram changes were temporary lasting for less than three months. Kazzi et al. (1987) reported an increase in "low dextrostix" occurring within an hour of delivery after the use of ritodrine. Musci et al. (1986) studied 200 women given the drug and found that fetuses exposed beginning at 30 weeks or less needed photo therapy for icterus more often than those treated after 30 weeks. Hypoglycemia was more common in infants exposed up to the day of delivery.

Imai et al. (1984) studied this drug in rats and rabbits. In rabbits, up to 750 mg per kg was given orally or up to 35 mg intravenously during organogenesis. In the oral studies growth retardation and embryolethality occurred at the maternal toxic dose of 750 mg per kg. At 750 mg per kg of 89 fetuses examined, there were 7 skeletal anomalies (mostly of digits), 6 ventricular septal defects and 8 other external defects. The intravenous treatment produced embryo lethality but no increase in defects. Maximum oral doses of 500 mg per kg for fertility and 1,000 mg per kg for studies during organogenesis and perinatal periods were used in the rat. They used 50 mg per kg intravenously for fertility studies and 110 mg per kg for studies during organogenesis. At the highest doses which were toxic to the dams, growth retardation was seen in the fetus. The higher doses inhibited uterine contraction in the rat and produced frequent maternal death which made it difficult to assess fetal function. No malformation increase occurred in the rat fetuses.

Imai, K.; Makita, T.; Nakajo, M.; Ohba, M.; Ikeda, S.; Ozawa, S.; Sakai, Y.; Yamamoto, K. and Hirasawa, K.: Reproduction of ritodrine hydrochloride studies in rats and rabbits. Clin. Report 18:6233–6281, 1984, 19:1351–1357, 2002–2018, 1984.

Kazzi, N.J.; Gross, T.L.; Kazzi, G.M. and Williams, T.G.: Neonatal complications following in utero exposure to intravenous ritodrine. Acta Obstet. Gynecol. Scand. 66:65–69, 1987.

Musci, M.N.; Abbasi, S.; Otis, C. and Bolognese, R.J.: Prolonged fetal ritodrine exposure and immediate neonatal outcome. J. Perinatol. 8:27–32, 1986.

Nuchpuckdee, P.; Brodsky, N.; Porat, R. and Hurt, H.: Ventricular septal thickness and cardiac function in neonates after in utero exposure. J. Pediatr. 109:687–691, 1986.

2139 RMI 12,936
17–β–Hydroxy–7–α–methylandrost–5–en–3–one

Kendle (1975) interrupted rat pregnancies by administering ten mg subcutaneously on day eight.

Kendle, K.E.: Some biological properties of RMI 12,936, a new synthetic antiprogestational steroid. J. Reprod. Fertil. 43:505–513, 1975.

2140 RMI 14,514

Gibson et al. (1981) gave pregnant rats up to 150 mg per kg orally during organogenesis or the entire gestation and found no adverse fetal or postnatal effects.

Gibson, J.P.; Larson, E.J.; Yarrington, J.T.; Hook, R.H.; Kariya, T. and Blohm, T.R.: Toxicity and teratogenicity studies with the hypolipidemic drug RMI 14,514 in rats. Fundamental and Applied Toxicology 1:19–25, 1981.

2141 Robaveron *KN–7 Prostate Extract*

Maeda et al. (1985) gave this extract of mature pig prostate to mice orally using 500 mg per kg before fertilization, during organogenesis and in the perinatal period. Some embryo–lethality occurred at the highest dose level and growth retardation was found postnatally. Rabbits given up to 5,000 mg per kg on days 6–18 had normal fetuses.

Maeda, H.; Yoshifune, S.; Mori, Y.; Sugiyama, K. and Tatsami, H.: Reproduction studies of KN–7 in mice and rabbits. Clin. Report 19:359–400, 601–608, 1985. Ax5;!

2142 Ronidazole

Greenaway et al. (1986) using in vitro culture of rat embryos found axial asymmetry and growth retardation at concentrations of 0.5 millimoles.

Greenaway, J.C.; Fantel, A.G. and Juchau, M.R.: On the capacity of nitroheterocyclic compounds to elicit an unusual axial asymmetry in cultured rat embryos. Toxicol. Appl. Pharmacol. 82:307–315, 1986.

2143 Ronnel *Fenchlorphos* CAS 299–84–3

This insecticide was administered by gavage in doses of 400, 600 and 800 mg per kg to rats on days 6 through 15 (Khera et al., 1982). Extra ribs were increased in the 600 and 800 mg groups but fetal weight was not decreased. Berge and Nafstad (1983) gave 100 mg per kg to pregnant foxes and found fetuses with facial clefts, hydrocephalus and renal defects. Reduction of the granular layer of the cerebellum was found.

Berge, G.N. and Nafstad, I.: Teratogenicity and embryotoxicity of orally administered fenchlorphos in Blue foxes. Act. Vet. Scand. 24:99–112, 1983.

Khera, K.S.; Whalen, C. and Angers, G.: Teratogenicity study on pyrethrum and rotenone (natural origin) and ronnel in pregnant rats. J. Toxicol. Environ. Health 10:111–119, 1982.

2144 Rotenone CAS 83–79–4

This pesticide is known to be a specific and irreversible inhibitor of the electron transport chain between flavoprotein and cytochromes. Rao and Chauhan (1971) exposed early chick embryo explants for 15 minutes to 1 microgm per ml and observed after explantation various degrees of growth inhibition and neural tube defect.

Khera et al. (1982) gavaged rats on days 6 through 15 with 2.5, 5 or 10 mg per kg. Maternal and fetal weights were reduced at 5 and 10 mg. Minor skeletal defects were found in the 5 mg group and resorptions were 46 percent in the

10 mg group. Spencer and Sing (1982) found maternal toxicity in rats fed diets with 750 or 1,000 ppm. No increased resorptions or malformations were seen in the rats treated with 800 ppm.

Khera, K.S.; Whalen, C. and Angers, G.: Teratogenicity study of pyrethrum and rotenone (natural origin) and ronnel in pregnant rats. J. Toxicol. Environ. Health. 10:111–119, 1982.

Rao, K.V. and Chauhan, S.P.S.: Teratogenic effects of rotenone on the early development of chick embryos in vitro. Teratology 4:191–198, 1971.

Spencer, F. and Sing, L.: Reproductive responses to rotenone during decidualized pseudo–gestation and gestation in rats. Bull. Environ. Contam. 28:360–368, 1982.

2145 Rowachol™

Hasegawa and Toda (1978) studied this combination of terpenoids in the rat. No adverse effects on reproduction or the fetus were seen with treatment orally during organogenesis or in the perinatal period.

Hasegawa, M. and Toda, T.: Teratological studies on Rowachol™ , remedy for Cholelithiasis. Effect of Rowachol administered to pregnant rats during organogenesis on pre– and post–natal development of their offspring. Oyo Yakuri 15:1109–1119, 1978.

2146 Roxatidine Acetate Hydrochloride

Usui et al. (1991) gave rats up to 40 mg per kg intravenously from before fertilization through pregnancy and lactation. No adverse reproductive events were found and the postnatal development was not changed.

Usui, T.; Eguchi, K.; Ogawa, C.; Sone, H.; Makino, M.; Yamamoto, T. and Suzuki, M.R.: Effects of intravenously administered roxatidine acetate hydrochloride on fertility and general reproductive performance in rats. Oyo Yakuri 42:439–447, 1994

2147 RU–24722

Esaki and Sasa (1984) gave up to 100 mg per kg daily to rabbits during organogenesis. No adverse fetal effects were found.

Esaki, K. and Sasa, H.: Teratogenicity of intragastric administration of RU–24722 in rabbits. Preclin. Rep. Cent. Inst. Exp. Anim. 10(1):27–36, 1984.

2148 Rubella Vaccines

The inadvertent vaccination of pregnant women with the rubella vaccines has not produced any part of the rubella syndrome in the offspring. Ebbin et al. (1973) studied the pregnancy outcome of 60 women immunized in the first trimester and found no adverse fetal effects. Transplacental passage of the Cendehill virus has been shown (Ebbin et al., 1972; Bolognese et al., 1973) and the virus was isolated from the fetus even 94 days after vaccination. Modlin et al. (1975) reviewed the effects of the vaccines.

In a report summarizing 119 exposures to RA–273 and 94 to Cendehill or HDV–77 vaccine, Bart et al. (1985) found no abnormalities compatible with the congenital rubella syndrome among 216 newborns. Two infants had asymptomatic glandular hypospadias.

Officials at the Center for Disease Control reported no ill effects from vaccination including the RA–273 vaccine (Anonymous, 1982).

Anonymous: Rubella vaccination during pregnancy. Morbidity and mortality. Sept. 10th Weekly Report 31:477–480, 1982.

Bart, S.W.; Stetler, H.C.; Preblund, S.R.; Williams, N.M.; Orenstein, W.A.; Bart, K.J.; Hinman, A.R. and Herrmann, K.L.: Fetal risk associated with rubella vaccine: An update. Rev. Infect. Dis. 7(3):95–102, 1985.

Bolognese, R.J.; Corson, S.L.; Fuccillo, D.A.; Sever, J.L. and Traube, R.: Evaluation of possible transplacental infection with rubella vaccination during pregnancy. Am. J. Obstet. Gynecol. 117:939–941, 1973.

Ebbin, A.J.; Wilson, M.G.; Chandor, S.B. and Wehrle, P.F.: Inadvertent rubella vaccination in pregnancy. Am. J. Obstet. Gynecol. 117:505–512, 1973.

Ebbin, A.J.; Wilson, M.G.; Wehrle, P.F.; Chin, J.; Emmons, R.W. and Lennette, E.H.: Rubella vaccine and pregnancy (letter). Lancet 2:481 only, 1972.

Modlin, J.F.; Brandling–Bennett, D.; Witte, J.J.; Campbell, C.C. and Meyers, J.D.: A review of five years experience with rubella vaccine in the United States. Pediatrics 55:20–29, 1975.

2149 Rubella Virus German Measles

The original congenital rubella syndrome described by Gregg (1941) consisted of the triad of congenital heart disease, deafness and cataracts. Since the identification of the virus (Alford et al., 1964), the syndrome was expanded to include intrauterine growth retardation, encephalitis, thrombocytopenia, radiographic changes of the long bones and persistence of the virus in the infant for a number of months after birth (Banatvala et al., 1965; Cooper and Krugman, 1967; Korones et al., 1965; Plotkin et al., 1965; Rudolph et al., 1965). Menser et al. (1967) reported a 25 year follow–up of 50 congenital rubella patients in Australia and found deafness (96 percent), cataracts (52 percent), small stature (50 percent) and congenital cardiovascular defects (22 percent). Ninety percent of these patients were of normal intelligence. One congenitally infected mother gave birth to a child with proven congenital rubella. Neurological abnormalities were described by Desmond et al. (1967). Studies of the inner ear pathology (Ward et al., 1968) and function (Keir, 1965) were reported. The pathologic findings were discussed by Tondury and Smith (1966) and by Naeye and Blanc (1965). Menser and Reye (1974) reviewed the pathology and discussed the several rare forms of late onset of the disease which may represent an immunopathological mechanism. Diabetes mellitus may be found more commonly in the adult survivors.

The incidence of rubella syndrome is higher in women exposed during the first 90 days of gestation, and after this period clinical manifestations are less common (Micheals and Mellin, 1960). Sever et al. (1965) reported from data collected prospectively from the collaborative study of cerebral palsy that 10 percent of women with clinical rubella in the first trimester had offspring with rubella syndrome diagnosed within the first month. Because of relative late registration of these mothers in the study program, it is reasonable to expect a higher incidence for the total first–trimester–exposure women. Of mothers with first trimester exposure but no clinical illness, 0.6 percent had a child with the congenital rubella syndrome. Increased attack rates among Japanese women are best explained by pre–existing antibodies in the population (Ueda et al., 1978).

Ueda et al. (1979) reported the correlation between type of defect and the time in gestation of the maternal rubella. Among 55 patients, all 13 with cataract were exposed within the first 60 days after the last normal menstrual period. The occurrence of intrauterine growth retardation in the same group of patients occurred only when the infection was within the first 100 days of gestation; it was not related to the type of defects that were present (Ueda et al., 1981). Miller et al. (1982) found that when the rash was in the first 12 weeks of pregnancy, 80 percent of the offspring had congenital defects. If the rash occurred between the 13th to 16th weeks, 54 percent were involved. Between 15–16 weeks there were 14 of which 7 had deafness alone. After the 16th week, none of the 63 exposed fetuses had defects. Follow–up studies extending to 17 years were reported on 363 exposed children and it appeared that if the mother's infection was after the 60th gestational day, growth rates returned to normal (Tokugawa et al., 1986).

The production of virologic model syndromes in monkeys (Parkman et al., 1965; Delahunt and Rieser, 1967), rabbits (Kono et al., 1969) and rats (Cotlier et al., 1968) were reported.

Alford, C.A.; Neva, F.A. and Weller, T.H.: Virologic and serologic studies on human products of conception after maternal rubella. N. Eng. J. Med. 271:1275–1281, 1964.

Banatvala, J.E.; Horstmann, D.M.; Payne, M.B. and Gluck, L.: Rubella syndrome and thrombocytopenic purpura in newborn infants. N. Eng. J. Med. 273:474–478, 1965.

Cooper, L.Z. and Krugmann, S.: Clinical manifestations of postnatal and congenital rubella. Arch. Ophthalmol. 77:434–439, 1967.

Cotlier, E.; Fox, J.; Bohigian, G.; Beaty, C. and Dupree, A.: Pathogenic effects of rubella virus on embryos and newborn rats. Nature 217:38–40, 1968.

Delahunt, C.S. and Rieser, N.: Rubella–induced embryopathies in monkeys. Am. J. Obstet. Gynecol. 99:580–588, 1967.

Desmond, M.M.; Wilson, G.S.; Melnick, J.L.; Singer, D.B.; Zion, T.E.; Rudolph, A.J.; Pineda, R.G.; Ziai, M. and Blattner, R.J.: Congenital rubella encephalitis. Course and early sequelae. J. Pediatr. 71:311–331, 1967.

Gregg, N.M.: Congenital cataract following German measles in the mother. Trans. Ophthalmol. Soc. Aust. 3:35–46, 1941.

Keir, E.H.: Results of rubella in pregnancy. II. Hearing defects. Med. J. Aust. 2:691–698, 1965.

Kono, R.; Hayakawa, Y.; Hibi, M. and Ishii, K.: Experimental vertical transmission of rubella virus in rabbits. Lancet 1:343–347, 1969.

Korones, S.B.; Ainger, L.E.; Monif, G.R.G.; Roane, J.A.; Sever, J.L. and Fuste, F.: Congenital rubella syndrome: New clinical aspects with recovery of virus from infants. J. Pediatr. 67:166–181, 1965.

Menser, M.A. and Reye, R.D.K.: The pathology of congenital rubella: A review written by request. Pathol. 6:215–222, 1974.

Menser, M.A.; Dods, L. and Harley, J.D.: A twenty–five year follow–up of congenital rubella. Lancet 2:1347–1350, 1967.

Micheals, R.H. and Mellin, G.W.: Prospective experience with maternal rubella and associated congenital malformations. Pediatrics 26:200–209, 1960.

Miller, E.; Cradock–Watson, J.E. and Pollack, T.M.: Consequences of confirmed maternal rubella at successive stages of pregnancy. Lancet 2:781–784, 1982.

Naeye, R.L. and Blanc, W.: Pathogenesis of congenital rubella. JAMA 194:1277–1283, 1965.

Parkman, P.D.; Phillips, P.E. and Meyer, H.M.: Experimental rubella virus infection in pregnant monkeys. Am. J. Dis. Child. 110:390–394, 1965.

Plotkin, S.A.; Oski, F.; Hartnett, E.M.; Hervada, A.R.; Friedman, S. and Gowing, J.: Some recently recognized manifestations of the rubella syndrome. J. Pediatr. 67:182–191, 1965.

Rowe, R.D.: Maternal rubella and pulmonary artery stenosis. Report of eleven cases. Pediatrics 32:180–185, 1963.

Rudolph, A.J.; Yow, M.D.; Phillips, C.A.; Desmond, M.M.; Blattner, R.J. and Melnick, J.L.: Transplacental rubella infection in newly born infants. JAMA 191:843–845, 1965.

Sever, J.L.; Nelson, K.B. and Gilkeson, M.R.: Rubella epidemic, 1964. Effect on 6,000 pregnancies. Am. J. Dis. Child. 110:395–407, 1965.

Tokugawa, K.; Ueda, K.; Fukkushige, J.; Koyanagi, T. and Hisanaga, S.: Congenital rubella syndrome and physical growth: A 17–year, prospective, longitudinal follow–up in the Ryukyu Islands. Rev. Infect. Dis. 8(6):874–883, 1986.

Tondury, G. and Smith, D.W.: Fetal rubella pathology. J. Pediatr. 68:867–879, 1966.

Ueda, K.; Hisanaga, S.; Nishida, Y. and Shepard, T.H.: Low–birth weight and congenital rubella syndrome: Effect of gestational age at time of maternal gestation. Clin. Pediatr. 20:730–733, 1981.

Ueda, K.; Nishida, Y.; Oshima, K. and Shepard, T.H.: Congenital rubella syndrome: Correlation of gestational age at time of maternal rubella with type of defect. J. Pediatr. 94:763–765, 1979.

Ueda, K.; Nishida, Y.; Oshima, K.; Yoshikama, H. and Nonaka, S.: An explanation for high incidence of congenital rubella in Ryukyu. Am. J. Epidemiol. 107:344–351, 1978.

Ward, P.H.; Honrubia, V. and Moore, B.S.: Inner ear pathology in deafness due to maternal rubella. Arch. Otolaryngol. 87:40–46, 1968.

2150 Rubeola *Measles*

In a prospective study of the offspring of 60 mothers with rubeola during pregnancy, Siegel and Fuerst (1966) found no increase in the incidence of congenital defects.

Siegel, M. and Fuerst, H.T.: Low birth weight and maternal disease. A prospective study of rubella, measles, mumps, chicken pox and hepatitis. JAMA 197:680–684, 1966.

2151 Rubratoxin B CAS 21794–01–4

Hood et al. (1973) studied this mold metabolite in pregnant mice. Treatment consisted of single intraperitoneal injections on one of days 6 through 12 of gestation. At doses of 0.6 mg per kg malformations were produced and generally above this dose embryo lethality was found. The malformations found were exencephaly, open eye and umbilical hernia. Evans and Harbinson (1977) found that the structural requirement for toxicity was an α–β unsaturated lactone ring.

Evans, M.A. and Harbinson, R.D.: Prenatal toxicity of rubratoxin B and its hydrogenated analog. Toxicol. Appl. Pharmacol. 39:13–22, 1977.

Hood, R.D.; Innes, J.E. and Hayes, A.W.: Effects of rubratoxin B on prenatal development in mice. Bull. Environ. Contamin. Toxicol. 10:200–207, 1973.

2152 Ruelene™ *Crufomate*

Rumsey et al. (1974) gave heifers 8.8 gm intravenously or 0.25 gm in the amnion on day 35 of gestation and found no changes in size or defects in the fetuses. Placental transfer studies were also done.

Rumsey, T.S.; Samuelson, G.; Bond, J. and Daniels, F.L.: Teratogenicity to 35–day fetuses, excretion patterns and placental transfer in beef heifers administered 4–tert–butyl–2–chlorophenyl methyl methyl–phosphoramidate (Ruelene™). J. Anim. Sci. 39:386–391, 1974.

2153 Rufocromomycin

Maraud et al. (1963) added 1 or 3 microgm to the vascular area of the chick embryo on the third day of incubation and produced defects of the eye and skeleton. Thyroxine reduced the teratogenicity of this antimitotic antibiotic.

Maraud, R.; Coulaud, H. and Stoll, R.: Sur l'action teratogene, chez l'embryon de poulet de la rufocromomycine associee ou non a la thyroxine. C. R. Soc. Biol. (Paris) 157:1566–1569, 1963.

2154 Saccharin CAS 81–07–2

Kline et al. (1978) found no increase in spontaneous abortions among women taking saccharin.

Lorke (1969) tested pregnant mice with saccharin and found no evidence of teratogenicity. The mice received up to 25 mg per kg daily from the 6th through the 15th days. Cyclamates up to 250 mg per kg was also given without producing fetal changes. Fritz and Hess (1968) reported no teratogenic effects in the rat fetus when the mother received 25 mg per kg from days 6 through 15 of gestation. Kroes et al.

(1977) also found no teratogenicity in long term studies in mice. In the male offspring of rats maintained on a diet of 7.5 percent saccharin, an increase in bladder neoplasms was found (Taylor et al., 1980). No increase was found with a diet containing 5 percent saccharin.

Lederer (1977) studied embryofetotoxicity of possible intermediates or contaminants of commercially prepared saccharin. Administered orally to rats at 0.1 percent of the diet, o–toluenesulfonamide was devoid of toxicity, o–sulfobenzoic acid increased the number of fetal resorptions slightly, but o–sulfamoylbenzoic acid, and especially NH–4 o–sulfobenzoic acid markedly increased resorptions.

The general subject of saccharin toxicity is reviewed in detail by Cranmer (1980).

Cranmer, M.F.: Saccharin, a report. American Drug Institute Inc. and Pathotox Publishers Inc., 1980.

Fritz, H. and Hess, R.: Prenatal development in the rat following administration of cyclamate, saccharin and sucrose. Experientia 24:1140–1141, 1968.

Kline, J.; Stein, Z.A.; Susser, M. and Warburton, D.: Spontaneous abortion and the use of sugar substitutes. Am. J. Obstet. Gynecol. 130:708–711, 1978.

Kroes, R.; Peters, P.W.J.; Berkvens, J.M.; Verschuuren, T.D. and Van Esch, G.J.: Long term toxicity and reproduction study (including a teratogenicity study) with cyclamate, saccharin and cyclohexylamine. Toxicology 8:285–300, 1977.

Lederer, J.: Problem of saccharin. Med. Nutr. 13:23–32, 1977.

Lorke, D.: Untersuchungen von Cyclamat und Saccharin auf embryotoxische und Teratogene Wirkung an der Maus. Arzneim. Forsch. 19:920–922, 1969.

Taylor, J.M.; Weinberger, M.A. and Friedman, L.: Chronic toxicity and carcinogenicity to the urinary bladder of sodium saccharin in the utero–exposed rat. Toxicol. Appl. Pharmacol. 54:57–75, 1980.

2155 Sairei–to Extract

An extract of this Chinese herb was given orally to rats during organogenesis in amounts of up to 2,000 mg per kg daily and no adverse fetal effects were found. (Kiwaki et al., 1989)

Kiwaki, S.; Ono, C.; Sakai, K.; Nakamura, T. and Oketani, Y.: A teratological evaluation of orally administered sairei–to extract in rats. Oyo Yakuri 38(4):255–270, 1989.

2156 Salbutamol

Gummerus and Haolnen carried out a control study of 200 women with multiple pregnancies. They received 4 mg five times daily for 6 weeks in the 3rd trimester. Neither adverse effects nor benefit were found.

Gummerus, M. and Halonen, O.: Prophylactic long–term oral tocolysis multiple pregnancies. British J. Obstet. Gynaecology 94:249–251, 1987.

2157 Salicylamide CAS 65–45–2

This analgesic drug was fed as 2 percent of the diet to rats from the 5th to 11th or 12th to eighteenth days and a high proportion of the fetuses had skeletal defects of the ribs, spine and extremities (Knight and Roe, 1978).

Knight, E. and Roe, D.A.: Effect of salicylamide and protein restriction on the skeletal development of the rat. Teratology 18:17–22, 1978.

2158 Salicylate–Methyl, Sodium and Acetyl Forms
Acetylsalicylic Acid Aspirin™ CAS 63–36–5

In humans, Turner and Collins (1975) studied 144 mothers taking salicylates regularly. The stillbirth rate was increased and the birth weight was reduced as compared to matched controls. No increase in congenital defects was noted. They confirmed the findings of Lewis and Schulman (1973) that therapeutic doses of salicylate are often associated with the post–maturity syndrome. Nelson and Forfar (1971) reported 8 anomalies in 458 pregnancies where salicylates were taken in the first 28 days of gestation. Although this was higher than the 3 malformations in 911 controls, the exposed group included achondroplasia and Down's syndrome, two defects of probable preconceptual origin.

Richards (1969) studied 833 pregnancies with anomalies in the offspring and matched these against pregnancies with normal offspring. For the women ingesting salicylates during the first trimester, the rate was 22 and the control 14, a difference which was significant at the 0.1 percent level. The author did not find an increase in the number of salicylate exposed malformation pregnancies during the second and third pregnancies. Crombie et al. (1970) queried 10,000 and found no excess of aspirin users among those who had malformed infants. Slone et al. (1976) found no increase in malformation rates among nearly 14,000 offspring of heavy or occasional aspirin users. Corby (1978) reviewed the effect of aspirin on the mother and fetus.

Rumack et al. (1981) studied intracranial hemorrhage in 108 infants born at 34 weeks gestation or earlier. The incidence of hemorrhage was 71 percent (12 of 17) among the offspring of mothers who took aspirin within a week of delivery. This was significantly more common than in the control group (31 of 71 infants). Ten of twenty infants exposed to acetaminophen had bleeding and this rate did not differ from the control. Stuart et al. (1982) studied 10 full term infants whose mothers took aspirin within 5 days of delivery and found none with bleeding tendancies. Their discharge hemoglobin was significantly less than 37 controls. Among the 37 controls there was only one with a bleeding problem. In the exposed group malondialdehyde levels were significantly reduced. Forty percent of the exposed infants had profuse petechiae over the presenting part. One had bleeding from the circumcision and two had hematuria. No increased clinical bleeding occurred in the group of offspring whose mothers took aspirin 5 to 10 days before delivery.

Streissguth et al. (1987) in a prospective study, found that heavy and moderate aspirin users had offspring with significant decreases in IQ at age 4. This association was present after correcting for conceivable confounding factors. The questioning of the mothers was during the first trimester and data on third trimester use which might affect platelet function in the fetus during delivery was not available. A large carefully controlled study from the collaborative perinatal project did not identify a decrease in IQ among the offspring of aspirin users (Klebanoff and Berendes, 1988). In fact, the offspring of users at age four were two points higher than the non–users. Other differences in the two studies did not give a clear–cut explanation for the contrasting results.

Koshakji and Schulert (1973) concluded that the teratogenic effect was due to salicylic acid, a hydrolysis product. They found that shifting or substituting functional groups on the aromatic ring eliminated teratogenic activity. Tanaka et al. (1973) administered 0.06 to 0.4 percent in the diet to rats on days 8–14. At 0.2 percent, there was fetal growth retardation and 2 of 160 fetuses had myeloschisis. No effect was seen at 0.1 percent.

Warkany and Takacs (1959) using methyl or sodium salicylate in the rat produced craniorachischisis, exencephaly, hydrocephaly, facial clefts, eye defects, gastroschisis and irregularities of the vertebrae and ribs. The methyl salicylate in doses of 0.1 to 0.5 ml was injected on the 9th, 10th and 11th gestational days. The sodium salicylate was injected on the same days in doses of 60 to 180 mg. These dose levels caused some maternal deaths. Since this work appeared, numerous workers have confirmed and extended it (for summary see Kalter, 1968).

Goldman and Yakovac (1963) showed that treatment on the 10th day with 300 mg per kg was non–teratogenic but 400 and 500 mg per kg caused defects in the rat fetus. Larsson and Eriksson (1966) gave mice 10 mg of sodium salicylate on various days of of gestation and found skeletal defects and hemorrhages of the extremities with a few cases of exencephaly. Trasler (1965) gave acetylsalicylic acid (with a trace of Tween 80) to mice on the 9th and 10th or 10th and 11th days of pregnancy and produced cleft lip, exencephaly, microcephaly and spina bifida. Butcher et al. (1972) administered 250 mg per kg orally to rats on the 8th, 9th and 10th days. Fetuses which were not malformed were cross fostered with control mothers and subsequently studied behaviorally. Learning behavior by maze testing showed a significant decrease in the treated fetuses.

Lapointe and Harvey (1964) used salicylamide (d–hydroxybenzamide) in the hamster and produced "cranial blisters" in the offspring. Wilson (1971) gave monkeys 250 mg per kg twice daily for three days at various times between the 18th to 26th days of gestation and produced two abortions, two small normal fetuses, one normal and two fetuses with malformations (one multiple and both with heart defects). At doses of 100 or 150 mg per kg per day on days 22 through 32 in the monkey, Wilson et al. (1977) observed two fetuses out of 41 with defects (cystic kidney and cranioschisis). They compared the pharmacodynamics in the rat and monkey and found higher concentrations and longer duration of concentrations in the rat. Gulienetti et al. (1962) found increased amniotic fluid volume in treated rats.

Goldman and Yakovac (1964, 1965) published a series of studies attempting to determine the mechanism of teratogenic action. Adrenalectomy did not change the abnormality rate but sodium carbonate and chloride and central nervous system depressants protected the fetuses. Larsson (1971) reviewed his work on the action of salicylate on the fetus.

Chebotar (1967) reported that both immobilization and ammonium chloride intensified while sodium bicarbonate reduced the teratogenic effects of sodium salicylate in the rat. Both ammonium chloride and immobilization caused the maternal salicylate level to increase.

Beall and Klein (1977) found that aspirin teratogenicity was increased in rats which were also food restricted. Salicylate added directly to culture at 600 microgram per ml of medium produced dilation of the rhombencephalon and mesencephalon of the rat embryos (McGarrity et al., 1978; Greenaway et al., 1982). Saglo (1982) reported that female Wistar rats delivered fetuses with numerical and structural chromosome aberrations after receiving one intragastric dose of 700 mg per kg salicylate. The same effect was observed after treatment of male rats with the same dose of this substance. Vasilenko et al. (1979) established that administration of this substance (one–tenth LD–50) to male rats during 1.5 months seriously damaged spermatogenesis and repeated inhalations (25 mg per cubic meter) during four months produced pathological changes in spermatogenic epithelium of offspring.

Beall, J.R. and Klein, M.F.: Enhancement of aspirin–induced teratogenicity by food restriction in rats. Toxicol. Appl. Pharmacol. 39:489–495, 1977.

Butcher, R.E.; Vorhees, C.V. and Kimmel, C.A.: Learning impairment from maternal salicylate treatment in rats. Nature New Biology 236:211–212, 1972.

Chebotar, N.A.: Peculiarities of action of sodium salicylate at various stages of embryogenesis in rats and the influence of certain shifts in the female organism on its teratogenic activity. Pharmacol. Toxicol. (Russian) 2:221–225, 1967.

Corby, D.G.: Aspirin in pregnancy: Maternal and fetal effects. Pediatrics 62:930–945, 1978.

Crombie, D.L.; Pinsent, K. and Slater, B.C.: Teratogenic drugs R.C.G.P. survey. Br. Med. J. 4:178–179, 1970.

Goldman, A.S. and Yakovac, W.C.: The enhancement of salicylate teratogenicity by maternal immobilization in the rat. J. Pharmacol. Exp. Ther. 142:351–357, 1963.

Goldman, A.S. and Yakovac, W.C.: Prevention of salicylate teratogenicity in immobilized rats by certain central nervous system depressants. Proc. Soc. Exp. Biol. Med. 115:693–696, 1964.

Goldman, A.S. and Yakovac, W.C.: Teratogenic action in rats of reserpine alone and in combination with salicylate and immobilization. Proc. Soc. Exp. Biol. Med. 118:857–862, 1965.

Greenaway, J.C.; Shepard, T.H.; Fantel, A.G. and Juchau, M.R.: Sodium salicylate teratogenicity in vitro. Teratology 26:167–171, 1982.

Gulienetti, R.; Kalter, H. and Davis, N.C.: Amniotic fluid volume and experimentally–induced congenital malformations. Biol. Neonate 4:300–309, 1962.

Kalter, H.: Teratology of the Central Nervous System. Chicago: University of Chicago Press, 153–154, 1968.

Klebanoff, M.A. and Berendes, H.W.: Aspirin exposure during the first 20 weeks of gestation and IQ at four years of age. Teratology 37:249–255, 1988.

Koshakji, R.P. and Schulert, A.R.: Biochemical mechanisms of salicylate teratology in the rat. Biochem. Pharmacol. 22:407–416, 1973.

Lapointe, R. and Harvey, E.B.: Salicylamide–induced anomalies in hamster embryos. J. Exp. Zool. 156:197–200, 1964.

Larsson, K.S.: Action of salicylate on prenatal development. In: Malformations Congenitales Des Mammiferes, H. Tuchmann–Duplessis (ed.), Paris: Masson, 171–186, 1971.

Larsson, K.S. and Eriksson, M.: Salicylate–induced fetal death and malformations in two mouse strains. Acta Paediatr. 55:569–576, 1966.

Lewis, R.B. and Schulman, J.D.: Influence of acetylsalicylic acid, an inhibitor of prostaglandin synthesis, on the duration of human gestation and labour. Lancet 2:1159, 1973.

McGarrity, C.; Samani, N.J. and Beck, F.: The in vivo and in vitro action of sodium salicylate on rat embryos. J. Anat. 127:646 only, 1978.

Nelson, M.M. and Forfar, J.O.: Associations between drugs administered during pregnancy and congenital anomalies of the fetus. Br. Med. J. 1:523–527, 1971.

Richards, I.D.G.: Congenital malformations and environmental influences in pregnancy. Br. J. Prev. Med. 23:218–225, 1969.

Rumack, C.M.; Guggenheim, M.A.; Rumack, B.H.; Peterson, R.G.; Johnson, M.L. and Braithwaite, W.R.: Neonatal intracranial hemorrhage and maternal use of aspirin. Obstet. Gynecol. 58(5):52S–56S, 1981.

Saglo, V.J.: Effect of sodium salicylate on the chromosome apparatus of rat somatic and embryonal cells. Farmakol. Toksikol. (USSR) 5:88–89, 1982.

Slone, D.; Heinonen, O.P.; Kaufman, D.W.; Siskind, V.; Monson, R.R. and Shapiro, S.: Aspirin and congenital malformations. Lancet 1:1373–1375, 1976.

Streissguth, A.P.; Treder, R.P.; Barr, H.M.; Shepard, T.H.; Bleyer, W.A.; Sampson, P.D. and Martin, D.C.: Aspirin and acetaminophen use by pregnant women and subsequent child IQ and attention decrements. Teratology 35:211–219, 1987.

Stuart, M.J.; Gross, S.J.; Elrad, H. and Graeber, J.E.: Effects of acetylsalicylic–acid ingestion on maternal and neonatal hemostasis. New Eng. J. Med. 307:909–912, 1982.

Tanaka, S.; Kawashima, K.; Nakaura, S.; Nagao, S.; Kuwamura, T.; Takanaka, A. and Omori, Y.: Studies on the teratogenicity of food additives (3). Teratogenic effect of dietary salicylic acid in rats. J. Food Hyg. Soc. Jpn. 14:549–557, 1973.

Trasler, D.G.: Aspirin–induced cleft lip and other malformation in mice. Lancet 1:606–607, 1965.

Turner, G. and Collins, E.: Fetal effects of regular salicylate ingestion during pregnancy. Lancet 2:338–339, 1975.

Vasilenko, N.M.; Manzhelay, E.S. and Gnezdilova, A.J.: Gonadotoxic action of acetylsalicylic acid. Farmakol. Toksikol. (USSR) 4:421–423, 1979.

Warkany, J. and Takacs, E.: Experimental production

of congenital malformations in rats by salicylate poisoning. Am. J. Pathol. 35:315–331, 1959.

Wilson, J.G.: Use of rhesus monkeys in teratological studies. Fed. Proc. 30:104–109, 1971.

Wilson, J.G.; Ritter, E.J. and Fradkin, R.: Comparative distribution and embryotoxicity of acetylsalicylic acid in pregnant rats and rhesus monkeys. Toxicol. Appl. Pharmacol. 41:67–68, 1977.

2159 Salmeterol Xinafoate

Itabashi et al. (**1993**) gave rats orally up to 10 mg per kg during organogenesis and found no adverse fetal effects or changes in postnatal behavior. The fetuses in the higher dose ranges were increased in body weight.

Itabashi, M.; Amano, Y.; Hashimoto, K.; Saitoh, T.; Sannai, S.; Ezaki, H.; Masuoka, M. and Tokado, H.: Reproduction study on oral administration of salmeterol xinafoate during the period of organogenesis in rata. Yakuri to Chiryo 21:2401–2423, 1993.

2160 Salsalate *Sasapyrine*

This is a salycilate used in rheumatic disease. Momma et al. (1984) found a mild constricting effect on the ductus arteriosis of the rat fetus when 100 mg per kg was given orally.

Momma, K.; Hagiwara, H. and Konishi, T.: Constriction of fetal ductus arteriosus by non–steroidal anti–inflammatory drugs: study of additional 34 drugs. Prostaglandins 28:527–536, 1984.

2161 Santicizer 141 and 148 *2–Ethylhexyldiphenyl Phosphate, Isodecyldiphenyl Phosphate*

Santicizer 141 was administered to rats by gavage on days 6–15 in amounts up to 3000 mg per kg daily. No teratogenic effects were seen. Similar studies with santicizer 148 but with treatment on days 6–19 were non–teratogenic (Robinson et al., 1986).

Robinson, E.C.; Hammond, B.G.; Johannsen, F.R.; Levinskas, G.J. and Rodwell, D.E.: Teratogenicity studies of alkylaryl phosphate ester plasticizers in rats. Fund. Appl. Toxicol. 7:138–143, 1986.

2162 Sarin *Isopropyl Methylphosphonofluoridate* CAS 107–44–8

This anticholinesterase inhibitor was tested in rats by Lu et al. (1984) by gavage at doses of up to 380 microgram per kg during organogenesis. No adverse fetal effects occurred even though maternal toxicity was present.

Lu, M.H.; Filler, R.; Bates, H.K.; LaBorde, J.B.; Bazare, J.; Gaylor, D.W. and Kimmel, C.A.: Teratogenicity evaluation of sarin in rats. (abs) Teratology 29:45A only, 1984.

2163 Sarkomycin CAS 489–21–4

Takaya (1965) injected rats with 5–11 mg per kg daily on the days 6 through 10 and produced 10 percent malformations in the offspring. The malformations included hydronephros and microophthalmia.

Takaya, M.: Teratogenic effects of antibiotics. J. Osaka City Med. Cent. 14:107–115, 1965.

2164 Sarpogrelate

Tanaka et al. (**1991**) gave this platelet coagulation inhibitor to rats before and during the first gestational week and found no adverse reproductive effects at 320 mg per kg. Hiraide et al. (**1991** A) found no teratogenicity after organogenesis studies but at 160 and 640 mg per kg there was a decrease in body weight. Perinatal and postnatal studies by Hiraide et al. (**1991** B) reported decreased weight and survival of pups exposed to 80 and 240 mg per kg.

Hiraide, Y.; Kashima, M.; Takahashi, M. and Tanaka, E.: Reproduction studies of sarpogrelate hydrochloride (MCI–9042) (II): Study on oral administration during the period of organogenesis in rats. Yakuri to Chiryo 19:s717–s729, 1991A.

Hiraide, Y.; Kashima, M.; Takahashi, M. and Tanaka, E.: Reproduction studies of sarpogrelate hydrochloride (MCI–9042) (III): Perinatal and postnatal study in rats by oral administration. Yakuri to Chiryo 19:s731–s740, 1991B.

Tanaka, E.; Toyooka, M.; Komatsu, K.; Umeshita, C.; Mizusawa, R. and Toshida, K.: Reproduction studies of sarpogrelate hydrochloride (MCI–9042) (I): Study on oral administration prior to and in the early stages of pregnancy in rats. Yakuri to Chiryo 19:s707–s716, 1991.

2165 Sauna Bathing

Saxen et al. (1982) studied the sauna habits of 100 women giving birth to children with neural defects and 202 giving birth to children with orofacial defects. Their habits did not differ from the control Finnish women. The authors pointed out that nearly every woman in Finland visited the sauna bath, and yet, the central nervous system defect rate in Finland is among the lowest.

Harvey et al. (1981) concluded from their studies that remaining in a hot tube at 39 degrees for at least 15 minutes and 41.1 degrees for least 10 minutes was unlikely to damage a human pregnancy. None of the volunteers were able to remain in the sauna for long enough to increase significantly body temperature. Ridge and Budd (**1990**) in contrast to the above found that women emersed in a 40 degree C bath did have temperature elevations to 39 degrees C without complaining of discomfort.

Harvey, M.A.S.; McRorie, M.M. and Smith, D.W.: Suggested limits to the use of hot tub and sauna in pregnant women. Canad. M.A.J. 125:50–53, 1981.

Ridge, B.R. and Budd, G.M.: How long is too long in a spa pool? Lancet 323:835 only, 1990.

Saxen, L.; Holmberg, P.C.; Nurminen, M. and Koosma, E.: Sauna and congenital defects. Teratology 25:309–313, 1982.

2166 SCE–2787

Sugitani et al. (**1992** A,B) and Nakatsu et al. (**1992**) gave this drug to rats before gestation, during organogenesis and in the perinatal and postnatal periods using up to

1000 mg per kg subcutaneously. They found no adverse teratogenic effect or effects on reproduction or postnatal development. Ooshima et al, (**1992**) gave rabbits up to 300 mg per kg during organogenesis and found no adverse fetal effects.

Nakatsu, T.; Takatani, O.; Kumada, S.; Sugitani, T.; Yoshizaki, H. and Kanamori, H.: Teratological study of SCE–2787 (HCl) in rats. Yakuri to Chiryo 20:s2723–s2739, 1992.

Ooshima, Y.; Horinouchi, A. and Myasukawa, Jh: Teratological study of SCE–2787 (HCl) in rabbits. Yakuri to Chiryo 20:s2741–s2747, 1992.

Sugitani, T.; Nakatsu, T.; Kanamori, H.; Yoshida, T.; Kumada, S. and Yoshizaki, H.: Reproduction study of SCE–2787 (HCl) in rats. Yakuri to Chiryo 20:s2707–s2721, 1992A.

Sugitani, T.; Nakatsu, T.; Yoshida. T.; Kanamori, H.; Yoshizaki, H. and Kumada, S.: Peri–and post–natal study of SCE–2787 (HCl) in rats. Yakuri to Chiryo 20:s2749–s2764, 1992B.

2167 SCH–12041 7–Chloro–1–(2, 2, 2–trifluoroethyl)–5–(phenyl 1,3–dihydro–2H,1, 4–dibenzodiazepin–2–one)

Beall (1972) reported that oral doses of up to 200 mg per kg during organogenesis did not produce adverse fetal effects in rats. Rabbits were also tested at a lower dose and no evidence of teratogenicity was found.

Beall, J.R.: Study of the teratogenic potential of diazepam and SCH 12041. Canad. Med. Assoc. J. 106:1061 only, 1972.

2168 Schizophrenia

Rieder et al. (**1975**) studied pregnancy outcome among 186 women who prospectively reported that the husband or wife had been hospitalized for a psychiatric problem. Among the test group there were 7.5% neonatal or fetal deaths as compared to 4.3% among the control. No pattern was identified and no association with medications was found.

Rieder, R.O.; Rosenthal, D.; Wender, P. and Blumenthal, H.: The offspring of schizophrenics. Arch. Gen. Psychiatry 32:200–211, 1975.

2169 Schizophyllan CAS 9050–67–3

This antitumor glycan was given by Ishizaki et al. (1982) subcutaneously to rats and rabbits during organogenesis. The highest dose in the rat was 50 mg per kg and in the rabbit 25 mg per day. No significant changes were produced in the fetuses of either species.

Ishizaki, O.; Daidohji, S.; Ohmori, Y. and Saito, G.: Reproductive studies of schizophyllan (SPG). Oyo Yakuri 23:935–951, 1982.

2170 Scopolamine CAS 51–34–3

George et al. (1990 A&B) gave this compound orally during organogenesis to rats and mice. Maternal toxicity was observed at 100 mg per kg and at 450 mg per kg shortened ribs were found in the rat study. In mice no fetal toxicity was

observed at 10 and 100 mg per kg and no teratogenicity was found at doses to 900 mg per kg.

George, J.D.; Price, C.J.; Marr, M.C.; Kimmel, C.A.; Schwetz, B.A. and Morrissey, R.E.: Teratologic evaluation of scopolamine hydrobromide (cas no 114–49–8) administered to cd rats on gestational days 6 to 15. National Toxicology Program NTP–87–145, 1990.

George, J.D.; Price, C.J.; Marr, M.C.; Kimmel, C.A.; Schwetz, B.A. and Morrissey, R.E.: Teratologic evaluation of scopolamine hydrobride (cas no 114–49–8) administered to cd–1 mice on gestational days 6 through 15. National Toxicology Program NTP–87–102, 1990.

2171 Scopoletin CAS 92–61–5

Discussed under Solanine

2172 Sebacic Acid

Greco et al. (**1990**) fed rats and rabbits 500 mg or 1,000 mg per kg respectively. No malformations were reported.

Greco, A.V.; Mingrone, G.; Mastromattei, E.A. Finotte, E. and Castogneto, M.: Toxicity of disodium sebacate. Drugs Exp Clin Res 16(10):531–536, 1990.

2173 Secalonic Acid D CAS 35287–69–5

Mayura et al. (1982) studied the fungal metabolite in pregnant rats using 15–25 mg subcutaneously on individual days 6–10 or 12 or 14. Anophthalmia, exencephaly and skeletal defects were found. Tracheo–esophageal fistula (43 percent) and renal agenesis (6.5 percent) occurred with treatment on day 10.

Mayura, A.; Hayes, A.W. and Berndt, W.O.: Teratogenicity of secalonic acid D in rats. Toxicology 25:311–322, 1982.

2174 Selenium CAS 7782–49–2

Robertson inquired about possible reproductive problems among women who are exposed to selenium in bacteriology laboratories where it is used as a component of culture medium. He did not find any evidence for reproductive problems but in his own laboratory among 5 pregnancies four spontaneous abortions occurred. Franke et al. (1936) applied 0.5 mg per kg (of egg) to the air cell of the chick and produced eye or beak defects in about 50 percent of the survivors. Ridgway and Karnofsky (1952) produced head and beak defects and cysts of the rump in chick embryos treated with 0.01 to 0.08 mg of selemious acid during the first 8 days of incubation. Schroeder et al. (1971) fed 3 ppm in the drinking water to mice and observed a significant increase in the number of runts produced, and by the third generation the treatment group became reduced in number. Yonemoto et al. (1983) produced decrased live births and fetal weight by giving 18 to 40 micromoles of sodium selenite on day 12 to mice.

Willhite et al. (**1990**) showed distribution of selenate and selenomethionine into the hamster embryo but no teratogenic activity at maternally toxic levels. Hawkes et al. (**1994**) found no deleterious effects of treatment of monkeys on days 20–50 with up to 300 microgm of selenium as L–selenomethionine per kg.

Franke, K.W.; Moxon, A.L.; Poley, W.E. and Tully, W.C.: Monstrosities produced by the injection of selenium salts into hens eggs. Anat. Rec. 65:15–22, 1936.

Hawkes, W.C.; Willhite, C.C.; Omaye, S.T.; Cox, D.N.; Choy, W.N. and Tarantal, A.F.: Selenium kinetics, placental transferr, and neonatal exposure in cynomolgus macaques (macaca fascicularis). Teratology 50:148–159, 1994.

Ridgway, L.P. and Karnofsky, D.A.: The effects of metals on the chick embryo: toxicity and production of abnormalities in development. Ann. N.Y. Acad. Sci. 55:203–215, 1952.

Robertson, D.S.F.: Selenium–a possible teratogen? The Lancet 1:518 only 1970.

Schroeder, H.A. and Mitchener, M.: Toxic effects of trace elements on the reproduction of mice and rats. Arch. Environ. Health 23:102–106, 1971.

Willhite, C.C.; Ferm, V.H. and Zeise, L.: Route-dependent pharmacokinetics, distribution, and placental permeability of organic and inorganic selenium in hamsters. Teratology 42:359–371, 1990.

Yonemoto, J.; Satoh, H.; Himeno, S. and Suzuki, T.: Toxic effect of sodium selenite on pregnant mice and modification of the effects by vitamin E or reduced glutathione. Teratology 28:333–340, 1983.

2175 Semicarbazide HCl CAS 563–41–7

Neuman et al. (1956) injected 2.0 mg of this chemical on the 6th day into the yolk and observed bent tarsometatarsal and tibiotarsal bones and malformed beaks in the 14–day–chick embryos. p–Hydrazinbenzoic acid and thiosemicarbazide produced similar defects, but 1–phenylsemicarbazide produced shortening of the leg bones with moderate edema and whitish areas in the liver. No defects were found with benzoic hydrazide. Stoll et al. (1970) were able to show that vitamin B_6 administration reduced the number of defects in the semicarbazide–treated chick and that desoxypyridoxamine enhanced this teratogen. De La Fuente (1986) reported brain damage, hemorrhage of liver and intestines and hydronephrosis in fetuses from maternal rats given 17 mg per kg subcutaneously.

De La Fuente, M.: Teratogenic effect of semicarbazide in Wistar rats. Biol. Neonate 49:150–157, 1986.

Neuman, R.E.; Maxwell, M. and McCoy, T.A.: Production of beak and skeletal malformations of chick embryo by semicarbazide. Proc. Soc. Exp. Biol. Med. 92:578–581, 1956.

Stoll, R.; Bodit, F. and Maraud, R.: Sur l'action teratogene de la semi–carbizide. Role de la vitamine B_6 et des amines biogenes. C. R. Soc. Biol. 164:1011–1013, 1970.

2176 Semlii Forest Virus Mutant ts22

Mabruk et al. (1988) injected this virus into pregnant mice on day 10 and found hemorrhages and fragility of the skin and skeletal defects (scoleosis and small mandibles) on day 18 fetuses.

Mabruk, M.J.E.M.F.; Flack, A.M.; Glasgow, G.M.; Smyth, J.M.B.; Folan, J.C.; Bannigan, J.G.; O'Sullivan, M.A.; Sheahan, B.J. and Atkins, G.J.: Teratogenicity of the semliki forest virus mutant ts22 for the foetal mouse: in-

duction of skeletal and skin defects. J. Gen. Virol. 69:2755–2762, 1988.

2177 Sendai Virus Parinfluenza 1 Virus

Coid and Wardman (1971) reported fetal wastage in rats infected with Sendai virus. No malformations were detected. Tuffrey et al. (1972) demonstrated the virus by immunofluorescence in fertilized mouse ova.

Coid, R. and Wardman, G.: The effect of para–influenza type 1 (Sendai) virus infection on early pregnancy in the rat. J. Reprod. Fertil. 24:39–43, 1971.

Tuffrey, M.; Zisman, B. and Barnes, R.D.: Sendai (parainfluenza 1) infection of mouse eggs. Br. J. Exp. Path. 53:638–640, 1972.

2178 Sennaglucosides

Mizutani et al. (1980) studied the reproductive effects of this vegetable laxative on rats. They gave up to 90 mg per kg daily either before mating and during the first 7 days of gestation or on days 7–17. There were no reproductive problems or adverse fetal effects found.

Mizutani, M.; Izutsu, M.; Hoshimoto, Y.; Nagao, T. and Matsuda, H.: Effects of sennaglucosides on reproductive function and fetal development and differentiation in rats. Kiso to Rinsho 14:380–396, 1980.

2179 Sennosides A and B

These stereoisomers are broken down in the large bowel to monoanthrone–a glycone structure which exerts a laxative action. Mengs (1986) gave rats and rabbits up to 100 and 20 mg per kg, respectively, during organogenesis and found no evidence of teratogenicity.

Mengs, U.: Reproductive toxicological investigations with sennosides. Arzneim Forsch 36–2:1355–1358, 1986.

2180 Serotonin 5–Hydroxytryptamine CAS 50–67–9

Reddy et al. (1963) injected rats with 0.5 to 5.0 mg daily during pregnancy and at the 1.5 mg level found abnormalities in 4 of 17 live fetuses. The congenital defects consisted of anophthalmia, hydrocephalus, exencephaly and omphalocele. Vacuolization of the myocardial cells was found also. They reported that a woman with a carcinoid tumor gave birth to three infants who died of neonatal respiratory distress and a fourth infant with undetermined multiple defects.

Poulson et al. (1963) injected 2 ml of the creatine sulfate form into mice on the 8th, 9th or 10th day of gestation and produced defects of the eye, limbs and tail. Some skull and central nervous system abnormalities were found also. From earlier work (Poulson et al., 1960), they proposed that the mechanism of action of this substance is through placental hemorrhagic changes. Marley et al. (1967) performed detailed equilibrium studies in the rat on days 9 and 10 and showed that 5–hydroxytryptamine greatly slowed the transfer of radioactive sodium to the embryo. Thompson and Gautieri (1969) injected mice subcutaneously with 5 or 10 mg per kg on days 7 through 12 and produced defects including hydrocephalus, exencephalus, hydronephrosis, renal agenesis and gastroschisis.

Marley, P.B.; Robson, J.M. and Sullivan, F.M.: Embryotoxic and teratogenic action of 5–hydroxytryptamine: Mechanism of action in the rat. Br. J. Pharmacol. Chemother. 31:494–505, 1967.

Poulson, E.; Botros, M. and Robson, J.M.: Effect of 5–hydroxytryptamine and iproniazid on pregnancy. Science 131:1102–1103, 1960.

Poulson, E.; Robson, J.M. and Sullivan, F.M.: Teratogenic effect of 5–hydroxytryptamine in mice. Science 141:717–718, 1963.

Reddy, D.V.; Adams, F.H. and Baird, C.D.C.: Teratogenic effects of serotonin. J. Pediatr. 63:394–397, 1963.

Thompson, R.S. and Gautieri, R.F.: Comparison and analysis of the teratogenic effects of serotonin, angiotensin–2 and bradykinin in mice. J. Pharm. Sci. 58:406–412, 1969.

2181 Serotonin Deficiency

Madani et al. (**1994**) reported that mouse embryos grown in serotonin–depleted serum showed oculo–neural malformations. They proposed that the deficiency accounts for malformations found in the offspring of phenylketonuric women.

Madani, M.; Launay, X.; Rey, F.; Mulliez, N.; Kolf, M.; Citadelle, D. and Roux, C.: Mechanism(s) of PKU embryopathy. (abs) Teratology 50:29A, 1994.

2182 Sertaconazole CAS 99592–32–2

Romero et al. (**1992**) gavaged rats and rabbits with up to 150 mg per kg during organogenesis or perinatally. The no effect dose was 100 mg per kg. No teratogenicity was found but in rabbits at 150 mg per kg an increase in fetal hepatic hemorrhage and enlargement was found.

Romero, A.; Grau, M.T.; Villamayor, F.; Sacristan, A. and Ortiz, J.A.: Reproduction toxicity of sertaconazole. Arzneim.–Forsch./Drug Res. 42:739–742, 1992.

2183 Setiptiline Maleate

Shibutani et al. (**1989** A) gave this anti–depressant orally to rats on days 7–17 in doses of up to 30 mg per kg. No teratogenicity was found but at the highest dose fetal lethality and growth retardation in males was found. At 10 mg per kg the dams failed to gain weight normaly. Rabbits treated during organogenesis with the same doses had increased resorptions but no teratogenicity. (Shibutani et al., **1989** B) Studies of fertility and postnatal studies in rats (Shibutani et al. **1989** C,D) were done and prolonged estrus was seen at doses over 30 mg per kg. Decreases in viability of the offspring were seen but cross–fostering indicated that the deaths were related to postnatal exposures to the drug. (Shibutani et al. **1990** E)

Shibutani, Y.; Kurokawa, M. and Inoue, K.: Toxicity studies of 1,2,3,4–tetrahydro–2–methyl–9h–dibenzo[3, 4: 6,7] cyxlohepta[1,2–c]pyridine maleate (mo–82820 (iv) teratogenicity study in rats. Iyakuhin Kenkyu 20:101–115, 1989.

Shibutani, Y.; Hamada, Y. and Obata, M.: Toxicity studies of 1,2,3,4–tetrahydro–2–methyl–9h–dibenzo[3, 4:6, 7]

cyclohepta [1,2–c]pyridine maleate (mo–8282) (v) teratogenicity study in rabbits. Iyakuhin Kenkyu 20:116–122, 1989.

Shibutani, Y.; Kurokawa, M. and Inoue, K.: Toxicity studies of 1,2,3,4–tetrahydro–2–methyl–9h–dibenzo[3, 4: 6, 7] cyclohepta[1,2–c]pyridine maleate (mo–8282) (iii) fertility study in rats. Iyakuhin Kenkyu 20:94–100, 1989.

Shibutani, M.; Kurokawa, M.; Inoue, K. and Hanada, Y.: Toxicity studies of 1,2,3,4–tetrahydro–2–methyl–9h–dibenzo [3, 4,: 6, 7]cyclohepta[1,2–c]pyridine maleate (mo–8282) (vi) perinatal and postnatal study in rats. Iyakuhin Kenkyu 20:123–135, 1989.

Shibutani, Y.; Hamada, Y. and Inoue, K.: Toxicity studies of 1,2,3,4–tetrahydro–2–methyl–9h–dibenzo[3, 4,: 6, 7] cyclohepta[1,2–c]pyridine maleate (mo–8282) (vii) cross–fostering study in rats. Iyakuhin Kenkyu 20:136–141, 1989

2184 Sfericase *Bacillus Sphaericus*

Koeda et al. (1978) fed this protease to pregnant mice and rabbits at maximum doses of 1,000 and 750 mg per kg daily, respectively. Some body weight reduction occurred in the mouse dams and fetuses at the highest doses. An increase in fetal mortality and decrease in dam weight occurred in the rabbit at the highest dose. No teratogenic effects were seen.

Koeda, T.; Moriguchi, M. and Hirano, F.: Effect of sfericase on reproductive performance of mice and rabbits. Oyo Yakuri 16:941–950, 1978.

2185 Shigella Toxin

Olsen and Storeng (1986) found that 0.01 picograms per ml inhibited development of mouse blastocysts in vitro.

Olsen, W.M. and Storeng, R.: Effect of shigella toxin on preimplantation of mouse embryos in vitro. Teratology 33:243–246, 1986.

2186 Shortwave

Discussed under Microwave Radiation

2187 Sho–saiko–to

Aso et al. (**1992**) studied this chinese medicinal prescription in rats during organogenesis. Oral doses of up to 2,500 mg per kg had no ill effect on the dams, fetuses or postnatal development.

Aso, S.; Horiwaki, S.; Mashiba, C.; Tatsumi, Y. and Suzuki, A.: Teratogenicity study of orally administered sho–saiko–to in rats. Oyo Yakuri 44:659–673, 1992.

2188 Silver CAS 7440–22–4

Ridgway and Karnofsky (1952) using $AgNO_3$ determined that the LD–50 was 0.10 mg in chick eggs. No defects were reported. Mankes et al. (1976) reported that insertion of a silver wire into the uterus of a rat did not exert contraceptive action.

Mankes, R.; Estel, G.; Abraham, R. and Skalko, R.: Reproductive effects of intrauterine devices in the rat. Toxicol. Appl. Pharmacol. 37:125–126, 1976.

Ridgway, L.P. and Karnofsky, D.A.: The effects of metals on the chick embryo: Toxicity and production of abnormalities in development. Ann. N.Y. Acad. Sci. 55:203–215, 1952.

2189 Silymarin CAS 22888–70–6

This antihepatotoxic substance from the plant Silybum marianum was tested in rats and rabbits by Hahn et al. (1968) and no evidence for teratogenicity was found. The pregnant rats received 1 gm per kg and the rabbits 100 mg per kg during the active period of organogenesis.

Hahn, G.; Lehmann, H–D.; Kurten, M.; Uebel, H. and Vogel, G.: Zur Pharmacologie und Toxikologie von Silymarin des Antihepatotoxischen Wirkprinzipes aus Silybum marianum (l.) Gaertn. Arzneim. Forsch. 18:698–704, 1968.

2190 Simazine

Dilley et al. (1977) exposed rats to up to 317 mg per square meter during days 7 through 14 and produced no prenatal fetal changes.

Dilley, J.V.; Chernoff, N.; Kay, D.; Winslow, N. and Newell, G.W.: Inhalation teratology studies of five chemicals in rats. Toxicol. Appl. Pharmacol. 41:196, 1977.

2191 Simvastatin CAS 79902–63–9

Wise et al. (1990 A) gave this inhibitor of cholesterol biosynthesis to rats on days 6–17 in doses of up to 12.5 mg per kg twice daily orally. No teratogenicity was found but at the highest dose fetal and maternal weight was reduced compared to controls. Studies of rabbits treated during organogenesis with doses of up to 10 mg per kg found no teratogenicity. (Wise et al. 1990 B) Fertility studies in rats at 12.5 mg per kg twice daily revealed no ill effects. (Wise et al. 1990 C) Perinatally administered drug did not interfer with postnatal behavior or fertility. (Minsker et al. 1990)

Minsker, D.H.; Robertson, R.T. and Bokelman, D.L.: Simvastatin (mk–0733): oral late gestation and lactation study in rats. Oyo Yakuri 39(2):169–179, 1990.
Wise, L.D.; Majka, J.A.; Robertson, R.T. and Bokelman, D.L.: Simvastatin (mk–0733): oral teratogenicity study in rats pre–and postnatal obersvation. Oyo Yakuri 39(2):143–158, 1990.
Wise, L.D.; Prahalada, S.; Robertson, R.T.; Bokelman, D.L.; Akutsu, S. and Fujii, T.: Simvastatin (mk–0733): oral teratogenicity study in rabbits. Oyo Yakuri 39(2)1590–167, 1990.
Wise, L.D.; Minsker, D.H.; Robertson, R.T.; Bokelman, D.L.; Akutsu, S. and Fujii, T.: Simvastatin (mk–0733): oral fertility study in rats. Oyo Yakuri 39(2):127–141, 1990.

2192 Sintamil

This antidepressant was studied in rats by Rao (1975). Doses of up to 60 mg per kg were gavaged daily before, during and at the end of gestation. No adverse effects were found.

Rao, R.R.: Effect of sintamil: A new dibenzoxazepine antidepressant on reproductive processes. Ind. J. Med. Res. 63:58–65, 1975.

2193 Sisomycin CAS 32385–11–8

Esaki et al. (1978) administered this antibiotic to pregnant mice on days 6 through 15 in doses up to 120 mg per kg and found no adverse fetal changes. Tanioka et al. (1978) gave up to 30 mg per kg to monkeys on days 23 through 36 and found no teratogenic changes. Taniguchi et al. (1984A,B) gave up to 20 and 15 mg per kg to rats and rabbits on days 7–17 or days 6–18, respectively, and found no adverse fetal effects.

Esaki, K.; Ohshio, K. and Yoshikawa, K.: Effects of subcutaneous administration of sisomycin on reproduction in the mouse. Experiments on drug administration during the development period of fetuses. (Japanese) CIEA Preclin. Rpt. 4:157–164, 1978.
Taniguchi, H.; Himeno, Y.; Hoshino, K.; Ichiki, T.; Ogata, H.; Inoue, H. and Kodama, R.: Reproduction studies of sisomycin. Teratogenicity in rats by intravenous administration. Yakuri to Chiryo 12:2759–2767, 1984A.
Taniguchi, H.; Ohtsuka, T.; Setoguchi, T.; Fujita, M.; Ichiki, T.; Ohkubo, M.; Ogata, H.; Noguchi, K. and Kodama, R.: Reproduction study of sisomycin intravenous administration. Teratogenicity in rabbits. Yakuri to Chiryo 12:2727–2757, 1984B.
Tanioka, Y.; Koizumi, H. and Inaba, K.: Teratogenicity test by intramuscular administration of sisomycin in rhesus monkeys. CIEA Preclin. Rpt. 4:57–71, 1978.

2194 Sjogrens Syndrome

In a study of 168 offspring of mothers with Sjogren's Syndrome, 7 children with congenital complete heart block were identified (Manthorpe and Manthorpe, 1992). The relative risk for this condition is about 500 and about 2% of the offspring are affected. Infertility and abortion were not increased in the mothers.

Manthorpe, T. and Manthorpe, R.: Congenital complete heart block in children of mothers with primary Sjogren's syndrome. The Lancet 340:1359–1360, 1992.

2195 SM–857 SE

11–Oxo–11H–pyrido(2,1–b)quinazoline–2–carboxylic Acid

Nishimura et al. (1988) studied SM 857 SE, an inhibitor of histamine release from the mast cell in the pregnant rat and rabbit. A the highest dose of 400 mg per kg given during organogenesis to the rat, maternal and fetal toxicity as well as microphthalmia and vertebral–costal defects were increased. No teratogenicity was found in the rabbit at 400 mg per kg during days 6–18.

Nishimura, M.; Kast, A.; Tsunenari, Y. and Kobayashi, S.: Teratogenicity of the antiallergic SM 857 SE in rats versus rabbits. Teratology 38:351–367, 1988.

2196 SM–1652

Tanaka et al. (1983) studied this cephalosporin antibiotic intravenously in the rat during organogenesis. No adverse effects were found at doses up to 2000 mg per kg.

Tanaka, Y.; Ohimakoshi, Y.; Kato, T. and Nakatani, H.: Teratology study of SM 1652 in rats. Clin. Report 17:1000–1004, 1983.

2197 SMA 1440–H Resin

Winek and Burgun (1977) treated pregnant rats with 4 ml per kg on days 1–7, 8–14 or 14–20. No adverse fetal effects were reported.

Winek, C.L. and Burgun, J.J.: Acute and subacute toxicology and safety evaluation of SM 1440–H resin. Clin. Toxicol. 10:255–260, 1977.

2198 Small Pox Vaccination

Vaccination against small pox during pregnancy has not been associated with increased defects in the human. Saxen et al. (1968) found no effect of a mass population vaccination on fetal wastage and congenital defects in the Finnish population. Over 300,000 persons were vaccinated during a 5–month period.

Green et al. (1966) reviewed experience with fetal vaccinia infection which may occur after vaccination of the mother during pregnancy.

Thalhammer (1957) reported that cataracts could be produced experimentally in the fetal mouse by maternal injection of the virus; however, no lesions could be found by Theiler (1966).

Green, D.M.; Reid, S.M. and Rhaney, K.: Generalized vaccinia in the human fetus. Lancet 1:1296–1298, 1966.

Saxen, L.; Cantell, K. and Hakama, M.: Relation between small pox vaccination and outcome of pregnancy. Am. J. Public Health 58:1910–1921, 1968.

Thalhammer, O.: Die Vakzine–virusembryopathie der weissen Maus. Wien. Z. Inn. Med. 38:4–72, 1957.

Theiler, K.: Gibt es eine Vakzine–virusembryopathie? Pathol. Microbiol. 29:825–836, 1966.

2199 Smilagenin

Keeler and Binns (1968) gave 1.2–1.5 gm on the 4th day of gestation to 3 ewes and found no craniofacial defects in the offspring.

Keeler, R.F. and Binns, W.: Teratogenic compounds of Veratrum californicum (Durand). Teratology 1:5–10, 1968.

2200 Snake Oil

Hashimoto et al. (1979) gave this oil extracted from the digestive tract of the cobra to mice orally from days 6 through 15 in amounts of 4,500 mg per kg and found no adverse fetal effects.

Hashimoto, T.; Takeuchi, K.; Nagase, M. and Akatuka, K.: Pharmacological studies of snake oil. Teratology test in mice. (Japanese) Clinical Report 13:808–814, 1979.

2201 Sobuzoxane

Kato et al. (1991 A,B,C) studied this drug in rats using up to 250 mg per kg before pregnancy and early pregnancy, during organogenesis and in the perinatal period. Oral doses did not produce changes in the fetuses or their behavior. Rabbits received 250 mg per kg during organogenesis and found no adverse fetal effects.

Kato, K.; Uehara, M.; Hosokawa, T.; Furuhashi, T.; Muramoto, A.; Ono, H.; Shiohara, Y.; Nobuhara, A.; Narita, T.

and Takase, M.: Reproduction studies of sobuzoxane (I) fertility study in rats. Iyakuhin Kenkyu 22: 626–643, 1991.

Kato, K.; Ushida, K.; Kodama, R.; Hosokawa, T.; Furuhashi, T.; Fukuda, S.; Ono, H.; Shiobara, Y.; Takehara, M.; Nobuhara, A. and Takase, M.: Reproduction studies of sobuzoxane (II) teratological study in rats. Iyakuhin Kenkyu 22:644–665, 1991.

Kato, I.; Ushida, K.; Takei, A.; Hosokawa, T.; Furuhashi, T.; Muramoto, A.; Ono, H.; Shiohara, Y.; Takehara, M.; Nobuhara, A. and Takase, M.: Reproduction studies of sobuzoxane (III) perinatal and postnatal study in rats. Iyakuhin Kenkyu 22:666–680, 1991.

Kato, I.; Itagaki, Y.; Kurihara, H.; Murase, E.; Hosokawa, T.; Furuhashi, T.; Fukuda, S.; Ono, H.; Shiohara, Y.; Takehara, M.; Nobuhara, A. and Takase, M.: Reproduction studies of sobuzoxane (IV) teratological study in rabbits. Iyakuhin Kenkyu 22:681–691, 1991.

2202 Sodium 5–(Acetylamino)–3,5–dideoxy–D–glycero–D–galacto–2–nonulosonate
KI-111

Toshihiro et al. (**1991**) exposed rats before fertilizaion and during the first week of gestation and found no reproductive alterations to 500 mg per m^3 for one hour daily. Similar studies during organogenesis were done and no teratogenicity or behavioral changes were found (Tanaka et al., **1991**). Tanaka et al. (**1991**) gave up to 90 mg per kg to rabbits and found no fetal ill effects. In the rat perinatal studies with the same exposure was not associated with abnormal postnatal behavior.

Tanaka, N.; Takeuchi, Y.; Hanafusa, T.; Mitomi, M. and Ogasawara, S.: Reproductive and developmental toxicity studies of sodium 5–(acetylamino)–3,5–dideoxy–d–glycero–d–galacto–2–nonulosonate (KI–111)–effect on teratogenicity in rats. The Clinical Report 25:873–813, 1991.

Tanaka, N.; Takeuchi, Y.; Hanafusa, T.; Mitomi, M. and Ogasawara, S.: Reproductive and developmental toxicity studies of sodium 5–(acetylamino)–3,5–dideoxy–d–glycero–d–glacto–2–nonulosonate (KI–111)–teratological effect on rabbits. The Clinical Report 25:833–840, 1991.

Tanaka, N.; Kumashiro, J.; Kakuta, K.; Hanafusa, T.; Mitomi, M. and Ogasawara, S.: Reproductive and developmental toxicity studies of sodium 5–(acetylamino)–3,5–dideoxy–d–glycero–d–galacto–2–nonulosonate (KI–111)–effect of perinatal and postnatal administration in rats. The Clinical Report 25:842–856, 1991.

Ui, T.; Hatanaka, K.; Miura, M.; Hanafusa, T.; Mitomi, M. and Ogasawara, S.: Reproductive and developmental toxicity studies of sodium 5–(acetylamino–3,5–dideosy–d–glycero–d–galacto–2–nonulosonate (KI–111)–effect on fertility and general reproductive performance of rats. The Clinical Report 25:799–872, 1991.

2203 Sodium Bicarbonate

Khera (**1991**) presented evidence that coadministration of sodium bicarbonate with salicylate or ethylene glycol reduced teratogenicity in the mouse. He postulated that maternal acidosis and other changes produced by salicylate and ethyleneglycol reduced fetal nutrition and materno–embryonic gaseous exchange and resulted in altered development.

Khera, K.S.: Chemically induced alterations in maternal homeostasis and histology of conceptus: Their etiologic significance in rat fetal anomalies. Teratology 44:259–297, 1991.

2204 Sodium Chloride CAS 7647–14–5

Nishimura and Miyamoto (1969) injected 1900 or 2500 mg per kg into mice on the 10th or 11th day of pregnancy and produced up to 18 percent skeletal defects. Clubfoot was the most frequently found defect. The type of defect was different than found with fasting.

Nishimura, H. and Miyamoto, S.: Teratogenic effects of sodium chloride in mice. Acta Anat. (Basel) 74:121–124, 1969.

2205 Sofalcone 2–Carboxymethoxy–4,4–bis(3–methyl–2–butenyloxy)–chalcone CAS 64506–49–6

Yamada et al. (1980) administered up to 1000 mg per kg daily orally before mating, during the first week or during all of gestation and found no adverse effects on the fetuses or offspring. Maternal weight gain was reduced at the highest dose.

Yamada, T.; Tanaka, Y.; Suzuki, H.; Nogariya, T.; Nakone, S.; Sasajima, M. and Ohzeki, M.: Reproductive studies of 2–carboxymethoxy–4, 4–bis(3–methyl–2–butenyloxy)–chalcone. Fertility perinatal and postnatal study in rats. Oyo Yakuri 19:515–553, 1980.

2206 Solanine CAS 20562–02–1

Solanine, a toxic alkaloid isolated from both deadly nightshade and from shoots of stored potatoes, was tested for teratogenicity in chick embryos by Nishie et al. (1971). Ten to 25 mg was administered on the 4th day of incubation to the chick and no increase in congenital malformations was found. The embryonic LD–50 was 19 mg. Kline et al. (1961) fed rats a diet containing 10 percent potato sprouts from the time pregnancy was indicated by increased weight gain. No defects were detected but the pups died before weaning. Ruddick et al. (1974) found no congenital defects in rats after they administered up to 25 mg per kg on days 8 through 11. Ruddick et al. (1974) also fed rats scopoletin and α–chaconine with negative teratogenic findings. Renwick et al. (1984) gavaged hamsters on day 8 with 200 mg per kg of the purified alkaloid and found 59 percent of the litters with defects including neural tube defects.

Kline, B.E.; Elbe, H.U.; Dahle, N.A. and Kupchan, S.M.: Toxic effects of potato sprouts and of solanine fed to pregnant rats. Proc. Soc. Exp. Biol. Med. 107:807–809, 1961.

Nishie, K.; Gumbmann, M.R. and Keyl, A.C.: Pharmacology of solanine. Toxicol. Appl. Pharmacol. 19:81–92, 1971.

Renwick, J.H.; Claringbold, W.D.B.; Earthy, M.E.; Few, J.D. and McLean, A.C.: Neural–tube defects produced

in Syrian hamsters by potato glycoalkaloids. Teratology 30:371–381, 1984.

Ruddick, J.A.; Harwig, J. and Scott, P.M.: Nonteratogenicity in rats of blighted potatoes and compounds contained in them. Teratology 9:165–168, 1974.

2207 Solstitialin A 13–Acetate Thistle, Yellow Star Cynaropicrin

Cheng et al. (1992) found that this extract of yellow star thistle was toxic to fetal rat brain in culture. Horses ingesting this plant develop a syndrome similar to Parkinsonism.

Cheng, D.H.K.; Costall, B.; Hamburger, M.; Hostettmann, K.; Naylor, R.J.; Wang, Y. and Jenner, P.: Toxic effects of solstitialin a 13–acetate and cynaropicrin from centaurea solstitialis l (asteraceae) in cell cultures of foetal rat brain. Neuropharmacology 31:271–277, 1992.

2208 Solvents, Organic also see Toluene, Xylene, Trichlorethylene, Methyl Ethyl Ketone, 2–Ethoxyethanol and Methyl Chloride

Holmberg (1979) found solvent exposure in 14 mothers of 132 offspring with neural tube closure defects or hydrocephaly. Since only two of the matched controls had exposure, he reported that there was a signifcant increase in exposure. The type of solvent was variable and included denatured alcohol, acetone, styrene, toluene, xylene, ethylene oxide, benzene and methylethylketone. Syrovadko and Malsheva (1977) studied 311 female enamelers who were exposed to 2–ethoxyethanol, chlorobenzene, tricresol and solvent naphtha. Congenital anomalies including heart defects and talipes were significantly increased (10.0 vs 3.9 percent). The control workers were from a regional hospital, silica plant and elsewhere.

Heidam (1983) studied abortions among women in industries where solvent exposures existed. She found that the increased abortion rate was accounted for by the increase in gravidity as compared to housewives who were the controls. Holmberg et al. (1982) studied solvent exposure in 388 pregnancies which resulted in oral clefts. Each index case was paired with a matched control. Fourteen mothers were exposed to solvents while among the controls only four gave such a history. Lacquer petrol was the most common solvent although xylene, acetone and toluene were also mentioned.

Olsen and Rachootin (1983) could not detect any decreased birth weight among 546 Danish mothers exposed to degreasers, lacquer, cutting lubricants and other solvents. Axelsson et al. (1984) studied 1160 pregnancies occurring in women employed in laboratory work. A slightly increased but not significant difference (RR 1.31) was found in the miscarriage rate. The perinatal mortality and defect rates were not different.

Eskenazi et al. (1988) interviewed over 4,000 mothers and selected 1059 who reported solvent exposure. Two industrial hygienists independently interviewed this group and agreed that 90 patients had significant solvent exposure (hydrocarbon solvents, halogenated solvents, ketones and aldehydes). These mothers were mostly laboratory technicians, art workers, assemblers or operating room workers. Forty–one of these were matched with controls and detailed neurodevelopmental studies were performed at age three and one–half. No significant differences were found.

Zielhuis et al. (1984), Hemminki and Vineis (1985) and Pradhan et al. (1988) reviewed the animal and human data on teratogenicity from organic solvents. Courtney et al. (1986) exposed mice to up to 1,500 mg per square meter on days 6–16. Some moderate changes in calcified centers of the paws and rib profiles were observed and some dilated renal pelves occurred. Hood and Ottley (1985) reviewed the animal teratogenicity studies including those with isomers of xylene and conclude that a selective teratogenic effect has yet to be presented.

Axelsson, G.; Lutz, C. and Rylander, R.: Exposure to solvents and outcome of pregnancy in university laboratory employees. Br. J. Ind. Med. 41:305–312, 1984.

Courtney, K.D.; Andrews, J.E.; Springer, J.; Menache, M.; Williams, T.; Dalley, L. and Graham, J.A.: A perinatal study of toluene in CD–1 mice. Fund Appl. Toxicol. 6:145–154, 1986.

Eskenazi, B.; Gaylord, L.; Bracken, M.B. and Brown, D.: In utero exposure to organic solvents and human neurodevelopment. Dev. Med. Child. Neurol. 30:492–501, 1988.

Heidam, L.Z.: Spontaneous abortions among factory workers. Scand J. Soc. Med. 11:81–85, 1983.

Hemminki, K. and Vineis, P.: Extrapolation of the evidence on teratogenicity of chemicals between humans and experimental animals. Teratogenesis, Carcinogenesis and Mutagenesis 5:251–318, 1985.

Holmberg, P.C.: Central–nervous–system defects in children born to mothers exposed to organic solvents during pregnancy. Lancet 2:177–179, 1979.

Holmberg, P.C.; Hernberg, S.; Kurppa, K.; Rantala, K. and Riala, R.: Oral clefts and organic solvent exposure during pregnancy. Int. Arch. Occup. Environ. Health 50:371–376, 1982.

Hood, R.D. and Ottley, M.S.: Developmental effects associated with exposure to xylene: A review. Drug Chem. Toxicol. 8(4):281–297, 1985.

Olsen, J. and Rachootin, P.: Organic solvents as possible risk factors for low birthweight. J. Occup. Med. 25:854–855, 1983.

Pradhan, S.; Ghosh, T.K. and Pradhan, S.N.: Teratological effects of industrial solvents. Drug Develop. Res. 13:205–212, 1988.

Syrovadko, O.N. and Malsheva, Z.U.: Work conditions and their effect on certain specific functions among women who are engaged in the production of enamel–insulated wire. (Russian) Gig Tr. Prof. Zabol. 4:25–28, 1977.

Zielhuis, R.L.; Stijkel, A.; Verberk, M.M. and Van de Poel–Bot, M.: Health Risks to Female Workers in Occupational Exposure to Chemical Agents. Springer–Verlag, Berlin, New York, 3–14, 1984.

2209 SOM–M *Triglycerides*

Sakimura et al. (1991 A,B,C) gave this intravenously to rats before and during organogenesis and in the perinatal and postnatal period. Doses of up to 35 mg per kg were not associated with adverse reproduction or fetal development. In the perinatal studies some inhibition of body weight increase was found in the offspring. Similar studies by Sakimura et al. (1991 D) in rabbits during organogenesis were non–teratogenic.

Sakimura, M.; Onishi, M.; Noda, Y.; Oneda, S.; Yoshizaki, K.; Baba. S.; Kamitani, T.: Fertility study of intravenously infused SOM–M emulsion in rats. Yakuri to Chriyo 19:s1691–s1699, 1991A.

Sakimura, M.; Onishi, M.; Noda, Y.; Nagata, R.; Yoshizaki, K.; Okazaki, K.; Kamitani, T. and Oneda, S.: Teratogenicity study of intravenously infused SOM–M emulsion in rats. Yakuri to Chriyo 19:s1701–s1715, 1991B.

Sakimura, M.; Onishi, M.; Noda, Y.; Nagata, R.; Yoshizaki, K.; Okazaki, K.; Kamitani, T. and Oneda, S.: Perinatal and postnatal study of intravenously infused SOM–M emulsion in rats. Yakuri to Chriyo 19:s1723–s1734, 1991C.

Sakimura, M.; Onishi, M.; Noda, Y.; Nagata, R.; Yoshizaki, K.; Okazaki, K.; Kamitani, T. and Oneda, S.: Teratogenicity stud of intravenously infused SOM–M emulsion in rabbits. Yakuri to Chriyo 19:s1717–s1722, 1991D.

2210 Soman *Phosphorofluoridic Acid Methyl–,1,2,2,trimethyl propyl ester*

This organophosphate anticholinesterase was given orally to rats and rabbits during orgoanogenesis in doses up to 165 microgm and 15 microgm respectively (Bates et al. 1990). Maternal toxicity occurred at the highest dose levels but no teratogenicity or fetal toxicity was found.

Bates, H.K.; LaBorde, J.B.; Dacre, J.C. and Young, J.F.: Developmental toxicity of soman in rats and rabbits. Teratology 42:15–23, 1990.

2211 Somatomedin C *Somazon IGFI*

Saegusa et al. (1992) studied rats and rabbits during gestation in subcutaneous doses of up to 2.0 mg per kg. Prenatal, postnatal and during organogenesis they found no adverse effects. Some increase in the growth of the dams was found. In the rabbit at the highest dose maternal and fetal mortality were increased as well as resorptions.

Saegusa, T.; Fujimoto, Y.; Hatakeyama, K.; Tanizaki, S.; Matsuo, S.; Shirai, K.; Ohara, K. and Noguchi, H.: Toxicity study of somazon–reproductive and developmental toxicity studies. The Clinical Report 26:2081–2091, 1992.

2212 Somatomedin Inhibitors

Balkan et al. (1988) added human serum fractions with somatomedin inhibiting activity to in vitro cultures of rat embryos and produced developmental abnormalities and growth decrease.

Balkan, W.; Rooman, R.P.; Hurst–Evans, A.; Phillips, L.S.; Goldstein, S.; Du Caju, M.V.L. and Sadler, T.W.: Somatomedin inhibitors from human serum produce abnormalities in mouse embryos in culture. Teratology 38:79–86, 1988.

2213 Somatropin *Somatotrophin Pituitary Growth Hormone* CAS 12629–01–5 CAS 9002–72–6

Hultquist and Engfeldt (1949), using two preparations from the anterior pituitary, produced enlarged rat fetuses. The preparations used, Phyol™ (80 to 120 units) or antuitrin G (120 to 180 units), were contaminated with other endocrine substances, and the effect of litter size and duration of gestation complicated the analysis of fetal growth. Clendinnen and Eayrs (1961) injected 3.2 mg of purified somatotropin into rats from the 7th to the 19th days of gestation and observed a significant increase in birth weight over the control animals. The duration of the gestation periods in the control and treated was not stated. The treated group of fetuses showed enhancement of cortically–mediated behavior, and this was supported by histologic studies showing hypertrophy of neurones. Zamenof et al. (1971) extended this work and showed that growth hormone reverses the adverse effect of starvation on brain development.

Watase et al. (**1993**) injected a recombinant in doses 12.5 IU per kg subcutaneously into males and females before conception and during the first week of pregnancy. The adult weights were increased and as well the number of implants. Fukanishi et al. (**1993**) treated rats similarly during organogenesis and found no ill effects. Perinatal studies by Watanabe et al. (**1993**) were not associated with adverse behavioral or reproductive effects. In rabbits given up to 1.25 IU during organogenesis no adverse effects were seen (Hanamoto et al., **1993**).

Croskerry and Smith (1975) presented convincing evidence that human somatotrophin prolongs pregnancy in the rat. They believed that this postmaturity along with an increase in maternal weight might explain the increases in brain growth and learning observed by others after growth hormone therapy during pregnancy.

Clendinnen, B.G. and Eayrs, J.T.: The anatomical and physiological effects of prenatally administered somatotrophin on cerebral development in rats. J. Endocrinol. 22:183–193, 1961.

Croskerry, P.G. and Smith, G.K.: Prolongation of gestation by growth hormone: A confounding factor in the assessment of its prenatal action. Science 189:648–650, 1975.

Fukunishi, K.; Watanabe, I.; Unno, T.; Nurimoto, S.; Yuki, Y.; Hiratani, H. and Kato, K.: Reproductive and developmental toxicity study of recombinant human growth hormone (JR–8810)–Teratology study in rats. The Clinical Report 27:5749–5773, 1993.

Hamamoto, S.; Moriwaki, M.; Kohata, K.; Yamanaka, H.; Yuki, Y.; Hiratani, H. and Kato, K.: Reproductive and developmental toxicity study of recombinant human growth hormone (JR–8810)–Teratological study in rabbits. The Clinical Report 27:5776–5787, 1993.

Hultquist, G.T. and Engfeldt, B.: Giant growth of rat fetuses produced experimentally by means of administration of hormones to the mother during pregnancy. Acta Endocrinol. 3:365–376, 1949.

Watanabe, I.; Fukunishi, K.; Unno, T.; Nurimoto, S.; Yuki, Y.; Hiratani, H. and Kato, K.: Reproductive and devel-

opmental toxicity study of recombinant human growth hormone (JR–8810)–Perinatal and postnatal study in rats. The Clinical Report 27:5789–5807, 1993.

Watase, T.; Hamamoto, S.; Kogami, H.; Yamanaka, H.; Yuki, Y.; Hiratani, H. and Kato, K.: Reproductive and developmental toxicity study of recombinant human growth hormone (JR–8810)–Fertility study in rats. The Clinical Report 27:5733–5748, 1993.

Zamenof, S.; Mathens, E.V. and Gravel, L.: Prenatal cerebral development: Effect of restricted diet, reversal by growth hormone. Science 174:954–955, 1971.

2214 Sorbitol

No increase in defects was found among rat offspring whose mother fed 2.5 to 10% sorbitol in their diets.

MacKenzie, K.M.; Hauck, W.N.; Wheeler, A.G. and Roe, F.J.C.: Three–generation reproduction study of rats ingesting up to 10% sorbitol in the diet—and a brief review of the toxicological status of sorbitol. Food Chem Toxicol 24:191–200, 1986.

2215 Sparfloxacin Trihydrate

Terada et al. (**1991** A) gave this antibiotic to adult rats in doses up to 500 mg per kg and found no adverse effect on reproduction. At 300 mg per kg orally Funabashi et al. (**1991** A) found decreased fetal weight at 300 mg per kg. An increase in ventricular septal defects (5.5 percent) was found in fetuses exposed to 300 mg per kg during organogenesis. After similar perinatal exposure behavior and reproduction were not altered (Funabashi et al., **1991** B). Terada et al. (**1991** B) gave rabbits up to 60 mg per kg and found no teratogenicity although the weight gain of the dams was reduced.

Funabashi, H.; Mukumoto, K.; Shigematsu, K.; Nishimura, K. and Ohnishi, K.: Reproductive and developmental toxicity studies of sparfloxacin (2)–teratogenicity study in rats. Jpn. Pharm. Therap. 19:1257–1274, 1991A.

Funabashi, H.; Mukumoto, K.; Imura, Y.; Nishimura, K. and Ohnishi, K.: Reproductive and developmental toxicity studies of sparfloxacin (3)–perinatal and postnatal study in rats. Jpn. Pharm. Therap. 19:1275–1287, 1991B.

Terada, Y.; Aoki, Y.; Mukumoto, K.; Shigematsu, K.; Nishimura, K. and Ohnishi, K.: Reproductive and developmental toxicity studies of sparfloxacin (1)–fertility study in rats. Jpn. Pharm. Therap. 10:1241–1255, 1991A.

Terada, Y.; Mukumoto, K.; Imura, Y.; Nishimura, K. and Onhishi, K.: Reproductive and developmental toxicity studies of sparfloxacin (4)–teratogenicity study in rabbits. Jpn. Pharm. Therap. 19:1289–1297, 1991B.

2216 Spermine CAS 71–44–3

Butros (1972) studied the effect of this polyamine on chick embryos. Injecting 2.6–10 mg per egg at 0–7 days of incubation, he observed a general arrest of development at 2.5 days. A striking finding was an over production of immature erythrocytes which at times packed the brain vesicles.

Butros, J.: Action of spermine on early chick development. I. Morphogenesis and histogenesis. Teratology 6:181–190, 1972.

2217 Spiclomazine *APY–606*
8(3–Chloro–10–phenothaz–inyl)–propyl, clospirazine

Hamada et al. (1970) gave rats and mice orally up to 100 and 200 mg respectively during organogenesis. At the highest doses some embryolethality and fetal growth inhibition were found. Some skeletal variations were found also. At 200 mg per kg and above in mice during the first 5 days of gestation, embryonic loss was found.

Hamada, Y.; Namba, T.; Okada, T. and Izaki, K.: Studies on psychotropic drugs. IX. Teratological studies of APY–606. Oyo Yakuri 4:497–504, 1970.

2218 Spray Adhesives

In 1973, Dr. J.R. Seely reported publicly that ten persons exposed to spray adhesives had significant increases in the number of chromosomal breaks and gaps in peripheral lymphocytes and two of these persons were infants with multiple anomalies. Reanalysis of the slides by an ad hoc committee failed to reach definitive conclusions (Anonymous, 1973). Separate studies of humans exposed to spray adhesives (see Murphy et al., 1975, for review) and epidemiologic studies failed to find that spray adhesives were associated with chromosomal damage or birth defects (Oakley et al., 1974).

No defects were found in fetal hamsters exposed from days 5–10 of gestation to inhalation of foil art adhesive (Murphy et al., 1975).

Anonymous: Adhesive spray studies are inconclusive so far. Prod. Safety Lett. 2:2 only, 1973.

Murphy, J.C.; Collins, T.F.X.; Black, T.N. and Osterberg, R.E.: Evaluation of teratogenic potential of a spray adhesive in hamsters. Teratology 11:243–246, 1975.

Oakley, G.P.; Nissim, J.E.; Hanson, J.W.; Boyce, J.M. and Roberts, M.: Epidemiologic investigations of possible teratogenicity of spray adhesives. (abs) Teratology 9:31a–32a, 1974.

2219 Stanolone *17–β–Hydroxy–3–androstanone*

Schultz and Wilson (1974) administered 4 mg per day on days 14–21 in the rat and did not produce masculinization of the female fetal internal genitalia.

Schultz, F.M. and Wilson, J.D.: Virilization of the wolffian duct in the rat fetus by various androgens. Endo. 94:979–986, 1994.

2220 Stanozolol

Kawashima et al. (1977) found masculinization of female rat fetuses exposed to 0.5 mg per kg.

Kawashima, K.; Nakaurra, S.; Nagao, S.; Tanaka, S.; Kuwamura, T. and Omorri, Y.: Virrilizing activities of various steroids in female rat fetuses. Endocrinol. Japon. 24:77–81, 1977.

2221 Staphlococcal Phage Lysate

Hirayama et al. (1980) gave this product subcutaneously to pregnant rats before gestation, during the first 7 days, during days 6 through 16 and during day 17 through 7 postnatal days. The maximum dose was 2.0 ml (400 microunits) per kg per day. No ill effects were found on fertility, the fetuses or postnatal development. Rabbits were given up to 0.5 ml daily during gestation with no ill effects being found in the fetuses.

Hirayama, H.; Wada, S.; Kimura, T.; Enokuya, Y.; Ohkuma, H. and Hikita, J.: Reproductive evaluation of staphlococcal phage lysate (SPL). Oyo Yakuri 20:487–499, 575–594, 595–608, 1980.

2222 Staurosporine

Fujinaga et al. (1994) found in whole embryo culture of rats that this protein kinase inhibitor did not prevent adrenergic–induced situs inversus. An embryonic syndrome consisting of mesencephalic cysts and abnormal migration of neural cells was found with concentrations of 0.5 and 1.0 micromolar.

Fujinaga, M.; Park, H.W.; Shepard, T.H.; Mirkes, P.E. and Baden, J.M.: Staurosporine does not prevent adrenergic–induced situs inversus, but causes a unique syndrome of defects in rat embryos grown in culture. Teratology 50:261–274, 1994.

2223 Stobadine

Ujhazy et al. (1994) gave this antiarrhythmic to mice intravenously on days 3, 6, 9 or 12 in doses of 1 or 3 mg per kg and found no teratogenicity. Oral doses of up to 122 mg per kg on days 4–16 caused some decrease in implantation and fetal weight but no increase in malformations.

Ujhazy, E.; Dubovicky, M.; Balonova, T. and Jansak, J.: Teratological assessment of stobadine after single and repeated administration in mice. Journal of Applied Toxicology 14:357–363, 1994.

2224 Streptomycin CAS 57–92–1

Two case reports questioned the association of streptomycin medication with congenital nerve deafness. Leroux (1950) reported a deaf infant following treatment of the mother with 30 gm of streptomycin during the 8th month. Kern (1962) reported deafness of an infant whose mother was given 20 gm of dihydrostreptomycin during the first 4 months of pregnancy. Robinson and Cambon (1964) reported two congenitally deaf children whose mothers were treated with one gm per day during the last four months in one case and during the 6th to 14th weeks in the other. Varpela et al. (1969) found normal hearing in 50 children who had been exposed in utero to dihydrostreptomycin or streptomycin. Conway and Birt (1965) examined 13 children exposed to the drugs in utero and found no disability but the caloric test in six and audiograms in four were abnormal.

Heinonen et al. (1977) found no congenital defect increase among 135 offspring exposed in the first four lunar months or among 335 exposed any time in pregnancy.

Suzuki and Takeuchi (1961) were unable to detect deafness in mice or rats exposed to 200 to 600 mg per kg during pregnancy. Boucher and Delost (1964) gave 25 or 250 mg per kg to mice on the 14th gestational day and detected growth failure for a month after birth.

Warkany (1979) reviewed the teratogenicity of this and other antituberculous drugs.

Boucher, D. and Delost, P.: Developpement post–natal des descendants issus de meres traitees par la streptomycine au cours de la gestation chez la souris. C. R. Soc. Biol. (Paris) 158:2065–2069, 1964.

Conway, N. and Birt, B.D.: Streptomycin in pregnancy: Effect on the fetal ear. Br. Med. J. 2:260–263, 1965.

Heinonen, O.P.; Slone, D. and Shapiro, S.: Birth Defects and Drugs in Pregnancy. Publishing Sciences Group Inc., Littleton, Mass, 1977.

Kern, G.: Zur Frage der Intrauterinen Streptomycin Schadigung. Schweiz. Med. Wochenschr. 92:77–79, 1962.

Leroux, M.L.: Existe–t–il une surdite congenitale acquise due a la streptomycine? Ann. Otolaryngol. 67:194–196, 1950.

Robinson, G.C. and Cambon, K.G.: Hearing loss in infants of tuberculous mothers treated with streptomycin during pregnancy. N. Eng. J. Med. 271:949–951, 1964.

Suzuki, Y. and Takeuchi, S.: Etude experimentale sur l'influence de la streptomycine sur l'appareil auditif du foetus apres administration de doses variees a la mere enceinte. Keio J. Med. 10:31–41, 1961.

Varpela, E.; Hietalahti, J. and Aro, M.J.T.: Streptomycin and dihydrostreptomycin medication during pregnancy and their effect on the child's inner ear. Scand. J. Resp. Dis. 50:101–109, 1969.

Warkany, J.: Antituberculous drugs. Teratology 20:133–138, 1979.

2225 Streptonigrin CAS 3930–19–6

Warkany and Takacs (1965) injected rats intraperitoneally with 0.25 mg per kg on the 9th, 10th or 11th day of gestation. Following treatment on the 10th day, 96 percent of the fetuses were abnormal. A wide variety of defects occurred: Omphalocele, exencephaly, hydrocephaly and defects of the eye and skeleton predominated. The axial skeleton was commonly involved and the authors described in detail one fetus with iniencephalus. Chaube et al. (1969) confirmed the findings and found that methyl ester streptonigrin and isopropylidine azastreptonigrin were equally teratogenic.

Chaube, S.; Kuffer, F.R. and Murphy, M.L.: Comparative teratogenic effects of streptonigrin (NSC–45383) and its derivatives in the rat. Cancer Chemother. Rep. 53:23–31, 1969.

Warkany, J. and Takacs, E.: Congenital malformations in rats from streptonigrin. Arch. Pathol. 79:65–79, 1965.

2226 Streptozotocin CAS 18883–66–4

Discussed under Diabetes

2227 Strontium ^{90}Sr CAS 7440–24–6

Nilsson and Henricson (1969) injected pregnant mice on the 11th or 16th gestational day with 20 microcuries of ^{90}Sr.

A general reduction of fetal oocytes was found. Finkel and Biskis (1969) concluded from dog experiments that the fetal dog was not more sensitive than the adult to either the lethal or the oncogenic effect of radiostrontium. Underdeveloped jaws, disproportionate growth of the long bones and many fractures were found in fetuses after administration of 1 millicurie per kg to the mother 6 days before delivery. Hiraoka (1961) injected 5 to 10 microcuries intraperitoneally into pregnant mice at various times during pregnancy and found an increase in skeletal defects.

Sternglass (1969) compared excess fetal death rate with the ^{90}Sr content of fetal bone and was of the opinion that a strong correlation existed between the two measurements. This assertion by Sternglass has been clearly refuted by Lindop and Rotblat (1969) who pointed out that he manipulated the data to fit his conclusions and many better explanations than radiation effects could account for the diminished decline in infant mortality.

Finkel, M.P. and Biskis, B.O.: Pathological consequences of radiostrontium administered to fetal and infant dogs. In: Radiation Biology of the Fetal and Juvenile Animal. M.R. Sikov and D.D. Mahlum (eds.), Oak Ridge, Tennessee: United States Atomic Energy Commission, 543–566, 1969.

Hiraoka, S.: The transplacental effects of the radio–strontium–90 upon the mouse embryos. Acta Anatomica Nipponica 36:161–171, 1961.

Lindop, P.J. and Rotblat, J.: ^{90}Strontium and infant mortality. Nature 244:1257–1260, 1969.

Nilsson, A. and Henricson, B.: Effect of ^{90}Sr on the ovaries of fetal mice. In: Radiation Biology of The Fetal and Juvenile Animal, M.R. Sikov and D.D. Mahlum (eds.), Oak Ridge, Tennessee: United States Atomic Energy Commission, 313–324, 1969.

Sternglass, E.J.: Evidence for low–level radiation effects on the human embryo and fetus. In: Radiation Biology of the Fetal and Juvenile Animal, M.R. Sikov and D.D. Mahlum (eds.), Oak Ridge, Tennessee: United States Atomic Energy Commission, 693–718, 1969.

2228 Styrene CAS 100–42–5

Murray et al. (1978) studied the effect of inhalation and gavage in the pregnant rat and rabbit during organogenesis. At doses of 600 ppm for 7 hours daily, no fetal changes were produced in either species. Rats were given up to 150 mg per kg without fetal effects. Barlow and Sullivan (1982) have summarized the reproductive toxicology of this substance. Kishi et al. (1995) exposed rats to 50 or 300 ppm for 6 hours daily on days 7–21. Although learning behavior was not affected, motor coordination was decreased at both levels.

Barlow, S.M. and Sullivan, F.M.: Reproductive Hazards of Industrial Chemicals. An Evaluation of Animal and Human Data. Academic Press Inc. (London), 1982.

Kishi, R.; Chen, B.Q.; Katakura, Y.; Ikeda, T. and Miyake, H.: Effect of prenatal exposure to styrene on the neurobehavioral development, activity, motor coordination, and learning behavior of rats. Neurotoxicology and Toxicology 17:121–130, 1995.

Murray, F.J.; John, J.A.; Balmer, M.F. and Schwetz, B.A.: Teratologic evaluation of styrene given to rats and rabbits by inhalation and by gavage. Toxicology 11:335–343, 1978.

2229 Styrene Oxide CAS 96–09–3

Kimmel et al. (1984) reviewed studies in rats and rabbits exposed to 100 and 50 ppm 7 hours per day during gestation. No evidence of teratogenicity was found although maternal toxicity and an increased fetal mortality was reported.

Sikov et al. (1986) exposed rats and rabbits to 100 ppm for 7 hours per day 5 days per week for 3 weeks. Preimplantation losses occurred in the rat. Delayed ossification occurred but no teratogenicity. Rabbits exposed to the same regime had increased resorptions at dose levels of 15 and 50 ppm. No teratogenicity was found.

Kimmel, C.A.; LaBorde, J.B. and Hardin, B.D.: Reproductive and developmental toxicology of selected epoxides. In: Toxicology and the Newborn, Chapter 13, S. Kacew and M.J. Reasor (eds.), Elsevier Sci. Publ. Amsterdam, 270–287, 1984.

Sikov, M.R.; Cannon, W.C.; Carr, D.B. and Miller, R.A.: Reproductive toxicology of inhaled styrene oxide in rats and rabbits. J. Appl. Toxicol. 6:155–164, 1986.

2230 Succinate Tartrates

Petersen and Daston (1989) found no adverse fetal effects in rat fetuses exposed up to 1.0 gm per kg via drinking water on days 6–15 of gestation.

Petersen, D.W.; Schardein, J.L. and Daston, G.P.: Evaluation of the developmental toxicity of succinate tartrates in rats. Food Chem Toxicol 27(4):249–254, 1989.

2231 Succinonitrile

Doherty et al. (1983) studied the effect of this aliphatic nitrile on the hamster giving up to 6.24 mmol per kg on day 8 of pregnancy. Neural tube defects and encephalocoeles occurred as well as stunting at doses above 0.147 mmol per kg and the effect was reversed by thiosulfate.

Doherty, P.A.; Smith, R.P. and Ferm, V.H.: Comparison of the teratogenic potential of two aliphatic nitriles in hamsters: Succinonitrile and tetramethylsucconitrile. Fund. Appl. Toxicol. 3:41–48, 1983.

2232 Sucrose CAS 57–50–1

Seta (1931) produced skeletal changes in the guinea pig fetus by feeding the mother 5 to 10 gm per kg during the later half of pregnancy. Hirata (1936) added 5 to 7 gm per kg of body weight to the diet of pregnant rabbits. Although the results in the first litters were almost absent, treatment was continued and the fetuses from the second and third litters were reported to have cataract, microphthalmia and hydrocephalus. Furukawa (1939) confirmed this work.

Hamel et al. (1986) fed mice a diet which supplied 64 percent of the calories from sucrose and compared it with an isocaloric control diet containing grain. The average litter size was reduced from 6.4 to 2.5 in the sucrose fed group.

All of the surviving pups from the sucrose diet died and at autopsy the cause was not evident.

Furukawa, S.: Experimental production of malformation. Nisshin Igaku (Japanese) 28:1119–1160, 1939.

Hamel, E.E.; Santisteban, G.A.; Ely, J.T.A. and Read, D.H.: Hyperglycemia and reproductive defects in non-diabetic gravidas: A mouse model test of a new theory. Life Sciences 39:1425–1428, 1986.

Hirata, M.: Experimental study on congenital hydrocephaly. Part I. Relationship between production of congenital hydrocephaly and maternal diet. Nisshin Igaku (Japanese) 25:1980–2004, 1936.

Seta, S.: Effect of maternal nutrition during pregnancy on the skeletal development of the fetus. Nisshin Igaku (Japanese) 21:486–504, 1931.

2233 Sucrose Polyester

Nolen et al. (1987) published a two–generation reproductive study in rats using up to 10 percent of this fat substitute in the diet. Small amounts of vitamin E and A were added to the test diet. No adverse effects were found. The weight of diet consumed was increased at the highest diet level.

Nolen, G.A.; Wood, F.E., Jr. and Dierckman, T.A.: A two–generation reproductive and developmental toxicity study of sucrose polyester. Food Chem. Toxicol. 25(1):1–8, 1987.

2234 Sudan Grasses

Prichard and Voss (1967) observed dystocia and ankylosis of fetal joints in the offspring of mares who grazed on hybrid Sudan grass pasture.

Prichard, J.T. and Voss, J.L.: Fetal ankylosis in horses associated with hybrid Sudan pasture. J.A.V.M.A. 150:871–873, 1967.

2235 Sufentanil

Fujinaga et al. (1988) administered this narcotic to rats giving 10, 50 or 100 mg per kg on days 5–20 via implanted osmotic minipumps. No adverse fetal effects were found.

Fujinaga, M.; Mazze, R.I.; Jackson, E.C. and Baden, J.M.: Reproductive and teratogenic effects of sufentanil and alfentanil in Sprague–Dawley rats. Anesth. Analg. 67:166–169, 1988.

2236 Suicide

Czeizel et al. (1988) studied 109 women who attempted suicide. The stillbirth rate (4 percent) was higher than controls but the miscarriage and congenital defect rate was not significantly increased. The majority of women took sedatives and tranquilizers. McElhatton et al. (1991) followed the pregnancies of 267 women who overdosed on aspirin (41), acetaminophen (147) codeine (11) or nonsteroidal anti inflammatories (19). The majority of the babies were normal and 9 percent ended in spontaneous abortion or fetal death. Eleven newborns of 195 continuing pregnancies had abnormalities. Two of the abnormalities were clearly unrelated to drug usage and 8 were most unlikely to be asociated. Gunnarskog and Kallen (1993) identified 424 infants born to Swedish mothers with chemical intoxications. Seventy women attempted suicide with psychoactive drugs and none had offspring with congenital defects.

Czeizel, A.; Sventesi, I.; Szekeres, I.; Molnar, G.; Glauber, A. and Bucski, P.: A study of adverse effects on the progeny after intoxication during pregnancy. Arch. Toxicol. 62:1–7, 1988.

Gunnarskog, J. and Kallen, A.J.B.: Drug intoxication during pregnancy: a study with central registries. Reproductive Toxicology 7:117–121, 1993.

McElhatton, P.R.; Sullivan, F.M. and Walton, L.: Analgesic overdose during pregnancy. (Abs) Teratology 44:17A, 1991.

2237 Sulbactam CAS 68373–14–8

Horimoto et al. (1984) gave this antibiotic intravenously to rats on days 7–17 and 17–21 as well as two weeks before mating and then the first seven days. Doses of up to 500 mg per kg daily produced no adverse reproductive effects.

Horimoto, M.; Sakai, T.; Ohtsuki, I. and Noguchi, Y.: Reproduction studies with sulbactam and combinations of sulbactam and cefoperazone in rats. Chemotherapy 32:108–115, 1984.

2238 Sulconazole Nitrate

Kobayashi et al. (1985) gave this antimycotic agent subcutaneously to rats and rabbits before fertilization, during organogenesis and in the rats on days 17–20. At doses of 10 mg per kg or more, decreased fertility occurred in both sexes of each species. Resorptions were increased at 60 mg and 30 mg in rats and rabbits treated during organogenesis. No teratogenicity was found in either species and postnatal behavioral changes were not found in the rat.

Kobayashi, T.; Ariyuki, F.; Higaki, K.; Shibano, T.; Kano, M.; Kitahara, S. and Nakagawa, H.: Reproductive studies in rats and rabbits given sulconazole nitrate (RS 44872). Oyo Yakuri 30(3):451–465, 1985.

2239 Sulfacetamide CAS 127–56–0; CAS 144–80–9; CAS 6209–17–2

No increase in defects was found among the offspring of 93 women taking this antibiotic in the first four lunar months (Heinonen et al., 1977).

Heinonen, O.P.; Slone, D. and Shapiro, S.: Birth Defects and Drugs in Pregnancy. Publishing Sciences Group Inc., Littleton, Mass, 1977.

2240 Sulfadiazine

Discussed under Sulfonamides

2241 Sulfadimethoxine *Sulphamoprine* also see *Sulphonamides* CAS 155–91–9

Paget and Thorpe (1964) found that dietary levels of 0.025 percent produced dental anomalies in the offspring of rats and mice but had no effect on rabbits.

This long–acting sulfonamide drug was given orally to rats on the 11th, 13th, 14th and 16th days of gestation, and

postnatally the offspring developed short snouts with positional defects of the incisors (Goultschin and Ulmansky, 1971). The dosage was 75 mg per kg daily.

Goultschin, J. and Ulmansky, M.: Skull and dental changes produced by sulfonamide in rats. Oral Surg. 31:290–294, 1971.

Paget, G.E. and Thorpe, E.: A teratogenic effect of a sulphonamide in experimental animals. Br. J. Pharmacol. 23:305–312, 1964.

2242 Sulfadoxine

This sulfonamide taken in combination with pyrimethamine was used by Daffos et al. (1988) to treat 6 pregnancies where fetal infections were diagnosed. No treatment–related fetal adverse effects were found and only 2 of 15 fetuses treated with sulfadiazine or sulfadoxine had choreoretinitis. Spiramycin was also used in therapy.

Daffos, F.: Prenatal management of 746 pregnancies at risk for congenital toxoplasmosis. N. Eng. J. Med. 318:271–275, 1988.

2243 Sulfaethylthiadiazole

Maren and Ellison (1972) injected rats subcutaneously with 250 mg per kg on days 10 and 11 and produced no increase in defects.

Maren, T.H. and Ellison, A.C.: The teratological effect of certain thiadiazoles related to acetazolamide and thiazide diuretics. Johns Hopkins Med. J. 130:95–104, 1972.

2244 Sulfaguanidine *Sulgin*™ CAS 57–67–0

This oral hypoglycemic was tested by Bariljak (1968) in rats using 3.0 gm per kg on the 10th day of gestation. Although a 10 percent fetal mortality occurred no congenital defects were noted.

Bariljak, I.R.: Comparison of antithyroidal and teratogenic activity of some hypoglycemic sulphonamides. Prob. Endocrinol. (Russian) 14(6):89–94, 1968.

2245 Sulfaguanol *N–[(4, 5–Dimethyl–2–oxazolyl) amidino]–sulfanilamide* CAS 27031–08–9

Kuhne et al. (1973) fed this antibiotic to pregnant rats and rabbits during organogenesis in amounts up to 250 mg per kg and found no fetal changes.

Kuhne, J.; Leuschner, F. and Neumann, W.: Untersuchungen zur Toxikologie von Sulfaguanol. Arzneim–Forsch 23:178–184, 1973.

2246 Sulfamethazine

Bates et al. (1989) fed rats up to 2400 ppm for three generations and then bred them. No adverse reproductive effects were seen.

Bates, H.K. and LaBorde, J.B.: Developmental toxicity evaluation of sulfamethazine in Fischer 344 (F-344) rats. Teratology 39:P79, 1989.

2247 Sulfamethoxazole *Gantanol*™ CAS 723–46–6

Sulfamethoxazole crosses the human placenta and reaches a peak at 10 hours. After a few gestational weeks, the concentration of sulfamethoxazole is lower in amniotic fluid and in the foetus than in maternal serum (Reid et al., 1975). Williams et al. (1969) treated 120 pregnant women and found no increase in defects in their offspring; only 10 of the women were treated before the 16th week. Heinonen et al. (1977) reported no increase in malformation rates in the offspring from 46 pregnancies when treatment was given in the first 4 lunar months.

Heinonen, O.P.; Slone, D. and Shapiro, S.: Birth Defects and Drugs in Pregnancy. Publishing Sciences Group Inc., Littleton, Mass., 1977.

Reid, D.W.J.; Caille, G. and Kaufmann, N.R.: Maternal and transplacental kinetics of trimethoprim and sulfamethoxazole, separately and in combination. Can. Med. Assoc. J. 112:67S–72S, 1975.

Williams, J.D.; Brumfitt, W.; Condie, A.P. and Reeves, D.S.: The treatment of bacteriuria in pregnant women with sulphamethoxazole and trimethoprim. A microbiological, clinical and toxicological study. Postgrad. Med. J. Suppl. 45:71–76, 1969.

2248 o–Sulfamoylbenzoic Acid CAS 632–24–6

Discussed under Saccharin

2249 Sulfamoyldapsone

2–Sulfamonyl–4,4–diaminodiphenylsulfone CAS 17615–73–5

Asano et al. (1975) gave pregnant rats up to 4,000 mg per kg orally from days 9 through 14 of gestation and found no significant increase in fetal defects. Post–natal studies were negative.

Asano, Y.; Susami, M.; Ariyuki, F. and Higaki, K.: The effects of administration of 2–sulfamonyl–4,4–diaminodiphenylsulfone (SDDS) on rat fetuses. (Japanese) Oyo Yakuri 9:695–707, 1975.

2250 Sulfanilamide also see *Sulphonamides* CAS 63–74–1

Bariljak (1968) gave 3000 mg per kg of body weight to rats on the 10th day of gestation and found no lethal or teratogenic properties.

Bariljak, I.R.: Comparison of antithyroidal and teratogenic activity of some hypoglycemic sulphonamides. Prob. Endocrinol. (Russian) 14(6):89–94, 1968.

2251 Sulfanilyurea *Urosulfan* CAS 547–44–4

Bariljak (1968) showed that urosulfan in doses of 1000 mg per kg of body weight on the 10th day of pregnancy caused death in 44 percent of implanted rat fetuses but did not manifest teratogenicity.

Bariljak, I.R.: Comparison of antithyroidal and teratogenic activity of some hypoglycemic sulphonamides. Prob. Endocrinol. (Russian) 14(6):89–94, 1968.

2252 Sulfasalazine CAS 599–79–1

Newman and Correy (1983) reported two infants exposed in utero to this medication. One had coarctation and

ventricular septal defect and the other was a twin with polycystic kidney (Potter 2a). The co–twin had hypoplastic lungs and absent kidneys. Mogadam et al. (1980) studied 531 pregnancies associated either with Crohn's disease or ulcerative colitis. In 244, neither corticoids nor sulphasalazine were used and in 287, the mothers had taken either or both of these drugs. In neither group was the malformation rate increased.

Mogadam, M.; Dobbins, W.O. and Korelitz, B.I.: The safety of corticosteroids and sulfasalazine in pregnancy associated with inflammatory bowel disease. Gastroenterol. 78:1224, 1980.

Newman, N.M. and Correy, J.F.: Possible teratogenicity of sulphasalazine. Med. J. Australia 1:528–529, 1983.

2253 Sulfisoxozole *Gantrisin™ Sulfafurazole* CAS 127–69–5

Sulfafurazole was shown to transfer into the amniotic fluid during early fetal life (Blum et al., 1975). Mellin (1964) found no increase in the defect rate when mothers were treated in the first trimester. Heinonen et al. (1977), who examined records of 796 offspring of mothers treated in the first 4 lunar months of pregnancy, also found no increase in malformation rate.

When mice and rats were administered 1000 mg per kg of sulfafurazole orally on days 7–12 and 9–14 of pregnancy, respectively. A significant increase in cleft palate and skeletal defects was found in offspring of both species, in addition mandibular defects were present in the rat fetuses (Kato and Kitagawa, 1973).

Blum, M.; Elian, I. and Ben–Tovim, R.: Transfer of antibiotics across the placenta in early pregnancy (Heb.). Harefuah 88:510–512, 1975.

Heinonen, O.P.; Slone, D. and Shapiro, S.: Birth Defects and Drugs in Pregnancy. Publishing Sciences Group Inc., Littleton, Mass., 1977.

Kato, T. and Kitagawa, S.: Production of congenital skeletal anomalies in the fetuses of pregnant rats and mice treated with various sulfonamides. Cong. Anomal. 13:17–23, 1973.

Mellin, G.W.: Drugs in the first trimester of pregnancy and the fetal life of Homo sapiens. Am. J. Obstet. Gynecol. 90:1169–1180, 1964.

2254 o–Sulfobenzoic Acid CAS 632–25–7

Discussed under Saccharin

2255 Sulfonamide CS–61

3,6–Dimethoxy–4–sulfanilamidopyridazine

Kato and Kitagawa (1974) studied this compound in pregnant rats and mice. At doses of 750 and 1000 mg per kg in the rat on days 9 through 14, a significant increase in absence of the kidney occurred in the fetuses. At the same dose range in the mouse, there was an increase in cleft palate.

Kato, T. and Kitagawa, S.: Effects of a new antibacterial sulfonamide (CS–61) on mouse and rat fetuses. Toxicol. Appl. Pharmacol. 27:20–27, 1974.

2256 Sulfuric Acid Aerosol CAS 7664–93–9

John et al. (1979) exposed mice and rabbits to up to 20 mg per square meter of air for 7 hours daily during major organogenesis and found no teratogenic effects.

Murray, F.J.; Schwetz, B.A.; Nitschke, K.D.; Crawford, A.A.; Quast, J.F. and Staples, R.E.: Embryotoxicity of inhaled sulfuric acid aerosol in mice and rabbits. J. Environ. Sci. Health C13:251–266, 1979.

2257 Sulfuryl Fluoride

Rabbits and rats were exposed during organogenesis to concentrations of up to 225 ppm for 6 hours daily. No evidence of teratogenicity was found but at the highest concentration in the rabbit the weights of the dams and fetuses were reduced.

Hanley, T.R., Jr.; Calhoun, L.L.; Kociba, R.J.; Greene, J.A.: The effects of inhalation exposure to sulfuryl fluoride on fetal development in rats and rabbits. Fundamental and Applied Toxicology. 13:79–86, 1989.

2258 Sulindac

Lione and Scialli (**1995**) reviewed the use and toxicity of this derivative of indomethacin. There is a smaller risk of gastrointestinal bleeding since this drug is not activated until after absorption. Momma and Takeuchi (**1985**) found that usual clinical doses caused constriction of the rat fetal ductus arteriosus.

Lione, A. and Scialli, A.R.: The developmental toxicity of indomethacin and sulindac. Reproductive Toxicology 9:7–20, 1995.

Momma, K. and Takeuchi, H.: Constriction of fetal ductus arteriosus by non–steroidal anti–inflammatory drugs. Prostaglandins 26:631–643, 1983b.

2259 Suloctidil CAS 54063–56–8

Ikeda et al. (1983) gave this drug to rats orally before fertilization, during organogenesis and in the perinatal periods. The maximum dose of 100 mg per kg produced no adverse reproductive effects.

Ikeda, Y.; Sukegawa, J. and Fujii, O.: Reproduction studies of suloctidil in rats. Iyakuhin Kenkyu 14:743–769, 1983.

2260 Sulphonamides *Sulfamethazine N–Sulphasalazine N–Sulphanilyacetamide Sulphapyridine*

There is no conclusive evidence that these compounds are teratogenic in man (Smithells, 1966). Heinonen et al. (1977) reported no increase in defects among 1,400 pregnancies exposed during the first four months. At four years, no IQ differences were found.

Landauer and Wakasugi (1968) summarized their extensive work with sulphonamides as teratogens in the chick embryo. Using acetazolamide, dichlorphenamide and methazolamide with doses of 0.5 to 2.0 mg given at 96 hours incubation they produced short upper beaks and syndactylism. p-Sulphamoylbenzoic acid was non–teratogenic.

For a more detailed description of the very specific teratogenic lesions found in mammals after administration of the strong carbonic anhydrase inhibiting sulfonamides, reference should be made to the entry under the heading, Acetazolamide.

Various sulphanilamides with bacteriostatic properties were shown by Ancel (1945) to be teratogenic in the chick. He showed that sulphanilamide, sulphapyridine and N–sulphanilylacetamide produced a syndrome of micromelia, syndactylism and parrot beak. Doses of 1 mg per egg at 96 hours of incubation were used by Landauer and Clark (1964) to produce the syndrome. They concluded that these agents interfered with purine synthesis or with functions of NAD–linked dehydrogenases, resulting in local deficiency of folic acid.

Paget and Thorpe (1964) reported that sulfadimethoxy pyrimidine (sulphamoprine) treatment led to skeletal and dental defects in rats. Bertazzoli et al. (1965) reported that sulfamoprine in doses of 50 mg per kg daily from the 11th through the 20th gestational days was associated with maloccluded incisors in the weaned rat offspring. Kalter and Warkany (1959) listed a number of references to studies where large doses of sulfonamides given to pregnant animals caused only fetal resorption. Using 8 separate sulfonamides, Kato and Kitagawa (1973a,b) administered orally 500–1,000 mg per kg per day to mice on the 7th to the 12th days and to rats on the 9th through the 14th days of gestation. In both species, cleft palate was produced by sulfamonomethoxine, sulfamethoxine, sulfamethomidine, sulfamethoxypyridazine, sulfisoxazole and sulfadiazine. Sulfanilamide and sulfisomidine did not produce defects. Using sulfamethopyrazine Suzuki et al. (1973) produced an increase in cleft palates in rat and mouse fetuses by giving mice 1,000 or 2,000 mg per kg per day of organogenesis and had similar results in the rat at 700 mg per kg. At 685 and 865 mg of sulfamethazine per kg on days 6–15, Wolkowski–Tyl et al. (1982) found cleft palate, hydroureter and hydronephrosis in rat fetuses.

Ancel, P.: L'achondroplasie. Sa realisation experimentale: Sa pathogene. Ann. Endocrinol. 6:1–24, 1945.

Bertazzoli, C.; Chieli, T. and Grandi, M.: Absence of tooth malformation in the offspring of rats treated with a long–acting sulphonamide. Experientia 21:151–152, 1965.

Kalter, H. and Warkany, J.: Experimental production of congenital malformations in mammals by metabolic procedure. Physiol. Rev. 39:69–115, 1959.

Kato, T. and Kitagawa, S.: Production of congenital anomalies in fetuses of rats and mice with various sulfonamides. Cong. Anom. 13:7–15, 1973a.

Kato, T. and Kitagawa, S.: Production of congenital skeletal anomalies in fetuses of pregnant rats and mice treated with various sulfonamides. Cong. Anom. 13:17–23, 1973b.

Landauer, W. and Clark, E.M.: On the teratogenic nature of sulfanilamide and 3–acetylpyridine in chick development. J. Exp. Zool. 156:313–322, 1964.

Landauer, W. and Wakasugi, N.: Teratological studies with sulfonamides. J. Embryol. Exp. Morphol. 20:261–284,

1968.

Paget, G.E. and Thorpe, E.: A teratogenic effect of sulphonamide in experimental animals. Br. J. Pharmacol. 23:305–312, 1964.

Heinonen, O.P.; Slone, D. and Shapiro, S.: Birth Defects and Drugs in Pregnancy. Publishing Sciences Group, Inc., Littleton, Mass., 1977.

Smithells, R.W.: Drugs and human malformations. In: Advances in Teratology. D.H.M. Woollam (ed.), New York: Academic Press, 1:251–278, 1966.

Suzuki, Y.; Wakita, Y.; Kondo, S.; Okada, F.; Suzuki, I.; Asano, F.; Matsuo, M. and Chiba, T.: Effects of sulfamethopyrazine administered to pregnant animals upon the development of their fetuses and neonates. Oyo Yakuri 7:1005–1019, 1973.

Wolkowski–Tyl, R.; Jones–Price, C.; Kimmel, C.A.; Ledoux, T.A.; Reel, J.R. and Langhoff–Paschke, L.: Teratologic evaluation of sulfamethazine in CD rats. (abs) Teratology 25:81A–82A, 1982.

2261 Sulpiride

Tuchmann–Duplessis (**1975**) administered 30 mg per kg parenterally from day 15 until delivery. No effect on delivery or sexual development of the offspring was found.

Tuchmann–Duplessis, H.: Influence of neuro–drugs on prenatal development. New Approaches to the Evaluation of Abnormal Development. Eds. D. Neubert and M.J. Merker, Thieme Stuttgard, 1975, pp 716–727.

2262 Sultopride CAS 53583–79–2

Inoue et al. (1984) studied this neuroleptic drug in mice, rats and rabbits. Oral doses which were up to 300 mg per kg in the rat and 600 mg per kg in the mouse did not produce malformation increases. Some resorptions and decreased fetal weights were seen at the higher doses.

Inoue, H.; Kawaguchi, Y.; Hayashi, T.; Takayama, S.; Niokawa, H.; Kaji, M.; Sato, K.; Genra, Y. and Yokoyama, Y.: Reproductive studies of sultopride hydrochloride in rats, mice and rabbits. Oyo Yakuri 28:663–685, 1984.

2263 Sumatriptan Succinate

3–[2–Dimethyl(amine)ethyl]–N–methyl–indol–5–methanesulphonamide Monosuccinate]

Ezaki et al. (**1993**) gave this anti–migraine drug to rats orally in doses of up to 1000 mg per kg during organogenesis and found no adverse fetal effects.

Ezaki, H.; Utusumi, K.; Hirata, M. and Tokado, H.: Reproductive study (seg II) on sumatriptan succinate in rats by oral route. Yakuri to Chiryo 21:2071–2091, 1993.

2264 Supidimide *CG–3033 EM–87*

This structural analog of thalidomide was tested in pregnant baboons (Hendrickx and Helm, 1980) and monkeys (Scott et al., 1980) at up to 40 and 15 mg per kg, respectively. Treatment was during organogenesis and no defects were produced.

Hendrickx, A.G. and Helm, F.CH.: Nonteratogenicity of a structural analog of thalidomide in pregnant baboons (Papio cynocephalus). Teratology 22:179–182, 1980.

Scott, W.J.; Wilson, J.G. and Helm, F.CH.: A metabolite of a structural analog of thalidomide lacks teratogenic effect in pregnant rhesus monkeys. Teratology 22:183–185, 1980.

2265 Suplatast Tosilate *IPD–1151T*

Aso et al. (**1992**) studied this anti–allergic medication in rats before and during early pregnancy at up to 1600 mg orally and found maternal and fetal toxicity at the highest doses. Yamakita et al. (**1992** A) gave up to 2700 mg per kg during rat organogenesis and found maternal and fetal toxicity above 900 mg per kg but no teratogenicity or postnatal behavioral changes. No treatment–related effects of perinatal dosing were found (Aso et al, **1992**). Oral administration to rabbits during organogenesis failed to produce teratogenicity (Yamakita et al., **1992** B).

Aso, S.; Sueta, S.; Kajiwara, Y.; Morita, S.; Horiwaki, S. and Yamakita, O.: Reproductive and developmental toxicity study of suplatast tosilate (IPD–1151T) (1): Fertility study in rats by oral administration. J. Tox. Sc. 17:141–154, 1992.

Aso, S.; Kajiwara, Y.; Anai, M.; Sueta, S.; Horiwaki, S. and Yamakita, O.: Reproductive and developmental toxicity study of suplatast tosilate (IPD–1151T) (4): Perinatal and postnatal study in rats by oral administration. J. Tox. Sc. 17:187–205, 1992.

Yamakita, O.; Shinomiya, M.; Koida, M.; Katayama, S.; Ikebuchi, K. and Yoshida, R.: Reproductive and developmental toxicity study of suplatast tosilate (IPD–1151T) (2): Teratological study in rats by orad administration J. Tox. Sc. 17:155–174, 1992.

Yamakita, O.; Shinomiya, M.; Kukokawa, M.; Koida, M.; Mizutani, T.; Nakagawa, M.; Manabe, H. and Sugimoto, S.: Reproductive and developmental toxicity study of suplatast tosilate (IPD–1151T) (3): Teratological study in rabbits by oral administration. J. Tox. Sc. 17:175–185, 1992.

2266 Suprofen *2–(p–(2–Thenoyl)phenyl)proprionic Acid* CAS 40828–46–4

Fukimura et al. (1983) studied this anti–inflammatory in pregnant rats and rabbits. In rats given 24 mg per kg daily before fertilization and during the first 7 days, during days 7–17 of gestation and in the perinatal period, the only adverse effect was a suppression of fetal weight. Delayed fetal ossification was found in the rabbit fetus at 200 mg per kg given during organogenesis.

Fujimura, H.; Hiramatsu, Y.; Tamura, Y. and Kobuka, S.: Reproduction studies of 2–[p–(2–thenoyl)phenyl] proprionic acid (Suprofen) in rats and rabbits. Oyo Yakuri 26:441–459 and 523–442, 1983.

2267 Surfactants *See Also Under Aklylbenzene Sulfonate*

Palmer et al. (1975) studied the surfactants linear alkylbenzene sulphonate (LAS), alcohol sulphate (AS) and the commercial light duty liquid detergent containing 17 percent LAS and 7 percent dodecyl ethoxy sulphate. Oral doses were given to mice, rats and rabbits during organogenesis. At the higher doses, 600 to 2400 mg per kg for LAS and AS, respectively, maternal toxicity occurred but no significant fetotoxicity or teratogenicity. The commercial liquid detergent was similarly not teratogenic although at maternal lethal doses fetal loss also occurred.

Palmer, A.K.; Readshaw, M.A. and Neuff, A.M.: Assessment of the teratogenic potential of surfactants. part 1: LAS, AS, CLD. Toxicology 3:91–106, 1975.

2268 Suxibuzone CAS 27470–51–5

This anti–inflammatory compound was studied in pregnant mice, rats and rabbits by Yoshida et al. (1980). In the rat at oral doses of 284 mg per kg, no adverse effects on fertility were found. Decreased fetal weight was found with doses of 142 mg per kg or above. At 284 mg per kg resorptions were increased. No external defects were found. In the mouse, a dose of 800 mg per kg was associated with decreased body weight and delayed ossification in the fetuses. Phenylbutazone was also studied in both rats and mice. In the rabbit, Yoshida et al. (1980) found an increase in resorptions at 36 mg per kg orally and at 142 mg decreased implantation was found.

Yoshida, R.; Asanoma, K.; Kurokawa, M. and Morita, K.: Reproductive studies of suxibuzone (4–butyl–4)β–carboxypropionyl–oxymethyl–1,2–diphenyl–3, 5–pyrazolidinedione. Oyo Yakuri 20:281–288, 289–298, 377–386, 387–392, 1980.

2269 SV–40 Virus

Farwell et al. (1980) reported an increase in central nervous system tumors among the offspring of mothers given poliomyelitis vaccine contaminated with SV–40 virus between 1955 and 1961. Of 15 children with meduloblastoma, 10 had been exposed to the contaminated vaccine and this was a statistically significant increase over matched controls. Suggestive association was made for gliomas. Heinonen et al. (1973,1977) listed 8 CNS tumors in the offspring of polio vaccinated mothers and this was 17.9 times the hospital standardized relative risk.

Farwell, J.W.; Dohrmann, G.J.; Marrett, L.D. and Meigs, J.W.: Effect of SV–40 virus–contaminated polio vaccine on the incidence and type of CNS neoplasm in children. A population–based study. Trans. Am. Neurolog. Ass. 104:1–4, 1980.

Heinonen, O.P.; Slone, D. and Shapiro, S.: Birth Defects and Drugs in Pregnancy. Publishing Sciences Group, Inc., Littleton, Mass., 1977.

Heinonen, O.P.; Shapiro, S.; Monson, R.R.; Hartz, S.C.; Rosenberg, L. and Slone, D.: Immunization during pregnancy against poliomyelitis and influenza in relation to childhood malignancy. Int. J. Epidemiol. 2:229–235, 1973.

2270 Swaddling

Swaddling of young infants has been shown to be related to congenital dislocation of the hips. A higher incidence of the defect occurring in infants born during the colder months was noted by Record and Edwards (1958) who postulated that during the colder months the infant was exposed to more swaddling which held the hips in adduction predisposing toward dislocation. The use of craddle boards by Indians has the same effect (Salter, 1968).

Record, R.G. and Edwards, J.H.: Environmental influences related to the aetiology of congenital dislocation of the hip. Br. J. Prev. Soc. Med. 12:8–12, 1958.

Salter, R.B.: Etiology, pathogenesis and possible prevention of congenital dislocation of the hip. Can. Med. Assoc. J. 98:933–945, 1968.

2271 Swine Fever Virus

Harding et al. (1966) produced some circumstantial evidence that swine fever led to cerebellar hypoplasia and spinal hypomyelinogenesis in newborn piglets. An association between the congenital defect and the presence of immunity to the agent was made in 32 of 33 herds studied.

Harding, J.D.J.; Done, J.T. and Darbyshire, J.H.: Congenital tremors in piglets and their relation to swine fever. Vet. Rec. 79:388–390, 1966.

2272 Syndactyly from Transgenic Insertion

Overbeek et al. (1986) introduced a chimeric gene containing Rous sarcoma virus linked with bacterial chloramphenicol acetyltransferase into mouse ova pronuclei and produced two strains of mice, one with dominant embryonic lethality and the other with a recessive trait of fused toes on all four feet.

Overbeek, P.A.; Lai, S–P.; Van Quill, K.R. and Westphal, H.: Tissue–specific expression in transgenic mice of a fused gene containing RSV terminal sequences. Science 231:1574–1577, 1986.

2273 Syphilis

The causative organism, Treponema pallidum, can be transmitted transplacentally and produces characteristic pathologic findings in the human fetus (Ingall and Norris, 1976; Grossman, 1977). Although Halter and Benirschke (1976) demonstrated the organisms in abortus material of less than 12 weeks, the pathologic picture does not appear before the 4th to 5th month of gestation, and may evolve primarily as a result of maturation of the fetal immune system. The infection spreads hematogenously from the placenta which is characteristically large and pale. The visceral and skin lesions resemble those seen in the postnatally acquired disease.

Grossman, J.: Congenital syphilis. Teratology 16:217–224, 1977.

Halter, C.A. and Benirschke, K.: Fetal syphilis in the first trimester. Am. J. Obstet. Gynecol. 124:705–711, 1976.

Ingall, D. and Norris, L.: Syphilis, Infectious Diseases of the Fetus and Newborn Infant, Chapter 9. J.S. Remington and J.O. Klein, (eds), Saunders Co., Philadelphia, 414–463, 1976.

2274 TAT–3–hydrochloride (N–(2–picolyl)N–phenyl–N–2–piperidinethyl)amine chloride)

Misutani et al. (1970) gave this antitussive to mice and rats in doses of up to 120 mg per kg daily during organogenesis and found no adverse effects.

Mizutani, M.; Ihara, T.; Kanamori, H.; Takatani, O. and Kaziwara, K.: The effect of TAT–3–Hydrochloride upon the development of the mouse and rat. J. Takeda Res. Lab 29:297–309, 1970.

2275 T–2 Toxin

This mold metabolite of Fusarium tricinetum given in amounts of 0.5 mg per kg intraperitoneally to mice on days 8 or 10, produced tail and limb abnormalities (Hood et al., 1978). Rousseaux and Schiefer (1987) confirmed the above findings and studied the effect of giving the toxin on single days to mice.

Hood, R.D.; Kuczuk, M.H. and Szczech, G.M.: Effects in mice of simultaneous prenatal exposure to ochratoxin A and T–2 toxin. Teratology 17:25–30, 1978.

Rousseaux, C.G. and Schiefer, H.B.: Maternal toxicity, embryolethality and abnormal fetal development in CD–l mice following one oral dose of T–2 toxin. J. Appl. Toxicol. 7:281–288, 1987.

2276 Tacrolimus FK5065

Saegusa et al. (1992) studied this macrolide antibiotic in pregnant rats and rabbits using oral doses of up to 3.2 mg per kg. No teratogenicity was found. Fewer implantations occurred in rats treated before and during early pregnancy. Abortion was increased in rabbits at 0.32 mg per kg and above.

Saegusa, T.; Ohara, K. and Noguchi, H.: Reproductive and developmental studies of tacrolimus (FK506) in rats and rabbits.

2277 Talinolol CAS 57460–41–0

Wendler and Schmidt (1975) gave one–half the LD–50 of this β adrenergic blocker to rats on days 6 through 14 and found no indication of adverse reproductive or fetal effects.

Wendler, D. and Schmidt, W.: Talinolol (Cordanum, 02–115) im teratologischen Versuch. Pharmazie 30:669–671, 1975.

2278 Talipexole Dihydrochloride

Matsuo et al. (1993) treated pregnant rats before fetilization and during early pregnancy, during organogenesis and in the perinatal period. Oral doses of up to 2.5 mg per kg were used. Maternal toxicity occurred at the higher doses as well as some fetal growth decrease. No teratogenicity or behavioral changes were found. Rabbits received up to 10 mg per kg during organogenesis and no ill effects found in the fetuses.

Matsuo, A.; Niggeschulze, A. and Kohei, H.: Oral reproductive and developmental toxicology of talipexole dihydrochloride. Oyo Yakuri/Pharmacometrics 46:9–27, 1993.

2279 Tamoxifen CAS 54965–24–1; CAS 10540–29–1

Ruiz–Velasco et al. (1979) reported no defects in the offspring of nine women whose ovulation was induced with tamoxifen. Five of 14 women so treated spontaneously aborted. This antineoplastic agent was given orally to rabbits in doses of 0.25 to 4.0 mg per kg daily (Esaki and Sakai, 1980). A high fetal loss was found at 4 mg per kg. Some fetal death increase was found at 0.5 and 2.0 mg per kg but no defect increase occurred with treatment on days 6–18. Iguchi et al. (1986) found various abnormalities in gonad and genitourinary tracts of female mice given 2 to 100 micrograms starting at birth. Follicular degeneration in the ovary and vaginal adenosis was found.

Esaki, K. and Sakai, Y.: Influence of oral administration of tamoxifen in the rabbit fetus. Preclin. Rep. Cent. Inst. Exp. Animal 6:217–238, 1980.

Iguchi, T.; Hirokawa, M. and Takasugi, N.: Occurrence of genital tract abnormalities and bladder hernia in female mice exposed neonatally to tamoxifen. Toxicology, 42:1–11, 1986.

Ruiz–Valesco, V. Rosas–Areco, J. and Matute, M.M.: Chemical inducers of ovulation: comparative results. Int. J. Fert. 24:61–64, 1979.

2280 Tandospirone Citrate SM–3997

Kannan et al. (1992) gave this 5–HT receptor agonist to rats and rabbits in oral doses of up to 80 mg per kg. Fertility and implantation were affected in the rat at 50 mg per kg. No adverse effects on development of the rat fetuses was found although postnatally the weight gain of the pups was decreased. In the studies of rabbits the fetal weight was reduced after 150 mg per kg during organogenesis but no teratogenicity was found.

Kanna, N.; Matsumoto, Y.; Okamoto, T.; Koda, A.; Kato, T. and Yamada, H.: Reproductive and developmental toxicity studies of SM–3997 in rats and rabbits. The Clinical Report 26:1803–1823, 1992.

2281 Taurine CAS 107–35–7

Takahashi et al. (1972) gave 4 gm per kg orally for 7 days starting on the 7th day in the pregnant mouse and found no alterations in the treated fetuses. Yamada et al. (1981) administered this compound orally to rats on days 7–17 in amounts of up to 3000 mg per kg and reported no adverse fetal effects.

Takahashi, H.; Kaneda, S.; Fukuda, K.; Fujihira, E. and Nakazawa, M.: Studies on the teratology and three generation reproduction of taurine in mice. Oyo Yakuri 6:535–540, 1972.

Yamada, T.; Nogariya, T.; Nakane, S. and Sasajima, M.: Reproductive studies of taurin. Kiso to Rinsho 15:4229–4240, 1981.

2282 Taxol

Scialli et al. (1992) found cardiovascular hemorrhages and decreased eye pigment in chick embryos exposed to 14 mmole and above.

Scialli, A.R.; DeSesso, J.M. and Goeringer, G.C.: Taxol toxicity in the developing chick (A). Teratology 45:478, 1992.

2283 Tazanolast *Butyl 3'(1 H–tetrazol–5–yl) oxanilate*

Morita et al. (**1989**) gave this antiallergic orally to rats on days 7–17 in amounts of up to 2100 mg per kg and found no adverse effects on the fetuses or offspring.

Morita, Y.; Kamiya, Y.; Mizurani, M.; Edanami, K.; Tsuruya, M.; Shimamura, K. and Yanakawa, H.: Teratological study of tazanolast (wp–833) in rats. Kiso to Rinsho 23:265–281, 1989.

2284 Tazobactam

Lochry et al. (**1991** gave rats intraperitoneally up to 320 mg per kg and found maternal and fetal toxicity at the highest dose. Fertility, organogenesis and postnatal studies did not reveal any unique reproductive toxicity.

Lochry, E.A.; Hoberman, A.M.; Filler, R.; Dougherty, W.J. and Traitor, C.E.: Fertility and general reproduction study of tazobactam administered intraperitoneally to crl:CD(D)BR rats alone or in combination with piperacillin. (Abs) Teratology 43:428, 1991.

2285 TD–2061

Komai et al. (**1989**) treated rats intravenously with up to 10 mg per kg during organogenesis with this tissue type plaminogen activator. No adverse fetal effects were found.

Komai, Y.; Ogura, H.; Hattori, M.; Inoue, S.; Kamada, K.; Isowa, K.; Ishimura, K. and Watanabe, T.: Reproduction study of td–2061 (ii)–teratogenicity study in rats by intravenous administration. Prog. Med. 9:421–435, 1989.

2286 TE–031 *A–56268*

Yamada et al. (1988) gave rabbits up to 125 mg per kg orally on days 6–18 and found no teratogenicity or fetotoxicity. Yamada et al. (**1988**) gave rats orally up to 160 mg per kg during organogenesis. This antibiotic was associated with increased fetal death at the highest dose but no teratogenicity was found and postnatal development was normal. Rabbits treated with up to 125 mg per kg on days 6–18 had normal offspring.

Yamada, T.; Nishiyama, T.; Ohno, H. and Nakane, S.: Reproduction studies of TE–031 (A–56268). (III) Administration study during the period of fetal organogenesis in rabbits. Chemotherapy 36:362–369, 1988.

Yamada, T.; Uchida, H.; Matsuzawa, N. and Nakane, S.: Reproduction studies of te–031(a–56268)(i)–fertility study in rats. Chemotherapy 36:334–344, 1988.

2287 Tecoram CAS 5836–23–7

Van Steenis and Van Logten (1971) inoculated chick eggs with 0.01 to 10 mg of the dithiocarbamate of tecoram on day 7 of incubation. With doses over 1.0 mg paralysis, shortening of extremities and muscular atrophy was found at the 19th day.

Van Steenis, G. and Van Logten, M.J.: Neurotoxic effect of dithiocarbamate tecoram on the chick embryo. Toxicol. Appl. Pharmacol. 19:675–686, 1971.

2288 Tegafur–Uracil *UFT* CAS 74578–38–4

Tegafur(1–(2–tetrahydrofuryl)5–fluorouracil) was mixed in a 1:4 molar ratio with uracil to produce an antineoplastic agent. Asanoma et al. (1980,1981) gave it orally to rabbits on days 6–18 of gestation in amounts of up to 12.96 mg per kg daily and found no significant fetal differences. At 81 mg orally in the rat during days 7–17 some fetal weight decrease, increase in resorptions and delay in fetal ossification was found. Postnatal function was not affected. Pre–mating dosing of the males and females did not alter fertility.

Asanoma, K.; Matsubara, T. and Morita, K.: Effect of UFT on reproduction. Oyo Yakuri 20:1001–1007, 1980; 22:85–107, 109–129, 1981.

2289 TEI–5103

Matsuzawa et al. (1988a, b) studied this anti–ulcer agent in rats. At 2000 mg per kg given before fertilization, no adverse effects were seen. Rats were treated orally on days 7–17 with up to 3000 mg per kg and no adverse fetal effects were seen. Perinatal studies (Sugawara et al., 1988) were done using 3000 mg per kg and no adverse effects were seen. Matsuzawa et al. (1988c) found no adverse effects of up to 900 mg per kg given to rabbits on days 6–18.

Matsuzawa, K.; Enjo, H. and Makita, T.: Toxicity studies of TEI–5103 (6). Fertility study of TEI–5103 in rats. Jpn. Pharmacol. Ther. 16:217–226, 1988a.

Matsuzawa, K.; Sugawara, S.; Enjo, H.; Kunii, M. and Makita, T.: Toxicity studies of TEI–5103 (7). Teratogenicity study of TEI–5103 in rats. Jpn. Pharmacol. Ther. 16:227–243, 1988b.

Matsuzawa, K.; Enjo, H.; Nishizawa, S. and Makita, T.: Toxicity studies of TEI–5103 (8). Teratogenicity study of TEI–5103 in rabbits. Jpn. Pharmacol. Ther. 16:245–253, 1988c.

Sugawara, S.; Matsuzawa, K.; Enjo, H.; Kunii, M. and Makita, T.: Toxicity studies of TEI–5103 (9). Perinatal and postnatal study of TEI–5103 in rats. Jpn. Pharmacol. Ther. 16:255–272, 1988.

2290 Tellurium CAS 13494–80–9

Garro and Pentschew (1964) fed pregnant rats diets containing 500 to 2,500 parts per million and observed a dose–related increase in postnatal hydrocephalus. Agnew and Curry (1972a) found the most vulnerable time for induction of hydrocephalus was day 9 and 10 in the rat and reported on the distribution of tellurium–127M in the pregnant rat (1972b). Duckett (1971) fed a diet with 3000 ppm throughout the rat gestation period. Although communicating hydrocephalus was present at birth, it became converted to an obstructive type within a few days. Of 237 newborn rats, 179 developed hydrocephalus within three days. Duckett (1971) presented evidence that the element was present in fetal brain tissue.

Perez–D'Gregorio and Miller (1988) gave 0–1,000 microM per kg subcutaneously to rats on days 15–19 and produced hydrocephalus, eye defects, small kidneys and other defects. Johnson et al. (1988) fed up to 5,000 ppm to rats and 5,250 ppm to rabbits during organogenesis and at the highest levels found increased fetal delay in ossification and malformations. The primary effect in the rat was dilation of the lateral ventricles. In both species, maternal toxicity occurred concurrent with fetal changes.

Agnew, W.F. and Curry, E.: Period of teratogenic vulnerability of rat embryo to induction of hydrocephalus by tellurium. Experientia 2:1444–1445, 1972a.

Agnew, W.F.: Transplacental uptake of tellurium–127M studied in whole–body radioautography. Teratology 6:331–338, 1972b.

Duckett, S.: The morphology of tellurium–induced hydrocephalus. Exp. Neurol. 31:1–16, 1971.

Garro, F. and Pentschew, A.: Neonatal hydrocephalus in the offspring of rats fed during pregnancy non–toxic amounts of tellurium. Arch. Psychiat. Neurol. 206:272–280, 1964.

Johnson, E.M.; Christian, M.S.; Hoberman, A.M.; DeMarco, C.J.; Kilpper, R. and Mermelstein, R.: Developmental toxicology investigation of tellurium. Fund. Appl. Toxicol. 11:691–702, 1988.

Perez–D'Gregorio, R.E. and Miller, R.K.: Teratogenicity of tellurium dioxide: Prenatal assessment. Teratology 37:307–316, 1988.

2291 Temafloxacin

Tarantal et al. (**1990**) gavaged moneys with 25, 50 or 100 mg per kg on days 20–50. At 100 mg per kg there was maternal toxicity. No adverse fetal effects were found but one fetus of the 30 treated had microphthalmia.

Tarantal, A.F.; Lasley, S.B. and Hendrickx, A.G.: Developmental toxicity of temafloxacin hydrochloride in the long–tailed macaque (macaca fascicularis). Teratology 42:233–242, 1990.

2292 Temocapril Hydrochloride

Takahashi et al. (**1991**) treated rats orally with this ACE inhibitor during organogenesis with doses of 5, 50 and 500 mg per kg. Maternal weight gain was suppressed at 500 mg per kg. Postnatal weight gain was decreased and survival rate on postnatal day 4 was decreased. At 50 mg per kg or more an increase in bloody pericardial cavities was seen. Tanase et al. (**1991**) gave rabbits 0.2, 1 or 5 mg per kg during organogenesis. Many of the dams died at the highest dose and there were increased abortions. No malformation or skeletal variation increases were found in the fetuses. Takahashi et al. (**1993** A) studied this ACE inhibitor in rats before fertilization and 7 days into pregnancy. Oral doses of up to 100 mg per kg caused no fetal or reproductive adverse effects. Takahashi et al. (**1993** B) gave up to 500 mg pe kg perinatally and postnatally. At the highest dose one third of the pups died with dilated renal pelves and urine–filled bladders. At 500 mg per kg the food intake of the pups was decreased.

Takahashi, M.; Kashima, M. and Hiraide, Y.: Reproductive and developmental toxicity study on temocapril hy-drochloride (CS–622) administered during the period of fetal organogenesis in rats. Jpn. Pharm. Therap. 19:3955–3971, 1991.

Takashi, M.; Sakurai, T.; Karasawa, N. and Hiraide, Y.: Reproductive and developmental toxicity study on temocapril hydrochloride (CS–622) administered prior to and in the early stages of pregnancy in rats. Yakuri to Chiryo 21:2523–2530, 1993A.

Takahashi, M.; Sakurai, T.; Nakasone, K. and Hiraide. Y.: Reproductive and developmental toxicity study of temocapril hydrochloride (CS–622) administered during the perinatal and lactation periods in rats. Yakuri to Chiryo 21:2531–2545, 1993B.

Tanase, H.; Asai, M. and Hirose, K.: Reproductive and developmental toxicity study on temocapril hydrychloride (CS–622) administered during the period of fetal organogenesis in rabbits. Jpn. Pharm. Therap. 19:3947–3953, 1991.

2293 Teniposide VM–26

This podophyllotoxin derivative was injected intraperitoneally into mice in amounts of 0.5 and 1.0 mg per kg on days 6, 7 or 8. Numerous malformations were found but dextrocardia, exencephaly and skeletal defects were most common (Sieber et al., 1978).

Sieber, S.M.; Whang–Peng, J.; Botkin, C. and Knutsen, J.: Teratogenic and cytogenic effects of some plant–derived antitumor agents (vincristine, colchicine, maytansine, VP–16–213 and VM–16) in mice. Teratology 18:31–48, 1978.

2294 Tenoxicam CAS 59804–37–4

Shimizu et al. (1984A,B,C) and Shiozaki et al. (1984) studied this drug in mice, rats and rabbits using maximum doses of 4.8, 12.0 and 32 mg per kg. The mouse study was on days 6–15 of gestation, the rat studies on days 7–17 and 17–21 and the rabbit study on days 6–18. Embryolethality was increased in the rabbit at 32 mg per kg and neonatal deaths increased in the rat at 0.5 mg per kg. No malformations were increased.

Shimizu, M.; Honma, M.; Takahashi, M. and Udaka, K.: Toxicity study of tenoxicam: Reproduction segment 2 study in mice. Yakuri to Chiryo 12:873–889, 1984A.

Shimizu, M.; Sato, C.; Inagaki, M.; Sato, M.; Noda, K. and Udaka, K.: Toxicity study of tenoxicam: Reproduction segment 3 study in rats. Yakuri to Chiryo 12:901–914, 1984B.

Shimizu, M.; Sato, C.; Inagaki, M.; Sato, M.; Noda, K. and Udaka, N.: Toxicity study of tenoxicam: Reproduction segment 2 study in rats. Yakuri to Chiryo 12:853–871, 1984C.

Shiozaki, H.; Nakagama, S.; Noda, K. and Udaka, K.: Toxicity study of tenoxicam: Reproduction segment 2 study in rabbits. Yakuri to Chiryo 12:891–900, 1984.

2295 Tenuazonic Acid CAS 610–88–8

This metabolite of alternaria tenus inhibits tumor growth but has little growth retarding effect in the chick embryo (Gitterman et al., 1964).

Gitterman, C.O.; Dulaney, E.L.; Kaczka, E.A.; Campbell, G.W.; Hendlin, D. and Woodruff, H.B.: The human tumor–egg host system. 3. Tumor–inhibitory properties of tenuazonic acid. Cancer Res. 24:440–443, 1964.

2296 Terbinafine

Watanabe et a; (**1990**) gave rats this antifungal orally on days 7–17 in amounts up to 260 mg per kg and found no adverse fetal effects. Watanabe et al. (**1993**) treated rabbits during organogenesis with oral doses of up to 200 mg per kg and found no teratogenesis although at the highest dose some of the dams did not gain weight normally.

Watanabe, M.; Miyashita, T.; Amano, N.; Asada, Y. and Nakazima, Y.: Teratogenicity study of terbinafine hydrochloridde in rats. Kiso to Rinsho 25:7925–7936, 1990.

Watanabe, M.; Amano, N. and Asada, Y.: Teratogenicity study of terbinafine hydrochloride in rabbits. The Clinical Report 27:5851–5861, 1993.

2297 Terbutaline

This adrenergic drug used as a tocolytic agent can cause fetal tachycardia. Hypoglycemia in the neonatal period has also been associated. Briggs et al. (1986) were unable to find long–term follow–up of exposed newborns.

Briggs, G.G.; Freeman, R.K. and Yaffe, S.J.: Drugs in Pregnancy and Lactation, 2nd ed., Williams and Wilkins, Baltimore, 422–423, 1986.

2298 Terfenadine *Seldane*[TM]

Schick et al. (**1994**) examined 134 offspring of women taking this antihistamine. No significant differences in outcome were found in comparison to a control group. This antihistamine was studied by Gibson et al. (1982) in the rat on days 7–16 at doses of up to 300 mg per kg daily and in the rabbit at doses of 500 mg per kg on days 7–19. No adverse fetal effects occurred.

Gibson, J.P.; Huffman, K.W. and Newborne, J.W.: Preclinical safety studies with terfenadine. Arzneimittelforschung 22:1179–1184, 1982.

Schick, B.; Hom, M.; Librizzi, R; Arnon, J. and Donnenfeld, A.: Terfenadine (Seldane)m exposure in early pregnancy. (abs) Teratology 49:417, 1994.

2299 Terguride

Kodama et al. (**1993** A) gave this anti–hyperprolactinemia drug orally to female rats in doses of up to 1.0 mg per kg before pregnancy and found no decrease in fertility. Kodama et al. (**1993** B) gave up to 0.1 mg per kg during the first 7 days of gestation and found inhibition of implantation at the highest dose. Kodama (**1993** C) in studies carried out during organogenesis found normal development and postnatal behavior of the pups born after exposure to 0.01 mg per kg. By starting on day 8 Kodama et al (**1993** D) were able to give up to 10 mg per kg without fetal ill effects. Perinatal and postnatal studies Kodama et al. (**1993** E) found decreased physical growth but behavior and reproduction of the pups were normal. Cross fostering suggested that the effects were due to decreased lactation (Kodama et al., **1993** F).

Kodama, N.; Kato, K.; Urabe, K. and Kageyama, A.: Toxicity study of terguride; fertility study in female rats. Yakuri to Chiryo 21:379–383, 1993A.

Kodama, N.; Kato, K.; Urabe, K. and Kageyama, A.: Toxicity study of terguride; preimplantation reproduction study in female rats. Yakuri to Chiryo 21:385–390, 1993B.

Kodama, N.; Kato, K.; Urabe, K. and Kageyama, A.: Toxicity study of terguride: teratogenicity study in rats. Yakuri to Chiryo 21:391–406, 1993C.

Kodama, N.; Kato, K.; Urabe, K. and Kageyama, A.: Toxicity study of terguride: teratogenicity study in rats. Yakuri to Chiryo 21:407–412, 1993D.

Kodama, N.; Kato, K.; Urabe, K.; and Kageyama, A.: Toxicity study of terguride: perinatal and postnatal study in rats. Yakuri to Chiryo 21:413–426, 1993E.

Kodama, N.; Kato, K.; Urabe, K. and Kageyama. A.: Toxicity study of terguride: foster mother study in rats. Yakuri to Chiryo 21:427–439, 1993F.

2300 Terpin Hydrate CAS 2451–09–8 CAS 80–53–5

Congenital defects were not increased among 1,762 women who took this medication during pregnancy (Heinonen et al., 1977).

Heinonen, O.P.; Slone, D. and Shapiro, S.: Birth Defects and Drugs in Pregnancy. Publishing Sciences Group Inc., Littleton, Mass., 1977.

2301 Tertiary Butyl Hydroperoxide

Sheveleva (**1976**) exposed rats during gestation to 226, 34 or 2 mg per cubic meter for 4 hrs each day. Maternal toxicity occurred at the two highest doses. At the hghest concentration fetal weight was reduced but no malformation increase was found.

Sheveleva, G.A.: Effect of tertiary butyl hydroperoxide on the body of the mother and fetal development (Russian). Gig. Tr. Prof. Zabol. 12:46–48, 1976.

2302 Testosterone CAS 58–22–0

Masculinization of the external genitalia of female infants was observed following maternal administration of testosterone, even in amounts small enough to have no effect in the mother (Van Wyk and Grumbach, 1968). Methyltestosterone and testosterone proprionate were associated with clitoromegaly with or without fusion of the labia minora. Moncrieff (1958) reported dosage levels of methyltestosterone as low as 6 mg daily. Hoffman et al. (1955) reported a masculinized fetus following administration to the mother of testosterone enanthate from the 4th to the 9th months. Grumbach and Ducharme (1960) summarized the human reports. Drawing from the more numerous examples of masculinization by the use of progestins, Grumbach et al. (1959) observed that labioscrotal fusion is more commonly found when treatment is given the mother before the 80th to 90th days of gestation. Clitoromegaly may result from treatment at any period.

An extensive literature on experimental production of masculinization led to the above–described human observations. Greene et al. (1939) carried out detailed studies in the rat; the mouse was used by Raynaud (1947); Bruner and Witschi (1946) utilized the hamster; and Jost's work (1953) in the rabbit is well known. Wells and Van Wagenen (1954) produced pseudohermaphroditism in the female monkey. Grumbach and Ducharme (1960) and Jost (1953) reviewed the subject.

To summarize briefly, the urogenital sinus and its derivatives and the external genitalia can be masculinized. Except in the hamster the wolffian ducts do not regress and form seminal vescicles, epididymides and to a variable extent, vasa deferentia. The mullerian ducts are not appreciably affected and inversion of ovarian development does not occur.

Bruner, J.A. and Witschi, E.: Testosterone–induced modifications of sexual development in female hamsters. Am. J. Anat. 79:293–320, 1946.

Greene, R.R.; Burrill, M.W. and Ivy, A.C.: Experimental intersexuality. The effect of antenatal androgens on sexual development of female rats. Am. J. Anat. 65:415–469, 1939.

Grumbach, M.M. and Ducharme, J.R.: The effects of androgens on fetal sexual development. Fertil. Steril. 11:157–180, 1960.

Grumbach, M.M.; Ducharme, J.R. and Moloshok, R.E.: On the fetal masculinizing action of certain oral progestins. J. Clin. Endocrinol. Metab. 19:1369–1380, 1959.

Hoffman, F.; Overzier, C. and Uhde, G.: Zur Frage der Hormonalen erzeugung Fotaler Zwittenbildungen beim Menschen. Geburtshilfe Frauenheilkd. 15:1061–1070, 1955.

Jost, A.: Problems of fetal endocrinology: The gonadal and hypophyseal hormones. Recent Progr. Horm. Res. 8:379–418, 1953.

Moncrieff, A.: Non-adrenal female pseudohermaphrodism associated with hormone administration in pregnancy. Lancet 2:267–268, 1958.

Raynaud, A.: Observations sur le developpement normal des ebauches de la glande mammaire des foetus males et femelles de souris. Ann. Endocrinol. 8:349–359, 1947.

Van Wyk, J. and Grumbach, M.M.: Disorders of sex differentiation. In: Textbook of Endocrinology, R.H. Williams (ed.), Philadelphia: W.B. Saunders, 537–612, 1968.

Wells, L.J. and Van Wagenen, G.: Androgen–induced female pseudohermaphroditism in the monkey (Macaca mulatta) anatomy of the reproductive organs. Carnegie Inst. Contrib. Embryol. 35:93–106, 1954.

2303 1,1,3,3–Tetrabutylurea *TBU* CAS 4559–86–8

Kennedy et al. (1987) applied 25, 50 or 100 mg to the shaven skin of rats on days 6–15. Resorptions were increased at the 100 mg but no structural defects were found in the fetuses.

Kennedy, G.L.; Lu, M.H. and McAlack, J.W.: Teratogenic evaluation of 1,1,3,3–Tetrabutylurea in the rat following dermal exposure. Food Chem. Toxicol. 25:173–176, 1987.

2304 Tetrahydrofuran

Mast et al. (**1992**) exposed rats and mice to up to 5,000 ppm for 6 hours per day on days 6–19 and 6–17 respectively. At the highest dose maternal lethality was increased in the mice. No increase in malformations occurred but at the highest doses fetal weights were reduced.

Mast, T.J.; Weigel, R.J.; Westerberg, R.B.; Schwetz, B.A. and Morrissey, R.E.: Evaluation of the potential for developmental toxicity in rats and mice following inhalation exposure of tetrahydrofuran. Fund. Appl. Toxicol. 18:255–265, 1992.

2305 Tetrasodium Biscitrato Ferrate

Okada et al. (**1988**) gave rabbits up tp 1,000 mg per kg orally on days 6–18 and found no fetal ill effects.

Okada, F.; Masubara, Y.; Gotoh, M.; Oksumi, I.; Kawaguchi, T.; Okuda, Y. and Nishimura, O.: Teratological study in rabbits of tetrasodium bicitrato ferrate (scf). Kiso to Rinsho 22:4681–4690, 1988.

2306 Tetrachloroacetone CAS 632–21–3

John et al. (1979) gave oral doses to mice and rabbits during active organogenesis and observed major malformations in mice at 15 and 50 mg per kg per day. A low incidence of malformations was found in rabbits at 1 to 10 mg per kg per day.

John, J.A.; Murray, F.J.; Murray, J.S.; Schwetz, B.A. and Staples, R.E.: Evaluation of environmental contaminants, tetrachloroacetone, hexachlorocyclopentadiene and sulfuric acid aerosol for teratogenic potential in mice and rabbits. (abs) Teratology 19:32A–33A, 1979.

2307 3,3,4,4–Tetrachloroazoxybenzene

Hassoun et al. (1984) studied this TCDD congener in mice using intraperitoneal doses of 6–8 mg per kg on days 10–12 of gestation. Cleft palate and hydronephrosis occurred in the offspring of C–57BL mice but no clefts occurred in a resistant strain (DBA).

Hassoun, E.; d'Argy, R.; Dencker, L. and Sandstrom, G.: Teratological studies of the TCDD congener 3,3,4,4–Tetrachloroazoxybenzene and sensitive and non–sensitive mouse strains: Evidence for a direct effect on embryonic tissues. Arch. Toxicol. 55:20–26, 1984.

2308 Tetrachlorobenzenes

Kacew et al. (1984) gavaged rats on days 6 through 15 with 50, 100 and 200 mg per kg with the 1,2,3,4–, 1,2,3,5– and 1,2,4,5–congeners. No increase in defects was observed. At the highest level of 1,2,4,5–the number of live fetuses was reduced.

Kacew, S.; Ruddick, J.A.; Parulekar, M.; Valle, V.E.; Chu, I. and Villeneuve, D.C.: A teratological evaluation and analysis of fetal tissue levels following administration of tetrachlorobenzene isomers to the rat. Teratology 29:21–27, 1984.

2309 3,4,3',4'–Tetrachlorobiphenyl CAS 32598–13–3

Wardell et al. (1982) gave rats orally up to 10 mg per kg on days 6–18 of gestation. At 3 and 10 mg per kg levels, there was an increase in bloody amniotic fluid and blood in the fetal intestinal tract. Fetal mortality increase and growth retardation were found at the 3 and 10 mg levels. McNulty (1985) fed 3150 or 630 micrograms per kg to rhesus monkeys in early pregnancy and all six aborted. Ronnback (**1991**) gave mice 1.5 to 15 mg per kg on day 13 and at the highest dose found reduction in germ cells.

McNulty, W.P.: Toxicity and fetotoxicity of TCDD, TCDF and PCB isomers in rhesus macaques (Macaca mulatta). Environ. Health Persp. 60:77–88, 1985.
Ronnback, C.: Effects of 3,3',4,4'–tetrachlorobiphenyl(TCB) on ovaries of foetal mice. Pharmacology and Toxicology 68:340–345. 1991.
Wardell, R.E.; Seegmiller, R.E. and Bradshaw, W.S.: Induction of prenatal toxicity in the rat by diethylstilbestrol, zeranol, 3,4,3,4–tetrachlorobiphenyl, cadmium and lead. Teratology 26:229–236, 1982.

2310 1,3,6,8–Tetrachlorodibenzo–p–dioxin CAS 33423–92–6

This isomer of 2,3,7,8–TCDD was given in doses of up to 3,000 mg per kg from day 8 through 15 of gestation. A slight decrease in weight and ossification at 3000 mg per kg was the only significant finding (Kamata, 1983).

Kamata, K.: Effect of 1,3,6,8–tetrachlorodibenzo–p–dioxin on the rat fetus. Oyo Yakuri 25:713–718, 1983.

2311 2,3,7,8–Tetrachlorodibenzo–p–dioxin TCDD CAS 1746–01–6

During the manufacture of 2,4,5–trichlorophenoxyacetic acid, a number of dioxins are formed, the most prevalent of which is TCDD. The concentration of TCDD in different batches of Agent Orange varied greatly but averaged about 2 parts per million of 2,4,5–T. The LD–50 ranges from 0.001 to 0.3 mg per kg. (Kociba and Schwetz, **1982**)

Following an industrial accident in Seveso, Italy, a number of studies were performed on the reproductive outcome of exposed women. Among 623 pregnancies, no increase in spontaneous abortion was found (Reggiani, 1980) and no increase in defects was observed (see summary by Friedman, 1984). May (1982) followed up on the reproductive outcome of 41 male employees who had chloracne after an industrial accident. Ten years later no adverse effect could be detected as compared to 31 non–exposed employees. Townsend et al. (1982) studied 737 conceptions after occupational exposure by 370 men to TCDD and other dioxins in a Midland, Michigan plant. No adverse reproductive effects were found when comparison was made with other non–exposed plant workers. Further epidemiologic publications may be found under Agent Orange.

After oral administration to various animals, mutagenicity studies reviewed by Friedman gave both positive and negative studies in test systems. There were over 14 animal teratogenicity reports based on studies in mice, rats,

rabbits and monkeys (Friedman, 1984). McNulty (1985) fed rhesus monkeys 1 microgram per kg in early pregnancy and 13 out of 16 aborted.

Sparschu et al. (1971) found that very small amounts of this material were embryotoxic to the rat embryo. They produced death, resorptions and fetal gastrointestinal hemorrhage with doses of 0.125 to 8.0 microgm per kg given orally on days 6 through 16 of gestation. Courtney and Moore (1971) using TCDD subcutaneously in amounts of 3 microgram per kg daily from day 6 through 15 of mouse gestation, produced cleft palate and renal malformations in the offspring. Cleft palates did not appear in the rat but renal defects were found with a dose of 0.5 microgram per kg daily. Renal defects were produced in the rabbit by Giavani et al. (1982) at a dose level of 0.25 mg per kg per day. Silkworth et al. (**1989**) found in a organic phase leachate containing over 100 chemicals from the Love Canal dump site that the teratogenicity was largely explained by the level of TCDD.

Neubert et al. (1973) reviewed the embryotoxic effects of this compound. Silbergeld and Mattison (1987) reviewed published data on this agent. Conture et al. (**1990**) summarized the literature on teratogenicity. Peterson et al. (**1993**) have reviewed in detail the developmental and reproductive toxicity of dioxins and related compounds.

Courtney, K.D. and Moore, J.A.: Teratology studies with 3,4,5–trichlorophenoxyacetic acid and 2,3,7,8–tetrachlorodibenzo–p–dioxin. Toxicol. Appl. Pharmacol. 20:396–403, 1971.
Couture, L.A.; Abbott, B.D. and Birnbaum, L.S.: A critical review of the developmental toxicity and teratogenicity of 2,3,7,8–tetrachlorodibenzo–p–dioxin: Recent advances toward understanding the mechanism. Teratology 42:619–627, 1990.
Friedman, J.M.: Does Agent Orange cause birth defects? Teratology 29:193–221, 1984.
Giavini, E.; Pratt, M. and Vismara, C.: Rabbit teratology study with 2,3,7,8–tetrachlorodibenzo–p–dioxin. Environ. Res. 27:74–78, 1982.
Kociba, R.J. and Schwetz, B.A.: Toxicity of 2,3,7,8–tetrachlorodibenzo–p–dioxin (TCDD). Drug Metab. Rev. 13:387–406, 1982.
May, G.: Tetrachlorodibenzo dioxin. A survey of subjects ten years after exposure. Br. J. Ind. Med. 39:128–135, 1982.
McNulty, W.P.: Toxicity and fetotoxicity of TCDD, TCDF and PCB isomers in rhesus macaques (Macaca mulatta). Environ. Health Persp. 60:77–88, 1985.
Neubert, D.; Zens, P.; Rothenwaller, A. and Merker, H–J.: A survey of the embryotoxic effects of TCDD in mammalian species. Environ. Health Persp. 5:67–69, 1973.
Peterson, R.E.; Theobald, H.M. and Kimmel, G.L.: Developmental and reproductive toxicity of dioxins and related compounds: cross–species comparisons. Critical Reviews in Toxicology 23(3):283–335, 1993.
Reggiani, G.: Acute human exposure to TCDD in Seveso, Italy. J. Toxicol. Environ. Health 6:27–43, 1980.
Silbergeld, E.K. and Mattison, D.R.: Experimental and clinical studies on the reproductive toxicology of 2,3,7,8–tetrachlorodibenzo–p–dioxin. Am. J. Ind. Med. 11:131–144,

1987.

Silkworth, J.B.; Cutler, D.S.; Antrim, L.; Houston, D.; Tumasonis, C. and Kaminsky, L.S.: Teratology of 2,3,7,8–tetrachlorodibenzo–p–dioxin in a complex environmental mixture from the love canal. Fundamental and Applied Toxicology 13:1–15, 1989.

Sparschu, G.L.: Dunn, F.L. and Rowe, V.K.: Study of the teratogenicity of 2,3,7,8–tetrachlordibenzo–p–dioxin in the rat. Food Cosmet. Toxicol. 9:405–412, 1971.

Townsend, J.D.; Bodner, K.M.; Van Peenen, P.F.D.; Olson, R.D. and Cook, R.R.: Survey of reproductive events of wives of employees exposed to chlorinated dioxins. Am. J. Epidemiol. 115:695–713, 1982.

2312 2,3,7,8–Tetrachlorodibenzofuran

Hassoun et al. (1984) injected this chemical intraperitoneally into mice on single days 10–13 of gestation. Dosage was 0.1 to 0.8 mg per kg. At the lowest dose on day 12, thymic hypoplasia was found in the fetuses. Cleft palate was also observed at this dose level. The findings were similar to those seen with 2,3,7,8–tetrachlorodibenzo–p–dioxin.

Hassoun, E.; d'Argy, R. and Dencker, L.: Teratogenicity of 2,3,7,8–tetrachlorodibenzofuran in the mouse. J. Toxicol. Environ. Health 14:337–351, 1984.

2313 Tetrachlorophenol

Sonawane et al. (1987) gavaged rats with 25, 100 or 200 mg per kg on days 6–15. At the highest dose preimplantation loss and decreased maternal weight gain occurred. No significant increase in defects was found.

Sonawane, B.R.; Price, C.J.; Rubenstein, R. and DeRosa, C.: Teratological evaluation of 2,3,4,6–tetrachlorophenol (TCP) in rats. Teratology 36:63A, 1987.

2314 Tetrachloroethylene *Perchloroethylene* CAS 127–18–4

Schwetz et al. (1975) exposed pregnant mice and rats to concentrations of 300 ppm. Both species were exposed for 7 hours daily periods on days 6 through 15 of gestation. No fetal toxicity or teratogenicity was found. Nelson et al. (1980) performed behavioral tests on the offspring of rats exposed to 100 ppm for 7 hours daily on days 14–20 of gestation and found no changes from the control pups. At exposure levels of 900 ppm, the maternal animals gained less weight and the offspring performed less well on neuromotor tests and had lower levels of brain acetylcholine and dopamine. Pair fed controls were not used.

Nelson, B.K.; Taylor, B.J.; Setzer, J.V. and Hornung, R.W.: Behavioral teratology of perchloroethylene in rats. J. Environ. Path. Toxicol. 3:233–250, 1980.

Schwetz, B.A.; Leong, B.K.J. and Gehring, P.J.: The effect of maternally inhaled trichloroethylene, perchloroethylene, methyl chloroform and methylene chloride on embryonal and fetal development in mice and rats. Toxicol. Appl. Pharmacol. 32:84–96, 1975.

2315 2,3,5,6–Tetrachloronitrobenzene

Courtney et al. (1976) gave 200 mg per kg orally to mice on days 7–18 and produced no increase in defects. The dams gained only 3.6 gm in body weight during the treatment.

Courtney, K.D.; Copeland, M.F. and Robbins, A.: The effects of pentachloronitrobenzene, hexachlorobenzene, and related compounds on fetal development. Toxicol. Appl. Pharmacol. 35:239–256, 1976.

2316 Tetracycline CAS 60–54–8

Baden (1970) reviewed the literature on tetracycline staining of deciduous teeth. Brownish staining of the teeth may result following administration after the 4th month of pregnancy. Generally only the deciduous teeth are involved although with administration close to term, the crowns of the permanent teeth may be stained.

Heinonen et al. (1977) found no increase in malformations among 341 infants exposed during the first four lunar months of pregnancy. Harley et al. (1964) reported that among eight infants with congenital cataracts three mothers had received tetracycline and a fourth oxytetracycline.

Although Fillippi (1967) reported cleft palate and shortened extremities in rat fetuses whose mothers received 5 mg daily from the 5th to 20th day, two other studies did not confirm the findings. Bevelander and Cohlan (1962) produced stunting but no defects in the rat fetus by injecting 40 to 80 mg per kg on the 10th through the 15th gestational day. Hurley and Tuchmann–Duplessis (1963) could not produce defects in the rat fetus. McColl et al. (1965) found only an increase in hydroureters in rat fetuses exposed to 500 mg daily.

Dolgova et al. (1987) showed that in rats, administration of tetracycline during preimplantation and postimplantation periods, as well as during placentation and organogenesis, has a negative effect on the structure and function of lymph nodes. Petrova (1984) studied the morphology of thymus in Wistar rats in antenatal and early postnatal periods after the action of tetracycline hydrochloride (therapeutic doses) in the preimplantation period (from day 1 through 6). She reported retardation of thymus development, alteration in the ultrastructure of epithelioreticulocytes and thymocytes, and a decreased number of small lymphocytes. Savitskaya (1984) studied ante–and postnatal development of tracheobronchial and mesenteric ganglia in Wistar rat embryos and offspring after the action of tetracycline hydrochloride (therapeutic doses) in preimplantation development. She reported anomalies in lymph nodes, distrophy and destruction of reticular cells.

Baden, E.: Environmental pathology of the teeth. In: Thomas Oral Pathology. R.J. Gorlin, and H.M. Goldman (eds), St. Louis: C.V. Mosby Co.; 6th ed., 189–191, 1970.

Bevelander, G. and Cohlan, S.Q.: The effect on the rat fetus of transplacentally acquired tetracycline. Biol. Neonate 4:365–370, 1962.

Dolgova, M.A. and Savitskaya, T.N.: The structure and role of mesenteric ganglia in preventing the action of antibiotics in prenatal development. In: Surgical Anatomy and Surgeries of the Vascular System in Children. (USSR) Leningrad, 70–77, 1987.

Fillippi, B.: Antibiotics and congenital malformations: evaluation of the teratogenicity of antibiotics. In: Advances in Teratology. D.H.M. Woollam (ed.), New York: Academic Press, 2:237–256, 1967.

Harley, J.D.; Farrar, J.F.; Gray, J.B. and Dunlop, I.C.: Aromatic drugs and congenital cataracts. Lancet 1:472–473, 1964.

Heinonen, O.P.; Slone, D. and Shapiro, S.: Birth Defects and Drugs in Pregnancy. Publishing Sciences Group Inc., Littleton, Mass., 1977.

Hurley, L.S. and Tuchmann–Duplessis, H.: Influence de la tetracycline sur la developpement pre– et post–natal du rat. C. R. Acad. Sci. (Paris) 257:302–304, 1963.

McColl, J.D.; Globus, M. and Robinson, S.: Effect of some therapeutic agents on the developing rat fetus. Toxicol. Appl. Pharmacol. 7:409–417, 1965.

Petrova, T.B.: The thymus structure in antenatal and early postnatal ontogeny after the action of tetracycline. Arkhiv Anatomiy, Gistologiy i Embriologiy (USSR) 2:85–92, 1984.

Savitskaya, T.N.: Development of lower tracheo-bronchial and mesenteric ganglia under the action of tetracycline hydrochloride. Arkhiv Anatomiy, Gistologiy i Embriologiy (USSR) 2:70–80, 1984.

2317　12–O–Tetradecanoyl–phorbol–13–acetate

Chen and Hales (**1994**) studied this tumor activator in whole rat embryo culture. Reduced prosencephalon, growth retardation and kinks in the body were found after culturing 6 hours in 50 or 100 nanomolar concentrations in the media. An abundance of embryonic E–cadherin mRNA was found after culture.

Chen, B. and Hales, B.F.: 12–O–tetradecanoyl–phorbol–13–acetate–induced rat embryo malformations in vitro are associated with an increased relative abundance of embryonic E–cadherin mRNA. Teratology 50:302–310, 1994.

2318　Tetradecyl Sulfate, Sodium

Heinonen et al. (1977) reported a two-fold increase in congenital defects among the offspring of 95 mothers injected with this agent for varicose veins in the first 4 lunar months. Among 606 women treated at any time in pregnancy no significant increase was found.

Heinonen, O.P.; Slone, D. and Shapiro, S.: Birth Defects and Drugs in Pregnancy, Publishing Sciences Group Inc., Littleton, Mass., 1977.

2319
Tetrahydro–3,5–dimethyl–4H–1,3,5–oxadiazine–4–thione

Stula and Krauss (1977) applied the chemical cutaneously in doses of 1500 mg per kg on day 12 and found no adverse fetal effects. One fetus exposed to 500 mg per kg on days 12 and 13 had subcutaneous hemorrhage.

Stula, E.F. and Krauss, W.C.: Embryotoxicity in rats and rabbits from cutaneous application of amide–type solvents and substituted ureas. Toxicol. Appl. Pharmacol. 41:35–55, 1977.

2320　N–(2–Tetrahydrofuryl) 5–fluorouracil CAS 17902–23–7

Morita et al. (1971) gave this anticancer agent intravenously to pregnant mice and rats once daily during active organogenesis. The highest dose in the rat was 150 mg per kg and 90 mg per kg in the mouse. Some fetal growth retardation and increased resorptions were found in both species but no teratogenic action was found.

Morita, K.; Watanabe, S.; Mizuno, T.; Takikawa, K. and Harima, K.: Teratogenic study of N–(2–tetrahydrofuryl)–5–fluorouracil. Oyo Yakuri 5:555–568, 1971.

2321
2–(1,2,3,4–Tetrahydro–1–naphthylamino)–2–imidazine HCl

Nakayama et al. (1965) found an increase in fetal mortality and skeletal anomalies in the offspring of mice treated with 12.5 or 25 mg per kg daily during organogenesis. An increase in rats with hydronephrosis was found at similar dosage.

Nakayama, Y.; Masaki, R.; Mesaki, R. and Kowa, Y.: 2–(1,2,3,4–Tetrahydro–1–naphthylamino)–2–imidazoline hydrochloride. Nihon Yakurigaku Zasshi (Folia Pharmacologica Japonica) 61:490–496, 1965.

2322　Tetrahydrophthalimide CAS 85–40–5

Verrett et al. (1969) injected this material into chick eggs and produced congenital defects in 4.8 percent of the survivors. Doses of 3–20 mg per kg of egg were used. The type of defect included micromelia of the legs, skull defects and ectopia of the viscera. The mortality rate was 38 percent at the 20 mg level. DMSO or absolute ethanol were used as vehicles.

Verrett, M.J.; Matchler, M.K.; Scott, W.F.; Reynaldo, E.F. and McLaughlin, J.: Teratogenic effects of captan and related compounds in the developing chicken embryo. Ann. N.Y. Acad. Sci. 160:334–343, 1969.

2323　(2″ R)–4'–O–Tetrahydropyranyladriamycin *THP*

Kurebe et al. (1986a) found no adverse reproductive effects by intravenous treatment of rats in daily doses of up to 0.1 mg per kg. Kurebe et al. (1986b) using up to 0.3 mg per kg during embryogenesis in the rat found reduced fetal weight and retarded ossification but no increase in defects. Rabbits given 0.1 mg per kg did not produce an increase in fetal defects. Kurebe et al. (1986c) gave rats 0.1 mg per kg perinatally and found no significant ill effects.

Kurebe, M. and Asaoka, H.: A study on the effect of (2″ R)–4'–O–tetrahydropyranyladriamycin, a new antitumor antibiotic, on reproduction. I. Its effect on the fertility of rats. Jpn. J. Antibiot. 2:463–476, 1986a.

Kurebe, M.; Asaoka, H.; Moriguchi, M.; Hata, T.; Okano, K. and Iko, M.: A study on the effect of (2″ R)–4'–O–tetrahydropyranyladriamycin, a new antitumor antibiotic, on reproduction. II. Its teratogenicity in rats and rabbits. Jpn. J. Antibiot. 2:477–506, 1986b.

Kurebe, M.; Asaoka, H.; Hata, T.; Watanabe, T. and Sawazaki, S.: A study on the effect of (2" R)–4'–O–tetrahydropyranyladriamycin, a new antitumor antibiotic, on reproduction. III. Its effects on perinatal and postnatal rats. Jpn. J. Antibiot. 2:507–524, 1986c.

2324 Tetrahydro–2–opyrimidinethiol

Discussed under Ethylenethiourea

2325 Tetramethylated Tetralin and Indane, Analogs of Retinoic Acid *RO 13–6307 RO 13–2389 RO 13–4306*

Howard et al. (1987) reported that false retinoids were teratogenic in the hamster. Oral doses on day 8 induced a syndrome of malformations identical to that induced by all–trans–retinoic acid. They were more potent embryotoxins than retinoic and doses of 0.3 to 0.6 mg per kg were teratogenic. Binding affinities for cRAPB correlated with the biological activity. Kochhar and Penner (**1992**) found in the mouse Ro 13–6307 to be 40 times more potent in teratogenicity than retinoic acid. The concentration in the fetus was only a fraction of that found with teratogenic retinoic acid.

Howard, W.B.; Willhite, C.C. and Sharma, R.P.: Structure–toxicity relationships of the tetramethylated tetralin and indane analogs of retinoic acid. Teratology 36:303–311, 1987.

Kochhar, D.M. and Penner, J.D.: Analysis of high dysmorphogenic activity of ro 13–6307, a tetramethylated tetralin analog of retinoic acid. Teratology 45:637–645, 1992.

2326 Tetramethylsuccinonitrile

Doherty et al. (1983) gave hamsters up to 0.147 mmol per kg on day 8 intraperitoneally. No teratogenicity was found. The crown–rump length of the embryos was reduced at 0.073 mM per kg.

Doherty, P.A.; Smith, R.P. and Ferm, V.H.: Comparison of the teratogenic potential of two aliphatic nitriles in hamsters: Succinonitrile and tetramethylsuccinonitrile. Fund. Appl. Toxicol. 3:41–48, 1983.

2327 1,1,3,3–Tetramethylthiourea also see *Ethylenethiourea*

Stula and Krauss (1977) exposed rats cutaneously to 250 mg on days 10 and 11 or 12 and 13 and found no adverse fetal effects. At 500 mg per kg on days 12 and 13, an increase in fetal subcutaneous hemorrhage was found. Teramoto et al. (1981) gave 250 mg per kg orally to rats on day 12 and to mice on day 10 and found no increase in the defect rate or resorptions in the rats. Skeletal defects and cleft palates were increased in the mouse fetuses.

Stula, E.F. and Krauss, W.C.: Embryotoxicity in rats and rabbits from cutaneous application of amide–type solvents and substituted ureas. Toxicol. Appl. Pharmacol. 41:35–55, 1977.

Teramoto, S.; Kaneda, M.; Aoyama, H. and Shirasu, Y.: Correlation between the molecular structure of n–alkylureas and n–alkylthioureas and their teratogenic properties. Teratology 23:335–342, 1981.

2328 1,1,3,3–Tetramethylurea

Teramoto et al. (1981) gave 250–1000 mg per kg orally to rats on day 12 and to mice on day 10 and found an increase in the defect rate and resorptions. Short tails were found in both species and the mouse fetuses had cleft palate and skeletal defects.

Teramoto, S.; Kaneda, M.; Aoyama, H. and Shirasu, Y.: Correlation between the molecular structure of n–alkylureas and n–alkylthioureas and their teratogenic properties. Teratology 23:335–342, 1981.

2329 Tetramisole *Levamisole Hydrochloride* CAS 5036–02–2 CAS 16595–80–5

Ohguro et al. (1982) gave rats and rabbits orally up to 240 and 90 mg per kg daily, respectively. Fertility and perinatal studies in the rat indicated some decrease in implantation at 120 and 240 mg levels and some decrease in viability over 60 mg per kg. No defect increases were found in either species but some growth retardation was found at 120 mg per kg in the rat.

Ohguro, Y.; Imamura, T.; Hara, T.; Nishikawa, S. and Miyazaki, E.: Study on safety of KW–2–LE–T (Levamisole HCl). Reproductive studies. Yakuri to Chiryo 10:3155–3167, 1982.

2330 Thalidomide α–*Phthalimidoglutarimide Distaval*™ *Kevadon*™ *Contergan*™ CAS 50–35–1

This sedative became the most notorious teratogen known to man. Although these events are becoming part of medical history, several teratologic principles were forcefully illustrated by observations made of the outbreak. The first point was that there existed extreme variability in species susceptibility to thalidomide and the second was that there was a very sharp relationship between the time of exposure and the presence and type of congenital defect. Further comments will be made about the epidemiology of the problem and about the relationship between configuration of the compound and its teratogenic action.

The type of defect observed in children has been well correlated with the time of treatment. German workers (Lenz and Knapp, 1962; Knapp et al., 1962) traced the date of prescription in 86 cases and of this group the approximate date of conception was known in 32 mothers. None of these mothers who gave birth to limb defective children were known to have taken the drug only before the 27th day or only after the 40th day. Accordingly, the critical period appeared to be no longer than 14 days. Furthermore, administration of the drug from the 27th to the 30th days was associated most often with only the arms affected while treatment during the 30th to 33rd days caused leg deformity with less involvement of the arms. These findings correlate with the appearance of the lower limb buds in the human embryo at about the 30th day. The previously quoted time of 27 to 28 days for appearance of the lower limbs increased by 2 or 3 days to the 30th day (Iffy et al., 1967; Schumacher, 1975; Jonsson, 1972).

Defects of the external ears are the earliest occurring of the thalidomide anomalies (approximately day 21 to day 27). Other anomalies associated with phocomelia are facial hemangioma, atresia of the esophagus or duodenum, tetralogy of Fallot and renal agenesis. Cleft palate is a rare complication and the central nervous system is not adversely affected. In a study of 86 adults with the syndrome Miller and Stromland (**1993**) found 5 with autism. Cranial nerve involvement was associated. Rodier (**1994**) pointed out that the gross deficiencies of the motor neurons in the facial, ambiguus, and hypoglossal brain nuclei found in a case of infantile autism agree with the timing of thalidomide embryopathy.

It is estimated that defective children resulted in about 20 percent of mothers who ingested the drug during the sensitive period. A single dose of this sedative has been documented to produce human defects. Speirs (1962) gave an interesting report of how an initial survey of the drug ingestion of mothers of phocomelic children gave no hint of a common agent. When thalidomide was later implicated eight of these ten mothers were shown to have ingested it.

Lenz (1962) and McBride (1961) were the first to report that thalidomide was the source of an almost world–wide epidemic of phocomelia. A number of general reviews of the subject appeared (Lenz, 1962; Taussig, 1962; Mellin and Katzenstein, 1962; Cahen, 1966). The epidemiologic aspects of the outbreak are covered by Taussig (1962). Lenz (1966) showed the very close correlation between thalidomide sales and incidence rates of phocomelia. Over 5,000 cases were known in West Germany while only 17 were found in the United States where the drug was not permitted on the market. A detailed clinical analysis of the Japanese cases was published (Kida, 1987). Lenz (**1992**) and Lipson (**1992**) have added interesting footnotes about the politics surrounding the discovery of thalidomide embryopathy.

Species variability: The mouse and rat embryo are relatively insensitive to thalidomide while the rabbit, monkey and man are sensitive. The comparable teratogenic doses on a mg per kg dose are approximately 1 mg for man and monkey, and 50 mg for the rabbit. Axelrod (1970) summarized the experimental work done in monkeys in which a single dose of 8 to 10 mg per kg on the 25th to 30th day produces the thalidomide syndrome. Salzgeber and Salaun (1965) produced defects in the extremity of the chick. Cahen (1966) tabulated the results of the treatment of many species tested.

The New Zealand rabbit has been shown to be teratologically the most sensitive small animal and doses of 250 mg per kg on days 8 to 10 are suitable to produce limb defects. The early pathology in rabbit limb buds indicated a defect of cartilage condensation; no increased cell necrosis was found (Vickers, 1967). Staples and Holtkamp (1963) reported that riboflavin therapy ameliorated the limb defects produced in the rabbit. McBride (1974) found neuronal degeneration and reductions in cell number in the dorsal root ganglia of newborn rabbits with thalidomide defects.

Cahen (1966) listed 14 separate publications dealing with the mouse; nearly all reported negative findings or else

a few defects which did not resemble the thalidomide syndrome. A similar outcome was found from studies in the rat which included many strains. Dyban and Akimova (1966) produced eye and palate defects in the rat with thalidomide when the mothers were deficient of riboflavin and folic acid. Parkhie and Webb (1983) injected Wistar rats with single doses (45 mg per kg) on days 10 through 12 and produced rib and vertebral defects but no phocomelia.

McBride (1977) summarized the evidence that the mechanism of action in the embryo is via changes in the peripheral nerves of the limbs. Evidence that an arene oxide metabolite may be associated with thalidomide toxicity in a cell culture system was presented (Gordon et al., 1981). Fabro (1981) reviewed the mechanisms of action that were proposed. Stephens (1988) listed and discussed 24 cellular or biochemical mechanisms of teratogenesis and concluded that none offered sufficient evidence to substantiate an acceptable mechanism.

Although much careful work has been done in an attempt to understand the mechanism of action, the reason for the chemical and species specificity remains an intriguing pharmacologic riddle (Jonsson, 1972; Schumacher, 1975). The work of Smith et al. (1965), Keberle et al. (1965) and Wuest et al. (1968) gave the numerous studies of compounds chemically related to thalidomide. The embryotoxic properties of the compound do not appear to be due to the simple chemical subunits such as phthalimide, phthalic acid, 3—aminoglutarimide and glutamine.

The compound with the carbonyl group in the phthalimide ring changed to a methylene has been shown to retain teratogenicity (Schumacher et al., 1972). Stockinger and Koch (1969), from studies of a new hypnotic, suggested that the aromatic phthalidimide variety of thalimide was responsible for teratogenicity. Jonsson et al. (1972) studied some derivatives of isoindolinone, benzisothiazoline and 4(3H) quinazolinone and did not find thalidomide teratogenicity.

Axelrod, L.R.: Drugs and nonhuman primate teratogenesis. In: Advances in Teratology. D.H.M. Woollam (ed), New York: Academic Press, 4:217–230, 1970.

Cahen, R.L.: Experimental and clinical chemoteratogenesis. Adv. Pharmacol. 4:263–349, 1966.

Dekker, A. and Mehrizi, A.: The use of thalidomide as a teratogenic agent in rabbits. Bull. Johns Hopkins Hosp. 115:223–230, 1964.

Dyban, A.P. and Akimova, I.M.: Significance of vitamin B complex and genetic factors in reaction to thalidomide in rat embryos. Arch. Anat. (Russian) 51(8):3–17, 1966.

Fabro, S.: Biochemical basis of thalidomide teratogenicity. In: The Biochemical Basis of Chemical Teratogenesis, Chapter 5, Mont R. Juchau (ed.), Elsevier–North Holland, New York, 157–178, 1981.

Gordon, G.B.; Spielberg, S.P.; Blake, D.A. and Balasubramanian, V.: Thalidomide teratogenesis: Evidence for a toxic arene oxide metabolite. Proc. Natl. Acad. Sci. 78:2545–2548, 1981.

Iffy, L.; Shepard, T.H.; Jakobovits, A.; Lemire, R.J. and Kerner, P.: The rate of growth in young human embryos of streeters horizons XIII to XXIII. Acta Anatomica 66:178–186, 1967.

Jonsson, N.A.; Mikiver, L. and Selberg, U.: Chemical structure and teratogenic properties II synthesis and teratogenic activity on rabbits of some derivatives of phthalimide, isoindoline–1–one, 1,2–benzisothiazoline–3–one1,1–dioxide and 4(3H)quinazolinone. Acta Pharm. Suec. 9:431–446, 1972.

Jonsson, N.A.: Chemical structure and teratogenic properties. 3. A review of available data on structure–activity relationships and mechanism of action of thalidomide analogues. Acta Pharm. Suecica 9:521–542, 1972.

Keberle, H.; Faigle, J.W.; Fritz, H.; Knusel, F.; Loustalot, P. and Schmid, K.: Theories on the mechanism of action of thalidomide. In: Embryopathic Activity of Drugs, J.M. Robson, F.M. Sullivan and R.L. Smith (eds.), Boston: Little Brown and Co., 210–233, 1965.

Kida, M.: Thalidomide Embryopathy in Japan. Kodansha, Ltd., Tokyo, Japan., 1987.

Knapp, K.; Lenz, W. and Nowack, E.: Multiple congenital abnormalities. Lancet 2:725 only, 1962.

Lenz, W.: Malformations caused by drugs in pregnancy. Am. J. Dis. Child. 112:99–106, 1966.

Lenz, W. and Knapp, K.: Thalidomide embryopathy. Arch. Environ. Health 5:100–105, 1962.

Lenz, W.: Thalidomide and congenital abnormalities. Lancet 1:271–272, 1962.

Lenz, W.: A personal perspective on the thalidomide tragedy. Teratology 46:417–418.

Lipson, A.H.: Thalidomide retrospective: what did the clinical teratologist learn? Teratology 46:411-413, 1992.

McBride, W.G.: Thalidomide and congenital abnormalities. Lancet 2:1358 only, 1961.

McBride, W.G.: Fetal nerve cell degeneration produced by thalidomide in rabbits. Teratology 10:283–292, 1974.

McBride, W.G.: The pathogenesis of thalidomide embryopathy. Adv. Stud. Birth Defects 1:113–117, 1977.

Mellin, G.W. and Katzenstein, M.: The saga of thalidomide. Neuropathy to embryopathy, with case reports of congenital anomalies. N. Eng. J. Med. 267:1184–1193, 1238–1244, 1962.

Miller, M.T. and Stromland, K.: Thalidomide embryopathy: an insight into autism? (abs) Teratology 47:387–388, 1993.

Parkhie, M. and Webb, M.: Embryotoxicity and teratogenicity of thalidomide in rats. Teratology 27:327–332, 1983.

Rodier, P.M.: An embryologic cause for autism. (Abs) Teratology 49:406, 1994.

Salzgeber, B. and Salaun, J.: Action de la thalidomide sur l'embryon de poulet. J. Embryol. Exp. Morphol. 13:159–170, 1965.

Schumacher, H.J.; Terapane, J.; Jordan, R.L. and Wilson, J.G.: The teratogenic activity of a thalidomide analogue, EM_{12} in rabbits, rats, and monkeys. Teratology 5:233–240, 1972.

Schumacher, H.J.: Chemical structure and teratogenic properties. In: Methods for Detection of Environmental Agents that Produce Congenital Defects. T.H. Shepard, J.R. Miller and M. Marois (eds.), Amsterdam: North Holland–American Elsevier, 65–77, 1975.

Smith, R.L.; Fabro, S.; Schumacher, H.J. and Williams, R.T.: Studies on the relationships between chemical structure and embryotoxic activity of thalidomide and related compounds. In: Embryopathic Activity of Drugs, J.M. Robson, F.M. Sullivan and R.L. Smith (eds), Boston: Little Brown and Co.; 194–209, 1965.

Speirs, A.L.: Thalidomide and congenital abnormalities. Lancet 1:303–305, 1962.

Staples, R.E. and Holtkamp, D.E.: Effects of parental thalidomide treatment on gestation and facial development. Exp. Mole. Path. Suppl. 2:81–106, 1963.

Stephens, T.D.: Proposed mechanisms of action in thalidomide embryopathy. Teratology 38:229–239, 1988.

Stockinger, L. and Koch, H.: Teratologische Untersuchung einer Neuen, dem Thalidomid Strukturell Nahestehenden Sedativ–hypnotisch Wirksamen Verbindung (K–2004). Arzneim. Forsch. 19:167–169, 1969.

Taussig, H.B.: A study of the German outbreak of phocomelia. The thalidomide syndrome. JAMA 180:1106–1114, 1962.

Vickers, T.H.: Concerning the morphogenesis of thalidomide dysmelia in rabbits. Br. J. Exp. Pathol. 48:579–592, 1967.

Wuest, H.M.; Fox, R.R. and Crary, D.D.: Relationship between teratogenicity and structure in the thalidomide field. Experientia 24:993–994, 1968.

2331 Thallium CAS 7440–28–0

Karnofsky et al. (1950) reported that chicks receiving 0.5 to 2.0 mg of thallium sulfate via the yolk sac on the 4th day of incubation developed an achondroplastic–like condition. Gibson and Becker (1970) administered the sulfate form intraperitoneally in amounts up to 10 mg per kg to pregnant rats during critical times of development. Some fetal weight reduction and a slight increase in hydronephrosis was reported.

Gibson, J.E. and Becker, B.A.: Placental transfer, embryo toxicity and teratogenicity of thallium sulfate in normal and potassium–deficient rats. Toxicol. Appl. Pharmacol. 16:120–132, 1970.

Karnofsky, D.A.; Ridgway, L.P. and Patterson, P.A.: Production of achondroplasia in the chick embryo with thallium. Proc. Soc. Exp. Biol. Med. 73:255–259, 1950.

2332 Thebaine CAS 115–37–7

Geber and Schramm (1975) found a 2 to 4 percent incidence of cranioschisis in hamster fetuses exposed to 140 and 193 mg per kg on day 8 of gestation. Maternal mortalities at these two dose levels were 10 and 75 percent.

Geber, W.F. and Schramm, L.C.: Congenital malformations of the central nervous system produced by narcotic analgesics in the hamster. Am. J. Obstet. Gynecol. 123:705–713, 1975.

2333 Theofibrate CAS 54504–70–0 *ML 1024*

1–(Theophyllin–7–yl)–ethyl–2–[2–(p–chlorophenoxy)–2–methylpropionate]

Metz et al. (**1977**) gave rats and rabbits up to 600 and 750 mg per kg respectively during major organogenesis. Oral dosing was used. Some reduction in fetal weight was found but no teratogenicity.

Metz, G.; Specker, M.; Sterner, W.; Heisler, E. and Grahwit, G.: 1–(Theophyllin–y–7yl)–ethyl–2–[p–chlorophenoxy)–2–met Forsch./Drug Res 27:1173–1177, 1977.

2334 Theobromine CAS 83–67–0

Fujii and Nishimura gave up to 600 mg per kg on day 12 to mice intraperitoneally and found cleft palates and difital defects. Tarka et al. (1986) fed rats diets of 0.135 percent theobromine and found delays in ossificaiton but no specific adverse effects when administration was given perinatally or during organogenesis. This amount is equivalent to approximately

Fujii, T. and Nishimura, H.: Teratogenic actions of some methylated xanthines in mice. Okajimas Fol. Anat. Jpn. methylatka, S.M., Jr.; Applebaum, R.S. and Borzelleca, J.F.: Tarka, S.M., Jr.; Applebaum, R.S. and Borzelleca, J.F.: Evaluation of the perinatal, postnatal and teratogenic effects of cocoa powder and theobromine in Spra defects among 76 women exposed to the drug.

2335 Theophylline CAS 58–55–9

Heinonen et al. (1977) found no increase in defects among 76 women exposed to the drug.

Lindstrom et al. (**1990**) studied mice and rats during organogeneses with oral doses of up to 359 or 396 mg per kg daily respectively. No significant increases in malformations were found. In mice some increase in resorptions occurred at the highest dose.

Morrissey et al. (1988) studied fertility in male and female mice and observed a decrease in the number of live pups at a dose level of 265 mg per kg. Fujii and Nishimura (1969) gave up to 600 mg per kg to mice on day 12 and found an increase in cleft palate, digital defects and micrognathia. Ishikawa et al. (1978) applied 1.0 ml of 0.02 M theophylline directly to the chorioallantoic membrane of stage 24 to 27 chick embryos and produced 63 percent aortic aneurysms, many with ventricular septal defects and some with truncus arteriosus. Robkin et al. (1974), at doses of 50 microgram per ml of medium, showed cardiac acceleration in day 11 explanted rat embryos.

Fujii, T. and Nishimura, H.: Teratogenic actions of some methylated xanthines in mice. Okajimas Fol. Anat. Jpn. 46:167–175, 1969.

Heinonen, O.P.; Slone, D. and Shapiro, S.: Birth Defects and Drugs in Pregnancy. Publishing Sciences Group Inc., Littleton, Mass., 1977.

Ishikawa, S.; Gilbert, E.F.; Bruyere, H.J. and Cheung, M.O.: Aortic aneurysm associated with cardiac defects in theophylline stimulated chick embryos. Teratology 18:23–30, 1978.

Lindstrom, P.; Morrissey, R.E.; George, J.D.; Price, C.J.; Marr, M.C.; Kimmel, C.A. and Schwetz, B.A.: The developmental toxicity of orally administered theophylline in rats and mice. Fundamental and Applied Toxicology 14:167–178, 1990.

Morrissey, R.E.; Collins, J.J.; Lamb, J.C.; Manus, A.G. and Gulati, D.K.: Reproductive effects of theophylline in mice and rats. Fund. Appl. Toxicol. 10:525–536, 1988.

Robkin, M.A.; Shepard, T.H. and Baum, D.: Autonomic drug effects on the heart rate of early rat embryos. Teratology 9:35–49, 1974.

2336 Thiabendazole CAS 148–79–8

Ogata et al. (1984) gavaged mice on days 7–15 with amounts of 700–2400 mg per kg. Fusion of vertebrae, cleft palate and skeletal anomalies were increased as was the maternal mortality. Reduction deformities of the limbs were also noted. Further studies (Yoneyama et al., 1985) found irreversible binding of the drug to macromolecules in the mouse fetus. Tanaka et al. (1982) fed rats up to one percent in the diet on days 7–17 of gestation. Fetal growth retardation occurred with maternal toxicity but no teratogenicity was found.

Ogata, A.; Ando, H.; Kubo, Y. and Hiraga, K.: Teratogenicity of thiabendazole in ICR mice. Food Chem. Toxicol. 22:509–520, 1984.

Tanaka, S.; Kawashima, K.; Nakaura, S.; Takanaka, A. and Omori, Y.: Effect of dietary administration of thiabendazole on pregnant rats and fetal development. J. Food Hyg. Soc. Jpn. 23:468–473, 1982.

Yoneyama, M.; Ogata, A. and Hiraga, K.: Irreversible in vivo binding of thiabendazole to macromolecules in pregnant mice and its relation to teratogenicity. Food Chem. Toxicol. 23:733–736, 1985.

2337 Thiadiazole

Discussed under 2–Ethylamino–1–3,4, thiadiazole

2338 Thiamine Deficiency

Nelson and Evans (1955), in an extensive analysis of the effect of thiamine deficiency on the pregnant rat, detected no increase in congenital defects in the offspring. Dietary intake which was reduced in the thiamine deficient animals was partly responsible for the increased death rate and small size of the fetuses. Pfaltz and Severinghaus (1956) reported that some rat fetuses from deficient mothers had hemorrhages, edema of the head and torso and sometimes exencephaly. Full details were not given.

Two chemically related antagonists, oxythiamine and neopyrithiamine, may be teratogenic and are discussed under their own separate headings.

Nelson, M.M. and Evans, H.M.: Relation of thiamine to reproduction in the rat. J. Nutr. 55:151–163, 1955.

Pfaltz, H. and Severinghaus, E.L.: Effects of vitamin deficiencies on fertility, course of pregnancy, and embryonic development in rats. Am. J. Obstet. Gynecol. 72:265–276, 1956.

2339 Thiamphenicol CAS 15318–45–3

Suzuki et al. (1973) studied this antibiotic drug in pregnant rats and mice by giving it by gavage during organogenesis. Complete resorptions resulted in the rats receiving more than 50 mg per kg per day and in mice with more than 1,000 mg per kg. Fetal mortality and growth retardation were found at the highest doses but no teratogenic action was found. Bass et al. (1978) studied respiration and replication in mitochondria from exposed embryos. Postnatal studies did not reveal differences between the treated and control groups. Nau et al. (1981) demonstrated transplacental passage in the early human pregnancy.

Bass, R.; Detlef, O.; Krowke, R. and Spielmann, H.: Embryonic development and mitochondrial function. 3. Inhibition of resorptions and ATP generation in rat embryos by thiamphenicol. Teratology 18:93–102, 1978.

Nau, H.; Welsch, F.; Ulbrich, B.; Bass, R. and Lange, J.: Thiamphenicol during the first trimester of human pregnancy: Placental transfer in vivo, placental uptake in vitro and inhibition of mitochondrial function. Toxicol. Appl. Pharmacol. 60:131–141, 1981.

Suzuki, Y.; Kondo, S.; Okada, F.; Suzuki, I.; Asano, F.; Matsuo, M. and Chiba, T.: Effects of thiamphenicol administered to the pregnant animals upon the development of their fetuses and neonates. (Japanese) Oyo Yakuri 7:41–51, 1973.

2340 Thiamphenicol Glycinate HCl CAS 2611–61–2

Suzuki et al. (1973) gave this drug intraperitoneally to pregnant rats and mice in maximum oral doses of 100 mg and 700 mg per kg per day, respectively. Fetal mortality was found increased in rats at the 100 mg level and in mice at the 700 mg level. Fetal growth inhibition also occurred at the higher doses, but no teratogenic activity occured. Both species were treated during active organogenesis. Although some increase in congenital defects occurred, the number was not significantly greater than the controls.

Suzuki, Y.; Okada, F.; Kondo, S.; Suzuki, I.; Asano, F.; Matuo, M. and Chiba, T.: Effects of thiamphenicol glycinate hydrochloride administered to the pregnant animals upon the development of their fetuses and neonates. (Japanese) Oyo Yakuri 7:859–870, 1973.

2341 Thiamylal Sodium

Tanimura (1965) injected mice intraperitoneally with 20–140 mg per kg once or twice on days 7–14 of gestation. At doses above 60 mg per kg on day 10, malformations of the foot joint were increased. Less common defects such as malformation of the tail and polydactyly were also seen.

Tanimura, T.: Effect of administration of thiamylal sodium to pregnant mice upon the development of their offspring. Kaibogaku Zasshi Acta Anat. Nippon. 40:323–328, 1965.

2342 β–2–Thienylalanine CAS 139–86–6

This structural analogue of phenylalanine inhibits differentiation of ectomesenchyme derived from cranial neural crest. It was added to organ culture medium of fusing palatal shelves in the concentration of 2 mm (Barrd and Verrusio, 1973) and the fusion was prevented.

Barrd, G. and Verrusio, A.C.: Inhibition of palatal fusion in vitro by β–2–thienylalanine. Teratology 7:37–48, 1973.

2343 Thiethylperazine

This phenothiazine was given by Szabo andd Brent (1974) to mice and rats at doses of 50 and 200 mg per kg daily respectively and increases in cleft palate occurred.

Szabo, K. and Brent, R.L.: Species differences in experimental teratogenesis by tranquillising agents. Lancet 1:565, 1974.

2344 Thimerosal *Merthiolate*

The relative risk for congenital anomalies among 56 children whose mothers used this antiseptic topically during the first 4 months was 2.69 (p ¡0.05) (Heinonen et al., 1977). Gasset et al. (1975) gave daily intraperitoneal injections of 1.0 ml of a 0.2% solution to rats on the 6–18th days of pregnancy and found no adverse reproductive effects. At a dose level of 2.0% there was increased fetal deaths.

Gasset, A.R.; Motokazu, I.; Ishii, Y. and Ramer, R.M.: Teratogenesis of ophthalmic drugs II teratogenicities and tissue accumulation of thimerosal. Arch. Opthalmol 93:52–55, 1975.

Heinonen, O.P.; Slone, D. and Shapiro, S.: Birth Defects and Drugs in pregnancy. John Wright Publishing Sciences Group, Inc., Littleton, Mass. 1977.

2345 Thimet

Dilley et al. (1977) exposed rats to vapors of up to 1.94 mg per square meter during days 7 through 14 and found increased fetal lethality and reduced fetal weight, but no increase in defects.

Dilley, J.V.; Chernoff, N.; Kay, D.; Winslow, N. and Newell, G.W.: Inhalation teratology studies of five chemicals in rats. Toxicol. Appl. Pharmacol. 41:196, 1977.

2346 1–Thiocarbamyl–2–imidazolidinone

Fantel et al. (1988) gave evidence that this breakdown product of niridazole is associated with teratogenicity.

Fantel, A.G.; Person, R.E.; Tracy, J.W. and Juchau, M.R.: Niridazole metabolism by rat embryos in vitro. Teratology 37:213–221, 1988.

2347 Thioguanine CAS 154–42–7

Thiersch (1957) injected intraperitoneally pregnant rats with 10 mg per kg. Complete resorption was found with treatment on the 7th and 8th days while unspecified types of malformation or runting were found after treatment on the 4th and 5th or 11th and 12th days.

Thiersch, J.B.: Effect of 2–6 diaminopurine (2–6 DP): 6 Chlorpurine (CLP) and thioguanine (THG) on rat litter in utero. Proc. Soc. Exp. Biol. Med. 94:40–43, 1957.

2348 Thiopental Sodium also see *Barbituric Acid* CAS 71–73–8

Tanimura et al. (1967) found no defects in the offspring of female mice receiving 100 mg per kg by injection on day

11 of pregnancy. Some reduction in fetal weight was found with doses of 50 mg or more per kg. Persaud (1965) injected 50 and 100 mg into pregnant rats on the 4th day of gestation and found no defective offspring.

Persaud, T.V.N.: Tierexperimentelle Untersuchungen zur Frage der teratogenen Wirkung von Barbituraten. Acta Biol. Med. Ger. 14:89–90, 1965.

Tanimura, T.; Owaki, Y. and Nishimura, H.: Effect of administration of thiopental sodium to pregnant mice upon the development of their offspring. Okajimas Folia Anat. Jpn. 43:219–226, 1967.

2349 Thiophanate ethyl 1,
2-Bis–(3–ethoxycarbonyl–thioureido)–benzene CAS 23564–06–9

Makita et al. (1970) gave up to 1000 mg per kg orally to pregnant mice during organogenesis and observed no increase in malformations although some growth retardation was recorded. A three–generation test with 3,000 PPM in the diet was negative.

Makita, T.; Hashimoto, Y. and Noguchi, T.: Toxicological evaluation of thiophanate. 2. Studies on the teratology and three generation reproduction of thiophanate in mice. Oyo Yakuri 4:23–30, 1970.

2350 Thiophanate–methyl Dimethyl 4, 4–O–Phenylene bis–(3-thioallophanate) CAS 23564–05–8

This systemic fungicide was given intraperitoneally to mice on days 1 through 15 of gestation in doses up to 1000 mg per kg. No teratogenic effect was found although fetal death occurred at the highest dose (Makita et al., 1973).

Makita, T.; Hashimoto, Y. and Noguchi, T.: Mutagenic, cytogenetic and teratogenic studies on thiophanate-methyl. Toxicol. Appl. Pharmacol. 24:206-215, 1973.

2351 Thiophosphamide Thiotepa

Korogodina and Kaurov (1984) gave different strains of mice 5 mg per kg on the 12th day of gestation and found increased malformations of the skeletal system especially ribs and caudal region of the spine. The 101/H mice were more susceptible to teratogenesis and mutagenesis than the CBA mice.

Korogodina, Y.V. and Kaurov, B.A.: Teratogenic effect of thiophosphamide on mice of different genetypes. Biull. Eksp. Biol. Med. 97:331–332, 1984.

2352 Thioridazine Mellaril™ CAS 50–52–2

Discussed under Phenothiazines

2353 Thiosemicarbazide also see Semicarbazide CAS 79–19–6

Neuman et al. (1956) and Bodit et al. (1966) produced beak abnormalities and skeletal defects of the legs in chick embryos. Bodit and his co–workers used 0.5–1.0 mg applied to the vascular area on the third day of incubation.

Bodit, F.; Stoll, R. and Maraud, R.: Action de l'hydroxyuree, de la semicarbazide sur le developpement de l'embryon de poulet. C. R. Soc. Biol. (Paris) 160:960–963, 1966.

Neuman, R.E.; Maxwell, M. and McCoy, T.A.: Production of beak and skeletal malformations of chick embryo by semicarbazide. Proc. Soc. Exp. Biol. Med. 92:578–581, 1956.

2354 Thio–TEPA Triethylene Thiophosphoramide CAS 52–24–4

Stevens and Fisher (1965) reported on one woman treated in the 7th month with two daily doses of 15 mg each. The infant was sectioned 5 weeks before term but progressed well and was a lusty child.

Murphy et al. (1958), using the rat, determined that the dose per kg which killed 50 percent of the fetuses on the 12th gestational day was 5.0 mg per kg. The maternal LD–50 was 1.75 times greater. Growth retardation, syndactyly and skeletal defects were frequent and encephaloceles were occasionally produced. They were unable to demonstrate teratogenicity by treating the 4–day chick embryo with 0.01 mg. Thiersch (1957) also reported that two daily injections of 5 mg per kg caused many resorptions if given on two consecutive days during the rat pregnancy. Many of the survivors were malformed. Korogodina and Kaurov (1984) gave 5 mg per kg to mice on day 12 of gestation and observed skeletal malformations in 95 and 61 percent of the day 18 fetuses from 101–H and CBA mice, respectively. Tanimura (1968) published an extensive study of the effect of this compound on the mouse embryo.

Korogodina, Y.V. and Kaurov, B.A.: Teratogenic action of thiophosphamide on mice of different genotypes. Byull. Eksper. Biol. Med. 97:331–332, 1984.

Murphy, M.L.; Delmoro, A. and Lacon, C.R.: The comparative effects of five polyfunctional alkalating agents on the rat fetus with additional notes on the chick embryo. Ann. N.Y. Acad. Sci. 68:762–782, 1958.

Stevens, F.R.T. and Fisher, H.M.: Pregnancy in leukemia. Aust. N.Z.J. Obstet. Gynaec. 5:38–40, 1965.

Tanimura, T.: Relationship of dosage and time of administration to teratogenic effects of thio–TEPA in mice. Sonderab. Okaj. Folia Anatom. Japon. 44:203–253, 1968.

Thiersch, J.B.: Effect of 2,4,6, triamin–S–triazine (TR), 2,4,6 tris (ethyleneimino)–S–triazine (TEM) and N,N, N–triethylenephosphoramide (TEPA) on rat litter in utero. Proc. Soc. Exp. Biol. Med. 94:36–40, 1957.

2355 Thiothixene Navane™ CAS 5591–45–7

Owaki (1969a) gave up to 90 mg per kg per day to pregnant mice from days 7 through 12 and observed no teratogenic or postnatal effects. Similar findings were reported from rabbits (Owaki et al., 1969b).

Owaki, Y.; Momiyama, H. and Yokoi, Y.: Teratological studies on thiothixene in mice. (Japanese) Oyo Yakuri 3:315–320, 1969a.

Owaki, Y.; Momiyama, H. and Yokoi, Y.: Teratological studies on thiothixene (navane) in rabbits. (Japanese) Oyo Yakuri 3:321–324, 1969b.

2356 Thiourea

Teramoto et al. (1981) gave 2000 mg per kg orally to rats on day 12 and 1000 mg per kg to mice on day 10 and found no increase in the defect rate but a slight increase in resorptions.

Teramoto, S.; Kaneda, M.; Aoyama, H. and Shirasu, Y.: Correlation between the molecular structure of n–alkylureas and n–alkylthioureas and their teratogenic properties. Teratology 23:335–342, 1981.

2357 Thiram *Tetramethyl–thiuram Disulfide* CAS 137–26–8

Roll (1971) reported production of cleft palate, curved long bones and micrognathia in two strains of mice. Treatment with doses over 250 mg per kg on the 12th and 13th days of gestation produced the highest incidence of defects. Robens (1969) gave hamsters 31 to 500 mg per kg orally on day 7 or 8. The material was suspended in DMSO. At the highest doses fetal lethality was very high. Neural tube, skeletal and other defects occurred. When the compound was given in carboxymethyl cellulose defects were rare at 125 mg per kg. Vasilos et al. (1978) showed this compound in subtoxic maternal doses induced subcutaneous hematomas and a decrease in fetal weight. It also caused embryonic resorptions.

Robens, J.F.: Teratologic studies of carbaryl, diazinon, norea, disulfiram and thiram in small laboratory animals. Toxicol. Appl. Pharmacol. 15:152–163, 1969.

Roll, R.: Teratologische Untersuchungen mit Thiram (TMTD) an zwei Mausestammen. Arch. Toxikol. 27:173–186, 1971.

Vasilos, A.F.; Anisimova, L.A.; Todorova, E.A. and Dmitrienko, V.D.: The reproductive function of rats in acute and chronic intoxication with tetramethylthiiuramdisulfide. Gig. Sanit. (USSR) 6:37–40, 1978.

2358 Thyroid Antibodies

Maternal autoimmunization to thyroid has been implicated (Blizzard et al., 1960) etiologically in athyrotic cretinism but the incidence of thyroid antibodies in the serum of the mothers of cretins was low at 25 percent (Chandler et al., 1962). The possibility exists that the antibodies resulted from an unidentified agent which affected both mother and fetus, destroying the fetal thyroid and damaging the maternal gland sufficiently to produce a maternal antibody to thyroid.

Blizzard, R.M.; Chandler, R.W.; Landing, B.H.; Pettit, M.D. and West, C.D.: Maternal autoimmunization to thyroid as probable cause of athyrotic cretinism. N. Eng. J. Med. 263:327–336, 1960.

Chandler, R.W.; Blizzard, R.M.; Hung, W. and Kyle, M.: Incidence of thyrocytotoxic factor and other antithyroid antibodies in the mothers of cretins. N. Eng. J. Med. 267:376–380, 1962.

2359 Thyroidectomy also see *Hypothyroidism*

Langman and Van Faassen (1955) partially thyroidectomized rats before pregnancy and found in subsequently produced fetuses about a 50 percent abnormality rate in the lenses. Besides retinal folding, various degrees of lenticular degeneration were present. Partial thyroidectomy early in pregnancy was not associated with defects. The postnatal effects following maternal thyroidectomy of rats were reported by Bakke et al. (1975). Persistently enlarged thyroids and elevated thyroid stimulating hormone levels were found in the male offspring.

Bakke, J.L.; Lawrence, N.L.; Robinson, S. and Bennett, J.: Endocrine studies of the untreated progeny of thyroidectomized rats. Pediatr. Res. 9:742–748, 1975.

Langman, J. and Van Faassen, F.: Congenital defects in the rat embryo after partial thyroidectomy of the mother animal: A preliminary report of eye defects. Am. J. Ophthalmol. 40:65–76, 1955.

2360 Thyrotropin–Releasing–Hormone *TRH* CAS 9002–71–5

Asano et al. (1974) injected this synthetic neuroendocrine substance intraperitoneally into pregnant rats and mice on days 7 through 16 and 6 through 15, respectively. Doses of 0.2 to 30 mg per kg were used. No teratogenicity or postnatal changes were found except for growth retardation in the rat offspring. A reduction in the weight of the fetal mouse thyroids was found.

Asano, Y.; Ariyuki, F. and Higaki, K.: Effects of administration of synthetic thyrotropin–releasing–hormone on mouse and rat fetuses. Oyo Yakuri 8:807–816, 1974.

2361 Thyroxine CAS 51–48–9

Giroud and Rothschild (1951) tube–fed 0.25 to 0.30 mg of thyroxine daily to rats and found cataracts in 38 percent of the offspring. Other types of defects were not seen. Stoll et al. (1966), using 3 to 24 microgm of L–thyroxine per egg on the 2nd, 3rd or 5th day of incubation, produced a monstrosity characterized by eversion of the body wall and sac from which the legs, wings and head protrude. Triodothyronine (0.1 microgm) produced a similar type of defect. A decrease in cleft palate in the mouse associated with thyroxine administration was reported by Woollam and Millen (1960). A possible explanation that the thyroxine causes an increased fetal loss rate in the affected but not the normal litter mates was offered by Juriloff and Fraser (1977).

Giroud, A. and Rothschild, B.: Repercussions de la thyroxine sur l'oeil du foetus. C. R. Soc. Biol. (Paris) 145:525–526, 1951.

Juriloff, D.M. and Fraser, F.C.: Differential mortality of the cleft lip embryos in response to maternal treatment with thyroxine. (abs) Teratology 15:18–19A, 1977.

Stoll, R.; Coulaud, H.; Faucounau, N. and Maraud, R.: Sur l'action teratogene des hormones thyroidiennes chez l'embryon de poulet. Les Mechanisms Morphogenetiques De La Strophosomie. Arch. Anat. Microsc. Morphol. Exp. 55:59–76, 1966.

Woollam, D.H.M. and Millen, J.W.: Influence of thyroxine on the incidence of hairlip in the "strong A" line of mice. Br. Med. J. 1:1253–1254, 1960.

2362 Tiapride Hydrochloride

Suzuki et al. (1985) studied this antipyschotic in rats using up to 500 mg during organogenesis or from day 17 until 3 weeks after birth. No teratogenicity was found but at 500 mg per kg there was slight retardation of postnatal growth.

Suzuki, T.; Ikeda, S.; Nishimura, N.; Hirakawa, T.; Fujii, T. and Fuke, H.: Reproduction studies of tiapride hydrochloride in rats. Clin. Report 19:1961–1976, 1985.

2363 Tiaprofenic Acid

5–Benzoyl–α–methyl–2–thiophene Acetic Acid CAS 33005–95–7

Esaki and Oshio (1980) in a series of papers studied the effects of this anti–inflammatory drug on breeding, prenatal and perinatal development in the mouse. Dosing up to 80 mg per kg orally produced some delay in fetal development but no other adverse effects were noted.

Hiramatsu et al. (1980) gave this antiflammatory agent orally to rabbits during organogenesis. At the highest dose (75 mg per kg), there was maternal toxicity, implantations were reduced and fetal ossification was delayed.

Esaki, K. and Oshio, K.: Effects of oral administration of tiaprofenic acid (RU–15060) on reproduction in the mouse. Preclin. Rep. Cent. Inst. Exp. Anim. 6:195–216, 1980.

Hiramatsu, Y.; Tamura, Y. and Koniba, S.: Teratological study of RU–15060 (5-benzoyl-α-methyl-2-thiophene). Yakuri to Chiryo 8:1773–1776, 1980.

2364 Tiaramide

4[5-Chloro-2-oxo–(3–benzothiazolinyl)acetyl]–1–piperazin Ethanol Hydrochloride CAS 32527–55–2

This anti–inflammatory compound was given to mice, rats and rabbits in maximum daily oral doses of 250, 1000 and 250 mg per kg (Watanabe et al., 1973). No defects or fetal effects of significance were found.

Watanabe, N.; Takashima, T.; Ito, N.; Fujii, T. and Miyazaki, K.: Toxicological and teratological studies of 4[(5–chloro–2–oxo–3–benzothiazolinyl)acetyl]–1–piperazine ethanol hydrochloride (tiaramide hydrochloride), an anti–inflammatory drug. Arzneim Forsch 23:504–508, 1973.

2365 Ticlopidine Hydrochloride

Thieno[3,2-c]pyridine,5-[(2-chlorophenyl)methyl]–4,5,6,7–tetrahydro–hydrochloride CAS 55142–85–3

Watanabe et al. (1980) gave up to 320 mg per kg orally to rats before and during pregnancy. Sedation and other side effects were noted in the adults but no adverse effects on reproduction were found. Some delay in ossification of the fetus was observed.

Watanabe, T.; Takagi, S.; Mochida, K.; Ohura, K.; Matsuhashi, K.; Morita, H. and Akimoto, T.: Reproductive studies of triclopidine hydrochloride fertility studies in rats. Iyakuhin Kenkyu 11:255–264, 1980.

2366 Tilidine *DL–trans–2–dimethylamino–1–phenyl–cyclohex–3–en–trans–2*

dimethylamino–1–phenyl–cyclohex–3–en–trans–1–carbonacidethyl–ester mydrochloride

Herrmann et al. (1970) gavaged rats and dogs with up to 100 or 25 mg per kg respectively and produced no increase in congenital defects. Treatment was given to the rats on days 6–15.

Herrmann, V.M.; Wiegleb, J. and Leuschner, F.: Toxikologische untersuchungen uber ein neues stark wirksames analgeticum. Arzneim.–Forsch. (Drug Res)20:983–990, 1970.

2367 Tilorone

Terry et al. (1992) studied this immunomodulator in rats using 250 and 500 mg per kg on day 10 or 400 mg per kg on days 17–19. In the early treatment the embryos examined at 72 hours after 500 mg per kg had an increased death rate. No growth changes were found in the fetuses from the late treatment group. A partial prevention of fetotoxicity was found after concurrent progesterone treatment.

Terry, R.D.; Marks, T.A.; Hamilton, R.D.; Pitts, T.W. and Renis, H.E.: Prevention of tilorone developmental toxicity with progesterone. Teratology 46:237–250, 1992.

2368 Timelastine

Freeman et al. (1989) gave rats up to 300 mg per kg on days 6–15 and found no maternal toxicity or reproductive ill effect. The histamine–H1 antagonist was given by gavage.

Freeman, S.J.; Irvine, L. and Walker, T.F.: Teratological evaluation in rat of SK&F 93944 administered alone or combined with pseudoephedrine HCl. Toxicologist 9:31, 1989.

2369 Timentin

This antibiotic consisting of ticarcillin and clavulanic acid was studied by Fujita et al. (1986 A) in rats during organogenesis. At intravenous doses of 150 mg per kg (Clavulanic) and 1200 mg per kg (ticarcillin) maternal toxicity occurred but no teratogenicity or postnatal ill effects were seen. Perinatal and fertility studies were also performed without adverse findings. Rabbits treated intravenously during organogenesis with up to 15o mgs (clavulanic acid) and 125 mgs (ticarcillin) had no fetal ill effects. (Fujita et al., 1989 B)

Fujita, K.; Nishioka, Y.; Kurakata, Y.; Koshima, Y.; Nakamura, A.; Ishida, S. and Baldwin, J.A.: Reproduction studies of potassium clavulanate and brl28500 (II) teratology studies in rats. Chemotherapy 34:142–151, 1986.

Fujita, K.; Nishioka, Y.; Kurakata, Y.; Koshima, Y.; Furukawa, S.; Morinaga, T.; Toteno, I. and Baldwin, J.A.: Reproduction studies of potassium clavulanate and brl28500 (IV) teratology studies in rabbits. Chemotherapy 34:161–166, 1986.

2370 Timepidium Bromide

1,1–Dimethyl–5–methoxy–3–(dithien–2–y-methylene) –piperidinium bromide CAS 35035–05–3

Fujisawa et al. (1973) gave this drug intraperitoneally and orally to pregnant mice and rats during organogenesis.

Orally the maximum doses were 160 mg per kg daily. The maximum intraperitoneal dose was 40 mg per kg per day in mice and 20 mg per kg per day in rats. No increase in defects was found and no postnatal differences were observed in the four–week postnatal study.

Fujisawa, Y.; Fujii, T.M. and Kowa, Y.: The effects of 1,1–dimethyl–5–methoxy–3–(dithien–2–y–methylene) piperidinium bromide (SA–504) upon offsprings of mice and rats administered maternally during critical period of pregnancy. (Japanese) Oyo Yakuri 7:1293–1304, 1973.

2371 Timiperone 4–(4–(2, 3–Dihydro–2–thioxo–1H–benzimidazol–l–yl)–1–piperidinyl)–1–(4–fluoro–phenyl)–1–butanone CAS 57648–21–2

Nakashima et al. (1981) found neither teratogenic effect nor changes in postnatal function in rat fetuses exposed to up to 10 mg per kg daily during organogenesis. Maternal and fetal weight reduction occurred. Similar treatment of rabbits revealed no significant differences from controls.

Nakashima, K.; Yamashita, N. and Morita, H.: Reproductive studies of timiperone. Iyakuhin Kenkyu 12:861–880, 1981.

2372 Timonacic Thiazolidine Carboxylic Acid CAS 444–27–9

Bertrand and Piton (1972) gave 25–250 mg per kg by mouth to pregnant mice, rats and rabbits and found no fetal effects. The rats and mice were treated between the 3rd and 15th day and the rabbits between the 2nd and 27th gestational days.

Bertrand, M. and Piton, Y.: Recherche du risque d'effet teratogene de l'acide thiazolidine carboxylique. Gazzetta Medica Italiana 131:268–271, 1972.

2373 Tin

Themrer et al. (1971) fed rats sodium pentafluorostannite up to 200 or up to 500 ppm of the sodium salt of tin during pregnancy and found no decrease in fertility or adverse fetal effects.

Themrer, R.C.; Mahoney, A.W. and Sarett, H.P.: Placental transfer of fluoride and tin in rats given various fluoride and tin salts. J. Nutr. 101:525–532, 1971.

2374 Tinidazole Ethyl [2–(2–methyl–5–nitro–1–imidazolyl) ethyl] sulfone CAS 19387–91–8

Owaki et al. (1974) gave this compound orally to pregnant mice and rats in maximum doses of 2,000 mg per kg per day of organogenesis and found no teratogenic effects.

Owaki, Y.; Momiyama, H.; Sakai, T. and Nabata, H.: Effects of tinidazole on the fetuses and their postnatal development in mice and rats. Oyo Yakuri 8:421–427, 1974.

2375 Tinoridine HCl 2–Amino–3–ethoxycarbonyl–6–benzyl–4,5,6,7–tetrahydrothieno–(2,3–c) pyridine Hydrochloride CAS 24237–54–5

This anti–inflammatory agent was given to pregnant mice on days 7–12. The maximum dose of 1000 mg per kg

produced no fetal changes. Rats received up to 700 mg per kg on days 9–14 and some growth retardation of the fetuses was found (Nanba et al., 1970).

Nanba, T.; Hamada, Y.; Izaki, K. and Imamura, H.: Studies on anti–inflammatory agents. XV. Toxicological studies of 2–amino–3–ethoxycarbonyl–6–benzyl–4,5,6,7–tetrahydrothieno–(2,3–c) pyridine hydrochloride. Yakugaku Zasshi 90:1447–1451, 1970.

2376 Tioconazole CAS 65899–73–2

Noguchi et al. (1982) gave this antifungal agent which is an imidazole derivative subcutaneously to rats. When 30 and 100 mg per kg were given before mating and during the first 7 days of gestation, a reduction in implantation and live fetuses was found. The same doses were associated after treatment on days 17–21 with decreased fetal survival and reduced lactation. No malformations were increased after treatment on days 7–17.

Noguchi, Y.; Tochibana, M.; Nabatake, H.; Iijima, M.; Horimoto, M.; Yamakawa, S.; Ishikawa, J. and Otsaki, I.: Preclinical safety evaluation of tioconazole. Yakuri to Chiryo 10:3849–3861, 1982.

2377 Tiopronin 2–Mercaptopropionyl Glycine

Fujimoto et al. (1979) used this sulfhydryl compound to prevent cleft palate induced in mice by methylmercuric chloride. They cite unpublished work that showed no teratogenicity.

Fujimoto, T.; Fuyuta, M.; Kiyofuji, E. and Hirata, S.: Prevention of tiopronin (2–mercaptopropionyl glycine) of methylmercuric chloride–induced teratogens and fetotoxic effects in mice. Teratology 20:297–302, 1979.

2378 Tiron Sodium 4,5–Dihydroxybenzene–1,3–disulfonate 4,5–Dihydroxybenzene–1,2–disulfonate sodium

Ortega et al. (1991) gave up to 3,000 mg per kg intraperitoneally to mice on dys 6–15 and found maternal toxicity and resorption increase at the highest dose. No effect was found at 1500 mg per kg.

Ortega, A, Sanchez, D.J.; Domingo, J.L.; Llobet, J.M. and Corbella, J.: Development toxicity evaluation of tiron (sodium 4, 5–dihydroxybenzene–1,3–disulfonate) in mice. Research Communications in Chemical Pathology and Pharmacology 73:97–106, 1991.

2379 Tiropramide

Shimazu et al. (1992 A) gave this bladder muscle relaxant to rats before and in early pregnancy in amounts of up to 100 mg per kg orally and found no decrease in fertility. Similar treatment during organogenesis was associated with an increase in thymic reminants in the neck but no other adverse effects (Shimazu et al, 1992 B). Perinatal studies were followed by normal behavior and reproduction in the offspring (Shimazu et al., 1992 C).

Shimazu, H.; Shiota, Y.; Katsumata, Y.; Fujioka, M.; Suzuki, K.; Watase, E.; Yasuda, E.; Ohta, M. and Ito, S.: Reproductive and developmental toxicity studies of tiropramide hydrochloride (I)–fertility study in rats by oral administration. Jpn. Pharmacol. Ther. 20:3037–3047, 1992A.

Shimazu, H.; Katsumata, Y.; Shiota, Y.; Fujioka, M.; Suzuki, K.; Ogawa, J.; Yasuda, E.; Ohta, M. and Ito, S.: Reproductive and developmental toxicity studies of tiropramide hydrochloride (II)–teratological study in rats by oral administration. Jpn. Pharmacol. Ther. 20:3049–3063, 1992B.

Shimazu, H.; Kattsumata, Y.; Shiota, Y.; Fujioka, M.; Suzuki, K.; Tsuchiya, Y.; Yasuda, E.; Ohta, M. and Ito, S.: Reproductive and developmental toxicity studies of tiropramide hydrochloride (III)–peri-natal and post-natal study in rats by oral administration. Jpn. Pharmacol. Ther. 20:3065–3078, 1992C.

2380 Tissue Plasminogen Activator *AK–124*

Aso et al. (**1988**) and Shibano et al. (**1988**) gave this tissue plasminogen activator intravenously to pregnant rats and rabbits in doses up to 100,000 AKU per kg daily during organogenesis and found no adverse effects on the offspring of either species.

Aso, S.; Sueta, S.; Ehara, H.; Kajiwara, Y.; Horiwaki, S.; Kanabayashi, T.; Wakabayashi, S. and Moriwaki, T.: Teratogenicity study of tissue plasminogen activator (ak–124) by intravenous administration during period of fetal organogenesis in rats. Yakuri to Chiryo 16:3633–3652, 1988.

Shibano, T.; Sakai, Y.; Kinoshita, K.; Yoneyama, S.; Kanabayashi, T.; Koga, H. and Nishigaki, K.: Teratogenicity study of tissue plasminogen activator (ak–124) by intravenous administration during the period of fetal organogenesis in rabbits. Yakuri to Chiryo 16:1403–1413, 1988.

2381 Titanocene Dichloride CAS 1271–19–8

Kopf–Maier and Erkenswick (1984) studied this antitumor agent in mice given 30 or 60 mg per kg on single days 8, 10, 12, 14 or 16. At both dose levels cleft palates were increased. Growth retardation and osseous delay also occurred.

Kopf–Maier, P. and Erkenswick, P.: Teratogenicity and embryotoxicity of titanocene dichloride in mice. Toxicology 33:171–181, 1984.

2382 Tobacco Stalks

Crowe and Swerczek (1974) fed aqueous tobacco leaf filtrates or tobacco stalks to sows and found congenital arthrogryposis in the offspring. The material was administered from the 4th through the 53rd day of gestation. The causative agent was not determined. Keeler and Crowe (1981) isolated anabasine from these stalks and fed N. glauca, a plant containing the chemical as its sole alkaloid, to cows and pigs and produced congenital defects.

Crowe, M.W. and Swerczek, T.W.: Congenital arthrogryposis in offspring of sows fed tobacco (Nicotiana tabacum). Am. J. Vet. Res. 35:1071–1073, 1974.

Keeler, R.F. and Crowe, M.W.: Congenital deformities in livestock induced by maternal Nicotiana ingestion. (abs) Teratology 23:44A, 1981.

2383 Tobramycin CAS 32986–56–4

This aminoglycoside was given subcutaneously to rats and rabbits in doses of 100 mg and 20 or 40 mg per kg, respectively. No effect on fetal development occurred (Welles et al., 1973). Akiyoshi (1978) studied ototoxicity in the guinea pig and found none in those treated in the first four weeks of gestation but in the last four weeks four of 38 newborns had hearing loss at 20,000 Hz. Doses of 100 mg per kg were used daily.

Akiyoshi, M.: Evaluation of ototoxicity of tobramycin in guinea pigs. J. Antimicrob Chemoth. 4(suppl. A):69–72, 1978.

Welles, J.S.; Emmerson, J.L.; Gibson, W.R.; Nickander, R.; Oben, N.V. and Anderson, R.C.: Preclinical toxicology studies of tobramycin. Toxicol. Appl. Pharmacol. 25:398–409, 1973.

2384 Tocoretinate

Ryo et al. (**1992**) gave rats up to 100 mg per kg before pregnancy and during early in the prenatal period and during the perinatal period and found no adverse effects on reprodiction. During organogenesis 2,000 mg per kg was given without ill effect in the fetuses. Postnatal behavior and reproduction were normal (Narita et al., **1992** and Marakami et al., **1992**). Tanaka et al. (**1992**) gave rabbits up to 1,000 per kg during organogenesis and found no ill effects.

Murakam, Y; Ogasawara, H.; Sakauchi, N.; Narirta, H.; Hanada, S.; Ohashi, M.; Noguchi, O.; Tanaka, N.; Misawa, N. and Inomata, N.: Reproductive and developmental toxicity study of tocoretinate (IV) administration to female rats during the perinatal and postnatal periods. Oyo Yakuri 43:329–339, 1992.

Narita, H.; Sakauchi, N.; Ohashi, M.; Murrakami, Y.; Hamada, S.; Ogasawarra, H.; Tanaka, N.; Noguchi, O.; Misawa, N. and Inomata, N.: Reproductive and developmental toxicity study of tocorretinate (II) teratological study in rats by oral administration. Oyo Yakuri 42:311–321, 1992.

Ryo, Ohta, Hashimoto, Y.; Mizutani, M.; Misawa, N. and Inomata, N.: Reproductive and developmental toxicity study of tocoretinate (I) effects on reproductive function and fertility in rats (segment I study). Oyo Yakuri 43:303–310, 1992.

Tanaka, N.; Murakami, Y.; Noguchi, S.; Narita, H.; Hamada, S.; Ohashi, M.; Misawa, N. and Inomata, N.: Reproductive and developmental toxicity study of tocoretinate (III) teratological study in rabbits by oral administration. Oyo Yakuri 43:323–327, 1992.

2385 Todralazine *N1–Ethoxy carbonyl–N2–hydrazinophthalazine HCl 621–BT, Hydralazine derivative* CAS 14679–73–3

This hypotensive agent was tested in mice and rats (Koyama et al., 1969). Oral doses of 200 mg per kg in mice and 100 mg per kg in rats produced minor skeletal defects. A few exencephalics and cleft palates were found in the mice.

Koyama, K.; Imamura, S.; Ohguro, Y. and Hatano, M.: Toxicological studies on N1–ethoxy carbonyl–N2–hydrazinophthalazine hydrochloride (621–BT). Yamaguchi Igaku 18:29–38, 1969.

2386 Tofisopam CAS 22345–47–7

This benzodiazepin tranquilizer was tested by Hayashi et al. (1981) in rats. They used oral doses of up to 100 mg per kg daily before conception and on days 7–17 and 17–21 of gestation. No adverse effects on reproduction was found.

Hayashi, Y.; Inoue, K.; Kasuys, S.; Tomita, K.; Ito, C. and Ohnishi, H.: Toxicological studies of tofisopam: Reproductive studies of tofisopam in rats. Iyakuhin Kenkyu 12:565–580, 1981.

2387 Tokishakuyaku–san

Aburada et al. (1982) fed up to 800 mg per kg daily to rats during days 7–17 and found no adverse fetal or maternal effects. This material is an old Chinese medicine.

Aburada, M.; Akiyama, Y.; Ichio, Y.; Nakamura, A. and Ichikawa, N.: Teratological study of tokishakuyaku–san in rats. Oyo Yakuri 23:981–997, 1982.

2388 Tolbutamide CAS 64–77–7

Sterne (1963) reviewed 34 pregnancies treated with oral hypoglycemics and reported 28 normal births and six stillbirths but no congenital defects.

Tuchmann–Duplessis and Mercier–Parot (1959) and Demeyer (1961) produced congenital defects of the eye in the offspring of treated rats. Demeyer (1961) used 300 mg per day on the 8th and 9th days. Smithberg and Runner (1963) injected mice intraperitoneally with one mg per gm of weight on the 9th gestational day and found exencephaly, rib and vertebral defects in many of the offspring. Lazarus and Volk (1963) injected 125 mg per kg twice daily into rabbits from the 7th through the 14th gestational day and found no evidence of teratogenicity. These authors reviewed the available negative evidence for teratogenicity in man. Smoak (1992) studied the effect of rat serum treated with tolbutamide on cultured mouse embryos. She found growth retardation and malformations in the tolbutamide serum and these were not corrected by adding glucose.

Demeyer, R.: Etude experimentale de la glycoregulation gravidique et de l' action teratogene des perturbations du metabolisme glucidique. Paris: Masson Et Cie, 175–183, 1961.

Lazarus, S.S. and Volk, B.W.: Absence of teratogenic effect of tolbutamide in rabbits. J. Clin. Endocrinol. 23:597–599, 1963.

Smithberg, M. and Runner, M.N.: Teratogenic effects of hypoglycemic treatments in inbred strains of mice. Am. J. Anat. 113:479–489, 1963.

Smoak, I.W.: Teratogenic effects of tolbutamide on early–somite mouse embryos in vitro. Diabetes Research and Clinical Practice 17:161–167, 1992.

Sterne, J.: Antidiabetic drugs and teratogenicity. Lancet 1:1165 only, 1963.

Tuchmann–Duplessis, H. and Mercier–Parot, L.: Influence de divers sulfamides hypoglcemiants sur le developpement de l'embryon. Etude experimentale chez le rat. Acad. Nat. De Medecine (Paris) 143:238–241, 1959.

2389 Tolciclate CAS 50838–36–3

This topical antifungal agent was given subcutaneously to rats and rabbits in maximal doses of 100 mg per kg daily by Harakawa et al. (1981). No ill effects were found on fetuses of either species. Fertility and postnatal studies also showed no ill effects.

Harakawa, T.; Suzuki, T.; Hayashizaki, A.; Nishimura, N.; Sano, Y.; Nishikawa, M.; Kato, M.; Sato, Y.; Iwasaki, K. and Kihara, T.: Reproductive studies of tolciclate. Kiso to Rinsho 15:2413–2425, 1981.

2390 Tolmetin CAS 26171–23–3

This non–steroid anti–inflammatory was given orally by Nishimura et al. (1977) to pregnant rabbits on days 6 through 18 in amounts of up to 100 mg per kg daily. At 100 mg per kg, the dam's weight gain was reduced and fetal mortality was increased. No teratogenic effect was found.

Nishimura, K.; Fukagawa, S.; Shigematsu, K.; Makumoto, K.; Terada, Y.; Sasaki, H.; Nanto, T. and Tatsumi, H.: Teratogenicity study of tolmetin sodium in rabbits. (Japanese) Iyakuhin Kenkyu 8:158–164, 1977.

2391 Tolnaftate

This antifungal agent was studied in mice and rats by Noguchi et al. (1966) at doses up to 2000 or 500 mg per kg in the mouse and rat respectively there was no increase in defect rates.

Noguchi, T.; Hashimoto, Y.; Makita, T. andd Tanimura, T.: Teratogenesis study of tolnaftate an antitrichophyton agent. Toxicol Appl Pharmacol 8:386–397, 1966.

2392 Tolobuterol

O–Chloro–α–(tert–butylaminomethyl)–benzylalcohol Hydrochloride

Tsuruzaki et al. (1977) tested this bronchodilator in pregnant rats and with oral doses of up to 75 mg per kg during organogenesis, produced no ill effects in the fetuses. Kawana et al. (1977) found no teratogenic effect of oral doses of 40 mg per kg in the rabbit.

Kawana, S.; Watanabe, G.; Ito, T.; Yamamoto, M. and Kamimura, K.: Effects of the bronchodilator C–78 on fetal development in rabbits. Acta Medica et Biologica 25:67–73, 1977.

Tsuruzaki, T.; Shita, T.: Yamasaki, M.: Kubo, N.; Yamamota, M. and Vemura, K.: Effects of O–chloro–α–(tert–butylaminomethyl)benzylalcohol on the reproductive function of rats. (Japanese) Clin. Rep. 11:439–453, 1977.

2393 Toluene CAS 108–88–3

Syrovadko (1977) reported a higher incidence of fetal asphyxia and low birth weight in the offspring of women working with "organosilicon" varnishes which increase the exposure to toluene and other agents. Hersh et al. (1985) reported three children with microcephaly, central nervous system dysfunction, and minor cranofacial and limb anomalies. All three were exposed in utero to regular recreational toluene sniffing. The facial features resembled those of fetal alcohol syndrome (FAS) and the fingernails were blunted. Hersh (1989) has reported two additional cases. Goodwin (1988) identified five women with toluene sniffing (0.5 to 2 cans per day) and all had renal tubular acidosis. Four of the five offspring were less than 2500 mg and two that were studied had renal acidosis. One child had features of FAS but there was no history of alcohol exposure. Another had deformed ears, ventricular septal defect, micrognathia and hydronephrosis. Donald et al. (**1991**) have reviewed and discussed human and animal literature. Growth retardation in exposed fetuses was common at higher exposure levels. Arnold et al. (**1994**) reviewed the clinical findings in 35 offspring of mothers abusing toluene. The signs were similar to the fetal alcohol syndrome but included blunt finger tips. Only one of the six interviewed mothers abused alcohol.

Ericson et al. (1984) studied 1161 Swedish laboratory workers. In a case control study of 26 early postnatal deaths or newborns with significant defects they found that two had toluene exposure, a rate similar to that among laboratory workers, with normal pregnancy outcomes. The overall rate for these two critical categories among laboratory workers was 3.5 % as compared to the general population with 2.6%. Axelsson et al. (1984) found no increase in spontaneous abortions among 140 laboratory workers exposed to toluene.

Hudak and Ungvary (1978) exposed rats to 1500 mg per square meter of air from day 1 through day 8 or 1000 mg per square meter for 8 hours daily from day 1 through day 21 and found no teratogenic effect. Some fetal growth retardation occurred at the higher dose which also killed some of the mothers. Mice were exposed during days 6–13 to 1500 mg per square meter of air with similar results. Studies in rabbits exposed on days 6–18 to up to 500 ppm revealed no maternal or fetal toxicity (Klimisch et al., **1992**). Nawrot and Staples (1979) gaage fed mice 1.0 mg per kg on days 6 through 15 and found increased cleft palate in the offspring.

Arnold, G.L.; Kirby, R.S.; Langendoerger, S. and Wilkins–Haug, L.: Toluene embryopaathy: clinical delineation and developmental follow-up. Pediatrics 93:216–220, 1994.

Axelsson, G.; Lutz, C. and Rylander, R.: Exposure to solvents and outcome of pregnancy in university laboratory employees. Br J Ind Med. 41:305–312, 1984.

Donald, J.M.; Hooper, K. and Hopenhayn–Rich, C.: Reproductive and developmental toxicity of toluene: a review. Environmental Perspectives 94:237–244, 1991.

Ericson, A.; Kallen, B.; Zetterstrom, R.; Eriksson, M. and Westerholm, P.: Delivery outcome of women working in laboratories during pregnancy. Arch. Environ. Health 39:5–10, 1984.

Goodwin, T.M.: Toluene abuse and renal tubular acidosis in pregnancy. Obstet. Gynecol. 71:715–717, 1988.

Hersh, J.H.; Podrich, P.E.; Rogers, G. and Weisskopf, B.: Toluene embryopathy. J. Pediatr. 106:922–927, 1985.

Hersh, J.H.: Toluene embryopathy: two new cases. J. Med Genetics 26:333–337, 1989.

Hudak, A. and Ungvary, G.: Embryotoxic effects of benzene and its methyl derivatives: Toluene, xylene. Toxicology 11:55–63, 1978.

Klimisch, H.J.; Hellwig, J. and Hofmann, A.: Studies on the prenatal toxicity of toluene in rabbits following ihalation exposure and proposal of a pregnancy guidance value. Arch. Toxicol. 66:373–381, 1992.

Nawrot, P.S. and Staples, R.E.: Embryo–fetal toxicity and teratogenicity of benzene and toluene in the mouse. (abs) Teratology 19:41A, 1979.

Syrovadko, O.N.: Working conditions and health status of women handling organosilicon varnishes containing toluene. Gig. Tr. Prof. Zabol. 12:15–19, 1977.

2394   2,5–Toluene Diamine Sulfate CAS 6369–59–1

Discussed under Hair Dyes

2395   o–Toluene Sulfonamide CAS 88–19–7

Discussed under Saccharin

2396   Torasemide

Ohta et al. (**1994**) gave this diuretic in 30 mg per kg orally to rats during organogenesis. Wavy ribs were increased at 6 and 30 mg doses. No other adverse effects were found and behavior including reproduction of the offspring was normal. In perinatal and postnatal studies the viability was reduced (Ohta et al., **1994** B). In the rabbit, Ohta et al. (**1994** C) found increased resorptions at 1 mg per kg given during organogenesis.

Ohta, T.; Kobayashi, Y.; Kato, M.; Koshiba, H. and Iwai, M.: Teratogenicity study of torasemide in rats. Yakuri to Chiryo 22:s1113–s1132, 1994.

Ohta, T.; Kobayashi, Y.; Kato, M.; Koshiba, H. and Iwai, M.: Perinatal and postnatal development study in rats. Yakuri to Chiryo 22:s1143–s1158, 1994B.

Ohta, T.; Nishiwaki, M.; Kato, M.; Koshiba, H. and Iwai, M.: Teratogenicity study of torasemide in rabbits. Yakuri to Chiryo 22:s1133–s1142, 1994C.

2397   Tordon 202c

This herbicide, a combination of picloram and 2,4–D was studied by Blakely et al. (1989) in male mice. Treatment was given in drinking water for 60 days prior to the day of mating and not to the dams. Fetal weight was decreased and abnormal fetuses were increased in the offspring of the group given 0.42 percent in the drinking water. The malformed fetuses were 4.7 percent in the 0.42 percent group and included ablepharon, cleft palate and unilateral agenesis of the testes. All the males in this high–treatment group died. However, in the highest dosage group (0.84 percent), there was no increase in malformed fetuses. When given with the diet during all of gestation (Blakely et al., 1989) as 0.42 percent or above, increased incidences of cleft palate, renal agenesis and hydronephrosis were found. Undescended testes and umbilical hernias were also increased. The maternal liver weight was also increased.

Blakely, P.M.; Kim, J.S. and Firneisz, G.D.: Effects of preconceptual and gestational exposure to tordon 202c on fetal growth and development in CD–1 mice. Teratology 39:547–554, 1989.

Blakely, P.M.; Kim, J.S. and Firneisz, G.D.: Effects of paternal subacute exposure to tordon 202c on fetal growth and development in CD–1 mice. Teratology 39:237–241, 1989.

2398 Toremifene

Hirsimaki et al. (1990) found that this non–steroidal anti–estrogen did not compromise pregnancy at 50 mg per kg orally in the rat. In the rabbit 10 mg and 50 mg per kg caused total litter loss.

Hirsimaki, Y.; Beltrame, D.; McAnulty, P.; Tesh, J. and Wong, L.: Preliminary investigations of the reproductive consequences of toremifene citrate treatment. Toxicologist 10:223, 1990.

2399 Tosufloxacin

Nakada et al. (1988) gave this antibiotic orally to rats in doses of up to 3,000 mg per kg on days 7–17 and found no teratogenicity or changes in fertility or postnatal behavior. At 500 and 3,000 mg per kg there was an increase in supernummerary cornary orifices.

Nakada, H.; Nakamura, S.; Komae, N.; Sanzen, T.; Nojima, Y.; Akasaka, M.; Nishio, Y. and Yoneda, T.: Reproduction studies of t–3262 in rats. Chemotherapy 36294–319, 1988.

2400 Toxaphene *Polychlorcamphene* CAS 8001–35–2

This chlororganic pesticide was administered per os to pregnant rats for two weeks at doses of 12 mg per kg daily (Badaeva, 1979). It was demonstrated that the substance in question decreased cholinesterase activity of fetal neural structures and hampered differentiation of cardiac neural elements. Chu et al. (1988) studied rats in a two generation study at dietary levels of 4.0 to 500 ppm. No adverse effects on reproduction were found but postnatal weight gain was reduced at the highest dose level. Martson and Shepelskaya (1980) gave this compound to rats from 6 through 15 days (40 mg per kg) or from day 1 through 20 (4 mg per kg) and to hamsters in the same doses from days 7 through 11 or on days 1 through 15. The compound showed slightly teratogenic and gonadotoxic activites.

Badaeva, L.N.: The effect of some pesticides on cholinesterase activity in the cardiac neural elements of pregnant animals and fetuses. Arkh. Anat. Gistol. Embryol. (USSR) 67,4:ll68–1171, 1979.

Chu, I.; Secours, V.; Villeneuve, D.C.; Valli, V.E.; Nakamura, A.; Colin, D.; Clegg, D.J. and Arnold, E.P.: Reproduction study of toxaphene in the rat. J. Environ. Sci. Health B23:101–126, 1988.

Martson, L.V. and Shepelskaya, N.R.: Study of the generative function in animals exposed to polycholorocamphene. Gig. Sanit. (USSR) 5:14–16, 1980.

2401 Toxic Waste Sites

Budnick et al. (1984) studied the residents of a Pennsylvania toxic waste site area and found no increase in birth defects among about 500 pregnancies.

Budnick, L.D.; Sokal, D.C.; Falk, H.; Logue, J.N. and Fox, J.M.: Cancer and birth defects near the Drake Superfund site, Pennsylvania. Arch. Environ. Health 39:409–413, 1984.

2402 Toxiferine

Shoro (1977) produced club foot in rat fetuses by intramuscular injection of 0.001 mg on days 16–19.

Shoro, A.A.: Intra–uterine growth retardation and limb deformities produced by neuromuscular blocking agents in the rat fetus. J. Anat. 123:341–350, 1977.

2403 Toxoplasmosis Gondii

Congenital toxoplasma gondii infection may be acquired through asymptomatic maternal infection. The incidence varies in the newborn population from 0.25 to 8 per 1000. The maternal infection is acquired by ingestion of raw meat (lamb or pork especially) or by oral contamination with infected cat feces. Pregnant women should be made aware of these environmental dangers. A large proportion of infected infants remain asymptomatic and develop normally. Sturcher et al. (1987) found 8 infants out of 10,000 newborns in Switzerland with persistent specific IgM, but none of these were symptomatic. Fuith et al. (1988) detected 20 women with evidence of recent toxoplasmosis from 8,289 pregnancies in Austria. All women were treated with pyrimethamine and salfadiazine and none of the infants had clinical signs of toxoplasmosis. Two infants had specific IgG but none had IgM. Hohlfeld et al. (1990) reported that the addition of spiramycin to a regime of pyrimethamine and sulfonomides significantly reduced the number of severe congenital toxoplasmosis cases. Berrebi et al. (1994) treated 163 mothers with acute toxoplasmosis before 28 weeks using 9 million international units of spiramycin orally. Three fetuses died in utero and 27 had proven congenital toxoplasmosis neonatally but were free from neurologic symptoms at 15 to 71 months of age. Wallon et al. (1994) briefly report on 540 women with acute toxoplasmosis who were treated with sulpha doxine–pyrimeth–amine or spiramycin. Neurologic changes were found by ultrasound in 5 cases and termination performed. Two infants had some visual impairment but none showed neurologic or mental defect.

Women who seroconvert in early pregnancy have a much higher chance of giving birth to infants with the clinical syndrome than do those converting in later pregnancy. Those with the syndrome may have hepatosplenomegaly, icterus, maculopapular rash, chorioretinitis and cerebral calcifications. Hydrocephalus or microcephalus may develop and lead to functional defects. There is no evidence that a woman with infection prior to pregnancy can infect her fetus. Berrebi and Kobuch (**1994**) drew mainly on sonographic evidence on microcephaly in making a poor prognosis. The subject was reviewed by Feldman (1968), Warkany (1971) and Remington and Desmonts (1990).

The role of toxoplasmosis in the etiology of repeated spontaneous abortion is controversial. Langer (1963) claims that a high proportion of habitual or repeated abortions, prematures and stillbirths in Germany are due to toxoplasmosis. Other studies have shown a much lower incidence. Kimball et al. (1971) in a prospective study of 5,000 New York women, found that toxoplasma antibodies were not associated with habitual abortion and could not culture toxoplasma from 260 spontaneous abortions. A significant increase from 31 to 38 percent positive dye tests was observed in the group of women who reported previous spontaneous abortions as compared to those without abortions.

Daffos et al. (1988) treated 15 pregnancies with spiramycin and pyrimethamine with either sulfadoxine or sulfadiazine. Two infants had choreoretinitis but all 15 had a normal neurologic status up to 30 months of age.

Berrebi, A.; Kobuch, W.E.; Bessieres, M.H.; Bloom, M.C.; Rolland, M.; Sarramon. M.F.; Roques, C. and Fournie, A.: Termination of pregnancy for maternal toxoplasmosis. The Lancet 344:36–39, 1994.

Berrebi, A. and Kobuch, W.E.: Toxoplasmosis in pregnancy. The Lancet 334:950, 1994.

Daffos, F.: Prenatal management of 746 pregnancies at risk for congenital toxoplasmosis. N. Eng. J. Med. 318:271–275, 1988.

Feldman, H.A.: Toxoplasmosis. Medical Progress 279:1371–1431, 1432–1437, 1968.

Fuith, L.C.; Reibnegger, G.; Honlinger, M. and Wachter, H.: Screening for toxoplasmosis in pregnancy. Lancet 2:1196, 1988.

Hohlfeld, P.; Daffos, F.; Thulliez, P.; Aufrant, C.; Couvreur, J.; MacAleese, J.; Descombey, D. and Forestier, F.: Fetal toxoplasmosis: outcome of pregnancy and infant follow-up after in utero treatment. J. Pediat. 115(5):765–769, 1989.

Kimball, A.C.; Kean, B.H. and Fuchs, F.: The role of toxoplasmosis in abortion. Am. J. Obstet. Gynecol. III:219–226, 1971.

Langer, H.: Repeated congenital infection with toxoplasma gondii. Obstet. Gynecol. 21:318–329, 1963.

Remington, J.S. and Desmonts, G.: Toxoplasmosis. In: Infectious Diseases of the Fetus and Newborn Infant, 3rd ed., J.S. Remington, J.O. Klein and W.B. Saunders (eds), Philadelphia, 1990.

Sturcher, D.; Berger, R. and Just, M.: Die konnatale Toxoplasmose in der Schweiz. Schweiz Med. Wschr. 117:161–168, 1987.

Wallon, M.; Gandihon, F.; Peyron, F. and Mojon, M.: The Lancet 344:541, 1994.

Warkany, J.: Congenital Malformations Notes and Comments. Chicago: Year Book Medical Publishers, 78–81, 1971.

2404 Tramadol
1–(m–Methoxyphenyl)–2–dimethylaminomethyl

cyclohexanol hydrochloride CG–315

Yamamoto et al. (**1972**) treated mice and rats orally and subcutaneously during organogenesis. No teratogenicity was found. Maximum dosage subcutaneously was 120 mg per kg in the mice and 60 mg per kg in the rats.

Yamamoto, H.; Kuchii, M.; Hayano, T. and Nishino, H.: A study on teratogenicity of both CG–315 and morphine in mice and rats. Oyo Yakuri 6:1055–1069, 1972.

2405 Trandolapril *RU44570*

Matsuura et al. (**1993** A) treated rats found decreased implantation at doses of 30-300 mg per kg but no adverse fetal effects in the offspring of adults treated prenatally and during the first 7 days of pregnancy. Matsuura et al. (**1993** B) gave up to 300 mg per kg orally during organogenesis to rats but found only slightly reduced weight at the highest dose. In perinatal studies at 30 and 300 mg per kg dilation of the renal pelves and increased water intake was found postnatally (Matsuura et al., **1993** C).

Matsuura, T.; Kurio, W.; Fujishima, H.; Kumagai, Y.; Narama, I.; Hiramatsu, Y. and Takabatake, E.: Fertility study of trandolapril (RU44570) in rats. J. Tox. Sc. 18:93–105, 1993A.

Matsuura, T.; Kurio, W.; Maeda, H.; Kumagai, Y.; Narama, I.; Hiramatsu, Y. and Takabatake, E.: Teratological study of trandolapril (RU44570) in rats. J. Tox. Sc. 18:107–132, 1993B.

Matsuura, T.; Kurio, W.; Fujishima, H.; Kumagai, Y.; Narama, I.; Hiramatsu, Y. and Takabatake, E.: Perinatal and postnatal study of trandolapril (RU44570) in rats. J. Tox. Sc. 18:133–159, 1993C.

2406 Tranexamic Acid
Trans–4–(aminomethyl)cyclohexane–carboxylic Acid CAS 1197–18–8

This antifibrinolytic agent was given orally to mice and rats during organogenesis in doses as high as 1500 mg per kg per day. No fetal effects or teratogenicity occurred (Morita et al., 1971).

Morita, H.; Tachizawa, H. and Okimoto, T.: Evaluation of the safety of tranexamic acid. 3. Teratogenic effects in mice and rats. Oyo Yakuri 5:415–420, 1971.

2407 Tranilast *N–(3–4–Dimethoxycinnamoyl)*
Anthranilic Acid CAS 53902–12–8

Nakazawa et al. (1978) found no effect of 600 mg per kg given orally to rats on days 7 through 17. No adverse effects on the rabbit fetus were observed after oral doses of 750 mg per kg on days 6 through 18 (Iwadare et al., 1978).

Iwadare, M.; Ooba, M.; Tsukamoto, T.; Imai, K. and Nakazawa, M.: Effects of N–(3–4–dimethoxycinnamoyl) anthranilic acid (N–5) administered orally on reproductive performance in rabbits. Teratogenicity test. (Japanese) Iyakuhin Kenkyu 9:187–193, 1978.

Nakazawa, M.; Ooba, M.; Imaki, K. and Iwadare, M.: Effects of N–(3–4–dimethoxycinnamoyl) anthranilic acid (N–5) administered orally on reproductive performance of rats.

2. Teratogenicity. (Japanese) Iyakyuhin Kenkyu 9:161–172, 1978.

2408 Transgenic Insertion

Discussed under Lens Ablation; Limb Defects; Pancreatic Ablation; Pituitary Ablation and Syndactyly.

2409 Tranylcypromine *2–Phenyl Cyclopropylamine*

Poulson and Robson (1963) gave 0.1 mg per day orally to pregnant mice on days 1–6, 4–7, 6–11 and 11–16 and produced no decrease in the number of litters. Maternal weight was reduced at the highest dose but no fetal changes occurred.

Poulson, E. and Robson, J.M.: The effect of amine oxidase inhibitors on pregnancy. J. Endocrin. 27:147–152, 1963.

2410 Trapidil *Trapymin*

Ito et al. (1976) gave up to 120 mg per kg on days 6–18. Maternal weight was reduced at the highest dose but no fetal changes occurred.

Ito, C.; Shibutani, Y.; Nakano, K. and Ohnishi, H.: Toxicological studies of trapymin. 5. Teratological studies of trapymin in rabbits. Iyakuhin Kenkyu 7:195–199, 1976.

2411 Trauma also see *Laparotomy Surgery*

There are only a small number of convincing case reports where trauma contributed to production of a human anomaly (Hinden, 1965). Brodsky (1983) received five reports of women who were operated upon during pregnancy. No increase in congenital defects among the offspring was determined but one study found an increased spontaneous abortion rate. Mazze and Kallen (1989) studied 5,405 Swedish women undergoing surgery during pregnancy and found no increase in stillbirths or malformations. Infants with birth weights below 1500 gm were increased with a relative risk of 2.2 (1.8–2.8 interval). Kallen and Mazze (1990) reviewed 2,252 Swedish births following first trimester operations and found 6 with neural tube defects. Five hundred and seventy–two had operations during gestation of weeks 4–5 and found 5 of the 6 cases with defects. They stated that if there is a causal relationship to any factor in surgery the risk "could be eight to nine times, or even more, the absolute risk for an NTD". Ford and Picker (**1989**) report a case of intrauterine brain hemorrhage and death following a car accident. Williams et al. (**1990**) studied 84 pregnancies in which significant abdominal trauma occurred. Abruption was detected in two and 28% had preterm labor most of which were successfully treated by tocolysis.

Brent and Franklin (1960) reported the effect of uterine vascular clamping in the rat. With clamping for 1.5 to 3 hours on the 9th day, they produced 16.7 percent malformations which included eye, kidney and vascular systems as well as anencephaly and omphalocele. Webster et al. (1987) clamped the fat near the uterine artery or handled the uterus for 5 minutes and produced fetal hemorrhages followed by reduction defects in the rat fetus. Handling the uterus through the abdominal wall did not produce defects.

Brodsky, J.B.: Anesthesia and surgery during early pregnancy and fetal outcome. Clin. Obstet. Gynecol. 26:449–457, 1983.

Brent, R.L. and Franklin, J.B.: Uterine vascular clamping: New procedure for the study of congenital malformations. Science 132:89–91, 1960.

Ford, R.M. and Picker, R.H.: Fetal head–injury following motor–vehicle accident; an unusual case of intrauterine death. Aust. Nz. J. O. 20(1):72–73, 1989.

Hinden, E.: External injury causing foetal deformity. Arch. Dis. Child. 40:80–81, 1965.

Kallen, B. and Mazze, R.I.: Neural tube defects and first trimester operations. Teratology 41:717–720, 1990.

Mazze, R.I. and Kallen, B.: Reproductive outcome after anesthesia and operation during pregnancy: a registry study of 5405 cases. Amer Journal of Obstetrics and Gnyecology, 161:1178–1185, 1989.

Webster, W.S.; Lipson, A.H. and Brown–Woodman, P.D.C.: Uterine trauma and limb defects. Teratology 35:253–260, 1987.

Williams, J.K.; McClain, L.; Rosemurgy, A.S. and Colorado, N.M.: Evaluation of blunt abdominal trauma in the third trimester of pregnancy: Maternal and fetal considerations. Obstetrics and Gynecology 75:33–37, 1990.

2412 Traxanox CAS 58712–69–9

This anti–allergic agent was tested in rats by Imanishi et al. (1983). For the fertility and perinatal studies maximum oral doses of 500 mg per kg were used and for the studies on days 7–17, 2500 mg were employed. No adverse effects on reproduction were found.

Imanishi, M.; Takeuchi, M. and Kato, Y.: Reproduction studies on traxanox sodium in rats. Iyakuhin Kenkyu 14:667–708, 1983.

2413 Trazodone *Desyrel™*

Rosa (**1994**) found only one defect among 112 offspring exposed. Barcellona (1970) gave rats and rabbits oral doses of up to 210 and 75 mg per kg, respectively. No teratogenicity was found.

Barcellona, P.S.: Investigations on the possible teratogenic effects of trazodone in rats and rabbits. Boll. Chim. Farm. 109:323–332, 1970.

Rosa, F.: Medicaid antidepressant pregnancy exposure outcomes. (abs) Reproductive Toxicology 8:444–445, 1994.

2414 Triacetyl–6–azauridine

Discussed under 6–Azauridine

2415 3 β,23,N–Triacetylveratramine

Keeler and Binns (1968) gave 0.9–1.1 gm on the 14th day of gestation to 8 ewes and found no craniofacial defects in the offspring.

Keeler, R.F. and Binns, W.: Teratogenic compounds of Veratrum californicum (Durand). Teratology 1:5–10, 1968.

2416 Triamcinolone *Aristocort™* CAS 124–94–7

Walker (1965) produced cleft palates in mice by administering daily as little as 1 microgm on days 11 through 14.

The difference in cleft–palate–producing doses between cortisone and triamcinolone was 200 times whereas their therapeutic effect in man differs by only about 6 times. Parker and Hendrickx (1983) produced craniofacial and brain malformations in subhuman primates. For the rhesus they used 10 mg per kg of triamcinolone acetonide injected on days 23, 25, 27, 29 and 31.

The work in the monkey was extended by Tarara et al. (1989) who used 100 times the human equivalent dose and found a spectrum of midline cranial dysraphism in the occipital region near the posterior fontanelle.

Parker, R.M. and Hendrickx, A.G.: Craniofacial and central nervous system malformations induced by triamcinolone acetonide in nonhuman primates: 2 Craniofacial pathogenesis. Teratology 28:35–44, 1983.

Tarara, R.P.; Cordy, D.R. and Hendrickx, A.G.: Central nervous system malformations induced by triamcinolone acetonide in nonhuman primates: Pathology. Teratology 39:75–84, 1989.

Walker, B.E.: Cleft palate produced in mice by human–equivalent dosage with triamcinolone. Science 149:862–863, 1965.

2417 Triazolam Halcion™ CAS 28911–01–5

Matsuo et al. (1979) gave this tranquilizer by gastric gavage to rats and rabbits in maximum total daily doses of 300 and 50 mg per kg respectively. Given during organogenesis, there was no teratogenic effect in either species. In the rabbit, doses of 3 and 50 mg per kg daily caused a decrease in implantations, viable fetuses and fetal weight.

Matsuo, A.; Kast, A. and Tsunenari, Y.: Reproduction studies of triazolam in rats and rabbits. (Japanese) Iyakuhin Kenkyu 10:52–67, 1979.

2418 Tribendimidin

Shao et al. (1988) studied this antihelmintic agent in rats giving oral doses of up to 200 mg per kg on days 8–10 of gestation. No evidence of teratogenicity was found.

Shao, B–R.; Zhan, C–Q.; Xu, Y–Q. and Ha, S–H.: Mutagenicity and teratogenicity tests on anthelmintic agent tribendimidin. Pharmaceutical Industry 19:112–115, 1988.

2419 Tribromoethanol CAS 75–80–9

This anesthestic was tested in explanted rat embryos and at anesthetic dosage, a reduction in embryonic growth as measured by protein increase was found (Kaufman and Steele, 1976).

Kaufman, M.H. and Steele, C.E.: Deleterious effect of an anesthetic on cultured mammalian embryos. Nature 260:782–783, 1976.

2420 Tricaprylin CAS 538–23–8

Ohta et al. (1970) gave 10 ml per kg to mice and observed no fetal effects. Studies with the pregnant rabbit were also negative for teratogenicity.

Ohta, K.; Matsuoka, Y.; Ichikawa, Y. and Yamamoto, K.: Toxicity, teratogenicity and pharmacology of tricaprylin. (Japanese) Oyo Yakuri 4:871–882, 1970.

2421 Trichlorfon O,O–Dimethyl–1–hydroxy–2, 2, 2–trichloroethylphosphonate; Dipterex™ Chlorophos Metrifonate Neguvon CAS 52–68–6

A cluster of defects was reported by Czeizel et al. (1993) in a small Hungarian town where eleven of fifteen had congenital defects. The types of defects varied but 4 had Down syndrome (two with evidence of meiosis II non–dysfunction). The local fish from a farm had a high content of trichlorfon (100 mg per kg).

Kronevi and Backstrom (1977) reported a possible connection between maternal treatment with this organophosphorous compound and congenital tremor with hypoplasia of the cerebellum in the piglet. Knox et al. (1978) fed 60 mg per kg to sows on two days between days 55 and 70 of pregnancy and produced cerebellar hypoplasia in the offspring.

This agent was given to pregnant hamsters and mice during active organogenesis (300 mg per kg daily) by Martson (1979). A low incidence of cleft palate and some skeletal defects in hamster fetuses were found. All of the abnormal fetuses occurred in hamsters treated on day 8. Neither embryotoxicity nor teratogenicity was found in mice. Courtney and Andrews (1980) found an increase in extra ribs in mice treated at 200 and 400 mg per kg. The route and time of administration was not given.

Courtney et al. (1986) gave mice this insecticide at 300 and 400 mg per kg on days 7–16 by intragastric route. Malformations were seen but these levels are maternally lethal. At 200 mg per kg no defects were seen but retarded ossification occurred. In rats using doses up to 200 mg per kg on days 7–19 or 8–20, enlarged renal pelves and minor skeletal changes were reported. Some maternal lethality was seen at the highest dose.

This compound was given by gavage or in the diet to rats on days 6 through 15 (Staples et al., 1976). With gavage at 145 mg per kg per day, malformation of the skull and brain occurred in the fetus as well as skeletal and palate defects. In the diet experiment, 75 mg per kg did not cause malformations. Martson and Voronina (1976) gave 80 mg per kg on day 9 and produced open eye and exencephaly in the rat fetus. Eight mg per kg throughout gestation caused no malformation increase.

Courtney, K.D. and Andrews, J.E.: Extra ribs indicate fetotoxicity and maldevelopment. (abs) Teratology 21:35A, 1980.

Courtney, K.D.; Andrews, J.E. and Springer, J.: Assessment of teratogenic potential of trichlorfon in mice and rats. J. Environ. Sci. Health B21(3):207–227, 1986.

Czeizel, A.E.; Elek, C.; Gundy, S.; Metneki, J.; Nemes, E.; Reis, A.; Sperling, K.; Timar, L.; Tusnady, and Viragh, Z.: Environmental trichlorfon and cluster of congenital abnormalities. The Lancet 341:539–542, 1993.

Knox, B.; Askaa, J.; Basse, A.; Bitsch, V.; Eskildsen, M.; Mandrap, M.; Ottosen, H.E.; Overby, E.; Pedersen, K.B. and Rasmussen, F.: Congenital ataxia and tremor with cerebellar hypoplasia in piglets borne by sows treated with Neguvon vet (metrifonate, trichlorfon) during pregnancy. Nord Vet–Med 30:538–545, 1978.

Kronevi, T. and Backstrom, L.: Kongenital tremor (Skaksjuka) hos gris. Sartryck ur Svensk Veterinartidning 21:837–841, 1977.

Martson, L.V.: Teratological studies on chlorophos in golden hamsters and white mice. Gigiena i Sanitariya (USSR) 7:70-72, 1979.

Martson, L.V. and Voronina, V.M.: Experimental study of the effect of a series of phosphoroorganic pesticides (dipterex and imidan) on embryogenesis. Environ. Health Persp. 13:121–125, 1976.

Staples, R.E.; Kellam, R.G. and Haseman, J.K.: Development toxicity in the rat after ingestion or gavage of organophosphate pesticides (dipterex, imidan) during pregnancy. Environ. Health Perspect. 13:133–140, 1976.

2422 Trichlormethiazide

This thiazide diuretic was used in 405 pregnancies and no increase in congenital defects observed (Heinonen, et al. **1977**). Only two of these women were treated in the first trimester.

Heinonen, O.P.; Slone, D. and Shapiro, S.: Birth Defects and Drugs in Pregnancy. Littleton, Massachusetts: John Wright–PSG, INC., p 441, 1977.

2423 Trichloroacetic Acid

Smith et al. (1988) fed rats this acid after neutralization on days 6–15 in amounts up to 1600 mg per kg per day. Resorptions were increased at the 800 mg level. An increase in heart and limb defects was found at the higher levels (300 mg and above). Maternal weight gain was reduced at the 800 mg level. Heinenon et al. (1977) reported on 405 women who were treated and found no increase in defects. Only two women were treated in the first trimester.

Heinonen, O.P.; Slone, D. and Shapiro, S.: Birth Defects and Drugs in Pregnancy. John Wright Publishing Sciences Group, Inc., Littleton, Mass. 1977.

Smith, M.K.; Randall, J.L. and Stober, J.A.: Teratogenic activity of trichloroacetic acid in the rat. Teratology 40:445–451, 1989.

2424 1,2,3–Trichloro–1,2–butadiene

The male rats were subjected to chronic inhalation of 1,2,3–trichloro–1,3–butadiene at the doses 1.77 mg/m^3, 5.9, 15.6 mg/m^3 for 2.5 months. This agent induced mutagenic as well as embryotoxic effects. The threshold concentration of mutagenic action (1.77 mg/m^3) was found to be 3 times lower than of embryotoxic and gonadotoxic action. (Petrosyan et al. **1989**)

Petrosyan, F.R.; Nalbandyan, T.L. and Gizhlaryan, M.S.: Indirect effects of 1,2,3–trichloro–1,3–butadiene on rats. Biol. Zh. Arm. 42(5):470–473, 1989.

2425 1,1,1–Trichloroethane *Methylchloroform* CAS 71–55–6

Schwetz et al. (1975) exposed pregnant mice and rats to this vapor in concentrations of 875 ppm. Both species were exposed for 7 hour daily periods on days 6 through 15 of gestation. No fetal toxicity or teratogenicity was found. Lane

et al. (1982) exposed mice to 1,000 mg per kg in drinking water and found no adverse effects on reproduction. York et al. (1982), using 2100 ppm exposure (6 hours per day) during the rat gestation, found some reduction in fetal weight and delay in osseous development. Their postnatal studies revealed no differences from controls. Deane et al. (**1989**) investigated a cluster of women exposed to a water source contaminated with trichloroethane. Over 250 exposed and unexposed women were interviewed. In the exposed the odds ration for spontaneous abortion was 2.3 (95% CI 1.3–4.2) and for congenital defects it was 3.1 (95% CI 1.1–10.4). Because of lack of precise exposure information they felt no causal inference could be made. No pattern of defect was found among the 13 infants. Swann et al. (**1989**) studied the same population and found 12 infants with congenital heart defect when 6 were expected. The temporal distribution of the cases did not match the exposure period suggesting that the solvent leaks was not related to the excess.

Deane, M.; Swan, S.H.; Harris, J.A.; Epstein, D.M. and Neutra, R.R.: Adverse pregnancy outcomes in relation to water contamination, Santa Clara county, California, 1980–1981. American Journal of Epidemiology 129:892–904, 1989.

Lane, R.W.; Riddle, B.L. and Borzelleca, J.F.: Effects of 1, 2–dichloroethane and 1,1,1–trichloroethane in drinking water on reproduction and development in mice. Toxicol. Appl. Pharmacol. 63:409–421, 1982.

Schwetz, B.A.; Leong, B.K.J. and Gehring, P.J.: The effect of maternally inhaled trichloroethylene, perchloroethylene, methyl chloroform and methylene chloride on embryonal and fetal development in mice and rats. Toxicol. Appl. Pharmacol. 32:84–96, 1975.

Swan, S.H.; Shaw, G.; Harris, J.A. and Neutra, R.R.: Congenital cardiac anomalies in relation to water contaminations, santa clara, county, california 1981–1983. Am. J. Epidemiol. 129:885–893, 1989.

York, R.G.; Sowry, B.M.; Hastings, L. and Manson, J.M.: Evaluation of teratogenicity and neurotoxicity with maternal inhalation exposure to methyl chloroform. J. Toxicol. Environ. Health 9:251–266, 1982.

2426 Trichloroethylene CAS 79–01–6

Schwetz et al. (1975) exposed pregnant mice and rats to 300 ppm. Both species were exposed to seven hour daily periods on days 6 through 15 of gestation. No fetal toxicity or teratogenicity was found. Dorfmueller et al. (1979) exposed female rats six hours daily for two weeks before pregnancy and for the first 20 days of gestation to 1800 ppm and found anomalies of skeletal and soft tissues which were considered to be indicative of developmental delay. Behavioral evaluation of the offspring showed no effects from the treatment. Sperm examination from mice exposed to 0.3 percent for four hours daily for five days revealed increased abnormalities after 28 days (Land et al., 1981). Dawson et al. (**1990**) continuously perfused the rat uterine cavity from day 7 to term with 15 or 1,500 ppm of trichloroethylene in saline. The control treated fetuses had 1.5% heart defects while at the two treatment levels 28 and 28% had heart defects. Defects of the valves and septa were noted and all examinations were carried out on coded specimens. Coberly et al. (**1992**) treated mice with up to 483 mg per kg while ova were traversing pronuclear stages of development. No changes in embryo cell proliferation were found.

Cosby and Dukelow (**1992**) treated mice by gavage with up to one-tenth the LD50 on various days before implantation and during organogenesis and found no adverse reproductive effects.

Coberly, S.; Oudiz, D.J.; Overstreet, J.W. and Wiley, L.M.: Effects of maternal exposure to trichloroethylene (tce) on cell proliferation in the mouse preimplantation embryo. Reproductive Toxicology 6:241–245, 1992.

Cosby, N.C. and Dukelow, W.R.: Toxicology of maternally ingested trichloroethylene (tce) on embryonal and fetal development in mice and of tce metabvolites on in vitro fertilization. Fundamental and Applied Toxicology 19:268–274, 1992.

Dawson, B.V.; Johnson, P.D.; Goldberg, S.J. and Ulreich, J.B.: Cardiac teratogenesis of trichlorethylene and dichloroethylene in a mammalian model. J. Am. Coll. Cardiol. 16:1304–1309, 1990.

Dorfmueller, M.A.; Henne, S.P.; York, R.G.; Bornschein, R.L. and Manson, J.M.: Evaluation of teratogenicity and behavioral toxicity with inhalation exposure of maternal rats to trichloroethylene. Toxicology 14:153–166, 1979.

Land, P.C.; Owen, E.L. and Linde, H.W.: Morphological changes in mouse spermatozoa after exposure to inhalational anesthetics during early spermatogenesis. Anesthesiology 54:53–66, 1981.

Schwetz, B.A.; Leong, B.K.J. and Gehring, P.J.: The effect of maternally inhaled trichorethylene, perchloroethylene, methyl chloroform and methylene chloride on embryonal and fetal development in mice and rats. Toxicol. Appl. Pharmacol. 32:84–96, 1975.

2427 2,4,5–Trichlorophenol CAS 95–95–4

Hood et al. (1979) administered this compound by gavage to pregnant mice in single doses of 800–900 mg per kg or multiple doses of 250–300 mg per kg and found no significant fetal effects.

Hood, R.D.; Patterson, B.L.; Thacker, G.T.; Sloan, G.C. and Szczech, G.M.: Prenatal effects of 2,4,5T, 2,4,5–trichlorophenol and phenoxyacetic acid in mice. J. Environ. Sci. Health C13:189–204, 1979.

2428 2,4,5–Trichlorophenoxyacetic Acid 2,4,5–T CAS 93–76–5

Since this chemical is a widely used herbicide, more than 18 cytogenetic studies in animals and of human cells have been published (see review by Friedman, 1984). Cytogenetic aberrations were found in human lymphocytes exposed in vitro (Fujita et al., 1975). Exposed humans sometimes had cytogenetic changes (see Friedman, 1984).

Nelson et al. (1979) reviewed and contributed new data on the effect of 2,4,5–T on the human. By analysis of 1,201 cases of cleft palate over 32 years in the rice growing area of Arkansas, they could establish no relationship between 2,4,5–T use per county and incidence of clefts. The rate of several congenital defects in Hungary did not change with the introduction and heavy use of 2,4,5–T (Thomas, 1980). Field and Kerr (1979) noted that the neural tube defect rate in Australia was parallel to the amount of herbicides used. Smith et al. (1981,1982) in New Zealand, and Townsend et al. (1982) found no increase in defects or abortions among wives of exposed workers.

At doses of 200 mg per kg at the time of ovulation, dominant lethality and chromosomal defects were found in albino rats by Chebotar (1980). In a study by Chebotar et al. (1978), female rats received by gavage 2,4,5–trichlorphenoxyacetic acid in the dose of 400 mg per kg, commercial sample of butyl ester of 2,4,5–T in the dose of 50, 100, 200 mg per kg and purified ester of 2,4,5–T in the dose of 800 mg per kg daily on the 9th through 14th day of gestation. Embryotoxic effect of 2,4,5–T was highest in the group of animals treated on the 9th day of gestation, and that of butyl ester of 2,4,5–T on the 10th day. Immobilization of pregnant rats (two hours daily) significantly increased teratogenic action of 2,4,5–T and butyl ester of 2,4,5–T. Nelson et al. (**1992**) could find no differences between the technical and analytical grade of this chemical as tested in mice. Dose responses for developmental toxicity in 4 strains of mice were given by Holson et al. (**1992**).

Emmerson et al. (1971) gave rats 1 to 24 mg by mouth on days 6 through 15 and rabbits 10 to 40 mg on days 6 through 18. No effects on litter size, fetal weight or incidence of congenital defects were found. Previous findings by Courtney et al. (1970) and Courtney and Moore (1971) about teratogenicity of 2,4,5–T may be explained by dioxin contaminants in the material. Roll (1971) reported cleft palate production in mice with a preparation contaminated with less than 0.1 ppm of the dioxane.

Most of the epidemiologic literature was discussed under Agent Orange.

Chebotar, N.A.: Cytogenetic and morphogenetic changes in oogenesis and embryogenesis induced by primethamine and 2,4,5–trichlorophenoxyacetic acid in meiotic oocytes in albino rats. (Russian) Genetika 16:1220–1227, 1980.

Chebotar, N.A.; Grebenic, L.A. and Kirilenko, A.I.: Effect of stress (immobilization) on embryotoxic and teratogenic action of 2,4,5–trichlorphenoxyacetic acid and its derivatives. Arkh. Anat. Gistol. Embryol. (USSR) 75,8:30–35, 1978.

Courtney, K.D.; Gaylor, D.W.; Hogan, M.D.; Falk, H.L.; Bates, R.R. and Mitchell, I.: Teratogenic evaluation of 2,4,5–T. Science 168:864–866, 1970.

Courtney, K.D. and Moore, J.A.: Teratology studies with 2,4,5–trichlorophenoxyacetic acid and 2,3,7,8–tetrachlorodibenzo–p–dioxin. Toxicol. Appl. Pharmacol. 20:396–403, 1971.

Emmerson, J.L.; Thompson, D.J.; Strebing, R.J.; Gerbig, C.G. and Robinson, V.B.: Teratogenic studies on 2,4,5–trichlorophenoxyacetic acid in the rat and rabbit. Food Cosmet. Toxicol. 9:395–404, 1971.

Field, B. and Kerr, C.: Herbicide use and incidence of neural tube defects. Lancet 1:1341–1342, 1979.

Friedman, J.M.: Does agent orange cause birth defects? Teratology 29:193-221, 1984.

Fujita, K.; Fujita, H.M. and Funazaki, Z.: Chromosomal abnormality caused by 2,4,5–T. J. Jpn. Assoc. Rural Med. 24:77–79, 1975.

Holson, J.F.; Gaines, T.B.; Nelson, C.J.; LaBorde, J.B.; Gaylor, D.W.; Sheehan, D.M. and Young, J.F.: Developmental toxicity of 2,3,5–trichlorophenoxyacetic acid (2,4,5–t) I. multireplicated dose–response studies in four inbred strains and one outbred stock of mice. Fundamental and Applied Toxicology 19:286–297, 1992.

Nelson, C.J.; Holson, J.F.; Green, H.G. and Gaylor, D.W.: Retrospective study of the relationship between agricultural use of 2,4,5–T and cleft palate occurrence in Arkansas. Teratology 19:377–384, 1979.

Nelson, C.J.; Holson, J.F.; Gaines, T.B.; LaBorde, J.B.; McCallum, W.F.; Wolff, G.L.; Sheehan, D.M. and Young, J.F.: Developmental toxicity of 2,4,5–trichlorophenoxyacetic acid (2,4,5–t) II. Multireplicated dose–response studies with technical and analytical grades of 2,4,5–t in four–way outcross mice. Fundamental and Applied Toxicology 19:298–306, 1992.

Roll, R.: Untersuchungen uber die Teratogene Wirkung von 2,4,5–T bei Mausen. Food. Cosmet. Toxicol. 9:671–676, 1971.

Smith, A.H.; Fisher, D.O.; Pearce, N. and Chapman, C.J.: Congenital defects and miscarriages among New Zealand 2,4,5–T sprayers. Arch. Environ. Health 37:197–200, 1982.

Smith, A.H.; Matheson, D.P.; Fisher, D.O.; Chapman, C.J.: Preliminary report of reproductive outcomes among pesticide applications using 2,4,5–T. N.Z. Med. J. 93:177–179, 1981.

Thomas, H.F.: 2,4,5–T use and congenital malformation rates in Hungary. Lancet 2:214–215, 1980.

Townsend, J.C.; Bodner, K.M.; Van Peenen, P.F.D.; Olson, R.D. and Cook, R.R.: Survey of reproductive events of wives of employees exposed to chlorinated dioxins. Am. J. Epidemiol. 115:695–713, fb82.

2429 1,2,3 Trichloropropane

Hardin et al. (1981) injected rats intraperitoneally with 37 mg per kg on days 1–15. Maternal toxicity but not fetal toxicity or teratogenicity was found.

Hardin, B.D.; Bond, G.P.; Sikov, M.R.; Andrew, F.D.; Beliles, R.P. and Niemeier, R.W.: Testing of selected workplace chemicals for teratogenic potential. Scand. J. Work Environ. Health 7(4):66–75, 1981.

2430 Trichosanthin

Chan et al. (1993) injected this protein from the tubers of Trichosanthes Kirilowii intraperitoneally to mice on day 8. At 5 mg per kg or above fetal weight was reduced and stunting, skeletal defects and exencephaly were increased.

Chan, W,Y,; Ng, T.B.; Wu, P.U. and Yeung, H.W.: Developmental toxicity and teratogenicity of trichosanthin, a
ribosome–inactivating protein, in mice. Teratogenesis, Carcinogenesis, and Mutagenesis 13:47–57, 1993.

2431 Triclabendazole

Yoshimura (1987) administered this antihelmintic to rats by gavage on days 8–15 in doses between 10 and 200 mg per kg. No resorption or defect rate increase was found but some fetal weight reduction was found over 100 mg per kg.

Yoshimura, H.: Teratogenic evaluation of triclabendazole in rats. Toxicology 43:283–287, 1987.

2432 Triclopyr *3,5,6–Trichloro–2–pyridylacetic Acid* CAS 55335–06–3

Hanley et al. (1984) studied this herbicide in the pregnant rat and rabbit. No teratogenicity was found in the rat with doses up to 200 mg per kg during days 6 through 15. At doses of 25 mg per kg in the rabbit, no teratogenicity was observed.

Hanley, T.R., Jr.; Thompson, D.J.; Palmer, A.K.; Beliles, R.P. and Schwetz, B.A.: Teratology and reproduction studies with triclopyr in the rat and rabbit. Fund. Appl. Toxicol. 4:872–882, 1984.

2433 Triclosan CAS 3380–34–5
2,4,4'–Trichloro–2'–hydroxydiphenylether

This antimicrobiol agent was studied in mice, rats and rabbits at doses up to 10 mg per kg. No tertaogenicity was found but at the highest dose ranges maternal and fetal toxicity occurred. Kawashima et al. (1987) gavaged rats on days 7–17 with up to 400 mg per kg. Maternal toxicity occurred at the highest dose but no fetal death or malformation increase was found.

DeSalva, S.J.; Kong, B.M. and Lin, Y.J.: Triclosan: a safety profile. Am. J. Dent. 2:185–196, 1989.

Kawashima, K.; Nakaura, S.; Yamaguchi, M.; Tanaka, S. and Takanaka, A.: Effects of triclosan on fetal developments of rats. Bull. Natl. Inst. Hyg. Sci. 105:27–32, 1987.

2434 Tridemorph *N–Tridecyl–2,6–dimethylmorpholine* CAS 24602–86–6

This fungicide which is 83 percent of Calixin™ was studied in mice and rats by Merkle et al. (1984). No embryotoxicity was found at levels of 27.5 mg and 20.6 mg per kg for mice and rats, respectively. They were not able to reproduce the work of Stenberg et al. (1981) who found teratogenic effects at 0.6 mg per kg. Merkle et al. (1984) found no teratogenicity at 3.9 mg per kg of Calixin™

Merkle, J.; Schulz, V. and Gelbke, H.P.: An embryotoxicity study of the fungicide Tridemorph and its commercial formulation Calixin. Teratology 29:259–269, 1984.

Stenberg, A.J.; Zaeva, G.N.; Rysina, T.Z. and Gavrilenko, E.V.: The teratogenic effect of the fungicide Calixin. Voprosi Pitanija 6:55–61, 1981.

2435 Tridiphane *2–(3,5 Dichlorophenyl)–2–(2,2,2–trichloroethyloxirane)*

Hanley et al. (1987) treated pregnant rats and mice with this herbicide. At 75 and 250 mg per kg maternal toxicity occurred in the mouse and when used on days 6–15, a number of animals had complete early resorptions. On days 8–15 treatment was associated with an increase in cleft palate. In rats treated on days 8–15, retardation of the fetal skeleton was found at 200 mg per kg.

Hanley, T.R., Jr.; John–Greene, J.A.; Hayes, W.C. and Rao, D.S.: Embryotoxicity and fetotoxicity of orally administered tridiphane in mice and rats. Fund. Appl. Toxicol. 8:179–187, 1987.

2436 Triethylene Glycol Dimethyl Ether

George et al. (1987) gavaged mice during organogenesis with up to 1000 mg per kg. At the highest dose, neural, craniofacial and skeletal defects were increased and at 500 mg per kg fetal growth retardation was seen. No adverse effects were seen at 250 mg per kg.

Further information is given under the heading, Ethylene Glycol–Monoethyl Ether.

George, J.D.; Price, C.J.; Kimmel, C.A. and Marr, M.C.: The developmental toxicity of triethylene glycol dimethyl ether in mice. Fund. Appl. Toxicol. 9:173–181, 1987.

2437 Triethylenemelamine *TEM* CAS 51–18–3

Sieber and Adamson (1975) summarized the findings in three women treated with TEM. Two received it early in pregnancy and one had a spontaneous abortion and the other had a normal infant. The third received 5 mg every 3–6 days from 5.5 months to term and gave birth to a normal infant.

Thiersch (1957) reported that this compound was embryolethal in the rat. Chaube and Murphy (1968) injected 0.3 to 0.6 mg per kg on the 11th or 12th gestational day of the rat and produced defects of the central nervous system, palate and skeleton. Jurand (1959) studied the embryopathogenesis of this drug in the mouse embryo and reported growth retardation, enlargement of the myocele, somite degeneration and other diverse defects. Kageyama and Nishimura (1961) described the multiple defects found in the term mouse fetus. Sanyal et al. (1981) exposed rat embryos in vitro to 1–5 micrograms per ml and retarded growth was found. Tanaka et al. (**1992**A&B) studied pregnant mice exposed to up to 12,000 mg per ml of drinking water during gestation. Hemorrhages and various brain abnormalities increased at levels of 6,000 and 12,000 mg per kg. Fetal copper was decreased.

Chaube, S. and Murphy, M.L.: The teratogenic effects of the recent drugs active in cancer chemotherapy. In: Advances in Teratology, D.H.M. Woollam (ed.), New York: Academic Press 3:181–237, 1968.

Jurand, A.: Action of triethanomelamine (TEM) on early and late stages of mouse embryos. J. Embryol. Exp. Morphol. 7:526–539, 1959.

Kageyama, M. and Nishimura, H.: Developmental anomalies in mouse embryos induced by triethylene melamine (T.E.M.). Acta Med. Univ. Kyoto 37:318–327, 1961.

Sanyal, M.K.; Kitchin, K.T. and Dixon, R.L.: Rat conceptus development in vitro. Comparative effects of alkalating

agents. Toxicol. Appl. Pharmacol. 57:14–19, 1981.

Sieber, S.M. and Adamson, R.H.: Toxicity of antineoplastic agents in man. Chromosomal aberrations, antifertility effects, congenital malformations and carcinogenic potential. In: Advances in Cancer Research, G. Klein and S. Weinhouse (eds.), Academic Press, 22:57–155, 1975.

Tanaka, H.; Yamanouchi, M.; Imai, S. and Hayashi, Y.: Low copper and brain abnormalities in fetus from triethylene tetramine dihydrochloride treated pregnant mouse. J. Nutr. Sci. Vitaminol. 38:545–554, 1992.

Tanaka, H.; Inomata, K. and Arima, M.: Teratogenic effects of triethylene tetramine dihydrochloride on the mouse brain. J. Nutr. Sci. Vitaminol. 39:177–188, 1993.

Thiersch, J.B.: Effect of 2,4,6, triamino–"S"–triazine (TR), 2,4,6, "Tris" (ethyleneimino)–"S"–triazine (TEM) and N,N',N"–triethylenephosphoramide (TEPA) on the rat litter in utero. Proc. Soc. Exp. Biol. Med. 94:36–40, 1957.

2438 Triethylenephosphoramide *TEPA*

Thiersch (1957) gave rats 5 mg per kg intramuscularly on days 4, 5 or 7, 8 or 11, 12. Stunting and defects of the face and cranium were found.

Thiersch, J.B.: Effect of 2,4,6,triamino–"S"–triazine (TR),2,4,6 "Tris" (ethyleneimino)–"S"–triazine (TEM) and N,N',N"–triethylenephosphoramide (TEPA) on rat litter in utero. Proc. Soc. Exp. Biol. Med. 94:36–40, 1957.

2439 Triethylenetetramine *TETA* CAS 112–24–3

Walsche (1982) reported no malformations among six infants of mothers treated with TETA for Wilson's disease.

Keen et al. (1983) fed rats 0.17, 0.83 and 1.66 percent of TETA in their diets during all of pregnancy. Resorptions were increased at all dosages and at the two highest levels, fetal abnormalities occurred. The abnormalities consisted of hemorrhage and edema. Fetal copper was reduced but zinc was increased. Tanaka et al. (**1992** A,B) studied pregnant mice exposed to up to 12,000 mg per ml of drinking water during gestation. Hemorrhages and various brain abnormalities increased at levels of 6,000 and 12,000 mg per kg. Fetal copper was decreased.

Keen, C.L.; Cohen, N.L.; Lonnerdal, B. and Hurley, L.S.: Teratogenesis and low copper status resulting from triethylenetetramine in rats. Proc. Soc. Exp. Biol. Med. 173:598–605, 1983.

Tanaka, H.; Yamanouchi, M.; Imai, S. and Hayashi, Y.: Low copper and brain abnormalities in fetus from triethylene tetramine dihydrochloride–treated pregnant mouse. J. Nutr. Sci, Vitaminol.38:545–554, 1992.

Tanaka, H.; Inomata, K. and Arima, M.: Teratogenic effects of triethylene tetramine dihydrochloride on the mouse brain. J. Nutr. Sci. Vitaminol. 39:177–188, 1993.

Walsche, J.M.: Treatment of Wilson's disease with trientine (triethylenetetramine) dihydrochloride. Lancet 1:643–647, 1982.

2440 Triethyl Phosphate

Gumbmann et al. (1968) reported that 1 percent in the diet "adversely affected" rat reproduction and 5 percent prevented reproduction. One–half percent had no effect on litter size.

Gumbmann, M.R.; Gagne, W.E. and Williams, S.N.: Short–term toxicity studies of rats fed triethyl phosphate in the diet. Toxicol. Appl. Pharmacol. 12:360–371, 1968.

2441 Trifluridine *FTdR*

This nucleotide analog used topically in the eye was studied by Itoi et al. (1975) who reported absence of teratogenicity in rabbits treated with intraocular doses which were ten times greater than doses of idoxuridine which produced teratogenicity.

Itoi, M.; Gefter, J.W.; Kaneo, N.; Ishii, Y.; Ramer, R.M. and Gasset, A.R.: Teratogenicities of ophthalmic drugs. Arch Ophth. 93:46–51, 1975.

2442 Trifluoperazine also see *Phenothiazines* *Stelazine*™ CAS 117–89–5

In a review of the literature on the teratogenicity of trifluoperizine, Moriarity (1963) reported the use of this drug in over 700 early pregnancies and could not find any increase in congenital malformations. Heinonen et al. (1977) reported the outcome of 42 pregnancies treated in the first four months with trifluoperizine. There was no significant increase in congenital defects.

Wheatley (1964) reported 59 exposures with a 1.7 percent abnormality rate which was not increased over the normal. Schrire (1963) reported that there were no skeletal defects in 478 women who were treated with the drug during pregnancy. Hall (1963) reported one child with a reduction defect in the arm. The mother was treated with Stelazine for 2–3 days around the 25th day of gestation.

In the literature dealing with animal experimentation, Vichi et al. (1968), using doses of .05 to .8 milligrams per mouse during several four–day periods in mid–pregnancy, produced an increase in cleft palate at the higher dose. Other defects were not mentioned. This dose, if converted to a per kilogram dose based upon an assumption that the mouse is about 20 gm, would be equivalent to 40 milligrams of the drug per kilogram per day. At this very high dose, the mouse might be very sleepy and therefore have partial starvation. Starvation in the mouse is particularly prone to producing cleft palate.

Hall, G.: A case of phocomelia of the upper limbs. Med. J. Austr. 1:449–450, 1963.

Heinonen, O.P.; Slone, D. and Shapiro, S.: Birth Defects and Drugs in Pregnancy. Publishing Sciences Group Inc., Littleton, Mass., 1977.

Moriarity, A.J.: Trifluoperizine and congenital malformations. Can. Med. Assoc. J. 88:97, 1963.

Schrire, I.: Trifluoperizine and foetal anomalies. Lancet 1:174, 1963.

Vichi, F.; Pierleoni, P.; Orlando, S. and Tollaro, I.: Palatoschisi indolte da trifluoperidolo nel topo. Sperimentale 118:245–250, 1968.

Wheatley, D.: Drugs and the embryo. Br. Med. J. 1:630 only, 1964.

2443 2–Trifluoroacetamido–1,3,4–thiadiazole–5–sulfonamide

Maren and Ellison (1972) gave 600 mg per kg subcutaneously on days 10 and 11 to rats and did not produce defects.

Maren, T.H. and Ellison, A.C.: The teratological effect of certain thiadiazoles related to acetazolamide with a note on sulfanilamide and thiazide diuretics. Johns Hopkins Med. J. 130:95–104, 1972.

2444 5–Trifluoromethyl–2–deoxyuridine CAS 70–00–8

Kury and Crosby (1967) injected the yolk sac of developing chick eggs once during the first 4 days of incubation with 0.1 to 3.0 microgm of this compound. Cleft palate, amelia and other skeletal defects as well as renal hypoplasias were produced.

Kury, G. and Crosby, R.J.: The teratogenic effect of 5–trifluoromethyl–2–deoxyuridine in chicken embryos. Toxicol. Appl. Pharmacol. 11:72–80, 1967.

2445 Trifluperidol *Psicoperidol*™ CAS 749–13–3

Vichi et al. (1968) produced cleft palates in the offspring of mice treated with 0.1 to 0.8 mg per day from the 10th through the 13th day of gestation.

Vichi, F.; Pierleoni, P.; Orlando, S. and Tollaro, I.: Palatoschisi indotte la trifluperidolo nel topo. Spermimentale 118:245–250, 1968.

2446 Trifluralin™ α,α,α–Trifluoro–2,6–dinitro–N,N–di N propyl–p–toluidine CAS 1582–09–8

This herbicide was given by Beck (1981) to mice on days 6–15 of gestation. The dose was 1.0 gm per kg. On day 60 postnatally alizarin red studies of the skeleton revealed an increase in 14th ribs and other minor skeletal defects.

Beck, S.L.: Assessment of adult skeletons to detect prenatal exposure to 2,4,5–T and Trifluralin in mice. Teratology 23:33–55, 1981.

2447 Trihexyphenidyl CAS 144–11–6

This anticholinergic was given to 9 mothers during the first trimester and no increase in major defects was observed (Heinonen et al., 1977).

Heinonen, O.P.; Slone, D. and Shapiro, S.: Birth Defects and Drugs in Pregnancy. Publishing Sciences Group Inc., Littleton, Mass., 1977.

2448 Trimebutine CAS 39133–31–8

Asano et al. (1982) studied this digestive tract antimotility agent in rats and rabbits at oral doses of up to 1000 mg per kg. No fertility decrease or change in postnatal function was found in rats and in neither species was there any change in the fetuses.

Asano, Y.; Fujisawa, K.; Ono, T.; Ariyuki, F. and Higaki, K.: Reproductive studies of trimebutine maleate in rats and rabbits. Kiso to Rinsho 16:633–650, 1982.

2449 Trimellitic Anhydride

Ryan et al. (1989) used this chemical intermediate with known immunologically–based clinical syndromes on rats and guinea pigs. Six hourly exposures to 500 microg per m^3 were used during organogenesis. Although the dams had lung hemorrhage, no teratogenicity was found. With challenge of the pups, no lung foci appeared.

Ryan, B.M.; Hatoum, N.S.; Zeiss, C.R. and Garvin, P.J.: Immuno–teratologic investigation of trimellitic anhydride (TMA) in the rat and guinea pig. Teratology 39:P114, 1989.

2450 Trimethadione

3,5,5–Trimethyl–2,4–oxazolidinedione CAS 127–48–0

German et al. (1970) reviewed 11 women treated during pregnancy with either trimethadione or paramethadione and found an increased incidence of congenital defects in the 11 exposed children. These defects included (4) cleft palate, (3) cardiac anomalies and other types of conditions. Two were normal. Zackai et al. (1975) described a specific phenotype in children born in three families where the mother took trimethadione. The syndrome included developmental delay, v–shaped eyebrows, low–set ears with anteriorly folded helix, high–arched palate and irregular teeth. Feldman et al. (1977) added four cases and summarized the clinical picture from 53 pregnancies.

Brown et al. (1979) administered 543 to 858 mg per kg to mice on days 8 through 10 or 11 through 13 and produced a high incidence of visceral and skeletal malformations. Aortic arch and vertebal defects were the most common. Intraperitoneal routes gave the same results as oral dosing. Fetal weight reduction occurred at dose levels of 35 mg per kg.

Brown, N.A.; Shull, G. and Fabro, S.: Assessment of the teratogenic potential of trimethadione in the CD–1 mouse. Toxicol. Appl. Pharmacol. 51:59–71, 1979.

Feldman, G.L.; Weaver, D.D. and Lovrien, E.W.: The fetal trimethadione syndrome. A. J. Dis. Child. 131:1389–1392, 1977.

German, J.; Kowal, A. and Ehlers, K.H.: Trimethadione and human teratogenesis. Teratology 3:349–362, 1970.

Zackai, E.H.; Melman, W.J.; Neiderer, B. and Hanson, J.W.: The fetal trimethadione syndrome. J. Pediatr. 87:280–284, 1975.

2451 Trimethobenzamide *Tigan*^{TM} CAS 138–56–7

This antinauseant given during the first 84 days in 193 pregnancies was associated with 2.6 percent serious congenital defects in the offspring studied at one year and in 5.8 percent at 5 years. The 5.8 percent incidence was borderline significant (p ¡ 0.05) as compared to non–treated control rates (3.2 percent). No pattern of malformation was identified (Milkovich and Van den Berg, 1976).

Milkovich, L. and Van den Berg, B.J.: An evaluation of the teratogenicity of certain antinauseant drugs. Am. J. Obstet. Gynecol. 125:244–248, 1976.

2452 Trimethoprim *Septra*^{TM} *Co–Trimoxozole* CAS

738–70–5

Williams et al. (1969) treated 120 pregnant women with combined sulfamethoxazole and trimethoprim and found no increase in defects in the offspring. Only 10 of the women were treated before 16 weeks of pregnancy. It was pointed out that trimethoprim has an antifolic acid activity. Czeizel (1990) in a Hungarian study of 5,667 paired (normal and defect) found some increase in exposure among offspring with cleft lip and hypospodias but the time of exposure did not coincide with the critical embryonic periods. Among the group with defects 2.31% were exposed while in the unexposed group 1.25 had defects. He concluded that the maternal disorders treated could explain the higher drug use. Colley et al. (1982) studied 211 offspring of mothers treated with trimethoprim–sulfamethoxazole during pregnancy and found no increase in congenital defects. First trimester use was employed in 83.

Helm et al. (1976) reported studies in the rat and rabbit using the drug in a 1 to 5 dose ratio with sulfamethoxazole. The rats received up to 600 mg per kg of the combination on days 8 through 15 and cleft palate, micrognathia and limb shortening were found in the fetuses. Doses of 180 mg per kg had no effect. In the rabbit treated on days 8 through 14 given 600 mg per kg, there were no malformations but the fetal loss was increased. Sulfamethoxazole alone in corresponding doses was not embryotoxic.

Colley, D.P. and Gibson, K.J.: Study of the use in pregnancy of cotrimoxazole sulfamethizole. Aust J Pharm 63:570–575, 1982.

Czeizel, A.: A case–control analysis of the teratogenic effects of co–trimoxazole. Reproductive Toxicology 4:305–313, 1990.

Helm, F.; Kretzschmar, R.; Leuschner, F. and Neumann, W.: Untersuchungen uber den Einfluss der Kombination Sulfamoxol Trimethoprim (CN 3123) und Fertilitat und Embryonalentwicklung an Ratten und Kaninchen. Arzneim Forsch. 26:643–651, 1976.

Williams, J.D.; Condie, A.P.; Brumfitt, W. and Reeves, D.S.: The treatment of bacteriuria in pregnant women with sulphamethoxazole and trimethoprim. Postgrad. Med. J. Suppl. 45:71–76, 1969.

2453 L–1–(3,4,5–Trimethoxybenzyl)–6, 7–dihydroxy–1,2,3,4–tetrahydroisoquinoline HCl

AQL–208

Kowa et al. (1968) studied this bronchodilator in rats and mice at maximum oral doses of 500 and 1000 mg per kg, respectively, during organogenesis. Maximum doses of 250 mg per kg were also given subcutaneously. No teratogenicity was found but the highest doses were associated with some fetal weight decrease and slight increases in fetal mortality in both species.

Kowa, Y.; Ariyuki, F.; Takashima, I. and Suma, M.: A teratological study of L–1–(3,4,5–trimethoxybenzyl)–6, 7–dihydroxy–1,2,3,4–tetrahydroisoquinoline HCl (AQL–208). Oyo Yakuri 2:383–396, 1968.

2454 3,4,5–Trimethoxy–N–(3–piperidyl)benzamide
KU–54

Sugimoto et al. (1984) studied this drug in pregnant rats, mice and rabbits. In fertility studies and organogenesis studies in the mouse and rat, no adverse fetal effects were seen at up to 800 mg per kg orally. Rabbits received up to 200 mg per kg on days 6–18 and no adverse fetal findings were reported. Perinatal studies using up to 300 mg per kg in the mouse did not reveal behavioral changes. At 800 mg per kg in the rat perinatal some growth retardation was found.

Sugimoto, T.; Suzuki, H.; Irikura, T.; Imai, S.; Abe, H.; Ichiba, S.; Tanase, H.; Miyajima, N.; Hosomi, J. and Imada, O.: Reproduction studies of 3,4,5–trimethoxy–N–(3–piperidyl)benzamide (KU 54) in the mouse, rat and rabbit. Clin. Report 18:33–90, 1984.

2455 Trimethylamine

Guest and Varma (**1991**) gave 2.5 or 5.0 millimoles per kg intravenously on days 1–17 to mice. At the highest dose 5 of 11 dams died and at both doses fetal weight was decreased. No significant increase in congenital defects was found.

Guest, I. and Varma, D.R.: Developmental toxicity of methylamines in mice. Journal of Toxicology and Environmental Health 32:319–330, 1991.

2456 1,1,3–Trimethyl–5–phenylbiuret

Yamakita et al. (**1989**) gave this antirheumatic drug to rats during short periods of organogenesis in doses of 900 or 1800 mg per kg and found ventricular septal defects to be highest when treatment was on day 9 (16–40%). Three strains of rat were affected.

Yamakita, O.; Wakasugi, N.; Tomita, T. and Ito, N.: Critical period and strain differences in ventricular septal defect induction in rats by transplacental treatment with 1,1,300–trimethyl–5–phenylbiuret (ST–281). Toxicologist 9:269, 1989.

2457 Trimethyl Phosphite

Mehlman et al. (1984) studied this organophosphorous alkylating agent in pregnant rats. At the highest dose 164 mg per kg gavaged on days 6–15, scoliosis, neural tube defects, cleft palates and agnathia occurred in the fetuses. No maternal toxicity was observed.

Mehlman, M.A.; Craig, P.H. and Gallo, M.A.: Teratological evaluation of trimethyl phosphite in the rat. Toxicol. Appl. Pharmacol. 72:119–123, 1984.

2458 Trimethyltin

Paule et al. (1986) injected rats intraperitoneally on day 7, 12 or 17 in doses of up to 9 mg per kg. Subtle changes in the fetal brain region of Ammon's horn of the hippocampus were found in those exposed on day 12 or 17. Findings were present at the highest dose which was toxic to the dams and increased fetal death.

Lipscomb et al. (1986) studied the equilibration of this organic tin preparation in the pregnant rat. On day 19 the fetal blood levels were 50 percent of those in the mother.

Lipscomb, J.C.; Paule, M.G. and Slikker, W., Jr.: The fetomaternal disposition of 14C-trimethyltin in the rat. NeuroToxicol. 7(2):581–590, 1986.

Paule, M.G.; Reuhl, K.; Chen, J.J.; Ali, S.F. and Slikker Jr., W.: Developmental toxicity of trimethyltin in the rat. Toxicol. Appl. Pharmacol. 84:412–417, 1986.

2459 4,4–10–β–Trimethyl–trans–decal–3–β–01 *TMD*

This inhibitor of oxidosqualene cyclase, a distal step in cholesterol synthesis, induced growth retardation and caudal neural tube closure in rat embryos grown in vitro in 50 microgram per ml of medium (Jin et al., 1985).

Jin, L–S.; Freeman, S.J. and Brown, N.A.: Actions of cholesterol synthesis inhibitors on whole rat embryos in culture. (abs) Teratology 31:43A, 1985.

2460 Trimetozine

N–(3,4,5–Trimethoxybenzoyl)tetrahydro–1,4–oxazine CAS 635–41–6

Saito et al. (1976) administered by mouth maximum doses of 1000 and 600 mg per kg daily to mice and rats, respectively. The medication was given during organogenesis. At the highest doses, maternal and fetal weight were reduced. Resorptions were increased in the rat at 600 mg per kg. Some increase in 14th ribs was present in mouse fetuses exposed to 500 mg per kg. Internal malformations were not studied.

Saito, K.; Hikito, S.; Komori, A.; Matsubara, T.; Okuda, T.; Terashima, A.; Matsubara, T.; Saito, H.; Yamamoto, H. and Moritoki, H.: Studies on the toxicity of N–(3,4,5–trimethoxybenzoyl)tetrahydro–1, 4–oxazine (trimetozine). 3. Teratogenicity test in mice and rats. Yakuri to Chiryo 4:252–274, 1976.

2461 Trimipramine Maleate

Jurand (1980) injected mice subcutaneously on day 9 with 85 mg per kg and examined them on day 13. They found exencephaly in 4.8 percent and a few other defects of the central nervous system. In 67 control litters, not a single defect of the nervous system was reported.

Jurand, A.: Malformations of the central nervous system induced by neurotropic drugs in mouse embryos. Develop. Growth Differ. 22(1):61–78, 1980.

2462 Tri–n–butyltin Acetate

Noda et al. (**1991**) treated rats orally with up to 16 mg per kg from days 7–17 of gestation. At the highest dose of the acetate form maternal and fetal toxicity occurred along with cleft palate, cervical ribs and rudimentary lumbar ribs. Itami et al. (**1990**) gave the chloride form to rats in doses of up to 25 mg per kg during days 7–15 of gestation. Maternal toxicity was seen at 9 mg per kg and above. Fetal toxicity occurred at the highest doses and some variations in fetal seleton were reported. Placental weights were significantly greater in the treated groups.

Noda, T., Morita, S.; Yamano, T.; Shimizu, M.; Nakamura, T.; Saitoh, M. and Yamada, A.: Teratogenicity study of tri–n–butyltin acetate in rats by oral administration. Toxicology Letters 55:109–115, 1991.

2463 Tri–n–butyltin Chloride

Itami et al. (**1990**) gave up to 15 mg per kg daily to rats on days 7–15. No teratogenicity was found but placental weights were increased with cystic degenetration of the spongiosa.

Itami, T.; Ema, M. and Kawasaki, H.: Increased placental weight induced by tributyltin chloride in rats (abs). Teratology 42:42A, 1990.

2464 Trinitro–RDX *(1,3,5–Triaza–cyclohexane)*

This explosive was studied by Angerhofer et al. (1986) who gavaged 2, 6, and 20 mg per kg on days 6–15 of the pregnant rat. At 20 mg per kg most of the dams died. At the other doses, reduced fetal weight was seen but no teratogenic effects.

Angerhofer, R.A.; Davis, G. and Balczewski, L.: Teratological assessment of trinitro–RDX in rats, study no. 75–51–0573–86, June 1985–January 1986. United States Army Environmental Hygiene Agency, Aberdeen Proving Ground, pp 1–M1, 1986.

2465 Triparanol *MER–29* CAS 78–41–1

Roux and Dupuis (1961) treated five rats with 1 gm per kg from the 7th through the 10th days of gestation. Three litters were completely resorbed. Of the other two litters, all fetuses had facial defects and rachischisis of the central nervous system was common. Later studies detailed various forms of holoprosencephaly and genitourinary defects (Roux, 1964).

Roux, C.: Action teratogene du triparanol chez l' animal. Arch. Franc. Pediat. 21:451–464, 1964.

Roux, C. and Dupuis, R.: Action teratogene du triparanol. C. R. Soc. Biol. (Paris) 155:2255–2257, 1961.

2466 Triphenyl Phosphate CAS 115–86–6

Welsh et al. (1987) fed rats up to 1.0 percent in the diet before reproduction and then during gestation. No toxicity or teratogenic effects were found.

Welsh, J.J.; Collins, T.F.X.; Whitby, K.E.; Black, T.N. and Arnold, A.: Teratogenic potential of triphenyl phosphate in Sprague–Dawley (Spartan) rats. Toxicology and Industrial Health 3:357–369, 1987.

2467 Triprolidine HCl and Pseudoephedrine HCl *Actifed*[TM]

Aselton et al. (1985) studied the offspring of 244 mothers who took this medication in the first trimester and found only three with malformations.

Aselton, P.A.; Jick, H.; Milunsky, A.; Hunter, J.R. and Stergachis, A.: First–trimester drug use and congenital disorders. Obstet. Gynecol. 65:451–455, 1985.

2468 Tris(2–chloroethyl) Phosphate

Kawashima et al. (1983) gavaged rats with up to 200 mg per kg on days 7–15. No adverse fetal effects were found and behavioral studies were not abnormal.

Kawashima, K.; Tanaka, S.; Nakaura, S.; Nagao, S.; Endo, T.; Onoda, K.–I.; Takanaka, A. and Omori, Y.: Effect of oral administration of tris(2–chloroethyl) phosphate to pregnant rats on prenatal and postnatal development. Bull. Natl. Inst. Hyg. Sci. 101:55–61, 1983.

2469 Tris(chloropropyl) Phosphate *TCPP* CAS 26248–87–3

Kawasaki et al. (1982) fed rats up to 1 percent of this flame retarder in the diet on days 0 through 20 of gestation and did not produce any adverse fetal effects.

Kawasaki, H.; Murai, T. and Kanoh, S.: Fetal toxicity of tris(chloropropyl) phosphate (TCPP) in rats. Oyo Yakuri 24:697–702, 1982.

2470 Tris(2,3–dibromopropyl) Phosphate

Kawashima et al. (1980) gavaged rats with up to 200 mg per kg on days 7–15. At the highest dosage, fetal retardation occurred along with a decrease in maternal weight. No teratogenicity or behavioral effects were found. Seabaugh et al (**1981**) in similar experiments found no teratogenic effects.

Kawashima, K.; Tanaka, S.; Nakaura, S.; Nagao, S.; Onoda, K.–I.; Kasuya, Y. and Omori, Y.: Effects of tris(2,3–dibromopropyl) phosphate on the prenatal and postnatal development of the rats. Bull. Natl. Inst. Hyg. Sci. 98:50–55, 1980.

Seabaugh, V.M.; Collins, T.F.X.; Hoheisel, C.A.; Bierbower, G.W. and McLaughlin, J.: Rat teratology study of orally administered tris–(2,3–dibromopropyl) phosphate. Fd. Cosmet. Toxicol. 19:67–72, 1981.

2471 Tris(1,3–dichloroisopropyl) Phosphate *TDCPP*

Tanaka et al. (1981) gavaged rats on days 7–15 with up to 400 mg per kg. No increase in fetal death or malformations was found at 200 mg per kg. Fetal death was increased at 400 mg per kg which was very toxic to the dams.

Tanaka, S.; Nakaura, S.; Kawashima, K.; Nagao, S.; Endo, T.; Onoda, K.–I.; Kasuya, Y. and Omori, Y.: Effect of oral administration of tris(1,3–dichloroisopropyl) phosphate to pregnant rats on prenatal and postnatal development. Bull. Natl. Inst. Hyg. Sci. 99:50–55, 1981.

2472 Trisodium Citrate

Nolen et al. (1972) gave 20 mg per kg in the drinking water to rats on days 6–15 of gestation. No adverse fetal effects occurred.

Nolen, G.A.; Bohne, R.L. and Buehler, E.U.: Effects of trisodium nitrilotriacetate, trisodium citrate and a trisodium nitrilotriacetate–ferric chloride mixture on cadmium and methyl mercury toxicity and teratogenesis in rats. Toxicol. Appl. Pharmacol. 23:238–250, 1972.

2473 Tritium *Thymidine* CAS 10028–17–8

During pregnancy rats were maintained at constant body activities of tritiated water ranging through 1 to 100 microcuries per ml of body water (Cahill and Yuile, 1970). Although gross defects were not seen, the offspring exposed to 10 microcuries or more were stunted and had microcephaly and marked reduction in the size of the testes or ovaries. Snow (1973) found that tritiated thymidine added to culture medium containing mouse ova was associated with a reduction in blastocyst cell number when concentrations over 0.01 microcuries were used.

Chandhuri et al. (1971) infused tritiated thymidine into pregnant rats from day 9 to term. Some chromosome abnormalities in the newborns were found after 288 microcuries per day. At 576 microcuries, growth retardation occurred. Exencephaly was present in groups exposed to 1152 microcuries or above. Pareek et al. (1989) exposed rats to 0.3 microCurie per ml of maternal drinking water from day 15 through 5 weeks of postnatal age. The brain weights were reduced at most periods of examination and reductions in glycogen phospholipid and cholesterol content were found.

Cahill, D.F. and Yuile, C.L.: Tritium: some effects of continuous exposure in utero on mammalian development. Radiat. Res. 44:727–737, 1970.

Chandhuri, J.P.; Haas, R.J.; Schreml, W.; Knorr–Gartner, H. and Fliedher, T.M.: Cytogenetic and teratological studies in rats continuously infused with various doses of tritiated thymidine during pregnancy. Teratology 4:395–404, 1971.

Pareek, S.; Jain, N. and Bhatia, A.L.: Tritium toxicity in postnatally developing mouse brain: neurochemical changes after continuous exposure. Nucl. Med. Biol. 16:347–350, 1989.

Snow, M.H.L.: Abnormal development of preimplantation mouse embryos grown in vitro with [^{3}H] thymidine. J. Embryol. Exp. Morph. 29:601–615, 1973.

2474 Tritolyl Phosphate *Tri–o–cresyl Phosphate TOCP* CAS 78–30–8

Tocco et al. (1987) gavaged rats with up to 350 mg per kg on days 6–18. No adverse fetal effects were found even at near lethal doses. Mele and Jensh (1977) produced no fetal changes in rat fetuses exposed via maternal intubation to 750 mg per kg on the 18th and 19th days of gestation.

Mele, J.M. and Jensh, R.P.: Teratogenic effects of orally administered tri–o–cresyl phosphate on Wistar albino rats. (abs) Teratology 15:32A only, 1977.

Tocco, D.R.; Randall, J.L.; York, R.G. and Smith, M.K.: Evaluation of the teratogenic effects of tri–ortho–cresyl phosphate in the Long–Evans hooded rat. Fund. Appl. Toxicol. 8:291–297, 1987.

2475 Triton–W.R. 1339 CAS 25301–02–4

Roussel and Tuchmann–Duplessis (1968) injected mice on the 6th through 8th gestational days with 200 mg per kg and produced 21 percent defective offspring. Neural tube closure defects, microphthalmos and omphaloceles were present. Concurrent administration of progesterone caused a doubling in the defect rate.

Roussel and Tuchmann–Duplessis (1970) suggested that the teratogenic action of the compound is through modification of lysosomal function in the yolk sac.

Roussel, C. and Tuchmann–Duplessis, H.: Dissociation des actions embryotoxique et teratogene du triton WR 1339 Chez La Souris. C. R. Acad. Sci. (Paris) (d) 266:2171–2174, 1968.

Roussel, C. and Tuchmann–Duplessis, H.: A propos des actions embryotoxique et teratogene du 'triton WR–1339' chez la souris: Influence de la vitamine A. C. R. Acad. Sci. (Paris) (d) 271:215–218, 1970.

2476 Trypan Blue CAS 72–57–1

This azo dye and its chemical–related compounds have been the subject of a good deal of teratogenic investigation since Gillman et al. (1948) first noted that trypan blue had teratogenic activity in mammals and that it was produced without entry of the dye into the embryo proper. Beck and Lloyd (1966) presented in detail the early work with trypan blue including the chemistry and relative teratogenicity of the other azo dyes and the types of defects obtained. Gillman et al. (1948) reported hydrocephalus, spina bifida, tail defects, eye defects and a few anomalies of other systems. Cleft lip and palate and skeletal defects were uncommon. Wilson (1959) later found 22 percent of the fetuses had cardiovascular defects. Fox and Goss (1958) extended the work and noted that transposition was common. Kreschover et al. (1957) described defects in dentition. Goldstein (1957) described the urogenital defects. The mechanism producing hydrocephalus was extensively studied (Warkany et al., 1958; Wilson, 1959; Gunberg, 1956; Vickers, 1961).

A dose of 50 mg per kg appears to be the optimum teratogenic dose. A characteristic observation with trypan blue is that with treatment after the 9th day of gestation defects are rare. This fact supported other evidence that the mechanism of action was dependent on disruption of yolk sac nutrition.

The mouse (Hamburgh, 1954), hamster (Ferm, 1958), chick (Beaudoin and Wilson, 1958) and guinea pig (Hoar and Salem, 1961) are all susceptible to the teratogenic activity of trypan blue. In the monkey, Wilson (1971) caused abortions with doses of 50 mg per kg with single or two daily doses between the 20th to 25th day.

An interesting series of studies produced evidence indicating a possible action of trypan blue on a nutritive function of the visceral yolk sac. The failure of trypan blue to act directly upon the embryo is generally held with the exception of Davis (1968) who demonstrated some localization in the hindgut possibly due to incorporation through the yolk stalk. Experiments with ring–labelled radioactive trypan blue did not give evidence of any embryonic incorporation of the ^{14}C (Wilson et al., 1963). The absence of teratogenic action after the initiation of chorio–allantoic placentation also indicated that yolk sac function was important in pathogenesis. The dye can be visualized in the cells of the visceral yolk sac.

Turbow (1965) studied the production of defects in embryos isolated from the mother in an in vitro system. Lloyd and Beck (1969) summarized their work on the role of lysosomes of the visceral yolk sac in trypan blue teratogenesis. They showed that both horseradish peroxidase and the protein–trypan blue complex are concentrated in lysosomes. Through disruption of the enzymatic digestive process in the yolk sac lysosome, trypan blue may interfere with normal embryonic nutritive processes. It seems likely that work on trypan blue will lead to more knowledge about the normal physiologic role of the yolk sac in the mammal. Lloyd (1990) has reviewed the cell physiology of pinocytosis and intralysosomal digestion of proteins in the yolk sac. His studies used 125^{v-1}I–labeled polyvinylpyrrolidone and bovine serum albumin to trace uptake and digestion with macromolecules in rat yolk sac cells.

Ema et al. (1985) studied postnatal function of rats treated on the 7th day of gestation with either 10 or 50 mg per kg. Open field tests were decreased in the offspring treated with the higher dose. Their reproduction was not affected.

Beaudoin, A.R. and Wilson, J.G.: Teratogenic effects of trypan blue on the developing chick. Proc. Soc. Exp. Biol. Med. 97:85–92, 1958.

Beck, F. and Lloyd, J.B.: The teratogenic effects of azo dyes. In: Advances in Teratology, D.H.M. Woollam (ed.), New York: Academic Press 1:131–193, 1966.

Davis, H.W. and Gunberg, D.L.: Trypan blue in the rat embryo. Teratology 1:125–134, 1968.

Ema, M.; Itami, T.; Kawasaki, H. and Kanoh, S.: Postnatal development of offspring from pregnant rats treated with trypan blue. (Japanese) Eiseishikenjohokohu 103:75–81, 1985.

Ferm, V.H.: Teratogenic effects of trypan blue on hamster embryos. J. Embryol. Exp. Morphol. 6:284–287, 1958.

Fox, M.H. and Goss, C.M.: Experimentally produced malformations of the heart and great vessels in rat fetuses. Transposition complexes and aortic arch abnormalities. Am. J. Anat. 102:65–92, 1958.

Gillman, J.; Gilbert, C.; Gillman, T. and Spence, I.: A preliminary report on hydrocephalus, spina bifida and other congenital anomalies in the rat produced by trypan blue. S. Afr. J. Med. Sci. 13:47–90, 1948.

Goldstein, D.J.: Trypan blue induced anomalies in the genito–urinary system of rats. S. Afr. J. Med. Sci. 22:13–22, 1957.

Gunberg, D.L.: Spina bifida and the Arnold–Chiari malformation in the progeny of trypan blue injected rats. Anat. Rec. 126:343–367, 1956.

Hamburgh, M.: The embryology of trypan blue induced abnormalities in mice. Anat. Rec. 119:409–427, 1954.

Hoar, R.M. and Salem, A.J.: Time of teratogenic action of trypan blue in guinea pigs. Anat. Rec. 141:173–181, 1961.

Kreschover, S.J.; Knighton, H.T. and Hancock, J.A.: Influence of systemically administered trypan blue on prenatal development of rats and mice. J. Dent. Res. 36:677–683, 1957.

Lloyd, J.B.: Cell physiology of the rat visceral yolk sac: a study of pinocytosis and lysosome function. Teratology 41:383–393, 1990.

Lloyd, J.B. and Beck, F.: Teratogenesis. In: Lysosomes in Biology and Pathology, J.T. Dingle and H.B. Fell (eds.), Amsterdam: North–Holland Publishing Co., 333–449, 1969.

Turbow, M.M.: Teratogenic effect of trypan blue on rat embryos cultivated in vitro. Nature 206:637 only, 1965.

Warkany, J.; Wilson, J.G. and Geiger, J.F.: Myeloschisis and myelomeningocele produced experimentally in the rat. J. Comp. Neurol. 109:35–64, 1958.

Wilson, J.G.: Experimental studies on the mechanism of teratogenic action of trypan blue. J. Chron. Dis. 10:111–130, 1959.

Wilson, J.G.: Use of rhesus monkeys in teratological studies. Fed. Proc. 30:104–109, 1971.

Wilson, J.G.; Shepard, T.H. and Gennaro, J.F.: Studies on the site of action of C(14)–labeled trypan blue. (abs) Anat. Rec. 145:300 only, 1963.

Vickers, T.H.: Concerning the mechanism of hydrocephalus in the progeny of trypan blue treated rats. Arch. Entw. Mech. Org. 153:255–261, 1961.

2477 Tryptophan CAS 73–22–3

Naidu (1974) administered 2.0 mg of this amino acid to 72–hour chick embryos and produced limb deformities, rumplessness and visceral defects. Meier and Wilson **1983** fed hamsters diets high in tryptophan (1.8 to 8.0%) and observed decrease litter size and birth weight. Matsueda and Niiyama (**1982**) fed pregnant rats diets with 5% tryptophan and found reduced fetal body and brain weights.

Matsueda, S. and Niiyama, Y.: The effects of excess aminoo acids on maintenance of pregnancy and fetal growth in rats. J. Nutr Sci Vitaminol 28:557–573, 1982.

Meier, A.H. and Wilson, J.M.: Tryptophan feeding adversely influences pregnancy. Life Sci 32:1193–1196, 1983.

Naidu, R.C.M.: Teratogenic effects of tryptophane on the developing chick embryo. Experientia 30:1462–1463, 1974.

2478 Tsumacide m–Tolyl–N–methylcarbamate

Yasuda (1972) fed this insecticide to rats in their diet during organogenesis at up to 4000 ppm and found no adverse effects in the fetuses at term or at 6 weeks postnatally.

Yasuda, M.: Teratologic evaluation of tsumacide in the rat. Botyu-Kagaku 37:161–165, 1972.

2479 Tuberculosis

Snider et al. (**1980**) reviewed the literature on tuberculosis treatment during pregnancy and recommendded isoniazid and ethambutol with the addition of rifampin if necessary. Streptomycin because of ototoxicity was not advised unless rifampin was contraindicated or unsatisfactory. No congenital defects occurred among 11 infants exposed to these drugs.

Snider, D.E.; Layde, P.M.; Johnson, M.W. and Lyle, M.A.: Treatment of tuberculosis during pregnancy. American Review of Respiratory Disease 122:65–79, 1980.

2480 D–Tubocurarine also see *Curare* CAS 57–95–4 *Pancuronium, Atracurium, Vanuronium*

Jago (1970) reported arthrogryposis in an infant whose mother was treated for tetanus with D–tubocurarine for 19 days starting on about the 55th day of gestation.

Chandramohan Naaidu (1974) injected the yolk sac of developing chicks with 2.0 mg of tryptophane at 72 hours of gestation and produced defects in 70 percent of the survivors. Limb deformities, rumplessness and visceral defects were found. Drachman and Coulombre (1962) produced clubfoot and arthrogryposis in chick embryos infused for 24 to 48 hour periods with 30 microgm per hour during the last quarter of incubation. This elegant experiment emphasized the important role that prenatal movement plays in development of joints. Baker (1960) reviewed the general lack of evidence for transplacental passage of this drug during parturition.

Rodriguez et al. (**1992**) carried out detailed bone studies of rat fetuses treated from day 17 to term. Contractures, growth retardation and shortened umbilical cords were found.

Jacobs (1971) administered 0.6 to 5 mg per kg intramuscularly to rats on day 13.5 of gestation. Although muscle flaccidity occurred in the fetuses, their palates closed normally. Shoro (1972) produced club foot in 8 percent of rat fetuses injected in the intrascapular region between the 17th and 19th days of gestation.

Fujinaga et al. (**1992**) studied the effect of d–tubocurarine, pancuronium, atracurium and vecuronium on rat embryos in culture. Growth retardation was found but only at concentrations 30 fold higher than levels in patients.

Baker, J.B.E.: The effects of drugs on the foetus. Pharmacol. Rev. 12:37–90, 1960.

Chandramohen Naidu, R.: Teratogenic effects of tryptophane on the development of the chick embryo. Experientia 30:1462–1463, 1974.

Drachman, D.B. and Coulombre, A.J.: Experimental club foot and arthrogryposis multiplex congenita. Lancet 2:523–526, 1962.

Fujinaga, M.; Baden, J.M. and Mazze, R.I.: Developmental toxicity of nondepolarizing muscle relaxants in cultured rat embryos. Anesthesiology 76:999-1003, 1992.

Jacobs, R.M.: Failure of muscle relaxants to produce cleft palate in mice. Teratology 4:25–30, 1971.

Jago, R.H.: Arthrogryposis following treatment of maternal tetanus with muscle relaxants. Arch. Dis. Child. 45:227–279, 1970.

Rodriguez, J.I.; Palacios, J.; Ruiz, A.; Sanchez, M.; Alvarez, I. and Demiguel, E.: Morphological changes in long bone development in fetal akinesia deformation sequence: an experimental study in curarized rat fetuses. Teratology 45:213–221, 1992.

Shoro, A.A.: Club–foot and intrauterine growth retardation produced by tubocurarine in the rat fetus. (abs) J. Anat. 111:506–508, 1972.

2481 Tumor Necrosis Factor *TNF PT–050*

Terada et al. (**1990** A) gave up to 50,000 JRU per kg daily on days 7–17 in the rat and observed no adverse fetal effects. At the higher dosages male fetal weight was increased. Terada (**1990** B), in rabits, similarly treated found increased fetal mortality at 500,000 and 50,000 JRU per kg but not teratogenicity.

Terada, Y.; Funabashi, H.; Imura, Y.; Nishimura, K. and Ohnishi, K.: Reproduction studies of pt–050 (recombinant human tnf)–teratogenicity study in rats–(cesarean section and natural delivery studies). Yakuri to Chiryo 18:449–473, 1990.

Terada, Y.; Kishi, H.; Aoki, Y.; Shigematus, K.; Mukumoto, K.; Funabashi, H.; Imura, Y.; Satoh, K.; Yoshioka, M.; Nishimura, K. and Ohnishi, K.: Reproduction studies of pt–050 (recombinant human tnf)–teratogenicity study in rabbits. Yakuri to Chiryo 18:497–506, 1990.

2482 Tungsten

Wide (1984) treated mice intravenously on days 3 or 8 with 25 millimoles and found no decrease in implantation but some increase in resorptions in the day 8 group. Bone maturation and weight in day 17 fetuses was similiar to the controls.

Wide, M.: Effect of short–term exposure to five industrial metals on the embryonic and fetal development of the mouse. Environ. Res. 33:47–53, 1984.

2483 Tween 60

Ema et al. (1988) fed rats up to 10 percent of their diet on days 7–14. No adverse effects on reproduction or the fetus were found. Previously published negative teratogenicity studies were cited.

Ema, M.; Itami, T.; Kawasaki, H. and Kanoh, S.: Teratology study of tween 60 in rats. Drug Chem. Toxicol. 11(3):249–260, 1988.

2484 Twinning

Kallen (1986) studied records of 15,427 infants who were twins. Hydrocephaly was increased and this was explained only in part by a higher rate of low–weight infants among twins. Some components of the VACTERL Syndrome were increased, notably esophageal and anal atresia and severe kidney malformations. Ten patients with esophageal atresia were found when the expected was 4.3. A two–fold increase in dislocation of the hip was reported. Nine of the 27 twins with multiple defects fell into the VACTERL constellation.

Kallen, B.: Congenital malformations in twins: A population study. Acta Genet. Med. Gemellol. 35:167–178, 1986.

2485 TYB–5220 *Erythropoietin, Human*

Hashimoto et al. (**1990**) gave this recombinant DNA agent intravenously to rats on days 7–17 at doses of 1.3 and 13 microgm per kg. Osseous retardation was found. Placental weight was increased at 13 microgm per kg. No teratogenicity or postnatal functional changes were found.

Hashimoto, Y.; Kaneko, Y.; Ishizuka, Y. and Mizutani, M.: Reproduction and developmental toxicity study of try–5220–Teratognicity study of tyb–5220 by intravenous injection in rats. Yakuri to Chiryo 18:s1175–s1191, 1990.

2486 U–600

Ito et al. (1981) studied this anti–ulcer agent which was extracted from bovine spleen. Rats and rabbits were given up to 400 mg per kg per day subcutaneously during organogenesis and no adverse effects were found in the fetuses. Treatment of the rat in late gestation was done and no behavioral changes were found.

Ito, R.; Kajiwara, S.; Mori, S.; Ondo, T.; Miyamoto, K. and Sugimoto, T.: Fertility study on a new antiulcer agent U–600 in rats. Yakubutsu Ryoho (Medical Treatment) 14:43–56, 1981.

2487 U–11, 100A

Morris et al. (1967) gave up to 5.0 mg per kg to rabbits during short periods of organogenesis and found no malformations. Implantation was inhibited when it was given earlier.

Morris, J.M.; Van Wagenen, G.; McCann, T. and Jacob, D.: Compounds interfering with ovum implantation and development. Fertil. Steril. 18:18–34, 1967.

2488 U–11, 555A

Morris et al. (1967) gave rabbits up to 20 mg per kg on days 10–20 and found no malformation increase in the fetuses. An antifertility effect was found when the drug was given during tubal transport.

Morris, J.M.; Van Wagenen, G.; McCann, T. and Jacob, D.: Compounds interfering with ovum implantation and development. Fertil. Steril. 18:18–34, 1967.

2489 U–67,590A

Ishida et al. (1993 A) gave this glucocortoid to rats and rabbits intravenously. Treatment at the highest dose (27 mg per kg) was associated with an increase in cleft palates. No effect of 9 mg per kg on fertility was seen (Ishida et al., 1993 B). Perinatal treatment did not alter fetal development although atrophy of maternal spleen, thymus and adrenals occurred. (Ishida et al., 1993 C). In the rabbit treatment of 0.09 and 0.9 mg per kg caused maternal toxicity with decreased body and placental weight but no fetal abnormalities.

Ishida, S.; Ikegaya, M.; Fujioka, M.; Shimazu, H. and Koike, S.: Fertility and general reproductive performance study in rats treated intravenously with U–67,590A. The Clinical Report 27:2911–2935, 1993A.

Ishida, S.; Ishikawa, Y.; Fujioka, M.; Shimazu, H. and Koike, S.: Teratological study in rats treated intravenously with U–67,590A. The Clinical Report 27:2937–2957, 1993B.

Ishida, S.; Shibuya, M.; Shiota, Y.; Simazu, H. and Koike, S.: Peri–and post–natal study in rats treated intravenously with U–67,570A. The Clinical Report 27:2959–2978, 1993C.

Nishimura, N.; Nanada, O.; Serizawa, K. and Koike, S.: Teratological study of U–67,570A in rabbits by intravenous administration. The Clinical Report 27:2979–2990, 1993.

2490 Ubiquinone–9 CAS 303–97–9

Nakazawa et al. (1969) gave this compound to rats and mice on days 9 through 14 and 7 through 12, respectively. Oral doses to 4,000 mg per kg daily in mice and 2,000 mg per kg per day in rats were used. Intravenous dosages were also used. No teratogenic action was observed.

Nakazawa, M.; Ohzeki, M.; Takahashi, N. and Tsuchida, T.: Toxicity tests of ubiquinone–9. 2. Teratogenicity. Oyo Yakuri 3:155–159, 1969.

2491 Ufenamate Butyl–2–((3–(trifluoromethyl)phenyl)amino)benzoate

Isuruzaki et al. (1979) studied this nonsteroidal anti–inflammatory agent in rats orally before mating and during the first 7 days, on days 7–17 and days 17–20 of gestation. No effects on reproduction were noted after giving up to 60 mg per kg. Up to 120 mg per kg was given during organogenesis without adverse fetal effects. Perinatal administration of up to 45 mg per kg produced no abnormal postnatal changes.

Isuruzaki, T.; Kubo, S.; Shimo, T.; Yamazaki, M.; Kato, H. and Yamamoto, M.: Reproductive studies of butyl–2–((3–trifluromethylphenyl)amino) benzoate (HF–264) in rats. Kiso to Rinsho 13:3279–3287, 3288–3301, 3302–3313, 1979.

2492 Ultrasound

A major body of data on the biological effects of ultrasound has been published and reviewed (Nyborg, 1977; Repacholi, 1981; Stewart and Stratmeyer, 1982; WHO, 1982; National Council on Radiation Protection, 1983 and National Institutes of Health, 1984). Reece et al. (1990) reviewed the literature and concluded that current data indicate that diagnostic levels are safe. Ziskin and Petitti (1989) found no evidence for adverse effects in reviewing epidemiologic studies of ultrasound. Brent et al. (1991) have reviewed the animal and human studies dealing with reproductive effects of ultrasound.

The strength of the ultrasound field is usually expressed as intensity. Intensity describes the spatial concentration of power in units of watts per square centimeter (w per cm squared) or milliwatts per square centimeter (mW per cm squared). The former unit being 1,000 times greater than the latter. The most important biologic effect of ultrasound is mediated by increased temperature in the tissue. Two of the main factors in heat production are intensity and time of exposure. Although cavitation can occur in plants where micro bubbles are present, no similar situation has been identified in animals. At the intensity and time presently used by ultrasound imaging in humans, conditions for production of increased heat are unlikely.

Carefully controlled studies in humans are almost impossible to perform because of the widespread and diagnostically beneficial use of ultrasound imaging. Analysis is further complicated because ultrasound tends to be used more in those pregnancies which are complicated. A study from Norway (Eik–Nes et al., 1984) used a randomly controlled prospective study to show no ill effect of ultrasound imaging. Kohorn et al. (1967) found no electroencephalographic changes in newborns exposed to 5 minutes of brain scanning. Ott et al. (1980) found no differences in a study of 298 newborns exposed as compared to 201 not exposed. Bernstein (1969) found no increase in complications among 720 newborns where Doppler examinations were used. The comparison was the general U.S. naval rate. Hellman et al. (1970) found no increase in defects among 1,079 exposed individuals. Falus et al. (1972), using developmental studies in 171 intrauterine–exposed children, could find no differences.

Scheidt et al. (1978) compared the results of amniocentesis with and without ultrasound. Those with added ultrasound (303) did not differ significantly in birth weight from those with amniocentesis alone (679) and at one year, no significant differences in hearing or development were found. An abnormal tonic neck reflex was reported in the ultrasound–exposed group at birth. Stark et al. (1984) reported a three–year hospital study of the long–term effects of ultrasound on children aged 7 to 12 years. The authors found "no biologic differences" from a group matched by seriousness of pregnancy complication. By combining the number of dyslexics from all three hospitals there were 79 individuals from the exposed group and 26 in the unexposed group.

Most experimental conditions in animals where adverse effects were found have intensity and times which should produce excess heating. Nyborg (1977) illustrated this in a number of animal experiments. At 100 mW per cm squared (spatial peak, temporal average), increased temperature might be observed only after 60 minutes exposure and at the critical area of the biological test. The intensities at the transducer surface of scanners varies from 0.1 to 200 mW per cm squared. Pulsed doppler varies from 350–700 mW per cm squared. Sikov et al. (1984) measured temperature increases within the uterus after different intensities and found that fetal damage was found only when temperature was increased. Sikov (1986) summarized the biologic effects of ultrasound in development. Child et al. (**1988**) exposed mice at 8 days post fertilization to maximum pulsed intensities up to 60 W/cm^2 for 5 minutes. No fetal changes were found on day 18. Bosward et al. (**1991**) in a preliminary report showed up to a 1.5 degree C rise in temperature of guinea pig brain adjacent to ultrasounded bone (230 mW).

Kimmel et al. (1989) exposed pregnant mice to pulsed ultrasound on gestational day 8. With spacial peak pulse average intensities of 90 W per cm squared for 6.5 microseconds some of the dams died but no adverse fetal effects were found on day 18. Fisher et al. (**1994**) scanned pregnant rats for 10 minutes on days 4–19 using up to 30 W per cm^2 (3MH$_3$ pulsed wave) and found no embryotoxicity. Vorhess, et al. (**1994**) found some postnatal neurobehavioral dysfunction in rats exposed 10 minutes daily to continuous high doses of 30 W/cm^2 but not to 10 or 20 W/cm^2 doses. They calculate that the temperature at 20 and 30 W/cm^2 rise was to 43.6

and 46.3 degrees C respectively in the embryos.

Over 40 studies of ultrasound in pregnant animals have been published. Although the physical dosage in many experiments could not be estimated, those experiments which produced congenital defects were associated with exposure conditions which could produce hyperthermia which alone is teratogenic (critically reviewed by National Institute of Health, 1984). Kimmel et al. (1983) exposed mice on day 8 to 1.0 W per cm squared for two minutes and found no increase in defects as compared to sham–treated animals. Mannor et al. (1972) exposed mice to doses of up to 1050 mW per cm squared for 60 minutes for varying number of days and found no increase in defects or decrease in postnatal fertility. O'Brien (1983) summarized his extensive work which showed a dose related effect of ultrasound on mouse fetal weight at doses of 0.5 to 5.5 W per cm squared for 10 to 300 seconds. Sikov (1986) reviewed the animal teratogenicity including postnatal effects and concluded that deleterious results are only found when doses far exceeding human exposure were used. Vorhees et al. (**1992**) treated rats on days 4–19 for 15 minutes daily to up to 30.0 W/cm^2 of continuous wave ultrasound. No evidence of embryotoxicity was found.

Sikov and Hildebrand (1982) observed some neuromuscular delay in rat fetuses exposed to 10.5 W per cm squared for 5 to 15 minutes. Kimmel et al. (1987) exposed mice to 1 MHz on day 8 for 10 minutes. The peak intensity (spacial peak, pulse average) was 90 W per cm square. No changes from the controls were found on day 18 and postnatal studies were not different from sham treated controls.

Tarantal and Hendrickx (1989A) scanned 16 monkeys multiple times weekly from day 21 through day 60 and once weekly from day 61–150. None of the physical characteristics of the fetuses at birth were different from the sham treated controls. Segmented neutrophies and monocytes were significantly reduced through the 17th postnatal day but no differences were present at 6 months of age. The same authors (1989B) found reduced body weight gain in the offspring during the first 3 months but these were not present at 6, 9 and 12 months. Neurobehavioral tests revealed transient increased muscle tone and more quiet activities in the exposed animals.

Lele (1985) summarized the bioeffects on human reproduction. Stewart et al. (1985) compiled data on all the biologic effects of ultrasound. Fisher and Stratmeyer (1987) and Stratmeyer and Fisher (1987) reviewed the animal and human effects of ultrasound, respectively.

Bernstein, R.L: Safety studies with ultrasonic doppler technic: A clinical followup of patients and tissue culture study. Obstet. Gynecol. 34:707–709, 1969.

Bosward, K.L.; Barnett, S.B.; Wood, A.K.W. and Edwards, M.J.: The effect of ossification of the guinea pig fetal skull on brain heating during exposure to pulsed ultrasound. Australian Teratology Abstracts Teratology 44:477, 1991.

Brent, R.L.; Jensh, R.P. and Beckman, D.A.: Medical sonography: Reproductive effects and risks. Teratology 44:123–146, 1991.

Child, S.Z.; Carstensen, E.L.; Gates, A.H. and Hall,

W.J.: Testing for the teratogenicity of pulsed ultrasound in mice. Ultrasound in Med. and Biol. 14:493–498, 1988.

Eik–Nes, S.; Okland, D.; Avre, J.C. and Ulstein, M.: Ultrasound screening in pregnancy: A randomized controlled trial. Lancet 1:347, 1984.

Falus, M.; Korany, G. and Sobel, M.: Followup studies on infants examined by ultrasound during the fetal age. Orv. Hetil. 113:2119–2121, 1972.

Fisher, B.R. and Stratmeyer, M.E.: Developmental and teratogenic effects of ultrasound. In: Ultrasound. M.H. Repacholi, M. Grandolfo and A. Rindi (eds.), Plenum Publishing Corporation, 195–217, 1987.

Fisher, J.E.; Acuff–Smith, K.ED.: Schilling. M.A.; Vorhees, C.V.; Meyer, R.A.; Smith, N.B. and O'Brien, W.D. Jr.: Tertologic evaluation of rats prenatally exposed to pulse-wave ultrasound. Teratology 49:150–155, 1994.

Hellman, L.M.; Duffus, G.M.; Donald, I. and Sunden, B.: Safety of diagnostic ultrasound in obstetrics. Lancet 1:1133–1135, 1970.

Kimmel, C.A.; Stratmeyer, M.E.; Galloway, W.D.; Brown, N.T.; LaBorde, J.B. and Bates, H.K.: Developmental exposure of mice to pulsed ultrasound. Teratology 54A, 1987.

Kimmel, C.A.; Stratmeyer, M.E.; Galloway, W.D.; LaBorde, J.B.; Brown, N. and Pinkavitch, F.: The embryotoxic effects of ultrasound exposure in pregnant ICR mice. Teratology 27:245–251, 1983.

Kimmel, C.A.; Stratmeyer, M.E.; Galloway, W.D.; Brown, N.T.; LaBorde. J.B. and Bates, H.K.: Developmental exposure of mice to pulsed ultrasound. Teratology 40:387–393, 1989.

Kohorn, E.I.; Pritchard, J.W. and Hobbins, J.C.: The safety of clinical ultrasound examination. Obstet. Gynecol. 29:272–274, 1967.

Lele, P.P.: Ultrasound bioeffects and human reproduction. In: Ultrasound Annual, R.C. Sanders and M.C. Hill (eds.), Raven Press, New York, 1985.

Mannor, S.M.; Serr, D.M.; Tamari, A.; Meshorer, A. and Frei, E.H.: The safety of ultrasound in fetal monitoring. Am. J. Obstet. Gynecol. 113:653–661, 1972.

National Council on Radiation Protection Biological - Effects of Ultrasound: Mechanisms and Clinical Implications NCRP No 74, Washington, D.C., 1983.

National Institute of Health: Diagnostic Ultrasound Imaging in Pregnancy. Report of a concensus. NIH Publication No. 84–667, U.S. Government Printing Office, Washington, D.C., 1984.

Nyborg, W.L.: Physical mechanisms for biologic effects of ultrasound. HEW Publication (FDA) 78–8062, 1977.

O'Brien Jr., W.D.: Dose–dependent effect of ultrasound on fetal weight in mice. J. Ultrasound Med. 2:1–8, 1983.

Ott, W.J.; Callaghan, G.M.; Ritter, H.A. and Ritter, P.J.: Routine real–time ultrasound in a low–risk obstetric population. J. Reprod. Med. 24:203–207, 1980.

Reece, E.A.; Assimaklpoulos, E.; Zheng, X; Hagay, Z. and Hobbins, J.C.: The safety of obstetric ultrasonography: concern for the fetus. Obstet. and Gynec. Review 76:139–145, 1990.

Repacholi, M.H.: Ultrasound: Characterizations and biological action, National Research Council of Canada, Publication No. NRCC 19244, Ottowa, 1981.

Scheidt, P.C.; Stanley, F. and Bryla, D.A.: One–year followup of infants exposed to ultrasound in utero. Am. J. Obstet. Gynecol. 131:743–748, 1978.

Sikov, M.R.: Effects of ultrasound on development. Part 2. Studies in mammalian species and overview. J. Ultrasound Med. 5:651–661, 1986.

Sikov, M.R.: Effect of ultrasound on development. Part 1. Introduction and studies in inframammalian species. Report of the Bioeffects Committee of the American Institute of Ultrasound in Medicine, October, 1986.

Sikov, M.R.; Collins, D.H. and Carr, D.B.: Measurement of temperature rise in prenatal rats during exposure of the exteriorized uterus to ultrasound. IEEE Transactions on Sonics and Ultrasonics Vol. SU–31, No. 5, (Sept) 497–503, 1984.

Sikov, M.R. and Hildebrand, B.P.: Effects of prenatal exposure to ultrasound in teratology testing. T.V.N. Persaud (ed), MTP Press Ltd, 267–291, 1982.

Stark, C.R.; Orleans, M.; Havercamp, A.D. and Murphy, J.: Short and long–term risks after exposure to diagnostic ultrasound in utero. Obstet. Gynecol. 63:194–200, 1984.

Stewart, H.D.; Stewart, H.F.; Moore, R.M. and Garry, J.: Compilation of reported biological effects data and ultrasound levels. J. Clin. Ultrasound 13:167–186, 1985.

Stratmeyer, M.E. and Fisher, B.R.: Exposure to medical ultrasound: Studies of human effects. In: Ultrasound, M.H. Repacholi, M. Grandolfo and A. Rindi (eds.), Plenum Publishing Corporation, 219–232, 1987.

Stewart, H.F. and Stratmeyer, M.E.: An overview of ultrasound: Theory, measurement, medical applications and biological effects. HHS Publication FDA 82–8290, U.S. Government Printing Office, Washington, D.C. 1982.

Tarantal, A.F. and Hendrickx, A.G.: Evaluation of the bioeffects of prenatal ultrasound exposure in the cynomolgus Macaque (Macaca fascicularis): I. Neonatal/infant observations. Teratology 39:137–147, 1989A.

Tarantal, A.F. and Hendrickx, A.G.: Evaluation of the bioeffects of prenatal ultrasound exposure in the cynomolgus Macaque (Macaca fascicularis): II. Growth and behavior during the first year. Teratology 39:149–162, 1989B.

Vorhees, C.V.; Acuff-Smith, K.D.; Weisenburger, W.P.; Meyer, R,A.; Smith, N.B. and O'Brien Jr., W.D.: A teratologic evaluation on continuous–wave, daily ultrasound exposure in unanesthetized pregnant rats. Teratology 44:667–674, 1991.

Vorhees, C.V.; Acuff–Smith, K.D.; Schilling, M.A.; Fisher, Jr., J.E.; Meyer, R.A.; Smith, N.B.; Ellis, D.S. and O'Brien, Jr., W.D.: Behavioral teratologic effects of prenatal exposure to continuous–wave ultrasound in unanesthetized rats. Teratology 50:238–249, 1994.

World Health Organization (WHO): Environmental Health Criteria 22 for Ultrasound. Geneva, Switzerland, 1982.

Ziskin, M.C. and Petitti, D.B.: Epidemiology of human exposure to ultrasound: a critical review. Ultrasound in

Med. and Biol. 14:91–96, 1988.

2493 Uracil CAS 66–22–8

Kosmachevskaya and Tichodeeva (1968) injected 0.25 to 4.0 mg of this chemical into chick eggs at 24 hours of incubation and found 24 percent of the surviving embryos deformed. Microphthalmia, rumplessness and axial skeletal defects were observed.

Kosmachevskaya, E.A. and Tichodeeva, I.I.: Peculiarities of action of some pyrimidine derivatives on different stages of chicken embryogenesis. Bull. Exp. Biol. (Russian) 3:109–113, 1968.

2494 Uracil Mustard also see Nitrogen Mustard CAS 66–75–1

Chaube and Murphy (1968) reported that 0.3 to 0.6 mg per kg injected into rats on the 11th or 12th gestational day produced congenital defects in the offspring. Exencephaly or encephalocele and skeletal defects were observed.

Chaube, S. and Murphy, M.L.: The teratogenic effects of the recent drugs active in cancer chemotherapy. In: Advances in Teratology, D.H.M. Woollam (ed), New York: Academic Press, 3:181–237, 1968.

2495 Uranium

Domingo et al. (1989) gavaged mice with 5 to 50 mg per kg of uranyl acetate dihydrate on days 6–15. Maternal and fetal toxicity were found at 10, 25, and 50 mg per kg. Cleft palate and skeletal variations were also increased at these dose levels.

Domingo, J.L.; Paternain, J.L.; Llobet, J.M. and Corbella, J.: The developmental toxicity of uranium in mice. Toxicology 55:143–152, 1989.

2496 Urea

Teramoto et al. (1981) gave 2000 mg per kg orally to rats on day 12 and to mice on day 10 and found no increase in the defect rates or resorptions.

Teramoto, S.; Kaneda, M.; Aoyama, H. and Shirasu, Y.: Correlation between the molecular structure of N-alkylureas and N-alkylthioureas and their teratogenic properties. Teratology 23:335–342, 1981.

2497 17–β–Ureido–1,4–androstadien–3–one

This steroid inhibitor was given subcutaneously to rats on days 13–21 in amounts of 10 or 30 mg per kg and the male fetuses were found to have significantly reduced anogenital distances (Goldman et al., 1976).

Goldman, A.S.; Eavey, R.D. and Baker, M.K.: Production of male pseudohermaphroditism in rats by two new inhibitors of steroid 17 α–hydroxylase and C–17–20 lyase. J. Endocrinol. 71:289–297, 1976.

2498 Urethane Ethylurethan Ethyl Carbamate CAS 51–79–6

Sinclair (1950) produced neural tube closure defects in the mouse fetus by injecting 15 mg on the 7th day of gestation. Nishimura and Kuginuki (1958) gave 1.5 gm per kg

daily to mice from the 9th through 12th days of gestation and found skeletal and palate defects. Hall (1953) demonstrated eye developmental defects in rats treated with this substance. Takaori et al. (1966) gave 1 gm per kg to rats on days 6 through 11 and produced a high incidence of skeletal defects. Gross defects were uncommon.

Ferm (1966) produced defects in hamsters by injecting 25 to 100 mg intravenously on days 9, 10 or 11. Neural tube defects and cardiac malformations were found. Itoh and Matsumoto (1984) found growth retardation in mouse embryos grown in vitro at 50 millimolar concentrations of urethane. Studies using methane during the later part of pregnancy in several strains of mice reported increased lung tumors, hepatomas and ovarian tumors in the offspring. Luebke et al. (1986) treated rats during organogenesis with 1.0 or 2.0 mg per kg and found that the offspring had suppressed humoral immune function. The reproductive toxicity of urethane was reviewed (Anonymous, 1974).

Anonymous: Urethane. IARC 7:111–140, 1974.

Ferm, V.H.: Severe developmental malformations. Arch. Path. 81:174–177, 1966.

Hall, E.K.: Developmental anomalies in the eye of the rat after various experimental procedures. Anat. Rec. 116:383–393, 1953.

Itoh, A. and Matsumoto, N.: Organ specific suceptibility to clastogenic effect of urethane: A trial of application of whole embryo culture in testing systems for clastogen. J. Toxicol. Sci. 9:175–192, 1984.

Luebke, R.W.; Riddle, M.M.; Rogers, R.R.; Rowe, D.G.; Garner, R.J. and Smialowicz, R.J.: Immune function in adult C57BL6J mice following exposure to urethan pre- or postnatally. J. Immunopharmacol. 8(2):243–257, 1986.

Nishimura, H. and Kuginuki, M.: Congenital malformations induced by ethyl–urethane in mouse embryos. Okajimas Folia Anat. Jpn. 31:1–10, 1958.

Sinclair, J.G.: A specific transplacental effect of urethane in mice. Tex. Rep. Biol. Med. 8:623–632, 1950.

Takaori, S.; Tanabe, K. and Shimamoto, K.: Developmental abnormalities of skeletal system induced by ethyl-urethan in the rat. Jpn. J. Pharmacol. 16:63–73, 1966.

2499 Urinary Tract Infection

Urinary tract infection has been reported by many to be associated with low birth weight and prematurity (Kass, 1960). McGrady et al. (1985) summarized the subject and reported that the relative risk for prematurity was increased to 2.04.

Kass, E.H.: Bacteriuria and pyelonephritis of pregnancy. Arch. Intern. Med. 105:42–46, 1960.

McGrady, G.A.; Daling, J.R. and Peterson, D.R.: Maternal urinary tract infection and adverse fetal outcomes. Am. J. Epidemiol. 121:377–381, 1985.

2500 Urokinase CAS 9039–53–6

Akutsu et al. (1974) used human purified urokinase in doses of 100,000 iu per kg in mice and rats during active organogenesis. Intraperitoneal doses did not produce any teratogenic action nor were differences produced in the off-spring during a three–week postnatal period.

Akutsu, T.; Ito, C.; Sakai, K.; Arigaya, Y.; Ohnishi, H. and Ogawa, N.: Studies on teratogenicity of urokinase in mice and rats. Oyo Yakuri 8:981–989, 1974.

2501 Ursodeoxycholic Acid CAS 128–13–2

Toyoshima et al. (1978) gave rats up to 200 mg per kg orally on the 7th through 17th days. No adverse fetal effects were found.

Toyoshima, S.; Fujita, H.; Sakurai, T.; Sato, R. and Kashima, M.: Reproduction studies of ursodeoxycholic acid in rats. II. Teratogenicity study. Oyo Yakuri 15:931–945, 1978.

2502 D–Usnic Acid CAS 7562–61–0

Wiesner and Yudkin (1955) gave 1.5 mg to mice immediately after copulation and prevented pregnancy. The litter size was reduced in mice who were maintained on the compound for long periods before pregnancy started.

Wiesner, B.P. and Yudkin, J.: Control of fertility by antimitotic agents. Nature 176:249–250, 1955.

2503 Uterine Clamping

Discussed under Trauma

2504 Vaccinia Virus

Heath et al. (1956) injected chick eggs with this virus at 48 hours of incubation and found microcephaly and axial abnormalities in the embryos.

Heath, H.D.; Shear, H.H.; Imagawa, D.T.; Jones, M.H. and Adams, J.M.: Teratogenic effects of herpes simplex, vaccinia, influenza–A (NWS) and distemper virus infections on early chick embryos. Proc. Soc. Exp. Biol. Med. 92:675–682, 1956.

2505 Valexon CAS 14816–18–3

This insecticide caused gonadotoxic effects in male rats (0.7 mg per kg daily during 10 weeks) but did not induce damage of female gonads (Shepelskaya, 1980). Treatment of pregnant rats at 90 mg per kg daily caused some congenital malformations in the offspring. Abnormalities in heart ganglia and reduced activity of cholinesterase occurred.

Shepelskaya, N.R.: The gonadotoxic effect of valexon in experiments. Gig. Sanit. (USSR) 7:77, 1980.

2506 Valnoctamide Nirvanil™

Tuchmann–Duplessis and Mercier–Parot (1965) gavaged rats, mice and rabbits during various periods of orrganogenesis. The maximum dose in the rat was 300 mg per kg and in the other two species 400 mg per kg. No teratogenic effect was found.

Tuchmann–Duplessis, H. and Mercier–Parot, L.: Difficultes d'interpretation rencontrees au cours de l'etude teratogene d'un neuro–sedatif. C.R. Soc. Biol (Paris)159:6–10, 1965.

2507 Valoron™ DL
trans–2–Dimethylamino–1–phenyl–cyclohex–3–ene–trans carbonic acid ethylester HCl

Herrmann et al. (1970) treated rats and rabbits orally with up to 25 and 50 mg per kg, respectively. No adverse fetal effects were found.

Herrmann, M.; Wiegleb, J. and Leuschner, F.: Toxikologische Untersuchunger uber ein neues stark Wirksames Analgeticum. Arzneim–Forsch. 20:983–990, 1970.

2508 Valproic Acid *2–En–Valproic Acid* CAS 99–66–1

Jager–Roman et al. (1986) summarized 14 pregnancies in which valproic acid exposure occurred and reported that there were four with major defects. The number of minor defects were four times higher than controls. Head circumferences were not reduced. Lindhout and Schmidt (1986) questioned 18 groups using valproic acid and found 6 of 393 women who used valproic acid (and another anticonvulsant) had infants with neural tube defects. Those women who used valproic acid alone numbered 120 and they gave birth to three with neural tube defects (2.5 percent). In the total group, the ratio of spina bifida to anencephaly was eleven. Robert (1988) summarized the international efforts to collect data on valproic acid exposure. The data suggested that monodrug therapy gave a higher risk of meningomyelocele than valproic acid with phenobarbital treatment. Kallen et al. (**1989**) analyzed 318 malformed infants born to mothers with epilepsy. A significant association between valproic acid and spina bifida was found but an increased non–significant association was found with carbamazepine. Cotarin and Zaidman (**1991**) have summarized the developmental toxicity in humans and experimental systems.

Nau et al. (1981) reported on the pharmacodynamics of the drug. They also reported that among 12 exposed human pregnancies there were two offspring with microcephaly and four with head circumferences less than the 10th percentile. Dalens et al. (1980) reported an infant with facial dysmorphology and congenital heart disease. The mother received 1000 mg daily during most of pregnancy. Clay et al. (1981) reported an infant with microcephaly, facial dysmorphogenesis and ventricular septal defect. The mother and child had neurofibromatosis and the mother received 750 mg daily during pregnancy.

Robert and Guibaud (1982) found that among 72 women giving birth to offspring with caudal neural tube defects, nine were taking valproic acid. The doses were usually over one gm daily. This would give an odds ratio of approximately 20 for the defect to occur among valproic acid using mothers. Jeavons (1982) found nine spina bifida defects among 196 pregnancies during which the mother took valproic acid or valproic acid and other antiepileptics. Ardinger et al. (1988) summarized the clinical findings of 19 children exposed in utero. Postnatal growth deficiency and microcephaly were found in two–thirds and developmental delay or neurologic abnormality was present in 71 percent of these patients collected retrospectively from multiple centers. Bertollini et al. (1987) studied 62 pregnancies where monotherapy was employed and found four congenital defects when two were expected. The defects were spina bifida, hydrocephaly, microophthalmia and micrognathia.

International studies are in progress (Bjerkedal et al., 1982). Lindhout (1985) reported ten new cases among valproic acid users in the Netherlands during an 18–month period and he estimated that their relative risk for neural tube defects was approximately 30–fold increased. Huot et al. (1987) added two cases and summarized the wide spectrum of findings in 46 patients.

Sucheston et al. (1979) administered this anticonvulsant to mice orally on days 7 through 12 of gestation at doses of 225 to 560 mg per kg daily. Defects occurred at all dose levels and consisted of neural closure defects and occasional limb defects. In a monograph published by the manufacturer (Abbott Laboratory, 1978), the teratogenic dose in mice, rats and rabbits was given as 150, 65 and 315 mg per kg. Cleft palate, skeletal as well as renal and other defects were found. Khera (1992) described changes in the uteran labyrinth after 600–1000 mg per kg given to rats on day 13. Sonoda et al. (1993) described a dose–responsive increase in cardiac malformations in the offspring of mice treated on days 7, 8 or 9 with 300 to 700 mg per kg intraperitoneally. Guest et al. (1994) found reduced DNA in rat embryos grown at concentrations of 0.1 mM or above.

Nau and Hendrickx (1987) defined the structural elements of the molecule that are responsible for teratogenesis. Nau and Scott (1987) studied the mechanism of teratogenicity in the mouse. The main breakdown product 2–en–valproic acid and 3–keto–valproic acid were found in the fetus. The roles of intracellular binding and plasma binding were discussed. Nau (1986) reported pharmacologic measures of valproic acid and its metabolite 2–en–valproic acid in the pregnant mouse. The metabolite was much weaker in teratogenicity, producing only 3.5 percent neural tube defects at 10 mg per kg given on day 8 intraperitoneally as compared to 63 percent when valproic acid was used. Hauck and Nau (1989) reported that the R enantiomer of 2–n–propyl–4–penetenoic acid was less teratogenic than the S form.

Hendrickx et al. (1988) found craniofacial and skeletal defects in monkeys treated with 20 to 600 mg per kg. The treatment was oral and varied between day 20–50. Pharmacokinetic studies showed that the embryos were exposed to about one–half of the free drug in maternal plasma. Binkerd et al. (1988) reported a dose–dependent increase in skeletal and facial defects as well as cardiac and urogenital defects in rat fetuses. The dosages used were 200–800 mg per kg orally on days 8–17. Turner et al (1990) studied abnormal neural tube closure in the mouse embryo and found disrupted neuroepithelium with loss of integrity of the apical and basal surfaces. The study used scanning electron and light microscopy, gel electrophoresis of proteins and immunoreactivity to laminin and fibronectin. The distribution of the last two chemicals was not altered by treatment. Hansen and Grafton (1991) studied the toxicity in whole rat embryo culture and found that folinic acid did not ameliorate the toxicity. Vorhees et al. (1991) showed at the metabolite of valproic acid trans–2–ene–valproic acid was not teratogenic in

the rat when given orally on days 7–18 in amounts of 300 mg per kg. Sonoda et al. (1991) gave a single dose of 600 mg per kg intraperitoneally on day 6, 7, 8, or 9 and found increased congenital hearts. Treatment on day 7 produced various types of congenital heart disease in 29% of the offspring.

Nosel and Klein (1992) found that methionine supplementation in the rat reduced the resorptions produced by valproate. The serum from these rats also reduced embryotoxicity in whole embryo culture with valproate.

Michejda and McCollough (1987) produced spina bifida with meningomyelocele using 170 mg per kg on days 24, 25 and 26 in the monkey (malatta). Denker et al. (1990) showed by radioautography that the drug is accumulated in high concentration in the neuroepithelium of the rat fetus. Ehlers et al. (1992) showed that spina bifida occulta was induced in mice given 300 mg per kg three times daily on day 9. Similar treatment with 450 mg per kg three times daily produced 1% spina bifida aperta.

Abbott Laboratory: Depakene, Valproic Acid. Drug Monograph, 1978.

Ardinger, H.H.; Atkin, J.F.; Blackston, R.D.; Elsas, L.J.; Clarren, S.K.; Livingston, S.; Flannery, D.B.; Pellock, J.M.; Harrod, M.J.; Lammer, E.J.; Majewski, F.; Schinzel, A.; Toriella, H.V. and Hanson, J.W.: Verification of the fetal valproate syndrome phenotype. Am. J. Med. Gen. 29:171–185, 1988.

Bertollini, R.; Kallen, B.; Mastroiacovo, P. and Robert, E.: Anticonvulsant drugs in monotherapy. Effect on the fetus. Eur. J. Epidemiol. 3:164–171, 1987.

Binkerd, P.E.; Rowland, J.M.; Nau, H. and Hendrickx, A.G.: Evaluation of valproic acid (VPA) developmental toxicity and pharmacokinetics in Sprague–Dawley rats. Fund. Appl. Toxicol. 11:485–493, 1988.

Bjerkedal, T.; Czeizel, A.; Goujard, J.; Kallen, B.; Mastroiacova, P.; Nevin, N.; Oakley, G.P. and Robert, E.: Valproic acid and spina bifida. Lancet 2:1096 only, 1982.

Clay, S.A.; McVie, R. and Chen, H.: Possible teratogenic effect of valproic acid. J. Pediatr. 99:828 only, 1981.

Cotariu, D. and Zaidman, J.L.: Minireview developmental toxicity of valproic acid. Life Sciences 48:1341–1350, 1991.

Dalens, B.; Raynaud, E–J. and Gaulme, J.: Teratogenicity of valproic acid. J. Pediatr. 97:332–333, 1980.

Denker, L.; Nau, H. and D'Argy, R.: Marked accumulation of valproic acid in embryonic neuroepithelium of the mouse during early organogenesis. Teratology 41:699–706, 1990.

Ehlers, K.; Sturje, H. Merker, H. and Nau, H.: Valproic acid–induced spina bifida: a mouse model. Teratology 45:145–154, 1992.

Guest, I.; Buttar, H.S.; Smith, S. and Varma, D.R.: Evaluation of the rat embryo culture system as a predictive test for human teratogens. Can. J. Physiol. Pharmacol. 72:57–62, 1994.

Hansen, D.K. and Grafton, T.F.: Lack of attenuation of valproic acid–induced effects by folinic acid in rat embryos in vitro. Teratology 43:575–582, 1991.

Hauck, R.S. and Nau, H.: Asymmetric–synthesis and enantioselective teratogenicity of 2–n–propyl–4–pentenoic acid (4–en–vpa), an active metabolite of the anticonvulsant drug, valproic acid. Tox. Lett. 49(1):41–48, 1989.

Hendrickx, A.G.; Nau, H.; Binkerd, P.; Rowland, J.M.; Rowland, J.R.; Cukierski, M.J. and Cukierski, M.A.: Valproic acid developmental toxicity and pharmacokinetics in the rhesus monkey: An interspecies comparison. Teratology 38:329–345, 1988.

Huot, C.; Gauthier, M.; Lebel, M. and Larbrisseau, A.: Congenital malformations associated with maternal use of valproic acid. Can. J. Neurol. Sci. 14:290–293, 1987.

Jager–Roman, E.; Deichl, A.; Jakob, S.; Hartmann, A–M.; Koch, S.; Rating, D.; Steldinger, R.; Nau, H. and Helge, H.: Fetal growth, major malformations and minor malfunctions in infants born to women receiving valproic acid. J. Pediatr. 108:997–1004, 1986.

Jeavons, P.M.: Sodium valproate and neural tube defects. Lancet 2:1282–1283, 1982.

Kallen, B.; Robert, E.; Mastroiacovo, P.; Martinez–Frias, M.L.; Castilla, E.E. and Cocchi, G.: Anticonvulsant drugs and malformations is there a drug specificity? Eur. J. Epidemiol. 5(1):31–36, 1989.

Khera, K.S.: Valproic acid induced placental and teratogenic effects in rats. Teratology 45:603–610, 1992.

Lindhout, D.: Valproate and spina bifida in the Netherlands. (abs) Teratology 32:15A, 1985.

Lindhout, D. and Schmidt, D.: In utero exposure to valproate and neural defects. Lancet 1:329–393, 1986.

Michejda, M. and McCollough, D.: New animal model for the study of neural tube defects. Z Kinderchirurgie 42 I:32–35, 1987.

Nau, H.: Transfer of valproic acid and its main active unsaturated metabolite to the gestational tissue: Correlation with neural tube defect formation in the mouse. Teratology 33:21–27, 1986.

Nau, H. and Hendrickx, A.G.: Valproic acid teratogenesis. ISI Atlas of Science: Pharmacology 52–56, 1987.

Nau, H.; Rating, D.; Koch, S.; Hauser, I. and Helge, H.: Valproic acid and its metabolites: Placental transfer, neonatal pharmacokinetics, transfer via mothers milk and clinical status in neonates of epileptic mothers. Pharmacol. Exp. Ther. 219:768–777, 1981.

Nau, H. and Scott, W.J.: Teratogenicity of valproic acid and related substances in the mouse: Drug accumulation and pH in the embryo during organogenesis and structure–activity considerations. Arch. Toxicol. Suppl. 11:128–139, 1987.

Nosel, P.G. and Klein, N.W.: Methionine decreases the embryotoxicity of sodium valproate in the rat: in vivo and in vitro observations. Teratology 46:499–507, 1992.

Robert, E.: Valproic acid as a human teratogen. Cong. Anom. 28:S71-S80, 1988.

Robert, E. and Guibaud, P.: Maternal valproic acid and congenital neural tube defects. Lancet 2:937 only, 1982.

Sonoda, T.; Ohdo, S.; Ohba, K.; Okishima, T. and Hayakawa, K.: Sodium valproate–induced cardiovascular abnormalities in the Jcl:ICR mouse fetus. Cong. Anom.
31:89–94, 1991.

Sonoda, T.; Ohdo, S.; Ohba, K.; Okishima, T. and Hayakawa, K.: Sodium valproate–induced cardiovascular abnormalities in the Jcl:ICR mouse fetus: peak sensitivity of gestational day and dose–dependent effect. Teratology 48:127–132, 1993.

Sucheston, M.E.; Hayes, T.G.; Paulson, R.B. and King, J.E.: Fetal malformations in valporate sodium treated CD–1 mice. (abs) Teratology 19:49A only, 1979.

Turner, S.; Sucheston, M.E.; De Philip, R.M. and Paulson, R.B.: Teratogenic effects on the neuroephithelium of the cd–1 mouse embryo exposed in utero to sodium valproate. Teratology 41:421–442, 1990.

Vorhees, C.V.; Acuff–Smith, K.D.; Weisenburger, W.P.; Minck, D.R.; Berry, J.S.; Setchell, K.D.R. and Nau, H.: Lack of teratogenicity of trans–2–ene–valproic acid compared to valproic acid in rats. Teratology 43:583–590, 1991.

2509 Vamicamide

Katsumata et al. (**1994**) treated pregnant rats and rabbits with this anticholinergic in oral doses of up to 100 mg per kg. No teratogenic or antifertility effects were seen. Postnatal development of the rats was normal.

Katsumata, Y.; Ishida, S.; Umemura, T.; Shimazu, H.; Saegusa, T.; Iwanami, K. and Mine, Y.: Toxicity study of vamicamide–reproductive and developmental toxicity studies. The Clinical Report 28:983–998, 1994.

2510 Vanadium CAS 7440–62–2

Roschin and Kazimov (1980) gave orthovanadate Na (0.85 mg per kg) intraperitoneally and subcutaneously to male rats and an alteration of spermatogenesis was found. Treating the pregnant rats with this substance caused increased preimplantation embryonic mortality. Wide (1984) found no decrease in implantation in mice treated with 1.5 millimoles intravenously on day 3 but those treated on day 8 had reduced skeletal maturation. Domingo et al. (**1993**) found that Tiron, a chelating agent, reduced vanadium fetotoxicity.

Domingo, J.L.; Bosque, M.A.; Luna, M. and Corbella, J.: Prevention by tiron (sodium 4,5–dihydroxybenzene–1,3–disulfonate) of vanadate–induced developmental toxicity in mice. Teratology 48:133–138, 1993.

Roschin, A.V. and Kazimov, M.A.: Effect of vanadium on the generative function of tested animals. Gig Tr. Prof. Zabol. (USSR) 5:49–51, 1980.

Wide, M.: Effect of short–term exposure to five industrial metals on the embryonic and fetal development of the mouse. Environ. Res. 33:47–53, 1984.

2511 Vanadium Pentoxide

Altamirano–Lozano et al. (**1993**) injected pregnant mice intraperitoneally with 8.5 mg per kg and found increased shortening of the limbs and weight reduction.

Altamirano–Lozano, M.; Alvarez–Barrera, L. and Roldan–Reyes, E.: Cytogenetic and teratogenic effects of vanadium pentoxide on mice. Med. Sci. Res. 21:711–713, 1993.

2512 Vancomycin

Reyes et al. (1989) studied the offspring of 10 women treated for varying period in the 2nd and 3rd trimester with 1 gm intrvenously every 12 hours. auditory brainstem response and renal function was normal in all.

Reyes, M.P.; Ostola, E.M.; Cabinian, A.E.; Schmitt, C. and Rintelmann, W.: Vancomycin during pregnancy: does it cause hearing loss or nephrotoxicity in the infant? Am. J. Obstet. Gynecol. 161:977–980, 1989.

2513 Varicella *Herpes zoster*

The largest prospective study of women with varicella during pregnancy found 9 cases (0.7%) of the varicella syndrome among 1,294 infected women. (Enders et al., **1994**) Their cases included mostly congenital scarring of skin and in 7 of the 9 limb defects. All were infected before the 20th week. In the 97 women who had varicella after prophylaxis with anti–varicella immunoglobulin there were no infants with the syndrome. No infants with the syndrome were found among 366 women with herpes zoster during pregnancy.

Siegel and Fuerst (1966) made a prospective study of the offspring of 135 mothers with varicella during pregnancy and found no increase in congenital defect rate. The incidence of low birth weight was not increased. Savage et al. (1973) reported an infant exposed in utero to varicella at the 9th week with reduction defect of the extremity, Horner's syndrome, and meningo–encephalitis with defective swallowing. The presence of a depigmented depressed skin lesion on the deformed limb resembled those described in two other cited cases. Jones et al. (1990) identified prospectively 108 women with varicella. Seventy were in their first trimester. None of the 91 offspring had the varicella syndrome. One had microtia, one had a preauricular tag and one had bowel obstruction. Jones et al. (**1994**) followed up 172 women who had varicella. The infection occurred in the first trimester in 132. Among the 146 liveborns there was one with cutaneous scars, Horner's Syndrome and blindness associated with a retinal scar. Five others had single malformations: nystagmius, unilaterral microtia, bilaterial double ureters, and thryoglossal duct cyst. Pastuszak et al (**1994**) based on follow up of 106 mothers with first trimester varicella found the increased risk was 2.2 percent. Only one varicella embryopathy was found.

A case report by McKendry and Bailey (1973) described a newborn whose mother had varicella at 11 weeks of pregnancy. The infant had focal skin and muscle defects, delayed growth and seizures. The clinical picture may include disorganization of the eye (Frey et al., 1977). A brief review of this congenital syndrome was given by Bai and John (1979). Trlifajova et al. (1986) looked for evidence of placental transfer of the virus in 20 pregnancies. In three the evidence was compatible with transfer. Six of the offspring had congenital disease but the syndromes did not fit a pattern and only one of the four tested had evidence of infection. The one positive case had cicatrical skin lesions, Horner's syndrome, bone and skin hypoplasia and psychomotor retardation.

The incidence of congenital defects following first trimester varicella is still unknown (Srabstein et al., 1974) but an estimate of the fetal attack rate of 20 percent was made. Further analysis and a summary of 4 studies including 461 pregnancies indicated that 3 (0.6%) new-borns were effected. (Preblund et al., 1986). When there is maternal varicella within 4–6 days of delivery, Herpes zoster immun serum should be given to protect the newborn from disseminated disease.

Alkalay et al. (1987) summarized the clinical features of 22 infants with the syndrome and suggested that three criteria be used to establish the association. They are: 1. evidence of maternal Varicella zoster infection; 2. presence of a congenital skin lesion corresponding to a derivative distribution; and 3. immunologic proof of persisting infection in the newborn. They reaffirmed the absence of any significant increase in defects among many prospective studies of women with the infection.

Alkalay, A.L.; Pomerance, J.J. and Rimoin, D.L.: Fetal varicella syndrome. J. Pediatr. 111:320–321, 1987.

Bai, P.V.A. and John, J.: Congenital skin ulcers following varicella in late pregnancy. J. Pediatr. 94:65–66, 1979.

Enders, G.; Miller, E.; Cradock–Watson, J.; Bolley, I. and Ridehalgh, M.: Consequencess of varicella and herpes zoster in pregnancy: prospective study of 1739 cases. The Lancet 343:1548–1551, 1994.

Frey, H.M.; Bialkin, G. and Gershon, A.A.: Congenital varicella: Case report of a serologically proved long–term survivor. Pediatrics 59:110–112, 1977.

Jones, K.L.; Johnson, K.A. and Chambers, T.A.: Prospective follow–up of offspring born to women infected with varicella during pregnancy. (abs) Teratology 41:569, 1990.

Jones, K.L.; Johnson, K.A. and Chambers, C.D.: Offspring of women infected with varicella durring pregnancy: a prospective study. Teratology 49:29–32, 1994.

McKendry, J.B.J. and Bailey, J.D.: Congenital varicella associated with multiple defects. Can. Med. Assoc. J. 108:66–67, 1973.

Pastuszak, A.L.; Levy, M.; Shick, B.; Zuber, C.; Feldkamp, M.; Fladstone, J.; Bar–Levy, F.; Jackson, E.; Mesohino, W. and Koren, G.: Pregnancy outcome following maternal varicella infection in the first 20 weeks of gestation: a controlled multicenter study. (Abst) Pediat. Res. 35:92A, 1994.

Preblund, S.R.; Cochi, S.L. and Orenstein, W.A.: Letter to the editor. New Eng. J. Med. 315:1416–1417, 1986.

Savage, M.D.; Moosa, A. and Gordon, R.R.: Maternal varicella infection as a cause of malformations. Lancet 1:352–354, 1973.

Siegel, M. and Fuerst, H.T.: Low birth weight and maternal virus diseases. A prospective study of rubella, measles, mumps, chicken pox and hepatitis. JAMA 197:680–684, 1966.

Srabstein, J.C.; Morris, N.; Larke, R.B.B.; Derek, J.D.; Castelmo, B.B. and Sum, E.: Is there a congenital varicella syndrome? J. Pediatr. 84:239–243, 1974.

Trlifajova, J.; Benda, R. and Benes, C.: Effect of maternal varicella–zoster virus infection on the outcome of pregnancy and analysis of transplacental virus transmission. Acta Virol. 30:249–255, 1986.

2514   Varnish Workers

Syrovadko and Malysheva (1977) studied reproduction among workers who produced enamel–insulated wire. Their main exposures were vinylflex and polyether varnishes. Hypermenstruation occurred in 47 percent and some of the offspring of workers were "reluctant" to breastfeed but controls were not discussed.

Syrovadko, O.N. and Malysheva, Z.U.: Work conditions and their effect on certain specific functions among women who are engaged in the production of enamel–insulated wire. Gig. Tr. Prof. Zabol. 4:25–28, 1977.

2515   Vasopressin *Pitressin* CAS 11000–17–2

Jost (1951) injected rabbit fetuses intraperitoneally after 17 days of gestation with 5 to 500 milli units of Parke–Davis pitressin. With the lowest dose, edema was produced and doses above 10 milli units were associated with hemorrhagic necrosis and amputations of the distal extremities. Davis and Robson (1970) injected the amniotic cavity of the mouse fetus on the 15th or 16th day and obtained lesions similar to those seen in the mouse. By direct observation they concluded that intensive arterial vasospasm occurred.

Davis, J. and Robson, J.M.: The effects of vasopressin, adrenaline and noradrenaline on the mouse fetus. Br. J. Pharmacol. 38:446 only, 1970.

Jost, A.: Sur le role de la vasopressine et de la corticostimuline (A.C.T.H.) dans la production experimentale de lesions des extremities foetales (hemorragies, necroses, amputations congenitales). C. R. Soc. Biol. (Paris) 145:1805–1809, 1951.

2516   Venezuelan Equine Encephalitis

London et al. (1977) injected this virus directly into the brain of fetal monkeys and at birth they had cataracts and hydrocephalus.

Human infected pregnancies were reviewed when hydroanencephaly was found (Wenger, 1977).

London, W.T.; Levitt, N.H.; Kent, S.G.; Wong, V.G. and Sever, J.L.: Congenital cerebral and ocular malformations induced in rhesus monkeys by Venezuelan equine encephalitis virus. Teratology 16:285–296, 1977.

Wenger, F.: Venezuelan equine encephalitis. Teratology 16:359–362, 1977.

2517   Venoglobulin–IH *Immunoglobulin, Heat Inactivated*

Komai et al. (1989) gave this heat–treated human immunoglobulin intravenously to rats on days 7–17 in doses of up to 1,000 mg per kg and found no effect on the fetuses or offspring. Rabbits received up to 200 mg per kg on days 6–18 withough fetal ill effects.

Komai, Y.: Teratology study of venoglobulin–ih in rats. Kiso to Rinsho 23:6689–6716, 1989.

2518   Venom, Snake

Viper venom was injected subcutaneously on various days of the mouse gestation and following treatment on the 7th and 8th days facial anomalies, anencephaly and heart anomalies were produced. Cleft palate in low incidence was produced with treatment on the 8th through 14th days. The dose was 200 microgm of powdered venom which was the LD–50 (Clavert and Gabriel–Robez, 1974).

A report of 30 mothers bitten by poisonous snakes found 6 spontaneous abortions and 7 fetal deaths (Dunnihoo et al., 1992). No malformations were noted but only 5 were in the first trimester when bitten. Three of the mothers died.

Clavert, J. and Gabriel–Robez, O.: The effects on mouse gestation and embryo development of an injection of viper venom (Vipera aspis). Acta Anat. 88:11–21, 1974.

Dunnihoo, D.R.; Rush, B.M.; Wise, R.B.; Brooks, G.G.; and Otterson, W.N.: Snake bite poisoning in pregnancy. The Journal of Reproductive Medicine 37:653–658, 1992.

2519   Veracevine

Keeler and Binns (1968) fed 0.6–0.7 gm to 2 ewes on the 14th day of gestation and found no cyclopia in the offspring.

Keeler, R.F. and Binns, W.: Teratogenic compounds of veratrum californicum (Durand). Teratology 1:5–10, 1968.

2520   Verapamil

Magee et al. (1994) found no significant increase in defects among 33 offspring of women identified prospectively during their first trimester. Lee and Nagele (1986) studied the effect of this calcium antagonist on explanted chick embryos and found that 15 microgram per ml of medium produced neural tube defects. Nishikawa et al. (1988) found that phenobarbital significantly potentiated the teratogenicity in chicks.

Lee, H. and Nagele, R.G.: Toxic and teratogenic effects of verapamil on early chick embryos. Evidence for the involvement of calcium in neural tube closure. Teratology 33:203–211, 1986.

Magee, L.A.; Conover, B.; Schick, B.; Sage, S.; Cook, L.; Raman–Williams, L. and Koren, G.: Exposure to calcium channel blockers in human pregnancy. A prospective controlled multicentre cohort study. (abs) Teratology 49:372, 1994.

Nishikawa, T.; Kajita, A.; Ando, M.; Takao, A.; Bruyere, H.J. and Gilbert, E.F.: Potentiating effects of verapamil on cardiovascular teratogenicity of phenobarbital in the chick embryos. Dev. Pharmacol. Ther. 11:322–327, 1988.

2521   Veratrosine

Discussed under Cyclopamine

2522   Veratrum Californicum

This plant was the dietary source of cyclopamine which caused outbreaks of cyclopia with cleft palate in the fetuses of sheep who fed on it during the second and third week of gestation. The plant grows at relatively high altitudes. A more detailed list of references may be found under cyclopamine. Kalter (1968) summarized the subject.

Kalter, H.: Teratology of the Central Nervous System. Chicago: University of Chicago Press, 144–146, 1968.

2523 Vidarabine CAS 5536–17–4

This antiviral compound was teratogenic when given intramuscularly in the rat (30 mg per kg) and in the rabbit (5 mg per kg) (Schardein et al., 1977). The defects included orofacial clefts, skeletal and renal malformations. Application of a 10 percent ointment topically to 5 or 10 percent of the body surface of the rabbit dam also produced defects. Of five exposed monkey pregnancies, one had an enlargement of the tail and abnormally shaped ribs.

In whole embryo culture studies was the most toxic of several antiviral compounds. Retardation and decreased protein content was found at around 10 micromolar concentrations (Stahlmann et al., **1993**).

Schardein, J.L.; Hertz, D.L.; Petrere, J.A.; Fitzgerald, J.E. and Kurtz, S.M.: The effect of vidarabine on the development of the offspring of rats, rabbits and monkeys. Teratology 15:231–242, 1977.

Stahlmann, R.; Klug, S.; Foerster, M. and Neubert, D.: Significance of embryo culture methods for studying the prenatal toxicity of virustatic agents. Reproductive Toxicology7:129–143, 1993.

2524 Video Display Terminals *VDT Magnetic Field*

Marcus (1990) has summarized seven epidemiologic studies some of which are detailed below and found only one of six to show an increase in defects. Among five studies of abortion rates two were increased among workers exposed for longer than 20 hours per week. In view of the lack of consistency and association of recall bias and other factors such as stress it is a "shame that we may be terrorizing a generation of women without a clear scientific imperative to do so", Scialli (1990). Kavet and Tell (**1991**) have also reviewed the field levels and reproductive toxicity studies.

Ericson and Kallen (1986A) reported pregnancy outcome of a high proportion of Swedish women for the years 1976–77 and 1980–81. They were divided into work categories with high, medium or low exposure to video displays. No significant increase in malformations in newborns or low birth weight could be associated with exposure. Ericson and Kallen (1986B) using the same group, did a case control study using 522 pregnancies which ended in abortion, malformations or low birth weight and compared them to 1,032 women whose pregnancies had none of the adverse outcomes. Stress and smoking were associated with video screen work, the effect of video screen work was analyzed after stratification for stress and smoking–no statistically significant effect of video screen work was seen but odds ratios were above 1. The analysis of 44 infants with birth defects showed no signs of specificity of type. They discuss the various recall biases which are possible in a retrospective study. Kurppa et al. (1985) studied 235 women exposed to video display terminals and found no difference in malformations as compared to 255 nonexposed.

Goldhaber et al. (1988) studied 1,583 women using video terminals. The odds ratio for spontaneous abortions for women using VDTs for more than 20 hours weekly was 1.8 but no significant increase was found for women working with them for less than 20 hours weekly. Six maternal variables (age, education, occupation, smoking, alcohol consumption and other characteristics) could not explain the increased risk but the authors questioned whether those with miscarriage might have over–reported their VDT exposure times while those with normal births may have underreported. Windham et al. (**1990**) studied 852 VDT users in California and found no association with spontaneous abortion rates, low birth weight or intrauterine growth retardation. Intrauterine growth retardation was somewhat elevated among women with greater VDT use.

A few clusters of abnormal reproductive outcomes have been noted among exposed mothers but they consisted of small numbers over short periods and no distinct reproducible syndrome of abnormalities was found (Petersen, 1983). X–ray, ultraviolet, infrared, microwave and radio frequency emissions from video display terminals have been measured and no hazard identified (Lazarus and Burke, 1982; Letourneau, 1981 and Weiss, 1983). McDonald et al. (1986) interviewed women regarding over 56,000 current pregnancies and 48,000 past pregnancies. A slight excess of abortions (7.5 vs 6.8 percent) was found among those exposed to video display units but the authors stated that it could be due to biased recall. No differences were found in the previous pregnancy study (22.0 vs 20.7 for control). In occupational groups ranked by percentage use of video display units, the abortion rate was the same irrespective of the amount of use. Abenhaim et al. (1988) critically reviewed seven epidemiological studies of VDT users and concluded that there was no increased risk for congenital anomalies, stillbirth, premature birth or spontaneous abortion.

Abnormalities in chick embryos exposed to 48 hours of 0.5 msec pulses with pulse repetition rates of 10 to 1000 Hz and magnetic field intensities of up to 12 mG were reported by Delgado et al. (1982). However, two groups, Durfee et al. (1975) and Maffeo et al. (1984) were unable to confirm their findings. The conditions used by Maffeo et al. were virtually equal to those used by Delgado et al. Berman (1990) has summarized the animal work with electromagnetic currents. One protocol carried out in six laboratories gave negative results in 4 and positive in 2. (Berman et al. **1990**) He indicated that without understanding the exposure variables and the mechanism of effect the limited evidence for teratogenicity in animals was not extrapolatable to humans.

Stuchly et al. (1988) exposed rats to magnetic fields of up to 66 microT for seven hours daily from two weeks before, until the end of pregnancy. They found no fetal ill effects from field intensities much greater than those to which VDT users are exposed.

Abenhaim, L.; Lert, F.; Kaminski, M.; Mamelle, N.; Spira, N. and Ayme, S.: Travail sur terminal et grossesse. Evaluation des risques par consensus. Rev. Epidemiol. Sante Publ. 36:235–245, 1988.

Berman, E.: The developmental effects of pulsed magnetic fields on animal embryos. Reproductive Toxicology 4:45–49, 1990.

Berman, E.; House, C.D.; Koch, B.A.; Koch, W.E.; Leal, J.; Lovtrup, S.; Mantiply, E.; Martin, A.H.; Martucci, G.I.; Mild, K.H.; Monahan, J.C.; Sandstrom, M.; Shamsaifar, K.; Tell, R.; Trillo, M.A.; Ubeda, A. and Wagner, P.: Development of chicken embryos in a pulsed magnetic field. Bioelectromagnetics 11:169–187, 1990.

Delgado, J.M.R.; Leal, J.; Monteagudo, J.L. and Gracia, M.G.: Embryological changes induced by weak, extremely low frequency electromagnetic fields. J. Anat. 134:533, 1982.

Durfee, W.K.; Glante, P.R. and Muthukrishnan, S.: Extremely low frequency electromagnetic fields in domestic birds. A MRDC report compilation of navy sponsored biomedical and ecological research reports, 1975.

Ericson, A. and Kallen, B.: An epidemiological study of work with video screens and pregnancy outcome: I. A registry study. Am. J. Ind. Med. 9:447–457, 1986A.

Ericson, A. and Kallen, B.: An epidemiological study of work with video screens and pregnancy outcome: II. A case-control study. Am. J. Ind. Med. 9:459–475, 1986B.

Goldhaber, M.K.; Polen, M.R. and Hiatt, R.A.: The risk of miscarriage and birth defects among women who use visual display terminals during pregnancy. Am. J. Ind. Med. 13(6):695–706, 1988.

Kavet, R. and Tell, R.A.: VDTs: Field levels, epidemiology, and laboratory studies. Health Physics 61:47–57, 1991.

Kurppa, K.; Holmberg, P.C.; Rantala, K.; Nurminen, T. and Saxen, L.: Birth defects and exposure to video display terminals during pregnancy. Scand. J. Work Environ. Health 11:353–356, 1985.

Lazarus, M.G. and Bourke, J.A.: Problems associated with use of visual display units by bank clerical staff. Med. J. Aust. 2:186 only, 1982.

Letourneau, E.G.: Are video display terminals safe? Can. Med. Assoc. J. 15:535–539, 1981.

Maffeo, S.; Miller, M.W. and Carstensen, E.L.: Lack of effect of weak low frequency electromagnetic field on chick embryogenesis. J. Anat. 139:613–618, 1984.

Marcus, M.: Epidemiologic studies of vdt use and pregnancy outcome. Reproductive Toxicology 4:51–56, 1990.

McDonald, A.D.; Cherry, N.M.; Delorme, C. and McDonald, J.C.: Visual display units and pregnancy: Evidence from the Montreal survey. J. Occup. Med. 8:1226–1231, 1986.

Petersen, R.: Bioeffects of microwaves: A review of current knowledge. J. Occup. Med. 25:103–110, 1983.

Scialli, A.R.: The history of concerns about vdts. Reproductive Toxicology 4:43-44, 1990.

Stuchly, M.A.; Ruddick, J.; Villeneuve, D.; Robinson, K.; Reed, B.; Lecuyer, D.W.; Tan, K. and Wong, J.: Teratological assessment of exposure to time–varying magnetic field. Teratology 38:461–466, 1988.

Weiss, M.: The video display terminal? Is there a radiation hazard? J. Occup. Med. 25:98–100, 1983.

Windham, G.C.; Fenster, L.; Swan, S.H. and Neutra, R.R.: Use of video display terminals during pregnancy and the risk of spontaneous abortion, low birthweight, or intrauterine growth retardation. American Journal of Industrial Medicine 18:675–688, 1990.

2525 Vinblastine *Vincaleukoblastine* CAS 865–21–4

Discussed under Vincristine

2526 Vinconate Hydrochloride *Methyl 3–ethyl–2,3,3a,4–tetrahydro–1H–indolo[3,2,1–de] [1,5]naphthyridine–6–carboxylate monohydrochloride OM–853*

Shimazu et al. (**1993**) gave up to 500 mg per kg to rats before pregnancy and during the first 7 days of gestation and found no adverse fetal effects. Ishida et al. (**1993**) gave the same treatment perinatally and postnatally and found a decrease in pup survival and body weight at the highest dose but no behavioral changes or teratogenicity. Shimazu et al. (**1992**) gave this drug to rats and rabbits at oral doses of up to 500 mg per kg during organogenesis and found no teratogenic effects.

Ishida, S.; Ikeya, M.; Fujioko, M.; Shioda, Y.; Tamura, K.; Katsumata, T.; Tamaki, Y. and Sato, Y.: Study on orad administration of OM–853 to rats during the period of perinatal and lactation. Yakuri to Chiryo 21:2495–3521, 1993.

Shimazu, H.; Ishida, S.; Ikeya, M.; Tamura, K.; Katsumata, T.; Tamaki, Y. and Sato, Y.: Study on oral administration of OM–853 to rats prior to and in the early stages of pregnancy. Yakuri to Chiryo 21:2479–2494, 1993.

Shimazu, H.; Ishida, S.; Ikeya, M.; Tamura, K.; Katsumata, T.; Shichida, O.; Serizawa, K.; Furuta, M.; Tamaki, Y. and Sato, Y.: Study on administration of OM–853 during the period of organogenesis in rats and rabbits. Yakuri to Chiryo 20:s2033–s2067. 1992.

2527 Vincristine CAS 57–22–7

At least 26 normal infants have been born to mothers treated with vinblastine (Freidman & Polifka, **1994**) and no pattern of malformation increase has been identified. Green et al. (**1991**) studied 30 offspring exposed to vinblastine or vincristine and other treatments and found 5 with congenital anomalies.

The vinca alkaloids derived from the common periwinkle plant Vinca rosea (linn.) include vincristine and vinblastine. These antineoplastic substances are teratogenic in hamsters, rats, rabbits and monkeys. In the rhesus monkey, Courtney and Valerio (1968) injected a single daily dose of 0.15 to 0.175 mg of vincristine per kg on the 27th or 29th day of pregnancy and produced one fetus with encephalocele and another with syndactyly.

Ferm (1963) injected hamsters intravenously with 0.25 mg per kg of vinblastine or 0.1 mg per kg of vincristine on the 8th gestational day and produced skeletal and eye defects as well as spina bifida and exencephaly. Demeyer (1964, 1965) used vinblastine (0.25 mg) and vincristine (0.05 to 0.075 mg) in the pregnant rat on day 9 of gestation and produced a high incidence of eye defects with some microcephaly and neural tube closure defects. Some of the fetuses had midfacial defects with incomplete development of the prosencephalon. He observed an accumulation of arrested metaphase figures in the treated embryos. Cohlan and Kitay (1965) observed an increase in the number of mitotic figures in rat fetuses exposed to vinblastine. Ohzu and Shoji

(1965), using 2.5 mg per kg daily in the mouse on the 11th through the 14th days, produced a few cleft palates. Morris et al. (1967) studied the teratogenesis of vinblastine in the rabbit. Spielmann et al. (1986) injected 0.25–1.0 mg per kg into mice intraperitoneally on day 2 and found increased lethality on day 17.

Cohlan, S.Q. and Kitay, D.: The teratogenic action of vincaleukoblastine in the pregnant rat. J. Pediatr. 66:541–544, 1965.

Courtney, K.D. and Valerio, D.A.: Teratology in the Macaca mulatta. Teratology 1:163–172, 1968.

Demyer, W.: Vinblastine–induced malformations of face and nervous system in two rat strains. Neurology (Minneap.) 14:806–808, 1964.

Demyer, W.: Cleft lip and jaw induced in fetal rats by vincristine. Archives of Anatomy 48:181–186, 1965.

Ferm, V.H.: Congenital malformations in hamster embryos after treatment with vinblastine and vincristine. Science 141:426 only, 1963.

Freidman. J.M. and Polifka, J.E.: The Teratogenic Effects of Drugs, a Resource for Clinicians (TERIS). Johns Hopkins Press, Baltimore, 1994, pp 650–657.

Green, D.M.; Zevon, M.A.; Lowrie, G.; Seigelstein, N. and Hall, B.: Congenital anomalies in children of patients who received chemotherapy for cancer in childhood and adolescence. N. Engl. J. Med. 325(3):141–146, 1991.

Morris, J.M.; Van Wagenen, G.; Hurteau, G.D.; Johnston, D.W. and Carlsen, R.A.: Compounds interfering with ovum implantation and development. I. Alkaloids and antimetabolites. Fertil. Steril. 18:7–17, 1967.

Ohzu, E. and Shoji, R.: Preliminary notes on abnormalities induced by velban in developing mouse embryos. Proceedings of the Japanese Academy 41:321–325, 1965.

Spielmann, H.; Kruger, C.; Tenschert, B. and Vogel, R.: Studies on the embryotoxic risk of drug treatment during the preimplantation period in the mouse. Arzneim–Forsch. 36:219–223, 1986.

2528 Vinpocetine *TCV–3B* CAS 42971–09–5

Furuhashi et al. (1983) studied this derivative of vinca minor alkaloid in rats at oral doses of up to 125 mg per kg on days 0–7, 7–17 and 17–21. Some increase in embryolethality and survival rate was seen at 125 mg but no other adverse fetal effects occurred. Lactation decreased and abnormal vaginal bleeding was seen in the dams. In rabbits treated on days 6–18 with up to 125 mg per kg no adverse fetal effects occurred but above 5 mg per kg the implanted pregnancies were reduced.

Furuhashi, T.; Tsuji, K.; Honda, T.; Takei, A.; Vehara, M.; Kato, I. and Nakayoshi, H.: Effect of TCV–3B on the rat and rabbit pregnancy. Yakuri to Chiryo 11:1113–1142, 3559–3575, 3585–3595, 1983.

2529 Vinyl Acetate CAS 108–05–4

Minor skeletal defects with smaller weights were found in rat fetuses exposed to as high as 1000 ppm on days 6–15. When 5000 ppm was added to drinking water on days 6–15, no adverse fetal effects were found in the rat offspring (Anonymous, 1983).

Anonymous: The Society of the Plastics Industry, Inc., Vinyl acetate: Oral and inhalation teratology studies in the rat. FYI–AX–1283–0278 (sequence F), Washington, D.C.; US Environmental Protection Agency, 1983.

2530 Vinyl Chloride *Polyvinyl Chloride* CAS 75–01–4

Although Infante et al. (1976a,b) reported increased rates of malformations in one city where a vinyl chloride plant was located, subsequent studies by Edmonds et al. (1978) found the parents of these children were not workers in the plant nor were they living closer to the manufacturing source than the controls. Infante et al. (1976b) found a significant increase in fetal loss among the wives after paternal exposure. They used workers in rubber plants as a "control" group. Sanotsky et al. (1980) did not find an increase in spontaneous abortions among the wives of vinyl chloride workers. Theriault et al. (1983) studied a community near a vinyl chloride plant and found an increased defect rate. There was no association with occupation or site of residence and the authors felt that an association could not be substantiated. Lindbohm et al. (1985) studied abortions among 44 plastics workers exposed to vinyl chloride polyurethane and styrene plastics. No significant increase was found.

Since this compound causes mutations in bacterial systems there have been a number of dominant lethal studies in animals. Negative findings in rats exposed to up to 1,000 ppm were reported by Short et al. (1977). Anderson et al. (1977), using inhalation doses of up to 30,000 ppm six hours daily for five days, produced no dominant lethal effects in the mouse.

John et al. (1977) exposed rodents and rabbits to inhalation of varying doses of 50 to 2500 ppm during major organogenesis and found no adverse fetal effects. Ungvary et al. (1978) exposed rats to 1500 ppm during pregnancy and produced increased fetal mortality but no malformations. Salnikova and Kitsovskaya (1980) exposed Wistar rats during the entire gestation by inhalation (in a dose of 4.8 mg per cubic meter). An alteration of blood vessel permeability, nervous system functional disturbance and other abnormalities in offspring were found. The dose of 35.3 mg per cubic meter produced a slight embryotoxic effect.

Anderson, D.; Hodge, M.C.E. and Purchase, I.F.H.: Dominant lethal studies with halogenated olefins vinyl chloride and vinylidene dichloride in male CD–1 mice. Environ. Health Persp. 21:71–78, 1977.

Edmonds, L.D.; Anderson, C.E.; Flynt, J.W. and James, L.M.: Congenital central nervous system malformations and vinyl chloride monomer exposure: A community study. Teratology 17:137–142, 1978.

Infante, P.F.; Wagoner, J.K.; McMichael, A.J.; Waxweiler, R.J. and Falk, H.: Genetic risks of vinyl chloride. Lancet 1:734–735, 1976a.

Infante, P.F.; Wagoner, J.K. and Waxweiler, R.J.: Carcinogenic, mutagenic and teratogenic risks associated with vinyl chloride. Mutation Research 41:131–142, 1976b.

John, J.A.; Smith, F.A.; Leong, B.K.J. and Schwetz,

B.A.: The effects of maternally inhaled vinyl chloride on embryonal and fetal development in mice, rats and rabbits. Toxicol. Appl. Pharmacol. 39:497–513, 1977.

Lindbohm, M.; Hemminki, K. and Kyyronen, P.: Spontaneous abortions among women employed in the plastics industry. Am. J. Ind. Med. 8:579–586, 1985.

Salnikova, L.S. and Kitsovskaya, I.A.: Effect of vinyl chloride on embryogenesis in the rat. Gig. Tr. Prof. Zabol. (USSR) 3:46–47, 1980.

Sanotsky, I.V.; Davtian, R.M. and Glushchenko, V.I.: Study of the reproductive function in men exposed to chemicals. Gig. Tr. Prof. Zabol. (USSR) 5:28–32, 1980.

Short, R.D.; Minor, J.L.; Winston, J.M. and Lee, C.C.: A dominant lethal study in male rats after repeated exposures to vinyl chloride or vinylidene chloride. J. Toxicol. Environ. Health 3:965–968, 1977.

Theriault, G.; Iturra, H. and Gingras, S.: Evaluation of the association between birth defects and exposure to ambient vinyl chloride. Teratology 27:359–370, 1983.

Ungvary, G.; Hudak, A.; Tetrai, E.; Lorincz, M. and Folly, G.: Effects of vinyl chloride exposure alone and in combination with trypan blue, applied systematically during all thirds of pregnancy on the fetuses of CFY rats. Toxicology 11:45–54, 1978.

2531 4–Vinylcyclohexene

Hooser et al. (**1991**) have reported the specific toxicity of this chemical for female primary germ cells in the developing mouse. Substitution of the phenyl group eliminated follicular loss. Grizzle et al. (**1994**) gavaged mice with up to 500 mg per kg before and during gestation with no ill effects in the offspring. The offspring were raised and were able to reproduce normally in spite of a 17% decrease in sperm and a 33% decrease in primordial germ cells.

Grizzle, T.B.; George, J.D.; Fail, P.A.; Seely, J.C. and Heindel, J.J.: Reproductive effects on 4–vinylcyclohexene in swiss mice assessed by a continuous breeding protocol. Fundamental and Applied Toxicology 22:122–129, 1994.

Hooser, S.B.; Parola, L.R.: douds, D.A.; Hoyer, P.B. and Sipes, I.G.: Reproductive toxicity of 4–vinylcyclohexene and 4–phenylcyclohexene in mice. Abstract Society of Toxicology Feb 26, 1991, p 110.

2532 Viomycin

Takaya (1965) gave 1400–2800 mg per kg subcutaneously on days 6–10 to the pregnant rat and produced no adverse effects.

Takaya, M.: Teratogenic effects of antibiotics. Osaka City Med. Cen. J. 14:107–115, 1965.

2533 Viriditoxin CAS 35483–50–2

Hood et al. (1976) injected mice intraperitoneally on days 7–10 with up to 3.5 mg per kg and observed no adverse effect on the fetuses.

Hood, R.D.; Hayes, A.W. and Scammell, J.G.: Effects of prenatal administration of vitricin and viriditoxin to mice. Food Cosmet. Toxicol. 14:175–178, 1976.

2534 Virilizing Tumors

Verhoeven et al. (1973) reviewed the effect of various masculinizing ovarian tumors on the genitalia of the fetus. Examples of masculinization of the female fetus have occurred with arrhenoblastomas, Lehdig cell tumors, adrenal rest tumors, leuteoma, granulosa cell tumor and Krukenberg tumors. The masculinization was especially common with arrhenoblastomas and Krukenberg tumors.

Verhoeven, A.T.M.; Mastboom, J.L.; Leusden, H.A.I.M. and Van der Velden, W.H.M.: Virilization in pregnancy coexisting with an (ovarian) mucinous cystadenoma: A case report and review of virilizing ovarian tumors in pregnancy. Obstet. Gynecol. Survey 28:597–622, 1973.

2535 Vitamins

Vitamin supplementation as it relates to neural tube defects is described under Folic Acid in this edition. Seller and Nevin (1984) studied a group of mothers who had previous infants with neural tube defects. Among 424 who were vitamin supplemented, three offspring had hydrocephalus as compared to one infant among the 543 unsupplemented group. The general rate of malformation did not differ significantly between the two groups. The vitamins used contained 0.36 mg of folic acid and 4,000 units of vitamin A. Mulinare et al. (1988) studied the use of vitamins in the 3 months before and during the first 3 months of pregnancy among 347 pregnancies with neural tube closure defects. An apparent protective effect was noted but the protection could have been associated with other characteristics of the women who used vitamins. Forty-five percent of the pregnancies with neural tube defects and 38 percent (1092 of 2829) of the controls did not use multiple vitamins.

Tolarova (1982) offered multiple vitamins and 10 mg folic acid to mothers who had given birth previously to infants with cleft palate and/or cleft lip. There was one recurrence in 84 who took the treatment and 15 among the 202 who did not. Fraser and Warburton (1964) in their vitamin supplement study were unable to decrease the recurrence rate. Aro et al. (1984) studied 453 pregnancies that were complicated by limb reduction and could find no evidence that fewer vitamins were taken as compared to a matched group. Dudas and Czeizel (**1992**) report no increase in defects in a randomized blind control of 1,200 women taking preconceptual vitamins containing 6,000 IU vitamin A.

Li et al. (**1993**) reported on 96 women with offspring having urinary tract abnormalities. They adjusted potential confounders. Compared to 325 controls the perivitamin use odds ratio was 0.2 (confidence interval 0.1–0.7). The ratio was 0.4 for women who used vitamins in the 2nd and third trimesters.

Aro, T.; Haapakoski, J.; Heinonen, O.P. and Saxen, L.: Lack in association between vitamin intake during early pregnancy and reduction limb defects. Am. J. Obstet. Gynecol. 150:433 only, 1984.

Dudas, I. and Czeizel, A.E.: Use of 6,000 iu vitamin a during early pregnancy without teratogenic effect. Teratology 45:335–336, 1992.

Fraser, F.C. and Warburton, D.: No association of emotional stress on vitamin supplementation during pregnancy to cleft lip or palate in man. Plastic Reconstructive Surgery 33:395–399, 1964.

Li, D.K.; Daling, J.R.; Hickok, D.E.; Mueller, B.A.; Fantel, A.G. and Weiss, N.S.: Perinatal vitamin use and congenital urinary tract anomalies. (abs) Teratology 47:392–393, 1993.

Mulinare, J.; Cordero, J.F.; Erickson, J.D. and Berry, R.J.: Periconceptional use of multivitamins and the occurrence of neural tube defects. JAMA 260:3141–3181, 1988.

Seller, M.J. and Nevin, N.C.: Periconceptual vitamin supplementation and the prevention of neural tube defects in south–east England and Northern Ireland. J. Med. Genet. 21:325–330, 1984.

Tolarova, M.: Periconceptual supplementation with vitamins and folic acid to prevent cleft lip. Lancet 2:217 only, 1982.

2536 Vitamin A Deficiency

The general lack of evidence that vitamin A deficiency plays asignificant role in human congenital abnormalties is discussed by Warkany (1971). Giroud and Tuchmann–Duplessis (1962) briefly illustrated an infant with anophthalmia born to a deficient mother.

Hale (1937), using a vitamin A deficient diet, demonstrated for the first time that a non–genetic factor could produce congenital defects. Since then, vitamin A deficiency has been a widely studied teratologic experimental model. The vitamin A deficient congenital syndrome in the rat consists of over 90 percent ocular and urogenital anomalies, 50 percent diaphragmatic hernias and 17 percent congenital heart defects (Wilson et al., 1953). The eye defects consist of overgrowth of connective tissue between the hyaloid vessels in place of the vitreous and coloboma with retinal eversion (Warkany and Schraffenberger, 1946).

Many of the heart defects simulate malformations in man and include incomplete interventricular septation and anomalies of the aortic arch (Wilson and Warkany, 1950). The work by Millen and Woollam (1956) and Lamming et al. (1954) on hydrocephalus in the offspring of vitamin A deficient rabbits was summarized by Kalter (1968). Increased intracranial pressure with secondary aqueductal compression appears to be part of the mechanism underlying the hydrocephalus. Keratinizing epithelial metaplasia, a classic sign of vitamin A deficiency in postnatal life, was not seen in the rat fetus before the 18th fetal day (Wilson and Warkany, 1947). Palludan (1966) described extensive work in the pig model. Kalter and Warkany (1959) summarized the general topic.

Giroud, A. and Tuchmann–Duplessis, H.: Malformations congenitales. Role des facteurs exogenes. Pathol. Biol. 10:119–151, 1962.

Hale, F.: Relation of maternal vitamin A deficiency to microphthalmia in pigs. Texas State J. Med. 33:228–232, 1937.

Kalter, H.: Teratology of the Central Nervous System. Chicago. University of Chicago Press, 35–40, 1968.

Kalter, H. and Warkany, J.: Experimental production of congenital malformations in mammals by metabolic procedure. Physiol. Rev. 39:69–115, 1959.

Lamming, G.E.; Woollam, D.H.M. and Millen, J.W.: Hydrocephalus in young rabbits associated with maternal vitamin A deficiency. Br. J. Nutr. 8:363–369, 1954.

Millen, J.W. and Woollam, D.H.M.: Effect of the duration of vitamin–A deficiency in female rabbits on incidence of hydrocephalus in their young. J. Neurol. Neurosurg. Psychiatry 19:17–20, 1956.

Palludan, B.: A–Vitaminosis in Swine: A Study on the Importance of vitamin A for Reproduction. Copenhagen: Munksgaard, 1966.

Warkany, J.: Congenital Malformations Notes and Comments. Chicago: Year Book Medical Publishers, 127–128, 1971.

Warkany, J. and Schraffenberger, E.: Congenital malformations induced in rats by maternal vitamin A deficiency. I. Defects of the eye. Arch. Ophthalmol. 35:150–169, 1946.

Wilson, J.G.; Roth, C.B. and Warkany, J.: An analysis of the syndrome of malformations induced by vitamin A deficiency. Effects of restoration of vitamin A at various times during gestation. Am. J. Anat. 92:189–217, 1953.

Wilson, J.G. and Warkany, J.: Epithelial keratinization as evidence of vitamin A deficiency. Proc. Soc. Exp. Biol. Med. 64:419–422, 1947.

Wilson, J.G. and Warkany, J.: Cardiac and aortic arch anomalies in the offspring of vitamin A deficient rats correlated with similar human anomalies. Pediatrics 5:708–725, 1950.

2537 Vitamin B Deficiency

Discussed under Riboflavin; Folic Acid; Nicotinamide or Pyridoxine Deficiency

2538 Vitamin B₁₂ Deficiency

A series of papers on the teratogenic effects of a diet deficient in folic acid and B$_{12}$ indicated that a low percentage of fetal hydrocephalus could be produced (Grainger et al., 1954). Using a Steenbock–Black rachitogenic diet supplemented with viosterol and riboflavin, 11.7 percent of rat offspring developed hydrocephalus, 14.4 had bone defects and 6.7 had eye defects. When the same diet was supplemented with B$_{12}$ virtually no defects were observed. Woodard and Newberne (1966) confirmed the findings using a better defined basal diet.

They reported finding hydronephrosis, cleft lip and neural closure defects. The hydrocephalus is thought to be due to intermittant obstruction of the cerebral aqueduct (Overholser et al., 1954). Kalter (1968) summarized this subject.

Grainger, R.B.; O'dell, B.L. and Hogan, A.G.: Congenital malformations as related to deficiencies of riboflavin and vitamin B$_{12}$, source of protein, calcium to phosphorous ratio and skeletal phosphorous metabolism. J. Nutr. 54:33–48, 1954.

Kalter, H.: Teratology of the Central Nervous System. Chicago: University of Chicago Press, 28–32, 1968.

Overholser, M.D.; Whitley, J.R.; O'dell, B.L. and Hogan, A.G.: The ventricular system in hydrocephalic rat brains produced by a deficiency of vitamin B_{12} or of folic acid in the maternal diet. Anat. Rec. 120:917–934, 1954.

Woodard, J.C. and Newberne, P.M.: Relation of vitamin B_{12} and one–carbon metabolism to hydrocephalus in the rat. J. Nutr. 88:375–381, 1966.

2539 Vitamin E CAS 59–02–9

Momose et al. (1972) administered vitamin E in amounts of 150 or 300 mg per kg per day to pregnant mice subcutaneously on days 6, 8 and 10 of gestation. Growth retardation and fetal survival were increased in the treated group. The incidence of cleft palate was increased. Studies with the rat (75 mg per day) were negative (Sato, 1973).

Momose, Y.; Akiyoshi, S.; Mori, K.; Nishimura, N.; Fujishima, H.; Imaizumi, S. and Agata, I.: On teratogenicity of vitamin E. Reports from the department of anatomy. Mie Perfectural University School of Medicine 20:27–35, 1972.

Sato, Y.: Study of developmental pharmacology on vitamin E. (Japanese) Folia Pharmacol. 69:293–298, 1973.

2540 Vitamin E Deficiency

Kalter (1968) reviewed the experimental literature dealing with the teratogenicity of a diet deficient in vitamin E. Rats were reared from about 3 weeks of age on a deficient diet. By supplementing the deficient diet with 2–4 mg of vitamin E once on the 10th to the 13th day, surviving offspring with congenital defects were produced. Exencephaly or hydrocephalus was found in about 30 percent of the fetuses (Cheng and Thomas, 1953; Cheng et al., 1960).

Hook et al. (1974) found no significant teratogenicity when they gavaged pregnant mice with 591 IU on days 7–11. An equivalent dose in the human would be about one million IU.

Steele et al. (1974) using the explanted rat embryo system were able to show that vitamin E acts directly on the conceptus. N,N–diphenyl–p–phenylenediamine had a similar effect to vitamin E but ethoxyquin was inactive in vitro.

Cheng, D.W.; Bairnson, T.A.; Rao, A.N. and Subbammal, S.: Effect of variations of rations on the incidence of teratogenicity in vitamin E–deficient rats. J. Nutr. 71:54–60, 1960.

Cheng, D.W. and Thomas, B.H.: Relationship of time of therapy to teratogeny in maternal avitaminosis E. Proceedings of the Iowa Academy of Science 60:290–299, 1953.

Hook, E.B.; Healy, K.M.; Niles, A.M. and Skalko, R.G.: Vitamin E: A teratogen or antiteratogen? Lancet 1:809 only, 1974.

Kalter, H.: Teratology of the Central Nervous System. Chicago: University of Chicago Press, 41–43, 1968.

Steele, C.E.; Jeffery, E.H. and Diplock, A.T.: The in vitro study of explanted rat embryos. J. Reprod. Fertil. 38:115–123, 1974.

2541 Vitamin K_2 *Menaquinone*

Suzuki et al. (1971) gave this vitamin to mice and rats on the 7th through the 14th days of gestation, respectively.

The maximum doses were 1,000 mg per kg orally and 100 mg per kg intraperitoneally. No growth inhibition, teratogenicity or postnatal effects were observed. Some delay in appearance of ossification centers was found in the mice fetuses.

Mikami et al. (1981) gave this agent orally to rats in amounts of up to 1000 mg per kg daily either before pregnancy and during the first 7 days or on days 7–21 of gestation. No adverse effects on fertility or fetal postnatal function were observed. Kosuge (1973) fed 15 mg and 250 mg per day to rats and at both levels, fetal resorptions were increased but no increase in defects was found. A marked retardation of skeletal ossification was observed.

Kosuge, Y.: Study of developmental pharmacology on vitamin D_3. Part 1. Effect of vitamin K_3 on the rat fetus. Folia Pharmacol. Japon 69:285–291, 1973.

Mikami, T.; Mochida, H.; Osumi, I.; Gioto, K. and Suzuki, Y.: Fertility study and perinatal and postnatal study of menatetrenone in rats. Kiso to Rinsho 15:1143–1159, 1981.

Suzuki, Y.; Yayanagi, K.; Okada, F. and Furuuchi, M.: Toxicological studies of menaquinone–4 on development of fetuses and offsprings in mice and rats. (Japanese) Oyo Yakuri 5:469–487, 1971.

2542 Warfarin CAS 81–81–2

Discussed under Coumadin derivatives

2543 Water, Drinking CAS 7732–18–5

Chernoff et al. (1979) and Staples et al. (1979) could find no fetal changes in mice drinking municipal water.

Chernoff, N.; Rogers, E.H.; Carver, B.; Kavlock, R. and Gray, E.: The fetotoxic potential of municipal drinking water in the mouse. Teratology 19:165–170, 1979.

Staples, R.E.; Worthy, W.C. and Marks, T.A.: Influence of drinking water: Tap versus purified on embryo and fetal development in mice. Teratology 19:237–244, 1979.

2544 Wesselbron Disease Virus

Coetzer et al. (1979) injected 15 pregnant cows subcutaneously or intravenously with this virus. Three aborted and the other fetuses were normal. One fetus injected directly had porencephaly and cerebellar hypoplasia.

Coetzer, J.A.W.; Theodoridis, A.; Herr, S. and Kritzinger, L.: Wesselbron disease: A cause of congenital porencephaly and cerebellar hypoplasia in calves. Onderstepoort J. of Vet. Res. 46:165–169, 1979.

2545 Western Equine Encephalitis

Medovy (1943) reported two infants who had encephalitis during the first week of life after the mothers had mild febrile illnesses. Both infants developed spastic diplegia with either cortical atrophy or microcephaly.

London et al. (1982) injected intracerebral virus into monkey fetuses and produced hydrocephalus in 12 of 16.

London, W.T.; Levitt, N.H.; Altshuler, G.; Curfman, B.L.; Kent, S.G.; Palmer, A.E.; Sever, J.L. and Houff, S.A.:

Teratological effects of western equine encephalitis virus on the fetal nervous system of Macaca mulatta. Teratology 25:71–79, 1982.

Medovy, H.: Western equinine encephalomyelitis in infants. J. Pediatr. 22:308–318, 1943.

2546 WIN–35833

2–(p(2,2–Dichlorocyclopropyl)phenoxy)–2–methyl propionic acid Ciprofibrate CAS 52214–84–3

Tuchmann–Duplessis et al. (1976) found no teratologic effects of this hypolipemic in rats. Postnatal effects such as postnatal thrombosis and increased liver size were found.

Tuchmann–Duplessis, H.; Hiss, H. and Legros, J.: Teratological and prenatal toxicity evaluations of a new hypolipemic agent. (abs) Teratology 14:376, 1976.

2547 Xanthinol Niacinate CAS 437–74–1

Taniguchi et al. (1974) gave this compound orally and subcutaneously to mice and rats from the 7th through the 14th day of gestation. In mice, the maximum oral dose was 5,000 and subcutaneous dose 1,250 mg per kg. In the rats, oral doses to 10,000 mg and subcutaneous doses to 1,250 mg per kg were used. No teratogenicity was found, but some growth retardation of the fetus was seen.

Taniguchi, S.; Yamada, A. and Morita, S.: Teratogenic studies of xanthinol nicotinate. Oyo Yakuri 8:1145–1156, 1974.

2548 Xenon CAS 7440–63–3

Discussed under Nitrous Oxide

2549 Xenytropium Bromide CAS 511–55–7

This anticholinergic was given intraperitoneally to rats and mice on the 7th through the 14th day of gestation and no teratogenicity or postnatal functional changes were found. The rats received up to 1000 mg per kg per day and the mice 500 mg per kg per day (Aono et al., 1970).

Aono, K.; Mizusawa, H.; Oketani, Y. and Masuda, N.: Teratogenic study of xenytropium bromide in mice and rats. Oyo Yakuri 4:725–739, 1970.

2550 Xorphanol

17–Cyclobutylmethyl–3–hydroxy–6–β–methylmorphinan methanesulfonate; TR 5379M CAS 7728–89–9

This narcotic antagonist was tested orally in rats on days 6 through 15. At doses of up to 250 mg per kg there were no adverse reproductive effects even though the dams did not gain weight normally (Porter et al., 1983).

Porter, M.C.; Hartnagel, R.E.; Clemens, G.R.; Kowalski, R.L.; Bare, J.J.; Halliwell, W.E. and Kitchen, D.N.: Preclinical toxicity and teratogenicity studies with the narcotic antagonist analgesic drug TR 5379M. Fund. Appl. Toxicol. 3:478–482, 1983.

2551 Xylene CAS 1330–20–7

Among nine pregnancies producing offspring with caudal regression syndrome, five mothers had exposure to fat solvents (Kucera, 1968). These included acetone, trichloroethylene, methylchloride and xylene. Subsequent observations showed less association with solvents (personal communication with author, 1977).

Kucera (1968) reported studies in chick embryos exposed for 60 to 240 minutes to a xylene atmosphere at developmental periods up to the 10 somite stage. A high malformation rate was found and nearly one–half of the defects were rumplessness, a defect resembling caudal regression syndrome.

In mice, Marks et al. (1982) found fetal deaths increased when they gavaged 3.6 ml of xylene mixtures per kg on days 5 through 14 of gestation. This dose was close to the LD–50 for the dams. Increased defects (mostly cleft palate) were found at dose levels of 2.4, 3.0 and 3.6 ml per kg per day. Rosen et al. (1986) exposed rats to up to 7000 mg per square meter during organogenesis and found no postnatal change in growth rate or acoustic startle response. Hudak and Ungvary (1978) exposed rats to 1000 mg per square meter of air during days 9 through 14 and found no teratogenic results although minor skeletal anomalies occurred. Ungvary and Tatrai (1985) exposed rats, mice and rabbits to ortho–, meta–, and paraxylene 1000 mg per meter squared during organogenesis and found growth and skeletal retardation but no defect increase. Hood and Ottley (1985) reviewed the evidence for developmental effects and discussed the o and p isomers.

Hood, R.D. and Ottley, M.S.: Developmental effects associated with exposure to xylene: A review. Drug and Chemical Toxicology 8:281–297, 1985.

Hudak, A. and Ungvary, G.: Embryotoxic effects of benzene and its methyl derivatives: Toluene, xylene. Toxicology 11:55–63, 1978.

Kucera, J.: Exposure to fat solvents: A possible cause of sacral agenesis in man. J. Pediatr. 72:857–859, 1968.

Marks, T.A.; Ledoux, T.A. and Moore, J.A.: Teratogenicity of a commercial xylene mixture in the mouse. J. Toxicol. Environ. Health 9:97–105, 1982.

Rosen, M.B.; Crofton, K.M. and Chernoff, N.: Postnatal evaluation of prenatal exposure to p–xylene in the rat. Toxicology Letters 34:223–229, 1986.

Ungvary, G. and Tatrai, E.: On the embryotoxic effects of benzene and its alkyl derivatives in mice, rats and rabbits. Arch. Toxicol. Suppl. 8:425–430, 1985.

2552 Xylometazoline HCl *Otrivin*™ CAS 526–36–3

Aselton et al. (1985) studied the offspring of 207 women exposed to this medication in the first trimester and found only five with congenital defects.

Aselton, P.A.; Jick, H.; Milunsky, A.; Hunter, J.R. and Stergachis, A.: First–trimester drug use and congenital disorders. Obstet. Gynecol. 65:451–455, 1985.

2553 β–D–Xylosides

Gibson and Doller (1978) using p–nitrophenyl–β–D–xyloside and 4–methylumbelliferyl–β–D–xyloside and their derivatives in doses of 5–10 mg into the amniotic sac of 9 day chick embryo produced generalized stunting of the skeleton.

Gibson, K.D. and Doller, H.J.: Beta–D–xylosides cause abnormality of growth and development in chick embryos. Nature 273:151–153, 1978.

2554 Xyloxemine Hydrochloride *2–[2–(Di–2,6–xylylmethoxy)ethoxy]–N,N–dimethylethylamine Hydrochloride*

Van Eeken and Mulder (**1967**) treated rats and rabbits during gestation with up to 35 and 10 mg per kg respectively. No evidence of teratogenicity was found. In rats receiving 300 mg per kg of diet there were no adverse effects found in three generations.

Van Eeken, C.J. and Mulder, D.: The influence of xyloxemine hydrochloride, 2–[2–(Di–2,6–xylylmethoxy)ethoxy]–N,N–dimethylethylamine hydrochloride, on reproduction in laboratory animals. Arch. Int. Pharmacodyn. 167:135–141, 1967.

2555 Yellow, Dye and Coloring No. 8 *Acid Yellow 73 Fluorescein Sodium* CAS 518–47–8

Burnett and Goldenthal (1986) treated rats and rabbits by gavage during organogenesis with doses of up to 1500 and 250 mg, respectively. No maternal toxicity or adverse fetal effects were found. McEnerney et al. (1977) injected pregnant rats intravenously with 1.4 ml of this solution on day 5, 6, 8, 13, 15 or 16 and observed no adverse fetal effects.

Burnett, C.M. and Goldenthal, E.I.: The teratogenic potential in rats and rabbits of D and C Yellow no. 8. Food Chem. Toxicol. 24:819–823, 1986.
McEnerney, J.K.; Wong, W.P. and Peyman, G.A.: Evaluation of the teratogenicity of fluorescein sodium. Am. J. Opthalmol. 84:847–850, 1977.

2556 YN–72 *D–Arbinofuranosyl–(E)–5–[2–bromovinyl] uracil*

Ishida et al. (**1990**) gave this antiviral orally to rats on days 7–17 in doses up to 200 mg per day. No adverse fetal effects were found. Umemura et al. (**1990**) treated rabbits with up to 500 mg per kg during organogenesis and found no adverse effects.

Ishida, S.; Ishikawa, Y.; Ikeya, M. and Fujita, M.: Teratological study in rats treated orally with yn–72. Kiso to Rinsho 24:977–993, 1990.
Umemura, T. Takigami, H. and Ishikura, T.: Teratological study in rabbits treated orally with yn–72. Kiso to Rinsho 24:995–1001, 1990.

2557 Yohimbine

Bovet–Nitti and Bovet (1959) gave rats 20 mg per kg intramuscularly on day 5 of pregnancy and found no ill effects on 8 pregnancies.

Bovet–Nitti, F. and Bovet, D.: Action of some sympatholytic agents on pregnancy in the rat. Proc Soc Exp Biol Med 100:555–557, 1959.

2558 Ytterbium CAS 7440–64–4

In the hamster, Gale (1975) found damage to the axial skeleton of the fetus when doses of 50 to 100 mg per kg were given intravenously on the 8th day. A few nervous system defects and ventral body wall openings were also found.

Gale, T.F.: The embryotoxicity of ytterbium chloride in golden hamsters. Teratology 11:289–296, 1975.

2559 Zaltoprofen

Shimazu et al. (**1990**) gave rats orally up to 33 mg per kg on days 7–17. The dams had digestive disorders but no ill effects were found in the fetuses. A slight decrease in weight gain and survival was found in postnatal studies of exposed pups.

Shimazu, H.; Matsuoka, T.; Ikeya, M.; Katahira, K.; Nishikawa, M.; Hirakawa, T.; Kuhara, K. and Ikka, T.: Reprodductive and developmental toxicity studies of cn–100 in rats. Kiso to Rinsho 24:6773–6784, 1990.

2560 Zatosetron Maleate

Byrd et al. (**1991**) gavaged rats and rabbits during organogenesis with up to 70 and 7.5 mg per kg respectively. Maternal and fetal toxicity were found at the higher doses but no selective toxicity was found.

Byrd, R.A.; Hoover, D.M. and Kelich, S.L.: Developmental toxicity of zatosetron maleate. (Abs) Teratology 43:458, 1991.

2561 Zearalenone CAS 17924–92–4

This is an estrogenic mycotoxin produced by fusarium. Ruddick et al. (1976) treated pregnant rats orally with 1 to 10 mg per kg. At the higher doses there was some fetal growth retardation and some delay or absence of ossification centers. No visceral defects occurred.

Ruddick, J.A.; Scott, P.M. and Harwig, J.: Teratological evaluation of zearalenone administered orally to the rat. Bull. Environ. Contam. Toxicol. 15:678–681, 1976.

2562 Zeniplatin

This anticancer agent was studied by Filler et al. (**1992**) in rats in doses of 0.03 to 1.2 mg per kg and in rabbits in doses of 0.04 to 0.08 mg per kg. An intravenous mode of administration was used. Growth retardation, gross and visceral defects were produced in both species.

Filler, R.; York, R.G. and Schardein, J.L.: Differential sensitivity to the developmental toxicity of zeniplatin (ZPt) in the rat and rabbit (A). Teratology 45:471, 1992.

2563 Zeranol CAS 26538–44–3

The anabolic agent was tested in rats with oral doses of up to 4.0 mg per kg on days 6–18 of gestation. No adverse fetal effects were found except for increased resorptions at the 4.0 mg level. Tibial length was not altered (Wardell et al., 1982).

Wardell, R.E.; Seegmiller, R.E. and Bradshaw, W.S.: Induction of prenatal toxicity in the rat by diethylstilbestrol, zeranol, 3,4,3,4–tetrachlorobiphenyl, cadmium and lead. Teratology 26:229–237, 1982.

2564 Zidovudine *AZT ZDU Azidothymidine*

Toltzis et al. (1991) administered 0.25–2.5 mg per ml of drinking water to mice from age 8 weeks and found no pregnancies. At 0.25 mg per ml of water increased resorptions were found. Fetal weights were not given. No "physical anomalies were noted". In vitro studies of the maturation of eggs indicated reduced development in concentrations of 1 or 10 microgm per ml.

Lopez–Anaya et al. (1990) showed that this experimental anti–HIV agent crosses the placental barrier. Ha et al. (1994) gave monkeys 1.5 mg per kg every 4 hours throughout pregnancy. Of the 12 term pregnancies studied no ultrasound differences or behavioral delays were found.

A prospective registry of exposed first trimester exposures is maintained by Burroughs Wellcome Co. with the Centers for Disease Control and NIH (800–722–9292 ext 8465). Of 41 exposures only one was associated with a defect; five ended in induced abortion and 35 were healthy. One had pectus excavatum and the other unilateral renal agenesis, thymic cyst and brain "inactivity" by scan. Sperling et al. (1992) found no teratogenic abnormalities among 12 infants exposed in the first trimester. Anemia and growth retardation was found in a minority of infants exposed throughout pregnancy. Wellcome Co,. Research Triangle Park. Dec. 17, 1992.

Eldridge, R.R.: Personal communication Burroughs

Ha, J.C.; Nosbisch, C.; Conrad, S.H.; Ruppenthal, G.C.; Sackett, G.P.; Abkowitz, J. and Unadkat, J.D.: Fetal toxicity of zidovudine (azzidothymidine) in macaca nemestrina: preliminary observations. Journal of Acquired Immune Deficiency Syndromes 7:154–157, 1994.

Lopez–Anaya, A.; Unadkat, J.D.; Schumann, L.A. and Smith, A.L.: Pharmacoinetics of zidovudine (azidothymidine) i transplacental transfer. Journal of Acquired Immune Deficiency Syndromes 3:959–964, 1990.

Sperling, R.S.; Stratton, P.; O'Sullivan, M.J.; Boyer, P.; Watts, D.H.; Lambert, J.S.; Hammill, H.; Livingston, E.G.; Gloeb, D.J.; Minkoff, H. and Fox, H.E.: A survey of zidovudine use in pregnant women with human immunodeficiency virus infection. The New England Journal of Medicine 326:857–861, 1992.

Toltzis, P.; Marx, C.M.; Kleinman, N.; Levine, E.M. and Schmidt, E.V.: Zidovudine–associated embryonic toxicity in mice. The Journal of Infectious Diseases 163:1212–1218, 1991.

2565 Zinc Chloride CAS 7646–85–7

Zimmerman (1984) measured zinc in the cord bloods of nine anencephalics and three meningomyelocele infants and found it significantly elevated above controls. Maternal zinc levels were normal but there was a shift in the distribution of zinc from α 2M to albumin.

Chang et al. (1977) injected 20 mg per kg intraperitoneally on day 8, 9, 10 or 11 of the mouse pregnancy and produced a delay and some malformations in fetal ossification. Ferm and Carpenter (1977), using 2 mg per kg intravenously on the eighth day in the hamster, did not find increased malformations.

Chang, C–H.; Mann, D.E. and Gautieri, R.F.: Teratogenicity of zinc chloride, 1,10–phenanthroline and a zinc–1,10–phenanthroline complex in mice. J. Pharm. Sci. 66:1755–1758, 1977.

Ferm, V.H. and Carpenter, S.J.: Teratogenic effect of cadmium and its inhibition by zinc. Nature 216:1123 only, 1967.

Zimmerman, A.W.: Hyperzincemia in anenecephaly and spina bifida: A clue to the pathogenesis of neural tube defects? Neurology 34:443–450, 1984.

2566 Zinc Deficiency

Sever and Emanuel (1973) pointed out that zinc deficiency is clinically prevalent in Egypt and that the incidence of CNS defects there is high (7.88 per 1,000 births). Cadvar et al. (1980) reported from Turkey that maternal serum Zn levels are significantly lower in those giving birth to anencephalics than in controls.

Hambridge et al. (1975) raised the question that zinc deficiency in patients with acrodermatitis enteropathica might account for two major defects occurring among seven pregnant patients with the disease. The two defects were anencephaly and fatal achondrogenesis. Ghosh et al. (1985) studied serum levels in 437 Chinese women and could not associate low levels with spontaneous abortions, low birth weight or malformations. Six samples from mothers with anencephalics were similar to controls. The serum levels dipped during pregnancy and did not correlate with hair zinc. Milunsky et al. (1992) analyzed zinc content of toenail clippings from second trimester women with a neural tube defects. An increase in the proportion of low levels was found in the effected mothers. Matched analyses controlling for folic acid and other variables gave an odds ration of 5.0 (95% confidence 1.1–21.6). Keen and Hurley (1987)Keen et al. (1993) have updated the evidence for human and animal teratogenicity. Jameson (1993) has reviewed the work showing zinc supplementation during pregnancy may reduce perinatal complications.

Warkany and Petering (1972) provided detailed descriptions of the central nervous system defects in the rat model. Hurley and Swenerton (1966) reared rats on a marginally deficient Zn diet and at the onset of gestation placed them on a Zn deficient diet. Nearly all of the surviving fetuses exhibited one or more congenital malformations. Cleft palate, skeletal defects, hydrocephalus (65 percent), eye, heart, lung and urogenital (49 percent) abnormalities were found. A reduction of Zn content in the fetuses was found. Hurley et al. (1971) after exposing rats to only a few days of Zn deficiency, produced fetal defects.

Purichia and Erway (1972) produced a reduction in the otoliths of rat fetuses whose mothers were maintained on deficient diets. Evidence that a three–day period of zinc deficiency can produce abnormal rat blastocysts and morulae was published by Hurley and Shrader (1975). Mieden et al. (1986) reversed the brain and general growth retardation of in vitro grown rat embryos by adding zinc to the deficient medium. Oteiza et al. (1988) gave evidence that zinc deficiency affects the microtubule assembly in maternal brains of rats. Keen et al. (1988) showed that zinc and metallothionein concentration in the liver of marginally deficient monkeys are useful in measuring long term deprivation. Welsh et al. (**1990**) studied the effect of textured vegetable protein from soybeans on pregnant rats. The defects observed were shown to relate to Zn deficiency. Supplementation of the diet with Zn prevented the defects. A number of chemicals (urethanes, ethanol, melphalan, arsenic and alpha–hederin) produce animal developmental toxicity by altering maternal and embryonic Zn metabolixm (Taubeneck et al., **1994**).

Cadvar, A.O.; Arcasoy, A.; Baycu, T. and Himmetoglu, O.: Zinc deficiency and anencephaly in Turkey. Teratology 22:14 only, 1980.

Ghosh, A.; Fong, L.Y.Y.; Liang, S.T.; Woo, J.S.K. and Wong, V.: Zinc deficiency is not a cause for abortion, congenital abnormality and small–for–gestational age infant in Chinese women. Br. J. Obstet. Gynecol. 92:886–891, 1985.

Hambridge, K.M.; Neldner, K.H. and Walravens, P.A.: Zinc, acrodermatitis enteropathica, and congenital malformations. Lancet 1:577–578, 1975.

Hurley, L.S.; Gowan, J. and Swenerton, H.: Teratogenic effects of short–term and transitory zinc deficiency in rats. Teratology 4:199–204, 1971.

Hurley, L.S. and Shrader, R.E.: Abnormal development of preimplantation rat eggs after three days of maternal dietary zinc deficiency. Nature New Biology 254:427–429, 1975.

Hurley, L.S. and Swenerton, H.: Congenital malformations resulting from Zn deficiency in rats. Proc. Soc.

Jameson, S.: Zinc Status in pregnancy: the effect of zinc therapy on perinatal mortality, prematurity, and placental ablation. Keen, C.L.; A. Benedich, A. and Willhite, C.C., eds. NY Acad of Science. Maternal Nutrition and Pregnancy Outcome 678:178–192, 1993.

Keen, C.L.; Golub, M.S.; Gershwin, M.E.; Lonnerdal, B. and Hurley, L.S.: Studies of marginal zinc deprivation in rhesus monkeys. III. Use of liver biopsy in the assessment of zinc status. Am. J. Clin. Nutr. 47:1041–1045, 1988.

Keen, C.L. and Hurley, L.S.: Effects of zinc deficiency on prenatal and postnatal development. NeuroToxicol. 8(3):379–388,

Keen, C.L.; Taubeneck, M.W.; Daston, G.P.; Rogers, J.M. and Gershwin, M.E.: Primary and secondary zinc deficiency as factors underlyikng abnormal cns development. C.L. Keen; A. Bendish and C.C. Willhite, eds. N.Y. Acad Sciences. Maternal Nutrition and Pregnancy Outcome 678:37–47, 1993.

Mieden, G.D.; Keen, C.L.; Hurley, L.S. and Klein, N.W.: Effects of whole rat embryos cultured on serum from zinc– and copper–deficient rats. J. Nutr. 116:2424–2431, 1986.

Milunsky, A.; Morris, J.S.; Jick, H.; Rothman, K.J.; Ulcickas, M.; Jick, S.S.; Shoukimas, P. and Willett, W.: Maternal zinc and fetal neural tube defects. Teratology 46:341–348, 1992.

Oteiza, P.I.; Hurley, L.S.; Lonnerdal, B. and Keen, C.L.: Marginal zinc deficiency affects maternal brain microtubule

assembly in rats. J. Nutr. 118:735–738, 1988.

Purichia, N. and Erway, L.C.: Effects of dichlorophenamide, zinc and manganese on otolith development in mice. Dev. Biol. 27:395–405, 1972.

Sever, L.E. and Emanuel, I.: Is there a connection between maternal zinc deficiency and congenital malformations of the central nervous system in man? Teratology 7:117–118, 1973.

Taubeneck, M.W.; Daston, G.P.; Rogers, J.M. and Keen, C.L.: Altered maternal zinc metabolism following exposure to diverse developmental toxicants. Reprod. Toxicol. 8:25–40, 1994.

Warkany, J. and Petering, H.G.: Congenital malformations of the central nervous system in rats produced by maternal zinc deficiency. Teratology 5:319–334, 1972.

Welsh, J.J.; Rader, J.I.; Collins, T.F.X.; Black, T.N.; Rorie, J.I. and Kopral, C.A.: Developmental effects and mineral interactions in rats fed textured vegetable protein. Teratology 42:67–78, 1990.

2567 Zinc Diethyldithiocarbamate

Nakaura et al. (1984) gavaged rats with up to 250 mg per kg on days 7–15. Even at maternally toxic doses, no teratogenicity or postnatal behavioral changes were found.

Nakaura, S.; Tanaka, S.; Kawashima, K.; Takanaka, A. and Omori, Y.: Effects of zinc diethyldithiocarbamate on the prenatal and postnatal development of the rats. Bull. Natl. Inst. Hyg. Sci. 102:55–61, 1984.

2568 Zinc Dimethyldithiocarbamate *Ziram*

Giavini et al. (**1983**) studied rat pregnancies exposed by gavage to 12.5 to 100 mg per kg during implantation or orqanogenesis. At all levels maternal weight was decreased. Fetal weight was decreased and at 100 mg per kg two fetuses with diaphragmatic hernias and one with ventricular septal defect were found among 33 examined. Major malformations were not increased at 50 mg per kg.

Giavini, E.; Vismaa, C. and Boccia, M.L.: Pre- and postimplantation embryotoxic effects of zinc dimethyldithiocarbamate (ziram) in the rat. Ecotoxicology and Environmental Safety 7:531–537, 1983.

2569 Zonisamide
1,2–Benzisoxazole–3–methanesulfonamide

Terada (1987) gavage fed rats with up to 60 mg per kg from day 17 of gestation to day 20 of lactation. No adverse effects were found in the pups.

Terada, Y.; Aoki, Y.; Mukumoto, K.; Imura, Y.; Yoshioka, M.; Nishimura, K. and Ohnishi, K.: Reproduction studies of zonisamide. 3. Perinatal and postnatal study in rats. Jpn. Pharmacol. Ther. 15:91–107, 1987.

2570 Zopiclone CAS 43200–80–2

Esaki et al. (1983) gavaged rats on days 7–17 with up to 250 mg per kg and produced no adverse fetal effects. Given during the perinatal period the surviving fetuses were decreased in numbers. Crab–eating monkeys were gavaged with 8 mg per day from day 23 to 36 and 4 normal fetuses and one abortion resulted (Tanioka et al., 1983).

Esaki, K.; Umemura, T.; Yamaguchi, K.; Takada, K. and Yanagita, T.: Reproduction studies of zopiclone in rats. Preclin. Rep. Cent. Inst. Anim. 9:127–156, 1983.

Tanioka, Y.; Koizumi, H. and Shibuya, A.: Teratogenicity study of zopiclone in crab–eating monkeys. Preclin. Rep. Cent. Inst. Anim. 9:157–174, 1983.

2571 Zotepine
2–Chloro–11,(2–dimethylaminoethoxy)–dibenzo[b,f]thiepine

Fukuhara et al. (1979) gave up to 16 mg per kg to rats and up to 64 mg per kg to rabbits during organogenesis and found no adverse fetal effects.

Fukuhara, K.; Emi, Y.; Furukawa, T.; Fujii, T.; Iwanami, K.; Watanabe, N. and Tsubura, Y.: Toxicological and teratological studies of 2–chloro–11–(2–dimethylaminoethoxy)dibenzo[b,f] thiepine, a new neuroleptic drug. Arzneim–Forsch. 29:1600–1606, 1979.

AUTHOR INDEX

Barwell, B.E., 1920
Basaran, A.H., 506
Basett, R.C., 1723
Basford, A.B., 1177
Baskar, R., 2108
Bass, R., 2339
Basse, A., 2421
Bassir, O., 1849
Bateman, D.A., 615
Bateman, J., 624
Bates, H., 59, 1926, 2001
Bates, H.K., 2162, 2210, 2246, 2492
Bates, R.R., 2428
Batra, B.K., 1569
Baudin, M., 1014
Bauer, A., 1422, 1599
Bauerfeind, J.C., 2131
Baum, C., 102, 606, 740, 1637, 1642
Baum, D., 2053, 2335
Baumann, C.A., 1713, 1842
Bavous, F., 601
Bawdon, R.E., 1886
Baxi, L., 2126
Baxley, M.N., 178
Baycu, T., 2566
Bazare, J., 2162
Bazeley, P.L., 687
Beach, J.E., 641
Beachler, D.W., 410
Beall, J.R., 740, 1158, 1173, 1809,
   2158, 2167
Beals, R.K., 1476
Beare, J.L., 2113
Beasley, M., 1920
Beasley, R.P., 1187
Beattie, A.D., 1402
Beattie, B., 562
Beaty, C., 2149
Beaudoin, A., 410
Beaudoin, A.R., 127, 178, 577, 628,
   860, 1150, 1307, 1720, 2476
Beaudry, P.H., 123
Beaufils, M., 871
Beazley, J., 2
Becci, P.J., 671
Becerra, J.E., 731
Becher, R., 2095
Bechter, R., 1590
Beck, F., 51, 1015, 1078, 1146, 1147,
   1320, 1720, 2158, 2476
Beck, J.M., 906
Beck, S.L., 384, 2446
Beck-Mannagetta, G., 2026
Becker, B.A., 257, 669, 673, 740, 868,
   1402, 1677, 1911, 2331
Becker, H.C., 59
Becker, M.H., 637
Becker, R.F., 572

Becking, G.C., 1945
Beckman, D.A., 912
Beckman, D.A., 397, 924, 2108, 2492
Beckman, L., 178, 1792
Beckwith, J.B., 2008
Bederka, J.P., 117, 118
Bedrick, A.D., 938
Behm, M.C., 1240
Behringer, R.R., 1851, 1988
Beilin, L.J., 1572
Bejar, R., 615
Belanger, K., 572
Beliles, R.P., 295, 372, 632, 981, 984,
   1203, 1503, 1563, 1619, 1760,
   2429, 2432
Bell, A., 1402
Bell, J.A., 687, 1024
Bellanti, J.A., 1868
Bellart, J., 740
Bellinger, D., 1402
Bellinger, D.C., 1402
Bellows, R.A., 636, 643
Bellville, J.W., 157
Beltrame, A., 1510
Beltrame, D., 1392, 2398
Beluhan, F.Z., 740
Belyayeva, A.P., 165
Ben-Tovim, R., 2253
Benacerraf, B., 1920
Benacerraf, B.R., 731
Benach, J.L., 1451
Bencova, E., 1010, 1027
Benda, R., 2513
Benedetti, J., 1198
Benedetti, T.J., 940
Benedict, J.H., 984
Benes, C., 2513
Benet, L.Z., 812
Beniashvili, D.Sh., 1603
Benirschke, K., 615, 640, 687, 923,
   2273
Benitz, K.F., 59
Benjamin, D., 1998
Benjamin, D.R., 1747
Benjamin, S.A., 614
Benke, P.J., 2122
Bennett, J., 2359
Bennett, P.H., 731
Bennett, R.R., 975
Bennett, W.R., 912
Benny, R., 1383
Bensen, P., 922
Benson, R., 637
Benzaken, C., 601
Beohar, V., 276
Beral, V., 59, 384
Berbey, B., 1552
Berdasco, N.M., 1738

Berencsi, G., 1402
Berendes, H., 190
Berendes, H.W., 59, 2158
Berge, G.N., 2143
Berger, F.M., 1499
Berger, O.G., 1402
Berger, R., 1920, 2403
Berget, A., 1637
Berglund, K., 572
Bergman, U., 740
Bergstrom, R.M., 1412, 1455
Berhane, Z., 1899
Berkovitch, M., 1785
Berkvens, J.M., 658, 2154
Berlin, Jr., C.M., 1920
Berman, E., 1644, 2524
Berman, L., 1874
Berman, W., 868
Bernal, E., 475
Bernard, J.B., 400
Bernardi, M.M., 778
Berndt, W.O., 581, 2173
Bernhard, W.G., 46
Bernhardt, I.B., 1279
Bernstein, A., 1410
Bernstein, H., 1111
Bernstein, R.L, 2492
Bernuzzi, V., 84
Berrebi, A., 2403
Berry, C.L., 1543
Berry, D.L., 176, 2122
Berry, H.K., 1920
Berry, J.S., 2508
Berry, R.J., 2535
Berry, S., 1198
Bertazzoli, C., 383, 2260
Berteau, P.E., 17, 48, 2047
Bertoli, D., 1184
Bertollini, R., 399, 868, 2508
Bertrand, M., 384, 2372
Besedina, E.I., 2006
Bessieres, M.H., 2403
Bestetti, A., 1414
Bethenod, M., 231
Betremieux, P., 130
Beuhler, B., 1920
Beuker, E.D., 22, 198
Bevelander, G., 1380, 2316
Bevencs, Gy., 1598
Beyer, B.K., 132, 566, 801
Beyer, P.E., 1913
Beyers, P.E., 655
Beyler, S.A., 1759
Bhandari, N.R., 276
Bhargava, A.K., 1422
Bhargava, A.S., 806
Bhat, N., 1532
Bhat, N.G., 296

Gal, I., 2023
Galbreath, C., 1476
Gale, T.F., 564, 2558
Galina, M.P., 1326
Gallo, M.A., 2457
Galloway, T.M., 1684
Galloway, W.D., 2492
Galvin, M.J., 1644
Gamagari, Z., 1476
Gamm, S.H., 381
Gandelman, R., 2022
Gandier, B., 1920
Gandihon, F., 2403
Ganter, P., 694, 1629, 1637
Gao, C., 614
Garau, A., 262
Garbis, H., 596
Garcia, J.D., 1510
Garcia, R.E., 556
Gardlund, A.T., 1510
Gardner, A., 1069
Gardner, K.A., 572
Gardner, L.I., 1327, 1454, 1476
Gardner, R.J.M., 878
Gare, D., 692
Gareis, F.J., 1327
Garg, G.P., 175, 421
Garg, S.K., 175, 421
Garibova, T.L., 1055
Garmasheva, N.L., 955
Garner, R.J., 2498
Garny, V., 1647
Garrah, M.S., 436
Garrett, R.J.B., 10
Garrettson, L.K., 410
Garro, F., 2290
Garry, J., 2492
Garvin, P.J., 2449
Garzia, G., 208
Gaskin, J.M., 1684
Gasparini, M., 2126
Gasset, A.R., 1292, 2344, 2441
Gastellu, J., 1864
Gastol-Lewinska, L., 2085
Gaston, J., 1279
Gates, A.H., 2492
Gatling, R.R., 942, 1220, 1780, 1925
Gaub, M.L., 1766
Gaudry, R., 962, 2032
Gaulme, J., 2508
Gaunt, I.F., 332, 422
Gauthier, M., 2508
Gauthier, R., 384
Gautier, M., 58, 1512
Gautieri, R.F., 158, 187, 198, 309,
    615, 1240, 1677, 2180, 2565
Gavaler, J.S., 573
Gavrilenko, E.V., 90, 2434

Gaworski, C.L., 578
Gaylor, D.W., 52, 2162, 2428
Gaylord, L., 2208
Gaynor, M.F., 1543
Gazarian, M., 1785
Geber, W.F., 132, 158, 617, 1454,
    1476, 1514, 1527, 1677, 1888, 2332
Gebhardt, D.O.E., 2042
Gee, T., 392
Geelen, J.A.G., 1279
Gefter, J.W., 1292, 2441
Gehring, P.J., 411, 533, 696, 769,
    1310, 1579, 1583, 1883, 2314,
    2425, 2426
Geiger, J.F., 1472, 2476
Geiger, J.M., 1014
Geissler, F., 50
Gelbke, H.P., 2434
Gellert, R.J., 695
Generoso, W.M., 987, 1576, 1771,
    2122
Genieser, N.B., 637
Gennaro, J.F., 2476
Genra, Y., 1686, 2262
George, E.L., 17, 493, 505, 506, 526
George, J. D., 799
George, J.D., 295, 617, 793, 794, 2170,
    2335, 2436, 2531
George, J.D. Price, C.J., 1572
Georges, A., 121, 229
Gerard, A., 1122
Gerard, H., 1122
Gerber, G.B., 1402
Gerbig, C.G., 1956, 2428
Gerges, S.E., 57
Gerhardsson, M., 740
Gerhart, J.M., 1800
Gerling, F.S., 937
German, J., 2450
German, M.A., 1277
Gershon, A.A., 1198, 2513
Gershon, R., 2120
Gershwin, M.E., 84, 2566
Gersony, W.M., 1684, 2053
Geschwind, S.A., 1179
Gessner, P.K., 615
Geva, D., 1476
Ghali, F.E., 1277
Ghidini, A., 1638
Ghosh, A., 2566
Ghosh, T.K., 2208
Ghysen, J., 210
Giannina, T., 801
Gianutsos, G., 1476
Giaquinto, C., 31
Giavini, E., 760, 790, 936, 1138, 2311,
    2568
Gibbon, J., 1279

Gibson, A.A.M., 1868
Gibson, G.T., 1792
Gibson, J.E., 669, 673, 856, 860, 1294,
    2331
Gibson, J.P., 242, 780, 896, 1008,
    1628, 2140, 2298
Gibson, K.D., 2553
Gibson, K.J., 2452
Gibson, R.R., 1540
Gibson, W.A., 1324
Gibson, W.B., 984
Gibson, W.R., 1028, 2383
Gidley, J.T., 1005, 1784
Gift, H.C., 157, 1766
Giknis, M., 483
Giknis, M.L.A., 197, 484, 2080
Gilbert, C., 292, 1578, 2476
Gilbert, D.L., 1026
Gilbert, E.F., 384, 942, 1360, 2024,
    2335, 2520
Gilbert, E.S., 2108
Gilbert, T., 1122
Gilbert, W.M., 615
Gililland, J., 49
Gilkeson, M.R., 2149
Gill, T.J., 1221
Gill, W.B., 801
Gillam, M.P., 615
Gillet, C., 58, 1512
Gillette, J.R., 2042
Gillman, J., 292, 1578, 2476
Gillman, T., 292, 1578, 2476
Gilloteaux, J., 615
Gilmore, D.H., 1868
Gilstrap III, L.C., 615, 1277, 1886
Gilstrap, L.C., 728
Gindoff, P.R., 572
Gingras, S., 2530
Ginsberg, J., 1184
Ginsberg, J.S., 1184
Ginsberg, M.D., 410
Ginsberg, N.A., 562
Ginsburg, B.E., 59
Ginsburg, J.S., 637
Ginter, L., 59
Gioto, K., 2541
Giovanelli, L., 2041
Giraldo, A., 1792
Giroud, A., 281, 635, 639, 1078, 1279,
    1852, 2361, 2536
Giroud, J., 384
Giroud, M., 639
Giroud, P., 639
Gitterman, C.O., 2295
Giuliani, P., 1414
Giurgea, C., 699
Giurgea, M., 699
Givelber, H.M., 1454

Greene, R.M., 744
Greene, R.R., 154, 155, 703, 801, 953, 2302
Greenfield, V.S., 1058
Greenfield, W., 157, 1766
Greengard, P., 1725
Greenspan, B.J., 1113
Greenwald, M., 615
Greenwood, M.R., 1510
Greer, D.L., 1727
Gregg, K.U., 572
Gregg, N.M., 2149
Gregory, E.R.W., 2113
Greindl, M.G., 699
Grether, J.K., 1465
Grice, H.C., 50
Griesbach, U., 221
Griffith, D.P., 15
Griffith, R.W., 319
Griffiths, P.D., 687
Grifo, A.P., 1528
Grilly, D.M., 1476
Grimm, J.K., 1276
Grin, N.V., 631
Grix, A.W., 2122
Grizzle, T.B., 2531
Grollman, A., 63, 545, 635, 714
Grollman, E.F., 63, 545, 635, 714
Gromisch, D.S., 615
Grosch-Worner, C., 31
Gross, A., 495
Gross, P., 1540
Gross, S.J., 2158
Gross, T.L., 2138
Grosse, F.R., 59, 315
Grossman, J., 2273
Grossman, S.J., 1040
Grote, W., 156, 408, 637, 1014, 1275
Gruber, B., 562
Grudy, R.R., 406
Grumbach, M.M., 801, 932, 962, 1620, 2302
Grundel, D., 773
Gruning, K., 562
Gu, X-Q., 1577
Gudernatsch, J.F., 2109
Guerino, G., 471
Guerrero, R., 706
Guerriero, F.J., 200, 1146
Guest, I., 615, 835, 1593, 2455, 2508
Guggenheim, M.A., 2158
Guibaud, P., 2508
Guibbert, D., 1569
Guines, A., 1039
Gulati, D.K., 286, 983, 2335
Gulati, S.K., 392
Gulienetti, R., 2158
Gumbmann, M.R., 2206, 2440

Gummerus, M., 2156
Gunberg, D.L., 2476
Gundersen, J., 801
Gunderson, V.M., 59
Gundy, S., 2421
Gunnarskog, J., 2236
Gunning, D.B., 2122
Gunther, T., 297, 1461, 1874
Gunzel, P., 806, 1141
Gupta, B.N., 1172, 1755, 1756, 1758
Gupta, C., 1910
Gupta, P.K., 927, 1946
Guram, M.S., 132
Gurd, M.R., 1534
Gurian, J.M., 2042
Gushow, T.S., 72, 514, 760, 937, 2057
Gustafsson, J.A., 596
Gustavson, J.C., 801
Gustavson. C.R., 801
Gutberlet, R., 1803
Gutova, M., 227
Gyurik, R.J., 56
Ha, J.C., 2564
Ha, S-H., 2093, 2418
Haapakoski, J., 2535
Haar, T.G., 981
Haas, R.J., 2473
Haberman, S., 1946
Habersang, S., 1295, 1835, 2119
Habuchi, M., 1321
Hackett, P.L., 349, 1402, 1994
Hackett, T.N., 1688
Hada, R., 1161
Haddad, R., 1560
Haddow, J., 1078
Haddow, J.E., 1118
Hadeed, A.J., 615
Hadley, J.A., 1644
Hagan, J.V., 1526
Hagay, Z., 2492
Hagbard, L., 731
Hage, M.L., 811
Hagele, M., 264, 678, 1487
Haggerty, G.C., 2029
Hagino, T., 1321
Hagita, K., 1321
Hagiwara, H., 2160
Hagopian, G.S., 1448, 1449, 1768
Haguro, S., 425, 1663
Hahn, G., 2189
Hahn, J.D., 678
Hahn, L.J., 1510
Hahn, M.A., 186
Hailey, F.J., 1748
Haines, J.S., 485
Hakama, M., 2198
Hakosalo, H., 1316
Hakosalo, J., 1316

Halasz, P., 940
Hale, F., 2536
Hale, L.J., 2098
Hale, R.L., 59
Hales, B.F., 33, 71, 381, 673, 1137, 1139, 1505, 1930, 2045, 2317
Halfmann, U., 1305
Haling, R.F., 1078
Hall, B., 392, 2527
Hall, E.K., 2498
Hall, G., 2442
Hall, I-H., 1784
Hall, I.H., 667
Hall, J.E., 1534
Hall, J.G., 637, 1184, 1541
Hall, L.L., 655, 1913
Hall, R., 1020
Hall, W.J., 2492
Halldorsson, S., 637
Hallesy, D.W., 12, 772
Hallett, P., 282
Halling, H., 1206
Halliwell, W.E., 2550
Halloran, M.W., 369
Halmesmaki, E., 59
Halonen, O., 2156
Halter, C.A., 2273
Hama, T., 1659
Hamabuchi, T., 918, 1062
Hamada, S., 2384
Hamada, H., 1302
Hamada, M., 287, 303, 379, 1387
Hamada, N., 1538
Hamada, S., 2384
Hamada, Y., 262, 420, 518, 560, 603, 1007, 1321, 1640, 1656, 1734, 2014, 2183, 2217, 2375
Hamajimma, Y., 357
Hamamoto, K., 812
Hamamoto, S., 2213
Hamar, D.W., 1967
Hambridge, K.M., 2566
Hamburger, J.I., 1327
Hamburger, M., 2207
Hamburger, S., 1644
Hamburgh, M., 922, 2476
Hamby, B.T., 320
Hameed, M.S., 381
Hamel, E.E., 2232
Hamilton, L., 494
Hamilton, R.D., 2367
Hamm, T.E., 1567
Hammers, B., 1748
Hammill, G.C., 1327
Hammill, H., 2564
Hammond, P.B., 1402
Hammond, B.G., 2161
Han, S., 1154, 2091

Hanada, S., 924, 2384
Hanada, Y., 2183
Hanafusa, T., 2202
Hanari, H., 901
Hanasono, G.K., 1690, 1955
Hanaway, J.K., 1454
Hancock, J.A., 1452, 2476
Handa, H., 1603
Handa, J., 682, 1603, 1770
Handa, S., 1006
Hande, M.P., 2108
Handschumacher, R.E., 213
Haneberg, B., 315
Haney, A.F., 801
Hanify, J.A., 52
Hanley Jr, T.R., 1738
Hanley, Jr., T.R., 980
Hanley, T.R., Jr., 72, 514, 554, 760,
    771, 937, 974, 1526, 2257, 2432,
    2435
Hanlon, D.P., 178
Hanninen, H., 1792
Hansborough, L.A., 1719
Hansen, D.K., 2120
Hansen, C.A., 615
Hansen, D.K., 1175, 1432, 2508
Hansen, E.V., 1776
Hansen, H.S., 1041
Hansen, L.A., 2122
Hansen, L.G., 1201
Hansen, O.M., 1045
Hansen, R., 1727
Hanser, P.L., 1644
Hanshaw, J.B., 687
Hansmann, M., 1868
Hanson, J.W., 59, 615, 868, 1118,
    1775, 2218, 2450, 2508
Hanson, K.J., 1967
Hara, H,, 952
Hara, H., 899, 909, 1153, 1332
Hara, K., 443, 738, 1479, 1682, 1813
Hara, M., 1818
Hara, T., 186, 191, 886, 1144, 1153,
    1322, 1374, 1375, 1376, 1384,
    1643, 2329
Hara, Y., 2021
Harada, H., 1683
Harada, S., 333, 433, 898, 1797, 1897
Harada, T., 239
Harada, Y., 1510
Harakawa, T., 2389
Harbinson, R.D., 868, 2151
Hardin, B.D., 295, 349, 372, 632, 794,
    847, 981, 984, 987, 1203, 1503,
    1563, 1567, 1579, 1619, 1704,
    1754, 1760, 2042, 2229, 2429
Harding, A.J., 878, 1277
Harding, J.D.J., 2271

Hardof, A., 2053
Hardwick, B.C., 631
Hardy, J.B., 572, 1316
Hardy, P., 516, 762
Hardy, R.P., 1646
Hardy, T.L., 199
Hare, J.W., 731
Haresaku, M., 356
Harima, K., 2320
Harinasuta, T., 1491
Haring, O.M., 408
Harlap, S., 59, 572, 1775, 1803
Harley, E., 731
Harley, E.E., 59
Harley, J.D., 2149, 2316
Harley, J.M.G., 596
Harlow, H.F., 1920
Harman, M.T., 185
Harmon, J.B., 89
Harmon, J.R., 2053
Harms, D., 1014
Harne, L.C., 1370
Haro, S., 231, 868
Haro, T., 1049
Harousseau, H., 59
Harpel, H.S., 1677
Harper, K.H., 494, 607, 777, 1300,
    1876
Harpey, J-P., 2091
Harran, D., 82
Harris, C., 266, 267, 541, 733, 1422
Harris, J.A., 1465, 2425
Harris, M., 782
Harris, M.M., 1088
Harris, M.W., 179, 749, 1895
Harris, R., 1078
Harris, S.B., 737, 1177, 1350, 1510,
    1766, 2057
Harrison, W.P., 178
Harrod, M.J., 2508
Harry, G.J., 1905
Harstad, T.W., 1886
Hart, C.W., 541
Hart, M.M., 695
Hartel, A., 922, 1301
Hartel, G., 922, 1301
Hartley, W.J., 1277
Hartman, H.A., 1517
Hartmann, A-M., 2508
Hartnagel, R.E., 1834, 2136, 2550
Hartnett, E.M., 2149
Hartwig, N.G., 1868
Hartz, S.C., 231, 502, 868, 1315,
    1499, 2269
Haruguchi, T., 1286
Harvey, D., 637
Harvey, E.A., 868
Harvey, E.B., 2158

Harvey, M.A.S., 868, 1277, 2165
Harvey, R.B., 732
Harvey, S., 912
Harwart, A., 1141
Harwig, J., 683, 1310, 2206, 2561
Hasan, H., 2023
Hasan, S.H., 393
Hasebe, M., 1103
Hasegawa, H., 214
Hasegawa, K., 414
Hasegawa, M., 1321, 2145
Hasegawa, N., 87, 472, 1236, 1694
Hasegawa, T., 237, 247
Hasegawa, Y., 104, 441, 443, 461, 474,
    478, 738, 943, 1479, 1682, 1813,
    1898, 2020, 2052
Haseman, J.K., 52, 664, 1939, 1944,
    2421
Hasewaga, Y., 1444
Hashiguchi, M., 420, 532, 1007, 1734
Hashimoto, K., 2159
Hashimoto, M., 719, 1009
Hashimoto, T., 426, 1160, 2200
Hashimoto, Y., 45, 46, 92, 454, 556,
    599, 834, 890, 1035, 1228, 1262,
    1336, 1401, 1597, 1665, 1692,
    2349, 2350, 2391, 2485
Hashimoto,Y., 1802
Hass, U., 1615
Hassanein, K., 572
Hassert, G.L., 475
Hassinger, D.W., 392
Hassoun, E., 2307, 2312
Hastings, L., 1579, 2425
Hasunuma, K., 446
Hata, M., 1378
Hata, T., 473, 625, 1481, 2323
Hatab, P., 801
Hatakeyama, K., 2211
Hatakeyama, Y., 354, 1052, 1323
Hatanaka, K., 2202
Hatano, M., 49, 169, 186, 250, 589,
    692, 812, 883, 1722, 1961, 2385
Hatch, E.E., 1476
Hatcher, R.P., 1905
Hatori, M., 2030
Hatoum, N.S., 2449
Hattori, M., 145, 169, 443, 478, 1379,
    1481, 1660, 1686, 2285
Hattori, N., 1846
Hauck, R.S., 2508
Hauck, W.N., 2214
Hauser, I., 2508
Hautefeuille, H., 297
Havelka, J., 494
Havercamp, A.D., 2492
Havlena, J., 1337, 1347
Hawkes, W.C., 2174

Ito, M., 446, 1479, 1481, 1929
Ito, N., 2364, 2456
Ito, R., 405, 719, 831, 931, 1716, 1894, 2010, 2486
Ito, S., 1483, 2379
Ito, T., 1153, 1921, 2392
Ito, Y., 331, 486, 516, 1107, 1767
Itoh, A., 2498
Itoh, I., 145, 1515
Itoh, M., 104, 474, 1813
Itoi, M., 1292, 2441
Itono, Y., 1752
Itou, I., 738
Itou, M., 738
Itou, S., 148, 1497
Itri, L.M., 2122
Iturra, H., 2530
Iuliucci, J.D., 1677
Ivankovic, S., 1000, 1248, 1923, 2029
Ivankovic, V.S., 1761
Ivanov, V.V., 256
Ivanova-Chemishanska, L., 228
Ivanova-Tchemishanska, L., 881
Ivarsson, S.A., 1277
Iverson, F., 990, 1569, 1764
Ivy, A.C., 154, 155, 703, 801, 953, 2302
Iwadare, M., 1373, 2407
Iwadate, K., 1097
Iwai, M., 634, 2396
Iwai, M., 2396
Iwaki, M., 1961
Iwaki, R., 828
Iwaki, T., 449
Iwama, A., 1228, 1290, 1649
Iwamoto, H.S., 812
Iwamoto, S., 354
Iwanami, K., 597, 1207, 2509, 2571
Iwasa, K., 1006
Iwasaki, K., 2389
Iwasaki, S., 158, 1770
Iwase, T., 634, 809, 1343
Izaki, K., 603, 2217, 2375
Izawa, M., 446, 1481
Izawa, Y., 686, 1850
Izumi, H., 713, 952
Izumi, T., 1057
Izumiyama, K., 79, 237, 1260, 1386, 1441, 1475
Izutsu, M., 1336, 2178
Jackh, R., 840, 1551, 1946
Jackson, A.J., 915
Jackson, B.A., 1025, 1651
Jackson, C.W., 2060
Jackson, D., 619, 764, 1101
Jackson, E., 2513
Jackson, E.C., 64, 2235
Jackson, M.R., 1024

Jackson, R.A., 244
Jackson, W.P.U., 552
Jacob, D., 1857, 2487, 2488
Jacob-Muller, U., 673, 678, 1487
Jacobs, M., 483
Jacobs, R.M., 547, 1112, 1301, 2480
Jacobs, R.T., 895
Jacobsen, L., 2108
Jacobsohn, D., 678
Jacobson, B.D., 962
Jacobson, J.L., 516
Jacobson, S.J., 1432
Jacobson, S.W., 516
Jacquet, P., 1402
Jacquillat, C., 694
Jacson, L.G., 562
Jaenisch, R., 624
Jaffe, I., 2126
Jaffe, N., 392
Jaffe, P., 1803
Jager-Roman, E., 2026, 2508
Jagiello, G., 1803
Jago, M.V., 1181
Jago, R.H., 2480
Jahn, V., 1071
Jain, A.K., 77
Jain, N., 2473
Jakob, S., 2508
Jakobovits, A., 2330
James, D.A., 196
James, L.F., 165, 178, 293, 631, 672, 1435
James, L.M., 52, 242, 2530
James, M.L., 156, 807
James, N., 1078
James, P., 920, 998
James, P.A., 199
James, R.W., 998
James, W.H., 596
Jameson, C.W., 179
Jameson, S., 2566
Janardhan, A., 402
Janerich, D.T., 1775, 1803
Janig, U., 1014
Janisch, W., 1000
Jankowiak, M.E., 867
Janos, G., 242
Jansak, J., 2223
Jansons, R.A., 1727
Janz, D., 868
Jarman, J.A., 1327
Jarrel, B.E., 676
Jarrell, J.F., 673
Jaschevatzy, O., 1966
Jasmin, G., 946
Jasperse, D., 59
Jaswal, O.B., 436
Jeavons, P.M., 2508

Jefferies, A.L., 89
Jeffery, E.H., 2540
Jeffries, J.A., 801
Jeffries, T.K., 1956
Jelinek, P.V., 602, 1892
Jenkins, D.H., 873, 2084
Jenkins, E., 637
Jenner, P., 2207
Jennings, V.H., 706
Jensen, B., 1045
Jensen, N.M., 256
Jensh, R.P., 2108, 2474, 2492
Jenson, R.P., 2108
Jerome, C.P., 727
Jessop, W.J.E., 1316
Jetten, A.M., 2122
Jewett, G.L., 696
Jewett, T., 615
Jeyasak, N., 874
Jiang, H., 2122
Jiang, H., 2122
Jick, H., 10, 242, 482, 596, 812, 859, 863, 910, 1078, 1277, 1287, 1313, 1642, 1775, 1790, 1904, 1934, 2073, 2122, 2467, 2552, 2566
Jick, S.S., 910, 1078, 1277, 2122, 2566
Jida, T., 939
Jimenez, L., 183
Jin, L-S., 2459
Jirasek, J., 213
Job-Spira, N., 392
Jochmann, G., 1599
Joesoef, M.R., 384
Johannesson, T., 1677
Johannsen, F.R., 17, 48, 833, 1094, 1466, 2047, 2161
Johansen, K.T., 1432
Johansson, B., 731
John, D.M., 998
John, D.M., 535, 1324
John, E.M., 59
John, J., 1526, 2513
John, J.A., 36, 72, 248, 514, 554, 600, 760, 763, 937, 974, 1205, 1310, 2228, 2306, 2530
John-Greene, J.A., 771, 1956, 2435
Johnson, A.O.K., 1188
Johnson, A.R., 369
Johnson, C.K., 1261
Johnson, D.E., 873
Johnson, D.H., 1325
Johnson, E.M., 1078, 1279, 1720, 2130, 2290
Johnson, F.M., 1000
Johnson, J., 2103
Johnson, J.A., 1792, 1893
Johnson, J.M., 2029
Johnson, K., 1432

485

Murakami, U., 188, 204, 737, 1285, 1402, 1510, 1600, 1644, 1674, 1929, 2108
Murakami, Y., 2384
Muramoto, A., 2201
Muranaka, R., 738, 1813
Muraoka, Y., 738
Murase, E., 952, 2201
Murata, M., 922
Murdock, R.M., 1402
Murofushi, A., 1789
Murofushi, K., 1021
Murphree, O.D., 2038
Murphy, S.J., 1241
Murphy, D.P., 731
Murphy, J., 2492
Murphy, J.C., 2218
Murphy, J.L., 845
Murphy, M.L., 15, 69, 209, 321, 322, 347, 492, 519, 540, 673, 682, 694, 736, 744, 976, 1045, 1060, 1061, 1064, 1065, 1067, 1068, 1171, 1252, 1253, 1267, 1268, 1483, 1508, 1658, 1708, 1751, 1993, 1996, 2029, 2225, 2354, 2437, 2494
Murphy, P.F., 701
Murrakami, Y., 2384
Murray, C.L., 49, 392
Murray, EDS, 1700
Murray, F.J., 36, 248, 766, 1205, 1310, 2228, 2256, 2306
Murray, J., 554, 1526
Murray, J.S., 763, 974, 1205, 2306
Murray, S.M., 1017
Murrofushi, K., 1021
Murthy, R.C., 381
Murti, C.R.K., 1422
Musci, M.N., 2138
Musselman, A., 868
Mutanen, P., 987, 1792
Mutchinick, O., 2032
Mutchler, M.K., 1079, 1948
Muthukrishnan, S., 2524
Muzutani, M., 707
Myasukawa, Jh, 2166
Myers, C.B., 304
Myers, G.J., 687
Myers, M.H., 392
Myers, R.E., 410, 1285
Myers, T.F., 1652
Myhre, J.L., 2035
Myhre, S.A., 2026
Mylecraine, L., 2096
Myon, J., 694, 1629
Myrianthopoulos, N.C., 59, 868, 1277
Naas, D., 1077
Nabata, H., 2374
Nabatake, H., 2376

Naber, E.C., 1713, 1842
Nabeshima, J., 356
Nachtomi, E., 763, 766
Nadas, A.S., 1684
Nadaskay, N., 1775
Nadeenko, V.G., 613, 1723
Naeye, R.L., 572, 618, 728, 1197, 1276, 1285, 1792, 2149
Nafstad, I., 2143
Nagae, Y., 239
Nagai, H., 215, 1978
Nagai, M., 2021
Nagai, N., 215
Nagai, T., 1228
Nagami, K., 1947
Nagano, A., 1460
Nagano, K., 984
Nagano, M., 569, 919, 1076
Nagao, H., 269
Nagao, S., 509, 669, 826, 993, 1084, 1106, 1453, 1489, 1703, 1781, 1837, 2032, 2054, 2124, 2158, 2220, 2468, 2470, 2471
Nagao, T., 235, 599, 918, 1336, 1390, 1603, 1692, 1893, 2178
Nagaoka, S., 1228
Nagaoka, T., 337, 576, 692, 812, 883, 1218, 1294, 1389, 1722
Nagari, H., 414
Nagasawa, H., 465
Nagase, M., 1287, 2200
Nagashima, Y., 573
Nagata, R., 2209
Nagele, R.G., 1854, 2520
Nagymajtenyi, L., 1402
Nahas, G., 1914
Nahmias, A.J., 1198
Nahrwold, M.L., 1766
Naidu, N.R.G., 1669
Naidu, R.C.M., 2477
Naimzada, M.K., 394
Nair, A., 276
Nair, R.S., 1094
Nair, T.B., 296
Naito, Y., 365, 1129, 1323, 2012
Naitoh, J., 275, 1184, 1844
Naitoh, Y., 294
Nakada, H., 19, 407, 442, 451, 1817, 1972, 2399
Nakagama, S., 2294
Nakagawa, H., 604, 905, 1612, 2238
Nakagawa, K., 1458
Nakagawa, KK, 1458
Nakagawa, M., 1740, 2265
Nakagawa, S., 1894
Nakagawa, T., 899, 952
Nakahara, N., 1207
Nakai, K., 384, 828, 1727

Nakai, S., 405
Nakai, T., 597, 2025
Nakaichi, M., 984
Nakaji, Y., 246
Nakajima, H., 1229
Nakajima, I., 1298
Nakajima, K., 1185
Nakajima, S., 1927
Nakajima, T., 1387, 1633
Nakajima, Y., 1228
Nakajo, M., 2138
Nakajoh, M., 1844
Nakame, S., 1818
Nakamori, K., 1770
Nakamoto, N., 2010
Nakamura, M., 2021
Nakamura, A., 890, 2369, 2387, 2400
Nakamura, E., 1629
Nakamura, H., 432, 903, 1324, 1413, 1679
Nakamura, K., 238, 269, 890, 1321
Nakamura, M., 1286, 1407, 1869, 2021
Nakamura, S., 19, 442, 451, 1972, 2399
Nakamura, T., 828, 1228, 1290, 2155, 2462
Nakamura, Y., 536, 1761
Nakane, S., 1236, 1652, 1818, 2281, 2286
Nakane, Y., 12, 399, 940, 2026
Nakanishi, Y., 1099
Nakano, K., 1642, 2410
Nakanowatari, J., 902
Nakao, H., 202, 1440
Nakashima, K., 1822, 2371
Nakasone, K., 2292
Nakata, M., 453
Nakatani, H., 1108, 2196
Nakatsu, I., 431
Nakatsu, T., 1, 140, 159, 1251, 2166
Nakatsuka, T., 12, 807, 889, 924, 1103
Nakaura, S., 59, 499, 509, 669, 702, 752, 826, 993, 1084, 1106, 1190, 1453, 1489, 1575, 1588, 1703, 1781, 1837, 2032, 2054, 2124, 2158, 2336, 2433, 2468, 2470, 2471, 2567
Nakaurra, S., 2220
Nakayama, A., 917
Nakayama, E., 984
Nakayama, T., 1163, 1507
Nakayama, Y., 1225, 2321
Nakayoshi, H., 222, 237, 438, 459, 1228, 1717, 2528
Nakazawa, M., 1428
Nakazawa, M., 339, 428, 464, 1228, 1428, 1791, 2281, 2407, 2490

Nakazawa, Y., 12, 940, 2026
Nakazima, M., 1021
Nakazima, Y., 303, 2296
Naki, H., 331
Nakone, S., 2205
Nalbandyan, T.L., 2424
Namba, H., 393
Namba, K., 599
Namba, T., 262, 603, 2217
Namerow, P.B., 615
Namikawa, R., 801
Namiki, M., 634
Namiki, M., 634
Namkung, M.J., 20, 248, 1590, 1911
Namoto, T., 431
Nanada, O., 2489
Nanama, I., 576
Nanba, T., 2375
Nanto, T., 1970, 2390
Napalkov, N.P., 125, 1603, 1763
Nara, H., 443
Nara. H., 443
Narama, I., 97, 99, 236, 294, 337, 365,
    623, 805, 996, 1129, 1294, 1323,
    1389, 1635, 1693, 2012, 2050, 2405
Naramo, I., 1218
Naranjo, E., 863
Naranjo, P., 863
Narirta, H., 2384
Narita, M., 1826
Narita, H., 2384
Narita, M., 1826
Narita, R., 1369
Narita, T., 2201
Narrod, S.A., 869, 1484
Nartosky, M.G., 320
Narzullaeva, M.S., 59
Nash, J., 384
Nash, J.E., 572, 1727
Nash, L., 990, 1655
Naso, B., 615
Nasu, Y., 1811
Nataksuka, T., 885
Nath, D., 77
National Institute of Health, 2492
Nau, H., 967, 968, 1013, 1566, 1680,
    1727, 2026, 2122, 2339, 2508
Naughton, M.D., 392
Naughton, M.J., 1793
Naulty, J.S., 1761
Naunton, R.F., 541
Navarro, J., 496
Nawrot, P.S., 248, 922, 1644, 2393
Naya, M., 243, 1067, 1144, 1322,
    1374, 1377
Nayak, B.N., 381, 1402
Naylor, R.J., 2207
Nazirov, B.D., 59

NcCarter, R.J., 731
Neal, J.V., 2108
Nebel, L., 556, 776, 956
Nebert, D.W., 256, 837
Nechkina, M.A., 769
Nechushkina, L.V., 302
Neda, K., 1074
Nedleman, H.L., 1402
Nee, J., 801
Needleman, H.L., 1058, 1402
Neel, J.V., 2108
Neerhof, M.G., 615
Neff, R.K., 731
Negishi, R., 140, 1413
Negri, S.R., 84
Nehez, M., 326, 1598
Nehlig, A., 384
Neiderer, B., 2450
Neill, C.A., 1803
Neldner, K.H., 2566
Nelson, B.K., 60, 352, 971, 984, 995,
    1214, 1358, 1533, 1566, 1887,
    2043, 2314
Nelson, C.J., 52, 2120, 2428
Nelson, G.W., 1261
Nelson, K.B., 2149
Nelson, K.G., 516
Nelson, M.M., 119, 1078, 1110, 1852,
    2158, 2338
Nelson, R.A., 1710
Nelson, R.C., 2131
Nelson, T., 725, 1803, 2087
Nelson, W.O., 1729
Nelson. J.A., 31
Nemec, M., 1077
Nemec, M.D., 223, 898
Nemec, M.J., 1191
Nemes, E., 2421
Nemoto, K., 472
Nera, E.A., 1510
Neu, R.L., 1454, 1476
Neubert, D., 42, 599, 915, 1122, 1603,
    2122, 2311, 2523
Neuff, A.M., 2267
Neugut, R.H., 1775
Neuman, F., 393
Neuman, R.E., 115, 2175, 2353
Neumann, F., 529, 678
Neumann, J., 746
Neumann, W., 1304, 2245, 2452
Neumerzhitskaja, L.V., 849
Neutra, R., 1465
Neutra, R.R., 2425, 2524
Neva, F.A., 2149
Nevin, N., 2508
Nevin, N.C., 399, 596, 1078, 2535
New, D.A.J., 1456, 1457
New, D.A.T., 161, 660, 1277, 1646

Newberne, J.W., 242, 780, 896, 925,
    1628
Newberne, P.M., 2538
Newbold, R.R., 801
Newborne, J.W., 2298
Newcombe, R.G., 572
Newell, G.W., 313, 2190, 2345
Newell-Morris, L.L., 1279, 2122
Newman, L.M., 1279
Newman, N.M., 2252
Newport, G.D., 1300, 2120
Newsome, W.H., 753, 990
Ng, C., 1969
Ng, S.K.C., 615
Ng, T.B., 1244, 1245, 1666, 2430
Ng, W.W., 147
Nguyen-Dinh, X., 1792
Nicholas, A.G., 1775
Nicholas, N.S., 2126
Nichols, H.P., 1091
Nichols, M.I., 1402
Nickander, R., 2383
Nickola, J.M., 1905
Nicod, I., 59
Nicolaides, K.H., 637
Nicolini, U., 1578, 1868
Niebyl, J.R., 1476
Niemeier, R.W., 295, 372, 632, 981,
    984, 1203, 1503, 1563, 1619, 1760,
    2429
Niemi, M-L., 987
Niemier, R.W., 984
Niemierko, A., 683
Niggeschulze, A., 330, 941, 2278
Nigli, M., 381, 1402
Nihill, M.R., 161
Niii, N., 909
Niimi, J., 719
Niino, K., 2070
Niiyama, Y., 2477
Nikashima, T., 379
Niki, R., 460
Nikiforov, B., 1750
Nikitin, A.I., 59, 248
Nikitina, S.S., 41
Nikonorow, M., 1946
Niles, A.M., 2540
Nilsson, A., 2227
Nilsson, L., 1828
Ninomiya, H., 465, 954
Ninomiya, M., 767
Niokawa, H., 2262
Nirubagam T., 425
Nishi, N., 437, 1791
Nishiawa, T., 145
Nishiazwa, S., 1850
Nishida, M., 1321
Nishida, Y., 2149

Ono, M., 1287
Ono, T., 399, 940, 2026, 2448
Onoda, K-I., 1106, 2468, 2470, 2471
Onodera, N., 893
Ooba, M., 2407
Oobayashi, H., 984
Oohira, A., 204, 872
Ooshima, Y., 1, 432, 731, 1291, 1396, 1413, 2166
Opitz, J.M., 315
Oppenheimer, A., 1476
Orav, E.J., 615
Orenstein, W.A., 2148, 2513
Orgata, E.S., 731
Orikawa, M., 19
Orlando, S., 2442, 2445
Orleans, M., 2492
Ormission, A., 59
Ornoy, A., 13, 556, 731, 912, 928, 1243, 1997
Ornstein, M., 242
Ortega, A, Sanchez, D.J., 2378
Ortega, A., 827
Ortiz, J.A., 2182
Osamu, N., 1439
Osamu, F., 809
Osamu, T., 1439
Osborn, R.A., 1277
Osborne, A.W., 2108
Osborne, C.A., 15
Oser, B.L., 16, 658
Oshchepksv, V.I., 1792
Oshima, K., 2149
Oshima, M., 719
Oshima, T., 1373
Oshima, Y., 337
Oshio, K., 79, 80, 1163, 1260, 1386, 2363
Oski, F., 2149
Osman, S.F., 485
Ostby, J., 1747
Ostby, J.S., 858
Oster, G., 289, 878, 1037
Osterberg, R.E., 2218
Osterburg, I, 1099
Osterburg, I., 120, 454, 1897
Ostergaard, A.H., 1045
Osterloh, V.G., 254
Ostola, E.M., 2512
Ostrea, E.M., 1197
Osuga, F., 883, 1218
Osuka, F., 692, 812
Osumi, I., 205, 444, 2541
Osumi, S., 1306
Osztovics, M., 7, 59, 878
Ota, T., 6, 1513, 1770, 2016, 2020
Otaka, T., 428
Otaka, T., 428, 1099

Otake, M., 2108
Otani, G., 2016
Otani, K., 87, 399, 940, 2026
Oteiza, P.I., 2566
Otis, C., 2138
Otome, S., 1874
Otsaki, I., 2376
Otsuka, T., 713, 1663
Otsuki, S., 12, 940, 2026
Ott, W.J., 2492
Otterson, W.N., 2518
Ottesen, E.A., 1039
Ottley, M.S., 2208, 2551
Otto, D.A., 1402
Ottoboni, A., 695
Ottolenghi, A.D., 664
Ottosen, H.E., 2421
Ottosen, L.O., 84
Ouchi, M., 78
Ouchi, S., 939
Ouda, T., 831
Oudiz, D.J., 2426
Ouellette, J.H., 1956
Ounsted, M.K., 1572
Oura, K., 1436
Oura, T., 1920
Ovadia, M., 130
Ovellete, E.M., 59
Overbeck, G.W., 615
Overbeek, P.A., 2272
Overby, E., 2421
Overholser, M.D., 2538
Overman, D.O., 1088
Overmann, G.J., 1180
Overstreet, J.W., 2426
Overzier, C., 2302
Owaki, Y., 893, 1980, 2079, 2348, 2355, 2374
Owen, E.L., 2426
Owen, N.V., 1527, 1690, 1893, 2052
Oyama, K., 1161
Oyama, M., 622
Oyama, T., 1603
Ozaki, K., 1229
Ozaki, M., 233, 1894
Ozawa, H., 984
Ozawa, S., 275, 1844, 2138
Ozeki, K., 2061
Ozeki, Y., 1109, 1826, 2062
Paccagnini, S., 1189
Pace, H.B., 1476
Pacheco, H., 1569
Packard, D.S., 745, 1292
Pacque, M., 1365
Padmanabhan, R., 381
Page, E.W., 1103
Paget, G.E., 2241, 2260
Paghini, G., 361

Pahnish, O.F., 636, 643
Palacios, J., 2480
Palazzolo, R.J., 1932
Palermo-Neto, J., 131
Pallin, I.M., 1915
Palludan, B., 185, 2536
Palm, P.E., 384
Palmer, A.E., 710, 2545
Palmer, A.K., 453, 494, 607, 777, 1300, 1405, 1422, 1608, 1876, 2267, 2432
Palmer, K.H., 1784
Palmer, M.S., 731
Palmer, R.H., 59
Palmiter, R.D., 1851, 1988
Pandit, P.B., 89
Panopio, L., 1655
Panopio, L.G., 1201
Pant, S.S., 637
Pantelakis, S.N., 2
Panter, K.E., 642, 1183
Papadimitriou, G.C., 2
Papiernik, E., 1646
Parano, E., 2122
Pardo, F., 706
Pardthaisong, T., 1487
Pareek, S., 2473
Parekh, A.J., 615
Parent, R.A., 33
Parer, R.M., 846
Pares, P.J., 1237
Paris, J., 871
Parish, R.C., 56
Parisi, T., 740
Park, H.W., 1280, 2222
Parker, C.A., 535, 704
Parker, J., 748
Parker, R.M., 846, 2086, 2416
Parker, S., 573
Parkhie, M., 2330
Parkhie, M.R., 1946
Parkinson, M.M., 282, 1075, 1802
Parkman, P.D., 2149
Parks, W.P., 31
Parmelee, A.H., 1326
Parola, L.R., 2531
Parr, R.W., 1232
Parra, J.E., 1792
Parrish, H.M., 137
Parrish-Johnson, J.C., 59
Parsells, J.L., 740
Parulekar, M., 2308
Pascal, S., 694, 1629
Paschke, L.A., 159
Paschke, L.L., 857, 1841
Pascoe-Mason, J.M., 673
Pashayan, H., 740
Pasquet, J., 694, 1629

Plasterer, M.R., 847, 1704, 1754
Platner, W.S., 22, 198, 1954
Platzek, T., 18, 599, 1603
Pleet, H.B., 1277
Plessinger, M.A., 615
Plotkin, S., 1198
Plotkin, S.A., 687, 2149
Plouin, P-F., 924
Plummer, G., 2108
Podrich, P.E., 2393
Poetsche, G., 1365
Poggel, H.A., 806
Pohland, R.C., 93
Poindexter, A.N. III, 1413
Poland, B.J., 1803
Polani, P.E., 1078, 2108
Polednak, A.P., 1775
Polen, M.R., 2524
Poley, W.E., 2174
Polifka, J.E., 2122
Polifka, J.E., 637, 2527
Polishuk, W.Z., 677
Polk, B.F., 731
Poll-The, B.T., 1920
Pollack, T.M., 2149
Pollock, J.J., 1803
Pomerance, J.J., 2513
Pons, J.C., 1646
Ponte, C., 1920
Popendiker, K., 842
Popick, F., 801
Poplack, D.G., 186, 392
Popov, B.V., 59
Popov, V.B., 7, 673, 2091
Popov, V.G., 59
Popova, N.V., 108
Poppe, S.M., 329, 1442
Poppen, N.K., 150
Populaire, P., 694, 1629
Porat, R., 2138
Porfirieva, R.P., 1903
Port, R., 637
Porter, J.B., 1775
Porter, J.F., 289
Porter, M.C., 2550
Poskanzer, D.C., 801
Posner, H.S., 82, 1484
Possamai, A.M., 2008
Poston, K.A., 612
Poswillo, D.E., 1277, 2008
Potashnik, G., 753
Potempska, A., 1920
Potter, S.S., 1418
Potturi, R.B., 615
Potworowski, M., 963
Poulson, E., 537, 1178, 1194, 1226,
    1233, 1337, 1661, 1865, 1906,
    2180, 2409

Powell, D., 1198
Powell, H.R., 1543
Powell, J.G., 1028, 1313
Powers, H., 1476
Pozanski, A.K., 868
Pozharsky, K.M., 2135
Poznakhirev, P.R., 1903
Pradhan, S., 2208
Pradhan, S.N., 2208
Prahalada, S., 242, 801, 807, 1040,
    1487, 1782, 1785, 2191
Pras, M., 619
Prati, M., 760
Pratt, M., 2311
Pratt, R.M., 2122
Pratten, M.K., 1078, 1320
Preblund, S.R., 2148, 2513
Pregliasco, P., 2122
Presburg, H.J., 1325
Preston-Martin, S., 1792
Preussmann, R., 8, 973, 1000, 1587,
    1627, 1923, 1995
Pribram, H.W., 2122
Price, C.J., 34, 159, 242, 295, 304,
    360, 617, 793, 794, 799, 857, 983,
    1123, 1841, 1946, 2170, 2313,
    2335, 2436
Price, K.R., 2008
Price, W.A., 1645
Prichard, J.T., 2234
Priest, R.G., 922
Principi, N., 1189
Prindle, R.A., 1285
Printz, R.H., 1510
Pritchard, A.L., 312
Pritchard, A.L., 1152, 1866
Pritchard, J.A., 1078
Pritchard, J.F., 1031
Pritchard, J.W., 2492
Pritts, I.M., 983, 985, 1592, 1746
Prober, C., 1198
Probst, K.S., 1768
Prochazka, J., 494
Proffit, W.R., 964
Proinova, V.A., 302
Proll, J., 510
Prytherch, J.P., 1748
Prywes, R., 1803
Puchkov, V.F., 7, 59, 123, 673, 1067
Pueschel, S.M., 1058
Puigdevall, J., 699
Pujol, M., 1353
Pulkkinen, M.O., 2023
Pumariega, A.J., 801
Purchase, I.F.H., 766, 2530
Purichia, N., 772, 2566
Purmalis, A., 1421
Purmalis, B.P., 588

Pushkin, Joan, 1876
Pushkina, N.N., 248, 1088
Pyatkin, E.L., 59
Pylant, F., 1792
Qamar, I.U., 89
Qazi, Q.H., 31
Qian, C., 1577
Qin, X-B., 1577
Quaife, C.J., 1851
Quaini, E., 637
Quast, J.F., 72, 937, 1918, 2256
Queenan, J.T., 706
Queralto, J.M., 740
Quesenberry, C.P., 231
Quiec, D., 284
Quimby, E.H., 1327
Quinton, R.M., 1415
Raab, D.M., 908
Rabe, A., 1402, 1560, 1920
Rabello, Y., 746
Rabinovitch, O., 619
Rabinowitz, M., 1402
Rabinowitz, M.B., 1402
Rachelefsky, G.S., 1300
Rachootin, P., 2208
Racowsky, C., 1727
Racz, K., 319
Rader, J.I., 2566
Rader, M., 922
Radovskaya, T.L., 613
Radow, B., 59
Radvany, R.A., 731
Rahimtula, A.D., 612
Rahm, U., 18, 42
Rahnevik, G., 596
Rahwan, R.G., 288
Raisys, V.A., 615
Rakalska, Z., 510, 1495
Raman-Williams, L., 2520
Ramazzotto, L.J., 1417
Ramcharan, S., 1775, 1803
Ramel, C., 1510
Ramer, R.M., 1292, 2344, 2441
Ramey, S.L., 615
Rampy, L.W., 248, 766
Randall, C.L., 59
Randall, E.J., 513
Randall, J.L., 317, 758, 1862, 1946,
    2423, 2474
Randall, J.R., 759
Randall, J.L., 758
Randel, R.D., 1150
Randels, S.P., 59
Ranganathan, S., 2128
Rantala, K., 1792, 2208, 2524
Rao, A.N., 2540
Rao, D.S., 2435
Rao, G.S., 1277

SUBJECT INDEX